Stochastic Dynamics, Filtering and Optimization

Stochastic processes and probability theory are widely used mathematical tools for modeling uncertainties of both epistemic and aleatory types. The calculus for diffusion processes, often used to model such noisy fluctuations, differs fundamentally from that of deterministic smooth functions. The theory of diffusive stochastic processes is also a ubiquitous ingredient in modern research strategies for a broad class of optimization problems including stochastic filtering. This book provides a balanced treatment of theory and applications in this area of importance.

It covers fundamentals of stochastic processes with applications to dynamical systems in science, engineering and recursive search algorithms. It discusses fundamental concepts and theory of stochastic processes, calculus, Ito-Taylor expansion and numerical integration of stochastic differential equations (SDEs) in detail. Topics such as Radon-Nikodym derivatives and Girsanov theorems with emphasis on Ito diffusion processes are comprehensively discussed. The text discusses advances in numerically integrating dynamical systems, non-linear stochastic filtering and generalized Bayesian updating theories. It covers many applications of stochastic filtering and global optimization. MATLAB codes for all the applications discussed here appear on the weblink www.cambridge.org/9781107182646

Debasish Roy is currently working as Professor, Computational Mechanics Lab, Department of Civil Engineering, Indian Institute of Science, Bangalore. His research interests include computational mechanics of non-classical continua, stochastic dynamical systems and optimization/inverse problems.

G. Visweswara Rao is an Engineering Consultant, Bangalore. His research interests include structural dynamics specific to earthquake engineering, non-linear and random vibration, and stochastic structural dynamics.

Stochastic Dynamics, Filtering and Optimization

Debasish Roy
G. Visweswara Rao

CAMBRIDGE
UNIVERSITY PRESS

CAMBRIDGE
UNIVERSITY PRESS

University Printing House, Cambridge CB2 8BS, United Kingdom
One Liberty Plaza, 20th Floor, New York, NY 10006, USA
477 Williamstown Road, Port Melbourne, vic 3207, Australia
4843/24, 2nd Floor, Ansari Road, Daryaganj, Delhi - 110002, India
79 Anson Road, #06–04/06, Singapore 079906

Cambridge University Press is part of the University of Cambridge.

It furthers the University's mission by disseminating knowledge in the pursuit of education, learning and research at the highest international levels of excellence.

www.cambridge.org
Information on this title: www.cambridge.org/9781107182646

First published 2017

Printed in India by Nutech Print Services, New Delhi

A catalogue record for this publication is available from the British Library

ISBN 978-1-107-18264-6 Hardback

Additional resources for this publication at www.cambridge.org/9781107182646

To our wives

Runa and Raji

Contents

Figures

Tables

Preface

This book is written with the express purpose of providing the readership with a reasonably self-contained treatise on stochastic dynamical systems whilst emphasizing on a class of applications that pervade across several disciplines and thus has a general appeal. The targeted audience are doctoral students, researchers and practitioners in science and engineering who wish to understand and exploit the principles of stochastic dynamics as a tool to address their scientific or applied problems. In view of the stated aim, what one should not expect is a treatment that might be deemed rigorous enough by researchers in pure and applied mathematics. Moreover, this book deals only with diffusive stochastic processes and does not address non-Markovian and / or non-diffusive stochastic dynamical systems.

With these broad goals in mind, the question as to whether there are similar other books may naturally arise. Alternatively, one may ask if there are any aspects of originality that we, the authors, could lay our claim to. Indeed there are a number of very well-written mathematical texts on stochastic processes and calculus, some of which also cover applications to such areas as finance (e.g., stock options), biology (e.g., birth–death processes) and estimation or control. Talking of applications, there are several mathematical texts on stochastic filtering problems, even though the focus therein may not so much be on the applied aspects covering higher dimensional filtering and identification problems—an area given some prominence in this book. There are even monographs dedicated entirely to the exploitation of the theory of stochastic processes and calculus to numerical integration of stochastic differential equations. Despite such plentiful and laudatory compilations, there is, to the best of our knowledge, no book or monograph on stochastic processes that simultaneously furnishes a reasonably in-depth treatment of the problem of global optimization based on stochastic search, thereby foregrounding the role of stochastic processes and calculus in the development of robust optimization schemes. Another novel feature is the use of change of measures as the driving refrain in most of the applications covered in this book-from numerical solutions to stochastic differential equations to filtering to global optimization.

The claimed self-contained structure of the book requires that the chapters be arranged to provide an appropriately sequenced and graded presentation of the subject. This we have tried to do, resulting in the following overall arrangement of the ten chapters constituting this volume. This brief look-up of the contents, provided below, will also help us prescribe to the potential reader a few guidelines on how to go about reading this book. Chapter 1 begins with a measure theoretic introduction to probability theory leading up to the notion of a probability density, interpretable as a Radon–Nikodym derivative. Concepts such as those of a random variable, their expectations, independence and transformation are also presented in this chapter. Following up in the second chapter, we continue to build up our arsenal of information on random variables by introducing conditional expectations with respect to sub-σ algebras and how these relate to change of probability measure. A natural consequence of this discourse is the generalized Bayes' rule. As a precursor to the applied and computational work undertaken from Chapter 5 onwards, we end Chapter 2 by giving an introduction to Monte Carlo simulations of random variables. Chapter 3 dwells upon stochastic processes that invariably identify with solutions to or evolutions of stochastic dynamical systems of practical import. Here we start with characterizing a stochastic process by all the finite dimensional distributions of the associated random variables and, in doing so, make use of Kolmogorov's extension theorem. Then we shift our focus to Brownian motion—a process of fundamental significance in modelling both input 'noises' and solutions of stochastic dynamical systems, and identify Brownian motion as a sub-class of a more general class of stochastic processes referred to as martingales. The theory of martingales is a powerful tool in the understanding and applications of stochastic processes and is repeatedly made use of in this book. Chapter 4 is devoted to an exposition of stochastic calculus (e.g., Ito's calculus) that formalizes the notion of stochastic differential equations (SDEs) whose solutions are stochastic processes representing time evolutions of functions of Brownian motion. Based on Ito's formula, we discuss such evolutions that typically involve stochastic (or Ito) integrals and contrast them with evolutions of differentiable functions, as via ordinary differential equations in classical calculus. While we do touch upon the Stratonovich form of stochastic integrals, this book consistently employs the Ito representation throughout. We introduce the backward Kolmogorov operator and Feynman–Kac formula for determining expectations of such time-evolving stochastic processes and wind up this chapter by presenting the Girsanov theorems that extend the concept of change of measures to Ito diffusion processes. With what we believe is a graded presentation of the basic probability theory and stochastic processes in the first four chapters, we start off with applications, a main motive behind this treatise, in Chapter 5 to discuss some numerical methods to integrate SDEs, often making use of the Ito–Taylor expansion—an asymptotic expansion derivable through repeated applications of the Ito formula. Explicit and implicit integration techniques, in their strong and weak forms, are considered. In the course of this presentation, we also try to highlight the relative difficulty or complexity in producing formally higher order numerical integrators for SDEs in comparison with those available for ordinary differential equations (ODEs). In Chapter 6, we introduce the idea of non-linear stochastic filtering as a rational route to solving state estimation and parameter identification problems in non-linear dynamical systems conditioned on the available, time evolving and noisy experimental data on a few, probably not all, system states. As a generic means of solution of the filtering problem, we use generalized Bayes' formula and thus derive the KS equation. Though beset with such numerical bottlenecks as circularity and higher dimensionality curse, the KS equation is important in that it formally governs evolutions of the required estimates. Indeed, by considering a few Monte Carlo or particle filtering schemes, we show that approximate empirical solutions to the filtering problem or the KS equation are obtainable. While many of these particle filtering schemes are not directly guided by the structure of the KS equation, we consider in Chapter 7 a family of Monte Carlo filters that are. Of specific interest are variants of the so-called KS filter wherein we are directly guided by the innovation integral in the KS equation in deriving the update terms in the numerical filtering schemes. In Chapter 8, we go beyond filtering problems and demonstrate the power of KS-type equations, which in essence enable obtaining sample realizations corresponding to a new transitional probability measure by additively correcting the realizations from an old transitional measure, in the context of numerical integration of SDEs. Specifically, we develop a weakly corrected Euler–Maruyama scheme and show, using a few illustrative examples, how it may be employed to improve the accuracy of the numerically integrated solution. Chapters 9 and 10, the last two chapters, are devoted to an exposition of global optimization schemes based on stochastic search. We initiate our presentation of the topic in Chapter 9 by describing a family of such global search schemes, covering a few that are relatively more formally grounded (e.g., simulated annealing, CMA–ES) and a few others that are biologically or sociologically motivated and best described as memetic schemes (e.g., genetic algorithms, particle swarms, differential evolutions). In Chapter 10, we detail a recently developed, formally grounded approach to global optimization based on change of measures, wherein updates to global extrema are often guided by KS-type equations. To the best of our understanding, this last scheme, whose acronym is COMBEO (Change Of Measure Based Evolutionary Optimization), probably turns out to be amongst the best performers across a fairly broad range of Np-hard optimization problems.

The reader will find this volume reasonably self-contained provided she has undergone a few basic maths courses (e.g., elements of linear algebra, probability theory, basic calculus) at the undergraduate level. Information on any additional mathematical topics, generally not covered at this level, is provided, albeit in a concise form, in a series of Appendices. Despite our sincere efforts, a book of such volume will have its share of typos and other errors. We would be grateful if our readers bring them to our notice.

Parts of the work in Chapters 8 and 10 are funded by a research grant (Grant # ERIP/ER/1201130/M/01/1509) from the Defence Research and Development Organization of India. The authors are grateful to Professor Karmeshu of the Jawaharlal Nehru University (New Delhi, India) for his very helpful suggestions in improving upon the presentation of the basic theory covered in the first couple of chapters. We also take this opportunity to express our thankfulness to Dr. Sreehari Kumar for his yeoman's service in formatting the documents.

Some guidelines on how to use this book

If used as a textbook for graduate students for a semester–long course on stochastic processes and calculus, the first four chapters of the book will come in particularly handy. Specifically, the instructor may decide to cover Chapter 1 in full, except perhaps a few theorems (e.g., Caratheodory's extension theorem in 1.3.1) and a few other aspects (e.g., the Nataf transformation in 1.11.4). Similarly, in covering Chapter 2, a few sections (e.g., Rosenblatt transformation in 2.2.5, the relatively more advanced topics in Monte Carlo simulations in 2.5) may be left out. In Chapter 3, one may consider omitting parts of 3.4 on martingales, some details on stopping times and stopped processes in 3.5 (e.g., 3.5.3–3.5.5), most of 3.6 (except for the statement of the martingale convergence theorem and a short discussion on its implications) and 3.7. The spectral representation of stochastic processes in 4.5 of Chapter 4 need not be covered in such a course. Parts of 4.6 (especially the technical aspects beyond the basic concepts), 4.7 and 4.8, a few portions in 4.9 (proofs of some theorems) and 4.10 may also be omitted. If time permits, the course may possibly end with an elementary coverage of the Euler–Maruyama method considered at the beginning of Chapter 5.

It is assumed that the researchers/practitioners, who wish to make good use of this book, have the benefit of exposure to a course, as above, on stochastic processes and calculus. In any case, it may be a good idea to have a quick recap of the first four chapters. Chapters 5 and 8 will be useful only to those needing relatively more accurate numerical or semi-analytical solutions to stochastic differential equations. Readers desirous of picking up the basics of the filtering theory and/or stochastic optimization, may go through Chapters 6, 7, 9 and 10 directly without requiring the matter provided in Chapters 5 and 8 as a prerequisite. However, an appreciation of the optimization methods dealt with in Chapters 9 and 10 will necessitate an understanding of the basics of the filtering theory covered in Chapters 6 and 7. The latter half of the book being aimed at researchers and practitioners, we have not included any exercises in Chapters 7, 8, 9 and 10.

Acronyms

ATP	Annealing-type parameter
ABS	Auxiliary bootstrap filter
ADP	Artificial diffusion parameter
ATP	Annealing type parameter
BS	Bootstrap filter
CDF	Cumulative distribution function
CMA	Covariance matrix adaptation
CLT	Central limit theorem
COMBEO	Change Of Measure Based Evolutionary Optimization
dof	Degree-of-freedom
DE	Differential equation
DiEv	Differential evolution
EnKF	Ensemble Kalman Filter
EnKS	Ensemble Kushner–Stratonovich
EM	Euler–Maruyama
EKF	Extended Kalman filter
fdd	finite dimensional distribution
FSM	Fundamental solution matrix
GA	Genetic algorithm
GCEM	Gisnov corrected Euler–Maruyama
GCLM	Girsanov corrected linearization method
GLM	Girsanov linearization method
IGSF	Iterated gain-based stochastic filter
i.i.d	Independent and identically distributed
KF	Kalman Filter
KS	Kushner–Stratonovich
LTL	Locally transversal linearization
MCMC	Markov chain Monte Carlo
MC	Monte Carlo
MSI	Multiple stochastic integrals
ODE	Ordinary differential equation
PDE	Partial differential equation
PF	Particle filter
PSD	Power spectral density
PSO	Particle swarm optimization
pdf	Probability density function
QV	Quadratic Variation
RHS	Right hand side
SIR	Sampling importance resampling
SIS	Sequential importance sampling
SMC	Sequential Monte Carlo
SA	Simulated annealing
SDE	Stochastic differential equation
UI	Uniformly integrable

General Notations

$a(t)$, $a(t,X)$, $a_i(t,\boldsymbol{X})$	drift terms in an SDE
$\boldsymbol{a}(t)$, $\boldsymbol{a}(t,\boldsymbol{X})$	drift vector
$B_t(\omega)$	Wiener process (Brownian motion)—often denoted in short by B_t
$\boldsymbol{B}_t(\omega)$	vector Brownian motion
$\mathcal{B}$, $\mathcal{B}(\mathbb{R})$	Borel σ-algebra
$cov(X,Y)$	covariance of random variables X and Y
$C_{XX}(t_i,t_j)$	cross-correlation of the stochastic process, X
C^1, C^2	continuously differentiable, twice continuously differentiable
$\mathcal{C}_0^\infty$	space of compactly supported smooth functions that are infinitely differentiable
$E[.]$	expectation operator with respect to P
$E_Q[.]$	expectation operator with respect to Q
$E[X/H]$	conditional expectation of X under H with respect to P
$f_X(x)$	probability density function, pdf
$F_X(x)$	probability (cumulative) distribution function, CDF
$f(y,t;x,s)$	transition pdf
$f(y;t\|x;s)$	transition CDF
$\mathcal{F}$	$\sigma - algebra$
$\mathcal{F}_t$	filtration (see Section 3.3.1 – Chapter 3)
$\mathcal{G}$	power set, 2^Ω

$\mathcal{H}^*$	class of stochastic processes satisfying the conditions in Eqs. (4.26), (4.45), and (4.47)
$\mathbf{I}$	identity matrix
$I(.)$	indicator function (see Eq. 1.13)
L_t	generator of the process $X(t)$, the backward Kolmogorov operator
L_t^*	adjoint of L_t
$\mathcal{M}(t)$	martingale process
N	number of time steps in the partition Π_n
N_e	ensemble size
N_p	particles size
$\mathcal{N}(m,\sigma)$	normal random variable parametered by m and σ
$\mathbb{N}$	set of natural numbers
P	probability measure on $\mathcal{F}$
P_B	probability measure
$P(E)$	probability of event, E
$P(y,t;x,s)$	conditional probability distribution function of a process at time t given that it is x at time $s \geq t$ and is $P(X(t) \leq y \| X(s) = x)$
$P \ll Q$	the measure P is absolutely continuous with respect to the measure Q
$\mathbb{R}$	set of real numbers (real line)
$\mathbb{R}^+$	non-negative real numbers
$\mathbb{R}^n$	n-dimensional Euclidean space
$\mathbb{R}^{n\times m}$	$n \times m$ matrices with real elements
t	time variable
$\mathbb{U}(.)$	unit step function (Heavyside step function) (Eq. 1.68)
$U(a,b)$	uniform distribution parametered by a and b
$Var(.)$	symbol for variance of a random variable
$X_t(\omega)$	scalar stochastic process
$\boldsymbol{X}_t(\omega)$	vector stochastic process

$[X,X](t)$	quadratic variation of the stochastic process X
$[X,X](t)$	quadratic covariation of stochastic processes X and Y
$\mathbb{Z}$	set of integers
$\Delta_n = \max_{1\leq j\leq n}(t_j - t_{j-1})$	of a partition Π_n of a time interval (see notation Π_n below)
$\delta(.)$	Dirac delta function (see Eq. (1.78) and Eq. 2.180)
δ_{ij}	Kronecker delta (see Eq. 2.167), equal to unity if $i = j$ and zero otherwise
$\Delta B_j = B(t_{j+1}) - B(t_j)$,	the Brownian increment
Π_n	partition of the time interval $t_0 - t_n$ into n parts
σ_X^2, σ_X	variance and standard deviation of the random variable X
$\sigma-$	sigma algebra
$\sigma(\mathcal{A})$	sigma algebra generated by $\mathcal{A}$
$\sigma(t)$, $\sigma(t,X)$, $\sigma_k(t)$, $\sigma_{ij}(t,\mathbf{X})$	diffusion terms in an SDE
$\boldsymbol{\sigma}(t)$, $\boldsymbol{\sigma}(t,\mathbf{X})$	diffusion vector
$(\Omega,\mathcal{F},P)$	probability space
$(\Omega,\mathcal{F},\mu)$	measure space
$(\Omega,\mathcal{F},\{\mathcal{F}_t\},P)$	filtered probability space
ζ_1, ζ_2	random permutations on the integer set $\{1,N_p\}/\{j\}$.
$\emptyset$	null set
$\varnothing$	characteristic function
$\Phi_Z(z)$ and $\Phi_Z(z)$	CDF and pdf of a standard normal random variable Z
$\|\cdot\|_p$	L^p norm
$\cup$	union operator
$\cap$	intersection operator

Chapter 1

Probability Theory and Random Variables

1.1 Introduction

Uncertainty or randomness appears to pervade across most natural, socio–economic and engineering phenomena—be it a simple game of chance, like tossing of a coin, or the complex analysis of an engineering system with or without uncertain (i.e., inadequately known) parameters under stochastic or random excitations. In this context, even without going into the scientific rigor, one could grossly appreciate the uncertainty associated with a seismic event or a nuclear accident probably leading to the failure of an engineering system during the plant operation and the catastrophic consequences there of. Note that, for engineering systems of importance or their components, the so called probability of failure against which the design calculations must be performed may be in the order of 10^{-6} or even less. A cut-off probability of, say, 10^{-6} may mean that not even one item shall fail out of a manufactured (or analyzed) lot of 10^6 products. Alternatively, one may also suppose that the item shall survive without failure, 10^6 events of the external calamity for which it is designed, given the projected frequency of occurrence of such an event.

Probability as a notion might have its origins in game theory. Earliest contributions in this regard probably date back to the works of Fermat, Pascal, Leibniz, Huygens, de Moivre, Bernoulli, and Bayes. Applications of probabilistic concepts had a conspicuous start in the nineteenth century due to Laplace, Chebyshev, and Markov. The applications included such diverse areas as mathematical statistics, psychology, medical science, and statistical mechanics. In the sequel, Kolmogorov [1933, 1950] pioneered both axiomatic and measure–theoretic approaches to the development of modern probability theory. Beyond its applications to natural sciences and engineering, the theory has today emerged as an indispensable tool in the modelling and analysis of financial markets [Berger 1985] as well. The stock market is known to fluctuate and shows either a decreasing or increasing trend whenever interest rates decline or increase. Knowing the fluctuations of the market price alone, one may have a gross idea about how the stock price responds in a probabilistic sense; e.g., the probability of price increase, known as prior probability, say, on a particular day. A more insightful behavior of the market price may be had by determining how the price increases or decreases when the interest rate is known to have increased or decreased. The particular formula yielding this desirable information, known as posterior probability, is called Bayes' rule. This is closely related to the important notion of conditional probability and is dealt with in Chapter 2 in more detail.

As a more specific example of a random phenomenon, some possible records (outcomes in terms of time histories) of wind velocity, say, on three consecutive days observed at the same time, are shown in Fig. 1.1. In this example, while the sample realizations vary with time, the uncertainty in the signal manifests itself in the apparent randomness in the magnitude of the wind velocity at any given time instant. It thus seems that, towards an appropriate characterization of such signals as random or stochastic processes, a collection (an ensemble) of such sample signals, as against a single sample history, would be needed.

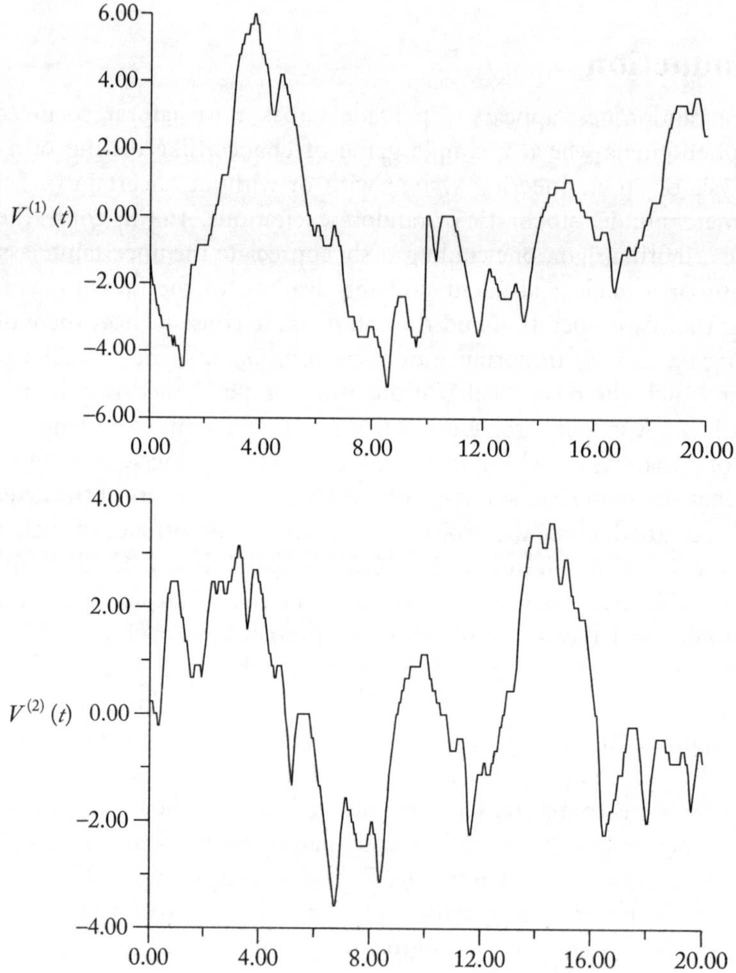

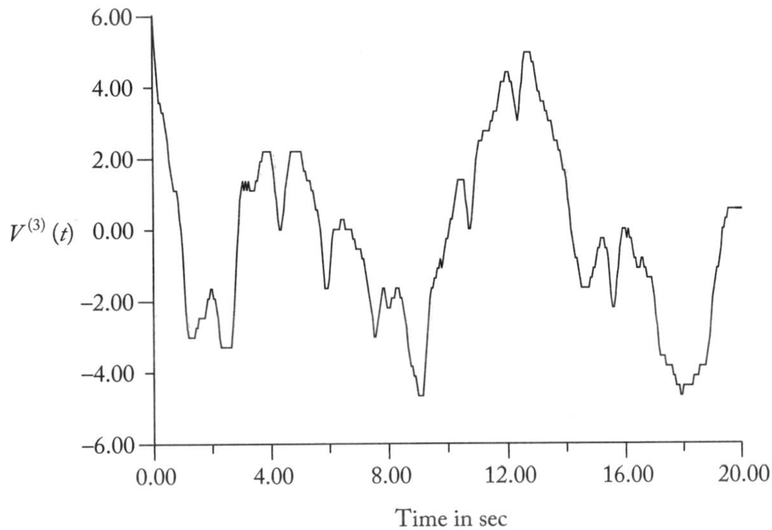

Fig. 1.1 Random signals; typical wind velocity (in m/sec) wave forms, $V^{(i)}(t)$, $i = 1, 2, 3$

This notion is made more precise whilst introducing stochastic processes in Chapter 3. Instances of such stochastic signals or excitations are omnipresent in nearly all disciplines of science and engineering, including diverse fields such as medical sciences, mathematical finance, and social sciences. The probability theory provides for a rational basis to arrive at workable mathematical models that make it possible to study the system performance within a random environment.

If we refer to an activity accompanying such uncertainty as a random experiment, a specific outcome could be among many such possible ones. While the experiments conducted repetitively may not yield any visibly repetitive arrangement amongst the possible outcomes (unlike a deterministic or non-random experiment), they might nevertheless possess a form of regularity in a statistical sense. Indeed, much of the machinery of the theory of probability or that of stochastic processes is aimed at revealing such regular features. Figure 1.2 depicts the familiar case of coin tossing experiment numerically simulated here by familiar simulation methods (see Section 2.5, Chapter 2 for details). We notice the regularity with which the number of heads reaches one half of the total number of tosses, especially with the total number of tosses becoming large, even as the occurrence of a head or otherwise in each toss remains unpredictable. This is the kind of regularity inherent in random phenomena that indeed helps us to make statistical inferences from the associated probabilistic models. In case our interest is in studying a system response in random environment, it eventually leads us to the response quantities in terms of a distinct pattern of reliable averages. The present chapter dwells on some of these averaged quantities and their role in the probabilistic models. The presentation focuses mainly on the modern probability theory [Kolmogorov 1950] which utilizes a measure–theoretic approach, treating the probability as a measure satisfying a definite set of axioms. Before we delve into these aspects in detail, it is also informative to have an appraisal of the erstwhile theories and approaches (towards treating uncertainties) mostly motivated by the games of chance (gambling, to wit).

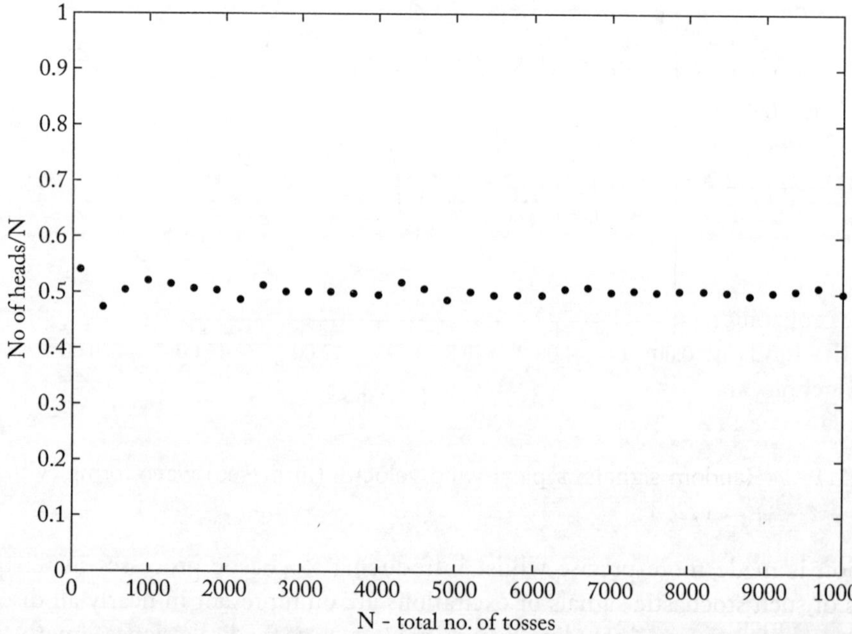

Fig. 1.2 Simulation of coin tossing; a random experiment and statistical regularity

Role of the classical and relative frequency concepts in probability theory

The early development of probability theory had its roots in the classical and relative frequency approaches. Before we proceed with the description of these two approaches, we need to define the sample space and events of a random experiment. The set of all possible outcomes of a random experiment is known as the sample space, denoted by Ω. Any subset of Ω is an event, E. Obviously if $E = \Omega$, the event is sure to occur. In the classical approach, all the possible outcomes of a random experiment are known or specifiable and the method assigns probability *a priori* to the occurrence of each of the outcomes on an equally-likely hypothesis. Hence, if N_E is the number of outcomes constituting the event E and N the total number of outcomes (cardinality of Ω), the method defines the probability $P(E)$ of the event E as:

$$P(E) = \frac{N_E}{N} \tag{1.1}$$

No experimentation is involved in the approach, which requires Ω to be finite. It mostly makes use of combinatorial mathematics to evaluate N and N_E. On the other hand, the relative frequency approach is founded on experimental observations and defines $P(E)$ as:

$$P(E) = \lim_{N \to \infty} \frac{N_E}{N} \tag{1.2}$$

That the random experiment can be repeated *ad infinitum* is inherent in the above concept. The definition in Eq. (1.2) appears to be more reasonable and intuitively satisfying than the classical approach, wherein the assignment of equal probabilities to all outcomes may be questionable [Papoulis 1991]. The sample space being restricted to a finite one in the classical approach is also an obvious limitation. Even though the last restriction is apparently absent in relative frequency approach, a significant drawback is the likely infeasibility of indefinitely repeating an experiment. Thus, in the rest of this chapter, we introduce the concepts leading to the development of modern probability theory [Kolmogorov 1933] based on a measure–theoretic approach. We first define axiomatic probability, random variable and their underlying framework rooted in measure theory. The fundamental notions of σ-algebra (e.g., Borel σ-algebra), independence of events, algebras and random variables are presented in the sequel followed by the description of conditional probability which is central to the probability theory.

1.2 Probability Space and Basic Definitions

The notions of an experiment, its outcomes, an event, and the sample space continue to remain fundamental to the probability theory. Simple yet much quoted examples of random experiments include tossing a coin and throwing a dice. The former has only two outcomes head or tail with the sample space $\Omega = \{\text{head}, \text{tail}\}$ and the latter yields $\Omega = \{f_1, f_2, f_3, f_4, f_5, f_6\}$ with only six outcomes representing the six faces of the dice. These two experiments are examples of a discrete sample space, wherein Ω is countable (finite or even countably infinite). Note that a set is countable if it is in one-to-one correspondence with $\mathbb{N}$, the set of natural numbers. Consider now the hypothetical experiment (see Fig. 1.1) of recording wind velocity at a specific location. The velocity V so obtained from an uncountable infinite set of outcomes (in this case, the velocity value is a positive real number) that define a continuous sample space, $\Omega = \mathbb{R}$. Here an event is a subset of the sample space. For example, in an experiment of throwing two dices, one can have the event E = sum of the faces showing on the two dice equals six and thus $E = \{(1,5), (2,4), (3,3), (4,2), (5,1)\}$. Before the axioms of probability are stated, it is necessary to define an algebra, the σ-algebra, and a measurable space.

Algebra and σ-algebra

A non-empty set $\mathcal{F}$ consisting of all observable subsets belonging to Ω (discrete or continuous) is an algebra if it is closed under complementation and finite unions of its subsets (including Ω). Thus $\mathcal{F}$ is an algebra if:

(i) $\Omega \in \mathcal{F}$ (1.3a)

(ii) $A \in \mathcal{F} \Longrightarrow A^c \in \mathcal{F}$, and (1.3b)

(iii) $A_1 \in \mathcal{F}, A_2 \in \mathcal{F} \Longrightarrow A_1 \cup A_2 \in \mathcal{F}$ (1.3c)

Note that the null set $\emptyset$ belongs to $\mathcal{F}$. Now, $\mathcal{F}$ is a σ-algebra if, along with (i) and (ii) in Eq. (1.3) it is closed under countable unions of $A_i s$,

$$A_i \in \mathcal{F},\ i = 1,2,.. \Longrightarrow \cup_{i=1}^{\infty} A_i \in \mathcal{F}. \tag{1.4}$$

An event is defined as any element $E \in \mathcal{F}$. Note that for a discrete sample space, algebra and σ-algebra are the same. The power set denoted by 2^{Ω} is the set all subsets of Ω and is a σ-algebra. For example, in the discrete sample space corresponding to the case of Bernoulli trials [Papoulis 1991, Ross 2002], Ω contains only two outcomes—success or failure (as head or tail in the tossing of a coin). Denoting, for convenience, the word success by the letter s and failure by f, we have $\Omega = \{s, f\}$ and σ-algebra, $\mathcal{F}$ is the power set of Ω containing $2^{\Omega} = 2^2 = 4$ elements: $\emptyset$, s, f, Ω.

In fact, the power set is the largest σ-algebra for a random experiment and $\{\Omega, \emptyset\}$ is the trivial (smallest) algebra. Further, we can see that the set $\{\Omega, \emptyset, A, A^c\}$ is a σ-algebra for every $A \subset \Omega$ and is known as the σ-algebra generated by A. The smallest σ-algebra containing every $A_i \subset \Omega$ $(i = 1,2,\ldots)$ and denoted by $\sigma\{A_i\}$, is also referred to as the σ-algebra generated by A_i. As an example, if $\Omega = \{f_1, f_2, f_3, f_4, f_5, f_6\}$ as in the dice experiment and if $A_1 = \{\{f_2\}, \{f_3\}\}$, there may be several σ-algebras containing $\{f_2\}$ and $\{f_3\}$. However, the intersection of all these σ-algebras yields the smallest σ-algebra generated by $\{\{f_2\}, \{f_3\}\}$ and is given by $\sigma\{\{f_2\}, \{f_3\}\} = \{\{f_2\}, \{f_3\}, \{f_2\}^c, \{f_3\}^c, \{f_2, f_3\}, \{f_2, f_3\}^c, \Omega, \phi\}\}$.

We now utilize the above notion in general to specify only the σ-algebra generated by those subsets of Ω about which information is available from a random experiment instead of, say, dealing with the power set, which is the largest σ-algebra and typically too large to be useful. Roughly, this also goes to explain why uncountable unions and intersections are not considered in the definition of σ-algebra, as it should apparently have been, particularly when the sample space is continuous. This important aspect, related to the measure theory, will be addressed again in a later section of this chapter. We also note in passing that unions or intersections of σ-algebras are σ-algebras too. Suppose $\{\sigma_j\}$ is a collection of σ-algebras on Ω. Let us prove that $\cap_j \sigma_j$ is also a σ-algebra.

Since σ_j is a σ-algebra, we have:

$$\text{i) } \Omega \in \sigma_j\ \forall\ j \Longrightarrow \Omega \in \cap_j \sigma_j \tag{1.5a}$$

$$\text{ii) } A \in \cap_j \sigma_j \Longrightarrow A \in \sigma_j \forall j \Longrightarrow A^c \in \sigma_j \forall j \Longrightarrow A^c \in \cap_j \sigma_j \tag{1.5b}$$

$$\text{iii) } \{A_m\} \in \cap_j\ \sigma_j \Longrightarrow A_m \in \sigma_j \forall\ m,\ j \Longrightarrow \cup_m A_m \in \sigma_j\ \forall\ j \Longrightarrow \cup_m A_m \in \cap_j \sigma_j \tag{1.5c}$$

Since the requirements postulated in Eqs. (1.3) and (1.4) are thus satisfied, the required result follows. Note that the words algebra and field are synonymous and so are σ-algebra and σ-field.

Continuing the discussion on continuous sample spaces, we now proceed to define a Borel σ-algebra which is the algebra associated with a topological space (Appendix A). In this context, if $\Omega = \mathbb{R}$ and A_1 a collection of open sets (Appendix A), $\sigma(A_1)$ is called the

Borel σ-algebra on Ω, and denoted by $\mathcal{B}(\mathbb{R})$. The Borel σ-algebra contains all the open sets in $\mathbb{R}$ and all those sets obtained by countable unions, intersections and complements of the open sets in A. The elements of $\mathcal{B}(\mathbb{R})$ are called Borel sets and it is the smallest σ-algebra containing the open sets in $\mathbb{R}$. We can show that the subsets (i) $[a,b]$, $a \leq b$, (ii) $(a,b]$, $a \leq b$, (iii) $[a,b)$, $a \leq b$ and (iv) any closed subset of $\mathbb{R}$ belong to $\mathcal{B}(\mathbb{R})$. In particular, the proofs for (i) and (iv) are given below:

Proof of i): $[a,b]$ is an intersection of open sets in $\mathbb{R}$ in the form:

$$[a,b] = \cap_{n=1}^{\infty}\left(a - \frac{1}{n}, b + \frac{1}{n}\right)$$

$$\Longrightarrow [a,b] \in \mathcal{B}(\mathbb{R}) \tag{1.6}$$

Proof of iv): If a closed set $A \in \mathbb{R}$, A^c is open $\Longrightarrow A^c \in \mathcal{B}(\mathbb{R})$. Hence, $(A^c)^c = A \in \mathcal{B}(\mathbb{R})$.

Proofs of (ii) and (iii) are left as an exercise to readers (also see Exercise 1.1). Here we need to note that different collections of sets belonging to $\mathbb{R}$ generate the same σ-algebra. Suppose that A_2 = set of all closed sets in $\mathbb{R}$ such as $[a,b]$, $a \leq b$ in item (i) above. Since any open and bounded interval can be written as the union of countable collection of closed and bounded intervals:

$$A_1 \subseteq \sigma(A_2) \Longrightarrow \sigma(A_1) \subseteq \sigma(A_2) \tag{1.7}$$

Equation (1.6) gives:

$$A_2 \subseteq \mathcal{B}(\mathbb{R}) \Longrightarrow \sigma(A_2) \subseteq \mathcal{B}(\mathbb{R}). \tag{1.8}$$

But, we have earlier established that $\mathcal{B}(\mathbb{R}) = \sigma(A_1)$ and as such Eq. (1.7) leads to $\mathcal{B}(\mathbb{R}) \subseteq \sigma(A_2)$. This fact coupled with the one in Eq. (1.8) yields the result $\mathcal{B}(\mathbb{R}) = \sigma(A_2)$. In fact by similar arguments, we can show that the σ-algebras generated by sets of the form $(a,b]$, $(-\infty,b)$, $[a,\infty)$ are also the same as $\mathcal{B}(\mathbb{R})$.

1.3 Probability as a Measure

Let us first define a measure and enumerate its properties.

If $\mathcal{F}$ is a σ-algebra on a sample space Ω and $\mu : \mathcal{F} \rightarrow [0,\infty]$, μ is an additive set function if for all $A, B \in \mathcal{F}$ with $A \cap B = \emptyset$,

$$\mu(A \cup B) = \mu(A) + \mu(B). \tag{1.9}$$

It is sub-additive if for all $A, B \in \mathcal{F}$,

$$\mu(A \cup B) \leq \mu(A) + \mu(B) \tag{1.10}$$

It is finitely additive for $A_i \in \mathcal{F}$ which are disjoint, if:

$$\mu\left(\cup_{i=1}^{N} A_i\right) = \sum_{i=1}^{N} \mu(A_i), N \in \mathbb{N} \tag{1.11}$$

μ is countably additive for all $A_i \in \mathcal{F}$ which are disjoint, if:

$$\mu\left(\cup_{i=1}^{\infty} A_i\right) = \sum_{i=1}^{\infty} \mu(A_i) \tag{1.12}$$

A countably additive set function μ is referred to as a measure. More formally we have the following definition for the measure:

A map $\mu : \mathcal{F} \to [0, \infty]$ is called a measure if $\mu(\emptyset) = 0$ and μ is countable additive (also known as σ-additive).

Counting and Dirac measures are the simplest examples of measure. The counting measure gives the cardinality of a finite set and in case of infinite sets assigns the value ∞. The Dirac measure indicates the measure of a set A as $\mu(A) = I_A(x)$ where $I_A(x)$ is an indicator function:

$$I_A(x) = 1, \quad x \in A$$

$$= 0, \quad x \notin A \tag{1.13}$$

Having defined a measure, we can make the following statements: (i) The triple $(\Omega, \mathcal{F}, \mu)$ is called the measure space, (ii) the events which are the elements of $\mathcal{F}$ are the measurable sets and (iii) $(\Omega, \mathcal{F}, \mu)$ is a finite measure space if $\mu(\Omega) < \infty$. Here we note that μ is non-decreasing. This is clear from the following:

For all $A, B \in \mathcal{F}$ with $A \subseteq B$, we have $B = A \cup (B \backslash A)$ where $B \backslash A = B \cap A^c$ and it follows from Eq. (1.9):

$$\mu(B) = \mu(A) + \mu(B \backslash A) \geq \mu(A) \tag{1.14}$$

In particular, $\mu(B \backslash A) = \mu(B) - \mu(A)$. Here note that σ-additivity (countable additivity) $\Longrightarrow$ additivity $\Longrightarrow$ sub-additivity. A measure has the following monotone-convergence properties with respect to limiting sequences:

i) Let $A_n \uparrow A$ (i.e., $A_n \subseteq A_{n+1} \forall n \in \mathbb{N}$ and $\lim_{n \to \infty} \cup A_n = A$). Then the monotone-convergence property of the measure is stated as: if $A_{n \in \mathbb{N}} \in \mathcal{F}$, $\mu(A_n) \uparrow \mu(A)$

Proof: Let $B_1 = A_1$ and $B_n = A_n \backslash A_{n-1}$, $n \geq 2$ so that the sets $B_{n \in \mathbb{N}}$ are disjoint. By σ-additive property of a measure, we have:

$$\mu(A_n) = \mu(B_1 \cup B_2 \cup \ldots \cup B_n) = \sum_{k \leq n} \mu(B_k) \uparrow \sum_{k \leq \infty} \mu(B_k) = \mu(A) \tag{1.15}$$

ii) Let $A_n \downarrow A$ (i.e., $A_{n+1} \subseteq A_n \ \forall n \in \mathbb{N}$ and $\lim_{n\to\infty} \cap A_n = A$). Then the second monotone-convergence property of the measure is stated as: If $A_{n\in\mathbb{N}} \in \mathcal{F}$ and $\mu(A_k) < \infty$ for some k, then $\mu(A_n) \downarrow \mu(A)$. In this case the proof follows the same steps as in (i) above except that we write $B_1 = A_1$ and $B_n = A_n \backslash A_{n+1}$, $n \in \mathbb{N}$ and apply the σ-additive property of a measure to obtain the result.

In a metric space (Appendix A) we are familiar with an associated metric which can indeed be regarded as a measure. The elementary notions of length, area and volume are particular cases of measure. If we consider the case $\Omega = \mathbb{R}$ and a bounded interval $A = (a, b)$ or $(a, b]$, or $[a, b)$ or $[a, b]$, then length of A is defined by $l(A) = |b - a|$. The length is additive since A can be decomposed into a disjoint union of a finite sequence $\{A_i\}_{i=1}^{m}$ of intervals. This leads to $A = \cup_{i=1}^{m} A_i \implies l(A) = \cup_{i=1}^{m} l(A_i)$. We can have similar consideration towards area and volume in $\mathbb{R}^2$ and $\mathbb{R}^3$ respectively and we find that these measures satisfy Eq. (1.12) and are all σ-additive. Measure is a generalization of a volume to all other abstract spaces including the Euclidean space. Another measure is the Lebesgue measure which is defined on σ-algebra over $\mathbb{R}^n$. This measure has important applications in integrations of functions associated in particular with the probability theory and will be discussed in a later section of this chapter.

Probability is also one such measure. Specifically if μ is normalized such that $\mu(\Omega) = 1$, μ is a probability measure. If, henceforth, we denote the probability measure by the letter P, the triplet $(\Omega, \mathcal{F}, P)$ is called the probability space. It is called a complete probability space if $\mathcal{F}$ contains all subsets of Ω with P-outer measure zero, i.e., with $P^*(A) := \inf\{P(F) : F \in \mathcal{F}, A \in \mathcal{F}\} = 0$. Any probability space can be made complete by adding to $\mathcal{F}$ all sets of P-outer measure zero and by extending P accordingly. The following properties constitute the axioms of probability:

i) $P(E) \geq 0, \forall E \in \mathcal{F},$ (1.16a)

ii) $P(\cup_{i=1}^{\infty} A_i) = \sum_{i=1}^{\infty} P(A_i) \forall A_1, A_2, \cdots \in \mathcal{F}$, with $A_i \cap A_j - \emptyset, i \neq j$ (1.16b)

iii) $P(\emptyset) = 0$ (1.16c)

P is a σ-additive function. It follows from the properties of a measure that if A_i, $i = 1, 2, 3, \ldots$ etc. are not disjoint, $P(\cup_{i=1}^{\infty} A_i) \leq \sum_{i=1}^{\infty} P(A_i)$. The events $(A_i \in \mathcal{F})$ are measurable sets. If Ω is a topological space, e.g., $\mathbb{R}$ with $\mathcal{B}$ as the Borel σ-algebra, P is known as the Borel measure. Note that for a non-empty subset $B \subset \Omega$, $(B, \mathcal{F}_B, P_B)$ is also a probability space with σ-algebra $\mathcal{F}_B = \sigma(B)$ and with sample space reduced from Ω to B and the new measure P_B satisfying $P_B(B) = 1$. This aspect will be further highlighted in Section 1.8 wherein the notion of independence of random variables is discussed.

For a discrete sample space, if the probability measure (point probability) of each outcome, $\omega \in \Omega$, is assigned such that $P(\{\omega\}) \in [0, 1]$ and $\sum_{\omega\in\Omega} P(\{\omega\}) = 1$, the probability measure of an event $E \in \mathcal{F}$ is given by:

$$P(E) = \sum_{\{\omega\}\in E} P(\{\omega\}), \text{by } \sigma\text{-additivity} \tag{1.17}$$

For finite Ω, the summation is straightforward. The number of terms in the summand is infinite for countably infinite Ω and the sum must be convergent. An example is the case of the experiment where one seeks the probability of success for the first time during a number of Bernoulli trials. Define the event $E = \{s, fs, ffs, fffs, ...\}$. If $P(s) = P(f) = 1/2$, then $P(E) = \sum_{r=1}^{\infty} \frac{1}{2^r}$. This infinite series converges to 1.

From the above discussion on discrete sample spaces, we infer that while the assignment of probability to all the elements of $\mathcal{F}$ could be extremely laborious, it might suffice to assign probability to a much smaller subset $\mathbb{G} \subset \mathcal{F}$, with $\mathcal{F} = \sigma(\mathbb{G})$, followed by an extension of the definition to the whole of $\mathcal{F}$. Particularly in the coin tossing experiment, $\mathbb{G} = \{heads, tails\}$. While an extension of this idea to continuous sample spaces (e.g., $\Omega = \mathbb{R}$) has great practical significance, it involves a difficulty too. To assign a probability measure in this case to all individual outcomes ω, e.g., real numbers in $\mathbb{R}$, is impossible since the sample space is, in the first place, uncountable and more importantly, the measure of any individual outcome $\omega \in \Omega$ is zero. In fact, for a continuous sample space, point measures carry no information and this is analogous to the property of Lebesgue measure defined on any topological space [Rudin 1976], say $\mathbb{R}$. This is further clarified as follows. The measures of an open set (a, b) and the corresponding closed set $[a,b]$ in $\mathbb{R}$ are the same, $|b-a|$. Moreover, for a continuous sample space, integration replaces summation (Eq. 1.17) in obtaining $P(E)$. Thus, if $\Omega \subset \mathbb{R}$ and x , $r \in \mathbb{R}$, the probability of the event $E = (x-r, x+r)$ (Fig. 1.3) is given by:

$$P(E) = \int_{x-r}^{x+r} dP \rightarrow 0 \text{ as } r \rightarrow 0 \tag{1.18}$$

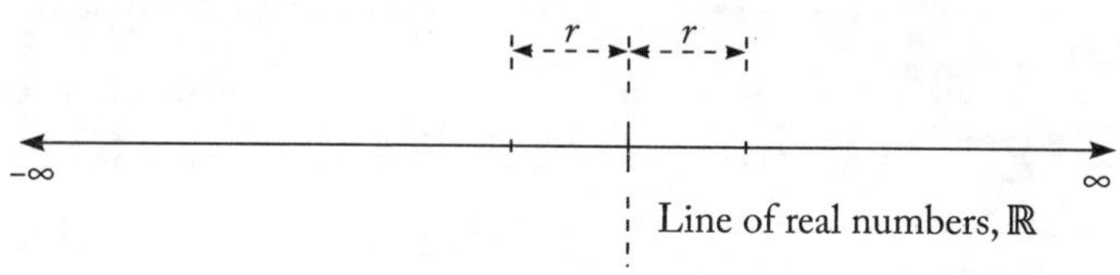

Fig. 1.3 Continuous sample space, $\Omega = \mathbb{R}$; zero probability of a point

1.3.1 Carathéodory's extension theorem

In measure theory, Carathéodory's extension theorem states that a countably additive set function on an algebra of sets can be extended to a measure on the σ-algebra generated by that algebra. In general, the continuous sample spaces of our interest being the topological spaces – in particular $\mathbb{R}^n$, it is customary to define a (probability) measure μ_0 on some basic class of sets, say, the intervals belonging to $\mathbb{R}^n$ that form an algebra. Since open sets are obtained as countable unions/intersections of these intervals, μ_0 can be extended to $\mathcal{B}(\mathbb{R}^n)$ by the σ-additive property of a measure. In fact, the concept of extension of a measure defined on an algebra over Ω to the σ-algebra is contained in Carathéodory's extension theorem which is stated below.

Let Ω be a set, $\mathcal{A}$ be an algebra on Ω, and $\mathcal{F} = \sigma(\mathcal{A})$ be the σ-algebra generated by $\mathcal{A}$. If $\mu_0 : \mathcal{A} \to [0,\infty]$ is a countably additive map then there exists a measure on $(\Omega, \mathcal{F})$ such that $\mu = \mu_0$ for all $A \in \mathcal{A}$.

Proof of the Carathéodory extension theorem

Step 1. We introduce a non-negative function known as an outer measure $\mu^* : 2^\Omega \to [0,\infty]$ with the properties:

i) $\mu^*(\emptyset) = 0$ (1.19a)

ii) It is monotonic, i.e., for $A, B \in 2^\Omega$ with $A \subset B, \mu^*(A) \le \mu^*(B)$ (1.19b)

iii) μ^* is countably sub-additive, i.e., $\mu^*\left(\cup_{j=1}^{\infty} A_j\right) \le \sum_{j=1}^{\infty} \mu^*\left(A_j\right) \forall A_j \in 2^\Omega$ (1.19c)

If $\mathcal{G}$ denotes 2^Ω, consider a set $A \subseteq \mathcal{G}$ and a collection of countable sets $A_{n \in \mathbb{N}} \subseteq \mathcal{A}$ covering the set A (see Appendix A for definition of a cover), i.e.,

$$A \subseteq \cup_{n \in \mathbb{N}} A_n \tag{1.20}$$

and define

$$\mu^*(A) = \inf \sum_n \mu_0(A_n) \tag{1.21}$$

The infimum in the above equation is taken over all $A_{n \in \mathbb{N}}$ in $\mathcal{A}$. Even though we are yet to show that μ^* in Eq. (1.21) is an outer measure on $(\Omega, \mathcal{G})$, with some abuse of notation, the symbol μ^* (earlier used to denote an outer measure) is used in this equation also. First, it is clear that $\mu^*(\emptyset) = 0$ and it is a monotonically increasing function. The countably sub-additive property (Eq. 1.19c) is proved in the following.

Suppose for $\{A_n\} \in \mathcal{G}$, $\mu^*(A_n) < \infty \; \forall n$. Given $\varepsilon > 0$, choose sequences $\{A_{nm}\}_{n,m \in \mathbb{N}}$ in $\mathcal{A}$ such that

$$A_n \subseteq \cup_m A_{nm},$$

$$\mu^*(A_n) + \varepsilon\, 2^{-n} > \sum_m \mu_0(A_{nm}) \tag{1.22}$$

Then with A defined as in Eq. (1.20), we have:

$$A \subseteq \cup_n \cup_m A_{nm} \tag{1.23}$$

It follows from the definition of an outer measure and Eq. (1.23) that:

$$\mu^*(A) \le \sum_n \sum_m \mu_0(A_{nm})$$

$$< \sum_n (\mu^*(A_n) + \varepsilon\, 2^{-n}) = \sum_n \mu^*(A_n) + \varepsilon \quad \text{(by Eq. 1.22)} \tag{1.24}$$

Since ε is arbitrary, we have:

$$\mu^*(A) \le \sum_n \mu^*(A_n) \tag{1.25}$$

This proves that μ^* is countably sub-additive and is an outer measure. Note that an outer measure as defined over the measurable space $(\Omega, \mathcal{G})$ is weaker than a measure in that it is countably sub-additive unlike a measure which is countably additive.

Here we also define μ^*- measurability (Caratheodory measurability) of sets. A set $A \in \mathcal{G}$ is μ^*- measurable if for all $B \in \mathcal{G}$, we have:

$$\mu^*(B) = \mu^*(A \cap B) + \mu^*(A^c \cap B) \tag{1.26}$$

Note that if A is μ^*- measurable, so is A^c. In the steps to follow, we utilize the notion of μ^*- measurability to show that a) the set, say, M of μ^*- measurable sets in $\mathcal{G}$ is a σ-algebra on which the outer measure μ^* is countably additive, b) $\mu^* = \mu_0$ on $\mathcal{A}$ and c) M contains $\mathcal{A}$ such that, $\mathcal{F} := \sigma(\mathcal{A}) \subseteq M$.

Step 2. To prove that M is a σ-algebra and μ^* is a measure restricted to M.

First let us show that M is an algebra. That is, if $A_1, A_2 \in M$, then $A_1 \cup A_2 \in M$. In other words, since A_1 and A_2 are μ^*- measurable, it suffices to show that $A_1 \cup A_2$ is μ^*- measurable.

Let $B \in \mathcal{G}$ be any set and we have by μ^*- measurability of A_2 (Eq. 1.26):

$$\mu^*(B \cap A_1{}^c) = \mu^*(B \cap A_1{}^c \cap A_2) + \mu^*(B \cap A_1{}^c \cap A_2^c) \tag{1.27}$$

We also have the set identity:

$$B \cap (A_1 \cup A_2) = (B \cap A_1) \cup (B \cap A_1{}^c \cap A_2) \tag{1.28}$$

By the countable sub-additive property of an outer measure, the above identity leads to:

$$\mu^*(B \cap (A_1 \cup A_2)) \le \mu^*(B \cap A_1) + \mu^*(B \cap A_1{}^c \cap A_2) \tag{1.29}$$

Now by μ^*- measurability of A_1, we have:

$$\mu^*(B) = \mu^*(B \cap A_1) + \mu^*(B \cap A_1^c)$$

$$= \mu^*(B \cap A_1) + \mu^*(B \cap A_1{}^c \cap A_2) + \mu^*(B \cap A_1{}^c \cap A_2^c) \text{ (by Eq. 1.27)}$$

$$\ge \mu^*(B \cap (A_1 \cup A_2)) + \mu^*(B \cap A_1{}^c \cap A_2^c) \text{ (by Eq. 1.29)}$$

$$= \mu^* (B\cap (A_1 \cup A_2)) + \mu^* (B\cap (A_1 \cup A_2)^c) \tag{1.30}$$

The above equation indicates that $A_1 \cup A_2$ is μ^*- measurable. Since $A_1{}^c$ and $A_2{}^c$ are also μ^*- measurable and we can write $(A_1 \cap A_2)^c = A_1{}^c \cup A_2^c$, it follows that intersection $A_1 \cap A_2$ of two measurable sets A_1 and A_2 is also μ^*- measurable. Hence M is an algebra.

For M to be a σ-algebra, it remains to be shown that for any countable collection of sets $A_{n\in\mathbb{N}}$ in M and for $A = \cup_n A_n$:

$$A \in M \text{ and } \mu^* (A) = \textstyle\sum_n \mu^*(A_n) \tag{1.31}$$

The first part of the above equation requires that M is closed under countable union of sets A_n in M. This is equivalent to showing that A $(= \cup_n A_n)$ is also μ^* measurable. The second part of Eq. (1.31) is on the requirement of showing μ^* to be countably additive on M and thus a measure restricted to M.

Letting $\overline{A}_1 = A_1, \overline{A}_2 = A_2 \backslash A_1 = A_2 \cap A_1^c, \ldots, \overline{A}_j = A_j \cap A_{j-1}^c, \ldots,$ we have a sequence of disjoint sets $\overline{A}_j$, $j = 1,2,\ldots n \in \mathbb{N}$ such that $A = \cup_n A_n = \cup_n \overline{A}_n$. Now for any $B \in \mathcal{G}$, μ^* measurability of A_1 gives:

$$\mu^* (B) = \mu^* (B \cap A_1) + \mu^* (B \cap A_1^c) \tag{1.32}$$

Utilizing the μ^* measurability of A_2, the second term on the right hand side of the above equation can be written as:

$$\begin{aligned} \mu^* (B \cap A_1^c) &= \mu^* (B \cap A_1^c \cap A_2) + \mu^* (B \cap A_1^c \cap A_2^c) \\ &= \mu^*\left(B \cap \overline{A}_2\right) + \mu^* (B \cap A_1^c \cap A_2^c) \end{aligned} \tag{1.33}$$

Here $\overline{A}_2 = A_1^c \cap A_2$. With the above identity for $\mu^* (B \cap A_1^c)$, Eq. (1.32) is rewritten as:

$$\mu^* (B) = \mu^*\left(B \cap \overline{A}_1\right) + \mu^*\left(B \cap \overline{A}_2\right) + \mu^* (B \cap A_1^c \cap A_2^c) \tag{1.34}$$

Continuing this process involved in Eqs. (1.33) and (1.34), we obtain:

$$\begin{aligned} \mu^* (B) &= \mu^*\left(B \cap \overline{A}_1\right) + \mu^*\left(B \cap \overline{A}_2\right) + \ldots + \mu^*\left(B \cap \overline{A}_n\right) + \mu^* (B \cap A_1^c \cap A_2^c \cap A_3^c \cap \ldots A_n^c) \\ &= \sum_j \mu^*\left(B \cap \overline{A}_j\right) + \mu^* (B \cap A_1^c \cap A_2^c \cap A_3^c \cap \ldots A_n^c) \end{aligned} \tag{1.35}$$

Here we also have:

$$B \cap A_1^c \cap A_2^c \cap A_3^c \cap \ldots A_n^c = B \cap A^c \Longrightarrow \mu^* (B \cap A_1^c \cap A_2^c \cap A_3^c \cap \ldots A_n^c) \leq \mu^* (B \cap A^c) \tag{1.36}$$

Substituting the above result in Eq. (1.35), we get:

$$\mu^*(B) \geq \sum_j \mu^*\left(B \cap \overline{A}_j\right) + \mu^*(B \cap A^c)$$
$$\geq \mu^*\left(B \cap \left(\cup_j \overline{A}_j\right)\right) + \mu^*(B \cap A^c)$$
$$= \mu^*\left(B \cap \left(\cup_j A_j\right)\right) + \mu^*(B \cap A^c)$$
$$= \mu^*(B \cap A) + \mu^*(B \cap A^c) \tag{1.37}$$

From Eqs. (1.35) and (1.36), we also have the set equality:

$$B = \cup_j\left(B \cap \overline{A}_j\right) + B \cap A^c = B \cap \cup_j \overline{A}_j + B \cap A^c = B \cap \cup_j A_j + B \cap A^c \tag{1.38}$$

By the countable sub-additive property of an outer measure, we have:

$$\mu^*(B) \leq \mu^*\left(B \cap \cup_j A_j\right) + \mu^*(B \cap A^c) \tag{1.39}$$

The two equations, Eqs. (1.37) and (1.39), yield the result:

$$\mu^*(B) = \mu^*(B \cap A) + \mu^*(B \cap A^c) \tag{1.40}$$

Hence A is μ^*-measurable and belongs to M showing that M is a σ-algebra.

Further, setting $B = A$ in Eq. (1.39), we get the result $\mu^*(A) = \sum_n \mu^*(A_n)$ corresponding to the second part of Eq. (1.31) showing that μ^* is indeed a measure on M.

Step 3. We show that μ^* extends μ_0, i.e., $\mu^*(A) = \mu_0(A)\ \forall A \in \mathcal{A}$.

Consider $A \in \mathcal{A}$. By definition of an outer measure (Eq. (1.19a)), it is obvious that:

$$\mu^*(A) \leq \mu_0(A) \tag{1.41}$$

Now, suppose that $A_n \in \mathcal{A}$ and $A \subseteq \cup_n A_n$. Also, let $B_1 = A_1$ and $B_n = A_n \backslash A_{n-1}, n \geq 2$ so that the sets $B_{n \in \mathbb{N}}$ are disjoint, $B_n \subseteq A_n$ and $\cup_n B_n = \cup_n A_n \supseteq A$. Since μ_0 is countably additive:

$$\mu_0(A) = \mu_0(\cup(A \cap B_n)) = \sum \mu_0(A \cap B_n) \tag{1.42}$$

This results in:

$$\mu_0(A) \leq \sum \mu_0(B_n)$$
$$\leq \sum \mu_0(A_n) \quad (\text{since } B_n \subseteq A_n) \tag{1.43}$$

Since the infimum over $\sum \mu_0(A_n)$ is the outer measure $\mu^*(A)$, we obtain from the above equation:

$$\mu_0(A) \le \mu^*(A) \tag{1.44}$$

From Eqs. (1.41) and (1.44), we have, $\mu^*(A) = \mu_0(A) \;\; \forall A \in \mathcal{A}$.

Step 4. We prove that M contains $\mathcal{A}$.

Let $A \in \mathcal{A}$ and $B \in \mathcal{G}$. To prove $\mathcal{A} \subseteq M$, all we need to show is that all $A \in \mathcal{A}$ are μ^*-measurable, i.e.:

$$\mu^*(B) = \mu^*(A \cap B) + \mu^*(A^c \cap B) \tag{1.45}$$

From the set identity $B = (A \cap B) + \ (A^c \cap B)$ and by the definition of an outer measure with its sub-additive property, we readily have:

$$\mu^*(B) \le \mu^*(A \cap B) + \mu^*(A^c \cap B) \tag{1.46}$$

To get the identity in Eq. (1.45), it is only required to show that:

$$\mu^*(B) \ge \mu^*(A \cap B) + \mu^*(A^c \cap B) \tag{1.47}$$

With the assumption that $\mu^*(B) < \infty$, we can find a sequence $\{B_{n \in \mathbb{N}}\}$ in $\mathcal{A}$ such that:

$$B = \cup_n B_n, \tag{1.48a}$$

$$\mu^*(B) + \varepsilon \ge \sum_n \mu^*(B_n) \text{ with } \varepsilon > 0 \tag{1.48b}$$

Then we have:

$$(B \cap A) \subseteq \cup_n (B_n \cap A), \tag{1.49a}$$

$$(B \cap A^c) \subseteq \cup_n (B_n \cap A^c) \tag{1.49b}$$

The above equation leads to:

$$\mu^*(B \cap A) \le \sum_n \mu^*(B_n \cap A) \text{ and } \mu^*(B \cap A^c) \le \textstyle\sum_n \mu^*(B_n \cap A^c)$$

$$\Longrightarrow \mu^*(B \cap A) + \mu^*(B \cap A^c) \le \sum_n \mu^*(B_n \cap A) + \sum_n \mu^*(B_n \cap A^c) = \sum_n \mu^*(B_n)$$

$$\Longrightarrow \mu^*(B \cap A) + \mu^*(B \cap A^c) \le \sum_n \mu^*(B_n)$$

$$\Longrightarrow \mu^*(B \cap A) + \mu^*(B \cap A^c) \leq \mu^*(B) + \varepsilon \tag{1.50}$$

Since ε is arbitrary, we obtain the required result (Eq. 1.47). Hence Eq. (1.45) is satisfied for all $A \in \mathcal{A}$ implying that M contains $\mathcal{A}$.

The proof of the Carathéodory's extension theorem is complete.

1.3.2 Uniqueness criterion for measures

The uniqueness of the extended measure requires some consistency conditions that are contained in Dynkin's Lemma. Before this Lemma is discussed, we need a few definitions particularly of a π-system and a λ-system.

π-system

A non-empty collection $\mathfrak{I}$ of subsets of Ω, which are closed under finite intersections is known as π-system. This implies:

$$A \cap B \in \mathfrak{I} \text{ for all } A,\ B \in \mathfrak{I} \tag{1.51}$$

λ-system

A λ-system is a collection $\mathcal{D}$ of subsets of Ω that satisfies the following properties:

i) $\Omega \in \mathcal{D}$ (1.52a)

ii) If $A, B \in \mathcal{D}$ and $A \subseteq B, B \backslash A \in \mathcal{D}$ (1.52b)

iii) If $\{A_n\}$ is an increasing sequence of subsets of Ω with $A_n \uparrow A$ (i.e., $A_n \subseteq A_{n+1}\ \forall n \in \mathbb{N}$ and $\lim_{n\to\infty} \cup A_n = A$), then $A \in \mathcal{D}$ (1.52c)

We note here that a system Σ which is both a π-system and λ-system is a σ-algebra. This is clarified as follows.

Let A, B and $A_1, A_2, \ldots, A_{n\in\mathbb{N}} \in \Sigma$. Then $A^c \in \Sigma$ by the property of λ-system. $A \cup B = \Omega \backslash (A^c \cap B^c) \in \Sigma$ by the properties of π-system and λ-system. Hence the finite unions $A_1 \cup A_2 \cup \cdots \cup A_n \in \Sigma$. Now, if $A_n = \cup A_j \uparrow A$, the countable unions $\cup A_j \in \Sigma$ by the property of a λ-system. Hence Σ is a σ-algebra.

Dynkin's Lemma

The lemma states that if $\mathfrak{I} \subset \mathcal{D}$, then $\mathcal{D}(\mathfrak{I}) \supseteq \sigma(\mathfrak{I})$. In other words, if a λ-system contains a π-system, it contains the σ-algebra generated by the π-system and the proof is as follows.

Proof: Without loss of generality, let $\mathcal{D}$ be the intersection of all λ-systems that contain $\mathfrak{I}$.

Consider the class D_1 of all sets $B, C \in \mathcal{D}$ such that $B \cap C \in \mathcal{D}$, i.e., $D_1 = \{B \in \mathcal{D} : B \cap A \in \mathcal{D},\ \forall A \in \mathcal{D}\}$. By this definition, D_1 is a π system.

i) Further, that D_1 contains $\mathfrak{I}$ is clear as follows. For all those $A, B \in \mathfrak{I} (\subset \mathcal{D})$, $B \cap A \in \mathfrak{I}$ since it is a π-system. Hence $B \cap A \in \mathcal{D} \Longrightarrow B \in D_1\ \ \forall A,\ B \in \mathfrak{I} \Longrightarrow D_1 \supset \mathfrak{I}$.

ii) D_1 is a λ-system as deduced from the following statements.

a) $\Omega \in D_1$ (satisfying Eq. 1.52a) (1.53a)

b) If $B_1, B_2 \in D_1$ and $B_1 \subseteq B_2$, then for any $A \in \mathcal{D}$, we have $(B_2 \backslash B_1) \cap A = (B_2 \cap A) \backslash (B_1 \cap A)$. Since $B_2 \cap A \in \mathcal{D}$ and $B_1 \cap A \in \mathcal{D}$, it is clear that $(B_2 \backslash B_1) \cap A \in \mathcal{D}$ so that:

$B_2 \backslash B_1 \in D_1$ (satisfying Eq. 1.52b) (1.53b)

c) If $B_{i \in \mathbb{N}} \in D_1$ and $B_n \uparrow B$, then for $A \in \mathcal{D}$ we have $(B_n \cap A) \uparrow (B \cap A)$ so that:

$B \cap A \in \mathcal{D}$ and hence $B \in D_1$ (satisfying Eq. 1.52c) (1.53c)

Finally we have that D_1 is a λ-system as well as a π-system. Since $\mathcal{D}$ is the smallest λ-system, we have $D_1 \supseteq \mathcal{D}$. But by definition $D_1 \subseteq \mathcal{D}$ and, therefore, $D_1 = \mathcal{D}$. This proves the Lemma, i.e., $\mathcal{D} \supseteq \sigma(\mathfrak{T})$.

Thus a λ-system generated by a π-system on a sample space Ω coincides with the σ-algebra generated by the π-system. We shall now proceed with the uniqueness theorem on extension of a measure and the relevant criterion.

The statement of the *uniqueness theorem* is as follows.

Consider a measurable space $(\Omega, \mathcal{S})$. Let $\mathfrak{T}$ be a π-system and $\mathcal{S} = \sigma(\mathfrak{T})$. If $\mu_1$and μ_2 are two measures on $\mathcal{S}$ such that $\mu_1(\Omega) = \mu_2(\Omega) < \infty$ and $\mu_1(A) = \mu_2(A) \;\; \forall A \in \mathfrak{T}$, then μ_1 and μ_2 are equal, i.e., $\mu_1(B) = \mu_2(B) \; \forall B \in \mathcal{S}$.

The proof is given below.

Proof: Define $\mathcal{D} = \{A \in \mathcal{S} : \mu_1(A) = \mu_2(A)\}$. Our aim is to show that $\mathcal{D} = \mathcal{S}$. By definition $\mathcal{D} \supset \mathfrak{T}$ (since $\mu_1(A) = \mu_2(A) \forall A \in \mathfrak{T}$). By Dynkin's lemma above, it suffices to verify that $\mathcal{D}$ is a λ-system. $\Omega \in \mathcal{D}$ follows from the hypothesis $\mu_1(\Omega) = \mu_2(\Omega)$. If $B, C \in \mathcal{D}$ and $B \in \mathcal{S}$, $\mu_1(C \backslash B) = \mu_1(C) - \mu_1(B) = \mu_2(C) - \mu_2(B) = \mu_2(C \backslash B) \Longrightarrow C \backslash B \in \mathcal{D}$. Finally, $B_n \in \mathcal{D}$ and $B_n \uparrow B$, then $\mu_1(B_n) = \mu_2(B_n) \Longrightarrow \mu_1(B) = \mu_2(B)$ by limiting properties of measures (see Eq. 1.15). Thus $\mathcal{D}$ is a λ-system containing a π-system and Dynkin's Lemma leads to the result that $\mathcal{D} = \mathcal{S} = \sigma(\mathfrak{T})$. The theorem implies that a measure defined on a π-system generating the σ-algebra is unique.

Example 1.1

An important application of uniqueness of extension of a measure is that a finite measure on the Borel subsets of a metric space is uniquely determined by its values on open sets. In particular, if the metric space is $\mathbb{R}$, there exists a unique Borel measure μ on $\mathbb{R}$ such that, for all $a, b \in \mathbb{R}$ with $a < b$, $\mu((a, b)) = b - a$. The measure μ is called Lebesgue measure on the measurable space $(\mathbb{R}, \mathcal{B}(\mathbb{R}))$. Note that the Lebesgue measure space (R, $\mathcal{B}(\mathbb{R})$, μ) is translation-invariant, i.e., if, for all $x \in \mathbb{R}$ and all $B \in \mathcal{B}(\mathbb{R}), \mu(B + x) = \mu(B)$.

Example 1.2

Here is a counter-example to illustrate the necessity of defining a measure on a π-system (or algebra) for unique extension to σ-algebra. Let $\Omega = (1,2,3,4)$ and we choose a system $E = (E_1, E_2)$ where $E_1 = (1, 2)$ and $E_2 = (2, 3)$ and a measure of $1/2$ is assigned to each. Here $\sigma(E)$ is the power set of Ω. Note that E is not a π-system. Two different measures μ_1 and μ_2 on Ω with the choice of $\mu_1 = (1/4, 1/4, 1/4, 1/4)$ and $\mu_2 = (1/2, 0, 1/2, 0)$ satisfy the above assignment on E_1 and E_2 and also $\mu_1(\Omega) = \mu_2(\Omega) = 1$. For uniqueness of extension, it is essential that E is a π-system (closed under intersections) and thus it must be $(\{1,2\}, \{2,3\}, \{2\})$.

1.4 Random Variables and Measurable Functions

In many cases, we may have readily discernible associations of outcomes ($\omega \in \Omega$) and events ($E \in \mathcal{F}$) with (real) numbers. However, in many others, outcomes may also be descriptive and non-numeric viz., $\Omega = \{\text{failure}, \text{no failure}\}$ as in the case, say, of the structural response exceeding a specified critical level or otherwise. However, a numerical or computational treatment of uncertainty requires that the outcomes or events be made to correspond one-to-one with real numbers via an appropriate mapping from Ω to $\mathbb{R}^n$.

In general, if $(\Omega, \mathcal{F})$ and (Σ, Υ) are measurable spaces, a map $X : \Omega \to \Sigma$ is called $\mathcal{F} - \Upsilon$ measurable if:

$$X^{-1}(A) = \{\omega \in \Omega : X(\omega) \in A\} \in \mathcal{F} \text{ for every } A \in \Upsilon \tag{1.54}$$

$X^{-1}(A)$ denotes the inverse image of A by X. If $(\Omega, \mathcal{F}, P)$ is a probability space and (Σ, Υ) is a measurable space, then a measurable function $X : \Omega \to \Sigma$ is referred to as Σ-valued random variable. The case with $\Sigma = \mathbb{R}$ and $\Upsilon = \mathcal{B}(\mathbb{R})$, the Borel σ-algebra, is of special interest and the function X is here known as a Borel measurable function or, simply, a random variable. Thus, in general, we define the (vector-valued) random variable to be a (invertible) mapping $X : \Omega \to \mathbb{R}^n$ such that if U is an open set in $\mathbb{R}^n$, then $E = X^{-1}(U) = \{\omega \in \Omega : X(\omega) \in U\} \in \mathcal{F} = \sigma(U)$. X is thus a function between two measurable spaces $(\Omega, \mathcal{F})$ and $(\mathbb{R}^n, \mathcal{B})$ where $\mathcal{B}$ is the σ-algebra generated by the subsets in $\mathbb{R}^n$.

Note that the above definition of a measurable function/random variable is valid for Ω being discrete or continuous. If X is a continuous (and bounded) mapping $\Omega \to \mathbb{R}^n$, it is measurable with respect to Borel σ-algebra $\mathcal{B}$. If the random variable is discrete, it maps outcomes of Ω to a countable set in $\mathbb{R}^n$. For $n = 1$ (the case of a scalar-valued random variable), $X(\omega_i \in \Omega) = x_{i \in \mathbb{N}} \in \mathbb{R}$, $i = 1, 2..m$ where m is total number of outcomes in Ω (i.e., the cardinality of Ω).

1.4.1 Some properties of random variables

Let $X : \Omega \to \mathbb{R}$ be a Borel measurable function.

i) With A, B, $A_{n \in \mathbb{N}} \in \mathcal{B}(\mathbb{R})$, X preserves all set operations, e.g., $X^{-1}(A^c) = \left(X^{-1}(A)\right)^c$, $X^{-1}(A \cap B) = X^{-1}(A) \cap X^{-1}(B)$, $X^{-1}(\cup_n A_n) = \cup_n X^{-1}(A_n)$ etc.

ii) If $C \subseteq \mathcal{B}(\mathbb{R})$ and $\sigma(C) = \mathcal{B}(\mathbb{R})$, then $X : \Omega \to \mathbb{R}$ is Borel measurable if and only if $X^{-1}(C) \subseteq \mathcal{F}$.

The if-part is obvious since if X is Borel measurable, then $X^{-1}(C) \subseteq \mathcal{F}$. Conversely, suppose that $X^{-1}(C) \subseteq \mathcal{F}$. Consider the family of sets:

$$E{:=}\left\{A \in \mathcal{B}(\mathbb{R}) : X^{-1}(A) \subseteq \mathcal{F}\right\} \tag{1.55}$$

Using the property (i) above, we can show that E is a σ-algebra on $\mathbb{R}$ containing C. Thus E coincides with $\mathcal{B}(\mathbb{R})$.

iii) The set $\left\{X^{-1}(A) : A \in \mathcal{B}(\mathbb{R})\right\}$ is a σ-algebra on Ω. Similarly, $\{X(A) : A \in \mathcal{F}\}$ is a σ-algebra on $\mathcal{B}(\mathbb{R})$. Proof is provided for the first part which can be restated as:

A function $X : \Omega \to \mathbb{R}$ is Borel measurable if and only if the set $\left\{X^{-1}(A) : A \in \mathcal{B}(\mathbb{R})\right\} \in \mathcal{F}$. The proof is given in the following statements.

a) If $X^{-1}(A) \in \mathcal{F}$ for each $A \in \mathcal{B}(\mathbb{R})$, then certainly $X^{-1}(U) \in \mathcal{F}$ for any open set U in $\mathbb{R}$ (since $\mathcal{B}(\mathbb{R}) = \sigma(U)$). So X is Borel measurable.

b) Conversely, suppose that $X^{-1}(U) \in \mathcal{F}$ for any open set U in $\mathbb{R}$. Let C denote the collection of subsets of $\mathbb{R}$ given by $C = \left\{B \subseteq \mathbb{R} : X^{-1}(B) \in \mathcal{F}\right\}$. Then C must be a σ-algebra which is clear from the following.

1. $X^{-1}(\mathbb{R}) = \Omega \in \mathcal{F} \Longrightarrow \mathbb{R} \in C$ (1.56a)
2. $X^{-1}(\mathbb{R} \backslash B) = \Omega \backslash X^{-1}(B)$, so if $B \in C$, then so is $\mathbb{R} \backslash B$ (1.56b)
3. If $B_1, B_2, \ldots, B_{n \in \mathbb{N}} \in C$, then

$$X^{-1}(\cup_n B_n) = \cup_n X^{-1}(B_n) \in \mathcal{F}$$

and so

$$\cup_n B_n \in C \tag{1.56c}$$

Thus, C is indeed a σ-algebra. But as per the hypothesis, C consists of all open sets in $\mathbb{R}$ which leads to $\mathcal{B}(\mathbb{R}) \subseteq C$. Hence for any $A \in \mathcal{B}(\mathbb{R})$, we have the result $X^{-1}(A) \in \mathcal{F}$. Proof of the second part of property (iii) above is left as an exercise to the reader.

Example 1.3

$\left\{X^{-1}(B) : B \in \mathcal{B}(\mathbb{R})\right\}$ is a σ-algebra on Ω – Case of a discrete random variable

Consider a coin tossing experiment involving two tosses. Treating the outcome of a head as success (denoted by s) and a tail as failure (denoted by f), we have

$\Omega = \{\omega_1 = (s,s), \omega_2 = (s,f), \omega_3 = (f,s), \omega_4 = (f,f)\}$. Let $\mathcal{F}$ be the power set 2^Ω. Suppose that our interest is in a success on the first toss and thus denote $A = (\omega_1, \omega_2)$. The σ-algebra generated by A is given by $\mathcal{F}_1 = \sigma(A) = (\phi, \Omega, A, A^c)$. Define a mapping $X_1 := X_1(\omega_1) = X_1(\omega_2) = 1.5$, $X_1(\omega_3) = X_1(\omega_4) = 0.5$. We observe that X_1 is a random variable on $\mathcal{F}_1$. Indeed, the set $\{\omega \in \Omega : X_1(\omega) = 1.5\} = (\omega_1, \omega_2) = A \in \mathcal{F}_1$. Also, $\{\omega \in \Omega : X_1(\omega) = 0.5\} = (\omega_3, \omega_4) = A^c \in \mathcal{F}_1$. If we consider another mapping $X_2 := X_2(\omega_1) = 1.25$, $X_2(\omega_2) = 0.75$, $X_2(\omega_3) = 0.75$, $X_2(\omega_4) = 0.25$, X_2 is not $\mathcal{F}_1$-measurable and hence not a random variable on $\mathcal{F}_1$. The reason is that the set $\{\omega \in \Omega : X_2(\omega) = 0.75\} = (\omega_2, \omega_3) \notin \mathcal{F}_1$. However, X_2 is a random variable on $\mathcal{F}$.

iv) If Ω is a topological space, any continuous function $X : \Omega \to \mathbb{R}$ is Borel measurable

Proof: A function between two topological spaces is continuous if and only if inverses of open sets are open [Rudin 1976]. Since the given function X is continuous, we have that whenever U is open in $\mathbb{R}$, the inverse image $X^{-1}(U)$ is open in Ω and generates the σ-algebra $\mathcal{B}(\Omega)$.

A few more useful properties of measurable functions are listed below; see references at the end of the chapter for proofs.

v) Compositions of any pair of measurable functions is measurable

vi) Sums and products of measurable functions are measurable

vii) A family of functions $X_i : \Omega \to \mathbb{R}, i \in \mathbb{N}$ are measurable if $\sigma\left(X_i^{-1}(A) : \forall A \in \mathcal{B}(\mathbb{R})\right) = \mathcal{F}_i$ where $\mathcal{F}_i$ is the σ-algebra generated by X_i, $i \in \mathbb{N}$.

For real-valued functions, it is often convenient to identify the one-dimensional range with the extended real line. Thus the function $X : \Omega \to [-\infty, \infty]$ is an extended real function and is Borel measurable if $\{\omega \in \Omega : X(\omega) \le a\} = X^{-1}((-\infty, a]) \in \mathcal{F}$ for all $a \in \mathbb{R}$. The Borel σ- algebra is generated by semi-closed intervals of the form$(-\infty, a], a \in \mathbb{R}$. In this context, the above function X is also measurable if:

i) $\{\omega \in \Omega : X(\omega) < a\} = X^{-1}((-\infty, a)) \in \mathcal{F}$ (1.57a)

ii) $\{\omega \in \Omega : X(\omega) > a\} = X^{-1}((a, \infty)) \in \mathcal{F}$ (1.57b)

iii) $\{\omega \in \Omega : X(\omega) \ge a\} = X^{-1}([a, \infty)) \in \mathcal{F}$ (1.57c)

Henceforth we denote random variables using upper case letters, e.g., $X(\omega)$ and the values that a random variable may take by lower case letters, e.g., x. In discrete probability space, the random variable X is measurable on $\mathcal{F}$ if the sets $\{\omega : X(\omega) = x\}$ are all measurable. Analogously, the measurability of X in a continuous probability space requires that the sets $\{\omega : a \le X(\omega) \le b\}$ are measurable.

Example 1.4

In a random experiment involving tossing of two dice, we have $\Omega = \{\omega = (i,j) : 1 \le i, j \le 6\}$. A random variable $X : \Omega \to \mathbb{R}$ may be defined as $X(\omega) = i + j$, with the range consisting of the integers $2, 3, 4, \ldots, 12$.

Example 1.5

If $X(\omega)$ is a random variable representing the time taken by a radioactive material to decay, it takes an uncountably infinite number of possible values. It is a continuous random variable with $\Omega = \{\omega : a \le X(\omega) \le b,\ a, b \in [0, \infty),\ a < b\}$.

1.5 Random Variables and Induced Probability Measures

Consider a probability space $(\Omega, \mathcal{F}, P)$ and suppose that $X : (\Omega, \mathcal{F}) \to (\Sigma, \Upsilon)$ is measurable. Then X induces an image measure $\mu_X = P \circ X^{-1}$ on Υ given by:

$$\mu_X(A) = P \circ X^{-1}(A) = P\left(X^{-1}(A)\right), \quad A \in \Upsilon \tag{1.58}$$

μ_X is indeed a probability measure on Υ which is clear from the following.

i) $\mu_X(\Sigma) = P \circ X^{-1}(\Sigma) = P(\Omega) = 1$ (1.59a)

ii) $\mu_X(A) = P \circ X^{-1}(A) \ge 0,\ \forall A \in \Upsilon$ (1.59b)

iii) If $A_{n \ge 1} \in \Upsilon$ are disjoint,

$$\begin{aligned}\mu_X(\cup_n A_n) &= P \circ X^{-1}(\cup_n A_n) \\ &= P\Big(\sum_n X^{-1}(A_n)\Big) \\ &= \sum_n P(X^{-1}(A_n)) \\ &= \sum_n P \circ X^{-1}(A_n) (\text{since } \left\{X^{-1}(A_n)\right\}_{n \ge 1} \text{ are disjoint in } \Omega)\end{aligned} \tag{1.59c}$$

Thus, if X is a random variable (discrete or continuous and, say, scalar-valued), then $P \circ X^{-1}$ is a probability measure induced on $\mathbb{R}$. The measurability of X makes it possible to assign probabilities to the events in $\mathcal{B}(\mathbb{R})$, once they are assigned to those in $\mathcal{F}$ via P.

Example 1.6

Suppose that, in the dice experiment of Example 1.4, $P(\omega) = \frac{1}{36}$ for each $\omega = (i,j): 1 \leq i,\ j \leq 6$, where the dice are identical and unbiased. Let us find $P \circ X^{-1}(i+j=4)$. This measure is given by $P(\{\omega : X(\omega) = 4\}) = P(\omega \in \{(1,3),$ $(2,2),(3,1)\}) = 1/12$.

1.5.1 σ-algebra generated by a random variable

In general, if X is a measurable function given by $X : (\Omega, \mathcal{F}) \rightarrow (\Sigma, \Upsilon)$, then the σ-algebra generated by X is defined as:

$$\sigma(X) = \sigma(X^{-1}(\Upsilon)) = \{[X \in B], B \in \Upsilon\} \tag{1.60}$$

Specifically, for a random variable X, the σ-algebra generated by X is defined as:

$$\sigma(X) = \sigma(X^{-1}(\mathcal{B}(\mathbb{R}))) = \{X \in B, B \in \mathcal{B}(\mathbb{R})\} \tag{1.61}$$

We shortly find that the concept of independence (see Section 1.9) of random variables indeed stems from the independence of σ-algebras generated by them.

Example 1.7

Consider a measurable space $(\Omega, \mathcal{F})$. An indicator function I_A (see Eq. 1.13) is a measurable function (random variable), if A is a measurable set (equivalently an event), i.e., $A \in \mathcal{F}$.

Suppose that $X = I_A$ for some $A \in \mathcal{F}$. It has the range $\{0,1\}$. In this case, the σ-algebra generated by X is given by:

$$\sigma(X) = \{[X \in B], B \in \mathcal{B}(\mathbb{R})\} = \{\emptyset, \Omega, A, A^c\} \tag{1.62}$$

The four elements in $\sigma(X)$ respectively correspond to the following cases of B: (i) $0 \notin B,\ 1 \notin B$, (ii) $0 \in B,\ 1 \in B$, (iii) $0 \notin B,\ 1 \in B$ and (iv) $0 \in B,\ 1 \notin B$.

1.6 Probability Distribution and Density Function of a Random Variable

The inverse measure $\mu_X = P \circ X^{-1}$ in Eq. (1.58) induced by a random variable (assumed to be scalar-valued for expositional convenience) $X : (\Omega, \mathcal{F}) \rightarrow (\mathbb{R}, \mathcal{B}(\mathbb{R}))$ is known as the distribution of X. In other words, the distribution of X is the probability measure μ_X on $\mathcal{B}(\mathbb{R})$, defined by:

$$\mu_X(A) = P\left(X^{-1}(A)\right),\ A \in \mathcal{B}(\mathbb{R}) \tag{1.63}$$

Since it is obvious that the set $E = \{(-\infty,x] : -\infty \leq x \leq \infty\}$ of semi-closed (left open and right closed) intervals is a π-system, the image measure is uniquely determined by its value on these intervals and is denoted by $P(X \leq x)$ given by:

$$P(X \leq x) = \mu_X((-\infty,x]) = P \circ X^{-1}((-\infty,x]) \tag{1.64}$$

Example 1.8

Let us consider the dice experiment with $\Omega = \{1, 2, 3, 4, 5, 6\}$, $\mathcal{F} = 2^{\Omega}$ and $P(\{\omega\}) = 1/6,\ \omega = 1,2,\ldots,6$. Suppose that the discrete random variable X on $(\mathbb{R}, \mathcal{B}(\mathbb{R}))$ is defined by $X(x) = x$. Since it is defined on a discrete sample space, its range would also be discrete. The set forming the range of the random variable has a finite number of values/elements. The distribution of X is given by:

$$\mu_X = \frac{1}{6}(\delta_1 + \delta_2 + \delta_3 + \delta_4 + \delta_5 + \delta_6) \tag{1.65}$$

Here $\delta_j,\ j = 1,2,\ldots,6$ is the Dirac measure (see Section 1.3 for the definition of a Dirac measure).

1.6.1 Probability distribution function

$P(X \leq x)$ in Eq. (1.64) is a function of x and is commonly referred to as the probability (cumulative) distribution function $F_X(x) : \mathbb{R} \to [0,1]$ of the random variable X. This function is denoted by CDF in the rest of this chapter and of the book. The CDF $F_X(x)$ has the following properties (see Appendix A for proofs):

i) $0 \leq F_X(x) \leq 1,\ \forall x \in \mathbb{R}$

ii) $F_X(x)$ is a non-decreasing function, i.e., $F_X(x) \leq F_X(y)$ whenever $x \leq y$

iii) $\lim_{x \to \infty} F_X(x) = 1$ and $\lim_{x \to -\infty} F_X(x) = 0$

iv) $F_X(x)$ is right continuous, i.e., for each $x \in \mathbb{R}$, $F_X(x) = \lim_{h \downarrow 0} F_X(x+h)$ (1.66)

Because of the property (iv), $F_X(x)$ is continuous except possibly for countably many jumps (see Appendix A for the definition of continuity of functions and the nature of discontinuities) where a jump is a discontinuity of the first kind. Further, since $F_X(x)$ is non-decreasing on $\mathbb{R}$, it follows that the function increases to a limiting value $\sup_{x \uparrow a} F_X(x)$ for any given $a \in \mathbb{R}$ and $F_X(a)$ is an upper bound to this limit. Thus if $F_X(a_-)$ denotes $\sup_{x \uparrow a} F_X(x)$, then $F_X(a_-) \leq F_X(a)$. The jump, if any, of $F_X(x)$ at $x = a$ is the difference, $F_X(a) - F_X(a_-)$. If $F_X(x)$ is continuous at $x = a$, the jump is zero. Otherwise the jump is equal to $P(X = a)$. It is important to note that the set of non-zero jumps of a distribution function $F_X(x)$ is countable (see Appendix A for proof). This is indeed true with any function of bounded variation (Appendix A). Here we note that since $F_X(x)$ is monotonic, it is a function of bounded variation. This is clear from Eq. (A.21) of Appendix A wherein we find that $\sum_{j=1}^{n} \left(F_X(a_j) - F_X(a_{j-1})\right) \leq 1$ and thus is of bounded variation over any partition of any interval $(a,b] \subset \mathbb{R}$, e.g., $a = a_0 < a_1 < \cdots < a_n = b$.

Example 1.9

Refer to Example 1.8, dealing with a discrete random variable. The probabilities (of occurrence) of the events $\{X(\omega) \le x_i\}, i = 1,2,\ldots,6$ are given by the CDF $F_X(x_i)$ in the form:

$$F_X(x_i) = P(-\infty \le X \le x_i) = \sum_{k=1}^{i} P(X = x_k)\mathbb{U}(X - x_k) \tag{1.67}$$

$\mathbb{U}(.)$ denotes a unit step function defined by:

$$\mathbb{U}(X-a) = 1, \quad X \ge a$$

$$= 0, \quad \text{otherwise} \tag{1.68}$$

Figure 1.4a shows the piecewise continuous function $F_X(x)$ in Eq. (1.67).

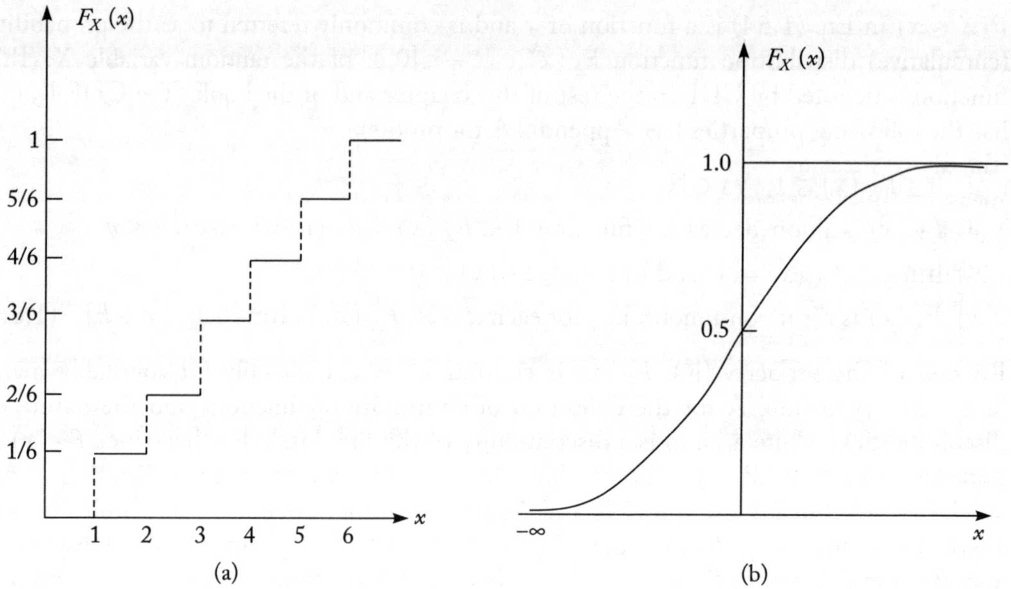

Fig. 1.4 CDFs for (a) a discrete random variable (the case of a piecewise continuous function with finite number of jumps – throw of a dice) and (b) a continuous random variable (the example of a normal or Gaussian probability measure - Example 1.10)

Example 1.10

Consider the probability space $(\Omega, \mathcal{F}, P)$ with $\Omega = \mathbb{R}$, $\mathcal{F} = \mathcal{B}(\mathbb{R})$ and the probability measure P constructed using the CDF, $F_X(x) = \frac{1}{2}(1 + \mathrm{erf}(x))$. Here $\mathrm{erf}(x) := \frac{1}{\sqrt{2\pi}} \int_0^x e^{-\frac{u^2}{2}} du$ is the error function whose values are typically available in standard tables. This CDF corresponds to a continuous random variable popularly referred to as Gaussian or normal and is shown in Fig. 1.4b.

1.6.2 Lebesgue–Stieltjes measure

If $\mathcal{A}$ is an algebra on $\mathbb{R}$, it consists of sets of the form:

$$A = \cup_{i=1}^{n} B_i \tag{1.69}$$

where each B_i is an interval of the types $(a,b), (a,b], [a,b)$ and $[a,b]$. The intervals on $\mathbb{R}$ are expressible in terms of the semi-closed (left open and right closed) intervals and from Eq. (1.64) we have:

$$\mu_X((a,b]) = P(X \le b) - P(X \le a) = F_X(b) - F_X(a) \tag{1.70}$$

Assuming that B_i in Eq. (1.69) are disjoint, we have:

$$\mu_X(\cup_{i=1}^{n} B_i) = \sum_{i=1}^{n} (F_X(b_i) - F_X(a_i)) \tag{1.71}$$

The above measure is known as Stieltjes measure or more generally Lebesgue–Stieltjes measure. The Lebesgue measure discussed earlier in Section 1.3 is a particular case of the Lebesgue–Stieltjes measure with $F_X(x) = x$. With $F_X(x)$ being right-continuous and non-decreasing, the Lebesgue–Stieltjes measure is countably additive. With $\sigma(\mathcal{A}) = \mathcal{B}(\mathbb{R})$, the additivity allows extension of the measure to $\mathcal{B}(\mathbb{R})$.

Now suppose that a function $G(x)$ satisfies the properties in Eq. (1.66). Then there exists a random variable X such that $G(x) = F_X(x)$ uniquely.

Proof: Since $G(x)$ satisfies the properties in Eq. (1.66), there exists a unique probability measure P which is the Lebesgue–Stieltjes measure on $(\mathbb{R}, \mathcal{B}(\mathbb{R}))$ as given in Eq. (1.70) or (1.71). Let us now specify that the random variable X is defined on the measure space $(\mathbb{R}, \mathcal{B}(\mathbb{R}), P)$. Without loss of generality, suppose that $X(x) = x$, $\forall\, x \in \mathbb{R}$ with the result that the inverse measure $\mu_X = P$. Hence we have:

$$F_X(a) = P(X(x) \le a \,:\, a, x \in \mathbb{R}) \tag{1.72}$$

$$= P(x \le a \,:\, a, x \in \mathbb{R})$$

$$= P((-\infty, a])$$

$$= G(a) \tag{1.73}$$

Thus every function satisfying the properties in Eq. (1.66) is the distribution function of a random variable.

1.6.3 Probability density function

Before we define a probability density function of a random variable, we need the notion of absolute continuity of a measure and its relation to that of the distribution function, $F_X(x)$. See Appendix A for the definition of absolute continuity of a function. Analogously, a measure μ_X on $\mathcal{B}(\mathbb{R})$ is said to be absolutely continuous with respect to Lebesgue measure λ (denoted as $\mu_X \ll \lambda$) if, for every $\varepsilon > 0$ there exists $\delta > 0$ such that $\mu_X(A) < \varepsilon$ for all Borel sets A of Lebesgue measure $< \delta$. For such a measure, there exists a Lebesgue integrable (Appendix A) unique (up to λ-null sets) function g which is known as the Radon–Nikodym derivative or density of the absolute continuous measure μ_X such that:

$$\mu_X(A) = \int_A g \, d\lambda, \ \forall A \in \mathcal{B}(\mathbb{R}) \tag{1.74}$$

In this connection, more inquisitive readers may like to go through the following *Radon-Nikodym theorem.*

1.6.4 Radon–Nikodym theorem

If a measure μ_X is absolutely continuous with respect to another measure ν and both measures are σ-finite, then μ_X has a density or Radon–Nikodym derivative with respect to ν which means that there exists a ν- measurable function $g = \frac{d\mu_X}{d\nu}$ such that for any ν-measurable set A, we have:

$$\mu_X(A) = \int_A g \, d\nu \tag{1.75}$$

Now, a finite measure μ_X on $\mathcal{B}(\mathbb{R})$ is related to the CDF $F_X(x)$ via Eq. (1.64) and thus it is absolutely continuous if $F_X(x)$ is absolutely continuous. If the absolute continuity holds, the Radon–Nikodym derivative of μ_X is equal to the derivative of $F_X(x)$ almost everywhere. If we define $f_X(x) = \frac{dF_X(x)}{dx} = F'_X(x)$, it denotes the probability density function of the random variable X. The density function always exists for a continuous random variable $X : \Omega \to \mathbb{R}$, since the fact that $P\left(\left\{\omega \in \Omega \,\middle|\, X(\omega) = x\right\}\right) = 0 \ \forall x \in \mathbb{R}$ renders $F_X(x)$ absolutely continuous. Thus, we have:

$$P(x \le X \le x + dx) = F_X(x + dx) - F_X(x) = \frac{dF_X(x)}{dx} dx = f_X(x) \, dx \tag{1.76}$$

Also Eq. (1.76) implies that:

$$P(a \le X \le b) = F_X(b) - F_X(a) = \int_a^b f_X(x)\,dx,\ F_X(x) = \int_{-\infty}^{x} f_X(u)du \quad (1.77\text{a,b})$$

and

$$\int_{-\infty}^{\infty} f_X(x)dx = F_X(\infty) = P(-\infty \le x \le \infty) = 1 \tag{1.77c}$$

Taking the case of a discrete random variable, for instance, one may still define a (generalized) probability density function $f_X(x)$ by using Dirac delta functions (point masses) $\delta(X-x)$ in the form:

$$f_X(x_i) = \sum_{k=1}^{n} P(X = x_k)\delta(X - x_i) \tag{1.78}$$

such that we have $F_X(x_i) = \int_{-\infty}^{x_i} f_X(u)du$ as defined in Eq. (1.67). This follows from the definition of the Dirac delta function $\delta(x)$ as a generalized function in the sense that $\int_{-\infty}^{\infty} g(x)\,\delta(x-a)\,dx = g(a)$. Henceforth, the acronym *pdf* is used to denote the probability density function of a random variable.

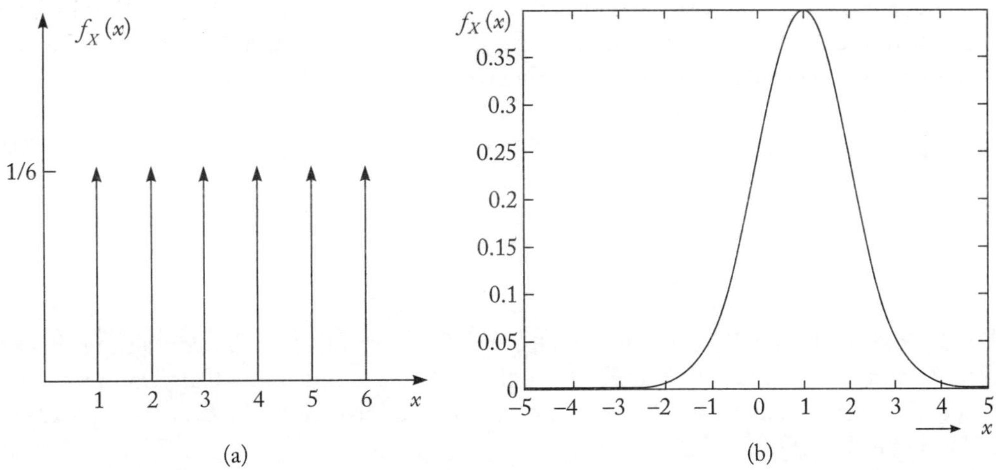

Fig. 1.5 Probability density function (pdf) (a) discrete random variable and (b) continuous random variable (normal or Gaussian random variable - see Example 1.12)

Example 1.11

Consider the discrete random variable in Example 1.8 whose CDF is given in Fig. 1.4a. Equation (1.78) gives the corresponding *pdf* which is shown in Fig. 1.5a. Each term

$\sum_{k=1}^{n} P(X = x_k)\delta(X - x_i)$ in this equation is shown as a vertical arrow in the figure with height equal to $P(X = x_k)$ whenever $i = k$.

Example 1.12

For the normal random variable whose CDF (Fig. 1.4b) is differentiable everywhere on $\mathbb{R}$, the *pdf* $f_X(x)$ is given by $\frac{dF_X(x)}{dx} = \frac{1}{\sqrt{2\pi}}e^{-\frac{x^2}{2}}$. Figure 1.5b shows the graph of the *pdf*.

1.7 Vector-valued Random Variables and Joint Probability Distributions

In forming a probability model for a joint random experiment involving more than one random variable, it is necessary first to introduce the notions of a product space, product σ-algebra and product measure. Specifically, if we consider two probability spaces $(\Omega_1, \mathcal{F}_1, P_1)$ and $(\Omega_2, \mathcal{F}_2, P_2)$, the product sample space Ω is given by the Cartesian product $\Omega_1 \times \Omega_2$:

$$\Omega = \{(\omega_1, \omega_2), \omega_1 \in \Omega_1, \omega_2 \in \Omega_2\} \tag{1.79}$$

Here obviously, $\omega_i, i = 1, 2$ are the individual outcomes of each of the two experiments. The σ-algebra of the joint experiment is the product σ-algebra $\mathcal{F} = \mathcal{F}_1 \times \mathcal{F}_2$ that includes all events of the forms $\{\omega_1 \in A_1\}$ and $\{\omega_2 \in A_2\}$ where $A_1 \in \mathcal{F}_1$ and $A_2 \in \mathcal{F}_2$. For the joint experiment, these events are more explicitly given by:

$$A_1 \times \Omega_2 = \{(\omega_1, \omega_2) : \omega_1 \in A_1, \omega_2 \in \Omega_2\} \tag{1.80a}$$

and

$$\Omega_1 \times A_2 = \{(\omega_1, \omega_2) : \omega_1 \in \Omega_1, \omega_2 \in A_2\} \tag{1.80b}$$

Thus $\mathcal{F}_1 \times \mathcal{F}_2$ is the smallest product σ-algebra of subsets of $\Omega_1 \times \Omega_2$ containing the sets $A_1 \times \Omega_2$ and $\Omega_1 \times A_2$ for all $A_i \in \mathcal{F}_i$. Note that $\mathcal{F}_1 \times \mathcal{F}_2$ is thus not a Cartesian product of $\mathcal{F}_1$ and $\mathcal{F}_2$. Further, since a σ-algebra is closed under intersection, $A_1 \times A_2 = (A_1 \times \Omega_2) \cap (\Omega_1 \times A_2) \in \mathcal{F}_1 \times \mathcal{F}_2$.

The product measure P is defined as $P := P_1 \times P_2$ on $\Omega_1 \times \Omega_2$. The collection of disjoint unions of sets (rectangles) $A_1 \times A_2$ forms an algebra $\mathcal{A}$ (also a π-system) and $\mathcal{F}$ is the σ-algebra generated by $\mathcal{A}$. Therefore, by Caratheodory extension theorem, the product measure P extends uniquely to a countably additive probability measure on the algebra $\mathcal{A}$.

As an illustration, let us consider the two-dimensional plane $\mathbb{R}^2$ which is the Cartesian product of $\mathbb{R}$ with itself. Each measurable space is $(\mathbb{R}, \mathcal{B}(\mathbb{R}), \lambda)$ with λ being the one-dimensional Lebesgue measure. The resulting product σ-algebra is $\mathcal{B}(\mathbb{R}) \times \mathcal{B}(\mathbb{R})$ and the product measure is the two-dimensional Lebesgue measure which we denote here by $\overline{\lambda}$ and is given by:

$$\overline{\lambda}([a,b]\times[c,d]) = \lambda([a,b]).\,\lambda([c,d]) = (b-a).(d-c) \tag{1.81}$$

We notice that the measure $\overline{\lambda}$ corresponds to the usual notion of an area in $\mathbb{R}^2$.

Extension of the above arguments to multi-dimensional measurable spaces is straightforward. If $(\Omega_i, \mathcal{F}_i, P_i)$ are the given probability spaces with $1 \le i \le n$, we construct the unique product probability measure $P = P_1 \times P_2 \times \cdots \times P_n$ on the product space $\Omega = \Omega_i \times \Omega_2 \times \cdots \times \Omega_n$ such that the product σ-algebra $\mathcal{F}$ is generated by measurable sets $A_1 \times A_2 \times \ldots A_n$ with $A_i \in \mathcal{F}_i$.

1.7.1 Joint probability distributions and density functions

With the vector random variable defined in Section 1.7, an n-dimensional random variable $\boldsymbol{X}(\omega \in \Omega) := (X_1(\omega) = x_1, X_2(\omega) = x_2, \ldots.., X_n(\omega) = x_n)$ is a measurable function between the measurable spaces $(\Omega, \mathcal{F})$ and $(\mathbb{R}^n, \mathcal{B}(\mathbb{R}^n))$. Considering such a random variable (often called a joint random variable) and with $\boldsymbol{x} := (x_1, x_2, \ldots., x_m)$, we have the joint CDF, $F_{\boldsymbol{X}}(\boldsymbol{x}) : \mathbb{R}^n \to [0,1]$ defined as:

$$F_{\boldsymbol{X}}(\boldsymbol{x}) = P(X_1 \le x_1, X_2 \le x_2, \ldots., X_n \le x_n) \tag{1.82}$$

If $F_{\boldsymbol{X}}(\boldsymbol{x})$ is sufficiently differentiable, we have the joint *pdf* as $\frac{\partial^n F_{\boldsymbol{X}}(\boldsymbol{x})}{\partial x_1 \partial x_2 \ldots \partial x_n}$. Alternatively, $F_{\boldsymbol{X}}(\boldsymbol{x})$ is given by the n-dimensional integral in terms of the *pdf* (analogous to Eq. 1.77b):

$$F_{\boldsymbol{X}}(\boldsymbol{x}) = \int_{-\infty}^{x_1}\int_{-\infty}^{x_2} \cdots \int_{-\infty}^{x_n} f_{\boldsymbol{X}}(x_1, x_2, \ldots., x_n)\, dx_1 dx_2 \ldots. dx_n \tag{1.83}$$

If $\boldsymbol{X}$ is a discrete (say, with m mass points) random variable, its joint *pdf* can be expressed in terms of the point probabilities as:

$$f_{\boldsymbol{X}}(\boldsymbol{x}_i) = \sum_{k=1}^{m} P(\boldsymbol{X} = \boldsymbol{x}_k)\, \delta(\boldsymbol{X} - \boldsymbol{x}_i), i = 1, 2 \ldots, m \tag{1.84}$$

$F_{\boldsymbol{X}}(\boldsymbol{x}_i)$ is a non-decreasing function and has properties analogous to the distribution function of a scalar-valued random variable (Eq. 1.66), viz., $F_{\boldsymbol{X}}(x_1,\ x_2, \ldots x_{i-1},\ -\infty,\ x_{i+1}, \ldots, x_n) = 0$ for any $i \le n$ and $F_{\boldsymbol{X}}(\infty,\ \infty, \ldots,\ \infty) = 1$.

1.7.2 Marginal probability distributions and density functions

$F_{X_k}(x_k)$, $k = 1, 2 \ldots, n$ are the marginal (i.e., individual) probability distribution functions corresponding to the kth scalar component of $\boldsymbol{X}$. These are projections (restrictions) of $F_{\boldsymbol{X}}(\boldsymbol{x})$ to one dimensional (sample) spaces and are obtained as:

$$F_{X_k}(x_k) = \int_{-\infty}^{\infty}\int_{-\infty}^{\infty} \cdots \int_{-\infty}^{x_k} \cdots. \int_{-\infty}^{\infty} f_{\boldsymbol{X}}(x_1, x_2, \ldots x_k .., x_n)\, dx_1 dx_2 \ldots dx_k .. dx_n, k = 1, 2 \ldots, n$$

$$= F_{\mathbf{X}}(\infty,\infty,\ldots,x_k,\infty,\ldots,\infty) \tag{1.85}$$

The marginal density functions $f_{X_k}(x_k)$, $k = 1,2\ldots,n$ follow from the existence of $\frac{\partial F_{X_k}(x_k)}{\partial x_k}$ and are given by the following $(n-1)$-dimensional integral with respect to dx_i, $i = 1,2,\ldots n$, $i \neq k$:

$$f_{X_k}(x_k) = \int_{-\infty}^{\infty}\int_{-\infty}^{\infty}\cdots\int_{-\infty}^{\infty} f_{\mathbf{X}}(x_1,\ldots x_k..,x_n)\,dx_1\ldots dx_{k-1}dx_{k+1}\ldots..dx_n \tag{1.86}$$

Similar to the definitions in Eqs. (1.85) and (1.86) above, one also obtains l-dimensional marginal joint distribution / density functions from $F_{\mathbf{X}}(\boldsymbol{x})$ by integrating out all scalar variables except the ones of interest.

1.8 Integration of Measurable Functions and Expectation of a Random Variable

If (Σ,Υ,μ) is a measure space and X a Υ-measurable function, we define the integral of X with respect to μ as:

$$\mu(X) = \int_{\Sigma} X d\mu \tag{1.87}$$

We now define a random variable based on a representation through simple functions of the form:

$$Y = \sum_k \alpha_k I_{A_k} \tag{1.88}$$

where $\alpha_k \in \mathbb{R}$, I_{A_k} is an indicator function (Eq. 1.13) and $A_k \in \Upsilon$. Since an indicator function is measurable (see Example 1.7), a simple function which is a linear combination of indicator functions (defined on measurable sets) takes only countably many values and is measurable. For these simple functions the μ- integral is defined as:

$$\mu(Y) = \sum_k \alpha_k \mu(A_k) \tag{1.89}$$

In fact, one can show that X is a random variable if and only if it is a simple function or a limit of simple functions (see Appendix A for proof). Let us now define the μ- integral of a non-negative measurable function Y as:

$$\mu(Y) = \sup\{\mu(Z) : Z \text{ simple}, Z \leq Y\} \tag{1.90}$$

Here note that if we set $X^+(x) := \max(X(x),0)$ and $X^-(x) := \max(-X(x),0)$, then X^+ and X^- are non-negative functions. Also, $X(x) = X^+(x) - X^-(x)$ and

$|X(x)| = X^+(x) + X^-(x)$. If X is a measurable function and $\mu(|X|) < \infty$, then we say X is μ-integrable similar to the definition $\int_\Sigma |X(s)|\, ds < \infty$ for integrability of functions in general. Further, $\mu(X) = \mu(X^+) - \mu(X^-)$.

1.8.1 Integration with respect to product measure and Fubini's theorem

Consider the measurable spaces $(\Omega_1, \mathcal{F}_1, \mu_1)$ and $(\Omega_2, \mathcal{F}_2, \mu_2)$ with the product sample space $\Omega = \{(\omega_1, \omega_2), \omega_1 \in \Omega_1, \omega_2 \in \Omega_2\}$ (see Eq. 1.79), the product σ-algebra $\mathcal{F} = \mathcal{F}_1 \times \mathcal{F}_2$ and μ_i, $i = 1, 2$ being finite measures.

If $X(\omega_1, \omega_2) : \Omega \to \mathbb{R}$ is $\mathcal{F}$-measurable, then with $\mu = \mu_1 \times \mu_2$, we have for all $\omega_2 \in \Omega_2$, the projection or the map $\omega_1 \mapsto X(\omega_1, \omega_2) : \Omega_1 \longrightarrow \mathbb{R}$ is $\mathcal{F}_1$-measurable on Ω_1. Similarly, for all $\omega_1 \in \Omega_1$, the map $\omega_2 \mapsto X(\omega_1, \omega_2) : \Omega_2 \longrightarrow \mathbb{R}$ is $\mathcal{F}_2$-measurable on Ω_2. Hence, analogous to the integral in Eq. (1.87), we define the following integrals (called iterated integrals):

$$\mathcal{J}_{X_1}(\omega_1) = \int_{\Omega_2} X(\omega_1, \omega_2)\, d\mu_2(\omega_2) \tag{1.91a}$$

and

$$\mathcal{J}_{X_2}(\omega_2) = \int_{\Omega_1} X(\omega_1, \omega_2)\, d\mu_1(\omega_1) \tag{1.91b}$$

Now Fubini's theorem states that if X is non-negative, then

$$\begin{aligned} \mu(X) &= \int_{\Omega_1} \mathcal{J}_{X_1}(\omega_1)\, d\mu_1(\omega_1) \\ &= \int_{\Omega_2} \mathcal{J}_{X_2}(\omega_2)\, d\mu_2(\omega_2) \end{aligned} \tag{1.92}$$

Fubini's theorem helps in interchanging the order of integration in multiple integrals whenever the integrand is non-negative and μ-measurable.

Example 1.13

Consider the measurable spaces $(\Omega, \mathcal{F})$ and $(\mathbb{R}^+ = [0, \infty), \mathcal{B}(\mathbb{R}^+))$ with $\mu_1 = P$ and $\mu_2 =$ Lebesgue measure on $\mathbb{R}^+$. Let the random variable $X(\omega)$ be an indicator function I_A where the set $A = \{(\omega, x) : 0 \le x \le X(\omega)\}$. Then the integrals in Eq. (1.91) are:

$$\mathcal{J}_{X_1}(\omega) = X(\omega) \text{ and } \mathcal{J}_{X_2}(x) = P(X \ge x) \tag{1.93}$$

Thus we obtain:

$$\mu(A) = \int_{\Omega} X(\omega)\, dP(\omega) = \int_{\mathbb{R}^+} P(\omega : X(\omega) \geq x)\, dx \tag{1.94}$$

Here dx is the Lebesgue measure. We shortly see in the next Section 1.8.3 that the last integral in Eq. (1.94) is the well-known result for expectation of a non-negative random variable.

Example 1.14

If we consider a Gaussian random variable with its *pdf* given by $f_X(x) = \frac{1}{\sqrt{2\pi}} e^{-\frac{x^2}{2}}$, $x \in \mathbb{R}$, Fubini's theorem helps in getting the established result $\int_{\mathbb{R}} f_X(x) dx = 1$. Let us have $\mathcal{J}_1 = \int_{\mathbb{R}} e^{-\frac{x^2}{2}}\, dx$. Taking the square of $\mathcal{J}_1$ gives:

$$\mathcal{J}_1{}^2 = \int_{\mathbb{R}} e^{-\frac{x^2}{2}}\, dx \,.\, \int_{\mathbb{R}} e^{-\frac{y^2}{2}}\, dy$$

$$= \int_{\mathbb{R}} \int_{\mathbb{R}} e^{-\frac{x^2+y^2}{2}}\, dx\, dy \tag{1.95}$$

Here we identify the two integrals Eq. (1.91) as $\mathcal{J}_{X_1}(\omega_1) = \int_{\mathbb{R}} e^{-\frac{x^2+y^2}{2}} dy$ and $\mathcal{J}_{X_2}(\omega_2) = \int_{\mathbb{R}} e^{-\frac{x^2+y^2}{2}} dx$. The integrands are non-negative and dx and dy are the Lebesgue measures. By Fubini's theorem (see Eq. 1.92), the double integral in Eq. (1.95) can be evaluated as an area integral:

$$\mathcal{J}_1{}^2 = \int_{\mathbb{R}} e^{-\frac{x^2+y^2}{2}}\, d(x,y) \tag{1.96}$$

By change of coordinates from Cartesian to polar, i.e., $x = r\cos\theta$ and $y = r\sin\theta$, we have $d(x,y) = r d\theta dr$ and:

$$\mathcal{J}_1{}^2 = \int_0^{\infty} \int_0^{2\pi} r e^{-\frac{r^2}{2}}\, d\theta dr \tag{1.97}$$

By interchanging the integrals, the above can be easily evaluated and found to be 2π. Hence, $\frac{1}{\sqrt{2\pi}} \int_{\mathbb{R}} e^{-\frac{x^2}{2}}\, dx = 1$ as required.

1.8.2 Monotone convergence theorem

In the context of integration with respect to a measure, the monotone convergence theorem, which is stated below, plays a central role.

If $\{X_n\}$ is a sequence of non-negative measurable functions such that $X_n \uparrow X$, then $\mu(X_n) \uparrow \mu(X) \leq \infty$, or alternatively we state:

$$\int_{\Omega} X_n(\omega)\,\mu(d\omega) \uparrow \int_{\Omega} X(\omega)\,\mu(d\omega) \tag{1.98}$$

The monotone convergence theorem above is closely related to the monotone convergence properties of measures described in Section 1.3 (see Eq. 1.15). For proof of the above theorem and the related results such as the Fatou Lemma and dominated-convergence theorem, see Appendix C.

We note that for all μ-integrable functions, say X, Y, the following results also hold:

i) $\mu(\alpha X + \beta Y) = \alpha\mu(X) + \beta\mu(Y),\ \alpha, \beta \in \mathbb{R}$ (Linearity) (1.99a)

ii) $X \leq Y \Longrightarrow \mu(X) \leq \mu(Y)$ (1.99b)

iii) $X = 0$ a.e. (almost everywhere) $\Longrightarrow \mu(X) = 0$ (1.99c)

1.8.3 Expectation of a random variable

If X is a random variable defined on a probability space $(\Omega, \mathcal{F}, P)$, the integral in Eq. (1.87) is called the expectation of X and is denoted by $E[X]$ which is the Lebesgue integral $\int_{\Omega} X dP = \int_{\Omega} X(\omega) dP(\omega)$ with respect to P. If the inverse measure μ_X of a real (scalar-valued) random variable X has the probability density $f_X(x)$ with respect to the Lebesgue measure (see Eq. 1.74), then we have:

$$E[X] = \int_{\Omega} X dP = \int_{\mathbb{R}} x dF_X(x) = \int_{\mathbb{R}} x f_X(x) dx \tag{1.100}$$

Distribution or density function may fully describe a random variable X and thus provide the probabilities of outcomes / events of the associated experiment. However, we often seek a few (or rather a sequence of) deterministic quantities (called moments) that, consistent with its distribution, also characterize X. These moments are of practical significance in engineering analysis and design. The first (order) moment, denoted by $E[X]$ as given by Eq. (1.100), is known the expectation or mean of X. Further, the expectation represents a random variable in a weighted average sense in that if one can perform the random experiment of sampling X indefinitely many times, then the average of the resulting numbers approaches $E[X]$.

The above definition applies to both discrete and continuous random variables except that, for discrete X, summation replaces the integration as:

$$E[X] = \sum_{k} x_k f_X(x_k) \tag{1.101}$$

Properties of expectation

By properties of Lebesgue integral and a probability measure, expectation of a random variable has the following properties:

i) X is integrable if $E[|X|] < \infty$.

ii) X is integrable if both $E[X^+]$ and $E[X^-]$ are finite. This is same as $E[|X|] = (E[X^+] + E[X^-]) < \infty$

iii) If X and Y are integrable, then $E[\alpha X + \beta Y] = \alpha E[X] + \beta E[Y]$

iv) If X and Y are such that $|X| \le Y$ and Y is integrable, then X is integrable and $E[|X|] \le E[Y]$

v) If $X \ge 0$, then $E[X] = 0$ if and only if $P(X = 0) = 1$

vi) If X is integrable, $\left|E[X]\right| \le E[|X|]$ (see Jensen's inequality in Section 2.4, Chapter 2)

vii) For non-negative random variable $X, E[X] = \int_0^\infty P(X > x)\,dx = \int_0^\infty (1 - F(x))\,dx$

For all non-negative Borel functions $g(x) : \mathbb{R} \to \mathbb{R}$ such that $\int_{\mathbb{R}} \left|g(X)\right| dP \le \infty$, the corresponding Lebesgue integral is given by:

$$E[g(X)] = \int_{\mathbb{R}} g(x) f_X(x) dx \tag{1.102}$$

The above equation is a generalization of the definition of expectation to include a function $g(X)$ of a random variable.

1.8.4 Higher order expectations of a random variable

The mth order moment of X is defined by $E[X^m] = \int_{-\infty}^{\infty} x^m f_X(x)\,dx$ in case of a continuous random variable and by $E[X^m] = \sum_{k=1}^{n} {x_k}^m f_X(x_k)$ in the discrete case. Using the above definition of $E[X]$, one can define the L^p norm $\|\mathbf{X}\|_p$ of a vector-valued random variable $\mathbf{X} : \Omega \to \mathbb{R}^n$ as:

$$\|\mathbf{X}\|_p = \left\{ \int_B |\mathbf{X}|^p dP \right\}^{1/p} \tag{1.103}$$

Here $|\boldsymbol{x}|^p = \sum_{i=1}^{n} |x_i|^p$. Note that for $1 \le p < \infty$, $X \in L^p(\Omega, \mathcal{F}, P)$ if $E[|X|^p] < \infty$. Moreover, if $X,\ Y \in L^2$, then by the monotonicity of norms, we have $X, Y \in L^1$.

The definition in Eq. (1.103) is not a norm in the strictest sense in that $\int_{\mathbb{R}} |\boldsymbol{x}|^p P(dB) = 0$ may have non-zero X as a solution, when the probability of occurrence of such non-zero values of X is zero. Somewhat loosely and in the context of deterministic (Borel measurable) functions, the above norm associates with L^p spaces which are Banach spaces (normed linear spaces). $L^2(\Omega, \mathcal{F}, P)$, in particular, could be loosely thought of as a Hilbert space with inner product $E[X.Y] = \int_{-\infty}^{\infty} xy f_{XY}(x,y) dx dy$ where, $X, Y \in L^2(\Omega, \mathcal{F}, P)$.

While a parallel with Banach spaces of deterministic real analysis could be drawn in this way, notions of such spaces in probability theory are typically not much made use of.

Central moments of a random variable

A random variable may also be characterized by its central moments—the moments centered about its mean. For instance, consider a vector random variable $\mathbf{X}(\omega \in \Omega) := (X_1(\omega) = x_1, X_2(\omega) = x_2, \ldots, X_n(\omega) = x_n)$ with $\boldsymbol{x} := (x_1, x_2, \ldots, x_n)$. Denoting the mean vector as $E[\mathbf{X}] := \boldsymbol{\mu}_{\mathbf{X}}$, the central moments are given by:

$$E\left[(\mathbf{X} - \boldsymbol{\mu}_{\mathbf{X}})^m\right] = \int_{-\infty}^{\infty} (\boldsymbol{x} - \boldsymbol{\mu}_{\mathbf{X}})^m f_{\mathbf{X}}(\boldsymbol{x})\, d\boldsymbol{x}, \text{ for continuous } \mathbf{X}$$

$$= \sum_{k=1}^{n} (\boldsymbol{x}_k - \boldsymbol{\mu}_{\mathbf{X}})^m f_{\mathbf{X}}(\boldsymbol{x}_k), \text{ for discrete } \mathbf{X} \tag{1.104}$$

Here $\mathbf{X}^m$ is defined as $\mathbf{X}^m := (X_1^m, X_2^m, \ldots, X_n^m)$.

Specifically, for $m = 2$, $E\left[\mathbf{X}^2\right]$ is known as the vector of the mean square values and $\sigma_{\mathbf{X}}{}^2 = E\left[(\mathbf{X} - \boldsymbol{\mu}_{\mathbf{X}})^2\right]$, the vector of the variances of $\mathbf{X}$. The variance is a measure of dispersion of the distribution of $\mathbf{X}$ about its mean. Note that $\boldsymbol{\sigma}_{\mathbf{X}}$ is known as the standard deviation of $\mathbf{X}$. By Chebyshev's inequality (Section 2.4, Chapter 2), $\sigma_{\mathbf{X}} = 0$ signifies that $\mathbf{X}$ is deterministic (non-random). Expanding $E\left[(\mathbf{X} - \boldsymbol{\mu}_{\mathbf{X}})^2\right]$, we have $\sigma_{\mathbf{X}}{}^2 = E\left[\mathbf{X}^2\right] - 2\boldsymbol{\mu}_{\mathbf{X}} E[\mathbf{X}] + \boldsymbol{\mu}_{\mathbf{X}}{}^2 = E\left[\mathbf{X}^2\right] - \boldsymbol{\mu}_{\mathbf{X}}{}^2$. Here one may define the covariance matrix $\boldsymbol{C}$ of $\mathbf{X}$ as:

$$\mathbf{C} = \begin{bmatrix} C_{11} \ C_{12} \ldots\ldots C_{1n} \\ C_{21} \ C_{22} \ldots\ldots C_{2n} \\ \ldots\ldots\ldots\ldots\ldots \\ C_{n1} \ C_{n2} \ldots\ldots C_{nn} \end{bmatrix} \tag{1.105}$$

where $C_{jk} = E\left[\left(X_j - \mu_{X_j}\right)\left(X_k - \mu_{X_k}\right)\right]$. If $\boldsymbol{C}$ is diagonal with $C_{jk} = 0,\ j \neq k$, we say that the (components of) vector random variable $\boldsymbol{X}$ is (are) uncorrelated and, in that case, the diagonal elements are merely the mean square values $C_{jj} = E\left[X_j{}^2\right] = \sigma_{X_j}{}^2$.

Example 1.15

Consider the probability space $(\Omega, \mathcal{F}, P)$ and a scalar normal random variable $X : \Omega \to \mathbb{R}$ with a probability distribution given by:

$$f_X(x) = \frac{1}{\sigma\sqrt{2\pi}} \exp\left(-\frac{1}{2}\left(\frac{x-m}{\sigma}\right)^2\right) \tag{1.106}$$

where $\sigma > 0$ and m are constants. Then $E[X] = m$ and variance $E\left[(X-m)^2\right] = \sigma^2$ as shown in the following steps.

The first moment (or mean) of X is given by Eq. (1.100). Therefore we have:

$$E[X] = \int_{\mathbb{R}} x f_X(x)\,dx = \frac{1}{\sigma\sqrt{2\pi}} \int_{\mathbb{R}} x \exp\left(-\frac{1}{2}\left(\frac{x-m}{\sigma}\right)^2\right) dx \tag{1.107}$$

With change of variable $z = \frac{x-m}{\sigma}$, the above integral takes the form:

$$E[X] = \frac{m}{\sqrt{2\pi}} \int_{-\infty}^{\infty} \exp\left(-\frac{1}{2}z^2\right) dz + \frac{1}{\sqrt{2\pi}} \int_{-\infty}^{\infty} z \exp\left(-\frac{1}{2}z^2\right) dz \tag{1.108}$$

The second integral is zero since the integrand is an odd function. From Example 1.14, it is seen that $\frac{1}{\sqrt{2\pi}} \int_{\mathbb{R}} e^{-\frac{z^2}{2}}\,dz = 1$ and hence we have the result $E[X] = m$.

The variance or the second central moment of X is given by:

$$\begin{aligned} E\left[(X-m)^2\right] &= \int_{\mathbb{R}} (x-m)^2 f_X(x)\,dx \\ &= \frac{1}{\sigma\sqrt{2\pi}} \int_{-\infty}^{\infty} (x-m)^2 \exp\left(-\frac{1}{2}\left(\frac{x-m}{\sigma}\right)^2\right) dx \end{aligned} \tag{1.109}$$

With the same change of variable $z = \frac{x-m}{\sigma}$, the last integral in Eq. (1.109) can be simplified as:

$$E\left[(X-m)^2\right] = \frac{\sigma^2}{\sqrt{2\pi}} \int_{-\infty}^{\infty} z^2 \exp\left(-\frac{1}{2}z^2\right) dz \tag{1.110}$$

Integration by parts with $u = z$ and $dv = z\exp\left(-\frac{1}{2}z^2\right) dz$ of the above integral results in:

$$\begin{aligned} E\left[(X-m)^2\right] &= \frac{\sigma^2}{\sqrt{2\pi}} \left[\left(-z\exp\left(-\frac{1}{2}z^2\right)\right)_{-\infty}^{\infty} + \int_{-\infty}^{\infty} \exp\left(-\frac{1}{2}z^2\right) dz \right] \\ &= \sigma^2 \end{aligned} \tag{1.111a}$$

The normal probability density function is a bell-shaped curve with range $\mathbb{R}$ (see Fig. 1.5b). We observe that

$$F_X(m) = P(X \le m) = \frac{1}{\sigma\sqrt{2\pi}} \int_{-\infty}^{m} \exp\left(-\frac{1}{2}\left(\frac{x-m}{\sigma}\right)^2\right) dx = 0.5 \tag{1.111b}$$

and thus the probability density curve is symmetric about m.

1.8.5 Characteristic and moment generating functions

Continuing with the vector random variable $\boldsymbol{X} : \Omega \to \mathbb{R}^{\boldsymbol{n}}$, the characteristic function of $\boldsymbol{X} = (X_1, X_2, \ldots, X_n)$ is defined by:

$$\begin{aligned} &\varnothing_{\mathrm{X}}(\lambda_1, \lambda_2, \ldots, \lambda_n) \\ &= E[\exp i(\boldsymbol{\lambda}, \boldsymbol{X})] = \int_{\mathbb{R}^n} e^{i(\lambda_1 x_1 + \lambda_2 x_2 + \ldots + \lambda_n x_n)} f_{\mathrm{X}}(\boldsymbol{x})\, d\boldsymbol{x} \end{aligned} \tag{1.112}$$

Here $(\boldsymbol{\lambda}, \boldsymbol{X})$ is inner product of $\boldsymbol{X}$ and the vector $\boldsymbol{\lambda} := (\lambda_1, \lambda_2, \ldots, \lambda_n)$. Also, $\boldsymbol{x} := (x_1, x_2, \ldots, x_n)$. The above definition implies that $\varnothing_{\mathrm{X}}(.)$ is the Fourier transform of the *pdf* of $\boldsymbol{X}$. We note that since $f_{\mathrm{X}}(\boldsymbol{x}) \geq 0$, $\varnothing_{\mathrm{X}}(\boldsymbol{\lambda})$ is maximum at $\boldsymbol{\lambda} = \boldsymbol{0}$. Thus with $\varnothing_{\mathrm{X}}(\boldsymbol{0}) = 1$, $\varnothing_{\mathrm{X}}(\boldsymbol{\lambda}) \leq \varnothing_{\mathrm{X}}(\boldsymbol{0})$. Given that Fourier transform is an isomorphism (1-1 mapping) from the space of square integrable functions to itself, it follows that $\varnothing_{\mathrm{X}}(\boldsymbol{\lambda})$ uniquely characterizes the *pdf*, provided that the latter is square integrable.

The characteristic function is real-valued and even, if the *pdf* is symmetric around the origin. For example, for $f_X(x) = \frac{1}{\sigma}\exp\left(-\frac{x^2}{2\sigma^2}\right)$ which is known as the zero mean normal random variable, $\varnothing_X(\lambda) = \exp\left(-\frac{\sigma^2\lambda^2}{2}\right)$. Further, the transformation in Eq. (1.112) is bijective and the bijection is continuous, i.e., if the sequence $F_{X_i}(x_i), i = 1,2,\ldots$ converges to the distribution $F_X(x)$, the corresponding sequence $\varnothing_{X_i}(\lambda_i)$ also converges to $\varnothing_X(\lambda)$, the characteristic function of $F_X(x)$. Convergence of random variables is discussed in Chapter 2 (Section 2.3).

Replacing $i\boldsymbol{\lambda}$ with $\boldsymbol{s} = (s_1, s_2, \ldots, s_n)$ in Eq. (1.112), we obtain the moment generating function $\psi(\boldsymbol{s}) = \int_{\mathbb{R}^n} e^{(\boldsymbol{s},\boldsymbol{x})} f_{\mathrm{X}}(\boldsymbol{x})\, d\boldsymbol{x}$ where $(\boldsymbol{s}, \boldsymbol{x}) = s_1x_1 + s_2x_2 + \ldots + s_nx_n$. Differentiation of $\psi(\boldsymbol{s})$ with respect to $\boldsymbol{s}$ to different orders yields information on the joint moments of the random variable $\boldsymbol{X}$. For example, with $n = 1$, we find that $\psi'(0) = \int_{-\infty}^{\infty} x_1 e^{s_1x_1} f_{X_1}(x_1)\, dx_1 = E[X_1]$ and thus, in general, the mth order moment is given by $E[\boldsymbol{X}^m] = \psi^m(\boldsymbol{0})$ where the index m in $\psi^m(.)$ indicates the mth order derivative of ψ.

1.9 Independence of Random Variables

The notion of independence has a significant role in probability theory and its applications. It is basically related to the idea that one can often repeat a (random) experiment whose outcomes neither influence nor are influenced by the outcomes of other experiments.

Start, as usual, with a probability space $(\Omega, \mathcal{F}, P)$.

1.9.1 Independence of events

The events A_i, $i = 1, 2, \ldots, n$ are independent if and only if

$$P\left(\cap_j A_j\right) = \prod_j P\left(A_j\right) \tag{1.113}$$

Example 1.16

Let $\Omega = [0,1], \mathcal{F} = \mathcal{B}([0,1])$ and $P =$ Lebesgue measure. If $A = [0,1/2]$ and $B = [1/4,3/4]$, we find that the intervals A and B are independent since $P(A \cap B) = 1/4 = P(A)P(B)$.

Here, note that if the events $A_i, i = 1,2,\ldots,n$, are independent, then $A_i{}^c, i = 1,2,\ldots,n$ and the indicator functions I_{A_i}, $i = 1,2,\ldots,n$ are also independent.

1.9.2 Independence of classes of events

Let us have classes of events C_i, $i = 1,2,..,m \subset \mathcal{B}$ (the σ-algebra). Consider the events $A_j, j = 1,2,\ldots,n$ with $A_j \in C_i$ for fixed i. The classes C_i (for different i's) are independent if A_j are independent.

1.9.3 Independence of σ-algebras

The σ-algebras $\mathcal{A}_{i \in I} \subseteq \mathcal{F}$ are independent if the elements of the set $\{A_{i \in I}\}$ are independent whenever $A_i \in \mathcal{A}_i$ for all i. To establish the independence of σ-algebras, it is useful to note the following theorem.

If $\mathfrak{I}_1$ and $\mathfrak{I}_2$ are π-systems contained in $\mathcal{F}$ and $P(A_1 \cap A_2) = P(A_1)P(A_2)$ for $A_1 \in \mathfrak{I}_1$ and $A_2 \in \mathfrak{I}_2$, then the σ-algebras $\sigma(\mathfrak{I}_1)$ and $\sigma(\mathfrak{I}_2)$ are independent.

Proof: Let us fix $A_1 \in \mathfrak{I}_1$ and define for $A \in \mathcal{F}$:

$$\mu_1(A) = P(A_1 \cap A) \text{ and } \mu_2(A) = P(A_1)P(A) \tag{1.114}$$

$\mu_1$and μ_2 are measures which are equal on the π-system $\mathfrak{I}_2$ with $\mu_1(\Omega) = \mu_2(\Omega) = P(A_1) < \infty$. By uniqueness of extension, for all $A_2 \in \sigma(\mathfrak{I}_2) \subseteq \mathcal{F}$, $\mu_1(A_2) = \mu_2(A_2)$. Therefore we have:

$$P(A_1 \cap A_2) = \mu_1(A_2) = \mu_2(A_2) = P(A_1)P(A_2), \forall A_2 \in \mathfrak{I}_2 \tag{1.115}$$

Similarly, if we fix $A_2 \in \mathfrak{I}_2$ and proceed with the same arguments with:

$$\mu_1'(A) = P(A_2 \cap A) \text{ and } \mu_2'(A) = P(A_2)P(A), \tag{1.116}$$

We can show that for all $A_1 \in \sigma(\mathfrak{I}_1) \subseteq \mathcal{F}, P(A_1 \cap A_2) = P(A_1)P(A_2)$. Hence the result follows.

1.9.4 Independence of random variables

A countable set of random variables $\{X_n\}_{n \in \mathbb{N}}$ is said to be independent if the σ-algebras generated by these random variables are independent. As an illustration, let us consider the following case of indicator functions which are simple functions and themselves random variables.

Example 1.17

Suppose that A_i, $i = 1, 2, \ldots, n$ are independent. Let I_{A_i}, $i = 1, 2, \ldots, n$ be indicator functions on A_i. For any A_i, we have the corresponding σ-algebra, $\sigma\left(I_{A_i}\right) = \left\{\emptyset, \Omega, A_i, A_i^c\right\}$. Since A_i's are independent, these σ-algebras are independent too and thus I_{A_i} are independent.

1.9.5 Independence in terms of CDFs

A countable set of random variables $\{X_n\}_{n \in \mathbb{N}}$ is said to be independent if and only if:

$$F_{X_1, X_2, \ldots, X_n}(x_1, x_2, \ldots, x_n) = F_{X_1}(x_1) F_{X_2}(x_2) \ldots F_{X_n}(x_n), \forall x_1, x_2, \ldots, x_n \in \mathbb{R} \quad (1.117a)$$

It follows from the above equation that:

$$f_{X_1, X_2, \ldots, X_n}(x_1, x_2, \ldots, x_n) = f_{X_1}(x_1) f_{X_2}(x_2) \ldots f_{X_n}(x_n), \forall x_1, x_2, \ldots, x_n \in \mathbb{R} \quad (1.117b)$$

Proof: Since the independence of random variables $\{X_n\}_{n \in \mathbb{N}}$ requires independence of corresponding σ-algebras, it is equivalent to the condition that for any collection of Borel sets $E_1, E_2, \ldots, E_n$:

$$P(X_1 \in E_1, X_2 \in E_2, \ldots, X_n \in E_n) = P(X_1 \in E_1) . P(X_1 \in E_1) \ldots P(X_1 \in E_1) \quad (1.118a)$$

The above condition leads to:

$$P\left(X_1^{-1}(E_1) \cap X_2^{-1}(E_2) \cap \ldots X_n^{-1}(E_n)\right) = \prod_i P\left(X_i^{-1}(E_i)\right) \quad (1.118b)$$

By setting in the above equation $E_1 = (-\infty, x_1]$, $E_2 = (-\infty, x_2]$ and $E_n = (-\infty, x_n]$ for any $x_1, x_2, \ldots, x_n \in \mathbb{R}$, we have the result.

It follows from Eq. (1.117b), the characteristic function of independent random variables is of the following form:

$$\emptyset_X(\lambda_1, \lambda_2, \ldots, \lambda_n)$$

$$= \int_{\mathbb{R}} e^{i(\lambda_1 x_1)} f_{X_1}(x_1) dx_1 \int_{\mathbb{R}} e^{i(\lambda_2 x_2)} f_{X_2}(x_2) dx_2 \ldots \int_{\mathbb{R}} e^{i(\lambda_n x_n)} f_{X_n}(x_n) dx_n$$

$$\Longrightarrow \emptyset_X(\lambda_1, \lambda_2, \ldots, \lambda_n) = \emptyset_X(\lambda_1) \emptyset_X(\lambda_2) \ldots \emptyset_X(\lambda_n) \quad (1.119)$$

Independence of discrete random variables

Specifically, a collection of discrete real-valued random variables $X_1, ..., X_n$ are independent if and only if:

$$P(X_1 = x_1, X_2 = x_2, \ldots, X_n = x_n) = \prod_{i=1}^{n} P(X_i = x_i) \tag{1.120}$$

1.9.6 Independence of functions of random variables

Suppose that $X_1, X_2 \ldots, X_n$ are independent real-valued random variables and that $g_1, g_2, \ldots, g_n$ are Borel functions. Then the random variables $Y_1 = g_1(X_1), Y_2 = g_2(X_2), \ldots, Y_n = g_n(X_n)$ are independent.

Proof: For any collection of Borel sets $E_1, E_2, \ldots, E_n$ in $\mathbb{R}$, we have:

$$P(Y_1 \in E_1, Y_2 \in E_2, \ldots, Y_n \in E_n)$$

$$= P(g_1(X_1) \in E_1, g_2(X_2) \in E_2, \ldots, g_n(X_n) \in E_n)$$

$$= P\left(X_1 \in g_1^{-1}(E_1), X_2 \in g_2^{-1}(E_2), \ldots, X_n \in g_n^{-1}(E_n)\right)$$

$$= P(Y_1 \in E_1) P(Y_2 \in E_2) \ldots P(Y_n \in E_n) \tag{1.121}$$

Hence the result.

1.9.7 Independence and expectation of random variables

Let X and Y be two independent random variables and assume that they are mean-square bounded, i.e., $X, Y \in L^2$. Then $XY \in L^1$ and $E[XY] = E[X]E[Y]$.

A simple elucidation of the above assertion is through the following arguments. For simplicity, let $|X| < M$ and $|Y| < N$ with $M, N \in \mathbb{R}$. We approximate both X and Y in terms of simple functions $\mathcal{V}$ and $\mathcal{W}$ respectively of the forms:

$$\mathcal{V}(\omega) = \sum_{i=1}^{m} \alpha_i I_{A_i}(\omega) \text{ and } \mathcal{W}(\omega) = \sum_{j=1}^{n} \beta_j I_{B_j}(\omega) \tag{1.122}$$

Here the events $A_i = X^{-1}([\alpha_i, \alpha_{i+1}))$ and $B_j = Y^{-1}([\beta_j, \beta_{j+1}))$ with $-M = \alpha_0 < \alpha_1 < \cdots < \alpha_m = M$ and $-N = \beta_0 < \beta_1 < \cdots < \beta_n = N$. Now we can express $E[X]$, $E[Y]$ and $E[XY]$ as:

$$E[X] \approx E[\mathcal{V}] = \sum_{i} \alpha_i P(A_i), \tag{1.123a}$$

$$E[Y] \approx E[W] = \sum_j \beta_j P(B_j) \tag{1.123b}$$

$$\begin{aligned} E[XY] \approx E[VW] &= \sum_i \sum_j \alpha_i \beta_j P\left(A_i \cap B_j\right) \\ &= \sum_i \sum_j \alpha_i \beta_j P(A_i) P(B_j) \\ &= \sum_i \alpha_i P(A_i) \sum_j \beta_j P(B_j) \cong E[X] E[Y] \end{aligned}$$

$$\Longrightarrow E[XY] = E[X] E[Y] \tag{1.123c}$$

The above equation also implies that if two random variables are independent, they are uncorrelated (zero covariance). That is, from Eq. (1.117b), we can also write $E[XY] = \int_{-\infty}^{\infty} \int_{-\infty}^{\infty} xy f_{XY}(x,y)\, dx dy = \int_{-\infty}^{\infty} x f_X(x)\, dx \int_{-\infty}^{\infty} y f_Y(y)\, dy = E[X] E[Y]$. Thus, we have the covariance of X and Y given by:

$$E[(X - E[X])(Y - E[Y])] = E[XY] - E[X] E[Y] = 0 \tag{1.123d}$$

However, if two random variables are uncorrelated, they are not necessarily independent (left as an exercise to the reader), unless they are Gaussian.

1.9.8 Additional remarks on independence of random variables

In all the cases above, an infinite number of events or σ-algebras or random variables are said to be independent if all finite collections of them are independent.

i) Even if, for every $A_i \in S_i, i = 1, 2, \ldots, n$, with $S_i \subset \Omega$, we have $P(A_1 \cap A_2 \cap \cdots \cap A_n) = P(A_1) P(A_2) \ldots P(A_n)$, the σ-algebras $\Sigma(S_i)$ need not be independent unless S_i are π-systems.
ii) Checking pair-wise independence is not enough to verify independence. For example, suppose that three events A_1, A_2 and A_3 are independent with $P(A_i = -1) = P(A_i = +1) = 1/2$. If $B_1 = A_1 \cap A_2, B_2 = A_2 \cap A_3$ and $B_3 = A_3 \cap A_1$, B_i are not independent but only pair-wise independent.
iii) If $X, Y \in L^2$ are two independent random variables, $XY \in L^1$. This may not be necessarily true if X and Y are not independent.

1.10 Some oft-used Probability Distributions

Several probability distribution and density functions, associated with a random variable X, are popularly used (probably more often than the others) in the literature. Choice of a distribution may however be facilitated through an understanding of the nature of uncertainties in the mathematical model itself or some physical phenomena associated with the model (some of which may not have been incorporated within the model and, therefore, may qualify as additional forms of uncertainty). Some of these distributions, such as binomial, Poisson (which are discrete), uniform, normal, and Rayleigh (which are continuous) are described in the following sections.

1.10.1 Binomial distribution

For ease of exposition, consider X to be a scalar-valued random variable. Binomial distribution is related to a random experiment consisting of a fixed n number of Bernoulli trials where the random variable X (or X_n, to be more specific) is the number of successes in these independent n trials. Introduce a parameter θ (a constant) representing the probability of a success in each Bernoulli trial. Here the sample space is the set $\{1,2,...,n\} \subset \mathbb{N}$ and the Borel σ-algebra $\mathcal{B}$ contains 2^n subsets (power set) of the sample space. Since $(1-\theta)$ denotes the probability of failure, the probability $P(X=r)$ is given by $\binom{n}{r}\,\theta^r\,(1-\theta)^{n-r}$ which is incidentally the $r+1^{\text{th}}$ term in the binomial expansion of $\{\theta+(1-\theta)\}^n$, which evaluates to unity. Thus we have $\sum_{r=0}^{n}\binom{n}{r}\,\theta^r\,(1-\theta)^{n-r}=1$. The *pdf* of the binomial distribution symbolically expressed as $b\,(r;n,\theta)$ can be written in the form:

$$f_X\,(r)=\sum_{k=0}^{n}\binom{n}{k}\,\theta^k\,(1-\theta)^{n-k}\delta(r-k) \tag{1.124}$$

It follows from the above equation that the CDF is given by $F_X\,(r)=\sum_{k=0}^{r}P\,(X=k)\,U(X-k)$. The mean of X is thus given by $E\,[X]=\mu_X=\sum_{k=0}^{n}k\binom{n}{k}\theta^k\,(1-\theta)^{n-k}=n\theta$ and the variance by $\sigma_X{}^2=E\left[X^2\right]-\mu_X{}^2=n\theta(1-\theta)$ (see Exercise 1.4).

1.10.2 Poisson distribution

Computation of probabilities from binomial distribution as $n\to\infty$ and $\theta\to 0$ is cumbersome. In this limiting case, together with the assumption of $n\theta$ $(=\lambda$, say) remaining finite, it can be shown that [Feller 1968]:

$$P\,(X=k)=\binom{n}{k}\,\theta^k\,(1-\theta)^{n-k}\approx\frac{e^{-\lambda}\,\lambda^k}{k!} \tag{1.125}$$

The above limiting approximation to $b(r;n,\theta)$ is known as Poisson distribution represented by a single parameter, λ. Poisson distribution finds wide application as a useful model for random experiments with rare events (θ being small). As an example, let us consider the case of a possible structural failure when the response (such as displacement or stress) histories (similar to those in Fig. 1.1) exceed a critical level, c. Here we take the number of critical level crossings in a time interval $(0,T)$ as the random variable, X_T. If the interval is divided into n equal parts of Δt such that, $n = \frac{T}{\Delta t}$, X_T follows a binomial distribution. Let us now assume that (i) the probability, θ of crossing the level c exactly once in a sufficiently small Δt is proportional to Δt, (ii) the level crossing events in any two sub-intervals are independent of each other and (iii) probability of critical level crossing more than once in Δt is negligibly small. In the limiting case of large n (i.e., large T) and small θ (e.g., if the level c is high), the above assumptions lead to a Poisson distribution (Eq. 1.125). As such, the probability of no failure within $(0, T)$ is equivalent to knowing the probability of the number of critical level crossings being zero. We can compute this probability as $P(X_T = 0) = e^{-\lambda}$, provided, of course, $\lambda\ (= n\theta)$ is known. Note that, for Poisson distribution, $E(X_T) = \sum_{k=0}^{n} k\frac{e^{-\lambda}\lambda^k}{k!} = \lambda$ and ${\sigma_X}^2 = \lambda$.

1.10.3 Normal distribution

A scalar-valued random variable $X : \Omega \to \mathbb{R}$ is normal (or Gaussian) if the *pdf* of X is given by Eq. (1.106). The parameters m and σ in the equation are indeed the mean $E[X]$ and standard deviation $\sqrt{E\left[(X - E[X])^2\right]}$ of X with the density being symmetric about m. The normal distribution is thus completely represented by the two parameters $m \in \mathbb{R}$ and $\sigma \in [0, \infty)$ and the normal random variable is usually written as $\mathcal{N}(m, \sigma)$. The distribution function $F_X(x)$ is given by:

$$F_X(x) = P(X \le x) = \frac{1}{\sigma\sqrt{2\pi}} \int_{-\infty}^{x} e^{-\frac{1}{2}\left(\frac{u-m}{\sigma}\right)^2} du \tag{1.126}$$

$F_X(x)$ needs to be evaluated numerically as no anti-derivative of $f_X(x)$ in Eq. (1.126) is analytically known.

A normal random variable remains normal under linear transformation (shown later). With X being a normal random variable and given $Y = aX + b\ (a, b \in \mathbb{R})$, Y is also normal with mean m_Y and variance ${\sigma_Y}^2$ given by:

$$m_Y = E[Y] = aE[X] + b = am + b \text{ and } {\sigma_Y}^2 = E[Y - m_Y]^2 = a^2\sigma^2 \tag{1.127}$$

Specifically, with change of variable, $Z = \frac{X-m}{\sigma}$, we have:

$$E[Z] = E\left[\frac{X - m}{\sigma}\right] = \frac{1}{\sigma}(E[X] - m) = 0, \tag{1.128a}$$

$$E[(Z - E[Z])^2] = E\left[Z^2\right] = E\left[\left(\frac{X - m}{\sigma}\right)^2\right] = \frac{1}{\sigma^2}E\left[(X - m)^2\right] = 1 \tag{1.128b}$$

Z is commonly referred to as the standard normal variable $\mathcal{N}(0, 1)$ with zero mean and unit variance. From Eq. (1.126), we get the distribution function $F_Z(z) = \frac{1}{\sqrt{2\pi}} \int_{-\infty}^{z} e^{-\frac{1}{2}u^2} du$.

Since $\frac{1}{\sqrt{2\pi}}\int_{-\infty}^{\infty}e^{-\frac{1}{2}z^2}dz = 1$ (as shown in Example 1.14), $F_Z(z)$ can also be written as $F_Z(z) = \frac{1}{\sqrt{2\pi}}\int_{-\infty}^{0}e^{-\frac{1}{2}z^2}dz + \frac{1}{\sqrt{2\pi}}\int_{0}^{z}e^{-\frac{1}{2}u^2}du = \frac{1}{2} + \text{erf}(z)$. Obviously, the *pdf* $f_Z(z)$ of the standard normal random variable is $\frac{1}{\sqrt{2\pi}}e^{-\frac{z^2}{2}}$. Further,

$$
\begin{aligned}
F_Z(z) &= P(Z \le z) \\
&= P\left(\frac{X-m}{\sigma} \le z\right) \\
&= P(X \le m + \sigma z) \\
&= F_X(m + \sigma z)
\end{aligned}
\tag{1.129}
$$

Alternatively, for $\mathcal{N}(m,\sigma)$, $F_X(x) = F_Z\left(\frac{x-m}{\sigma}\right) = \frac{1}{2} + \text{erf}\left(\frac{x-m}{\sigma}\right)$. Note that the notations $\Phi_Z(z)$ and $\Phi_Z(z)$ are more commonly used to respectively denote the distribution and density functions of the standard normal random variable, Z.

Example 1.18

The characteristic function for $X \sim \mathcal{N}(m,\sigma)$ is given by $\exp(im\lambda - \frac{\sigma\lambda^2}{2})$ which is derived as follows.

Let us first find the characteristic function of $Z \sim \mathcal{N}(0,1)$. Here we have:

$$
\begin{aligned}
\emptyset_Z(\lambda) &= E[\exp(i\lambda Z)] \\
&= \int_{\mathbb{R}} e^{i\lambda z} f_Z(z)\,dz \\
&= \frac{1}{\sqrt{2\pi}}\int_{-\infty}^{\infty} e^{i\lambda z}e^{-\frac{z^2}{2}}dz \\
&= \frac{1}{\sqrt{2\pi}}\int_{-\infty}^{\infty} e^{-\left(\frac{z^2-2i\lambda z}{2}\right)}dz \\
&= \frac{1}{\sqrt{2\pi}}e^{-\frac{\lambda^2}{2}}\int_{-\infty}^{\infty} e^{-\frac{(z-i\lambda)^2}{2}}dz \\
&= e^{-\frac{\lambda^2}{2}}
\end{aligned}
\tag{1.130}
$$

Now, to get the characteristic function $\emptyset_X(\lambda) = E(\exp((i\lambda X))$ of a normal random variable $\mathcal{N}(m,\sigma)$ with non-zero mean, we utilize the transformation $X = m+\sigma Z$ and obtain:

$$\emptyset_X(\lambda) = E\left[\exp\left(i\lambda\left(m+\sigma Z\right)\right)\right]$$

$$= e^{i\lambda m}\, E\left[i\sigma\lambda Z\right] = e^{i\lambda m}\, \emptyset_Z(\sigma\lambda) = e^{i\lambda m}\, e^{-\frac{\sigma^2\lambda^2}{2}}$$

$$\Longrightarrow \emptyset_X(\lambda) = \exp\left(im\lambda - \frac{\sigma^2\lambda^2}{2}\right) \tag{1.131}$$

Characteristic function of a vector normal random variable

Similar to the above case of a single normal random variable, we can derive the characteristic function of the vector normal random variable $\boldsymbol{X} = (X_1, X_2, \ldots..X_n)$ in the form:

$$\emptyset_X(\boldsymbol{\lambda}) = E\left[\exp\left(i\boldsymbol{\lambda}\boldsymbol{X}^T\right)\right] = \exp\left(-\frac{1}{2}\boldsymbol{\lambda}\boldsymbol{C}\boldsymbol{\lambda}^T + i\boldsymbol{\lambda}\boldsymbol{m}^T\right) \tag{1.132}$$

where, $\boldsymbol{\lambda} = (\lambda_1, \lambda_2, \ldots, \lambda_n)$. In terms of $\boldsymbol{m} = (m_1, m_2, \ldots m_n)$ and the symmetric positive semi-definite covariance matrix, $\boldsymbol{C}$, the joint density function of $\boldsymbol{X}$ follows from Fourier inversion and is given by:

$$f_X(\boldsymbol{x}) = \frac{1}{(2\pi)^{n/2}\sqrt{\Delta}} \exp\left\{-\frac{1}{2}(\boldsymbol{X}-\boldsymbol{m})\,\boldsymbol{C}^{-1}(\boldsymbol{X}-\boldsymbol{m})^T\right\} \tag{1.133}$$

where, Δ stands for the determinant of the matrix $\boldsymbol{C}$.

The example below shows an application of characteristic function in deriving a more general result concerning joint normal random variables.

Example 1.19

The vector random variable $\boldsymbol{X} = (X_1, X_2, \ldots..X_n)$ is jointly normal if only if $Y = a_1X_1 + a_2X_2 + \cdots + a_nX_n$ is normal for any $a_j \in \mathbb{R},\ j = 1,2,\ldots,n$.

First assume that $\boldsymbol{X}$ is (jointly) normal. With $\boldsymbol{A} = (a_1, a_2, \ldots a_n)$, use the notation $\boldsymbol{A}\,\boldsymbol{X}^T = \sum_{j=1}^{n} a_jX_j$. The characteristic function of Y is:

$$\emptyset_Y\left(y = \boldsymbol{A}\,\boldsymbol{X}^T\right)$$

$$= E\left[\exp\left(i\lambda\boldsymbol{A}\boldsymbol{X}^T\right)\right] = \exp\left(-\frac{1}{2}\sum_{j,k}^{n}\lambda^2 a_j a_k \sigma_{jk}^2 + i\sum_{j}^{n}\lambda a_j m_j\right)$$

$$= \exp\left(-\frac{1}{2}\lambda^2 \sum_{j,k}^{n} a_j a_k \sigma_{jk}^2 + i\lambda \sum_{j}^{n} a_j m_j\right)$$

$$= \exp\left(-\frac{1}{2}\lambda^2 \boldsymbol{A}\boldsymbol{C}\boldsymbol{A}^T + i\lambda \boldsymbol{A}\boldsymbol{m}^T\right) \tag{1.134}$$

Here $m_j = E(X_j)$ and $\sigma_{jk}^2 = C_{jk} = E\left[\left(X_j - m_j\right)(X_k - m_k)\right]$, the elements of the covariance matrix $\boldsymbol{C}$. Eq. (1.134) shows that Y is normal with $E[Y] = m_Y = \boldsymbol{A}\boldsymbol{m}^T = \sum_j^n a_j m_j$ and $\sigma_Y^2 = \boldsymbol{A}\boldsymbol{C}\boldsymbol{A}^T = \sum_{j,k}^n a_j a_k C_{jk}$.

Conversely, suppose that $Y = \sum_{i=1}^n a_i X_i$ is normal with mean, m_Y and variance, σ_Y^2. By the definition of a characteristic function of an $\mathcal{N}(m,\sigma)$ random variable, we have:

$$E[\exp i\lambda Y] = E[\exp(i\lambda(a_1X_1 + a_2X_2 + \ldots.. + a_nX_n))]$$

$$= \exp\left(-\frac{1}{2}\lambda^2\sigma_Y^2 + i\lambda m_Y\right) \tag{1.135}$$

Here:

$$m_Y = \sum_j^n a_j E\left[X_j\right] \text{ and}$$

$$\sigma_Y^2 = E\left[(Y - m_Y)^2\right] = E\left[\left(\sum_{j=1}^n a_j X_j - \sum_j^n a_j E\left[X_j\right]\right)^2\right]$$

$$= E\left[\left(\sum_{j=1}^n a_j\left(X_j - m_j\right)\right)^2\right]$$

$$= \sum_{j,k}^n a_j a_k\, E\left[\left(X_j - m_j\right)(X_k - m_k)\right] \tag{1.136}$$

On substitution of m_Y and σ_Y^2 in Eq. (1.135) we find that the form of the characteristic function is consistent with the one given in Eq. (1.132) of a vector random variable. Hence $\boldsymbol{X}$ is normal.

Following Eq. (1.133), we have with $n = 2$ the *pdf* of a 2-dimensional joint (and correlated) normal random variable as:

$$f_{X_1,X_2}(x_1,x_2) = \frac{1}{2\pi\sigma_1\sigma_2\sqrt{(1-\rho^2)}}\exp\left[-\frac{1}{2(1-\rho^2)}\left\{\left(\frac{x_1-m_1}{\sigma_1}\right)^2 -2\rho\left(\frac{x_1-m_1}{\sigma_1}\right)\left(\frac{x_2-m_2}{\sigma_2}\right)+\left(\frac{x_2-m_2}{\sigma_2}\right)^2\right\}\right] \quad (1.137)$$

Incidentally, the marginal densities of the random variables X_1 and X_2 are given by (Appendix A):

$$f_{X_1}(x_1) = \frac{1}{\sigma_1\sqrt{2\pi}}\exp\left\{-\frac{1}{2}\left(\frac{x_1-m_1}{\sigma_1}\right)^2\right\} \quad (1.138a)$$

and

$$f_{X_2}(x_2) = \frac{1}{\sigma_2\sqrt{2\pi}}\exp\left\{-\frac{1}{2}\left(\frac{x_2-m_2}{\sigma_2}\right)^2\right\} \quad (1.138b)$$

The above Eqs. (1.138) indicate that m_1 and m_2 are the expectations of the random variables X_1 and X_2 with σ_1^2 and σ_2^2 being their respective variances. It is shown in Appendix A that ρ in Eq. (1.137) stands for $\frac{\sigma_{12}}{\sigma_1\sigma_2}$ where, $\sigma_{12} = E[(X_1-m_1)(X_2-m_2)]$, the cross-covariance between X_1 and X_2. ρ is known as the correlation coefficient and, as its name indicates, it is a quantitative measure of correlation between the two random variables. Note that $|\rho| \leq 1$ which signifies that $\sigma_{12} \leq \sigma_1\sigma_2$. This is clarified in the following steps.

$$E\left[\{a(X-m_1)+(Y-m_2)\}^2\right] = a^2\sigma_1^2 + 2a\sigma_{12} + \sigma_2^2 \quad (1.139)$$

The right hand side expression in the above equation is a positive quadratic in a, its discriminant $\sigma_{12}^2 - \sigma_1^2\sigma_2^2 \leq 0$ and hence $\left|\frac{\sigma_{12}}{\sigma_1\sigma_2}\right| \leq 1$.

Normal random variables if independent are uncorrelated too. Whilst this is true for all random variables, the converse is also true for normal random variable which may not be the case with others. For example, if $\rho = 0$ in Eq. (1.137), we find that:

$$f_{X_1,X_2}(x_1,x_2) = \frac{1}{2\pi\sigma_1\sigma_2}\exp\left[-\frac{1}{2}\left\{\left(\frac{x_1-m_1}{\sigma_1}\right)^2+\left(\frac{x_2-m_2}{\sigma_2}\right)^2\right\}\right]$$

$$\Longrightarrow f_{X_1,X_2}(x_1,x_2) = \frac{1}{\sqrt{2\pi}\sigma_1}e^{-\frac{1}{2}\left(\frac{x_1-m_1}{\sigma_1}\right)^2}\frac{1}{\sqrt{2\pi}\sigma_2}e^{-\frac{1}{2}\left(\frac{x_2-m_2}{\sigma_2}\right)^2} \quad (1.140)$$

The above equation shows that the normal random variables X_1 and X_2 are independent (See Eq. 1.117b).

Example 1.20

Consider $\{Z_i\}$, $i = 1,2,\ldots,n$, a set of n independent standard normal random variables with the joint *pdf* as:

$$f_{Z_1,Z_2\ldots Z_n}(z_1,z_2,\ldots,z_n) = \frac{1}{(2\pi)^{\frac{n}{2}}}\exp\left[-\frac{1}{2}\left\{{z_1}^2 + {z_2}^2 + \cdots + z_n^2\right\}\right] \tag{1.141}$$

We find the distribution of random variable, $W = {Z_1}^2 + {Z_2}^2 + \cdots + Z_n^2$.

First, if $W_1 = {Z_1}^2$, it follows from transformation of random variables (Section 1.11) that $f_{W_1}(w_1) = \frac{1}{\sqrt{2\pi}} w_1^{-\frac{1}{2}} \exp(-\frac{w_1}{2})$, for $w_1 \geq 0$ and zero otherwise. Since Z_i, $i = 1,2,\ldots,$ n are independent, ${Z_1}^2$, $i = 1,2,\ldots,n$ are also independent. Therefore, with the help of Dirac delta function one has the result [Ross 2002]:

$$f_W(w) = \int_{\mathbb{R}^n} \delta\left(w - \left({z_1}^2 + {z_2}^2 + \cdots + z_n^2\right)\right) f_{Z_1,Z_2\ldots Z_n}(z_1,z_2,\ldots,z_n)\, dz_1 dz_2 \ldots dz_n$$

$$= \frac{1}{2^{\frac{n}{2}}} \frac{w^{\frac{n}{2}-1}}{\Gamma(\frac{n}{2})} \exp(-\frac{w}{2}),\ w \geq 0$$

$$= 0, \quad \text{otherwise} \tag{1.142}$$

where, $\Gamma(x)$ is the Gamma function with $\Gamma(x+1) = x\Gamma(x)$ and $\Gamma\left(\frac{1}{2}\right) = \sqrt{\pi}$.

W in Eq. (1.142) is known as the chi-square distribution with n degrees of freedom and familiarly denoted as $\chi^2(n)$. Further $\mu_{\chi^2} = E[\chi^2(n)] = n$ and $E\left[(\chi^2 - \mu_{\chi^2})^2\right] = 2n$. The chi-square distribution is widely used in statistics for estimation of variances in the form of chi-square tests [Pearson 1900].

With $n = 3$, Eq. (1.142) leads to the well-known Maxwell–Boltzmann distribution [Laurendeau 2005] for velocity of a gas molecule, which is fundamental to kinetic theory of gases. Suppose that V_1, V_2 and V_3 represent the three components of the velocity of a gas molecule at temperature T and are normally distributed with zero mean and variance $\sigma^2 = \frac{K_B \mathrm{T}}{m_g}$ where m_g is the mass of a gas molecule and K_B is known as Boltzmann's constant. Maxwell–Boltzmann distribution of $V = \sqrt{V_1^2 + V_2^2 + V_3^2}$ is obtained as:

$$f_V(v) = \frac{1}{\sigma} f_{\chi^2}\left(\frac{v}{\sigma}\right) = 4\pi \sqrt{\left(\frac{m_g}{4\pi K_B T}\right)^3} v^2 \exp(-\frac{m_g v^2}{2K_B T}) \tag{1.143}$$

Maxwell–Boltzmann distributions for momentum and kinetic energy follow from that of the velocity V.

Normal random variables are commonly encountered in engineering applications. For instance, measurement errors in experiments (or noise as is commonly called) are often modelled by normal distribution. Even a discrete random variable like binomial, $b(r;n,\theta)$ is approximated by a normal distribution when n is large and $n\theta(1-\theta) \gg 1$. This result is due to DeMoivre–Lapalce limit theorem [Feller, 1968]. With n large, a binomial random variable may be well approximated by a Poisson variable for $\lambda(= n\theta)$ small and by a normal variable for λ large. Moreover, the normal distribution also derives its importance from the central limit theorem (Section 2.3.3, Chapter 2) which implies that sum of a large number of independent and identically distributed random variables is approximately normal. This justifies a normal approximation to a binomial random variable, if the latter is considered to be a sum of n independent Bernoulli random variables.

In structural analysis involving random external loading, a response quantity could be regarded as a transformation of the input random parameters (e.g., the loading) through the governing equations of motion and hence the response (at a given time) would admit characterization as a normal random variable if the input parameters are themselves normal and the governing equations of motion linear.

1.10.4 Uniform distribution

Let a random variable X take values in the real interval $[a,b]$, $b > a$ such that its CDF is:

$$F_X(x) = 0, \quad x < a$$

$$= \frac{x-a}{b-a}, \quad a \le x \le b$$

$$= 1, \quad x > b \tag{1.144}$$

$F_X(x)$ is differentiable everywhere except at $x = a$ and $x = b$. This gives the *pdf* of X as:

$$f_X(x) = \frac{1}{b-a}, \quad x \in (a,b)$$

$$= 0, \text{ otherwise} \tag{1.145}$$

Such a random variable (Fig. 1.6) is said to be uniformly distributed in $[a,b]$ and denoted by $U(a,b)$. A standard uniform variate is $U(0,1)$ with uniform *pdf* in $(0,1)$. A significant application of the uniform distribution is in simulation studies wherein it is extensively used to generate random numbers with other (specified) distributions. Generation of (pseudo-) random numbers uniformly distributed in $[0,1]$ is relatively easy with present-day computing machines and random numbers with a specified probability distribution (other than uniform) are obtained via an appropriate transformation (Section 1.11) and by Monte Carlo (MC) simulation (Section 2.5, Chapter 2).

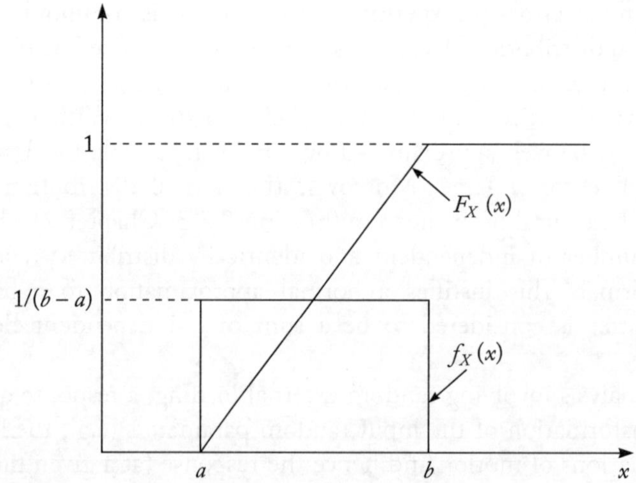

Fig. 1.6 Uniform random variable, distribution and density functions

1.10.5 Rayleigh distribution

A scalar-valued random variable $X : \Omega \to \mathbb{R}_+$ is Rayleigh distributed if the *pdf* and CDF of X are given by:

$$f_X(x) = \frac{x}{\sigma^2} e^{-\frac{x^2}{2\sigma^2}} \tag{1.146a}$$

$$F_X(x) = 1 - e^{-\frac{x^2}{2\sigma^2}}, \ x \geq 0$$

$$= 0, \ x < 0 \tag{1.146b}$$

For this distribution, we have the mean $= \sigma\sqrt{\frac{\pi}{2}}$ and variance $= \frac{(4-\pi)\sigma^2}{2}$.

Example 1.21

The Rayleigh distribution is functionally related to the normal random variable in that if X and Y are independent zero mean normal random variables with common variance σ^2, then $Z = \sqrt{X^2 + Y^2}$ follows Rayleigh distribution (Fig. 1.7).

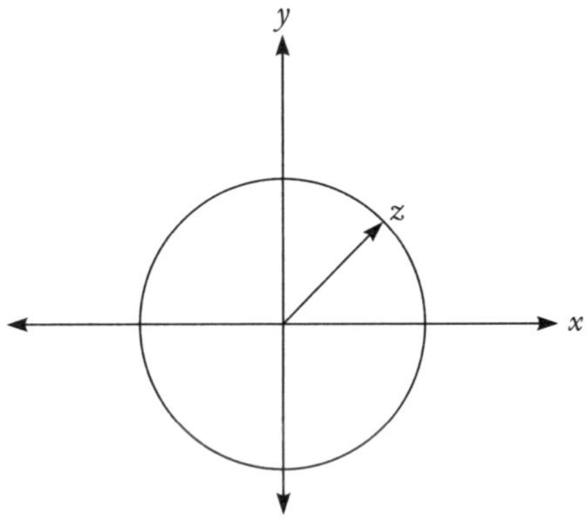

Fig. 1.7 Rayleigh distribution derived from two normal random variables X and Y

Let X and Y be independent and identically distributed (i.i.d.) random variables distributed as $\mathcal{N}(0,\sigma)$. To find the distribution of the random variable, $Z = \sqrt{X^2 + Y^2}$, we proceed as follows:

$$F_Z(z) = P(Z \le z) = P\left(\sqrt{X^2 + Y^2} \le z\right) = \int_{\Omega} f_{XY}(x,y)\, dxdy \tag{1.147}$$

Here $dxdy = d\Omega$ is the region in the x–y plane that satisfies the condition, $\sqrt{x^2 + y^2} \le z$. Hence, the requirement is to find the set of points in the x–y plane such that the events $\{\omega : Z(\omega) = z\}$ and $\{\omega : \{X(\omega), Y(\omega)\} \in d\Omega\}$ are equal.

Using polar coordinates $x = r\cos\theta$ and $y = r\sin\theta$, we have $dx\, dy = r\, dr\, d\theta$ and $\sqrt{x^2 + y^2} \le z \Rightarrow r \le z$. Noting that, $f_{XY}(x,y) = \frac{1}{2\pi\sigma^2} e^{-\frac{x^2+y^2}{2\sigma^2}}$, we have:

$$F_Z(z) = \frac{1}{2\pi\sigma^2} \int_0^{2\pi} d\theta \int_0^z r e^{-\frac{r^2}{2\sigma^2}} dr = \left(1 - e^{-\frac{z^2}{2\sigma^2}}\right) \mathbb{U}(z) \tag{1.148}$$

$\mathbb{U}(z)$ is the unit step function (Eq. 1.68). The *pdf* is given by:

$$f_Z(z) = \frac{z}{\sigma^2} e^{-\frac{z^2}{2\sigma^2}}, \text{ for } z \ge 0 \text{ and zero otherwise} \tag{1.149}$$

1.11 Transformation of Random Variables

Suppose $Y = g(X)$ is a Borel measurable function of the random variable, X; $g(X)$ is again a random variable such that the distribution function of the new random variable is given by:

$$F_Y(y) = P(Y \le y) = P(g(X) \le y) = P\left(X \le g^{-1}(y)\right) = P(X \in R_y) \tag{1.150}$$

where, R_y is the region containing the values of X for which $g(x) \leq y$. For strictly monotonic functions (either increasing or decreasing), because of the one-to-one correspondence between X and Y, $F_Y(y)$ is easily derivable in terms of $F_X\left(x = g^{-1}(y)\right)$ and the *pdf* of Y is given by [Papoulis 1991, Ang and Tang 1975]:

$$f_Y(y) = f_X(x)\left|\frac{dx}{dy}\right| \tag{1.151}$$

If $g(X)$ is not a monotonic function, the *pdf* of Y is obtained as:

$$f_Y(y) = \sum_i f_X(x_i)\left|\frac{dx}{dy}\right|_{x=x_i} \tag{1.152}$$

where the summation is over all the real roots $x_i = g^{-1}(y)$. The above result is equally applicable to the case of discrete random variables. It is noteworthy that the transformations may be usefully employed in the simulation of random variables with a specified probability distribution (see Section 2.5, Chapter 2– Monte Carlo simulation of random variables).

1.11.1 Transformation involving a scalar function of vector random variables

If $\boldsymbol{X} = (X_1, X_2, \ldots, X_n)$ is a vector random variable, the CDF of the scalar random variable $Y = g(\boldsymbol{X})$ is given by $F_Y(y) = P(g(\boldsymbol{x}) \leq y)$ where the probability covers all the points $\boldsymbol{x} := (x_1, x_2, \ldots, x_n)$ in $\mathbb{R}^n$ such that, $g(\boldsymbol{x}) \leq y$. Hence knowing, for example, the *pdf* of $\boldsymbol{X}$, we obtain the CDF of Y as:

$$F_Y(y) = \int_{g(\boldsymbol{x}) \leq y} f_{\boldsymbol{X}}(\boldsymbol{x})d\boldsymbol{x} \tag{1.153a}$$

and the *pdf* of Y as:

$$f_Y(y) = \frac{dF_Y(y)}{dy} = \frac{d}{dy}\left(\int_{g(\boldsymbol{x}) \leq y} f_{\boldsymbol{X}}(\boldsymbol{x})d\boldsymbol{x}\right) \tag{1.153b}$$

Example 1.22

If X and Y are independent random variables with *pdf*s given by $f_X(x) = \alpha e^{-\alpha x}\mathbb{U}(x)$ and $f_Y(y) = \beta e^{-\beta y}\mathbb{U}(y)$ respectively, let us find the *pdf* of $Z = X + Y$.

Directly from Eq. (1.153b) with $g(\boldsymbol{x}) = x + y$, we have:

$$f_Z(z) = \frac{d}{dz}\left(\int_{x+y \leq z} f_{XY}(x,y)dxdy\right) = \frac{d}{dz}\int_0^{\infty}\int_0^{z-y} f_{XY}(x,y)dxdy \tag{1.154}$$

By Lebnitz rule,

$$\left(\frac{\partial}{\partial z}\int_{a(z)}^{b(z)} f(x,y)\,dx = \frac{\partial b}{\partial z} f(b(y),y) - \frac{\partial a}{\partial z} f(a(y),y) + \int_{a(z)}^{b(z)} \frac{\partial}{\partial z} f(x,y)\,dx\right),$$

we simplify Eq. (1.154) as:

$$f_Z(z) = \int_0^\infty f_{XY}(z-y,y)\,dy \tag{1.155a}$$

By writing $f_Z(z) = \frac{d}{dz}\left(\int_{x+y\le z} f_{XY}(x,y)dxdy\right) = \frac{d}{dz}\int_0^\infty \int_0^{z-x} f_{XY}(x,y)dydx$, one may alternatively obtain:

$$f_Z(z) = \int_0^\infty f_{XY}(x,z-x)dx \tag{1.155b}$$

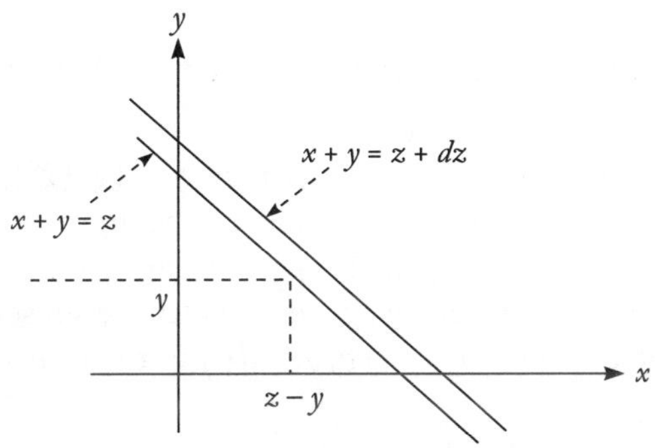

Fig. 1.8 Convolution integral in Eq. (1.155)

Clearly, the two integrals in the above equations are convolution integrals and we can obtain the solution from any one of them. If we take the first integral (Fig. 1.8) and substitute for $f_{XY}(x,y) = f_X(x)f_Y(y)$ (since X and Y are independent), we obtain:

$$\begin{aligned}
f_Z(z) &= \alpha\beta \int_0^\infty e^{-\alpha(z-y)}\, e^{-\beta y}\mathbb{U}(z-y)\,\mathbb{U}(y)dy \\
&= \alpha\beta \int_0^z e^{-\alpha(z-y)}\, e^{-\beta y}dy \\
&= \alpha\beta e^{-\alpha z} \int_0^z e^{(\alpha-\beta)y}dy \\
&= \frac{\alpha\beta}{\alpha-\beta}\left(e^{-\beta z} - e^{-\alpha z}\right)\mathbb{U}(z)
\end{aligned} \tag{1.156}$$

Alternatively, we can derive the above result more elegantly by making use of the characteristic functions. The characteristic function of the random variable $Z = X + Y$ is given by:

$$\varnothing_Z(\lambda) = E\left[e^{i\lambda(X+Y)}\right] = E\left[e^{i\lambda X}\right]E\left[e^{i\lambda Y}\right]$$

$$\Longrightarrow \varnothing_Z(\lambda) = \alpha\beta \int_0^\infty e^{i\lambda x}e^{-\alpha x}dx \int_0^\infty e^{i\lambda y}e^{-\beta y}dy$$

$$= \alpha\beta \frac{1}{(\alpha - i\lambda)(\beta - i\lambda)} \tag{1.157}$$

$f_Z(z)$ results from the inverse Fourier transform of $\varnothing_Z(\lambda)$. When we write $\frac{1}{(\alpha-i\lambda)(\beta-i\lambda)} = \frac{1}{\beta-\alpha}\left(\frac{1}{\alpha-i\lambda} - \frac{1}{\beta-i\lambda}\right)$, the inverse transform yields the same result as in Eq. (1.156)

1.11.2 Transformation involving vector functions of random variables

Consider the vector random variable $\mathbf{Y} = \mathbf{g}(\mathbf{X})$ where $\mathbf{g} = (g_1(\mathbf{X}), g_2(\mathbf{X}), \ldots, g_n(\mathbf{X}))$ and $\mathbf{X} = (X_1, X_2, \ldots, X_n)$, the vector random variable in the argument of $\mathbf{g}$. Given the joint *pdf* of $\mathbf{X}$, the task to find the joint *pdf* of $\mathbf{Y}$ is essentially a similar one attempted earlier through Eqs. (1.150) and (1.151). That is, if we consider the subset $E_{\mathbf{X}} = (\mathbf{x}, \mathbf{x} + d\mathbf{x})$ and the corresponding mapped element $E_{\mathbf{Y}} = (\mathbf{y}, \mathbf{y} + d\mathbf{y})$ in the transformed space, then:

$$P(E_{\mathbf{Y}}) = P(E_{\mathbf{X}}) \Longrightarrow f_{\mathbf{Y}}(\mathbf{y})\,d\mathbf{y} = f_{\mathbf{X}}(\mathbf{x})\,d\mathbf{x} \tag{1.158}$$

$d\mathbf{x}$ and $d\mathbf{y}$ are infinitesimal hypervolumes (Lebesgue measures). The above equation leads us to:

$$f_{\mathbf{Y}}(\mathbf{y}) = f_{\mathbf{X}}(\mathbf{x}) \frac{d\mathbf{x}}{d\mathbf{y}} \tag{1.159}$$

The ratio $\frac{d\mathbf{y}}{d\mathbf{x}}$ of the two hypervolumes can be shown to be the Jacobian of the transformation which is the determinant of the matrix of derivatives defined by:

$$|J| = \begin{bmatrix} \frac{\partial g_1}{\partial x_1} & \frac{\partial g_1}{\partial x_2} & \cdots & \frac{\partial g_1}{\partial x_n} \\ \frac{\partial g_2}{\partial x_1} & \frac{\partial g_2}{\partial x_2} & \cdots & \frac{\partial g_2}{\partial x_n} \\ & \cdots & & \\ & \cdots & & \\ & \cdots & & \\ \frac{\partial g_n}{\partial x_1} & \frac{\partial g_n}{\partial x_2} & \cdots & \frac{\partial g_n}{\partial x_n} \end{bmatrix} \tag{1.160}$$

Noting that $\frac{dx}{dy} = |J|^{-1}$, we have:

$$f_Y(y) = f_X(x)\,|J|^{-1} \tag{1.161a}$$

In the general case of the infinitesimal hypervolume dy mapping into more than one infinitesimal hypervolume $dx_k, k = 1, 2, ..r$, we have:

$$f_Y(y) = \sum_{k=1}^{r} f_X(x_k)\,|J_k|^{-1} \tag{1.161b}$$

In case the dimension of the vector $\boldsymbol{Y}$ is less than that of $\boldsymbol{X}$, the above derivation of $f_Y(y)$ can still be followed on identical lines after suitably augmenting the vector $\boldsymbol{Y}$ with auxiliary random variables from the vector $\boldsymbol{X}$ itself (see Exercise 1.17).

1.11.3 Diagonalization of covariance matrix and transformation to uncorrelated random variables

It is often required (including cases involving simulation studies) to transform a set of correlated random variables into one of uncorrelated ones since it is easy to do mathematical manipulations with the latter. The covariance matrix, $\boldsymbol{C} \in \mathbb{R}^{nxn}$ in Eq. (1.133) is a symmetric positive semi-definite matrix. This property is made use of to transform $\boldsymbol{X}$ to a vector $\boldsymbol{Y}$ of uncorrelated random variables $(Y_1, Y_2, \ldots Y_n)$. Here one can use the transformation $\boldsymbol{Y} = \boldsymbol{V}^T\boldsymbol{X}$ where $\boldsymbol{V}$ is the matrix of orthonormalized eigenvectors of $\boldsymbol{C}$ with $\boldsymbol{V}^T\boldsymbol{V} = \boldsymbol{I}$ where $\boldsymbol{I}$ is the identity matrix. Diagonalization of $\boldsymbol{C}$ leads to $\boldsymbol{V}^T\boldsymbol{C}\,\boldsymbol{V} = \boldsymbol{D}$ with $\boldsymbol{D}$ being a diagonal matrix containing the eigenvalues of $\boldsymbol{C}$. $\boldsymbol{D}$ is also the covariance matrix of $\boldsymbol{Y}$. The diagonal elements give the correlations (mean square values) $E\left[Y_j^2\right]$ of the scalar random variables of $\boldsymbol{Y}$. The following example illustrates the diagonalization of the matrix $\boldsymbol{C}$ and the related transformation to uncorrelated random variables.

Example 1.23

Consider the case of a bi-variate normal random variable $\boldsymbol{X} = (X_1, X_2)$ with the joint *pdf* given by Eq. (1.137). We transform $\boldsymbol{X}$ to uncorrelated random variables.

Solution

For the sake of simplicity in exposition, let us assume $m_1 = m_2 = 0$ and $\sigma_1 = \sigma_2 = \sigma$ so that:

$$f_{X_1,\,X_2}(x_1, x_2) = \frac{1}{2\pi\sigma\sqrt{(1-\rho^2)}}\exp\left[-\frac{1}{2\sigma^2(1-\rho^2)}\left\{x_1^2 - 2\rho x_1 x_2 + x_2^2\right\}\right] \tag{1.162}$$

Comparing Eqs. (1.133) and (1.162), we identify that $\frac{1}{\sigma^2(1-\rho^2)}\begin{bmatrix} a & b \\ c & d \end{bmatrix} = \boldsymbol{C}^{-1}$. Let us now write the quadratic inside the curly braces on right hand side of the last equation as:

$$x_1^2 - 2\rho x_1 x_2 + x_2^2 = (x_1 \ \ x_2)\begin{bmatrix} a & b \\ c & d \end{bmatrix}(x_1 \ \ x_2)^T \tag{1.163}$$

With $\boldsymbol{C}$ being symmetric, Eq. (1.163) gives the values of $a = 1,\ b = c = -\rho$ and $d = 1$. With these values, the covariance matrix $\boldsymbol{C}$ can be obtained as:

$$\boldsymbol{C} = \begin{bmatrix} \sigma^2 & \rho\sigma^2 \\ \rho\sigma^2 & \sigma^2 \end{bmatrix} = \sigma^2\begin{bmatrix} 1 & \rho \\ \rho & 1 \end{bmatrix} \tag{1.164}$$

Diagonalization of $\boldsymbol{C}$ first requires the computation of the eigenvalues $\lambda_i, i = 1,2$ and orthonormalized eigenvectors $\boldsymbol{v}_i, i = 1,2$ of the matrix. Let $\rho = 0.5$. Eigenvalues of the 2×2 matrix are presently given by $\lambda_{1,2} = 0.5\sigma^2, 1.5\sigma^2$ corresponening to the respective eigenvectors $\boldsymbol{v}_{1,2} = \left(\frac{1}{\sqrt{2}} \ -\frac{1}{\sqrt{2}}\right)^T$ and $\left(\frac{1}{\sqrt{2}} \ \frac{1}{\sqrt{2}}\right)^T$. Writing $\boldsymbol{V} = [\boldsymbol{v}_1 \ \ \boldsymbol{v}_2]$, we can verify that $\boldsymbol{V}^T\boldsymbol{V} = \mathbf{I}$. Thus, the transformation which we need for diagonalization of $\boldsymbol{C}$ is given by:

$$\boldsymbol{Y} = \boldsymbol{V}^T\boldsymbol{X} = \begin{bmatrix} \frac{1}{\sqrt{2}} & \frac{1}{\sqrt{2}} \\ -\frac{1}{\sqrt{2}} & \frac{1}{\sqrt{2}} \end{bmatrix}^T \boldsymbol{X} \tag{1.165}$$

The transformed covariance matrix is accordingly given by:

$$\boldsymbol{C}' = \boldsymbol{V}^T\boldsymbol{C}\boldsymbol{V} = \sigma^2\begin{bmatrix} \frac{1}{\sqrt{2}} & \frac{1}{\sqrt{2}} \\ -\frac{1}{\sqrt{2}} & \frac{1}{\sqrt{2}} \end{bmatrix}^T \begin{bmatrix} 1 & 0.5 \\ 0.5 & 1 \end{bmatrix}\begin{bmatrix} \frac{1}{\sqrt{2}} & \frac{1}{\sqrt{2}} \\ -\frac{1}{\sqrt{2}} & \frac{1}{\sqrt{2}} \end{bmatrix}$$

$$= \sigma^2\begin{bmatrix} \frac{1}{2} & 0 \\ 0 & \frac{3}{2} \end{bmatrix} = \boldsymbol{D} \tag{1.166}$$

$\boldsymbol{C}'$ is diagonal indicating that the transformed random variables Y_1 and Y_2 are indeed uncorrelated and their variances are $\sigma_1^2 = 0.5\sigma^2$ and $\sigma_2^2 = 1.5\sigma^2$ respectively. With $\Delta = 3\sigma^2/4$ the joint *pdf* of $\boldsymbol{Y} = (Y_1,\ Y_2)^T$ is therefore given by:

$$f_{\boldsymbol{Y}}(\boldsymbol{y}) = \frac{1}{2\pi\sqrt{\Delta}}\exp\left(-\frac{1}{2}\boldsymbol{y}^T\boldsymbol{C}'^{-1}\boldsymbol{y}\right) = \frac{1}{\sqrt{3}\,\sigma\pi}\exp\left(-\frac{1}{\sigma^2}\left(y_1^2 + \frac{y_2^2}{3}\right)\right) \tag{1.167}$$

1.11.4 Nataf transformation

In reliability and risk analyses, one often requires transforming correlated non-normal random variables X_i, $i = 1,2,...,n$ to independent normal ones. If the original joint CDF $F_{\boldsymbol{X}}(\boldsymbol{x})$ is fully available along with the correlation data, Rosenblatt transformation (see Section 2.2.5, Chapter 2) is generally used. However, in most of the practical applications, only the individual (marginal) CDFs may be accessible. In such cases, Nataf [Nataf 1962] transformation is used, mainly to transform the correlated random variables $\boldsymbol{X}$ to correlated standard normal random variables $\boldsymbol{Z}$ and then, if needed, to transform the latter to uncorrelated ones according to the procedure already described in the previous section. Let the probability measures for the random variables $\boldsymbol{X}$ and $\boldsymbol{Z}$ be absolutely continuous with respect to each other. Thus consider the transformation of the marginal distributions $F_{X_i}(x_i)$ to standard normal distributions $F_{Z_i}(z_i)$ where each Z_i is $\mathcal{N}(0,1)$. If $\boldsymbol{\rho}$ is the specified correlation matrix of $\boldsymbol{X}$ and $\boldsymbol{\rho}'$ the unknown correlation matrix of $\boldsymbol{Z}$, Nataf transformation is given by:

$$F_{Z_i}(z_i) = F_{X_i}(x_i),\ i = 1,2,....,n \tag{1.168}$$

The transformation for the joint *pdf*s of $\boldsymbol{X}$ and $\boldsymbol{Z}$ takes a form similar to the one in Eq. (1.161a), i.e., $f_{\boldsymbol{Z}}(\boldsymbol{z}) = f_{\boldsymbol{X}}(\boldsymbol{x})\,|J|^{-1}$ where J is the Jacobian matrix (Frechet derivative) $\frac{d\boldsymbol{z}}{d\boldsymbol{x}}$. Since only marginal distributions are involved in Nataf transformation, the determinant $|J|$ can be shown to be equal to $\prod_{i=1}^{n}\frac{f_{Z_i}(z_i)}{f_{X_i}(x_i)}$ with $z_i = F_{Z_i}^{-1}\left\{F_{X_i}(x_i)\right\}$. Thus, for any pair of random variables $\left(X_i, X_j\right)$, we have:

$$f_{Z_iZ_j}\left(z_i, z_j; \rho'_{ij}\right) = f_{X_iX_j}\left(x_i, x_j; \rho_{ij}\right)\frac{f_{X_i}(x_i)\,f_{X_j}\left(x_j\right)}{f_{Z_i}(z_i)\,f_{Z_j}\left(z_j\right)} \tag{1.169}$$

Here $z_i = F_{Z_i}^{-1}\left\{F_{X_i}(x_i)\right\}$ and $z_j = F_{Z_j}^{-1}\left\{F_{X_j}\left(x_j\right)\right\}$. The next phase involved in the transformation is the evaluation of the correlation matrix $\boldsymbol{\rho}'$ from $\boldsymbol{\rho}$. For any $\left(X_i, X_j\right)$, we have:

$$\rho_{ij} = \iint_{-\infty}^{\infty}\left(\frac{x_i - m_{x_i}}{\sigma_{x_i}}\right)\left(\frac{x_j - m_{x_j}}{\sigma_{x_j}}\right) f_{X_iX_j}\left(x_i, x_j\right) dx_i dx_j \tag{1.170}$$

Here m_{x_i} and σ_{x_i} are the known mean and standard deviation of the component random variables X_i, $i = 1,2,\ldots,n$. In view of Eq. (1.169), we rewrite the last equation as:

$$\rho_{ij} = \iint_{-\infty}^{\infty}\left(\frac{x_i - m_{x_i}}{\sigma_{x_i}}\right)\left(\frac{x_j - m_{x_j}}{\sigma_{x_j}}\right) f_{Z_iZ_j}\left(z_i, z_j; \rho'_{ij}\right)\left(\frac{f_{Z_i}(z_i)\,f_{Z_j}\left(z_j\right)}{f_{X_i}(x_i)\,f_{X_j}\left(x_j\right)}\right) dx_i dx_j \tag{1.171}$$

The above equation shows an implicit integral relationship between ρ_{ij} and ρ'_{ij}. With $f_{Z_iZ_j}\left(z_i, z_j; \rho'_{ij}\right)$ given by Eq. (1.137) or more precisely by (1.162), one elegant way of obtaining ρ'_{ij} for a given ρ_{ij} is to first solve Eq. (1.171) for ρ_{ij} for different assumed values of $\rho'_{ij} \in (-1,1)$. Then one easily obtains ρ'_{ij} for the specified value of ρ_{ij} by interpolation.

Example 1.24

Let X_1 and X_2 be given as random variables having uniform and log normal distributions respectively with $m_{x_1} = 0.1$, $m_{x_2} = 100$ and $\sigma_{x_1} = 0.03$, $\sigma_{x_2} = 10$. Assume that $\rho_{12} = 0.4$ with $\rho_{11} = \rho_{22} = 1.0$. Knowing the marginal distributions of X_1 and X_2, we can evaluate ρ'_{12} from Eq. (1.171). Solving this equation, we obtain $\rho'_{12} = 0.4104$ for the given value of $\rho_{12} = 0.4$.

Example 1.25

Let us compute the reliability index [Maymon 1998] of a simple system consisting of a straight bar under an axial load T which follows a normal distribution. Suppose that the cross-sectional area A of the bar is also normal and that the yield stress S_y of the material has a Weibull distribution. The specified statistics (in appropriate units) of the three random variables is:

mean:	$\mu_{S_y} = 2500$	$\mu_T = 4166$	$\mu_A = 2$
standard deviation:	$\sigma_{S_y} = 75$	$\sigma_T = 125$	$\sigma_A = 0.04$

We need to evaluate the reliability index and the probability of failure of the system (see Appendix A for short notes on the topic needed to appreciate this example). With $\mathbf{X} = \left(S_y, T, \mathrm{A}\right)$, the limit state function in this case is $g\left(\mathbf{X}\right) = g\left(S_y, T, \mathrm{A}\right) = S_y - \frac{T}{\mathrm{A}}$. The marginal distributions are shown in Fig. 1.9. In particular, the Weibull distribution of S_y is given by:

$$F_{S_y}\left(s_y\right) = 1 - \exp(-\alpha {s_y}^{\beta}),\ \ s_y \geq 0 \tag{1.172a}$$

$$f_{S_y}\left(s_y\right) = \alpha\beta {s_y}^{\beta-1} \exp(-\alpha {s_y}^{\beta}),\ \ s_y \geq 0$$

$$= 0,\quad \text{otherwise} \tag{1.172b}$$

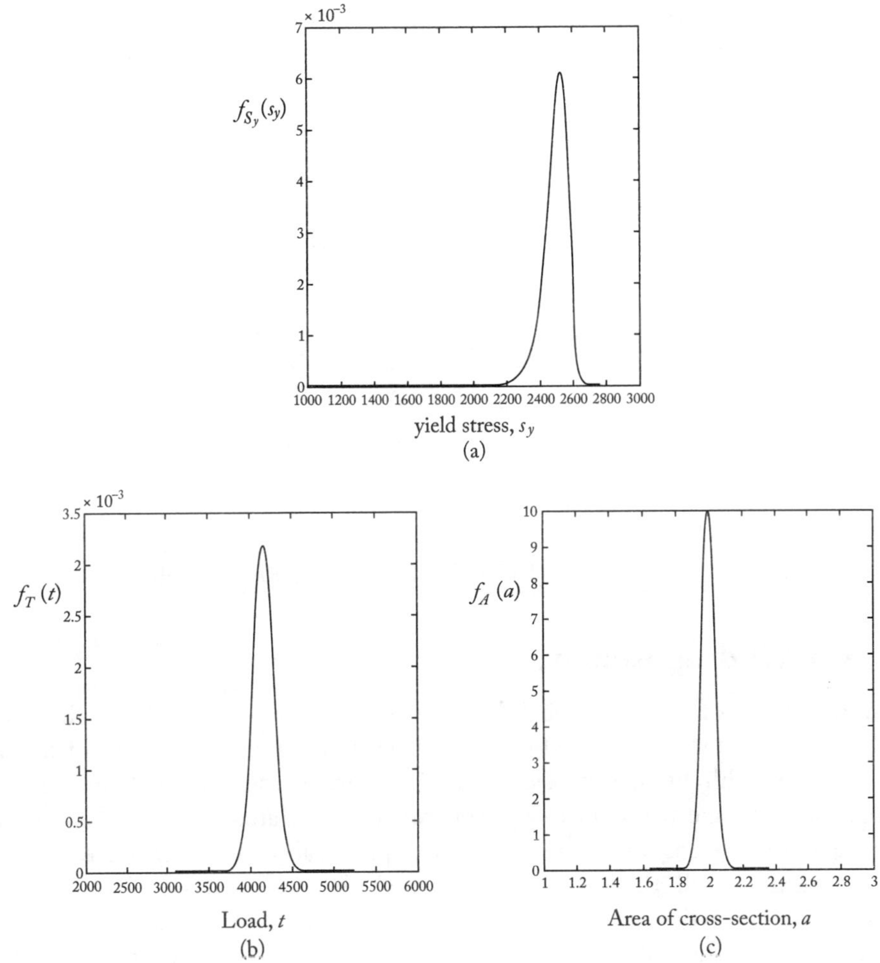

Fig. 1.9 Specified marginals in Example 1.25; (a) yield stress S_y (Weibul), (b) load T (normal) and (c) area of cross-section A (normal)

The two parameters α and β of the Weibull distribution are related to moments μ_{S_y} and σ_{S_y} as:

$$\mu_{S_y} = \frac{1}{\alpha^{1/\beta}} \Gamma\left(1 + \frac{1}{\beta}\right)$$

$$\sigma_{S_y} = \frac{1}{\alpha^{2/\beta}} \left\{ \Gamma\left(1 + \frac{2}{\beta}\right) - \Gamma^2\left(1 + \frac{1}{\beta}\right) \right\} \tag{1.173}$$

Here $\Gamma(.)$ is the gamma function. Knowing μ_{S_y} and σ_{S_y}, α and β are obtained as 8.18211×10^{-144} and 42.03862, respectively. Following Rackwitz–Fiessler (Table A.1

in Appendix A) procedure [1978], the reliability index is obtained as $\beta_F = 3.1755$ by an iterative process and the results are given in Table 1.1. The design point is given by $S_y = 2186.3$, $T = 4304.7$ and $A = 1.9689$. The probability of failure is given by $\Phi(-\beta_F)$ = 0.00075, which corresponds to an associated reliability of 99.925% under the given load condition.

Table 1.1 Results for Example 1.25 by Rackwitz–Fiessler iterative procedure (Table A.1 of Appendix A)

Iteration no.	S_y	T	A	β_F
1	2302.2	4450.2	1.9330	3.8638
2	2206.0	4330.2	1.9629	3.2369
3	2188.5	4307.5	1.9683	3.1768
4	2186.3	4304.7	1.9689	3.1755
5	2186.1	4304.4	1.9690	3.1755

1.12 Concluding Remarks

The basic principles of modern probability theory have been introduced in this Chapter. Definition of a measure as a countably additive set function and treatment of probability as a measure are highlighted. Extension of a probability measure defined on an algebra on the sample space to the whole of σ-algebra and the associated uniqueness are explained via Carathéodory's extension theorem. This has practical significance since the need to assign a measure to all the elements in a σ-algebra is by passed by the theorem, particularly when we deal with σ-algebras of continuous sample spaces such as the topological spaces. The notion of a random variable which is central to the probability theory is presented next. The essential idea is that a random variable is a mapping or function between two measurable spaces $(\Omega, \mathcal{F})$ and $(\mathbb{R}^n, \mathcal{B})$ where $\mathcal{B}$ is the σ-algebra generated by subsets in $\mathbb{R}^n$. Equivalence of the *pdf* of a random variable and Randon–Nikodym derivative of the associated probability measure is touched upon. This is followed by the interpretation of expectation of a random variable as a Lebesgue integral with respect to the probability measure and use of characteristic or moment generating function in evaluating expectations. Independence of random variables is elucidated from the perspectives of events, distributions and σ-algebras. Having dwelt on a few oft-used probability distributions and their properties, the topic on transformation of (scalar or vector) random variables is discussed in view of its myriad practical applications. In the next chapter, we continue to dwell on the theme of this chapter to discuss some additional basic concepts of probability theory such as conditional probability and related topics—conditional expectation and Bayes' rule—and also Monte Carlo simulation of random variables.

Exercises

1. Consider tossing of two fair coins. Then $\Omega = \{HH, HT, TH, TT\}$. $\mathcal{F}_0 = 2^{\Omega}$, the power set of Ω, carries all the information about the outcomes of both the tosses. $\mathcal{F}_1 = \{\emptyset, \{HH,HT\}, \{TH,TT\}, \Omega\}$ carries information only about the first toss, $\mathcal{F}_2 = \{\emptyset, \{HH,TH\}, \{HT,TT\}, \Omega\}$ about the second toss and $\mathcal{F}_3 = \{\emptyset, \{HH,TT\}, \{HT,TH\}, \Omega\}$ about whether the outcome on both the tosses is the same or different. Suppose that we define a random variable:

$$X(\omega) = 0, \text{ if } \omega = HH \text{ or } HT$$

$$= 1, \text{ if } \omega = TH \text{ or } TT.$$

   Then X is measurable in $\mathcal{F}_0$ and $\mathcal{F}_1$ but not in $\mathcal{F}_2$and $\mathcal{F}_3$. State the reasons.

2. Show that the following subsets of $\mathbb{R}$ belong to $\mathcal{B}(\mathbb{R})$.

   (i) $(-\infty, b)$ for any $b \in \mathbb{R}$
   (ii) (a, ∞) for any $a \in \mathbb{R}$
   (iii) $(-\infty, b]$ for any $b \in \mathbb{R}$
   (iv) $[a, \infty)$ for any $a \in \mathbb{R}$

3. Let $(\Omega, \mathcal{F}, \mu)$ be a measure space with $\mu(\Omega) = 1$, and let $A \in \mathcal{F}$. Show that the set of all $B \in \mathcal{F}$ which satisfy $\mu(A \cap B) = \mu(A)\mu(B)$, is a λ-system.

4. Find the mean and variance of a (i) binomial distribution considering it to be the sum of n independent Bernoulli trials, with p being the probability of success in a trial and (ii) Poisson distribution with parameter λ.

5. Given two independent Poisson random variables X and Y with respective parameters λ_1 and λ_2, find the distribution of $X + Y$, e.g., show that it is Poisson with parameter $\lambda_1 + \lambda_2$.

6. Given that X_i, $i = 1, 2, \ldots, n$, are jointly normal and the variables have zero mean and covariance matrix C. Show that (i) $X = a_1 X_1 + a_2 X_2 + \cdots + a_n X_n$ with $a_i \in \mathbb{R}$, $i = 1, 2, \ldots, n$, has the characteristic function $\emptyset_X(\lambda) = E\left[e^{j\lambda X}\right] = \exp\left(-\frac{1}{2}\lambda^2 C\right)$. Also show that (ii) $E[X_1 X_2 X_3 X_4] = C_{12}C_{34} + C_{13}C_{24} + C_{14}C_{23}$ where $C_{jk} = E\left[X_j X_k\right]$. (hint for (ii): expand $E\left[e^{j\lambda X}\right]$ and $\exp\left(-\frac{1}{2}\lambda^2 C\right)$ separately and equate the coefficients of the term that contains $a_1 a_2 a_3 a_4$ in the two expressions)

7. Find the characteristic functions of (a) the binomial distribution (b) the Poisson distribution. Prove that both the distributions approach the standard normal distribution as one of the parameters tends to ∞ under suitable normalization.

8. If $A_1, A_2, \ldots, A_n$ are independent events, prove that $A_1^c, A_2^c, \ldots, A_n^c$ and $I_{A_1}, I_{A_2}, \ldots, I_{A_n}$ are independent.

9. Consider independent and identically distributed random variables $X_1, X_2, \ldots, X_n$ each having mean μ and variance σ^2. Let $\overline{X} = \sum_{j=1}^{n} \frac{X_j}{n}$ be the sample mean. Find the variance of $\overline{X}$. Also find $E[Y]$ where $Y = \sum_{j=1}^{n} \frac{(X_j - \overline{X})^2}{n-1}$. [Ans. $var\left[\overline{X}\right] = \frac{\sigma^2}{n}$ and $E[Y] = \sigma^2$].

10. Random variables X_i, $i = 1, 2, \ldots, n$ are independent and uniformly distributed in the interval $[0,1]$ with means m_i and variances $\sigma_i^2, i = 1, 2, \ldots, n$. Compare the density of $X = X_1 + X_2 + \cdots + X_n$ with the normal approximation $f_X(x) = \frac{1}{\sigma\sqrt{2\pi}} \exp\left(-\frac{1}{2}\left(\frac{x-m}{\sigma}\right)^2\right)$ with $m = m_1 + m_2 + \cdots + m_n$ and $\sigma^2 = \sigma_1^2 + \sigma_2^2 + \cdots + \sigma_n^2$ for $n = 2, 3$ and 4 (analytically and by simulation – see Chapter 2 for MC simulation methods).

11. Let X be a random variable with cumulative distribution function F strictly increasing on the range of X. Let $Y = F(X)$. Show that Y is uniformly distributed in the interval [0, 1].

12. Let X follow an exponential distribution with pdf $f_X(x) = \lambda \exp(-\lambda x)$, $x > 0$. If $Y = X^{1/\alpha}$ for $\alpha > 0$, then Y has a (two-parameter) Weibull distribution. Show that $f_Y(y) = \alpha\lambda y^{\alpha-1} \exp(-\lambda y^{\alpha})$, $y > 0$ (see Example 1.25). Weibull distribution is often used in lifetime (survival) analysis: see Cox and Oakes 1984, Kalbfleish and Prentice 1980 in the Bibliography.

13. The density function of a Cauchy random variable is given by:

$$f_X(x) = \frac{1}{\pi(1+x^2)}, \; -\infty < x < \infty$$

Show that $1/X$ is also a standard Cauchy random variable.

14. The arrival times T_1 and T_2 of two flights are uniformly distributed in a specified interval spanning an hour. Thus $\Omega = \{(t_1, t_2) : 0 \le t_1 \le 60; 0 \le t_2 \le 60\}$. If the expected arrival times are separated at least by 10 minutes, determine the probability that the flights arrive within an interval of 10 minutes (i.e., $P\{(t_1, t_2) \in \Omega; |t_1 - t_2| \le 10\}$).

15. Let X and Y be uniformly distributed in $[0,1]$. Find the distributions of $\min(X, Y)$ and $\max(X, Y)$. Are they independent?

16. Let $\mathbf{X} = (X_1, X_2, \ldots, X_n)$ be a random vector consisting of n independent random variables with $X_i \sim \mathcal{N}(0,1)$, $i = 1, 2, \ldots, n$. Let $\Xi \in \mathbb{R}^{n \times n}$ be a given positive semi-definite symmetric matrix, and $\mathfrak{M} \in \mathbb{R}^n$, a given vector. Show that there exists an affine transformation $T : \mathbb{R}^n \to \mathbb{R}^n$ such that the random vector $T(\mathbf{X})$ is Gaussian with $T(\mathbf{X}) \sim N\left(\mathfrak{M}, \Xi^{1/2}\right)$.

17. Given $f_{XY}(x,y) = \frac{1}{2\pi\sigma^2} \exp\left(-\frac{1}{2\sigma^2}(x^2 + y^2)\right)$, find the pdf of $U = \sqrt{(X^2 + Y^2)}$.
(Hint: Use an auxiliary random variable $V = X$ and first derive $f_{UV}(u,v)$ and then the marginal *pdf* of U).

Notations

A	area of cross-section
A, A_1, A_2, A_m, A_n	sets, events
A^c	complement of A
B, B_1, B_k, B_n	sets, events
$b(r;n,\theta)$	binomial distribution parmetered by r, n and θ
$\mathcal{A}$	algebra
$\mathcal{D}$	collection of sets
E, E_1, E_2	events
erf(.)	error function, $\frac{1}{\sqrt{2\pi}}\int_0^x e^{-\frac{u^2}{2}}du$
$\mathcal{G}$	sigma algebra
i, j	integers
$\mathfrak{T}_1$ and $\mathfrak{T}_2$	π-systems
$\|J\|$	Jacobian
K_B	Boltzmann's constant
m	mean (of a random variable)
N	total number of outcomes in the sample space, Ω (cardinality)
N_E	number of elements in the event E
S	cvcnt
S_y	yield stress
T	temperature
T	load
$V^{(i)}(t)$	random signals parametered via sample number `i' and time t
V_1, V_2, V_3	velocities of a gas molecule (random variables)
W, W_1	random variables
X, Y, X_1, X_2	scalar random variables
$\boldsymbol{X}, \boldsymbol{Y}, \boldsymbol{Z}$	vector random variables
β_F	reliability index

$\Gamma(.)$	gamma function
$\delta_1, \delta_2, \delta_3, \delta_4, \delta_5, \delta_6$	Dirac measures (see Section 1.2 for definition)
θ	a parameter (in a probability distribution)
$\mu, \mu^*, \mu_0, \mu_1, \mu_2, \mu_X$	measures
ρ	correlation coefficient
$\boldsymbol{\rho}$	correlation matrix
σ_1^2, σ_2^2	variances
σ_{12}	covariance between two random variables, e.g., $E[(x_1 - m_1)(x_2 - m_2)]$
(Σ, Υ)	measure space
$\chi^2(n)$	chi-square distribution with n degrees of freedom
ψ	moment generating function
$\omega \in \Omega$	outcome or element or sample point of sample space Ω
$\omega_1, \omega_2, \omega_3, \omega_4$	outcomes of Ω
Ω	probability sample space
$\mathcal{F}_1, \mathcal{F}_2, \mathcal{F}_B$	sigma algebras

Chapter 2

Random Variables: Conditioning, Convergence and Simulation

2.1 Introduction

The concept of conditional probability is central to probability theory and is closely related to the notion of independence of random variables (Chapter 1). It is the probability of an event A conditioned on or given the occurrence of another event B. The event B may belong to a sub-σ-algebra that may or may not have an overlap with the sub-σ-algebra to which the event A belongs. Convention has it that the conditional probability be denoted by $P(A|B)$, which is defined as the likelihood (i.e., a ratio of probabilities or a ratio of measures or a Radon–Nikodym derivative: $P(A \mid B) = \frac{P(A \cap B)}{P(B)}$ —see Eq. (2.2) below) that event A also occurs, known that B had already occurred. One may also interpret it as a way of describing A when the statistical information used to arrive at this description is restricted to the sub-σ-algebra generated by the subsets of B. This may remind us of the problem of the best representation of a function or a map through projection on a subspace, such as the Galerkin method [Elishakoff and Ren 2003, Roy and Rao 2012]. As a somewhat simplistic example on conditional probability, one may consider the problem of whether a flight gets delayed at a particular time on a day. As is known, delay or no delay depends on many parameters; e.g., the weather condition with a known probability of fair or otherwise, the departure time on the day and such. Figure 2.1 pictorially shows the use of partial information on weather condition to estimate the probability of a flight landing on time.

$$\begin{aligned} P(A) &= P(A \cap G) + P(A \cap B) \\ &= P(A|G)P(G) + P(A|B)P(B) \end{aligned} \tag{2.1}$$

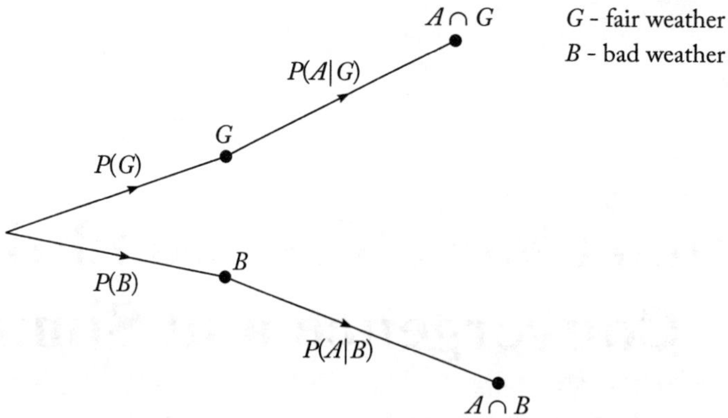

Fig. 2.1 Conditional probabilities and computation of total probability

As another example, if one considers the stock price in a financial market, it invariably depends on other variables (e.g., interest rates etc.) influencing its fluctuations and consideration of dependence through conditional probabilities is a necessity in drawing inferences. Similar is the case with assessment of a seismic hazard [Field 2007] which is largely influenced by the data collected on fault patterns, slip rates, return periods and recurrence of past events in a given region. This is suggestive of updating an existing hypothesis by using the new or extra information or evidence as they become available. In fact, the posterior probability obtainable by statistical inferences, using Bayes' rule (see the Introduction to Chapter 1) is nothing but a conditional probability. Viewed over a wider perspective, Bayes' rule has varied applications—the concept is introduced in this chapter and its applications to dynamical systems of practical interest elaborated in subsequent chapters. In engineering science, its relevance is well-known in signal processing and filtering [Kailath 1981, Candy 2009, Haykin 1996], pattern recognition [Bishop 2006], robotics [Thrun *et al.* 2005], and machine learning [Mitchell 1997]. Modelling of a dynamical system is fraught with uncertainties and the initial system states may be conjectured or predicted with a prior probability. Information by measurements (generally collected using, say, sensors attached to the system) which, by themselves, may be noisy and hence uncertain, are used in updating the system states using recursive Bayesian estimation.

Another intriguing idea in the modern probability theory is a change of measure over a probability space, $(\Omega, \mathcal{F}, P)$. This aspect was briefly touched upon in Chapter 1 (Radon–Nkodym theorem in Section 1.6). The *pdf* of a random variable is in fact a change of measure from P on $\mathcal{B}(\mathbb{R})$ to Lebesgue measure dx in the form $dP = dF_X(x) = f_X(x)dx$. A valid probability measure, say Q, over the same probability space, yet different from P, may simplify the mathematical model and facilitate better understanding of the physical phenomena. The concept has wide appeal in mathematical finance [Hubalek and Schachermayer 2001, Glasserman 2003] related to insurance or stock market price dynamics and the associated risk management. Specific to engineering applications, the concept plays an important role in computing conditional expectations (e.g., generalized Bayes' formula—Section 2.2.3 of this Chapter), importance sampling methods and related simulation studies. Given the importance of numerical simulation of stochastic dynamical systems and the role played by Monte Carlo (MC) methods in such studies, we also consider in this Chapter, MC simulation of random variables—scalar or vector, uncorrelated or correlated. Use of MC methods in the evaluation of multi-dimensional integrals that play so significant a role in reliability studies, system identification (Chapter 6 and 7) and stochastic optimization (Chapters 9 and 10) problems, is also highlighted in the sequel.

In statistical inferences of diverse fields—from population statistics to engineering and science research, one always finds the distribution of an average tending to a normal distribution. This is indeed implied by the central limit theorem (CLT), which is a fundamental result in probability theory. In simple words, the theorem contends that the mean of a large sample converges to a normal, irrespective the original distribution. The proof of the theorem relies on the concept of convergence of random variables and their distributions. Different convergence criteria are discussed in this Chapter followed by the proof of CLT. Figure 2.2 shows the sum of n uniform and Rayleigh random variables approaching a normal distribution as n increases.

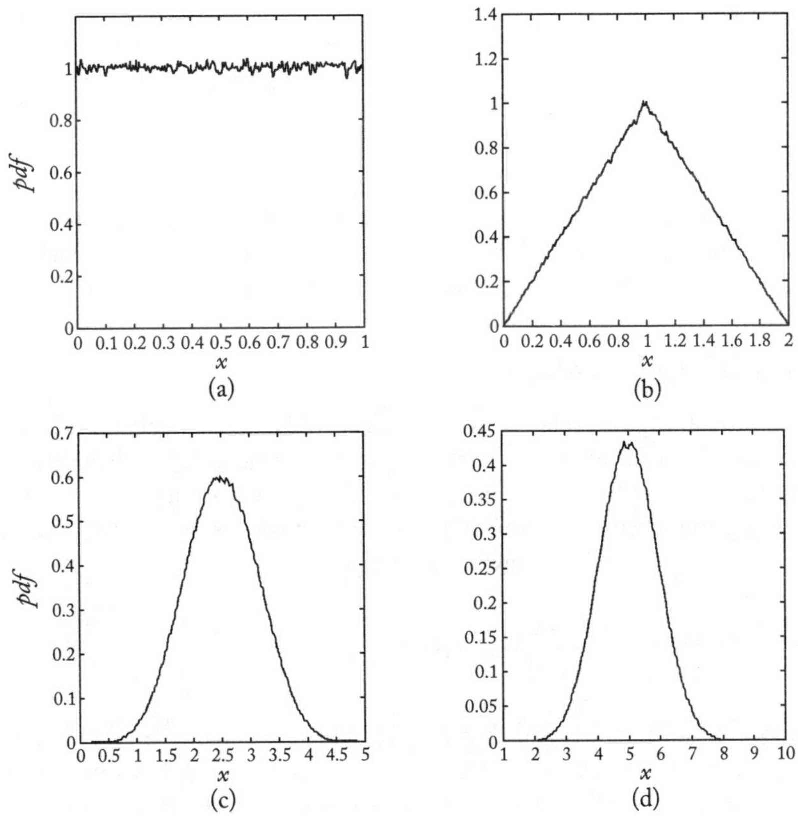

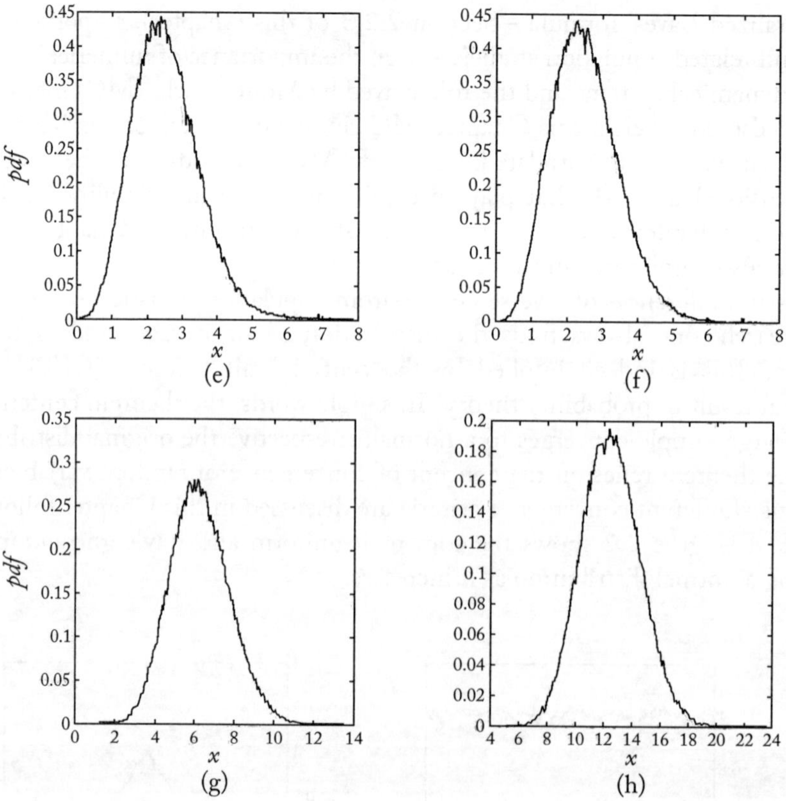

Fig. 2.2 Central limit theorem; (a–d) convergence of uniform random variable for $n = 1,2,5$ and 10, respectively; (e–h) convergence of Rayleigh random variable for $n = 1,2,5$ and 10, respectively

2.2 Conditional Probability

Conditional probability is related to the probability of a subset of events when information on the occurrence of another subset (not necessarily disjoint) is available. Thus, given $(\Omega, \mathcal{F}, P)$ and two elements, $A, B \subset \mathcal{F}$, the probability of A conditioned on B (i.e., with B having occurred being a given hypothesis) is known as the conditional probability $P(A|B)$ or $P_B(A)$. It may be defined as:

$$P_B(A) = P(A|B) = \frac{P(A \cap B)}{P(B)}, P(B) > 0 \tag{2.2}$$

From the above definition, we find that $P_B(A)$ is proportional to $P(A \cap B)$. For a given B, $P(B)$ can be considered to be a normalizer so that $0 \leq P_B(A) \leq 1$. Conditioning restricts the sample space to $B \subset \Omega$ leading to a new probability space $(B, \mathcal{F}_B, P_B)$ with $\mathcal{F}_B$ being the sub-σ-algebra generated by all subsets of B. Thus the conditional probability is always associated with a reduced sample space and, in this sense, P_B is a restriction to $\mathcal{F}_B$ of P. In case the conditional probability $P_B(A)$ equals the unconditional probability, $P(A)$, then the event A is statistically independent of the event B (see Section 1.9, Chapter 1). This also implies that the joint probability is given by:

$$P(A \cap B) = P(A)P(B) \tag{2.3}$$

Now consider a random variable (discrete or continuous) $X : \Omega \to \mathbb{R}$ with $A, B \subset \mathbb{R}$ being open intervals. We define conditional distribution function as:

$$F_X(x|B) := \frac{P(\{X \leq x\} \cap B)}{P(B)} \tag{2.4}$$

The conditional probability distribution function satisfies all the properties of $F_X(x)$ as stated in Eq. (1.66) of Chapter 1 with probabilities replaced by conditional probabilities. The conditional density function $f_X(x|B)$, for instance, is the derivative (if it exists) of $F_X(x|B)$:

$$f_X(x|B) = \frac{dF_X(x|B)}{dx} = \lim_{\Delta x \to 0} \frac{F_X(x+\Delta x|B) - F_X(x|B)}{\Delta x}$$

$$= \lim_{\Delta x \to 0} \frac{P(x < X \leq x + \Delta x | B)}{\Delta x}. \tag{2.5}$$

Example 2.1

Let us consider an event A describing the life of a component beyond, say t units, conditioned on the event B corresponding to the information that it has already survived for a certain period, say, b units. Here the life of the system, say T, is a continuous random variable, $T : \Omega \to [0, \infty) \subset \mathbb{R}$, and we are interested in $P(A = \{T > t\})$ given the event $B = \{T > b\}$. Now $P(A|B) = P(\{T > t\}|\{T > b\})$ and it is given in terms of the conditional distribution function $F_T(t|B)$ as:

$$P(T > t|B) = 1 - F_T(t|B) = 1 - \frac{P(T \leq t \cap T > b)}{P(T > b)} = 1 - \frac{F_T(t) - F_T(b)}{1 - F_T(b)}$$

$$= \frac{1 - F_T(t)}{1 - F_T(b)}, t > b$$

$$= 1, t \leq b \tag{2.6}$$

$F_T(t)$ stands for the CDF of the random variable T.

Conditional probabilities involving two or more (scalar-valued) random variables can be defined analogously. Let $X, Y : \Omega \to \mathbb{R}$ be two random variables defined on the probability space $(\Omega, \mathcal{F}, P)$. With $A = \{X \leq x\}$ and $B = \{Y \leq y\}$, the conditional probability, $P(\{X \leq x\}|\{Y \leq y\})$ is the conditional distribution function $F_X\left(x|\{Y \leq y\}\right)$ given by:

$$F_X\left(x|Y \leq y\right) = \frac{P(X \leq x \cap Y \leq y)}{P(Y \leq y)} = \frac{F_{XY}(x,y)}{F_Y(y)} = \frac{\int_{-\infty}^{y}\int_{-\infty}^{x} f_{XY}(u,v)\,du\,dv}{\int_{-\infty}^{y}\int_{-\infty}^{\infty} f_{XY}(x,v)\,dx\,dv} \tag{2.7}$$

It follows from Eq. (2.7) that if $F_{XY}(x,y)$ is differentiable, one can define the conditional density function, $f_X\left(x|Y \leq y\right) = \frac{\partial F_{XY}(x,y)/\partial x}{F_Y(y)}$. In particular, to find $f_X\left(x|Y = y\right)$, we first consider $F_X\left(x|y \leq Y \leq y + \Delta y\right)$ which is given by:

$$F_X\left(x|y \leq Y \leq y + \Delta y\right) = \frac{F_{XY}(x, y + \Delta y) - F_{XY}(x,y)}{F_Y(y + \Delta y) - F_Y(y)}$$

$$= \frac{\int_{-\infty}^{\infty}\int_{y}^{y+\Delta y} f_{XY}(x,v)\,dv\,dx}{f_Y(y)\,\Delta y}. \tag{2.8}$$

Differentiating with respect to x, we have:

$$f_X\left(x|\{y \leq Y \leq y + \Delta y\}\right) = \frac{\int_{y}^{y+\Delta y} f_{XY}(x,z)\,dz}{f_Y(y)\,\Delta y} = \frac{f_{XY}(x,y)\,\Delta y}{f_Y(y)\,\Delta y} \tag{2.9}$$

Hence we obtain $f_X\left(x|y\right)$ as:

$$f_X\left(x|y\right) = \lim_{\Delta y \to 0} f_X\left(x|y \leq Y \leq y + \Delta y\right) = \frac{f_{XY}(x,y)}{f_Y(y)} \tag{2.10}$$

Independence of random variables follows easily from the above equation and we say that X and Y are independent (also see Eq. 1.117b of Chapter 1) if:

$$f_X\left(x|y\right) = f_X(x) \Rightarrow f_{XY}(x,y) = f_X(x)\,f_Y(y) \tag{2.11}$$

2.2.1 Conditional expectation

Consider a random variable X defined on $(\Omega, \mathcal{F}, P)$ such that $E(|X|) < \infty$ (i.e., X is integrable). Let $\mathcal{G} \subset \mathcal{F}$ be a sub-σ-algebra, then the random variable $E[X|\,\mathcal{G}\,]$ is referred to as a conditional expectation of X given $\mathcal{G}$ satisfying the following properties:

(i) $E[X|\mathcal{G}]$ is $\mathcal{G}$-measurable (2.12a)

(ii) $\int_G E[X|\mathcal{G}]dP = \int_G XdP \ \forall\ G \in \mathcal{G}$ (2.12b)

In order to verify that $E[X|\mathcal{G}]$ exists and is $P_{\mathcal{G}}$- unique, with $P_{\mathcal{G}}$ denoting the measure restricted to $\mathcal{G}$, one may appeal to the Radon–Nikodym theorem (Section 1.6.4, Chapter 1). Without loss of generality, assume X to be a strictly positive random variable (if not, then one may decompose $X = X^+ - X^-$ and apply the arguments individually to the two positive components). Then, we may define a positive measure $\mu_{\mathcal{G}}$ as:

$$\mu_{\mathcal{G}}(G) = \int_G XdP \ \forall\ G \in \mathcal{G} \tag{2.13}$$

Clearly, $\mu_{\mathcal{G}}$ and $P_{\mathcal{G}}$ are absolutely continuous measures with respect to each other and by invoking the Radon–Nikodym theorem, we have a $P_{\mathcal{G}}$-measurable unique function $\Lambda = \frac{d\mu_{\mathcal{G}}}{dP_{\mathcal{G}}}$ such that for any $P_{\mathcal{G}}$- measurable set G, we have:

$$\mu_{\mathcal{G}}(G) = \int_G \Lambda\ dP_{\mathcal{G}} \tag{2.14}$$

However, given the definition in Eq. (2.12b) and the uniqueness of Λ, it follows that:

$$E\left[X\middle|\mathcal{G}\right] = \Lambda = \frac{d\mu_{\mathcal{G}}}{dP_{\mathcal{G}}} \tag{2.15}$$

The above equation implies that as $P_{\mathcal{G}}$ is restricted to a sub-σ-algebra $\mathcal{G} \subset \mathcal{F}$, the restriction has the conditional expectation $E[X|\mathcal{G}]$ as its Radon–Nikodym derivative. In the special case of all the σ-algebras above being Borel (so that X is a real-valued random variable measurable in $\mathcal{B}(\mathbb{R})$ with $(\mathbb{R}, \mathcal{B}(\mathbb{R}), P)$ being the complete probability space), one may consider yet another random variable Y so that $\sigma(Y) = \mathcal{G} \subset \mathcal{B}(\mathbb{R})$. In this case, one may define $E[X|Y] := E[X|\mathcal{G}]$. Thus $d\mu_{\mathcal{G}}$ in Eq. (2.15) can be written with respect to the Borel sets in $\mathcal{G}$ by (i) letting $dP = P(x \le X \le x + dx, Y \le y)$ and (ii) integrating the weighted sum XdP over $\mathcal{G}$. Thus we obtain $d\mu_{\mathcal{G}}$ as:

$$d\mu_{\mathcal{G}} = \int_{\mathcal{G}} XP\left(x \le X \le x + dx, Y \le y\right) \tag{2.16}$$

Since $P\left(x \le X \le x + dx, Y \le y\right) = F_{XY}(dx, y)$ where F_{XY} denotes the joint distribution of X, Y, we can rewrite the above equation as:

$$d\mu_{\mathcal{g}} = \int_{\mathcal{g}} x f_{XY}(x,y)\,dx \tag{2.17}$$

Recall that, if F_{XY} is sufficiently differentiable, f_{XY} denotes the joint density function of X and Y. Regarding the measure $dP_{\mathcal{g}}$ in Eq. (2.15), it follows from the fact $\sigma(Y) = \mathcal{g}$ that

$$dP_{\mathcal{g}}(Y \le y) = dF_Y(y) = f_Y(y)\,dy = \int_{\mathcal{g}} f_{XY}(x,y)\,dx \tag{2.18}$$

With $d\mu_{\mathcal{g}}$ and $dP_{\mathcal{g}}$ defined as above, we arrive at the classical definition typically adopted for conditional expectations in many texts [Papoulis 1991, Ash 2008]:

$$E[X|Y] := \int_{-\infty}^{\infty} x f_{X|Y}(x|y)\,dx = \int_{-\infty}^{\infty} \frac{x f_{XY}(x,y)}{f_Y(y)}\,dx = \frac{\int_{-\infty}^{\infty} x f_{XY}(x,y)\,dx}{\int_{-\infty}^{\infty} f_{XY}(x,y)\,dx} \tag{2.19}$$

Note that $f_{X|Y}(x|y) = \frac{f_{XY}(x,y)}{f_Y(y)}$ is generally referred to as the conditional density of X given Y.

Properties of conditional expectation [*Breiman* 1968]

(i) If X takes on positive as well as negative values, then $E[X|Y] = E[X^+|Y] - E[X^-|Y]$

(ii) $E[X|Y]$ is a function of Y, say $h(Y)$. If $h(Y)$ is integrable, we have:

$$E[h(Y)] = E[E[X|Y]] = \int_{-\infty}^{\infty}\left(\int_{-\infty}^{\infty} x f_{XY}(x|y)\,dx\right) f_Y(y)\,dy$$
$$= \int_{-\infty}^{\infty}\int_{-\infty}^{\infty} x f_{XY}(x,y)\,dx\,dy = \int_{-\infty}^{\infty} x f_X(x)\,dx = E[(X)] \tag{2.20}$$

(iii) If $X \ge 0$, then we have $E[X|\,\mathcal{g}] \ge 0$ almost surely (a.s.)

(iv) If $0 \le X_n \uparrow X$, then monotone convergence theorem (see Section 1.8.2, Chapter 1) applies and we have $E[X_n|\,\mathcal{g}] \uparrow E[X|\,\mathcal{g}]$ a.s. Fatou's lemma and dominated convergence theorem are similarly applicable for the conditional expectation operator.

(v) $E[(aX + bY)|\,\mathcal{g}] = aE[X|\,\mathcal{g}] + b\,E[Y|\,\mathcal{g}]$ (linearity)

(vi) If $\hbar$ is a sub-σ-algebra of $\mathcal{g}$, then $E[E[X|\,\mathcal{g}]|\hbar] = E[X|\hbar]$ (tower property)

(vii) If Y is $\mathcal{g}$-measurable (or $Y \subset \mathcal{g}$) and bounded, then $E[XY|\,\mathcal{g}] = YE[X|\,\mathcal{g}]$ a.s.

(viii) If X is $\mathcal{G}$-measurable, $E[X|\mathcal{G}] = X$

(ix) If Y is independent of X, then $E[X|Y] = E[X]$

Analogous to the definition of conditional expectation, conditional variance can be defined. Given $\mathcal{G} \subset \mathcal{F}$ a sub-σ-algebra, we define the conditional variance of X given $\mathcal{G}$ (denoted by $\sigma^2_{X|\mathcal{G}}$ as the random variable $E[(X - E(X \mid \mathcal{G}))^2) \mid \mathcal{G}]$. With respect to a random variable Y ($\mathcal{G}$-measurable), the conditional variance, $\sigma^2_{X|Y} = E\left[\left(X - m_{X|Y}\right)^2 |Y)\right] = \int_{-\infty}^{\infty} \left(x - m_{X|Y}\right)^2 f_X\left(x|Y = y\right)dx$. Here $m_{X|Y}$ denotes $E[X|Y]$.

2.2.2 Change of measure

The definition of a density function $f_X(x)$ in Chapter 1 (Section 1.6) and of conditional expectation in the present section is closely related to the concept of change of measure. In both the cases, two positive measures which are equivalent (i.e., absolutely continuous with respect to each other) are uniquely linked through a Radon–Nikodym derivative. More formally, if Q and P are two measures on a probability space, say discrete, and $Q \ll P$ (the notation $\ll$ denoting the absolute continuity of Q with respect to P), then Radon–Nikodym theorem ensures the existence and P—a.s. uniqueness of a non-negative measurable function (random variable) Λ such that:

$$Q(\omega) = \Lambda(\omega)\, P(\omega), \forall \omega \in \Omega \tag{2.21}$$

If X is a discrete random variable defined on $\mathcal{F}$, the expectations with respect to P and Q are given by:

$$E_P[X] = \sum_i X_i P(\omega_i), E_Q[X] = \sum_i X_i Q(\omega_i) \tag{2.22}$$

In view of the relationship in Eq. (2.21), $E_Q[X]$ can be written as:

$$E_Q[X] = \sum_i X_i \Lambda(\omega_i)\, P(\omega_i) \Longrightarrow E_Q[X] = E_P[\Lambda X] \tag{2.23}$$

We also have $Q(\Omega) = \sum_i Q(\omega_i) = 1 = \sum_i \Lambda(\omega_i) P(\omega_i) = E_P[\Lambda]$. Thus for a change of measure $Q \to P$, we find Λ with $E_P[\Lambda] = 1$ and thereby define $Q(\omega) = \Lambda(\omega)\, P(\omega)$.

In the case of a continuous sample space, say $\mathbb{R}$, we have (compare with Eq. 2.21):

$$dQ(\omega) = \Lambda(\omega)\, dP(\omega) \tag{2.24}$$

Here the notations dP and dQ are kept on for sake of consistency but otherwise, one can instead use the respective distribution functions as well. Now, if $f_X^P(x)$ and $f_X^Q(x)$ are the probability density functions (with respect to the Lebesgue measure) corresponding to Q and P respectively, the relation between these probability densities is given by:

$$f_X^Q(x) = \Lambda(x) f_X^P(x) \tag{2.25}$$

From the above equation, $Q(\Omega) = \int_{\mathbb{R}} f_X^Q(x)\,dx = 1 = \int_{\mathbb{R}} \Lambda(x) f_X^P(x)\,dx = E_P[\Lambda(x)]$.

Following examples illustrate the above ideas regarding change of probability measures as applied to random variables.

Example 2.2

Consider the familiar coin tossing experiment with $\Omega = \{\omega_1 = h, \omega_2 = t\}$ with h denoting the occurrence of the head and t of the tail. Let P be the probability measure if the coin is fair, with $P(h) = 1/2$ and $P(t) = 1/2$. If the coin is biased, we have an associated measure Q given by $Q(h) = q_1$ and $Q(t) = q_2 = 1 - q_1$. With this change of measure one can define $Q(\omega_i) = \Lambda(\omega_i) P(\omega_i),\ i = 1, 2$. In this simple case, we have $\Lambda(h) = 2q_1$ and $\Lambda(t) = 2q_2$. We observe that $\Lambda(\omega_i) > 0$ and also that $Q \ll P$.

Example 2.3

Consider two Gaussian measures P, Q which are respectively $\mathcal{N}(0,1)$ and $\mathcal{N}(\mu,1)$. With the corresponding *pdf*s known, the exercise to find Λ is trivial in this case too. We have by Eq. (2.25):

$$\frac{1}{\sqrt{2\pi}} \exp\left\{-\frac{1}{2}(x-\mu)^2\right\} = \exp\left(\mu x - \frac{\mu^2}{2}\right) . \frac{1}{\sqrt{2\pi}} \exp\left\{-\frac{1}{2}x^2\right\} \tag{2.26}$$

It thus follows that:

$$\Lambda(x) = \exp\left(\mu x - \frac{\mu^2}{2}\right) \tag{2.27}$$

We can verify that:

$$E_P[\Lambda] = \frac{1}{\sqrt{2\pi}} \int_{-\infty}^{\infty} \exp\left(\mu x - \frac{\mu^2}{2}\right) \exp\left\{-\frac{1}{2}x^2\right\} dx = \exp\left(-\frac{\mu^2}{2}\right) E_P\left(e^{\mu x}\right) = 1 \tag{2.28}$$

Note that $E_P\left[e^{\mu x}\right]$ is the moment generating function of $X \sim \mathcal{N}(0,1)$ and is equal to $\exp\left(\frac{\mu^2}{2}\right)$. Using Λ, the moment generating function with respect to the measure Q is obtained as:

$$E_Q\left(e^{\lambda X}\right) = E_P\left(\Lambda e^{\lambda X}\right) = E_P\left(e^{(\lambda+\mu)X - \frac{\mu^2}{2}}\right) = e^{-\frac{\mu^2}{2}} E_P\left(e^{(\lambda+\mu)X}\right) = e^{\lambda\mu + \lambda^2/2} \tag{2.29}$$

Example 2.4

Consider once more the measures P, Q as in the last example with $\mu = 6$. The problem is the estimation of $P(X < 0), X \sim Q$, by simulation. Specifically, if $x_i := X(\omega_i)$ denotes the ith realization of the random variable X, then we have the required probability with respect to Q as:

$$P(X < 0) \cong \frac{1}{n}\sum_{i=1}^{n} I_{x_i<0} \tag{2.30}$$

for sufficiently large n. Since the above probability is very low, being of $O(10^{-10})$, one expects $n \gg 10^{10}$ if one were to have a few realizations of X to be negative for which the indicator function on the RHS of the above equation would evaluate to 1. In other words, a straightforward numerical computation of the above probability would necessitate at least tens of billions of such realizations to be obtained. On the other hand, a substantial reduction in the computational effort can be effected through a change of measure from Q to $P = \mathcal{N}(0,1)$, so that one would typically have about half of the realizations negative. With $\Lambda(x) = \exp\left(\mu x - \frac{\mu^2}{2}\right)$, we may therefore write:

$$\begin{aligned} P(X < 0) &\cong \frac{1}{n}\sum_{i=1}^{n} \Lambda(x_i) I(x_i < 0) \\ &= \frac{1}{n}\sum_{i=1}^{n} \exp(6x_i - 18)\, I(x_i < 0) \\ &= e^{-12}\frac{1}{n}\sum_{i=1}^{n} \exp 6(x_i - 1)\, I(x_i < 0) \end{aligned} \tag{2.31}$$

with x_i now denoting realizations from $\mathcal{N}(0,1)$ distribution. This example illustrates the use of change of measure in simulation for estimation of probability of rare events.

2.2.3 Generalized Bayes' formula and conditional probabilities

Conditional expectations can be computed using Radon–Nikodym derivative Λ defined on $\mathcal{F}$ using a change of measure. Thus consider two measures P and Q defined on $\mathcal{F}$ and let $\mathcal{G}$ be any sub-σ-algebra of $\mathcal{F}$. Let $Q \ll P$ with $dQ = \Lambda dP$ and X be Q-integrable ($E_Q[|X|] < \infty$). Then ΛX is P-integrable, since $E_P[|\Lambda|] = 1$ from Eq. (2.25) and $E_P[|\Lambda X|] = E_Q[|X|] < \infty$. Generalized Bayes' formula gives the following relation for conditional probabilities over P and Q:

$$E_Q\left[X|\mathcal{G}\right] = \frac{E_P\left[\Lambda X|\mathcal{G}\right]}{E_P\left[\Lambda|\mathcal{G}\right]} \tag{2.32}$$

Proof: For any bounded $\mathcal{G}$-measurable random variable Z:

$$E_Q[ZX] = E_Q\left[ZE_Q\left[X|\mathcal{G}\right]\right] \tag{2.33}$$

The right hand side of Eq. (2.32) being $\mathcal{G}$-measurable, we have:

$$\begin{aligned}
E_Q\left[Z\frac{E_P\left[\Lambda X|\mathcal{G}\right]}{E_P\left[\Lambda|\mathcal{G}\right]}\right] &= E_P\left[\Lambda Z\frac{E_P\left[\Lambda X|\mathcal{G}\right]}{E_P\left[\Lambda|\mathcal{G}\right]}\right] \\
&= E_P\left[ZE_P\left[\Lambda|\mathcal{G}\right]\frac{E_P\left[\Lambda X|\mathcal{G}\right]}{E_P\left[\Lambda|\mathcal{G}\right]}\right] \\
&= E_P[ZE_P[\Lambda X|\mathcal{G}]] \\
&= E_P[Z\Lambda X] \\
&= E_Q[ZX] = E_Q[ZE_Q[X|\mathcal{G}]] \quad \text{(from Eq. 2.33)}
\end{aligned}$$

$$\Longrightarrow \frac{E_P\left[\Lambda X|\mathcal{G}\right]}{E_P\left[\Lambda|\mathcal{G}\right]} = E_Q\left[X|\mathcal{G}\right], Q \quad \text{a.s.} \tag{2.34}$$

In Eqs. (2.33) and (2.34), one can as well use indicator function I_Z in place of Z.

The generalized Bayes' formula will be used extensively when we study the filtering and optimization problems in Chapters 6–7 and 9–10, respectively.

2.2.4 Conditional expectation as the least mean square error estimator

If X is a square integrable random variable (i.e., $E\left[X^2\right] < \infty$) defined on $\mathcal{F}$ and $\mathcal{G} \subseteq \mathcal{F}$, then $E[X|\mathcal{G}]$ is the unique (up to almost sure equivalence) random variable Y that minimizes the mean square error between X and any $\mathcal{G}$-measurable approximation to X.

The result (proved below) can be geometrically interpreted by considering $E[(X-Y)^2]$ as the square of the L^2-distance between X and Y in the Hilbert space $L^2(\Omega, \mathcal{F}, P)$. The conditional expectation operator $E[.|\mathcal{G}]$ applied to a random variable X results in a random variable closest to X, picked from the square integrable measurable functions in the linear space $L^2(\Omega, \mathcal{G}, P) \subset L^2(\Omega, \mathcal{F}, P)$. In other words, $E[.|\mathcal{G}]$ is an orthogonal projection operator on to the subspace $L_2\left(\Omega, \mathcal{G}, P\right)$ and is unique.

Proof: Let us denote $Z := E[X|\mathcal{G}]$. If Y is $\mathcal{G}$-measurable with $E\left[Y^2\right] < \infty$), then:

$$E\left[(X-Y)^2\right] = E\left[((X-Z)-(Y-Z))^2\right]$$
$$= E\left[(X-Z)^2\right] + E\left[(Y-Z)^2\right] - 2E\left[(X-Z)(Y-Z)\right] \quad (2.35)$$

With $W = Y - Z$ also being $\mathcal{G}$-measurable, we can have:

$$E[WX] = E\left[E\left[WX|\mathcal{G}\right]\right] = E\left[WE\left[X|\mathcal{G}\right]\right] = E[WZ] \quad (2.36)$$

Noting that $E\left[(WX)^2\right] \le \left(E\left[W^2\right]E\left[X^2\right]\right) \le \infty$ by Cauchy–Schwartz inequality [Appendix B] we obtain:

$$E[(X-Z)(Y-Z)] = E[(X-Z)W] = 0 \quad (2.37)$$

In view of the above, Eq. (2.35) simplifies to:

$$E\left[(X-Y)^2\right] = E\left[(X-Z)^2\right] + E\left[(Y-Z)^2\right] \quad (2.38)$$

Letting $E\left[(Y-Z)^2\right] = \Delta^2 \ge 0$, it is clear that Z (as an element of $L_2\left(\Omega, \mathcal{G}, P\right)$) is closest to X. To prove the uniqueness of the result let Z^* be another $\mathcal{G}$-measurable function that minimizes $E\left[(X-Y)^2\right]$, i.e., $E\left[(X-Z^*)^2\right] = E\left[\left(X - E\left[X|\mathcal{G}\right]\right)^2\right]$. Similar to Eq. (2.38), we can again show that $E\left[(X-Y)^2\right] = E\left[(X-Z^*)^2\right] + E\left[(Y-Z^*)^2\right]$ and for Z^* to qualify as being closest to X, we must have $Y = Z^*$, i.e., $E[X|\mathcal{G}] = Z^* = Z$ a.s.

2.2.5 Rosenblatt transformation

The Rosenblatt transformation [Rosenblatt 1952] is discussed in this section since it uses the concept of conditional probabilities in transforming a vector of statistically dependent random variables with a specified joint *pdf*, say $f_{\boldsymbol{X}}(\boldsymbol{x})$, into one of independent standard normal variables $\boldsymbol{Z}$. The procedure involves the following successive conditioning of random variables:

$$z_1 = \Phi_{Z_1}{}^{-1}\left(F_{X_1}(x_1)\right)$$

$$z_2 = \Phi_{Z_2}{}^{-1}\left(F_{X_2|X_1}(x_2|x_1)\right)$$

$$z_3 = \Phi_{Z_3}{}^{-1}\left(F_{X_3|X_2,X_1}(x_3|x_2,x_1)\right)$$

$$\ldots\ldots\ldots\ldots\ldots\ldots\ldots\ldots\ldots\ldots$$

$$z_n = \Phi_{Z_n}{}^{-1}\left(F_{X_n|X_{n-1},X_{n-2},\ldots,X_1}(x_n|x_{n-1},x_{n-2},\ldots,x_1)\right) \tag{2.39}$$

Given $f_{\mathbf{X}}(\mathbf{x})$, the conditional distributions in the above equation are obtainable from successive use of Eqs. (1.86) (of Chapter 1) and (2.10). We find from the equation that $\prod_i F_{Z_i}(z_i) = F_{X_1}(x_1) F_{X_2|X_1}(x_2|x_1) \ldots F_{X_n|X_{n-1},X_{n-2},\ldots,X_1}(x_n|x_{n-1},x_{n-2},\ldots,x_1)$, whence the independence of Z_i is apparent.

Since Z_i, $i = 1,2,\ldots,n$ are all standard normal variables, $F_{Z_i}(z_i)$ are identical for all i and are given by $\frac{1}{2} + \text{erf}(z_i)$ (see Section 1.10.3, Chapter 1). Thus, given a joint *pdf* of correlated non-normal random vector $\mathbf{X}$, Eq. (2.39) helps in transformation to independent standard normal variables and also provides a means to generate random sequence corresponding to each Z_i that may find use in simulation studies (see Section 2.5, Monte Carlo simulation). Also note that the Rosenblatt transformation (and hence its performance) depends on the ordering of the input random variables X_i.

Example 2.5

We use the Rosenblatt transformation to evaluate the reliability index of a system [Ang and Tang 1984] consisting of a uniform beam with section modulus S_m and subjected to a bending moment B_m at the pertinent section. The yield strength S_y of the material and S_m follow a lognormal distribution and B_m a Type I asymptotic distribution (see Eq. 2.47 in the text below). The following are the specified statistics (in appropriate units) of the three random variables:

mean :	$\mu_{S_y} = 40$	$\mu_{S_m} = 50$	$\mu_{B_m} = 1000$
standard deviation:	$\sigma_{S_y} = 5$	$\sigma_{S_m} = 2.5$	$\sigma_{B_m} = 200$

The variables S_y and S_m are partially correlated with correlation coefficient $= 0.4$.

Solution

We refer to Appendix A for short notes on the topic of system reliability and Appendix B for evaluating the reliability index and the associated probability of failure by Rosenblatt transformation. With $\mathbf{X} = \left(S_y, S_m, B_m\right)$, the limit state function in this example is $g(\mathbf{X}) = g\left(S_y, S_m, B_m\right) = S_y S_m - B_m$. Since S_y and S_m are correlated non-normals, we follow the iterative procedure [Hohenbichler and Rackwitz 1981] enumerated in Table B.1 of Appendix B, to find the reliability index. To this end, we use the Rosenblatt transformation to first transform the correlated non-normals S_y and S_m into independent (uncorrelated) standard normals at each iteration.

The Rosenblatt transformation if applied to S_y and S_m, yields (see Eq. 2.39):

$$z_1 = \Phi^{-1}\left[F_{S_y}\left(s_y\right)\right] \tag{2.40a}$$

$$z_2 = \Phi^{-1}\left[F_{S_m|S_y}\left(s_m \left| s_y\right.\right)\right] \tag{2.40b}$$

Since S_y is lognormal, $F_{S_y}\left(s_y\right) = \Phi\left(\frac{\ln s_y - \lambda_{s_y}}{\zeta_{s_y}}\right)$ and $f_{S_y}\left(s_y\right) = \Phi\left(\frac{\ln s_y - \lambda_{s_y}}{\zeta_{s_y}}\right)$ where ζ_{s_y} and λ_{s_y} are related to μ_{S_y} and σ_{S_y} by:

$$\zeta_{s_y} \simeq \frac{\sigma_{S_y}}{\mu_{S_y}}, \lambda_{s_y} = \ln \mu_{S_y} - \frac{1}{2}\left(\zeta_{s_y}\right)^2 \tag{2.41}$$

Equation (2.40a) gives:

$$z_1 = \Phi^{-1}\left[\Phi\left(\frac{\ln s_y - \lambda_{s_y}}{\zeta_{s_y}}\right)\right] = \frac{\ln s_y - \lambda_{s_y}}{\zeta_{s_y}} \tag{2.42}$$

Further, $F_{S_m|S_y}\left(s_m \left| s_y\right.\right)$ in Eq. (2.40b) is:

$$\begin{aligned} F_{S_m|S_y}\left(s_m \left| s_y\right.\right) &= \int_{-\infty}^{s_m} f_{S_m|S_y}\left(u \left| s_y\right.\right) du \\ &= \int_{-\infty}^{s_m} \frac{f_{S_m,S_y}\left(u, s_y\right)}{f_{S_y}\left(s_y\right)} du \end{aligned} \tag{2.43}$$

Since S_y and S_m are lognormals, $f_{S_m,S_y}\left(s_m, s_y\right)$, the joint *pdf* of S_y and S_m, is given by:

$$f_{S_y,S_m}\left(s_y, s_m\right)$$

$$= \frac{1}{2\pi s_y s_m \zeta_{s_y} \zeta_{s_m} \sqrt{1-r^2}}$$

$$\left[\exp\left\{-\frac{1}{2}\left(\frac{\ln s_y - \lambda_{s_y}}{\zeta_{s_y}}\right)^2 + r\left(\frac{\ln s_y - \lambda_{s_y}}{\zeta_{s_y}}\right)\left(\frac{\ln s_y - \lambda_{s_y}}{\zeta_{s_y}}\right) - \frac{1}{2}\left(\frac{\ln s_y - \lambda_{s_y}}{\zeta_{s_y}}\right)^2\right\}\right]$$

$$= \frac{1}{2\pi s_y s_m \zeta_{s_y} \zeta_{s_m} \sqrt{1-r^2}}$$

$$\left[\exp\left\{ -\frac{1}{2}\left(\frac{\ln s_y - \lambda_{s_y}}{\zeta_{s_y}}\right)^2 - \frac{1}{2}\left(\frac{\ln s_m - \lambda_{s_m} - r\frac{\zeta_{sm}}{\zeta_{sy}}\left(\ln s_y - \lambda_{s_y}\right)}{\zeta_{s_m}\sqrt{1-r^2}}\right)^2 \right\} \right] \tag{2.44}$$

Equations (2.40b), (2.43) and (2.44) lead to:

$$F_{S_m|S_y}\left(s_m \middle| s_y\right) = \int_{-\infty}^{s_m} f_{S_m|S_y}\left(u \middle| s_y\right) du$$

$$= \int_{-\infty}^{s_m} \frac{1}{\sqrt{2\pi} u \zeta_{s_m} \sqrt{1-r^2}} \left[\exp\left\{ -\frac{1}{2}\left(\frac{\ln u - \lambda_{s_m} - r\frac{\zeta_{sm}}{\zeta_{sy}}\left(\ln s_y - \lambda_{s_y}\right)}{\zeta_{s_m}\sqrt{1-r^2}}\right)^2 \right\} \right] du$$

$$= \Phi\left(\frac{\ln s_m - \lambda_{s_m} - r\frac{\zeta_{sm}}{\zeta_{sy}}\left(\ln s_y - \lambda_{s_y}\right)}{\zeta_{s_m}\sqrt{1-r^2}}\right)$$

$$\Longrightarrow z_2 = \Phi^{-1}\left(\Phi\left(\frac{\ln s_m - \lambda_{s_m} - r\frac{\zeta_{sm}}{\zeta_{sy}}\left(\ln s_y - \lambda_{s_y}\right)}{\zeta_{s_m}\sqrt{1-r^2}}\right)\right)$$

$$\Longrightarrow z_2 = \frac{\ln s_m - \lambda_{s_m} - r\frac{\zeta_{sm}}{\zeta_{sy}}\left(\ln s_y - \lambda_{s_y}\right)}{\zeta_{s_m}\sqrt{1-r^2}} \tag{2.45}$$

z_3, the independent standard normal corresponding to the non-normal variable B_m is:

$$z_3 = \Phi^{-1}\left[F_{B_m}(b_m)\right] \tag{2.46}$$

$F_{B_m}(b_m)$, the CDF, and $f_{B_m}(b_m)$, the *pdf*, correspond to Type I asymptotic (distribution) and are given by:

$$F_{B_m}(b_m) = \exp\left(-\exp\left(-\alpha(b_m - u)\right)\right) \tag{2.47a}$$

$$f_{B_m}(b_m) = \alpha \exp(-\alpha(b_m - u) - \exp(-\alpha(b_m - u))) \tag{2.47b}$$

where, the parameters α and u are related to μ_{B_m} and σ_{B_m} as $\alpha = \frac{\pi}{\sqrt{6}} \frac{1}{\sigma_{B_m}}, u = \mu_{B_m} - \frac{0.577}{\alpha}$.

Following Hohenbichler and Rackwitz iterative procedure, the reliability index is obtained as $\beta_F = 2.662$ and the results are given in Table 2.1. The design point is given by $S_y = 33.7, S_m = 47.8$ and $B_m = 1612.0$. The probability of failure is given by $\Phi(-\beta_F) = 0.0039$, which corresponds to a reliability of 99.61% under the given loading.

Table 2.1 Results for Example 2.5 by Hohenbichler and Rackwitz iterative procedure (Appendix B)

Iteration no.	S_y	S_m	B_m	β_F
1	40.0	50.0	1000.0	0.187
2	29.8	46.2	1328.2	2.8275
3	33.2	47.5	1597.9	2.7117
4	33.8	47.8	1614.4	2.662
5	33.7	47.8	1612.0	2.662

2.3 Convergence of Random Variables

Some essential concepts (definitions) of convergence of a sequence of random variables are described below. The stronger convergence criteria are given first, followed by progressively weaker ones. These criteria are useful in studying the limits of random variables. One such limit through monotone convergence is already given in Chapter 1 (Section 1.8.2) with regard to a measure.

2.3.1 Convergence of a sequence of random variables

Consider a sequence of random variables $\{X_1, X_2, \ldots.., X_n, \ldots\}$.

(i) Convergence almost surely (a.s.)

The sequence $\{X_n\}$ is said to converge to X almost surely (a.s.), if there exists a measurable set A with $P(A) = 0$ such that $\lim_{n\to\infty} X_n(\omega) \to X(\omega)$ for every $\omega \in A^c$.

(ii) Convergence in probability

If $P\left(\omega : \left|X_n(\omega) - X(\omega)\right| \geq \varepsilon\right) \to 0$ for every $\varepsilon > 0$ as $n \to \infty$, then the sequence $\{X_n\}$ is said to converge to X in probability.

(iii) Convergence in the mean square sense

If $E\left[|X_n - X|^2\right] \to 0$ as $n \to \infty$, then the sequence $\{X_n\}$ is said to converge to X in the mean square sense (or L_2 sense). In general, the sequence $\{X_n\}$ is said to converge to X in L_p if for any n, $E\left[|X_n|^p\right] < \infty$ and $E\left[|X_n - X|^p\right] \to 0$ as $n \to \infty$.

(iv) Convergence in distribution (weak convergence)

If $F_n(x)$ and $F(x)$ are the distribution functions of X_n and X respectively such that $F_n(x) \to F(x)$ as $n \to \infty$ for every point of continuity of $F(x)$, then the sequence $\{X_n\}$ is said to converge to X in distribution. Convergence in distribution is equivalent to convergence of expectations / characteristic functions (moment generating functions) of X_n to those of X.

Convergence criteria and inter-relationships

First, we need to put in place a few definitions and the related concepts on limits of events or random variables.

Consider a sequence of events, $(A_n : n \in \mathbb{N})$. Let us define (A_n, infinitely often), written in short as (A_n, i.o.), and (A_n, eventually), written in short as (A_n, ev). The first is also known as Lim sup A_n and the second one as Lim inf A_n

a) Definition: $(A_n, \text{i.o.}) := \limsup A_n := \bigcap_m \bigcup_{n \geq m} A_n$

$$= \{\omega : \text{ for every } m, \ \exists\, n(\omega) \geq m \text{ such that } \omega \in A_n(\omega)\}$$

$$= \{\omega : \ \omega \in A_n \text{ for infinitely many } n\} \tag{2.48}$$

b) Definition: $(A_n, \text{ev}) := \liminf A_n := \bigcup_m \bigcap_{n \geq m} A_n$

$$= \{\omega : \text{ for some } m(\omega), \ \omega \in A_n(\omega) \ \ \forall n \geq m(\omega)\}$$

$$= \{\omega : \ \omega \in A_n \text{ for all large } n\} \tag{2.49}$$

Proposition: $P(\liminf A_n) \geq \liminf P(A_n)$ (known as Fatou's Lemma)

Let $G_m := \bigcap_{n \geq m} A_n$. If $G := \liminf A_n$, then by the monotone convergence property (Section 1.3, Chapter 1) as $G_m \uparrow G, P(G_m) \uparrow P(G)$. This leads to $P(G_m) \leq \inf_{n \geq m} P(A_n)$. Hence:

$$P(G) \leq \lim\{\inf_{n \geq m} P(A_n)\} =: \liminf P(A_n) \tag{2.50}$$

Proposition: $P(\limsup A_n) \geq \limsup P(A_n)$; (Reverse Fatou's Lemma)

Let $G_m := \bigcup_{n \geq m} A_n$. If $G := \limsup A_n$, then by the monotone convergence property as $G_m \downarrow G$, $P(G_m) \downarrow P(G)$. This leads to $P(G_m) \geq \sup_{n \geq m} P(A_n)$. Hence:

$$P(G) \geq \lim\left\{ \sup_{n \geq m} P(A_n)\right\} =: \ \limsup P(A_n) \tag{2.51}$$

Now, we claim and prove that convergence almost surely and convergence in L_p imply convergence in probability that in turn implies convergence in distribution.

(i) First, we show that $X_n \to X$ *a.s.* $\Longrightarrow X_n \to X$ in probability.

For any $\varepsilon > 0$, there exists an $N(\omega, \varepsilon) \in \mathbb{N}$ so that $\left|X_n(\omega) - X(\omega)\right| \le \varepsilon$ for all $n > N(\omega, \varepsilon)$ with probability one. Therefore, $\left|X_n(\omega) - X(\omega)\right| > \varepsilon$ happens only for finitely many n's, i.e., $P(|X_n - X| > \varepsilon \text{ i.o.}) = 0$. Invoking reverse Fatou's lemma as above, we have lim sup $P(|X_n - X| > \varepsilon) \le P(|X_n - X| > \varepsilon \text{ i.o.}) = 0$.

(ii) Here we show that $X_n \to X$ in $L_p \Longrightarrow X_n \to X$ in probability.

First note that $P\left(\left|X_n(\omega) - X(\omega)\right| > \varepsilon\right) = P((X_n(\omega) - X(\omega))^p > \varepsilon^p)$. Now by Chebyshev inequality (Section 2.4) $((X_n(\omega) - X(\omega))^p > \varepsilon^p) \le \frac{E[|X_n - X|^p]}{\varepsilon^p}$. Given that $X_n \to X$ in L_p, we have $E[|X_n - X|^p] \to 0$ as $n \to \infty \Longrightarrow$ $P\left(\left|X_n(\omega) - X(\omega)\right| > \varepsilon\right) \to 0$ as $n \to \infty$ and hence the convergence in probability.

(iii) Finally, we show that $X_n \to X$ in probability $\Longrightarrow X_n \to X$ in distribution

We prove that $E[f(X_n)] \to E[f(X)]$ as $n \to \infty$ where $f(x)$ is a bounded and continuous function. Let $\left|f(x)\right| \le K/2$ for some $K < \infty$. $f(x)$ being continuous, for any $\varepsilon > 0$, $\exists\, \delta > 0$ such that $\left|f(x) - f(y)\right| < \varepsilon \Longrightarrow \left|x - y\right| < \delta$. Now let us consider:

$$\left|E[f(X_n) - f(X)]\right| \le E\left[\left|f(X_n) - f(X)\right|\right] \text{ (Jensen's inequality – Section 2.4)}$$

$$= E\left[\left(I_{|f(X_n)-f(X)|>\varepsilon} + I_{|f(X_n)-f(X)|\le\varepsilon}\right)\left|f(X_n) - f(X)\right|\right] \tag{2.52}$$

Here

$$E\left[I_{|f(X_n)-f(X)|\le\varepsilon}\left|f(X_n) - f(X)\right|\right] \le \varepsilon$$

and

$$E\left[I_{|f(X_n)-f(X)|>\varepsilon}\left|f(X_n) - f(X)\right|\right] \le KE\left[I_{|f(X_n)-f(X)|>\varepsilon}\right]$$

$$= KP\left(\left|f(X_n) - f(X) > \varepsilon\right|\right) \tag{2.53}$$

Thus, from the above two equations, we obtain:

$$\left|E[f(X_n) - f(X)]\right| \le \varepsilon + KP\left(\left|f(X_n) - f(X)\right| > \varepsilon\right)$$

$$\le \varepsilon + KP(|X_n - X| > \delta) \text{ by continuity of } f(X) \tag{2.54}$$

Since $X_n \to X$ in probability, $P(|X_n - X| > \delta) \to 0$ as $n \to \infty$. Finally, $|E[f(X_n) - f(X)]| \le \varepsilon \implies \lim\sup |E[f(X_n) - f(X)]| \le \varepsilon$. Since this is true for any $\varepsilon > 0$, $E[f(X_n)] \to E[f(X)]$ and hence the result. See Appendix C for additional details and proofs on Fatou's lemma and monotone convergence.

2.3.2 Law of large numbers

It may often be of interest to determine the asymptotic nature of probabilities of events, empirically estimated, for instance, from actual observations / experiments. The law of large numbers is concerned with the convergence of these estimates. There exist two versions of the law of large numbers, e.g., the weak and strong laws. While both of these refer to the estimate of a sample average from a large number of available data, they differ only in the mode of convergence of the estimate.

Strong law of large numbers

For the sequence of independent random variables X_j, $j = 1,2,..,$ each with finite mean m, and $E\left[\left(X_j - m\right)^4\right] \le K \in [0,\infty)$, the strong law states that:

$$P\left(\lim_{n\to 8} \frac{X_1 + X_2 + \ldots. + X_n}{n} = m\right) = 1 \tag{2.55}$$

Proof: Let $Y_j = X_j - m$ with $E\left[Y_j\right] = 0$. If $S_n = \sum_j Y_j$, the law states that $P\left(\frac{S_n}{n} \to 0\right) = 1$ or $\frac{S_n}{n} \to 0$ (*a.s.*). Now, we have $E\left[S_n{}^4\right] = E\left(\sum_j Y_j\right)^4$ and we can show that:

$$E\left[S_n{}^4\right] = \sum_k Y_k{}^4 + 6\sum_k \sum_{j<k} Y_j{}^2 Y_k{}^2 \tag{2.56}$$

Note that, in arriving at the above equation, independence of Y_j resulting in the identities $E\left[Y_i Y_j{}^3\right] = E\left[Y_i Y_j{}^2 Y_k\right] = E\left[Y_i Y_j Y_k Y_l\right] = 0$ for distinct i, j, k and l, has been made use of. Further, we have:

$$\left(E\left[Y_j{}^2\right]\right)^2 \le E\left[Y_j{}^4\right] \le K, \quad \forall j \tag{2.57a}$$

and

$$E\left[Y_j{}^2 Y_k{}^2\right] = E\left[Y_j{}^2\right] E\left[Y_k{}^2\right] \le K, \quad j \ne k \tag{2.57b}$$

In view of the above equation, Eq. (2.56) simplifies to:

$$E\left[S_n{}^4\right] \le nK + 3n(n-1)K \le 3Kn^2 \tag{2.58}$$

Finally, we obtain:

$$E\left[\sum_n\left(\frac{S_n}{n}\right)^4\right]\le 3K\sum_n\frac{1}{n^2}<\infty\Longrightarrow\sum_n\left(\frac{S_n}{n}\right)^4<\infty\ \text{a.s.}\ \Longrightarrow\frac{S_n}{n}\to 0\ \text{a.s.}\quad(2.59)$$

For the non-zero-mean X_j, the law thus leads to the conclusion that $\frac{S_n}{n}\to m$ a.s.

Weak law of large numbers

Let $X_j,\ j=1,2,..,$ be a sequence of independent random variables, each having a finite common mean, $E\left[X_j\right]=m$. Then, for any $\varepsilon>0$:

$$P\left(\left|\frac{X_1+X_2+\ldots.+X_n}{n}-m\right|\ge\varepsilon\right)\to 0\ \text{ as }\ n\to\infty\quad(2.60)$$

Thus, in contrast to the strong law where the convergence is a.s., the weak law implies convergence in *probability*. One can make use of Chebyshev's inequality (Section 2.4) to prove the weak law and the proof is left as an exercise to the reader.

The weak law of large numbers implies that for a specified large number n of trials of a random experiment, the sample mean $\frac{X_1+X_2+\ldots.+X_n}{n}$ stays near the value m and may differ, though infrequently, by large values. On the other hand, the strong law of large numbers states that, with probability 1, the difference of the sample mean from m may be made smaller than any given positive ε for sufficiently large n.

2.3.3 Central limit theorem (CLT)

Given n independent random variables $X_j,\ j=1,2,..,n$, with X_j having mean m_j and variance σ_j^2, define $Y=\sum_{j=1}^n X_j$ to be another random variable with mean $m=\sum_{j=1}^n m_j$ and variance $\sigma^2=\sum_{j=1}^n\sigma_j^2$. The central limit theorem states that, under certain conditions, the distribution $F_Y(y)$ of Y approaches a normal distribution with mean m and variance σ^2 as $n\to\infty$. The theorem provides an illustration for convergence in distribution. The theorem implies that if X_j is a sequence of continuous random variables, the density function of Y approaches the normal *pdf* $f_Y(y)=\frac{1}{\sigma\sqrt{2\pi}}\exp\left(-\frac{1}{2}\left(\frac{y-m}{\sigma}\right)^2\right)$.

In other words, if $Z=\frac{Y-m}{\sigma}$, the CLT states that $F_Z(z)$ tends to $\mathcal{N}(0,1)$ as $n\to\infty$.

Proof: Assume that $\varnothing_j(\lambda)$ and $\varnothing_Y(\lambda)$ are the characteristic functions of X_j and Y. In the particular case of X_j being identically distributed, the derivation is relatively easy and proceeds as follows. Since the random variables X_j are independent, we have:

$$\varnothing_Y(\lambda)=\prod_{j=1}^n\varnothing_j(\lambda)=\left(\varnothing_j(\lambda)\right)^n\quad(2.61)$$

Here λ is the parameter in the definition of the characteristic function (see Eq. 1.112 of Chapter 1). Thus, the characteristic equation of $Z = \frac{Y-m}{\sigma}$ is given by:

$$\varnothing_Z(\lambda) = \left[\exp\left(-\frac{i\lambda m}{\sigma\sqrt{n}}\right)\varnothing_j\left(\frac{\lambda}{\sigma\sqrt{n}}\right)\right]^n$$

$$= \exp\left(-\frac{i\lambda m}{\sigma\sqrt{n}}\right)\left(E\left[\exp\frac{i\lambda Y}{\sigma\sqrt{n}}\right]\right)^n$$

$$= \exp\left(-\frac{i\lambda m}{\sigma\sqrt{n}}\right)\left\{1 + E[Y]\frac{i\lambda}{\sigma\sqrt{n}} + \frac{E\left[Y^2\right]}{2}\left(\frac{i\lambda}{\sigma\sqrt{n}}\right)^2 + \ldots..\right\}^n$$

$$= \left[\exp\left(-\frac{i\lambda m}{\sigma\sqrt{n}}\right)\left\{1 + \frac{i\lambda m}{\sigma\sqrt{n}} + \frac{\left(m^2+\sigma^2\right)}{2}\left(\frac{i\lambda}{\sigma\sqrt{n}}\right)^2 + \ldots..\right\}\right]^n$$

$$= \left[1 - \frac{\lambda^2}{2n} + o\left(\frac{\lambda^2}{2n}\right)\right]^n \tag{2.62a}$$

where, $o\left(\frac{\lambda^2}{2n}\right)$ is the remainder in the above expansion that tends to zero as $n \to \infty$ faster than the second term $\frac{\lambda^2}{2n}$. Thus,

$$\varnothing_Z(\lambda) \to \exp\left(-\frac{\lambda^2}{2}\right) \text{ as } n \to \infty \text{ (using } \lim_{n\to\infty}\left(1+\frac{1}{n}\right)^n = e) \tag{2.62b}$$

The last result shows that $\varnothing_Z(\lambda)$ is the characteristic function of an $\mathcal{N}(0,1)$ random variable and hence the result.

In case X_j are not identically distributed, it is necessary that each random variable has finite variance and has, in the sense described below, small probability of X_j being large. This requirement, referred to as Lindeberg condition [Thomasian 1969], is given by:

$$\lim_{n\to\infty}\frac{1}{\sigma^2}\sum_{j=1}^{n}\int_{|X_j-m_j|>\varepsilon\sigma}\left(x_j-m_j\right)^2 f_{X_j}\left(x_j\right)dx_j = 0, \text{ for every } \varepsilon > 0 \tag{2.63}$$

We can find a violation of the above condition if X_j is Cauchy distributed $\left(f_{X_j}(x_j) = \frac{a}{\pi(x^2+a^2)}\right)$. Here, the variance is infinite and as such $Y = \sum_{j=1}^{n} X_j$ is not a normal distribution in the limit and indeed tends to a Cauchy distribution itself.

2.3.4 Random walk and central limit theorem

Consider a suspended particle in a fluid experiencing random displacement due to bombardment of the fluid molecules. If n is the number of such bombardments in time T and every such bombardment imparts to the particle an independent and identically distributed displacement X_j, then the position of the particle at time T is given by:

$$Y = \sum_{j=1}^{n} X_j \tag{2.64}$$

With the assumption that each bombardment contributes only a small displacement to the suspended particle and with $E\left[X_j\right] = 0$ and $E\left[{X_j}^2\right] = \sigma^2$ finite, suppose that one is interested to know the limiting behaviour of Y as $n \to \infty$. In this limit, the central limit theorem gives the distribution of Y as Gaussian with zero mean and variance $= n\sigma^2$. In fact, the random variable Y represents a random walk model. The limiting motion of the particle constitutes a stochastic process and is known as the Brownian motion or a Wiener process, a detailed account of which is given in Chapter 3.

2.4 Some Useful Inequalities in Probability Theory

Markov's inequality

If X is a non-negative random variable on $(\Omega, \mathcal{F}, P)$, then for any $\alpha > 0$, $P(X \geq \alpha) \leq \frac{E[X]}{\alpha}$.

Proof: Define the event $A = \{\omega \in \Omega : X(\omega) \geq \alpha\}$. Now, we have:

$$X(\omega) \geq \alpha I_A \Longrightarrow E[X] \geq \alpha E[I_A] = \alpha P(X \geq \alpha)$$

$$\Longrightarrow P(X \geq \alpha) \leq \frac{E[X]}{\alpha} \tag{2.65}$$

As a corollary, if X is a random variable and $g(X)$ a non-negative function, we have:

$$P(g(X) \geq \alpha) \leq \frac{E[(g(X))]}{\alpha} \tag{2.66}$$

In particular, if $g(X) = |X|$, the inequality takes the form:

$$P(|X| \geq \alpha) \leq \frac{E[|X|]}{\alpha} \tag{2.67}$$

Markov's inequality provides us with a bound on the probability of events at the tail of a probability distribution. Further, it implies that if the expectation of a non-negative random variable is finite, the random variable may take large values with almost zero probability. As an example, if $\alpha = 1000E[|X|]$ then $P(|X| \geq \alpha) \leq 0.001$. Indeed, Markov's inequality enables arriving at the above estimate without a recourse to the probability distribution of the random variable.

Chebyshev's inequality

Given both the mean m and variance σ^2 of a random variable X, Chebyshev's inequality helps in obtaining a sharper bound than Markov's inequality. The inequality is of the form:

$$P(|X - m| \geq \varepsilon) \leq \frac{\sigma^2}{\varepsilon^2} \tag{2.68}$$

Proof: We apply Markov's inequality to the (non-negative) random variable $(X - m)^2$ with $\alpha = \varepsilon^2$ to obtain:

$$P\left((X - m)^2 \geq \varepsilon^2\right) \leq \frac{E\left[(X - m)^2\right]}{\varepsilon^2} = \frac{\sigma^2}{\varepsilon^2} \tag{2.69a}$$

Since probability of the event $\left((X - m)^2 \geq \varepsilon^2\right)$ is the same as that of $(|X - m| \geq \varepsilon)$, we get $P(|X - m| \geq \varepsilon) \leq \frac{\sigma^2}{\varepsilon^2}$. If $\varepsilon = k\sigma$ is used, we have:

$$P(|X - m| \geq k\sigma) \leq \frac{1}{k^2} \tag{2.69b}$$

In a general case, the random variable $|X - m|^n$ takes positive values and hence with $\alpha = \varepsilon^n$, we have:

$$P(|X - m|^n \geq \varepsilon^n) \leq \frac{E[|X - m|^n]}{\varepsilon^n} \tag{2.69c}$$

As an example, given $X \sim \mathcal{N}(m, \sigma)$, Eq. (2.69b) gives an estimate of $P(|X - m| \geq \sigma) \leq 1$, $P(|X - m| \geq 2\sigma) \leq 0.25$ and $P(|X - m| \geq 3\sigma) \leq 0.111$. However, noting that $P(|X - m| \geq k\sigma) = 1 - \int_{-k\sigma}^{k\sigma} (x - m) f_X(x)\, dx$, one can exactly compute the above probabilities as $P(|X - m| \geq \sigma) = 0.317$, $P(|X - m| \geq 2\sigma) = 0.045$ and $P(|X - m| \geq 3\sigma) = 0.003$. Thus, for a normal random variable, while 68% of the probability mass is concentrated within a band of one standard deviation around the mean, 95% is within two standard deviations and 99% is within three standard deviations. The estimates by Chebyshev's inequality in Eq. (2.69c) may thus be conservative. Chebyshev's inequality may also be used to prove the weak law of large numbers (see Section 2.3.2).

Jensen's inequality

Let $g: I \to \mathbb{R}$ be a convex function (Appendix B) on an open interval $I = (a,b)$ in $\mathbb{R}$. If X is a random variable with range (a,b), then:

$$g(E(X)) \le E(g(X)) \tag{2.70a}$$

For instance, if $g(x) = |x|$, then:

$$|E(X)| \le E(|X|) \tag{2.70b}$$

Proof: Referring to Fig. 2.3, use the notation $\Delta_{c,d} = \frac{g(d)-g(c)}{d-c}$ and $\Delta_{d,e} = \frac{g(e)-g(d)}{e-d}$ with $c,d,e \in I$ and $a < c < d < e < b$. Since $g(x)$ is convex, $\Delta_{c,d} \le \Delta_{d,e}$. Specifically, we can find an m such that:

$$\sup_{c \uparrow d} \Delta_{c,d} \le m \le \inf_{e \downarrow d} \Delta_{d,e} \tag{2.71}$$

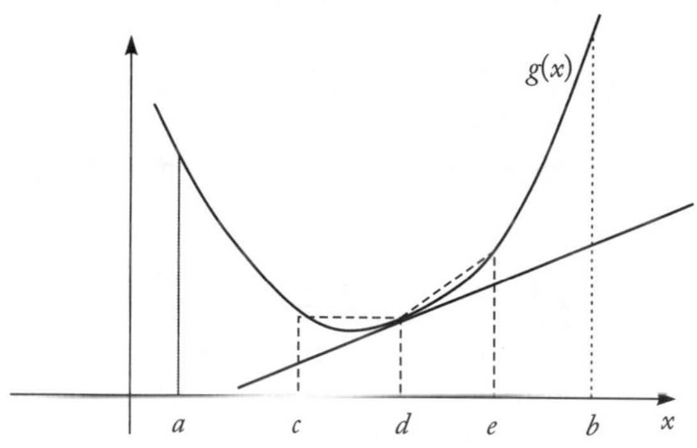

Fig. 2.3 A convex function and Jensen's inequality

Thus for every $d \in I$, one has:

$$g(x) \ge m(x-d) + g(d) \tag{2.72}$$

Physically, the above equation only asserts that $g(x)$ always lies above the tangent (when it exists) at d. In any case, with $d = E(X)$, we have:

$$g(X) \ge m(X - E(X)) + g(E(X)) \tag{2.73}$$

Taking expectations on both sides of the above inequality, we obtain Jensen's inequality in Eq. (2.70a).

As an example, take $X = \mathcal{N}(m,\sigma)$ and $g(X) = e^X$. g is a convex function and Jensen's inequality gives:

$$E(g(X)) = E\left(e^X\right) \geq g(E(X)) = g(m) = e^m \tag{2.74}$$

In fact, $E\left(e^X\right) = e^{m+0.5\sigma^2}$. If $g(X) = X^2$, Jensen's inequality shows the fundamental result that the variance of X is non-negative, i.e., $E\left[X^2\right] \geq (E[X])^2$.

Schwarz inequality

The Schwarz inequality is similar to and a consequence of the classical Cauchy–Schwarz inequality (Appendix B). The inequality states that If $X,\ Y \in L^2$ are non-negative random variables, then:

$$\left|E(XY)\right| \leq E(|XY|) \leq \|X\|_2 \, \|Y\|_2 \tag{2.75}$$

where, $\|.\|_p$ is defined as $\|X\|_p := (E(|X|^p))^{1/p}$.

Proof: Since $-|XY| \leq XY \leq |XY|$, the first inequality in Eq. (2.75) follows either from:

$$-E(|XY|) \leq E(XY) \leq E(|XY|) \Longrightarrow \left|E(XY)\right| \leq E(|XY|) \tag{2.76}$$

or, from Jensen's inequality. For the second inequality, define $\alpha = |X|/\|X\|_2, \beta = |Y|/\|Y\|_2$ so that:

$$\left(\frac{|X|}{\|X\|_2}\right)^2 + \left(\frac{|Y|}{\|Y\|_2}\right)^2 \geq \frac{2\ |X||Y|}{\|X\|_2\|Y\|_2} = 2\ \frac{|XY|}{\|X\|_2\|Y\|_2} \tag{2.77}$$

Applying expectation on both sides, we get:

$$\frac{1}{2}\left(\frac{E\left(|X|^2\right)}{(\|X\|_2)^2} + \frac{E\left(|Y|^2\right)}{(\|Y\|_2)^2}\right) \geq \frac{E|XY|}{\|X\|_2\|Y\|_2} \Longrightarrow 1 \geq \frac{E|XY|}{\|X\|_2\|Y\|_2}, \tag{2.78}$$

which yields the second inequality in Eq. (2.75).

Holder's inequality readily follows from Schwarz inequality and the former is of the form:

$$E|XY| \leq \|X\|_p\|Y\|_q \tag{2.79}$$

where $\frac{1}{p} + \frac{1}{q} = 1$.

2.5 Monte Carlo (MC) Simulation of Random Variables

Simulation studies by MC methods by far remain the most popular and versatile in solving a variety of problems in all branches of engineering and applied science [Fishman 1995, Rubinstein and Kroese 2007, Saha and Roy 2007, Kroese et al 2011]. In the absence of explicit analytical (closed-form) solutions to most dynamical systems under stochastic forcing, MC simulations often provide the only recourse to acceptable numerical solutions. Of late, these simulation methods are being effectively used to integrate stochastic dynamical systems (i.e., systems of stochastic differential equations, *SDE*s), wherein a crucial step is the numerical generation of appropriate sequences of random variables. Thus, simulation on a computer is essentially a process of generating samples, i.e., sequences of realizations of real-valued random variables according to specified probability distributions and the MC methods in turn use them in computing sample statistics of the system response processes.

Of related interest is the principle that exploits MC simulations in the evaluations of deterministic (possibly multidimensional) integrals (see Section 2.5.4 below) that often arise in and beyond studies related to stochastic dynamics. For instance, in reliability studies, the main focus is often on numerically evaluating the probability of failure from deterministic integrals of the form $\int \ldots \int_{G(\boldsymbol{x}) \le 0} f_X(\boldsymbol{x})\, d\boldsymbol{x}$ where $G(\boldsymbol{x})$ is known as the limit state function. In the stochastic filtering theory dealing with system state estimation and/or parameter identification problems (covered in Chapters 6 and 7), numerical evaluations of similar integrals are frequently needed in determining the state/parameter estimates.

Numerical simulations of random variables are generally accomplished using pseudo-random numbers. The word pseudo implies the use of a deterministic algorithm to generate the so-called random numbers (realizations) on a computer. To start with, a random number generator actually outputs integers with a definite number of bits (of a word size in a computer) that are converted to floating point real numbers $\in [0,1]$ appropriately inside the algorithm. It is clear that the finite precision arithmetic of computers in retaining the number of bits invariably results in the sequence of random numbers being periodic. Consequently, to ensure a closer resemblance to the true randomness in the sequence generated, the period must be long enough to produce an adequately large array of such disparate (non-repeating) numbers needed for a specific application. See references [Knuth 1997, Matsumoto and Nishimura 1998, Gentle 2003] for the essential properties and other pertinent details of random number generators.

2.5.1 Random number generation—uniformly distributed random variable

The pseudo-random numbers are usually generated according to any one of the following equations:

$$X_{j+1} = AX_j (\text{mod } M) \tag{2.80a}$$

or,

$$X_{j+1} = AX_j + B(\text{mod } M) \tag{2.80b}$$

The first of the equations above is known as a multiplicative congruential generator and the second a mixed congruential generator. Here A, B and M are non-negative integers. Both the above forms are congruential generators in that $C \equiv D(mod\ M)$ is read "C is congruent to D modulo M". This relation yields $C = D - K \times M$ where K is the largest positive integer less than D/M. X_0 is the starting value which is known as the seed of the random number generator. By the recurrence formula in Eq. (2.80b), we can show at the end of n iterations that:

$$X_n = A^n X_0 + B\,\frac{A^{n-1}-1}{A-1}\ (\text{mod } M) \tag{2.81}$$

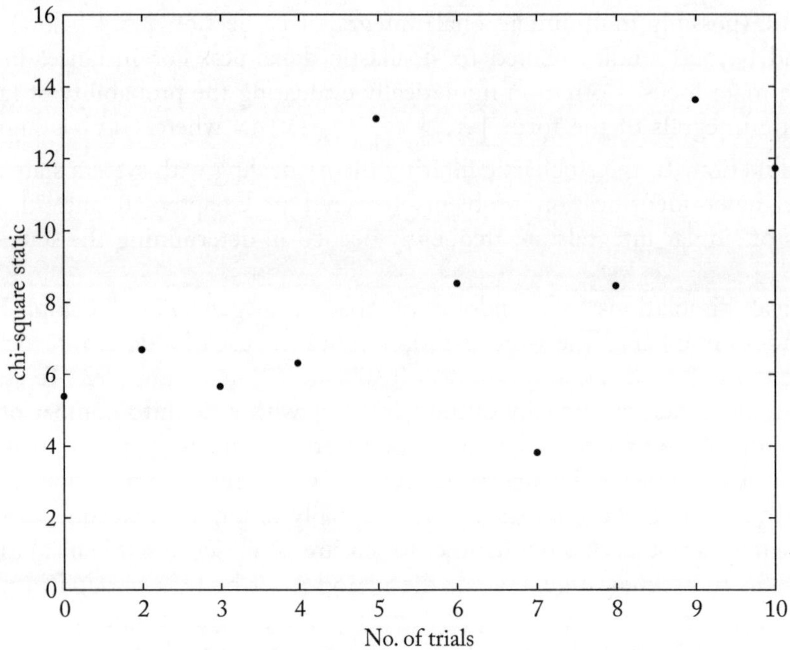

Fig. 2.4 Simulation of uniformly distributed sampled data and chi-square goodness of fit test

The period of the pseudo-number generator is derivable from the above equation by forcing $X_n = X_0$. Note that if the number of bits associated with a word size in a computer is m, the largest possible period (M) is found to be $M = 2^m$.

From X_n generated for each n, we obtain the uniformly distributed pseudo-random number $U_n = X_n/M$. To assess the fitness of the sampled data, (i.e., how much the observed distribution compares with the theoretical distribution $U(m,\sigma)$), we use the chi-square, χ^2, goodness of fit test [Papoulis 1991]. The comparison is through a chi-square statistic:

$$\chi^2 = \sum_{k=1}^{n} \frac{\left(f_k^O - f_k^T\right)^2}{f_k^T} \tag{2.82}$$

Here f_k^O and f_k^T are respectively the observed and theoretical frequencies (of the number of sampled data that fall in the kth interval). These frequencies exactly agree if $\chi^2 = 0$. Figure 2.4 shows the value of χ^2 for a number of independent trials (ten in the present case) of generating 1000 uniformly distributed numbers in each trial (on a 36 bit computer). In each trial, the data is sorted in the ascending order so as to fit into m intervals of 100 numbers each. The value ${\chi^2}_{0.95}(m-1)$ so calculated (with $m = 10$) is less than the critical value for 0.95 confidence probability (as defined and listed in standard Tables) for all the trials.

2.5.2 Simulation for other distributions

Random numbers with a specified probability distribution (other than uniform) are obtained by an appropriate transformation. Considering two scalar random variables X and Y, we can write:

$$F_Y(y) = F_X(x) \Longrightarrow y = F_Y^{-1}(F_X(x)) \tag{2.83}$$

In order to make sense of the last identity involving the inverse of F_Y, it is assumed that $F_X \ll F_Y$.

Let $X \sim U(0,1)$, i.e., the CDF is $F_X(x) = x \in [0,1]$. Using the transformation, $F_Y(y) = F_X(x) = x$, we obtain the cumulative probability for the random variable y as:

$$P(Y \leq y) = P\left({F_Y}^{-1}(X) \leq y\right) = P(X \leq F_Y(y)) = F_X(F_Y(y)) = F_Y(y) \tag{2.84}$$

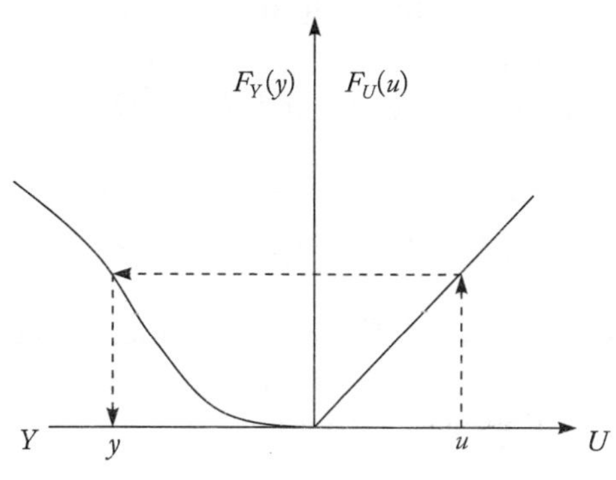

Fig. 2.5 Generation of realizations for Y of specified $F_Y(y)$ by a transformation using $U(0,1)$

Inverse method

Thus, with a set of numbers $x_1, x_2, \ldots.$ (generated as realizations of the random variable $X \sim U(0,1)$), the corresponding set for the random variable Y is obtained by the inverse transformation:

$$y_i = F_Y^{-1}(x_i), \; i = 1,2\ldots \tag{2.85}$$

The computer generation of a random variable Y is illustrated graphically in Fig. 2.5. For example, given an exponential distribution with $f_Y(y) = e^{-y}$, the CDF is $F_Y(y) = 1 - e^{-y}$ and the transformation is $F_Y(y) = x$ that leads to $y = F_Y^{-1}(x) = -\ln(1-x)$. Note that one may use $y = -\ln(x)$ as well. Similarly, realizations for a normal random variable Y with $F_Y(y) = \frac{1}{2}(1 + \text{erf}(y))$ may also be obtained.

The inverse transformation method is advantageous when $F_Y^{-1}(x)$ is known explicitly in terms of x. In other cases, one has to take recourse to obtaining the solution by any one of the numerical methods, e.g., bisection, secant (also known as regula falsi) or Newton–Raphson methods. The computations could be tedious in the inverse method and thus it may not always be efficient.

In the particular case of a normal random variable, an alternative and efficient way is via the Box–Muller method, which allows for the simultaneous generation of two independent normal random variables Y and Z.

Box–Muller method

Suppose that we have $f_{YZ}(y,z) = \frac{1}{2\pi}\exp\left\{-\frac{1}{2}\left(y^2+z^2\right)\right\}$. The Box–Muller method involves a two-stage transformation, the first of which involves a representation in polar coordinates, i.e., $Y = R\cos\Theta$ and $Z = R\sin\Theta$. From Eq. (1.161a) of Chapter 1, we have, $f_{YZ}(y,z) = f_{R\Theta}(r,\theta)\,|J|^{-1}$ with $|J| = r$. The joint *pdf* $f_{R\Theta}(r,\theta)$ also involves two independent random variables and factorizes trivially as $f_\Theta(\theta) = \frac{1}{2\pi}$ and $f_R(r) = r\exp\left\{-\frac{1}{2}r^2\right\}$. Here R represents a Rayleigh distributed random variable (see Eq. 1.146a of Chapter 1). Now, starting with two uniformly distributed and independent random variables U and V, we readily obtain the realizations for R and Θ. Specifically, $\Theta = 2\pi U$ as it is uniform. Finally, for R, we use the 1-D transformation $F_R(r) = 1 - \exp\left\{-\frac{1}{2}r^2\right\} = V$. Having generated R and Θ, we transform back to Y and Z using the relations:

$$y_i = r_i\cos\theta_i \;\text{ and }\; z_i = r_i\sin\theta_i \; i = 1,2,\ldots \tag{2.86}$$

The empirical marginals as well as their CDFs for the numerically generated normal random variables Y and Z via the Box–Muller method are shown in Fig. 2.6. Figure 2.7 shows similar results for R, which is Rayleigh distributed.

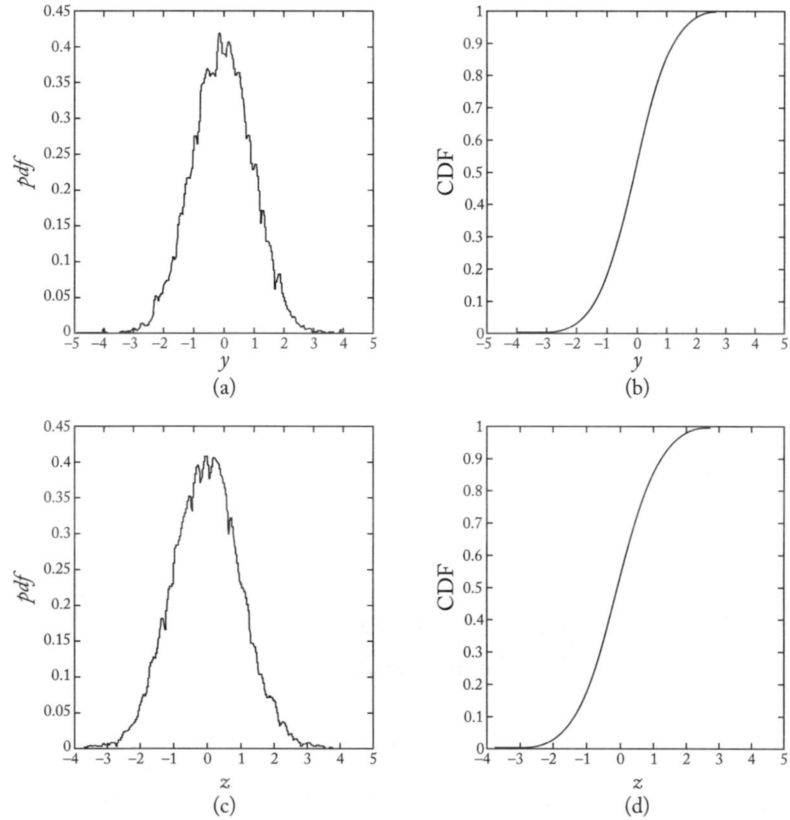

Fig. 2.6 MC simulation; generation of normal random variables by Box–Muller method, (a) *pdf* of Y, (b) CDF of Y, (c) *pdf* of Z, (d) CDF of Z

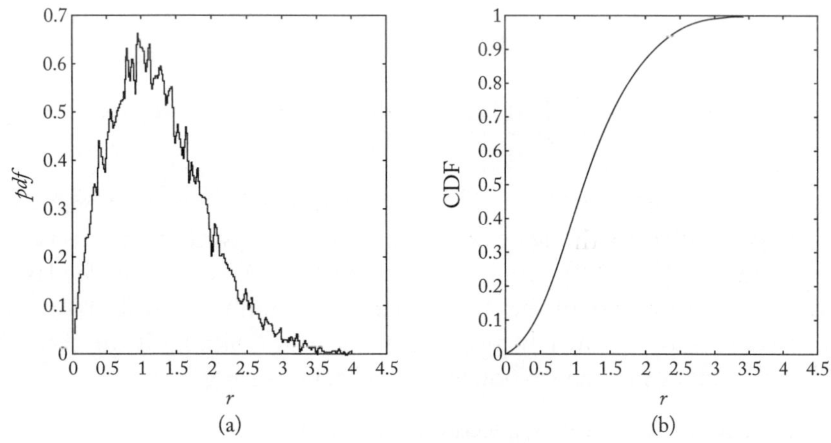

Fig. 2.7 MC simulation; generation of Rayleigh random variable R by normal random variables obtained from Box–Muller method, (a) *pdf* of R and (b) CDF of R

Rejection method

The rejection method is based on the following property of joint *pdf* s.

Theorem: Let X be a random variable with density $f_X(x)$ on $\mathbb{R}$. With $U(0,1)$ being an independent uniform random variable, $(X,\ \alpha f_X(x)U)$ is uniformly distributed on $A = \{(x,u) : x \in \mathbb{R}, 0 \le u \le \alpha f_X(x)\}$ where $\alpha = area(A) > 0$ is a constant. Conversely, if (X,U) is a random vector in $\mathbb{R}^2$ and is uniformly distributed on A, then X has density $f_X(x)$ on $\mathbb{R}$.

Proof: Let $B \subset A$. Then:

$$P((X,\alpha f_X(x)U) \in B) = \iint_B f_{XU}(x,u)\,du dx = \iint_B \left(\frac{1}{\alpha f(x)}\right) f(x)\,du\,dx$$
$$= \frac{1}{\alpha}\iint_B du\,dx = \frac{\iint_B du\,dx}{\iint_A du\,dx} \tag{2.87}$$

The required probability is proportional to $\iint_B du\,dx$ which is the area of B. This being the property of a uniform distribution, the first part, $(X,\ \alpha f_X(x)U)$, is uniformly distributed. Thus the theorem is proved.

To prove the converse, consider all Borel sets, $C \subset \mathbb{R}$. Then, we have:

$$P(X \in C) = P((X,U) \in D = \{(x,u) : x \in C,\ 0 \le u \le \alpha f_X(x)\})$$

$$= \frac{\iint_D du\,dx}{\iint_A du\,dx} \text{ (as a ratio of areas, since the joint } pdf \text{ is uniformly distributed)}$$

$$= \frac{\int_C \left(\int_0^{\alpha f_X(x)} du\right) dx}{\iint_A du\,dx} = \frac{1}{\alpha}\int_C \alpha f_X(x)dx = \int_C f_X(x)dx \text{ as required} \tag{2.88}$$

The rejection method starts with $h(x)$, a chosen *pdf* from which realizations can be more readily generated relative to the *pdf* $g(x)$ of the targeted distribution. It assumes that $g(x) \le \alpha h(x)\,\forall x$ (see Fig. 2.8) with α being a constant. According to the last theorem just proved, for a uniform *pdf* in the $X - Y$ plane, the number of realizations within an interval $(0, \alpha h(x))$ is asymptotically proportional to $h(x)$. Similar is the case with $g(x)$ and we follow the following steps to obtain the realizations for $g(x)$.

(i) Generate random sequence $\{x_i\}$ according to $h(x)$

(ii) Generate random sequence $\{u_i\}$ uniformly distributed over $(0, \alpha h(x_i))$. The point $(x_i,\ u_i)$ will be from a uniform distribution in the area below $\alpha h(x)$.

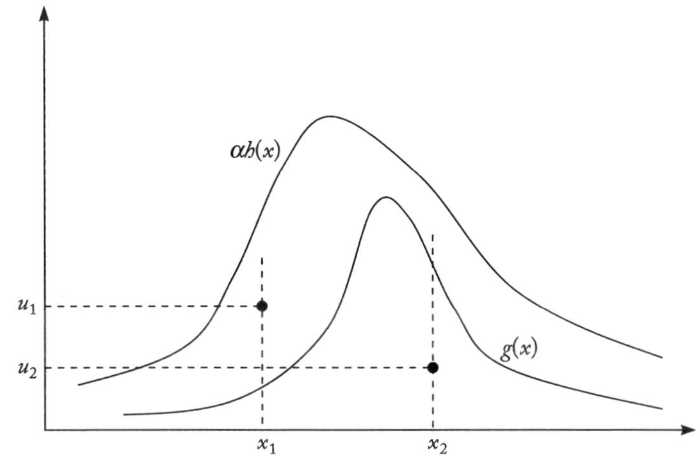

Fig. 2.8 Rejection method; generation of random numbers according to a general *pdf* $g(x)$

(iii) If $u_i \leq g(x_i)$, then accept x_i as a realization from $g(x)$. The point (x_i, u_i) is positioned in the area below the $g(x)$ curve (e.g., (x_2, u_2) in Fig. 2.8).

(iv) If $u_i > g(x_i)$, then reject x_i (e.g., (x_1, u_1) in Fig. 2.8) and return to the step (i).

The point (x_1, u_1) in Fig. 2.8 is rejected and (x_2, u_2) accepted, i.e., x_2 is an acceptable realization from $g(x)$. For each iteration k, the acceptance probability, $p = P(u_k \leq g(x_k))$, is given by:

$$p = \int_{-\infty}^{\infty} \int_{0}^{g(x_k) \geq \alpha h(x_k)} \frac{1}{\alpha} du dx = \frac{1}{\alpha} \int_{-\infty}^{\infty} g(x_k) dx_k = \frac{1}{\alpha} \tag{2.89}$$

Thus, the iterations follow a geometric distribution with *pdf* $f_K(k) = p(1-p)^{k-1}$ whose mean is $\frac{1}{p} = \alpha$. Hence the average number of times step i) is performed before the first acceptance is α.

Example 2.6

Consider generation of the Beta distribution with *pdf* $g(x) = 20x(1-x)^3$, $0 < x < 1$. Let $h(x)$ denotes the *pdf* for $U(0,1)$. To find the constant α such that $g(x) \leq \alpha h(x)$, we maximize $\frac{g(x)}{h(x)} = 20x(1-x)^3$. This leads to $x = \frac{1}{4}$ and we choose $\alpha = \frac{135}{64}$. In the present case, both steps (i) and (ii) of the rejection method involve generation of two uniformly distributed random sequences $\{x_i\}$ and $\{u_i\}$. Steps (iii) and (iv) determine the acceptance or otherwise of u_i being a valid realization from $g(x)$. Figure 2.9 shows the *pdf* and CDF of the generated random sequence for the Beta distribution by the rejection method. Total number of iterations (i.e., the number of times step (i) is executed) is 50000 and the number of successful realizations in the present example is nearly 24000, thus yielding an approximate mean rate of success = 2 ($\cong \alpha$). Comparisons with the theoretical *pdf* and CDF are also shown in Fig. 2.9.

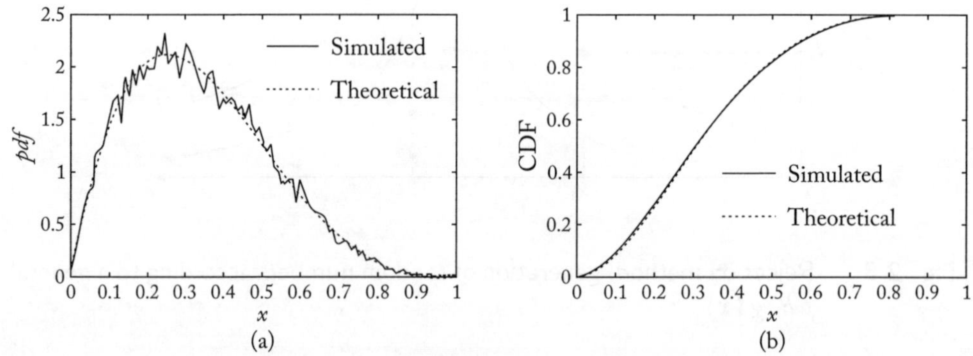

Fig. 2.9 Empirical approximation to the Beta distribution by the rejection method; Example 2.6; (a) *pdf* and (b) CDF

2.5.3 Simulation of joint random variables—uncorrelated and correlated

We first consider the generation of joint normal random variables, specifically a normal bi-variate (Eq. 1.137 of Chapter 1). In fact, the Box–Muller method described in Section 2.5.2 yields standard normals X and Y that are independent and hence uncorrelated. Figure 2.10a shows the (almost) zero correlation between the two random variables generated by the method. If it is needed to consider during simulation, nonzero correlation ($\rho \neq 0$) between the two normals, one can generate the corresponding random numbers from X and Y as follows:

$$\widehat{X}_i = X_i$$

$$\widehat{Y}_i = \rho X_i + \sqrt{(1-\rho^2)}\,Y_i, \quad i = 1,2,\ldots,n \tag{2.90}$$

Figures 2.10b and c show the correlated random variables $\widehat{X}$ and $\widehat{Y}$ with the former figure corresponding to a positive correlation and the latter to a negative correlation.

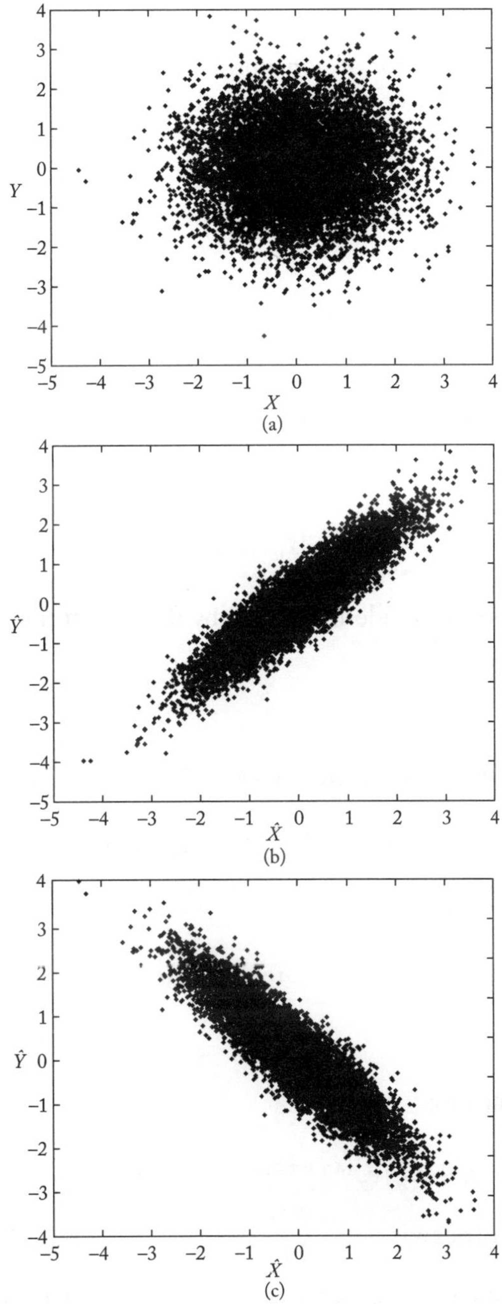

Fig. 2.10 MC simulation and Box–Muller method; (a) uncorrelated normal bivariate (X and Y); (b) positively correlated ($\rho = 0.9$) normal bivariate ($\widehat{X}$ and $\widehat{Y}$) and (c) negatively correlated ($\rho = -0.9$) normal bivariate ($\widehat{X}$ and $\widehat{Y}$)

Simulation of vector non-normal random variables

Let $\boldsymbol{X}$ be an n-dimensional vector of non-normal random variables with a specified joint *pdf* $f_{\mathbf{X}}(\boldsymbol{x})$. Assume that the ($n$-tuples of) realizations corresponding to n uniformly distributed independent random variables, U_i, $i = 1,2,\ldots,n$, are available. Following Rosenblatt transformation (Eq. 2.39), we can generate the random vector $\boldsymbol{X}$ via the following equalities:

$$u_1 = \Phi_{u_1}{}^{-1}\left(F_{X_1}(x_1)\right)$$

$$u_2 = \Phi_{u_2}{}^{-1}\left(F_{X_2|X_1}(x_2|x_1)\right)$$

$$u_3 = \Phi_{u_3}{}^{-1}\left(F_{X_3|X_2,X_1}(x_3|x_2,x_1)\right)$$

$$\ldots\ldots\ldots\ldots\ldots\ldots\ldots\ldots\ldots\ldots$$

$$u_n = \Phi_{u_n}{}^{-1}\left(F_{X_n|X_{n-1},X_{n-2},\ldots,X_1}(x_n|x_{n-1},x_{n-2},\ldots,x_1)\right) \tag{2.91}$$

The simulation of non-normal random variables by Rosenblatt transformation is illustrated by the following example with $n = 2$.

Example 2.7

Let the joint CDF of X and Y be specified as:

$$F_{XY}(x,y) = 1 - e^{-x} - e^{-y} + e^{-(x+y+xy)}, \quad x,y \geq 0$$
$$= 0 \quad \text{otherwise} \tag{2.92a}$$

Given that the two random variables are correlated with the correlation coefficient $\rho = -0.4$.

Solution

The joint *pdf* is obtained as:

$$f_{XY}(x,y) = (x + y + xy)\,e^{-(x+y+xy)}, \quad x,y \geq 0$$
$$= 0 \quad \text{otherwise} \tag{2.92b}$$

One obtains the marginal *pdf*s as $f_X(x) = e^{-x}$, $x \geq 0$ and $f_Y(y) = e^{-y}$, $y \geq 0$ and both X and Y are exponentially distributed. Further, $F_X(x) = 1 - e^{-x}$, $x \geq 0$ and the conditional CDF $F_{Y|X}(y|x)$ is obtained as:

$$F_{Y|X}(y|x) = \frac{\int_0^y f_{XY}(x,z)\,dz}{f_X(x)} = 1 - (1 + y)\,e^{-(y+xy)} \tag{2.93}$$

Following the procedure in Eq. (2.91), we generate the realizations x and y using two uniform random variables u_1 and u_2 in $[0,1]$ recursively as:

$$x = F_X^{-1}(u_1) \tag{2.94a}$$

and

$$y = F_{Y|X}^{-1}(u_2|u_1) \tag{2.94b}$$

The second equation above is solved for y given x (the latter obtained from the first equation) by bisection method. Each of the sequences generated for x and y contains 5000 realizations. The marginal CDFs and *pdf* s obtained from the simulated sequences are shown in Fig. 2.11. The correlation coefficient $\rho = \frac{\sigma_{XY}}{\sigma_X \sigma_X}$ is also evaluated using the generated samples and found to be -0.4054.

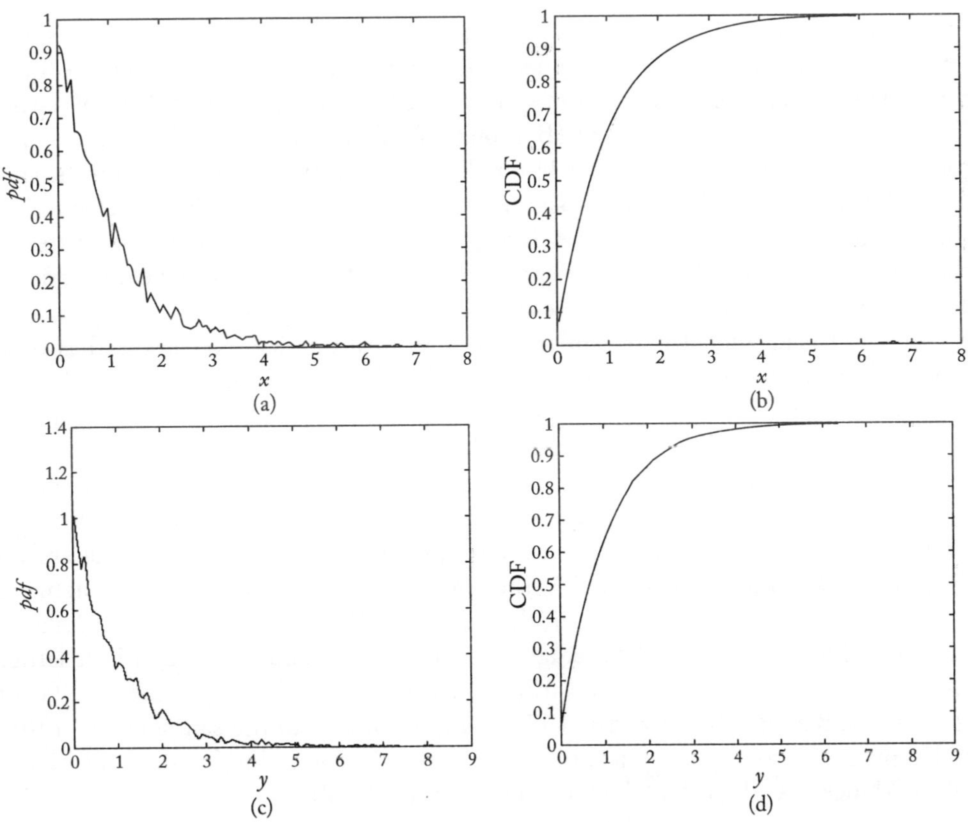

Fig. 2.11 Generation of correlated joint non-normal (exponential) random variables using Rosenblatt transformation; Example 2.7, MC simulation results, (a) *pdf* and (b) CDF of X and (c) *pdf* and (d) CDF of Y, simulated correlation coefficient, $\rho = -0.4054$

In an analogous manner, one can also use Nataf transformation (Section 1.11.4, Chapter 1) to generate correlated $\boldsymbol{X}$ if the available information is partial and only in terms of marginal *pdf* s and correlations.

Example 2.8

We consider 2 dimensional Ising model [Ising 1925, Tobochnik and Gould 2008] for identifying the thermo-dynamical phase transitions in a ferromagnet using MC simulation.

Solution

Brief notes on the Ising model and ferromagnetism is provided in Appendix B to get the readers familiarized with the basics underlying the phenomenon. There we have also detailed the Metropolis algorithm which is implemented for the MC simulation of the Ising model. The Ising model consists of a simplified visualization (see Fig. B.2 of Appendix B) of electron spins in an atom / molecule arranged on a 2-dimensional lattice. The model studies the effect of temperature on magnetization of a ferromagnetic material. The electron spins located at the nodes of an $N \times N$ square lattice are represented by discrete (Bernoulli) random variables $S_i, i = 1, 2, \ldots, N^2$ each with sample space $\Omega = \{-1, +1\}$ for all i. The negative (-) sign denotes the spin axis pointing downwards and the + sign upwards. Ferromagnetism is characterized by the alignment of spins in the same direction, up or down. As such, two states of thermal equilibrium exist for ferromagnetic materials for temperature below a critical value T_c where a phase transition takes place.

The 2-D Ising model is one of the few statistical physics models that has been solved analytically [Peierls 1936, Onsager 1944]. In this example, results are obtained numerically by MC simulation based on Metropolis algorithm (steps enumerated in Table B.2 of Appendix B). According to this algorithm, for each temperature, iterations start with a randomly selected initial (spin) configuration. At each iteration, flip a spin (i.e., swap -1 with +1 or vice versa) with probability $= \frac{1}{N^2}$ and accept or reject using an acceptance probability based on the Boltzmann distribution (Eq. B.8 of Appendix B). The interaction parameter J and the Boltzmann constant β are both taken as unity in the numerical simulations. Figure 2.12 shows the effect of temperature on the average magnetization M and energy E. Finite lattice size effects in the simulation of various quantities of interest may be significant [Fisher and Barber 1972, Landau 1976]. Improvements to the sampling algorithm and applications to other physical phenomena may be found in [Landau and Binder 2000, Manekar *et al.* 2001, Chaikin and Lubensky 2007].

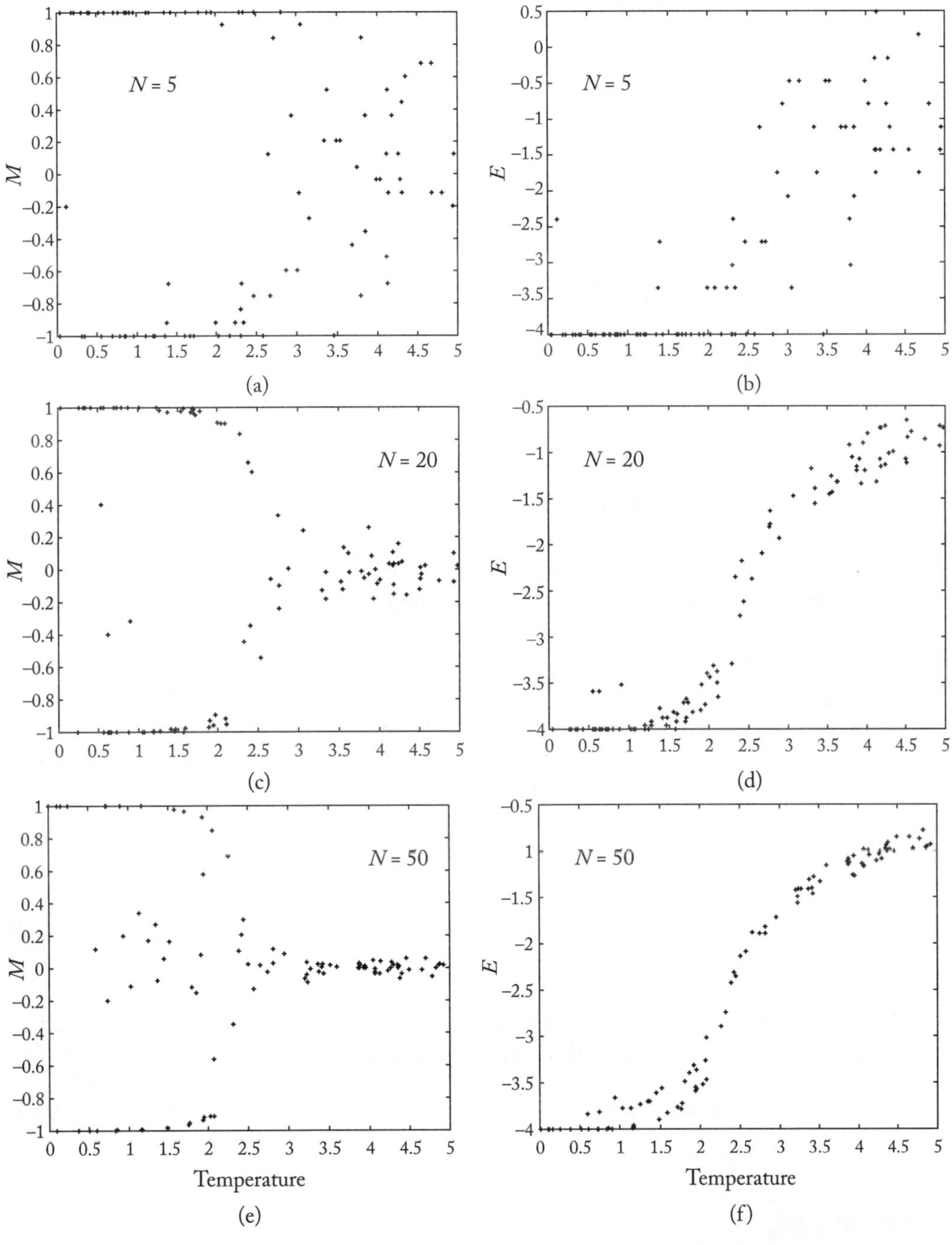

Fig. 2.12 A two-dimensional Ising statistical physics model for ferromagnetism; Example 2.8, (a), (c) and (e) average magnetization vs temperature and (b), (d) and (f) average energy vs temperature, maximum iterations $= 1000 \times N^2$

2.5.4 Multidimensional integrals by MC simulation methods

Consider the numerical evaluation of the multidimensional definite integral for $d \geq 1$:

$$I = \int_{(a,b)^d} g(\boldsymbol{x})\, d\boldsymbol{x} \tag{2.95}$$

$\Upsilon := (a,b)^d$ denotes a d-dimensional hypercube, in which a generic point is given by $\boldsymbol{x} = \left(x^{(1)}, x^{(2)}, \ldots, x^{(d)}\right)$, and $g : \mathbb{R}^d \to \mathbb{R}$. The MC simulation method aims at an approximate evaluation of the integral by relating it to the mathematical expectation of a function of a random vector $\boldsymbol{X}$ with joint *pdf* $h(\boldsymbol{x})$. Thus the integral I is written as:

$$I = \int_{(a,b)^d} \left\{ \frac{g(\boldsymbol{x})}{h(\boldsymbol{x})} \right\} h(\boldsymbol{x}) d\boldsymbol{x} = E\left[r(\boldsymbol{X})\right] \quad \text{with } r(\boldsymbol{x}) = \frac{g(\boldsymbol{x})}{h(\boldsymbol{x})} \tag{2.96}$$

Here it is assumed that $h(\boldsymbol{x}) > 0$ whenever $g(\boldsymbol{x}) \neq 0$ and $h(\boldsymbol{x}) \geq 0$ whenever $g(\boldsymbol{x}) = 0$. In such a case, $h(\boldsymbol{x})$ is said to be tolerant to $g(\boldsymbol{x})$. $h(\boldsymbol{x})$ is often referred to as the sampling distribution. In case the domain of integration is finite, one can conveniently assume $h(\boldsymbol{x})$ to be uniform. Thus, if V denotes the volume of the domain, one may adopt $h(\boldsymbol{x}) = \frac{1}{V}$, a constant over the domain, and the integral I is then given by:

$$I = VE[g(\boldsymbol{X})] \tag{2.97}$$

Without any loss of generality, we may assume that $V = 1$, which is indeed true if the domain of integration is over $(0,1)^d$. We now estimate I via $E[g(\boldsymbol{X})]$ by generating a sequence $\{\boldsymbol{x}_j = \left(x_j{}^{(1)}, x_j{}^{(2)}, \ldots, x_j{}^{(d)}\right) : j = 1, \ldots, N\}$ of i.i.d. pseudo-random realizations of $\boldsymbol{X}$ (with *pdf* $h(\boldsymbol{x})$) and computing the arithmetic mean given by:

$$I_N = \frac{1}{N} \sum_{j=1}^{N} g(\boldsymbol{x}_j) \tag{2.98}$$

By law of large numbers, the arithmetic mean converges to the mathematical expectation $E[g(\boldsymbol{x})]$ and so to I as $N \to \infty$. Thus, for any $\varepsilon > 0$ we have:

$$\lim_{N \to \infty} P\left(|I_N - I| > \varepsilon\right) = 0 \tag{2.99}$$

Example 2.9

Our interest here is to obtain the probability of failure P_f for the same system considered in Example 2.5 by direct simulation of the multiple integral (see Eq. A.40 of Appendix A) defining P_f. The equation is reproduced below.

$$P_f = \int_{g(\boldsymbol{X})<0} f(\boldsymbol{X})\, d\boldsymbol{X} \tag{2.100}$$

where, $\boldsymbol{X} = \left(S_y, S_m, B_m\right)$ is the system state vector of random variables. $g(\boldsymbol{X})$ is the performance function and $f(\boldsymbol{X})$ the joint *pdf* of $\boldsymbol{X}$. The failure surface is $g(\boldsymbol{X}) = 0 \Longrightarrow S_y S_m - B_m = 0$.

Solution

The integral in Eq. (2.100) is recast in the form:

$$P_f = \int_{\mathbb{R}^3} I_A(\boldsymbol{X}) f(\boldsymbol{X})\, d\boldsymbol{X} \tag{2.101}$$

where, $A = \{g(\boldsymbol{X}) < 0\}$ and $I_A(\boldsymbol{X})$ is the indicator function equal to 1 when $\boldsymbol{X} \in A$ and zero when $\boldsymbol{X} \notin A$. In the last equation, the integration domain is over the whole sample space $\mathbb{R}^3$ of $\boldsymbol{X}$ unlike the integral in Eq. (2.100). Hence the integral is given by the expectation of the indicator function with respect to the probability measure corresponding to $f(\boldsymbol{X})$. The Rosenblatt transformation in Eq. (2.39) gives the relationship between the correlated lognormals S_y and S_m and independent standard normals Z_1 and Z_2. Similarly, standard normal Z_3 is related to B_m by Eq. (2.46). These equations thus help in generating realizations of random variables S_y, S_m and B_m during simulation. The probability of failure is thus obtained by:

$$P_f = E\left[I_A(\boldsymbol{X})\right] = \frac{1}{N}\sum_{i=1}^{N} I_A(\boldsymbol{X} = \boldsymbol{x}_i), \quad \boldsymbol{x}_i = \left(s_{y,i}, s_{m,i}, b_{m,i}\right) \tag{2.102}$$

$\boldsymbol{x}_i$ is the ith realization of $\boldsymbol{X}$ obtained using Equations (2.42), (2.45) and (2.46). Table 2.2 shows the computed probability of failure obtained through 10 independent MC runs, each involving $N = 5 \times 10^5$ realizations. The table also shows the correlation coefficient r_{sim} between lognormals S_y, S_m computed from the realizations.

Table 2.2 MC simulation results; probability of failure for the system in Example 2.5

Independent MC run	1	2	3	4	5
P_f	0.0041	0.0039	0.0039	0.0039	0.0041
r_{sim}	0.3869	0.3872	0.3944	0.3885	0.3897
Independent MC run	**6**	**7**	**8**	**9**	**10**
P_f	0.0039	0.0041	0.0038	0.004	0.004
r_{sim}	0.387	0.3917	0.3903	0.3934	0.4

Estimate of error in simulation

For finite N, the error in the estimate of the integral is given by:

$$e_N = \left| \frac{1}{N} \sum_{j=1}^{N} g(\boldsymbol{x}_j) - I \right|$$

$$= \left| \frac{\sigma_g}{\sqrt{N}} \left(\frac{\sum_{j=1}^{N} g(\boldsymbol{x}_j) - NI}{\sigma_g \sqrt{N}} \right) \right| \tag{2.103}$$

where $\sigma_g{}^2 = E[(g(\boldsymbol{X}) - I)^2]$ is the variance of the random variable $g(\boldsymbol{X})$. Recognizing that $\sum_{j=1}^{N} g(\boldsymbol{x}_j)$ is the sum (of i.i.d. realizations) with mean NI and variance $N\sigma_g{}^2$ and letting $z = \frac{\sum_{j=1}^{N} g(\boldsymbol{x}_j) - NI}{\sigma_g \sqrt{N}}$, we have by the central limit theorem:

$$z = \frac{\sum_{j=1}^{N} g(\boldsymbol{x}_j) - NI}{\sigma_g \sqrt{N}} \rightarrow \mathcal{N}(0,1) \text{ as } N \rightarrow \infty \tag{2.104}$$

Thus the error $e_N = \frac{\sigma_g}{\sqrt{N}} \mathcal{N}(0,1)$ asymptotically converges with $O(\frac{1}{\sqrt{N}})$, which is independent of the dimension d. Note that e_N is also affected by the variance $\sigma_g{}^2$ and the sampling distribution $h(\boldsymbol{x})$. The last observation provides a means to reduce the error by suitably designing the simulation methods that reduce σ_g and hence achieve higher accuracy. Techniques that aim at reducing the variance $\sigma_g{}^2$ are known as variance reduction methods [Saha and Roy 2007] and some of these are described below.

Variance reduction methods

For continuity in exposition, we reproduce here the integral in Eq. (2.96):

$$I = E_h[r(\boldsymbol{X})] = \int_{\Upsilon} r(\boldsymbol{x})\, h(\boldsymbol{x})\, d\boldsymbol{x} \tag{2.105}$$

$E_h[.]$ is the expectation with respect to the *pdf* $h(x)$.

Integration on sub-domain (partial averaging) method

It is often possible that the integral can be evaluated analytically over a sub-domain $\Upsilon' \subset \Upsilon$. This gives rise to the method of integration on sub-domain (or partial averaging) by which the variance of the MC estimate can be significantly reduced. In this method, the integral I is expressed as:

$$I = I' + \int_{\Upsilon_1 := \Upsilon \setminus \Upsilon'} r(\boldsymbol{x}) h(\boldsymbol{x}) d\boldsymbol{x} \tag{2.106}$$

The method assumes that $I' = \int_{\Upsilon'} r(\boldsymbol{x})h(\boldsymbol{x})d\boldsymbol{x}$ is known analytically. If we define $c = \int_{\Upsilon'} h(\boldsymbol{x})d\boldsymbol{x}$ and a random variable $\boldsymbol{Y}$ with a *pdf* $h'(\boldsymbol{y}) = h(\boldsymbol{y}) / (1-c)$ over Υ_1, the integral I is expressible as:

$$I = I' + E_{h'}[r'(\boldsymbol{Y})]r \quad \text{with} \quad r'(\boldsymbol{y}) = (1-c)\,r(\boldsymbol{y}) \tag{2.107}$$

The above assertion is justified as follows.

$$\begin{aligned} I &= I' + E_{h'}[r'(\boldsymbol{Y})] = I' + E_{h'}[(1-c)\,r(\boldsymbol{Y})] \\ &= I' + (1-c)\,E_{h'}[r(\boldsymbol{Y})] \\ &= I' + (1-c)\int_{\Upsilon_1} r(\boldsymbol{y})\,h'(\boldsymbol{y})\,d\boldsymbol{y} \\ &= I' + \int_{\Upsilon_1} r(\boldsymbol{y})\,h(\boldsymbol{y})\,d\boldsymbol{y} \end{aligned} \tag{2.108}$$

With I' already known, it is only needed to evaluate the integral $\int_{\Upsilon_1} r(\boldsymbol{y})\,h(\boldsymbol{y})\,d\boldsymbol{y}$ in Eq. (2.108) which thus represents partial averaging. The improved accuracy by the method (accruing through reduced sample variance) is illustrated in the following steps.

For the random variable $r(\boldsymbol{X})$, the variance is given by:

$$\begin{aligned} \sigma_r{}^2 &= E_h\left[(r(\boldsymbol{X}) - I)^2\right] = \int_{\Upsilon} r^2(\boldsymbol{x})h(\boldsymbol{x})d\boldsymbol{x} - I^2 \\ &= \int_{\Upsilon_1} r^2(\boldsymbol{x})h(\boldsymbol{x})d\boldsymbol{x} + \int_{\Upsilon'} r^2(\boldsymbol{x})\,h(\boldsymbol{x})d\boldsymbol{x} - I^2 \end{aligned} \tag{2.109}$$

Noting that $E_{h'}[r'(\boldsymbol{Y})] = I - I'$, the variance with respect to the new random variable $r'(\boldsymbol{Y})$ defined on Υ_1 is given by:

$$\begin{aligned} \sigma_{r'}{}^2 &= E_{h'}\left[(r'(\boldsymbol{Y}) - (I - I'))^2\right] = E_{h'}[\{(1-c)\,r(\boldsymbol{Y}) - (I - I')\}^2] \\ &= E_{h'}\left[\left\{(1-c)\,r(\boldsymbol{Y}) - \int_{\Upsilon_1} r(\boldsymbol{y})\,h(\boldsymbol{y})\,d\boldsymbol{y}\right\}^2\right] \quad \text{(from Eq. 2.108)} \\ &= (1-c)^2\int_{\Upsilon_1} r^2(\boldsymbol{y})\,h'(\boldsymbol{y})\,d\boldsymbol{y} - \left(\int_{\Upsilon_1} r(\boldsymbol{y})\,h(\boldsymbol{y})\,d\boldsymbol{y}\right)^2 \end{aligned}$$

$$= (1-c)\int_{\Upsilon_1} r^2(y)h(y)\,dy - \left(\int_{\Upsilon_1} r(y)h(y)\,dy\right)^2 \tag{2.110}$$

From the last two equations, we have:

$$(1-c)\sigma_r^{\,2} - \sigma_{r'}^{\,2} = (1-c)\int_{\Upsilon'} r^2(x)h(x)dx - (1-c)I^2 + \left(\int_{\Upsilon_1} r(y)h(y)\,dy\right)^2$$

$$= (1-c)\int_{\Upsilon'} r^2(x)h(x)dx - (1-c)I^2 + (I-I')^2 \tag{2.111}$$

Noting that $\int_{\Upsilon'} r^2(x)h(x)dx - \frac{I'^2}{c} = \int_{\Upsilon'}\left(r(x) - \frac{I'}{c}\right)^2 h(x)dx$, Eq. (2.111) can be simplified as:

$$(1-c)\sigma_r^{\,2} - \sigma_{r'}^{\,2} = (1-c)\left\{\int_{\Upsilon'}\left(r(x) - \frac{I'}{c}\right)^2 h(x)dx + \frac{I'^2}{c}\right\} - (1-c)I^2 + (I-I')^2$$

$$= (1-c)\int_{\Upsilon'}\left(r(x) - \frac{I'}{c}\right)^2 h(x)dx + \left(cI^2 - 2II' + \frac{I'^2}{c}\right)$$

$$= (1-c)\int_{\Upsilon'}\left(r(x) - \frac{I'}{c}\right)^2 h(x)dx + \left(\sqrt{c}I - \frac{I'}{\sqrt{c}}\right)^2 \geq 0 \tag{2.112}$$

That $\sigma_{r'}^{\,2} < \sigma_r^{\,2}$ is evident from the last equation.

Example 2.10

Consider the integral $I = \iint_{(0,1)^2} e^{-\frac{(x-0.5)^2+(y-0.5)^2}{2}} dxdy$. The integral can be analytically evaluated over $\Upsilon' \subset \Upsilon = (0,1)^2$ where Υ' may, for example, be the sector $x \leq 0.75$, $y \leq 0.75$, $x \geq 0$ and $y \geq 0$ and thus $I' = \iint_{\Upsilon'} e^{-\frac{(x-0.5)^2+(y-0.5)^2}{2}} dxdy = 0.4156$. Let $h(x,y)$ be a uniformly distributed *pdf* over Υ and hence we readily have $c = \iint_{\Upsilon'} h(x,y)\,dxdy = 0.4418$. By MC simulation we perform the partial averaging corresponding to the integral $I - I' = \int_{\Upsilon_1} r(y)h(y)\,dy$ (see Eq. 2.108). Here $y = (x,y)$. Figure 2.13 shows the result on the integral estimates and variances from ten such simulations. The results from full averaging are also plotted in the figure.

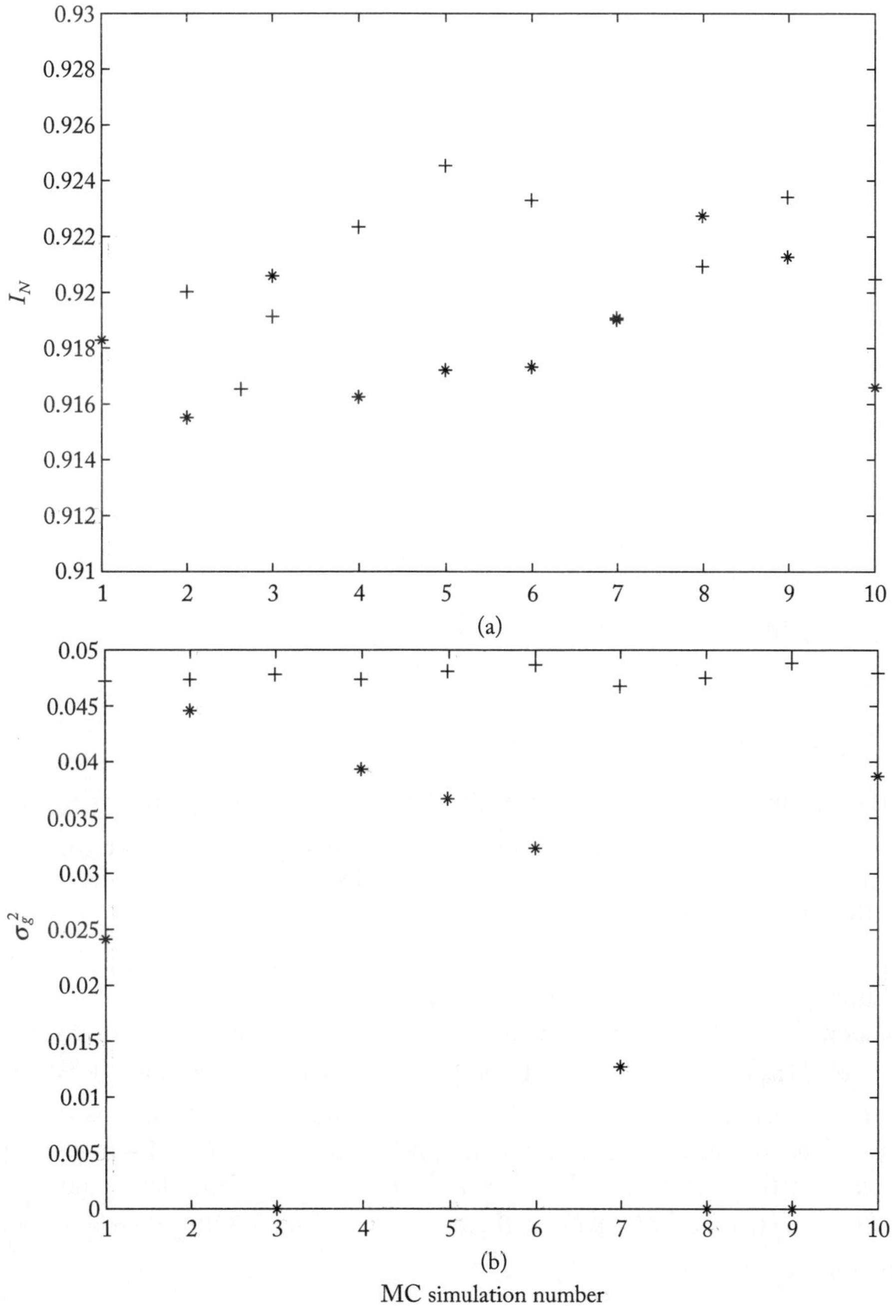

Fig. 2.13 MC simulation and partial averaging method for Example 2.10; (a) estimate of the integral and (b) variance of the estimate, + standard simulation, * simulation by partial averaging

Method of Antithetic variables

Referring to the general case of simulation according to Eq. (2.96), we have $I_N = \frac{1}{N}\sum_{j=1}^{N} r(\boldsymbol{x}_j)$. The sequence $\left\{r(\boldsymbol{x}_j)\right\}$ is generated from the i.i.d. realizations $\boldsymbol{x}_j$ from the *pdf* $h(\boldsymbol{x})$. Here, one way to reduce the variance ${\sigma_r}^2$ of the summand is to use $\boldsymbol{x}_j$ having desirable correlations so as to have cancellations in the sum expressing the variance. With $r(\boldsymbol{x}_j)$ denoted by r_j, the strategy is explained in the following steps.

$$
\begin{aligned}
I_N &= \frac{1}{N}\sum_{j=1}^{N} r_j = \frac{2}{N}\sum_{j=1}^{N/2} \frac{r_{2j-1}+r_{2j}}{2} \\
&= \frac{2}{N}\sum_{j=1}^{N/2} s_j \quad \text{with} \quad s_j = \frac{r_{2j-1}+r_{2j}}{2}
\end{aligned} \tag{2.113}
$$

If ${\sigma_s}^2$ denotes the sample variance of the set $\{s_j\}$, we have:

$$
\begin{aligned}
{\sigma_s}^2 &= \frac{1}{4}\left\{{\sigma_r}^2 + {\sigma_r}^2 + 2cov\,(s_j, s_{j-1})\right\} \\
&= \frac{1}{2}\left\{{\sigma_r}^2 + cov\,(r_{2j}, r_{2j-1})\right\}
\end{aligned} \tag{2.114}
$$

If $\boldsymbol{x}_j$ has negative pair-wise correlations, then $cov\,(r_{2j}, r_{2j-1}) < 0$ so that ${\sigma_s}^2 < \frac{1}{2}{\sigma_r}^2$. In particular, if $h(\boldsymbol{x})$ is uniformly distributed, say, over $(0,1)^d$ and $\boldsymbol{u}_i$ are the corresponding realizations by simulation, we choose $\boldsymbol{x}_{2j-1} = \boldsymbol{u}_j$ and $\boldsymbol{x}_{2j} = \mathbf{1} - \boldsymbol{u}_j$ where 1 is the vector of 1's of the same dimension as $\boldsymbol{u}_j$. Here $\mathbf{1} - \boldsymbol{u}_j$ is also a realization from uniform distribution over $(0,1)^d$ and moreover is negatively correlated with $\boldsymbol{u}_j$. If $\mathbb{K}(\boldsymbol{x})$ is a non-uniform distribution (corresponding to a non-uniform *pdf* $\Bbbk(\boldsymbol{x})$), the realizations are generated by the inverse technique so that $\boldsymbol{x} = \mathbb{K}^{-1}(\boldsymbol{u})$. It follows that $\boldsymbol{x}_{2j-1} = \mathbb{K}^{-1}\left(\boldsymbol{u}_j\right)$ and $\boldsymbol{x}_{2j} = \mathbb{K}^{-1}\left(\mathbf{1} - \boldsymbol{u}_j\right)$ have the same distribution. $\mathbb{K}(\boldsymbol{x})$, being a distribution function, is monotonically increasing and so is $\mathbb{K}^{-1}(\boldsymbol{u})$. Now, if r is a monotone function (increasing or decreasing), $r(h^{-1}(\boldsymbol{u}))$ and $r(h^{-1}(\mathbf{1}-\boldsymbol{u}))$ also remain monotone functions of $\boldsymbol{u}$. It follows that since $\boldsymbol{u}$ is an independent random vector, $cov\,\left(r\left(\mathbb{K}^{-1}(\boldsymbol{u})\right), r(\mathbb{K}^{-1}(\mathbf{1}-\boldsymbol{u}))\right) \le 0$ [Ross 2002, Dimov 2008] thereby leading to a reduction in ${\sigma_s}^2$.

Example 2.11

Consider the integral $I = \int_{-4}^{4} \frac{\exp(-x^2)}{\left(|x^2-1|+0.01\right)^{1/2}}\,dx$. $h(x)$ is a uniformly distributed *pdf* over $\Upsilon = (-4, 4)$ and 1000 independent realizations are used in each estimation, wherein the simulation procedure outlined above (Eq. 2.113) is followed. Figure 2.14 shows ten such estimates (through ten independent MC runs). The results obtained via the standard MC simulation (without using antithetic variables) are also shown in the figure.

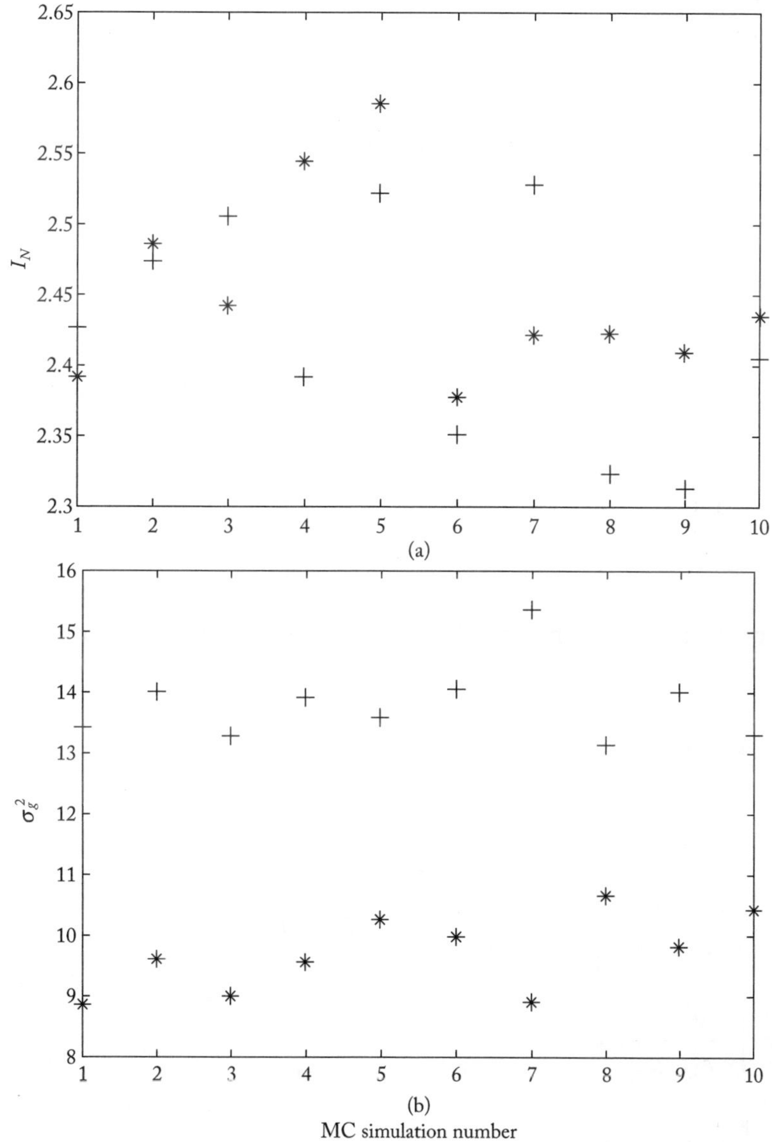

Fig. 2.14 MC simulation and antithetic variables method for Example 2.11; (a) estimate of the integral and (b) variance of the estimate, + standard simulation, $*$ simulation with antithetic variables

Importance sampling method

In MC simulation, the choice of the sampling density is crucial in that the regions (subsets of Υ) that contribute dominantly to the integral I need to be densely sampled. More often, the estimation also requires simulation of rare events, wherein an inappropriate sampling may even lead to unbounded variance. The importance sampling method principally focuses on providing guidelines on the selection of the sampling distribution obtaining accurate estimates, i.e., unbiased ones with reduced sample variance. Consider estimating the integral $I = \int_{\Upsilon} g(\boldsymbol{x})\, d\boldsymbol{x}$ with the random variable $\boldsymbol{X}$ realized from the *pdf* $f_{\boldsymbol{X}}(\boldsymbol{x})$ (with the subscript $\boldsymbol{X}$ omitted below). The estimation takes the form:

$$I = E_f\left[\frac{g(\boldsymbol{X})}{f(\boldsymbol{X})}\right] \tag{2.115}$$

Assume that $f(\boldsymbol{x})$ is tolerant to $g(\boldsymbol{x})$. It can be shown [Dimov 2008] that if $f(\boldsymbol{x}) \propto |g(\boldsymbol{x})|$ i.e., $f(\boldsymbol{x}) = c\,|g(\boldsymbol{x})|$ with c being a constant, the estimate has the minimum variance given by:

$$\sigma_g{}^2{}_{\min} = \left(\int_{\Upsilon} |g(\boldsymbol{x})|\, d\boldsymbol{x}\right)^2 - I^2 \tag{2.116}$$

Proof: The proportionality constant c is obtained (from the condition of normalization of a *pdf*) as:

$$c = 1 \Big/ \int_{\Upsilon} |g(\boldsymbol{x})|\, d\boldsymbol{x} \tag{2.117}$$

Now, the variance is given by:

$$\begin{aligned}
\sigma_g{}^2 &= E_f\left[\left(\frac{g(\boldsymbol{X})}{f(\boldsymbol{X})}\right)^2\right] - I^2 = \int_{\Upsilon}\left(\frac{g(\boldsymbol{x})}{f(\boldsymbol{x})}\right)^2 f(\boldsymbol{x})\, d\boldsymbol{x} - I^2 \\
&= \int_{\Upsilon} \frac{g^2(\boldsymbol{x})}{c\,|g(\boldsymbol{x})|}\, d\boldsymbol{x} - I^2 \\
&= \frac{1}{c}\int_{\Upsilon} |g(\boldsymbol{x})|\, d\boldsymbol{x} - I^2 \\
&= \left(\int_{\Upsilon} |g(\boldsymbol{x})|\, d\boldsymbol{x}\right)^2 - I^2 \quad \text{(by Eq. 2.117)}
\end{aligned} \tag{2.118}$$

It remains to be shown that ${\sigma_g}^2$ in the above equation is indeed the minimum. To this end, let us consider any other tolerant *pdf* $h(\boldsymbol{x})$. Thence it follows:

$$\left(\int_{\Upsilon} |g(\boldsymbol{x})|\, d\boldsymbol{x}\right)^2 = \left(\int_{\Upsilon} |g(\boldsymbol{x})|\; h^{-1/2} h^{1/2}\, d\boldsymbol{x}\right)^2 \tag{2.119}$$

By Cauchy–Schwarz inequality, the above equation can be expressed as:

$$\left(\int_{\Upsilon} |g(\boldsymbol{x})|\, d\boldsymbol{x}\right)^2 \leq \int_{\Upsilon} \frac{g^2(\boldsymbol{x})}{h}\, d\boldsymbol{x}.\ \int_{\Upsilon} h\, d\boldsymbol{x} \leq \int_{\Upsilon} \frac{g^2(\boldsymbol{x})}{h}\, d\boldsymbol{x} \tag{2.120}$$

Recognizing that $\int_{\Upsilon} \frac{g^2(\boldsymbol{x})}{h}\, d\boldsymbol{x} - I^2$ is the variance when $h(\boldsymbol{x})$ is the sampling density, we have the desired result (${\sigma_g}^2 = {\sigma_g}^2_{\ min}$ if $(\boldsymbol{x}) = c\, \left|g(\boldsymbol{x})\right|$). In the particular case with $g(\boldsymbol{x})$ not changing sign in Υ, we find that the corresponding $f(\boldsymbol{x})$ leads to zero variance (${\sigma_g}^2 = 0$). That is, the integral estimate is obtained via just one sample alone (with probability 1). However, this is a hypothetical supposition since in this case one has $c = \frac{1}{I_N}$, which requires I_N, the very estimate for which the MC simulation is being employed, to be known a priori. Nevertheless, the observation guides us towards a proper choice for $f(\boldsymbol{x})$ that may lead to significant variance reduction.

Example 2.12

Consider the integral $I = \int_{\mathbb{R}} x^4\, e^{x^2/4}\ I_{\{x \leq 2\}} \cdot \left(\frac{e^{-x^2/2}}{\sqrt{2\pi}}\right) dx$ where $I_{\{x \leq 2\}}$ is an indicator function. Notice that $I = E_f[g(x)]$ where $g(x) = x^4\, e^{x^2/4}\ I_{\{x \leq 2\}}$ and $f(x) = \frac{1}{\sqrt{2\pi}} e^{-x^2/2}$ is the normal *pdf* corresponding to $X \sim \mathcal{N}(0,1)$. Let the importance sampling density $h(x)$ be chosen as $\mathcal{N}(\mu, 1)$ and I be estimated from $E_h[\frac{g(x)f(x)}{h(x)}]$. Here $\frac{g(x)f(x)}{h(x)}$ is given by:

$$\frac{g(x)\,f(x)}{h(x)} = \frac{x^4\, e^{x^2/4}\ I_{\{X \leq 2\}}\left(\frac{e^{-x^2/2}}{\sqrt{2\pi}}\right)}{\frac{e^{-(x-\mu)^2/2}}{\sqrt{2\pi}}} = x^4\, e^{x^2/4}\ I_{\{x \leq 2\}} e^{-\mu x + \mu^2/2} \tag{2.121}$$

Thus the estimate I_N (by important sampling) for the integral $I = E_h\left[X^4\, e^{X^2/4}\ I_{\{X \leq 2\}}.e^{-\mu X + \mu^2/2}\right]$ is obtained by simulating X as a random variable from $\mathcal{N}(\mu, 1)$ with $\mu = 2$. The results, obtained for ten independent MC runs, are shown in Fig. 2.15. The result of standard MC simulation in the Figure pertains to that obtained from $X \sim \mathcal{N}(0,1)$.

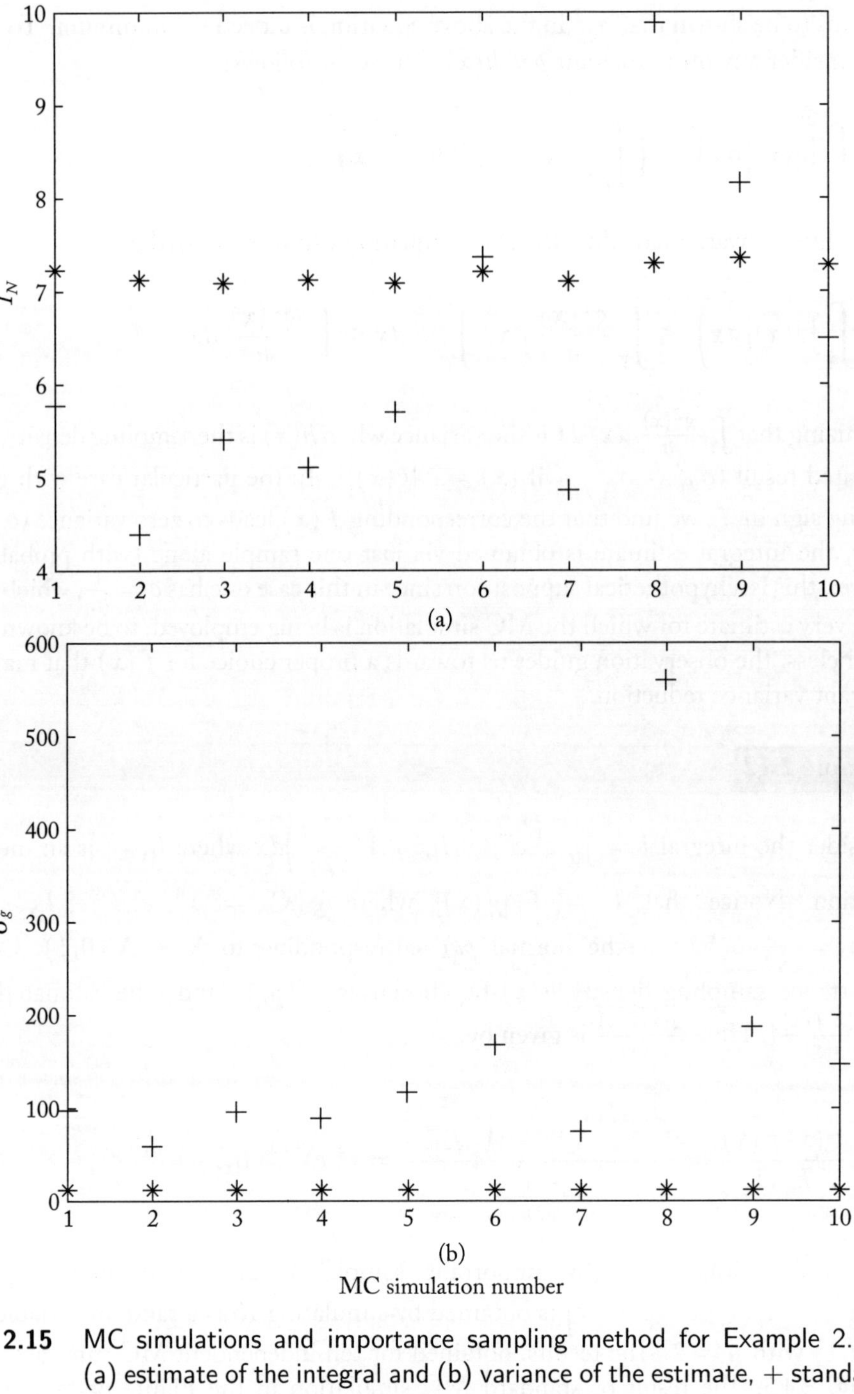

Fig. 2.15 MC simulations and importance sampling method for Example 2.12; (a) estimate of the integral and (b) variance of the estimate, + standard simulation, * simulation by important sampling

Choice of the sampling pdf in the importance sampling method

Consider the integral $I = E_f\left[g\left(\boldsymbol{X}\right)\right] = \int_{\Upsilon} g\left(\boldsymbol{x}\right) f\left(\boldsymbol{x}\right) d\boldsymbol{x}$ estimated by the random variable $\boldsymbol{X}$ generated from the *pdf* $f(\boldsymbol{x})$. Suppose that $h(\boldsymbol{x})$ is another sampling *pdf* with support identical to that of $f\left(\boldsymbol{x}\right)$. In this case, the integral takes the form:

$$I = \int_{\Upsilon} \frac{g\left(\boldsymbol{x}\right) f\left(\boldsymbol{x}\right)}{h\left(\boldsymbol{x}\right)} h\left(\boldsymbol{x}\right) d\boldsymbol{x} = E_h\left[g^*(\boldsymbol{X})\right]; g^*\left(\boldsymbol{x}\right) = \frac{g\left(\boldsymbol{x}\right) f\left(\boldsymbol{x}\right)}{h\left(\boldsymbol{x}\right)} \tag{2.122}$$

The respective variances with respect to the two sampling *pdf*s are obtained as:

$$\sigma_g{}^2 = \int_{\Upsilon} g\left(\boldsymbol{x}\right)^2 f(\boldsymbol{x}) d\boldsymbol{x} - I^2 \quad \text{and} \tag{2.123a}$$

$$\sigma_{g^*}{}^2 = \int_{\Upsilon} g^*\left(\boldsymbol{x}\right)^2 h(\boldsymbol{x}) d\boldsymbol{x} - I^2 = \int_{\Upsilon} \frac{g\left(\boldsymbol{x}\right)^2 f\left(\boldsymbol{x}\right)^2}{h(\boldsymbol{x})} d\boldsymbol{x} - I^2 \tag{2.123b}$$

The difference in the two variances is given by:

$$\sigma_g{}^2 - \sigma_{g^*}{}^2 = \int_{\Upsilon} g\left(\boldsymbol{x}\right)^2 \left(1 - \frac{f\left(\boldsymbol{x}\right)}{h\left(\boldsymbol{x}\right)}\right) f(\boldsymbol{x}) d\boldsymbol{x} \tag{2.124}$$

For variance reduction, the right hand side of the above equation needs to be positive. Thus, the main consideration in the choice of $h(\boldsymbol{x})$ is to have $\frac{f(\boldsymbol{x})}{h(\boldsymbol{x})} < 1$ whenever $g\left(\boldsymbol{x}\right) f\left(\boldsymbol{x}\right)$ in the integral I is large, i.e., to have $f\left(\boldsymbol{x}\right)$ to be dominated by $h(\boldsymbol{x})$ in the sub-domain(s) of Υ where $g\left(\boldsymbol{x}\right) f\left(\boldsymbol{x}\right)$ has larger absolute values. In regions of relatively smaller $g\left(\boldsymbol{x}\right) f\left(\boldsymbol{x}\right)$, one may have $\frac{f(\boldsymbol{x})}{h(\boldsymbol{x})} > 1$, but this condition is of less relevance since contributions to the integral from these regions are insubstantial.

One choice for a good sampling density is to have $g(\boldsymbol{x}) f\left(\boldsymbol{x}\right)$ and $h(\boldsymbol{x})$ attaining maximum values at the same $\boldsymbol{x}$. Whilst this may generally lead to many choices for $h(\boldsymbol{x})$, it is particularly useful if the importance sampling density follows the same family of distributions as $f(\boldsymbol{x})$. For instance, in Example 2.12 with $d = 1$, $h\left(x\right)$ is chosen to be normal, similar to $f(x)$. Thence, $h\left(x\right)$ attains its maximum at $x = \mu = E_h\left[X\right]$. The maximum of $g\left(x\right) f\left(x\right)$ is given by:

$$\max\, x^4\, e^{x^2/4}\; I_{\{x \le 2\}} \left(\frac{e^{-x^2/2}}{\sqrt{2\pi}}\right) = \max_{x \le 2} x^4\, e^{-x^2/4} = \sqrt{8} \tag{2.125}$$

The graphs of $g\left(x\right) f(x)$, $f(x)$ and $h(x)$ are plotted in Fig. 2.16a. The resulting variance reduction in estimating the integral is reported in Fig. 2.16b, when X is sampled from $h\left(x\right)$ with $\mu = \sqrt{8}$. Compare this with the earlier result, obtained via a normal density with $\mu = 2$ and reproduced in Fig. 2.16b for a ready reference.

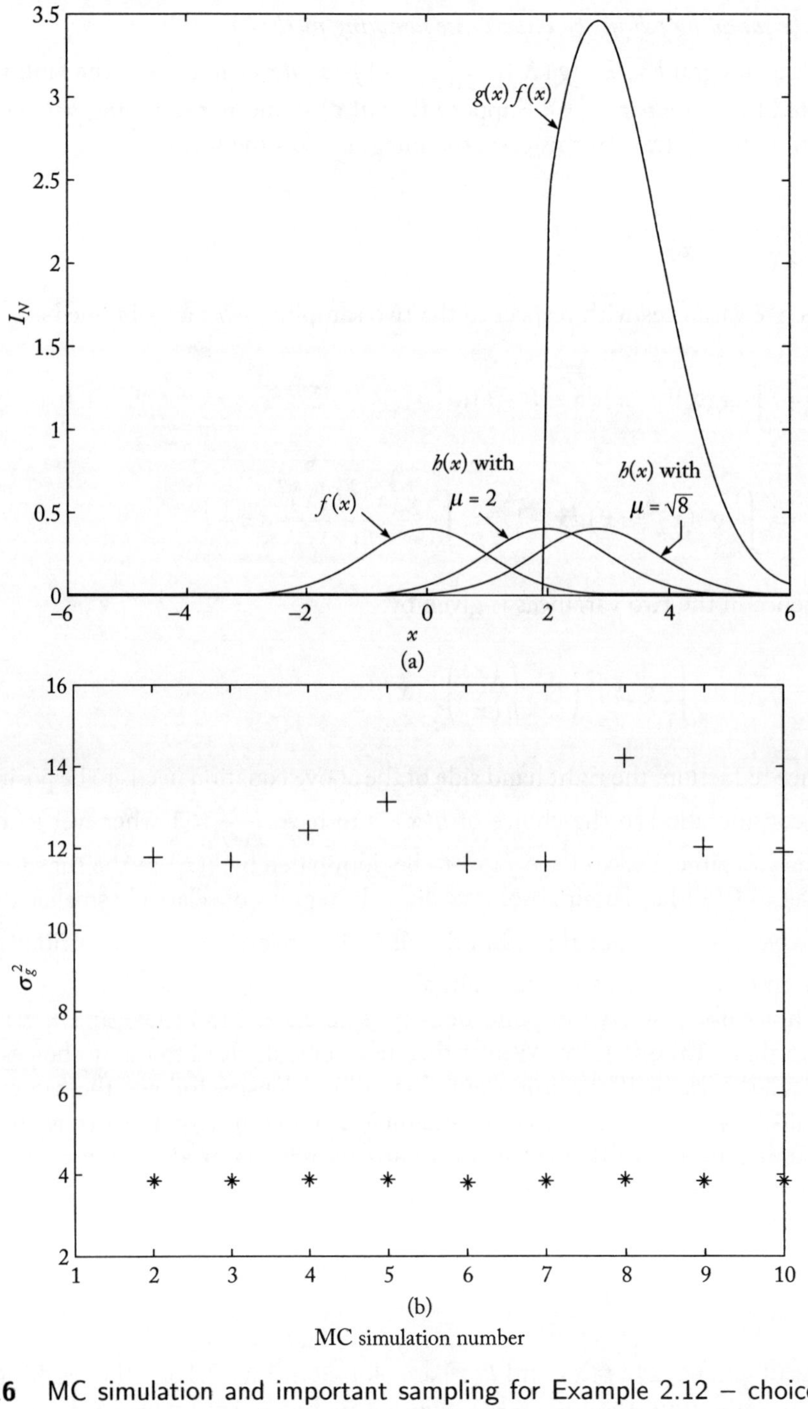

Fig. 2.16 MC simulation and important sampling for Example 2.12 – choice of sampling density; (a) plots of $g(x), f(x)$ and $h(x)$ and (b) variance reductions with $\mu = 2$ (marked +) and $\mu = \sqrt{8}$ (marked *)

2.5.5 Rao–Blackwell theorem and a general approach to variance reduction techniques

A broader approach to and an insightful understanding of variance reduction is provided by the Rao–Blackwell theorem given below. Toward this, it is convenient to first bring in the notion of a statistic. Specifically, given a $\mathcal{B}(\mathbb{R})^n$-measurable random variable X with a finite ensemble of n i.i.d. realizations (data) $B_X^n = \{X_1, X_2, \ldots, X_n,\}^T$ (with each X_i drawn from μ_X, the given distribution of X_i, a statistic $S(X_1, \ldots, X_n,)$ is a measurable function $\mathcal{B}(\mathbb{R})^n \to \mathcal{B}(\mathbb{R})^n$ of the elements of B_X^n. If $S^{-1}(\mathcal{B}(\mathbb{R})^n) := \{S^{-1}(G) : G \in \mathcal{B}(\mathbb{R})^n\}$, then $S^{-1}(\mathcal{B}(\mathbb{R})^n)$ is also a σ-algebra and is indeed the smallest σ-algebra on $\mathbb{R}^n$ in which S is measurable. Note that B_X^n defines a vector random variable with the family of product measures $\mu_X \times \ldots \times \mu_X = \mu_X{}^n$ (See Section 1.7, Chapter 1 for the definition of a product measure), parametrized by n. For example, given the data B_X^n, one statistic is $S = \frac{1}{N}\sum_{i=1}^{N\le n} X_{j_i}$ where, $j_i \in [1,n]$ and $j_i \neq j_k$ for $i \neq k$. Another possible statistic is $S = \max_i X_i$. In fact, the statistic $S = \frac{1}{N}\sum_{i=1}^{N} g(X_i)$ has been used in this sub-section (of MC simulations) from which the integral I (in Eq. 2.95 or 2.96) was estimated. In the last case, $B_g^n = \{g(X_i)\}$ was the data. This notion of a statistic is more commonly used in parameter estimation problems. We now define a sufficient sub-σ-algebra and hence a sufficient statistic.

Sufficient sub-σ-algebra

Given a family $\boldsymbol{\mu}$ (e.g., $\{\mu_X^n : n = 1,2\ldots\}$) of probability measures on $\mathcal{B}(\mathbb{R})^n$, a sub-$\sigma$-algebra $\mathcal{A} \subset \mathcal{B}(\mathbb{R})^n$ is called sufficient for the family $\boldsymbol{\mu}$ if and only if for every $B \in \mathcal{B}(\mathbb{R})^n$ there is an $\mathcal{A}$-measurable function $g_B \ge 0$ such that $g_B = \mu(B|\mathcal{A})$ μ-almost surely for every $\mu \in \boldsymbol{\mu}$. A statistic $S : (\mathbb{R}^n, \mathcal{B}(\mathbb{R})^n) \to (\mathbb{R}^n, \mathcal{B}(\mathbb{R})^n)$ is termed sufficient if and only if $S^{-1}(\mathcal{B}(\mathbb{R})^n)$ is sufficient.

Pair-wise sufficiency of a sub-σ-algebra

A sub-σ-algebra $\mathcal{A}$ is pair-wise sufficient if and only if for every pair μ_1, $\mu_2 \in \boldsymbol{\mu}$, the likelihood ratio (Radon–Nikodym derivative) $\frac{d\mu_1}{d\mu_2}$ is $\mathcal{A}$-measurable.

More generally, let a family $\boldsymbol{\mu}$ of probability measures be parametrized as $\boldsymbol{\mu} = \{\mu^\theta\}$ where $\theta \in \Theta$ denotes a real vector of parameters. Let every member of $\boldsymbol{\mu}$ be absolutely continuous to another probability measure ν on $(\mathbb{R}^d, \mathcal{B}(R^d))$. Then, we can define the set of Radon–Nikodym derivatives $\left\{\Lambda(\theta, x) = \frac{d\mu^\theta}{d\nu}(x)\right\}$, $x \in \mathbb{R}^d$. The following factorization theorem is due to Neyman [1935] and Halmos and Savage [1949].

Factorization theorem

As above, let $(\mathbb{R}^n, \mathcal{B}(\mathbb{R})^n)$ be a measurable space, $\mathcal{A} \subset \mathcal{B}(\mathbb{R})^n$ a sub-σ-algebra and $\boldsymbol{\mu}$ a non-empty family of probability measures on $\mathcal{B}(\mathbb{R})^n$. If every element of $\boldsymbol{\mu}$ is absolutely continuous to a probability measure ν, the following statements are equivalent:

i) $\mathcal{A}$ is sufficient for $\boldsymbol{\mu}$.

ii) $\exists$ a $\mathcal{B}(\mathbb{R})^n$-measurable function $h \geq 0$ such that $\forall\ \mu^\theta \in \boldsymbol{\mu}$, and a $\mathcal{A}$-measurable function g_θ with the Radon–Nikodym derivative $\frac{d\mu_X{}^\theta}{d\nu} = f_\theta\ h$. Here $h \in L^1(\mathbb{R}^n, \mathcal{B}(\mathbb{R})^n, \nu)$.

iii) $\mathcal{A}$ is pair-wise sufficient for $\boldsymbol{\mu}$.

An immediate consequence of the above theorem can be rephrased in the following equivalent statements.

Let $S(x)$ be a statistic on the sample space $(\mathbb{R}^n, \mathcal{B}(\mathbb{R})^n)$ and $\boldsymbol{\mu}$, ν be as defined in the last theorem. Then we have:

i) S is sufficient for $\boldsymbol{\mu}$

ii) the likelihood functions (Radon–Nikodym derivatives) can be written as:

$$\Lambda(\theta,x) = f(\theta, S(x))\,h(x) \tag{2.126}$$

where, for each θ, $f(\theta,.)$ is a $\mathcal{A}$-measurable scalar-valued and $h(x)$ a scalar-valued measurable function on $(\mathbb{R}^n, \mathcal{B}(\mathbb{R})^n)$.

iii) for any $\mu^{\theta_1}, \mu^{\theta_2} \in \boldsymbol{\mu}$, the likelihood function $\Lambda(\theta_1,\theta_2,x) = \frac{\mu^{\theta_2}}{\mu^{\theta_1}}(x)$ is a measurable function of $S(x)$.

The above equivalence statements remain valid even if ν is any σ-finite measure. It is important to note that the factor $h(x)$ in Eq. (2.126) is independent of θ. The following examples provide illustrations to this factorization.

Example 2.13

Consider a sequence of i.i.d. random variables $\{X_1, X_2, \ldots, X_n\}$ generated via Bernoulli trials such that each $X_i = 1$ with probability p and $X_i = 0$ with probability $q = 1 - p$. Here, $\Theta = \{p\}$ is a singleton. Then, by the factorization theorem, we have the likelihood function as:

$$\Lambda(p,x) = \prod_{i=1}^{n} p^{x_i}\,(1-p)^{1-x_i}$$

$$= p^{\sum_{i=1}^n x_i}\,(1-p)^{n-\sum_{i=1}^n x_i} \tag{2.127}$$

The factorization into the form as in Eq. (2.126) is obvious from the last equation. That is, we identify:

$$f(\theta, S(x)) = p^{\sum_{i=1}^{n} x_i} (1-p)^{n-\sum_{i=1}^{n} x_i}, h(x) = 1, \;\; S(x) = \sum_{i=1}^{n} x_i \tag{2.128}$$

$\sum_{i=1}^{n} x_i$ is thus a sufficient statistic for the unknown probability measure p of the Bernoulli trials.

Example 2.14

Let $\{X_1, X_2, \ldots, X_n\}$ be a sequence of independent normal random variables with the unknown parameters m, the mean and σ^2, the variance. Thus $\Theta = \{m, \sigma^2\}$. In this example, the likelihood ratio (the joint density of $X_1, X_2, \ldots, X_n$ with respect to the Lebesgue measure) is given by:

$$\begin{aligned}\Lambda\left(m, \sigma^2, x\right) &= \frac{1}{(2\pi\sigma^2)^{n/2}} \exp\left(-\frac{1}{2\sigma^2} \sum_{i=1}^{n} (x_i - m)^2\right) \\ &= \frac{1}{(2\pi\sigma^2)^{n/2}} \exp\left(-\frac{1}{2\sigma^2}\left(nm^2 - 2m\sum_{i=1}^{n} x_i + \sum_{i=1}^{n} {x_i}^2\right)\right)\end{aligned} \tag{2.129}$$

In this example also, we observe that $h(x) = 1$.

$$f(\Theta, S(x)) = \frac{1}{(2\pi\sigma^2)^{n/2}} \exp\left\{-\frac{1}{2\sigma^2}\left(nm^2 + S(x)(-2m,\ 1)^T\right)\right\} \tag{2.130}$$

where, the superscript T stands for transposition. Each element of the vector function $S(x) = \left(\sum_{i=1}^{n} x_i,\ \sum_{i=1}^{n} {x_i}^2\right)$ is a sufficient statistic for $\Theta = \{m, \sigma\}$.

Sufficient statistic is not unique for a given family of distributions. In the above two examples, the sample mean $\frac{1}{n}\sum_{i=1}^{n} x_i$ and sample variance $\frac{1}{n}\sum_{i=1}^{n} {x_i}^2$ (the latter applicable only for the last example) are also sufficient statistics. If $g(x)$ is an $\mathcal{A}$-measurable function on the realizations $X_1, X_2, \ldots, X_n$, then $\frac{1}{n}\sum_{i=1}^{n} g(x_i)$ and $\frac{1}{n}\sum_{i=1}^{n} g({x_i}^2)$ are also sufficient statistics. Note that such functions have been used in estimating the integral I in Eq. (2.95) or (2.96). We are now in a stage to state Rao–Blackwell theorem.

Rao–Blackwell theorem

Let $I_N = I_N(X_1, X_2, \ldots, X_N)$ be an estimator for I with $E\left[{I_N}^2\right] < \infty$. Suppose that $S(x)$ is a sufficient statistic for I and let $I^* = E[I_N \mid S] = E[I_N \mid \mathcal{A}]$. Then for all I, we have:

$$E[I^* - I]^2 \leq E[I_N - I]^2 \tag{2.131}$$

Proof:

$$E[I^* - I]^2 = E\,[E\,[I_N \mid \mathcal{A}] - I]^2$$

$$= E\,[E\,[I_N - I|\mathcal{A}]]^2 \le E\left[E\left[(I_N - I)^2|\mathcal{A}\right]\right] = E[I_N - I]^2 \tag{2.132}$$

The last inequality follows from Jensen's inequality i.e., if X is a random variable, $E[X - E(X)]^2 = E[X^2] - (E[X])^2 \ge 0$. The equality holds if I_N is a function of $S\,(x)$, in which case $E\,[I_N \mid \mathcal{A}] = I_N$.

Note that this theorem is equally valid for any convex function of $(I^* - I)$. For instance, $E\,[|I^* - I|] \le E\,[|I_N - I|]$. Thus reduction of sample variance of an estimate (e.g., evaluating an integral as in Examples 2.9 – 2.12) is ensured whenever it is conditioned on a sufficient statistic, if available.

2.6 Concluding Remarks

Our attempt, in the current and the preceding chapters, has been on providing the readers with a self-contained introduction to the basic concepts in the modern probability theory. Despite using elements of measure theory as the basic building block, the language we employ lacks the pristine rigor of a purely mathematical text; it is rather targeted to a readership whose primary interest is in using these notions in the contexts of applied scientific or engineering research. Having thus begun with the rudiments of random variables—moments, characteristic functions and transformations in Chapter 1, the present chapter goes in for a further exposition of the subject with regard to the crucial topics—conditioning, Bayes' formula and MC simulation. Using an interlacing of the theory with a number of illustrative examples, we have aimed at emphasizing the diverse possible applications of the theory. With this, we are now ready to develop the basic machinery to define stochastic processes (including the all-important Brownian motion), the associated σ-algebra and build up a framework through which such processes could be characterized.

Exercises

1. Consider two indicator variables:

$$I_A\,(x) = 1, \quad x \in A$$

$$= 0, \quad \text{otherwise}$$

and

$$I_B(x) = 1, \quad x \in B$$

$$= 0, \quad \text{otherwise}$$

Show that the indicator variables are positively correlated, uncorrelated and negatively correlated according as $P(A|B)$ is greater than, equal to or less than $P(A)$.

2. Let Ω be the set of events consisting of an infinite sequence of independent coin tosses. Let $\{X_n\}$ be the sequence of independent random variables with $X_n = 1$, if n^{th} toss is a head and $= 0$, if n^{th} toss is a tail. Given that the probability of each these outcomes is $1/2$, and a random variable $X = \sum_{n=1}^{\infty} \frac{X_n}{2^n}$, show that $\Omega_X = [0,1]$ and find the corresponding probability measure $P((a,b]) = P(a < X \leq b)$, $0 \leq a < b \leq 1$.

3. Let $Q << P$ and $P << R$. Prove that $Q << R$ and that $\frac{dQ}{dR} = \frac{dQ}{dP}\frac{dP}{dR}$.

4. Let $X \sim b(r;n,p)$ (binomial distribution) and $Y \sim \mathcal{N}(0,1)$. Let $Z = X$ with probability $1/2$ and $Z = Y$ with probability $1/2$. Let the distributions of X, Y and Z be P_X, P_Y and P_Z. Show that $P_X \ll P_Z$ and $P_Y \ll P_Z$. Also find $\frac{dP_X}{dP_Z}$ and $\frac{dP_Y}{dP_Z}$.

5. Denote by P the standard normal distribution and by Q the exponential distribution with rate 1. Show that $Q << P$, compute $\frac{dQ}{dP}$ and also show that P is not absolutely continuous with respect to Q.

6. A random variable X denotes the time interval that passes before an emission of a radioactive isotope takes place. It follows an exponential *pdf* given by:

$$f_X(x) = \lambda e^{-\lambda x}, \quad x \geq 0$$

$$= 0, \quad \text{otherwise}$$

Given that there has been no emission for t seconds after a clock is started, find the probability that we need to wait for s more seconds for an emission to took place. [Hint: The required probability is $P(A|B)$ where $A = \{X \geq t\}$ is the event that no emission has taken place in t seconds and $B = \{X \geq t + s\}$, the event that emission has not taken place for another s seconds. One can show that $P(A \mid B) = e^{-s}$ which implies that this probability is independent of the previous waiting time t. This is known as the *memoryless* property of exponential and geometric distributions.]

7. Given two independent Poisson random variables X and Y with respective parameters λ_1 and λ_2, find the conditional distribution $P(X|X+Y=n)$. Show that it is a binomial distribution $b\left(r;n,\frac{\lambda_1}{\lambda_1+\lambda_2}\right)$.

8. Given two continuous random variables X_1 and X_2 with joint *pdf* $f_{X_1,X_2}(x_1,x_2)$, show that: $E_{f_{X_1}}[g(X_1)] = E_{f_{X_1|X_2}}[g(X_1)|X_2 = x_2]$ (law of iterated expectation).

9. $X_1, X_2, \ldots$ are independent and identically distributed (iid) discrete random variables with probability distributions $\{P(X_i = k) = a_k, k = 0, 1, \ldots\}$, $i = 1, 2, \ldots$. Let $Y_n = X_1 + X_2 + X_n$, $n = 1, 2, \ldots$ with $Y_0 = 0$. Show that $\{Y_n\}$ is a Markov chain with homogeneous transition probabilities; (see Appendix B for some details on Markov chains).

10. If X is a random variable and an event $A : \{X \le d\}$ with $-\infty < d < \infty$, obtain expressions for the conditional probability distribution and density functions, i.e., $F_X(x|X \le d)$ and $f_X(x|X = d)$. Deduce from the result that both the distribution and density functions are never smaller than $F_X(x)$ and $f_X(x)$ respectively.

11. Using the fact that for any real constants $\lambda, c > 0$, $e^{\lambda(x-c)} \ge u(x-c)$ where $u(.)$ is a unit step function and $\mathcal{M}_X(\lambda)$ a moment generating function,

    a) prove Chernoff's inequality
    $$P(X \ge c) \le e^{-\lambda x}\mathcal{M}_X(\lambda) \tag{E2.1}$$

    b) for $X \sim \mathcal{N}(0,1)$, find the minimum bound for $P(X \ge c)$.

    [Hint: for b) minimize the right hand side of (E2.1) above with respect to the parameter c].

12. Generate realizations of a uniform random variable in the interval $[0,1]$ by writing an appropriate program and continue adding the numbers until the first time the sum is greater than 1. If the exercise is repeated a number of times, estimate the expected number of such exercises in order that the sum first exceeds 1.

13. Obtain the conditional *pdf* $f_{X|Y}(x|y)$ for two jointly Gaussian zero-mean random variables X and Y with a non-singular covariance matrix. The variances of X and Y are respectively denoted by σ_1^2 and σ_2^2 and the correlation coefficient by r. Show that given $Y = y$, the conditional *pdf* of X is Gaussian with mean $r\sigma_1\sigma_2 y$ and variance $\sigma_1^2(1-r^2)$.

14. Gambler's ruin problem: Two players A and B have initial financial capital of a units and b units respectively and play a game until one of them is bankrupt. At the end of each game, the loser pays one unit of currency to the winner. The probability of success for A is p and that of B is q. $r = 1 - p - q$ is the probability of a tie. If X_n represents the capital of the player A at the end of the nth game, find the distribution of X_n at the end of, say, n games. Assume $a = 10, b = 5, p = 0.2, q = 0.4, n = 100$.

    (Hint: X_n represents a 1-dimensional random walk starting at $X_0 = a$ with the absorbing barriers of the capital equal to zero and $a + b$. The first barrier corresponds to the bankruptcy of A and the second to that of B. The 1-dimensional random walk is the simplest case of a discrete parameter Marko chain (which is known as a discrete stochastic process – discussed in Chapter 3). The states are discrete and are, in this specific case, given by

$\{0,1,\ldots,a+b=13\}$. If $\boldsymbol{P}$ denotes the transition probability matrix and $\boldsymbol{p}$ the vector of probabilities of the states at end of a game, then $\boldsymbol{p}_n = \boldsymbol{P}^T \boldsymbol{p}_{n-1} = \left(\boldsymbol{P}^T\right)^n \boldsymbol{p}_0$. See Appendix B for an introductory account on Markov chains.

Notations

A_n	sequence of events with $n \in \mathbb{N}$
B_m	bending moment (Example 2.5)
B_X^n	$=\{X_1, X_2, \ldots, X_n\}$, a vector random variable
e_N	error estimate (Eq. 2.103)
$E_P[X]$	expectation with respect to the probability measure P
$f_{X\|Y}(x\|y)$	conditional *pdf*
$F_n(x)$ and $F(x)$	distribution functions of X_n and X respectively
$F_{X\|Y}(x\|y)$	conditional CDF
$g(x), h(x)$	continuous *pdf*s
$g(\boldsymbol{X})$	performance function (Example 2.9)
I, I'	finite integral
I_N	integral as a finite sum (Eq. 2.98)
$\mathbb{K}(x)$, $\mathbb{k}(x)$	CDF and *pdf* of random variable X
$m_{X\|Y}$	conditional mean
P_f	probability of failure (Example 2.9)
$P_{\mathcal{G}}$	measure
$P_B(A)$	conditional probability $P(A\|B)$ (Eq. 2.2)
S	statistic – a measurable function (Section 2.5.5 - *Rao–Blackwell Theorem*
S_m	section modulus (Example 2.5)
S_y	yield strength of the material (Example 2.5)
X, $X_{1\le j\le n}$, Y	random variables
U, V	random variables
β_F	reliability index
Λ	Radon–Nikodym derivative (Eq. 2.14 and 2.15)

$\Lambda(\omega)$	Radon–Nikodym derivative in the case of discrete random variable
$\Lambda(x)$	Radon–Nikodym derivative in the case of continuous random variable
$\mu_{\mathcal{G}}$	measure
$\mu = \{\mu_X^n : n = 1, 2\ldots\}$	family of measures
$\sigma_{X\|Y}^2$	conditional variance
$\mathcal{F}_B$	sigma algebra

Chapter **3**

An Introduction to Stochastic Processes

3.1 Introduction

Given a complete probability space $(\Omega, \mathcal{F}, P)$, a stochastic process is a parametrized family of random variables $X = \{X_t(\omega) : t \geq 0,\ t \in \mathfrak{T}, \omega \in \Omega\}$, where, t is often a scalar (real) parameter defined on an index set $\mathfrak{T}$ (finite or countable or even uncountable) such that the random variable $X_t(\omega)$ is $\mathcal{F}$-measurable for every $t \in \mathfrak{T}$. While the indexing parameter t generally refers to time, discrete or continuous, and takes values in $\mathbb{R}^+$, it may sometimes be a vector referring, for instance, to one or more spatial (or spatiotemporal) coordinate dimensions. In the latter case the process $X_t(\omega)$ is better known as a random field, e.g., one used to describe the aerodynamic pressure fluctuations on a 2-dimensional panel as a function of the two spatial coordinates. The focus in this treatise is however on stochastic processes with t interpreted only as time. Formally, then, a stochastic process may be looked upon as a two-parameter mapping $X_t(\omega) = X(\omega, t) : \Omega \times \mathbb{R}^+ \to \mathbb{R}^n$ $(n \geq 1)$. While for a fixed $\omega^* \in \Omega$, $X(\omega^*, t)$ is often called a path (or realization or trajectory) of the stochastic process, for a fixed $t^* \in T$, $X(\omega, t^*)$ is a random variable in $(\Omega, \mathcal{F}, P)$.

Thus, a stochastic process may be treated either as an ensemble of (possibly infinitely many) trajectories evolving in time or as a collection of random variables sampled at (possibly infinitely many) time instants. A graphical interpretation of the idea is provided in Fig. 1.1 (which is reproduced here as Fig. 3.1 as a ready reference), wherein three different paths of the underlying stochastic process, the wind velocity, are indicated. Each path may represent a realization collected (observed) over a time interval $[0, T = 20\ s]$ on different occasions (e.g., on different days) at any fixed spatial location. Figure 3.2 shows three simulated paths of a Brownian motion, a quintessential stochastic process of great significance in the developments to follow and briefly touched upon in Chapter 2 as a limiting random walk model.

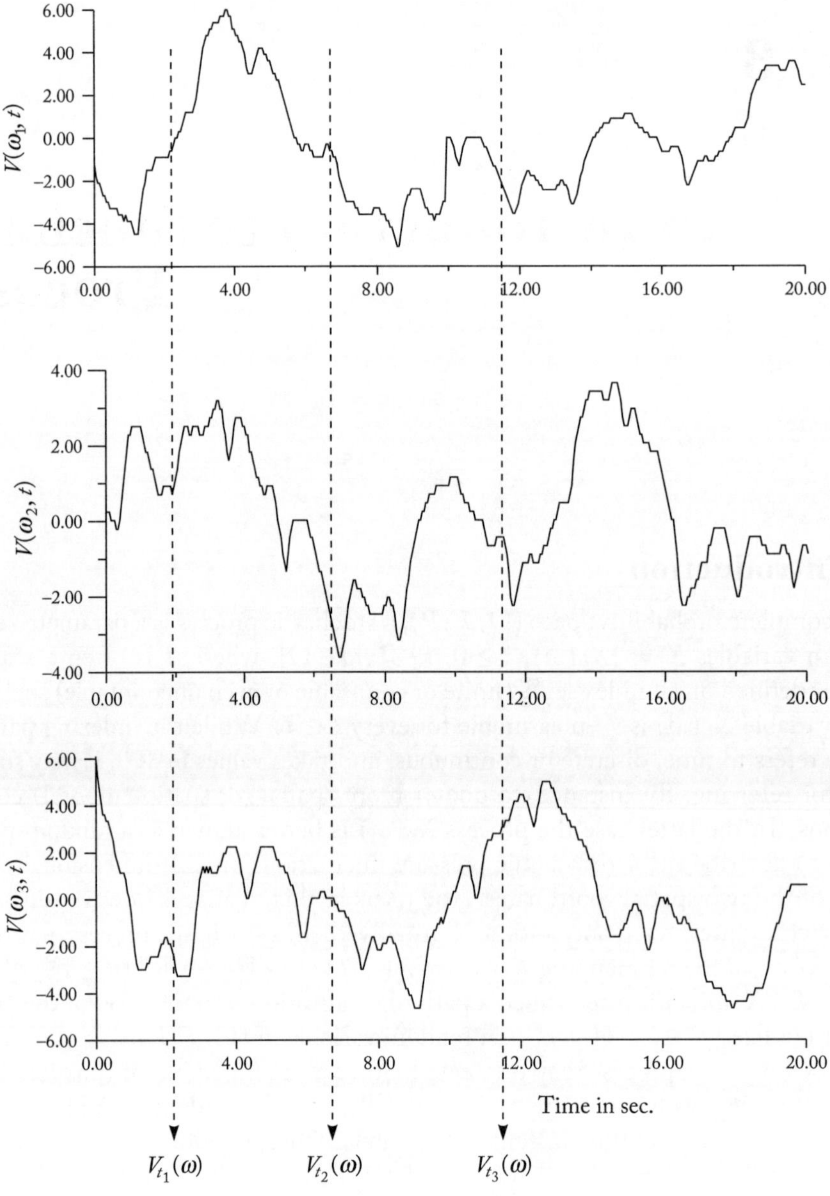

Fig. 3.1 Stochastic signals; typical wind velocity (in m/sec.) trajectories, $V(\omega_j, t), j = 1, 2$ and 3 and random variables $V_{t_i}(\omega),\ i = 1, 2$ and 3

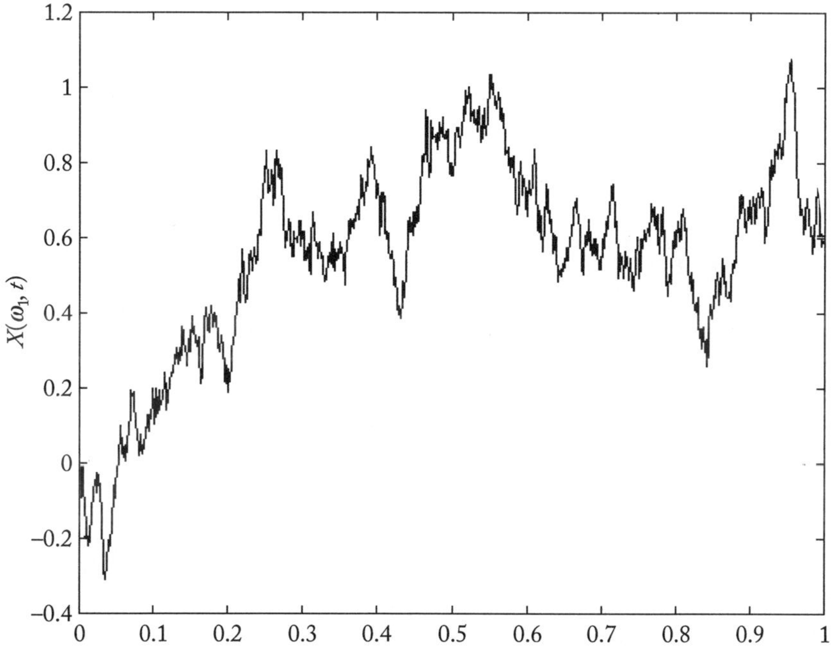
1.2
1
0.8
0.6
0.4
0.2
0
−0.2
−0.4
$X(\omega_1, t)$
0
0.1
0.2
0.3
0.4
0.5
0.6
0.7
0.8
0.9
1

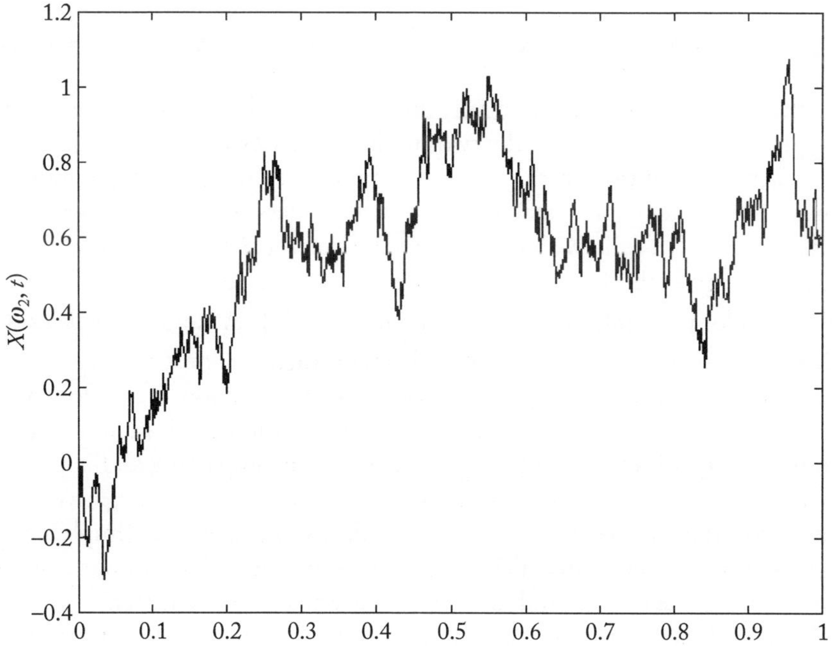
1.2
1
0.8
0.6
0.4
0.2
0
−0.2
−0.4
$X(\omega_2, t)$
0
0.1
0.2
0.3
0.4
0.5
0.6
0.7
0.8
0.9
1

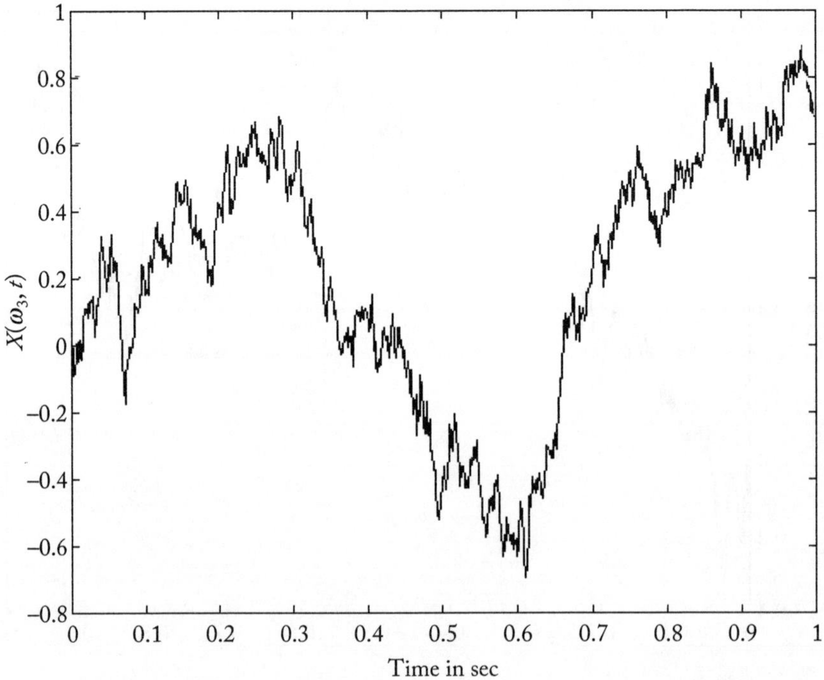

Fig. 3.2 Brownian motion as a stochastic process; paths $X(\omega_j, t), j = 1, 2,$ and 3

Pending a clearer exposition and based only on a cursory look at the paths plotted in the last two figures, it is tempting to qualify the two stochastic processes as continuous in t. Setting aside a satisfactory resolution of this issue for the nonce (which requires more advanced tools than we have developed so far), one may nevertheless proceed to describe a discrete-time stochastic process as one with the (ordered) index set $\mathfrak{T}$ finite or countable, e.g., $\mathfrak{T} = \mathbb{N} = \{1, 2...\}$. In this case, the stochastic process $X(\omega, t)$ is accordingly a finite/countable collection of random variables $\left\{X_{t_i}(\omega) := X_i(\omega),\ i \in \mathbb{N}\right\}$. A random walk as defined in Section 2.3.4, Chapter 2 is an example of a discrete stochastic process.

Having thus far defined a stochastic process as a collection of random variables, a pertinent question that arises is whether an understanding of random variables, as developed in Chapters 1 and 2, should be adequate in encompassing all the properties of a (discrete) stochastic process. The answer to this question is in the negative as one must also account for the precise role played by ordered index set $\mathfrak{T}$ (implying gradually increasing time) in characterizing such properties—an aspect that did not come into play while characterizing the (joint) random variables in the last two chapters.

Indeed, the utilitarian part of studying stochastic processes is that they are routinely used to model various phenomena [Gardner 1990, Nigam and Narayanan 1994, Yates and Goodman 1999] such as noise in electrical/communication systems, seismic, wind and sea wave forces partly characterized by variations that are intrinsically unpredictable. Modelling such phenomena has implications in engineering design and beyond, including such diverse fields as financial market pricing [Black and Scholes 1973, Karatzas and Shreve 1998], statistical/quantum mechanics [Nagasawa 2000], continuum mechanics admitting long-range interaction effects [Sarkar et al. 2015, Roychowdhury et al. 2016] and geophysics [Ewald 2002]. Uncertainties may also be present in system model parameters rendering the associated mathematical operator itself random. Thus, looking ahead, the theory of stochastic processes must be brought in to quantify the response or precisely characterize the models of dynamical systems under noisy time-varying inputs or with imprecisely known parameters. This in turn requires studying dynamical system equations involving stochastic (noisy) forcing terms—often referred to as stochastic differential equations (SDEs). Given that time-dependent conservation laws and operators for dynamical systems often lead to time-continuous solution paths in the absence of noisy inputs (and parameter uncertainties), the issue of extrapolating the notion of continuity for solutions of SDEs also demands to be resolved. This done, an appropriate characterization of such solutions (e.g., their uniqueness, stability and enumerable expressions describing them within an appropriate probabilistic framework) must be sought. This brings into play the so-called stochastic calculus—a counterpart of the well-known Newtonian calculus employed to cover solutions of ordinary differential equations (ODEs). This chapter is all about a progressive exposition of some of these basic ideas *en route* to building up the tools for various applications in the later chapters.

3.2 Stochastic Process and its Finite Dimensional Distributions

A stochastic process is determined by the set of its finite dimensional distributions (*fdd*s), i.e., by seeking probabilities of the form $P(X(t_1) \in \mathcal{F}_1, X(t_2) \in \mathcal{F}_2, \ldots, X(t_k) \in \mathcal{F}_k)$ for any choices of $0 \le t_1 \le t_2 \le \ldots\ldots t_k \le T$, where, $\mathcal{F}_i, i = 1, 2.., k$ are Borel sets in $\mathbb{R}^n$. In other words, a stochastic process is typically characterized by all its finite dimensional joint distributions of thc form:

$$F_X(x_1, x_2, \ldots\ldots x_m; t_1, t_2, \ldots, t_m) = P(X(t_1) \le x_1, X(t_2) \le x_2, \ldots, X(t_m) \le x_m) \quad (3.1)$$

for $0 \le t_1 \le t_2 \le \ldots\ldots t_m \le T$. If this distribution function is differentiable, we have the joint density function $f_X(x_1, x_2, \ldots\ldots x_m; t_1, t_2, \ldots, t_m) = \frac{\partial^m F_X(x_1, x_2, \ldots\ldots x_m; t_1, t_2, \ldots, t_m)}{\partial x_1 \partial x_2, \ldots\ldots \partial x_m}$. Thus, for a given stochastic process, we can look for all such finite dimensional joint distributions associated with the random variables that in turn correspond to the snapshots of the stochastic process at different time instants. Kolmogorov's extension theorem provides the basis for proving the reverse. It states that given a family of distributions of finite dimensional random variables that is consistent that (see below), there exists a stochastic process satisfying the distributions.

Kolmogorov's extension theorem

According to this theorem, the consistency requirement [Øksendal 2003] states that for all $t_1, t_2, \ldots, t_k \in [0, T]$, $k \in \mathbb{N}$, if $\gamma_{1,2,\ldots,k} := \gamma_{t_1, t_2, \ldots, t_k}$ is a probability measure on $\mathbb{R}^{nk}$ such that (i) $\gamma_{1,2,\ldots,k}(\mathcal{F}_1 \times \mathcal{F}_2 \times \cdots \times \mathcal{F}_k) = \gamma_{\sigma(1),\sigma(2),\ldots,\sigma(k)}(\mathcal{F}_{\sigma(1)} \times \mathcal{F}_{\sigma(2)} \times \cdots \times \mathcal{F}_{\sigma(k)})$ for all permutations σ (σ being a permutation operator on $1,2...k$) and (ii) $\gamma_{1,2,\ldots,k}(\mathcal{F}_1 \times \mathcal{F}_2 \times \cdots \times \mathcal{F}_k) = \gamma_{1,2,\ldots,k,k+1,\ldots,k+l}(\mathcal{F}_1 \times \mathcal{F}_2 \times \cdots \times \mathcal{F}_k \times \mathbb{R}^n \times \ldots\ldots \times \mathbb{R}^n)$ for all $l \in \mathbb{N}$, then there exists a probability space $(\Omega, \mathcal{F}, P)$ and a stochastic process $X(\omega, t)$ on $\Omega; X_t : \Omega \to \mathbb{R}^n$ such that $\gamma_{1,2,\ldots,k}(\mathcal{F}_1 \times \mathcal{F}_2 \times \cdots \times \mathcal{F}_k) = P(X_{t_1} \in \mathcal{F}_1, X_{t_2} \in \mathcal{F}_2, \ldots, X_{t_k} \in \mathcal{F}_k)$ for all $t_k \in [0,T]$, $k \in \mathbb{N}$, and for all Borel sets $\mathcal{F}_i$, $i = 1,2\ldots,k$.

The first consistency condition of the theorem implies that the joint probability information is invariant with respect to a permutation of its arguments and the second one that higher dimensional distributions contain all lower dimensional (marginal) ones.

It may be possible to construct a stochastic process by its *fdd*s (as per Kolmogorov's extension theorem). However, the process determined by these distributions may not be unique. It may seem that having the *fdd*s is all that matters and that this interpretation potentially makes the theory of stochastic processes, particularly of those continuous in time, simpler. On the contrary, we note that the continuity of the sample paths of the process is not exclusively determined by the knowledge of these distributions. The following example illustrates specifically the continuity or otherwise of two processes for which all the *fdds* are identical.

Example 3.1

Consider two processes X and Y with $X(t) = 0 \ \forall \ t \in [0,1]$ and $Y(t) = 0$ for $t \neq \tau$ and $Y(t) = 1$ for $t = \tau$ where τ is uniformly distributed over $[0,1]$. Here, we find that $P(X(t) = 0) = P(Y(t) = 0) = 1$ and $P(X(t) \neq 0) = P(Y(t) \neq 0) = 0$ so that the 1-dimensional distributions are identical. Similarly, all the *fdd*s are also the same. However, the sample paths of $X(t)$ are continuous in $t \in [0,1]$ whilst every sample path of $Y(t)$ has a discontinuity of the first kind [Appendix A] in the form of a jump at $t = \tau$.

3.2.1 Continuity of a stochastic process

We want to employ the theory of stochastic processes in modeling dynamical systems and, in the process, wish to study the dynamics in continuous time, i.e., for $\forall t \in [0,T]$. In this connection, recalling from Chapter 2 the different kinds of convergence of random variables, one may define several forms of continuity for a stochastic process. For example:

Mean square continuity

A stochastic process is mean square continuous at any time $t_k \in [0,T]$, if $t \to t_k$ implies $E\left[\left|X(\omega,t) - X(\omega,t_k)\right|^2\right] \to 0$. This type of continuity is also referred to as L^2 continuity. One can as wll define continuity in the L^p sense for any $p \in [1,\infty]$.

Continuity in probability

A stochastic process is continuous in probability at any time $t_k \in [0,T]$, *if* $t \to t_k$ implies $P(\omega \in \Omega \mid (X(\omega,t) - X(\omega,t_k) = 0)) = 1$.

Neither L^p continuity nor continuity in probability ensures path-wise continuity (defined below) of a stochastic process.

Path-wise continuity

A stochastic process is path-wise continuous at any $t_k \in [0,T]$, if, for almost all $\omega \in \Omega$, $t \to t_k$ implies $X(\omega,t) \to X(\omega,t_k)$. Thus, a process is continuous if, for almost all $\omega \in \Omega$, $X(\omega,.)$ is a continuous function. Path-wise continuity implies L^p continuity.

A weaker path-wise continuity is provided, for instance, by a right continuous function with left limits (a càdlàg process). More specifically, a function $x(t)$ is càdlàg if for every $t_k \in [0,T]$, $t \downarrow t_k$ implies $x(t) \to x(t_k)$ where $x(t_k)$ is the limit from the right and for $t \uparrow t_k$, $\lim_{t \uparrow t_k} x(t)$ exists but is not equal to $x(t_k)$ (see Fig. 3.3).

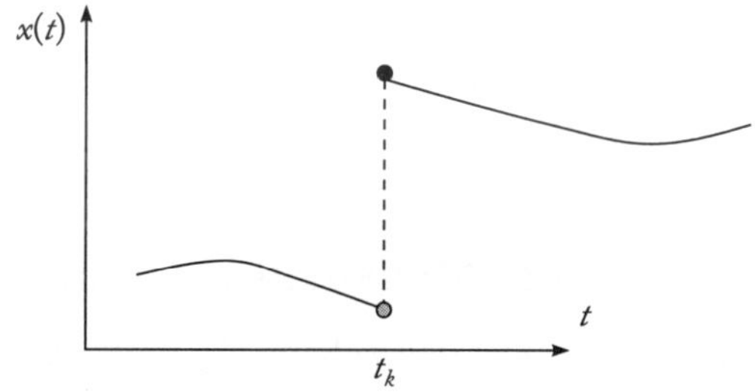

Fig. 3.3 A càdlàg function

A stochastic process $X(\omega,t)$ is càdlàg if almost all its sample paths are càdlàg. Such a notion of right continuity, incorporated within a stochastic process, is very helpful in view of its features and keeping in mind several applications to be described in due course. As noted earlier, while the set of *fdd*s may correspond to a stochastic process by dint of Kolmogorov's extension theorem, such a process need not be unique. In other words, there could be several versions of the stochastic process, possibly with varying continuity properties, each satisfying the *fdd*s. Amongst these versions, the easiest to work with could be the one with good regularity features. Indeed, the ideal thing to look for would be if a continuous version of the stochastic process exists.

3.2.2 Version/modification of a stochastic process

Two stochastic processes X and Y are versions or modifications of each other if:

$$P(\{\omega \in \Omega \mid X(t_k,\omega) = Y(t_k,\omega)\}) = 1, \forall t_k \in [0,T], k \in \mathbb{N} \tag{3.2}$$

In Example 3.1, $P(X(t) = Y(t)) = 1$, $\forall t \in [0,T]$ and hence the two processes X and Y are versions of each other having the same *fdd*s. Note that the path properties of the two versions may differ from each other. However, this provides a choice to select and, if possible, work with the version with continuous paths. The condition for existence of the versions of a stochastic process that are continuous is stated by Kolmogorov's continuity theorem.

Kolmogorov's continuity theorem

If, for all $T > 0$, there exist positive constants, α, β and C such that:

$$E\left[|X_t - X_s|^{\alpha}\right] \leq C\,(t-s)^{1+\beta}, 0 \leq s < t \leq T \tag{3.3}$$

then there exists a version of the stochastic process $X : \mathbb{R}^+ \times \Omega \to \mathbb{R}^n$ with continuous sample paths. Refer to Stroock and Varadhan [1979] for a proof of the theorem. Similarly, the condition for the existence of versions with paths having jump discontinuities (called regular paths) is stated in the following part of the theorem.

If, for all $T > 0$, there exist positive constants, $\alpha_1, \alpha_2, \beta$ and C such that:

$$E\left[|X_u - X_s|^{\alpha_1}|X_t - X_u|^{\alpha_2}\right] \leq C\,(t-s)^{1+\beta}, 0 \leq s \leq u \leq t \leq T \tag{3.4}$$

then there exists a version of the stochastic process X with sample paths having jump discontinuities only (i.e., discontinuities of the first kind).

3.3 Stochastic Processes–Measurability and Filtration

Consider the stochastic process $X(\omega, t)$ defined on $(\Omega, \mathcal{F}, P)$. Since $X(\omega, t)$ is a function of two variables, it is measurable if the function:

$$X : (\Omega \times [0,T], \mathcal{F} \otimes \mathcal{B}([0,T])) \to (\mathbb{R}^n, \mathcal{B}(\mathbb{R}^n)) \tag{3.5}$$

is jointly measurable. Here $\mathcal{F} \otimes \mathcal{B}([0,T] = \sigma(\mathcal{F}, \mathcal{B}([0,T])$ is the product σ-algebra. However, not all stochastic processes are measurable. If $E := \{\omega : X_t(\omega) \leq a \in \mathbb{R}\}$, then the event $E_1 = \cap_{\substack{t \in [0,T],\\ t \text{ rational}}} E$ is measurable since the corresponding set of random variables $\{X_t(\omega) : t \text{ rational}\}$ is measurable. On the other hand, the event of the type $E_2 = \cap_{t \in [0,T]} E$ is an intersection of uncountable number of sets and it is far from easy to decide on the measurability of such events. It is here necessary [Itô and McKean Jr. 1996] to have a restriction in the form of separability of a stochastic process so that it can be completely described despite the countable nature of the associated σ-algebra (e.g., through all the *fdd*s). Specifically, a stochastic process $X_t(\omega), t \in S \subset \mathbb{R}$ (S being possibly uncountable) defined on a probability space $(\Omega, \mathcal{B}(S), P)$ is separable if there exists a countable set $S_1 \subset S$ such that for any open interval $I \subset \mathbb{R}$, one has

$$P\left(\cap_{t \in I \cap S}(X_t \in A) \setminus \cap_{t \in I \cap S_1}(X_t \in A)\right) = 0 \tag{3.6}$$

for any $A \in \mathcal{B}(S)$.

From the point of view of numerous applications involving stochastic processes (that follow in later Chapters), it suffices to deal with only separable stochastic processes and in turn only with the separable versions/modifications of the processes. Prior to that, however, we would need a few more definitions.

3.3.1 Filtration and adapted processes

On a probability space, $(\Omega, \mathcal{F}, P)$, a filtration is an increasing sequence $\{\mathcal{F}_t\}, t \in \mathbb{N}$ or $t \in [0, T]$ or $t \in [0, \infty)$ of sub-σ-algebras of $\mathcal{F}$, i.e., $\mathcal{F}_t \subset \mathcal{F}, \forall t$ and for $s < t$ we have $\mathcal{F}_s \subset \mathcal{F}_t$. The quadruple $(\Omega, \mathcal{F}, \{\mathcal{F}_t\}, P)$ is known as the filtered probability space. $\mathcal{F}_t$ being a σ-algebra generated by inverse images of sets of the form $\{X_s(\omega) \leq a \in \mathbb{R}\}$ (for all $s \leq t$) contains the history of the process X_s up to the current time t. Now, a stochastic process $X = \{X_t(\omega)\}$ is said to be adapted to the filtration $\{\mathcal{F}_t\}$ if the random variable $X_t(\omega)$ is $\mathcal{F}_t$-measurable for all t. Specifically, if X is càdlàg and needs to be adapted to the filtration $\{\mathcal{F}_t\}$, then the filtered probability space must also satisfy the usual condition that filtration is right continuous, i.e., $\mathcal{F}_t = \mathcal{F}_{t+}$ where $\mathcal{F}_{t+} = \cap_{s>t} \mathcal{F}_s$ implying that information known immediately after t is also included in $\mathcal{F}_t$. However, for a strictly continuous stochastic process, such a condition need not be imposed.

In practice, we will often have a countable sequence of random variables, $\{X_{t_j}, j = 0, 1, \ldots, n, \ldots\}$, $n \in \mathbb{N}$, representing a stochastic process (in a time-discrete form). Then, the filtration generated by these random variables (of the discrete process) is given by $\mathcal{F}_n = \sigma(X_{t_0}, X_{t_1}, \ldots, X_{t_n})$ and the process is $\mathcal{F}_n$-adapted by construction. A discrete stochastic process is referred to as $\mathcal{F}_n$-predictable if $X_n(\omega) := X_{t_n}(\omega)$ is $\mathcal{F}_{n-1}$ measurable for every n.

We have thus far described some basic properties of a stochastic process and set up a stage to characterize/classify these processes, which in turn are helpful in modeling of input noises to dynamical systems.

3.3.2 Some basic stochastic processes

Before we proceed with a description of some of the oft-used stochastic processes, it is necessary to characterize them from the viewpoint of their statistics [Lin 1967, Lawler 1995, Nigam 1983, Ross 1995, Brzeźniak and Zastawniak 1999, Grimmettvand Stirzaker 2001, Cramér and Leadbetter 1967, 2004]. Given the nth order distribution function $F(x_1, x_2, \ldots, x_2; t_1, t_2, \ldots, t_n)$ of the associated random variables $\{X_{t_i}\}, i = 1, \ldots, n$ of a stochastic process $X(\omega, t)$, one can, in principle, determine the relevant statistical properties (say, the cross moments of any order $\leq n$). Here, the usual suffix that accompanies the definition of a distribution function is dropped for convenience. Indeed, in many applications, it is often of interest to know only the first and second order moments associated with such random variables forming the time-discrete version of $X(\omega, t)$. Definitions of moments of single/joint random variable/s have already been covered in Chapter 1 and only the additional aspects needing special attention vis-à-vis a stochastic process are described hereupon.

Auto-and cross-correlations and covariances

Let us define $X_i(\omega) := X_{t_i}(\omega)$. The second order statistics $E\left[X_i(\omega)\ X_j(\omega)\right], i, j \in [1, n]$ is specifically known as the auto-correlation $R_{XX}(t_i, t_j)$ defined as:

$$R_{XX}(t_i,t_j) = E\left[X_i(\omega)\ X_j(\omega)\right] = \int_{\mathbb{R}}\int_{\mathbb{R}} x_i x_j f(x_i,x_j;t_i,t_j)dx_i dx_j \tag{3.7}$$

Similarly, the auto-covariance is defined as:

$$C_{XX}(t_i,t_j) = E\left[(X_i(\omega)-\mu_X(t_i))\ \left(X_j(\omega)-\mu_X(t_j)\right)\right]$$

$$= \int_{\mathbb{R}}\int_{\mathbb{R}} (x_i-\mu_X(t_i))\left(x_j-\ \mu_X(t_j)\right)f(x_i,x_j;t_i,t_j)dx_i dx_j$$

$$= \ R_{XX}(t_i,t_j)-\mu_X(t_i)\ \mu_X(t_j) \tag{3.8}$$

$\mu_X(t_i)$ and $\mu_X(t_j)$ are the first order statistics (mean values) of the process given by $\int_{\mathbb{R}} x_i f(x_i;t_i)dx_i$ and $\int_{\mathbb{R}} x_j f(x_j;t_j)dx_j$ respectively. If $t_i = t_j$ in the definitions (3.7) and (3.8), they respectively yield the mean square value $E\left[X_i{}^2(\omega)\right]$ and the variance $E\left[(X_i(\omega)-\mu_X(t_i))^2\right]$ for the random variable $X_i(\omega)$. The auto-covariance is positive definite, i.e., for any $\alpha_i,\alpha_j \in \mathbb{R}$, one has:

$$\sum_{1\le i,j\le n} \alpha_i\alpha_j C_{XX}(t_i,t_j) \ge 0 \tag{3.9}$$

Proof:

$$\sum_{1\le i,j\le n} \alpha_i\alpha_j C_{XX}(t_i,t_j) = \sum_{1\le i,j\le n} \alpha_i\alpha_j E\left[(X_i(\omega)-\mu_X(t_i))\ \left(X_j(\omega)-\mu_X(t_j)\right)\right]$$

$$= E\left[\left(\sum_{i=1}^{n} \alpha_i(X_i(\omega)-\mu_X(t_i))\right)^2\right] \ge 0$$

One may also define cross-correlation and cross-covariance between two discrete-time stochastic processes X and Y as:

$$R_{XY}(t_i,t_j) = E\left[X_i(\omega)\ Y_j(\omega)\right] = \int_{\mathbb{R}}\int_{\mathbb{R}} x_i y_j f(x_i,y_j;t_i,t_j)dx_i dy_j \tag{3.10a}$$

$$C_{XY}(t_i,t_j) = E\left[(X_i(\omega)-\mu_X(t_i))\ \left(Y_j(\omega)-\mu_Y(t_j)\right)\right]$$

$$= \int_{\mathbb{R}} \int_{\mathbb{R}} (x_i - \mu_X(t_i)) \left(y_j - \mu_Y(t_j)\right) f(x_i, y_j; t_i, t_j) dx_i dy_j \tag{3.10b}$$

These statistical quantities are relevant in determining the input and output relationships of dynamical systems. Passing in the limit to continuous-time processes $X_t(\omega)$ and $Y_s(\omega)$, these quantities may be thought of as functions of two time variables, viz. $R_{XY}(t,s)$ and $C_{XY}(t,s)$.

3.3.3 Stationary stochastic processes

A stochastic process $X_t(\omega)$ is called stationary if for any k time instants $t_1, \ldots, t_k$ and any time shift τ, the random vectors $\left\{X_{t_1}, X_{t_2}, \ldots, X_{t_k}\right\}$ and $\left\{X_{t_1+\tau}, X_{t_2+\tau}, \ldots, X_{t_k+\tau}\right\}$ have the same probability distributions. Specifically, this property is referred to as stationarity of order k. The process X is covariance (weakly or wide-sense) stationary if $k = 2$. It follows that, for a stationary process, the probability density function is invariant in time-translations, i.e., for any $\tau > 0$, we have:

$$f(x_1, x_2, \ldots, x_k; t_1, t_2, \ldots, t_k) = f(x_1, x_2, \ldots, x_k; t_1 + \tau, t_2 + \tau, \ldots, t_k + \tau) \tag{3.11}$$

Thus, for a weakly stationary process, the covariance $C_{XX}(s,t)$ (also known as the autocovariance function given by $[(X_s - \mu_s)(X_t - \mu_t)]$) is a function of only the time difference $t - s$. It follows that the autocorrelation function, $R_{XX}(s,t) (= E[X_s X_t])$ is also a function of the time difference only. Note that stationarity of order k implies stationarity of all lower orders. Thus weak stationarity implies that $E[X_t]$ is constant (since, for $k = 1$, the probability distribution of X is invariant of t). Also observe that $C_{XX}(s,t) = E[X_s X_t] - \mu_{X(s)}\mu_{X(t)} = R_{XX}(s,t) - \mu_s \mu_t$. The autocorrelation function is maximum at $\tau = 0$. This is derivable from the following:

$$E\left[(X_t \pm X_{t+\tau})^2\right] \geq 0$$

$$\Longrightarrow E\left[{X_t}^2 \pm 2X_t X_{t+\tau} + {X_{t+\tau}}^2\right] \geq 0$$

$$= R_{XX}(0) \pm 2R_{XX}(\tau) + R_{XX}(0) = 2\ (R_{XX}(0) \pm R_{XX}(\tau)) \geq 0$$

$$\Longrightarrow R_{XX}(0) \geq R_{XX}(\tau) \geq 0\ \forall\ \tau \tag{3.12}$$

$R_{XX}(\tau)$ is an even function of τ as $E[X_t X_{t+\tau}] = E[X_t X_{t-\tau}]$ and it follows that $\left.\frac{dR_{XX}(\tau)}{d\tau}\right|_{\tau=0} = 0$ provided the derivative exists. For the derivative process $\dot{X}_t(\omega) = \frac{dX_t(\omega)}{dt}$ (if it exists), the autocorrelation function $R_{\dot{X}\dot{X}}(t, t+\tau)$ is obtained as:

$$\frac{dR_{XX}(\tau)}{d\tau} = \frac{d\{E[X_t X_{t+\tau}]\}}{d\tau} = E\left[X_t \dot{X}_{t+\tau}\right] = E[X_{t-\tau}\dot{X}_t]$$

$$\Rightarrow \frac{d^2 R_{XX}(\tau)}{d\tau^2} = \frac{dE[X_{t-\tau}\dot{X}_t]}{d\tau} = -E\left[\dot{X}_{t-\tau}\dot{X}_t\right] = -E[\dot{X}_t \dot{X}_{t+\tau}] \tag{3.13}$$

From the last equation, we have the autocorrelation function of a derivative process as $R_{\dot{X}\dot{X}}(t,t+\tau) = -\frac{\partial^2 R_{XX}(\tau)}{\partial \tau^2}$. As an example, if we have $X_t(\omega)$ representing a displacement process, the second order derivative of its autocorrelation function (with respect to the time shift τ) is the autocorrelation function of the velocity but for a change of sign. Also observe that $\frac{dE[\dot{X}_t]}{dt} = 0$ since $E[\dot{X}_t]$ is a constant for a (weakly) stationary process. A weakly stationary process will be described henceforth as a stationary process for the sake of convenience.

3.3.4 Wiener process/ Brownian motion

Wiener process [Wiener 1923, Karatzas and Shreve 1991, Bhattacharya and Waymire 2009] is a stochastic process having numerous practical applications, especially in modelling noise terms in the equations for dynamical systems. Based upon an intuitive understanding, we typically expect a noise process $W(t)$ to satisfy the following:

a) $W(t)$ should be a zero-mean process for all t
b) $W(t)$ should be stationary
c) $W(t_1)$ and $W(t_2)$ are independent random variables for $t_1 \neq t_2$

Unfortunately, however, a process $W(t)$ satisfying the above three prescriptions does not have continuous, or even measurable, paths. Formally, $W(t)$ is often referred to as a white noise and may be characterized as a generalized process, in the sense that it can be interpreted as a probability measure on the space of tempered distributions over $[0,\infty)$, and not as a measure on $\mathbb{R}^+$. However, given a time discretization $0 = t_0 < t_1 < \cdots < t_k = t$ of a time interval $[0,t]$ of interest, one may construct the discrete process $B(t_{l+1}) = B(t_l) + W(t_l)(t_{l+1} - t_l)$, $l \in [0,k-1]$ such that the non-overlapping increments $B(t_{l+1}) - B(t_l)$ and $B(t_{m+1}) - B(t_m)$ are statistically independent for $l \neq m$ (based on our intuitive suggestion in c). Moreover, a) would imply that these increments in $B(t_l)$ are zero-mean random variables. The wide sense stationarity may be enforced by allowing $E\left[(B(t_{l+1}) - B(t_l))^2\right]$ to be a function of the time difference $\Delta t_l = t_{l+1} - t_l$ alone for every l. These are indeed the properties leading to the definition of a Wiener process (named after Wiener 1923) as a stochastic process $B_t(\omega)$, where a version with continuous paths exists as $\Delta t_l \to 0$. With the added assumption that $B_t(\omega)$ is Gaussian, it also provides a convincing mathematical model for the Brownian motion—the temperature induced random motion of pollen particles in fluid. To construct Brownian motion that represents, say, the position of a pollen particle at any time t, one specifies the probability measure $\gamma_{1,2,\ldots,k}(\mathcal{F}_1 \times \mathcal{F}_2 \times \cdots \times \mathcal{F}_k)$ that closely replicates the observed behaviour and satisfies the consistency conditions in Kolmogorov's extension theorem. As stated in Section 2.3.4, Chapter 2, a Wiener process can be thought of as the limiting form of a random walk model $B_n := B(t_n = n\Delta t) = \sum_{j=1}^{n} X_j$, with n being the number of bombardments by the fluid molecules on the pollen particle in time t_n and Δt, the time between two successive bombardments. X_j is the random distance travelled by the particle due to each bombardment. As $n \to \infty$, the central limit theorem assures a limiting *fdd* corresponding to the finite number of random variables $\{B_1, B_2, \ldots B_n\}$. Derivation of the limiting distribution is shown in the following.

Random walk model and the limiting distribution

The random walk [Weiss 1994, Hughes 1996] is a discrete time stochastic process. A convenient way to construct a random walk model is by means of a simple coin tossing experiment conducted at every Δt seconds and involving independent Bernoulli trials with $p =$ probability of a head or tail at each coin toss. These trials correspond a sequence of independent random variables $\{X_j\}, j \geq 1$ defined on a probability space $(\Omega, \mathcal{F}, P)$ with $P(X_j(\omega = \text{'head'}) = 1) = 1/2$ and $P(X_j(\omega = \text{'tail'}) = -1) = 1/2$. Let each occurrence of a head be followed by one step to the right and a tail by one step to the left on the real line. Suppose by the nth toss with $t_n = n\Delta t = t$, we have k heads (leading to k steps to the right) and $(n-k)$ steps to the left. If each step is of length σ units, the random walk (towards right) is described by $B_n = (2k - n)\sigma$ with the associated Binomial measure given by:

$$P(B_n = m\sigma) = \binom{n}{k}\frac{1}{2^n} \quad \text{with } m = (2k - n) \tag{3.14}$$

The random variables X_j are independent with $E[X_j] = 0$ and $E[X_j{}^2] = \sigma^2$. This leads to $E[B_n] = 0$ and $E\left[B_n{}^2\right] = n\sigma^2 = \frac{t}{\Delta t}\sigma^2$. In the limit as $n \to \infty$ (or equivalently as $\Delta t \to 0$), the probability measure approaches the Gaussian (by DeMoivre–Lapalce limit theorem) and the probability distribution takes the form of the standard normal distribution:

$$P(B_n \leq m\sigma) \to F_Z\left(z = m/\sqrt{n}\right) \tag{3.15}$$

where, $Z \sim \mathcal{N}(0,1)$. In the limit as $\Delta t \to 0$, we assume that $\sigma \sim o(\sqrt{\Delta t}) \to 0$ so that $E\left[B_t{}^2\right]$ remains finite and $\to \alpha t$ where $\alpha = \frac{\sigma^2}{\Delta t}$. Now, with $\mathscr{b} = m\sigma$, we have $\frac{m}{\sqrt{n}} = \frac{\mathscr{b}}{\sqrt{\alpha t}}$ and the 1-dimensional probability density of the limiting Wiener process $B_t(\omega)$ is given by:

$$f_B(\mathscr{b}) = \frac{1}{\sqrt{2\pi\alpha t}}\exp\left(-\mathscr{b}^2/2\alpha t\right), \text{ i.e., } B \sim \mathcal{N}(0, \sqrt{\alpha t}) \tag{3.16}$$

Wiener process as a Gaussian stochastic process with independent increments

Since the random variables, $\left\{B_i := B_{t_i}\right\}, i \in \mathbb{N}$, $(0 = t_0 < t_1 < t_2 \ldots)$ are zero mean normals, the incremental random variables $B_1 - B_0$, $B_2 - B_1$, etc., are also zero mean normals and, in addition, are independent. Thus, a Wiener process is a Gaussian stochastic process with independent increments. This is expected since the underlying random walk model B_n has independent increments, i.e., for $m, n \in \mathbb{Z}$, $B_{m+n} - B_m$ is independent of $\sigma(B_0, B_1, \ldots, B_m)$, the σ-algebra generated by all the previous random variables $B_0, B_1, \ldots, B_m$. Assuming that $B_0 = 0$ and $t_m < t_n$, we thus have for the Wiener process:

$$E\left[(B_m - B_0)(B_n - B_m)\right] = E\left[B_m\right] E\left[B_n - B_m\right] = 0 \tag{3.17}$$

From Eq. (3.17), the correlation between the random variables B_m and B_n is obtained as:

$$E\left[B_m B_n\right] = E\left[{B_m}^2\right] = \alpha t_m \tag{3.18}$$

Suppose if $t_m > t_n$, the last two steps lead to $E\left[B_m B_n\right] = E\left[{B_n}^2\right] = \alpha t_n$. Thus, in general, $E\left[B_t B_s\right] = t \wedge s$; here $t \wedge s$ is a notation for $\min(t, s)$. That $B_t(\omega)$ has independent increments can be directly shown by using the fact that normal random variables are independent if and only if they are uncorrelated, i.e.:

$$E\left[\left(B_i - B_{i-1}\right)\left(B_j - B_{j-1}\right)\right]$$

$$= E\left[B_i B_j\right] - E\left[B_{i-1} B_j\right] - E\left[B_i B_{j-1}\right] + E\left[B_{i-1} B_{j-1}\right],$$

$$= \alpha\left(t_i - t_{i-1} - t_i + t_{i-1}\right) = 0, \text{with } t_i < t_j \tag{3.19}$$

Now we are ready to have a formal definition of the Wiener process (the standard Brownian motion).

3.3.5 Formal definition of a Wiener process

A stochastic process, $B_t(\omega)$, $t \geq 0$ adapted to the filtration $\mathcal{F}_t$, is called an $\mathcal{F}_t$-measurable Wiener process if:

1. $B_0 = 0$ a.s. (3.20a)
2. The random variable $B_t - B_s \sim \mathcal{N}(0, \sqrt{t-s})$ for all $s \leq t$ (3.20b)
3. For $0 \leq s \leq t$, the random variable $B_t - B_s$ is independent of $\mathcal{F}_s$. In other words, it is a stochastic process with independent increments (3.20c)

The finite dimensional Gaussian distribution that characterizes a Wiener process satisfies the consistency conditions of Kolmogorov's extension theorem and thus there exists a stochastic process $B_t(\omega)$ satisfying the properties 1 – 3 above. Since a Wiener process is Gaussian, the finite dimensional distribution of the random variables $\{B(t_j) := B_j\}, j = 1,2,\ldots,k$, is jointly normal. Typically, the characteristic function of the vector $\{B_j\}$ is given by:

$$E[\exp i(\lambda,\{B_j\})] = \exp\left(-\frac{1}{2}\sum_{l=1}^{k}\sum_{m=1}^{k}\lambda_l C_{lm}\lambda_m + i\sum_{j=1}^{k}\lambda_j\mu_j\right) \tag{3.21}$$

Here $i = \sqrt{-1}$, $\boldsymbol{\lambda} := (\lambda_1, \lambda_2,\ldots,\lambda_k)\in\mathbb{R}^k$, $\mu_j = E[B_j]$ and $C_{lm} = E[(B_l - \mu_l)(B_m - \mu_m)]$. Specifically, the symmetric positive semi-definite covariance matrix $\boldsymbol{C} = [C_{lm}] \in \mathbb{R}^{k\times k}$ is given by:

$$\boldsymbol{C} = \begin{bmatrix} t_1 & t_1 & \ldots & t_1 \\ t_1 & t_2 & \ldots & t_2 \\ \cdot & \cdot & & \cdot \\ \cdot & \cdot & & \cdot \\ \cdot & \cdot & & \cdot \\ t_1 & t_2 & \ldots & t_k \end{bmatrix} \tag{3.22}$$

3.3.6 Other properties of a Wiener process

A Wiener process has stationary independent increments

The Wiener process, in addition to being a process with independent increments (property 3 in Eq. 3.20), has stationary increments. This is true since $B_{t+h}-B_{s+h}$ has the same *fdd*s as B_t-B_s for all $s \le t$, $h > 0$. Hence, with $\Delta B_t(h) := B_{t+h}-B_t$, the variance $E\left[\Delta B_t(h)^2\right] = h$ is linear in the time lag h. Note that, with $E\left[{B_t}^2\right] = t$, the Wiener process $B_t(\omega)$ is not stationary.

Sample paths of Wiener process are continuous

From the property 2 in Eq. (3.20), we can show that the sample paths of (a version of the) Wiener process are continuous.

Proof: To show that $B_t(\omega)$ has a version that is continuous in time or has continuous sample paths, we use property 2, i.e., every increment $\Delta B_s(t-s) = B_t - B_s \sim \mathcal{N}(0,\sqrt{t-s})$. Thus, there exists a continuous version of the Wiener process.

For all $s \le t$, it readily follows that $E\left[(B_t - B_s)^4\right] = 3(t-s)^2$. Now, referring to Kolmogorov's continuity theorem and the condition in Eq. (3.3), we find that the criterion is satisfied for $\alpha = 4$ and $\beta = 1$.

Weiner process is not differentiable

Proof: To show this, consider $\frac{\Delta B_t(h)}{h} = \frac{B_{t+h} - B_t}{h}$ with h being an interval in $\mathbb{R}$. $\frac{\Delta B_t(h)}{h}$ is a zero-mean random variable with $E\left[\left(\frac{\Delta B_t(h)}{h}\right)^2\right] = \frac{1}{h}$. If $\lim_{h \to 0} \frac{\Delta B_t(h)}{h}$ converges in some sense (see Chapter 2 on convergence of a sequence of random variables) to a limit, then the sequence of the characteristic functions $E[e^{i\lambda \frac{\Delta B_t(h)}{h}}]$ converges to a limit which must be continuous in λ. We know that for a zero-mean normal random variable Z, the characteristic function $\Phi_Z(\lambda) = E\left[e^{i\lambda z}\right] = e^{-\frac{\lambda^2 \sigma_z^2}{2}}$ where σ_z^2 is the variance of Z. Hence, we have:

$$\lim_{h \to 0} E\left[e^{i\lambda \frac{\Delta B_t(h)}{h}}\right] = \lim_{h \to 0} e^{\frac{-\lambda^2}{2h}} = 1 \text{ if } \lambda = 0$$
$$= 0, \quad \text{otherwise} \tag{3.23}$$

We find from Eq. (3.23) that the characteristic function in the limit is not continuous in λ, and hence the derivative of the Weiner process does not converge.

A non-rigorous approach to the derivative of a Wiener process

Recalling the intuitive description, provided at the beginning of this section for a white noise process $W(t)$, we may define the white noise process as a generalized derivative of a Wiener process. Since $B_t(\omega)$ is nowhere differentiable, $W(t)$ is not a derivative in the usual sense and that is the reason why the word generalized is used. We first define, though not rigorously, a generalized derivative, $\dot{B}$ and show it to be a stochastic process with $E[\dot{B}] = 0$ and $E\left[\dot{B}_t \dot{B}_s\right] = \delta(t-s)$, where $\delta(.)$ is the Dirac delta function. The latter is strictly not a function, rather a distribution or a generalized function having the following property:

$$\int_{-\infty}^{\infty} g(u)\delta(u)\, du = g(0) \tag{3.24}$$

where, $g(t)$ is a (deterministic), strictly positive test function $\in C_0^\infty$, the space of compactly supported smooth functions that are infinitely differentiable. Now, loosely invoking the notion of the weak derivative, consider integrating $\int_0^t g(u)\dot{B}_u du$ (even though, this integral is strictly not a Lebesgue integral, but a stochastic integral whose rigorous treatment through stochastic calculus would be laid out in Chapter 4) by parts as:

$$\int_0^t g(u)\dot{B}_u du = g(t)B_t - \int_0^t \dot{g}(u)B_u du \tag{3.25}$$

Note again that the above formula for integration by parts, borrowed directly from deterministic calculus, is not strictly valid. Nevertheless, taking expectation on both sides of the last equation, we have:

$$E\left[\int_0^t g(u)\,\dot{B}_u du\right] = 0 \Rightarrow E\left[\dot{B}_t\right] = 0 \tag{3.26}$$

Further, let $\psi(t) = E\left[\int_0^t \int_0^t g(u)\,g(u')\,\dot{B}_u \dot{B}_{u'} du du'\right]$ and we have:

$$\psi(t) = \int_0^t \int_0^t g(u)\,g(u')\,E\left[\dot{B}_u \dot{B}_{u'}\right] du du'$$

$$= E\left[\int_0^t g(u)\,\dot{B}_u du \int_0^t g(u')\,\dot{B}_{u'} du'\right]$$

$$= E\left(g(t)\,B_t - \int_0^t \dot{g}(u)\,B_u du\right)^2, \text{(from Eq. (3.25))}$$

$$= g^2(t)E\left[B_t{}^2\right] - 2g(t)\int_0^t \dot{g}(u)\,E[B_t B_u]du + \int_0^t\int_0^t \dot{g}(u)\,\dot{g}(u')E\,[B_u B_{u'}]\,du du'$$

$$= tg^2(t) - 2g(t)\int_0^t \dot{g}(u)\,u du + \int_0^t \int_0^t \dot{g}(u)\,\dot{g}(u')\min(u,u')\,du du'$$

$$\Rightarrow \frac{d\psi(t)}{dt} = \{g^2(t) + 2tg(t)\dot{g}(t)\} - 2\left\{\dot{g}(t)\int_0^t \dot{g}(u)u\, du + tg(t)\dot{g}(t)\right\}$$

$$+2\dot{g}(t)\int_0^t \dot{g}(u)u du$$

$$= g^2(t)$$

$$\Rightarrow \psi(t) = \int_0^t g^2(u)du = \left[\int_0^t \int_0^t g(u)\,g(u')\,E\left[\dot{B}_u \dot{B}_{u'}\right] du du'\right] \tag{3.27}$$

However, we can also write $\int_0^t g^2(u)du$ as:

$$\int_0^t g^2(u)du = \int_0^t \int_0^t g(u)\,g(u')\,\delta(u-u')du du' \tag{3.28}$$

Comparing Eqs. (3.27) and (3.28), we infer:

$$E\left[\dot{B}_u \dot{B}_{u'}\right] = \delta(u-u') \tag{3.29}$$

Thus, allowing for a grossly improper usage of integration as above, we still see that the derivative process $\dot{B}_t$ is a zero mean Gaussian process with wide sense stationarity. These properties are the same as those ascribed earlier to the so-called white noise process, $W(t)$. Its autocorrelation function, which is incidentally a Dirac delta function, shows that $W(t_1)$ and $W(t_2)$ for $t_1 \neq t_2$ are uncorrelated, i.e., the process is completely memoryless.

A Wiener process and time-scaling

If B_t is a Wiener process, then the scaled process $\overline{B}_t := \alpha^{-\frac{1}{2}} B(\alpha t)$ with $\alpha \in \mathbb{R}$, is also a Wiener process.

Proof: Firstly, $\overline{B}_t$ starts at 0 a.s. and is Gaussian similar to B_t. Next, the incremental process $\overline{B}_t - \overline{B}_s = \alpha^{-\frac{1}{2}} (B_{\alpha t} - B_{\alpha s})$ is also Gaussian and is independent of $\mathcal{F}_s$, $s < t$. Thus $\overline{B}_t$ has independent increments. Finally, the variance of $\overline{B}_t - \overline{B}_s$ is given by:

$$var\left(\overline{B}_t - \overline{B}_s\right) = var\left(\alpha^{-\frac{1}{2}} (B_{\alpha t} - B_{\alpha s})\right) = \alpha \frac{(t-s)}{\alpha} = t - s, \forall s \le t \tag{3.30}$$

This shows the scaling invariance of a Wiener process.

Wiener process and time-shifting

If B_t is a Wiener process, then the time shifted process $\hat{B}_t := B_{t+T} - B_T = \Delta B_T(t)$ is also a Wiener process, where $T > 0$ is a constant.

Proof: The expected value of $\hat{B}_t$ is given by:

$$E\left[\hat{B}_t\right] = E\left[\Delta B_T(t)\right] = 0 \tag{3.31}$$

Being just a translation of B_t, it is obvious that increments of $\hat{B}_t$ are independent. The variance of the increment $\hat{B}_t - \hat{B}_s$ is given by:

$$\begin{aligned} var\left(\hat{B}_t - \hat{B}_s\right) &= var\left((B_{t+T} - B_T) - (B_{s+T} - B_T)\right) \\ &= var\left(B_{t+T} - B_{s+T}\right) = t - s, \forall s \le t \end{aligned} \tag{3.32}$$

Hence the result.

Brownian bridge

A Brownian bridge, say over $t \in [0,1]$, is a stochastic process of considerable usefulness and is defined as:

$$\tilde{B}_t := B_t - tB_1 \tag{3.33}$$

Since $B_0 = 0$ a.s., $\tilde{B}_0 = 0$. From the properties of the Wiener process, B_t, we have $E\left[\tilde{B}_t\right] = E\left[B_t - tB_1\right] = 0$. Also, with $0 \le s \le t \le 1$, the correlation function of $\tilde{B}$ is given by:

$$\begin{aligned} E\left[\tilde{B}_t\tilde{B}_s\right] &= E\left[\left(B_t - tB_1\right)\left(B_s - sB_1\right)\right] \\ &= E\left[B_tB_s\right] - tE\left[B_1B_s\right] - sE\left[B_tB_1\right] + tsE\left[{B_1}^2\right] \\ &= s - ts - st + ts = s(1-t) \end{aligned} \tag{3.34}$$

From Eq. (3.33), it is clear that the paths of a Brownian bridge are continuous. The Brownian bridge like the standard Brownian motion starts at zero but in addition ends also at zero with probability 1. The paths are tied up (pinned) at both ends thus justifying the nomenclature bridge. As the correlation structure indicates, a Brownian bridge is not a Wiener process. Figure 3.4 shows a few simulated paths of a Brownian bridge.

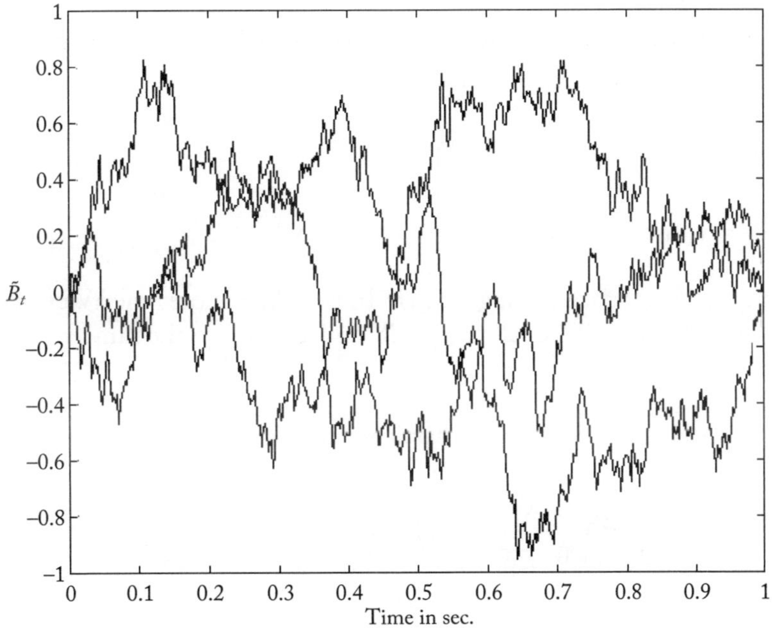

Fig. 3.4 Sample paths of a Brownian bridge

Time inversion property of the Wiener process

If B_t is a Wiener process, then the stochastic process defined by:

$$\begin{aligned} \breve{B}_t &= tB(1/t), t \ge 0 \\ &= 0, t = 0 \end{aligned} \tag{3.35}$$

is also a Wiener process.

Proof:

(i) $E[\breve{B}_t] = tE[B(1/t)] = 0, \quad t > 0.$

(ii) The covariance function of $\breve{B}_t$ is obtained as:

$$cov\left(\breve{B}_t\, \breve{B}_{t+h}\right) = t(t+h)\ cov\,(B(1/t)B(1/(t+h)))$$

$$= t(t+h)\min(\frac{1}{t}, \frac{1}{t+h}) = t \tag{3.36}$$

Further, the *fdd*s of the above process are Gaussian and the sample paths are continuous. So $\breve{B}_t$ is also a Wiener process.

Space homogeneity of Wiener process

A stochastic process is space homogeneous in that its *fdd*s remain unchanged with a shift in space, i.e., with $\left\{X\left(t_j\right) := X_j\right\}, j = 1, 2, \ldots, k$:

$$P(X_1 \le x_1, X_2 \le x_2, \ldots, X_k \le x_k | X_0 = 0)$$

$$= P(X_1 \le x_1 + x, X_2 \le x_2 + x, \ldots, X_k \le x_k + x | X_0 = x) \tag{3.37}$$

A Wiener process B_t is space homogeneous. Thus, if B_t^x denotes the Wiener process that starts at x, then $B_t^x = x + B_t^0$ i.e. all *fdd*s of B_t^x and $x + B_t^0$ are the same.

Quadratic variation of a Wiener Process

The Quadratic Variation (QV) of a stochastic process $X(t)$ is defined as:

$$[X, X](a, b) = \lim_{\substack{\Delta_n \to 0 \\ n \to \infty}} \sum_{j=1}^{n} \left(X\left(t_j\right) - X\left(t_{j-1}\right)\right)^2 \tag{3.38}$$

where, $\Delta_n = \max_{1 \le j \le n}\left(t_j - t_{j-1}\right)$ corresponds to a partition Π_n of the time interval as $a = t_0 < t_1 < \ldots < t_n = b$ and the limit is taken over all shrinking partitions of $[a, b]$. The QV of a Wiener process $[B, B](0, t)$ can be shown to be finite and is equal to t.

Proof: Letting $Q_n = \sum_{j=1}^{n}\left(B\left(t_j\right) - B\left(t_{j-1}\right)\right)^2$, we have:

$$E[Q_n] = E\left[\sum_{j=1}^{n}\left(B\left(t_j\right) - B\left(t_{j-1}\right)\right)^2\right]$$

$$= \sum_{j=1}^{n} E\left[\left(B\left(t_j\right) - B(t_{j-1})\right)^2\right]$$

$$= \sum_{j=1}^{n} \left(t_j - t_{j-1}\right) = t - 0 = t \tag{3.39}$$

Since the fourth moment of $\mathcal{N}(0,\sigma)$ is $3\sigma^4$, we have:

$$var\left(Q_n - t\right) = var\left(\sum_{j=1}^{n} \left(B\left(t_j\right) - B(t_{j-1})\right)^2 - t\right) = \sum_{j=1}^{n} var\left(B\left(t_j\right) - B(t_{j-1})\right)^2$$

$$= \sum_{j=1}^{n} 2\left(t_j - t_{j-1}\right)^2 \leq 2t \max\left(t_j - t_{j-1}\right) = 2t\Delta_n \tag{3.40}$$

Thus, clearly, $\lim_{\Delta_n \to 0} var\left(Q_n - t\right) = 0$. This shows that $Q_n - t$ is non-random as $n \to \infty$ and $[B,B](0,t) = \lim_{n\to\infty} Q_n = t$ in $L^2(P)$. Note that even though the sums involved in the definition of Q_n are random, the limit is non-random.

Another measure of a Wiener process is the total variation. For example we define for any real function $g(t)$ the total variation (TV) over a closed interval $[a,b]$ as:

$$TV\left(g;a,b\right) = \sup_{n\geq 1} \sup_{\{t_{j-1},t_j\}\in\wp(n,a,b)} \sum_{j=1}^{n} \left|g\left(t_j\right) - g(t_{j-1})\right| \tag{3.41}$$

$\wp(N,a,b)$ is the set of all partitions of the interval $[a,b]$ in the from $a = t_0 < t_2 < \cdots < t_N = b$. Clearly for real $g(t)$, $TV\left(g;a,b\right) = g(b) - g(a)$. A Wiener process is a.s. of infinite variation for any $a < b$ (see Handel 2007 for a proof). Accordingly, the Wiener process must oscillate very rapidly and infinite TV can be interpreted as a measure of the irregularity of its sample paths.

Wiener process is a Markov process

Markov processes are an important class of stochastic processes and the Markovian property is crucial in characterizing solutions to SDEs. A stochastic process $X(t)$ is Markov if the conditional distribution of $X\left(t+s\right)$ given $\mathcal{F}_t$ is the same as the conditional distribution of $X\left(t+s\right)$ given X_t for any increment $s > 0$. In other words,

$$P(X\left(t+s\right) \leq x|\mathcal{F}_t) = P(X\left(t+s\right) \leq x|X_t) \tag{3.42}$$

Thus, for a Markov process, a future state is dependent only on the present and not on the past. A Wiener process $B(t)$ possesses the Markovian property. This is proved below.

Consider the moment generating function of $B(t+s)$ given $\mathcal{F}_t$ in the form:

$$E\left[e^{\lambda B(t+s)} \,\middle|\, \mathcal{F}_t\right] = E\left[e^{\lambda(B(t+s)-B(t)+B(t))} \,\middle|\, \mathcal{F}_t\right]$$

$$= e^{\lambda B(t)}\, E\left[e^{\lambda(B(t+s)-B(t))} \,\middle|\, \mathcal{F}_t\right] \tag{3.43}$$

Since a Wiener process is one with independent increments, $e^{\lambda(B(t+s)-B(t))}$ is independent of $\mathcal{F}_t$. Hence we have:

$$E\left[e^{\lambda B(t+s)} \,\middle|\, \mathcal{F}_t\right] = e^{\lambda B(t)} E\left[e^{\lambda(B(t+s)-B(t))}\right] \tag{3.44}$$

Since Wiener increments are Gaussian, the moment generating function of $B(t+s)-B(t)$ is $e^{\lambda^2 s/2}$ and the last equation takes the form:

$$E\left[e^{\lambda B(t+s)} \,\middle|\, \mathcal{F}_t\right] = e^{\lambda B(t)} e^{\lambda^2 s/2} = e^{\lambda B(t)}\, E\left[e^{\lambda(B(t+s)-B(t))} \,\middle|\, B_t\right] = E\left[e^{\lambda B(t+s)} \,\middle|\, B_t\right] \tag{3.45}$$

This proves that B_t is Markovian. A Markov process is characterized by the transition probability function defined as:

$$P(y,t;x,s) = P(X(t) \le y | X(s) = x) \tag{3.46}$$

$P(y,t;x,s)$ is the conditional probability distribution function of the process at time t given that it is x at time $s < t$. For a Wiener process, since the increment $B(t) - B(s) \sim \mathcal{N}\left(0, \sqrt{t-s}\right)$, the transition probability function is given by:

$$P(y,t;x,s) = P\left(B(t) \le y \,\middle|\, B(s) = x\right) = \frac{1}{\sqrt{2\pi(t-s)}} \int_{-\infty}^{y} e^{-\frac{(u-x)^2}{2(t-s)}}\, du \tag{3.47}$$

with the corresponding transition probability density function $f(y,t;x,s) = \frac{1}{\sqrt{2\pi(t-s)}} e^{-\frac{(y-x)^2}{2(t-s)}}$. Further, $P(y,t;x,s) = P(y,t-s;x,0) = P(B(t-s) \le y | B(0) = x)$. For fixed t and x, $P(y,t;x,0)$ has the 1-dimensional density function $f(t,x,y)$ as:

$$f(t,x,y) = \frac{1}{\sqrt{2\pi t}} \exp\left(-\frac{(y-x)^2}{2t}\right) \tag{3.48}$$

It is also clear from the above that a Wiener process is time-homogeneous, i.e., its distributions is unaltered with a shift in time and so is the case with all its *fdd*s.

Wiener process in higher dimensions

We treat a Wiener process in higher dimensions, say, taking values in $\mathbb{R}^d$ $(d \geq 2)$, as a random vector function $B(t) = B^1(t), B^2(t), \ldots, B^d(t)$ where, $\{B^k(t)\}$, $k = 1, 2, \ldots, d$, is a vector of independently evolving 1-dimensional Wiener processes. Similar to a 1-dimensional Wiener process, $B(t)$ is a Markov Gaussian process homogeneous both in space and time. The corresponding transition probability density is given by, (Eq. 3.48):

$$f(t, \boldsymbol{x}, \boldsymbol{y}) = \frac{1}{\sqrt{2\pi t}} \exp\left(-\frac{|\boldsymbol{y} - \boldsymbol{x}|^2}{2t}\right), \boldsymbol{y} \in \mathbb{R}^d \text{ and given } \boldsymbol{x} \in \mathbb{R}^d \tag{3.49}$$

$|\boldsymbol{y} - \boldsymbol{x}|$ is the length of the vector $(\boldsymbol{y} - \boldsymbol{x})$.

A Wiener process is a martingale

A martingale $\mathcal{M}(t)$ is a special type of $\mathcal{F}_t$-adapted L^1 (i.e. $E\left[|\mathcal{M}(t)|\right] < \infty$) stochastic process satisfying the condition:

$$E\left[\mathcal{M}(t) \mid \mathcal{F}_s\right] = \mathcal{M}(s) \text{ a.s. for every } s \leq t \tag{3.50a}$$

$\mathcal{M}(t)$ is said to be a super-martingale with respect to the filtration $\{\mathcal{F}_t\}$ if:

$$E\left[\mathcal{M}(t) \mid \mathcal{F}_s\right] \leq \mathcal{M}(s) \text{ a.s. for every } s \leq t \tag{3.50b}$$

and a sub-martingale if:

$$E\left[\mathcal{M}(t) \mid \mathcal{F}_s\right] \geq \mathcal{M}(s) \text{ a.s. for every } s \leq t \tag{3.50c}$$

A martingale process is then both a super-martingale and a sub-martingale. Also, from the definition, we see that if $X(t)$ is a super-martingale, $-X(t)$ is a sub-martingale. To have an intuitive understanding of the above definitions, it is more convenient to fall back on a discrete stochastic process, e.g., the random walk described in Section 3.3.4. Since each Bernoulli random variable X_j is independent, we have that X_{j+1} is independent of the σ-algebra $\mathcal{F}_j = \sigma(X_0, X_1, \ldots, X_j)$ and hence:

$$E\left[X_{j+1} | \mathcal{F}_j\right] = E\left[X_{j+1}\right] \tag{3.51}$$

It follows that:

$$\begin{aligned} E\left[B_{j+1} | \mathcal{F}_j\right] &= E\left[(X_0 + X_1 + \cdots + X_{j+1}) | \mathcal{F}_j\right] \\ &= E\left[(X_0 + X_1 + \cdots + X_j) | \mathcal{F}_j\right] + E\left[(X_{j+1}) | \mathcal{F}_j\right] \end{aligned}$$

$$= X_0 + X_1 + \cdots + X_j + E\left[X_{j+1}\right]$$

$$= B_j + E\left[X_{j+1}\right], j \in \mathbb{N} \tag{3.52}$$

Thus, if $E\left[X_k\right] = 0$ for all $k = 1, 2, \ldots$, we get the result:

$$E\left[B_{j+1} | \mathcal{F}_j\right] = B_j$$

or, more generally

$$E\left[B_{j+k} | \mathcal{F}_j\right] = B_j, \ \forall k \geq 1 \tag{3.53}$$

In fact, each Bernoulli trial is assumed to be fair and has equal probability $p = \frac{1}{2}$ of having a head or a tail (equivalently, the chance of moving σ units either to right or left also has probability of $\frac{1}{2}$ each). Thus, $E\left[X_k\right] = 0$ and hence the random walk B_n is a discrete martingale which we infer from Eqs. (3.50a) and (3.53). Suppose $p = \frac{1}{3}$, then we find that $E\left[X_k\right] = \sigma\left(\frac{1}{3}\right) + \sigma\left(-\frac{2}{3}\right) = \left(-\frac{1}{3}\right)\sigma$ resulting in $E\left[B_{j+1} | \mathcal{F}_j\right] < B_j$ and therefore B_n is a super-martingale. In essence, a martingale represents a fair game of chance and the other two the opposite.

An $\mathcal{F}_t$-adapted Wiener process B_t is a martingale with respect to $\mathcal{F}_t$ and the proof follows.

Proof: The Wiener process is integrable with $E[B(t)] = 0 < \infty$. With $0 \leq s < t$, we have:

$$E\left[B_t \mid \mathcal{F}_s\right] = E\left[B_t - B_s + B_s \mid \mathcal{F}_s\right]$$

$$= E\left[(B_t - B_s) \mid \mathcal{F}_s\right] + E\left[B_s \mid \mathcal{F}_s\right]$$

$$= E\left[B_t - B_s\right] + B_s \tag{3.54}$$

Note that $\mathcal{F}_t = \sigma(B_s : s \leq t)$ and hence $B_t - B_s$ is independent of $\mathcal{F}_s$ leading to $E\left[(B_t - B_s) \mid \mathcal{F}_s\right] = E\left[B_t - B_s\right] = 0$. Therefore, we obtain:

$$E\left[B_t \mid \mathcal{F}_s\right] = B_s \tag{3.55}$$

Thus, a Wiener process is a martingale with continuous sample paths.

As will be shown later [Levy's characterization of Brownian motion—Chapter 4], a continuous martingale is a Brownian motion if its quadratic variation over an interval $[0, t]$ is equal to t.

$B^2(t) - t$ *is a martingale*

Proof: $B^2(t)$ is integrable with $E\left[B^2(t)\right] = t < \infty$. Further, we have:

$$B^2(t+s) = (B(t+s) - B(t) + B(t))^2$$

$$= (B(t+s) - B(t))^2 + 2\,B(t)\,(B(t+s) - B(t)) + B^2(t) \quad (3.56)$$

$$E\left[B^2(t+s) \mid \mathcal{F}_t\right] = E\left[\left((B(t+s) - B(t))^2 + 2\,B(t)\,(B(t+s) - B(t)) + B^2(t)\right)|\mathcal{F}_t\right]$$

$$= E\left[(B(t+s) - B(t))^2 \mid \mathcal{F}_t\right] + 2E\left[B(t)\,(B(t+s) - B(t))|\mathcal{F}_t\right]$$

$$+ E\left[B^2(t)|\mathcal{F}_t\right]$$

$$= s + 0 + B^2(t)$$

$$= B^2(t) + s \quad (3.57)$$

Equation (3.57) is obtained using the properties of Brownian increments, e.g., $(B(t+s) - B(t))$ is independent of $\mathcal{F}_t$, has mean zero and variance s. Subtracting $(t+s)$ from both sides of the last equation, we have the result.

Remark: The above result suggests a means to characterize a Brownian motion, i.e., if a process $X(t)$ is a continuous martingale such that $X(t)^2 - t$ is also a martingale, then $X(t)$ is Brownian. That this is indeed true may be seen through Levy's characterization of a Brownian motion (see Appendix D).

$e^{\lambda B(t) - \frac{\lambda^2}{2}t}$ is a martingale for any real λ

Proof: Since $B(t) \sim \mathcal{N}(0,t)$, the moment generating function is given by $E\left[e^{\lambda B(t)}\right] = e^{t\lambda^2/2} < \infty$ and thus $e^{\lambda B(t) - \frac{\lambda^2}{2}t}$ is integrable with $E\left[e^{\lambda B(t) - \frac{\lambda^2}{2}t}\right] = 1 < \infty$. Now consider $E\left[e^{\lambda B(t+s)}|\mathcal{F}_t\right]$ in the form:

$$E\left[e^{\lambda B(t+s)}|\mathcal{F}_t\right] = E\left[e^{\lambda(B(t+s) - B(t) + B(t))}|\mathcal{F}_t\right] \quad (3.58)$$

Here also we use the properties: $B(t)$ is $\mathcal{F}_t$-measurable ($\Longrightarrow e^{\lambda B(t)}$ is $\mathcal{F}_t$-measurable), the increment $B(t+s) - B(t)$ is independent of $\mathcal{F}_t$ and has zero mean and variance equal to s. Thus, we obtain:

$$E\left[e^{\lambda B(t+s)}|\mathcal{F}_t\right] = e^{\lambda B(t)}\ E\left[e^{\lambda(B(t+s)-B(t))}\right] = e^{\lambda B(t)} e^{\frac{\lambda^2}{2}s} \tag{3.59}$$

Multiplying both sides of Eq. (3.59) by $e^{-\frac{\lambda^2}{2}(t+s)}$, we find that $e^{\lambda B(t)-\frac{\lambda^2}{2}t}$ is a martingale, which is strictly positive and popularly known as an exponential martingale.

3.4 Martingales: A General Introduction

Much of the development in the theory of martingales is due to Levy [1937] and Doob [1953] and has now become central to a modern treatment of stochastic processes [Williams 1991, Rogers and Williams 2000]. It plays a key role in stochastic analysis and becomes exceedingly important in various applications in science and engineering as progressively exposed in the ensuing chapters of this book. As much of the stochastic analysis is about sequences of real-valued random variables and their convergence, we find that the inequality and convergence theorems on martingales provide an elegant approach towards understanding these issues including the proof of Radon–Nikodym theorem (earlier stated in Section 1.6.4, Chapter 1). Doob's decomposition and martingale representation and convergence theorems are also relevant and insightful in this context. In fact, the laws of large numbers and central limit theorem are derivable more easily using martingale convergence. We now formalize these concepts and build up the necessary tools to demonstrate the power of the martingale theory. Since it is relatively easy to comprehend the intuitive reasoning whilst drawing inspiration from the strategies used in games of chance, we often adopt this route for an understanding of the properties of martingales.

3.4.1 Doob's decomposition theorem

Any discrete stochastic process X_n, which is $\mathcal{F}_n$-adapted with $X_n \in L^1$, can be decomposed into a martingale part and a predictable part, i.e., we have by Doob's decomposition theorem:

$$X_n = X_0 + A_n + \mathcal{M}_n \text{ a.s.} \tag{3.60}$$

where, $\{A_n\}$ is $\mathcal{F}_n$-predictable (see below for the definition) with $A_0 = 0$ a.s. and $\mathcal{M}_n$ is an $\mathcal{F}_n$-martingale with $\mathcal{M}_0 = 0$ a.s. Also, the decomposition in the above equation is unique.

Proof: Define the predictable process $A_n = \sum_{j=1}^{n} E\left[\left(\left(X_j - X_{j-1}\right)|\mathcal{F}_{j-1}\right)\right]$. Then we have:

$$\mathcal{M}_n = X_n - X_0 - A_n = \sum_{j=1}^{n}\{X_j - X_{j-1} - E[(X_j - X_{j-1})|\mathcal{F}_{j-1}]\} \tag{3.61}$$

Indeed, from Eq. (3.61), $\mathcal{M}_0 = 0$. Further, $E\left[(\mathcal{M}_k - \mathcal{M}_{k-1})|\mathcal{F}_{k-1}\right] = 0$ proving that $\mathcal{M}_n$ is an $\mathcal{F}_n$-martingale and hence the validity of the decomposition. To prove the uniqueness of the decomposition, suppose that $\overline{\mathcal{M}}_n$ and $\overline{A}_n$ are another martingale and predictable process (respectively) satisfying $X_n = X_0 + \overline{A}_n + \overline{\mathcal{M}}_n$ a.s. This gives $\mathcal{M}_n - \overline{\mathcal{M}}_n + A_n - \overline{A}_n = 0$ a.s. Since A_n and $\overline{A}_n$ are predictable (i.e., $\mathcal{F}_{n-1}$-measurable), so is $A_n - \overline{A}_n$. This leads to:

$$A_n - \overline{A}_n = E\left[\left(A_n - \overline{A}_n\right)|\mathcal{F}_{n-1}\right] = E\left[\left(\overline{\mathcal{M}}_n - \mathcal{M}_n\right)|\mathcal{F}_{n-1}\right] = 0 \tag{3.62}$$

Thus, $A_n = \overline{A}_n$ a.s. $\Longrightarrow \mathcal{M}_n = \overline{\mathcal{M}}_n$ a.s. proving the uniqueness of the decomposition.

In the above decomposition, if the predictable part A_n is strictly decreasing (increasing), then the process X_n is super-martingale (sub-martingale). A predictable process is also known as a previsible process. The significance of the pre-visibility of the process can be intuitively understood by thinking of a stake to be decided by a player on game n. In this context, A_n must be decided based on the knowledge available on the games up to and including $n-1$. This aspect will be revisited when we describe a martingale transform.

An application of Doob's decomposition

If $\mathcal{M}_n$ is an L^2 zero-mean martingale $\left(E\left[\mathcal{M}_n^2\right] < \infty\ \forall n\right)$, then there exists an a.s. unique predictable increasing process A_n with $A_0 = 0$ such that $\mathcal{M}_n^2 - A_n$ is a zero-mean martingale. Further, the random variable $A_n - A_{n-1}$ is a version of conditional variance of $\mathcal{M}_n$ given $\mathcal{F}_{n-1}$.

Proof: By using Jensen's inequality (Chapter 2) and the property of a martingale, one has:

$$(E[\mathcal{M}_{n+1}])^2 \leq E[\mathcal{M}_{n+1}^2]$$

$$\Longrightarrow E\left[\mathcal{M}_{n+1}^2|\mathcal{F}_n\right] \geq \left(E\left[\mathcal{M}_{n+1}|\mathcal{F}_n\right]\right)^2 = \mathcal{M}_n^2, \forall n \tag{3.63}$$

Thus $\mathcal{M}_n^2$ is a sub-martingale. By Doob's decomposition theorem, $\mathcal{M}_n^2 (\in L^1)$ can be decomposed into a martingale and a unique increasing predictable process A_n with $A_0 = 0$. It then readily follows that $\mathcal{M}_n^2 - A_n$ is a martingale. About the statement on conditional variance, consider the increments of the predictable process A_n given by:

$$A_n - A_{n-1} = E\left[\mathcal{M}_n^2 - \mathcal{M}_{n-1}^2|\mathcal{F}_{n-1}\right]$$

$$= E\left[(\mathcal{M}_n - \mathcal{M}_{n-1})^2|\mathcal{F}_{n-1}\right] + 2E\left[\mathcal{M}_{n-1}\ (\mathcal{M}_n - \mathcal{M}_{n-1})\,|\mathcal{F}_{n-1}\right] \tag{3.64}$$

Since $\mathcal{M}_n$ is a zero-mean martingale (with $\mathcal{M}_{n-1}$ being $\mathcal{F}_{n-1}$-measurable), the second term on the right hand side of the above equation vanishes and we have:

$$E\left[\mathcal{M}_n^2 - \mathcal{M}_{n-1}^2|\mathcal{F}_{n-1}\right] = E\left[(\mathcal{M}_n - \mathcal{M}_{n-1})^2|\mathcal{F}_{n-1}\right] \tag{3.65}$$

From Eq. (3.65), $A_n - A_{n-1}$ can be written as:

$$A_n - A_{n-1} = E\left[(\mathcal{M}_n - E[\mathcal{M}_n|\mathcal{F}_{n-1}])^2|\mathcal{F}_{n-1}\right]$$

$$= var\ ((\mathcal{M}_n - \mathcal{M}_{n-1})|\mathcal{F}_{n-1})$$

$$\Longrightarrow A_n = \sum_{j=1}^{n}\left(A_j - A_{j-1}\right) = \sum_{j=1}^{n} var\ \left((\mathcal{M}_j - \mathcal{M}_{j-1})|\mathcal{F}_{j-1}\right) \tag{3.66}$$

The predictable process A_n is called the conditional variance process of the square integrable martingale $\mathcal{M}_n$.

From Eq. (3.65), it follows that Pythagoras' theorem holds for square integrable martingales, i.e.:

$$E\mathcal{M}_n^2 = E\mathcal{M}_0^2 + \sum_{j=1}^{n} E\left[\left(\mathcal{M}_j - \mathcal{M}_{j-1}\right)^2\right] \tag{3.67}$$

Remark: Note that the increments of a square integrable martingale are orthogonal, i.e., with $u \leq v \leq s \leq t$, $\mathcal{M}_t - \mathcal{M}_s$ is orthogonal to $L^2(\mathcal{F}_s)$. Thus $E[(\mathcal{M}_v - \mathcal{M}_u)(\mathcal{M}_t - \mathcal{M}_s)] = 0$. The following expression for $\mathcal{M}_n$ is therefore a sum of orthogonal terms:

$$\mathcal{M}_n = \mathcal{M}_0 + \sum_{j=1}^{n}\left(\mathcal{M}_j - \mathcal{M}_{j-1}\right) \tag{3.68}$$

Doob's decomposition may not be true for processes continuous in time. However, in most such cases where the continuous process is governed by an SDE (considered in Chapter 5), it may be reduced to a time-discrete process via numerical integration. Accordingly Doob's decomposition still remains valid for the time-discrete approximant to the continuous process. In this sense, the decomposition as in Eq. (3.60) is remarkably insightful.

3.4.2 Martingale transform

Consider an $\mathcal{F}_n$-adapted martingale $\mathcal{M}_n$ and an $\mathcal{F}_{n-1}$-measurable (predictable) process, A_n. We define the martingale transform $(A.\mathcal{M})_n$ as:

$$(A.\mathcal{M})_n := \sum_{j=1}^{n} A_j\left(\mathcal{M}_j - \mathcal{M}_{j-1}\right) \tag{3.69}$$

Specifically $(A.\mathcal{M})_n$ is the martingale transform of $\mathcal{M}_n$ by A_n and it is again a martingale provided that, $(A.\mathcal{M})_n \in L^1$.

Proof:

$$E\left[(A.\mathcal{M})_{n+1}|\mathcal{F}_n\right] = E\left[(A.\mathcal{M})_n|\mathcal{F}_n\right] + E\left[A_{n+1}\ (\mathcal{M}_{n+1} - \mathcal{M}_n)\,|\mathcal{F}_n\right]$$

$$= E\left[(A.\mathcal{M})_n|\mathcal{F}_n\right] + A_{n+1}\ E\left[(\mathcal{M}_{n+1} - \mathcal{M}_n)\,|\mathcal{F}_n\right] \tag{3.70}$$

Clearly, the second term on the right hand side of Eq. (3.70) is zero and thus we obtain:

$$E\left[(A.\mathcal{M})_{n+1}|\mathcal{F}_n\right] = E\left[(A.\mathcal{M})_n|\mathcal{F}_n\right] = (A.\mathcal{M})_n \tag{3.71}$$

Thus, the martingale transform is also a martingale (provided of course $(A.\mathcal{M})_n$ is integrable). We find that the integrability of $(A.\mathcal{M})_n$ is also ensured if A_n is integrable:

$$E\left[\left|(A.\mathcal{M})_n\right|\right] = E\left|\sum_{j=1}^{n} A_j\ \left(\mathcal{M}_j - \mathcal{M}_{j-1}\right)\right|$$

$$\leq \sum_{j=1}^{n} E\left|A_j\ \left(\mathcal{M}_j - \mathcal{M}_{j-1}\right)\right|$$

$$\leq 2K \sum_{j=1}^{n} E\left|\mathcal{M}_j\right| < \infty \quad \text{letting} \quad E\left[A_n\right] \leq K \in \mathbb{R} \tag{3.72}$$

This shows that $(A.\mathcal{M})_n \in L^1$, the requirement for it to be a martingale.

The martingale transform may be better understood in the context of familiar betting strategies, e.g., a gambling exercise or stock pricing. As we know already, a martingale is constant on average and thus in terms of martingale difference sequence $\{\mathcal{M}_{n+1} - \mathcal{M}_n\}$ we have $E\left[(\mathcal{M}_{n+1} - \mathcal{M}_n)\,|\mathcal{F}_n\right] = 0$. Now, if one is to stake a bet, i.e., A_n units before start of the nth game, based on the previous outcomes up to and including the $n-1^{\text{th}}$ game, then the amount gained (or lost) is $A_n\ (\mathcal{M}_n - \mathcal{M}_{n-1})$, where, $\mathcal{M}_n$ denotes the available capital at the end of the nth game. Thus, the martingale transform $(A.\mathcal{M})_n$ represents the amount won (or lost) up to the game n. Since $(A.\mathcal{M})_n$ is also a martingale, it again shows that one has no chance in a fair game to get benefitted on the average, by any type of betting strategy.

We can similarly define sub-martingale and super-martingale transforms. For example, if X_n is an $\mathcal{F}_n$-adapted sub-martingale and A_n is an $\mathcal{F}_{n-1}$-measurable (predictable) process, then $(A.X)_n$ is the sub-martingale transform. However, the necessary condition for both these transforms to exist is that A_n must be bounded. One important aspect about these transforms is that they are the discrete analogues of stochastic integrals of the type $\int A dX$, when the integrand A is predictable and the process $X(t)$ a diffusion process (Chapter 4).

3.4.3 Doob's upcrossing inequality

This inequality is needed to prove the martingale convergence theorem. Given a stochastic process X_j with $j \in [1, n]$, we define the number of upcrossings of the interval $[a, b]$ with $a < b$ as the number of times the process, starting from below a, ends up above b (see Fig. 3.5). The number of such upcrossings, denoted by $U_n(a,b)$, is a random variable on Ω. For martingales, Doob's upcrossing inequality (stated below) yields a simple bound on $U_n(a,b)$.

Thus, if $X_j = \mathcal{M}_j$ is a martingale, then:

$$E\left[U_n(a,b)\right] \leq \frac{E\left[(\mathcal{M}_n - a)^-\right]}{(b-a)} \tag{3.73}$$

Here $(\mathcal{M}_n - a)^- = \max(0, -(\mathcal{M}_n - a))$

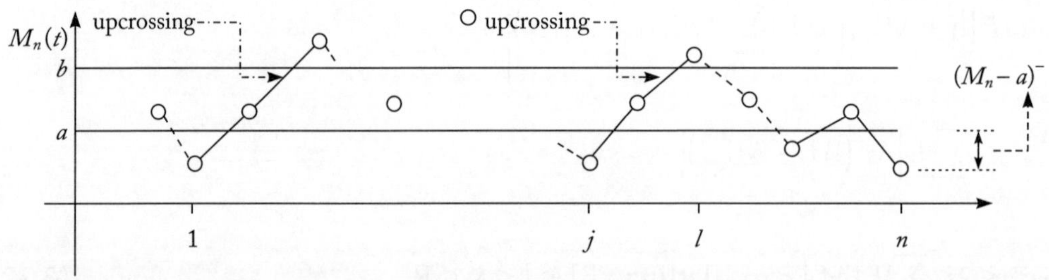

Fig. 3.5 A sample path of $\mathcal{M}$ and an illustration of upcrossings

Proof: We define a process A_j with $A_0 = 0$ and $A_j = 1$ if either $A_{j-1} = 1$ and $\mathcal{M}_{j-1} \leq b$ or if $A_{j-1} = 0$ and $\mathcal{M}_{j-1} < a$. Set $A_j = 0$ otherwise. Clearly, A_j is a predictable process since $A_j = I_{(A_{j-1}=1)} I_{(\mathcal{M}_{j-1} \leq b)} + I_{(A_{j-1}=0)} I_{(\mathcal{M}_{j-1}<a)}$, $1 < j \leq n$ and thus is $\mathcal{F}_j$-predictable ($I_{(.)}$ being an indicator function). Moreover, it is bounded too. Now we prove the following:

$$(A.\mathcal{M})_n \geq (b-a)\, U_n(a,b) - (\mathcal{M}_n - a)^- \tag{3.74}$$

$(A.\mathcal{M})_n$ is the martingale transform of $\mathcal{M}$ by A. It represents the sum of all increments $\mathcal{M}_j - \mathcal{M}_{j-1}$ for which $A_j = 1$ (dark lines on the sample path of $\mathcal{M}$ shown in Fig. 3.5). In contrast to the betting strategy discussed in Section 3.4.2, we presently place a stake only when $\mathcal{M}_j$ falls below a and stops staking when once $\mathcal{M}_j$ exceeds b. Thus, $(A.\mathcal{M})$ increases by at least by $(b-a)$ units during each completed upcrossing. Between upcrossing periods, $(A.\mathcal{M})$ remains constant. If the final time n falls within an upcrossing period, $\mathcal{M}$ may decrease before the next upcrossing is completed and hence $(A.\mathcal{M})$ can decrease by at most $(\mathcal{M} - a)^-$ units. The inequality in Eq. (3.74) is therefore proved. $(A.\mathcal{M})_n$ is a martingale, $E\left[(A.\mathcal{M})_n\right] = 0$ and hence we obtain the result:

$$(b-a)E\left[U_n(a,b)\right] \leq E\left[(\mathcal{M}_n - a)^-\right]$$

$$\Longrightarrow E\left[U_n(a,b)\right] \le E\left[(\mathcal{M}_n - a)^-\right]/(b-a) \tag{3.75}$$

It may seem that when $\mathcal{M}_n$ fluctuates, one continues to be successful in her stakes, gaining $(b-a)$ units for each upcrossing. However, since the transformed process is again a martingale and finally averages to zero, any predictable strategy must fail and in fact Doob's upcrossing inequality just provides that upper bound on $(b-a)U_n(a,b)$ in an averaged sense. Further, if $\mathcal{M}_n$ is bounded, it should not fluctuate forever and thus should converge to a random variable which is the essence of martingale convergence theorem.

3.4.4 Martingale convergence theorem

If $\mathcal{M}_n$ is an $\mathcal{F}_n$-martingale and bounded such that either a) $\sup_n E\left[|\mathcal{M}_n|\right] < \infty$ or b) $\sup_n E\left[\mathcal{M}_n{}^+\right] < \infty$ or c) $\sup_n E\left[\mathcal{M}_n{}^-\right] < \infty$, then there exists an $\mathcal{F}$-measurable random variable $\mathcal{M}_\infty \in L^1$ such that $\lim_{n\to\infty}\mathcal{M}_n \to \mathcal{M}_\infty$ a.s. where $\mathcal{F} := \mathcal{F}_\infty = \sigma(\mathcal{F}_n : n = 1,2,\ldots)$.

Proof:

Step 1. Without loss of generality, let us assume that $E\left[\mathcal{M}_n\right] = 0$. Since $\mathcal{M}_n = \mathcal{M}_n{}^+ - \mathcal{M}_n{}^-$, $E\left[\mathcal{M}_n{}^+\right] = E\left[\mathcal{M}_n{}^-\right]$. Also, $|\mathcal{M}_n| = \mathcal{M}_n{}^+ + \mathcal{M}_n{}^-$. Thus, the integrability conditions a) to c) are equivalent.

Step 2. $(\mathcal{M}_n - a)^- \le |\mathcal{M}_n - a| \le |\mathcal{M}_n| + |a|$. Hence, Doob's upcrossing inequality may be written as:

$$E\left[U_n(a,b)\right] \le \left(|a| + E\left[|\mathcal{M}_n|\right]\right)/(b-a)$$

$$\le \left(|a| + \sup_n E\left[|\mathcal{M}_n|\right]\right)/(b-a) < \infty \tag{3.76}$$

By monotone convergence (Appendix C), $E\left[U_\infty(a,b)\right] = \lim_{n\to\infty} E\left[U_n(a,b)\right] < \infty$. It is thus clear that $\mathcal{M}_n$ can only have a finite number of upcrossings $U_\infty(a,b)$.

Step 3. In general, as $n \to \infty$, $\mathcal{M}_n(\omega)$ may either converge to a finite value or converge to $-\infty$ or ∞ or have no limit and fluctuate forever (with distinct limits superior and inferior). In the last case, there must exist two rational numbers a and b such that $\mathcal{M}_n(\omega)$crosses $[a,b]$ (Fig. 3.5) infinite number of times. However, from Step 2 we have:

$$P\left(\exists a,b \in \mathbb{Q},\ a < b,\ \mathcal{M}_n(\omega) \text{ crossing } [a,b] \text{ i. o.}\right)$$

$$\le \sum_{a,b\in\mathbb{Q}} P(\mathcal{M}_n(\omega) \text{ crossing } [a,b] \text{ i. o.}) = P\left(U_\infty(a,b) = \infty\right) = 0 \tag{3.77}$$

Thus, $\mathcal{M}_n$ must converge a.s. to some $\mathcal{M}_\infty$.

Step 4. By Fatou's lemma [Williams 1991; also see Appendix C]:

$$E[|\mathcal{M}_\infty|] \le \liminf_n E[|\mathcal{M}_n|] < \infty \Longrightarrow \mathcal{M}_\infty \in L^1 \tag{3.78}$$

Conditional expectation re-visited using martingale convergence theorem

Recall that given two random variables X and Y on $(\Omega,\mathcal{F},P)$, we defined, in Chapter 2, the conditional expectation $E[X|Y]$. We attempt here to interpret the conditional expectation in a more general setting via the above martingale convergence theorem. To this end, we first express Y as the limit of a non-decreasing sequence of simple random variables:

$$Y_n = \sum_{k=1}^{n} \alpha_k I_{A_k} \quad \text{(see Eq. (1.88) of Chapter 1)} \tag{3.79}$$

where, $\alpha_k \in \mathbb{R}$, I_{A_k} is an indicator function and $A_k \in \mathcal{F}$. Assume further, without loss of generality, that Y_n is a non-negative random variable. Even otherwise, one can express $Y_n = Y_n{}^+ - Y_n{}^-$ and apply the following arguments separately to the positive components. Now define $\mathcal{M}_n = E[X \mid Y_1, Y_2, \ldots, Y_n]$. We claim that $\mathcal{M}_n$ is a martingale with respect to $\mathcal{F}_n = \sigma(Y_1, Y_2, \ldots, Y_n)$. To verify this, we first see that $\mathcal{M}_n$ is adapted. Further, using the tower property, we have for $m \le n$:

$$E[\mathcal{M}_n|\mathcal{F}_m] = E[E[X \mid \mathcal{F}_n]|\mathcal{F}_m] = E[X \mid \mathcal{F}_m] = \mathcal{M}_m \tag{3.80}$$

Thus, $\mathcal{M}_n$ is an $\mathcal{F}_n$- martingale. Further, by Jensen's inequality:

$$|\mathcal{M}_n| = \left|E[X \mid \mathcal{F}_n]\right| \le E[|X| \mid \mathcal{F}_n]$$

$$\Longrightarrow E|\mathcal{M}_n| \le E[E[|X| \mid \mathcal{F}_n]] = E[|X|] < \infty \tag{3.81}$$

assuming that X is integrable. Thus $\mathcal{M}_n$ is bounded. The martingale convergence theorem thus applies and the result is that $\mathcal{M}_n$ converges to $\mathcal{M}_\infty = E[X|Y]$ a.s. Thus, we may define $E[X \mid Y] = \lim_{n\to\infty} \mathcal{M}_n$.

Radon–Nikodym derivative re-visited using martingale convergence theorem

If Q and P are two measures defined on the measurable space $(\Omega,\mathcal{F})$ and $Q \ll P$, the Radon–Nikodym derivative as defined in Chapter 1 is a non-negative measurable function and is the density of Q with respect to P. In the context of random variables, the Radon–Nikodym theorem ensures the existence and uniqueness of the derivative (denoted by Λ) such that $Q(\omega_i) = \Lambda(\omega_i)\ P(\omega_i)$, $\forall \omega_i \in \Omega$ in the discrete case and $dQ(\omega) = \Lambda(\omega)\ dP(\omega)$, $\omega \in \Omega$ in the continuous case. In either case, the Radon–Nikodym derivative relates the two measures by $\frac{dQ}{dP} = \Lambda$ with $E_P[\Lambda] = 1$ and necessarily corresponds to a change of measure, an important tool in stochastic analysis. We here provide an alternative characterization of Λ based on a stochastic process defined over a separable σ-algebra.

Separable σ*-algebra*

A σ-algebra is separable if there exists a filtration $\{\mathcal{F}_n\}$ of discrete σ-algebras $\mathcal{F}_n \in \mathcal{F}$ such that $\mathcal{F} = \sigma(\mathcal{F}_n : n \in \mathbb{N})$. In other words, $\mathcal{F}$ is separable if $\mathcal{F} = \sigma\{A_n\}$ for a countable collection of events $A_n \in \mathcal{F}$. In general, most σ-algebras that we deal with in practice are separable. For instance, suppose $\{X_t\}_{t\in[0,\infty)}$ is a stochastic process with continuous sample paths. Then $\mathcal{F} = \sigma\ \{X_t : t \in [0,\infty)\}$ is separable. Owing to its continuity, the process is completely characterizable via a dense set of times (say, dyadic rationals). Thus it suffices to sample the process in terms of the sequence $\{X_{k\Delta t}\}$ (for a given $t > 0$) by defining $\Delta t = t2^{-n}$ (with $n \to \infty$) and thus construct the associated σ-algebra as $\mathcal{F}_t = \{X_{k\Delta t} : k \in [0,2^n]\}$ with respect to which X_s is measurable for $s \le t$.

Radon–Nikodym theorem

Consider a probability space $(\Omega,\mathcal{F},P)$ with $\mathcal{F} = \sigma(A_n : n \in \mathbb{N})$ for some collection of subsets $\{A_n\}$ of Ω. Let Q be an another finite probability measure on $\mathcal{F}$. Suppose $Q \ll P$; then by Radon–Nikodym theorem, there exists a unique non-negative $\mathcal{F}$-measurable function $\Lambda \in L^1(\Omega,\mathcal{F},P)$ with $E_P[\Lambda] = 1$ and $Q(A) = E_P[\Lambda I_A]\ \forall\ A \in \mathcal{F}$. The random variable Λ is the Radon–Nikodym derivative of Q with respect to P. Further, Λ is a uniformly integrable (UI) martingale and $\Lambda_n \to \Lambda_\infty$ a.s. in $L^1(P)$. The notion of uniform integrability needs to be introduced first before the proof for the Radon–Nikodym theorem is presented.

3.4.5 Uniform integrability

Recall that a random variable X is integrable, i.e., $X \in L^1$, if $E[|X|] < \infty$. Then we have:

$$\lim_{k\to\infty} E[|X| I(|X| > k)] = 0 \tag{3.82}$$

Indeed, for a random variable, the converse is also true.

Proof: In fact, if X is integrable, then $\lim_{k\to\infty} |X|\ I_{|X|>k} = 0$ and $|X|\ I_{|X|>k} \le |X|$. Thus, by dominated convergence, we have the result. To prove the converse, we can write $E[|X|] = E\left[|X|\ I_{|X|\le k}\right] + E\left[|X|\ I_{|X|>k}\right]$. The first term is bounded by k. Further, with the information given in Eq. (3.82), one can choose k large enough for the second term to be finite, so that $X \in L^1$.

The condition in Eq. (3.82) corresponds to uniform integrability (or just integrability) of a random variable. However, in the context of a stochastic process (or a sequence of random variables), while uniform integrability implies integrability, integrability need not ensure uniform integrability (see Example 3.2 below). A sequence C of random variables X_n is called uniformly integrable (UI), if given $\varepsilon > 0$, there exists k in $[0,\infty)$ such that $E\left[|X_n| I_{|X_n|>k})\right] < \varepsilon,\ \forall X_n \in C$.

Example 3.2

Consider $\Omega = [0,1]$ with $\mathcal{F} = \mathcal{B}[0,1]$. Choose $A_n = (0, \frac{1}{n})$ and $X_n = nI_{A_n}$. With Lebesgue measure $\mu(A_n) = \frac{1}{n}$, it is clear that $E[X_n] = 1,\ \forall n$ so that $X_n \in L^1$. Now, for any $k > 0$ and $n > k$, we have $E\left[|X_n| I_{|X_n|>k}\right] = n\ \mu(A_n) = 1$. Here $X_n \to 0$ but $E[X_n] \nrightarrow 0$ and hence $\{X_n\}$ is not UI.

The definition of UI generalizes for a discrete stochastic process $\{X_n : n \in \mathbb{N}\}$ that is nothing but a collection of UIrandom variables. For a continuous stochastic process $X(t)$, it is UI over $0 \le t \le T$, if $E\left[\left|X(t)\right| I_{|X(t)|>k}\right]$ converges to zero as $k \to \infty$ uniformly in t. This implies:

$$\lim_{k\to\infty} \sup_t E\left[\left|X(t)\right| I_{|X(t)|>k}\right] = 0 \text{ or } \sup_t E\left[\left|X(t)\right|\right] < \infty \tag{3.83}$$

The supremum in Eq. (3.83) is over $[0,T]$ or $[0,\infty]$ if the process is considered over a finite or an infinite interval respectively.

Some sufficient conditions for uniform integrability

(a) If the process $X(t)$ is dominated by an integrable random variable Y, i.e., $\left|X(t)\right| \le Y$ with $E[Y] < \infty$, then it is UI. In particular, if $E[\sup_t \left|X(t)\right|] < \infty$, then $X(t)$ is UI.

Proof:

$$E\left[\left|X(t)\right| I_{|X(t)|>k}\right] \le E[Y I_{|Y|>k}] \to 0 \text{ as } k \to \infty \tag{3.84}$$

For instance, if $Y = \sup_t \left|X(t)\right| \ge \left|X(t)\right|$, the above condition is satisfied and the result applies.

(b) If, for some positive increasing convex function $g(x)$ on $[0,\infty)$, we have $\lim_{x\to\infty} \frac{g(x)}{x} = \infty$, and $\sup_{t\le T} E[g(\left|X(t)\right|)] < \infty$, then $X(t),\ t \le T$ is UI.

Proof:

$$\sup E\left[\left|X(t)\right| I_{|X(t)|\ge k}\right] \le \sup_{y\ge k} \frac{y}{g(y)} \sup_t E\left[g(\left|X(t)\right|\right] = 0,\ k \to \infty \tag{3.85}$$

The second condition (b) above, particularly holds if $g(x) = x^r,\ r > 1$. Whilst it amounts to checking uniform integrability using moments, it is evident that boundedness in L^1 does not imply uniform integrability.

Proof of Radon–Nikodym theorem:

In the proof that follows, the expectations are with respect to P unless stated otherwise.

Step 1. Since $Q \ll P$, for any $A \in \mathcal{F}$, $P(A) = 0 \Longrightarrow Q(A) = 0$. This implies that given $\varepsilon > 0$, there exists $\delta > 0$, such that for $A \in \mathcal{F}$:

$$P(A) < \delta \Longrightarrow Q(A) < \varepsilon \tag{3.86}$$

Step 2. Let $\{A_{n,k}\}$ be the partition of Ω that generates $\mathcal{F}_n = \sigma(F_1, F_2, \dots, F_n)$. Each $A_{n,k}$ is of the form $B_1 \cap B_2 \cap \cdots \cap B_n$ with B_j being either F_j or $F_j{}^c$. We define a function $\Lambda_n(\omega) : \Omega \to [0, \infty)$ as:

$$\begin{aligned} \Lambda_n(\omega) &= Q(A_{n,k})/P(A_{n,k}), \text{ if } P(A_{n,k}) > 0 \\ &= 0, \text{ if } P(A_{n,k}) = 0 \end{aligned} \tag{3.87}$$

Clearly, $\Lambda_n(\omega)$ is $\mathcal{F}_n$-measurable and belongs to $L^1(P)$. Further, since any set $A \in \mathcal{F}_n$ is the union of (some of the) $A_{n,k}$s, we have :

$$Q(A) = E[\Lambda_n I_A] \tag{3.88}$$

Step 3. We can say that $\Lambda_n = \frac{dQ}{dP}$ on $(\Omega, \mathcal{F}_n)$. That $\Lambda = (\Lambda_n : n \in \mathbb{Z}^+)$ is a martingale is shown below.

Let $\{A_{n,j}\}$ and $\{A_{m,k}\}$ be the partitions of Ω generating $\mathcal{F}_n$ and $\mathcal{F}_m$ respectively. With $n > m$, every $A_{m,k}$ is a union of sets $A_{n,j}$ for suitable choices of j. Therefore:

$$\begin{aligned} E[\Lambda_n | A_{m,k}] &= \frac{1}{P(A_{m,k})} \sum_{j:\, A_{n,j} \subset A_{m,k}} E\left[\Lambda_n I_{A_{n,j}}\right] \\ &= \frac{1}{P(A_{m,k})} \sum_{j:\, A_{n,j} \subset A_{m,k}} Q(A_{n,j}) \\ &= \frac{Q(A_{m,k})}{P(A_{m,k})} = \Lambda_m \end{aligned} \tag{3.89}$$

Step 4. Since Λ is thus a non-negative bounded martingale, the martingale convergence theorem applies and $\Lambda_\infty := \lim \Lambda_n$ exists a.s.

Let $\varepsilon > 0$. Then by absolute continuity of measures, there exist $k \in (0, \infty)$ and δ such that $P(\Lambda_n > k) < \delta$. Specifically, choose k such that $k^{-1} Q(\Omega) < \delta$. Then, by Markov's inequality, we have:

$$P(\Lambda_n > k) \le k^{-1} E[\Lambda_n] = k^{-1} Q(\Omega) < \delta \tag{3.90}$$

Then, it can be shown that

$$Q(\Lambda_n > k) = E\left[\Lambda_n I_{\Lambda_n > k}\right] < \varepsilon \tag{3.91}$$

We prove the inequality above by assuming a contradiction and proving it to be false. Thus, if the inequality is not true, we can find, for some ε_0, a sequence $\{A_n \in \mathcal{F}_n\}$ such that $P(A_n) < 2^{-n}$ and $E[|\Lambda_n| I_{A_n}] \geq \varepsilon_0$. Let $B := \lim\sup A_n$. Referring to Borel-Cantelli Lemma (Appendix C), we have $P(B) = 0$. However, the reverse Fatou Lemma shows that $E[|\Lambda_n| I_B] \geq \varepsilon_0$ leading to contradiction.

Λ is thus a UI martingale and $\Lambda_n \to \Lambda$ P-a.s. in L^1. As $Q \ll P$, we also have that $\Lambda_n \to \Lambda$ Q-a.s. From Eq. (3.88), we have $E[\Lambda I_A] = Q(A)$ on the π-system $\bigcup_{n\geq 1} \mathcal{F}_n$ so that they agree on $\mathcal{F}$. This result is owing to the theorem on uniqueness of measures (Section 1.3.2, Chapter 1).

Step 5. We need to show that $E[\Lambda] = 1$ so that one finally has $\Lambda = \frac{dQ}{dP}$. Since Λ is a martingale, we have $E[\Lambda I_\Omega] = Q(\Omega) = E_Q[1] = 1$.

Step 6. It remains to prove that Λ is unique. If possible, let there be Λ and $\overline{\Lambda}$ both $\mathcal{F}$-measurable and $E[\Lambda I_A] = E\left[\overline{\Lambda} I_A\right] = Q(A)\ \forall A \in \mathcal{F}$. Let $A_1 = \left(\omega : \Lambda(\omega) > \overline{\Lambda}(\omega)\right)$ and $A_2 = \left(\omega : \Lambda(\omega) < \overline{\Lambda}(\omega)\right)$ such that $A_1, A_2 \in \mathcal{F}$. Since by assumption, $E\left[\left(\Lambda - \overline{\Lambda}\right) I_A\right] = 0$, $P(A_1) = P(A_2) = 0$ and thus $\Lambda = \overline{\Lambda}$ a.s.

The proof is complete.

The martingale theory, whilst providing an intuitive proof of Radon–Nikodym theorem, also yields a close inter-relation between the Radon–Nikodym derivative and conditional expectation as explained in the following. By martingale convergence theorem, we have (earlier in this section) obtained the limit for the conditional expectation $E[X \mid Y] = \mathcal{M}_\infty = \lim_{n\to\infty} \mathcal{M}_n$ where $\mathcal{M}_n = E[X \mid \mathcal{F}_n]$ and $\mathcal{F}_n = \sigma(Y_1, Y_n, \ldots, Y_n)$. Since the choice of the sequence $\{Y_n\}$ is non-unique, the uniqueness issue of the above limit must be addressed.

Let X be a non-negative random variable with $E[X] = 1$. These assumptions are made to define valid probability measures on Ω. Even otherwise, one can express $X = X^+ - X^-$ and also rescale $X \in L^1$ to have $E[X] = 1$. We now have for each $\mathcal{F}_n$:

$$E[X \mid \mathcal{F}_n] = \sum_j E\left[X | A_j\right] I_{A_j} = \sum_j \frac{E\left[X I_{A_j}\right]}{P\left[A_j\right]} \tag{3.92}$$

where, $\{A_j\}$ is a partition of Ω that generates $\mathcal{F}_n$. If we define $Q(A) = P(X I_A) = E[X I_A]$ to be another probability measure on Ω, then Eq. (3.92) can be rewritten as:

$$E[X \mid \mathcal{F}_n] = \sum_j \frac{Q\left[A_j\right]}{P\left[A_j\right]} I_{A_j} \tag{3.93}$$

We identify the conditional expectation $E[X \mid \mathcal{F}_n]$ as a Radon–Nikodym derivative $= \frac{dQ_{\mathcal{F}_n}}{dP_{\mathcal{F}_n}}$– where $P_{\mathcal{F}_n}$ and $Q_{\mathcal{F}_n}$are measures restricted to $\mathcal{F}_n$. By Radon–Nikodym theorem, we find that, in the limit as $n \to \infty$, the derivative converges and thus $E[X \mid \mathcal{F}_n] \to E[X \mid \mathcal{F}] = \frac{dQ_{\mathcal{F}}}{dP_{\mathcal{F}}}$ which is unique.

The above conclusions may also be summarized by Kolmogorov's definition of conditional expectation which reads as: Let $X \in L^1$ and $\mathcal{F}$ be a σ-algebra. Then, $E[X \mid \mathcal{F}]$ is the unique $\mathcal{F}$-measurable random variable that satisfies $E[XI_A] = E[E[X \mid \mathcal{F}]\, I_A]$ for all $A \in \mathcal{F}$.

Strong law of large numbers re-visited via martingale convergence

Before proving this law, we first define the notion of a downward filtration followed by Levy's downward martingale theorem on the convergence of reverse martingales. A sequence $\{\mathcal{G}_{-n}\}$ of σ-algebras satisfying:

$$\mathcal{G}_{-(n+1)} \subseteq \mathcal{G}_{-n} \subseteq \mathcal{G}_{-(n-1)} \subseteq \cdots \subseteq \mathcal{G}_{-1} \tag{3.94}$$

is called a downward filtration. Also define $\mathcal{G}_{-\infty} := \cap_n \mathcal{G}_{-n}$.

Levy's downward martingale theorem

For $\gamma \in L^1$, define the reverse martingale $\mathcal{M}_{-n} := E[\gamma | \mathcal{G}_{-n}]$. Then $\mathcal{M}_{-\infty} := \lim \mathcal{M}_{-n}$ exists a.s. and in L^1. Further, $\mathcal{M}_{-\infty} = E[\gamma | \mathcal{G}_{-\infty}]$ a.s.

The proof of the above theorem [Williams 1991] is similar to the forward martingale convergence theorem proved earlier in this section.

Now recall the law of large numbers (e.g., Section 2.3.2, Chapter 2) restated below: *Let $X_1, X_2, \ldots, X_n, \ldots$ be independent and identically distributed (i.i.d.) random variables with $E[|X_j|] \leq \infty$, $\forall j$. Given $E[X_j] = \mu\ \forall j$ and the same distribution $F_X(x)$, write the partial sums S_n as:*

$$S_n = X_1 + X_2 + \cdots + X_n \tag{3.95}$$

Then, $L^1 \ni \frac{S_n}{n} \to \mu$ a.s.

Proof: Define:

$$\mathcal{G}_{-n} := \sigma(S_n, S_{n+1}, \ldots) = \sigma(S_n, X_{n+1}, X_{n+2} \ldots) \tag{3.96}$$

Note that $\mathcal{G}_{-n}$ is a decreasing sequence of σ-algebras with increasing n. Since the tail σ-algebra $\mathcal{T}_n = \sigma(X_{n+1}, X_{n+2} \ldots)$ is independent of $\sigma(X_1, S_n)$ (which is a sub-σ-algebra of $\mathcal{F}_n := \sigma(X_1, X_2, \ldots, X_n)$ with $\mathcal{F}_\infty := \mathcal{F}$), we have:

$$E[X_1 \mid \mathcal{G}_{-n}] = E[X_1 \mid S_n] \tag{3.97}$$

It is also possible to write:

$$E\left[X_1 I_{S_n \in F \in \mathcal{F}}\right] = \int \dots \int_{S_n \in F} x_1\, F_{X_1}(dx_1) F_{X_2}(dx_2) \dots F_{X_n}(dx_n)$$

$$= E\left[X_2 I_{S_n \in F}\right] = \dots = E\left[X_n I_{S_n \in F}\right] \tag{3.98}$$

Therefore, we obtain:

$$E\left[X_1 \mid S_n\right] = E\left[X_2 \mid S_n\right] = \dots = E\left[X_n \mid S_n\right]$$

$$= \frac{E[(X_1 + X_2 + \dots + X_n) | S_n]}{n} = \frac{S_n}{n} \text{ a.s.} = E\left[X_1 \mid \mathcal{G}_{-n}\right] \tag{3.99}$$

For a given $m > 0$, define $\mu_m := \limsup \frac{X_{m+1} + X_{m+2} + \dots + X_{m+n}}{n}$. Clearly, μ_m is a bounded reverse martingale since, for $m < k$, $E\ [\mu_m | \mathcal{T}_k] = \mu_k$. Now, by Levy's downward theorem, we have:

$$\mu_\infty := \lim_{m \to \infty} \mu_m = \lim_{n \to \infty} \frac{S_n}{n} \tag{3.100}$$

μ_m is measurable with respect to the tail σ-algebra $\mathcal{T}_m$. By Kolmogorov's $0-1$ law (Appendix C), $P(\mu_m = a) = 1$ for some $a \in \mathbb{R}$. However, we also have:

$$a = E[\mu_m] = \lim E\left[\frac{S_n}{n}\right] = \mu \tag{3.101}$$

The proof is complete.

3.5 Stopping Time and Stopped Processes

Stopping time is another important notion in the theory of stochastic processes. It is a random variable τ indicating the time when an observable event happens. Specifically, considering a discrete stochastic process $\{X_n, n \geq 0\}$, such events are described by the associated filtration F_n. Based on a stopping time, one can also define a stopped process. We will see that if $\{X_n\}$ is a martingale, even the stopped processes are martingales and finally these notions lead to Doob's optional stopping theorem. One of the many applications of stopping time is in solving a class of parabolic boundary value problems (BVPs) based on the theory of diffusion processes (see Chapter 4).

3.5.1 Stopping time

If $\{X_n\}$ is on $(\Omega,\mathcal{F},P)$, we define an $\mathcal{F}_n$- stopping time as a random time $\tau = \Omega \to \{0,1,2,\ldots,\infty\}$ such that

$$\{\tau \le n\} = \{\omega \in \Omega : \tau(\omega) \le n\} \in \mathcal{F}_n = \sigma\left\{X_j, j = 1,2,\ldots,n\right\} \tag{3.102a}$$

for every n. Equivalently, we also have:

$$\{\tau = n\} = \{\omega \in \Omega : \tau(\omega) = n\} \in \mathcal{F}_n \tag{3.102b}$$

As τ represents time, $\tau \le n$ or $\tau = n$ are notations that may sometimes be replaced by $\tau \le t_n$ and $\tau = t_n$ respectively; certain liberty will be taken here while making use of such variations depending upon the context.

Example 3.3

Let X_n be adapted to the filtration $\{\mathcal{F}_n\}$. Let $b \in \mathbb{R}$ and define $\tau = \inf\{n : X_n \ge b\}$. Since $\{\tau \le n\} = \{X_0 \ge b\} \bigcup \{X_1 \ge b\} \bigcup \ldots \bigcup \{X_n \ge b\}$ and the events $\{X_k \ge b\} \in \mathcal{F}_k \subseteq \mathcal{F}_n$, we have that $\{\tau \le n\} \in \mathcal{F}_n$ and τ is $\mathcal{F}_n$-measurable.

The definitions in Eqs. (3.102a,b) are indeed equivalent since, by Eq. (3.102a), we have $\{\tau = n\} = \{\tau \le n \backslash \tau \le n-1\} \in \mathcal{F}_n$. On the other hand, Eq. (3.102b) implies $\{\tau \le n\} = \bigcup_{0 \le j \le n}\{\tau = j\}$. Intuitively, τ is a non-negative random variable and is the time when one can decide to stop further observations of X_n. Thus, the event $\{\tau \le n\}$ is completely determined by the information known up to time n i.e., $\{X_1, X_2, \ldots, X_n\}$ and is accordingly $\mathcal{F}_n$-measurable. Thus, stopping times are random times, whose occurrence can be determined without prior knowledge of the future. A random time which is generally not a stopping time is the instant when X reaches its maximum on a certain interval. The definition of stopping time also implies that if, for a certain outcome ω, the event never occurs, the value of the stopping time for this ω is $+\infty$.

Example 3.4

Suppose that $\{X_n\}$ is adapted to $\mathcal{F}$. $\tau = \inf\{n \ge 0 : X_n = \alpha\}$ is called the first passage time (time of first entry) of the process into the state $X = \alpha$. This is also known as the first hitting time of α.

Example 3.5

Suppose that A is a collection of states, for example $A = \{(\alpha,\beta,\gamma) \in \mathbb{R}\}$ and that τ is the first passage time into the set A, i.e., $\tau = \inf\{n \ge 0 : X_n \in A\}$. For instance, in a betting game, a gambler can choose to stop the game when she has either amassed an amount Q or gone bankrupt $(Q = 0)$. Defining $A = \{0, Q\}$, here $\{\tau = 0\} = \{X_0 \in A\}$

and $\{\tau = n\} = \{X_0 \notin A, X_1 \notin A, \ldots, X_{n-1} \notin A, X_n \in A\}$. Note that $\{\tau > n\} = \{X_0 \notin A, X_1 \notin A, \ldots, X_{n-1} \notin A, X_n \notin A\}$ is not a stopping time.

Clearly $T = \max\{n : n \leq l \in \mathbb{N}; X_n \in A\}$ is not a stopping time as one cannot ascertain whether n has reached the maximum value based on the available history; it requires future information so T is not $\mathcal{F}_n$ measurable. In this context, we recall that the number of upcrossings of an interval $[a, b]$ by a bounded martingale $\mathcal{M}_n$ is finite. The random time at which $\mathcal{M}_n$ last crosses b is not a stopping time since one has to look into the future to ascertain that it never crosses b again.

3.5.2 Stopped processes

Given a stochastic process X_n, if $\tau < \infty$ is a stopping time, then $X_\tau := X_{\tau(\omega)}(\omega)$ is a random variable and $X_n^\tau := X_{n\wedge\tau(\omega)}(\omega)$ is known as the stopped process. Here $n \wedge \tau(\omega)$ denotes the minimum of n and $\tau(\omega)$ and thus:

$$\begin{aligned}(n \wedge \tau)(\omega) &= n, \text{ if } n \leq \tau \\ &= \tau(\omega), \text{ if } n > \tau\end{aligned} \tag{3.103}$$

Hence, the stopped process is given by:

$$\begin{aligned}X_n^\tau &= X_n, \quad n < \tau < \infty \\ &= X_\tau, \quad n \geq \tau\end{aligned} \tag{3.104}$$

If τ is not bounded, then we set the random variable $X_{\tau(\omega)}(\omega) = 0$. It is very important to distinguish a stopped process X_n^τ from the random variable X_τ. Corresponding to the above definition of a stopped process, the filtration $\mathcal{F}_\tau$ is given by:

$$\mathcal{F}_\tau = \{(\tau \leq n) \in \mathcal{F}_n, \forall n\} \tag{3.105}$$

Here $\mathcal{F}_\infty = \sigma\{\mathcal{F}_n, n \in \mathbb{N}\}$. $\mathcal{F}_\tau$ is the σ-algebra of all the events that occur before τ and X_τ is $\mathcal{F}_\tau$-measurable.

Example 3.6

In this example, we consider a continuous stochastic process $B_t(\omega) : [0, \infty) \times \Omega \to \mathbb{R}$, a one-dimensional standard Brownian motion with $B_0 = 0$ and filtration $\mathcal{F} = \{\mathcal{F}_0, \mathcal{F}_1, \ldots, \mathcal{F}_n\}$. Here τ is a stopping time with respect to this filtration, if for each $t \in [t_0, t_1, \ldots, t_n]$, the event $\{\tau \leq t\} \in \mathcal{F}_t$. For instance, the first hitting time for the region $\{x \in \mathbb{R}; x \geq a\}$, $\tau = \inf\{t > 0 : B_t(\omega) \geq a\}$, is a stopping time. Then, the stopped Brownian motion is given by $B_t^\tau = B_t$ for $t < \tau$ and $B_t^\tau = a$ otherwise.

Stopped martingales are martingales

Observe that $I_{n\le\tau}$ is a predictable process since

$$\{\omega : n \le \tau(\omega)\} = \Omega \backslash \{\omega : \tau(\omega) \le n-1\} \in \mathcal{F}_{n-1}, \tag{3.106a}$$

we can write any stopped process in terms of $I_{n\le\tau}$ as:

$$X_{n\wedge\tau} = X_0 + \sum_{j=1}^{n} \left(X_j - X_{j-1}\right) I_{j\le\tau} \tag{3.106b}$$

If, in particular, X is a martingale (super- or sub-martingale) denoted by $\{\mathcal{M}_n\}$, then $\mathcal{M}_{n\wedge\tau}$ is the martingale transform $(A.\mathcal{M})_n$ with $\{A_n\} \equiv \{I_{n\le\tau}\}$. The predictable process A_n is bounded and $X_{n\wedge\tau}$ is thus a martingale (super- or sub-martingale). Moreover, in the case of a martingale, we have the basic stopping result:

$$E\left[\mathcal{M}_{n\wedge\tau}\right] = E\left[\mathcal{M}_0\right], \ \forall n \tag{3.107}$$

For super- and sub-martingales, the equality in Eq. (3.107) is replaced by $\le$ and $\ge$ respectively. Thus, even if one were to adopt a predictable stopping strategy during betting, the last result confirms the earlier assertion that $\mathcal{M}_{n\wedge\tau}$ (being a martingale) always corresponds to a fair game.

3.5.3 Doob's optional stopping theorem

For a stopped martingale, the result in Eq. (3.107) may not imply $E\left[\mathcal{M}_\tau\right] = E\left[\mathcal{M}_0\right]$, where, $\mathcal{M}_\tau$ is the random variable corresponding to the stopping time $\tau(\omega)$ that will be different form path to path (i.e., $\mathcal{M}_\tau \ne \mathcal{M}_{n\wedge\tau}$ unless $n \ge \tau(\omega)$ for almost all ω). For example, the random walk model (discrete) B_n in Eq. (3.53) is a martingale with $B_0 = 0$. Now if $\tau := \inf\,\{n : B_n = 1\}$, then with $P(\tau < \infty) = 1$ [see Appendix C for a proof], we have $1 = E\left[B_\tau\right] \ne E\left[B_0\right] = 0$ even though $E\left[B_{n\wedge\tau}\right] = E\left[B_0\right], \ \forall n$. However, under certain conditions on either τ or $\mathcal{M}_n$, it is possible to ensure that the stopping time does not change the expected value, which would be a much stronger result. The optional stopping theorem (stated and proved below) provides these required constraints to have $E\left[\mathcal{M}_\tau\right] = E\left[\mathcal{M}_0\right]$.

Optional stopping theorem

Given $\mathcal{M}_n$ a martingale and $\tau < \infty$ a stopping time, the equality $E\left[\mathcal{M}_\tau\right] = E\left[\mathcal{M}_0\right]$ holds if either (a) $\tau < k$ a.s. for $k \in \mathbb{N}$ or (b) $|\mathcal{M}_n| \le \alpha$ a.s. for some $\alpha \in [0,\infty)$ and for all n or (c) $|\mathcal{M}_n - \mathcal{M}_{n-1}| \le \beta$ a.s. for some $\beta \in [0,\infty)$ and for all n and $E[\tau] < \infty$.

Proof: In the case of (a), the proof is straightforward, i.e., given $\tau < k$:

$$E\left[\mathcal{M}_\tau\right] = E\left[\mathcal{M}_{k\wedge\tau}\right] = E\left[\mathcal{M}_0\right] \tag{3.108}$$

For (b), since, as $n \to \infty$, $n \wedge \tau = \tau$ $(< \infty)$, $\mathcal{M}_{n\wedge\tau} \to \mathcal{M}_\tau$ a.s. Also, $|\mathcal{M}_\tau| \le \alpha < \infty$ by hypothesis. By dominated convergence, $E[\mathcal{M}_{n\wedge\tau}] \to E[\mathcal{M}_\tau] \Longrightarrow E[\mathcal{M}_\tau] = E[\mathcal{M}_0]$.

For (c), Eq. (3.106b) gives:

$$|\mathcal{M}_{n\wedge\tau}| \le |\mathcal{M}_0| + \sum_{j=1}^{n} I_{j\le\tau} \left|\mathcal{M}_j - \mathcal{M}_{j-1}\right|$$

$$\le |\mathcal{M}_0| + \beta\ (n \wedge \tau)$$

$$\le |\mathcal{M}_0| + \beta\tau \tag{3.109}$$

Since the extreme right hand side of Eq. (3.105) is integrable (as $E[\beta\tau] < \infty$) by hypothesis, one can let $n \to \infty$ and thus get the result using dominated convergence.

3.5.4 A super-martingale inequality

Suppose that $\{\mathcal{M}_n\}$ is a non-negative super-martingale. We have, by the martingale convergence theorem, $\lim_{n\to\infty}\mathcal{M}_n \to \mathcal{M}_\infty$. Let $c > 0$ be a threshold such that $\mathcal{M}_\infty < c$ with some non-zero probability. It may happen that any sample path of $\mathcal{M}$ may exceed c before finally converging to $\mathcal{M}_\infty < c$. In this scenario, interest may lie in knowing the probability of sample paths of $\mathcal{M}$ exceeding c. Using the martingale theory and the notion of stopping time, this probability can be obtained in terms of the following inequality:

$$P\left(\sup_n \mathcal{M}_n \ge c\right) \le \frac{E[\mathcal{M}_0]}{c} \tag{3.110}$$

Proof: First consider a finite interval $0 \le n \le K$ to find $P\left(\sup_{n\le K<\infty}\mathcal{M}_n \ge c\right)$. Expressing $\sup_{n\le K}\mathcal{M}_n(\omega)$ by the stopped process $\mathcal{M}_{\tau(\omega)\wedge K}(\omega)$ where $\tau = \inf\{n : \mathcal{M}_n \ge c\}$ is stopping time, we have:

$$\left\{\omega : \sup_{n\le K}\mathcal{M}_n(\omega) \ge c\right\} = \left\{\omega : \mathcal{M}_{\tau(\omega)\wedge K}(\omega) \ge c\right\} \tag{3.111}$$

Clearly, if $\sup_{n\le K}\mathcal{M}_n \ge c$, then $\tau \le K$ and $\mathcal{M}_{\tau\wedge K} \ge c$ and if $\sup_{n\le K}\mathcal{M}_n < c$, then $\tau > K$ and $\mathcal{M}_{\tau\wedge K} < c$. Now, by Markov's inequality we obtain:

$$P\left(\sup_{n\le K}\mathcal{M}_n \ge c\right) = P\left(\mathcal{M}_{\tau\wedge K} \ge c\right) \le \frac{E[\mathcal{M}_{\tau\wedge K}]}{c} \le \frac{E[\mathcal{M}_0]}{c} \tag{3.112}$$

The last inequality on the right relates to the super-martingale property (see Eq. 3.107). Finally as $K \to \infty$, we get the result in Eq. (3.110) by invoking the monotone convergence theorem.

It is evident from the above result that the probability of crossing a threshold c can be controlled and made arbitrarily small by restricting the initial condition $\mathcal{M}_0$ close to zero a.s.

3.5.5 Optional stopping theorem for UI martingales

Before the theorem is revisited under UI conditions, let us establish some additional results on UI martingales and stopping times.

If $\mathcal{M}_n$ is a UImartingale, then (a) $\mathcal{M}_\infty := \lim \mathcal{M}_n$ exists a.s. and is in L^1. Also, (b) for every n, one has:

$$\mathcal{M}_n = E\left[\mathcal{M}_\infty | \mathcal{F}_n\right] \text{a.s.} \tag{3.113}$$

Proof:

(a) Since $\mathcal{M}_n$ is UI, it is in L^1. It follows from martingale convergence theorem that $\mathcal{M}_\infty := \lim \mathcal{M}_n$ exists a.s. That $\mathcal{M}_\infty$ is in L^1 is proved in the following steps (Williams 1991).

Step 1. Take $c \in [0,\infty)$ and define a function $g_c : \mathbb{R} \to [-c,c]$ as:

$$\begin{aligned} g_c(x) &= c, \quad \text{if } x > c \\ &= x, \quad \text{if } |x| \le c \\ &= -c, \quad \text{if } x < -c \end{aligned} \tag{3.114}$$

Step 2. Now, given $\varepsilon > 0$, and by the UIproperty of $\mathcal{M}_n$, we can set c such that:

$$E\left[\left|g_c(\mathcal{M}_n) - \mathcal{M}_n\right|\right] < \frac{\varepsilon}{3}, \ \forall n \text{ and } E\left[\left|\mathcal{M}_\infty - g_c(\mathcal{M}_\infty)\right|\right] < \frac{\varepsilon}{3} \tag{3.115}$$

Since $\left|g_c(x) - g_c(y)\right| \le \left|x - y\right|$, it is clear that $g_c(\mathcal{M}_n) \to g_c(\mathcal{M}_\infty)$ a.s. By bounded convergence theorem, we can choose n_0 such that for $n \ge n_0$,

$$E\left[\left|g_c(\mathcal{M}_\infty) - g_c(\mathcal{M}_n)\right|\right] < \frac{\varepsilon}{3} \tag{3.116}$$

By triangle inequality, we have:

$$\begin{aligned} &E\left[\left|(g_c(\mathcal{M}_n) - \mathcal{M}_n) + (\mathcal{M}_\infty - g_c(\mathcal{M}_\infty)) + (g_c(\mathcal{M}_\infty) - g_c(\mathcal{M}_n))\right|\right] \\ &\le E\left[\left|g_c(\mathcal{M}_n) - \mathcal{M}_n\right|\right] + E\left[\left|\mathcal{M}_\infty - g_c(\mathcal{M}_\infty)\right|\right] + E\left[\left|g_c(\mathcal{M}_\infty) - g_c(\mathcal{M}_n)\right|\right] < \varepsilon \end{aligned}$$

$$\Longrightarrow E\left[|\mathcal{M}_\infty - \mathcal{M}_n|\right] < \varepsilon \Longrightarrow \mathcal{M}_n \to \mathcal{M}_\infty \in L^1 \tag{3.117}$$

(b) For $\mathcal{F} \subset \mathcal{F}_n$ and $r \geq n$, the martingale property gives:

$$E\left[\mathcal{M}_r|\mathcal{F}\right] = E\left[\mathcal{M}_n|\mathcal{F}\right] \tag{3.118}$$

However, we have:

$$\left|E\left[\mathcal{M}_r|\mathcal{F}\right] - E\left[\mathcal{M}_\infty|\mathcal{F}\right]\right| \leq E\left[|\mathcal{M}_r - \mathcal{M}_\infty||\mathcal{F}\right] \text{ (Jensen's inequality)}$$

$$\leq E\left[|\mathcal{M}_r - \mathcal{M}_\infty|\right] \text{ (integration of a positive quantity over a larger domain)} \tag{3.119}$$

It follows from the last two equations that, as $r \to \infty$ (with $E\left[|\mathcal{M}_\infty - \mathcal{M}_r|\right] \to 0$):

$$E\left[\mathcal{M}_\infty|\mathcal{F}\right] = E\left[\mathcal{M}_n|\mathcal{F}\right] = \mathcal{M}_n \tag{3.120}$$

This proves the result.

The UI property of martingales gives another interesting result that follows as an extension of the previous one.

If $\mathcal{M}_n$ is a UI martingale, then for any stopping time τ:

$$E\left[\mathcal{M}_\infty|\mathcal{F}_\tau\right] = \mathcal{M}_\tau \text{ a.s.} \tag{3.121}$$

Proof: From the previous result in Eq. (3.113), we have for $k \in \mathbb{N}$:

$$E\left[\mathcal{M}_\infty|\mathcal{F}_k\right] = \mathcal{M}_k \text{ a.s.} \tag{3.122a}$$

and

$$E\left[\mathcal{M}_k|\mathcal{F}_{\tau\wedge k}\right] = \mathcal{M}_{\tau\wedge k} \text{ a.s.} \tag{3.122b}$$

By the tower property of conditional expectations and the UI property, it follows:

$$E\left[\mathcal{M}_\infty|\mathcal{F}_{\tau\wedge k}\right] = \mathcal{M}_{\tau\wedge k} \text{ a.s.} \tag{3.123}$$

If $F \in \mathcal{F}_\tau$, then $F \cap \{\tau \leq k\} \in \mathcal{F}_{\tau\wedge k}$ and, using Eq. (3.123):

$$E\left[\mathcal{M}_\infty|F \cap \{\tau \leq k\}\right] = E\left[E\left[\mathcal{M}_\infty|\mathcal{F}_{\tau\wedge k}\right]|F \cap \{\tau \leq k\}\right]$$

$$= E\left[\mathcal{M}_{\tau\wedge k}|F \cap \{\tau \leq k\}\right] = E\left[\mathcal{M}_\tau|F \cap \{\tau \leq k\}\right] \tag{3.124}$$

For the case when $\mathcal{M}_\infty \geq 0$, $\mathcal{M}_n = E[\mathcal{M}_\infty|\mathcal{F}_n] \geq 0$ a.s., i.e., $\mathcal{M}_n$ is a positive martingale. Now, as $k \to \infty$ in the last equation, we get by using monotone convergence:

$$E[\mathcal{M}_\infty|F \cap \{\tau < \infty\}] = E[\mathcal{M}_\tau|F \cap \{\tau < \infty\}] \tag{3.125}$$

Noting that $F = \{F \cap \{\tau < \infty\} \bigcup F \cap \{\tau = \infty\}\}$, and that the identity $E[\mathcal{M}_\infty| F \cap \{\tau = \infty\}] = E[\mathcal{M}_\tau|F \cap \{\tau = \infty\}]$ is trivial, we have:

$$E[\mathcal{M}_\infty|F] = E[\mathcal{M}_\tau|F] \Longrightarrow E[\mathcal{M}_\infty|\mathcal{F}_\tau] = E[\mathcal{M}_\tau|\mathcal{F}_\tau] = \mathcal{M}_\tau \tag{3.126}$$

The general case of $\mathcal{M}_n$ follows from its splitting into $\mathcal{M}_n{}^+$ and $\mathcal{M}_n{}^-$.

We can make use of the above result to prove the following.

If $\mathcal{M}_n$ is a UI martingale, and τ and ν are two stopping times with $\nu \leq \tau$ a.s., then:

$$E[\mathcal{M}_\tau|\mathcal{F}_\nu] = \mathcal{M}_\nu \text{ a.s.} \tag{3.127}$$

Proof: Equation (3.126) gives $E[\mathcal{M}_\infty|\mathcal{F}_\nu] = \mathcal{M}_\nu$ and $E[\mathcal{M}_\infty|\mathcal{F}_\tau] = \mathcal{M}_\tau$. Now by the tower property of conditional expectations, we get:

$$E[\mathcal{M}_\tau|\mathcal{F}_\nu] = E[E[\mathcal{M}_\infty|\mathcal{F}_\tau]|\mathcal{F}_\nu] = E[\mathcal{M}_\infty|\mathcal{F}_\nu] = \mathcal{M}_\nu \tag{3.128}$$

Optional stopping theorem revisited under UI conditions

The optional stopping theorem can be stated under UI conditions in the two following forms.

(i) If $\mathcal{M}_n$ is a UI martingale, and τ a stopping time, then $E[|\mathcal{M}_\tau|] < \infty$ and $E[\mathcal{M}_\tau] = E[\mathcal{M}_0]$

(ii) Given $\mathcal{M}_n$ a martingale and $\tau < \infty$ a stopping time, the equality $E[\mathcal{M}_\tau] = E[\mathcal{M}_0]$ holds if the stopped martingale $\mathcal{M}_{n\wedge\tau}$ is UI.

Proof for (i): Since $\mathcal{M}_n$ is a UI, it is earlier shown that (a) $\mathcal{M}_\infty := \lim \mathcal{M}_n$ exists a.s. and is in L^1 and also (b) $E[\mathcal{M}_\infty|\mathcal{F}_\tau] = \mathcal{M}_\tau$. Hence, $E[|\mathcal{M}_\tau|] < \infty$ follows (as $\mathcal{F}_\tau \subseteq \mathcal{F}$). Proving the second part $E[\mathcal{M}_\tau] = E[\mathcal{M}_0]$ is also straightforward. To this end, we utilize the result in Eq. (3.127). Take $\nu = 0$ (deterministic stopping time) and we get:

$$E[\mathcal{M}_\tau|\mathcal{F}_0] = \mathcal{M}_0 \text{ a.s.}$$

$$\Longrightarrow E[E[\mathcal{M}_\tau|\mathcal{F}_0]] = E[\mathcal{M}_0]$$

$$\Longrightarrow E[\mathcal{M}_\tau] = E[\mathcal{M}_0] \tag{3.129}$$

Proof for (ii): Suppose that $\mathcal{M}_{n\wedge\tau}$ is UI. It follows that it converges a.s. and since $\tau < \infty$, $\mathcal{M}_\tau = \lim_{n\to\infty}\mathcal{M}_{n\wedge\tau}$ exists a.s. and is in L^1. Further, we have:

$$E[\mathcal{M}_0] = E[\mathcal{M}_{0\wedge\tau}] = E[E[\mathcal{M}_\tau|\mathcal{F}_0]] = E[\mathcal{M}_\tau] \tag{3.130}$$

Interestingly, the UI constraint on $\mathcal{M}_{n\wedge\tau}$ in the last result is closely related to the conditions contained in the optional stopping theorem stated and proved in Section 3.5.3. This relationship is brought out in the following statement.

Given $\mathcal{M}_n$ a martingale and $\tau < \infty$ a.s. a stopping time, the stopped martingale $\mathcal{M}_{n\wedge\tau}$ is UI if (a) τ is bounded, or (b) $\mathcal{M}_n$ is a UI martingale or (c) $E[\tau] < \infty$ and there exists a non-zero constant $c \in \mathbb{R}$ such that $E[|\mathcal{M}_n - \mathcal{M}_{n-1}||\mathcal{F}_{n-1}] \le c$ a.s. for all n.

Proof:

(a) Let $\tau \le k(\in \mathbb{N})$ a.s. Then, $|\mathcal{M}_{n\wedge\tau}| \le |\mathcal{M}_1| + |\mathcal{M}_2| + \cdots + |\mathcal{M}_k| \in L^1$ and so the family $\{\mathcal{M}_{n\wedge\tau}\}$ is dominated by a random variable in L^1 and hence is UI (see sufficient conditions for uniform integrability of a sequence of random variables – Section 3.4.5).

(b) Since $\mathcal{M}_n$ is a UI martingale, there exists a random variable $Y \in L^1$ such that $\mathcal{M}_n = E[Y|\mathcal{F}_n]$ a.s. for all n. Now, we write the stopped martingale in the form:

$$\mathcal{M}_{n\wedge\tau} = \sum_{k=0}^{n} \mathcal{M}_k\, I_{\{\tau=k\}} + \mathcal{M}_n I_{\{\tau>n\}} \tag{3.131}$$

Then, we get

$$\mathcal{M}_{n\wedge\tau} = \sum_{k=0}^{n} E[Y|\mathcal{F}_k]\; I_{\{\tau=k\}} + E[Y|\mathcal{F}_n]\; I_{\{\tau>n\}} \tag{3.132}$$

Define a convex positive increasing function $g(x) : [0,\infty) \to [0,\infty)$ with$g(0) = 0$ and $E[g(Y)] < \infty$. Applying the convex function to both sides of the last equation and using Jensen's inequality gives:

$$g(|\mathcal{M}_{n\wedge\tau}|) \le \sum_{k=0}^{n} g\left(\left|E[Y|\mathcal{F}_k]\right|\right) I_{\{\tau=k\}} + g\left(\left|E[Y|\mathcal{F}_n]\right|\right) I_{\{\tau>n\}}$$

$$\le \sum_{k=0}^{n} E[g(|Y|)|\mathcal{F}_k]\, I_{\{\tau=k\}} + E[g(|Y|)|\mathcal{F}_n]\, I_{\{\tau>k\}}$$

$$= \sum_{k=0}^{n} E\left[g(|Y|)\, I_{\{\tau=k\}}|\mathcal{F}_k\right] + E\left[g(|Y|)\, I_{\{\tau>k\}}|\mathcal{F}_n\right]$$

$$\Longrightarrow E\left[g\left(|\mathcal{M}_{n\wedge\tau}|\right)\right] \le E\left[g\left(|Y|\right)\right] < \infty, \forall\ n \tag{3.133}$$

Hence $\mathcal{M}_{n\wedge\tau}$ is UI (see Eq. 3.85).

(c) As in (a), in this proof also, the given hypothesis leads to one of the sufficient conditions for UI, i.e., that $\mathcal{M}_{n\wedge\tau}$ is dominated by a random variable in L^1.

By martingale transform, Eq. (3.106), one has:

$$\begin{aligned}
\mathcal{M}_{n\wedge\tau} &= \mathcal{M}_0 + \sum_{j=1}^{n}\left(\mathcal{M}_{j\wedge\tau} - \mathcal{M}_{j-1\wedge\tau}\right) \\
&= \mathcal{M}_0 + \sum_{j=1}^{\infty}\left(\mathcal{M}_{j\wedge\tau} - \mathcal{M}_{j-1\wedge\tau}\right)I_{j\le n} \\
&= \mathcal{M}_0 + \sum_{j=1}^{\infty}\left(\mathcal{M}_{j} - \mathcal{M}_{j-1}\right)I_{j\le n}I_{\tau\ge j}
\end{aligned}$$

$$\Longrightarrow |\mathcal{M}_{n\wedge\tau}| \le |\mathcal{M}_0| + \sum_{j=1}^{\infty}\left|\mathcal{M}_j - \mathcal{M}_{j-1}\right|I_{\tau\ge j} \tag{3.134}$$

The right hand side of the last relation is independent of n and it is enough to show that $\sum_{j=1}^{\infty} E\left[\left|\mathcal{M}_j - \mathcal{M}_{j-1}\right| I_{\tau\ge j}\right] < \infty$. To this end, we utilize the fact that $\{\tau \ge j\} \in \mathcal{F}_{j-1}$ (See Eq. (3.106a)) for all $j \in \mathbb{N}$. Writing $\mathcal{M}_j - \mathcal{M}_{j-1} = E\left[\mathcal{M}_j - \mathcal{M}_{j-1}|\mathcal{F}_{j-1}\right]$, we have:

$$\begin{aligned}
\sum_{j=1}^{\infty} E\left[\left|\mathcal{M}_j - \mathcal{M}_{j-1}\right| I_{\tau\ge j}\right] &= \sum_{j=1}^{\infty} E\left[\left|E\left[\mathcal{M}_j - \mathcal{M}_{j-1}|\mathcal{F}_{j-1}\right]\right| I_{\tau\ge j}\right] \\
&\le \sum_{j=1}^{\infty} E\left[E\left[\left|\mathcal{M}_j - \mathcal{M}_{j-1}\right||\mathcal{F}_{j-1}\right] I_{\tau\ge j}\right] \\
&\le c\sum_{j=1}^{\infty} P\left(\tau \ge j\right) = cE[\tau] < \infty
\end{aligned} \tag{3.135}$$

From Eqs. (3.134) and (3.135), it is evident that $\mathcal{M}_{n\wedge\tau}$ is dominated by a bounded deterministic quantity $cE[\tau]$ and hence is UI.

Example 3.7

Gambler's ruin and computation of ruin probabilities using optional stopping theorem

Solution

This example is the familiar gambler's ruin problem generally expressed in terms of a simple random walk model. If a gambler has an initial capital $a > 0$ and the opponent $b > 0$, gambler's ruin occurs when all her capital is lost and when the game eventually stops. Let $S_n = \sum_{j=1}^{n} X_j$ denote her gain (over a, the initial capital) at the end of game n with $P\left(X_j = 1\right) = P\left(X_j = -1\right) = 1/2$ and $S_0 = 0$. If τ is the stopping time when the amount gained by the gambler is equal to $-a$ (her ruin) or b (the opponent's ruin), then the ruin probabilities of the two players add up to unity, i.e., $P(S_\tau = -a) + P(S_\tau = b) = 1$. For the simple random walk, the stopping time τ as defined above is finite. i.e., $P(\tau < \infty) = 1$. This is the property of recurrence of states of a 1-dimensional random walk and hence is applicable to 1-dimensional Brownian motion. Continuing further, the stopped martingale $S_{n\wedge\tau}$ is bounded by b and is a UI martingale. The conditions for applying the optional stopping theorem hold and therefore lead to:

$$0 = E[S_0] = E[S_\tau] = -aP(S_\tau = -a) + bP(S_\tau = b) \tag{3.136}$$

The ruin probabilities are thus obtained as $P(S_\tau = -a) = \frac{b}{a+b}$ and $P(S_\tau = b) = \frac{a}{a+b}$.

3.6 Some Useful Results for Time-continuous Martingales

While the material presented in the previous section mostly pertains to discrete martingales, either the theorems or other results are applicable/extendable to continuous processes as well. The concepts of uniform integrability, martingale convergence and optional stopping theorems hold and their proofs follow on similar lines except for a few subtleties, specific to time-continuous nature of these processes.

Uniform integrability

Under the sufficient conditions for uniform integrability (see Section 3.4.5), it is clear that for a positive increasing convex function $g(x)$ defined on $[0,\infty)$ with $\lim_{x\to\infty} \frac{g(x)}{x} = 0$, if $\sup_{t\leq T} E\left[g(|X(t)|)\right] < \infty$, then X is UI. In particular, a square integrable (i.e., $\sup_{t\leq T} E\left[|X^2(t)|\right] < \infty$ with $g(X) = X^2$) martingale is UI. For example, a Brownian motion $B(t)$ on a finite interval $0 \leq t \leq T$ is a square integrable martingale since $E\left[B^2(t)\right] = t \leq T < \infty$ and hence it is UI. Similar is the case with the square integrable martingale $B^2(t) - t$.

3.6.1 Doob's and Levy's martingale theorem

If Y is an integrable random variable, i.e., $E[Y] < \infty$ and $\mathcal{M}(t) := E[Y|\mathcal{F}_t]$, then $\mathcal{M}(t)$ is a UI martingale.

Proof: Since $E[\mathcal{M}_t|\mathcal{F}_s] = E[E[Y|\mathcal{F}_t]|\mathcal{F}_s] = E[Y|\mathcal{F}_s] = \mathcal{M}_s$, clearly $\mathcal{M}$ is a martingale. In the rest of the proof, it suffices to consider Y to be a non-negative random variable (since a general case follows from the decomposition of Y into Y^+ and Y^-). Let us next show that $\widehat{\mathcal{M}} = \sup_{t\leq T}\mathcal{M}(t) < \infty$. If this result were not true, then $\mathcal{M}_{t_n} \uparrow \infty$ as $t_n \uparrow \infty$. But this leads to $E\left[\mathcal{M}_{t_n}\right] \uparrow \infty$ by monotone convergence theorem and it is a contradiction since $E\left[\mathcal{M}_{t_n}\right] = E[Y] < \infty$. Now $E[\mathcal{M}(t)I(\mathcal{M}(t) > n)] = E[YI(\mathcal{M}(t) > n)]$. Since $\{\mathcal{M}(t) > n\} \subseteq \{\widehat{\mathcal{M}} > n\}$, we have:

$$E[YI(\mathcal{M}(t) > n)] \leq E\left[YI\left(\widehat{\mathcal{M}} > n\right)\right] \tag{3.137}$$

Thus $E[\mathcal{M}(t)I(\mathcal{M}(t) > n)] \leq E\left[YI\left(\widehat{\mathcal{M}} > n\right)\right]$. Since $E\left[YI\left(\widehat{\mathcal{M}} > n\right)\right]$ is independent of t, it leads to:

$$\sup_{t\leq T} E[\mathcal{M}(t)I(\mathcal{M}(t) > n)] \leq E\left[YI\left(\widehat{\mathcal{M}} > n\right)\right] \tag{3.138}$$

As Y is integrable and $\widehat{\mathcal{M}}$ finite, $\sup_{t\leq T} E[\mathcal{M}(t)I(\mathcal{M}(t) > n)] \to 0$ as $n \to \infty$ so that $\mathcal{M}(t)$ is a UI martingale. Recall that the result is equally valid in the discrete case, i.e., $\{\mathcal{M}_n\}$ is a UI martingale, when $\mathcal{M}_n = E[Y|\mathcal{F}_n]$ where, $Y \in L^1(\Omega, \mathcal{F}, P)$.

The above result provides a means to construct UI martingales. In addition, a corollary to the above theorem is that any martingale on a finite interval $0 \leq t \leq T < \infty$ is UI since it is closed by $\mathcal{M}(T)$. Further, the property $\mathcal{M}(t) := E[Y|\mathcal{F}_t]$ also holds for UI martingales on an infinite interval since a random variable namely $\mathcal{M}_\infty$ exists a.s. and is in L^1 owing to the martingale convergence theorem.

3.6.2 Martingale convergence theorem

Consider a martingale $\mathcal{M}_t$. If any of the following conditions: (a) $\sup_t E[|\mathcal{M}_t|] < \infty$, or (b) $\sup_t E\left[(\mathcal{M}_t)^+\right] < \infty$, or (c) $\sup_t E\left[(\mathcal{M}_t)^-\right] < \infty$ is satisfied, then there exists an $\mathcal{F}_\infty$-measurable function (random variable) $\mathcal{M}_\infty \in L^1$ such that $\mathcal{M}_t \to \mathcal{M}_\infty$ a.s.

Proof: To prove the theorem, it is only necessary to extend Doob's upcrossing lemma to the continuous case and show that the number of upcrossings in an interval $[a,b]$ by the sample paths of the process is finite. To this end, let $U_T(a,b)$ be the number of upcrossings of $[a,b]$ with $a < b$ by $\mathcal{M}_t$ within a time interval $[0,T]$. We construct a discrete time process $\mathcal{M}_{t_{n,k}}$ from $\mathcal{M}_t$ by considering a sequence of time instants $t_{n,k} = n\Delta t$, $n = 0,1,...,2^k$ with $\Delta t = T2^{-k}$ and $k \in \mathbb{N}$. If $U_T^k(a,b)$ denotes the number of upcrossings of $[a,b]$ by the discrete martingale $\mathcal{M}_{t_{n,k}}$, we have, by Doob's upcrossing inequality (Eq. 3.75), $E\left[U_T^k(a,b)\right] \leq E\left[\left(\mathcal{M}_{t_{n,k}} - a\right)^-\right]/(b-a)$. Since $\mathcal{M}_t$ has continuous sample paths, and the interval $[0,T]$ is compact, the sample paths are uniformly continuous on $[0,T]$ and thus $U_T^k(a,b) \uparrow U_T(a,b)$. This leads to the assertion that

$U_T(a,b)$ is finite and $U_T(a,b) = U_T^k(a,b)$ for sufficiently large k. Now, the steps followed in Section 3.4.4 whilst proving the martingale convergence theorem in the discrete case, very well apply to the continuous case and the final result thus follows.

3.6.3 Optional stopping theorem

A random time τ is a stopping time if for any $t > 0$, the sets $\{t \le \tau\} \in \mathcal{F}_t$. A martingale stopped at τ is the stopped process $\mathcal{M}_{t\wedge\tau}$ which is again a martingale; this is analogous to what we have already seen for discrete processes. It follows that the optional stopping theorem and super-martingale inequality equally hold in the continuous case (including the theorems under UI conditions). The continuous counterparts of the basic stopping result $E[\mathcal{M}_{t\wedge\tau}] = E[\mathcal{M}_0]$ on stopped martingales and the optional stopping theorem are further discussed here in the context of finding the distributions of stopping times, especially for the Wiener process. An important stopping time is the hitting time or the first passage time defined by $\tau_a(\omega) = \inf\{t : X_t \; \mathcal{R} \; a\}$ where $\mathcal{R}$ may be one of the relations $<, \le, \ge, >$.

First passage times of a Wiener process

If B is a Wiener process, $\tau_a := \inf\{t : B_t = a\}$ is the first passage time—or the hitting time—of the level a for the first time. Clearly τ_a is a stopping time. It can be shown (Appendix C) that $\lim_{t\to\infty} \sup B_t = \infty$ and $\lim_{t\to\infty} \inf B_t = -\infty$ a.s. and thus the sample paths of a Wiener process are unbounded. Further, as B_t is continuous, τ_a is finite a.s for each $a \in \mathbb{R}$. Here a few useful properties of stopped Wiener processes are described via some selected examples.

Example 3.8

Let $a < 0 < b$ with $a, b \in \mathbb{R}$. If $\tau_a = \inf\{t : B_t = a\}$ and $\tau_b = \inf\{t : B_t = b\}$ are respective stopping times (first passage times) of a Wiener process B_t with $B_0 = 0$, then one can show that:

$$P(\tau_a < \tau_b) = \frac{b}{b-a}, \quad P(\tau_b < \tau_a) = \frac{-a}{b-a} \tag{3.139}$$

Solution

The sample paths of B_t are unbounded and hence they eventually leave the interval $[a,b]$. This implies that if $\tau = \min(\tau_a, \tau_b)$, then $P(\tau < \infty) = 1$ [Williams 1991]. τ can be interpreted as the time to exit the interval $[a,b]$. Consequently, we have:

$$P(\tau_a < \tau_b) + P(\tau_b < \tau_a) = 1 \tag{3.140}$$

Moreover, $B_{t\wedge\tau}$ is a bounded martingale and B_τ is either a or b according as $\tau = \tau_a$(i.e., $\tau_a < \tau_b$) or $\tau = \tau_b$(i.e., $\tau_b < \tau_a$). Now, by the optional stopping theorem:

$$0 = E[B_0] = E[B_\tau] = aP(\tau_a < \tau_b) + bP(\tau_b < \tau_a) \tag{3.141}$$

The last two equations yield $P(\tau_a < \tau_b) = \frac{b}{b-a}$ and $P(\tau_b < \tau_a) = \frac{-a}{b-a}$.

One can recognize that this example is a continuous counterpart of the gambler's ruin problem. Note that if B_t is not bounded below (i.e., allowing $a\downarrow -\infty$) or above (i.e., allowing $b\uparrow\infty$), the stopped martingale $\mathcal{M}_{n\wedge\tau}$ is not UI and $E[B_0] \neq E[B_\tau]$.

Example 3.9

In continuation of the above example, one can show that $E[\tau] = |ab|$

Solution

For this result, use the martingale $\hat{B} = B^2 - t$ and apply the optional stopping theorem to get:

$$0 = E\left[\hat{B}_0\right] = E\left[\hat{B}_{t\wedge\tau}\right] = E\left[B^2_{t\wedge\tau}\right] - E[t\wedge\tau] \Longrightarrow E\left[B^2_{t\wedge\tau}\right] = E[t\wedge\tau] \quad (3.142)$$

As $t\to\infty$, the left hand side $E\left[B^2_{t\wedge\tau}\right]$ of Eq. (3.142) is bounded in the limit by $E\left[B^2_\tau\right]$ and $E[t\wedge\tau]$ converges to $E[\tau]$ by monotone convergence. Also $E\left[B^2_\tau\right]$ is given by:

$$E\left[B^2_\tau\right] = a^2 P(\tau_a < \tau_b) + b^2 P(\tau_b < \tau_a) = -ab = |ab| \quad (3.143)$$

Hence we obtain $E[\tau] = |ab|$.

Example 3.10

We show that the first passage time τ_a has the *pdf*:

$$f_{\tau_a}(t) = \frac{|a|}{\sqrt{2\pi t^3}} e^{-a^2/2t}, t\geq 0 \quad (3.144)$$

Solution

Step 1. Let us first show that the Laplace transform of τ_a is:

$$\mathcal{L}_{\tau_a}(s) := E\left[e^{-s\tau_a}\right] = e^{-|a|\sqrt{2s}}, s\geq 0 \quad (3.145)$$

Case (i) Let $a > 0$. Given the exponential martingale $\hat{B} = e^{\lambda B(t) - \frac{\lambda^2}{2}t}$ and with $\lambda \geq 0$, the stopped martingale $\hat{B}_{t\wedge\tau_a}$ is bounded by (as $t\to\infty$):

$$\hat{B}_{t\wedge\tau_a} \to \hat{B}_{\tau_a} = e^{\lambda a - \frac{\lambda^2\tau_a}{2}} \leq e^{\lambda a} \quad (3.146)$$

Hence the bounded martingale $\hat{B}_{t\wedge\tau_a}$ is UI. Applying the optional stopping theorem, one has:

$$E\left[\hat{B}_{\tau_a}\right] = E\left[e^{\lambda a - \frac{\lambda^2 \tau_a}{2}}\right] = E\left[\hat{B}_0\right] = 1$$

$$\Longrightarrow E\left[e^{-\frac{\lambda^2}{2}\tau_a}\right] = e^{-\lambda a} \tag{3.147}$$

If $\lambda := \sqrt{2s}$, then $\mathcal{L}_{\tau_a}(s) = E\left[e^{-s\tau_a}\right] = e^{-a\sqrt{2s}}$.

Case (ii) If $a < 0$, we start with $-B$ which is also a Weiner process and get $\mathcal{L}_{\tau_a}(s) = E\left[e^{-s\tau_a}\right] = e^{a\sqrt{2s}}$.

Thus, for $a \in \mathbb{R}$, $\mathcal{L}_{\tau_a}(s) = e^{-|a|\sqrt{2s}}$.

Step 2. In view of the uniqueness of the transform, we finally show that the Laplace transform of the right hand side of Eq. (3.144) is indeed $e^{-|a|\sqrt{2s}}$.

The distribution function $F_{\tau_a}(t)$ is given by $F_{\tau_a}(t) = P(\tau_a \le t)$. Identifying by the definition of a stopping time that $P(\tau_a \le t) = P(B_t \ge |a|)$ and given that the *fdd* of B_t is normal with zero mean for all t, we have:

$$F_{\tau_a}(t) = P(\tau_a \le t) = 2P(B_t \ge a) = 2\int_a^\infty \frac{1}{\sqrt{2\pi t}}\, e^{-\frac{u^2}{2t}}\, du \tag{3.148}$$

By definition of Laplace transform of $f_{\tau_a}(t)$, one has:

$$\mathcal{L}_{\tau_a}(s) = \int_0^\infty e^{-st} f_{\tau_a}(t)\, dt = \int_0^\infty e^{-st} \frac{dF_{\tau_a}(t)}{dt}\, dt \tag{3.149}$$

Integrating by parts the integral on the extreme right hand side of Eq. (3.149) and substituting for $F_{\tau_a}(t)$ yield:

$$\mathcal{L}_{\tau_a}(s) = \mathcal{L}\left(f_{\tau_a}(t)\right) = F_{\tau_a}(t)\, e^{-st}\Big|_0^\infty + s\int_0^\infty e^{-st} F_{\tau_a}(t) dt = s\int_0^\infty e^{-st} F_{\tau_a}(t) dt$$

$$= 2s\int_0^\infty e^{-st} \int_a^\infty \frac{1}{\sqrt{2\pi t}}\, e^{-\frac{u^2}{2t}}\, du\, dt \tag{3.150}$$

With the parameter $s > 0$ fixed, let $g(a) = \mathcal{L}_{\tau_a}(s)$. The integrand is non-negative and, by Fubini's theorem, the order of integration can be changed to obtain:

$$g(a) = 2s\int_a^\infty \left(\int_a^\infty e^{-st} \frac{1}{\sqrt{2\pi t}}\, e^{-\frac{u^2}{2t}}\, dt\right) du \tag{3.151}$$

Since $\int_a^\infty e^{-st}\frac{1}{\sqrt{2\pi t}}\,dt = \frac{1}{\sqrt{2\pi t}}\Gamma\left(\frac{1}{2}\right) < \infty$ where $\Gamma(.)$ is the gamma function, $e^{-st}\frac{1}{\sqrt{2\pi t}}$ is integrable and dominates the integrand. Therefore the inner integral in the last equation yields a function continuous in u. Now, differentiating the right hand side of the last equation once with respect to a gives:

$$g'(a) = -2s\int_0^\infty e^{-st}\frac{1}{\sqrt{2\pi t}}\,e^{-\frac{a^2}{2t}}\,dt \tag{3.152}$$

With a similar argument on the integrability of the integrand in Eq. (3.152), one gets:

$$g''(a) = 2s\int_0^\infty e^{-st}\frac{a}{\sqrt{2\pi t^3}}\,e^{-\frac{a^2}{2t}}\,dt = 2s\int_0^\infty e^{-st}f_{\tau_a}(t)\,dt,\ \ a > 0$$

$$= 2s\,\mathcal{L}_{\tau_a}(s) = 2s\,g(a) \tag{3.153}$$

The two roots of the characteristic polynomial of the second order DE: $g''(a) - 2s\,g(a) = 0$ are $\lambda_{1,2} = \pm\sqrt{2s}$ and the general solution is given by $g(a) = Ae^{a\sqrt{2s}} + Be^{-a\sqrt{2s}}$. Equation (3.150) gives $g(0) = A + B = 1$ and $g(\infty) = 0$ and we get $g(a) = \mathcal{L}_{\tau_a}(s) = e^{-a\sqrt{2s}}$. If Step 2 is repeated with $-a = b > 0$ one obtains $g(b) = e^{-b\sqrt{2s}}$. Clearly, for any $a \in \mathbb{R}$, $\mathcal{L}_{\tau_a}(s) = \mathcal{L}\left(f_{\tau_a}(t)\right) = e^{-|a|\sqrt{2s}}$.

Combining the results of Steps 1 and 2 gives the required result in Eq. (3.144).

Strong Markov property of a Wiener process

The strong Markov property is similar to the Markov property (Eq. 3.42 of Section 3.3.6) and it differs only in that it is defined with respect to a stopping time instead of a fixed time. If τ is a stopping time, the strong Markov property of a Wiener process is defined for $t \geq 0$ as:

$$P\left(B(\tau + t) \leq a|\mathcal{F}_\tau\right) = P\left(B(\tau + t) \leq a|B_\tau\right)\ \text{a.s.} \tag{3.154}$$

It follows [Rogers and Williams 2000] that the process:

$$\overline{B}(t) = B(\tau + t) - B(\tau) \tag{3.155}$$

is a Wiener process starting at zero (standard Brownian motion) and is independent of $\mathcal{F}_\tau$.

Reflection principle

If B is a Wiener process, it is easy to see that $-B$ is also a Wiener process. The process $-B$ is a reflection of B about the time axis and is the simplest example of the reflection principle of a Wiener process. In general if τ is a stopping time and we define a process $\hat{B}$ (Fig. 3.6) as:

$$\hat{B}(t) = B(t), \ t \leq \tau$$
$$= 2B(\tau) - B(t), \ t \geq \tau \tag{3.156}$$

then, $\hat{B}$ can be shown to be a standard Brownian motion [Mörters and Peres 2010]. It can be justified as follows. The reflected path for $t \geq \tau$ is $\hat{B}(t) - B(\tau) = -(B(t) - B(\tau))$ which is also a Brownian motion (by the strong Markov property). Now $\hat{B}$, constructed

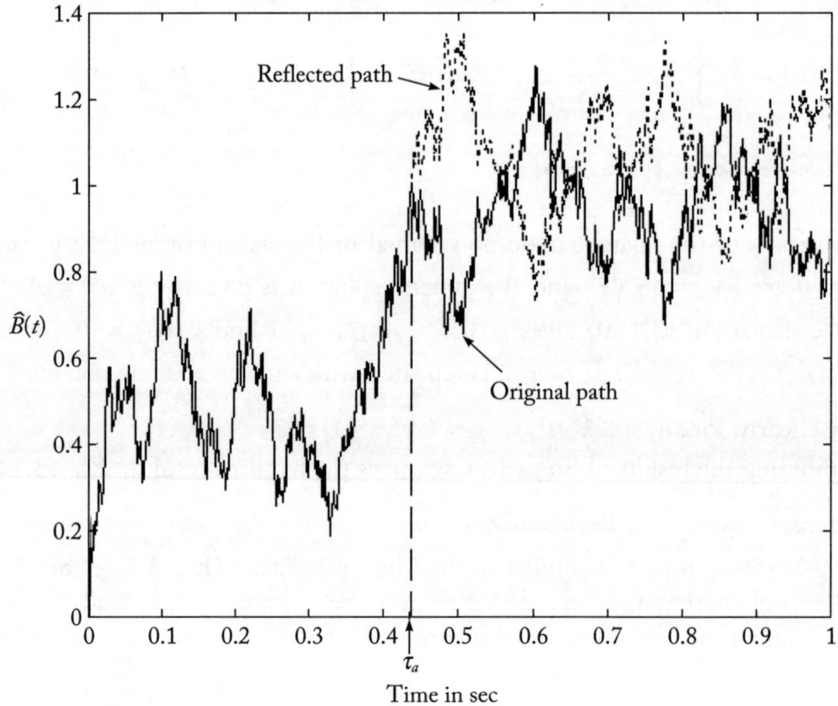

Fig. 3.6 Reflection principle; Brownian path reflected at $\tau_a = \inf\{t : B_t = 1\}$

from two Brownian motions - one the original B for $t \leq \tau$ and another Brownian motion after the stopping time is also a Brownian motion.

Example 3.11

If $M_t = \max_{0 \leq s \leq t} B_s$ be the maximum of the Brownian motion in the interval $[0, t]$, then one may show that for any $x > 0$, $P(M(t) \geq x) = 2P(B(t) \geq x) = 2(1 - \Phi(\frac{x}{\sqrt{t}}))$ where $\Phi(x) = \frac{1}{\sqrt{2\pi t}} \int_{-\infty}^{x} e^{-\frac{u^2}{t}} du$ stands for the standard normal distribution function.

Solution

To prove the above equalities, first consider $P(M(t) \geq x)$. By the definition of stopping time and from Eq. (3.144):

$$P(M(t) \geq x) = P(\tau_x \leq t) = \int_{-\infty}^{t} \frac{x}{\sqrt{2\pi u^3}} e^{-x^2/2u}\, du \tag{3.157}$$

Change of variable $u = \frac{x^2 t}{v^2}$ gives $du = -\frac{2x^2 t}{v^3} dv$ and Eq. (3.157) can be written as:

$$P(M(t) \geq x) = P(\tau_x \leq t) = 2\int_{x}^{\infty} \frac{1}{\sqrt{2\pi t}} e^{-v^2/2t}\, dv = 2\left(1 - \int_{-\infty}^{x} \frac{1}{\sqrt{2\pi t}} e^{-v^2/2t}\, dv\right)$$

$$= 2\left(1 - \Phi\left(\frac{x}{\sqrt{t}}\right)\right) = 2P(B(t) \geq x) \tag{3.158}$$

An alternative proof is as follows. With $\{\tau_x \leq t\} \supset \{B(t) \geq x\}$, we trivially have:

$$P(B(t) \geq x) = P(\tau_x \leq t \cap B(t) \geq x) \tag{3.159}$$

With $B(\tau_x) = x$ and for $t \geq \tau_x$, one has:

$$P(B(t) \geq x) = P(\tau_x \leq t \cap \{B(t) - B(\tau_x) \geq 0\})$$

$$= P(\{\tau_x \leq t\} \cap \{B(\tau_x + (t - \tau_x)) - B(\tau_x) \geq 0\})$$

$$= P\left(\{\tau_x \leq t\} \cap \overline{B}(t - \tau_x) \geq 0\right) \tag{3.160}$$

where, $\overline{B}(t - \tau_x) = B(\tau_x + (t - \tau_x)) - B(\tau_x)$. By the strong Markov property, $\overline{B}(t - \tau_x)$ is a standard Brownian motion which is independent of $\mathcal{F}_{\tau_x}$. Hence for $t \geq \tau_x$:

$$P\left(\overline{B}(t - \tau_x) \geq 0 | \mathcal{F}_{\tau_x}\right) = \frac{1}{2} \tag{3.161}$$

Equation (3.160) takes the form:

$$P(B(t) \geq x) = \frac{1}{2} E\left[I_{\{\tau_x \leq t\}}\right] = \frac{1}{2} P(\tau_x \leq t) = \frac{1}{2} P(M(t) \geq x) \tag{3.162}$$

which gives the result.

One may similarly obtain the distribution $P(m(t) \leq x)$ of the minimum $m(t) = \min_{0\leq s\leq t} B(s)$ of a Brownian motion using the identity $\min_{0\leq s\leq t} B(s) = \max_{0\leq s\leq t}(-B(s))$. Thus for $x < 0$:

$$P(m(t) \leq x) = P\left(\max_{0\leq s\leq t}(-B(s)) \geq -x\right)$$

$$= 2P(B(t) \geq -x) = 2P(B(t) \leq x) \tag{3.163}$$

As a cross-check, note from Eq. (3.158) that $P(\tau_x < \infty) = 2(1 - \Phi(0)) = 1$. Moreover, for a standard Brownian motion with any fixed $x \neq 0$, $E[\tau_x] = \infty$. This is obtained by integrating the tail $P(\tau_x > t)$ as:

$$E[\tau_x] = \int_0^\infty P(\tau_x > t)\, dt$$

$$= \int_0^\infty (1 - P(\tau_x \leq t))\, dt$$

$$= \int_0^\infty 1 - 2P(B(t) \geq x) dt$$

$$= \int_0^\infty \left(1 - \frac{2}{\sqrt{2\pi}} \int_{\frac{x}{\sqrt{t}}}^\infty e^{-u^2/2}\, du\right) dt$$

$$= \frac{2}{\sqrt{2\pi}} \int_0^\infty \left(\int_0^{\frac{x}{\sqrt{t}}} e^{-u^2/2}\, du\right) dt \tag{3.164}$$

By changing the order of integration on the right hand side integral of Eq. (3.164), we obtain:

$$E[\tau_x] = \frac{2}{\sqrt{2\pi}} \int_0^\infty \int_0^{\frac{x^2}{u^2}} e^{-u^2/2}\, dt\, du$$

$$= \frac{2}{\sqrt{2\pi}} x^2 \int_0^\infty \frac{1}{u^2} e^{-u^2/2}\, du$$

$$\geq \frac{2}{\sqrt{2\pi}} x^2 \int_0^1 \frac{1}{u^2} e^{-u^2/2}\, du$$

$$\geq \frac{2}{\sqrt{2\pi}} x^2 e^{-1/2} \int_0^1 \frac{1}{u^2} du = \infty \tag{3.165}$$

Example 3.12

Using the reflection principle, we find the joint probability density of (B_t, M_t).

Solution

Reflect B at $\tau_b = \inf\{t : B_t = b\}$ to get $\hat{B} = BI_{t \leq \tau} + (2b - B) I_{t \geq \tau}$ for $a \leq b,\ b > 0$. Then we have:

$$\begin{aligned}
P(B_t \leq a, M_t \geq b) &= P(B_t \leq a, \tau_b \leq t) && (\text{since } \{M_t \geq b\} = \{\tau_b \leq t\}) \\
&= P\left(2b - \hat{B}_t \leq a, \tau_b \leq t\right) && (\text{since for } \tau_b \leq t, \hat{B}_t = 2b - B_t) \\
&= P\left(\hat{B}_t \geq 2b - a, \tau_b \leq t\right) \\
&= P(B_t \geq 2b - a, \tau_b \leq t) && (\text{since } \tau_b \text{ is the same for } B \text{ and } \hat{B} \text{ due to the reflection principle}) \\
&= P(B_t \geq 2b - a) && (\text{since } \{\tau_b \leq t\} \supset \{B_t \geq 2b - a\}) \\
&= 1 - \Phi\left(\frac{2b - a}{\sqrt{t}}\right)
\end{aligned} \tag{3.166}$$

Thus:

$$P(B_t \leq a, M_t \geq b) = \int_{-\infty}^{a} \int_{b}^{\infty} f_{B,M}(u, v) du dv = 1 - \Phi\left(\frac{2b - a}{\sqrt{t}}\right) \tag{3.167}$$

The joint density $f_{B,M}(a, b) = \sqrt{\frac{2}{\pi}} \frac{(2b-a)}{t^{3/2}} e^{-\frac{(2b-a)^2}{t}}$ is obtained by differentiation.

Recurrence and transience of Brownian motion

Consider a one-dimensional ($d = 1$) Brownian motion B starting at zero with a stopping time $\tau_a = \inf\{t : t \geq 0,\ B_t = a\}$. τ_a is finite with probability 1, i.e., $P(\tau_a < \infty) = 1$. This is the recurrence property of a Brownian motion. Since $\lim_{t\to\infty} \sup B_t = \infty$ and $\lim_{t\to\infty} \inf B_t = -\infty$ a.s., it must be that $B(t)$ crosses zero i.o. The zero level and indeed by space–homogeneity, any level is recurrent for 1-dimensional $B(t)$.

Specifically, for $d = 1$, the Brownian motion is point recurrent, (i.e., $\{t \geq 0 :\ B(t) = x\}$ is unbounded a.s. for every $x \in \mathbb{R}$). The recurrence property is a class property and differs for $B(t)$ of higher dimensions. For $d = 2$, it is neighborhood recurrent, i.e., $\{t \geq 0 : |B(t)| \leq \varepsilon\}$ is unbounded a.s. for every $x \in \mathbb{R}^d$ and $\varepsilon > 0$. For $d \geq 3$, it is transient, i.e., a.s. $|B(t)| \to \infty$ as $t \to \infty$.

3.7 Localization and Local Martingales

Stopping times are useful in localizing certain properties of stochastic processes. A stochastic process X is said to locally satisfy some property if there exists a sequence of stopping times τ_n called a localizing sequence such that $\tau_n \uparrow \infty$ as $n \uparrow \infty$ and each stopped process $X_{t\wedge\tau_n}$ satisfies this property. For example, a non-negative and increasing process X is locally integrable if $E\left[X_{t\wedge\tau_n}\right] < \infty, \forall n$.

Another useful localization is with regard to the localizing the martingale property.

3.7.1 Definition of a local martingale

An adapted process $X(t)$ is a local martingale if there exists a localizing sequence τ_n such that $\tau_n \uparrow \infty$ as $n \uparrow \infty$ and for each n the stopped process $X_{t\wedge\tau_n}$ is a UI martingale in t.

Any martingale $\mathcal{M}$ is a local martingale. Also, the uniform integrability property holds locally for any martingale since a martingale convergent in L^1 is UI. This is clear with the choice of a localizing sequence $\tau_n = n$ leading to $\mathcal{M}_{n\wedge t} = \mathcal{M}_n$ for $t > n$. On the other hand, local martingales may not be martingales. However, if a local martingale is dominated by an integrable random variable, it is a UI martingale. This is proved in the following.

Proof: Suppose that $\mathcal{M}(t),\ 0 \leq t < \infty$ is a local martingale such that $|\mathcal{M}(t)| < Y$ where Y is a random variable with $E[Y] < \infty$. Let τ_n be a localizing sequence. We need to show that $\mathcal{M}$ is a UI martingale.

Step 1. $\mathcal{M}$ is integrable, since $E\left[|\mathcal{M}(t)|\right] \leq E[Y] < \infty$. As τ_n is a localizing sequence, $\mathcal{M}_{t\wedge\tau_n}$ is a UI martingale in t. Thus, for $s < t$, we have by martingale property of $\mathcal{M}_{t\wedge\tau_n}$:

$$E\left[\mathcal{M}_{t\wedge\tau_n}|\mathcal{F}_s\right] = \mathcal{M}_{s\wedge\tau_n}, \text{ for any } n \tag{3.168}$$

As $\lim_{n\to\infty}\mathcal{M}_{t\wedge\tau_n} \to \mathcal{M}_t$, it is clear by dominated convergence of conditional expectation that $\lim_{n\to\infty}E\left[\mathcal{M}_{t\wedge\tau_n}|\mathcal{F}_s\right] \to E\left[\mathcal{M}_t|\mathcal{F}_s\right]$. Further, $\lim_{n\to\infty}\mathcal{M}_{s\wedge\tau_n} \to \mathcal{M}_s$ so that by taking limits on both sides of Eq. (3.168), it follows that $\mathcal{M}(t)$ is a martingale.

Step 2. Given that $\mathcal{M}(t)$ is dominated by the random variable Y, the uniform integrability of $\mathcal{M}(t)$ also is established (Section 3.6.1 - Doob's and Levy's martingale theorem).

Note that a local martingale may fail to be a martingale even though it is integrable. This aspect assumes importance when we discuss stochastic filtering methods in Chapter 6. The filtering theory extensively utilizes the notion of change of measures, wherein the Radon–Nikodym derivative, which is a stochastic exponential and hence strictly positive, is only a local martingale. As we shall see there, it is not a martingale, rather a super-martingale. Indeed, such local martingales will be extensively made use of even in Chapters 7–10. One may show that a local martingale $\mathcal{M}$ is a UI martingale if and only if it is of the Dirichlet class, which is defined as a family $\{X(\tau)\}$ of UI processes with τ being a stopping time; see Klebaner [2005] for a proof.

3.8 Concluding Remarks

The chapter has dealt with the basic theory of stochastic processes, interpreted via Kolmogorov's extension theorem as an indexed collection of random variables with finite dimensional distributions. As an extension of the σ-algebra for a random variable, we have discussed the notion of filtration for a stochastic process. Of specific interest has been a particular stochastic process, the Brownian motion that is of fundamental significance in modeling both input noises to stochastic dynamical systems and their solutions. Brownian motion is a sub-class of a more general class of stochastic processes referred to as martingales. The theory of martingales is central to a modern treatment of stochastic processes. The inequality and convergence theorems on martingales, like those of random variables (covered in Chapter 2), provide an alternative elegant approach towards understanding the laws of large numbers, conditional expectations and even the CLT. Doob's decomposition and martingale representation and convergence theorems relevant in this context have been stated with proofs. The notion of martingales, as a special class of Markov processes, is of great interest in myriad applications, some of which are covered later on in this book. Of similar or related interest are the notions of stopping time, stopped processes, localization and local martingales and these have also been briefly covered in this chapter.

The governing equations of motion in stochastic dynamics being stochastic differential equations (SDEs), the next chapter attempts an exposition of stochastic calculus (e.g., Ito's calculus) that formalizes the notion of SDEs whose solutions are the stochastic processes representing evolutions of functions of Brownian motion.

Exercises

1. Show that if the stochastic processes $X(t)$ and $Y(t)$, with index set R^d, are continuous and versions of one another, then they are indistinguishable.

2. Let $X(t)$ be a weakly stationary process with covariance function, $R_{XX}(u) : \mathbb{R} \to \mathbb{R}$. Show that mean-square continuity of the process is equivalent to the covariance function, $R_{XX}(u)$, being continuous at 0.

3. For a Poisson process $X(t)$, show that the autocorrelation function is given by:

$$R_X(t_1, t_2) = \lambda t_1 (1 + \lambda t_2), \quad t_1 \leq t_2$$

$$= \lambda t_2 (1 + \lambda t_1), \quad t_2 \leq t_1$$

4. Let B be a Brownian motion and fix $t > 0$. Define the process $\hat{B}$ by:

$$\hat{B}_s = B_{s\wedge t} - (B_s - B_{s\wedge t}) = B_s, \ \ s \le t$$

$$= 2B_t - B_s, \ \ s > t$$

Show that $\hat{B}$ is again a BM.

5. If X is a martingale and g is convex function such that $g(X_n) \in L^1$, show that $Y = g(X)$ is a sub-martingale. (As an example, if X is a martingale, then $|X|$ is a sub-martingale.) Note that if g is concave, $g(X)$ is a super-martingale.

6. Let $\{X_n\}$ be a martingale. If $E\left[|X_n|^p\right] < \infty$, then prove that $|X_n|^p$, $p = 1,2,3,\ldots$ is a sub-martingale. (Hint: Use Jensen's inequality)

7. Let $\hat{B}_t = \alpha t + B_t$ where B_t is standard Brownian motion and α, a real constant. The process $\hat{B}_t$ is a Brownian motion with a drift. With a fixed $\lambda > 0$, $M_t := e^{\theta \hat{B}_t - \lambda t}$ is a martingale if and only if $\theta = \pm\sqrt{\alpha^2 + 2\lambda} - \alpha$.

8. Let B_t be a standard Brownian motion. Show that $Y_t = \exp\left(\sigma B_t - \frac{\sigma^2}{2} t\right)$, $t \ge 0$ is a martingale.

9. Let $Y_1, Y_2, \ldots$ be a sequence of independent non-negative random variables satisfying $E[Y_k] = 1$ for all $k \in \mathbb{N}$. Define $X_0 = 1$, $X_n = \prod_{j=1}^{n} Y_j$ and $\mathcal{A}_n = \sigma(Y_1, Y_2, \ldots, Y_n)$. Show that X_n is a martingale with respect to $\mathcal{A}_n$.

10. Let $\{X_n, n \ge 0\}$ be a given sequence of integrable and independent zero-mean random variables with corresponding σ-algebras $\{\mathcal{F}_n, n \ge 0\}$. Show that the sequence of partial sums $S_n = X_1 + X_2 + \cdots + X_n$ with $S_0 = 0$ is a martingale.

11. Let $\{X_n, n \ge 1\}$ be i.i.d. random variables such that $P(X_n = 1) = p$ and $P(X_n = -1) = 1 - p$ with $0 < p < 1$. Then $S_n = \left(\frac{1-p}{p}\right)^{X_1 + X_2 + \cdots + X_n}$ with $S_0 = 1$ is a martingale with respect to $\mathcal{F}_n = \sigma\{X_1, X_2, \ldots, X_n\}$.

12. Let $\{X_n\}$ be a martingale. Prove that $i)$ if $\{X_n\}$ is predictable, then it is constant in times. i.e., $X_n = X_0$ a.s. for all n; ii) if X_n is independent of $\mathcal{F}_{n-1}$, then X_n is constant, i.e., $X_n = E[X_0]$ for all n.

13. Suppose that $X_1, X_2, \ldots$ be i.i.d. random variables, each uniformly distributed in $[-1, 1]$. Let $S_n = \sum_{i=1}^{n} X_i$. If Y is a martingale with respect to $\mathcal{F}_n = \sigma(S_1, S_2, \ldots, S_n)$, show that there exists a predictable process A such that $Y = Y_0 + A.S$.

14. Suppose that X be a Poisson process with rate λ. Then $X_t - \lambda t$ is a martingale.

15. Let B_t a standard Brownian motion and define $X_t := c + \mu t + \sigma B_t$ with c, μ, $\sigma \in \mathbb{R}$. Prove that X_t is a martingale if and only if $\mu = 0$. Show that if $\mu \ne 0$, then ${X_t}^2$ is neither a submartingale nor a supermartingale.

16. Let M_n be a positive martingale and A_n a positive decreasing sequence such that A_n is $\mathcal{F}_{n-1}$ measurable, for any $n = 1,2,\ldots$ Then sequence $S_n := M_n + A_n$ is a supermartingale.

17. If σ and τ are stopping times with respect to the filtration $\mathcal{F}_t$, show that $\sigma \wedge \tau$ and $\sigma \vee \tau$ are also stopping times and determine the associated σ-algebras.

18. If $\{\tau_n\}$ is an increasing or decreasing sequence of stopping times then $\tau := \lim_{n\to\infty} \tau_n$ is a stopping time.

19. Show that if σ and τ are stopping times with respect to the filtration F_t then $\sigma + \tau$ is a stopping time as well.

20. Consider the Markov chain with state space $\Omega \subset \mathbb{Z}$, and the transition matrix given by $P_{ij} = p_i$, if $j = i+1$ with $p_1 = 1$; $P_{ij} = q_i$ if $j = i-1$ with $q_0 = 0$; $P_{ij} = 0$ otherwise. Take $i,j = 1,2,\ldots.$ p_i and q_i are positive real numbers satisfying $p_i + q_i = 1$ for all i (see Appendix B for introductory notes on Markov chains). Thus denote

$$\mathbf{P} = \left[P_{ij}\right] = \begin{bmatrix} 0 & 1 & 0 & 0 & \ldots \\ q_1 & 0 & p_1 & 0 & \ldots \\ 0 & q_2 & 0 & p_2 & \ldots \\ & & \cdot & & \\ & & \cdot & & \\ & & \cdot & & \end{bmatrix},$$

If the Markov chain has a stationary distribution $\pi = \left(\ldots, \pi_{j-1}, \pi_j, \pi_{j+1}, \ldots\right)$, show that:

$$\pi_j = p_{j-1}\pi_{j-1} + q_{j+1}\pi_{j+1}.$$

Assume that the Markov chain is initially at origin at time zero. (Hint: Show that $\sum_{n=1}^{\infty} \frac{p_1 p_2 \cdots p_{n-1}}{q_1 q_2 \cdots q_{n-1} q_n}$ converges so that the Markov chain has a stationary distribution.

21. Let X be an adapted process and $\tau_1, \tau_2, \ldots,$ τ_k stopping times such that $X_{t\wedge\tau_1}$, $X_{t\wedge\tau_2}, \ldots, X_{t\wedge\tau_k}$ are stopped martingales. Show that $X_{t\wedge(\tau_1\vee\tau_2\vee\cdots\vee\tau_k)}$ is a martingale. (Hint: for $k = 2$, write $X_{t\wedge(\tau_1\vee\tau_2)}$ in terms of X_{τ_1}, X_{τ_2} and $X_{\tau_1\wedge\tau_2}$)

22. Show that the exponential martingale $X(t) = \exp(B(t) - t/2)$ of a Wiener process is not uniformly integrable.

23. Consider a sequence of i.i.d. random variables ξ_i, $i = 1,2,\ldots$ with $P(\xi_i = 1) = P(\xi_i = -1) = \frac{1}{2}$. Let $S_0 = 0$ and for $n \geq 1$, $S_n = \sum_{i=1}^{n} \xi_i$. Then S_n is not UI.

Notations

$\{A_n\}$	$\mathcal{F}_n$-predictable process (or previsible process)
$(A.\mathcal{M})_n$	martingale transform, $\sum_{j=1}^{n} A_j(\mathcal{M}_j - \mathcal{M}_{j-1})$
B_n	Random walk
$\{\mathcal{G}_{-n}\}$	decreasing sequence of σ-algebras (Eq. 3.94) with increasing n
$\mathcal{M}_{n\Lambda\tau}$	stopped process, stopped martingale
S_n	partial sum of random variables (Eq. 3.95)
$TV(g;a,b)$	total variation of a real function $g(t)$ over a closed interval $[a,b]$
$U_n(a,b)$	number of up-crossings of the interval $[a,b]$, with $a < b$ by a stochastic process
$V_{t_i}(\omega), i = 1,2$ and 3	typical stochastic signals (Fig. 3.1)
$W(t)$	white noise process
$X(\omega,t), X_t(\omega)$	stochastic process. a parametrized family of random variables, $t \geq 0, t \in T, \omega \in \Omega$
X_n^{τ}	stopped process, $X_{n\Lambda\tau}(\omega)$
$\gamma_{t_1,t_2,\ldots,t_k}$	probability measure on $\mathbb{R}^{nk}$
$\mu_X(t_i)$	first order statistics (mean value) of the stochastic process, X
τ	stopping time
τ_a	first passage time – or the hitting time – of the level a for the first time $=$ inf $\{t : X_t(t) = a\}$

Chapter 4

Stochastic Calculus and Diffusion Processes

4.1 Introduction

Basic concepts on stochastic processes, supported by the underlying probability theory, have been attended to in the last chapter. The current focus is on stochastic differential equations (SDEs), whose integral representations typically contain integrals where the integrand and integrator may be stochastic processes. The interest is, more specifically, in evolving an understanding of the so called stochastic integrals (as contrasted with integrals of the Riemann–Stieltjes type), wherein the integrators involve stochastic processes (e.g., the Wiener process) that do not have finite variations. The topic is central to solving SDEs that arise in varied applications of stochastic dynamics—the main theme of this book. The exposition leads in the sequel to an understanding of stochastic calculus [Karatzas and Shreve 1991, Björk 2001, Klebaner 2005] that, as one finds, fundamentally differs from the Newtonian calculus dealing with deterministic functions, or more generally, those with (piecewise) continuous paths of finite variations. For example, if $\mathfrak{I} := \int_0^T X(s)\,dB(s)$ where X and B are stochastic processes, and the latter in particular is a Wiener process, then we find that $\mathfrak{I}$ is not the classical integral in Stieltjes sense. The reason is that a Wiener process is not differentiable and its total variation not finite (see Section 3.3.6, Chapter 3). Thus, the integral involves paths of unbounded variation so that a path-wise definition of the integral in the Stieltjes sense cannot be given.

The reader may be familiar with integrals such as:

$$I(a,b) = \int_a^b X(t)\,dt \quad \text{(Riemann type)} \tag{4.1a}$$

$$= \int_a^b X(t)\,dY(t) \quad \text{(Riemann – Stieltjes type)} \tag{4.1b}$$

These integrals can be defined as limits of approximating summations (e.g., see Eq. 4.2 below), provided $X(t)$ is continuous (or discontinuous with a finite number of jumps) and Y a function of finite variation. With any partition Π_N of the interval $[a,b]$ given by $a = t_0 < t_1 < \cdots < t_N = b$ and with $\Delta_N = \max_{0 \le i \le N-1} (t_{i+1} - t_i)$, the Riemann integral is approximated by the sequence:

$$I_N = \sum_{i=0}^{N-1} X(t_i)(t_{i+1} - t_i) \tag{4.2}$$

and $I_N \to I$ as X is sampled with an increasingly fine partition, i.e., as $\Delta_N \to 0$. In the case of the Riemann–Stieltjes integral (or simply the Stieltjes integral), if Y is differentiable, one can have $\int_a^b X(t)\, dY(t) = \int_a^b X(t)\, \dot{Y}(t)\, dt$ and represent the integral by the sum:

$$I(a,b) = \lim_{N\to\infty} \sum_{i=0}^{N-1} X(t_i)\, \dot{Y}(t_i)(t_{i+1} - t_i) \tag{4.3}$$

If $\dot{Y}$ does not exist and yet increments in Y are finite, the above representation may be modified as:

$$I(a,b) = \lim_{N\to\infty} \sum_{i=0}^{N-1} X(t_i)(Y(t_{i+1}) - Y(t_i)) \tag{4.4}$$

If $X(t)$ is a stochastic process, approximating the integral in Eq. (4.1a) through the sequence $\{I_N\}$ in Eq. (4.2) may not often pose problems. For instance, this is so if a continuous version of $X(t)$ exists. If, for the integral in Eq. (4.1b), both the integrand and the integrator are stochastic processes, it is logical to proceed with the representation in Eq. (4.3) provided $\dot{Y}$ exists. This is clearly not the case if $Y = B$, a Wiener process, which is not differentiable. It seems plausible to proceed in this case with the type of approximating sum in Eq. (4.4) that involves increments of a Wiener process, viz.:

$$\mathfrak{I} = \lim_{N\to\infty} \sum_{i=0}^{N-1} X(t_i)(B(t_{i+1}) - B(t_i)) \tag{4.5}$$

Here, $0 = t_0 < t_1 < \cdots < t_N = T$. Nevertheless, one may still have misgivings on the above supposition till it is proved that the limit on the right hand side exists in some sense. More pertinently, the issue of much needed invariance of the limit with respect to the partition Π_N of the interval $[0,T]$ and the question of a possible extension of the integral to an infinite one must be resolved. In practical applications, dynamics of physical systems is generally described in terms of DEs. Subjected to adequate smoothness of the functions involved, for a (scalar) function $X(t)$ in the DE $\dot{Z}(t) = X(t)\dot{Y}(t)$, the fundamental theorem of calculus [Rudin 1987] assures the following identity:

$$Z(t) = Z(0) + \int_0^t X(s)\dot{Y}(s)ds \tag{4.6}$$

Imagine if $Y(t)$ were not differentiable (as is the case with a Wiener process), both the DE and the identity in Eq. (4.6) fail to hold in the above forms. A more appropriate representation for the integral in the last equation could be $\int_0^T X(s)dY(s)$. However, in such a case, $Y(t)$ may not be a process of finite variation and thus a proper interpretation of the existence of such an integral must be laid out.

4.2 Stochastic Integral

Specifically, our goal is to properly define the stochastic integral, $\mathfrak{I}(X) = \int_0^T X(s)dB(s)$. To this end, first consider $X(t)$ a non-random (deterministic) process given by a simple function, i.e., a piece-wise constant function with jumps at a finite number of time instants t_i (Fig. 4.1) and also independent of $B(t)$. Thus, with a partition Π_N of the interval $[0,T]$ and constants $c_j, j = 1,\ldots,N-1$, we have:

$$X(t) = c_0 I_0(t) + \sum_{j=0}^{N-1} c_j I_{\left(t_j, t_{j+1}\right]}(t) \tag{4.7}$$

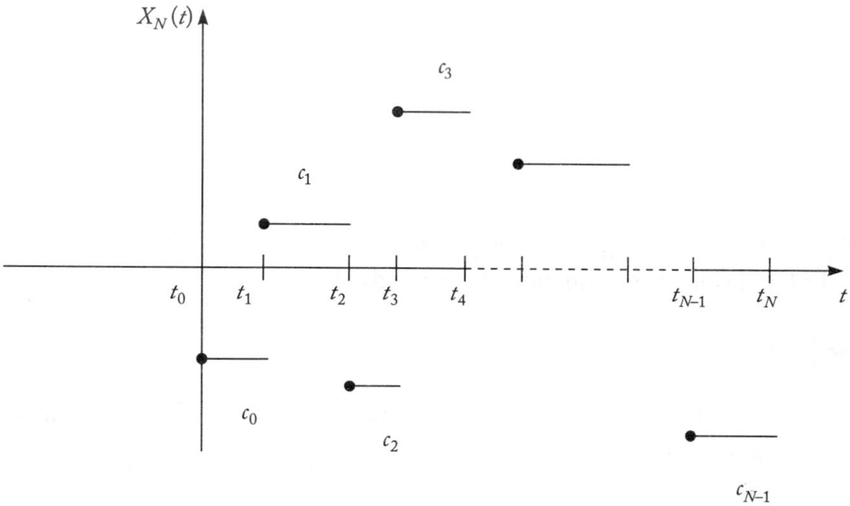

Fig. 4.1 $X(t)$ as a simple function as in Eq. (4.7)

Note that the function is left continuous with $c_j = X(t_j)$ and that $I_0(t) = 1$ for $t = 0$ and zero for $t \neq 0$. With $B_{t_j} = B(t_j)$, the integral $\mathfrak{T}$ is now defined as a sum in the form:

$$\mathfrak{T}(X) = \sum_{j=0}^{N-1} c_j \left(B_{t_{j+1}} - B_{t_j}\right) \tag{4.8}$$

By the independence of increments $B_{t_{j+1}} - B_{t_j}$ of the Wiener process, the integral $\mathfrak{T}$ is a Gaussian random variable with zero mean and variance given by:

$$var(\mathfrak{T}) = E\left[\left(\int_0^T X(s)\,dB(s)\right)^2\right] = var \sum_{j=0}^{N-1} c_j \left(B_{t_{j+1}} - B_{t_j}\right)$$

$$= \sum_{j=0}^{N-1} c_j^2 var\left(B_{t_{j+1}} - B_{t_j}\right) = \sum_{j=0}^{N-1} c_j^2 \left(t_{j+1} - t_j\right) \tag{4.9}$$

Clearly, the stochastic integral can be defined for more general (deterministic) functions by taking limits of simple functions as in Eq. (4.7). The identity $E[(\int_0^T X(s)dB(s))^2] = \sum_{j=0}^{N-1} c_j^2(t_{j+1} - t_j)$ in Eq. (4.9) signifies one of the important properties—the isometry of the stochastic integral.

4.2.1 Stochastic integral of a discrete stochastic process

With the integrand $X(t)$ taken to be a time–discrete stochastic process, we replace the constants $c_j, j = 1, N-1$ in Eq. (4.7) by random variables, $X_{t_j} = X(t_j)$. In terms of the simple functions as in this equation, $X(t)$ is expressed as:

$$X(t) = X_0 I_0 + \sum_{j=0}^{N-1} X_{t_j} I_{(t_j, t_{j+1}]}(t) \tag{4.10}$$

where, $X_{t_0} := X_0$. In defining $X(t)$, the particular choice of $X_{t_j} = X(t_j)$ in the summation on the right hand side of the equation above needs elaboration.

Example 4.1

Consider the following example of constructing the stochastic integral with another process $\overline{X}$ expressed as

$$\overline{X}(t) = \overline{X}_0 I_0 + \sum_{j=0}^{N-1} \overline{X}_{t_{j+1}} I_{(t_j, t_{j+1}]}(t) \tag{4.11}$$

where, $\overline{X}_{t_{j+1}} = \overline{X}\left(t_{j+1}\right)$ is now taken at the end of interval $(t_j, t_{j+1}]$.

Solution

With $\Delta B_j = B_{t_{j+1}} - B_{t_j}$, we have:

$$\mathfrak{T} = \sum_{j=0}^{N-1} X_{t_j} \Delta B_j \tag{4.12a}$$

$$\overline{\mathfrak{T}} = \sum_{j=0}^{N-1} \overline{X}_{t_{j+1}} \Delta B_j \tag{4.12b}$$

Since X_{t_j} is $\mathcal{F}_{t_j}$-measurable and hence independent of ΔB_j, it follows that:

$$E\left[\mathfrak{T})\right] = \sum_{j=0}^{N-1} E\left[X_{t_j} \Delta B_j\right] = 0 \tag{4.13a}$$

$$var\ (\mathfrak{T}) = E\left[\left(\int_0^T X(s)\,dB(s)\right)^2\right] = \sum_{j=0}^{N-1} E\left[X_{t_j}^2\right]\left(t_{j+1} - t_j\right) \tag{4.13b}$$

provided $E[X_{t_j}^2] < \infty$. However, for $\overline{\mathfrak{T}}$ in Eq. (4.12b) $\overline{X}_{t_{j+1}}$ is not independent of ΔB_j and the moment information for the integral cannot be so easily evaluated. Therefore, the two integrals should be different from each other. Parallel to Eq. (4.9), Eq. (4.13b) is referred to as the isometry of the stochastic integral $\mathfrak{T}$.

Example 4.2

Another example that is perhaps more demonstrative of the above result may be constructed with two discrete versions of the continuous Wiener process B representing the integrand of $\int_0^T B(s)dB(s)$

Solution

With X_{t_j} and $\overline{X}_{t_{j+1}}$ in Eq. (4.12) replaced by B_{t_j} and $B_{t_{j+1}}$ respectively, we have the following two representations:

$$\mathfrak{T} = \sum_{j=0}^{N-1} B_{t_j} \Delta B_j \tag{4.14a}$$

$$\overline{\mathfrak{T}} = \sum_{j=0}^{N-1} B_{t_{j+1}} \Delta B_j \tag{4.14b}$$

The first moments of the two summands are given by:

$$E[\mathfrak{T}] = \sum_{j=0}^{N-1} E\left[B_{t_j} \Delta B_j\right] = 0 \tag{4.15a}$$

$$\begin{aligned} E\left[\overline{\mathfrak{T}}\right] &= \sum_{j=0}^{N-1} E\left[B_{t_{j+1}} \Delta B_j\right] \\ &= \sum_{j=0}^{N-1} E\left[\left(B_{t_{j+1}} - B_{t_j}\right)^2 + B_{t_j} \Delta B_j\right] \\ &= \sum_{j=0}^{N-1} \left(E\left[\left(\Delta B_j\right)^2\right] + E\left[B_{t_j} \Delta B_j\right]\right) \end{aligned} \tag{4.15b}$$

Since B_{t_j} is independent of ΔB_j and $E\left[\Delta B_j\right] = 0$, $E\left[B_{t_j} \Delta B_j\right] = 0$ leading to:

$$E\left[\overline{\mathfrak{T}}\right] = \sum_{j=0}^{N-1} E\left[\left(\Delta B_j\right)^2\right] = T \tag{4.16}$$

As already noted, the two stochastic integrals are different.

In principle, one can create an infinite sequence $\{X'\}$ of such stochastic processes $X'(t) = X'_0 I_0 + \sum_{j=0}^{N-1} X'_{t'} I_{(t_j, t_{j+1}]}(t)$ corresponding to the choice of $t' \in \left[t_j, t_{j+1}\right] \forall j$ and thus define sequence of integrals $\mathfrak{T}' = \sum_{j=0}^{N-1} X_{t'} \left(B_{t_{j+1}} - B_{t_j}\right)$. But, the above examples show that, unlike in a Riemann–Stieltjes integral, the choice of t' matters in defining a stochastic integral. Specifically, the integral $\mathfrak{T}$ for $t' = t_j$ is called the Ito integral [Ito 1951]. Note that the associated stochastic process $X(t)$ is thus rendered adapted (Section 3.3.1, Chapter 3) with respect to the filtration $\mathcal{F}_t$ generated by the Wiener process B_t. Thus, for the simple adapted process, the Ito integral is given by:

$$\mathfrak{T}(X) = \sum_{j=0}^{N-1} X_{t_j} \left(B_{t_{j+1}} - B_{t_j}\right) \tag{4.17}$$

Since X_{t_j} are random, the integral need not be Gaussian.

Note that if $t' = \left(t_j + t_{j+1}\right)/2$ (the mid point in $\left[t_j, t_{j+1}\right]$), it leads to another integral representation known as Stratonovich integral [Stratonovich, 1966].

4.2.2 Properties of Ito integral of simple adapted processes

Some significant properties of the Ito integral are:

(a) Linearity

Given $c, d \in \mathbb{R}$ and if X and Y are simple adapted processes, then:

$$\int_0^T (cX(s) + dY(s))\, dB(s) = c\int_0^T X(s)\, dB(s) + d\int_0^T Y(s)\, dB(s) \tag{4.18}$$

Since a linear combination of simple processes is again simple, the linearity property of an Ito integral immediately follows.

(b) Zero–mean property

By the Cauchy–Schwartz inequality, one has:

$$E\left[\left|X_{t_j}\left(B_{t_{j+1}} - B_{t_j}\right)\right|\right] \le \sqrt{E\left[X_{t_j}^2\right] E\left[\left(B_{t_{j+1}} - B_{t_j}\right)^2\right]} < \infty$$

$$\Longrightarrow E\left[\left|\sum_{j=0}^{N-1} X_{t_j}\left(B_{t_{j+1}} - B_{t_j}\right)\right|\right] \le \sum_{j=0}^{N-1} E\left[\left|X_{t_j}\left(B_{t_{j+1}} - B_{t_j}\right)\right|\right] < \infty \tag{4.19}$$

The above inequality shows that the Ito integral possesses a mean. Now, since a Wiener process is a martingale and X_{t_j} is $\mathcal{F}_{t_j}$-measurable, it leads to:

$$E\left[X_{t_j}\left(B_{t_{j+1}} - B_{t_j}\right)|\mathcal{F}_{t_j}\right] = X_{t_j} E\left[\left(B_{t_{j+1}} - B_{t_j}\right)|\mathcal{F}_{t_j}\right] = 0$$

$$\Longrightarrow E\left[X_{t_j}\left(B_{t_{j+1}} - B_{t_j}\right)\right] = 0 \Longrightarrow E\left[\int_0^T X(s)\, dB(s)\right] = 0 \tag{4.20}$$

(c) Ito isometry

$$E\left[\left(\int_0^T X(s)\, dB(s)\right)^2\right] = \int_0^T E\left[(X(s))^2\right] ds \tag{4.21}$$

To prove this identity, we note that

$$E\left[\left(\int_0^T X(s)\,dB(s)\right)^2\right] = E\left[\left(\sum_{j=0}^{N-1} X_{t_j}\left(B_{t_{j+1}} - B_{t_j}\right)\right)^2\right]$$

$$= \sum_{j=0}^{N-1} E\left[X_{t_j}^2\left(B_{t_{j+1}} - B_{t_j}\right)^2\right] + 2\sum_{j<k} E\left[X_{t_j}X_{t_k}\left(B_{t_{j+1}} - B_{t_j}\right)\left(B_{t_{k+1}} - B_{t_k}\right)\right] \quad (4.22)$$

By the martingale property of a Wiener process, the first sum on the right hand side of the above equation takes the form:

$$\sum_{j=0}^{N-1} E\left[X_{t_j}^2\left(B_{t_{j+1}} - B_{t_j}\right)^2\right] = \sum_{j=0}^{N-1} E\left[E\left[X_{t_j}^2\left(B_{t_{j+1}} - B_{t_j}\right)^2 |\mathcal{F}_{t_j}\right]\right]$$

$$= \sum_{j=0}^{N-1} E\left[X_{t_j}^2 E\left[\left(B_{t_{j+1}} - B_{t_j}\right)^2 |\mathcal{F}_{t_j}\right]\right]$$

$$= \sum_{j=0}^{N-1} E\left[X_{t_j}^2 E\left[\left(B_{t_{j+1}} - B_{t_j}\right)^2\right]\right]$$

$$= \sum_{j=0}^{N-1} E\left[X_{t_j}^2\right]\left(t_{j+1} - t_j\right) \quad (4.23)$$

Similarly the second sum on the right hand side of Eq. (4.22) can be written for $j < k$ as:

$$\sum_{j<k} E\left[X_{t_j}X_{t_k}\left(B_{t_{j+1}} - B_{t_j}\right)\left(B_{t_{j+1}} - B_{t_j}\right)\right]$$

$$= \sum_{j<k} E\left[E\left[X_{t_j}X_{t_k}\left(B_{t_{j+1}} - B_{t_j}\right)\left(B_{t_{k+1}} - B_k\right)|\mathcal{F}_{t_k}\right]\right]$$

$$= \sum_{j<k} E\left[X_{t_j}X_{t_k}\left(B_{t_{j+1}} - B_{t_j}\right)E\left[\left(B_{t_{k+1}} - B_{t_k}\right)|\mathcal{F}_{t_k}\right]\right] = 0 \quad (4.24)$$

Combining the results in Eqs. (4.23) and (4.24), we get:

$$E\left[\left(\int_0^T X(s)\,dB(s)\right)^2\right] = \sum_{j=0}^{N-1} E\left[X_{t_j}^2\right]\left(t_{j+1} - t_j\right) \tag{4.25}$$

Thus the right hand side above reduces to $\int_0^T E\left[(X(s))^2\right]ds$ yielding the Ito isometry. For instance, if X is in $L^2(P)$, then Ito isometry is well defined: $\int_0^T E\left[(X(s))^2\right]ds < \infty$.

4.2.3 Ito integral for continuous processes

Having established the properties of Ito integrals of simple functions, we use them to extend the definition of the integral to a much broader class of functions. It is a set, say, $\mathcal{H}$ of functions (stochastic processes) $X(\omega, t)$ on the product space $\Omega \times [0, \infty)$ such that:

i) $X_t(\omega)$ is $\mathcal{B}[0,\infty) \times \mathcal{F}(\Omega)$-measurable (4.26a)

ii) $X_t(\omega)$ is $\mathcal{F}_t$ adapted (4.26b)

iii) $E\int_0^T X_t^2(\omega)dt < \infty$ (4.26c)

$\mathcal{H}$ may be given the structure of a Hilbert space with an inner product defined by the covariance:

$$\langle X, Y\rangle_{L^2(P)} = E[XY]$$

with the norm given by:

$$\|X\|_{L^2(P)} = E\left[\int_0^T X^2(t,\omega)\,dt\right] \tag{4.27}$$

(Note however that $\|X\|_{L^2(P)}$ can be zero even when $X_t(\omega) \neq 0$ over a subset of null P–measure). Since the space of simple functions is dense in $\mathcal{H}$, there exists a sequence $\left\{X^{(N)} = \sum_{j=0}^{N-1} X_{t_j} I_{(t_j, t_{j+1}]}(t)\right\}$ of simple functions with $X_{t_j} = X(t_j)$, so that for any function $X \in \mathcal{H}$, we have:

$$\left\|X - X^{(N)}\right\|_{L^2(P)} = E\left[\int_0^T \left(X - X^{(N)}\right)^2 dt\right] \to 0 \text{ as } N \to \infty \tag{4.28}$$

Now, we define the Ito integral $\mathfrak{I}$ for $X_t(\omega) \in \mathcal{H}$ as:

$$\mathfrak{I}(X) = \int_0^T X(s)\,dB(s) := \lim_{N\to\infty} \int_0^T X_s^{(N)}(\omega)dB(s) = \lim_{N\to\infty} \mathfrak{I}(X^N) \tag{4.29}$$

By Ito isometry in Eq. (4.21), $\mathfrak{I}$ forms a Cauchy sequence in $L^2(P)$ and hence the limit in the last equation exists as an element in $L^2(P)$. Moreover, the limit does not depend on the choice of the simple functions $X^{(N)}$ so long as Eq. (4.28) holds.

Clearly, $\mathfrak{I} : \mathcal{H} \to L^2$ is linear. If $g \in \mathcal{H}$, $0 \le S < T$ and $c \in \mathbb{R}$, then we have the following properties of the Ito integral for which the proofs are straightforward:

(i) $\int_0^T f\, dB(s) = \int_0^S f\, dB(s) + \int_S^T f\, dB(s)$ for almost all (a.a.) ω (4.30a)

(ii) $\int_0^T \left(cf + g\right) dB(s) = c\int_0^T f\, dB(s) + \int_0^T g\, dB(s)$ a.a. ω (4.30b)

(iii) $E\left[\int_0^T f\, dB(s)\right] = 0$ (4.30c)

In addition, by the definition of $f \in \mathcal{H}$, it is also clear that $\int_0^T f\, dB(s)$ is $\mathcal{F}_T$-measurable. Another useful property of the integral is the following.

If $f_t(\omega) \in \mathcal{H}$ and $f^{(N)}(\omega) \in \mathcal{H}$ for $N = 1, 2, \ldots$ and $E[\int_0^T (f - f^{(N)})^2 dt] \to 0$ as $N \to \infty$, then:

$$\int_0^T f^{(N)} dB(s) \to \int_0^T f\, dB(s) \tag{4.31}$$

in $L^2(P)$ as $N \to \infty$. The following example demonstrates the above property.

Example 4.3

We show that the Ito integral $\int_0^T B\, dB(s) = \frac{1}{2}\left(B_T^2 - T\right)$.

Solution

Assume that B is standard, i.e., $B_0 = 0$. Choose the simple functions $B_T^{(N)}(\omega)$ such that:

$$B_T^{(N)}(\omega) = \sum_{j=0}^{N-1} B_j\, I_{(t_j, t_{j+1}]}(t) \tag{4.32}$$

where, $B_j = B(t_j)$. Clearly B, $B_T^{(N)} \in \mathcal{H}$. We also have:

$$E\left[\int_0^T \left(B - B_T^{(N)}\right)^2 ds\right] = E\left[\sum_{j=0}^{N-1} \int_{t_j}^{t_{j+1}} \left(B - B_j\right)^2 ds\right] = \sum_{j=0}^{N-1} \int_{t_j}^{t_{j+1}} \left(s - t_j\right) ds$$

$$= \frac{1}{2}\sum_{j=0}^{N-1}\left(t_{j+1}-t_j\right)^2 \to 0 \text{ as } N \to \infty \tag{4.33}$$

Here $N \to \infty$ is equivalent to $\Delta t_j = \left(t_{j+1}-t_j\right) \to 0$. From Eq. (4.31), it follows that:

$$\int_0^T BdB(s) = \lim_{\Delta t_j \to 0} \int_0^T B_T^{(N)} dB(s)$$

$$= \lim_{\Delta t_j \to 0} \sum_{j=0}^{N-1} B_j \Delta B_j \quad \text{(From Eq. 4.32)} \tag{4.34}$$

Denoting $\Delta B_j^2 = B_{j+1}^2 - B_j^2$, we have:

$$\Delta B_j^2 = \left(B_{j+1}-B_j\right)^2 + 2B_j\left(B_{j+1}-B_j\right)$$

$$= \left(\Delta B_j\right)^2 + 2B_j\Delta B_j$$

$$\Longrightarrow B_j\Delta B_j = \frac{1}{2}\left(\Delta B_j^2 - \left(\Delta B_j\right)^2\right) \tag{4.35}$$

since $B_0 = 0$, $\sum_{j=0}^{N-1} \Delta B_j^2 = B_T^2$. Further $\sum_{j=0}^{N-1} \left(\Delta B_j\right)^2 \to T$ in $L^2(P)$ as $\Delta t_j \to 0$ (see Eq. 4.16). Therefore, Eq. (4.35) gives:

$$\lim_{\Delta t_j \to 0} \sum_{j=0}^{N-1} B_j \Delta B_j = \int_0^T BdB(s) = \frac{1}{2}\left(B_T^2 - T\right) \tag{4.36}$$

This is the desired result.

The example above shows that the Ito integral behaves differently from an ordinary integral. The second term $-T/2$ in the last result (on the extreme right hand side of Eq. 4.36) would be absent if the rules of standard calculus were applied to the integral. Given that an Ito integral has zero mean, we find that the expectation of the right hand side of Eq. (4.36) becomes zero only with the appearance of the term $-T/2$, since $E[B_T^2] = T$.

4.3 Ito Processes

So far, with $X_t(\omega) \in \mathcal{H}$, an Ito integral $\mathfrak{I} = \int_0^T X(s)\,dB(s)$ is defined for a fixed terminal time T. Since $\mathfrak{I}$ is a random variable for any fixed terminal time $t \leq T$, it immediately follows that $\int_0^t X(s)\,dB(s)$ as a function of the upper limit t can be considered as a stochastic process-generally known as an Ito process. It is shown below that there exists a time continuous version of the Ito integral $\mathfrak{I}_t := \mathfrak{I}(t,\omega) = \int_0^t X(s)\,dB(s)$ with continuous sample paths.

Consider the sequence $\left\{X^{(N)}(t,\omega)\right\}$ of simple functions $X^{(N)}(t,\omega) = \sum_j X_j(\omega)\, I_{\left(t_j^{(N)},t_{j+1}^{(N)}\right]}(t)$ with $X_j = X(t_j^{(N)})$, such that, for any function $X \in \mathcal{H}$, we have:

$$E\left[\int_0^T \left(X - X^{(N)}\right)^2 dt\right] \to 0 \text{ as } N \to \infty \tag{4.37}$$

where, $\left\{t_j^{(N)}\right\}$ is a partition Π_N of $[0,\ T]$ with $\Delta_N = \max_{1\leq j\leq N-1}\left(t_{j+1}^{(N)} - t_j^{(N)}\right) \to 0$ as $N \to \infty$. Define:

$$\mathfrak{I}^{(N)}(t,\omega) = \int_0^t X^{(N)}(s,\omega)\,dB(s) \tag{4.38}$$

$\mathfrak{I}^{(N)}(t,\omega)$ is continuous for all N (since B has continuous paths). Also, $\mathfrak{I}^{(N)}(t,\omega)$ has the important property that it is an $\mathcal{F}_t$–martingale.

Proof: For $t < s$, we have:

$$\begin{aligned}
E\left[\mathfrak{I}^{(N)}(s,\omega)|\mathcal{F}_t\right] &= E\left[\left(\int_0^t X^{(N)}dB + \int_t^s X^{(N)}dB\right)|\mathcal{F}_t\right] \\
&= \int_0^t X^{(N)}dB + E\left[\sum_{t\leq t_j^{(N)}\leq t_{j+1}^{(N)}\leq s} X_j^{(N)}\Delta B_j|\mathcal{F}_t\right], \Delta B_j = B_{j+1} - B_j \\
&= \int_0^t X^{(N)}dB + \sum_{t\leq t_j^{(N)}\leq t_{j+1}^{(N)}\leq s} E\left[E\left[X_j^{(N)}\Delta B_j|\mathcal{F}_{t_j^{(N)}}\right]|\mathcal{F}_t\right] \\
&= \int_0^t X^{(N)}dB + \sum_{t\leq t_j^{(N)}\leq t_{j+1}^{(N)}\leq s} E\left[X_j^{(N)}E\left[\Delta B_j|\mathcal{F}_{t_j^{(N)}}\right]|\mathcal{F}_t\right]
\end{aligned} \tag{4.39}$$

With $E[\Delta B_j|\mathcal{F}_{t_j^{(N)}}] = 0$, we get:

$$E\left[\mathfrak{T}^{(N)}(s,\omega)|\mathcal{F}_t\right] = \int_0^t X^{(N)} dB = \mathfrak{T}^{(N)}(t,\omega) \tag{4.40}$$

Thus $\mathfrak{T}^{(N)}(t,\omega)$ is an $\mathcal{F}_t$–martingale.

Now we prove that $\mathfrak{T}^{(N)}(t,\omega)$ uniformly converges for $t \in [0,T]$ and for a.a. ω to a limit that has continuous sample paths.

Proof: Clearly, $\mathcal{M}_{N,\mathfrak{m}}(t,\omega) := \mathfrak{T}^{(N)}(t,\omega) - \mathfrak{T}^{(\mathfrak{m})}(t,\omega)$ is an $\mathcal{F}_t$–martingale. By the martingale property, we have:

$$P\left(\sup_{0\le t\le T}|\mathcal{M}_{N,\mathfrak{m}}(t,\omega)| > \varepsilon\right) \le \frac{E\left[\left|\mathcal{M}_{N,\mathfrak{m}}(\tau,\omega)\right|^2\right]}{\varepsilon^2}$$

$$= \frac{1}{\varepsilon^2}E\left[\int_0^T \left(X^{(N)} - X^{(\mathfrak{m})}\right)^2 ds\right] \to 0, \text{ as } N,\mathfrak{m} \to \infty \tag{4.41}$$

So we can choose a suitable subsequence $N_k \uparrow \infty$ such that:

$$P\left(\sup_{0\le t\le T}\left|\mathfrak{T}^{(N_{k+1})}(t,\omega) - \mathfrak{T}^{(N_k)}(t,\omega)\right| > \frac{1}{2^k}\right) < \frac{1}{2^k} \tag{4.42}$$

By the Borel–Cantelli lemma (Appendix C), we have:

$$P\left(\sup_{0\le t\le T}\left|\mathfrak{T}^{(N_{k+1})}(t,\omega) - \mathfrak{T}^{(N_k)}(t,\omega)\right| > \frac{1}{2^k} \text{ for infinitely many } k\right) = 0 \tag{4.43}$$

Thus there exists a $k_1(\omega)$, such that for a.a. ω:

$$\sup_{0\le t\le T}\left|\mathfrak{T}^{(N_{k+1})}(t,\omega) - \mathfrak{T}^{(N_k)}(t,\omega)\right| \le \frac{1}{2^k} \tag{4.44}$$

for $k \ge k_1(\omega)$. This shows that $\mathfrak{T}^{(N_k)}(t,\omega)\left(= \int_0^t X^{(N_k)} dB\right)$ uniformly converges for $t \in [0,T]$ and for a.a. ω, to some process J_t with continuous sample paths a.s. Since it is already known that $\mathfrak{T}^{(N_k)}(t,\omega) \to \mathfrak{T}_t = \int_0^t X\, dB$ in $L^2(P)$ for all t, we have $\mathfrak{T}_t = J_t$ a.s. for all $t \in [0,T]$.

4.3.1 Larger class of integrands for Ito integral

The integrand $X(t,\omega)$ can be defined for a larger class $\mathcal{H}^*$ than $\mathcal{H}$ in that the $\mathcal{F}_t$-measurability condition in Eq. (4.26b) can be relaxed as:

$$X_t(\omega) \text{ is } \mathcal{F}_t^* - \text{adapted} \tag{4.45}$$

where, $\{\mathcal{F}_t^*\}$ is the increasing family of σ-algebras corresponding to the integrator which may not be a Brownian motion but nevertheless a martingale $\mathcal{M}(t,\omega)$ with respect to $\mathcal{F}_t^*$. Clearly $E\left[(\mathcal{M}_s - \mathcal{M}_t)|\mathcal{F}_t^*\right] = 0$ for all $s > t$ and the Ito integral $\mathfrak{I}_t = \int_0^t X(s)\, d\mathcal{M}(s)$ remains a martingale, and proof for the existence of a continuous version of the integral follows as before. For instance, this generalization is also helpful in characterizing Ito integrals defined with respect to a vector Brownian motion.

Example 4.4

Definition of a multi-dimensional Ito integral

Suppose that $\boldsymbol{B}(t,\omega)$ is an n-dimensional Brownian motion vector and $\mathcal{F}_t^* = \mathcal{F}_t^{(n)}$ is the σ-algebra generated by $B_1(s,\omega), B_2(s,\omega), \ldots, B_n(s,\omega)\, s \le t$. Each $B_k(t,\omega)$ is a martingale with respect to $\mathcal{F}_t^*$ and the Ito integral for the $\mathcal{F}_t^*$-adapted integrand $X(t,\omega)$ is defined as $\mathfrak{I}_t^k = \int_0^t X(s)\, dB_k$.

Now consider an $m \times n$ matrix of integrands, $\boldsymbol{X} = \left[X_{jk}(t,\omega)\right]$ where each $X_{jk} \in \mathcal{H}^*$. We can now define the multi-dimensional Ito integral (vector) as:

$$\int_0^t \boldsymbol{X} d\boldsymbol{B} = \int_0^t \begin{bmatrix} X_{11} & X_{12} & \ldots & X_{1n} \\ X_{21} & X_{22} & \ldots & X_{2n} \\ . & . & & . \\ X_{m1} & X_{m2} & \ldots & X_{mn} \end{bmatrix} \begin{Bmatrix} dB_1 \\ dB_2 \\ . \\ dB_n \end{Bmatrix} \tag{4.46}$$

In addition to the above extension of the Ito integral to integrands in $\mathcal{H}^*$, if the condition $E\int_0^T X_t^2(\omega)dt < \infty$ in Eq. (4.26c) is relaxed to:

$$P\left(\int_0^T X_t^2(\omega)dt < \infty\right) = 1 \tag{4.47}$$

then we may define the Ito integral as a limit in probability:

$$\mathfrak{I}_t = \int_0^t X(s,\omega)\, dB(s,\omega) = \lim_{N\to\infty} \int_0^t X^{(N)}(s,\omega)\, dB(s) \tag{4.48}$$

If $\mathcal{H}^*$ includes this extended class of integrands, one can show that there exist simple functions $X^{(N)}(t,\omega) \in \mathcal{H}^*$ for all t such that $\int_0^t \left|X - X^{(N)}\right|^2 ds \to 0$ in probability, followed by a continuous version of the integral [McKean 1969, Friedman 1975]. Here one may remark that the Ito integral is no longer a martingale, but a local martingale [Karatzas and Shreve 1991] (see Section 3.7, Chapter 3). In the further discussion, let $\mathcal{H}^*$ stand for the new class of functions with the relaxed conditions in Eq. (4.45) and (4.47).

4.4 Stochastic Calculus

Having defined stochastic integration via Ito integrals, the latter lead us to a differential calculus, referred to as stochastic calculus. An insightful treatment of dynamical systems under stochastic loading is possible via a systematic application of stochastic calculus of which the Ito calculus is the most widely used and the one adopted in this book. The primary tool here is Ito's formula which is an analogous version of the fundamental theorem of ordinary calculus suitably modified for stochastic differentials. For the Ito integral $X_t = \int_0^t \sigma(s)\, dB(s)$ with the integrand $\sigma(t) \in \mathcal{H}^*$, the stochastic differential of the integral is given by:

$$dX(t) = \sigma(t)\, dB(t) \tag{4.49}$$

Here X_t is assumed to be a one-dimensional (Ito) process. The above equation is amongst the simplest forms of an SDE. An Ito process $X(t)$ may in general be expressed as:

$$X(t) = X_0 + \int_0^t a(s)\, ds + \int_0^t \sigma(s)\, dB(s) \tag{4.50}$$

with X_0 denoting the measurable initial state of $X(t)$. The last integral on the right hand side above is the Ito integral and the preceding one a non-stochastic integral (despite $a(t)$ being possibly random). The corresponding SDE is of the form:

$$dX(t) = a(t)\, dt + \sigma(t)\, dB(t) \tag{4.51}$$

$a(t)$ is known as the drift term and $\sigma(t)$ the diffusion term. From physical considerations, the drift part of the SDE constitutes the classical dynamic equilibrium balance equations of a system (e.g., energy or momentum in the context of mechanical oscillators). The diffusion part represents the external noise term that arises from the stochastic environment to which the system is subjected. In fact, a more general form of the SDE in the last equation could be:

$$dX(t) = a(t,X)\, dt + \sigma(t,X)\, dB(t) \tag{4.52a}$$

An alternative integral form would be given by:

$$X(t) = X_0 + \int_0^t a(s,X(s))\, ds + \int_0^t \sigma(s,X(s))\, dB(s) \tag{4.52b}$$

The noise coefficient σ in the above equation is system-dependent, i.e., dependent on $X(t)$ also. In this case, the noise is referred to as multiplicative. Otherwise it is known as additive (Eq. 4.51). $\sigma(t,X)$ is often called the diffusion coefficient.

4.4.1 Integral representation of an SDE

In various applications of science and engineering, it is customary to treat the noise term in the SDE as one involving a white noise process $W(t)$ (see Section 3.3.4, Chapter 3) in the form:

$$\dot{X}(t) = a(t,X) + \sigma(t,X)\,W(t) \tag{4.53}$$

An intuitive description of a true noise process $W(t)$ appears to be best served through the following properties:

i) $W(t_1)$ is independent of $W(t_2)$ for any $t_1 \neq t_2$ (4.54a)

ii) $W(t)$ is stationary, and (4.54b)

iii) $E[W(t)] = 0 \;\forall\, t$ (4.54c)

Preference for white noise driven models in science and engineering may be traced to the fact that the white noise indeed conforms to all the three properties enumerated above. However, since $W(t)$ cannot have continuous paths, the existence of $W(t)$ as the graph of a function is hardly justifiable. Further, in reality a white noise is not a stochastic process in that its covariance $E[W_s W_t] = \delta(t-s)$ is not a proper function. The latter is generally known as a distribution or a generalized function. In Section 3.3.6, Chapter 3, a non-rigorous way of treating white noise as a generalized derivative of Wiener process had been presented. This prompts one to seek for an alternative representation of the SDE in Eq. (4.53) wherein $W(t)$ is suitably replaced by using $B(t)$, the Wiener process.

We write the following discrete version of the SDE (Eq. 4.53):

$$X_{j+1} = X_j + a\left(t_j, X_j\right)\Delta t_j + \sigma\left(t_j, X_j\right) W_j \Delta t_j \tag{4.55}$$

where, $X_j = X\left(t_j\right)$, $W_j = W(t_j)$ and $\Delta t_j = t_{j+1} - t_j$. Writing $\Delta Q_j = Q_{j+1} - Q_j = W_j \Delta t_j$ and passing to the limit as $\Delta t_j \to 0$, we wish to see if the process $Q(t)$ makes sense. From the properties in Eq. (4.54) of $W(t)$, it is clear that $Q(t)$ should have stationary independent increments with zero mean. But we know that a version of such a process with continuous paths is the Wiener process (Brownian motion) $B(t)$. Therefore it follows that $Q(t) = B(t)$. Eq. (4.55) may now be written as:

$$X_{j+1} = X_0 + \sum_{k=0}^{j} \{a(t_k, X_k)\Delta t_k + \sigma(t_k, X_k)\Delta B_k\} \tag{4.56}$$

As $\Delta t_j \to 0$, we can write a continuous form of the above equation as:

$$X(t) = X_0 + \int_0^t a(s, X(s))\, ds + \int_0^t \sigma(s, X(s))\, dB(s) \tag{4.57}$$

which is the same as Eq. (4.52b). Existence of an Ito integral has already been proved in Section 4.2 and the integral defined for a wide class of functions $\sigma(t, X)$. We thus have an alternative and a more appropriate integral representation for the SDE (4.53).

4.4.2 Ito's formula

In stochastic calculus, Ito's formula plays the counterpart of the conventional chain rule. For example, if $x(t) : [0, \infty) \to \mathbb{R}$ is continuous and of bounded variation and $g(x)$ a continuously differentiable function of x, then the fundamental theorem of conventional calculus gives:

$$g(x(t)) = g(x(0)) + \int_0^t g'(x(s))\, dx(s) \tag{4.58a}$$

$$\Longrightarrow dg(x(t)) = g'(x(t))\, dx(t) \tag{4.58b}$$

where, $g' = dg/dx$. Ito's formula is a stochastic analogue of the above rule when $x(t)$ is a stochastic (Ito) process.

One-dimensional Ito's formula for Wiener process

If $B(t)$ is a Wiener process on $[0, T]$ and $g(x)$ a twice differentiable continuous function on $\mathbb{R}$, then for any $t \le T$, Ito's formula is given by:

$$g(B(t)) = g(0) + \int_0^t g'(B(s))\, dB(s) + \frac{1}{2}\int_0^t g''(B(s))\, ds \tag{4.59}$$

Note that, in contrast to Eq. (4.58a), Ito's formula contains an additional term containing the second order derivative of $g(x)$.

Proof: In order to prove this formula, we need the following important result on $B(t)$. Consider a partition Π_N of the interval $[0, t]$ such that $\Delta_N = \max_{j \in [0, N-1]} \left\{ \left(t_{j+1} - t_j \right) \right\}$. Then, one can show that:

$$\sum_{j=0}^{N-1} \left(B\left(t_{j+1}\right) - B(t_j) \right)^2 = t, \text{ in } L^2(P) \tag{4.60}$$

in the limit as $\Delta_N \to 0$. First observe that $\sum_{j=0}^{N-1} h_j = \int_0^t ds = t$ (where $h_j = t_{j+1} - t_j$). Now, in order to prove the above equality (Eq. 4.60) with $\Delta B_j = B(t_{j+1}) - B(t_j)$, we take the expectation of the squared difference (between the left and right hand sides of the equality) leading to:

$$E\left[\sum_{j=0}^{N-1}\left(\Delta B_j{}^2 - h_j\right)\right]^2 = \sum_{j=0}^{N-1} E\left[\left(\Delta B_j{}^2 - h_j\right)\right]^2$$

$$+\sum_{j\neq k}\sum_{k} E\left[\left(\Delta B_j{}^2 - h_j\right)\left(\Delta B_k{}^2 - h_k\right)\right] \tag{4.61a}$$

Since $\Delta B_j{}^2 - h_j$ and $\Delta B_k{}^2 - h_k$ are independent for $j \neq k$, and $\sum_{j=0}^{N-1} E\left[\Delta B_j{}^2 - h_j\right] = 0 \; \forall j$, the double sum on the right hand side of the above equation vanishes. The first sum can be written as:

$$\sum_{j=0}^{N-1} E\left[\left(\Delta B_j{}^2 - h_j\right)\right]^2 = \sum_{j=0}^{N-1}\left(E\left[\Delta B_j{}^4\right] - 2E\left[\Delta B_j{}^2 h_j\right] + h_j^2\right)$$

$$= \sum_{j=0}^{N-1}\left(3h_j^2 - 2h_j^2 + h_j^2\right) = 2\sum_{j=0}^{N-1} h_j^2 \tag{4.61b}$$

Thus we get the result:

$$E\left[\sum_{j=0}^{N-1}\left(\Delta B_j{}^2 - h_j\right)\right]^2 = 2\sum_{j=0}^{N-1} h_j^2 \leq 2\Delta_N \sum_{j=0}^{N-1} h_j$$

$$= 2\Delta_N t \;\to 0 \text{ as } \Delta_N \;\to 0 \tag{4.61c}$$

Thus we get the following result:

$$\left(\sum_{j=0}^{N-1}\left(\Delta B_j\right)^2\right) - t \xrightarrow{L^2(P)} 0 \tag{4.62}$$

More general variants of this identity, viz.

$$\lim_{\Delta_N \to 0} \sum_{j=0}^{N-1} \psi_{t_j} \left(\Delta B_j\right)^2 = \int_0^t \psi(s) ds \tag{4.63a}$$

or,

$$\lim_{\Delta_N \to 0} \sum_{j=0}^{N-1} \psi\left(B\left(t_j\right)\right)\left(\Delta B_j\right)^2 = \int_0^t \psi\left(B(s)\right)\, ds \tag{4.63b}$$

may also be readily established in the $L^2(P)$ setting following the above procedure. For example, to prove Eq. (4.63a), we first observe that $\lim_{\Delta_N \to 0} \sum_{j=0}^{N-1} \psi_{t_j} \Delta t_j = \int_0^t \psi(s) ds$. Now the expectation of the squared difference (between the left and right hand sides of the equality in Eq. (4.63a) leads to:

$$E\left[\left(\sum_{j=0}^{N-1} \psi_{t_j} \left(\Delta B_j\right)^2 - \sum_{j=0}^{N-1} \psi_{t_j} \Delta t_j\right)^2\right] = E\left[\left(\sum_{j=0}^{N-1} \psi_{t_j} \left(\left(\Delta B_j\right)^2 - \Delta t_j\right)\right)^2\right]$$
$$= \sum_{j,k} E\left[\psi_{t_j} \psi_{t_k} \left(\left(\Delta B_j\right)^2 - \Delta t_j\right)\left(\Delta B_k{}^2 - \Delta t_k\right)\right] \tag{4.63c}$$

Since, for $j \neq k$, $\psi_{t_j} \psi_{t_k} \left(\left(\Delta B_j\right)^2 - \Delta t_j\right)$ and $\left((\Delta B_k)^2 - \Delta t_k\right)$ are independent and $\sum_{j=0}^{N-1} E[\Delta B_j{}^2 - \Delta t_j] = 0\ \forall j$, the corresponding terms on the extreme right hand side of the last equation vanish. For $j = k$, we have:

$$\sum_j E\left[\psi_{t_j}^2 \left(\left(\Delta B_j\right)^2 - \Delta t_j\right)^2\right] = \sum_j E\left[\psi_{t_j}^2\right] E\left[\left(\Delta B_j\right)^4 - 2\left(\Delta B_j\right)^2 \Delta t_j + \left(\Delta t_j\right)^2\right]$$
$$= \sum_j E\left[\psi_{t_j}^2\right]\left(3\left(\Delta t_j\right)^2 - 2\left(\Delta t_j\right)^2 + \left(\Delta t_j\right)^2\right)$$
$$= 2\sum_j E\left[\psi_{t_j}^2\right]\left(\Delta t_j\right)^2 \to 0 \tag{4.63d}$$

as $\Delta t_j \to 0$. This gives $\sum_{j=0}^{N-1} \psi_{t_j} \left(\Delta B_j\right)^2 - \sum_{j=0}^{N-1} \psi_{t_j} \Delta t_j \xrightarrow{L^2(P)} 0$ leading to the result in Eq. (4.63a).

The proof of Ito's formula now starts with the identity:

$$g(B(t)) = g(0) + \sum_{j=0}^{N-1} \left(g\left(B\left(t_{j+1}\right)\right) - g\left(B\left(t_j\right)\right)\right) \tag{4.64}$$

Applying Taylor's expansion to $g\left(B\left(t_{j+1}\right)\right)$, we get:

$$g\left(B\left(t_{j+1}\right)\right) = g\left(B(t_j)\right) + g'(B(t_j))\Delta B_j + \frac{1}{2}g''\left(\theta_j\right)\left(\Delta B_j\right)^2 \tag{4.65}$$

where, $\theta_j \in \left(B(t_j), B\left(t_{j+1}\right)\right)$. Substituting the above in Eq. (4.64), we get:

$$g(B(t)) = g(0) + \sum_{j=0}^{N-1} g'(B(t_j))\Delta B_j + \frac{1}{2}\sum_{j=0}^{N-1} g''\left(\theta_j\right)\left(\Delta B_j\right)^2 \tag{4.66}$$

As $\Delta_N \to 0$, the first sum on the right hand side above converges to the Ito integral $\int_0^t g'(B(s))\, dB(s)$ and the second sum yields $\frac{1}{2}\int_0^t g''(B(s))ds$ (see the identity in Eq. 4.63b). Thus, we obtain Ito's formula in Eq. (4.59). The SDE form of Ito's formula also immediately follows as:

$$dg(B(t)) = g'(B(t))\, dB(t) + \frac{1}{2}g''(B(t))dt \tag{4.67}$$

Example 4.5

Prove that $\int_0^t B(s)\, dB(s) = \frac{1}{2}B^2(t) - \frac{1}{2}t$

Solution

The identity was proved in Example 4.3 using simple functions. Here it is a direct application of Ito's formula with $g(B(t)) = B^2(t)$. The integral form of Ito's formula yields $g(B(t)) = g(B(0)) + 2\int_0^t B(s)\, dB(s) + \int_0^t ds$ with $B(0) = 0$:

$$B^2(t) = 2\int_0^t B(s)\, dB(s) + t,$$

$$\Longrightarrow \int_0^t B(s)\, dB(s) = \frac{1}{2}\left(B^2(t) - t\right) \tag{4.68}$$

Example 4.6

Obtain SDE form of Ito's formula for $g\left(B\left(t\right)\right)=e^{B(t)}$

Solution

By using the SDE form of Ito's formula (Eq. 4.67) we have:

$$d\left(e^{B(t)}\right)=e^{B(t)}dB\left(t\right)+\frac{1}{2}\,e^{B(t)}dt \tag{4.69}$$

where we identify $\frac{1}{2}\,e^{B(t)}$ as the drift coefficient and $e^{B(t)}$ as the diffusion coefficient. In other words, the stochastic process $X(t)=e^{B(t)}$ satisfies the SDE:

$$dX\left(t\right)=X\left(t\right)dB\left(t\right)+\frac{1}{2}X\left(t\right)dt \tag{4.70}$$

Before writing down Ito's formula for a general Ito process, it is convenient to introduce the notions of quadratic variation of $X(t)$ and quadratic covariation of stochastic processes $X\left(t\right)$ and $Y\left(t\right)$.

Quadratic variation: For a partition Π_N of $[0,t]$ into N intervals as before, the quadratic variation of $X(t)$ is defined as:

$$[X,X]\left(t\right)=\lim_{\Delta_N\to 0}\sum_{j=0}^{N-1}\left(X\left(t_{j+1}\right)-X\left(t_j\right)\right)^2 \tag{4.71a}$$

Quadratic covariation: For a partition Π_N of $[0,t]$, the quadratic covariation of stochastic processes $X(t)$ and $Y(t)$ is defined as:

$$[X,Y]\left(t\right)=\lim_{\Delta_N\to 0}\sum_{j=0}^{N-1}\left(X\left(t_{j+1}\right)-X\left(t_j\right)\right)\left(Y\left(t_{j+1}\right)-Y\left(t_j\right)\right) \tag{4.71b}$$

Example 4.7

We obtain the quadratic variation of Brownian motion $B(t)$

Solution

From Example 4.3, it is readily observed that the quadratic variation of $B(t)$ is given by:

$$[B,B](t) = \lim_{\Delta_N \to 0} \sum_{j=0}^{N-1} \left(B\left(t_{j+1}\right) - B(t_j)\right)^2 = \lim_{\Delta_N \to 0} \sum_{j=0}^{N-1} \left(\Delta B_j\right)^2 = t \quad (4.72)$$

While one may expect the above quadratic variation to be a random variable, it turns out to be (characterizable as) a deterministic quantity.

Example 4.8

To find the quadratic variation of a deterministic process

Solution

Let $X(t)$ be a continuous and at least once differentiable deterministic process. Then, using a deterministic Taylor expansion in the definition of quadratic variation, it is not difficult to see that:

$$[X,X](t) = 0 \quad (4.73)$$

It is interesting to note that the quadratic variation of a differentiable and bounded process (often encountered in deterministic dynamics of many engineering/physical systems) over any finite time interval is zero. This may be contrasted with the behavior of $B(t)$ and in turn with solutions to SDEs, for which the quadratic variation is a characteristic shared by the diffusion terms in the SDEs, as made precise in the next example.

Example 4.9

Find the quadratic variation of the Ito integral process $I(t) = \int_0^t \sigma(s)\, dB(s)$

Solution

Through a direct application of the definition of quadratic variation (alternatively from the proof of Ito's isometry), it follows that:

$$[I,I](t) = \int_0^t \sigma^2(s)\, ds \quad (4.74)$$

Example 4.10

To find the quadratic covariation of two stochastic processes $X(t)$ and $Y(t)$ where both are continuous and $Y(t)$ is of finite variation

Solution

By definition, the covariation (Eq. (4.71b)) is given by:

$$[X,Y](t) = \lim_{\Delta_N \to 0} \sum_{j=0}^{N-1} \left(X\left(t_{j+1}\right) - X(t_j)\right)\left(Y\left(t_{j+1}\right) - Y(t_j)\right)$$

$$\leq \lim_{\Delta_N \to 0} \max_j \left|Y\left(t_{j+1}\right) - Y(t_j)\right| \sum_{j=0}^{N-1} \left(X\left(t_{j+1}\right) - X(t_j)\right) \tag{4.75}$$

$Y(t)$ being continuous is uniformly continuous in the interval $[0,t]$. It follows that $\lim_{\Delta_N \to 0} \max_j \left|Y\left(t_{j+1}\right) - Y(t_j)\right| = 0$ and therefore their covariation $[X,Y](t)$ is zero.

Example 4.11

We obtain the quadratic variation of the Ito integral process $X(t) = X_0 + \int_0^t a(s)\,ds + \int_0^t \sigma(s)\,dB(s)$

Solution

The integral $\int_0^t a(s)\,ds$ is a continuous function of finite variation in t. In contrast, the Ito integral $\int_0^t \sigma(s)\,dB(s)$ is continuous, but not of bounded variation. Now, by the definition of quadratic variation, we have:

$$[X,X](t) = \lim_{\Delta_N \to 0} \sum_{j=0}^{N-1} \left(X\left(t_{j+1}\right) - X(t_j)\right)^2 \tag{4.76}$$

In the limit $\Delta_N \to 0$ as $N \to \infty$

$$[X,X](t) = \left[\int_0^t a(s)\,ds + \int_0^t \sigma(s)\,dB(s), \int_0^t a(s)\,ds + \int_0^t \sigma(s)\,dB(s)\right](t)$$

$$= \left[\int_0^t a(s)\,ds, \int_0^t a(s)\,ds\right](t) + 2\left[\int_0^t a(s)\,ds, \int_0^t \sigma(s)\,dB(s)\right](t)$$

$$+ \left[\int_0^t \sigma(s)\,dB(s), \int_0^t \sigma(s)\,dB(s)\right](t) \tag{4.77}$$

From the result in Example 4.10, the quadratic covariation of a continuous function with another of finite variation is zero. Thus the first two terms on the right hand side of the above equation vanish yielding the result:

$$[X,X](t) = \left[\int_0^t \sigma(s)\,dB(s), \int_0^t \sigma(s)\,dB(s)\right](t)$$

$$= \int_0^t \sigma^2(s)\,ds \tag{4.78}$$

Example 4.12

Find the quadratic covariation $[X,Y](t)$ from the following SDEs:

$$dX(t) = a_X(t)\,dt + \sigma_X(t)\,dB(t) \tag{4.79a}$$

$$dY(t) = a_Y(t)\,dt + \sigma_Y(t)\,dB(t) \tag{4.79b}$$

Solution

Definition of quadratic covariation gives:

$$d[X,Y](t) = dX(t)\,dY(t) = \sigma_X(t)\,\sigma_Y(t)\,(dB(t))^2 = \sigma_X(t)\,\sigma_Y(t)\,dt \tag{4.80}$$

The above result implies that $[X,Y](t) = \int_0^t \sigma_X(s)\,\sigma_Y(s)\,ds$.

Ito's formula for one-dimensional processes

We are now ready to state Ito's formula for a scalar function $g(t,X(t))$, where $X(t)$ is an Ito process given by:

$$X(t) = X_0 + \int_0^t a(s)\,ds + \int_0^t \sigma(s)\,dB(s), \quad 0 \le t \le T \tag{4.81a}$$

$$\Longrightarrow dX(t) = a(t)\,dt + \sigma(t)\,dB \tag{4.81b}$$

where, X_0 is $\mathcal{F}_0$–measurable and $a(t)$ and $\sigma(t)$ are $\mathcal{F}_t$–adapted stochastic processes such that $a(t)$ is integrable and $\sigma(t) \in \mathcal{H}^*$. $X(t)$ is thus the solution of the SDE (4.81b). Specifically, if $g(t,X(t))$ is twice continuously differentiable on $[0,\infty) \times \mathbb{R}$, i.e., $g \in C^2([0,\infty) \times \mathbb{R})$, we have Ito's formula:

$$dg(t,X_t) = \frac{\partial g}{\partial t}(t,X_t)\,dt + \frac{\partial g}{\partial x}(t,X_t)\,dX_t + \frac{1}{2}\frac{\partial^2 g}{\partial x^2}(t,X_t)\,(dX_t)^2 \tag{4.82}$$

Here the derivative of $g(t,X(t))$ with respect to X are supposed to be evaluated at $x = X_t$. For instance $\frac{\partial g}{\partial x}(t,X_t) = \left.\frac{\partial g(t,x)}{\partial x}\right|_{x=X_t}$.

Proof: Substituting Eq. (4.81b) in Eq. (4.82), the equivalent integral form of Ito's formula may be written as:

$$g(t,X_t) = g(0,X_0) + \int_0^t \left(\frac{\partial g}{\partial s}(s,X_s) + a_s \frac{\partial g}{\partial x}(s,X_s) + \frac{1}{2}\sigma_s^2 \frac{\partial^2 g}{\partial x^2}(s,X_s) \right) ds$$

$$+ \int_0^t \sigma_s \frac{\partial g}{\partial x}(s,X_s)\, dB_s \tag{4.83}$$

where, $a_s = a(s)$ and $\sigma_s = \sigma(s)$. Also, $(dX_t)^2 = d[X,X](t) = \sigma_t^2 dt$ is used in writing the above integral form. In simplifying $d[X,X](t)$, only terms of $O(dt)$ are retained. Hence, since dB is of $O\left(\sqrt{dt}\right)$, the following convention is made use of:

$$dt.dt = 0, \quad dt.dB = 0, \quad dB.dt = 0 \ \text{ and } dB.dB = t \tag{4.84}$$

By Taylor's expansion of $g(t,X(t))$ with respect to $g(0,X_0 := X(0))$, one has:

$$g(t,X(t)) = g(0,X_0) + \sum_j \Delta g\left(t_j,X_j\right)$$

$$= g(0,X_0) + \sum_j \frac{\partial g}{\partial t}\Delta t_j + \sum_j \frac{\partial g}{\partial x}\Delta X_j + \frac{1}{2}\sum_j \frac{\partial^2 g}{\partial t^2}\left(\Delta t_j\right)^2$$

$$+ \sum_j \frac{\partial^2 g}{\partial t \partial x}\Delta t_j \Delta X_j + \frac{1}{2}\sum_j \frac{\partial^2 g}{\partial x^2}\left(\Delta X_j\right)^2 + \sum_j R_j \tag{4.85}$$

where, $X_j := X\left(t_j\right)$, $\Delta g\left(t_j,X_j\right) := g\left(t_{j+1},X_{j+1}\right) - g\left(t_j,X_j\right)$, $\Delta t_j := t_{j+1} - t_j$, $\Delta X_j := X_{t_{j+1}} - X_{t_j}$. R_j is the remainder in Taylor's expansion and is $O\left(\left|\Delta t_j\right|^2 + \left|\Delta X_j\right|^2\right)\ \forall j$. The derivatives of $g(t,X(t))$ are evaluated at $\left(t_j,X_j\right)$. One can simplify the above expansion by representing $a(t)$ and $\sigma(t)$ through elementary functions. First, as $\Delta t_j \to 0$:

$$\sum_j \frac{\partial g}{\partial t}\Delta t_j \to \int_0^t \frac{\partial g}{\partial t}(s,X_s)ds$$

and

$$\sum_j \frac{\partial g}{\partial x} \Delta X_j \to \int_0^t \frac{\partial g}{\partial x}(s, X_s)\, dX_s$$

$$= \int_0^t \frac{\partial g}{\partial x}(s, X_s)\,(a_s dt + \sigma_s dB_s) \quad \text{(from Eq. 4.81b)}$$

$$\Longrightarrow \sum_j \frac{\partial g}{\partial x} \Delta X_j \to \int_0^t a_s \frac{\partial g}{\partial x}(s, X_s)\, ds + \int_0^t \sigma_s \frac{\partial g}{\partial x}(s, X_s)\, dB_s \tag{4.86}$$

Further, with $\Delta X_j = a_j \Delta t_j + \sigma_j \Delta B_j$ (where, $a_j = a(t_j)$, $\sigma_j = \sigma(t_j)$), we have:

$$\sum_j \frac{\partial^2 g}{\partial x^2} (\Delta X_j)^2 = \sum_j \frac{\partial^2 g}{\partial x^2} a_j^2 (\Delta t_j)^2 + 2 \sum_j \frac{\partial^2 g}{\partial x^2} a_j \sigma_j \Delta t_j \Delta B_j + \sum_j \frac{\partial^2 g}{\partial x^2} \sigma_j^2 (\Delta B_j)^2 \tag{4.87}$$

Clearly, as $\Delta t_j \to 0$, the first term on the right hand side of the above equation vanishes. Similar is the case with the second term (a zero–mean random variable) since:

$$E\left[\left(\sum_j \frac{\partial^2 g}{\partial x^2} a_j \sigma_j \Delta t_j \Delta B_j\right)^2\right] = \sum_j E\left[\left(\frac{\partial^2 g}{\partial x^2} a_j \sigma_j\right)^2\right] (\Delta t_j)^3 \to 0 \text{ in } L^2(P) \text{ as } \Delta t_j \to 0 \tag{4.88}$$

In deriving the above result, we have used the fact that ΔB_j is independent of $\frac{\partial^2 g}{\partial x^2} a_j \sigma_j$ in each term of the summation over j and $E\left[\Delta B_j^2\right] = \Delta t_j \ \forall j$. For the last term $\sum_j \frac{\partial^2 g}{\partial x^2} \sigma_j^2 (\Delta B_j)^2$ in Eq. (4.87), it follows from Eq. (4.63) that:

$$\sum_j \frac{\partial^2 g}{\partial x^2} \sigma_j^2 (\Delta B_j)^2 \to \int_0^t \frac{\partial^2 g}{\partial s^2} \sigma^2 ds \tag{4.89}$$

in $L^2(P)$ as $\Delta t_j \to 0$. Similar arguments yield that the remainder $\sum_j R_j \to 0$ as $\Delta t_j \to 0$. Thus Ito's formula (Eq. (4.83)) is proved. In differential form, it can be stated as:

$$dg(t, X_t) = \left(\frac{\partial g}{\partial t}(t, X_t) + a_t \frac{\partial g}{\partial x}(t, X_t) + \frac{1}{2}\sigma_t^2 \frac{\partial^2 g}{\partial x^2}(t, X_t)\right) dt + \sigma_t \frac{\partial g}{\partial x}(t, X_t)\, dB_t \tag{4.90}$$

It is clear from Ito's formula that $g(t, x)$ need only be C^1 with respect to t and C^2 with respect to x.

Example 4.13

Let us apply Ito's formula for $g(X(t)) = \log X(t)$ where $X(t)$ is the solution of the SDE $dX(t) = \frac{1}{2}X(t)\,dt + X(t)\,dB(t)$.

Solution

Noting that $\frac{d}{dx}(\log x) = 1/x$ and $\frac{d}{dx^2}(\log x) = -\frac{1}{x^2}$, we get via Ito's formula:

$$dg(X(t)) = d(\log X(t)) = \frac{1}{X(t)}dX(t) - \frac{1}{2X^2(t)}X^2(t)\,dt$$

$$(\text{by using } a(t) = \frac{1}{2}X(t), \sigma(t) = X(t))$$

$$= \left(dB(t) + \frac{1}{2}dt\right) - \frac{1}{2}dt = dB(t) \tag{4.91}$$

Thus, Ito's formula gives:

$$d(\log(X(t))) = dB(t) \tag{4.92}$$

Thus the solution process $X(t)$ is obtained as:

$$\Rightarrow \log(X(t)) = \log(X(0)) + B(t)$$

$$\Rightarrow X(t) = X(0)\,e^{B(t)} \tag{4.93}$$

Issues of existence and uniqueness of a solution to an SDE will be addressed in a later section of this chapter.

Example 4.14

Find the quadratic covariation $[g(B), B](t)$ by using Ito's formula

Solution

Let $g(B(t))$ be twice differentiable in its argument. We intend to find $[g(B(t)), B(t)](t)$ via Ito's formula of $g(B(t))$.

$$dg(B) = \frac{dg}{dB}dB + \frac{1}{2}\frac{d^2g}{dB^2}(dB)^2 = \frac{dg}{dB}dB + \frac{1}{2}\frac{d^2g}{dB^2}dt \tag{4.94}$$

Thus, we may obtain:

$$d[g(B),B](t) = dg(B(t))\,dB(t)$$

$$= \frac{dg}{dB}(dB(t))^2 + \frac{1}{2}\frac{d^2g}{dB^2}dB(t)\,dt$$

$$= \frac{dg}{dB}dt \quad (\text{since} \quad (dB)^2 = dt \text{ and } dBdt = 0) \tag{4.95}$$

Thus:

$$[g(B),B](t) = \int_0^t \frac{dg}{dB}ds \tag{4.96}$$

Example 4.15

If $g(t)$ is an increasing differentiable function and $X(t) = B(g(t))$, we can show that the quadratic variation of $X(t)$ is given by:

$$[X,X](t) = g(t) \tag{4.97}$$

Solution

Let $g(t) = \tau_t$, a strictly increasing function of time. Since there is one-to-one correspondence between t and τ_t, we can as well define the quadratic variation of $X(t) = B(\tau_t)$ over a shrinking partition Π_N of $[0, \tau_t]$:

$$[X,X](\tau_t) = [B(\tau_t), B(\tau_t)](\tau_t)$$

$$= \lim_{\Delta_N \to 0} \sum_{j=0}^{N-1} \left(B(\tau_{j+1}) - B(\tau_j)\right)^2$$

$$= \lim_{\Delta_N \to 0} \sum_{j=0}^{N-1} \Delta B_j^2 = \tau_t = g(t) \tag{4.98}$$

4.4.3 Ito's formula for higher dimensions

First consider Ito's formula for a function $g(t,X,Y)$ of two variables, which is C^1 in t and C^2 in the last two arguments. Let X and Y be Ito processes governed by SDEs (4.79). As before, Ito's formula is obtainable from a Taylor expansion as:

$$dg(t,X_t,Y_t) = \frac{\partial g(t,X_t,Y_t)}{\partial t}dt + \frac{\partial g(t,X_t,Y_t)}{\partial x}dX_t + \frac{\partial g(t,X_t,Y_t)}{\partial y}dY_t$$

$$+\frac{1}{2}\frac{\partial^2 g(t,X_t,Y_t)}{\partial x^2}(dX_t)^2 + \frac{\partial^2 g(t,X_t,Y_t)}{\partial x \partial y}(dX_t)(dY_t) + \frac{1}{2}\frac{\partial^2 g(t,X_t,Y_t)}{\partial y^2}(dY_t)^2 \quad (4.99)$$

Here $(dX_t)^2 = dX_t dX_t = d[X,X](t) = \sigma_X^2(t)dt$ and similarly $(dY_t)^2 = \sigma_Y^2(t)dt$ and $(dX_t)(dY_t) = d[X,Y](t) = \sigma_X(t)\sigma_Y(t)dt$. The proof is similar to the one-dimensional case and it leads finally to Ito's formula for a function of two variables as:

$$dg(t,X_t,Y_t) = \left(\frac{\partial g(t,X_t,Y_t)}{\partial t} + a_X(t)\frac{\partial g(t,X_t,Y_t)}{\partial x} + a_Y(t)\frac{\partial g(t,X_t,Y_t)}{\partial y}\right)dt$$

$$+\frac{1}{2}\left(\sigma_X^2(t)\frac{\partial^2 g(t,X_t,Y_t)}{\partial x^2} + 2\sigma_X(t)\sigma_Y(t)\frac{\partial^2 g(t,X_t,Y_t)}{\partial x \partial y} + \sigma_Y^2(t)\frac{\partial^2 g(t,X_t,Y_t)}{\partial y^2}\right)dt$$

$$+\left(\sigma_X(t)\frac{\partial g(t,X_t,Y_t)}{\partial x} + +\sigma_Y(t)\frac{\partial g(t,X_t,Y_t)}{\partial x}\right)dB_t \quad (4.100)$$

It is clear that the formula involves both the quadratic variations $[X,X](t)$ and $[Y,Y](t)$ and the quadratic covariation $[X,Y](t)$. Extension of the Ito's formula to dimensions > 2 is straightforward.

Indeed, we could consider the most general case where $\boldsymbol{B}(t) = (B_1(t),\ B_2(t),\ldots,\ B_n(t))^T$ is an n-dimensional vector of independent one-dimensional Wiener processes driving the m-dimensional SDE considered below. Let $\boldsymbol{X}(t)$ be an m-dimensional Ito process vector driven by $\boldsymbol{B}(t)$, i.e.:

$$d\boldsymbol{X}(t) = \boldsymbol{a}(t)dt + \boldsymbol{\sigma}(t)d\boldsymbol{B}(t) \quad (4.101)$$

Here $\boldsymbol{a}(t) \in \mathcal{H}^*$ is the drift vector and $\boldsymbol{\sigma}(t) \in \mathcal{H}^*$ the matrix of diffusion coefficients given by:

$$\boldsymbol{a}(t) = \begin{Bmatrix} a_1(t) \\ a_2(t) \\ \cdot \\ \cdot \\ \cdot \\ a_m(t) \end{Bmatrix}, \sigma(t) = \begin{bmatrix} \sigma_{11}(t) & \sigma_{12}(t) & \ldots & \sigma_{1n}(t) \\ \sigma_{21}(t) & \sigma_{22}(t) & \ldots & \sigma_{2n}(t) \\ & & \cdot & \\ & & \cdot & \\ & & \cdot & \\ \sigma_{m1}(t) & \sigma_{m2}(t) & \ldots & \sigma_{mn}(t) \end{bmatrix}, d\boldsymbol{B}(t) = \begin{Bmatrix} dB_1(t) \\ dB_2(t) \\ \cdot \\ \cdot \\ \cdot \\ dB_n(t) \end{Bmatrix} \quad (4.102)$$

It is assumed that $a_j(t)$ and $\sigma_{ij}(t)$ satisfy the conditions:

$$P\left(\int_0^t |a_i(s)|\,ds < \infty \text{ for all } t \geq 0\right) = 1, 1 \leq i \leq m \tag{4.103a}$$

$$P\left(\int_0^t (\sigma_{ij}(s))^2\,ds < \infty \text{ for all } t \geq 0\right) = 1, 1 \leq i \leq m, 1 \leq j \leq n \tag{4.103b}$$

such that each component Ito process dX_i is given by:

$$dX_i = a_i(t)\,dt + \sum_{j=1}^{n} \sigma_{ij}(t)\,dB_j(t), 1 \leq i \leq m \tag{4.104}$$

Now if $\boldsymbol{g}(t,\boldsymbol{X}_t) = (g_1(t,\boldsymbol{X}_t), g_2(t,\boldsymbol{X}_t), \ldots, g_p(t,\boldsymbol{X}_t))$ is a C^2 function such that: $[0,\infty) \times \mathbb{R}^m \to \mathbb{R}^p$, then Ito's formula for each member function $g_l(t,\boldsymbol{X}_t)$ is given by:

$$dg_l(t,\boldsymbol{X}_t) = \frac{\partial g_l(t,\boldsymbol{X}_t)}{\partial t}dt + \sum_{i=1}^{m} \frac{\partial g_l(t,\boldsymbol{X}_t)}{\partial x_i}dX_i + \frac{1}{2}\sum_{i=1}^{m}\sum_{k=1}^{m} \frac{\partial^2 g_l(t,\boldsymbol{X}_t)}{\partial x_i \partial x_k}dX_i dX_k \tag{4.105}$$

where, $1 \leq l \leq p$. With $dB_i dt = dt dB_i = 0$ and $dB_i dB_j = \delta_{ij} dt$ where δ_{ij} is the Dirac delta function (Section 1.6.4, Chapter 1), the formula above can be expressed in terms of the drift and diffusion coefficients as:

$$dg_l(t,\boldsymbol{X}_t) = \left(\frac{\partial g_l(t,\boldsymbol{X}_t)}{\partial t} + \sum_{i=1}^{m} a_i \frac{\partial g_l(t,\boldsymbol{X}_t)}{\partial x_i} + \frac{1}{2}\sum_{i=1}^{m}\sum_{k=1}^{m}\sum_{j=1}^{n} \sigma_{ij}\sigma_{kj} \frac{\partial^2 g_l(t,\boldsymbol{X}_t)}{\partial x_i \partial x_k}\right)dt$$

$$+ \sum_{i=1}^{m}\sum_{j=1}^{n} \sigma_{ij} \frac{\partial g_l(t,\boldsymbol{X}_t)}{\partial x_i} dB_j \tag{4.106}$$

It is interesting to observe that for any differentiable vector process $\boldsymbol{X}(t)$, one has $d[X_i, X_j](t) = 0$ for all i, j and hence Ito's formula in Eq. (4.105) reduces to the classical calculus-based relation for the total derivative:

$$\frac{dg_l(t,\boldsymbol{X}_t)}{dt} = \frac{\partial g_l(t,\boldsymbol{X}_t)}{\partial t} + \sum_{i=1}^{m} \frac{\partial g_l(t,\boldsymbol{X}_t)}{\partial x_i}\frac{dX_i}{dt} \tag{4.107}$$

Example 4.16

Consider two scalar valued stochastic processes $X(t)$ and $Y(t)$, then we have the following product rule in Ito calculus:

$$d(XY) = XdY + YdX + d[X,Y](t) = XdY + YdX + dX(t)\,dY(t) \quad (4.108)$$

Solution

To prove the above equality, choose a scalar function $g(X,Y) = XY$. By Ito's formula (Eq. 4.99) for the function of two variables, we have:

$$dg(X,Y) = \frac{\partial g(X,Y)}{\partial x} dX + \frac{\partial g(X,Y)}{\partial y} dY$$

$$+\frac{1}{2}\left(\frac{\partial^2 g(X,Y)}{\partial x^2}(dX)^2 + 2\frac{\partial^2 g(X,Y)}{\partial x \partial y}(dX)(dY) + \frac{\partial^2 g(X,Y)}{\partial y^2}(dY)^2\right)$$

$$\Longrightarrow dg(X,Y) = d(XY) = YdX + XdY + dXdY$$

$$\left(\text{since } \frac{\partial^2 g(X,Y)}{\partial x^2} = 0 = \frac{\partial^2 g(X,Y)}{\partial y^2} \text{ and } \frac{\partial^2 g(X,Y)}{\partial x \partial y} = 1\right)$$

$$\Longrightarrow d(XY) = YdX + XdY + d[X,Y](t) \quad (4.109)$$

Product rule and integration by parts

From the product rule in Eq. (4.109), we can deduce the following formula for integration by parts:

$$\int_0^t X(s)\,dY(s) = \int_0^t (d(XY) - YdX - d[X,Y](t))$$

$$= (X(t)Y(t) - X(0)Y(0)) - \int_0^t Y(s)\,dX(s) - \int_0^t dX(s)\,dY(s) \quad (4.110)$$

This formula can be established alternatively by considering the covariation of two Ito processes X and Y. Over decreasing partitions of $[0,t]$, one has:

$$[X,Y](t) = \lim_{\Delta_N \to 0} \sum_{j=0}^{N-1} \left(X(t_{j+1}) - X(t_j)\right)\left(Y(t_{j+1}) - Y(t_j)\right)$$

$$= \lim_{\Delta_N \to 0} \sum_{j=0}^{N-1} \left(X\left(t_{j+1}\right) Y\left(t_{j+1}\right) - X(t_j) Y(t_j) \right) - \lim_{\Delta_N \to 0} \sum_{0}^{t} X\left(t_j\right) \left(Y\left(t_{j+1}\right) - Y(t_j) \right)$$

$$- \lim_{\Delta_N \to 0} \sum_{j=0}^{N-1} Y(t_j) \left(X\left(t_{j+1}\right) - X(t_j) \right)$$

$$= \left(X(t) Y(t) - X(0) Y(0) \right) - \lim_{\Delta_N \to 0} \sum_{0}^{t} X\left(t_j\right) \left(Y\left(t_{j+1}\right) - Y(t_j) \right)$$

$$- \lim_{\Delta_N \to 0} \sum_{j=0}^{N-1} Y(t_j) \left(X\left(t_{j+1}\right) - X(t_j) \right) \tag{4.111}$$

The last two summation terms on the right hand side of the above equation converge a.s. to the Ito integrals $\int_0^t X(s) \, dY(s)$ and $\int_0^t Y(s) \, dX(s)$ respectively. Thus we have:

$$[X,Y](t) = X(t) Y(t) - X(0) Y(0) - \int_0^t X(s) \, dY(s) - \int_0^t Y(s) \, dX(s) \tag{4.112}$$

By re-arranging the terms in the above identity and with $[X,Y](t) = \int_0^t dX(s) \, dY(s)$, we arrive at the same formula as in Eq. (4.110) for integration by parts. In differential form, obviously, the formula gives the same stochastic (Ito's) product rule as in Eq. (4.109).

Note that if one of the two processes is continuous and of finite variation, then $d[X,Y](t) = 0$ and the product rule in Eq. (4.109) and the integration by parts rule in Eq. (4.110) reduce to the well-known rules of standard calculus.

An alternative form of quadratic variation of a process

Equation (4.112) leads to the following alternative expression for the quadratic variation of a stochastic process $X(t)$:

$$[X,X](t) = X^2(t) - X^2(0) - 2 \int_0^t X(s) \, dX(s) \tag{4.113}$$

Specifically for Wiener process $B(t)$ with $B(0) = 0$, the above formula gives:

$$[B,B](t) = B^2(t) - 2 \int_0^t B(s) \, dB(s) \tag{4.114}$$

With $[B,B](t) = t$, the above result is same as the one obtained in Example 4.3 where the Ito integral $\int_0^t B(s) \, dB(s)$ is shown to be $\frac{1}{2}\left(B^2(t) - t\right)$.

Example 4.17

We derive the equations determining the response moments of an SDE corresponding to a single degree of freedom (SDOF) mechanical oscillator. Here the equation of motion, in its conventional form, is given by:

$$\ddot{X}(t) + 2\xi\overline{\omega}\dot{X}(t) + \overline{\omega}^2 X(t) = \sigma_s(t) W(t) \tag{4.115}$$

$\overline{\omega}$ is known as the natural frequency of the oscillator and ξ the damping ratio. $W(t)$ is a white noise process with intensity $\sigma_s(t)$.

Solution

While the above oscillator equation may find relevance in many fields of engineering and science, it is specifically referred to here in the context of the dynamics of a structural system [Roy and Rao 2012]. It plays an important role as a building block in constructing the free or forced vibration solution of the system, irrespective of whether the system is continuous or discrete. A discrete system, typically represented by a family of coupled SDOF oscillators, is commonly referred to as a multi degree of freedom (MDOF) system. See Appendix D for brief notes on continuous and discrete structural systems driven by stochastic inputs. For the SDOF system represented by Eq. (4.115), $X(t)$ denotes the displacement, $\dot{X}(t) = \frac{dX}{dt}$ the velocity and $\ddot{X}(t) = \frac{d^2X}{dt^2}$ the acceleration processes. With $W(t) = \dot{B}(t)$ (see discussion in Section 4.4.1), Eq. (4.115) may be more appropriately recast into two coupled first order SDEs:

$$dX(t) = Y(t)\,dt \tag{4.116a}$$

$$dY(t) = -\left(2\xi\overline{\omega}Y(t) + \overline{\omega}^2 X(t)\right)dt + \sigma_s(t)\,dB(t) \tag{4.116b}$$

Of general interest is the evaluation of system response moments such as $E\left[X^2(t)\right]$, $E\left[Y^2(t)\right]$ and $E[X(t)Y(t)]$. To this end, we use Ito's formula to obtain SDEs for the solution processes. $X^2(t)$, $Y^2(t)$ and $X(t)Y(t)$. Comparing the above coupled equations with Eq. (4.102), one has:

$$\boldsymbol{a}(t,\boldsymbol{X}) = \begin{Bmatrix} a_1 \\ a_2 \end{Bmatrix} = \begin{Bmatrix} Y(t) \\ -2\xi\overline{\omega}Y(t) - \overline{\omega}^2 X(t) \end{Bmatrix}, \boldsymbol{\sigma}(t,\boldsymbol{X}) = \begin{bmatrix} \sigma_{11} \\ \sigma_{21} \end{bmatrix} = \begin{bmatrix} 0 \\ \sigma_s(t) \end{bmatrix},$$

$$d\boldsymbol{B}(t) = \begin{Bmatrix} 0 \\ dB(t) \end{Bmatrix}, \boldsymbol{X} = \begin{Bmatrix} X \\ Y \end{Bmatrix} \tag{4.117}$$

Now choose the scalar-valued functions $g_1(t,\mathbf{X}) = X^2(t)$, $g_2(t,\mathbf{X}) = X(t)Y(t)$ and $g_3(t,\mathbf{X}) = Y^2(t)$. Identifying $n = 1$, $m = 2$ and $p = 3$, we determine the following derivatives:

$$\frac{\partial g_1}{\partial x} = 2X(t), \ \frac{\partial g_1}{\partial y} = 0, \ \frac{\partial g_2}{\partial x} = Y(t), \ \frac{\partial g_2}{\partial y} = X(t), \ \frac{\partial g_3}{\partial x} = 0,$$

$$\frac{\partial g_3}{\partial y} = 2Y(t) \tag{4.118a}$$

and

$$\frac{\partial^2 g_1}{\partial x^2} = 2, \ \frac{\partial^2 g_1}{\partial x \partial y} = 0 = \frac{\partial^2 g_1}{\partial y \partial x}, \ \frac{\partial^2 g_1}{\partial y^2} = 0,$$

$$\frac{\partial^2 g_2}{\partial x^2} = 0, \ \frac{\partial^2 g_2}{\partial y \partial x} = 1 = \frac{\partial^2 g_2}{\partial x \partial y}, \ \frac{\partial^2 g_2}{\partial y^2} = 0 \tag{4.118b}$$

$$\frac{\partial^2 g_3}{\partial x^2} = 0, \ \frac{\partial^2 g_3}{\partial y \partial x} = 0 = \frac{\partial^2 g_3}{\partial x \partial y}, \ \frac{\partial^2 g_3}{\partial y^2} = 2 \tag{4.118c}$$

Using Eq. (4.106), we can now write the following SDEs:

$$dg_1 = d\left(X^2(t)\right) = a_1 \frac{\partial g_1}{\partial x} dt = 2X(t)\,Y(t)\,dt \tag{4.119a}$$

$$dg_2 = d\left(X(t)\,Y(t)\right) = \left(a_1 \frac{\partial g_2}{\partial x} + a_2 \frac{\partial g_2}{\partial y}\right) dt + \sigma_{21} \frac{\partial g_2}{\partial y} dB(t)$$

$$= \left\{Y^2(t) + X(t)\left(-2\xi\overline{\omega}Y(t) - \overline{\omega}^2 X(t)\right)\right\} dt + \sigma_s(t)\,X(t)\,dB(t)$$

$$= \left\{-\overline{\omega}^2 X^2(t) - 2\xi\overline{\omega}X(t)\,Y(t)\ \ Y^2(t)\right\} dt + \sigma_s(t)\,X(t)\,dB(t) \tag{4.119b}$$

and

$$dg_3 = d\left(Y^2(t)\right) = \left(a_2 \frac{\partial g_3}{\partial y} + \frac{1}{2}\sigma_{21}^2 \frac{\partial^2 g_3}{\partial y^2}\right) dt + \sigma_{21} \frac{\partial g_3}{\partial y} dB(t)$$

$$= \left\{\left(-2\xi\overline{\omega}Y(t) - \overline{\omega}^2 X(t)\right)\left(2Y(t)\right) + \frac{1}{2}\sigma_s^2(t)(2)\right\} dt + 2Y(t)\,\sigma_s(t)\,dB(t)$$

$$= \left(-4\xi\overline{\omega}Y^2(t) - 2\overline{\omega}^2 X(t)\,Y(t) + \sigma_s^2(t)\right) dt + 2Y(t)\sigma_s(t)\,dB(t) \tag{4.119c}$$

Note that unlike the SDE for g_1, those for g_2 and g_3 have Ito integral terms, which evolve as zero–mean martingales. To find the second moment statistics for the SDOF oscillator, we can take expectations of Eqs. (4.119a), (4.119b), and (4.119c) to arrive at the following system of ODEs:

$$\frac{d}{dt}\left(E\left[X^2(t)\right]\right) = 2E\left[X(t)\,Y(t)\right] \tag{4.120a}$$

$$\frac{d}{dt}\left(E\left[X(t)\,Y(t)\right]\right) = -\overline{\omega}^2 E\left[X^2(t)\right] - 2\xi\overline{\omega}E\left[X(t)\,Y(t)\right] + E\left[Y^2(t)\right] \tag{4.120b}$$

$$\frac{d}{dt}\left(E\left[Y^2(t)\right]\right) = -4\xi\overline{\omega}E\left[Y^2(t)\right] - 2\overline{\omega}^2 E\left[X(t)\,Y(t)\right] + \sigma_s^2(t) \tag{4.120c}$$

Equations (4.120 a–c) constitute a closed system of linear ODEs in the second moments which may be solved (analytically or through numerical integration), provided known initial conditions are supplied. In the special case of the SDOF system response being stationary (e.g., when $\sigma_s^2(t) = \sigma_s^2$, a constant), we must have $\frac{d}{dt}E\left[X^2\right] = \frac{d}{dt}E\left[XY\right] = \frac{d}{dt}E\left[Y^2\right] = 0$. Here the above set of ODEs reduces to 3 linear algebraic equations in as many unknowns. Solving this algebraic system yields the so-called stochastic steady state values:

$$E\left[X^2\right] = E\left[Y^2\right]/\overline{\omega}^2, E\left[Y^2\right] = \sigma_s^2/4\xi\overline{\omega}, E\left[XY\right] = 0 \tag{4.121}$$

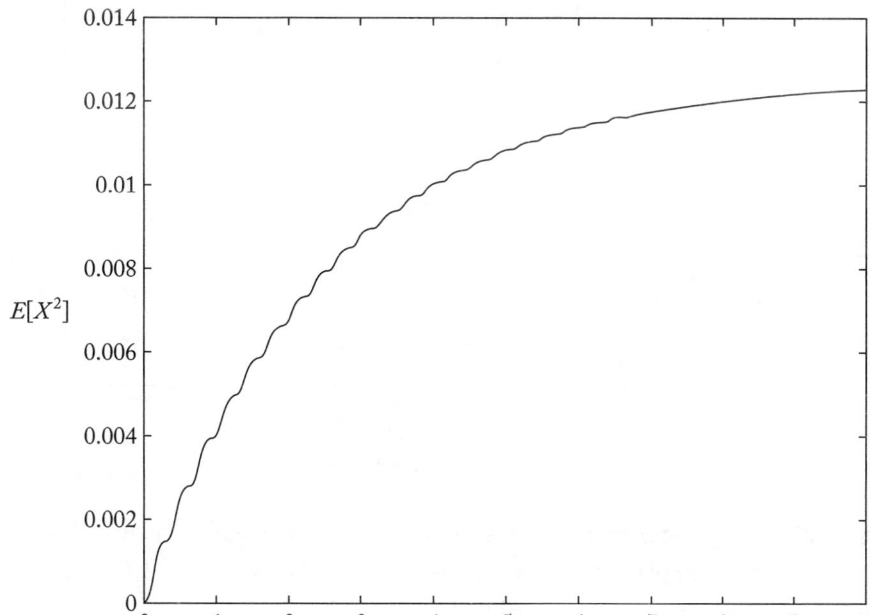

(a)

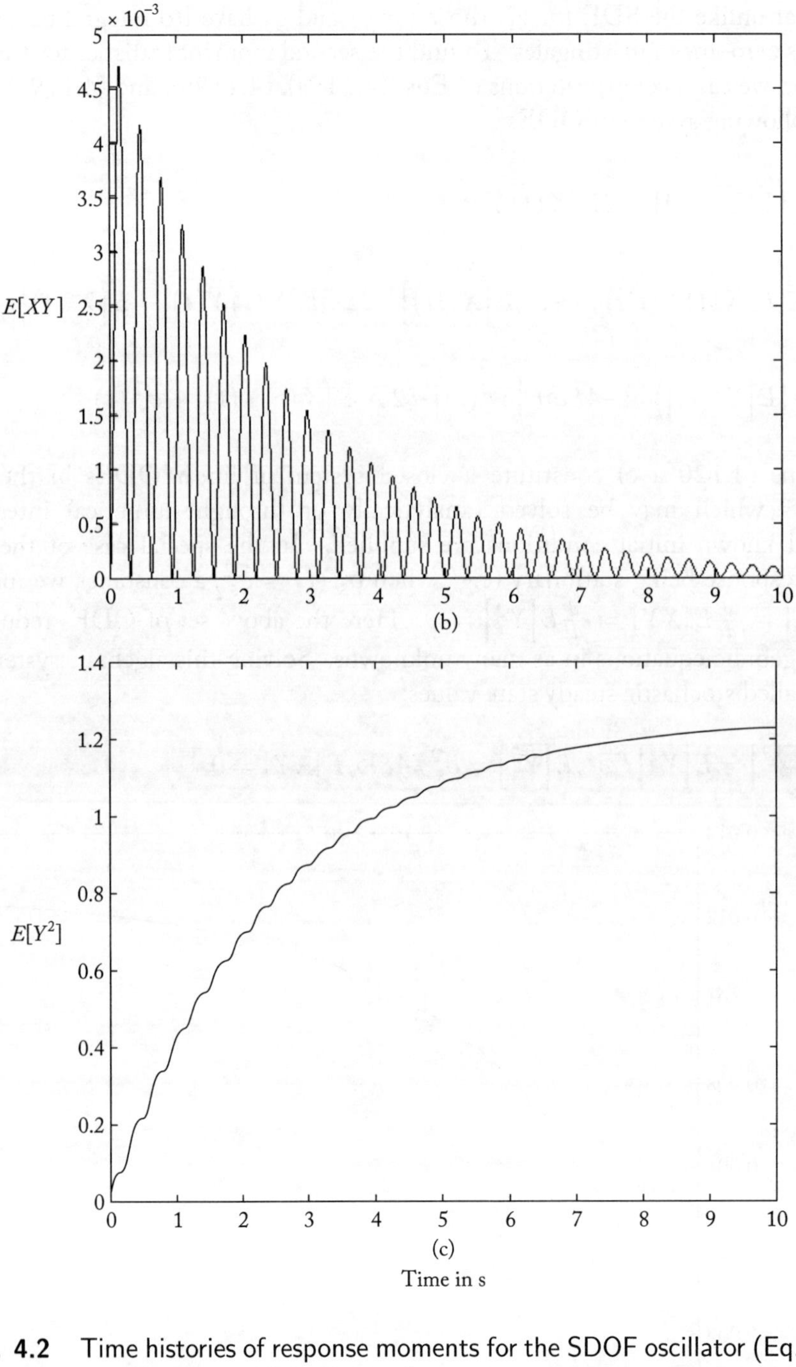

Fig. 4.2 Time histories of response moments for the SDOF oscillator (Eq. 4.120) of Example 4.17; $\overline{\omega} = 10\frac{\text{rad}}{s}$, $\xi = 0.02$ and $\sigma_s = 1.0$, (a) $E\left[X^2\right]$, (b) $E\left[XY\right]$, and (c) $E\left[Y^2\right]$

In order to determine the (transient) evolution of the above moments (i.e., before the so-called stochastic steady state is attained), the ODEs in Eq. (4.120) are numerically integrated by a 4th order Runge–Kutta algorithm (subject to identically zero initial conditions). The time histories of the three moments of the oscillator are shown in Fig. 4.2. Parameter values used for this purpose are given by $\overline{\omega} = 10\text{rad/sec.}$, $\xi = 0.02$ and $\sigma_s = 1.0$. In Appendix D, an analytical solution (see Fig. D.1) is obtained, for instance, for $E\left[X^2(t)\right]$ and can be found to be close to the numerical result shown in Fig. 4.2(a).

4.4.4 Dynamical system of higher dimension and application of Ito's formula

Consider an m-dimensional dynamical system (e.g., an MDOF mechanical oscillator) driven by an n-dimensional Brownian vector $\boldsymbol{B}(t)$. The SDEs may be expressed in the general form:

$$dX_i(t) = a_i(t,\boldsymbol{X})\,dt + \sum_{j=1}^{n} \sigma_{ij}(t,\boldsymbol{X})\,dB_j(t), i = 1,2\ldots,m \tag{4.122}$$

Here $\boldsymbol{X} = \{X_1(t), X_2(t), \ldots, X_m(t)\}^T$ denotes the response vector (which, for mechanical oscillators, contains both the displacement and velocity components). Thus, for an MDOF system, m is usually an even number with $\frac{m}{2}$ coupled SDOF oscillators corresponding to $\frac{m}{2}$ displacement components (see Eq. (4.116a). In this context, if $\mathcal{F}_t$ denotes the filtration generated by the Brownian vector $\boldsymbol{B}(s)$, $s \le t$, $\sigma_{ij}(t,\boldsymbol{X})$ (if multiplicative) are adapted to $\mathcal{F}_t$. For linear dynamical systems, the drift coefficient $a_i(t,\boldsymbol{X})$ must be linear in $\boldsymbol{X}$, even though Ito's formula in Eq. (4.105) has the same form for non-linear drift terms and for multiplicative diffusion coefficients $\sigma_{ij}(t,\boldsymbol{X})$ as well. An additional fact that needs to be noted is on the quadratic covariation of the scalar components of a vector $\boldsymbol{B}(t)$, which we state below.

Quadratic covariation of independent Brownian components

For any two independent Brownian motion components $B_k(t)$ and $B_l(t)$, $k,l \in [1,n]$ and $k \neq l$, we claim that $[B_k, B_l](t) = 0$ in $L^2(\mathrm{P})$. To see this, once more consider a partition Π_N of $[0,t]$ and, using the definition of quadratic covariation, construct the sum:

$$Q_n = \sum_{j=0}^{N-1} \left(B_k(t_{j+1}) - B_k(t_j)\right)\left(B_l(t_{j+1}) - B_l(t_j)\right) \tag{4.123}$$

The independence of $B_k(t)$ and $B_l(t)$ implies that $E[Q_n] = 0$. Moreover, since variance of Q_n is the sum of variances of the individual terms (which follows via the independence of Brownian increments), we get:

$$var\ Q_n = \sum_{j=0}^{N-1} E\left[\left(B_k\left(t_{j+1}\right) - B_k\left(t_j\right)\right)^2\right] E\left[\left(B_l\left(t_{j+1}\right) - B_l\left(t_j\right)\right)^2\right]$$

$$= \sum_{j=0}^{N-1} \left(t_{j+1} - t_j\right)^2 \le \Delta_N t_N; \Delta_N := \max_j \left(t_{j+1} - t_j\right) \tag{4.124}$$

Hence, Var $Q_N = E\left[Q_N{}^2\right] \to 0$ as $\Delta_N \to 0$, showing that the claim is valid in $L^2(P)$. Now if $g(\boldsymbol{X})$ is a scalar C^2 function of its vector argument so that $g(\boldsymbol{X})$ is an Ito diffusion process, then by Ito's formula for higher dimensions the SDE for $g(\boldsymbol{X})$ is given by $dg(\boldsymbol{X}) = \sum_{j=1}^{m} \frac{\partial g(\boldsymbol{X})}{\partial x_i} dX_i(t) + \sum_{i,j=1}^{m} \frac{1}{2} \frac{\partial^2 g(\boldsymbol{X})}{\partial x_i \partial x_j} dX_i(t) dX_j(t)$ (see Eq. 4.105).

Note that the zero quadratic covariation of Brownian components must be used whilst evaluating $dX_i(t)\, dX_j(t) = d\left[X_i, X_j\right](t)$. For example, consider the component processes X_1, X_2 driven by independent Brownian motion components $B_1(t)$ and $B_2(t)$ respectively. If $g(X_1, X_2) = X_1 X_2$, then we have by Ito's formula:

$$dg(X_1, X_2)(t) = d(X_1 X_2) = X_1(t)\, dX_2(t) + X_2(t)\, dX_1(t) + dX_1 dX_2$$

$$= X_1(t)\, dX_2(t) + X_2(t)\, dX_1(t) \tag{4.125}$$

The above is true since $d[X_1, X_2](t) = dX_1 dX_2 = 0$ which is implied by the fact that, $d[B_1, B_2](t) = 0$. The result here is identical to the product rule in conventional calculus.

Response moments for higher dimensional dynamical systems and Dynkin's formula

In Example 4.17, equations for second order response moments for a linear SDOF oscillator under white noise excitation were derived directly using Ito's formula. We now formalize this procedure in the context of an m-dimensional SDE given by Eq. (4.122). Thus, considering a scalar function $g(t, \boldsymbol{X})$ which is at least twice differentiable, we recast Ito's formula as:

$$dg(t, \boldsymbol{X}(t)) = \left(\frac{\partial g(t, \boldsymbol{X}(t))}{\partial t} + L_t(g(t, \boldsymbol{X}(t)))\right) dt + \frac{\partial g(t, \boldsymbol{X}(t))}{\partial \boldsymbol{x}} \sigma(t)\, d\boldsymbol{B}(t) \tag{4.126a}$$

where the operator L_t is defined as:

$$L_t g = \sum_{i=1}^{m} a_i(t, \boldsymbol{X}(t)) \frac{\partial g}{\partial_{x_i}} + \frac{1}{2} \sum_{i,j=1}^{m} \sum_{l=1}^{n} \sigma_{il}(t, \boldsymbol{X}(t)) \sigma_{jl}(t, \boldsymbol{X}(t)) \frac{\partial^2 g}{\partial_{x_i} \partial_{x_j}} \tag{4.126b}$$

Note that, in Eq. (4.126a), $\frac{\partial g(t,\boldsymbol{X}(t))}{\partial \boldsymbol{x}} = \left\{\frac{\partial g}{\partial x_1}, \frac{\partial g}{\partial x_1}, \ldots\ldots, \frac{\partial g}{\partial x_m}\right\}$ is a vector, σ is an $m \times n$ dimensional diffusion matrix and $\boldsymbol{B}(t)$ is an n-dimensional Brownian motion vector. The integral form of Eq. (4.126a) may be written as:

$$g(t,\boldsymbol{X}(t)) = g(0,\boldsymbol{X}(0)) + \int_0^t \left(\frac{\partial g(s,\boldsymbol{X}(s))}{\partial s} + L_s(g(s,\boldsymbol{X}(s)))\right) ds$$

$$+ \int_0^t \frac{\partial g(s,\boldsymbol{X}(s))}{\partial \boldsymbol{x}} \sigma(s)\, d\boldsymbol{B}(s) \tag{4.127}$$

where the last term (Ito integral) on the RHS of the above equation is a zero–mean martingale (as $\left|\frac{\partial g}{\partial \boldsymbol{x}}\right|$ is bounded owing to the stipulated continuity of g with respect to the components of $\boldsymbol{X}$). By taking expectations on both sides of Eq. (4.127) (with $\boldsymbol{X}(0)$ being non-random), we arrive at what is referred to as Dynkin's formula:

$$E[g(t,\boldsymbol{X}(t))] = g(0,\boldsymbol{X}(t)) + E\left[\int_0^t \left(L_s g(s,\boldsymbol{X}(s)) + \frac{\partial g(s,\boldsymbol{X}(s))}{\partial s}\right) ds\right] \tag{4.128}$$

Example 4.18

We solve the SDE corresponding to an MDOF mechanical oscillator in terms of the response moments.

Solution

Consider an MDOF system as shown in Fig. 4.3. It is a 2-DOF system represented by the displacement vector $(X_1, X_2)^T$ and excited by zero mean white noise processes $W_1(t)$ and $W_2(t)$ as shown in the figure.

Defining the two velocities $Y_1(t) := \dot{X}_1(t)$ and $Y_2(t) := \dot{X}_2(t)$, we can write the incremental form of SDEs for the MDOF system as follows:

$$dX_1(t) = Y_1(t)\, dt \tag{4.129a}$$

$$dY_1(t) = \frac{1}{m_1}(-c_1 Y_1(t) + c_2 Y_2(t) - k_1 X_1(t) + k_2 X_2(t))\, dt + \sigma_2(t)\, dB_1(t)/m_1 \tag{4.129b}$$

$$dX_2(t) = Y_2(t)\, dt \tag{4.129c}$$

$$dY_2(t) = \frac{1}{m_2}(c_3 Y_1(t) - c_4 Y_2(t) + k_3 X_1(t)$$

$$-k_4 X_2(t))\, dt\, \sigma_4(t)\, dB_2(t)/m_2 \tag{4.129d}$$

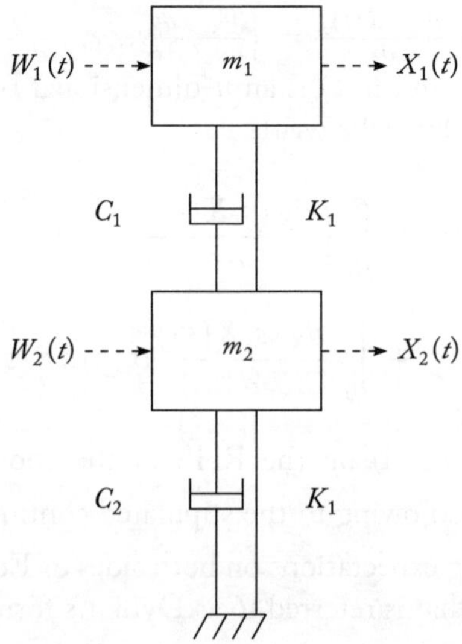

Fig. 4.3 MDOF system (Example 4.18) under stochastic input; masses: $m_1 = 1.0$, $m_2 = 2.0$, stiffnesses: $K_1 = 1200$, $K_2 = 1800$, damping values: $C_1 = 5.0$, $C_2 = 2.0$

Here $k_1 = K_1 = 1200.0$, $k_2 = K_1 = 1200.0$, $k_3 = K_1 = 1200.0$ and $k_4 = K_1 + K_2 = 3000.0$. Moreover, $c_1 = C_1 = 5.0$, $c_2 = C_1 = 5.0$, $c_3 = C_1 = 5.0$ and $c_4 = C_1 + C_2 = 7.0$. The units of these stiffnesses are in N/m and those of the damping coefficients are in N-s/m. The above is a four-dimensional ($m = 4$) system of SDEs and the Brownian motion vector $(B_1, B_2)^T$ is two-dimensional ($n = 2$). Thus, formally, $W_1(t) \equiv \sigma_2 \dot{B}_1$ and $W_2(t) \equiv \sigma_4 \dot{B}_2$. $\sigma(t)$, the 4×2 matrix of diffusion coefficients, is $\begin{bmatrix} 0 & \frac{\sigma_2}{m_1} & 0 & 0 \\ 0 & 0 & 0 & \frac{\sigma_4}{m_2} \end{bmatrix}^T$ where, $\sigma_2 = 1.0$ and $\sigma_4 = 0.5$ are the assumed intensities of the two white noise processes $W_1(t)$ and $W_2(t)$ respectively. With $\mathbf{X}(t) = \{X_1(t), Y_1(t), X_2(t), Y_2(t)\}^T$, the vector of drift coefficients is given by:

$$\boldsymbol{a}(t, \mathbf{X}(t)) = \left\{ \begin{array}{c} Y_1(t) \\ \frac{1}{m_1}\left(-c_1 Y_1(t) + c_2 Y_2(t) - k_1 X_1(t) + k_2 X_2(t)\right) \\ Y_2(t) \\ \frac{1}{m_2}\left(c_3 Y_1(t) - c_4 Y_2(t) + k_3 X_1(t) - k_4 X_2(t)\right) \end{array} \right\} \tag{4.130}$$

If the SDEs above are subject to zero initial conditions, it immediately follows that the first order moments, e.g., $E[X_1]$, $E[Y_1]$, $E[X_2]$, $E[Y_2]$ are identically zero. Now one may utilize either Dynkin's formula in Eq. (4.128) (which directly yields the ODE-s for the moments) or Ito's formula in Eq. (4.105) (which yields the associated SDEs and thus requires the additional step of taking expectations on both sides) to solve for the second order moments. Towards this, we have to choose appropriate scalar functions

$g(t,\boldsymbol{X})$, which in this case are the second order monomials in terms of the response variables, i.e., $g = {X_1}^2, {Y_1}^2, X_1Y_1, {X_2}^2, {Y_2}^2, X_2Y_2, X_1X_2, Y_1Y_2, X_1Y_2$ and X_2Y_1 and write Ito's (or Dynkin's) formula for each such g in turn. First let $g = {X_1}^2$ and noting that g is twice differentiable in $\boldsymbol{X}$, we have:

$$\frac{\partial g}{\partial \boldsymbol{x}} = \left(\frac{\partial g}{\partial x_1}, \frac{\partial g}{\partial y_1}, \frac{\partial g}{\partial x_2}, \frac{\partial g}{\partial y_2}\right) = (2X_1, 0, 0, 0) \text{ and} \tag{4.131a}$$

$$\left[\frac{\partial^2 g}{\partial \boldsymbol{x}_i \boldsymbol{x}_j}\right] = \begin{bmatrix} \frac{\partial^2 g}{\partial {x_1}^2} & \frac{\partial^2 g}{\partial x_1 \partial y_1} & \frac{\partial^2 g}{\partial x_1 \partial x_2} & \frac{\partial^2 g}{\partial x_1 \partial y_2} \\ \frac{\partial^2 g}{\partial y_1 \partial x_1} & \frac{\partial^2 g}{\partial {y_1}^2} & \frac{\partial^2 g}{\partial y_1 \partial x_2} & \frac{\partial^2 g}{\partial y_1 \partial y_2} \\ \frac{\partial^2 g}{\partial x_2 \partial x_1} & \frac{\partial^2 g}{\partial x_2 \partial y_1} & \frac{\partial^2 g}{\partial x_2^2} & \frac{\partial^2 g}{\partial x_2 \partial y_2} \\ \frac{\partial^2 g}{\partial y_2 \partial x_1} & \frac{\partial^2 g}{\partial y_2 \partial y_1} & \frac{\partial^2 g}{\partial y_2 \partial x_2} & \frac{\partial^2 g}{\partial {y_2}^2} \end{bmatrix} = \begin{bmatrix} 2 & 0 & 0 & 0 \\ 0 & 0 & 0 & 0 \\ 0 & 0 & 0 & 0 \\ 0 & 0 & 0 & 0 \end{bmatrix} \tag{4.131b}$$

Ito's formula in Eq. (4.105) gives the SDE for ${X_1}^2(t)$ as:

$$d(X_1^2(t)) = 2X_1Y_1dt \tag{4.132}$$

If $g = {Y_1}^2$, then:

$$\frac{\partial g}{\partial \boldsymbol{x}} = \left(\frac{\partial g}{\partial x_1}, \frac{\partial g}{\partial y_1}, \frac{\partial g}{\partial x_2}, \frac{\partial g}{\partial y_2}\right) = (0,\ 2Y_1, 0, 0) \text{ and} \left[\frac{\partial^2 g}{\partial \boldsymbol{x}_i \partial \boldsymbol{x}_j}\right] = \begin{bmatrix} 0 & 0 & 0 & 0 \\ 0 & 2 & 0 & 0 \\ 0 & 0 & 0 & 0 \\ 0 & 0 & 0 & 0 \end{bmatrix} \tag{4.133}$$

The SDE for ${Y_1}^2$ is:

$$\begin{aligned} d({Y_1}^2(t)) &= \frac{2Y_1}{m_1}(-c_1Y_1(t) + c_2Y_2(t) - k_1X_1(t) + k_2X_2(t))dt + \frac{1}{2}\left(2\frac{{\sigma_2}^2}{m_1^2}\right)dt \\ &\quad + 2Y_1(t)\frac{\sigma_2}{m_1}dB_1(t) \\ &= \frac{2Y_1}{m_1}(-c_1Y_1(t) + c_2Y_2(t) - k_1X_1(t) + k_2X_2(t))dt + \frac{{\sigma_2}^2}{m_1^2}dt \\ &\quad + 2Y_1(t)\frac{\sigma_2}{m_1}dB_1(t) \end{aligned} \tag{4.134}$$

If $g = X_1 Y_1$, we have:

$$\frac{\partial g}{\partial x} = \left(\frac{\partial g}{\partial x_1}, \frac{\partial g}{\partial y_1}, \frac{\partial g}{\partial x_2}, \frac{\partial g}{\partial y_2}\right) = (Y_1, X_1, 0, 0) \text{ and } \left[\frac{\partial^2 g}{\partial x_i \partial x_j}\right] = \begin{bmatrix} 0 & 1 & 0 & 0 \\ 1 & 0 & 0 & 0 \\ 0 & 0 & 0 & 0 \\ 0 & 0 & 0 & 0 \end{bmatrix} \tag{4.135}$$

The SDE for $X_1 Y_1$ is:

$$d(X_1 Y_1) = \left[{Y_1}^2 + \frac{X_1}{m_1}(-c_1 Y_1(t) + c_2 Y_2(t) - k_1 X_1(t) + k_2 X_2(t)) \right] dt + X_1(t) \frac{\sigma_2}{m_1} dB_1(t) \tag{4.136}$$

Similar steps yield the SDEs for the other response variables as:

$$d({X_2}^2(t)) = 2 X_2 Y_2 dt \tag{4.137}$$

$$d({Y_2}^2(t)) = \frac{2Y_2}{m_2}(c_3 Y_1(t) - c_4 Y_2(t) + k_3 X_1(t) - k_4 X_2(t))\, dt + \frac{{\sigma_4}^2}{m_2^2} dt + 2Y_2(t) \frac{\sigma_4}{m_2} dB_2(t) \tag{4.138}$$

$$d(X_2 Y_2) = \left[{Y_2}^2 + \frac{X_2}{m_2}(c_3 Y_1(t) - c_4 Y_2(t) + k_3 X_1(t) - k_4 X_2(t)) \right] dt + X_2(t) \frac{\sigma_4}{m_2} dB_2(t) \tag{4.139}$$

$$d(X_1 X_2) = (X_2 Y_1 + X_1 Y_2) dt \tag{4.140}$$

$$d(Y_1 Y_2) = \left[\tfrac{Y_2}{m_1}(-c_1 Y_1(t) + c_2 Y_2(t) - k_1 X_1(t) + k_2 X_2(t)) \right] dt + \left[\frac{Y_1}{m_2}(c_3 Y_1(t) - c_4 Y_2(t) + k_3 X_1(t) - k_4 X_2(t)) \right] dt + Y_2(t) \frac{\sigma_2}{m_1} dB_1(t) + Y_1(t) \frac{\sigma_4}{m_2} dB_2(t) \tag{4.141}$$

$$d(X_1 Y_2) = \left[Y_1 Y_2 + \frac{X_1}{m_2}(c_3 Y_1(t) - c_4 Y_2(t) + k_3 X_1(t) - k_4 X_2(t)) \right] dt + X_1(t) \frac{\sigma_4}{m_2} dB_2(t) \tag{4.142}$$

$$d(X_2Y_1) = \left[Y_1Y_2 + \frac{X_2}{m_1}(-c_1Y_1(t) + c_2Y_2(t) - k_1X_1(t) + k_2X_2(t))\right]dt$$
$$+ X_2(t)\frac{\sigma_2}{m_1}dB_1(t) \tag{4.143}$$

Taking expectation on both sides of SDEs (4.132), (4.134), (4.136) and (4.137)–(4.143), we obtain the following system of coupled ODEs for the second order statistics of the MDOF system (as an exercise, the reader may try obtaining them directly using Dynkin's formula):

$$\frac{dE\left[X_1^2(t)\right]}{dt} = 2E[X_1Y_1], \tag{4.144}$$

$$\frac{dE\left[Y_1^2(t)\right]}{dt} = \frac{2}{m_1}\left(-c_1E\left[Y_1^2\right] - k_1E[X_1Y_1] + c_2\,E[Y_1Y_2] + k_2\,E[Y_1X_2]\right)$$
$$+ \frac{{\sigma_2}^2}{m_1^2}, \tag{4.145}$$

$$\frac{dE[X_1Y_1]}{dt} = E\left[{Y_1}^2\right] + \frac{1}{m_1}\left(-c_1E[X_1Y_1] - k_1E\left[X_1^2\right] + c_2E[X_1Y_2] + k_2E[X_1X_2]\right), \tag{4.146}$$

$$\frac{dE\left[X_2^2(t)\right]}{dt} = 2E[X_2Y_2], \tag{4.147}$$

$$\frac{dE\left[Y_2^2(t)\right]}{dt} = \frac{2}{m_2}\left(-c_4E\left[Y_2^2\right] - k_4E[X_2Y_2] + c_3\,E[Y_1Y_2] + k_3\,E[X_1Y_2]\right) + \frac{{\sigma_4}^2}{m_2^2}, \tag{4.148}$$

$$\frac{dE[X_2Y_2]}{dt} = E\left[{Y_2}^2\right] + \frac{1}{m_2}\left(-c_4E[X_2Y_2] - k_4E\left[X_2^2\right] + c_3E[X_2Y_1] + k_3E[X_1X_2]\right), \tag{4.149}$$

$$\frac{dE[X_1X_2])}{dt} = E[X_2Y_1] + E[X_1Y_2], \tag{4.150}$$

$$\frac{dE[Y_1Y_2]}{dt} = \frac{1}{m_1}\left(-c_1E[Y_1Y_2] - k_1E[X_1Y_2] + c_2E\left[Y_2{}^2\right] + k_2E[X_2Y_2]\right)$$
$$+\frac{1}{m_2}\left(-c_4E[Y_1Y_2] - k_4E[X_2Y_1] + c_3e\left[Y_1{}^2\right] + k_3E[X_1Y_1]\right), \quad (4.151)$$

$$\frac{dE[X_1Y_2]}{dt} = E[Y_1Y_2] + \frac{1}{m_2}\left(c_3E[X_1Y_1] - c_4E[X_1Y_2] + k_3E\left[X_1{}^2\right] - k_4E[X_1X_2]\right) \quad (4.152)$$

$$\frac{dE[X_2Y_1]}{dt} = E[Y_1Y_2] + \frac{1}{m_1}\left(-c_1E[X_2Y_1] + c_2E[X_2Y_2] - k_1E[X_1X_2] + k_2E\left[X_2{}^2\right]\right) \quad (4.153)$$

The above ODEs are solved (presently subject to identically zero initial conditions) to get the histories of the second order moments some of which are shown in Figs. 4.4 and 4.5.

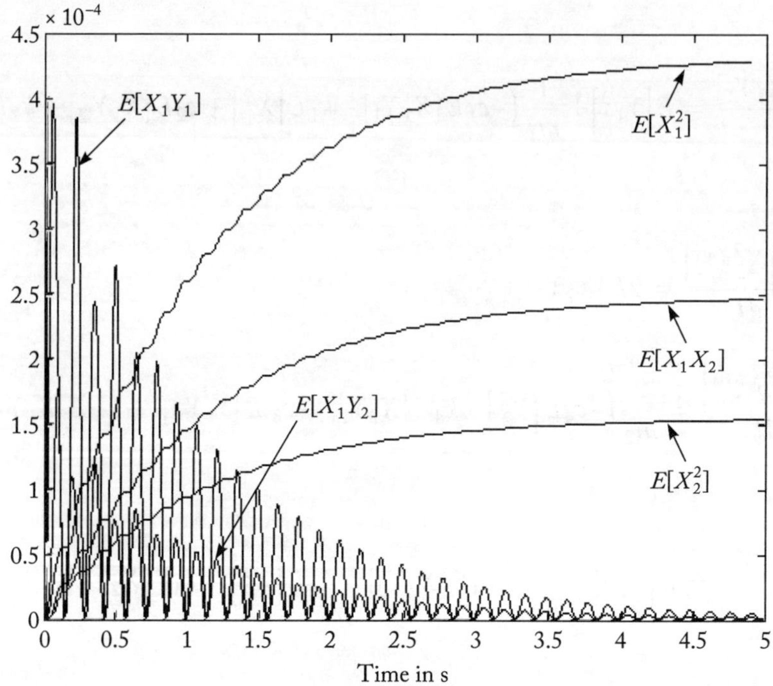

Fig. 4.4 Time histories of response moments for the MDOF oscillator (in Example 4.18) by Ito's formula: $E[X_1^2]$, $E[X_2^2]$, $E[X_1X_2]$, $E[X_1Y_1]$ and $E[X_2Y_2]$

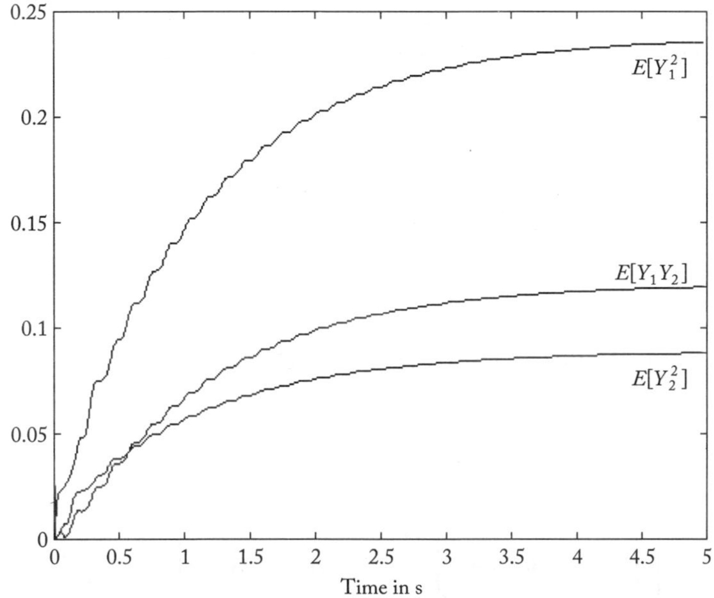

Fig. 4.5 Time histories of response moments for the MDOF oscillator (in Example 4.18) by Ito's formula; $E[Y_1^2]$, $E[Y_2^2]$, $E[Y_1 Y_2]$

4.5 Spectral Representations of Stochastic Signals

We digress a bit to briefly discuss the spectral representation of stochastic processes. In the sequel, we wish to find the response moments of a dynamical system (say, the linear SDOF oscillator described by Eq. 4.115) in frequency domain. For a stationary stochastic process $X(t)$ (see Section 3.3.3, Chapter 3 for a definition of stationarity), one has the Wiener–Khintchine relationship [Chatfield 2004] between the auto-correlation function $R_{XX}(\tau) = E[X(t)X(t+\tau)]$ and the so-called power spectral density (PSD), $s_{XX}(\lambda)$, defined by the following Fourier transform pairs:

$$R_{XX}(\tau) = \int_{-\infty}^{\infty} e^{i\lambda\tau} S_{XX}(\lambda) d\lambda \tag{4.154a}$$

and

$$s_{XX}(\lambda) = \frac{1}{2\pi} \int_{-\infty}^{\infty} e^{-i\lambda\tau} R_{XX}(\tau) d\tau \tag{4.154b}$$

It is useful to write the following spectral representation of a stochastic process, stationary or otherwise [Priestley 1965, 1967]:

$$X(t) = \int_{-\infty}^{\infty} \psi_t(\lambda)\, dz(\lambda) \tag{4.155}$$

$z(\lambda)$ is an orthogonal stochastic process in λ if $X(t)$ is stationary, i.e., the increment $dz(\lambda)$ is a random variable, for each λ, such that these increments are uncorrelated:

$$E\left[dz(\lambda)\,dz(\lambda')\right] = 0, \quad \lambda \neq \lambda' \tag{4.156}$$

Moreover, $\psi_t(\lambda) \in \mathbb{H}$, a suitable function space. For a stationary process, $\mathbb{H}$ is simply the family of complex exponentials, i.e., for each t:

$$\psi_t(\lambda) = e^{i\lambda t} \tag{4.157}$$

Thus, with $X(t)$ given by Eq. (4.155), the autocorrelation function $R_{XX}(\tau)$ of a stationary process is given by:

$$\begin{aligned} R_{XX}(\tau) &= E\left[X(t)\,X^*(t+\tau)\right] = E\left[X^*(t)\,X(t+\tau)\right] \\ &= E\left[\int_{-\infty}^{\infty} e^{-i\lambda t} dz^*(\lambda) \int_{-\infty}^{\infty} e^{i\lambda'}(t+\tau) dz(\lambda')\right] \\ &= \int_{-\infty}^{\infty} e^{i\lambda\tau} E\left[\left|dz(\lambda)\right|^2\right] \end{aligned} \tag{4.158}$$

The superscript * stands for complex conjugate. The last result is obtained by utilizing Eq. (4.156). While $dz(\lambda) \approx O(\sqrt{d\lambda})$ (recall the order $\sqrt{dt}$ increment of a Wiener process) and is not differentiable at $\lambda = 0$, $\left|dz(\lambda)\right|^2 \approx O(d\lambda)$ so we can write:

$$E\left[\left|dz(\lambda)\right|^2\right] = dZ(\lambda) = s_{XX}(\lambda)d\lambda \tag{4.159}$$

$dZ(\lambda)$ defines the incremental power spectrum for a stationary process and is the average incremental power at each frequency λ over all sample paths of $X(t)$. The density $s_{XX}(\lambda) = dZ(\lambda)/d\lambda$ is thus the PSD of $X(t)$. From Eqs. (4.158) and (4.159), we obtain, for a stationary process, the Fourier transform relationships in Eq. (4.154). Since $R_{XX}(0) = E\left[X^2(t)\right]$, the mean square response is obtained as:

$$E\left[X^2(t)\right] = \int_{-\infty}^{\infty} s_{XX}(\lambda)d\lambda \tag{4.160}$$

4.5.1 Non-stationary process and evolutionary power spectrum

For a non-stationary process, $X(t)$, the spectral representation by Eq. (4.155) is not strictly valid, since increments of $z(\lambda)$ may not be orthogonal. Here the increments $dz(\lambda)$ may be correlated leading to the definition of a two-dimensional PSD [Cramer 1960, Shinozuka and Jan 1972, Shinozuka 1972]:

$$s_{XX}(\lambda, \upsilon)\, d\lambda d\upsilon = E\left[dz(\lambda)\, dz^*(\upsilon)\right] \tag{4.161}$$

However, this representation entails certain assumptions about the smoothness of $s_{XX}(\lambda, \upsilon)$ so as to derive a satisfactory physical interpretation. Instead, we here follow the notion of an evolutionary power spectrum (EPS) mooted by Priestley [Priestley 1965] for a non-stationary stochastic process, rendering the spectral representation, as in Eq. (4.155), to retain the uncorrelated nature of the random amplitudes. Further, it allows for the following more physically justifiable interpretation of the model. If we define $s_{XX}(t, \lambda)$ - a time varying spectrum—to describe the local energy / power distribution of $X(t)$ in the neighbourhood of t, then it provides a similar interpretation as the power spectrum of a stationary process. The only difference lies in that, while the latter characterizes the process over all time, the former specifies the λ variation of power around each time instant t.

Thus, with the spectral representation in Eq. (4.155) accepted for a non-stationary stochastic process also, the issue that remains to be resolved is to arrive at a suitable family of functions $\mathbb{H}$ to define $\psi_t(\lambda)$. Recognizing that the general representation of a non-stationary signal (e.g., one with distinctly distinguishable temporal regimes) may require working simultaneously in both time and frequency and hence must include not only a time-oscillatory part but also an adequate scope for dependence on frequency, $\psi_t(\lambda)$ may be taken as an amplitude modulated wave in the form:

$$\psi_t(\lambda) = A_t(\lambda) e^{i\theta(\lambda)t} \tag{4.162}$$

$A_t(\lambda)$ is deterministic and is known as the modulating function. It may be interpreted as the envelope of $\psi_t(\lambda)$. If $\theta(\lambda)$ is a single valued function of λ, then, by suitably redefining the amplitude $A_t(\lambda)$, one is able to replace $\theta(\lambda)$ by λ so that Eq. (4.155) for a non-stationary stochastic process may be rewritten as:

$$X(t) = \int_{-\infty}^{\infty} A_t(\lambda) e^{i\lambda t} dz(\lambda) \tag{4.163}$$

With $A_t(\lambda) = 1$, the above representation reduces to that of a stationary process (Eq. 4.155) and therefore applies to any general oscillatory stochastic process $\{X(t)\}$. With the incremental power corresponding to a frequency λ at a given t being $|A_t(\lambda)|^2 d\mu(\lambda)$ where $d\mu(\lambda) = E[dz(\lambda)\, dz^*(\lambda)] = E\left[|\mathrm{dz}(\lambda)|^2\right]$ is an incremental measure, the evolutionary power spectrum (incremental form) is defined as:

$$dS_{XX}(t, \lambda) = |A_t(\lambda)|^2 d\mu(\lambda). \tag{4.164}$$

The autocorrelation function and the mean square value of $X(t)$ are respectively given by:

$$R_{XX}(t_1,t_2) = \int_{-\infty}^{\infty} A_{t_1}(\lambda)A_{t_2}^*(\lambda)e^{i\lambda(t_1-t_2)}d\mu(\lambda) \tag{4.165a}$$

$$R_{XX}(t,t) = E\left[X^2(t)\right] = \int_{-\infty}^{\infty} |A_t(\lambda)|^2 d\mu(\lambda) = \int_{-\infty}^{\infty} dS_{XX}(t,\lambda) \tag{4.165b}$$

The uncorrelated nature of $dz(\lambda)$ is utilized in obtaining the above result. If $d\mu(\lambda)$ is differentiable, we can define the evolutionary PSD of $X(t)$ as:

$$s_{XX}(t,\lambda) = |A_t(\lambda)|^2 s_{XX}(\lambda) \tag{4.166}$$

where $s_{XX}(\lambda) = \frac{d\mu(\lambda)}{d\lambda}$.

Example 4.19

We derive the expressions for the response moments in frequency domain of the SDOF oscillator under stationary and non-stationary inputs.

Solution

Case (i) Stationary input

Consider the SDE $\ddot{X}(t) + 2\xi\overline{\omega}\dot{X}(t) + \overline{\omega}^2 X(t) = \mathfrak{g}(t)$ with the input process $\mathfrak{g}(t)$ assumed to be stationary with mean μ_G (time invariant) and the autocorrelation function $R_{GG}(\tau := \tau_2 - \tau_1)$. We refer to Appendix D, wherein expressions for the mean and autocorrelation function are derived in time domain, for the response of an SDOF oscillator under the stationary input. With $h(t)$ being the impulse response function of the oscillator (Eq. D.5 of Appendix D), the response of the oscillator is given by the convolution integral $X(t) = \int_0^t \mathfrak{g}(\tau)h(t-\tau)d\tau$. This leads to the following expressions for the mean and autocorrelation functions of $X(t)$ (see Eq. D.16 of Appendix D):

$$\mu_X(t) = \mu_G \int_{-\infty}^{\infty} h(t-\tau)d\tau \tag{4.167a}$$

and

$$R_{XX}(t_1,t_2) = \int_{-\infty}^{\infty}\int_{-\infty}^{\infty} R_{GG}(\tau_2-\tau_1)h(t_1-\tau_1)h(t_2-\tau_2)d\tau_1 d\tau_2 \tag{4.167b}$$

The last equation shows that the response is non-stationary even if the input is stationary. However, for large times ($t_1, t_2 \to \infty$) but with finite $t_2 - t_1 = \tau$, we observe that $R_{XX}(t_1,t_2)$ may attain stationarity (the so-called stochastic steady state) and become $R_{XX}(t_2-t_1)$, a function of $\tau = t_2 - t_1$. In the last case, we can further obtain the response moment in frequency domain in terms of $S_{GG}(\lambda)$, the input PSD. Since $R_{GG}(\tau) = \int_{-\infty}^{\infty} S_{GG}(\lambda)e^{i\lambda\tau}d\lambda$ (by Wiener–Khintchine relationship in Eq. (4.154a)), we have from Eq. (4.167b):

$$R_{XX}(t_1,t_2) = \int_{-\infty}^{\infty}\int_{-\infty}^{\infty}\left(\int_{-\infty}^{\infty} S_{GG}(\lambda)e^{i\lambda\tau}d\lambda\right)h(t_1-\tau_1)h(t_2-\tau_2)\,d\tau_1 d\tau_2 \quad (4.168)$$

The transformations $t_1 - \tau_1 = \alpha_1$ and $t_2 - \tau_2 = \alpha_2$ modify the above integral as:

$$R_{XX}(t_1,t_2) = \int_{-\infty}^{\infty}\int_{-\infty}^{\infty}\left(\int_{-\infty}^{\infty} S_{GG}(\lambda)e^{i\lambda(t_2-t_1+\alpha_1-\alpha_2)}d\lambda\right)h(\alpha_1)h(\alpha_2)\,d\alpha_1 d\alpha_2$$

$$= \int_{-\infty}^{\infty}\int_{-\infty}^{\infty}\int_{-\infty}^{\infty} S_{GG}(\lambda)e^{i\lambda(t_2-t_1)}h(\alpha_1)\, e^{i\lambda\alpha_1}\, h(\alpha_2)e^{-i\lambda\alpha_2}d\alpha_1 d\alpha_2 d\lambda$$

$$= \int_{-\infty}^{\infty} S_{GG}(\lambda)e^{i\lambda(t_2-t_1)}\left(\int_{-\infty}^{\infty} h(\alpha_1)\, e^{i\lambda\alpha_1}d\alpha_1\right)\left(\int_{-\infty}^{\infty} h(\alpha_2)e^{-i\lambda\alpha_2}d\alpha_2\right)d\lambda$$

$$= \int_{-\infty}^{\infty}\left[S_{GG}(\lambda)H^*(\lambda)H(\lambda)\right]e^{i\lambda(t_2-t_1)}d\lambda \quad (4.169)$$

Since the impulse response function $h(t) = \frac{e^{-\xi\overline{\omega}t}}{m}\overline{\omega}_d \sin\overline{\omega}_d t$, each of the integrals of type $\int_{-\infty}^{\infty} h(\alpha_2)e^{-i\lambda\alpha_2}d\alpha_2$ gives the Fourier transform $H(\lambda)$ of $h(t)$. $H(\lambda)$ is known as the complex frequency response function. With $\overline{\omega}_d = \overline{\omega}\sqrt{1-\xi^2}$, we therefore have:

$$H(\lambda) = \int_{-\infty}^{\infty} h(\alpha)e^{-i\lambda\alpha}d\alpha = \frac{1}{\left(\lambda^2-\overline{\omega}^2\right)+i(2\xi\lambda\overline{\omega})} \quad (4.170)$$

$H^*(\lambda)$ in Eq. (4.169) is the complex conjugate of $H(\lambda)$. Comparison of Eqs. (4.169) and (4.154a) shows that $S_{GG}(\lambda)H^*(\lambda)H(\lambda)$ is indeed the PSD, $S_{XX}(\lambda)$, of the output (displacement) response. It implies that $S_{XX}(\lambda)$ represents the steady state PSD of the oscillator output given by:

$$S_{XX}(\lambda) = S_{GG}(\lambda)H^*(\lambda)H(\lambda) = S_{GG}(\lambda)\left|H(\lambda)\right|^2 \quad (4.171)$$

The mean square response is given by:

$$E\left[X^2(t)\right] = R_{XX}(0) = \int_{-\infty}^{\infty} |H(\lambda)|^2 S_{GG}(\lambda)\, d\lambda \tag{4.172}$$

The output autocorrelation $R_{XX}(\tau)$ can be obtained, if needed, by taking an inverse Fourier transform of $S_{XX}(\lambda)$.

Case (ii) Non-stationary input

If $\mathfrak{g}(t)$ denotes the input non-stationary process and $X(t)$ the output displacement response process of the SDOF oscillator, then the response mean and autocorrelation are given by:

$$\mu_X(t) = \int_{-\infty}^{\infty} \mu_G(\tau)\, h(t-\tau)\, d\tau \tag{4.173a}$$

and

$$R_{XX}(t_1,t_2) = \int_{-\infty}^{\infty}\int_{-\infty}^{\infty} R_{GG}(\tau_1,\tau_2)\, h(t_1-\tau_1)\, h(t_2-\tau_2)\, d\tau_1 d\tau_2 \tag{4.173b}$$

where, $\mu_G(t)$ and $R_{GG}(t_1,t_2)$ are respectively the mean and the autocorrelation of the input process $\mathfrak{g}(t)$. With the spectral representation as in Eq. (4.163) of a non-stationary process, $R_{GG}(t_1,t_2)$is given by Eq. (4.165a) so the output autocorrelation function $R_{XX}(t_1,t_2)$ in Eq. (4.173b) takes the form:

$$R_{XX}(t_1,t_2) = \int_{-\infty}^{\infty}\int_{-\infty}^{\infty}\int_{-\infty}^{\infty} e^{i\lambda(\tau_1-\tau_2)} A_{\tau_1}(\lambda) A^*_{\tau_2}(\lambda) h(t_1-\tau_1)\, h(t_2-\tau_2)$$

$$d\tau_1 d\tau_2 d\mu(\lambda) \tag{4.174}$$

Transformations of variables $t_1 - \tau_1 = \alpha_1$ and $t_2 - \tau_2 = \alpha_2$ in the last equation leads to:

$$R_{XX}(t_1,t_2) = \int_{-\infty}^{\infty}\int_{-\infty}^{\infty}\int_{-\infty}^{\infty} e^{i\lambda(t_1-t_2)} A_{t_1-\alpha_1}(\lambda) h(\alpha_1) e^{-i\lambda\alpha_1} A^*_{t_2-\alpha_2}(\lambda) e^{i\lambda\alpha_2} h(\alpha_2)$$

$$d\alpha_1 d\alpha_2 d\mu(\lambda)$$

$$= \int_{-\infty}^{\infty} e^{i\lambda(t_1-t_2)} \left\{ \int_{-\infty}^{\infty} A_{t_1-\alpha_1}(\lambda) h(\alpha_1) e^{-i\lambda\alpha_1}\, d\alpha_1 \right.$$

$$\int_{-\infty}^{\infty} A^{*}_{t_2-\alpha_2}(\lambda)e^{i\lambda\alpha_2}h(\alpha_2)d\alpha_2\Big\} d\mu(\lambda)$$

$$= \int_{-\infty}^{\infty} e^{i\lambda(t_1-t_2)}H(t_1,\lambda)H^{*}(t_2,\lambda)d\mu(\lambda) \tag{4.175}$$

Here $H(t,\lambda)$ is given by:

$$H(t,\lambda) = \int_{-\infty}^{\infty} A_{t-\tau}(\lambda)e^{-i\lambda\tau}h(\tau)d\tau \tag{4.176}$$

$H(t,\lambda)$ may be interpreted as a generalized frequency response function (transfer function) of the SDOF oscillator. Equation (4.175) implies (see Eq. 4.163) that the non-stationary response process $X(t)$ has the spectral representation:

$$X(t) = \int_{-\infty}^{\infty} H(t,\lambda)\,e^{i\lambda t}dz(\lambda) \tag{4.177}$$

Equation (4.175) gives the mean square response $R_{XX}(t,t)$ under non-stationary input with $t_1 = t_2 = t$ as:

$$R_{XX}(t,t) = \int_{-\infty}^{\infty} \left|H(t,\lambda)\right|^2 s_{GG}(\lambda)\,d\lambda \tag{4.178a}$$

From which the evolutionary PSD of the non-stationary output can be taken as:

$$s_{XX}(t,\lambda) = \left|H(t,\lambda)\right|^2 s_{GG}(\lambda) \tag{4.178b}$$

Example 4.20

As an application of the concept of evolutionary power spectrum, we investigate the dynamics of a structural system [Nigam and Narayanan 1994], say an aircraft crudely approximated by the following SDOF oscillator excited by a zero mean non-stationary gust $Z(t)$:

$$\ddot{X}(t) + \gamma_g\dot{X}(t) = -\gamma_g Z(t) \tag{4.179}$$

Solution

Here γ_g is a parameter influenced by the surface area exposed to the gust and the lift force phenomenon associated with a typical fluid-structure interaction problem (Blevins 1990). The non-stationary process $Z(t)$ is taken as:

$$Z(t) = A(t)\mathfrak{g}(t) \tag{4.180}$$

where, $\mathfrak{g}(t)$ is a stationary process with spectral representation $\int_{-\infty}^{\infty} e^{i\lambda t} dz(\lambda)$ and $E\left[|dz(\lambda)|^2\right] = s_{GG}(\lambda)d\lambda$. Here with $\psi_t(\lambda) = A(t)e^{i\lambda t}$, the modulating function $A(t)$ is taken as a non-negative real valued function in time as [Howell and Lin 1971]:

$$A(t) = c\left(e^{-\alpha_1 t} - e^{-\alpha_2 t}\right) \tag{4.181}$$

with $c, \alpha_1, \alpha_2 \in \mathbb{R}^+$, and $\alpha_2 > \alpha_1$. Further, c is a normalizing constant so that $\sup_t A(t) = 1$. The PSD $S_{GG}(\lambda)$ of $\mathfrak{g}(t)$ has the form of the Dryden spectrum [Zipfel 2000]:

$$S_{GG}(\lambda) = \frac{\sigma_v^2 \ell}{2\pi v_s}\left(\frac{1 + (3\lambda\zeta)^2}{1 + (\lambda\zeta)^2}\right) \tag{4.182}$$

where, σ_v^2 is the variance of the gust velocity, v_s the speed of the aircraft. ℓ a parameter describing the length scale of turbulence [Blevins 1990] and $\zeta = \ell/v_s$. Following Eq. (4.139), the spectral representation of $Z(t)$ is now of the form:

$$Z(t) = A(t)\int_{-\infty}^{\infty} e^{i\lambda t} dz(\lambda) \tag{4.183}$$

Thus the evolutionary spectrum of $Z(t)$ is $dS_{ZZ}(t,\lambda) = A_t(\lambda)^2 S_{GG}(\lambda)d\lambda$. With this specific form of the modulating function $A(t)$, the input process $Z(t)$ is known as the uniformly modulated process since the spectral components vary over time in exactly the same way, i.e., for two frequencies λ_1, λ_2 and time instants t_1, t_2:

$$\frac{dS_{ZZ}(t_1,\lambda_1)}{dS_{ZZ}(t_2,\lambda_1)} = \frac{dS_{ZZ}(t_1,\lambda_2)}{dS_{ZZ}(t_2,\lambda_2)} \tag{4.184}$$

Now, in terms of the velocity, $V = \frac{dX}{dt}$, the dynamical Eq. (4.179) takes the form:

$$\dot{V}(t) + \gamma_g V(t) = -\gamma_g Z(t) \tag{4.185}$$

The impulse response $h(t)$ with respect to the velocity response is given by:

$$h_v(t) = e^{-\gamma_g^t}, \quad t > 0$$

$$= 0, \quad \text{otherwise} \tag{4.186}$$

$H_v(t,\lambda)$, the generalized transfer function of the oscillator (for velocity response) is:

$$H_v(t,\lambda) = \int_0^t A(t-\tau)e^{-i\lambda\tau}h_v(\tau)\,d\tau \tag{4.187}$$

Substituting for $A(t)$ and $h_v(t)$, the integral in the above equation can be determined in closed form as:

$$H_v(t,\lambda) = \gamma_g e^{-(\gamma_g+i\lambda)t}\left(\frac{1}{\alpha_2-\gamma_g-i\lambda} - \frac{1}{\alpha_1-\gamma_g-i\lambda}\right)$$

$$+\gamma_g\left(\frac{e^{-\alpha_1 t}}{\alpha_1-\gamma_g-i\lambda} - \frac{e^{-\alpha_2 t}}{\alpha_2-\gamma_g-i\lambda}\right) \tag{4.188}$$

Now, the velocity response in terms of its autocorrelation can be obtained from Eq. (4.175). Using the above expression for $H_v(t,\lambda)$, results are obtained for the mean square velocity response $E\left[V^2(t)\right] = R_{VV}(t,t)$ from Eq. (4.178a) by numerically evaluating the integral $\int_{-\infty}^{\infty}\left|H_v(t,\lambda)\right|^2 S_{GG}(\lambda)\,d\lambda$ over $\lambda \in [0,10]$ in rad/s. for each time t varied in the range 0 – 5 s. The other parameters used in the computation are: $\ell = 500$ ft, $v_s = 200$ ft/s, $\sigma_v = 1$ ft/s, $\gamma_g = 1.11$ sec^{-1} and $\alpha_1 = 0.0$. Figure 4.6a shows the evolution of $E\left[V^2(t)\right]$ for $\alpha_2 = 0.5,\ 1,\ 2.5,$ 5.0 and 10.0. It is also of interest to determine the acceleration response of the dynamical system. This requires the impulse response function $h_a(t)$ for the acceleration response, which is given by:

$$h_a(t) = \delta(t) - \gamma_g e^{-\gamma_g t}\mathbb{U}(t) \tag{4.189}$$

where, $\delta(t)$ is the Dirac delta function. $\mathbb{U}(t)$ is the Heaviside unit step function which is unity for $t > 0$ and zero otherwise. Similar to $H_v(t,\lambda)$, one can obtain $H_a(t,\lambda) = \int_0^t A(t-\tau)e^{-i\lambda\tau}h_a(\tau)\,d\tau$, the generalized transfer function for acceleration response, in closed form. Utilizing this function, the mean square response for acceleration $R_{aa}(t,t) = E\left[\dot{V}^2(t)\right]$ is evaluated by numerical quadrature as in the case of velocity response. These are shown in Fig. 4.6b for different values of α_2.

It can be observed from the figure that as the modulating function $A(t)$ approaches unity, the mean square acceleration tends to that obtained from $\mathfrak{g}(t)$ if it were stationary. This may as well be verified (left as an exercise to the reader) from a computation of $E\left[\dot{V}^2(t)\right]$ using Eq. (4.172).

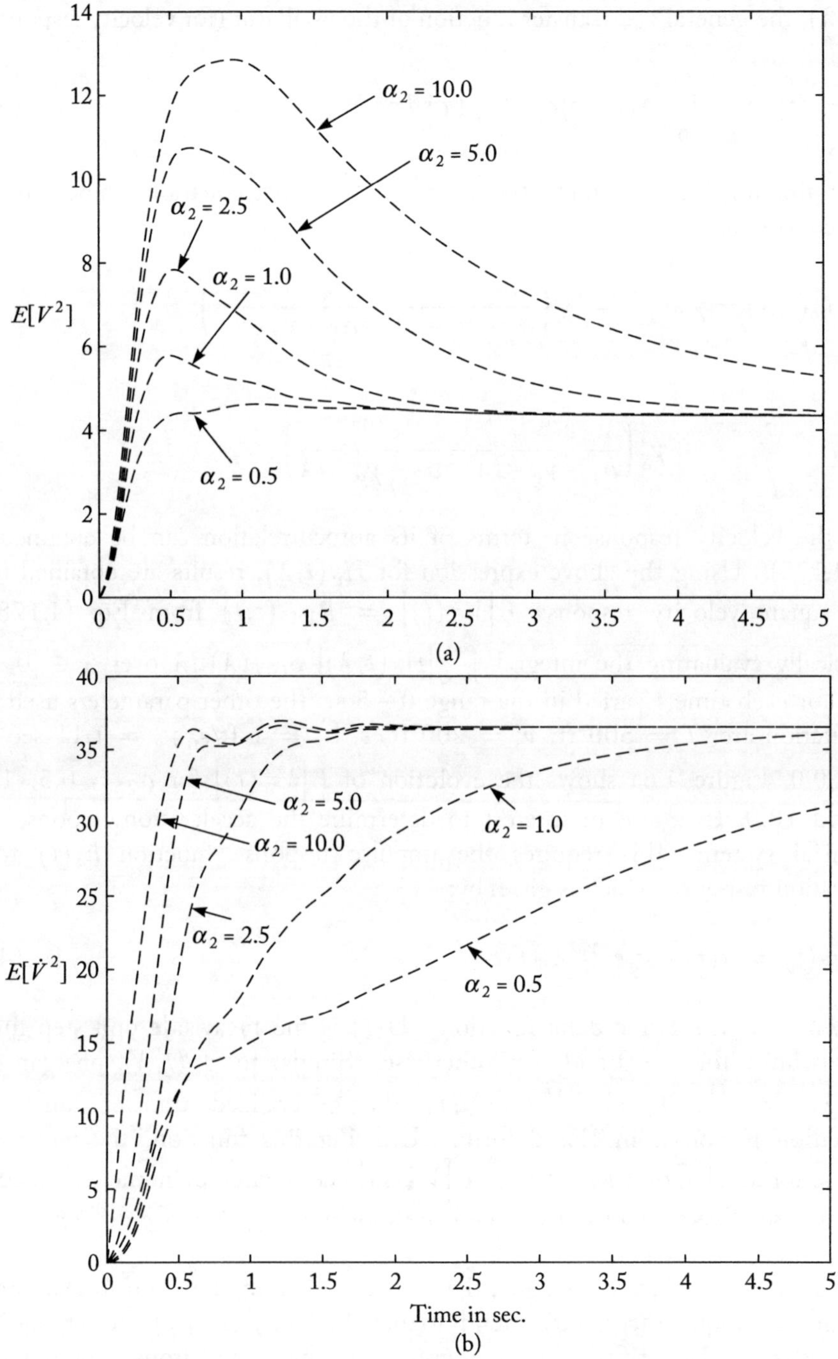

Fig. 4.6 Mean square response (Example 4.20) under non-stationary gust for different β with the modulating function $A(t) = 1 - e^{-\alpha_2 t}$; (a) velocity and (b) acceleration

Spectral representation of seismic forces

Similar to the wind excitation considered in the above example, non-stationary stochastic models of other environmental forces (e.g., tsunamis) arising from seismic events [Kiureghian and Crempien 1989, Zhang *et. al.*, 1991], near-shore ocean waves [Nigam and Narayanan 1994] and even the rail and road undulations [Yadav and Nigam 1978, Harrison and Hammond 1986] follow spectral representation such as the one in Eq. (4.163)

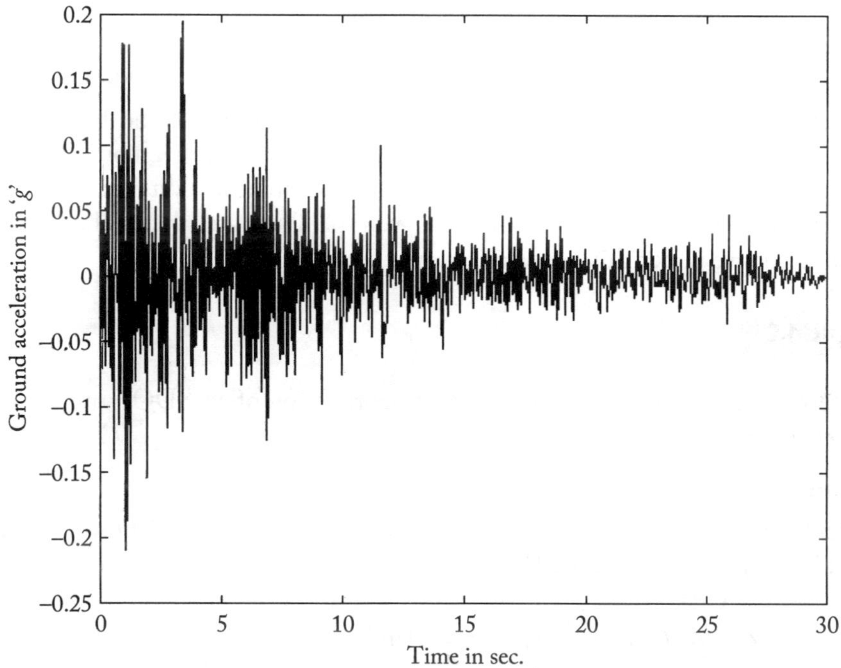

Fig. 4.7 An earthquake accelerogram; El Centro 1940-earthquake (vertical) component

The earthquake forces, thanks to their characteristic nature of releasing energy within short time intervals, are typically non-stationary stochastic processes. These forces normally pick up at the start (Fig. 4.7), reach their peak soon afterwards (known as the strong motion period) followed by the final decay at the tail spread out over a longer duration.

It is common in earthquake engineering practice to adopt the uniformly modulated process model $A(t)\,\mathfrak{g}(t)$(Eq. 4.180) for ground motion. Here $\mathfrak{g}(t) = \int_{-\infty}^{\infty} e^{i\lambda t} dz(\lambda)$ is a stationary stochastic process with a specified PSD, $S_{GG}(\lambda)$. In the seismic analysis of vibrating mechanical/structural systems, one popular choice for the modulating function $A(t)$ is the non-negative, real-valued function $\mathrm{c}\,(e^{-\alpha_1 t} - e^{-\alpha_2 t}), \alpha_2 > \alpha_1 \geq 0,\ \mathrm{c} > 0$, due to Shinozuka and Sato [1967]. Further, the stationary process model for $\mathfrak{g}(t)$ is generally constructed from the knowledge of the spectral properties of the site soil conditions. In this context, the following form of PSD known as Kanai–Tajimi spectrum [Kanai 1957, Tazimi 1960] is extensively used to represent $\mathfrak{g}(t)$:

$$S_{GG}(\lambda) = S_0 \frac{1 + 4\xi_g^2(\lambda/\varpi_g)^2}{\left(1 - (\lambda/\varpi_g)^2\right)^2 + 4\xi_g^2(\lambda/\varpi_g)^2} \tag{4.190}$$

It can be recognized from the above equation that $S_{GG}(\lambda)$is the output PSD of a filtered white noise model where the two parameters ξ_g and ϖ_g are respectively the damping ratio and the natural frequency of the filter whose equation is:

$$\ddot{U} + 2\xi_g\varpi_g\dot{U} + \varpi_g^2 U = -W(t) \tag{4.191}$$

The model assumes that the ground motion (in acceleration) at the bed rock level is represented by $W(t)$, a white noise process with a constant PSD given by S_0. This implies that the stationary part of the input acceleration to the vibrating system at the soil level is given by $\ddot{\mathfrak{g}}(t) = \ddot{U}(t) + W(t)$. The following example illustrates the method of analyzing the system response under this non-stationary model.

Example 4.21

We determine the time varying response autocorrelation of an SDOF oscillator using the above filtered white noise model.

Solution

The oscillator equation under the support motion due to the acceleration input at its base is given by [Roy and Rao 2012]:

$$\ddot{Z}(t) + 2\xi\varpi\dot{Z}(t) + \varpi^2 Z(t) = -A(t)\ddot{\mathfrak{g}}(t) \tag{4.192}$$

Here $Z(t)$ is the relative displacement response of the oscillator with respect to the ground motion. Knowing the impulse response $h(t) = \frac{e^{-\xi\varpi t}}{\varpi_d}\sin\varpi_d t$ with $\varpi_d = \varpi\sqrt{1-\xi^2}$, the generalized transfer function of the oscillator for $z(t)$ is given by:

$$H(t,\lambda) = \int_0^t A(t-\tau)e^{-i\lambda\tau}h(\tau)\,d\tau \tag{4.193}$$

The evolutionary PSD $S_{ZZ}(t,\lambda)$of the response follows from Eq. (4.178b) and the mean square response $E\left[Z^2(t)\right] = R_{ZZ}(t,t)$from Eq. (4.178a). Here it is possible to determine $H(t,\lambda)$ in closed form by a direct evaluation of the integral in Eq. (4.193). This results in:

$$H(t,\lambda)=\frac{c}{\overline{\omega}_d}\left[\frac{e^{-\alpha_1 t}}{D_1^2}\left\{\omega_d\left(1-e^{(\alpha_1-\xi\overline{\omega}-i\lambda)t}\cos\overline{\omega}_d t\ \right)\right.\right.$$

$$\left.+(\alpha_1-\xi\overline{\omega}-i\lambda)\,e^{(\alpha_1-\xi\overline{\omega}-i\lambda)t}\sin\overline{\omega}_d t\right\}+\frac{e^{-\alpha_2 t}}{D_2^2}\left\{\overline{\omega}_d\left(1-e^{(\alpha_2-\xi\overline{\omega}-i\lambda)t}\cos\overline{\omega}_d t\right)\right.$$

$$\left.\left.+(\alpha_2-\xi\overline{\omega}-i\lambda)\,e^{(\alpha_2-\xi\overline{\omega}-i\lambda)t}\sin\overline{\omega}_d t\right\}\right] \tag{4.194}$$

where, $D_1^2=\overline{\omega}_d^2+(\alpha_1-\xi\overline{\omega}-i\lambda)^2$ and $D_2^2=\overline{\omega}_d^2+(\alpha_2-\xi\overline{\omega}-i\lambda)^2$. Figure 4.8 shows the evolution of $E\left[Z^2(t)\right]$ over a time interval of 0 – 30 s with the oscillator parameters as in Example 4.16. The filter parameters $\overline{\omega}_g$ and ξ_g are taken as 5π rad/s and 0.6, respectively [Nigam and Narayanan 1994]. The parameters of $A(t)$ (in Eq. 4.181) are chosen as $c=1$, $\alpha_1=0.13$ and α_2=0.45. Without loss of generality, it is assumed that the intensity S_0 of the white noise is unity. The integral in Eq. (4.178a) is evaluated by numerical quadrature over the frequency range (-50, 50) in Hz.

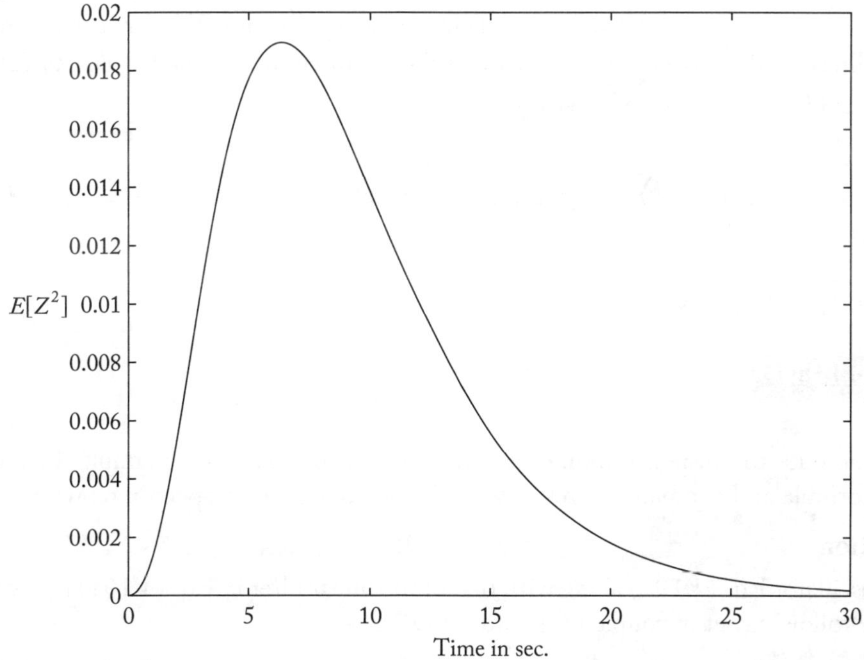

Fig. 4.8 Mean square response of an SDOF oscillator (Example 4.21) under non-stationary support motion by spectral representation approach: $A(t)=e^{-\alpha_1 t}-e^{-\alpha_2 t}$; $\overline{\omega}_g=5\pi$ rad/s, $\xi_g=0.6$, $\overline{\omega}=10$ rad/s, $\xi=0.02$, $\alpha_1=0.13$ and $\alpha_2=0.45$

4.5.2 Some interesting aspects of evolutionary power spectrum

The spectral representation in Eq. (4.155) of a stochastic process, stationary or otherwise, is not unique. For a stationary process, the choice of the harmonic functions for $\psi_t(\lambda)$ as in Eq. (4.157) is an obvious one, even though other choices may be possible. For a non-stationary process, the function space $\mathbb{H}$ containing functions of the form $\psi_t(\lambda) = A_t(\lambda)e^{i\theta(\lambda)t}$ is one amongst the possible options. Thus, $\psi_t(\lambda)$ is an amplitude modulated sine wave with frequency $\theta(\lambda)$. As a function of t, $\psi_t(\lambda)$ is an oscillatory function, provided that for some $\theta(\lambda)$, the function $A_t(\lambda)$ is of the form:

$$A_t(\lambda) = \int_{-\infty}^{\infty} e^{itu} dF_\lambda(u) \tag{4.195}$$

with $\left|dF_\lambda(u)\right|$ having an amplitude maximum at $u = 0$. It implies that the Fourier transform of $A_t(\lambda)$ is largely concentrated around zero, i.e., it is a slowly varying function of time and can be considered as an envelope of $\psi_t(\lambda)$.

It is also interesting to note that the spectral representation $X(t) = \int_{-\infty}^{\infty} \psi_t(\lambda)\, dz(\lambda)$ in Eq. (4.155) is similar to an Ito integral. In the definition of the Ito integral, $\mathfrak{I} = \int_0^t X(s)dB(s)$, the square of the orthogonal (independent) increment dB defines the time measure dt whereas the square of the increment dz corresponds to the measure $d\mu(\lambda)$. Thus $z(\lambda)$ is a Brownian motion analogue in frequency domain. We can as well prove (on similar lines as adopted earlier in the case of the Ito integral $\mathfrak{I}$) that $\int_{-\infty}^{\infty} \psi_t(\lambda)\, dz(\lambda)$ is a limit in $L^2(P)$ of the sums of the form:

$$X(t) = \lim_{\sup \Delta\lambda_j \to 0} \sum_{\lambda_j=-\infty}^{\infty} \psi_t(\lambda_j)\left(z(\lambda_{j+1}) - z(\lambda_j)\right) \tag{4.196}$$

where, $\Delta\lambda_j = \lambda_{j+1} - \lambda_j$.

Example 4.22

We compute the response moments of the SDOF oscillator in Example 4.21 using Ito's formula and compare the result with the one obtained by spectral representation.

Solution

The oscillator Eq. (4.192) along with that of the linear filter in Eq. (4.191) is rewritten as the following set of coupled first order SDEs:

$$dX_1(t) = X_2(t)\, dt \tag{4.197a}$$

$$dX_2(t) = -\left(2\xi\varpi X_2(t) + \varpi^2 X_1(t)\right)dt - A(t)\left(dX_4(t) + dB(t)\right) \tag{4.197b}$$

$$dX_3(t) = X_4(t)\,dt \tag{4.197c}$$

$$dX_4(t) = -\left(2\xi_g\overline{\omega}_g X_4(t) + \overline{\omega}_g^2 X_3(t)\right)dt + dB(t) \tag{4.197d}$$

Here $\mathbf{X}(t) := \{X_1(t), X_2(t), X_3(t), X_4(t)\}^T$ is the 4-dimensional Ito process vector such that $X_1(t) = Z(t)$, $X_2(t) = \dot{Z}(t)$, $X_3(t) = U(t)$ and $X_4(t) = \dot{U}(t)$. Also, $W(t)\,dt = dB(t)$ is utilized in writing Eq. (4.197). We have the vector of drift coefficients and matrix of diffusion coefficients (with $m = 4$ and $n = 1$) as:

$$\boldsymbol{a}(t, \mathbf{X}(t)) = \begin{Bmatrix} X_2(t) \\ -\left(2\xi\overline{\omega}X_2(t) + \overline{\omega}^2 X_1(t)\right) + A(t)\left(2\xi_g\overline{\omega}_g X_4(t) + \overline{\omega}_g^2 X_3(t)\right) \\ X_4(t) \\ -\left(2\xi_g\overline{\omega}_g X_4(t) + \overline{\omega}_g^2 X_3(t)\right) \end{Bmatrix} \tag{4.198a}$$

and

$$\sigma(t) = \begin{bmatrix} 0 \\ -A(t) \\ 0 \\ 1 \end{bmatrix} \tag{4.198b}$$

Now, we choose scalar functions $g(\mathbf{X})$ as $g = {X_1}^2, {X_2}^2, X_1X_2, {X_3}^2, {X_4}^2, X_3X_4, X_1X_3, X_2X_4, X_1X_4$ and X_2X_3. With the number of these functions $p = 10$, one can write the SDEs for these functions by Ito's formula (Eq. 4.106) and finally obtain, by the procedure outlined earlier, ODEs for the p response moments in the following form:

$$\frac{dE\left[X_1^2(t)\right]}{dt} = 2E[X_1X_2] \tag{4.199}$$

$$\frac{dE\left[X_2^2(t)\right]}{dt} = -4\xi\overline{\omega}E\left[X_2^2\right] - 2\overline{\omega}^2 E[X_1X_2] + A(t)\left(2\xi_g\overline{\omega}_g E[X_2X_4] + \overline{\omega}_g^2\, E[X_1X_3]\right) + {\sigma_{21}}^2 \tag{4.200}$$

$$\frac{dE[X_1X_2]}{dt} = E\left[{X_2}^2\right] - 2\xi\overline{\omega}E[X_1X_2] - \overline{\omega}^2 E\left[X_1^2\right] + A(t)\left(2\xi_g\overline{\omega}_g E[X_1X_4] + \overline{\omega}_g^2\, E[X_1X_3]\right) \tag{4.201}$$

$$\frac{dE\left[X_3^2(t)\right]}{dt} = 2E\left[X_3X_4\right] \tag{4.202}$$

$$\frac{dE\left[X_4^2(t)\right]}{dt} = -4\xi_g\overline{\omega}_gE\left[X_4^2\right] - 2\overline{\omega}_g^2E\left[X_3X_4\right] + {\sigma_{41}}^2 \tag{4.203}$$

$$\frac{dE\left[X_3X_4\right]}{dt} = E\left[X_4^2(t)\right] - 2\xi_g\overline{\omega}_gE\left[X_3X_4\right] - \omega_g^2E\left[X_3^2\right] \tag{4.204}$$

$$\frac{dE\left[X_1X_3\right])}{dt} = E\left[X_2X_3\right] + E\left[X_1X_4\right] \tag{4.205}$$

$$\frac{dE\left[X_2X_4\right]}{dt} = -2\xi\overline{\omega}E\left[X_2X_4\right] - \overline{\omega}^2E\left[X_1X_4\right] + A(t)\left(2\xi_g\overline{\omega}_gE\left[X_4^2(t)\right]\right.$$

$$\left.+\overline{\omega}_g^2E\left[X_3X_4\right]\right) - 2\xi_g\overline{\omega}_gE\left[X_2X_4\right] - \omega_g^2E\left[X_2X_3\right] + \sigma_{21}\sigma_{41} \tag{4.206}$$

$$\frac{dE\left[X_1X_4\right]}{dt} = E\left[X_2X_4\right] - 2\xi_g\overline{\omega}_gE\left[X_1X_4\right] - \overline{\omega}_g^2E\left[X_1X_3\right] \tag{4.207}$$

$$\frac{dE\left[X_2X_3\right]}{dt} = E\left[X_2X_4\right] - 2\xi\overline{\omega}E\left[X_2X_3\right] - \overline{\omega}^2E\left[X_1X_3\right]$$

$$+A(t)\left(2\xi_g\overline{\omega}_gE\left[X_3X_4\right] + \overline{\omega}_g^2E\left[X_3^2\right]\right) \tag{4.208}$$

These coupled ODEs can be solved (subject to identically zero initial conditions) to get the histories of the various second order moments of the SDOF oscillator. Figure 4.9 shows particularly the time history of $E\left[X_1^2(t)\right] = E\left[Z^2(t)\right]$ obtained using the same parameters as in Example 4.21. The time history closely matches the corresponding result obtained by the spectral representation of the non-stationary input (see Fig. 4.8).

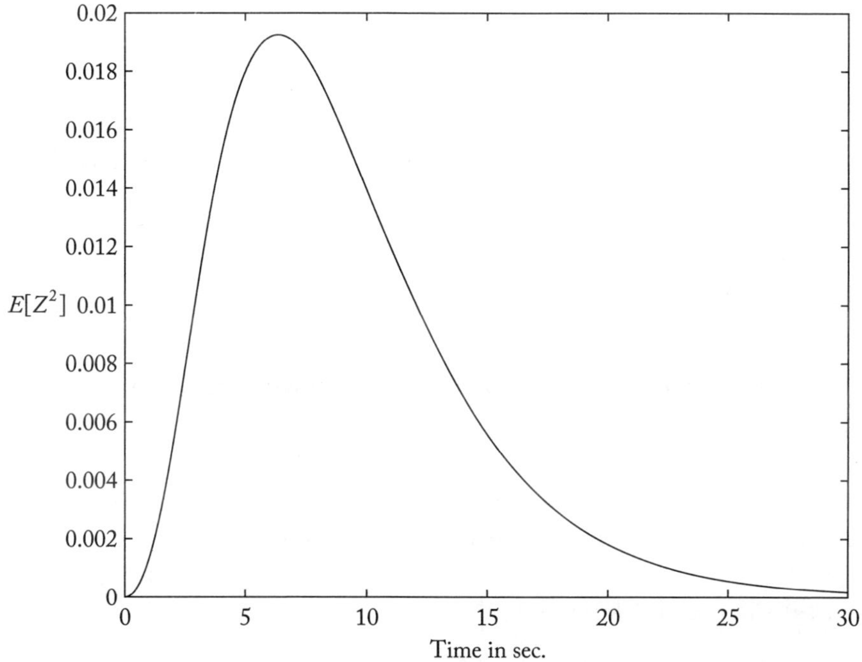

Fig. 4.9 Mean square response of an SDOF oscillator (Example 4.22) under non-stationary support motion by Ito's formula: $a(t) = e^{-\alpha_1 t} - e^{-\alpha_2 t}$, $\alpha_1 = 0.13$, $\alpha_2 = 0.45$; $\overline{\omega}_g = 5\pi$ rad/s, $\xi_g = 0.6$, $\overline{\omega} = 10$ rad/s and $\xi = 0.02$

4.6 Existence and Uniqueness of Solutions to SDEs

Consider the following m-dimensional SDE driven by an n-dimensional Wiener process, $\boldsymbol{B}(t), t \geq 0$:

$$d\boldsymbol{X}(t) = \boldsymbol{a}(t,\boldsymbol{X})\,dt + \boldsymbol{\sigma}(t,\boldsymbol{X})\,d\boldsymbol{B}(t), \quad \boldsymbol{X}(0) = \boldsymbol{X}_0 \tag{4.209}$$

The prescribed initial condition $\boldsymbol{X}_0$ may be a non-random vector or a random vector, independent of the σ-algebra $\mathcal{F}_\infty^n$ generated by $\boldsymbol{B}(s), s \geq 0$ with $E[|\boldsymbol{X}_0|^2] < \infty$. With the coefficients $\boldsymbol{a}(t,\boldsymbol{x}) : [0,\infty] \times \mathbb{R}^m \rightarrow \mathbb{R}^m$ and $\boldsymbol{\sigma}(t,\boldsymbol{x}) : [0,\infty] \times \mathbb{R}^m \rightarrow \mathbb{R}^{m \times n}$, a stochastic process $\boldsymbol{X}(t)$ is a unique and time-continuous solution to the above SDE if:

(i) $\boldsymbol{a}$ and $\boldsymbol{\sigma}$ are measurable functions, uniformly continuous in $t \in [0,T]$ and Lipschitz in $\boldsymbol{x}$, i.e.:

$$\left|\boldsymbol{a}(t,\boldsymbol{x}) - \boldsymbol{a}(t,\boldsymbol{y})\right| + \left|\boldsymbol{\sigma}(t,\boldsymbol{x}) - \boldsymbol{\sigma}(t,\boldsymbol{y})\right| \leq K\left|\boldsymbol{x} - \boldsymbol{y}\right|, \boldsymbol{x},\ \boldsymbol{y} \in \mathbb{R}^m \text{ and } 0 \leq t \leq T, \tag{4.210}$$

for some positive constant K and

(ii) the coefficients satisfy a linear growth condition, i.e.:

$$\left|\boldsymbol{a}\left(t,\boldsymbol{x}\right)\right|+\left|\sigma\left(t,\boldsymbol{x}\right)\right|\leq C(1+|\boldsymbol{x}|),\quad \boldsymbol{x}\in\mathbb{R}^m \text{ and } 0\leq t\leq T \tag{4.211}$$

for some positive constant C. The solution is adapted to the filtration $\mathcal{F}_t^0$ generated by $\boldsymbol{B}\left(s\right), s\leq t$ and $\boldsymbol{X}_0$ with:

$$E\int_0^T |\mathbf{X}_t|^2 dt < \infty \tag{4.212}$$

We may recall that in the case of an ODE $\frac{dx}{dt}=u(t,x)$ with $x(t_0)=x_0$, a solution exists if $u(t,x)$ is continuous and bounded in a rectangle $R: |x-x_0|<a, \ \ |t-t_0|<s$. In addition, if $\frac{\partial u}{\partial x}$ is continuous and bounded, the solution is unique. However, the uniqueness is guaranteed if $u(t,x)$ is only Lipschitz continuous (which is a weaker condition than the requirement of $\frac{\partial u}{\partial x}$ being continuous), i.e.:

$$\left|u\left(t,x_2\right)-u\left(t,x_1\right)\right|\leq K\left|\left(x_2-x_1\right)\right| \tag{4.213}$$

for some positive constant K. Similar to the above equation, Eq. (4.210) is the Lipschitz condition equivalent for the SDE (4.209) to have a unique solution.

For example, if we consider the one-dimensional equation $dX\left(t\right)=3X^{2/3}dt$ with $X_0=0$, it has more than one solution since $a(t,x)=3x^{2/3}$ is not Lipschitz continuous at $x=0$. Note that $X\left(t\right)=0\ \forall t$ is one solution and also for any $t_c>0$ it has another solution given by:

$$\begin{aligned} X\left(t\right) &= 0, \qquad t<t_c \\ &= \left(t-t_c\right)^3, \quad t\geq t_c \end{aligned} \tag{4.214}$$

The linear growth condition in Eq. (4.211) ensures that the solution $X\left(t\right)$ remains finite $\forall t$ and does not explode. The equation $dX\left(t\right)=X^2dt$ with $X_0=1$ has the solution:

$$X\left(t\right)=\frac{1}{t-1},\quad 0\leq t<1 \tag{4.215}$$

In this case $a\left(t,x\right)=x^2$ does not satisfy the linear growth condition and we find that the above solution, even though unique, is not globally valid but tends to ∞ for any interval containing $t=1$.

With the above two conditions (in Eqs. 4.210 and 4.211) imposed on $\boldsymbol{a}$ and σ, existence and uniqueness of $\boldsymbol{X}(t)$ can be proved. The steps to prove existence follow closely the familiar method of Picard iterations employed for ODEs. The method considers the integral form of the SDE as a non-linear map:

$$\pi(\boldsymbol{Y}_t) = \boldsymbol{X}_0 + \int_0^t \boldsymbol{a}(t,\boldsymbol{Y}_s)\,ds + \int_0^t \sigma(t,\boldsymbol{Y}_s)\,dB(s) \tag{4.216}$$

It generates the sequence $\boldsymbol{Y}_t^1 = \pi(\boldsymbol{Y}_t)$, $\boldsymbol{Y}_t^2 = \pi\left(\boldsymbol{Y}_t^1\right)$, ... and it is shown by successive approximations that $\boldsymbol{Y}_t^r$ is a Cauchy sequence in $L^2(P)$ and hence converges to a $\mathcal{F}_t^0$ adapted process $\boldsymbol{X}_t$. Thus $\boldsymbol{X}_t$ can be regarded as a fixed point of the mapping π, i.e., $\pi^r(\boldsymbol{Y}_t) \to \pi(\boldsymbol{X}_t)$ as $r \to \infty$. See Rogers and Williams [1990], Handel [2007] for details.

4.6.1 Locally Lipschitz condition and unique solution to SDE

For the SDE $dX(t) = 3X(t)^{2/3}dt$, it was seen that the Lipschitz condition is not satisfied at $x = 0$. But, the property holds otherwise in any compact set in $\mathbb{R}\backslash\{0\}$. Therefore, if the coefficients $\boldsymbol{a}(t,\boldsymbol{x})$ and $\sigma(t,\boldsymbol{x})$ in Eq. (4.209) are even locally Lipschitz continuous, one may still have a unique solution to the SDE within an interval of interest.

A function $\mathcal{g}(\boldsymbol{x}) : \mathbb{R}^m \to \mathbb{R}^m$ is locally Lipschitz if for any $c < \infty$, there exists a $K_c < \infty$ such that $\left|\mathcal{g}(\boldsymbol{x}) - \mathcal{g}(\boldsymbol{y})\right| \le K_c\left|\boldsymbol{x} - \boldsymbol{y}\right|$ for all $|\boldsymbol{x}|, |\boldsymbol{y}| \le c$.

If $\boldsymbol{a}(t,\boldsymbol{x})$ and $\sigma(t,\boldsymbol{x})$ satisfy the locally Lipschitz property and the linear growth condition with $\boldsymbol{X}_0$ being $L^2(P)$, then the SDE (4.209) has a unique solution in the time interval $[0,\tau]$ where τ is a stopping time (Section 3.5, Chapter 3). τ is also known as the explosion time and may be ∞ with $P(\tau < \infty) = 1$.

4.6.2 Strong and weak solutions

The unique solution to SDE (4.209), under the conditions in Eqs. (4.210) and (4.211), is known as the strong solution. A strong solution $\boldsymbol{X}(t)$ is the one obtained path-wise for a given realization of $\boldsymbol{B}(t)$ on a given probability space $(\Omega, \mathcal{F}, P)$ satisfying the SDE. Thus it is a functional of the driving process and is $\mathcal{F}_t^0$-adapted. It is associated with a path-wise uniqueness in that all possible realizations of the solution process satisfy the SDE. See Example 4.24 below for the derivation of a strong solution to an SDE popularly known as Ornstein–Uhlenbeck process.

However, strong solutions to a given SDE may not always exist. In such cases, weak solutions derived under less stringent conditions on $\boldsymbol{a}(t,\boldsymbol{x})$ and $\sigma(t,\boldsymbol{x})$ may be meaningful. Weak solutions must be interpreted in distribution and thus the path-wise solution to each individual realization of $\boldsymbol{B}(t)$ does not make any sense. To weakly solve an SDE with the given coefficients $\boldsymbol{a}(t,\boldsymbol{x})$ and $\sigma(t,\boldsymbol{x})$, we aim at constructing a Brownian motion $\widehat{\boldsymbol{B}}(t)$ and a process $\widehat{\boldsymbol{X}}(t)$ on some probability space $(\Omega, \mathcal{F}, P)$ such that:

$$\widehat{\boldsymbol{X}}(t) = \widehat{\boldsymbol{X}}(0) + \int_0^t \boldsymbol{a}\left(t,\hat{X}_s\right)ds + \int_0^t \sigma\left(t,\hat{X}_s\right)d\hat{B}(s) \tag{4.217}$$

where the equality above is valid only in distribution. In other words, $\widehat{\boldsymbol{X}}(t)$ or more aptly, the pair of processes $\left(\widehat{\boldsymbol{B}}(t), \widehat{\boldsymbol{X}}(t)\right)$ satisfies the SDE (4.209) and is called a weak solution. A strong solution by definition is also a weak solution and the uniqueness of the former implies the uniqueness of the latter. The converse, however, is not necessarily true.

Example 4.23

We consider the one-dimensional SDE:

$$dX(t) = \text{sign}(X_t)\, dB(t), \quad X_0 = 0 \tag{4.218}$$

where

$$\begin{aligned} \text{sign}(x) &= +1, \quad x \geq 0 \\ &= -1, \quad x < 0 \end{aligned} \tag{4.219}$$

Solution

This SDE is popularly known as Tanaka's equation. Here $\sigma(t, X) = \sigma(X) = \text{sign}(X_t)$ and is not Lipschitz continuous. It can be shown that the SDE has no strong solution [Øksendal 2003]. To find the weak solution to the SDE, we invert the roles of $X(t)$ and $B(t)$. In fact, since $(\text{sign}\, x)^2 = 1$, Eq. (4.218) can be rewritten with the roles of $B(t)$ and $X(t)$ reversed as:

$$dB(t) = \text{sign}\,(X_t)\, dX(t) \tag{4.220}$$

We take a probability space $(\Omega, \mathcal{F}, P)$ equipped with a Brownian motion, $X(t)$. Consider the filtration $\mathcal{F}_t^X$ generated by $X(t)$ and completed under P. Now, we define the process $\hat{B}(t)$ as:

$$\hat{B}(t) = \int_0^t \text{sign}(X_t)\, dX(t) \tag{4.221}$$

$\hat{B}(t)$ is a martingale which is $\mathcal{F}_t^X$-adapted with quadratic variation $\left[\hat{B}, \hat{B}\right](t) = t$. Hence it is an F_t^X–Brownian motion. The pair $\left(X, \hat{B}\right)$ solves the Tanaka equation.

4.6.3 Linear SDEs

Given an m-dimensional SDE:

$$d\boldsymbol{X}(t) = \boldsymbol{a}(t, \boldsymbol{X})\, dt + \boldsymbol{\sigma}(t, \boldsymbol{X})\, d\boldsymbol{B}(t), \quad \boldsymbol{X}(0) = \boldsymbol{X}_0 \tag{4.222}$$

with the coefficients $\boldsymbol{a}(t,\boldsymbol{X})$ and $\boldsymbol{\sigma}(t,\boldsymbol{X})$ linear in $\boldsymbol{X}$, these equations may sometimes be solved exactly and in closed-form. This is, for instance, true if $\boldsymbol{a}(t,\boldsymbol{X}) = \boldsymbol{A}\boldsymbol{X}$ (where $\boldsymbol{A}$ is a constant matrix with real elements) or if $\boldsymbol{\sigma}(t,\boldsymbol{X}) = \boldsymbol{\sigma}(t)$ is an additive diffusion coefficient matrix. In such a case, both the conditions in Eqs. (4.210) and (4.211) are readily satisfied and thus strong solutions exist and are unique. For example, the solution to the 1-diemsional SDE $dX(t) = adt + \sigma dB(t)$ with a and σ being scalar constants and $X(0) = X_0$ is given by:

$$X(t) = X_0 + at + \sigma B(t) \tag{4.223}$$

The above solution is known as an arithmetic Brownian motion with a drift.

Example 4.24

Consider the SDE:

$$dX(t) = -aX(t)dt + \sigma dB(t), \qquad X(0) = X_0 \tag{4.224}$$

Solution

Here a and σ are non-negative constants. To solve the equation, we consider the process $Y(t) = e^{at}X(t)$ and use Ito's formula for $g(e^{at}X(t))$. Since the differential of the quadratic covariation $d[e^{at}, X(t)](t) = d(X)d(e^{at}) = ae^{at}dt\ dX(t) = 0$, we have by Ito's product rule:

$$dY(t) = d\left(e^{at}\right)X + e^{at}dX(t)$$

$$= ae^{at}Xdt + e^{at}(-a\,Xdt + \sigma dB(t))$$

$$= \sigma e^{at}dB(t)$$

$$\Longrightarrow Y(t) = Y(0) + \int_0^t \sigma e^{as}dB(s)$$

$$\Longrightarrow X(t) = e^{-at}X_0 + \int_0^t \sigma e^{a(s-t)}dB(s), \text{ since } X(0) = Y(0) \tag{4.225}$$

The solution, $X(t)$, is known as the Ornstein–Uhlenbeck process. This is a functional of $X(0)$ and $(B_s)_{s\leq t}(t)$ and is a strong solution. We can verify this as follows.

Proof: Applying Ito's product rule to $X(t) = e^{-at}\left(X_0 + \int_0^t \sigma e^{as} dB(s)\right)$, we have:

$$dX(t) = d\left(e^{-at}\left(X_0 + \int_0^t \sigma e^{as} dB(s)\right)\right)$$

$$= d\left(e^{-at}\right)\left(X_0 + \int_0^t \sigma e^{as} dB(s)\right) + e^{-at} d\left(X_0 + \int_0^t \sigma e^{as} dB(s)\right)$$

$$+ d\left[e^{-at}, \left(X_0 + \int_0^t \sigma e^{as} dB(s)\right)\right](t) \tag{4.226a}$$

The last term on the right hand side is the quadratic covariation of e^{-at} and $X_0 + \int_0^t \sigma e^{as} dB(s)$ and thus vanishes. Simplification of Eq. (4.226a) finally leads to:

$$dX(t) = \left(-ae^{-at} dt\right)\left(X_0 + \int_0^t \sigma e^{as} dB(s)\right) + e^{-at}\left(\sigma e^{at} dB(t)\right)$$

$$= -aX(t)\, dt + \sigma dB(t) \tag{4.226b}$$

This verifies that the solution process in Eq. (4.225) satisfies (path-wise) the SDE in Eq. (4.224).

Following a similar procedure, the strong solution for the SDE:

$$dX(t) = (b - a\,X)\, dt + \sigma dB(t), \qquad X(0) = X_0, \tag{4.227}$$

may be obtained as:

$$X(t) = e^{-at}\, X_0 + \frac{b}{a}\left(1 - e^{-at}\right) + \int_0^t \sigma e^{a(s-t)} dB(s) \tag{4.228a}$$

The last solution process $X(t)$ is known as the mean reverting Ornstein–Uhlenbeck process. Solutions to both Eqs. (4.224) and (4.227) are unique. To prove this, let $\hat{X}(t)$ be another solution with the same initial condition. Then, in either case, we have:

$$d\left(X(t) - \hat{X}(t)\right) = -a\left(X - \hat{X}\right) dt, \; X(0) - \hat{X}(0) = 0 \tag{4.228b}$$

The unique solution of the above equation is $X(t) - \hat{X}(t) = 0$ and hence the result.

General linear one-dimensional SDEs

The linear SDE (4.222) may be further generalized to the following form:

$$dX(t) = (a(t) + b(t)X(t))\, dt + (\gamma(t) + \sigma(t)X(t))\, dB(t), \; X(0) = X_0 \tag{4.229}$$

The coefficients a, b, γ and σ are continuous in time. The solution to the above SDE is obtained via the following steps.

Step 1. Before attempting to find $X(t)$, consider the case with $a = 0$ and $\gamma = 0$ leading to the modified SDE of the form:

$$dY(t) = b(t)Y(t)dt + \sigma(t)Y(t)dB(t), \quad Y(0) = Y_0 \tag{4.230}$$

The above SDE can be rewritten as:

$$dY(t) = Y(t)dZ(t), \quad Y(0) = Y_0 \tag{4.231}$$

where,

$$dZ(t) = b(t)dt + \sigma(t)dB(t) \tag{4.232}$$

The solution $Y(t)$ of Eq. (4.231) is known as the stochastic exponential of $Z(t)$ and is given by:

$$Y(t) = \mathcal{E}(Z)(t) := Y_0 e^{Z(t) - Z_0 - \frac{1}{2}[Z,Z](t)} \tag{4.233}$$

This solution is unique and this may be verified, as shown in the proof below, by Ito's formula applied to $g(V) = e^{V(t)}$ with $V(t) = Z(t) - Z_0 - \frac{1}{2}[Z,Z](t)$.

Proof: Note that $Z(t)$ is continuous and is of finite variation and thus $[Z,Z](t)$ is also of finite variation. This implies that $[Z,[Z,Z]](t) = 0$. Now, Ito's formula on $g(V)$ gives:

$$dg(V) = de^{V(t)} = e^{V(t)}dV(t) + \frac{1}{2}e^{V(t)}d[V,V](t)$$

$$\Longrightarrow d(\mathcal{E}(Z)(t)) = dY(t) = Y_0 dg(V) = Y_0\left\{e^{V(t)}dV(t) + \frac{1}{2}e^{V(t)}d[V,V](t)\right\}$$

$$= Y_0\left\{e^{V(t)}\left(dZ(t) - \frac{1}{2}d[Z,Z](t)\right) + \frac{1}{2}e^{V(t)}d[Z,Z](t)\right\}$$

$$= Y_0 e^{V(t)}dZ(t)$$

$$\Longrightarrow dY(t) = Y(t)dZ(t) \tag{4.234}$$

which is same as the SDE in Eq. (4.231).

Now the solution $Y(t)$ (Eq. 4.233) can be written, using the integral form of SDE (4.232) for $Z(t)$, as:

$$Y(t) = Y_0 \exp\left(\int_0^t b(s)\,ds + \int_0^t \sigma(s)\,dB(s) - \frac{1}{2}\int_0^t \sigma^2(s)\,ds\right)$$
$$= Y_0 \exp\left(\int_0^t \left(b(s) - \frac{1}{2}\sigma^2(s)\right)ds + \int_0^t \sigma(s)\,dB(s)\right) \tag{4.235}$$

In writing the above equation $[Z,Z](t) = \int_0^t \sigma^2(s)\,ds$ is utilized.

Step 2. We now assume that the solution to the SDE in Eq. (4.229) is of the form:

$$X(t) = Y(t)\,U(t), \quad X(0) = X_0 \tag{4.236}$$

where, $Y(t)$ is given by Eq. (4.235). The process $U(t)$ is not determined as yet. Assume that $Y(0) = 1$ and $U(0) = X_0$ consistent with the known initial condition $X(0) = X_0$. We write the stochastic differential for $U(t)$ with unknown coefficients $\alpha(t)$ and $\beta(t)$ as:

$$dU(t) = \alpha(t)\,dt + \beta(t)\,dB(t) \tag{4.237}$$

$\alpha(t)$ and $\beta(t)$ are obtained by comparing $dX(t)$ in Eq. (4.229) with $d(Y(t)\,U(t)) = U(t)\,dY(t) + Y(t)\,dU(t) + d[Y,U](t)$. This leads to:

$$\alpha(t)\,Y(t) = a(t) - \sigma(t)\,\gamma(t), \tag{4.238a}$$

$$\beta(t)\,Y(t) = \gamma(t) \tag{4.238b}$$

Knowing $Y(t)$ from Eq. (4.235), $\alpha(t)$ and $\beta(t)$ can be expressed in terms of the known coefficients. $U(t)$ is in turn obtained from Eq. (4.237). The solution $X(t)$ is finally obtained as:

$$X(t) = Y(t)\left(X_0 + \int_0^t \frac{a(s) - \sigma(s)\,\gamma(s)}{y(s)}\,ds + \int_0^t \frac{\gamma(s)}{y(s)}\,dB(s)\right) \tag{4.239}$$

Note that if $b(t) = 0$, then $Y(t) = Y_0 + \int_0^t \sigma(s)\,Y(s)dB(s)$ is an Ito integral and hence should be an exponential martingale. The precise condition for this to hold is given by the so-called Novikov's condition [Øksendal 2003] discussed below.

It is indeed the following Kazamaki condition that is sufficient for $Y(t)$ to be a martingale:

$$E\left[\exp\left(\frac{1}{2}\int_0^t \sigma(s)\,dB(s)\right)\right] < \infty, \text{ for all } t \le T \tag{4.240a}$$

However, the above condition is implied by Novikov's condition (which is a stronger condition based on Ito's isometry):

$$E\left[\exp\left(\frac{1}{2}\int_0^T \sigma^2(s)\,ds\right)\right] < \infty \tag{4.240b}$$

Example 4.25

Suppose that a linear SDE is of the form:

$$dX(t) = \alpha(t)X(t)dt + dB(t), \qquad X(0) = X_0 \tag{4.241}$$

The SDE is known as the Langevin equation. The solution to the SDE can be obtained as follows.

Solution

Method 1. Considering the process $Z(t) = e^{-\int_0^t \alpha(s)ds}X(t)$ and using Ito's product rule, we have:

$$dZ(t) = e^{-\int_0^t \alpha(s)ds}dX(t) - \alpha(t)e^{-\int_0^t \alpha(s)ds}X(t)\,dt$$

$$= e^{-\int_0^t \alpha(s)ds}dB(t) \quad \text{(by Eq. 4.241)}$$

$$\Longrightarrow d\left(e^{-\int_0^t \alpha(s)ds}X(t)\right) = e^{-\int_0^t \alpha(s)ds}dB(t) \tag{4.242}$$

Integrating the above equation on both sides, we get:

$$e^{-\int_0^t \alpha(s)ds}X(t) = X_0 + \int_0^t e^{-\int_0^u \alpha(s)ds}dB(u)$$

$$\Longrightarrow X(t) = e^{\int_0^t \alpha(s)ds}\left(X_0 + \int_0^t e^{-\int_0^u \alpha(s)ds}dB(u)\right) \tag{4.243}$$

Method 2. The solution may also be obtained by comparing the SDE (4.241) with the general linear SDE (4.229) and identifying $b(t) = \alpha(t)$, $\gamma(t) = 1$ and $a(t) = \sigma(t) = 0$. The general solution in (4.239), adapted for this case, gives the same expression as in Eq. (4.243).

4.6.4 Markov property of solutions to SDEs

If $X(t)$ is a unique solution of an SDE, it possesses the Markov property (Section 3.3.6, Chapter 3) which, for $s < t$, is given by:

$$P\big(X(t) \le y \,\big|\, \mathcal{F}_s\big) = P\big(X(t) \le y \,\big|\, X(s)\big) \text{ a.s.} \tag{4.244}$$

Consider a one-dimensional SDE $dX(t) = a(t,X)\,dt + \sigma(t,X)\,dB(t)$. For a small time interval Δ, we find that $X(t+\Delta) - X(t) \approx a(t,x)\Delta + \sigma(t,x)(B(t+\Delta) - B(t))$ where $x = X(t)$. Since a Wiener process is a Markov process, $B(t+\Delta) - B$ is independent of the past and therefore it is intuitively clear that $X(t)$ is also a Markov process.

$F(y;t|x;s) = P\big(X(t) \le y \,\big|\, X(s)\big)$ on the right hand side of Eq. (4.244) is known as the transition probability distribution function. If $f(y;t|x;s) := \frac{\partial F}{\partial y}$ exists, then it is the transition *pdf* for the Markov process. Further, we have with $t_0 < t_1 < \cdots < t_{N-1} < t_N$:

$$f(x_N;t_N,x_{N-1};t_{N-1},\ldots,x_0;t_0) = f(x_N;t_N,x_{N-1};t_{N-1},\ldots,x_1;t_1|x_0;t_0)f(x_0;t_0)$$

$$= f(x_N;t_N,x_{N-1};t_{N-1},\ldots,x_2;t_2|x_1;t_1,x_0;t_0)f(x_1;t_1|x_0;t_0)f(x_0;t_0)$$

$$= f(x_N;t_N,x_{N-1};t_{N-1},\ldots,x_2;t_2|x_1;t_1)f(x_1;t_1|x_0;t_0)f(x_0;t_0) \tag{4.245}$$

By following successive conditioning with respect to $(x_k;t_k)$, $k > 1$, we finally obtain the joint *pdf* as:

$$f(x_N;t_N,x_{N-1};t_{N-1},\ldots,x_0;t_0)$$

$$= f(x_N;t_N \mid x_{N-1};t_{N-1})\ldots f(x_1;t_1|x_0;t_0)f(x_0;t_0) \tag{4.246}$$

Thus, only the transition *pdf* $f(x_i;t_i \mid x_{i-1};t_{i-1})$ and the initial density are needed to describe the joint *pdf* of a Markov process.

Example 4.26

We find the transition *pdf* $f(y;t|x;s)$ for the solution process of the SDE:

$$dX(t) = bX(t)dt + \sigma X(t)dB(t), X(0) = X_0 \tag{4.247}$$

Solution

We first find the solution $X(t)$ by comparing the above SDE with the general linear SDE (4.229) and identifying $b(t) = b,\ \gamma(t) = 0,\ a(t) = 0$ and $\sigma(t) = \sigma$ with b

and σ being scalar constants. Referring to the general solution in Eq. (4.239), we get the solution as:

$$X(t) = X_0 e^{\left(b-\frac{1}{2}\sigma^2\right)t+\sigma B(t)}$$

$$\Longrightarrow X(t) = X_s e^{\left(b-\frac{1}{2}\sigma^2\right)(t-s)+\sigma(B(t)-B(s))} \tag{4.248}$$

The transition probability distribution function is obtained as:

$$\begin{aligned} F(y;t|x;s) &= P\left(X(t) \le y \,\middle|\, X(s)\right) \\ &= P\left(X_s e^{\left(b-\frac{1}{2}\sigma^2\right)(t-s)+\sigma(B(t)-B(s))} \le y|X(s)\right) \\ &= P\left(x e^{\left(b-\frac{1}{2}\sigma^2\right)(t-s)+\sigma(B(t)-B(s))} \le y|X(s)\right) \end{aligned} \tag{4.249}$$

Since $B(t) - B(s)$ is independent of $X(s)$, $F(y;t|x;s)$ may be written as:

$$\begin{aligned} F(y;t|x;s) &= P\left(x e^{\left(b-\frac{1}{2}\sigma^2\right)(t-s)+\sigma(B(t)-B(s))} \le y)\right) \\ &= P\left(e^{\left(b-\frac{1}{2}\sigma^2\right)(t-s)+\sigma(B(t)-B(s))} \le \frac{y}{x})\right) \\ &= P\left((B(t)-B(s)) \le \frac{\ln\left(\frac{y}{x}\right) - \left(b-\frac{1}{2}\sigma^2\right)(t-s)}{\sigma}\right) \end{aligned} \tag{4.250}$$

Since $B(t) - B(s) \sim \mathcal{N}(0, t-s)$, we have:

$$F(y;t|x;s) = \Phi\left(\frac{\ln\left(\frac{y}{x}\right) - \left(b-\frac{1}{2}\sigma^2\right)(t-s)}{\sigma\sqrt{t-s}}\right) \tag{4.251}$$

The transition *pdf* is therefore given by:

$$f(y;t|x;s) = \frac{1}{\sigma\sqrt{2\pi(t-s)}} e^{-\frac{1}{2}\left(\frac{\ln\left(\frac{y}{x}\right)-\left(b-\frac{1}{2}\sigma^2\right)(t-s)}{\sigma\sqrt{t-s}}\right)^2} \tag{4.252}$$

Path-wise continuity and transition probability

The transition *pdf* characterizes the path-wise continuity of a stochastic process. The concept of path-wise continuity is captured by the following definition.

Given any x, y satisfying $|y-x| > \epsilon$ and $t < 0$, define the limit:

$$\lim_{h\to 0} \frac{f(y;t+h|x;t)}{h} = g(y|x,t) \tag{4.253}$$

If $g(y|x,t) = 0$, then the process is said to be path-wise continuous. In other words, the process is said to be path-wise continuous if, for any $\epsilon > 0$:

$$\int_{|y-x|>\epsilon} f(y;t+h|x;t)\,dy \approx o(h) \tag{4.254}$$

The above equation implies that the probability of a finite fraction of paths larger than ϵ from x tends to zero faster than h.

Example 4.27

We verify the path-wise continuity of a Brownian motion via the definition in the last equation.

Solution

Since $B_t(\omega)$ is Gaussian, we have $f(y;t+h|x;t) = \frac{1}{\sqrt{2\pi h}} e^{-\frac{(y-x)^2}{2h}}$. Equation (4.254) gives:

$$\int_{|y-x|>\epsilon} f(y;t+h|x;t)\,dy = \frac{1}{\sqrt{2\pi h}} \int_{|y-x|>\epsilon} e^{-\frac{(y-x)^2}{2h}}\,dy$$

$$= \frac{1}{\sqrt{2\pi h}} \left(\int_{-\infty}^{x-\epsilon} e^{-\frac{(y-x)^2}{2h}}\,dy + \int_{x+\epsilon}^{\infty} e^{-\frac{(y-x)^2}{2h}}\,dy \right)$$

$$= \frac{2}{\sqrt{2\pi h}} \int_{\epsilon}^{\infty} e^{-\frac{u^2}{2h}}\,du$$

$$= \frac{1}{\sqrt{\pi}} \int_{\epsilon^2/2h}^{\infty} e^{-z} z^{-\frac{1}{2}}\,dz \text{ (by transformation } z = \frac{u^2}{2h})$$

$$\leq \frac{1}{\epsilon}\sqrt{\frac{2h}{\pi}} \int_{\epsilon^2/2h}^{\infty} e^{-z}\,dz \ \ (\text{since } z \geq \epsilon^2/2h \Longrightarrow z^{-\frac{1}{2}} \leq \frac{\sqrt{2h}}{\epsilon})$$

$$= \frac{1}{\epsilon}\sqrt{\frac{2h}{\pi}}\, e^{-\epsilon^2/2h} \to 0 \text{ as } h \to 0 \tag{4.255}$$

Hence the result.

4.7 Backward Kolmogorov Equation—Revisiting Evaluation of Expectations

As elucidated through Examples 4.17 and 4.18, expectations of a function (or a functional) of solution to an SDE—vector or scalar—can be evaluated by Ito's formula. These expectations may be obtained more elegantly via backward or forward Kolmogorov equations which are deterministic PDEs. These PDEs characterize, in a sense, the continuum properties of the underlying stochastic dynamics. Indeed, the second order partial differential operator L_t in Eq. (4.126b) is known as the backward Kolmogorov operator associated with the m-dimensional SDE (4.122). The operator is rewritten below:

$$L_t = \sum_{i=1}^{m} a_i(t,x)\frac{\partial}{\partial x_i}\frac{1}{2}\sum_{i,j=1}^{m}\sum_{k=1}^{n}\sigma_{ik}(t,x)\,\sigma_{jk}(t,x)\frac{\partial^2}{\partial x_i \partial x_j} \qquad (4.256)$$

4.7.1 Backward Kolmogorov equation

For expositional simplicity, consider a one-dimensional SDE with the coefficients $a(t,X_t)$ and $\sigma(t,X_t)$ satisfying Lipschitz continuity and linear growth conditions in Eqs. (4.210) and (4.211). Let $u(t,x)$ and $g(x)$ be bounded functions such that $u(t,x)$ is C^1 in t and C^2 in x and $g(x)$ is C^2. With $0 \le t \le T$, the backward Kolmogorov equation is given by the following PDE:

$$\frac{\partial u(t,x)}{\partial t} + L_t u(t,x) = 0 \qquad (4.257)$$

with the terminal condition $u(T,x) = g(x)$.

If the coefficients $a(t,\ x)$ and $\sigma(t\ x)$ are bounded and continuous and (i) $\sigma^2(t,\ x) \ge k > 0$, (ii) $a(t,\ x)$ and $\sigma^2(t,\ x)$ satisfy Holder condition, i.e., for $s,\ t > 0,\ \alpha \ge 0$ (not necessarily an integer), $K \in \mathbb{R}$ and $x, y \in \mathbb{R}$:

$$\left|a(t,y) - a(s,x)\right| + \left|\sigma^2(t,y) - \sigma^2(s,x)\right| \le \mathrm{K}\left(\left|y-x\right|^{\alpha} + |t-s|^{\alpha}\right) \qquad (4.258)$$

then the transition pdf $f(y;T|x;t)$ is the fundamental solution of the backward Kolmogorov PDE, which is unique and strictly positive. In other words, as $f(y;T|x;t)$ is the Green's function for the PDE, for any bounded continuous function $g(x)$ on $\mathbb{R}$ we may write:

$$u(t,\ x) = \int_{\mathbb{R}} g(y) f(y;T|x;t)\,dy \qquad (4.259)$$

We claim that the conditional expectation $E\left[g(X_T)\middle|X_t = x\right]$ satisfies Eq. (4.257) with $\lim_{t\uparrow T} u(t,x) = g(x)$. The last equation is significant in that the conditional expectation

$E\left[g(X_T)\middle|X_t = x\right]$ is given by the solution of the backward PDE with the terminal boundary condition, $u(T,x) = g(x)$. For a rigorous proof, we must first show that $u(t,X_t)$ is a martingale. This in turn requires the following two results.

Result 1. A general result on the backward Kolmogorov PDE is that if $u(t,x)$, in addition to being C^1 in t and C^2 in x, has bounded first derivative in x, then:

$$M_u(t) := u(t,X_t) - \int_0^t \left(\frac{\partial u(s,X_s)}{\partial t} + L_s u(s,X_s)\right) ds \tag{4.260}$$

is a martingale.

Proof: The integral form of Ito's formula for $u(t,X)$ is given by Eq. (4.127):

$$u(t,X_t) = u(0,X_0) + \int_0^t \left(\frac{\partial u(s,X_s)}{\partial s} + L_s u(s,X_s)\right) ds + \int_0^t \frac{\partial u(s,X_s)}{\partial x}\sigma(s,X_s)\,dB(s)$$

$$\Longrightarrow M_u(t) = u(0,X_0) + \int_0^t \frac{\partial u(s,X_s)}{\partial x}\sigma(s,X_s)\,dB(s) \tag{4.261}$$

By hypothesis, $\frac{\partial u(s,x)}{\partial x}$ is bounded for all $X(s) = x$ and s and hence $\left(\frac{\partial u(s,x)}{\partial x}\right)^2 < K_1 \in \mathbb{R}$. This leads to:

$$\int_0^t E\left[\frac{\partial u(s,X_s)}{\partial x}\sigma(s,X_s)\right]^2 ds \le \mathrm{K}_1 \int_0^t E\left[\sigma^2(s,X_s)\right] ds \tag{4.262}$$

The linear growth condition $|a(t,x)| + |\sigma(t,x)| \le C(1+|x|)$ implies $\sigma^2(s,X_s) \le C^2(1 + \sup_{s\le T} X^2(s))$ and hence:

$$\int_0^t E\left[\frac{\partial u(s,X_s)}{\partial x}\sigma(s,X(s))\right]^2 ds \le 2K_1C^2T\left(1 + E\left[\sup_{s\le t} X^2(s)\right]\right) \tag{4.263}$$

Further, from the existence and uniqueness result on the solution of an SDE and the assumption, $E\left[X^2(0)\right] < \infty$, one has $E\left[\sup_{s\le T} X^2(s)\right] < K_2\left(1 + E\left[X^2(0)\right]\right) < \infty$ where K_2 depends only on C and T. It follows that $\int_0^t E\left[\frac{\partial u(s,X_s)}{\partial x}\sigma(s,X_s)\right]^2 ds$ is finite and hence by isometry and martingale property of Ito integrals we have:

$$E\left[\int_0^t \frac{\partial u(s,X_s)}{\partial x}\sigma(s,X_s)\,dB(s)\right]^2 = \int_0^t E\left[\frac{\partial u(s,X_s)}{\partial x}\sigma(s,X_s)\right]^2 ds < \infty \tag{4.264}$$

Thus $\mathcal{M}_u(t)$ in Eq. (4.261) is a martingale with mean $u(0, X_0)$.

Result 2. As a corollary from the above result, if $u(t,x)$ satisfies the backward Kolmogorov PDE, then $u(t, X_t)$ is a martingale. For example, if $X(t) = B(t)$, then $L_t = \frac{1}{2}\frac{\partial^2}{\partial x^2}$ and we find that $u(t,x) = e^{x-\frac{t}{2}}$ satisfies the backward Kolmogorov PDE $\frac{\partial u(t,x)}{\partial t} + \frac{1}{2}\frac{\partial^2 u(t,x)}{\partial x^2} = 0$. Thus, by *Result* 1 above $\mathcal{M}_u(t) = u(t, X_t) = e^{B(t)-\frac{t}{2}}$ is the exponential martingale of a Wiener process.

With the above two results, we are ready to prove the following.

Let $u(t, X_t)$ solve the backward PDE:

$$L_t u + \frac{\partial u(t,x)}{\partial t} = 0 \text{ with } u(T,x) = g(x) \tag{4.265}$$

Then, subject to adequate smoothness of $u(t,x)$ as prescribed earlier, we have the following identity:

$$u(t,x) = E\left[g(X_T) \middle| X_t = x\right] \tag{4.266}$$

Proof: To see that the above assertion is true, we first note via the martingale property of $u(t, X_t)$ that:

$$E\left[u(T, X_T) \middle| F_t\right] = u(t,x) \tag{4.267}$$

Since on the boundary, $u(T, X_T) = g(X_T)$, we have:

$$E\left[g(X_T)\,\middle| F_t\right] = u(t,x) \tag{4.268}$$

By the Markov property of $X(t)$, we have:

$$E\left[g(X_T)\,\middle| F_t\right] = E\left[g(X_T) \middle| X_t = x\right] = u(t,x) \tag{4.269}$$

Thus the result in Eq. (4.266) follows. Note that the terminal time may be replaced by a stopping time τ. In that case, the result in Eq. (4.266) takes the form:

$$u(t,x) = \mathrm{E}\left[g(X_{T\wedge\tau}) \middle| X_t = x\right] \tag{4.270a}$$

In fact, a more general form of Ito's formula, with τ being an $\mathcal{F}_t$ stopping time, may be written as:

$$g(X_{t\wedge\tau}) = g(X) + \int_0^{t\wedge\tau} g(X_s)\,ds + \int_0^{t\wedge\tau} \frac{\partial g(X_s)}{\partial x} \sigma(s, X_s)\,dB(s) \tag{4.270b}$$

Note that the Ito integral may be written as

$$\int_0^{t\wedge\tau} \frac{dg(X_s)}{dx}\sigma(s,X_s)\,dB(s) = \int_0^t I(s\le\tau)\ \frac{\partial g(X_s)}{\partial x}\sigma(s,X_s)\,dB(s) \tag{4.270c}$$

Clearly $I(s\le\tau)$ is a bounded and an $\mathcal{F}_t$ adapted process and hence the Ito integral makes sense (i.e., the integrand appearing in the integral on the RHS of the last equation belongs to $\mathcal{H}^*$). Similarly, Dynkin's formula in Eq. (4.128) may also be expressed using a stopping time.

4.7.2 Inhomogeneous backward Kolmogorov PDE

Consider the inhomogeneous backward Kolmogorov PDE:

$$\frac{\partial u(t,x)}{\partial t} + L_t u(t,x) = -h(x) \text{ with } u(T,x) = g(x) \tag{4.271}$$

In this case, we have that $u(t,X_t) + \int_0^t h(X_s)ds$ is a martingale (see Eq. 4.260). Thus, with the introduction of the source term, $h(x)$, the result in Eq. (4.266) takes the form (proof left as an exercise):

$$u(t,x) = E\left[\left(g(X_t) + \int_0^t h(X_s)ds\right)\middle| X(t) = x\right] \tag{4.272}$$

4.7.3 Adjoint differential operator and forward Kolmogorov PDE

The second order differential operator L_t^* given by ($\mathbf{X}\in\mathbb{R}^m$):

$$L_t^* g = -\sum_{i=1}^m \frac{\partial}{\partial x_i}(a_i(t,\mathbf{x})g) + \frac{1}{2}\sum_{i,j=1}^m\sum_{k=1}^n \frac{\partial^2}{\partial x_i\partial x_j}\left(\sigma_{ik}(t,\mathbf{x})\sigma_{jk}(t,\mathbf{x})g\right) \tag{4.273}$$

is known as the adjoint of L_t (Eq. 4.256). The forward Kolmogorov PDE (also known as Fokker–Planck equation) is given by:

$$\frac{\partial u(t,\mathbf{x})}{\partial t} = L_t^* u(t,\mathbf{x}),\quad u(0,\mathbf{x}) = g(\mathbf{x}) \tag{4.274}$$

The Fokker–Planck operator L_t^* (Eq. 4.273) is adjoint with respect to the L^2-inner product of L_t in that for all sufficiently smooth functions $g(\mathbf{x})$ and $h(\mathbf{x})$ that tend to zero sufficiently fast as $|\mathbf{x}|\to\infty$, one has:

$$\int_{\mathbb{R}^m} g(\mathbf{x})\,L_t^* h(\mathbf{x})\,d\mathbf{x} = \langle L_t^* h, g\rangle = \langle h, L_t g\rangle = \int_{\mathbb{R}^m} h(\mathbf{x})\,L_t g(\mathbf{x})\,d\mathbf{x} \tag{4.275}$$

For the forward equation, the coefficients $\boldsymbol{a}(t,\boldsymbol{x})$ and $\boldsymbol{\sigma}(t,\boldsymbol{x})$ are assumed to satisfy certain additional conditions in addition to those specified for the backward PDE. Specifically, the coefficients must also have bounded and Holder continuous partial derivatives up to the second order with respect to $\boldsymbol{x}$. The derivation of the Fokker–Planck equation is given in Appendix D. The transition *pdf* $f(y;t|x;s)$, $s < t$, is also the fundamental solution to the Fokker–Planck PDE.

Example 4.28

We derive the transition *pdf* for the one-dimensional Ornstein–Uhlenbeck SDE $dX(t) = -aX(t)dt + \sigma dB(t)$.

Solution

Identifying $a(t,x) = -ax$ and $\sigma(t,x) = \sigma$, the transition *pdf* $f(y;t|x;s)$ satisfies the backward and forward Kolmogorov equations respectively given by $(t > s)$:

$$\frac{\partial f}{\partial s} + ax\frac{\partial f}{\partial x} + \frac{1}{2}\sigma^2\frac{\partial^2 f}{\partial x^2} = 0, \lim_{s\uparrow t} f\left(y;t \mid x;s\right) = \delta(x-y) \tag{4.276a}$$

$$\frac{\partial f}{\partial t} + a\frac{\partial(yf)}{\partial y} - \frac{1}{2}\sigma^2\frac{\partial^2 f}{\partial y^2} = 0, \ \lim_{t\downarrow s} f(y;t|x;s) = \delta(y-x) \tag{4.276b}$$

Note that Eq. (4.276a) is a PDE in the backward variables (x,s) while Eq. (4.276b) is in terms of the forward variables (y,t). The boundary conditions are f, $\frac{\partial f}{\partial x} \to 0$ as $x \to \pm\infty$ for the backward PDE and f, $\frac{\partial f}{\partial y} \to 0$ as $y \to 0$ for forward PDE. It can be verified that $f(y;t|x;s)$ is in both the cases given by:

$$f(y,t|x,s) = \sqrt{\frac{a}{\pi\sigma^2\left(1-e^{-2a(t-s)}\right)}}\exp\left(-\frac{a}{\sigma^2}\frac{\left(y-xe^{-a(t-s)}\right)^2}{\left(1-e^{-2a(t-s)}\right)}\right) \tag{4.277}$$

and satisfies both the PDEs. In the stationary limit, we have $\frac{\partial f}{\partial t} = 0$, whence the forward Kolmogorov PDE (4.276b) simplifies as:

$$\frac{a\partial(yf)}{\partial y} + \frac{\sigma^2}{2}\frac{\partial^2 f}{\partial y^2} = 0 \tag{4.278}$$

Integrating the above PDE once with respect to y, we have:

$$ayf + \frac{1}{2}\sigma^2\frac{\partial f}{\partial y} = C_1 \tag{4.279}$$

$C_1 = 0$ because of the boundary conditions. Integrating the PDE (4.279) once more, we get the stationary *pdf* as:

$$f(y;t) = C_2 e^{-\frac{a}{\sigma^2}y^2} \tag{4.280}$$

Using $\int_{-\infty}^{\infty} f\,dy = 1$, which yields $C_2 = \frac{1}{\sigma}\sqrt{\frac{a}{\pi}}$, the solution is:

$$f(y;t) = \frac{1}{\sigma}\sqrt{\frac{a}{\pi}} e^{-\frac{a}{\sigma^2}y^2} \tag{4.281}$$

The stationary mean and variance of the process $X(t)$ can be identified from the transition *pdf* as $E[X] = 0$ and $var(X) = \frac{\sigma^2}{2a}$.

In order to cross check the above results, we can obtain the mean and variance of the Ornstein–Uhlenbeck process by direct application of Ito's formula and expectation operator on the corresponding SDEs. For the SDE $dX(t) = -aX(t)\,dt + \sigma dB(t)$, we have:

$$\begin{aligned} &E[dX(t)] = -aE[X(t)] \\ &\Longrightarrow dE[X(t)] = -aE[X(t)], E[X(0)] = E[X_0] \\ &\Longrightarrow E[X(t)] = -E[X_0]\,e^{-at} \end{aligned} \tag{4.282}$$

The above equation shows that the mean tends to zero as $t \longrightarrow \infty$.

In obtaining the variance of the process, choose $g(X) = X^2$ and apply Ito's formula. This results in:

$$d\left[X^2(t)\right] = \left(-2aX^2(t) + \sigma^2\right)dt + 2\sigma X(t)\,dB(t) \tag{4.283}$$

Taking expectations on both sides of the above equation, we get:

$$\frac{dE\left[X^2(t)\right]}{dt} = \sigma^2 - 2aE\left[X^2(t)\right], \quad E\left[X^2(0)\right] = E\left[X_0^2\right] \tag{4.284}$$

Solution to the ODE in Eq. (4.284) is given by:

$$E\left[X^2(t)\right] = e^{-2at}\left(E\left[X_0^2\right] + \frac{\sigma^2}{2a}\left(e^{2at} - 1\right)\right) \tag{4.285}$$

The variance of the process is obtained as:

$$var(X) = E\left[X^2(t)\right] - \left(E[X(t)]\right)^2$$

$$= e^{-2at}\left(E\left[X_0^2\right] + \frac{\sigma^2}{2a}\left(e^{2at} - 1\right)\right) - e^{-2at}\left(E\left[X_0\right]\right)^2$$

$$= \frac{\sigma^2}{2a}\left(1 - e^{-2at}\right) + e^{-2at} var(X_0) \qquad (4.286)$$

As $t \longrightarrow \infty$, the variance reaches the stationary value $\frac{\sigma^2}{2a}$ as before.

4.7.4 Generator L_t

The backward Kolmogorov second order partial differential operator L_t is known as the generator of $X(t)$, solution to the time homogeneous SDE $dX(t) = -a(X(t))dt + \sigma(X(t))dB(t)$. An SDE is called time-homogeneous if the drift and diffusion coefficients do not have explicit time dependence. L_t^* is its adjoint as defined in Eq. (4.273). $L_t u(x)$ can be considered as the directional derivative of a function $u(x)$, averaged over the paths generated by the diffusion equation. Thus;

$$L_t u(x) = \lim_{\Delta t \to 0} \frac{E\left[u\left(X_{t+\Delta t}\right)|X_t = x\right] - u(x)}{\Delta t}$$

$$= \frac{E\left[du\left(X_t\right)|X_t = x\right]}{dt} \qquad (4.287)$$

But by Ito's formula, we have:

$$du\left(X_t\right) = a(X_t)\frac{\partial u\left(X_t\right)}{\partial x}dt + \frac{1}{2}\sigma^2(X_t)\frac{\partial^2 u\left(X_t\right)}{\partial x^2}dt + \sigma(X_t)\frac{\partial u\left(X_t\right)}{\partial x}dB \qquad (4.288)$$

By taking conditional expectations on both sides and noting that the last term on the RHS of the last equation is a zero mean martingale (Eq. 4.261), we have the required result:

$$\frac{E\left[du\left(X_t\right)|X_t = x\right]}{dt} = a(x)\frac{\partial u(x)}{\partial x} + \frac{1}{2}\sigma^2(x)\frac{\partial^2 u(x)}{\partial x^2} = L_t u(x) \qquad (4.289)$$

It is readily noticed that the generator (backward Kolmogorov operator) of a Brownian motion is $L_t = \frac{1}{2}\frac{\partial^2}{\partial x^2}$, a Laplacian. The fundamental solution or the transition density $f(y,t|x,s)$ of the corresponding backward Kolmogorov PDE is given by $f(y,t|x,s) = \frac{1}{\sqrt{2\pi(t-s)}}e^{-\frac{1}{2}\frac{(y-x)^2}{(t-s)}}$. In this case, the adjoint operator L_t^* is the same as L_t so that L_t is self-adjoint. Note that the transition density above is a function of $(t-s)$ only (and not of t and s separately). Indeed, this is generally true for any time homogeneous SDE, as elaborated in the following remark.

Remarks: For the time homogeneous SDE $dX(t) = a(X(t))dt + \sigma(X(t))dB(t)$, the transition density depends on $(t-s)$ only, i.e., $f(y,t|x,s) = f(y,t-s|x,0)$. The following arguments corroborate this statement.

Let $s > 0$ denote an initial time and $X(s) = x$ the initial condition. Let such a process be denoted by $X^{s,x}(s+t)$. For $t > 0$, one has the following integral form of the SDE:

$$\text{(a)} \quad X^{s,x}(s+t) = x + \int_s^{s+t} a(X^{s,x}(\tau))\,d\tau + \int_s^{s+t} \sigma(X^{s,x}(\tau))dB(\tau)$$

Let $Y(t) = X^{s,x}(s+t)$ and $B_1(t) = B(s+t) - B(s),\ t \geq 0$. Then $B_1(t)$ is a Brownian motion and $Y(t)$ satisfies:

$$\text{(b)} \quad Y(t) = x + \int_0^t a(Y(\tau))\,d\tau + \int_0^t \sigma(Y(\tau))dB_1(\tau), Y(0) = x$$

Putting $s = 0$ in the expression for $X^{s,x}(s+t)$ gives:

$$\text{(c)} \quad X^{0,x}(t) = x + \int_0^t a\Big(X^{0,x}(\tau)\Big)d\tau + \int_0^t \sigma(X^{0,x}(\tau))dB(\tau) \text{ and } X^{0,x}(0) = x$$

Therefore $Y(t)$ and $X^{0,x}(t)$ satisfy the same SDE showing that both are governed by the same (transition) distribution, which depends only on the time interval of evolution and not on the initial or terminal times. Thus, by the arguments above, we may simply write the transition density as $f^{t-s}(y,x)$. For an interval t, if we define the operation $F^t u := \int_y F^t(x,dy)\,u(y) = \int_y f^t(y,x)\,u(y)\,dy$, then we have, over non-overlapping intervals t and s, $F^{t+s}u = F^t F^s u$ (the semi-group property). Here F^t denotes transition distribution for the time-homogeneous process. Also note that $\frac{d}{dt}F^t u|_{t=0} = L_t u$, so that one may formally write $F^t \equiv \exp(tL_t)$ around $t = 0$ and that $u(X(t))$ is a martingale.

Of specific interest are the functions $u(x)$ satisfying $L_t u(x) = 0$. $u(x)$ is generally referred to as a harmonic function and plays an important role in determining the exit time probabilities of diffusion processes (e.g., see Example 4.32). Using $L_t u(x) = 0$, it may be readily checked that a solution to the first derivative of $u(x)$ is obtainable as $u'(x) = C\exp(-\int_{x_0}^{x} \frac{2a(s)}{\sigma(s)}ds)$, where C is an arbitrary constant of integration. Clearly, for $C > 0$, $u'(x)$ is positive and, for $C < 0$, $u'(x)$ is negative. Thus $u(x)$ is monotone for $C \neq 0$. Consequently, omitting the two arbitrary constants of integration, $u(x)$ may be expressed as the double integral $u(X) = \int_{X_0}^{X} \exp\left(-2\int_{X_0}^{z} \frac{a(y)}{\sigma^2(y)}dy\right) dz$. Using the fact that $L_t u(x) = 0$, we arrive at the identity $F^t u = \exp(tL_t)\,u = u$ for small t. Thus u is invariant under the transition kernel. If we define the time shift operator $\theta_t : X_s \to X_{t+s}$ and shift invariant functions $\mathfrak{H} \circ \theta_t = \mathfrak{H}$ (the latter denoting perhaps an exit or terminal or coffin state of X_t as $t \to \infty$), then one can construct a sigma–algebra $\mathcal{I}$ generated by such functions $\mathfrak{H}$. It turns out that a bounded harmonic function u is closely related to $\mathcal{I}$ and this is exemplified by Doob's h-transform (see Appendix D for an introduction; however the material therein should be read only after going through the Girsanov change of measures in Section 4.10).

4.7.5 Feynman–Kac formula

Consider a more general, inhomogeneous backward Kolmogorov PDE:

$$\frac{\partial u(t,x)}{\partial t} + L_t u\,(t,x) = h\,(t,x)\,u(t,x) \text{ with } u\,(T,x) = g(x) \tag{4.290}$$

For given bounded functions $h\,(t,x)$ and $g(x)$, if a solution to the PDE exists, it is unique and is given by:

$$u\,(t,x) = E\left[e^{-\int_t^T h(s,X_s)ds} g(X_T)|X\,(t) = x\right] \tag{4.291}$$

Proof: By Ito's formula:

$$du\,(t,X_t) = \left(\frac{\partial u\,(t,X_t)}{\partial t} + L_t u(t,X_t)\right)dt + \sigma\,(t,X_t)\,\frac{\partial u\,(t,X_t)}{\partial x}dB\,(t) \tag{4.292}$$

Note that $\int_0^t \sigma\,(t,X_t)\,\frac{\partial u(t,X_t)}{\partial x}dB\,(t)$ is a zero mean martingale, say, $\mathcal{M}(t)$ (see Eq. 4.261). Denoting thus the last term on the right hand side of Eq. (4.292) by $d\mathcal{M}(t)$ and substituting $\frac{\partial u(t,X_t)}{\partial t} + L_t u\,(t,X_t)$ by $h\,(t,x)\,u\,(t,x)$, we recognize Eq. (4.292) as an SDE of the Langevin type (Eq. 4.241). The solution between t and T (with $T \geq t$) may then be written as (see Eq. 4.243):

$$u\,(T,X_T) = e^{\int_t^T h(s,X_s)ds}\left(u\,(t,X_t) + \int_t^T e^{-\int_t^u h(s,X_s)ds}d\mathcal{M}\,(u)\right) \tag{4.293}$$

With $u\,(T,X_T) = g\,(X_T)$, we obtain:

$$g\,(X_T)\,e^{-\int_t^T h(s,X_s)ds} = u\,(t,X_t) + \int_t^T e^{-\int_t^u h(s,X_s)ds}d\mathcal{M}\,(u) \tag{4.294}$$

The last term on the right hand side is the integral of a bounded function with respect to a martingale and hence is itself a martingale with zero mean. Taking conditional expectation given $X\,(t) = x$, we get the required result.

4.8 Solution of PDEs via Corresponding SDEs

One useful and interesting consequence of the previous results on backward Kolmogorov PDE and the Feynman–Kac formula is in having a convenient/alternative numerical method for computing solutions of deterministic ODEs and PDEs (of the parabolic type) via the probabilistic representation of solutions of SDEs. The idea is better explained by means of an example.

Example 4.29

Consider the PDE:

$$\frac{\partial u}{\partial t}+ax\frac{\partial u}{\partial x}+\frac{1}{2}(\sigma x)^2\frac{\partial^2 u}{\partial x^2}=cu, 0\le t\le T, u(T,x)=x^2 \tag{4.295}$$

Here a, σ and c are positive constants. We obtain the solution to the PDE by solving the corresponding SDE.

Solution

Comparing with Eq. (4.290), one has $L_t = ax\frac{\partial}{\partial x}+\frac{1}{2}(\sigma x)^2\frac{\partial^2}{\partial x^2}$.

The SDE corresponding to L_t above is:

$$dX(t) = aX(t)\,dt+\sigma X(t)\,dB(t),\quad X(0)=X_0 \tag{4.296}$$

Solution to this SDE is known as the geometric Brownian motion. Referring to Eqs. (4.247) and (4.248) of Example 4.24, we have the solution as:

$$X(t) = X_0 e^{\left(a-\frac{1}{2}\sigma^2\right)t+\sigma B(t)} \tag{4.297}$$

It follows by the Feynman–Kac formula that:

$$u(t,x) = E\left[e^{-c(T-t)}X^2(T)\,|X_t = x\right] = e^{-c(T-t)}E\left[X^2(T)\,|X_t=x\right] \tag{4.298}$$

From Eq. (4.297), one has:

$$X(t) = X_t e^{\left(a-\frac{1}{2}\sigma^2\right)(T-t)+\sigma(B(T)-B(t))} \tag{4.299a}$$

and

$$E\left[X^2(T)\,|X_t=x\right] = x^2 e^{\left(2a+\sigma^2\right)(T-t)} \tag{4.299b}$$

This leads to the solution to the PDE as:

$$u(t,x) = x^2 e^{\left(2a+\sigma^2-c\right)(T-t)} \tag{4.300}$$

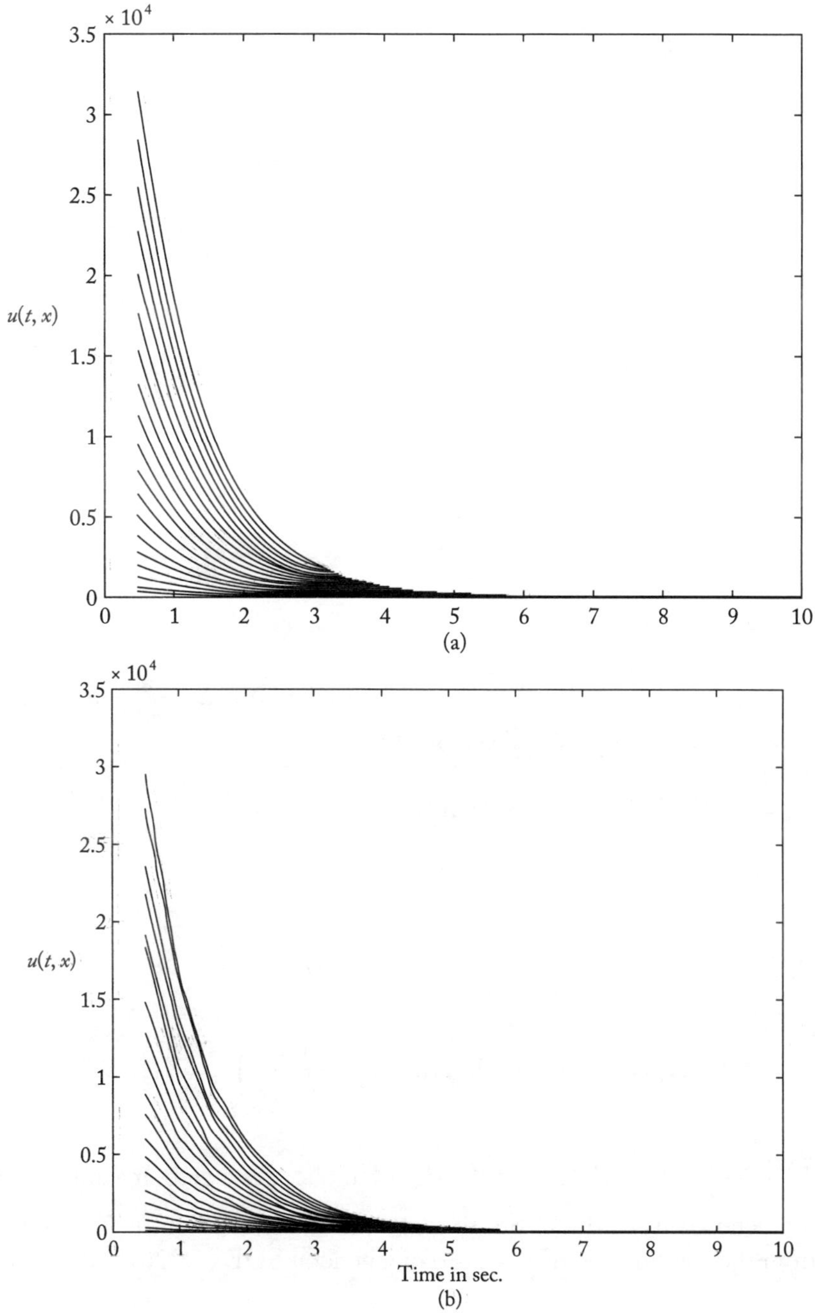

Fig. 4.10 Probabilistic solution to the PDE (4.295); Example 4.29, $a = 0.75$, $\sigma = 0.3$, $c = 0.5$ and $T = 10$, (a) closed form solution for $u(x,t)$ as in Eq. (4.300), and (b) solution by MC simulation with 10000 samples using Eq. (4.298)

The solution to the PDE is also obtained numerically by Monte Carlo (MC) simulation. This requires numerically integrating the SDE (4.296) by Euler–Maruyama method (described later in Chapter 5) to obtain $X(t)$. $u(t,x)$ is then obtained by averaging over all the samples using Eq. (4.298). Here, the MC simulation is performed with 10000 samples of (numerical approximations to) $X(t)$ with parameters $a = 0.75, \sigma = 0.3$, $c = 0.5$ and $T = 10$.

The range of x, the spatial variable, is taken as $(0-1.0)$ and that of t as $(0-10)$ s. The solution $u(t,x)$ is obtained for the cases (a) and (b) (see Fig. 4.10) at 20×20 equidistant grid points over an $x-t$ uniform grid confined in the range as indicated above. Figure 4.10 shows both the numerical and closed form solutions. Each of the curves in the figures corresponds to a trajectory of $u(t,x)$ for a fixed x.

4.8.1 Solution to elliptic PDEs

Consider the general elliptic PDE:

$$L_t w(\boldsymbol{x}) - r(\boldsymbol{x}) w(\boldsymbol{x}) = \hbar(\boldsymbol{x}), \quad \boldsymbol{x} \in \Sigma \tag{4.301a}$$

$$w(\boldsymbol{x}) = g(\boldsymbol{x}), \quad \boldsymbol{x} \in \delta\Sigma \tag{4.301b}$$

Here $L_t = \sum_{i=1}^{m} a_i(\boldsymbol{X}) \frac{\partial}{\partial x_i} + (1/2)\sum_{i,j=1}^{m}\sum_{l=1}^{n} \sigma_{il}(\boldsymbol{X})\sigma_{jl}(\boldsymbol{X}) \frac{\partial^2}{\partial x_i \partial x_j}$ is the generator of $\boldsymbol{X}(t)$, solution to the m-dimensional vector SDE:

$$d\boldsymbol{X}(t) = \boldsymbol{a}(\boldsymbol{X})\,dt + \boldsymbol{\sigma}(\boldsymbol{X})\,d\boldsymbol{B}(t), \boldsymbol{X}(0) = \boldsymbol{x} \tag{4.302}$$

By Ito's formula:

$$\begin{aligned} dw(\boldsymbol{X}_t) &= L_t w(\boldsymbol{X}_t)dt + \left(\frac{\partial w(\boldsymbol{X}_t)}{\partial \boldsymbol{x}}\right)^T \boldsymbol{\sigma}(\boldsymbol{X}_t)\,d\boldsymbol{B}(t) \\ &= \left(r(\boldsymbol{X}_t)w(\boldsymbol{X}_t) + \hbar(\boldsymbol{X}_t)\right)dt + \left(\frac{\partial w(\boldsymbol{X}_t)}{\partial \boldsymbol{x}}\right)^T \boldsymbol{\sigma}(\boldsymbol{X}_t)\,d\boldsymbol{B}(t) \end{aligned} \tag{4.303}$$

The term $\left(\frac{\partial w(\boldsymbol{X}_t)}{\partial \boldsymbol{x}}\right)^T \boldsymbol{\sigma}(\boldsymbol{X}_t)\,d\boldsymbol{B}(t)$ on the right hand side is a zero–mean martingale. Let this term be denoted by $d\mathcal{M}(t)$. Starting at time t, the solution $w(\boldsymbol{X}_{t'})$ with $t' \geq t$ is obtained from the solution (see Eq. 4.239) of a general SDE (4.229) and is given by:

$$\begin{aligned} w(\boldsymbol{X}_{t'}) = \exp\left(\int_t^{t'} r(\boldsymbol{X}_s)\,ds\right) \Bigg(w(\boldsymbol{X}_t) &+ \int_t^{t'} \hbar(\boldsymbol{X}_s)\,\exp\left(-\int_t^{s} r(\boldsymbol{X}_z)\,dz\right) ds \\ &+ \int_t^{t'} \exp\left(-\int_t^{s} r(\boldsymbol{X}_z)\,dz\right) d\mathcal{M}(s) \Bigg) \end{aligned} \tag{4.304}$$

Now, if τ is the first exit time of $\boldsymbol{X}(t)$ (Fig. 4.11) from the boundary $\delta\Sigma$, it is indeed a stopping time; $\tau = \inf\{t > 0 : \boldsymbol{X}(t) \in \delta\Sigma\}$. As $t \to \infty$, $t \wedge \tau = \tau$ a.s. and we have $w(\boldsymbol{X}_\tau) = g(\boldsymbol{X}_\tau)$. Thus, with $t' = \tau$ in (4.304), we have:

$$g(\boldsymbol{X}_\tau) = \exp\left(\int_t^\tau r(\boldsymbol{X}_s)\, ds\right) \left(\begin{array}{c} w(\boldsymbol{X}_t) + \int_t^\tau \hbar(\boldsymbol{X}_s) \exp\left(-\int_t^s r(\boldsymbol{X}_z)\, dz\right) ds \\ + \int_t^\tau \exp\left(-\int_t^s r(\boldsymbol{X}_z)\, dz\right) d\mathcal{M}(s) \end{array} \right) \quad (4.305)$$

Taking expectations conditioned on $\boldsymbol{X}(t) = \boldsymbol{x}$ (the initial condition) on both sides of the above equation and rearranging the terms, one finally obtains:

$$w(\boldsymbol{x}) = E\left[\exp\left(-\int_t^\tau r(\boldsymbol{X}_s)\, ds\right) g(\boldsymbol{X}_\tau) | \boldsymbol{X}(t) = \boldsymbol{x}\right]$$

$$-E\left[\int_t^\tau \hbar(\boldsymbol{X}_s) \exp\left(-\int_t^s r(\boldsymbol{X}_z)\, dz\right) ds | \boldsymbol{X}(t) = \boldsymbol{x}\right] \quad (4.306)$$

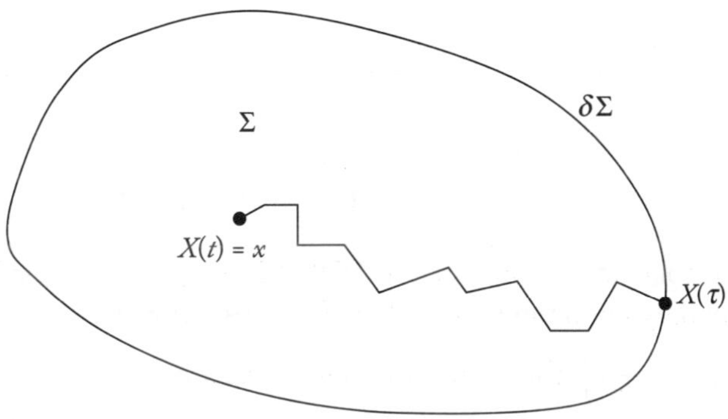

Fig. 4.11 The first exit time τ when $\boldsymbol{X}(t) \in \delta\Sigma$ for a schematic realization of MC simulation

Example 4.30

We solve the following two-dimensional Dirichlet–Poisson equation by the probabilistic approach described above:

$$\nabla^2 v(x,y) + \overline{\omega}^2 v(x,y) = \mathbb{K} \text{ with } \Sigma = (-a,a) \times (-b,b) \text{ and } x, y, \mathbb{K}, \overline{\omega}, a, b \in \mathbb{R} \quad (4.307a)$$

$$v(x,y) = 0 \text{ on the boundary } \delta\Sigma \quad (4.307b)$$

Here $\nabla^2 = \frac{\partial^2}{\partial x^2} + \frac{\partial^2}{\partial y^2}$ is the Laplacian.

Solution

The above PDE is popularly known as the (inhomogeneous) Helmholtz equation which finds interesting applications in the study of many physical problems. In structural engineering, it represents a vibrating membrane with edges clamped.

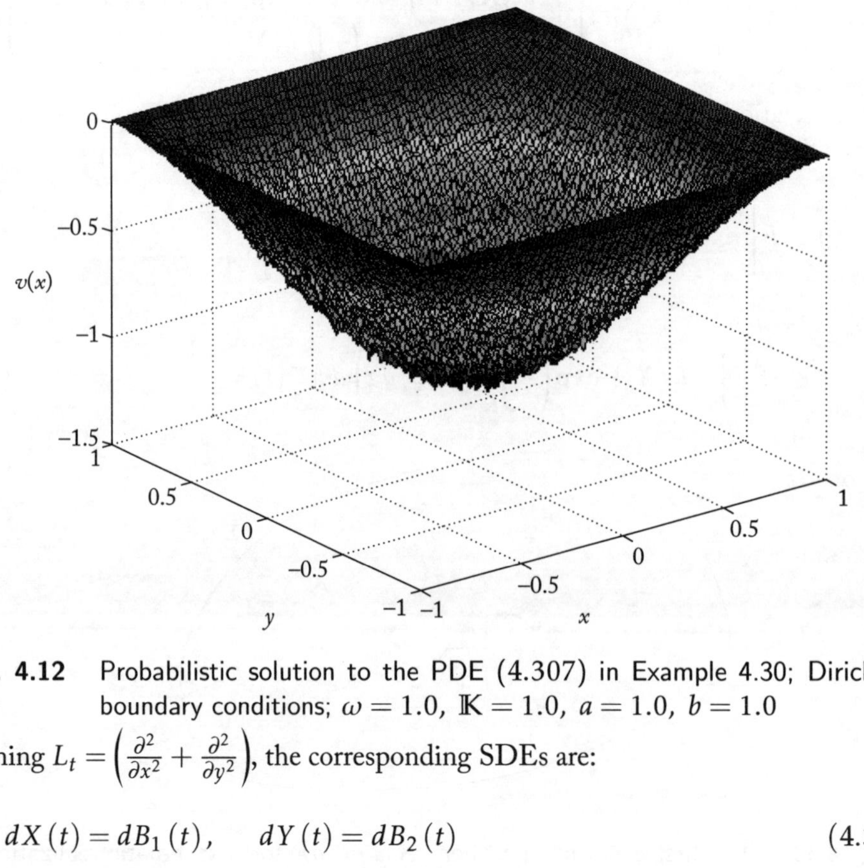

Fig. 4.12 Probabilistic solution to the PDE (4.307) in Example 4.30; Dirichlet boundary conditions; $\omega = 1.0$, $\mathbb{K} = 1.0$, $a = 1.0$, $b = 1.0$

Defining $L_t = \left(\frac{\partial^2}{\partial x^2} + \frac{\partial^2}{\partial y^2}\right)$, the corresponding SDEs are:

$$dX(t) = dB_1(t), \qquad dY(t) = dB_2(t) \tag{4.308}$$

$B_i(t), i = 1,2$ are Wiener processes. With $\boldsymbol{X} = (X,Y)^t$ and $\boldsymbol{x} = (x,y)^t$, the solution $v(\boldsymbol{x})$ is given by Eq. (4.306) which simplifies in the present case to:

$$v(\boldsymbol{x}) = -\mathbb{K}E\left[\int_t^{\tau} \exp\left(-\overline{\omega}^2(s-t)\right) ds\right]$$

$$= -\frac{\mathbb{K}}{\overline{\omega}^2}\left(1 - E\left[e^{-\overline{\omega}^2(\tau-t)}\right]\right) \tag{4.309}$$

For the numerical work, $\Sigma \cup \delta\Sigma$ is divided into a 50×50 grid and the solution obtained at each grid point by MC simulation. Starting at $\boldsymbol{X}(t) = \boldsymbol{x}$, the SDEs (4.308) are solved, the exit time τ at the boundary $\delta\Sigma$ captured (with linear interpolation wherever necessary, assuming thereby that such an approximation does not seriously impair the accuracy of the MC estimate) for each simulation and $v(\boldsymbol{x})$ obtained from (4.309). Figure 4.12 shows the solution of the PDE so obtained. The ensemble size is chosen to be 5000 for evaluation of $v(\boldsymbol{x})$ for every $\boldsymbol{x}$ with the parameters $\overline{\omega}$ and $\mathbb{K}$ prescribed as 1.0 and 1.0 respectively.

Example 4.31

We derive Jarzynski's equality - well-known in statistical physics [Jarzynski 1997, Hummer and Szabo 2001, Liphardt *et al.* 2002, Ge and Jiang 2008] - using the Feynman–Kac formula. The equality reads as:

$$\exp(-\beta_e \Delta F) = E[\exp(-\beta_e W_e)] \tag{4.310}$$

where $\Delta F := F_e(T) - F_e(0)$ is the change in free energy of a system (typically a mesoscopic thermodynamic fragment subject to temperature driven stochastic fluctuations), confined to a constant temperature heat bath, between two thermodynamic equilibria at time instants $t = 0$ and $t = T$. W_e is the external work done on the system to drive it, in a path-dependent manner, from the initial to the final state. $\beta_e = \frac{1}{k_B \chi}$ is known as the inverse temperature with k_B being the Boltzmann's constant and χ, the (absolute) temperature.

Solution

Jarzynski's equality in Eq. (4.310) relates the reversible ΔF to the path-averaged irreversible work W_e done *en route* to the journey of the system (from one equilibrium to another) along paths that may go far away from equilibrium. Such non-equilibrium paths generate, on an average, positive entropy—involving exchange of energy with the heat bath. By the second law of thermodynamics, the external work done on the system satisfies:

$$E[W_e] \geq \Delta F \tag{4.311}$$

The equality above is reached only for strictly reversible processes that drive the system from one thermodynamic equilibrium to another. On the other hand, the inequality in Eq. (4.311) between the average work and free energy change applies to an irreversible process connecting two isothermal equilibrium states (at absolute temperature χ) of an arbitrary system. $D = E[W_e] - \Delta F$ is the dissipation that relates to an increase in entropy in an irreversible process. Jarzynski's equality (4.310) may be viewed as a fluctuation theorem [Crooks 1999] or a non-trivial generalization over the second law of thermodynamics for non-equilibrium cases. With this short background, we proceed to outline the proof (adopted from Ge and Jiang [2008]) of Jarzynski's equality in the following.

Step 1. Consider the statistical thermodynamics of a driven system governed by the m-dimensional Langevin SDE:

$$dX(t) = a(t,X)\,dt + \sigma(t,X)\,dB(t) \tag{4.312}$$

Here $X \in \mathbb{R}^m$, $a : [0,+\infty) \times \mathbb{R}^m \to \mathbb{R}^m$ and $\sigma : [0,+\infty) \times \mathbb{R}^m \to \mathbb{R}^{m\times n}$. B is an n-dimensional standard Brownian motion. Assume that the coefficients $a(t,X)$ and $\sigma(t,X)$ are bounded and continuous with respect to both x and t and satisfy the conditions stated in Section 4.7.1 including the Holder condition. Now consider the Fokker–Planck equation corresponding to the SDE (4.312): $\frac{\partial \vartheta_t(x)}{\partial t} = L_t^* \vartheta_t(x)$ (see Eq. 4.274) and define a fictitious, possibly time-dependent pdf $\pi_t(x)$ that satisfies $L_t^* \pi_t(x) = 0$. Note that $\pi_t(x)$, referred to as the quasi–equilibrium (Boltzmann) density, is not the stationary solution to the last Fokker–Planck equation, i.e., $\pi_t(x) \neq \lim_{t\to\infty} \vartheta_t(x)$. However, $\pi_t(x)$ is an invariant distribution for the transition probability density $f_{X|Y}(x;t|y;t-t')$ in the sense that $\pi_t(x) = \int_{\mathbb{R}^n} f_{X|Y}(x;t|y;t-t')\,\pi_{t-t'}(y)\,dy$, $t' \in (0,t]$. Let π_t be given by:

$$\pi_t(x) = \frac{e^{-\beta_e H(t,x)}}{Z_\beta} \tag{4.313}$$

where, $H(t,x)$ is the Hamiltonian and $Z_\beta = \int_{\mathbb{R}^m} e^{-\beta_e H(t,x)} dx$ is the normalization constant, known as the partition function. Defining a continuous function $F_e(t)$ to denote the free energy, we express the Hamiltonian H of the system as:

$$H(t,x) = F_e(t) - \beta_e^{-1} \ln \pi_t(x) \tag{4.314}$$

This gives $\ln \pi_t(x) = -\beta_e H(t,x) - \ln \int_{\mathbb{R}^n} e^{-\beta_e H(t,x)} dx$ and one has:

$$F_e(t) = \beta_e^{-1} \ln \int_{\mathbb{R}^n} e^{-\beta_e H(t,x)} dx \tag{4.315}$$

We have the total variation $\Delta H = H(T,X_t) - H(0,X_0) = \int_0^T \frac{\partial H(s,X)}{\partial s} ds + \int_0^T \frac{\partial H(s,X)}{\partial x_s} dX_s$. The first integral $\int_0^T \frac{\partial H(s,X)}{\partial s} ds$ is the total work done W_e and the second one $\int_0^T \frac{\partial H(s,X)}{\partial x_s} dX_s$, the heat Q exchanged with the environment. Here ΔH, W_e and Q are treated as random variables, consistent with the foundations of stochastic thermodynamics. It follows that:

$$W_e = \int_0^T \frac{\partial\left(F_e(s) - \beta_e^{-1} \ln \pi_s(X)\right)}{\partial s} ds$$

$$= \Delta F - \int_0^T \frac{\partial\left(\beta_e^{-1} \ln \pi_s(\boldsymbol{X})\right)}{\partial s} ds \tag{4.316}$$

The second term on the RHS of the last equation denotes the dissipative part W_d of the work:

$$W_d = W_e - \Delta F = -\int_0^T \frac{\partial\left(\beta_e^{-1} \ln \pi_s(\boldsymbol{X})\right)}{\partial s} ds \tag{4.317}$$

Step 2. Let us denote:

$$u(t,\boldsymbol{x}) = E\left[\exp\left(\int_t^T \frac{\partial(\ln \pi_s(\boldsymbol{X}))}{\partial s} ds\right) | \boldsymbol{X}_t = \boldsymbol{x}\right] \tag{4.318}$$

Here $u(T,\boldsymbol{x}) = 1$. Note that the expectation in the above equation is taken conditioned on the event $\{\boldsymbol{X}_t = \boldsymbol{x}\}$. With the multi-dimensional diffusion process $\boldsymbol{X}_t$ satisfying the SDE (4.312), consider the inhomogeneous backward Kolmogorov PDE (see Eq. 4.290):

$$\frac{\partial u(t,\boldsymbol{x})}{\partial t} + L_t u(t,\boldsymbol{x}) = -\frac{\partial(\ln \pi_t(\boldsymbol{x}))}{\partial t} u(t,\boldsymbol{x}), \quad u(T,\boldsymbol{x}) = 1 \tag{4.319}$$

L_t is the infinitesimal generator (Eq. 4.256), Now, by Feynman–Kac formula and as proved in Section to PDE (4.6.5), $u(t,\boldsymbol{x}) = E\left[\exp \int_t^T \frac{\partial(\ln \pi_t(\boldsymbol{X}))}{\partial x} ds | \boldsymbol{X}_t = \boldsymbol{x}\right]$ is the solution to PDE (4.319).

Step 3. Let $G(t) = \int_{\mathbb{R}^n} \pi_t(\boldsymbol{x}) u(t,\boldsymbol{x}) d\boldsymbol{x}$. Note that $G(t)$ denotes $E[\exp(-\beta_e W_d)]$ where the average is taken along entire paths over t and T including t. Assume that $G(t) \leq +\infty$ for $t \in [0,T]$ and also $G(t)$ and $\int_{\mathbb{R}^n} \frac{\partial \pi_t(\boldsymbol{x}) u(t,\boldsymbol{x})}{\partial t} d\boldsymbol{x}$ are uniformly convergent for $t \in [0,T]$. Now, consider $\frac{dG(t)}{dt}$ and write as:

$$\frac{dG(t)}{dt} = \int_{\mathbb{R}^n} \left[\frac{\partial \pi_t(\boldsymbol{x})}{\partial t} u(t,\boldsymbol{x}) + \pi_t(\boldsymbol{x}) \frac{\partial u(t,\boldsymbol{x})}{\partial t}\right] \mathrm{dx}$$

$$= \int_{\mathbb{R}^n} \left[\frac{\partial \pi_t(\boldsymbol{x})}{\partial t} u(t,\boldsymbol{x}) + \pi_t(\boldsymbol{x})\right.$$

$$\left.\left\{-\frac{\partial(\ln \pi_t(\boldsymbol{x}))}{\partial t} u(t,\boldsymbol{x}) - L_t u(t,\boldsymbol{x})\right\}\right] dx; \text{(from Eq. (4.319))}$$

$$= \int_{\mathbb{R}^n} \left[\frac{\partial \pi_t(\boldsymbol{x})}{\partial t} u(t, \boldsymbol{x}) + \pi_t(\boldsymbol{x}) \left\{ -\frac{1}{\pi_t(\boldsymbol{x})} \frac{\partial \pi_t(\boldsymbol{x})}{\partial t} u(t, \boldsymbol{x}) - L_t u(t, \boldsymbol{x}) \right\} \right] dx$$

$$= \int_{\mathbb{R}^n} [-\pi_t(\boldsymbol{x}) L_t u(t, \boldsymbol{x})] dx \tag{4.320}$$

Integrating the RHS by parts gives:

$$\frac{dG(t)}{dt} = \int_{\mathbb{R}^n} [-u(t, \boldsymbol{x}) L_t^* \pi_t(\boldsymbol{x})] dx + \text{boundary terms; (from Eq. 4.275)} \tag{4.321}$$

One can see that the boundary terms identically vanish provided that the following conditions are met for each i, j:

$$\lim_{\|x\| \to \infty} \pi_t(\boldsymbol{x})\, a_i(t, \boldsymbol{x})\, u(t, \boldsymbol{x}) = 0$$

$$\lim_{\|x\| \to \infty} \pi_t(\boldsymbol{x})\, \sigma_{ij}(t, \boldsymbol{x}) \frac{\partial u(t, \boldsymbol{x})}{\partial x_j} = 0$$

$$\lim_{\|x\| \to \infty} \frac{\partial \pi_t(\boldsymbol{x})}{\partial x_i} \sigma_{ij}(t, \boldsymbol{x})\, u(t, \boldsymbol{x}) = 0 \tag{4.322}$$

Thus, we have:

$$\frac{dG(t)}{dt} = 0, \forall t \geq 0$$

$$\Longrightarrow G(t) \equiv 1, \text{ since } G(T) = 1 \tag{4.323}$$

It also follows that $G(0) = 1$, proving Jarzynski's equality. A more general form of Jarzynski's equality may be stated if the terminal time T is not deterministic but a stopping time $\tau \geq t$, a.s.

4.8.2 Exit time distributions from solutions of PDEs

The exit times or the hitting times are described in Section 3.5, Chapter 3. We can obtain certain interesting properties of the exit time τ via a probabilistic approach based on the solution of the (herein one-dimensional) boundary value problem:

$$L_t u(x) = -1, \quad x \in \Sigma = (a, b), b > a \tag{4.324a}$$

$$u(x) = 0, \quad x \in \delta\Sigma = \{a, b\} \tag{4.324b}$$

From the general result in Eq. (4.260) on backward Kolmogorov PDE, we can establish the martingale property of $\mathcal{M}(t \wedge \tau)$ given by:

$$\mathcal{M}(t \wedge \tau) = u(X_{t\wedge\tau}) - \int_0^{t\wedge\tau} \left(\frac{\partial u(X_s)}{\partial s} + L_s u(X_s) \right) ds \tag{4.325}$$

Consequently, writing Ito's formula for $u(X_{t\wedge\tau})$ and conditioning on $X(0) = x$, one gets:

$$E\left[\left(u(X_{t\wedge\tau}) - \int_0^{t\wedge\tau} (L_s u(X_s))\, ds \right) | X(0) = x \right] = u(x) \tag{4.326}$$

Using Eq. (4.324a), we have:

$$E[u(X_{t\wedge\tau}) | X(0) = x] + E[t \wedge \tau] = u(x)$$

$$\Longrightarrow E[t \wedge \tau] = u(x) - E[u(X_{t\wedge\tau}) | X(0) = x] \tag{4.327}$$

As $t \to \infty$, $t \wedge \tau \to \tau$ and $u(X_{t\wedge\tau}) \to u(X_\tau)$. Also, $X(t \wedge \tau) \in \Sigma$ for all t and $u(x)$ is bounded on $[a, b]$. Hence, $E[u(X_{t\wedge\tau}) | X(0) = x]$ is also bounded and therefore, $E[\tau] < \infty$. Further, $E[u(X_{t\wedge\tau}) | X(0) = x] \to E[u(X_\tau) | X(0) = x]$ by dominated convergence. Since $u(X_\tau) = 0$ on $\delta\Sigma$, $E[u(X_\tau) | X(0) = x] = 0$ yielding the result:

$$E[\tau] = u(x) \tag{4.328}$$

Thus, the mean exit time of the solution to the SDE is finite and is given by the solution of the PDE (4.324a).

Example 4.32

Consider the one-dimensional PDE:

$$L_t u(x) = 0, \quad x \in \Sigma = (a, b) \tag{4.329}$$

where, $L_t = \frac{1}{2}\sigma^2(x) \frac{\partial^2}{\partial x^2} + \alpha(x) \frac{\partial}{\partial x}$. If τ_a and τ_b are exit times of the interval respectively when $X(t) = a$ and $X(t) = b$ for the first time, then we show that the probability of the process $X(t)$ reaching b before it reaches a is given by

$$P(\tau_b < \tau_a) = \frac{u(x) - u(a)}{u(b) - u(a)} \tag{4.330}$$

Solution

Since $L_t u = 0$, u is harmonic and Eq. (4.326) gives for $\tau = \tau_a$ or τ_b (as $t \to \infty$):

$$E\left[u\left(X_{\tau}\right)|X\left(0\right)=x\right]=u\left(x\right) \tag{4.331}$$

As observed earlier, the exit times τ_a and τ_b are finite. If $\tau = \tau_a$ with probability $P\left(\tau_a < \tau_b\right)$, and $\tau = \tau_b$ with probability $P\left(\tau_b < \tau_a\right) = 1 - P\left(\tau_a < \tau_b\right)$, then Eq. (4.331) implies that:

$$u\left(a\right)\left(1-P\left(\tau_b<\tau_a\right)\right)+u\left(b\right)P\left(\tau_b<\tau_a\right)=u(x) \tag{4.332}$$

Rearranging the terms in the above equation, one obtains the result in Eq. (4.330). It is thus possible to evaluate the exit probability through the harmonic function, $u\left(x\right)$, the solution to Eq. (4.329). Indeed from Eq. (4.329), $\frac{1}{2}\sigma^2\left(x\right)\frac{\partial u'(x)}{\partial x}+\alpha\left(x\right)u'\left(x\right)=0$ and therefore one has $\frac{\partial u'(x)}{\partial x}=-\frac{2\alpha(x)}{\sigma^2(x)}u'\left(x\right)$. This leads to:

$$u'\left(x\right)=C\exp\left(\int_{x_0}^{x}\left(-\frac{2\alpha\left(y\right)}{\sigma^2\left(y\right)}\right)dy\right) \tag{4.333a}$$

It follows that:

$$u\left(x\right)=\int_{x_0}^{x}\exp\left(-2\int_{x_0}^{z}\frac{\alpha y}{\sigma^2\left(y\right)}dy\right)dz \tag{4.333b}$$

without explicitly quoting the associated constants of integration.

4.9 Recurrence and Transience of a Diffusion Process

The recurrence and transience properties of a Brownian motion were briefly introduced in Section 3.6.3, Chapter 3. In this section, we further consider these interesting properties in the context of diffusion processes in general. Suppose that $X\left(t\right)$ is a diffusion process with the drift $\alpha(x)$ and diffusion coefficient $\sigma(x)$ satisfying the conditions for the existence and uniqueness of a solution starting at any point x. In an n-dimensional case, a diffusion is recurrent if, for any starting point $\boldsymbol{x}\in\mathbb{R}^n$, the process visits a ball $b_{\epsilon}(\boldsymbol{y})$ around any point $\boldsymbol{y}\in\mathbb{R}^n$ of radius ϵ, however small, with unit probability, i.e., $P_X(\boldsymbol{X}(t)\in b_{\epsilon}(\boldsymbol{y})$ for some $t\in[0,\infty))=1$. $\boldsymbol{X}\left(t\right)$ is said to be transient if, for any starting point $\boldsymbol{x}\in\mathbb{R}^n$ within a ball howsoever large, the process leaves the ball and never returns.

Referring to a one-dimensional case, the recurrence property of $X\left(t\right)$ is equivalent to having, for any $x,y\in\mathbb{R}$, $P_X\left(\tau_y<\infty\right)=1$ where τ_y is the hitting time of level y starting at x. On the other hand, $X\left(t\right)$is transient if $P_X\left(\tau_y<\infty\right)<1$. To elaborate further, let us refer to Example 4.31 wherein Eq. (4.330) gives $P_X\left(\tau_b<\tau_a\right)$ with the generator $L_t=\frac{1}{2}\sigma^2\left(x\right)\frac{\partial^2}{\partial x^2}+\alpha(x)\frac{\partial}{\partial x}$. Now, if $y>x$, then:

$$P_X\left(\tau_y < \infty\right) = \lim_{a \to -\infty} P_X\left(\tau_y < \tau_a\right) = \lim_{a \to -\infty} \frac{u(x) - u(a)}{u(y) - u(a)} \tag{4.334}$$

where $u(x)$ is the solution given in Eq. (4.333b). If $\lim_{a \to -\infty} u(a) = \infty$ then $P_X\left(\tau_y < \infty\right) = 1$ and any point y is recurrent. Similarly, if $y < x$, then:

$$P_X\left(\tau_y < \infty\right) = \lim_{b \to \infty} P_X\left(\tau_y < \tau_b\right) = \lim_{b \to \infty} \frac{u(x) - u(b)}{u(y) - u(b)} \tag{4.335}$$

If $\lim_{b \to \infty} u(b) = \infty$, then $P_X\left(\tau_y < \infty\right) = 1$ and the point y is again recurrent. In case $\lim_{a \to -\infty} P_X\left(\tau_y < \tau_a\right)$ or $\lim_{b \to \infty} P_X\left(\tau_y < \tau_b\right)$ reaches a finite value, we have $P_X\left(\tau_y < \infty\right) < 1$ and it corresponds to the transient property. Incidentally, for $X(t) = B(t)$, $\alpha(x) = 0$ and $\sigma(x) = 1$ so that $u(x) = x$ (with the lower limit $x_0 = 0$ in the integral in Eq. (4.333b). Therefore, as can be inferred from Eqs. (4.334) and (4.335), a one-dimensional Brownian motion is always recurrent.

As a corollary to the above arguments for recurrence or otherwise of a one-dimensional diffusion process with $L_t u = \frac{1}{2}\sigma^2(x)\frac{\partial^2 u}{\partial x^2} + \alpha(x)\frac{\partial u}{\partial x}$, one can define:

$$I_1 = \int_{-\infty}^{x_0} \exp\left(-2\int_{x_0}^{y} \frac{\alpha(z)}{\sigma^2(z)} dz\right) dy, I_2 = \int_{x_0}^{\infty} \exp\left(-2\int_{x_0}^{y} \frac{\alpha(z)}{\sigma^2(z)} dz\right) dy \tag{4.336}$$

and state that $X(t)$ is recurrent if and only if both I_1 and I_2 are infinite and transient if any of I_1 or I_2 is finite (Klebaner 2005).

For higher dimensional processes, there may exist methods [Bhattacharya 1978, Khasminskii 1980] to test for the property of recurrence or transience. In the absence of strict conditions – necessary and sufficient - these methods are based on the combined use of Ito's formula and martingale properties. For instance, Brownian motion is recurrent for $n \le 2$ and transient for $n > 2$ [Øksendal 2003].

4.10 Girsanov's Theorem and Change of Measure

Change of measure of random variables and its relationship with the Radon–Nikodym derivative have been discussed in Section 2.2.2, Chapter 2. Suppose that $X \sim \mathcal{N}(0,1)$ is a random variable defined on the probability space $(\Omega, \mathcal{F}, P)$. Let $\Lambda = \exp\left(\mu X - \mu^2/2\right)$ be a random variable with $\Lambda \ge 0$ and $E[\Lambda] = 1$. It is clear that Λ can be used as a Radon–Nikodym derivative to effect a change of probability measure, say from P to Q , on $(\Omega, \mathcal{F})$. That is, with $\Lambda = \frac{dQ}{dP}$ and for any measurable set $A \in \mathcal{F}$, one has $Q(A) = E[\Lambda I_A]$ (where the notation $E[.]$ is used for the expectation operator with respect to P). More formally, if $g(x)$ is a bounded function, then with $E_Q[.]$ denoting the expectation operator with respect to Q, we have:

$$E_Q[g] = E[\Lambda g] = E\left[g(X)\exp\left(\mu X - \mu^2/2\right)\right]$$

$$= \int_{\mathbb{R}} g(x)\exp\left(\mu x - \mu^2/2\right)\frac{1}{\sqrt{2\pi}}\exp\left(-\frac{x^2}{2}\right)dx$$

$$= \frac{1}{\sqrt{2\pi}}\int_{\mathbb{R}} g(x)e^{-\frac{1}{2}(x-\mu)^2}dx \tag{4.337}$$

Thus, under the measure Q, the random variable $X \sim \mathcal{N}(\mu,1)$ acquires a mean μ. If the reverse transformation is effected with $\Lambda^{-1} = \frac{dP}{dQ}$, the change of measure removes the mean from X. The consequent advantage in simulation studies (for example of rare events) has been demonstrated in Example 2.4. Girsanov's theorem extends this concept of change of measures to Ito's diffusion processes.

4.10.1 Girsanov's theorem

Consider an Ito process $\boldsymbol{X}(t) \in \mathbb{R}^m$ satisfying the SDE $d\boldsymbol{X}(t) = \boldsymbol{\alpha}(t,\boldsymbol{X}_t)dt + \boldsymbol{\sigma}(t,\boldsymbol{X}_t)d\boldsymbol{B}(t)$; $t \geq 0$ with $\boldsymbol{B}(t) \in \mathbb{R}^n$, $\boldsymbol{\alpha}(t,\boldsymbol{X}_t) \in \mathbb{R}^m$ and $\boldsymbol{\sigma}(t,\boldsymbol{X}_t) \in \mathbb{R}^{m\times n}$. Suppose that there exist processes $\boldsymbol{\beta}(t,\boldsymbol{X}_t) \in \mathcal{H}^{*m}$ and $\boldsymbol{\gamma}(t,\boldsymbol{X}_t) \in \mathcal{H}^{*n}$ such that:

$$\boldsymbol{\sigma}(t,\boldsymbol{x})\boldsymbol{\gamma}(\boldsymbol{x},t) = \boldsymbol{\alpha}(t,\boldsymbol{x}) - \boldsymbol{\beta}(t,\boldsymbol{x}) \tag{4.338}$$

With $\boldsymbol{B}(t)$ a P –Brownian motion, we effect a change of measure to Q with $\Lambda = \frac{dQ}{dP}$ defined as:

$$\Lambda_t = \exp\left(-\int_0^t \boldsymbol{\gamma}(s,\boldsymbol{X}_s).d\boldsymbol{B}(s) - \frac{1}{2}\int_0^t \boldsymbol{\gamma}^2(s,\boldsymbol{X}_s)ds\right), t \leq T \tag{4.339}$$

$\boldsymbol{\gamma}(s,\boldsymbol{X}_s).d\boldsymbol{B}(s) = (\boldsymbol{\gamma}(s,\boldsymbol{X}_s))^T d\boldsymbol{B}(s)$ stands for the scalar product between the vectors $\boldsymbol{\gamma}(s,\boldsymbol{X}_s)$ and $d\boldsymbol{B}(s)$. Similarly $\boldsymbol{\gamma}^2(s,\boldsymbol{X}_s) = \boldsymbol{\gamma}(s,\boldsymbol{X}_s).\boldsymbol{\gamma}(s,\boldsymbol{X}_s)$. Girsanov's theorem states that if Λ_t is a martingale with respect to $\mathcal{F}_t^m$ and P, then the process $\widehat{\boldsymbol{B}}(t)$ given by:

$$\widehat{\boldsymbol{B}}(t) = \int_0^t \boldsymbol{\gamma}(s,\boldsymbol{X}_s)ds + \boldsymbol{B}(t) \tag{4.340}$$

is a Brownian motion with respect to $Q \ll P$, where Q is a probability measure on $\mathcal{F}_t^n$. Then the process $\boldsymbol{X}(t)$satisfies the SDE:

$$d\boldsymbol{X}(t) = \boldsymbol{\beta}(t,\boldsymbol{X}_t)dt + \boldsymbol{\sigma}(t,\boldsymbol{X}_t)d\widehat{\boldsymbol{B}}(t) \tag{4.341}$$

Before proving the above general result, we consider the particular case of $\mathbf{X}(t)$ being a P–Brownian motion and find that, by Girsanov's theorem, $\mathbf{X}(t)$ develops into a Q–Brownian motion with an additional drift term.

4.10.2 Girsanov's theorem for Brownian motion

With $m = 1$ and $n = 1$, let $X(t) = B(t)$. Then $\alpha(t, X_t) = 0$ and $\sigma(t, X_t) = 1$. Hence $\gamma(t) = -\beta(t, X_t)$. We examine the case $\beta(t, X_t) = \mu$, a constant. Then Λ_t in Eq. (4.339) takes the form:

$$\Lambda_t = \exp\left(\mu B_t - \frac{1}{2}\mu^2 t\right), \quad t \le T \tag{4.342}$$

By the Grisanov's theorem, we need first to prove that $Q = \int_\Omega \Lambda_t dP$ is a probability measure and next that $\widehat{B}(t) = B(t) - \mu t$ is a Q–Brownian motion.

Proof:

a) To show that Q is a probability measure on $\mathcal{F}_T$

$$dQ = \Lambda_t dP \Longrightarrow Q = \int_\Omega \Lambda_t dP \quad \text{is strictly positive} \tag{4.343a}$$

and

$$Q(\Omega) = E[\Lambda_t] = E\left[e^{\mu B_t - \frac{1}{2}\mu^2 t}\right] = e^{-\frac{1}{2}\mu^2 t} E e^{\mu B_t} \tag{4.343b}$$

As $B(t) \sim \mathcal{N}(0,t), E\left[e^{\mu B_t}\right] = e^{\frac{1}{2}\mu^2 t} \Longrightarrow E[\Lambda_t] = 1 = Q(\Omega)$.

b) To show that $\widehat{B}(t) = B(t) - \mu t$ is a Q–Brownian motion

We follow Levy's characterization [Klebaner 2005, Øksendal 2003, see also Appendix D] of a Brownian motion by which a process is a Brownian motion if it is a martingale with quadratic variation t. Since μt is a differentiable function, the quadratic variation of $\widehat{B}(t)$ is:

$$\left[\widehat{B}_t, \widehat{B}_t\right](t) = [B_t - \mu t,\ B_t - \mu t](t) = [B_t, B_t](t) = t \tag{4.344}$$

In fact, a process has the same quadratic variation under a change of measure from P to $Q \ll P$. We further notice that $\Lambda_t = \exp\left(\mu B_t - \frac{1}{2}\mu^2 t\right)$ is a martingale (an exponential martingale—Section 3.3.6, Chapter 3) under P. Now letting $Z(t) = \Lambda_t \widehat{B}(t)$, we have by Ito's formula:

$$dZ_i(t) = \Lambda_t d\widehat{B}(t) + \widehat{B}(t)\, d\Lambda_t + d\widehat{B}(t)\, d\Lambda_t$$

$$= \Lambda_t(-\mu dt + dB_t) + \widehat{B}(t)\,\Lambda_t(\mu dB_t)$$

$$+ (-\mu dt + dB_t)\,\Lambda_t(\mu dB_t)$$

$$= \Lambda_t dB_t\left(1 + \mu \widehat{B}_t\right) \tag{4.345}$$

Thus, $Z(t)$ is a P-martingale. Following the generalized Bayes' formula (Section 2.2.3, Chapter 2) on conditional expectation, one has:

$$E_Q\left[\widehat{B}_t | \mathcal{F}_s\right] = \frac{E\left[\Lambda_t \widehat{B}_t | \mathcal{F}_s\right]}{E\left[\Lambda_t | \mathcal{F}_s\right]}$$

$$= \frac{E\left[Z_t | \mathcal{F}_s\right]}{\Lambda_s}$$

$$= \frac{Z(s)}{\Lambda_s} = \widehat{B}(s) \text{ (since } Z(t) \text{ is a } P- \text{ martingale)} \tag{4.346}$$

$\widehat{B}(t)$ is therefore a martingale as claimed and as such it is a Q–Brownian motion by Levy's characterization.

This preliminary exposition of Girsanov's theorem shows that by a change of probability measure from P to an equivalent Q, a P–Brownian motion (say, of zero mean) can be changed to a Q–Brownian motion with a non-zero mean. One generalization of this result is the following.

4.10.3 Girsanov's theorem—Version 1

Consider again the Ito process $\boldsymbol{X}(t) \in \mathbb{R}^n$ satisfying the SDE:

$$d\boldsymbol{X}(t) = \boldsymbol{\alpha}(t, \boldsymbol{X}_t)\, dt + d\boldsymbol{B}(t), \boldsymbol{X}(0) = 0, 0 \le t \le T \tag{4.347}$$

$\boldsymbol{B}(t)$ is an n-dimensional standard Brownian motion with respect to P. Define Λ_t as:

$$\Lambda_t = \exp\left(-\int_0^t \alpha(s, \boldsymbol{X}_s).d\boldsymbol{B}_s - \frac{1}{2}\int_0^t \alpha^2(s, \boldsymbol{X}_s)\, ds\right) \tag{4.348}$$

If Λ_t is a martingale with respect to P and $\mathcal{F}_t^n$, then $\mathbf{X}(t)$is an n-dimensional standard Q-Brownian motion, $0 \leq t \leq T$, where Q is a probability measure on $\mathcal{F}_t^n$ and is given by:

$$dQ(\omega) = \Lambda_T dP(\omega) \tag{4.349}$$

Proof: Λ_t is a martingale if Novikov's condition (see Eq. 4.240b) is satisfied:

$$E\left[\exp\left(\frac{1}{2}\int_0^t \alpha^2(s,\mathbf{X}_s)\,ds\right)\right] < \infty \tag{4.350}$$

This can be verified by first letting $\Lambda_t = e^{Y(t)}$ with $Y(t) = -\int_0^t \alpha(s,\mathbf{X}_s).d\mathbf{B}_s - \frac{1}{2}\int_0^t \alpha^2(s,\mathbf{X}_s)\,ds$ and using Ito's formula:

$$d\Lambda_t = e^{Y(t)}dY(t) + \frac{1}{2}e^{Y(t)}(dY(t))^2$$

$$= \Lambda_t\left(\left(-\alpha(t,\mathbf{X}_t).d\mathbf{B}_t - \frac{1}{2}\alpha^2(t,\mathbf{X}_t)\,dt\right) + \frac{1}{2}\left(-\alpha(t,\mathbf{X}_t).d\mathbf{B}_t - \frac{1}{2}\alpha^2(t,\mathbf{X}_t)\,dt\right)^2\right)$$

$$= \Lambda_t\left(-\alpha(t,\mathbf{X}_t).d\mathbf{B}_t - \frac{1}{2}\alpha^2(t,\mathbf{X}_t)\,dt + \frac{1}{2}\alpha^2(t,\mathbf{X}_t)\,dt\right)$$

$$= -\alpha(t,\mathbf{X}_t)\Lambda_t.d\mathbf{B}(t)$$

$$\Longrightarrow \Lambda_t = -\int_0^t \alpha(s,\mathbf{X}_s)\Lambda_s.d\mathbf{B}_s \tag{4.351}$$

Hence Λ_t is a martingale. Therefore, for any bounded $\mathcal{F}_t^n$-measurable function, g, we have:

$$\int_\Omega g(\omega)\Lambda_t dP = E[g(\omega)\Lambda_t] = E[E[g(\omega)\Lambda_t|\mathcal{F}_t^n]]$$

$$= E[g(\omega)E[\Lambda_t|\mathcal{F}_t^n]] = E[g(\omega)\Lambda_t]$$

$$= \int_\Omega g(\omega)\Lambda_t dP \tag{4.352}$$

It also follows that:

$$Q(\Omega) = E_Q[I_\Omega] = E[\Lambda_T I_\Omega] = 1 \tag{4.353}$$

Thus, Q is a probability measure. To prove that $\boldsymbol{X}(t)$ is an n-dimensional Q–Brownian motion, we again invoke Levy's characterization of a Brownian motion. To this end, we first show that $\boldsymbol{X}(t)$ is a martingale with respect to Q. Letting $\boldsymbol{Z}(t) = \Lambda_t \boldsymbol{X}(t)$ and using Ito's formula, we have:

$$dZ_i(t) = \Lambda_t dX_i(t) + X_i(t)\, d\Lambda_t + dX_i(t)\, d\Lambda_t$$

$$= \Lambda_t \left(\alpha_i(t, \boldsymbol{X}_t)\, dt + dB_i(t)\right) + X_i(t)\, \Lambda_t \sum_{k=1}^{n} \left(-\alpha_k(t, \boldsymbol{X}_t)\, dB_k(t)\right)$$

$$+ \left(\alpha_i(t, \boldsymbol{X}_t)\, dt + dB_i(t)\right) \left(\Lambda_t \sum_{k=1}^{n} \left(-\alpha_k(t, \boldsymbol{X}_t)\, dB_k(t)\right)\right)$$

$$= \Lambda_t \left(dB_i(\mathrm{t}) - X_i(t) \sum_{k=1}^{n} \alpha_k(t, \boldsymbol{X}_t)\, dB_k(t) \right), \quad 1 \le i \le n \tag{4.354}$$

Thus $Z_i(t)$ is a martingale with respect to P. Now, using the generalized Bayes' rule as in Eq. (4.346), one obtains:

$$E_Q\left[X_i(t) | \mathcal{F}_s\right] = \frac{E\left[\Lambda_t X_i(t) | \mathcal{F}_s\right]}{E\left[\Lambda_t | \mathcal{F}_s\right]}$$

$$= \frac{E\left[Z_i(t) | \mathcal{F}_s\right]}{\Lambda_s}$$

$$= \frac{Z_i(s)}{\Lambda_s} = X_i(s) \tag{4.355}$$

Thus $\boldsymbol{X}(t)$ is a martingale with respect to Q. It remains now to show that $\left[X_i, X_j\right](t) = \delta_{ij} t$ a.s. for all $i, j \le n$. This is equivalent to showing that $X_i(t) X_j(t) - \delta_{ij} t,\ i, j \le n$ is a martingale with respect to Q (i.e., $X_i{}^2 - t,\ i \le n$ is a martingale and $X_i(t) X_j(t)$, $i, j \le n,\ i \ne j$ is a martingale). With $X_{i,t} := X_i(t)$ and $B_{i,t} := B_i(t)$, let us first define $Y_i(t) = \Lambda_t \left(X_{i,t}{}^2 - t\right)$ and use Ito's formula. Noting that:

$$d\left(X_{i,t}{}^2 - t\right) = 2X_{i,t}\left(\alpha_i(t, \boldsymbol{X}_t)\, dt + dB_{i,t}\right) + \left(\frac{1}{2}\right) 2\left(dX_{i,t}\right)^2 - dt$$

$$= 2X_{i,t}\left(\alpha_i(t, \boldsymbol{X}_t)\, dt + dB_{i,t}\right) \tag{4.356}$$

and following the arguments in Eqs. (4.354) and (4.355) one has:

$$dY_i(t) = \Lambda_t d\left(X_{i,t}{}^2 - t\right) + \left(X_{i,t}{}^2 - t\right) d\Lambda_t + d\left(X_{i,t}{}^2 - t\right) d\Lambda_t$$

$$= \Lambda_t \left(2X_{i,t}\left(\alpha_i(t,\mathbf{X}_t)\,dt + dB_{i,t}\right)\right) + \left(X_{i,t}{}^2 - t\right)\Lambda_t \sum_{k=1}^{n}\left(-\alpha_k(t,\mathbf{X}_t)\,dB_{k,t}\right)$$

$$+ \left(2X_{i,t}\left(\alpha_i(t,\mathbf{X}_t)\,dt + dB_{i,t}\right)\right)\left(\Lambda_t \sum_{k=1}^{n}\left(-\alpha_k(t,\mathbf{X}_t)\,dB_{k,t}\right)\right)$$

$$= \Lambda_t \left(2X_{i,t}\left(\alpha_i(t,\mathbf{X}_t)\,dt + dB_{i,t}\right)\right) + \left(X_{i,t}{}^2 - t\right)\Lambda_t \sum_{k=1}^{n}\left(-\alpha_k(t,\mathbf{X}_t)\,dB_{k,t}\right)$$

$$-2X_{i,t}\Lambda_t \alpha_i(t,\mathbf{X}_t)\,dt$$

$$= 2\Lambda_t X_{i,t} dB_i(t) - \Lambda_t\left(X_{i,t}{}^2 - t\right)\sum_{k=1}^{n}\alpha_k(t,\mathbf{X}_t)\,dB_{k,t} \tag{4.357}$$

Hence $Y_i(t) = \Lambda_t\left(X_{i,t}{}^2 - t\right)$ is a martingale. On similar lines, $X_i(t)X_j(t) - \delta_{ij}t,\ i,j \leq n,\ i \neq j$ can be shown to be a martingale. Thus by Levy's characterization, $\mathbf{X}(t)$ is a Q–Brownian motion.

Example 4.33

Consider the scalar SDE:

$$dX(t) = dt + dB(t), X(0) = x \tag{4.358}$$

We examine the effect of change of measure by Girsanov's theorem.

Solution

Equation (4.348) gives Λ_t as:

$$\Lambda_t = \exp\left(-B_t - \frac{t}{2}\right) \tag{4.359}$$

and

$$dQ(\omega) = \exp\left(-B_t - \frac{t}{2}\right) dP(\omega) \tag{4.360}$$

The solution to the SDE in Eq. (4.348) is:

$$X(t) = x + t + B(t)$$

$$\Longrightarrow E[X_t] = x + t \tag{4.361}$$

By Girsanov's theorem, $X(t)$is a Q–Brownian motion and hence, $E_Q[X(t)] = x$. This can be verified as follows.

$$\begin{aligned} E_Q[X_t] &= \int_\Omega X_t \Lambda_t dP(\omega) = \int_\Omega (x + t + B_t) \exp\left(-B_t - \frac{t}{2}\right) dP(\omega) \\ &= \frac{1}{\sqrt{2\pi t}} \int_{\mathbb{R}} (x + t + u) \exp\left(-u - \frac{t}{2}\right) \exp\left(-\frac{u^2}{2t}\right) du \\ &= x + t + \frac{1}{\sqrt{2\pi t}} \int_{\mathbb{R}} u e^{\left(-\frac{(u+t)^2}{2t}\right)} du \\ &= x \end{aligned} \tag{4.362}$$

The result in the last step is due to the fact that the integral in the penultimate step can be simplified and shown to be $-t$.

4.10.4 Girsanov's theorem—the general version

We are now ready to consider the general version of Girsanov's theorem stated at the beginning of this section and asserted via Eqs. (4.340) and (4.341). In this case, Novikov's condition rendering Λ_t a martingale is given by:

$$E\left[\exp\left(\frac{1}{2}\int_0^T \gamma^2(s, \boldsymbol{X}_s)\, ds\right)\right] < \infty \tag{4.363}$$

With $\boldsymbol{B}(t)$ a P–Brownian motion, the change of measure to $Q \ll P$ yields $\Lambda_t = \frac{dQ}{dP}\, \widehat{\boldsymbol{B}}(t)$ in Eq. (4.340) is now a Q–Brownian motion. The proof for this assertion closely follows the same procedural steps of version 1 of the theorem. Hence, the process $\boldsymbol{X}(t)$ under Q is obtained by substituting $d\widehat{\boldsymbol{B}}(t) - \gamma(t, \boldsymbol{X}_t)\, dt$ for $d\boldsymbol{B}(t)$ in the original SDE of $\boldsymbol{X}(t)$ resulting in Eq. (4.341).

Example 4.34

Consider the SDE:

$$d\boldsymbol{X}(t) = \begin{pmatrix} dX_1(t) \\ dX_2(t) \end{pmatrix} = \begin{pmatrix} 0 \\ 1 \end{pmatrix} dt + \begin{bmatrix} 1 & 3 \\ -1 & -2 \end{bmatrix} \begin{pmatrix} dB_1(t) \\ dB_2(t) \end{pmatrix},\ t \le T \tag{4.364}$$

We find the probability measure, $Q \ll P$, such that:

$$\widehat{\boldsymbol{B}}(t) = \begin{pmatrix} \widehat{B}_1(t) \\ \widehat{B}_2(t) \end{pmatrix} = \begin{pmatrix} -3t \\ t \end{pmatrix} + \begin{pmatrix} B_1(t) \\ B_2(t) \end{pmatrix} \tag{4.365}$$

Solution

If we assume $\boldsymbol{\beta}(t, \boldsymbol{X}_t) = (0,0)^T$ in Eq. (4.338); then:

$$\sigma(t,\boldsymbol{X})\gamma(t,\boldsymbol{X}) = \alpha(t,\boldsymbol{X}) - \boldsymbol{\beta}(t,\boldsymbol{X})$$

$$\gamma = \begin{bmatrix} 1 & 3 \\ -1 & -2 \end{bmatrix}^{-1} \begin{pmatrix} 0 \\ 1 \end{pmatrix} = \begin{pmatrix} -3 \\ 1 \end{pmatrix} \tag{4.366}$$

$dQ = \Lambda_t dP$ with Λ_t given by:

$$\Lambda_t = \exp\left(- \int_0^t (-3 \quad 1) \begin{pmatrix} dB_1(t) \\ dB_2(t) \end{pmatrix} - \frac{1}{2} \int_0^t 10 ds \right)$$

$$= \exp(3B_1(t) - B_2(t) - 5t) \tag{4.367}$$

Therefore, we get the Q–Brownian motion $\widehat{\boldsymbol{B}}$ as:

$$\widehat{\boldsymbol{B}} = \begin{pmatrix} \widehat{B}_1(t) \\ \widehat{B}_2(t) \end{pmatrix} = \int_0^t \gamma ds + \begin{pmatrix} B_1(t) \\ B_2(t) \end{pmatrix}$$

$$= \int_0^t \begin{pmatrix} -3 \\ 1 \end{pmatrix} ds + \begin{pmatrix} B_1(t) \\ B_2(t) \end{pmatrix} = \begin{pmatrix} -3 \\ 1 \end{pmatrix} t + \begin{pmatrix} B_1(t) \\ B_2(t) \end{pmatrix} \tag{4.368}$$

as in Eq. (4.365). The SDE representation of $\boldsymbol{X}(t)$ in terms of $\widehat{\boldsymbol{B}}$ is given by:

$$d\boldsymbol{X}(t) = \begin{pmatrix} dX_1(t) \\ dX_2(t) \end{pmatrix} = \sigma d\widehat{\boldsymbol{B}}(t) = \begin{bmatrix} 1 & 3 \\ -1 & -2 \end{bmatrix} \begin{pmatrix} d\widehat{B}_1(t) \\ d\widehat{B}_2(t) \end{pmatrix}, \; t \le T \tag{4.369}$$

4.11 Martingale Representation Theorem

It was shown earlier (Section 4.3) that $\int_0^t X(s,\omega)\, dB(s) (t \le T)$ is a martingale with respect to $\mathcal{F}_t$ and P, provided that $X(s,\omega) \in \mathcal{H}^*$. The martingale representation theorem states the converse:

If $\mathcal{M}_t$ is a continuous square-integrable martingale with respect to the Brownian filtration, $\mathcal{F}_t$, then there exists a unique stochastic process $X(s,\omega) \in \mathcal{H}^*$ such that for all $t \geq 0$:

$$\mathcal{M}_t = E[\mathcal{M}_0] + \int_0^t X(s,\omega)\, dB(s) \tag{4.370}$$

For example, if $\mathcal{M}_t = B_t^2 - t$, we have by Ito's formula (see Eq. 4.68):

$$B_t^2 - t = 2\int_0^t B_s dB_s \tag{4.371}$$

Here we find that, $X(t,\omega) = 2B_t$. Similar is the case with $\gamma_t = \exp\left(\mu B_t - \frac{1}{2}\mu^2 t\right)$ which is an exponential martingale with $d\gamma_t = \gamma_t dB_t$. In this case, $X(t,\omega) = \gamma_t$. The martingale representation theorem is indeed a consequence of Ito representation theorem which states that every square-integrable $\mathcal{F}$–measurable random variable $\mathbb{F}$ can be written as an Ito integral. More formally, the theorem reads as:

If $\mathbb{F} \in L^2(\mathcal{F}_t, P)$, then there exists a unique stochastic process $X(t,\omega) \in \mathcal{H}^*$ such that:

$$\mathbb{F}(\omega) = E[\mathbb{F}] + \int_0^T X(t,\omega)\, dB(t) \tag{4.372}$$

Proof (for Ito representation theorem): The uniqueness aspect follows from the Ito isometry. If there exists another possible representation for $\mathbb{F}$, say, in terms of $Y(t,\omega) \in \mathcal{H}^*$ as:

$$\mathbb{F}(\omega) = E[\mathbb{F}] + \int_0^T Y(t,\omega)\, dB(t) \tag{4.373}$$

We have:

$$0 = \int_0^T (X(t,\omega) - Y(t,\omega))\, dB(t)$$

$$\Longrightarrow 0 = E\left[\left(\int_0^T (X(t,\omega) - Y(t,\omega))\, dB(t)\right)^2\right] = \int_0^T E\left[(X(t,\omega) - Y(t,\omega))^2\right] dt$$

(by isometry)

$$\Longrightarrow X(t,\omega) = Y(t,\omega) \text{ for almost all } (t,\omega) \in [0,T] \times \Omega \tag{4.374}$$

Before proving the existence part of the theorem, we need to have the result [Øksendal 2003] that the set of random variables $\{u(B_{t_1}, B_{t_2}, \ldots, B_{t_n}) \in C_0^\infty(\mathbb{R}^n),$ $t_i \in [0, T],\ n = 1, 2, \ldots\},\ T > 0$ is dense in $L^2(\mathcal{F}_T^n, P)$ (see Appendix D for a proof). In other words, there exists a sequence of bounded continuous functions $u_n = u(B_{t_1}, B_{t_2}, \ldots, B_{t_n})$ on $\mathbb{R}^n$ such that $u_n \to \mathbb{F}$ in $L^2(\mathcal{F}_t^n, P)$. Consequently, we start with the assumption that $\mathbb{F} = u(B_{t_1}, B_{t_2}, \ldots, B_{t_n})$ with $0 \le t_1 \le t_2 \le \cdots \le t_n \le T$ and $u_n \in C_0^\infty(\mathbb{R}^n)$ and attempt to find $X(t, \omega)$ in Eq. (4.372).

Step 1. Guided by the earlier discussion on the probabilistic solution of PDEs using the generator (backward Kolmogorov operator, L_t), we here consider the degenerate case of the SDE $dX(t) = dB(t)$ and formulate a sequence of homogeneous PDEs. We define the functions, $v_k(t, B_{t_1}, \ldots, B_{t_{k-1}}, B_t),\ k = 1, 2, \ldots, n,\ t \in [t_{k-1}, t_k]$ as solutions of the following sequence of PDEs with the terminal condition for $k = n$ as $v_n(t_n, B_{t_1}, \ldots, B_{t_n}) = \mathbb{F}$. Starting first with the last interval, $t_{n-1} < t < t_n$, one can proceed backwards as follows:

$$\frac{\partial v_n}{\partial t} + \frac{1}{2}\frac{\partial^2 v_n}{\partial x_n^2} = 0, \quad t_{n-1} < t < t_n$$

$$v_n(t_n, x_1, \ldots, x_n) = u(x_1, x_2, \ldots, x_n) = u(B_{t_1}, B_{t_2}, \ldots, B_{t_n}) = \mathbb{F} \tag{4.375}$$

For $k = n-1, n-2, \ldots, 1$, the PDEs are:

$$\frac{\partial v_k}{\partial t} + \frac{1}{2}\frac{\partial^2 v_k}{\partial x_k^2} = 0, \quad t_{k-1} < t < t_k$$

$$v_k(t_k, x_1, \ldots, x_k) = v_{k+1}(t_k, x_1, \ldots, x_k, x_k) \tag{4.376}$$

Following Eq. (4.266), we obtain the solutions to the PDEs as:

$$v_k(t, x_1, \ldots, x_k) = E\left[v_{k+1}\left(t_k, x_1, \ldots, x_k, B_{t_k - t}\right)\right], k = n-1, \ldots, 2, 1 \tag{4.377}$$

Here $B_{t_k - t} = x_k$ at $t = t_k$.

Step 2. Assuming $v_k(t, x_1, \ldots, x_k),\ k = 1, 2, \ldots, n$ to be once continuously differentiable with respect to t and twice with respect to the rest of the arguments, one has for $t \in [t_{k-1}, t_k]$ by Ito's formula:

$$v_k\left(t, B_{t_1}, \ldots, B_{t_{k-1}}, B_t\right)$$

$$= v_k\left(t_{k-1}, B_{t_1}, \ldots, ,B_{t_{k-1}}, B_{t_{k-1}}\right) + \int_{t_{k-1}}^{t} \frac{\partial v_k}{\partial x_k}\left(s, B_{t_1}, \ldots, ,B_{t_{k-1}}, B_s\right) dB(s)$$

$$+ \int_{t_{k-1}}^{t} \left(\frac{\partial v_k}{\partial s} + \frac{1}{2}\frac{\partial^2 v_k}{\partial x_k^2}\right)\left(s, B_{t_1}, \ldots, ,B_{t_{k-1}}, B_s\right) ds$$

$$= v_k\left(t_{k-1}, B_{t_1}, \ldots, ,B_{t_{k-1}}, B_{t_{k-1}}\right) + \int_{t_{k-1}}^{t} \frac{\partial v_k}{\partial x_k}\left(s, B_{t_1}, \ldots, ,B_{t_{k-1}}, B_s\right) dB(s) \quad (4.378)$$

(in view of Eq. 4.376)

Step 3. From the last equation and Eq. (4.375):

$$\mathbb{F} = u\left(B_{t_1}, \ldots, B_{t_n}\right) = v_n\left(t_n, B_{t_1}, \ldots, B_{t_{n-1}}, B_{t_n}\right)$$

$$= v_n\left(t_{n-1}, B_{t_1}, \ldots, B_{t_{n-1}}, B_{t_{n-1}}\right) + \int_{t_{n-1}}^{t_n} \frac{\partial v_n}{\partial x_n}\left(s, B_{t_1}, \ldots, ,B_{t_{n-1}}, B_s\right) dB(s) \quad (4.379a)$$

$$= v_{n-1}\left(t_{n-1}, B_{t_1}, \ldots, B_{t_{n-2}}, B_{t_{n-1}}\right) + \int_{t_{n-1}}^{t_n} \frac{\partial v_n}{\partial x_n}\left(s, B_{t_1}, \ldots, ,B_{t_{n-1}}, B_s\right) dB(s) \quad (4.379b)$$

(from the terminal condition of the PDEs in Eq. 4.376)

Continuing thus, one can write the first term on the right hand side of the above equation as:

$$v_{n-1}\left(t_{n-1}, B_{t_1}, \ldots, B_{t_{n-2}}, B_{t_{n-1}}\right)$$

$$= v_{n-2}\left(t_{n-2}, B_{t_1}, \ldots, B_{t_{n-3}}, B_{t_{n-2}}\right) + \int_{t_{n-2}}^{t_{n-1}} \frac{\partial v_{n-1}}{\partial x_{n-1}}\left(s, B_{t_1}, \ldots, ,B_{t_{n-2}}, B_s\right) dB(s) \quad (4.380)$$

This simplifies Eq. (4.379b) as:

$$\mathbb{F} = v_n\left(t_n, B_{t_1}, \ldots, B_{t_{n-1}}, B_{t_n}\right)$$

$$= v_{n-2}\left(t_{n-2}, B_{t_1}, \ldots, B_{t_{n-3}}, B_{t_{n-2}}\right) + \int_{t_{n-2}}^{t_{n-1}} \frac{\partial v_{n-1}}{\partial x_{n-1}}\left(s, B_{t_1}, \ldots, ,B_{t_{n-2}}, B_s\right) dB(s)$$

$$+ \int_{t_{n-1}}^{t_n} \frac{\partial v_n}{\partial x_n}\left(s, B_{t_1}, \ldots, ,B_{t_{n-1}}, B_s\right) dB(s) \quad (4.381)$$

Repeating the recursive procedure in Step 3, we finally find that the Ito representation in Eq. (4.372) holds with:

$$X(t,\omega) = \frac{\partial v_k}{\partial x_k}\left(t,\ B_{t_1},\ \ldots,\ ,B_{t_{k-1}},B_t\right),\ \text{for } t_{k-1} < t < t_k \tag{4.382}$$

4.11.1 Proof of martingale representation theorem

By Ito representation theorem proved above, we have for $F = \mathcal{M}_t$

$$\begin{aligned}\mathcal{M}_t &= E[\mathcal{M}_t] + \int_0^t X^{(t)}(s,\omega)\,dB_s \\ &= E[\mathcal{M}_0] + \int_0^t X^{(t)}(s,\omega)\,dB_s \quad \text{(by martingale property)}\end{aligned} \tag{4.383}$$

$X^{(t)}(s,\omega) \in L^2(\mathcal{F}_t, P)$ is unique (the superscript stands for the upper temporal limit). With $0 \le t_1 < t_2$, we have;

$$\begin{aligned}\mathcal{M}_{t_1} &= E\left[\mathcal{M}_{t_2}|\mathcal{F}_{t_1}\right] = E[\mathcal{M}_0] + E\left[\int_0^{t_2} X^{(t_2)}(s,\omega)\,dB_s|\mathcal{F}_{t_1}\right] \\ &= E[\mathcal{M}_0] + \int_0^{t_1} X^{(t_2)}(s,\omega)\,dB_s\end{aligned} \tag{4.384}$$

However, one has:

$$\mathcal{M}_{t_1} = E[\mathcal{M}_0] + \int_0^{t_1} X^{(t_1)}(s,\omega)\,dB_s \tag{4.385}$$

It implies that:

$$0 = E\left[\left(\int_0^{t_1}\left(X^{(t_2)}(s,\omega) - X^{(t_1)}(s,\omega)\right)dB_s\right)^2\right] = \int_0^{t_1} E\left[\left(X^{(t_2)}(s,\omega) - X^{(t_1)}(s,\omega)\right)^2\right]ds$$

(by isometry)

$$\Longrightarrow X^{(t_1)}(t,\omega) = X^{(t_2)}(t,\omega) \quad \text{a.s. for } (t,\omega) \in [0,t_1] \times \Omega \tag{4.386}$$

Therefore, one can define $X(t,\omega)$ as:

$$X(t,\omega) = X^{(T)}(t,\omega),\, t \in [0,\ T] \tag{4.387}$$

This yields the required result:

$$\mathcal{M}_t = E[\mathcal{M}_0] + \int_0^t X^{(t)}(s,\omega)\, dB_s$$

$$= E[\mathcal{M}_0] + \int_0^t X(s,\omega)\, dB_s, \forall t \geq 0 \tag{4.388}$$

4.12 A Brief Remark on the Martingale Problem

Before concluding this chapter, a word on the martingale problem as conceptualized by Stroock and Varadhan [1972, 1969a,b] would perhaps be in order, especially as we make passing references to this important concept in some of the chapters to follow. The question here is whether a class of martingales given by Eq. (4.261) would suffice to characterize diffusion. In this context, first recall that the notions of independence and uncorrelatedness are equivalent for Gaussian random variables (see Section 1.9.7, Chapter 1). On the other hand, the perspicacious reader may imagine, based on the structure of an Ito's SDE, that a diffusion process should admit a local approximation by a Gaussian random variable. It therefore seems natural, at least in hindsight, to try and characterize diffusion processes in terms of martingales. This is precisely what Stroock and Varadhan have done. The issue at stake is thus to seek the existence and uniqueness of the probability measures, $\{P_s^{\boldsymbol{x}} : \boldsymbol{X}(s) = \boldsymbol{x} \in \mathbb{R}^m\}$, parametrized with respect to the initial condition and describing the associated diffusion processes. If it exists, $P_s^{\boldsymbol{x}}$ is the solution to the martingale problem. Before stating the results by Stroock and Varadhan, we first define the martingale problem.

Martingale problem: Let $\boldsymbol{X}(t)$ be defined over $(\Omega,\ \mathcal{F}_t^s)$ and governed by the SDE: $d\boldsymbol{X}(t) = \boldsymbol{a}(t,\boldsymbol{X})dt + \boldsymbol{\sigma}(t,\boldsymbol{X})\, d\boldsymbol{B}(t)$. $\mathcal{F}_t^s$ denotes the filtration up to t starting from $s \leq t$. Let $\boldsymbol{b} := \sigma\sigma^T \in \mathbb{R}^{m\times m}$ and $\boldsymbol{a} \in \mathbb{R}^m$. For $s \geq 0$, $\boldsymbol{x} \in \mathbb{R}^m$, a solution to the martingale problem for L_t or for $(\boldsymbol{a},\boldsymbol{b})$ starting from $(s,\boldsymbol{x})$ is a probability measure $P_s^{\boldsymbol{x}}$ on $(\Omega,\ \mathcal{F}_\infty^s)$ such that

(i) $P_s^{\boldsymbol{x}}(\{\omega : \omega(r) = \boldsymbol{x},\ 0 \leq r \leq s\}) = 1$ and

(ii) for each bounded scalar-valued function $g(t,\boldsymbol{X}(t)) \in C^{1,2}([0,\ \infty)\times\mathbb{R}^m)$
$g(t,\boldsymbol{X}(t)) - g(s,\boldsymbol{x}) - \int_s^t \left(\frac{\partial}{\partial r} + L_r\right) g(r,\boldsymbol{X}(r))\, dr$, $t \geq s$ is a $P_s^{\boldsymbol{x}}$-martingale

If for each $s \geq 0$, $\boldsymbol{x} \in \mathbb{R}^m$, there is only one solution to the above then we say that the martingale problem for L_t is well-posed.

Since descriptions based on the characteristic function or the probability distribution, are equivalent, condition (ii) above may be rephrased as follows.

$\exp\left[(\lambda,\boldsymbol{X}(t) - \boldsymbol{X}(s) - \int_s^t \boldsymbol{a}(r,\boldsymbol{X}(r))\, dr) - \frac{1}{2}\int_s^t (\lambda,\boldsymbol{b}(r,\boldsymbol{X}(r))\,\boldsymbol{\lambda})dr\right]$ is a $P_s^{\boldsymbol{x}}$-martingale where $\boldsymbol{\lambda} \in \mathbb{R}^m$.

We quote below the main result (theorem) by Stroock and Varadhan.

Let $\boldsymbol{a}(t,\boldsymbol{X}(t))$ be a bounded measurable function and $\boldsymbol{\sigma}(t,\boldsymbol{X}(t))$ a bounded continuous function such that the nondegeneracy (strict ellipticity) condition i.e., $(\boldsymbol{\lambda},\boldsymbol{b}(t,\boldsymbol{y})\boldsymbol{\lambda}) \geq C|\boldsymbol{\lambda}|^2$, $C > 0$, $\boldsymbol{y} \in \mathbb{R}^m$ holds for all t. Then for each $s \geq 0$, $\boldsymbol{x} \in \mathbb{R}^m$, there is a unique probability measure $P_s^{\boldsymbol{x}}$ on $(\Omega,\ \mathcal{F}_\infty^s)$ solving the martingale problem for $(\boldsymbol{a},\boldsymbol{\sigma})$. Also $(s,\boldsymbol{x}) \mapsto P_s^{\boldsymbol{x}}$ is continuous in the sense of weak convergence of probability measures. Moreover, under $\{P_s^{\boldsymbol{x}} : s \geq 0, \boldsymbol{x} \in \mathbb{R}^m\}$, the process $\boldsymbol{X}(t)$ is a continuous strong Markov process.

The proof of this theorem is beyond the scope of this book. However, interested readers may like to see Strook and Varadhan [1969a,b] and Ramasubramanian [2003].

4.13 Concluding Remarks

We have introduced the elements of stochastic calculus for diffusion processes, thereby providing the backbone material for an understanding of stochastic differential equations and their applications in the chapters to follow. Starting with the description of a stochastic integral, we have progressively built up the subject so as to highlight the essential differences from the conventional calculus. A striking contrast in developing the stochastic calculus lies in a proper interpretation of the stochastic integral, treated as a stochastic process, with regard to its non-differentiability and infinite total variation. Ito's interpretation is herein adopted as it has the advantage that local martingales are closed under the Ito integral. Ito's formula—a stochastic analogue of integration by parts—and its use in deriving SDEs and obtaining strong or weak solutions have been dwelt upon. In this context, existence and uniqueness of solutions to SDEs are also discussed. Another interesting aspect touched upon is the relation between SDEs and the solution of deterministic PDEs. Specifically, topics such as Kolmogorov's backward and forward operators as well as the Feyman–Kac formula have been laid down, including the insightful approaches that they enable in characterizing solutions to a class of deterministic ODEs and parabolic PDEs. Numerical exploitations of these strategies using MC simulations have been demonstrated with a few illustrative examples. Girsanov's theorem on the change of measures, as extended to Ito's processes, has been considered in detail. With our primary refrain being on the concept of change of measures and its myriad applications involving stochastic dynamical systems, we believe that a well-rounded appreciation of Girsanov's theorem is of the essence.

Having come thus far, we are now ready to begin studying SDEs and methods for their solutions in the strong and weak senses, with emphasis on the latter. This is what we take up in Chapter 5.

Exercises

1. Using Ito's formula, prove that: $\int_0^t B_s^2 dB_s = \frac{1}{3}B_t^3 - \int_0^t B_s ds$.
2. Let X and Y be two independent one-dimensional Brownian motions. For a partition Π_N of time interval $[t_0, t_N]$ with $\Delta_N = \max_j (t_j - t_{j-1})$, show that $\lim_{\Delta_N \to 0} \sum_{i=0}^{N-1} (X_{t_{i+1}} - X_{t_i})(Y_{t_{i+1}} - Y_{t_i}) = 0$ in $L^2(P)$.
3. Calculate the variances of the stochastic integrals: a) $\int_0^t \sqrt{|B_s|} dB_s$, b) $\int_0^t (B_s + s)^2 dB_s$
4. Consider the SDE: $dX_t = \mu dt + \sigma dB_t$, $X_0 = x$. Solve for X_t by writing Ito's formula for $Y = X_t - \mu t$. Show that X_t is Gaussian and find its mean and covariance.
5. Suppose that $X(t)$ satisfies $dX(t) = X^2(t)dt + X(t)dB(t)$ with $X(0) = 1$. Show that $X(t) = \exp\left(\int_0^t (X(s) - 1/2)ds + B(t)\right)$ is the solution.
6. Consider $dX_t = \mu(\theta - X_t)dt + \sigma dB_t$ (mean reverting, process). Find the stationary *pdf* using the Kolmogorov PDE (backward or forward), mean and covariance for large t.

   (Ans. mean $= \theta$, variance $= \frac{\sigma^2}{2\mu}$ for large t. See also Example 4.24).
7. Consider $dX_t = -\frac{X_t}{1-t}dt + dB_t$ with $X_0 = X_1 = 0$. Show that the solution to the SDE is the Brownian bridge $X_t = (1-t)\int_0^t \frac{1}{1-s} dB_s$, $t \in [0,1]$.
8. Consider the Cox–Ingersoll–Ross model for interest rates (of stock prices) :

   $$dX_t = \mu(\theta - \beta X_t)dt + \sigma\sqrt{X_t} dB_t$$

   Derive DEs for the first two moments by Ito's formula. Solve for the moments in closed form and show as $t \to \infty$ that $E[X_t] = \frac{\theta}{\beta}$ and $var(X_t) = \frac{\theta\sigma^2}{2\mu\beta^2}$.
9. Derive the Wiener–Khintchine relationship (Eqs. 4.154a and 4.154b) between the autocorrelation function and the power spectral density (PSD) of a stationary process (see Yates and Goodman 2005 in Bibliography).
10. Use the Wiener–Khintchine relationships in Eqs. (4.154a) and (4.154b) to derive the PSD of the response to the following linear DEs:

    i) $\dot{z}(t) + \alpha z(t) = f(t)$, $\alpha > 0$
    ii) $\ddot{z} + \beta\dot{z}(t) + \alpha z(t) = f(t)$, $\alpha, \beta > 0$ and $\beta^2 < \alpha$

    Assume that $f(t)$ is a white noise with $R_{ff}(\tau) = I\delta(\tau)$; see Appendix D for brief notes on linear systems under stochastic inputs.
11. Consider an SDOF oscillator governed by the SDE:

    $$\ddot{x}(t) + 2\xi\omega\dot{x}(t) + \omega^2 x(t) = f(t)$$

Derive the forward Kolmogorov *pdf* (see Appendix D for brief notes on the Fokker–Plank equation) when $f(t)$ is a white noise with $R_{ff}(\tau) = I\delta(\tau)$. Show that the stationary solution for the (transition) *pdf* is given by $\frac{1}{2\pi\sigma_x\sigma_{\dot{x}}}\exp\left\{-\frac{1}{2}\left(\frac{x^2}{\sigma_x^2}+\frac{\dot{x}^2}{\sigma_{\dot{x}}^2}\right)\right\}$ where $\sigma_x^2 = \frac{\pi I}{2\xi\omega^3}$ and $\sigma_{\dot{x}}^2 = \frac{\pi I}{2\xi\omega}$.

12. Consider the degenerate case of $dX(t) = dB(t)$ and define $\tau_a = \inf\{t \geq 0 : B_t = a\}$. In Example 3.10, it has been shown that the first passage time τ_a has the *pdf* $f_{\tau_a}(t) = \frac{|a|}{\sqrt{2\pi t^3}}e^{-a^2/2t}$, $t \geq 0$. Introduce a drift term μt in the SDE and show by a change of measures (using Girsanov's theorem) that the *pdf* of $\widehat{\tau}_a$, the first hitting time of the level a of the Brownian motion with drift, is $f_{\widehat{\tau}_a}(t) = \frac{|a|}{\sqrt{2\pi t^3}}e^{-(a-\mu t)^2/2t}$, $t \geq 0$.

13. Find by MC simulation the approximate probabilities that a Wiener process starting at zero exceeds levels $K = 1, 2, 3, 4$ in the time interval $[0, 1]$.

(Hint: Use change of measures).

Ans.

K	$E^Q[\hat{B}(t) \geq 0]$, where $\hat{B}(t) = B(t) - K$	$E^P[B(t) \geq K]$	Exact value $= \sqrt{\frac{2}{\pi}}\int_K^\infty \exp\left(-\frac{x^2}{2}\right)dx$
1	0.33	0.31	0.32
2	0.042	0.034	0.046
3	2.9E-3	2E-3	2.7E-3
4	5.8E-5	0	6.3E-5

Observation: For $K = 4$ and above, direct simulation under P (3rd column) shows that the level K is not exceeded even with an ensemble size of 500.

14. Let $X_1, X_2, \ldots$ be an i.i.d. sequence of Bernouilli random variables with probability of success equal to p and $F_n = \sigma(X_1, \ldots, X_n)$, $n \geq 1$. Let $\mathcal{M}$ be a martingale adapted to the generated filtration. Show that the martingale representation property holds: i.e., there exists a constant m and a predictable process Y such that $\mathcal{M}_n = m + (Y \,.\, S)_n$, $n \geq 1$, where $S_n = \sum_{k=1}^{n}(X_k - p)$.

Notations

$a(t)$, $a(t,X)$, $a_i(t,\boldsymbol{X})$	drift terms in an SDE
$A_t(\lambda)$	modulating function (Eq. 4.162)
B_1, B_2, $\widehat{\boldsymbol{B}}$	Brownian motion (Example 4.18)
c_1, c_2, c_3, c_4	damping coefficients (SDE 4.129)

$\varepsilon(Z)(t)$	stochastic exponential of the process $Z(t)$
$\mathfrak{f}, \mathfrak{g}$	stochastic processes $\in \mathcal{H}$
$\mathfrak{g}(t)$	excitation process to the SDOF oscillator in Example 4.19
$g(\boldsymbol{X})$	a scalar C^2 function of the vector process $\boldsymbol{X}$ and an Ito diffusion process
$g(t,x)$, $g(t,X(t))$	scalar function of a stochastic process X
$\boldsymbol{g}(t,\boldsymbol{X}_t)$	vector function of stochastic processes $\boldsymbol{X}$
$h_j = t_{j+1} - t_j$,	the time step
$h(t)$	impulse response of a SDOF oscillator (Eq. D.5 of Appendix D)
$h_v(t)$, $h_a(t)$	impulse response function with respect to velocity and acceleration respectively (in Example 4.20)
$H(t,x)$	Hamiltonian (Eq. 4.313)
$H(\lambda)$	complex frequency function (Fourier transform of $h(t)$) (Eq. 4.170)
$H_v(t,\lambda)$, $H_a(t,\lambda)$	transfer functions with respect to velocity and acceleration respectively (Example 4.20)
$\mathcal{H}$	class of stochastic processes satisfying the properties in Eq. (4.26)
$\mathcal{H}^*$	class of stochastic processes with relaxed conditions in Eqs. (4.45) and (4.47)
$\mathbb{H}$	family of complex eponentials
$I(a,b)$	finite integral over limits a and b
$\mathfrak{I}$, $\bar{\mathfrak{I}}$, I_N	summation approximating the integral $I(a,b)$ using a partition Π_N (Eqs. 4.2 and 4.12)
$\mathfrak{I}(X)$	stochastic integral, $\int_0^T X(s)dB(s)$
$\mathfrak{I}_t = \mathfrak{I}(t,\omega)$	Ito process, $\int_0^t X(s)dB(s)$
k_1, k_2, k_3, k_4	stiffness coefficients (SDE 4.129)
k_B	Boltzmann's constant
ℓ	parameter describing length scale of turbulence in Example 4.20

$m_1,\ m_2$	masses of the MDOF oscillator (Fig. 4.3)
$m,\ n$	integers
P_s^x	probability measure
Q_n	quadratic variation (Eq. 4.123)
$R_{XX}(\tau)$	auto-correlation of a stationary stochastic process X
$R_{XX}(t_i,t_j)$	auto-correlation of the stochastic process, X
$S_{XX}(\lambda)$	power spectral density (PSD) of a stationary process $X(t)$
$S_{XX}(t,\lambda)$	time dependent spectrum (evolutionary PSD –Eq. 4.178b)
$u(t,x)$	stochastic process in Eq. (4.257)
$U(t)$	stochastic processes in Eq. (4.236)
$v(t)$	velocity response process of the oscillator in Example 4.20
v_s	speed parameter in Example 4.20
$v(x,y)$	solution to the vibrating membrane in Example 4.30
W_d	dissipative part of the work done on the system
W_e	external work done on the system in Example 4.31
$W_1(t), W_2(t)$	white noise processes in Example 4.18
$X(t), X(t,\omega)$	stochastic processes
$\mathbf{X}(t)$	vector $\{X_1(t),\ X_2(t),\ Y_1(t),\ Y_2(t)\}^T$ of response processes in Example 4.18
$\{X^{(N)}(t,\omega)\}$	sequence of simple functions with each $X^{(N)}(t,\omega) = \Sigma_j X_j(\omega) I_{(t_j^{(N)},t_{j+1}^{(N)}]}(t)$
$Y(t), Y(t,\omega)$	stochastic processes
$z(\lambda)$	orthogonal stochastic process (Eq. 4.155)
$Z(t)$	stochastic process
Z_β	normalization constant (Eq. 4.313)
α_1, α_2	real constants
$\alpha(t)$	drift coefficient
$\boldsymbol{\alpha}(t,\mathbf{X}_t)$	vector of drift coefficients
β_e	real constant known as the inverse temperature (Example 4.31)

$\beta(t)$	diffusion coefficient
$\boldsymbol{\beta}(t,\boldsymbol{X}_t)$	vector stochastic process
ΔF	$:= F(T)-F(0)$, the change in free energy of a system at time instants $t=0$ and $t=T$ (in Eq. 4.310)
γ_g	parameter influenced by the surface area exposed to the gust (Example 4.20)
$\boldsymbol{\gamma}(t,\boldsymbol{X}_t)$	vector stochastic process
$\delta\Sigma$	boundary in $x-y$ plane (Example 4.30)
ζ	parameter of the SDOF oscillator in Eq. (4.182)
λ	spectral frequency (Eq. 4.154)
$\mu_X(t)$	time-dependent mean of a stochastic process X
ξ, $\overline{\omega}$	damping ratio and the natural frequency of the filter In Eqs. (4.115) and (4.192)
ξ_g, $\overline{\omega}_g$	damping ratio and the natural frequency of the filter In Eq. (4.191)
$\overline{\omega}_d$	damped natural frequency for the oscillator in Eq. (4.192)
$\pi_t(\boldsymbol{x})$	*pdf*
Σ	domain (Example 4.30 and 4.31)
σ_v	parameter of the SDOF oscillator in Eq. (4.182)
σ_2, σ_4	intensities of the two white noise processes in Eq. (4.129)
$\sigma_s(t)$	intensity of the white noise process in Example 4.17
τ	first exit time of $X(t)$ from the boundary $\delta\Sigma$ - the stopping time (see Fig. 4.11)
χ	absolute temparature
$\psi_t(\lambda)$	a complex exponential $\in \mathbb{H}$
$\nabla^2 = \frac{\partial^2}{\partial x^2} + \frac{\partial^2}{\partial y^2}$,	the Laplacian
$\mathcal{F}_t^s$	filtration up to t starting from $s \leq t$

Chapter 5

Numerical Solutions to Stochastic Differential Equations

5.1 Introduction

As is true with deterministic systems, obtaining closed form solutions—strong or weak—to SDEs is understandably a difficult task, except in a few cases (enumerated in Chapter 4). The familiar numerical integration methods such as Euler and other finite difference schemes [Simmons 1991, Carrier and Pearson 1991, Roy and Rao 2012] which are effective for ODEs may find applicability in solving SDEs also. However the issue of their precise implementation in the presence of diffusion (noise) terms assumes importance and may call for certain fundamental modifications to be effected. This invariably raises questions on how to comprehend the quality of numerical solutions and assess the order of convergence and the local / global error estimates of these integration methods for SDEs. Here we recall the stipulated conditions (Chapter 4) under which an SDE has a unique strong or weak solution. Some applications may particularly need computation of strong solutions (for example, whilst computing the response of a dynamical system being actively controlled) and the fidelity of a numerical method generating the path-wise solution is quantified by the strong order of convergence. In most other cases, a weak solution (e.g., one characterized in terms of expectations) may only be of interest and the accuracy of the associated integration method is quantified by the weak order of convergence. Further, with the solution characterized through the associated probability distribution, weak solvers, as can be expected, may be simpler and also adequate with the advantage of less computational effort. A universal framework for computing such weak solutions is provided through a Monte Carlo (MC) approach. The following error criteria distinguish the two types of convergence—strong and weak—of a numerical scheme as applied to an SDE.

Convergence criteria

Consider a probability space $(\Omega, \mathcal{F}, P)$ with increasing filtration $\{\mathcal{F}_t,\ 0 \le t \le T\}$ and an m-dimensional SDE given by:

$$dX(t) = a(t, X)\,dt + \sigma(t, X)\,dB(t), 0 \le t \le T, X(0) = X_0 \tag{5.1}$$

Here $X_t(\omega) = \{X_{j,t}(\omega) : 1 \le j \le m\}^T : \mathbb{R}_+ \times \Omega \to \mathbb{R}^m$ is the state vector of scalar stochastic processes and $B_t \in \mathbb{R}^n$ is an n-dimensional vector of independently evolving one-dimensional Brownian motions that drive the m-dimensional SDE. $a = \{a_j : 1 \le j \le m\}^T : \mathbb{R}_+ \times \mathbb{R}^m \to \mathbb{R}^m$ is the (possibly non-linear) drift vector process and $\sigma = \{\sigma_{jk} : 1 \le j \le m,\ 1 \le k \le n\} : \mathbb{R}_+ \times \mathbb{R}^m \to \mathbb{R}^{m \times n}$ is the diffusion coefficient matrix. X_0 is the vector of $\mathcal{F}_0$–measurable random variables representing the initial condition of the state vector. Let Π_N be a partition of the interval $[0, T]$ given by $t_0 = 0 < t_1 < t_2 < \cdots < t_N = T$ with $\Delta t = \max_{0 \le j \le N-1}(t_{j+1} - t_j)$. If $X_j^{(\Delta t)}$ denotes an approximation to $X(t_j)$ corresponding to a discrete time instant $t = t_j$ in the partition, the strong convergence of order β is defined by the mean square error criterion:

$$\varepsilon_s = \left(E\left[\left| X(t_j) - X_j^{(\Delta t)} \right|^2 \right] \right)^{1/2} \le C(\Delta t)^\beta \tag{5.2a}$$

where the constant $C \in \mathbb{R}$ is independent of Δt and j but depends on X_0. On the other hand, the weak order of convergence γ is defined by:

$$\varepsilon_w = \left| E\left[g\left(X(t_0 + T)\right)\right] - E\left[g\left(X^{(\Delta t)}(t_0 + T)\right)\right] \right| \le C(\Delta t)^\gamma \tag{5.2b}$$

Here g is a smooth test function. It is obvious that whilst ε_s represents a measure of path-wise deviation or closeness at a time instant, ε_w reflects a measure of convergence with respect to a moment. Further elaboration on the order of accuracy of a numerical scheme is provided in Section 5.2 of this chapter.

It is perhaps familiar to the reader that numerical integration schemes developed for solving deterministic ODEs are usually derived from the classical Taylor expansion [Simmons and Krantz 2006, Roy and Rao 2012]. Ito–Taylor expansion that is based on an iterated Ito's formula is the stochastic analogue of the classical one that helps to construct the integration schemes [Kloeden and Platen 1992] for SDEs. Similar to the deterministic case, truncation of terms in the Ito–Taylor expansion decides the order of convergence of a numerical scheme. However, truncations of an identical count of the conventional Taylor and Ito–Taylor expansions result in different orders of convergence – this will be made clearer in the subsequent sections of this chapter.

The chapter starts with a description of the Euler–Maruyama (EM) method which is the simplest method to numerically evaluate an approximate solution to an SDE and is the stochastic counterpart of deterministic Euler scheme. The simplicity of the method not with standing, its limitations are highlighted with regard to the requirement of sufficiently small time–step sizes (especially for the explicit variant of the scheme). It is followed by an exposition of implicit versions of the method. The Ito–Taylor expansion for both one- and multi-dimensional stochastic processes is introduced at this stage. This facilitates not only a more insightful understanding of the EM method, but also the possibility of developing higher order numerical integration schemes, e.g., the Milstein method and the stochastic Newmark method [Roy 2006], as well as the estimation of their orders of accuracy. Specific applications of the expansion towards obtaining numerical solutions of higher dimensional SDEs are illustrated using MC simulation methods. The Chapter also includes the local linearization techniques [Roy 2001, Saha and Roy 2007] suitably designed to handle non-linear SDEs in obtaining strong and weak solutions under both additive and multiplicative excitations.

5.2 Euler-Maruyama (EM) Method for Solving SDEs

Before describing the Ito–Taylor expansion and its role in constructing different numerical integration schemes for SDEs, we first present the most basic and popularly used numerical scheme, the EM method [Maruyama 1955, Kloeden and Platen 1992] in its both explicit and implicit forms. The presentation is so organized that the EM-based time stepping scheme has an intuitive appeal, even without an understanding of the Ito–Taylor expansion. We recall that most such numerical schemes are based on one-step recursive approximations to the true solution of an SDE. Thus, for instance, it suffices to describe the EM method through the associated one-step time marching map.

Let $\boldsymbol{X}_{t_j,\boldsymbol{x}}(t)$, $j \geq 0$, which is $\mathcal{F}_t$-measurable, denote the true solution of the SDE (5.1) corresponding to a time instant $t > t_j$, starting from the initial condition, $\boldsymbol{x} = \boldsymbol{X}(t_j)$. Here it is assumed that $E\left[\left|\boldsymbol{X}(t_0)\right|^2\right] < \infty$. Similarly let $\left(\hat{\boldsymbol{X}}_{t_j,\hat{\boldsymbol{x}}}(t_k), \mathcal{F}_{t_k}\right)$ be a numerical approximation to the solution at $t_k \geq t_j$ corresponding to a partition Π_N and with $\hat{\boldsymbol{x}} = \hat{\boldsymbol{X}}(t_j)$. A one-step approximation $\hat{\boldsymbol{X}}_{t_j,\hat{\boldsymbol{x}}}(t_{j+1})$, $t_0 \leq t_j \leq t_{j+1} = t_j + h_j \leq t_0 + T$, $h_j \leq t_{j+1} - t_j$ has the generic form:

$$\hat{\boldsymbol{X}}_{t_j,\hat{\boldsymbol{x}}}\left(t_{j+1}\right) = \hat{\boldsymbol{X}}_{j+1} = \hat{\boldsymbol{x}} + \boldsymbol{g}\left(t_j, \hat{\boldsymbol{x}}, h_j, \Delta\boldsymbol{B}_j\right), \hat{\boldsymbol{x}} = \hat{\boldsymbol{X}}_j, \Delta\boldsymbol{B}_j = \boldsymbol{B}_{j+1} - \boldsymbol{B}_j \tag{5.3}$$

In particular, the one-step explicit map for the EM scheme is given by:

$$\hat{\boldsymbol{X}}_{j+1} = \hat{\boldsymbol{x}} + \boldsymbol{a}\left(t_j, \hat{\boldsymbol{x}}\right)h_j + \boldsymbol{\sigma}\left(t_j, \hat{\boldsymbol{x}}\right)\Delta\boldsymbol{B}_j \tag{5.4}$$

Thus for the EM method, $\boldsymbol{g}\left(t_j, \hat{\boldsymbol{x}}, h_j, \Delta\boldsymbol{B}_j\right) = \boldsymbol{a}\left(t_j, \hat{\boldsymbol{x}}\right)h_j + \boldsymbol{\sigma}\left(t_j, \hat{\boldsymbol{x}}\right)\Delta\boldsymbol{B}_j$. While $\hat{\boldsymbol{X}}_{j+1} = \hat{\boldsymbol{X}}_{t_j,\hat{\boldsymbol{x}}_j}\left(t_{j+1}\right)$ yields the one-step approximation, $\hat{\boldsymbol{X}}_{j+1} = \hat{\boldsymbol{X}}_{t_0,\boldsymbol{x}_0}\left(t_{j+1}\right)$ corresponds to the approximation over the interval, $[t_0, t_{j+1}]$. Equation (5.4) recursively yields the numerical approximations $\left(\hat{\boldsymbol{X}}_k, \mathcal{F}_{t_k}\right), k = 0, 1, \ldots, N, t_N = t_0 + T$.

5.2.1 Order of convergence of EM method

It is known [Gikhman and Skorokhod 1972, Kloeden and Platen 1992, Higham 2001] that the order of accuracy of the EM method is rather poor, with a strong convergence of $O(\sqrt{\Delta t})$ and weak convergence of $O(\Delta t)$. In general, the following theorem [Milstein 1988, 1995] plays a fundamental role in establishing the strong order of convergence of a numerical scheme. The theorem essentially utilizes the relation between the global and local convergence criteria where the former corresponds to the error that accrues over an interval and the latter to that over a one-step approximation. We first assume that the drift and diffusion terms satisfy the Lipschitz and linear growth conditions in Eqs. (4.210) and (4.211) of Chapter 4, so that the SDE (5.1) has a unique t-continuous strong solution.

5.2.2 Statement of the theorem for global convergence

Suppose that the orders of errors in the mean and the mean square of the one-step approximation, vis-a-vis the true solution, are respectively given by:

$$E\left[\left|\boldsymbol{X}_{t_j,\boldsymbol{x}}(t_{j+1})-\hat{\boldsymbol{X}}_{t_j,\boldsymbol{x}}\left(t_{j+1}\right)\right|\right]\le K\left(1+|\boldsymbol{x}|^2\right)^{1/2}h^{p_1} \tag{5.5a}$$

and

$$\left(E\left[\left|\boldsymbol{X}_{t_j,\boldsymbol{x}}(t_{j+1})-\hat{\boldsymbol{X}}_{t_j,\boldsymbol{x}}\left(t_{j+1}\right)\right|^2\right]\right)^{1/2}\le K\left(1+|\boldsymbol{x}|^2\right)^{1/2}h^{p_2} \tag{5.5b}$$

Here we have replaced h_j by h (e.g., uniform time–step size) for sake of expositional ease. Now if

$$p_2\ge 1/2 \text{ and } p_1\ge p_2+1/2, \tag{5.6a}$$

then for any partition Π_N with $k=0,1,2,\ldots,N$, the following inequality holds for global convergence over the interval $[t_0,t_k]$:

$$\left(E\left[\left|\boldsymbol{X}_{t_0,\boldsymbol{x}_0}(t_k)-\hat{\boldsymbol{X}}_{t_0,\boldsymbol{x}_0}(t_k)\right|^2\right]\right)^{1/2}\le K\left(1+E\left[|\boldsymbol{x}_0|^2\right]\right)^{1/2}h^{p_2-1/2} \tag{5.6b}$$

i.e., the order of accuracy p of the scheme using a one-step approximation can be expressed as $p=p_2-1/2$. Here the real constant K, used generically, is independent of h and k and also $\boldsymbol{x}_0$.

Proof: Only an outline of the proof of the theorem is provided here; we refer to Milstein [Milstein 1995] for full details. Starting with the global error $\boldsymbol{X}_{t_0,\boldsymbol{x}_0}(t_{k+1})-\hat{\boldsymbol{X}}_{t_0,\boldsymbol{x}_0}(t_{k+1})$ on a sample path, we have:

$$\boldsymbol{X}_{t_0,\boldsymbol{x}_0}(t_{k+1})-\hat{\boldsymbol{X}}_{t_0,\boldsymbol{x}_0}(t_{k+1})=\boldsymbol{X}_{t_k,\boldsymbol{X}(t_k)}(t_{k+1})-\hat{\boldsymbol{X}}_{t_k,\hat{\boldsymbol{X}}_k}(t_{k+1}) \tag{5.7}$$

We rewrite the above equation as:

$$\mathbf{X}_{t_0,\mathbf{x}_0}(t_{k+1}) - \hat{\mathbf{X}}_{t_0,\mathbf{x}_0}(t_{k+1}) = \left(\mathbf{X}_{t_k,\mathbf{X}(t_k)}(t_{k+1}) - \mathbf{X}_{t_k,\hat{\mathbf{X}}_k}(t_{k+1})\right) + \left(\mathbf{X}_{t_k,\hat{\mathbf{X}}_k}(t_{k+1}) - \hat{\mathbf{X}}_{t_k,\hat{\mathbf{X}}_k}(t_{k+1})\right) \tag{5.8}$$

The first bracketed entity on the right hand side of the above equation corresponds to the error in the true solution due to the error in the initial condition at t_k. The second entity is the error between the true and numerical solutions over a time step. It follows that:

$$\begin{aligned}
E\left[\left|\mathbf{X}_{t_0,\mathbf{x}_0}(t_{k+1}) - \hat{\mathbf{X}}_{t_0,\mathbf{x}_0}(t_{k+1})\right|^2\right] &= E\left[\left(\mathbf{X}_{t_k,\mathbf{X}(t_k)}(t_{k+1}) - \mathbf{X}_{t_k,\hat{\mathbf{X}}_k}(t_{k+1})\right)^2\right] \\
&+ E\left[\left(\mathbf{X}_{t_k,\hat{\mathbf{X}}_k}(t_{k+1}) - \hat{\mathbf{X}}_{t_k,\hat{\mathbf{X}}_k}(t_{k+1})\right)^2\right] \\
&+ 2E\left[\left(\mathbf{X}_{t_k,\mathbf{X}(t_k)}(t_{k+1}) - \mathbf{X}_{t_k,\hat{\mathbf{X}}_k}(t_{k+1})\right)\left(\mathbf{X}_{t_k,\hat{\mathbf{X}}_k}(t_{k+1}) - \hat{\mathbf{X}}_{t_k,\hat{\mathbf{X}}_k}(t_{k+1})\right)\right] \\
&= E\left[E\left[\left(\mathbf{X}_{t_k,\mathbf{X}(t_k)}(t_{k+1}) - \mathbf{X}_{t_k,\hat{\mathbf{X}}_k}(t_{k+1})\right)^2 | \mathcal{F}_k\right]\right] \\
&+ E\left[E\left[\left(\mathbf{X}_{t_k,\hat{\mathbf{X}}_k}(t_{k+1}) - \hat{\mathbf{X}}_{t_k,\hat{\mathbf{X}}_k}(t_{k+1})\right)^2 | \mathcal{F}_k\right]\right] \\
&+ 2E\left[E\left[\left(\mathbf{X}_{t_k,\mathbf{X}(t_k)}(t_{k+1}) - \mathbf{X}_{t_k,\hat{\mathbf{X}}_k}(t_{k+1})\right)\left(\mathbf{X}_{t_k,\hat{\mathbf{X}}_k}(t_{k+1}) - \hat{\mathbf{X}}_{t_k,\hat{\mathbf{X}}_k}(t_{k+1})\right) | \mathcal{F}_k\right]\right]
\end{aligned} \tag{5.9}$$

In further simplifications to follow, we take recourse to the following results: (refer to Milstein [Milstein 1995] for proofs of these results):

a) If we write $\mathbf{X}_{t_k,\mathbf{X}(t_k)}(t_{k+1}) - \mathbf{X}_{t_k,\hat{\mathbf{X}}_k}(t_{k+1}) = \mathbf{X}(t_k) - \hat{\mathbf{X}}_k + \mathbf{Y}$ where $\mathbf{Y}$ is a $\mathcal{F}_{k+1}$-measurable random variable, then

$$E\left[\left|\mathbf{X}_{t_k,\mathbf{X}(t_k)}(t_{k+1}) - \mathbf{X}_{t_k,\hat{\mathbf{X}}_k}(t_{k+1})\right|^2\right] \le \left|\mathbf{X}(t_k) - \hat{\mathbf{X}}_k\right|^2 (1 + Kh) \tag{5.10a}$$

and

$$E\left[\mathbf{Y}^2\right] \le K\left|\mathbf{X}(t_k) - \hat{\mathbf{X}}_k\right|^2 h \tag{5.10b}$$

b) For all natural numbers $k = 0, 1, \ldots, N \in \mathbb{N}$, the following inequality holds:

$$E\left[\hat{\mathbf{X}}_k^{\,2}\right] \le K\left(1 + E|\mathbf{X}_0|^2\right) \tag{5.11}$$

c) If, for all natural numbers $k = 0,1,\ldots,N$ and arbitrary N, one has: $\epsilon_{k+1} = \epsilon_k(1+Ah) + Bh^p$, $A,B \geq 0$, $h = T/N$ and $\epsilon_k \geq 0$, then:

$$\epsilon_k \leq e^{At}\epsilon_0 + \frac{B}{A}\left(e^{At}-1\right)h^{p-1}, p \geq 1 \tag{5.12}$$

Referring to the first term on the right hand side of Eq. (5.9) and using the conditional version of the proposition in Eq. (5.10a), we have:

$$E\left[E\left[\left(\mathbf{X}_{t_k,\mathbf{X}(t_k)}(t_{k+1}) - \mathbf{X}_{t_k,\hat{\mathbf{X}}_k}(t_{k+1})\right)^2 |\mathcal{F}_k\right]\right] \leq E\left[\left|\mathbf{X}(t_k) - \hat{\mathbf{X}}_k\right|^2\right].(1+Kh) \tag{5.13}$$

By the conditional version of Eq. (5.5b) and the result in Eq. (5.11), the second term on the right hand side of Eq. (5.9) can be simplified as:

$$E\left[\,E\left[\left(\mathbf{X}_{t_k,\hat{\mathbf{X}}_k}(t_{k+1}) - \hat{\mathbf{X}}_{t_k,\hat{\mathbf{X}}_k}(t_{k+1})\right)^2 |\mathcal{F}_k\right]\right]$$

$$\leq K\left(1 + E\left|\hat{\mathbf{X}}_k\right|^2\right)h^{2p_2} \leq K\left(1 + E|\mathbf{X}_0|^2\right)h^{2p_2} \tag{5.14}$$

The last term on the right hand side of Eq. (5.9) can be simplified as follows. We first write $\mathbf{X}_{t_k,\mathbf{X}(t_k)}(t_{k+1}) - \mathbf{X}_{t_k,\hat{\mathbf{X}}_k}(t_{k+1}) = \mathbf{X}(t_k) - \hat{\mathbf{X}}_k + \mathbf{Y}$ as in Eq. (5.10a) and thus:

$$E\left[E\left[\left(\mathbf{X}_{t_k,\mathbf{X}(t_k)}(t_{k+1}) - \mathbf{X}_{t_k,\hat{\mathbf{X}}_k}(t_{k+1})\right)\left(\mathbf{X}_{t_k,\hat{\mathbf{X}}_k}(t_{k+1}) - \hat{\mathbf{X}}_{t_k,\hat{\mathbf{X}}_k}(t_{k+1})\right)|\mathcal{F}_k\right]\right]$$

$$= EE\left[\left(\mathbf{X}(t_k) - \hat{\mathbf{X}}_k\right)\left(\mathbf{X}_{t_k,\hat{\mathbf{X}}_k}(t_{k+1}) - \hat{\mathbf{X}}_{t_k,\hat{\mathbf{X}}_k}(t_{k+1})\right)|\mathcal{F}_k\right]$$

$$+EE\left[\mathbf{Y}\left(\mathbf{X}_{t_k,\hat{\mathbf{X}}_k}(t_{k+1}) - \hat{\mathbf{X}}_{t_k,\hat{\mathbf{X}}_k}(t_{k+1})\right)|\mathcal{F}_k\right]$$

$$= E\left[\left(\mathbf{X}(t_k) - \hat{\mathbf{X}}_k\right)E\left[\left(\mathbf{X}_{t_k,\hat{\mathbf{X}}_k}(t_{k+1}) - \hat{\mathbf{X}}_{t_k,\hat{\mathbf{X}}_k}(t_{k+1})\right)|\mathcal{F}_k\right]\right]$$

$$+EE\left[\mathbf{Y}\left(\mathbf{X}_{t_k,\hat{\mathbf{X}}_k}(t_{k+1}) - \hat{\mathbf{X}}_{t_k,\hat{\mathbf{X}}_k}(t_{k+1})\right)|\mathcal{F}_k\right] \tag{5.15}$$

By the conditional version of hypothesis (5.5a) in conjunction with Eq. (5.11), the first term on the right hand side of the above equation has the following bound:

$$E\left[\left|\mathbf{X}(t_k) - \hat{\mathbf{X}}_k\right|E\left[\left|\mathbf{X}_{t_k,\hat{\mathbf{X}}_k}(t_{k+1}) - \hat{\mathbf{X}}_{t_k,\hat{\mathbf{X}}_k}(t_{k+1})\right||\mathcal{F}_k\right]\right]$$

$$\le \left(E\left[\left|\boldsymbol{X}(t_k) - \hat{\boldsymbol{X}}_k\right|^2\right]\right)^{1/2} K\left(1 + E|\boldsymbol{X}_0|^2\right)^{1/2} h^{p_1} \tag{5.16}$$

The second term on the right hand side of Eq. (5.15) can be written as:

$$EE\left[\left|\boldsymbol{Y}\left(\boldsymbol{X}_{t_k,\hat{\boldsymbol{X}}_k}(t_{k+1}) - \hat{\boldsymbol{X}}_{t_k,\hat{\boldsymbol{X}}_k}(t_{k+1})\right)\right| |\mathcal{F}_k\right]$$

$$\le \left(EE\left[\boldsymbol{Y}^2|\mathcal{F}_k\right]\right)^{1/2}\left(EE\left[\left(\boldsymbol{X}_{t_k,\hat{\boldsymbol{X}}_k}(t_{k+1}) - \hat{\boldsymbol{X}}_{t_k,\hat{\boldsymbol{X}}_k}(t_{k+1})\right)^2 |\mathcal{F}_k\right]\right)^{1/2} \tag{5.17}$$

By the conditional version of Eq. (5.5b) combined with the results in Eqs. (5.10b) and (5.11), the last inequality reduces to:

$$EE\left[\left|\boldsymbol{Y}\left(\boldsymbol{X}_{t_k,\hat{\boldsymbol{X}}_k}(t_{k+1}) - \hat{\boldsymbol{X}}_{t_k,\hat{\boldsymbol{X}}_k}(t_{k+1})\right)\right| |\mathcal{F}_k\right]$$

$$\le K\left(E\left[\left|\boldsymbol{X}(t_k) - \hat{\boldsymbol{X}}_k\right|^2\right]\right)^{1/2} h^{1/2}\left(1 + E|\boldsymbol{X}_0|^2\right)^{1/2} h^{p_2}$$

$$= K\left(E\left[\left|\boldsymbol{X}(t_k) - \hat{\boldsymbol{X}}_k\right|^2\right]\right)^{1/2}\left(1 + E|\boldsymbol{X}_0|^2\right)^{1/2} h^{p_2+1/2} \tag{5.18}$$

With $\epsilon_k^2 = E\left[\left|\boldsymbol{X}(t_k) - \hat{\boldsymbol{X}}_k\right|^2\right]$, Eqs. (5.9), (5.13), (5.14), (5.16), and (5.18) lead to:

$$E\left[\left|\boldsymbol{X}_{t_0,\boldsymbol{x}_0}(t_{k+1}) - \hat{\boldsymbol{X}}_{t_0,\boldsymbol{x}_0}(t_{k+1})\right|^2\right] = \epsilon_{k+1}^2$$

$$\le \epsilon_k^2(1 + Kh) + K\left(1 + E|\boldsymbol{X}_0|^2\right)h^{2p_2} + K\epsilon_k\left(1 + E|\boldsymbol{X}_0|^2\right)^{1/2} h^{p_1}$$

$$+ K\epsilon_k\left(1 + E|\boldsymbol{X}_0|^2\right)^{1/2} h^{p_2+1/2} \tag{5.19}$$

If $p_1 \ge p_2 + 1/2$, we have for $h < 1$:

$$\epsilon_{k+1}^2 \le \epsilon_k^2(1 + Kh) + K\left(1 + E|\boldsymbol{X}_0|^2\right)h^{2p_2} + K\epsilon_k\left(1 + E|\boldsymbol{X}_0|^2\right)^{1/2} h^{p_2+1/2} \tag{5.20}$$

Using the inequality, $K\epsilon_k\left(1 + E|\boldsymbol{X}_0|^2\right)^{1/2} h^{p_2+1/2} \le \frac{\epsilon_k^2 k}{2} + K\left(1 + E|\boldsymbol{X}_0|^2\right)h^{2p_2}$, the last equation takes the form:

$$\epsilon_{k+1}^2 \le \epsilon_k^2(1 + Kh) + K\left(1 + E|\boldsymbol{X}_0|^2\right)h^{2p_2} \tag{5.21}$$

With $\epsilon_0^2 = 0$ and the result in Eq. (5.12), we finally obtain:

$$\epsilon_{k+1}^2 \le K\left(1 + E|\mathbf{X}_0|^2\right)h^{2p_2-1} \tag{5.22}$$

which proves the theorem and gives the result in the Eq. (5.6).

Using the above theorem, we now show that for the EM scheme (Eq. 5.4) $p_1 = 2$ and $p_2 = 1$ yielding the strong order of convergence as $p = p_2 - 1/2 = 1/2$. Subtracting the EM solution $\hat{\mathbf{X}}_{t,x}(t+h)$ from the integral form of the SDE (5.1), we get:

$$\mathbf{X}_{t,x}(t+h) - \hat{\mathbf{X}}_{t,x}(t+h) = \int_t^{t+h} \left(\boldsymbol{a}(s,\mathbf{X}_{t,x}(s)) - \boldsymbol{a}(t,\boldsymbol{x})\right) ds$$

$$+ \int_t^{t+h} \left(\boldsymbol{\sigma}(s,\mathbf{X}_{t,x}(s)) - \boldsymbol{\sigma}(t,\boldsymbol{x})\right) d\mathbf{B}(s) \tag{5.23}$$

The errors in the mean and the mean square of the one-step approximation in relation to the true solution are respectively given by:

$$E\left[\mathbf{X}_{t,x}(t+h) - \hat{\mathbf{X}}_{t,x}(t+h)\right] = E\left[\int_t^{t+h} \left(\boldsymbol{a}(s,\mathbf{X}_{t,x}(s)) - \boldsymbol{a}(t,\boldsymbol{x})\right) ds\right] \tag{5.24}$$

and

$$E\left[\left(\mathbf{X}_{t,x}(t+h) - \hat{\mathbf{X}}_{t,x}(t+h)\right)^2\right] = E\left[\left(\int_t^{t+h} \left(\boldsymbol{a}(s,\mathbf{X}_{t,x}(s)) - \boldsymbol{a}(t,\boldsymbol{x})\right) ds\right)^2\right]$$

$$+2E\left[\int_t^{t+h} \left(\boldsymbol{a}(s,\mathbf{X}_{t,x}(s)) - \boldsymbol{a}(t,\boldsymbol{x})\right) ds \int_t^{t+h} \left(\boldsymbol{\sigma}(s,\mathbf{X}_{t,x}(s)) - \boldsymbol{\sigma}(t,\boldsymbol{x})\right) d\mathbf{B}(s)\right]$$

$$+E\left[\left(\int_t^{t+h} \left(\boldsymbol{\sigma}(s,\mathbf{X}_{t,x}(s)) - \boldsymbol{\sigma}(t,\boldsymbol{x})\right) d\mathbf{B}(s)\right)^2\right]$$

$$\le E\left[\left(\int_t^{t+h} \left(\boldsymbol{a}(s,\mathbf{X}_{t,x}(s)) - \boldsymbol{a}(t,\boldsymbol{x})\right) ds\right)^2\right]$$

$$+2\left(E\left[\int_t^{t+h} \left(\boldsymbol{a}(s,\mathbf{X}_{t,x}(s)) - \boldsymbol{a}(t,\boldsymbol{x})\right)^2 ds\right]\right)^{1/2} \left(\int_t^{t+h} E\left[\left(\boldsymbol{\sigma}\left(s,\mathbf{X}_{t,x}(s)\right)\right.\right.\right.$$

$$\left.\left.\left. -\boldsymbol{\sigma}(t,\boldsymbol{x})\right)^2\right] d\mathbf{B}(s)\right)^{1/2}$$

$$+\int_t^{t+h} E\left[\left(\sigma\left(s,\boldsymbol{X}_{t,x}(s)\right)-\sigma(t,\boldsymbol{x})\right)^2\right]ds$$

(by Cauchy–Scwartz inequality and isometry)

$$\leq 2E\left[\left(\int_t^{t+h}\left(\boldsymbol{a}\left(s,\boldsymbol{X}_{t,x}(s)\right)-\boldsymbol{a}(t,\boldsymbol{x})\right)ds\right)^2\right]+2\int_t^{t+h} E\left[\left(\sigma\left(s,\boldsymbol{X}_{t,x}(s)\right)-\sigma(t,\boldsymbol{x})\right)^2\right]ds$$

$$(\text{since } 2\left(x^2+y^2\right)>(x-y)^2) \tag{5.25}$$

Assuming that the derivatives of the drift and diffusion coefficients with respect to t exist and vary at most linearly with $\boldsymbol{x}$ as $|\boldsymbol{x}|\to\infty$, one has the inequalities of the form:

$$\left|\boldsymbol{a}\left(s,\boldsymbol{X}_{t,x}(s)\right)-\boldsymbol{a}(t,\boldsymbol{x})\right|\leq K\left(1+|\boldsymbol{x}|^2\right)^{1/2}(s-t)+K\left|\boldsymbol{X}_{t,x}(s)-\boldsymbol{x}\right| \tag{5.26}$$

Further, if we write $\int_t^{t+h}\left(\boldsymbol{a}\left(s,\boldsymbol{X}_{t,x}(s)\right)-\boldsymbol{a}(t,\boldsymbol{x})\right)ds = \int_t^{t+h}1.\left(\boldsymbol{a}\left(s,\boldsymbol{X}_{t,x}(s)\right)-\boldsymbol{a}(t,\boldsymbol{x})\right)ds$ then by Cauchy–Schwartz inequality:

$$\left|\int_t^{t+h}\left(\boldsymbol{a}\left(s,\boldsymbol{X}_{t,x}(s)\right)-\boldsymbol{a}(t,\boldsymbol{x})\right)ds\right|^2\leq h\int_t^{t+h}\left(\boldsymbol{a}\left(s,\boldsymbol{X}_{t,x}(s)\right)-\boldsymbol{a}(t,\boldsymbol{x})\right)^2 ds \tag{5.27}$$

Moreover, based on the growth conditions on the drift and diffusion terms ensuring uniqueness of solutions in the strong sense, we may assume that (see Milstein [Milstein 1995] for a proof):

$$E\left[\left|\boldsymbol{X}_{t,x}(t+h)-\boldsymbol{x}\right|^2\right]\leq K\left(1+E\left[|\boldsymbol{x}|^2\right]\right)h \tag{5.28}$$

Now with the results in Eq. (5.26) the first term on the RHS of Eq. (5.25) is bounded by $\left(1+|\boldsymbol{x}|^2\right)h^3$. Using a similar inequality as in Eq. (5.26) for $\sigma\left(s,\boldsymbol{X}_{t,x}(s)\right)-\sigma(t,\boldsymbol{x})$ and the result in Eq. (5.28), the second term can be shown to be bounded by $K\left(1+|\boldsymbol{x}|^2\right)h$. Thus the order of the mean square error is $p_2=1$. Similarly we find that $p_1=2$ supported by the following arguments. If we assume that the derivatives $\frac{\partial a_i}{\partial x_j},\frac{\partial^2 a_i}{\partial x_j\partial x_k},i,j,k=1,2,\ldots$ are uniformly bounded, then:

$$\boldsymbol{a}\left(s,\boldsymbol{X}_{t,x}(s)\right)-\boldsymbol{a}(t,\boldsymbol{x})=\boldsymbol{a}(s,\boldsymbol{x})+\frac{\partial\boldsymbol{a}(t,\boldsymbol{x})}{\partial\boldsymbol{x}}\left(\boldsymbol{X}_{t,x}(s)-\boldsymbol{x}\right)-\boldsymbol{a}(t,\boldsymbol{x})+R \tag{5.29}$$

where the remainder $\boldsymbol{R}\leq K\left(\boldsymbol{X}_{t,x}(s)-\boldsymbol{x}\right)^2$. We have from the SDE (5.1):

$$\int_t^{t+h} d\boldsymbol{X} = \boldsymbol{X}_{t,x}(t+h) - \boldsymbol{x} = \int_t^{t+h} \boldsymbol{a}(u,\boldsymbol{X}_{t,x}(u))\,du + \int_t^{t+h} \sigma(u,\boldsymbol{X}_{t,x}(u))dB(u) \tag{5.30}$$

We also have from the Lipschitz condition on the drift coefficient:

$$\boldsymbol{a}(u,\boldsymbol{X}_{t,x}(u)) \le |\boldsymbol{a}(u,0)| + \left(|\boldsymbol{a}(u,\boldsymbol{X}_{t,x}(u))| - |\boldsymbol{a}(u,0)|\right) \le K + K|\boldsymbol{X}_{t,x}(u)| \tag{5.31}$$

From the last two equations, one gets:

$$E[\boldsymbol{X}_{t,x}(t+h) - \boldsymbol{x}] \le E\left[\int_t^{t+h} \boldsymbol{a}(u,\boldsymbol{X}_{t,x}(u))\,du\right] = \left(K + K|\boldsymbol{X}_{t,x}(u)|\right)h$$
$$\Longrightarrow \int_t^{t+h} E[\boldsymbol{X}_{t,x}(u) - \boldsymbol{x}]\,du \le K\left(1+|\boldsymbol{x}|^2\right)^{1/2} h^2 \tag{5.32}$$

Equations (5.29)–(5.32) indicate that the error in the mean of the one-step approximation in Eq. (5.24) is $O(h^2)$ and $p_1 = 2$ found to be $> p_2 + \frac{1}{2}$. Thus with $p_2 = 1$, the EM method is of order $p = p_2 - 1/2 = 1/2$. Suppose that the noise term SDE (5.1) is additive, i.e.:

$$d\boldsymbol{X}(t) = \boldsymbol{a}(t,\boldsymbol{X})\,dt + \sigma(t)\,d\boldsymbol{B}(t), 0 \le t \le T, \boldsymbol{X}(0) = \boldsymbol{X}_0 \tag{5.33}$$

The diffusion term being additive, the second term in Eq. (5.25) can be shown to be $O\left(h^3\right)$ as is the case with the first term. Thus from Eq. (5.5b), $p_2 = 3/2$. With p_1 remaining as 2, we still have, $p_1 \ge p_2 + 1/2$, yielding the strong order of convergence for EM method with additive noises as $p = 1$.

Example 5.1

We apply the EM method to a 1-dimensional SDE with $a(t,X) = \lambda X$ and $\sigma(t,X) = \mu X$ where λ and μ are real constants. Thus it is a linear SDE (commonly known as Black–Scholes SDE) given by:

$$dX(t) = \lambda X dt + \mu X dB(t) \tag{5.34}$$

Solution

As derived in Example 4.26 of Chapter 4, the exact solution to the above SDE is given by:

$$X(t) = X_0 \exp\left(\left(\lambda - \frac{1}{2}\mu^2\right)t + \mu B_t\right)$$

$$\Longrightarrow X_{t_0,X_0}\left(t_{j+1}\right) = X_0 \exp\left(\left(\lambda - \frac{1}{2}\mu^2\right)t_{j+1} + \mu B_{j+1}\right) \tag{5.35}$$

With a uniform time step size $h = \frac{T}{N}$, a one-step approximation by EM scheme Eq. (5.4) may be written as:

$$\hat{X}_{j+1} - \hat{X}_j = \Delta\hat{X}_j = \lambda\hat{X}_j h + \mu\hat{X}_j \Delta B_j \tag{5.36}$$

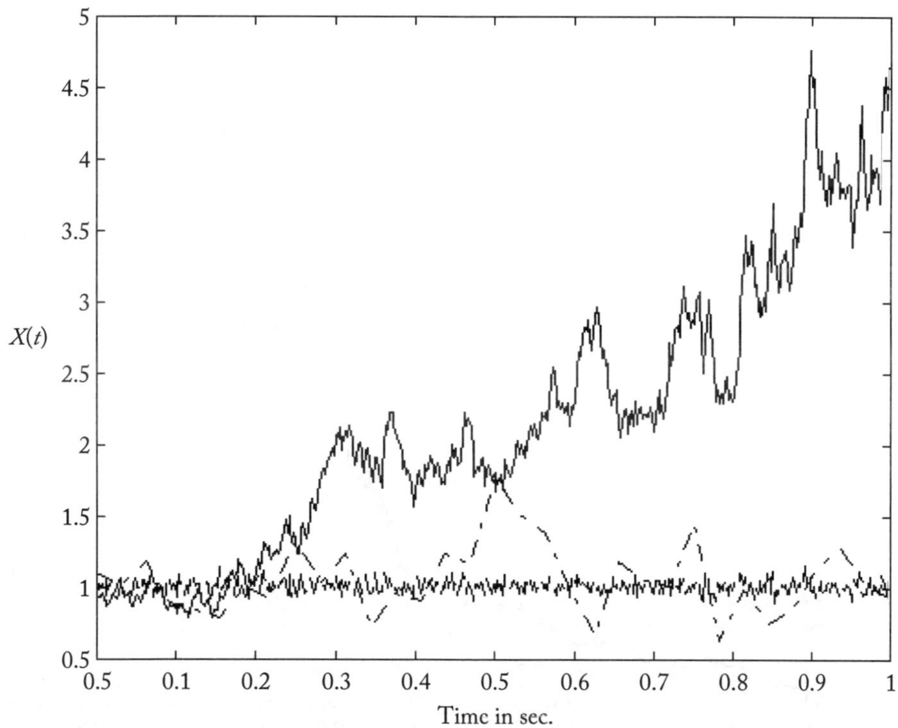

Fig. 5.1 Solution of Black–Scholes SDE (5.34); $\lambda = 2$, $\mu = 1$, $T = 1$, $x(0) = 1.0$, dark-line – true solution (Eq. 5.35), dashed line-solution by Eq. (5.37) with no convergence for $h = 2^{-5}$, dotted-line – solution by Eq. (5.37) with no convergence even with $h = 2^{-9}$

Denoting $\hat{X}_j = x$, the solution to the above difference equation may be approximately written in the form:

$$\hat{X}_{j+1} = \hat{X}_{t_j,x}\left(t_{j+1}\right) = x \exp\left(\lambda + \frac{\mu}{h}\Delta B_j\right) h \tag{5.37}$$

Since $E\left[e^{\mu \Delta B_j}\right] = e^{(\mu^2/2)h}$, the error in the mean is:

$$E\left[X_{t_j,x}(t_{j+1}) - \hat{X}_{t_j,x}\left(t_{j+1}\right)\right] = x\left(e^{\lambda h} - e^{\left(\lambda+\mu^2/2\right)h}\right) = O(h) \tag{5.38}$$

Thus $p_1 = 1$. It immediately follows from the conditions $p_1 \geq p_2 + 1/2$ and $p = p_2 - 1/2$ implied in the last theorem on convergence that the above (approximate) EM solution in Eq. (5.37) may not converge. Figure 5.1 shows such EM solutions that fail to converge to the true solution. In fact, the mean of true solution is $E\left[X(t)\right] = e^{\lambda t} E\left[X_0\right]$ and that of the approximate solution is $E\left[\hat{X}(t)\right] = e^{\left(\lambda+\mu^2/2\right)t} E\left[X_0\right]$.

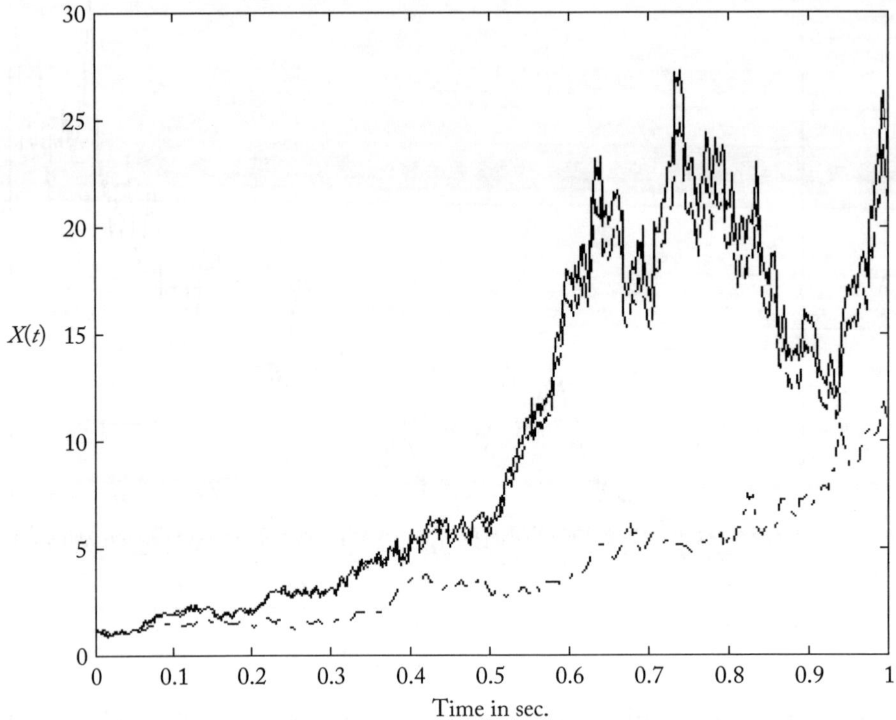

Fig. 5.2 Solution to Black–Scholes SDE (5.34) by EM method (Eq. 5.36); $\lambda = 2$, $\mu = 1$, $T = 1$, $x(0) = 1.0$; dark-line – true solution (Eq. 5.35) with $h = 2^{-9}$; dashed-line – EM solution with $h = 2^{-9}$; dash-dot line – EM solution with $h = 2^{-8}$

On the other hand, the EM solution (Eq. 5.36), when implemented numerically, shows that the strong order of convergence p is $1/2$. Note that β in Eq. (5.2a) is the same as $p\left(= p_2 - \frac{1}{2}\right)$ in Eq. (5.6b). With each sample path of Brownian motion simulated with a uniform step size of $h = 2^{-9}$ s, we apply the EM method for the SDE over the interval $[0,1]$. Figure 5.2 shows the EM solution along with a sample path of the true solution in Eq. (5.35). Two typical sample paths of the EM solution obtained with two different step sizes $h = 2^{-9}$ and $h = 2^{-8}$ s are also shown in the figure. The solution with $h = 2^{-9}$ almost converges to the true solution. Both strong and weak orders of convergence of the EM method are computed by the simulation. With N_e denoting the ensemble size, the EM solution is obtained for each sample and the sample mean of the global error between the true and numerical solutions at $t = T$ is computed for different choices of $h = 2^{-9}, 2^{-8}, 2^{-7}, 2^{-6}$ and 2^{-5} s. This gives ε_s corresponding to the strong order of convergence and the graph ε_s vs. h is shown in Fig. 5.3. The figure contains the results for three independent MC simulations with $N_e = 3000$. To examine the weak order of convergence, we consider $g(X) = X$, the identity function and the error ε_w between the sample–averaged true and EM solutions at $t = T$ is computed. Figure 5.4 shows the graphs of ε_w vs. h obtained for three independent MC runs with $N_e = 3000$.

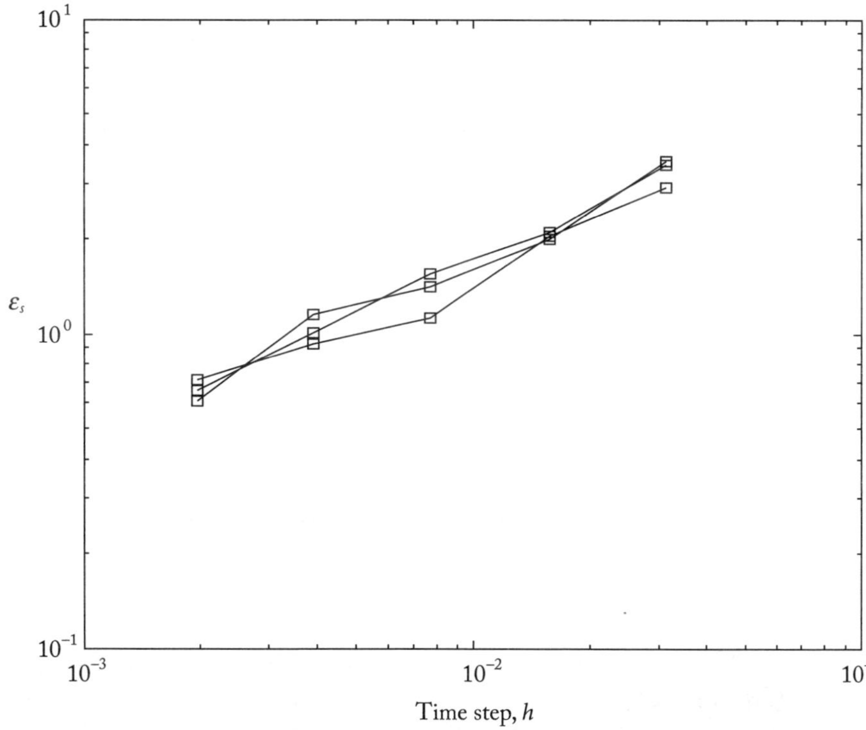

Fig. 5.3 Black–Scholes SDE (5.34); EM solution – strong order of convergence (Eq. (5.39a)), $\lambda = 2$, $\mu = 1$, $T = 1$, $x(0) = 1$; result from three independent MC simulation runs with sample size $N_e = 3000$

The error graphs in Figs 5.3 and 5.4 are drawn on log–log scale. From the definition of the orders of convergence, we have $\varepsilon_s \leq c\ h^{\beta}$ and $\varepsilon_w \leq d\ h^{\gamma}$ and the inequalities can be expressed as approximations on log–log scale as:

$$\log\varepsilon_s \approx \log c + \beta \log h \tag{5.39a}$$

and

$$\log\varepsilon_w \approx \log d + \gamma \log h \tag{5.39b}$$

A power law fit to the simulation results establishes the orders of convergence β and γ to be approximately equal to $\frac{1}{2}$ and 1 respectively as expected. In fact the strong and weak orders of convergence so obtained from the three independent MC runs are: $\beta = 0.588, 0.524$, and 0.578 and $\gamma = 0.974, 0.984$, and 1.08.

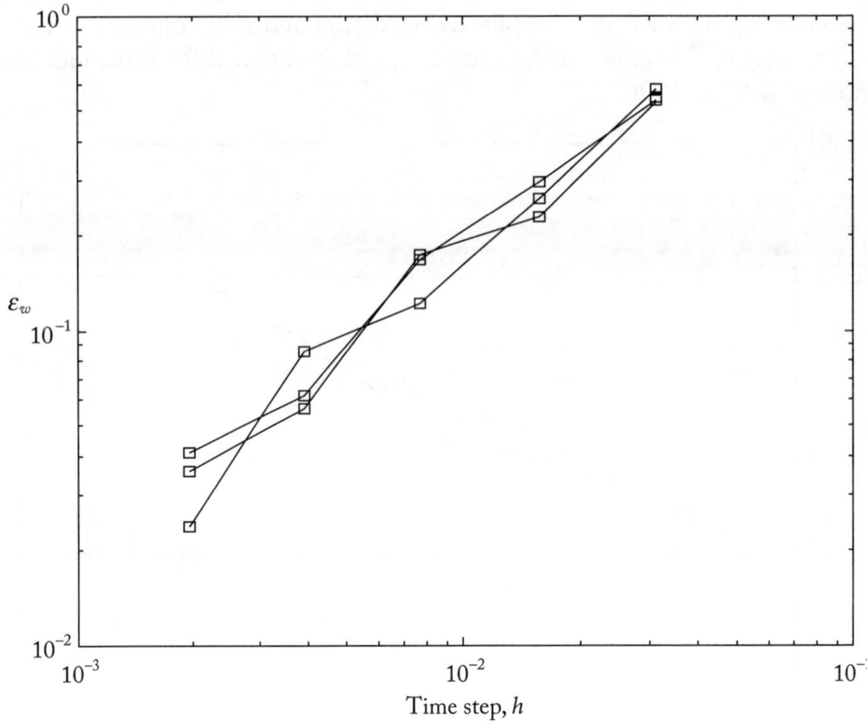

Fig. 5.4 Black–Scholes SDE (5.34); EM solution – weak order of convergence (Eq. (5.39b)), $\lambda = 2,\ \mu = 1,\ T = 1,\ x(0) = 1$; results from three independent MC simulation runs with sample size $N_e = 3000$

Example 5.2

In this example we consider the case of additive noise driving a non-linear (in the drift term) mechanical oscillator commonly known as the Duffing oscillator. Under $(\Omega, \mathcal{F}, P)$, the system dynamics of the oscillator may be formally written as:

$$\ddot{X} + c\,\dot{X} + kX + \upsilon X^3 = \mathcal{P}\cos 2\pi t + \sigma\,\dot{B}(t) \tag{5.40}$$

c and k are the damping and stiffness coefficients and υ denotes the drift non-linearity parameter. The first term on the RHS of the above equation stands for the deterministic forcing term which is harmonic. σ is the additive noise intensity, herein assumed to be time–independent.

Solution

Defining the displacement and velocity components by $X_1(t) = X(t)$ and $X_2(t) = \dot{X}_1(t)$ respectively, we have the state–space representation as:

$$dX_1 = X_2 dt \tag{5.41a}$$

$$dX_2 = \left(-cX_2 - kX_1 - \upsilon\, X_1^3 + \mathcal{P}\cos 2\pi t\ \right)dt + \sigma\, dB(t) \tag{5.41b}$$

The last equation represents the SDE–form of the oscillator in incremental form. The corresponding EM-based time marching map is:

$$X_{1,i+1} = X_{1,i} + X_{2,i}\ h \tag{5.42a}$$

$$X_{2,i+1} = X_{2,i} + \left(-cX_{2,i} - kX_{1,i} - \upsilon X_{1,i}^3 + \mathcal{P}\cos 2\pi t_i\ \right)h + \sigma\,\Delta B_i \tag{5.42b}$$

With Brownian trajectories simulated using a uniform step size of h, we apply the EM method for the SDE over the interval $[0, 5]$s. It is found that the explicit EM scheme fails to yield the approximate strong solution for larger step sizes. Figure 5.5 shows one such divergent solution obtained with $h = 0.05$. However, for adequately smaller step sizes ($h = 0.001$ or 0.0001) the EM solution is convergent. In fact the solutions with these two step sizes are visibly indistinguishable from one another as shown in the same figure. The response quantities reported in the figure are $E[X(t)]$ and $E\left[\dot{X}(t)\right]$ sample–averaged with $N_e = 100$. The noise intensity σ is taken as 0.01. In simulations, the parameters, c, k, υ and $\mathcal{P}$ are taken as 4, 100, 100 and 4, respectively.

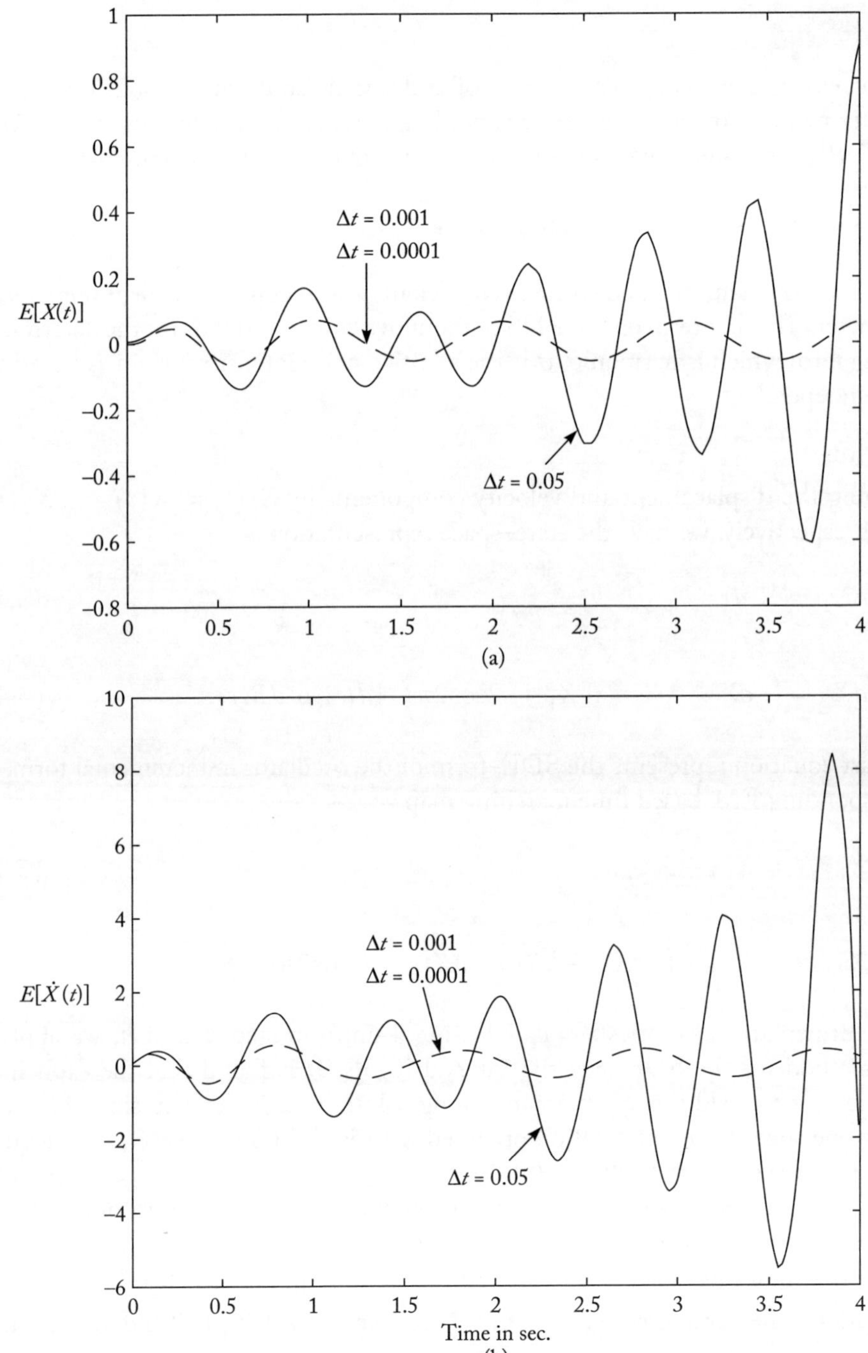

Fig. 5.5 Explicit EM solution for a non-linear Duffing oscillator – SDE in Eq. (5.40); $c = 4$, $k = 100$, $\upsilon = 100$, $\mathcal{P} = 4$, $\sigma = 10$, $N_e = 100$, (a) sample mean displacement and (b) sample mean velocity

5.3 An Implicit EM Method

An implicit version of the EM method is the stochastic counterpart of the so-called backward Euler scheme [Roy and Rao 2012] for deterministic ODEs. The scheme as applied to SDE (3.1) is given by the mapping:

$$\boldsymbol{X}_{j+1} = \boldsymbol{X}_j + \boldsymbol{a}\left(t_{j+1}, \boldsymbol{X}_{j+1}\right)\Delta t + \sigma\left(t_j, \boldsymbol{X}_j\right)\Delta \boldsymbol{B}_j \tag{5.43}$$

Note that only the drift coefficient is made implicit in the above equation whilst the diffusion coefficient (if it is multiplicative) term remains explicitly discretized. The drift-implicit EM scheme is shown [Higham *et al.* 2002] to converge to strong solution of the SDE with an order of convergence of $1/2$ in the root mean square sense (Eq. 5.5). The improvement of the EM solution via this so-called semi-implicit scheme is noticeable in Fig. 5.6. The result corresponds to the non-linear Duffing oscillator of Eq. (5.40). Specifically, the implicit scheme yields stable yet fairly accurate numerical solution to SDE (5.40) even with $h = 0.05$ for which the explicit solution was earlier observed to be divergent (this divergent solution is also included in this figure). The numerical solution with $h = 0.05$ also lies close to that obtained with a very small step size of $h = 0.0001$ showing the improved performance of the implicit scheme. In the absence of exact solutions, the EM simulation obtained with this small time step size may be considered to be as good as the true solution to the SDE.

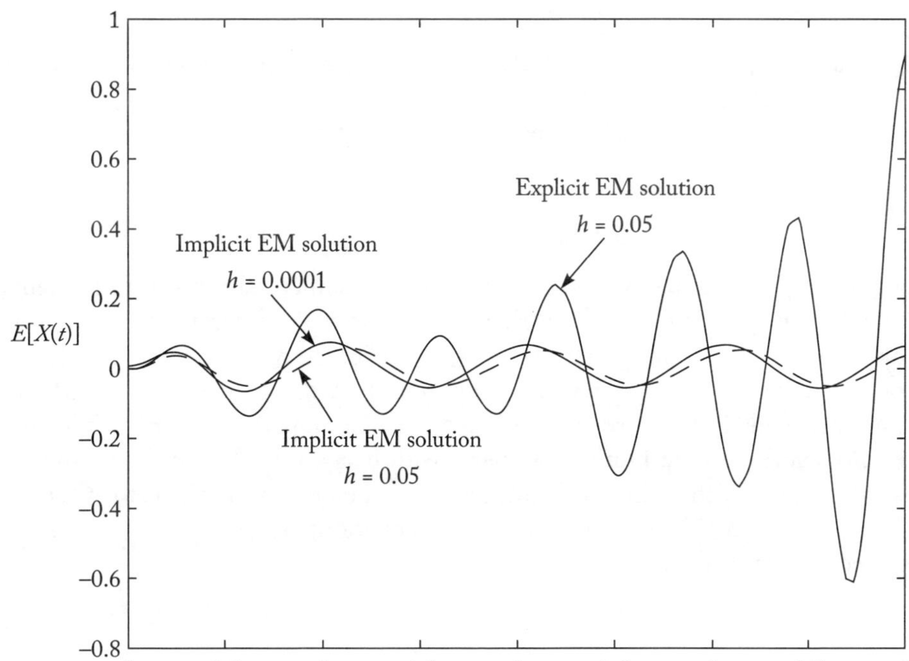

(a)

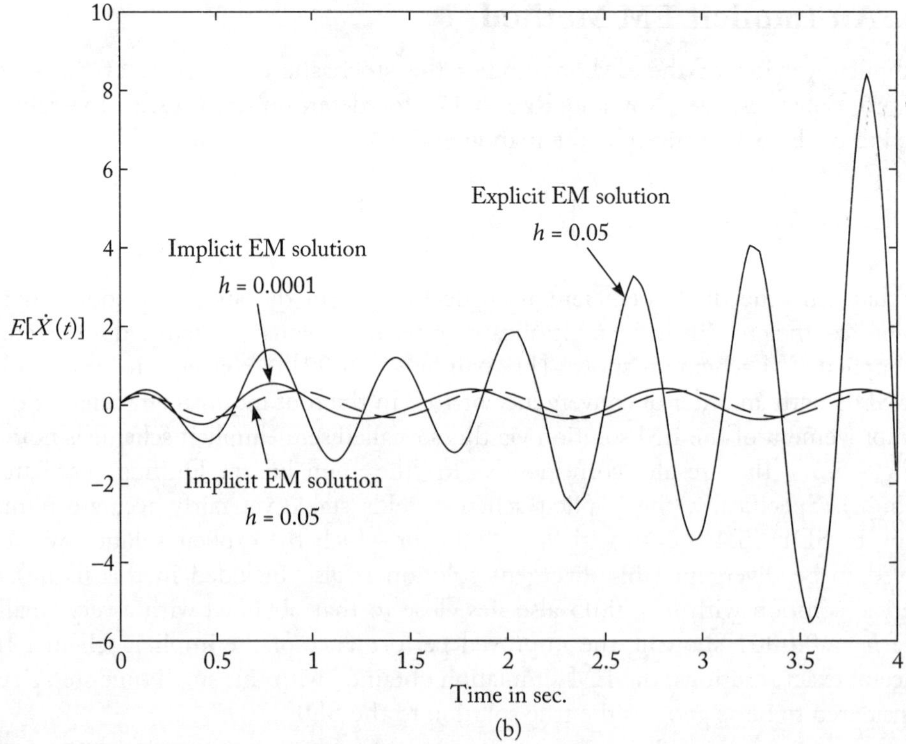

Fig. 5.6 Implicit EM method (Eq. 5.43); solution to the non-linear Duffing oscillator – SDE in Eq. (5.40); $c = 4,\ k = 100,\ \upsilon = 100,\ \mathcal{P} = 4$ and $\sigma = 10$; (a) sample mean displacement and (b) sample mean velocity

5.4 Further Issues on Convergence of EM Methods

The last Example 5.2 demonstrates the need for sufficiently small time–step sizes, especially for the explicit form of the EM method, to yield accurate numerical solution. The imperative usage of a small step size may be attributed to an unstable propagation, in time, of the numerical error. A typical manifestation of this instability is in the response becoming unbounded as $t \to \infty$ (with h fixed). This is apparent (see Fig. 5.5) from the explicit EM solution to the Duffing oscillator with $h = 0.05$. We recall here the stability criteria associated with numerical integration schemes as applied to ODEs in a deterministic set up. One considers a test equation, say, a 2-dimensional time–homogeneous linear ODE:

$$dx = Axdt \tag{5.44}$$

where, A is known as the amplification matrix having eigenvalues $\lambda_{1,2}$. Let $Re(\lambda_1) < Re(\lambda_2) < 0$ where $Re(.)$ denotes the real part of a complex number. Clearly this condition on the eigenvalues ensures that the exact solution is asymptotically stable, i.e., $\lim_{t\to\infty}\boldsymbol{x}(t) \to \mathbf{0}$. If the explicit EM (or rather Euler) method is applied to the test equation, it gives the following one-step approximation:

$$\boldsymbol{x}_{j+1} = \boldsymbol{x}_j + \boldsymbol{A}\boldsymbol{x}_j h \tag{5.45}$$

If the initial condition $\boldsymbol{x}_0$ is expressed as:

$$\boldsymbol{x}_0 = c_1\boldsymbol{v}_1 + c_2\boldsymbol{v}_2 \tag{5.46}$$

where, $\boldsymbol{v}_i, i = 1,2$ are the eigenvectors of $\boldsymbol{A}$ and $c_1, c_2 \in \mathbb{R}$, then one has the Euler solution as:

$$\boldsymbol{x}_{j+1} = (1+\lambda_1 h)^{j+1} c_1\boldsymbol{v}_1 + (1+\lambda_2 h)^{j+1} c_2\boldsymbol{v}_2 \tag{5.47}$$

The condition for the scheme to be stable is obviously $|1+\lambda_1 h| \le 1$. Here $|1+\lambda_1 h|$ is the spectral radius ρ of the matrix $\mathbf{I} + \boldsymbol{A}h$. If the above condition on ρ is not satisfied, the numerical solution will fail to converge and thus result in a (possibly sharp) increase in the error—local or global—leading to instability during computations. For an illustration, consider the mechanical oscillator in Example 5.2 without the noise term and the non linearity in the drift. The amplification matrix, its eigenvalues and the spectral radius are then given by:

$$\boldsymbol{A} = \begin{bmatrix} 0 & 1 \\ -k & -c \end{bmatrix}, \lambda_{1,2} = \frac{-c \pm i\sqrt{4k-c^2}}{2}, \ \rho = |1+\lambda_1 h| = 1 - ch + kh^2 \tag{5.48}$$

Since one commonly has $c \ll k$, $\lambda_{1,2}$ are in general complex. Now the condition $|1+\lambda_1 h| \le 1$ on the spectral radius in conjunction with Eq. (5.48) implies that with $h \le \frac{c}{k}$, the deterministic solution remains bounded and converges. The strict inequality $h < \frac{c}{k}$ provides the condition for asymptotic or absolute stability and the numerical scheme is then termed as A–stable.

In order to extend the discussion to the case of SDEs [Milstein 1995], we refer back to the SDE (5.1) purely additive and constant, i.e., $\sigma(t, \boldsymbol{X}) = \sigma$, and that the drift term is linear time-invariant, i.e., $\boldsymbol{a}(t, \boldsymbol{X}) = \boldsymbol{A}\boldsymbol{X}$. Then, analogous to the test ODE (5.44), we have the following test SDE:

$$d\boldsymbol{X} = \boldsymbol{A}\boldsymbol{X}dt + \sigma d\boldsymbol{B}(t) \tag{5.49}$$

Assume that $Re(\lambda_1) < Re(\lambda_2) < 0$ where $\lambda_{1,2}$ are the eigenvalues of $\boldsymbol{A}$ which are possibly complex and $\sigma \in \mathbb{R}^+$. The explicit EM map is given by:

$$\boldsymbol{X}_{j+1} = \boldsymbol{X}_j + \boldsymbol{A}\boldsymbol{X}_j h + \sigma\Delta\boldsymbol{B}_j \tag{5.50}$$

Since $\Delta \boldsymbol{B}_j$ is an independent Gaussian increment, it can be represented by the vector random variable $\zeta h^{1/2}$ whose scalar components are independent and $\mathcal{N}(0, \sqrt{h})$. This enables rewriting the explicit scheme as:

$$\boldsymbol{X}_{j+1} = \boldsymbol{X}_j + \boldsymbol{A}\boldsymbol{X}_j h + \sigma \zeta_j h^{1/2} \tag{5.51}$$

If $\sigma = 0$, it is the deterministic case discussed above and the EM solution is asymptotically stable. Otherwise ($\sigma \neq 0$) with $E\left[X_0^2\right] < \infty$, the EM solution has also second order moments that are uniformly bounded in j. It is therefore clear that for the SDE (5.49), the stability characteristics of the explicit EM scheme may be inferred from those of the associated deterministic solution. Similar is the case with the implicit EM scheme except that the scheme is always A-stable. To see this, first note that the solution by an implicit Euler (so-called backward Euler) scheme to the test Eq. (5.44) is given by:

$$\boldsymbol{X}_{j+1} = \boldsymbol{X}_j + \boldsymbol{A}\boldsymbol{X}_{j+1} h$$

$$\Longrightarrow \boldsymbol{X}_{j+1} = [\boldsymbol{I} - \boldsymbol{A}h]^{-1} \boldsymbol{X}_j = [\boldsymbol{I} - \boldsymbol{A}h]^{-(j+1)} \boldsymbol{X}_0$$

$$= (1-\lambda_1 h)^{-(j+1)} c_1 \boldsymbol{v}_1 + (1-\lambda_2 h)^{-(j+1)} c_2 \boldsymbol{v}_2 \tag{5.52}$$

In the above representation of $\boldsymbol{X}_{j+1}$, the decomposition of $\boldsymbol{x}_0$ in Eq. (5.46) is utilized. The test equation represents an asymptotically stable system with $Re(\lambda_1) < Re(\lambda_2) < 0$. Equation (5.52) points out that the implicit scheme is always A–stable (unlike the explicit scheme) in a deterministic setup, irrespective of the step size. The only consideration for the choice of h in this case may be based on convergence. As is the case with the explicit EM scheme, these observations on the stability of the implicit scheme may also be directly extrapolated to SDEs with additive noise. This follows from similar arguments as those in Eqs. (5.49)–(5.51). For further discussion on stability considerations for SDEs, refer to Saito and Mitsui [1996], Higham [1998] and Burrage et al. [2000].

5.5 An Introduction to Ito–Taylor Expansion for Stochastic Processes

The Taylor expansion is a popular tool for direct integration of DEs governing deterministic dynamical systems. An Ito–Taylor expansion is a stochastic analogue and is the basis for the development of integration techniques to numerically solve SDEs. The Wiener process (driving an SDE), even though continuous, has an unbounded variation over any given time interval. Further, the increments along a Wiener process path change by $O(h^{\frac{1}{2}})$ over a time interval of h. These non-smooth aspects associated with SDEs render the integration techniques quite different from their counterparts for ODEs. Before we introduce the Ito–Taylor expansion and its role in the numerical solutions of stochastic dynamical systems, first recall the derivation of the familiar Taylor's expansion in the deterministic case. Thus consider a (scalar) ODE of the form:

$$\frac{dx}{dt} = a(t,x) \tag{5.53}$$

Assume that $a(t,x)$ is sufficiently smooth and varies as an at most linear function of x as $x \to \infty$ so that a unique solution $x(t)$ exists. Now, for any sufficiently smooth function, $g(t,x(t))$, one has:

$$\frac{dg(t,x(t))}{dt} = Lg(t,x(t)) \tag{5.54}$$

where,

$$L = \frac{\partial}{\partial t} + a(t,x)\frac{\partial}{\partial x} \tag{5.55}$$

and we have:

$$g(t+h,x(t+h)) = g(t,x(t)) + \int_t^{t+h} Lg(s,x(s))ds \tag{5.56}$$

With $g(t,x(t)) = x(t)$, it follows that:

$$x(t+h) = x(t) + \int_t^{t+h} a(s,x(s))ds \tag{5.57}$$

Applying the formula in Eq. (5.56) to $a(t,x)$, the last equation becomes:

$$\begin{aligned} x(t+h) &= x(t) + \int_t^{t+h} \left(a(t,x(t)) + \int_t^s La(s_1,x(s_1))\,ds_1 \right) ds \\ &= x(t) + a(t,x(t)) \int_t^{t+h} ds + \int_t^{t+h} \int_t^s La(s_1,x(s_1))\,ds_1 ds \end{aligned} \tag{5.58}$$

Iterating once more by using Eq. (5.56) on $La(s_1,x(s_1))$ in the double integral on the RHS of the last equation, we get:

$$\begin{aligned} x(t+h) &= x(t) + h\,a(t,x(t)) + \frac{h^2}{2} La(t,x(t)) \\ &\quad + \int_t^{t+h} \int_t^s \int_t^{s_1} L^2 a(s_2,x(s_2))\,ds_2 ds_1 ds \end{aligned} \tag{5.59}$$

Repeated iterations with the help of Eq. (5.56) thus result in the familiar Taylor expansion of $x(t+h)$ in a neighborhood of t in powers of h and in different derivatives of $a(t,x(t))$:

$$x(t+h) = x(t) + ha(t,x(t)) + \frac{h^2}{2} La(t,x(t)) + \cdots + \frac{h^m}{m!} L^{m-1} a(t,x(t)) + R \quad (5.60a)$$

R is the remainder term given by:

$$R = \int_t^{t+h} \int_t^{s} \int_t^{s_1} \ldots \int_t^{s_{m-1}} L^m a(s_m, x(s_m)) ds_m \ldots ds_2 ds_1 ds \quad (5.60b)$$

The Taylor expansion in Eq. (5.60a) is the basis for generating explicit numerical integration methods to solve deterministic ODEs. It is clear that the truncated expansion excluding R corresponds to an explicit method with an accuracy of the mth order.

5.6 Derivation of Ito–Taylor Expansion

Extending the above arguments to an SDE as in Eq. (5.1), one obtains a corresponding expansion of the vector stochastic process $\boldsymbol{X}(t)$ in a neighbourhood of t, which is referred to as the stochastic Taylor expansion [Gikhma and Skorokhod 1972, Kloeden and Platen 1992]. The expansion is either Stratonovich or Ito–Taylor expansion depending on the interpretation of a stochastic integral–Stratonovich or Ito type (Chapter 4). Here discussion is limited to the derivation of an Ito–Taylor expansion with $\boldsymbol{X}(t) \in \mathbb{R}^m$, $\boldsymbol{a}(t,\boldsymbol{x}) \in \mathbb{R}^m$, $\boldsymbol{B}(t) \in \mathbb{R}^n$ and $\boldsymbol{\sigma}(t,\boldsymbol{x}) \in \mathbb{R}^{m \times n}$. We consider a sufficiently smooth scalar-valued function $g(t,\boldsymbol{x})$and obtain, by Ito's formula for $t_0 \le t \le s = t+h \le t_0 + T$, an expression analogous to Eq. (5.56) as:

$$g(t+h, \boldsymbol{X}(t+h))$$

$$= g(t,\boldsymbol{X}(t)) + \int_t^{t+h} L_0 g(s_1, \boldsymbol{X}(s_1)) ds_1 + \sum_{k=1}^{n} \int_t^{t+h} L_{1k} g(s_1, \boldsymbol{X}(s_1)) dB_k(s_1) \quad (5.61)$$

where the operators L_0 and L_{1k} are given by:

$$L_0 = \frac{\partial}{\partial t} + \sum_{j=1}^{m} a_j \frac{\partial}{\partial x_j} + \frac{1}{2} \sum_{k=1}^{n} \sum_{i=1}^{m} \sum_{j=1}^{m} \sigma_{ik} \sigma_{jk} \frac{\partial^2}{\partial x_i \partial x_j} \quad (5.62a)$$

$$L_{1k} = \sum_{j=1}^{m} \sigma_{jk} \frac{\partial}{\partial x_j}, \; k = 1,2,\ldots,n \quad (5.62b)$$

With $t \le s_1 \le s = t + h$, repeated applications of Ito's formula (Eq. 5.61) on $L_0 g(s_1, \boldsymbol{X}(s_1))$ and $L_{1k} g(s_1, \boldsymbol{X}(s_1))$ yield:

$$L_0 g(s_1, \boldsymbol{X}(s_1))$$

$$= L_0 g(t, \boldsymbol{X}(t)) + \int_t^{s_1} L_0{}^2 g(s_2, \boldsymbol{X}(s_2))\, ds_2$$

$$+ \sum_{k=1}^{n} \int_t^{s_1} L_{1k} L_0 g(s_2, \boldsymbol{X}(s_2))\, dB_k(s_2) \tag{5.63a}$$

and

$$L_{1k} g(s_1, \boldsymbol{X}(s_1)) = L_{1k} g(t, \boldsymbol{X}(t)) + \int_t^{s_1} L_0 L_{1k} g(s_2, \boldsymbol{X}(s_2))\, ds_2 + \sum_{j=1}^{n}$$

$$\int_t^{s_1} L_{1j} L_{1k} g(s_2, \boldsymbol{X}(s_2))\, dB_j(s_2) \tag{5.63b}$$

Substituting the above equations into Eq. (5.61), we have:

$$g(t+h, \boldsymbol{X}(t+h)) = g(t, \boldsymbol{X}(t)) + L_0 g(t, \boldsymbol{X}(t)) \int_t^{t+h} ds_1$$

$$+ \int_t^{t+h} \int_t^{s_1} L_0{}^2 g(s_2, \boldsymbol{X}(s_2))\, ds_2 ds_1$$

$$+ \sum_{k=1}^{n} \int_t^{t+h} \int_t^{s_1} L_{1k} L_0 g(s_2, \boldsymbol{X}(s_2))\, dB_k(s_2) ds_1$$

$$+ \sum_{k=1}^{n} L_{1k} g(t, \boldsymbol{X}(t)) \int_t^{t+h} dB_k(s_1)$$

$$+ \sum_{k=1}^{n} \int_t^{t+h} \int_t^{s_1} L_0 L_{1k} g(s_2, \boldsymbol{X}(s_2))\, ds_2 dB_k(s_1)$$

$$+ \sum_{k=1}^{n} \sum_{j=1}^{n} \int_t^{t+h} \int_t^{s_1} L_{1j} L_{1k} g(s_2, \boldsymbol{X}(s_2))\, dB_j(s_2) dB_k(s_1) \tag{5.64}$$

By continuing the application of Ito's formula on ${L_0}^2 g(s_2, \mathbf{X}(s_2))$, $L_{1k} L_0 g(s_2, \mathbf{X}(s_2))$, $L_0 L_{1k} g(s_2, \mathbf{X}(s_2))$ and $L_{1j} L_{1k} g(s_2, \mathbf{X}(s_2))$ in the above equation, we obtain:

$$\begin{aligned} g(t+h, \mathbf{X}(t+h)) = g(t, \mathbf{X}(t)) + L_0 g(t, \mathbf{X}(t)) \int_t^{t+h} ds_1 \\ + {L_0}^2 g(t, \mathbf{X}(t)) \int_t^{t+h} \int_t^{s_1} ds_2 ds_1 \\ + \sum_{k=1}^{n} L_{1k} L_0 g(t, \mathbf{X}(t)) \int_t^{t+h} \int_t^{s_1} dB_k(s_2) ds_1 \\ + \sum_{k=1}^{n} L_{1k} g(t, \mathbf{X}(t)) \int_t^{t+h} dB_k(s_1) \\ + \sum_{k=1}^{n} L_0 L_{1k} g(t, \mathbf{X}(t)) \int_t^{t+h} \int_t^{s_1} ds_2 dB_k(s_1) \\ + \sum_{k=1}^{n} \sum_{j=1}^{n} L_{1j} L_{1k} g(t, \mathbf{X}(t)) \int_t^{t+h} \int_t^{s_1} dB_j(s_2) dB_k(s_1) \\ + R \end{aligned} \tag{5.65}$$

R is the remainder term in the expansion and is given by:

$$\begin{aligned} R = \int_t^{t+h} \int_t^{s_1} \int_t^{s_2} {L_0}^3 g(s_3, \mathbf{X}(s_3)) ds_3\, ds_2 ds_1 \\ + \sum_{k=1}^{n} \int_t^{t+h} \int_t^{s_1} \int_t^{s_2} L_{1k} {L_0}^2 g(s_3, \mathbf{X}(s_3)) dB_k(s_3) ds_2 ds_1 \\ + \sum_{k=1}^{n} \int_t^{t+h} \int_t^{s_1} \int_t^{s_2} L_0 L_{1k} L_0 g(s_3, \mathbf{X}(s_3)) ds_3\, dB_k(s_2) ds_1 \\ + \sum_{k=1}^{n} \sum_{j=1}^{n} \int_t^{t+h} \int_t^{s_1} \int_t^{s_2} L_{1j} L_{1k} L_0 g(s_3, \mathbf{X}(s_3))\, dB_j(s_3) dB_k(s_2)\, ds_1 \end{aligned}$$

$$+\sum_{k=1}^{n}\int_{t}^{t+h}\int_{t}^{s_1}\int_{t}^{s_2} L_0{}^2 L_{1k} g\left(s_3, \boldsymbol{X}\left(s_3\right)\right) ds_3 ds_2 dB_k(s_1)$$

$$+\sum_{k=1}^{n}\sum_{j=1}^{n}\int_{t}^{t+h}\int_{t}^{s_1}\int_{t}^{s_2} L_{1j} L_0 L_{1k} g\left(s_3, \boldsymbol{X}\left(s_3\right)\right) dB_j\left(s_3\right) ds_2\, dB_k\left(s_1\right)$$

$$+\sum_{k=1}^{n}\sum_{j=1}^{n}\int_{t}^{t+h}\int_{t}^{s_1}\int_{t}^{s_2} L_0 L_{1j} L_{1k} g\left(s_3, \boldsymbol{X}\left(s_3\right)\right) ds_3\, dB_j\left(s_2\right) dB_k\left(s_1\right)$$

$$+\sum_{k=1}^{n}\sum_{j=1}^{n}\sum_{i=1}^{n}\int_{t}^{t+h}\int_{t}^{s_1}\int_{t}^{s_2} L_{1i} L_{1j} L_{1k} g\left(s_3, \boldsymbol{X}\left(s_3\right)\right) dB_i\left(s_3\right) dB_j\left(s_2\right) dB_k\left(s_1\right) \quad (5.66)$$

Removing the remainder term R from the Ito–Taylor expansion in Eq. (5.65), we observe that the formula contains several integrals with constant integrands. To recognize the usefulness of the expansion in identifying different numerical integration schemes of increasing order of accuracy, it is instructive to know the order of smallness of the different terms in the expansion. Before we proceed further, note that $\int_t^{t+h} ds$ is $O(h)$ and hence such a term corresponds to an order of smallness 1. Based on the same logic $\int_t^{t+h} dB(s)$ is taken as a term with an order of smallness $1/2$. Now, as applied to the truncated expansion (i.e., excluding R) in Eq. (5.65), the order of smallness of the terms that are retained, ranges from zero to 2. The first term on the RHS is of the zeroth order. The order of each term in $\sum_{k=1}^{n} L_{1k} g\left(t, \boldsymbol{X}\left(t\right)\right) \int_t^{t+h} dB_k\left(s_1\right)$ is $1/2$. Terms of the type $L_{1k} L_{1j} g \int_t^{t+h} \int_t^{s_1} dB_j(s_2) dB_k\left(s_1\right)$ and $L_0 g \int_t^{t+h} ds_1$ are of order 1 and the terms $L_0 L_{1k} g \int_t^{t+h} \int_t^{s_1} dB_k(s_2) ds_1$ and $L_{1k} L_0 g \int_t^{t+h} \int_t^{s_1} ds_2 dB_k\left(s_1\right)$ are of order $3/2$. There is also the term $L_0{}^2 g \int_t^{t+h} \int_t^{s_1} ds_2 ds_1$ of order 2.

5.6.1 One-step approximations—explicit integration methods

For a simple exposition to one-step approximations derivable from the above Ito–Taylor expansion, we consider without a loss of generality a scalar SDE driven by a vector Brownian motion $\boldsymbol{B}(t)$ given by:

$$dX\left(t\right) = a\left(t, X\right) dt + \sigma\left(t, X\right) d\boldsymbol{B}\left(t\right) \quad (5.67)$$

Here the diffusion matrix $\sigma\left(t, X\right)$ is a row vector and $d\boldsymbol{B}\left(t\right)$ a column vector. With $\sigma_k(t, X)$ denoting the diffusion coefficient associated with the kth Brownian motion B_k and with $g\left(t, X\left(t\right)\right) = X(t)$, we have $L_0 X = a$, $L_{1k} X = \sigma_k$. The Ito–Taylor expansion in Eq. (5.65) takes the form:

$$
\begin{aligned}
X(t+h) = X(t) &+ \sum_{k=1}^{n} \sigma_k \int_t^{t+h} dB_k(s_1) + a \int_t^{t+h} ds_1 \\
&+ \sum_{k=1}^{n} \sum_{j=1}^{n} \left(L_{1j}\sigma_k\right) \int_t^{t+h} \int_t^{s_1} dB_j(s_2) dB_k(s_1) \\
&+ \sum_{k=1}^{n} \left(L_0\sigma_k\right) \int_t^{t+h} \int_t^{s_1} ds_2 dB_k(s_1) + \sum_{k=1}^{n} \left(L_{1k}a\right) \int_t^{t+h} \int_t^{s_1} dB_k(s_2) ds_1 \\
&+ \sum_{k=1}^{n} \sum_{j=1}^{n} \sum_{i=1}^{n} \left(L_{1i}L_{1j}\sigma_k\right) \int_t^{t+h} \left(\int_t^{s_1} \left(\int_t^{s_2} dB_i(s_3) \right) dB_j(s_2) \right) dB_k(s_1) \\
&+ (L_0 a) \int_t^{t+h} \int_t^{s_1} ds_2 ds_1 + R
\end{aligned} \tag{5.68}
$$

The remainder term R is given by:

$$
\begin{aligned}
R = &\int_t^{t+h} \left(\int_t^{s_1} \left(\int_t^{s_2} {L_0}^2 a(s_3, X(s_3)) ds_3 \right) ds_2 \right) ds_1 \\
&+ \sum_{k=1}^{n} \int_t^{t+h} \left(\int_t^{s_1} \left(\int_t^{s_2} {L_0}^2 \sigma_k(s_3, X(s_3)) ds_3 \right) ds_2 \right) dB_k(s_1) \\
&+ \sum_{k=1}^{n} \int_t^{t+h} \left(\int_t^{s_1} \left(\int_t^{s_2} L_0 L_{1k} a(s_3, X(s_3)) ds_3 \right) dB_k(s_2) \right) ds_1 \\
&+ \sum_{k=1}^{n} \sum_{j=1}^{n} \int_t^{t+h} \left(\int_t^{s_1} \left(\int_t^{s_2} L_{1j} L_0 \sigma_k(s_3, X(s_3)) dB_j(s_3) \right) ds_2 \right) dB_k(s_1) \\
&+ \sum_{k=1}^{n} \int_t^{t+h} \left(\int_t^{s_1} \left(\int_t^{s_2} L_{1k} L_0 a(s_3, X(s_3)) dB_k(s_3) \right) ds_2 \right) ds_1 \\
&+ \sum_{k=1}^{n} \sum_{j=1}^{n} \int_t^{t+h} \left(\int_t^{s_1} \left(\int_t^{s_2} L_0 L_{1j} \sigma_k(s_3, X(s_3)) ds_3 \right) dB_j(s_2) \right) dB_k(s_1)
\end{aligned}
$$

$$+\sum_{k=1}^{n}\sum_{j=1}^{n}\int_t^{t+h}\left(\int_t^{s_1}\left(\int_t^{s_2} L_{1j}L_{1k}a\left(s_3,X\left(s_3\right)\right)dB_j\left(s_3\right)\right)dB_k\left(s_2\right)\right)ds_1$$

$$+\sum_{k=1}^{n}\sum_{j=1}^{n}\sum_{i=1}^{n}\int_t^{t+h}\left(\int_t^{s_1}\left(\int_t^{s_2}\left(\int_t^{s_3} L_0L_{1i}L_{1j}\sigma_k\left(s_4,X\left(s_4\right)\right)ds_4\right)\right.\right.\right.$$

$$\left.\left.\left.\times dB_i\left(s_3\right)\right)dB_j\left(s_2\right)\right)dB_k\left(s_1\right)+\sum_{k=1}^{n}\sum_{j=1}^{n}\sum_{i=1}^{n}\sum_{l=1}^{n}\int_t^{t+h}$$

$$\left(\int_t^{s_1}\left(\int_t^{s_2}\left(\int_t^{s_3} L_{1l}L_{1i}L_{1j}\sigma_k\left(s_4,X\left(s_4\right)\right)dB_l\left(s_4\right)\right)\right.\right.\right.$$

$$\left.\left.\left.\times\, dB_i\left(s_3\right)\right)dB_j\left(s_2\right)\right)dB_k\left(s_1\right) \tag{5.69}$$

In writing the expansion in Eq. (5.68), Ito's formula has again been applied to the last term on the RHS of Eq. (5.66) so that all the terms of order $1\frac{1}{2}$ are present in the truncated expansion. The reason for including this last term will be evident while discussing the order of convergence of these one-step approximations (in the section to follow). One then obtains the following possible one-step approximations:

$$X^{(1)}\left(t+h\right)=X\left(t\right)+\sum_{k=1}^{n}\sigma_k\int_t^{t+h}dB_k\left(s_1\right)+a\int_t^{t+h}ds_1 \tag{5.70a}$$

$$X^{(2)}\left(t+h\right)=X^{(1)}\left(t+h\right)+\sum_{k=1}^{n}\sum_{j=1}^{n}\left(L_{1j}\sigma_k\right)\int_t^{t+h}\int_t^{s_1}dB_j\left(s_2\right)dB_k\left(s_1\right) \tag{5.70b}$$

$$X^{(3)}\left(t+h\right)=X^{(2)}\left(t+h\right)$$

$$+\sum_{k=1}^{n}\left(L_0\sigma_k\right)\int_t^{t+h}\int_t^{s_1}ds_2dB_k\left(s_1\right)$$

$$+\sum_{k=1}^{n}\left(L_{1k}a\right)\int_t^{t+h}\int_t^{s_1}dB_k\left(s_2\right)ds_1+\sum_{k=1}^{n}\sum_{j=1}^{n}\sum_{i=1}^{n}\left(L_{1i}L_{1j}\sigma_k\right)$$

$$\int_t^{t+h}\left(\int_t^{s_1}\left(\int_t^{s_2} dB_i(s_3)\right)dB_j(s_2)\right)dB_k(s_1) \tag{5.70c}$$

$$X^{(4)}(t+h) = X^{(3)}(t+h) + (L_0 a)\int_t^{t+h}\int_t^{s_1} ds_2 ds_1 \tag{5.70d}$$

These approximations correspond to explicit integration schemes which may be contrasted with their deterministic counterparts [Gard 1988], specifically in assessing their orders of accuracy. To determine the global order of accuracy p (Eq. 5.6) of each of the above one-step approximations, we need to determine the orders p_1 and p_2 of the local error in the mean and the mean square (Eqs. 5.5a,b) of each approximation. The error in each of the one-step approximations is $\left|X(t+h) - X^{(j)}(t+h)\right|$, $j = 1,2,3,4$, where, $X(t+h)$ is the true solution represented by the Ito–Taylor expansion (including R) in Eq. (5.68). For example, for the one-step approximation in Eq. (5.70a), which is the popular explicit EM method, the error (denoted by $\mathfrak{E}$) is given by:

$$\begin{aligned}
\mathfrak{E}^{(1)} &= X(t+h) - X^{(1)}(t+h) \\
&= \sum_{k=1}^{n}\sum_{j=1}^{n}\left(L_{1j}\sigma_k\right)\int_t^{t+h}\int_t^{s_1} dB_j(s_2)dB_k(s_1) \\
&\quad + \sum_{k=1}^{n}(L_0\sigma_k)\int_t^{t+h}\int_t^{s_1} ds_2 dB_k(s_1) \\
&\quad + \sum_{k=1}^{n}(L_{1k}a)\int_t^{t+h}\int_t^{s_1} dB_k(s_2)ds_1 \\
&\quad + \sum_{k=1}^{n}\sum_{j=1}^{n}\sum_{i=1}^{n}\left(L_{1i}L_{1j}\sigma_k\right)\int_t^{t+h}\left(\int_t^{s_1}\left(\int_t^{s_2} dB_i(s_3)\right)dB_j(s_2)\right)dB_k(s_1) \\
&\quad + (L_0 a)\int_t^{t+h}\int_t^{s_1} ds_2 ds_1 + R
\end{aligned} \tag{5.71}$$

It may be verified that $\mathfrak{E}^{(j)}, j = 1,2,3,4$ for each one-step approximation contains the following two types of multiple stochastic integrals (MSIs):

$$I^{(1)}_{j_1 j_2 \cdots j_k}(h) = \int_t^{t+h} dB_{j_k}(s_1)\int_t^{s_1} dB_{j_{k-1}}(s_2)\ldots\int_t^{s_{k-1}} dB_{j_1}(s_k) \tag{5.72a}$$

$$I^{(2)}_{j_1 j_2 \ldots j_k}(g,h) = \int_t^{t+h} dB_{j_k}(s_1) \int_t^{s_1} dB_{j_{k-1}}(s_2)\ldots$$

$$\int_t^{s_{k-1}} L_{j_k} L_{j_{k-1}} \ldots L_{j_1} g(s_{k-1}, X(s_{k-1}))dB_{j_1}(s_k) \quad (5.72b)$$

The first type of integral (with the integrand being trivially 1) appears in truncated Ito–Taylor expansions, i.e., when the remainder term R is excluded. The second one appears only in R. The integrals are random variables which are independent of $\mathcal{F}_t$ with j_k assigned any of the alpha numerals $0, i, j\ldots$. Here $j_k = 0$ denotes integral with respect to s, i.e., with $dB_0(s) := ds$. Clearly, $E\left[I^{(1)}_{j_1 j_2 \ldots j_k}(h)\right] = 0$ if at least one $j_i \neq 0$ and $E\left[I^{(1)}_{j_1 j_2 \ldots j_k}(h)\right] = O(h^k)$ if all $j_i = 0,\ i = 1,2,\ldots,k$. Similarly, $E\left[I^{(2)}_{j_1 j_2 \ldots j_k}(g,h)\right] = 0$ if at least one $j_i \neq 0$.

The following two results [Milstein 1995] help in evaluating the orders of the MSIs in the mean square sense (see Appendix E):

(i) $$\left(E\left[I^{(1)}_{j_1 j_2 \ldots j_k}(h)\right]^2\right)^{1/2} = O\left(h^{\sum_{i=1}^{k} \frac{(2-\bar{j}_i)}{2}}\right) \quad (5.73)$$

where $\bar{j}_i = 0$ if $j_i = 0$

$= 1$ if $j_i \neq 0$

(ii) $$\left(E\left[I^{(2)}_{j_1 j_2 \ldots j_k}(g,h)\right]^2\right)^{1/2} \leq K\left(1 + E\left[\left|X(t)\right|^2\right]\right)^{1/2} h^{\sum_{i=1}^{k} \frac{(2-\bar{j}_i)}{2}} \quad (5.74)$$

provided that, $\left|L_{j_k} L_{j_{k-1}} \ldots L_{j_1} g(t,x)\right| \leq K\left((1+|x|^2\right)^{1/2}$

Eq. (5.74) indicates that the order of smallness of $I^{(2)}_{j_1 j_2 \ldots j_k}(g,h)$ is the same as that of $I^{(1)}_{j_1 j_2 \ldots j_k}(h)$, subjected to adequate regularity of g. Since the integrals $I^{(2)}_{j_1 j_2 \ldots j_k}(g,h)$ appear in the remainder term R (Eq. 5.69), the result shows that each term in R has order of smallness of at most 2. Based on the mean and mean square values of the two integrals, we are now ready to determine the order of accuracy p of the one-step approximations in Eq. (5.70).

For the EM scheme in Eq. (5.70a), the order p_1 of the local error is decided by the mean of the term $(L_0 a)\int_t^{t+h}\int_t^{s_1} ds_2 ds_1$ in Eq. 5.71). Since $E\left[\int_t^{t+h}\int_t^{s_1} ds_2 ds_1\right] = E\left[I^{(1)}_{00}(h)\right] = O(h^2)$, we have $p_1 = 2$. The order p_2 of the local error is determined by the mean square value of integrals appearing as component terms in the sum $\sum_{k=1}^{n}$ $\sum_{j=1}^{n}\left(L_{1j}\sigma_k\right)\int_t^{t+h}\int_t^{s_1} dB_j(s_2)dB_k(s_1)$. Since $\left(E\left[\int_t^{t+h}\int_t^{s_1} dB_j(s_2)dB_k(s_1)\right]^2\right)^{\frac{1}{2}} =$

$\left(E\left[I_{jk}^{(1)}(h)\right]^2\right)^{\frac{1}{2}} = O(h^{\sum_{i=1}^{2}\left(\frac{2-\bar{j}_i}{2}\right)}) = O(h)$, clearly $p_2 = 1 \geq 1/2$. Thus $p_1 = 2 \geq p_2 + 1/2$. Using Eq. (5.6), we find that the explicit EM scheme is globally convergent with order of convergence, $p = p_2 - 1/2 = 1/2$. We are already familiar with this basic result, which was proved in Section 5.2.

For the numerical scheme in Eq. (5.70b), p_1 is decided by the term $(L_0 a)\int_t^{t+h}\int_t^{s_1} ds_2 ds_1$ (same as in the EM scheme) and therefore, $p_1 = 2$. The order p_2 is determined by any of the components in the sums $\sum_{k=1}^{n}(L_0\sigma_k)\int_t^{t+h}\int_t^{s_1} ds_2 dB_k(s_1)$ or $\sum_{k=1}^{n}(L_{1k}a)\int_t^{t+h}\int_t^{s_1} dB_k(s_2)ds_1$. For example, we have $\left(E\left[\int_t^{t+h}\int_t^{s_1} ds_2 dB_k(s_1)\right]^2\right)^{\frac{1}{2}}$ $=\left(E\left[I_{0k}^{(1)}(h)\right]^2\right)^{\frac{1}{2}} = O(h^{\sum_{i=1}^{2}\left(\frac{2-\bar{j}_i}{2}\right)}) = O(h^{\frac{3}{2}})$ since $\bar{j}_1 = 0$ and $\bar{j}_2 = 1$. Thus $p_2 = \frac{3}{2}$, finally leading to $p = 1$.

For the scheme in Eq. (5.70c), p_1 and p_2 are both decided by the mean and mean square value of the term $(L_0 a)\int_t^{t+h}\int_t^{s_1} ds_2 ds_1$. This gives $p_1 = 2$ and $p_2 = 2$. To satisfy the condition $p_1 \geq p_2 + 1/2$ for convergence Eq. (5.6a), we must however take $p_2 = \frac{3}{2}$, which gives the order of convergence $p = 1$.

For the scheme in Eq. (5.70d), $\mathfrak{E} = R$ and p_1 is decided by the expectation of the term $\int_t^{t+h}\int_t^{s_1}\int_t^{s_2} L_0{}^2 a(s_3, X(s_3))\, ds_3 ds_2 ds_1$ (since the mean of all the other terms in R is zero). Thus, if the integrands in R satisfy the condition in Eq. (5.74), $E[R] = O(h^3)$ and $E\left[R^2\right] = O(h^4)$ yielding $p_1 = 3$ and $p_2 = 2$. Hence $p = \frac{3}{2}$.

A generalized rule to determine p

One can generalize the evaluation of p as elucidated below.

Case (a): In the first case, the truncation corresponding to a one-step approximation may include all terms of integral order up to m inclusive. Then, the error $\mathfrak{E}$ has terms of order $\geq m + \frac{1}{2}$. The terms of order $m + \frac{1}{2}$ have their mean zero and hence p_1 is decided by the terms of order $m + 1$. Therefore, $E[\mathfrak{E}] = O\left(h^{m+1}\right) \Longrightarrow p_1 = m + 1$. We get p_2 from the terms of order $m + \frac{1}{2}$ with $\left(E\left[\mathfrak{E}^2\right]\right)^{\frac{1}{2}} = O\left(h^{\frac{2m+1}{2}}\right) \Longrightarrow p_2 = m + \frac{1}{2}$. This gives the order of convergence $p = m$. Note that the condition in Eq. (5.74) must be satisfied by the terms of order $m + \frac{1}{2}$ (if the integrand is of the type appearing in the definition of $I_{j_1 j_2 \cdots j_k}^{(2)}(g, h)$). The one step approximation in Eq. (5.70b) is an example, wherein the approximation contains terms of order $m = 1$ inclusive. The error $\mathfrak{E}$ contains three terms of order $m + \frac{1}{2} = \frac{3}{2}$ and one term of order $m + 1 = 2$. Hence the numerical scheme has order of convergence, $p = m = 1$.

Case (b): In the second case, the one-step approximation may include all terms of half-integral order up to $m + \frac{1}{2}$ inclusive. Then, the error $\mathfrak{E}$ has terms of order $\geq m + 1$.

One amongst these terms specifically contains an integral of the type $I^{(1)}{}_{j_1 j_2 \cdots j_k}(h)$ with all $j_i = 0,\ i = 1,2,..,k$ and thus $E[\mathfrak{E}] = O(h^{m+1}) \Longrightarrow p_1 = m+1$ and $(E[\mathfrak{E}^2])^{\frac{1}{2}} = O(h^{2m+2}) \Longrightarrow p_2 = m+1$. Enforcing $p_2 = m + \frac{1}{2}$ so as to satisfy the convergence condition in Eq. (5.6a) yields $p = m$. This is the case with the numerical scheme in Eq. (5.70c) where the terms in the approximation are of order $\leq \frac{3}{2}$ with $m = 1$, finally resulting in an order of convergence $p = m = 1$.

It is possible to effect an improvement in the order of convergence of the general result in case (b) above and so of the associated numerical integration scheme. Suppose that the truncation of the Ito–Taylor expansion for the scheme has all terms of half integral order up to $m + \frac{1}{2}$ inclusive, as well as the term of order of $m+1$ having non-zero mean, i.e., $I^{(1)}_{j_1 j_2 \cdots j_k}(h) = L_0 a(s_m, X(s_m)) \int_t^{t+h} ds_1 \int_t^{s_1} ds_2 \ldots \int_t^{s_{k-2}} ds_m \int_t^{s_{k-2}} ds_{m+1}$. Then $p_1 = m+2$ since it is decided by a similar term in $\mathfrak{E}$ with a non-zero mean and of order $m+2$. The parameter p_2 remains at $m+1$, thus yielding $p = m + \frac{1}{2}$ instead of m. Here, of course, the corresponding integrands must satisfy the condition in Eq. (5.74). An example to this improvement in p can be seen for the scheme in Eq. (5.70d) over the one in Eq. (5.70c). Even the EM method in Eq. (5.70a) is an improved version of an otherwise non-convergent one-step approximation (corresponding to a possible truncation of the expansion in Eq. 5.68) given by:

$$X^{(0)}(t_{i+1}) = X(t_i) + \sum_{k=1}^{n} \sigma_k \int_t^{t+h} dB_k(s_1) \tag{5.75}$$

This truncation belongs to case (b) above with $m = 0$. It is readily seen in this case that $p_1 = p_2 = 1$ so that $p = 0$ signifying that the method is non-convergent. Now, adding the term $a \int_t^{t+h} ds_1$ to the truncation, as is the case with EM, one arrives at a scheme with an order of convergence $\frac{1}{2}$.

5.7 Implementation Issues of the Numerical Integration Schemes

As noted in the preceding section, the stochastic Taylor expansion in Eq. (5.68) is the basis for deriving explicit stochastic integration methods. The EM method (Eq. (5.70a)) with terms up to $O(h)$ (but not including all of them) retained in the expansion has $O(h^{\frac{1}{2}})$ global order of convergence. The strong Taylor scheme of order $O(h)$ given in Eq. (5.70b) is due to Milstein [1974]. Other strong Taylor schemes of $O(h^{\frac{3}{2}})$ and $O(h^2)$ have been proposed, among others, by [Wagner and Platen 1978, Kloeden and Platen 1992, and Milstein 1995]. However, the difficulty in using higher order Taylor approximations is the arduous task of numerically finding the MSIs. If, for example, we consider the explicit method in Eq. (5.70d), it contains the maximum number of MSIs to be evaluated. They are particularly of the type $I^{(1)}_{j_1 j_2 \cdots j_k}(h)$ with $k \leq 3$.

5.7.1 Evaluation of MSIs

For the numerical evaluation of MSIs, consider n independently evolving scalar Wiener processes, $B_j(t)$, $j = 1,2,...,n$. For an introductory illustration, methods of numerical evaluation of a few MSIs are only provided. For MSIs involving single integrals, one immediately gets:

$$I_0^{(1)}(h) = \int_t^{t+h} ds = h,\ I_j^{(1)}(h) = \int_t^{t+h} dB_j(s) = B_j(t+h) - B_j(t) \tag{5.76a}$$

Defining $\Delta_t B_j(h) := \int_t^{t+h} dB_j(s) = B_j(t+h) - B_j(t)$, one has $\Delta_t B_j(h) = h^{\frac{1}{2}} Z_j \sim \mathcal{N}\left(0, h^{\frac{1}{2}}\right)$ where Z_j is the standard normal random variable $\mathcal{N}(0,1)$.

For $I_{00}^{(1)}(h)$, we have the direct result:

$$I_{00}^{(1)}(h) = \int_t^{t+h} \left(\int_t^{s_1} ds_2 \right) ds_1 = \frac{h^2}{2!} \tag{5.76b}$$

For evaluating the integrals $I_{j_1 j_2}^{(1)}(h)$ and $I_{j_1 j_2 j_3}^{(1)}(h)$, where, $j_1, j_2, j_3 \neq 0$, the procedure may not be so direct. For example, to find $I_{jk}^{(1)}(h) = \int_t^{t+h} \left(\int_t^{s_1} dB_j(s_2) \right) dB_k(s_1)$, where, B_j and B_k are independent Weiner processes, we use the following result due to Ito's formula:

$$\begin{aligned}
&I_{jk}^{(1)}(h) + I_{kj}^{(1)}(h) = I_j^{(1)}(h) I_k^{(1)}(h) \\
&\qquad \Longrightarrow \int_t^{t+h} \left(\int_t^{s_1} dB_j(s_2) \right) dB_k(s_1) + \int_t^{t+h} \left(\int_t^{s_1} dB_k(s_2) \right) dB_j(s_1) \\
&\qquad = \int_t^{t+h} dB_j(s) \int_t^{t+h} dB_k(s) \\
&\qquad = \left(\Delta_t B_j(h) \right) \left(\Delta_t B_k(h) \right)
\end{aligned} \tag{5.77}$$

If these two integrals occur together in pairs, the above formula can be conveniently utilized. Otherwise, by writing $I_{jk}^{(1)}(h) = \int_t^{t+h} \left(B_j(s_1) - B_j(t) \right) dB_k(s_1)$, the task reduces to modelling the set of variables B_j, $j = 1,2,\ldots,n$ and the integrals of the type $\int_t^{t+h} B_j(s)\, dB_k(s)$ denoted by I_{jk}. It now follows:

$$I_{jk} = \int_t^{t+h} B_j(s)\, dB_k(s) = \sum_{i=1}^{l} \int_{t_{i-1}}^{t_i} B_j(s)\, dB_k(s)$$

$$\cong \sum_{i=1}^{l} B_j(t_{i-1})\left(B_k(t_i) - B_k(t_{i-1})\right) \text{ by the rectangle rule of integration} \tag{5.78}$$

The MSI $I_{jj}^{(1)}(h) = \int_t^{t+h}\left(\int_t^{s_1} dB_j(s_2)\right)dB_j(s_1)$ may be readily evaluated by the following procedure:

$$I_{jj}^{(1)}(h) = \int_t^{t+h}\left(\int_t^{s_1} dB_j(s_2)\right)dB_j(s_1) = \int_t^{t+h}\left(B_j(s_1) - B_j(t)\right)dB_j(s_1)$$

$$= \int_t^{t+h} B_j(s_1)\,dB_j(s_1) - \int_t^{t+h} B_j(t)\,dB_j(s_1) \tag{5.79}$$

Since, by Ito's formula, $dB_j^2(s_1) = 2B_j(s_1)\,dB_j(s_1) + dt$, the result for the integral is obtained as:

$$I_{jj}^{(1)}(h) = \int_t^{t+h}\left(\int_t^{s_1} dB_j(s_2)\right)dB_j(s_1)$$

$$= \frac{1}{2}\int_t^{t+h}\left(dB_j^2(s_1) - dt\right) - \int_t^{t+h} B_j(t)\,dB_j(s_1)$$

$$= \frac{1}{2}\int_t^{t+h}\left(dB_j^2(s_1) - dt\right) - B_j(t)\int_t^{t+h} dB_j(s_1)$$

$$= \frac{1}{2}\left(B_j^2(t+h) - B_j^2(t) - h\right) - B_j(t)\left(B_j(t+h) - B_j(t)\right)$$

$$= \frac{1}{2}\left(\left(I_j^{(1)}(h)\right)^2 - h\right) \tag{5.80}$$

Here note that in case of SDEs with additive noises, the MSI $I_{jk}^{(1)}(h)$ is absent, since $L_{1j}\sigma_k = 0$. In other words, the Milstein scheme in Eq. (5.70b) reduces to the EM scheme.

Next we consider evaluating the two integrals $I_{j0}^{(1)}(h) = \int_t^{t+h}\int_t^{s_1} dB_j(s_2)\,ds_1$ and $I_{0j}^{(1)}(h) = \int_t^{t+h}\int_t^{s_1} ds_2\,dB_j(s_1)$. Here we make use of the following identity:

$$I_{j0}^{(1)}(h) + I_{0j}^{(1)}(h) = \int_t^{t+h} ds_1 \int_t^{t+h} dB_j(s_1) = h\left(\Delta_t B_j(h)\right)$$

$$\Longrightarrow I_{j0}^{(1)}(h) = h\left(\Delta_t B_j(h)\right) - I_{0j}^{(1)}(h) \tag{5.81}$$

Denoting the random variable $I_{j0}^{(1)}(h)$ by ζ and $I_{0j}^{(1)}(h)$ by ξ, we find that these integrals are normal random variables. Their moments are obtained as follows:

$$E[\zeta] = \int_t^{t+h} \left(B_j(s_1) - B_j(t)\right) ds_1 = 0 \tag{5.82}$$

$$var(\zeta) = E\left[(\zeta - E[\zeta])^2\right] = E\left[(\zeta)^2\right] = E\left[\left(h\left(\Delta_t B_j(h)\right) - \xi\right)^2\right]$$

$$= E\left[h^2\left(\Delta_t B_j(h)\right)^2 - 2h\left(\Delta_t B_j(h)\right)\xi + \xi^2\right] \tag{5.83}$$

Using the identities:

$$E\left[\left(\Delta_t B_j(h)\right)^2\right] = h \tag{5.84a}$$

$$E\left[\left(\Delta_t B_j(h)\right)\xi\right] = E\left[\int_t^{t+h} dB_j(s_1) \int_t^{t+h}\int_t^{s_1} ds_2 dB_j(s_1)\right]$$

$$= E\left[\int_t^{t+h} dB_j(s_1) \int_t^{t+h}\int_t^{s_1} ds_2 dB_j(s_1)\right]$$

$$= \int_t^{t+h}\int_t^{s_1} ds_2 dt = \int_t^{t+h} (s_1 - t) dt \tag{5.84b}$$

$$E\left[\xi^2\right] = E\left[\int_t^{t+h}\int_t^{s_1} ds_2 dB_j(s_1) \int_t^{t+h}\int_t^{s_1} ds_2 dB_j(s_1)\right]$$

$$= \int_t^{t+h}\left(\int_t^{s_1} ds_2\right)^2 dt = \int_t^{t+h} (s_1 - t)^2 dt, \tag{5.84c}$$

one may simplify Eq. (5.83) to:

$$var(\zeta) = h^3 - 2h\int_t^{t+h} (s_1 - t)\, dt + \int_t^{t+h} (s_1 - t)^2 dt = \frac{1}{3}h^3 \tag{5.85}$$

Covariance of ζ with $\left(\Delta_t B_j(h)\right)$ is given by:

$$cov\left(\zeta,\left(\Delta_t B_j(h)\right)\right) = E\left[\zeta\left(\Delta_t B_j(h)\right)\right]$$

$$= E\left[\left(h\left(\Delta_t B_j(h)\right) - \xi\right)\left(\Delta_t B_j(h)\right)\right] \quad (\text{ from Eq. 5.81})$$

$$= E\left[\left(h\left(\Delta_t B_j(h)\right)^2 - \left(\Delta_t B_j(h)\right)\xi\right)\right]$$

$$= h^2 - \int_t^{t+h} (s_1 - t)\, dt \; = \frac{h^2}{2} \quad (\text{from Eq. (5.84b)}) \tag{5.86}$$

Thus, knowing the moment characteristics of ζ, one may simulate the pair of correlated normal random variables ζ and $\left(\Delta_t B_j(h)\right)$ as follows:

$$\Delta_t B_j(h) = h^{\frac{1}{2}} Z_{j1}, \qquad \zeta = \frac{1}{2} h^{\frac{3}{2}}\left(Z_{j1} + \frac{1}{\sqrt{3}} Z_{j2}\right) = I_{j0}^{(1)}(h) \tag{5.87}$$

Z_{j1} and Z_{j2} are independent standard normal random variables that can be generated, for example, by the Box–Muller method (Chapter 2). Following Eq. (5.81), we also obtain ξ as a normal random variable given by:

$$\xi = h\left(\Delta_t B_j(h)\right) - \zeta = \frac{1}{2} h^{\frac{3}{2}}\left(Z_{j1} - \frac{1}{\sqrt{3}} Z_{j2}\right) = I_{0j}^{(1)}(h) \tag{5.88}$$

Brief notes on the evaluation of MSIs are provided in Appendix E. For more details, the reader may refer to Milstein [1995], and Kloeden and Platen [1992].

Example 5.4

We apply the higher order numerical schemes in Eq. (5.70) to two non-linear oscillators: (i) the non-linear Duffing oscillator in Example 5.2 driven by an additive noise and a deterministic harmonic load and (ii) the Duffing-Van der Pol oscillator (see Eq. 5.91 below) driven by a multiplicative noise.

Solution

Case (i): For the non-linear Duffing oscillator in Example 5.2, $m = 2$ and $n = 1$. The drift vector $\boldsymbol{a}(t, \boldsymbol{X}(t))$ and diffusion coefficient $\boldsymbol{\sigma}(t, \boldsymbol{X}(t))$ (presently a vector) are identified for the example problem as:

$$\boldsymbol{a}(t,\boldsymbol{X}(t)) = \begin{Bmatrix} a_1(t,\boldsymbol{X}(t)) \\ a_2(t,\boldsymbol{X}(t)) \end{Bmatrix} = \begin{Bmatrix} X_2(t) \\ -cX_2 - kX_1 - \upsilon X_1^3 + \mathcal{P}\cos 2\pi t \end{Bmatrix} \tag{5.89a}$$

$$\boldsymbol{\sigma}(t) = \begin{Bmatrix} \sigma_1(t) \\ \sigma_2(t) \end{Bmatrix} = \begin{Bmatrix} 0 \\ \sigma \end{Bmatrix} \tag{5.89b}$$

With $L_0 = \frac{\partial}{\partial t} + \sum_{j=1}^m a_j \frac{\partial}{\partial x_j} + \frac{1}{2}\sum_{i=1}^m \sum_{j=1}^m \sigma_i \sigma_j \frac{\partial^2}{\partial x_i \partial x_j}$, $L_{11} = \sum_{j=1}^m \sigma_j \frac{\partial}{\partial x_j}$, we have $L_0 a_1 = a_2$, $L_{11} a_1 = \sigma$, $L_0 a_2 = -ca_2 - ka_1 - 3\upsilon a_1 {x_1}^2$, $L_{11} a_2 = -c\sigma$, $L_0 \sigma = 0$, $L_{11}\sigma = 0$ and $L_{11}L_{11}\sigma = 0$. By the higher order scheme in Eq. (5.70d), the approximate numerical solution $\hat{\boldsymbol{X}}(t) = \left(\hat{X}_1(t), \hat{X}_2(t)\right)^T$ is obtained by the following truncated Ito–Taylor expansion:

$$\hat{X}_{1,i} = \hat{X}_{1,i-1} + h\hat{X}_{2,i-1} + \sigma I_{10}^{(1)}(h) + a_2 I_{00}^{(1)}(h) \tag{5.90a}$$

$$\begin{aligned} \hat{X}_{2,i} = \hat{X}_{2,i-1} &+ h\left(-c\hat{X}_{2,i-1} - k\hat{X}_{1,i-1} - \upsilon\left(\hat{X}_{1,i-1}\right)^3 + \mathcal{P}\cos 2\pi t_{i-1}\right) + \sigma\left(\Delta_t B_1(h)\right) \\ &- c\sigma I_{10}^{(1)}(h) + \left(-ca_2 - ka_1 - 3\upsilon a_1\left(\hat{X}_{1,i-1}\right)^2\right) I_{00}^{(1)}(h) \end{aligned} \tag{5.90b}$$

Also $I_{00}^{(1)}(h) = \frac{h^2}{2}$ and the MSIs $I_{10}^{(1)}(h)$ and $I_{01}^{(1)}(h)$ are determined by Eqs. (5.87) and (5.88). The solution to the oscillator by the EM method has earlier been found to diverge for a step size ≥ 0.05 s (see Fig. 5.5). Since the noise is additive, the Milstein method (Eq. 5.70b) has no added advantage over the EM method (as pointed out earlier). Here the solution is obtained by the higher order scheme in Eq. (5.70d) and as represented by Eqs. (5.90) with a step size of 0.05 s. The equations are solved by MC simulation with $N_e = 1000$. Figure 5.7 shows the sample mean solutions for displacement and velocity. The solution obtained by the explicit EM method with a step size of 0.0001 s (which was earlier taken as the reference solution $\boldsymbol{X}_{t_0,x_0}(t_k)$ of the dynamical system – see Fig. 5.5) is also included in Fig. 5.7.

Case (ii): A Duffing–Van der Pol oscillator is governed by the SDE [Kloeden and Platen 1992]:

$$dX_1 = X_2 dt \tag{5.91a}$$

$$dX_2 = \left(X_1\left(a - X_1^2\right) - X_2\right)dt + \sigma X_1 dB \tag{5.91b}$$

$a \in \mathbb{R}$ and B is a 1-dimensional standard Brownian motion. $\sigma \geq 0$ is the strength of the multiplicative noise.

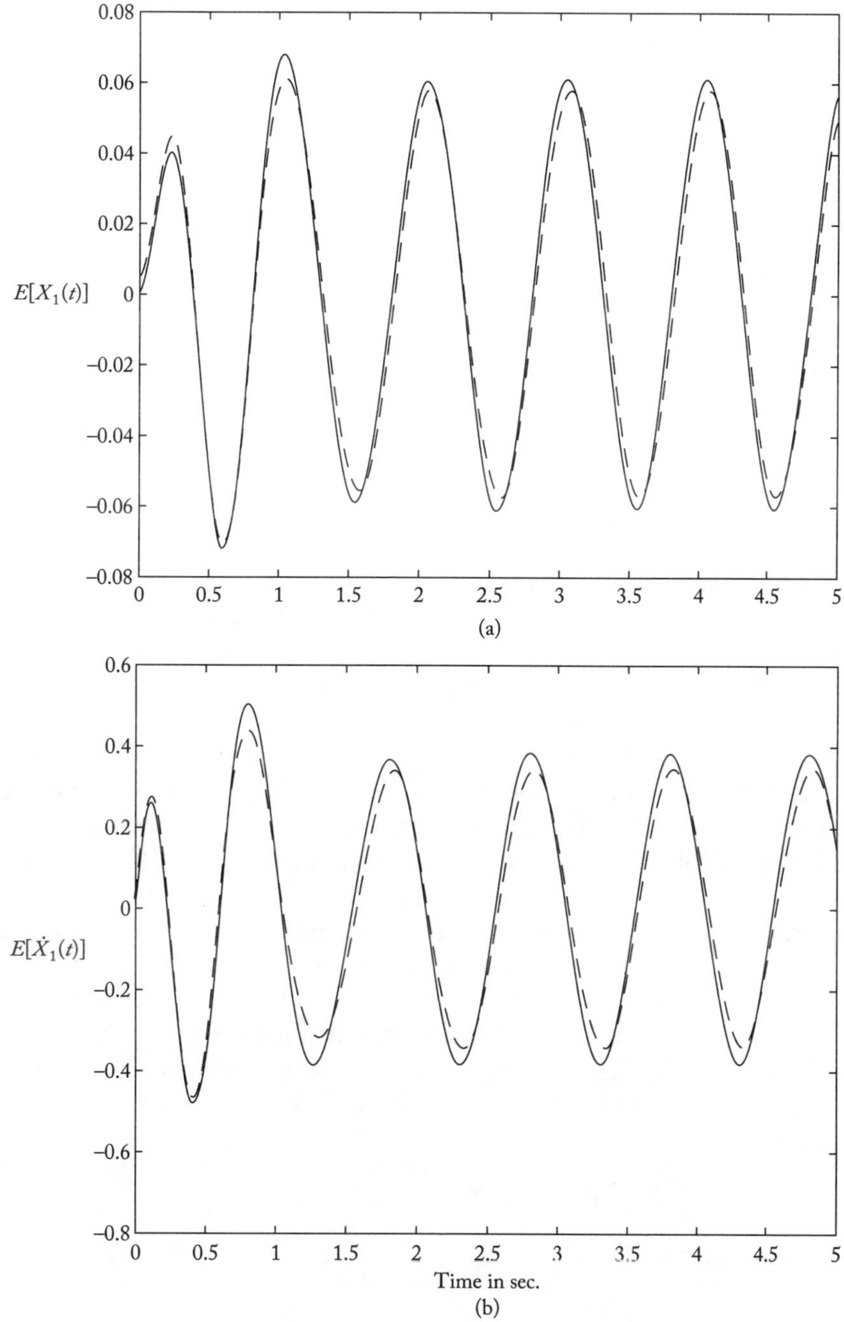

Fig. 5.7 Numerical solution to the non-linear Duffing oscillator of Example 5.2; $c = 4$, $k = 100$, $v = 100$, $\mathcal{P} = 4$, $\sigma = 0.01$, $N_e = 1000$, (a) sample mean displacement and (b) sample mean velocity; dashed-line – by higher order scheme (Eq. 5.70d) with $h = 0.01$ s and dark-line – by explicit EM method with $\Delta t = 0.0001$ s

In this case, the approximate numerical solution $\hat{\boldsymbol{X}}(t) = \left(\hat{X}_1(t), \hat{X}_2(t)\right)^T$ given by the higher order scheme in Eq. (5.70d) has the following form of time–discrete map:

$$\hat{X}_1(t_i) = \hat{X}_1(t_{i-1}) + h\hat{X}_2(t_{i-1}) + \sigma\hat{X}_1(t_{i-1}) I_{10}^{(1)}(h) + a_2 I_{00}^{(1)}(h) \tag{5.92a}$$

$$\begin{aligned}\hat{X}_2(t_i) = {} & \hat{X}_2(t_{i-1}) + ha_2 + \sigma\hat{X}_1(t_{i-1})(\Delta_t B_1(h)) \\ & -\sigma\hat{X}_1(t_{i-1}) I_{10}^{(1)}(h) + \sigma\hat{X}_2(t_{i-1}) I_{01}^{(1)}(h) \\ & +\left[a_1\left\{a - \left(\hat{X}_1(t_{i-1})\right)^2\right\} - a_2\right] I_{00}^{(1)}(h)\end{aligned} \tag{5.92b}$$

where, $a_1 = \hat{X}_2(t_{i-1})$ and $a_2 = \hat{X}_1(t_{i-1})\left(a - \left(\hat{X}_1(t_{i-1})\right)^2\right) - \hat{X}_2(t_{i-1})$. Figure 5.8 shows the trajectories in the phase plane staring with different initial conditions $x_0 \in (-4, 4)$ and $\dot{x} = 0$. The parameter $a = 1.0$ (not the diffusion coefficient) is kept constant in the simulations. Figure 5.8a shows the trajectories without noise. The two possible steady states $(\pm 1, 0)$ in the phase plane – the equilibrium points characterizing an unforced oscillator – are shown in the figure that have their specific regions of attraction in the initial condition space. The effect of noise is shown in Figs. 5.8b–e. Approximate solution by the higher order scheme (Eq. c) is included in the figures (Figs 5.8b–c). The solution by the higher order scheme (Eq. 5.70d) is assumed to be reliable with respect to the time step used $(h = 0.001)$. Noise with low intensity $(\sigma = 0.2)$ does induce stochasticity into the trajectories but they asymptotically approach one of the two attractors depending on the initial condition (Figs 5.8b and 5.8d). However, with higher noise level $(\sigma = 0.5)$ the trajectories may switch over from one steady state to the other with a random dwell time in each steady state (Figs 5.8c and 5.8e). This is perceivable from the long time behavior of the oscillator shown in Figs. 5.8f–g, where one sample each is obtained with two different initial conditions $(x_0 = -2,\ \dot{x}_0 = 0)$ and $(x_0 = -4,\ \dot{x}_0 = 0)$ by the higher order scheme in Eq. (5.70d).

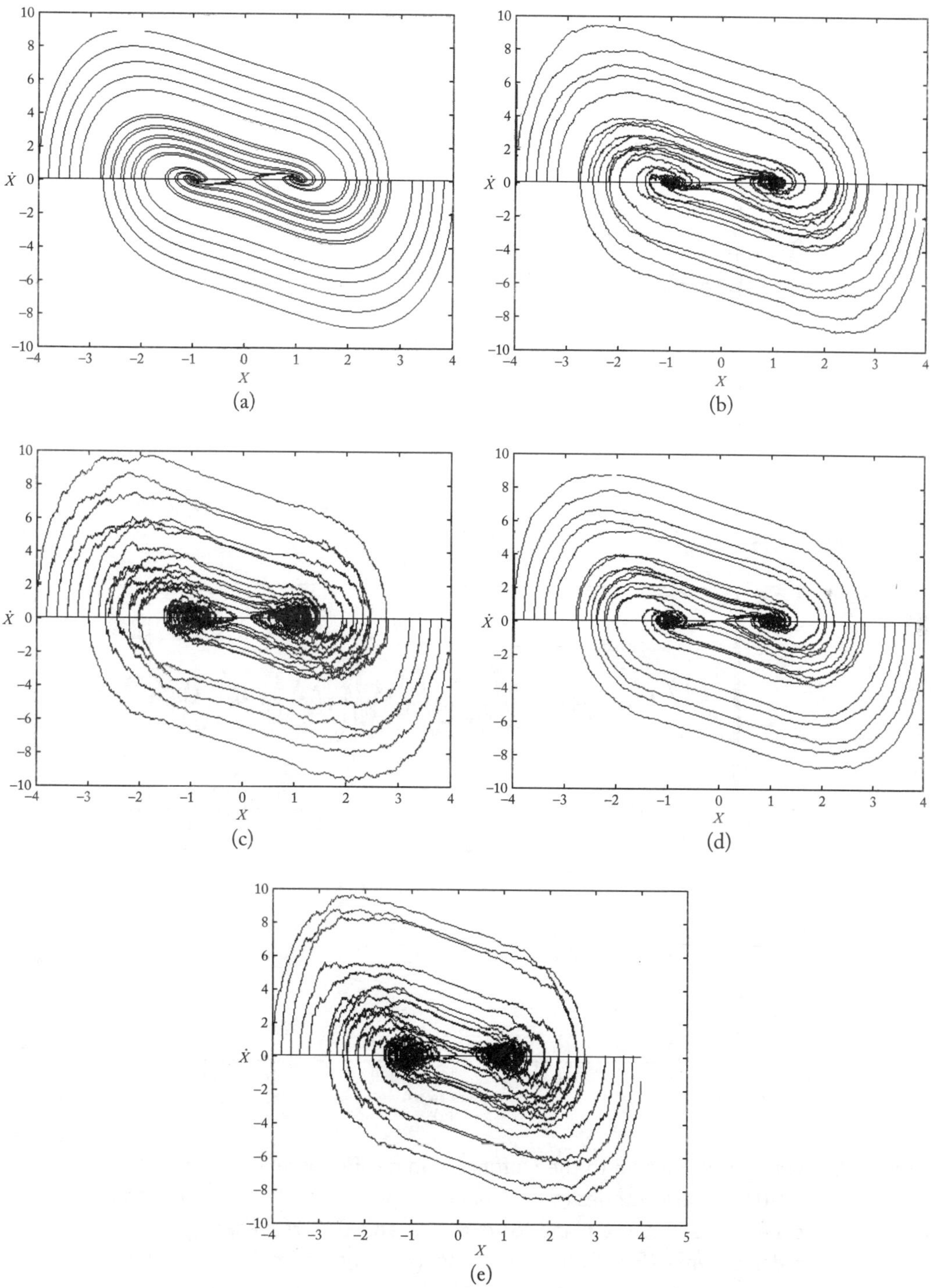
$\dot{X}$
X
(a)
(b)
(c)
(d)
(e)

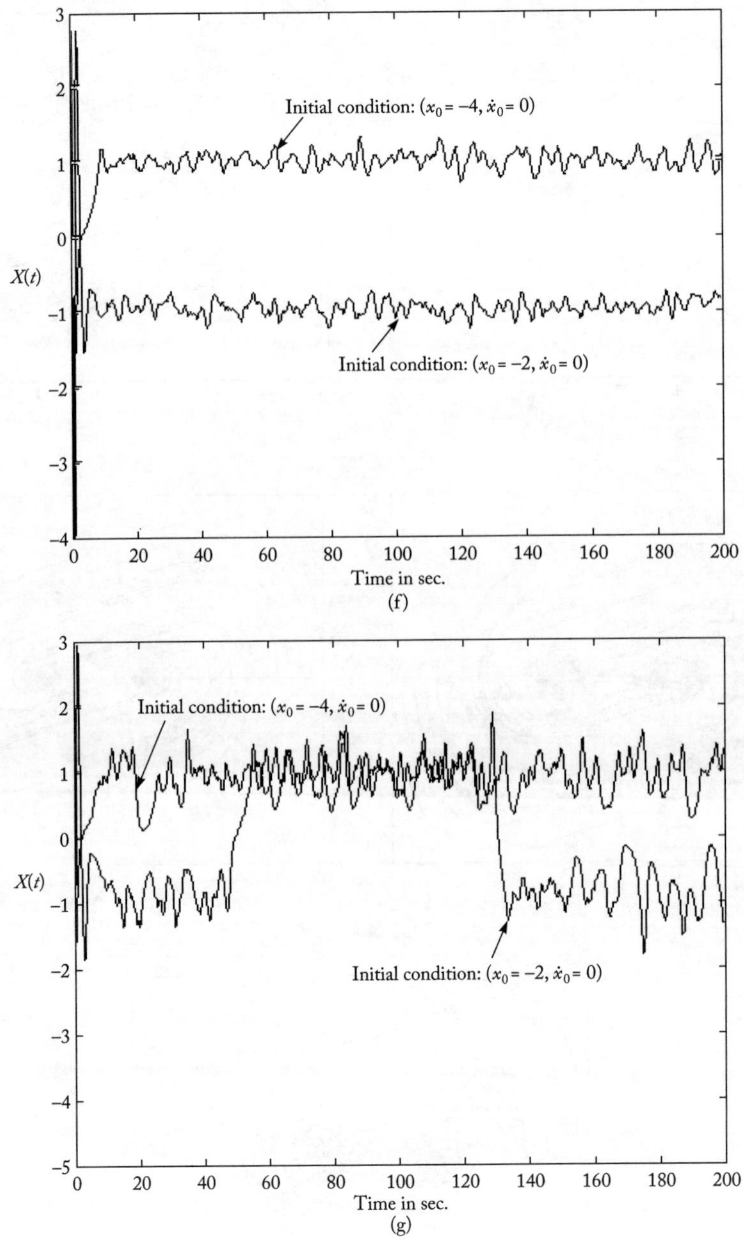

Fig. 5.8 Numerical solution to the Duffing–Van der Pol oscillator; $\alpha = 1.0$, $h = 0.001$; (a) trajectories in $x-\dot{x}$ plane without noise ($\sigma = 0$); (b) and (c) trajectories in $x-\dot{x}$ plane for $\sigma = 0.2$ and 0.5, respectively, by higher order scheme (Eq. 5.70c); (d) and (e) trajectories in $x-\dot{x}$ plane for $\sigma = 0.2$ and 0.5, respectively, by higher order scheme (Eq. 5.70d); and (f) and (g) long time trajectories by higher order scheme (Eq. 5.70d) with $\sigma = 0.2$ and 0.5, respectively

5.8 Stochastic Implicit Methods and Ito–Taylor Expansion

Consider the following Ito–Taylor expansion for $\boldsymbol{X}(t)$ of the m-dimensional SDE (5.1):

$$\boldsymbol{X}(t+h) = \boldsymbol{X}(t) + \int_t^{t+h} \boldsymbol{a}(s_1, \boldsymbol{X}(s_1))\, ds_1 + \sum_{k=1}^{n} \int_t^{t+h} \sigma_k(s_1, \boldsymbol{X}(s_1))\, dB_k(s_1) \tag{5.93}$$

Unlike in the deterministic case, one may introduce implicitness for both the drift and diffusion coefficients. To this end, we use Ito's formula in Eq. (5.61) to express $g(t, \boldsymbol{X}(t))$ in terms of $g(t+h, \boldsymbol{X}(t+h))$ as:

$$g(t, \boldsymbol{X}(t)) = g(t+h, \boldsymbol{X}(t+h)) - \int_t^{t+h} L_0 g(s_1, \boldsymbol{X}(s_1))\, ds_1 - \sum_{k=1}^{n} \int_t^{t+h} L_{1k} g(s_1, \boldsymbol{X}(s_1))\, dB_k(s_1) \tag{5.94}$$

with $t \le s_1 \le s = t+h$. Consequently, when the above implicit formula is used for both $\boldsymbol{a}(s_1, \boldsymbol{X}(s_1))$ and $\boldsymbol{\sigma}(s_1, \boldsymbol{X}(s_1))$ in Eq. (5.93), one has:

$$\begin{aligned}
\boldsymbol{X}(t+h) &= \boldsymbol{X}(t) + h\, \boldsymbol{a}(t+h, \boldsymbol{X}(t+h)) - \int_t^{t+h} \int_{s_1}^{t+h} L_0 \boldsymbol{a}(s_2, \boldsymbol{X}(s_2)) ds_2 ds_1 \\
&- \sum_{k=1}^{n} \int_t^{t+h} \int_{s_1}^{t+h} L_{1k} \boldsymbol{a}(s_2, \boldsymbol{X}(s_2)) dB_k(s_2) ds_1 \\
&+ \sum_{k=1}^{n} \left(\sigma_k(t+h, \boldsymbol{X}(t+h)) (\Delta_t B_k(h)) - \int_t^{t+h} \int_{s_1}^{t+h} L_0 \sigma_k(s_2, \boldsymbol{X}(s_2))\, ds_2 dB_k(s_1) \right) \\
&- \sum_{k=1}^{n} \sum_{j=1}^{n} \int_t^{t+h} \int_{s_1}^{t+h} L_{1j} \sigma_k(s_2, \boldsymbol{X}(s_2))\, dB_j(s_2) dB_k(s_1)
\end{aligned} \tag{5.95}$$

However, an attempt to apply the implicit formula, particularly to the diffusion coefficient, results in a solution which may grow without bounds [Milstein 1995, Burrage et al. 2004]. For examplc, consider the following scalar SDE with multiplicative noise:

$$dX = aXdt + \sigma X dB(t) \tag{5.96}$$

The implicit Ito–Taylor expansion in Eq. (5.95) gives:

$$X(t+h) = X(t) + haX(t+h) + \sigma X(t+h)(\Delta_t B(h)) + R \tag{5.97}$$

R is the remainder. The stochastic term containing $X(t+h)$ multiplied by the Wiener increment $\Delta_t B(h) = h^{\frac{1}{2}} \mathcal{N}(0,1)$ may assume large values during simulation depending on the generated normal random variable. To avoid such situations, it is necessary to be cautious in introducing implicitness in the expressions occurring in stochastic integrals. A more preferred option in the literature is thus to adopt semi-implicit methods wherein the drift coefficients are evaluated implicitly and the diffusion coefficients explicitly. The concept is elaborated in this section with prime focus on the construction of stochastic semi-implicit methods. In this context, it is interesting to derive a general two-parameter family of implicit methods similar to the ones that are well known [Newmark 1959, Hilber et al. 1977, Chung and Hulbert 1993, Roy and Rao 2012] in solving deterministic ODEs.

Now, referring back to the expansion in Eq. (5.93) and applying Ito's formula (Eq. 5.61) twice on $\boldsymbol{a}(s_1, \boldsymbol{X}(s_1))$, we obtain:

$$\begin{aligned}\boldsymbol{X}(t+h) &= \boldsymbol{X}(t) + h\boldsymbol{a}(t,\boldsymbol{X}(t)) + L_0\boldsymbol{a}(t,\boldsymbol{X}(t))\frac{h^2}{2} \\ &+ \sum_{k=1}^{n} L_{1k}\boldsymbol{a}(t,\boldsymbol{X}(t)) \int_t^{t+h}\int_t^{s_1} dB_k(s_2)ds_1 \\ &+ \sum_{k=1}^{n}\int_t^{t+h} \sigma_k(s_1,\boldsymbol{X}(s_1))\,dB_k(s_1) + \boldsymbol{R}_1 \end{aligned} \tag{5.98}$$

The remainder $\boldsymbol{R}_1$ is given by:

$$\begin{aligned}\boldsymbol{R}_1 &= \int_t^{t+h}\left(\int_t^{s_1}\left(\int_t^{s_2} L_0{}^2\boldsymbol{a}(s_3,\boldsymbol{X}(s_3))ds_3\right)ds_2\right)ds_1 \\ &+ \sum_{k=1}^{n}\int_t^{t+h}\left(\int_t^{s_1}\left(\int_t^{s_2} L_{1k}L_0\boldsymbol{a}(s_3,\boldsymbol{X}(s_3))dB_k(s_3)\right)ds_2\right)ds_1 \\ &+ \sum_{k=1}^{n}\int_t^{t+h}\left(\int_t^{s_1}\left(\int_t^{s_2} L_0L_{1k}\boldsymbol{a}(s_3,\boldsymbol{X}(s_3))ds_3\right) dB_k(s_2)\right)ds_1 \\ &+ \sum_{k=1}^{n}\sum_{j=1}^{n}\int_t^{t+h}\left(\int_t^{s_1}\left(\int_t^{s_2} L_{1j}L_{1k}\boldsymbol{a}(s_3,\boldsymbol{X}(s_3))dB_j(s_3)\right)dB_k(s_2)\right)ds_1 \end{aligned} \tag{5.99}$$

We introduce a real parameter α and write $\boldsymbol{a}(t,\boldsymbol{X}(t))$ as the sum:

$$\boldsymbol{a}(t,\boldsymbol{X}(t)) = \alpha\boldsymbol{a}(t,\boldsymbol{X}(t)) + (1-\alpha)\,\boldsymbol{a}(t,\boldsymbol{X}(t)) \tag{5.100}$$

Substituting the backward Ito–Taylor expansion in Eq. (5.94) for $\boldsymbol{a}(t,\boldsymbol{X}(t))$ in the second term of the above sum, one gets:

$$(1-\alpha)\boldsymbol{a}(t,\boldsymbol{X}(t)) = (1-\alpha)\left\{\begin{array}{c} \boldsymbol{a}(t+h,\boldsymbol{X}(t+h)) - \int_t^{t+h} L_0\boldsymbol{a}(s,\boldsymbol{X}(s))\,ds \\ -\sum_{k=1}^n \int_t^{t+h} L_{1k}\boldsymbol{a}(s,\boldsymbol{X}(s))\,dB_k(s) \end{array}\right\} \quad (5.101)$$

Applying (forward) Ito's formula on $L_0\boldsymbol{a}(s,\boldsymbol{X}(s))$ and $L_{1k}\boldsymbol{a}(s,\boldsymbol{X}(s))$ in the integrals of the second and third terms on the RHS of the above equation yields:

$$(1-\alpha)\boldsymbol{a}(t,\boldsymbol{X}(t)) = (1-\alpha)\left\{\begin{array}{c} \boldsymbol{a}(t+h,\boldsymbol{X}(t+h)) - hL_0\boldsymbol{a}(t,\boldsymbol{X}(t)) \\ -\sum_{k=1}^n L_{1k}\boldsymbol{a}(t,\boldsymbol{X}(t)) \int_t^{t+h} dB_k(s) + \boldsymbol{R}_2 \end{array}\right\} \quad (5.102)$$

The remainder $\boldsymbol{R}_2$ is given by:

$$\begin{aligned} \boldsymbol{R}_2 = &-\int_t^{t+h}\left(\int_t^{s_1} {L_0}^2\boldsymbol{a}(s_2,\boldsymbol{X}(s_2))\,ds_2\right)ds_1 \\ &-\sum_{k=1}^n \int_t^{t+h}\left(\int_t^{s_1} L_{1k}L_0\boldsymbol{a}(s_2,\boldsymbol{X}(s_2))\,dB_k(s_2)\right)ds_1 \\ &-\sum_{k=1}^n \int_t^{t+h}\left(\int_t^{s_1} L_0L_{1k}\boldsymbol{a}(s,\boldsymbol{X}(s))\,ds_2\right)dB_k(s_1) \\ &-\sum_{k=1}^n\sum_{j=1}^n \int_t^{t+h}\left(\int_t^{s_1} L_{1j}L_{1k}\boldsymbol{a}(s,\boldsymbol{X}(s))dB_j(s_2)\right)dB_k(s_1) \end{aligned} \quad (5.103)$$

Substituting the identity in Eq. (5.100) along with Eq. (5.102) for $\boldsymbol{a}(t,\boldsymbol{X}(t))$ in the second term on the RHS of Eq. (5.98), we obtain:

$$\begin{aligned} \boldsymbol{X}(t+h) = &\ \boldsymbol{X}(t) + \alpha h\boldsymbol{a}(t,\boldsymbol{X}(t)) + (1-\alpha)h\boldsymbol{a}(t+h,\boldsymbol{X}(t+h)) \\ &+(2\alpha-1)L_0\boldsymbol{a}(t,\boldsymbol{X}(t))\frac{h^2}{2} + \sum_{k=1}^n L_{1k}\boldsymbol{a}(t,\boldsymbol{X}(t))\int_t^{t+h}\int_t^{s_1} dB_k(s_2)ds_1 \\ &-(1-\alpha)h\sum_{k=1}^n L_{1k}\boldsymbol{a}(t,\boldsymbol{X}(t))\int_t^{t+h} dB_k(s) \end{aligned}$$

$$+\sum_{k=1}^{n}\int_{t}^{t+h}\sigma_k\left(s_1,\boldsymbol{X}\left(s_1\right)\right)dB_k\left(s_1\right)+\ \boldsymbol{R}_1+\left(1-\alpha\right)h\boldsymbol{R}_2 \tag{5.104}$$

Now a second parameter ß is introduced to express $L_0\boldsymbol{a}\left(t,\boldsymbol{X}\left(t\right)\right)$ in the fourth term on the RHS of the above equation as:

$$L_0\boldsymbol{a}\left(t,\boldsymbol{X}\left(t\right)\right)=ßL_0\boldsymbol{a}\left(t,\boldsymbol{X}\left(t\right)\right)+\left(1-ß\right)L_0\boldsymbol{a}\left(t,\boldsymbol{X}\left(t\right)\right) \tag{5.105}$$

$L_0\boldsymbol{a}\left(t,\boldsymbol{X}\left(t\right)\right)$ in the second term of the RHS above is expanded using backward Ito–Taylor expansion (Eq. 5.94). Substitution in Eq. (5.104) finally gives:

$$\begin{aligned}\boldsymbol{X}\left(t+h\right)&=\boldsymbol{X}\left(t\right)+\alpha h\boldsymbol{a}\left(t,\boldsymbol{X}\left(t\right)\right)+\left(1-\alpha\right)h\boldsymbol{a}\left(t+h,\boldsymbol{X}\left(t+h\right)\right)\\
&+\sum_{k=1}^{n}L_{1k}\boldsymbol{a}\left(t,\boldsymbol{X}\left(t\right)\right)\int_{t}^{t+h}\int_{t}^{s_1}dB_k(s_2)ds_1\\
&-\left(1-\alpha\right)h\sum_{k=1}^{n}L_{1k}\boldsymbol{a}\left(t,\boldsymbol{X}\left(t\right)\right)\int_{t}^{t+h}dB_k(s)\\
&+\left(2\alpha-1\right)ßL_0\boldsymbol{a}\left(t,\boldsymbol{X}\left(t\right)\right)\frac{h^2}{2}+\left(2\alpha-1\right)\left(1-ß\right)L_0\boldsymbol{a}\left(t+h,\boldsymbol{X}\left(t+h\right)\right)\frac{h^2}{2}\\
&+\sum_{k=1}^{n}\int_{t}^{t+h}\sigma_k\left(s_1,\boldsymbol{X}\left(s_1\right)\right)dB_k\left(s_1\right)\\
&+\boldsymbol{R}_1+\left(1-\alpha\right)h\boldsymbol{R}_2+\left(2\alpha-1\right)\left(1-ß\right)\frac{h^2}{2}\boldsymbol{R}_3\end{aligned} \tag{5.106}$$

The remainder term $\boldsymbol{R}_3$ arising out of a backward expansion of $L_0\boldsymbol{a}\left(t,\boldsymbol{X}\left(t\right)\right)$ is given by:

$$\boldsymbol{R}_3=-\int_{t}^{t+h}{L_0}^2\boldsymbol{a}\left(s_1,\boldsymbol{X}\left(s_1\right)\right)ds_1-\sum_{k=1}^{n}\int_{t}^{t+h}L_{1k}L_0\boldsymbol{a}\left(s_1,\boldsymbol{X}\left(s_1\right)\right)dB_k(s_1) \tag{5.107}$$

With $\boldsymbol{R}=\boldsymbol{R}_1+\left(1-\alpha\right)h\boldsymbol{R}_2+\left(2\alpha-1\right)\left(1-ß\right)\frac{h^2}{2}\boldsymbol{R}_3$, we find that $E\left[\boldsymbol{R}\right]=O(h^3)$ and $\left(E\left[\left(\boldsymbol{R}^2\right)\right]\right)^{\frac{1}{2}}=O(h^2)$. Thus discarding $\boldsymbol{R}$ from Eq. (5.106) yields a two-parameter family of implicit one step approximations with an order of accuracy equal to $3/2$.

5.8.1 Stochastic Newmark method—a two-parameter implicit scheme for mechanical oscillators

The deterministic Newmark integration method, a two-parameter implicit scheme, is well known [Belytschko and Hughes 1983, Miranda *et al.* 1989, Zienkiewicz and Taylor 1991, and Roy and Rao 2012] in solving linear as well as non-linear mechanical oscillators that govern the equations of motion of structural dynamical systems. The stochastic counterparts to this well-known integration method are developed in references [Roy and Dash 2002, Roy 2003] based on implicit Ito–Taylor expansions of the vector field. A natural starting point to describe the stochastic Newmark method is the following MDOF dynamical system:

$$\boldsymbol{M}\ddot{\boldsymbol{X}} + \boldsymbol{C}\big(\boldsymbol{X},\dot{\boldsymbol{X}}\big)\dot{\boldsymbol{X}} + \boldsymbol{K}\big(\boldsymbol{X},\dot{\boldsymbol{X}}\big)\boldsymbol{X} = \sum_{k=1}^{n} \boldsymbol{Q}_k(t,\boldsymbol{X},\dot{\boldsymbol{X}})\dot{B}_k + \mathcal{P}(t) \tag{5.108}$$

$\boldsymbol{X} = \left\{X_{1j} : j = 1,2,..,m\right\}^T \in \mathbb{R}^m$ is the displacement vector and $\dot{\boldsymbol{X}} = \left\{\dot{X}_{1j} : j = 1,2,..,m\right\}^T = \left\{X_{2j} : j = 1,2,..,m\right\}^T \in \mathbb{R}^m$, velocity vector. $\boldsymbol{M}$ is a constant mass matrix. $\boldsymbol{C}$ and $\boldsymbol{K}$ are $m \times m$ (state-dependent for non-linear systems) damping and stiffness matrices respectively. $\boldsymbol{Q}_k(t,\boldsymbol{X},\dot{\boldsymbol{X}})$is the kth column vector of $m \times n$ diffusion matrix and $\boldsymbol{B}(t)$ is an n-dimensional vector of independently evolving zero–mean Wiener processes with $B_k(0) = 0$ and $E\left[\left|B_k(t) - B_k(s)\right|^2\right] = t - s,\ t > s.$ $\mathcal{P}(t) = \{\mathcal{P}_j(t) : j = 1,2,\ldots,m\}$ is the external deterministic force vector. The dynamical system in the last equation may be more appropriately recast as the following system of $2m$ first order SDEs in an incremental form in the state space:

$$dX_{1j} = a_{1j}(t,\boldsymbol{X}_1,\boldsymbol{X}_2)\,dt \tag{5.109a}$$

$$dX_{2j} = a_{2j}(t,\boldsymbol{X}_1,\boldsymbol{X}_2)\,dt + \sum_{k=1}^{n} \overline{Q}_{jk}(t,\boldsymbol{X}_1,\boldsymbol{X}_2)dB_k(t)\ \ j = 1,2,...,m \tag{5.109b}$$

where, $\boldsymbol{X}_1 = \boldsymbol{X}$ and $\boldsymbol{X}_2 = \dot{\boldsymbol{X}} = \dot{\boldsymbol{X}}_1$. The drift vector and diffusion coefficient matrix are recognized from the above equation as:

$$\boldsymbol{a}_j(t,\boldsymbol{X}_1,\boldsymbol{X}_2) = \left\{\begin{array}{c} a_{1j}(t,\boldsymbol{X}_1,\boldsymbol{X}_2) \\ a_{2j}(t,\boldsymbol{X}_1,\boldsymbol{X}_2) \end{array}\right\}$$

$$= \left\{\begin{array}{c} X_{2j} \\ -\sum_{k=1}^{m}\overline{C}_{jk}(\boldsymbol{X}_1,\boldsymbol{X}_2)X_{2k} - \sum_{k=1}^{m}\overline{K}_{jk}(\boldsymbol{X}_1,\boldsymbol{X}_2)X_{1k} + \overline{\mathcal{P}_j}(t) \end{array}\right\},$$

$$j = 1,2,...,m \tag{5.110a}$$

$$\sigma_j\left(t,X_1,X_2\right)=\left\{\begin{array}{c}\sigma_{1j}\left(t,X_1,X_2\right)\\ \sigma_{2j}\left(t,X_1,X_2\right)\end{array}\right\}$$

$$=\left[\begin{array}{cccccc}0 & 0 & . & . & . & 0\\ \overline{Q}_{j1} & \overline{Q}_{j2} & . & . & . & \overline{Q}_{jn}\end{array}\right]_{2\times n},j=1,2,...,m \tag{5.110b}$$

$$\overline{C}=M^{-1}C,\overline{K}=M^{-1}K,\overline{\mathcal{P}}(t)=M^{-1}\mathcal{P}(t),\overline{Q}=M^{-1}Q \tag{5.110c}$$

As will be observed shortly, an explicit inversion of the matrix M may not be necessary. For existence and boundedness, it is assumed that a_j and σ_j are continuous and satisfy:

$$\left|a_j\left(t,\overline{X}\right)-a_j\left(t,\overline{Y}\right)\right|+\sum_{k=1}^{n}\left|Q_{jk}\left(t,\overline{X}\right)-Q_{jk}\left(t,\overline{Y}\right)\right|\leq K\left\|\overline{X}-\overline{Y}\right\|,\ \ \forall j\in[1,m] \tag{5.111}$$

Here $\overline{X},\overline{Y}\in\mathbb{R}^{2m}$ with overbars denoting $2m$ dimensional augmentation of solutions vectors that include both displacement and velocity vector components. With a partition Π_N and step size $h_i=t_i-t_{i-1}$, let $\overline{X}_{t_{i-1},\overline{x}}(t)$, $i\geq 0$, denote the $\mathcal{F}_t$–measurable solution of SDE (5.109) for $t\in(t_{i-1},t_i]$, starting from the initial condition $\overline{X}(t_{i-1})$. We assume that $E\left[\left|\overline{X}\left(t_0\right)\right|^2\right]<\infty$. It is now required to replace the non-linear system of SDEs by a suitable (stochastic) New mark map over the time interval $T_i=(t_{i-1},t_i]$ given the initial condition vector, $\overline{X}\left(t_{i-1}\right)$. For convenience of discussion, a uniform step size $h_i=h,\ \forall i$ is assumed.

Consider Eq. (5.109) and expand each element of the vectors $X_1\left(t_{i-1}+h\right)=X\left(t_{i-1}+h\right)$ and $X_2\left(t_{i-1}+h\right)=\dot{X}_1(t_i=t_{i-1}+h)$ in Ito–Taylor expansions around $X_1(t_{i-1})=X_{1,i-1}$ and $X_2(t_{i-1})=X_{2,i-1}$ respectively. Using Ito's formula in Eq. (5.68) and with $L_0X_{2j}=a_{2j}$ and $L_{1k}X_{2j}=\sigma_{2jk}$, the jth displacement component $X_{1j}(t_i)$ is expanded as:

$$X_{1j}\left(t_i\right)=X_{1j}\left(t_{i-1}\right)+\int_{t_{i-1}}^{t_i}X_{2j}\left(s_1\right)ds_1$$

$$=X_{1j}\left(t_{i-1}\right)+hX_{2j}\left(t_{i-1}\right)+\int_{t_{i-1}}^{t_i}\int_{t_{i-1}}^{s_1}a_{2j}\left(s_2,\overline{X}_{s_2}\right)ds_2ds_1$$

$$+\sum_{k=1}^{n}\int_{t_{i-1}}^{t_i}\int_{t_{i-1}}^{s_1}\sigma_{2jk}\left(s_2,\overline{X}_{s_2}\right)dB_k\left(s_2\right)ds_1 \tag{5.112}$$

Applying Ito's formula to a_{2j} and σ_{2jk} around $\left(t_{i-1}, \overline{\boldsymbol{X}}_{t_{i-1}}\right)$, one gets:

$$X_{1j}(t_i) = X_{1j}(t_{i-1}) + hX_{2j}(t_{i-1}) + \frac{h^2}{2} a_{2j}\left(t_{i-1}, \overline{\boldsymbol{X}}_{t_{i-1}}\right)$$

$$+\sum_{k=1}^{n} \sigma_{2jk}\left(t_{i-1}, \overline{\boldsymbol{X}}_{t_{i-1}}\right) \int_{t_{i-1}}^{t_i} \int_{t_{i-1}}^{s_1} dB_k(s_2)\, ds_1 + R_{1j} \tag{5.113}$$

R_{1j} is the remainder given by:

$$R_{1j} = \int_{t_{i-1}}^{t_i} \int_{t_{i-1}}^{s_1} \int_{t_{i-1}}^{s_2} L_0 a_{2j}\left(s_3, \overline{\boldsymbol{X}}_{s_3}\right) ds_3\, ds_2\, ds_1$$

$$+\sum_{k=1}^{n} \int_{t_{i-1}}^{t_i} \int_{t_{i-1}}^{s_1} \int_{t_{i-1}}^{s_2} L_{1k} a_{2j}\left(s_3, \overline{\boldsymbol{X}}_{s_3}\right) dB_k(s_3)\, ds_2\, ds_1$$

$$+\sum_{k=1}^{n} \int_{t_{i-1}}^{t_i} \int_{t_{i-1}}^{s_1} \int_{t_{i-1}}^{s_2} L_0 \sigma_{2jk}\left(s_3, \overline{\boldsymbol{X}}_{s_3}\right) ds_3\, dB_k(s_2)\, ds_1$$

$$+\sum_{k=1}^{n}\sum_{j=1}^{n} \int_{t_{i-1}}^{t_i} \int_{t_{i-1}}^{s_1} \int_{t_{i-1}}^{s_2} L_{1j} \sigma_{2jk}\left(s_3, \overline{\boldsymbol{X}}_{s_3}\right) dB_j(s_3)\, dB_k(s_2)\, ds_1 \tag{5.114}$$

Analogous to the deterministic Newmark technique, drift–implicitness is introduced in the expansion of $X_{1j}(t_i)$ by a real parameter α and this yields:

$$X_{1j}(t_i) = X_{1j}(t_{i-1}) + hX_{2j}(t_{i-1}) + \alpha\frac{h^2}{2} a_{2j}\left(t_{i-1}, \overline{\boldsymbol{X}}_{t_{i-1}}\right) + (1-\alpha)\frac{h^2}{2} a_{2j}\left(t_{i-1}, \overline{\boldsymbol{X}}_{i-1}\right)$$

$$+\sum_{k=1}^{n} \sigma_{2jk}\left(t_{i-1}, \overline{\boldsymbol{X}}_{t_{i-1}}\right) \int_{t_{i-1}}^{t_i} \int_{t_{i-1}}^{s_1} dB_k(s_2)\, ds_1 + R_{1j} \tag{5.115}$$

Writing $X_{1j}(t_i) := X_{1j,i}$, $X_{2j}(t_i) := X_{2j,i}$ etc., the fourth term on the RHS above is expanded by a backward Ito–Taylor expansion (see Eq. 5.94) giving:

$$X_{1j,i} = X_{1j,i-1} + hX_{2j,i-1} + \alpha\frac{h^2}{2} a_{2j}\left(t_{i-1}, \overline{\boldsymbol{X}}_{i-1}\right) + (1-\alpha)\frac{h^2}{2} a_{2j}\left(t_i, \overline{\boldsymbol{X}}_i\right)$$

$$+\sum_{k=1}^{n}\sigma_{2jk}\left(t_{i-1},\overline{X}_{i-1}\right)\int_{t_{i-1}}^{t_i}\int_{t_{i-1}}^{s_1}dB_k(s_2)ds_1+\left(R_{1j}-(1-\alpha)\frac{h^2}{2}R_{2j}\right)\quad(5.116)$$

The remainder R_{2j} is given by:

$$R_{2j}=\int_{t_{i-1}}^{t_i}L_0a_{2j}\left(s_1,\overline{X}_{s_1}\right)ds_1+\sum_{k=1}^{n}\int_{t_{i-1}}^{t_i}L_{1k}a_{2j}\left(s_1,\overline{X}_{s_1}\right)dB_k(s_1)\quad(5.117)$$

Similarly, an explicit Ito–Taylor expansion for the velocity component $X_{2j,i}$ gives:

$$X_{2j,i}=X_{2j,i-1}+\int_{t_{i-1}}^{t_i}a_{2j}\left(s_1,\overline{X}_{s_1}\right)ds_1+\sum_{k=1}^{n}\int_{t_{i-1}}^{t_i}\sigma_{2jk}\left(s_1,\overline{X}_{s_1}\right)dB_k(s_1)$$

$$=X_{2j,i-1}+ha_{2j}\left(t_{i-1},\overline{X}_{i-1}\right)+\sum_{k=1}^{n}\sigma_{2jk}\left(t_{i-1},\overline{X}_{i-1}\right)\int_{t_{i-1}}^{t_i}dB_k(s_1)+R_{3j}\quad(5.118)$$

R_{3j} is the remainder given by:

$$R_{3j}=\int_{t_{i-1}}^{t_i}\int_{t_{i-1}}^{s_1}L_0a_{2j}\left(s_2,\overline{X}_{s_2}\right)ds_2ds_1$$

$$+\sum_{k=1}^{n}\int_{t_{i-1}}^{t_i}\int_{t_{i-1}}^{s_1}L_{1k}a_{2j}\left(s_2,\overline{X}_{s_2}\right)dB_k(s_2)ds_1$$

$$+\sum_{k=1}^{n}\int_{t_{i-1}}^{t_i}\int_{t_{i-1}}^{s_1}L_0\sigma_{2jk}\left(s_2,\overline{X}_{s_2}\right)ds_2dB_k(s_1)$$

$$+\sum_{k=1}^{n}\sum_{j=1}^{n}\int_{t_{i-1}}^{t_i}\int_{t_{i-1}}^{s_1}L_{1j}\sigma_{2jk}\left(s_2,\overline{X}_{s_2}\right)dB_j(s_2)dB_k(s_1)\quad(5.119)$$

At this stage a second implicitness parameter $ß\in\mathbb{R}$ is introduced and $a_{2j}\left(t_{i-1},\overline{X}_{i-1}\right)$ in the second term on the RHS of Eq. (5.118) is expressed via the identity, $a_{2j}\left(t_{i-1},\overline{X}_{i-1}\right)=ßa_{2j}\left(t_{i-1},\overline{X}_{i-1}\right)+(1-ß)a_{2j}\left(t_{i-1},\overline{X}_{i-1}\right)$. In the second term of this sum, $a_{2j}\left(t_{i-1},\overline{X}_{i-1}\right)$ is expanded by a backward Ito–Taylor expansion and substituted in Eq. (5.118) to obtain:

$$X_{2j,i}=X_{2j,i-1}+ßha_{2j}\left(t_{i-1},\overline{X}_{i-1}\right)+(1-ß)ha_{2j}\left(t_i,\overline{X}_i\right)$$

$$+\sum_{k=1}^{n}\sigma_{2jk}\left(t_{i-1},\overline{X}_{i-1}\right)\int_{t_{i-1}}^{t_i} dB_k(s_1)+\left(R_{3j}-(1-ß)\,hR_{4j}\right) \tag{5.120}$$

Here the remainder term R_{4j} is given by:

$$R_{4j}=\int_{t_{i-1}}^{t_i} L_0 a_{2j}\left(s_1,\overline{X}_{s_1}\right)ds_1+\sum_{k=1}^{n}\int_{t_{i-1}}^{t_i} L_{1k}a_{2j}\left(s_1,\overline{X}_{s_1}\right)dB_k(s_1) \tag{5.121}$$

Thus a stochastic Newmark map for the jth scalar displacement X_{1j} is represented by Eq. (5.116) without the remainder terms. Similarly, a stochastic Newmark map for the jth scalar velocity X_{2j} is obtained via Eq. (5.120) whilst excluding the remainder terms. Pre-multiplying these equations on both sides by the mass matrix $\boldsymbol{M}$, one obtains the final forms of the Newmark map as:

$$\boldsymbol{M}\boldsymbol{X}_i=\boldsymbol{M}\boldsymbol{X}_{i-1}+h\boldsymbol{M}\dot{\boldsymbol{X}}_{i-1}-\alpha\frac{h^2}{2}\left(\boldsymbol{C}(\overline{\boldsymbol{X}}_{i-1})\dot{\boldsymbol{X}}_{i-1}+\boldsymbol{K}(\overline{\boldsymbol{X}}_{i-1})\boldsymbol{X}_{i-1}-\mathcal{P}_{i-1}\right)$$

$$+(1-\alpha)\frac{h^2}{2}\left(\boldsymbol{C}(\overline{\boldsymbol{X}}_i)\dot{\boldsymbol{X}}_i+\boldsymbol{K}(\overline{\boldsymbol{X}}_i)\boldsymbol{X}_i-\mathcal{P}_i\right)+\sum_{k=1}^{n}\boldsymbol{Q}_k\left(t_{i-1},\overline{\boldsymbol{X}}_{i-1}\right)I_{k0}^{(1)} \tag{5.122a}$$

$$\boldsymbol{M}\dot{\boldsymbol{X}}_i=\boldsymbol{M}\dot{\boldsymbol{X}}_{i-1}+ßh\left(\boldsymbol{C}(\overline{\boldsymbol{X}}_{i-1})\dot{\boldsymbol{X}}_{i-1}+\boldsymbol{K}(\overline{\boldsymbol{X}}_{i-1})\boldsymbol{X}_{i-1}-\mathcal{P}_{i-1}\right)$$

$$+(1-ß)h\left(\boldsymbol{C}(\overline{\boldsymbol{X}}_i)\dot{\boldsymbol{X}}_i+\boldsymbol{K}(\overline{\boldsymbol{X}}_i)\boldsymbol{X}_i-\mathcal{P}_i\right)+\sum_{k=1}^{n}\boldsymbol{Q}_k\left(t_{i-1},\overline{\boldsymbol{X}}_{i-1}\right)I_{k}^{(1)} \tag{5.122b}$$

Starting from any initial condition, $\{\overline{\boldsymbol{X}}_0\}=\{\boldsymbol{X}_0{}^T,\dot{\boldsymbol{X}}_0^T\}^T$, Eqs. (5.122) may be recursively solved for $\overline{\boldsymbol{X}}_i=\{\boldsymbol{X}_i{}^T,\dot{\boldsymbol{X}}_i^T\}^T$ at each time instant t_i, as a set of $2m$ coupled non-linear algebraic equations. The stochastic Newmark method is expected to have 'numerical stability' (no over-flows during computation due to large step sizes) owing to its implicit construction.

Error estimates

If the true solution of SDE (5.109) is represented by $\hat{\boldsymbol{X}}(t)$, the orders of the mean error $\left\|E\left[\hat{\boldsymbol{X}}-\overline{\boldsymbol{X}}\right]\right\|$ and mean square error $E\left[\left\|\hat{\boldsymbol{X}}-\overline{\boldsymbol{X}}\right\|^2\right]$ are determined by the first two statistical moments of the MSIs in the remainders in Eqs. (5.116) and (5.120). For the stochastic Newmark map for displacements, it is evident that $p_1=3$ and $p_2=2$ yielding the global error order $p=\frac{3}{2}$. Similarly for the velocity map (5.120), the local errors are obtained as

$p_1 = 2$ and $p_2 = 1$. Thus the global error order is $\frac{1}{2}$. Note that if the $O\left(h^2\right)$ term $U_1 := \alpha \frac{h^2}{2} a_{2j}\left(t_{i-1}, \overline{X}_{i-1}\right) + (1-\alpha)\frac{h^2}{2} a_{2j}\left(t_i, \overline{X}_i\right)$ is left out of the expansion for the displacement variable, p_1 and p_2 would both be 2. Since the inequality (Eq. 5.6a) $p_1 \geq p_2 + \frac{1}{2}$ is not satisfied, one must take $p_2 = \frac{3}{2}$ leading to an order of convergence equal to 1. In fact, the truncated expansion in Eq. (5.116) is not complete in $O\left(h^2\right)$.

If the term $U_2 := \sum_{k=1}^{n} \sum_{j=1}^{n} \int_{t_{i-1}}^{t_i} \int_{t_{i-1}}^{s_1} \int_{t_{i-1}}^{s_2} L_{1j} \sigma_{2jk}\left(s_3, \overline{X}_{s_3}\right) dB_j(s_3) dB_k(s_2) ds_1$ of R_{1j} (see Eq. 5.114) is also included in the expansion, one finds that $p_1 = 3$ and $p_2 = \frac{5}{2}$ yielding higher order of convergence equal to 2 but at the expense of an enhanced computational effort and the complexity of numerically treating higher order MSIs. If both the terms U_1 and U_2 are left out of the expansion, p would be 1 which is 0.5 less than that achieved while retaining U_1 alone which, unlike U_2, can easily be evaluated.

Example 5.5

Let us consider the Duffing oscillator in Example 5.2. The solution is now attempted by the stochastic Newmark method.

Solution

Referring to the oscillator Eq. (5.40), one has the drift and diffusion coefficients given by Eq. (5.89). With $m = 2$ and $n = 1$, the Newmark maps for displacement and velocity are respectively given by:

$$X_{1,i} = X_{1,i-1} + hX_{2,i-1} + \alpha \frac{h^2}{2}\left(-cX_{2,i-1} - kX_{1,i-1} - \upsilon\, X_{1,i-1}^3 + \mathcal{P}\cos 2\pi t_{i-1}\right)$$

$$+(1-\alpha)\frac{h^2}{2}\left(-cX_{2,i} - kX_{1,i} - \upsilon X_{1,i}^3 + \mathcal{P}\cos 2\pi t_i\right) + \sigma I_{10}^{(1)} \quad (5.123a)$$

$$X_{2,i} = X_{2,i-1} + ßh\left(-cX_{2,i-1} - kX_{1,i-1} - \upsilon\, X_{1,i-1}^3 + \mathcal{P}\cos 2\pi t_{i-1}\right)$$

$$+(1-ß)h\left(-cX_{2,i} - kX_{1,i} - \upsilon\, X_{1,i}^3 + \mathcal{P}\cos 2\pi t_i\right) + \sigma I_1^{(1)} \quad (5.123b)$$

The same data is assumed for the oscillator parameters as in Example 5.2, (i.e., $c = 4$, $k = 100$, $\upsilon = 100$, $\sigma = 0.01$ and $\mathcal{P} = 4$). Since the system is non-linear for any $\alpha, ß \neq 1$, Eqs. (5.123) constitute a set of coupled non-linear algebraic equations at each time step and may be solved via a standard Newton–Raphson method. With an ensemble of $N_e = 1000$, the equations are solved by MC simulation and Fig. 5.9 shows the sample mean solutions

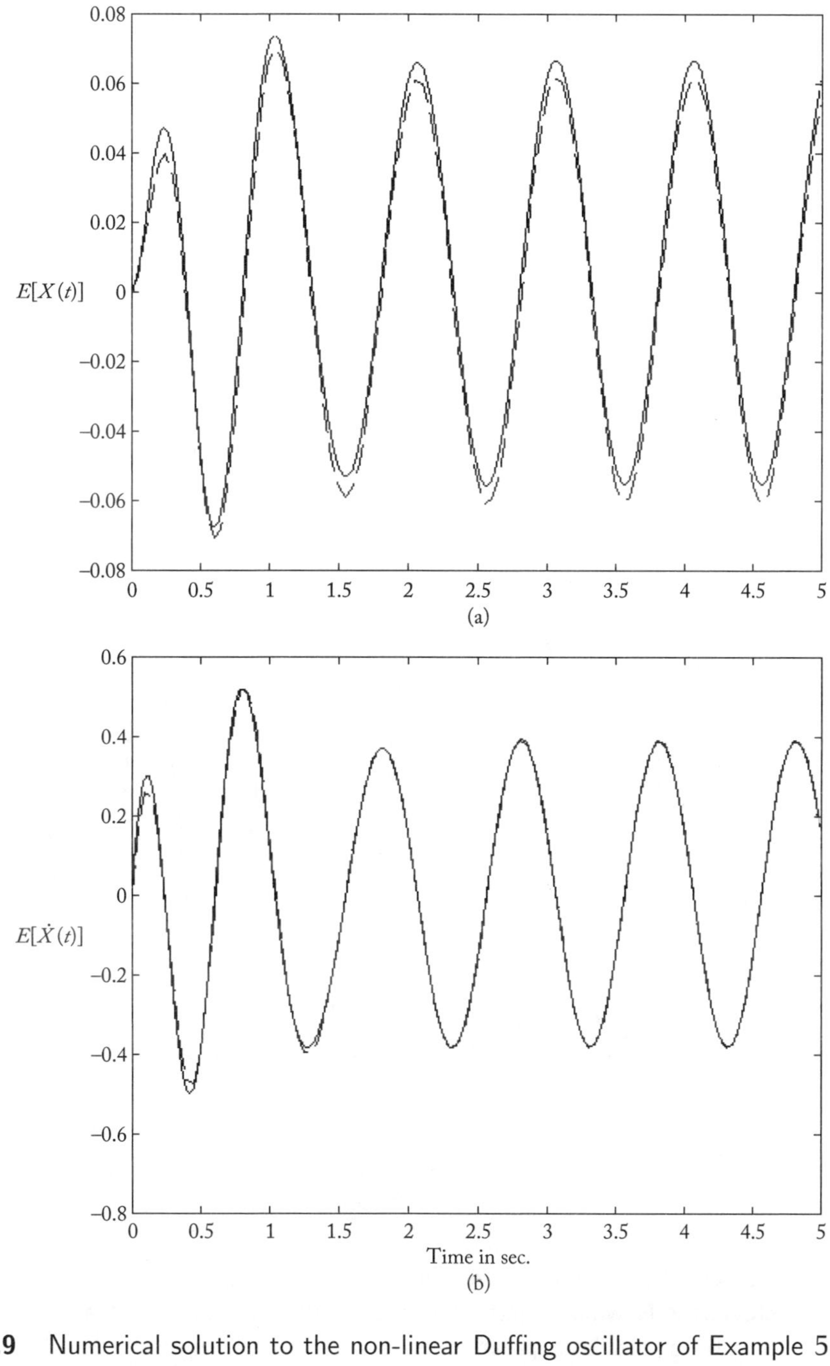

Fig. 5.9 Numerical solution to the non-linear Duffing oscillator of Example 5.2; $c = 4$, $k = 100$, $\upsilon = 100$, $\mathcal{P} = 4$, $\sigma = 0.01$ and $N_e = 1000$, (a) sample mean displacement and (b) sample mean velocity; dashed-line – solution by stochastic Newmark method with $h = 0.05$ s and $\alpha, ß = 0.5$; dark-line – reference solution by EM method with $h = 0.0001$ s

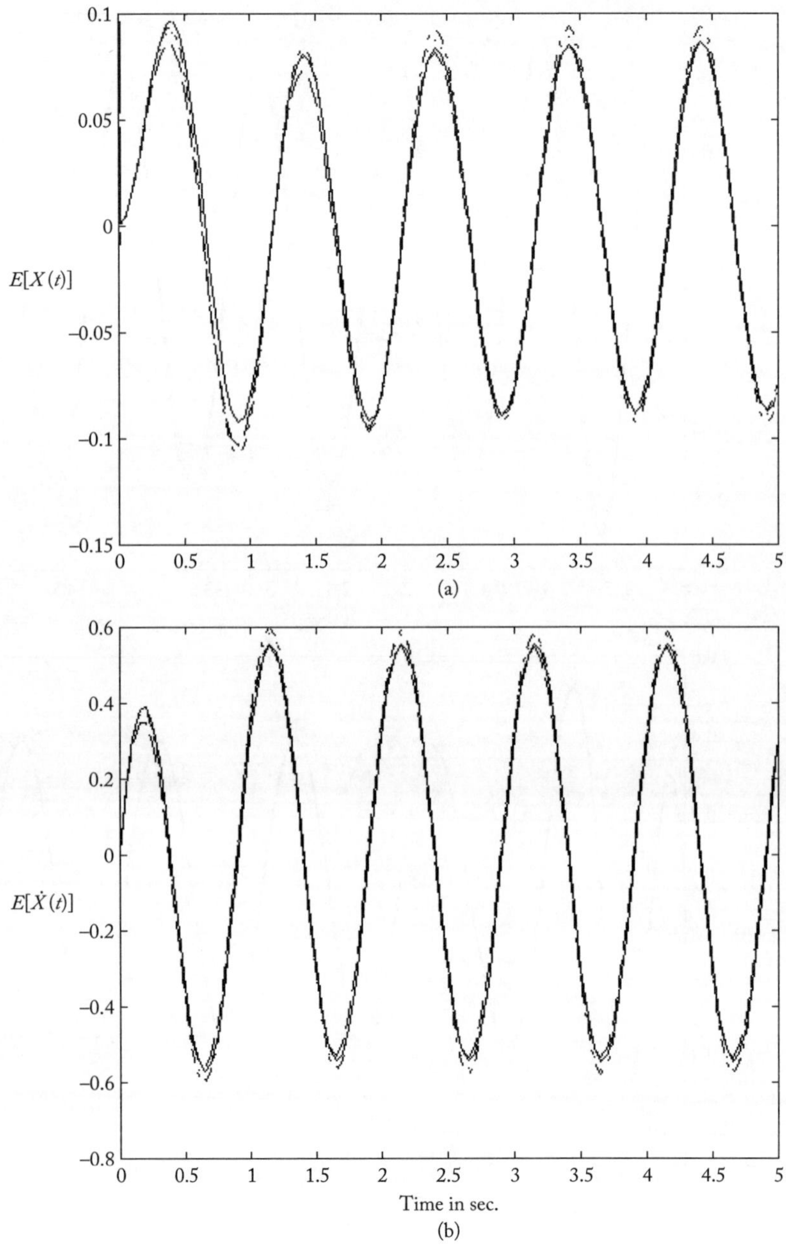

Fig. 5.10 Numerical solution to the non-linear Duffing oscillator of Example 5.2 by stochastic Newmark method; effect of different values of α and β with $h = 0.05$; oscillator parameters: $c = 4$, $k = 100$, $\upsilon = 100$, $\mathcal{P} = 4$, $\sigma = 0.01$, $N_e = 1000$, (a) sample mean displacement and (b) sample mean velocity; dark line – $\alpha = 0.5$, $\nu = 0.5$; dashed line – $\alpha = 0.5$, $ß = 0.25$; dotted-line – $\alpha = 0.25$, $ß = 0.5$; dash-dotted line – $\alpha = 0.75$, $ß = 0.5$

$E[X(t)]$ and $E\left[\dot{X}(t)\right]$ with $\alpha = ß = 0.5$. A time step size $h = 0.05$ is consistently adopted in the numerical results. To demonstrate the effect of the implicitness parameters α and ß, the displacement and velocity solutions are plotted in Fig. 5.10 for different choices of these parameters in $[0,1]$. The mean displacement and velocity trajectories of the present example problem do not sensitively depend on the choice of the parameters α and ß.

5.9 Weak One-step Approximate Solutions of SDEs

The strong and weak order of convergence of one-step approximations are distinguished by the respective requirements that the solution is close to the exact solution path-wise or in probability distribution. The main focus in weak approximations is thus on getting more information on the probability measure of $X(t)$ than on the sample trajectories. For a large dimensional and possibly non linear stochastic dynamical system under Gaussian white noise or filtered white noise, it is however difficult, if not impossible, to find the multi-dimensional joint transition probability distribution involving all the response variables. Nevertheless, for many problems of interest in science and engineering, it may often suffice to determine, with reasonable accuracy, the first few statistical moments of the response vectors. Obtaining weak approximate solutions (of an SDE) is thus the preferred option in such cases. These solutions are the truncated Ito–Taylor expansions derived as per the weak convergence criterion in Eq. (5.2b) as against the strong order of accuracy defined in Eq. (5.2a).

Let $X(t)$ and $\overline{X}(t)$ be respectively the exact and approximate numerical solutions to a 1-dimensional SDE. $\overline{X}(t)$ is said to be close to $X(t)$ in a weak sense, if $E\left[g(\overline{X}(t)\right]$ is close to $E[g(X(t)]$ for a reasonably large class of functions g. The weak approximations are basically marked by the following features.

a) Weak approximations with a desired order of accuracy involve fewer terms in the truncation compared to the corresponding strong schemes of the same order. For example, the EM method was numerically shown (in Example 5.1 of Section 5.2) to have a weak order of convergence of 1. The Milstein scheme (Eq. 5.70b) which has a strong order of convergence equal to 1 has additional terms over the EM method. These terms contain the integrals $I_{jk}^{(1)}$, $j,k = 1,2,..n$ whose evaluation needs extra computational effort.

b) Strong convergence implies weak convergence.

Suppose $\left(E\left[\left|\overline{X}(t) - X(t)\right|^2\right]\right)^{\frac{1}{2}} = O(h^{\beta})$; then it implies $E\left[g\left(\overline{X}(t)\right) - g(X(t))\right] = O(h^{\beta})$ for every function g which is Lipschitz. Since $E\left[\left|g\left(\overline{X}(t)\right)\right|\right] - E\left[\left|g(X(t))\right|\right] \leq \left(E\left[\left|g\left(\overline{X}(t)\right) - g(X(t))\right|\right]\right)$, the assertion follows.

Similar to the fundamental theorem on strong order of convergence (Section 5.2), one has the following theorem on weak approximations. Before stating the theorem, define a class of functions $\mathbb{F} = \{\mathbb{f}(\boldsymbol{x})\}$ such that there exist constants $K > 0, \kappa > 0$ satisfying the following inequality for all $\boldsymbol{x} \in \mathbb{R}^n$:

$$\left|\mathbb{f}(\boldsymbol{x})\right| \leq K(1 + |\boldsymbol{x}|^{\kappa}) \tag{5.124}$$

The function $\mathbb{f}$ may depend not only on $\boldsymbol{x}$ but also on time t. Then $\mathbb{f}(\boldsymbol{x})$ is said to belong to $\mathbb{F}$ with respect to $\boldsymbol{x}$, if an equality of the form in Eq. (5.124) holds uniformly in any time interval containing t.

5.9.1 Statement of the weak convergence theorem

With $\overline{\boldsymbol{X}}_0 = \boldsymbol{X}_0 = \boldsymbol{X}(t_0)$, consider the one-step approximation in the generic form Eq. (5.3) reproduced below:

$$\overline{\boldsymbol{X}}_{t_j,\overline{x}}\left(t_{j+1}\right) = \overline{\boldsymbol{X}}_{j+1} = \overline{\boldsymbol{x}} + g\left(t_j, \overline{\boldsymbol{x}}, h_j, \Delta \boldsymbol{B}_j\right),\ \overline{\boldsymbol{x}} = \overline{\boldsymbol{X}}_j(\omega),$$

$$\text{for a given } \omega,\ j = 0,1,\ldots \tag{5.125}$$

Define $\boldsymbol{\Delta} = \left\{\Delta^{(i_j)}\right\} := \boldsymbol{X} - \boldsymbol{x} = \boldsymbol{X}_{t,x}(t+h) - \boldsymbol{x}, \overline{\boldsymbol{\Delta}} = \left\{\overline{\Delta}^{(i_j)}\right\} := \overline{\boldsymbol{X}} - \boldsymbol{x} = \overline{\boldsymbol{X}}_{t,x}(t+h) - \boldsymbol{x}, i_j \in [1, m]$. More specifically $\Delta^{(i_j)}$ ($\overline{\Delta}^{(i_j)}$) denotes the $i_j{}^{\text{th}}$ component of the vector $\boldsymbol{\Delta}(\overline{\boldsymbol{\Delta}})$.

Suppose that

(a) the drift and diffusion coefficients are Lipschitz continuous and belong to $\mathbb{F}$ together with their partial derivatives of order up to $2\beta + 2$ (β is an integer) with respect to $\boldsymbol{x}$,

(b) the one-step approximation is such that:

$$\left|E\left[\prod_{j=1}^{\wp} \Delta^{(i_j)} - \prod_{j=1}^{\wp} \overline{\Delta}^{(i_j)}\right]\right| \leq K(\boldsymbol{x})\, h^{\beta+1}, \wp = 1, 2,$$

$$\ldots, 2\beta + 1, i_j \in [1, m]\ K(\boldsymbol{x}) \in \mathbb{F} \tag{5.126a}$$

$$E\left[\prod_{j=1}^{2\beta+2} \left|\overline{\Delta}^{(i_j)}\right|\right] \leq K(\boldsymbol{x})\, h^{\beta+1}, K(\boldsymbol{x}) \in \mathbb{F}, \tag{5.126b}$$

(c) for sufficiently large $L > 0, E\left|\overline{\boldsymbol{X}}_k\right|^{2L}, k = 0, 1, \ldots, n$ exist and are uniformly bounded, and

(d) a function $\mathbb{f}(\boldsymbol{x})$ together with partial derivatives up to order $2\beta + 2$ with respect to $\boldsymbol{x}$ belong to $\mathbb{F}$.

Then for all n and $k = 0, 1, \ldots, n$, the following inequality holds:

$$\left| E\left[\mathbb{f}(\boldsymbol{X}_{t_0,\boldsymbol{x}_0}(t_k)\right] - E\left[\mathbb{f}(\overline{\boldsymbol{X}}_{t_0,\boldsymbol{x}_0}(t_k)\right]\right| \leq K(\boldsymbol{x})h^{\beta} \tag{5.127}$$

The above implies that the one-step approximation has a weak order of accuracy β.

One may refer to Milstein [1995] for a proof of the theorem. We illustrate in the following an application of the theorem in deriving the weak order for a specific case of one-step approximation. To this end, we consider the following one-step approximation of the SDE (5.1):

$$\overline{\boldsymbol{X}}(t+h) = \overline{\boldsymbol{X}}(t) + \sum_{k=1}^{n} \sigma_k I_k^{(1)} + \boldsymbol{a} I_0^{(1)} + \sum_{k=1}^{n}\sum_{j=1}^{n} \left(L_{1j}\sigma_k\right) I_{jk}^{(1)}$$

$$+ \sum_{k=1}^{n} \left(L_0 \sigma_k\right) I_{0k}^{(1)} + \sum_{k=1}^{n} \left(L_{1k}\boldsymbol{a}\right) I_{k0}^{(1)} + \left(L_0 \boldsymbol{a}\right) I_{00}^{(1)} \tag{5.128}$$

Note that all the MSIs of the type $I_{j_1 j_2}^{(1)}$ (containing double stochastic integrals) are included in the above approximation. However, recall that if one were to include terms containing the integrals $I_{ijk}^{(1)}$ in the above approximation, the strong order of accuracy would have been $\frac{3}{2}$ (Eq. 5.70d). In contrast, the scheme in Eq. (5.128) yields a weak order of 2, even without the third level MSIs. This may be seen as follows. The remainder $\boldsymbol{R}$ for the scheme is (see Eq. 5.69):

$$\boldsymbol{R} = \sum_{k=1}^{n}\sum_{j=1}^{n}\sum_{i=1}^{n} L_{1i} L_{1j} \sigma_k I_{ijk}^{(1)}(h) + \boldsymbol{R}_1 \tag{5.129a}$$

$$\boldsymbol{R}_1 = \sum_{k=1}^{n}\sum_{j=1}^{n}\sum_{i=1}^{n}\sum_{l=1}^{n} I_{lijk}^{(2)}\left(L_{1l} L_{1i} L_{1j} \sigma_k, h\right)$$

$$+ \sum_{k=1}^{n}\sum_{j=1}^{n} I_{j0k}^{(2)}\left(L_{1j} L_0 \sigma_k, h\right) + \sum_{k=1}^{n}\sum_{j=1}^{n} I_{0jk}^{(2)}\left(L_0 L_{1j} \sigma_k, h\right)$$

$$+ \sum_{k=1}^{n}\sum_{j=1}^{n} I_{jk0}^{(2)}\left(L_{1j} L_{1k} \boldsymbol{a}, h\right) + \sum_{k=1}^{n} I_{0k0}^{(2)}\left(L_0 L_{1k} \boldsymbol{a}, h\right)$$

$$+\sum_{k=1}^{n} I_{k00}^{(2)}\left(L_{1k}L_0\boldsymbol{a},h\right)+\sum_{k=1}^{n} I_{00k}^{(2)}\left({L_0}^2\boldsymbol{\sigma}_k,h\right)$$

$$+\sum_{k=1}^{n}\sum_{j=1}^{n}\sum_{i=1}^{n} I_{0ijk}^{(2)}\left(L_0L_{1i}L_{1j}\boldsymbol{\sigma}_k,h\right)+I_{000}^{(2)}\left({L_0}^2\boldsymbol{a},h\right) \tag{5.129b}$$

Splitting the remainder $\boldsymbol{R}$ into $\boldsymbol{R}_1$ plus the term containing the MSI $I_{ijk}^{(1)}(h)$ is only meant for easy manipulation later while determining the weak order. Some intermediate results required for this exercise are first provided.

Proposition 1: Assuming that $\boldsymbol{a}$ and $\boldsymbol{\sigma}_k$ are Lipshitz continuous and belong to $\mathbb{F}$ together with their partial derivatives up to the sixth order with respect to $\boldsymbol{x}$, we show that the following inequalities hold for the approximation in Eq. (5.128):

$$\left|E\left[\boldsymbol{R}_1\right]\right| \le K(\boldsymbol{x})\,h^3,\ K(\boldsymbol{x})\in\mathbb{F} \tag{5.130a}$$

$$\left|E\left[\boldsymbol{R}_1^2\right]\right| \le K(\boldsymbol{x})\,h^4,\ K(\boldsymbol{x})\in\mathbb{F} \tag{5.130b}$$

$$\left|E\left[\boldsymbol{R}_1.I_k^{(1)}\right]\right| \le K(\boldsymbol{x})\,h^3,\ K(\boldsymbol{x})\in\mathbb{F} \tag{5.130c}$$

Proof: From the expression of $\boldsymbol{R}_1$ in Eq. (5.129), it follows that:

$$\left\|E\left[\boldsymbol{R}_1\right]\right\| \le \left\|E\left[I_{000}^{(2)}\left({L_0}^2\boldsymbol{a},h\right)\right]\right\| \tag{5.131}$$

Since ${L_0}^2\boldsymbol{a}\in\mathbb{F}$, one can find an even number $2L$ such that:

$$\left\|{L_0}^2\boldsymbol{a}\left(s,\boldsymbol{X}(s)\right)\right\| \le K\left(1+\left\|\boldsymbol{X}(s)\right\|^{2L}\right) \tag{5.132}$$

Substituting the above in Eq. (5.131) and noting that $E\left\|\boldsymbol{X}(s)\right\|^{2L}$ is bounded by a quantity $K\left(1+\|\boldsymbol{x}\|^{2L}\right)$, the first inequality in Eq. (5.130a) immediately follows. The terms in $\boldsymbol{R}_1$ containing the integrals of the type $I_{j0k}^{(2)}$, $I_{0jk}^{(2)}$, $I_{jk0}^{(2)}$ and $I_{ijkl}^{(2)}$ have the least order of $O(h^2)$. Thus the second inequality in Eq. (5.130b) follows by squaring $\boldsymbol{R}_1$, taking norm of each term and then the expectation. Proving the inequality in Eq. (5.130c) however requires some manipulations. Applying Ito's formula to the integrals in the first four terms of $\boldsymbol{R}_1$, one obtains terms of order two or higher with respect to h. Expectations of these terms multiplied by $I_k^{(1)}$ are zero. The remaining terms in $\boldsymbol{R}_1$ are at least of order of smallness $\frac{5}{2}$. Since $I_k^{(1)} = O(h^{\frac{1}{2}})$, expectation of the modulus of $I_k^{(1)}$ multiplying any of these terms can be shown to be $\leq K(\boldsymbol{x})h^3$ with $K(\boldsymbol{x}) \in \mathbb{F}$ by using the Cauchy–Schwarz inequality. Thus the inequality in Eq. (5.130c) holds.

Proposition 2: The following identities hold:

$$E\left[I_{ijk}^{(1)}\right] = 0,\; E\left[I_{ijk}^{(1)} I_r^{(1)}\right] = 0,\; E\left[I_{ijk}^{(1)} I_r^{(1)} I_s^{(1)}\right] = 0$$

and

$$E\left[I_{ijk}^{(1)} I_{rs}^{(1)}\right] = 0,\; i,j,k,r,s = 1,2,...,n \tag{5.133}$$

Proof: It is easy to see that the first, third and fourth identities are true because of an odd number of Weiner processes in the integral. The second identity is also satisfied for $r \neq i,\;\; r \neq j, r \neq k$. In the other cases we introduce a change of variables to prove the identity. To this end, let $t = 0$ without any loss of generality and:

$$dX(s) = B_i(s)dB_j(s), X(0) = 0 \tag{5.134a}$$

$$dY(s) = X(s)\,dB_k(s), Y(0) = 0 \tag{5.134b}$$

Then, $E\left[I_{ijk}^{(1)} I_r^{(1)}\right] = Y(h)\,B_r(h)$. Now if $r = i$, Ito's formula gives:

$$d(Y(s)\,B_r(s)) = B_r(s)\,X(s)\,dB_k(s) + Y(s)\,dB_r(s) + X(s)\,\delta_{ik} ds \tag{5.135}$$

δ_{ik} is the Kronecker delta. Taking expectations on both sides gives:

$$dE[Y(s)\,B_r(s)] = E[X(s)\,\delta_{ik} ds] \tag{5.136}$$

Since $E[X(s)] = 0$, we get $E[I_{ijk}^{(1)} I_r^{(1)}] = 0$. Other cases of $r = j$ and $r = k$ can be proved similarly.

With the help of the above two propositions, we may find that conditions in Eq. (5.126) of the weak convergence theorem are fulfilled, specifically with $\beta = 2$. Thus, for Lipschitz $\boldsymbol{a}$ and $\boldsymbol{\sigma}_k$, inequalities in Eq. (5.126) take the form:

$$\left\| E\left[\prod_{j=1}^{\mathcal{P}} \Delta^{(i_j)} - \prod_{j=1}^{\mathcal{P}} \overline{\Delta}^{(i_j)}\right]\right\| \leq K(\boldsymbol{x})\,h^3, i_j \in [1,m],$$
$$\mathcal{P} = 1,2,..,2\beta+1 = 5, K(\boldsymbol{x}) \in \mathbb{F} \tag{5.137}$$

and

$$E\left[\prod_{j=1}^{q}\left\|\overline{\Delta}^{(i_j)}\right\|\right] \leq K(x)\,h^3 \text{ with } q = 2\beta + 2 = 6 \tag{5.138}$$

Here, $\Delta = \overline{\Delta} + \boldsymbol{R}$. $E\left[\Delta - \overline{\Delta}\right] = E[\boldsymbol{R}] = E[\boldsymbol{R}_1]$. The result in Eq. (5.137) for $\wp = 1$ readily follows from Eq. (5.130a) of Proposition 1 above. With $\wp = 2$, $E[\Delta^{(i_j)^2} - \overline{\Delta}^{(i_j)^2}]$ contains terms with R_{1i_j} or $I_{ijk}^{(1)}$ as a factor. Terms with R_{1i_j} as a factor will have either $I_k^{(1)}$ or a term of order at least one as the other factor. In the former case, we utilize Eq. (5.130c) to find the order of the term as three. In the second case, any term of R_{1i_j} has order of at least two and the other factor has order of at least one. By the Cauchy–Schwarz inequality combined with Eq. (5.130b), one can again infer the order of the term to be at least three. Now, for terms with $I_{ijk}^{(1)}$ as a factor, one possibility is that they contain expressions of the form $I_{ijk}^{(1)} I_r^{(1)}$ or $I_{ijk}^{(1)} I_{rs}^{(1)}$ or $I_{ijk}^{(1)} h$. By the result in Proposition 2, expectations of such terms are zero. There are other terms containing $\left(I_{ijk}^{(1)}\right)^2$ with order three. Thus Eq. (5.137) is satisfied for $\wp = 2$. In a similar manner, the cases corresponding to $\wp = 3, 4, 5$ may be proved. The assertion in Eq. (5.138) easily follows from the fact that each term in $\overline{\Delta}$ is at least $O(h^{\frac{1}{2}})$.

We finally find that the conditions (a–d) of the weak convergence theorem are satisfied by the one-step approximation in Eq. (5.138) for $\beta = 2$. Thus the one-step approximation is proved to have an order of accuracy two in the weak sense.

5.9.2 Modelling of MSIs and construction of a weak one-step approximation

Another significant aspect of one-step weak approximations is the possibility for speedy numerical simulation of the MSIs by an alternate set of random variables. For Gaussian white noise inputs, the MSIs in the one-step approximations—strong or weak—are zero-mean Gaussian random variables. The difficulty in evaluating these MSIs during simulations is obvious and is more arduous in the case of higher order Taylor approximations. With the replacement of MSIs by the more easily evaluable random variables, we show below that the inequality in Eq. (5.127) still holds for the same β as obtained with original MSIs.

Referring to the one-step weak approximation in Eq. (5.128), just shown to be of order two, one can rewrite the equation by replacing the MSIs by an approximate set of random variables in the form:

$$\widetilde{\boldsymbol{X}}(t+h) = \widetilde{\boldsymbol{X}}(t) + \sum_{k=1}^{n} \sigma_k \xi_k h^{\frac{1}{2}} + \boldsymbol{a}h + \sum_{k=1}^{n}\sum_{j=1}^{n}\left(L_{1j}\sigma_k\right)\xi_{jk}h$$

$$+\sum_{k=1}^{n}(L_0\sigma_k)\xi_k h^{\frac{3}{2}}+\sum_{k=1}^{n}(L_{1k}\boldsymbol{a}-L_0\sigma_k)\zeta_k h^{\frac{3}{2}}+\frac{1}{2}(L_0\boldsymbol{a})h^2 \quad (5.139)$$

ξ_j, ξ_{jk} and ζ_k constitute a set of zero–mean random variables which may be non-uniquely established. The following proposition lists the conditions, to be satisfied by the moments of ξ_j, ξ_{jk} and ζ_k, so that the one-step approximation (Eq. 5.139) satisfies the weak convergence theorem for the same β as it does with the original MSIs.

Proposition 3: With Lipschitz boundedness of the drift and diffusion coefficients, assume that ξ_j, ξ_{jk} and ζ_k have finite moments up to order six (inclusive) and that the following conditions hold:

$$E\left[\xi_k h^{\frac{1}{2}}\right]=E\left[I_k\right]=0, E\left[\xi_{jk}h\right]=E\left[I_{jk}\right]=0, E\left[\zeta_k h^{\frac{3}{2}}\right]=E\left[I_{k0}\right]=0 \quad (5.140)$$

$$E\left[\xi_j\xi_k h\right]=E\left[I_j I_k\right]=\delta_{jk}h, E\left[\xi_i\xi_{jk}h^{\frac{3}{2}}\right]=E\left[I_i I_{jk}\right]=0 \quad (5.141)$$

$$E\left[\xi_j\zeta_k h^2\right]=E\left[I_j I_{k0}\right]=\delta_{jk}\frac{h^2}{2}, E\left[\xi_{ij}\xi_{kl}h^2\right]=E\left[I_{ij}I_{kl}\right]=\frac{h^2}{2}, \text{ if } i=k, j=l$$

$$=0, \text{ otherwise} \quad (5.142)$$

$$E\left[\xi_{ij}\zeta_k h^{\frac{5}{2}}\right]=E\left[I_{ij}I_{k0}\right]=0, E\left[\xi_i\xi_j\xi_k h^{\frac{3}{2}}\right]=E\left[I_i I_j I_k\right]=0 \quad (5.143)$$

$$E\left[\xi_i\xi_j\xi_{kl}h^2\right]=E\left[I_i I_j I_{kl}\right]=\frac{h^2}{2}, \text{ if } i=k, j=l \text{ or } i=l, j=k \text{ and } k\neq l$$

$$=h^2, \text{ if } i=j=k=l$$

$$=0, \text{ otherwise} \quad (5.144)$$

$$E\left[\xi_i\xi_j\zeta_k h^{\frac{5}{2}}\right]=E\left[I_i I_j I_{k0}\right]=0, E\left[\xi_i\xi_{jk}\xi_{lr}h^{\frac{5}{2}}\right]=E\left[I_i I_{jk} I_{lr}\right]=0 \quad (5.145)$$

$$E\left[\xi_i\xi_j\xi_k\xi_l h^2\right]=E\left[I_i I_j I_k I_l\right]=h^2 \text{ if } i,j,k,l \text{ form two pairs of equal numbers}$$

$$=3h^2 \text{ if } i=j=k=l$$

$$= 0, \quad \text{otherwise} \tag{5.146}$$

$$E\left[\xi_i \xi_j \xi_k \xi_{lr} h^{\frac{5}{2}}\right] = E\left[I_i I_j I_k I_{lr}\right] = 0, E\left[\xi_i \xi_j \xi_k \xi_l \xi_r h^{\frac{5}{2}}\right] = E\left[I_i I_j I_k I_l I_r\right] = 0 \tag{5.147}$$

Note that in the above statements, the superscript '(1)' in the MSIs is avoided for the sake of convenience.

Proof: If the inequalities in Eqs. (5.137) and (5.138) are rewritten so as to include the deviation, $\widetilde{\boldsymbol{\Delta}} = \widetilde{\boldsymbol{X}} - \boldsymbol{x}$, then one has:

$$\left\| E\left[\prod_{j=1}^{\mathscr{P}} \Delta^{(i_j)} - \prod_{j=1}^{\mathscr{P}} \widetilde{\Delta}^{(i_j)}\right]\right\| \le K(\boldsymbol{x}) h^3, i_j \in [1, m], \mathscr{P} = 1, 2, .., 5, K(\boldsymbol{x}) \in \mathbb{F} \tag{5.148}$$

Further, one has:

$$E\left[\prod_{j=1}^{q} \left\|\widetilde{\Delta}^{(i_j)}\right\|\right] \le K(\boldsymbol{x}) h^3 \text{ with } q = 6 \tag{5.149}$$

Now, the relations assumed in Proposition 3 among ξ_j, ξ_{jk} and ζ_k suffice for the conditions in Eqs. (5.148) and (5.149) to hold. It is simple to see that the condition in Eq. (5.148) is satisfied, if the left parts of the equalities in Eqs. (5.140)–(5.147) hold. Since the order of smallness of the terms in $\widetilde{\Delta}$ is at least $\frac{1}{2}$ with respect to h, the inequality in Eq. (5.149) also readily follows.

It remains to prove the right parts of Eqs. (5.140)–(5.147), which indeed provide the means to construct ξ_j, ξ_{jk} and ζ_k during simulation. Proof of the right part of (5.140) is direct. For $E\left[\xi_j \xi_k h\right] = \delta_{jk} h$ in (5.141), the result is straightforward since $E\left[I_j I_k\right] = h$ for $j = k$ and is zero otherwise. We find in general that most of these results on the random variables can be proved from the considerations of oddness in the number of Wiener processes and independence except in some special cases. For instance, to show $E\left[\xi_i \xi_{jk} h^{\frac{3}{2}}\right] = 0$ in the second part of Eq. (5.141), we use change of variables. That is, with $v_k = -B_k$, $k = 1, 2, .., n$, one has:

$$E\left[\xi_i \xi_{jk} h^{\frac{3}{2}}\right] = E\left[I_i I_{jk}\right] = E\left[\int_t^{t+h} dB_i(s) \int_t^{t+h} \int_t^{s_1} dB_j(s_1) dB_k(s)\right]$$

$$= -E\left[\int_t^{t+h} dv_i(s) \int_t^{t+h} \int_t^{s_1} dv_j(s_1) dv_k(s)\right] = -E\left[I_i I_{jk}\right] \tag{5.150}$$

which proves the result. For the right part of Eq. (5.142), note that $E\left[\xi_j\zeta_k h^2\right] = 0$ for $j \neq k$ because of the independence of $B_j(t)$ and $B_k(t)$. For $j = k$, one has:

$$\xi_j\zeta_k h^2 = I_j I_{j0} = \int_t^{t+h} dB_j(s) \int_t^{t+h}\int_t^{s_1} dB_j(s_1)ds \qquad (5.151)$$

Let $dX(s) = dB_j(s),\ dY(s) = X(s)ds$ with initial conditions $X(0) = Y(0) = 0$. Then by Ito's formula, $d\left(I_j I_{j0}\right) = d\left(X(s)Y(s)\right) = Y(s)\,dB_j(s) + X^2(s)\,ds$. Hence:

$$E\left[I_j I_{j0}\right] = \int_t^{t+h} X^2(s)\,ds = \int_t^{t+h} B_j^2(s)\,ds = \int_t^{t+h} s ds = \frac{h^2}{2} \qquad (5.152)$$

Similarly one can prove $E\left[\xi_{ij}\xi_{kl}h^2\right] = \delta_{ik}\delta_{jl}\frac{h^2}{2}$ by letting $dX(s) = B_i(s)\,dB_j(s)$, $dY(s) = B_k(s)\,dB_l(s)$ with initial conditions $X(0) = Y(0) = 0$, applying Ito's formula to $d\left(I_{ij}I_{kl}\right) = d(XY)$ and finally taking expectation. Here $t = 0$ is assumed without loss of generality. The right parts of Eq. (5.143) can be verified by a change of variables. To evaluate $E\left[\xi_i\xi_j\xi_{kl}h^2\right]$ in Eq. (5.144), we have:

$$E\left[\xi_i\xi_j\xi_{kl}h^2\right] = E\left[I_i I_j I_{kl}\right] = E\left[\int_0^h dB_i(s)\int_0^h dB_j(s)\int_0^h\int_0^{s_1} dB_k(s_2)dB_l(s_1)\right]$$

$$= E\left[B_i(h)\,B_j(h)\int_0^h B_k(s_1)dB_l(s_1)\right] \qquad (5.153)$$

Let $dY(s) = B_k(s)\ dB_l(s)$ with $Y(0) = 0$. By Ito's formula:

$$d\left(B_iB_jY\right) = B_iB_jB_k\ dB_l + B_iY dB_j + B_jY dB_i$$

$$+Y\delta_{ij}ds + B_iB_k\delta_{jl}ds + B_jB_k\delta_{il}ds$$

$$\Longrightarrow d\left(E\left[B_iB_jY\right]\right) = \delta_{jl}E\left[B_iB_k\right]ds + \delta_{il}E\left[B_jB_k\right]ds \qquad (5.154)$$

which yields the result.

Example 5.6

We examine the suitability of the following choice for the random variables ξ_j, ξ_{jk} and ζ_k in the weak one-step approximation of order two (Eq. 5.139). Symmetric independent random variable pairs $\varsigma_j, \xi_j, j = 1, 2, \ldots, n$ are chosen such that:

$$P\left(\xi_j = 0\right) = \frac{2}{3}, P\left(\xi_j = \pm\sqrt{3}\right) = \frac{1}{6} \tag{5.155a}$$

$$P\left(\varsigma_j = -1\right) = \frac{1}{2}, \mathrm{P}\left(\varsigma_j = 1\right) = \tfrac{1}{2} \tag{5.155b}$$

and

$$\zeta_k = \frac{1}{2}\xi_k \tag{5.156a}$$

$$\xi_{jk} = \frac{1}{2}\xi_j\xi_k - \frac{1}{2}r_{jk}\varsigma_j\varsigma_k, \text{ where } r_{jk} = 1 \quad \text{for } j \geq k \text{ and } = -1 \text{ for } j < k \tag{5.156b}$$

Solution

It is easy to verify the relations in Eqs. (5.140)–(5.147) for the present choice of the random variables, ξ_j, ξ_{jk} and ζ_k. Firstly, the expectation of the random variables is zero as required. Also $E\left[\xi_i^2\right] = 1$, $E\left[\xi_i^3\right] = E\left[\xi_i^5\right] = E\left[\varsigma_j\right] = E\left[\varsigma_i^3\right] = 0$, $E\left[\xi_i^4\right] = 3$ and $E\left[\varsigma_i^4\right] = 1$. It follows from independence of the random variables that $E\left[\xi_j\xi_k\right] = \delta_{jk}$, $E\left[\xi_i\xi_{jk}\right] = 0$ and $E\left[\xi_j\zeta_k\right] = \frac{1}{2}\delta_{jk}$. The identity $E\left[\xi_{ij}\xi_{kl}\right] = \frac{1}{2}$ only if $i = k, j = l$ follows from the following arguments. First note that:

$$E\left[\xi_{ij}\xi_{kl}\right] = \frac{1}{4}\Big(E\left[\xi_i\xi_j\xi_k\xi_l\right] - r_{ij}E\left[\varsigma_i\varsigma_j\xi_k\xi_l\right] - r_{kl}E\left[\varsigma_k\varsigma_l\xi_i\xi_j\right] + E\left[\varsigma_i\varsigma_j\varsigma_k\varsigma_l\right]\Big) \tag{5.157}$$

a) If $i = k$ and $j = l$, the result is true for all the possibilities of $i < j$ or $i > j$ or $i = j$. For example, in the first case ($i < j$), we have $k < l$ also. Hence $r_{ij} = r_{kl} = -1$, $E\left[\xi_i\xi_j\xi_k\xi_l\right] = E\left[\varsigma_i\varsigma_j\varsigma_k\varsigma_l\right] = 1$, $E\left[\varsigma_i\varsigma_j\xi_k\xi_l\right] = E\left[\varsigma_k\varsigma_l\xi_i\xi_j\right] = 0$ yielding $E\left[\xi_{ij}\xi_{kl}\right] = \frac{1}{2}$.

b) For all other cases, $E\left[\xi_{ij}\xi_{kl}\right]$ turns out to be zero. For example, let $i \neq k$. In case $j \neq i$ and $j \neq k$, then the expectations on the RHS of Eq. (5.157) are all zero yielding the result. The other possibilities may be $i \neq k$ along with $j = i$ or $j = k$. i) If $i \neq k$, $j = i$ and $k = l$, then $r_{ij} = r_{kl} = 1$ and all the expectations on the RHS are 1 giving the result. If $i \neq k$, $j = i$ and $k \neq l$, then each of the four expectations is zero and hence the result. ii) For $i \neq k$, $j = k$, one can similarly verify the result.

Example 5.7

Consider the Black–Scholes SDE $dX(t) = \lambda X dt + \mu X dB(t)$. We examine the strong and weak orders of convergence by applying the respective higher order (explicit) schemes in Eqs. (5.70d) and (5.128). Both the schemes involve MSIs.

Solution

For the SDE under consideration, $m = n = 1$, $a(X) = \lambda X$ and $\sigma(X) = \mu X$. Also, $L_0 a = \lambda^2 X, L_{11} a = \lambda\mu X$, $L_0\sigma = \lambda\mu X$, $L_{11}\sigma = \mu^2 X$, $L_{11}L_{11}\sigma = \mu^3 X$. In line with Eq. (5.70d), the higher order (strong) one-step approximation is given by:

$$X_i = X_{i-1} + \mu X_{i-1} I_1^{(1)} + \lambda X_{i-1} h + \mu^2 X_{i-1} I_{11}^{(1)} + \lambda\mu X_{i-1} I_{01}^{(1)} + \lambda\mu X_{i-1} I_{10}^{(1)}$$

$$+\mu^3 X_{i-1} I_{111}^{(1)} + \lambda^2 X_{i-1} \frac{h^2}{2} \tag{5.158a}$$

The MSIs $I_{11}^{(1)}, I_1^{(1)}, I_{01}^{(1)}$ and $I_{10}^{(1)}$ are respectively given by Eqs. (5.76), (5.80), (5.87) and (5.88). $I_{111}^{(1)}$ can be found (Appendix E) to be $\left(\frac{1}{6}\left(I_1^{(1)}\right)^2 - 3h\right) I_1^{(1)}$.

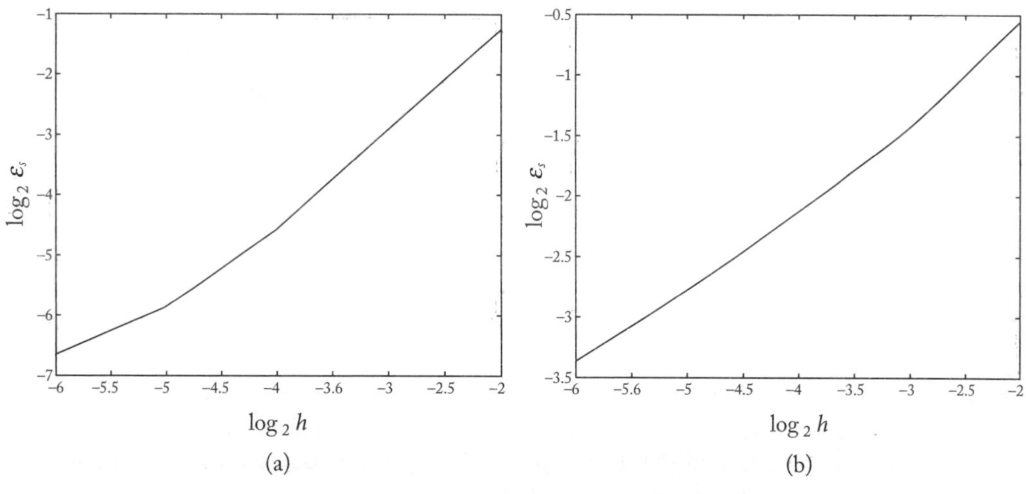

Fig. 5.11 Black–Scholes SDE; strong order of convergence; error plots from the higher order numerical scheme in Eq. (5.70d); (a) $\lambda - 2$, $\mu = 0.01$ and (b) $\lambda = 2$, $\mu = 0.1$, MC simulation with 20 iterations of $N_e = 100$, $X_0 = 1.0$

Figure 5.11 shows the numerical strong order of convergence. The weak order of convergence is similarly obtained for the SDE by the numerical scheme in Eq. (5.128) and shown in Fig. 5.12. This weak one-step approximation is the same as the one in Eq. (5.158a) above except that the term having the MSI $I_{111}^{(1)}$ is absent. $x_0 = 1.0$ is consistently adopted in the numerical simulations. Twenty different MC runs with an ensemble size of $N_e = 100$ are performed to estimate the global errors. Solutions are obtained for two sets of λ and μ. These are (i) $\lambda = 2, \mu = 0.01$ and (ii) $\lambda = 2, \mu = 0.1$. Figures 5.11 and 5.12 contain the sample–average plots (over 20 MC runs) of $\log_2 \varepsilon_s$ and $\log_2 \varepsilon_w$ versus $\log_2 h$ (ε_s and ε_w are defined in Eq. 5.2). Thus, the global errors for the jth MC run are obtained by computing $\varepsilon_s^{(j)} = \left(E\left[\left|\mathbf{X}_{t_0,x_0}(T) - \overline{\mathbf{X}}_{t_0,x_0}(T)\right|^2\right]\right)^{\frac{1}{2}}$ and $\varepsilon_w^{(j)} = \left|E\left[\mathbf{X}_{t_0,x_0}(T)\right] - E\left[\overline{\mathbf{X}}_{t_0,x_0}(T)\right]\right|$, $j = 1, 2, ..., 20$. The simulations are carried out for 5 different step sizes of $h = 2^{-6}$, 2^{-5}, 2^{-4}, 2^{-3} and 2^{-2} s. A linear fit to the mean graph by regression yields the strong order of convergence $p \cong 1.4$ (Fig. 5.11a) with $\mu = 0.01$ as against the theoretical estimate of 1.5. If the noise intensity is higher with $\mu = 0.1$, numerically estimated p reduces further to around 0.7 (Fig. 5.11b). A similar observation also holds for the weak convergence. It is found to be of order $\cong 1.9$ (Fig. 5.12a) with $\mu = 0.01$, close to the theoretical value of 2 and is $\cong 1.5$ with $\mu = 0.1$ (Fig. 5.12b).

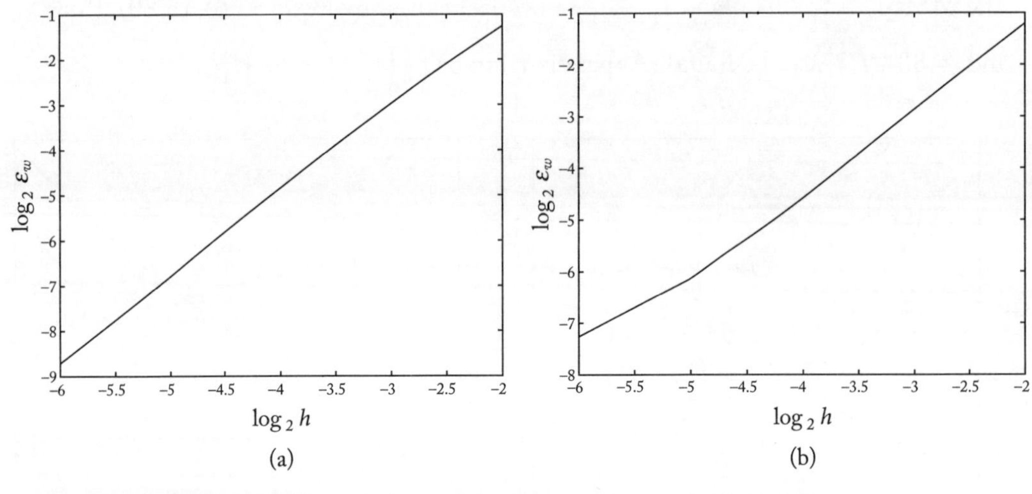

Fig. 5.12 Black–Scholes SDE; weak order of convergence; error plots from the higher order numerical scheme in Eq. (5.128a) with MSIs; (a) $\lambda = 2, \mu = 0.01$ and (b) $\lambda = 2, \mu = 0.1$; MC simulation with 20 iterations of $N_e = 100$ samples; $X_0 = 1.0$

Example 5.8

For the Black–Scholes SDE, we compare the weak numerical solutions obtained by the one-step approximations (i) with MSIs (Eq. 5.128) and (ii) equivalent random variables (Eq. 5.139).

Solution

MC simulations are performed with both the numerical schemes using the same initial condition ($x_0 = 1.0$). An ensemble size of $N_e = 100$ is used in approximating $E[X(t)]$. The time step is chosen as 2^{-7} s. For the Black–Scholes SDE, the one-step approximation in Eq. (5.139) takes the form:

$$X_i = X_{i-1} + \mu X_{i-1}\xi_1 h^{\frac{1}{2}} + \lambda X_{i-1}h + \mu^2 X_{i-1}\xi_{11}h + \lambda\mu X_{i-1}\left(\xi_1 - \zeta_1\right)h^{\frac{3}{2}}$$

$$+\lambda\mu X_{i-1}\zeta_1 h^{\frac{3}{2}} + \lambda^2 X_{i-1}\frac{h^2}{2} \quad (5.158b)$$

The random variables ξ_k and ς_k generated according to Eq. (5.155) are used in simulating ζ_k and ξ_{jk} via Eq. (5.156).

Figure 5.13 shows the solution to the SDE by the two numerical schemes. The true solution, simulated according to Eq. (5.35), is also included in the figure. The figure indicates a close match between the solutions obtained with MSIs and equivalent random variables.

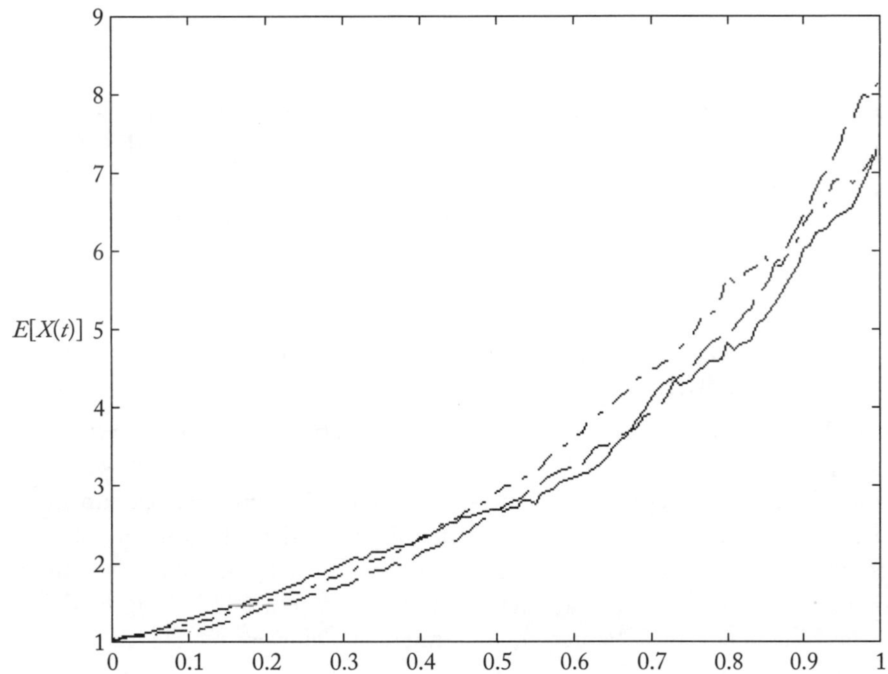

(a)

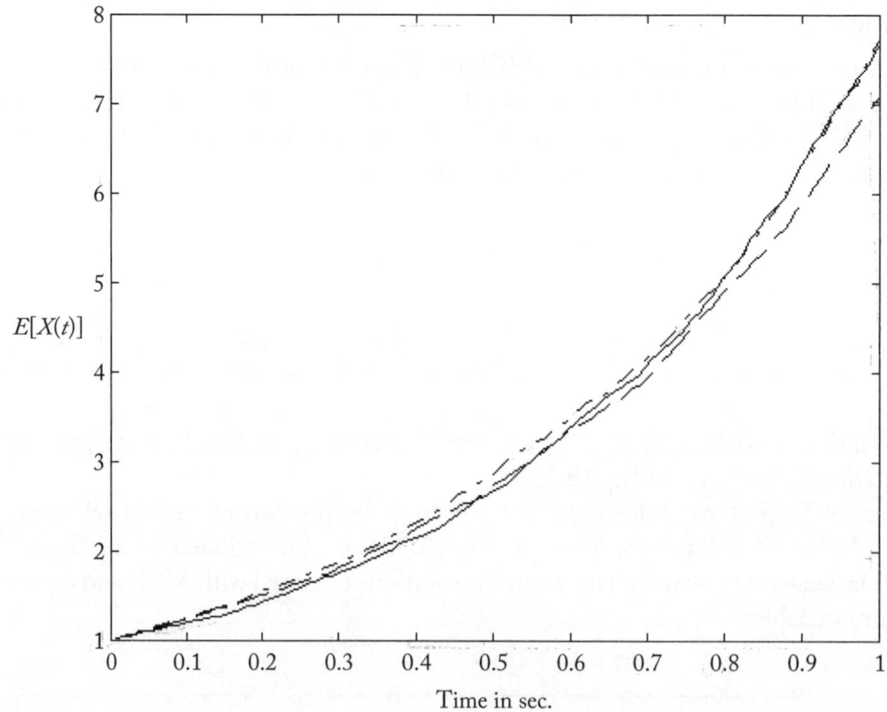

(b)

Fig. 5.13 Black–Scholes SDE; mean solution from weak one-step approximations with MSIs (Eq. 5.128) and equivalent random variables (Eq. 5.139); MC simulation with an ensemble size of $N_e = 100$, $x_0 = 1.0$ and $h = 2^{-7}$ s; (a) $\lambda = 2$, $\mu = 1.0$ and (b) $\lambda = 2, \mu = 0.5$; dark line – true solution, dashed-line—weak one-step solution with MSIs, dash-dot line – weak solution with equivalent random variables

5.9.3 Stochastic Newmark scheme using weak one-step approximation

Stochastic Newmark method has been presented in Section 5.8.1 with focus on obtaining strong sample path solutions of MDOF stochastic dynamical systems driven by (filtered) white noise inputs. As a computationally faster alternative, one may extend the implicit method to determine weak solutions in terms of the statistical moments of the system response vectors. In this context, one may note that closed form stationary solutions of the moment equations may be obtainable for linear systems [Lin 1967, Yang 1986]. Further, as illustrated in Chapter 4, a numerical solution to the moments in the non-stationary regime may also be possible for linear systems. One may also find a few efforts in the literature to derive approximate analytical solutions of non-linear stochastic dynamical systems. The related methods [Manohar 1995] include equivalent linearization [Spanos 1981, Roberts and Spanos 1990], perturbation [Crandall 1973], moment closure [Ibrahim 1978, Iyengar and Dash 1978], stochastic averaging [Iwan and Spanos

1978, Ibrahim 1985, Zhu 1988] and phase space linearization [Roy 2000a, 2000b]. But these techniques may be inapplicable to strongly non-linear systems especially those driven by multiplicative noise or of large dimensions. In view of this, numerical techniques based on direct MC simulation are often preferred to arrive at the required moment estimates. This in turn requires an accurate yet efficient direct integration of SDEs which is indeed possible via weak one-step approximations. Further, unlike the path-wise version of the Newmark method of Section 5.8.1, the weak adaptation of the method involves considerable simplification wherein the MSIs can be replaced by a set of random variables with much simplified (e.g., discrete) probability distributions.

Weak stochastic Newmark method (WSNM)

The dynamical system represented by the system of $2m$ first order SDEs in Eq. (5.109) written in the state space variables $\boldsymbol{X}, \dot{\boldsymbol{X}} \in \mathbb{R}^m$ is considered for describing the WSNM. The same mean square boundedness of the initial condition and the continuity and the Lipschitz growth conditions on the drift and diffusion coefficients are assumed. Denoting the full state vector as $\overline{\boldsymbol{X}}(t) = \left\{\boldsymbol{X}^T, \dot{\boldsymbol{X}}^T\right\}^T \in \mathbb{R}^{2m}$ and the local initial condition vector as $\overline{\boldsymbol{X}}(t_{i-1}) := \overline{\boldsymbol{X}}_{i-1}$, the Newmark map over the time interval $T_i = (t_{i-1}, t_i]$ is given by Eq. (5.122). The requirement is that the truncations in Eqs. (5.122a,b) corresponding to the displacement $\boldsymbol{X}(t_i)$ and velocity $\dot{\boldsymbol{X}}(t_i)$ (along with the associated remainders) satisfy the conditions of the weak convergence theorem. The arguments in support of this claim follow exactly in an analogous manner to those put forth for the one-step approximations (of weak solution) in Eqs. (5.128) and (5.139). While these are not repeated here, some elaboration may be needed on the error estimates of the solution vector, $\overline{\boldsymbol{X}}(t)$, particularly when the MSIs involved in the Eq. (5.122) are replaced by equivalent random variables.

Error estimates for WSNM

Consider the case with $\boldsymbol{M} = \boldsymbol{I}$, the identity matrix, in Eq. (5.122) without a loss of generality. Then, the solution maps are the same as those in Eqs. (5.116) and (5.120). Define $\boldsymbol{R}_d := \left(\boldsymbol{R}_1 - (1-\alpha)\frac{h^2}{2}\boldsymbol{R}_2\right)$ and $\boldsymbol{R}_v := (\boldsymbol{R}_3 - (1-ß)\,h\boldsymbol{R}_4)$ where the remainders $\boldsymbol{R}_1$ and $\boldsymbol{R}_2$ are given by Eqs. (5.114) and (5.117) and $\boldsymbol{R}_3$ and $\boldsymbol{R}_4$ are given by Eqs. (5.119) and (5.121), respectively. The subscripts '*d*' and '*v*' in $\boldsymbol{R}_d$ and $\boldsymbol{R}_v$ indicate that the remainders correspond respectively to the displacement and velocity maps. For the weak Newmark map expressed in terms of random variables (equivalent to MSIs), we denote the displacement and velocity components by $\widetilde{\boldsymbol{X}}_i = \left\{\widetilde{\boldsymbol{X}}_{1,i}^T, \widetilde{\boldsymbol{X}}_{2,i}^T\right\}^T$, and the numerical solution is given by:

$$\tilde{X}_{1j,i} = \tilde{X}_{1j,i-1} + h\tilde{X}_{2j,i-1} + \alpha\frac{h^2}{2}a_{2j}\left(t_{i-1}, \widetilde{\boldsymbol{X}}_{i-1}\right) + (1-\alpha)\frac{h^2}{2}a_{2j}\left(t_i, \widetilde{\boldsymbol{X}}_i\right)$$

$$+ \sum_{k=1}^{n} \sigma_{2,jk}\left(t_{i-1}, \widetilde{\boldsymbol{X}}_{i-1}\right)\zeta_k h^{\frac{3}{2}} \tag{5.159}$$

$$\tilde{X}_{2j,i} = \tilde{X}_{2j,i-1} + ßha_{2j}\left(t_{i-1}, \widetilde{\boldsymbol{X}}_{i-1}\right) + (1-ß)ha_{2j}\left(t_i, \widetilde{\boldsymbol{X}}_i\right)$$

$$+\sum_{k=1}^{n} \sigma_{2,jk}\left(t_{i-1}, \widetilde{\boldsymbol{X}}_{i-1}\right)\xi_k h^{\frac{1}{2}} \tag{5.160}$$

ξ_k and ζ_k are the zero mean random variables equivalent to $I_k^{(1)}$ and $I_{k0}^{(1)}$ respectively and are modelled according to Eqs. (5.155a) and (5.156a). Specific to the WSNM, the inequalities in Proposition 1 (Eq. 5.130) take the following form:

For displacement:

$$\left|E\left[\boldsymbol{R}_d\right]\right| \leq K(\boldsymbol{x})\,h^3,\; K(\boldsymbol{x}) \in \mathbb{F} \tag{5.161a}$$

$$\left|E\left[\boldsymbol{R}_d^2\right]\right| \leq K(\boldsymbol{x})\,h^4,\; K(\boldsymbol{x}) \in \mathbb{F} \tag{5.161b}$$

$$\left|E\left[I_k^{(1)}\boldsymbol{R}_d\right]\right| \leq K(\boldsymbol{x})\,h^3,\; K(\boldsymbol{x}) \in \mathbb{F} \tag{5.161c}$$

For velocity:

$$\left|E\left[\boldsymbol{R}_v\right]\right| \leq K(\boldsymbol{x})\,h^2,\; K(\boldsymbol{x}) \in \mathbb{F} \tag{5.162a}$$

$$\left|E\left[\boldsymbol{R}_v^2\right]\right| \leq K(\boldsymbol{x})\,h^2,\; K(\boldsymbol{x}) \in \mathbb{F} \tag{5.162b}$$

$$\left|E\left[I_k^{(1)}\boldsymbol{R}_v\right]\right| \leq K(\boldsymbol{x})\,h^2,\; K(\boldsymbol{x}) \in \mathbb{F} \tag{5.162c}$$

Knowing the expressions for $\boldsymbol{R}_d$ and $\boldsymbol{R}_v$ and following the arguments in the proof of Proposition 1 of Section 5.9.1, one can easily establish the above inequalities for both the displacement and velocity vectors. Consequently,the following inequalities (see Eqs. (5.126a,b) in the weak convergence theorem) are satisfied by the weak Newmark map for the displacement and velocity vectors ($i_j \in [1,m]$):

$$\left\|E\left[\prod_{j=1}^{p}\Delta_d{}^{(i_j)} - \prod_{j=1}^{p}\widetilde{\Delta}_d^{(i_j)}\right]\right\| \leq K(\boldsymbol{x})\,h^3, p = 1,2,..,5, K(\boldsymbol{x}) \in \mathbb{F} \tag{5.163a}$$

$$\left\| E\left[\prod_{j=1}^{p} \Delta_v{}^{(i_j)} - \prod_{j=1}^{p} \widetilde{\Delta}_v^{(i_j)}\right]\right\| \leq K(\boldsymbol{x})\, h^2, p = 1,2,..,5, K(\boldsymbol{x}) \in \mathbb{F} \tag{5.163b}$$

and

$$E\left[\prod_{j=1}^{q} \left\|\widetilde{\Delta}_d^{(i_j)}\right\|\right] \leq K(\boldsymbol{x})\, h^6 \text{ with } q = 6 \tag{5.164a}$$

$$E\left[\prod_{j=1}^{q} \left\|\widetilde{\Delta}_v^{(i_j)}\right\|\right] \leq K(\boldsymbol{x})\, h^3 \text{ with } q = 6 \tag{5.164b}$$

Here $\boldsymbol{\Delta}_d = \boldsymbol{X}_d - \boldsymbol{x}_d$ and $\boldsymbol{\Delta}_v = \boldsymbol{X}_v - \boldsymbol{x}_v$ denote the increments in the exact solutions and $\widetilde{\boldsymbol{\Delta}}_d = \widetilde{\boldsymbol{X}}_d - \boldsymbol{x}_d$ and $\widetilde{\boldsymbol{\Delta}}_v = \widetilde{\boldsymbol{X}}_v - \boldsymbol{x}_v$ the increments in the weak approximate solutions of displacement and velocity vectors. Here $\boldsymbol{X}_d$ and $\boldsymbol{X}_v$ more precisely stand for $(\boldsymbol{X}_d)_{t,\boldsymbol{x}_d}(t+h)$ and $(\boldsymbol{X}_v)_{t,\boldsymbol{x}_v}(t+h)$ etc. The inequalities in Eq. (5.163) follow from the proofs of Proposition 2 of Section 5.9.1 and Proposition 3 of Section 5.9.2. Underlying this assertion lies the statistical equivalence of the MSIs $I_k^{(1)}$ and $I_{k0}^{(1)}$ with $\xi_k$and ζ_k. The first part of Eq. (5.164) is proved from the fact that each term of $\widetilde{\boldsymbol{\Delta}}_d^{(i_j)}$ (see Eq. 5.159) is at least $O(h)$. The second part pertaining to the velocity solution follows from Eq. (5.160) wherein each term of $\widetilde{\boldsymbol{\Delta}}_v^{(i_j)}$ is at least $O(h^{\frac{1}{2}})$. Based on the preceding discussion, we have the following assertion for the order of accuracy of WSNM:

Let the conditions in Eqs. (5.161 – 5.162) and (5.163 – 5.164) hold and $g(\overline{\boldsymbol{X}}) \in \mathbb{F}$ be a function of displacement and velocity vectors such that all its partial derivatives up to order 6 exist and belong to the class $\mathbb{F}$ as well. In case g is a function of displacement alone, then one has:

$$\left\| E\left[g(\overline{\boldsymbol{X}})\right] - E\left[g(\widetilde{\overline{\boldsymbol{X}}})\right]\right\| \leq K(\boldsymbol{x})h^3 \tag{5.165a}$$

and for all other cases:

$$\left\| E\left[g(\overline{\boldsymbol{X}})\right] - E\left[g(\widetilde{\overline{\boldsymbol{X}}})\right]\right\| \leq K(\boldsymbol{x})h^2 \tag{5.165b}$$

Thus for the SDE (5.109) corresponding to the MDOF oscillator under consideration, the weak order of accuracy β is 2 for displacement and 1 for velocity. The interested reader may also refer to Roy [2003] for details of the proof.

Example 5.9

Consider the Duffing oscillator of Example 5.2. We solve the corresponding 2-dimensional SDE (in state space form) by WSNM.

Solution

With $m = 2$ and $n = 1$, the weak Newmark maps for displacement and velocity are respectively given by:

$$X_{1,i} = X_{1,i-1} + hX_{2,i-1} + \alpha\frac{h^2}{2}\left(-cX_{2,i-1} - kX_{1,i-1} - \upsilon\, X_{1,i-1}^3 + \mathcal{P}\cos 2\pi t_{i-1}\right)$$

$$+(1-\alpha)\frac{h^2}{2}\left(-cX_{2,i} - kX_{1,i} - \upsilon\, X_{1,i}^3 + \mathcal{P}\cos 2\pi t_i\right) + \sigma\zeta_k h^{\frac{3}{2}} \quad (5.166a)$$

$$X_{2,i} = X_{2,i-1} + ßh\left(-cX_{2,i-1} - kX_{1,i-1} - \upsilon\, X_{1,i-1}^3 + \mathcal{P}\cos 2\pi t_{i-1}\right)$$

$$+(1-ß)h\left(-cX_{2,i} - kX_{1,i} - \upsilon\, X_{1,i}^3 + \mathcal{P}\cos 2\pi t_i\right) + \sigma\xi_k h^{\frac{1}{2}} \quad (5.166b)$$

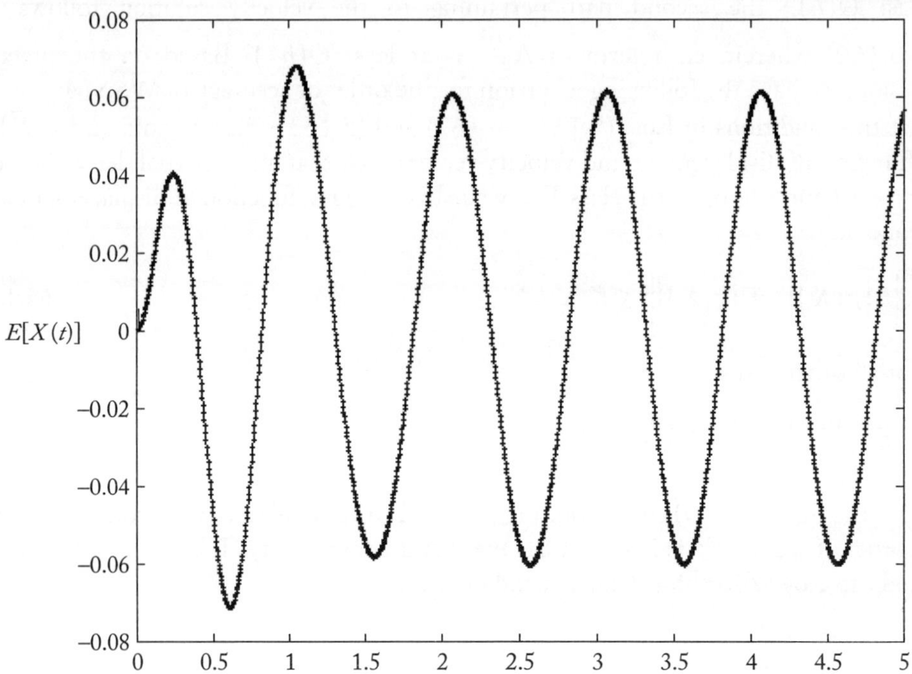

(a)

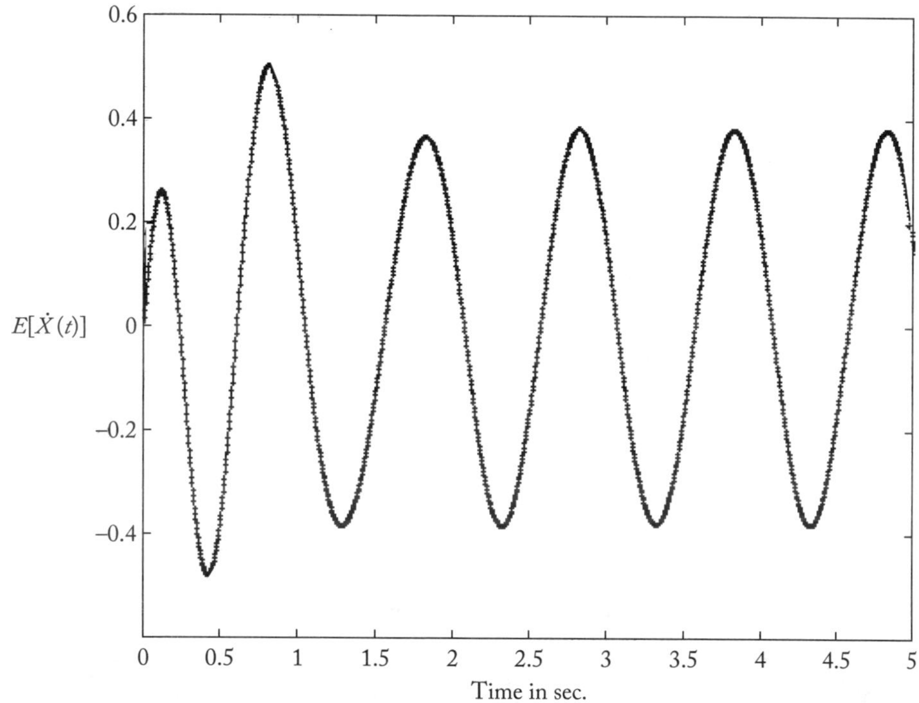

(b)

Fig. 5.14 Weak stochastic Newmark method; numerical solution to the Duffing oscillator of Example 5.2, $c = 4$, $k = 100$, $\upsilon = 100$, $\mathcal{P} = 4$, $\sigma = 0.01$, $h = 0.05$ s, $\alpha = ß = 0.5$, and $N_e = 100$, (a) sample mean displacement and (b) sample mean velocity; dark-line – solution by WSNM with equivalent random variables; dotted-line – WSNM with MSIs

The implicitness parameters α and ß are taken to be 0.5 each. The coupled non-linear algebraic equations at each time step are solved by the standard Newton–Raphson method. An ensemble of $N_e = 100$ is utilized in the MC simulation with a time step of 0.05 s. Figure 5.14 shows the sample mean solutions $E[X(t)]$ and $E\left[\dot{X}(t)\right]$. The solution with MSIs (Eq. 5.123) is also included in the Figure. A close match between the two solutions is obvious from the Figure.

Higher order variants of the Newmark scheme are possible [Roy and Dash 2005] by incorporating more terms in the associated stochastic Taylor expansions, i.e., by iterating the error terms in the lower order schemes with the help of Ito's formula. It may however result in the daunting task of modelling MSIs of third and higher levels. Need for simulation of far less random variables for the lower order methods often renders these schemes more attractive from a computational point of view, particularly for dynamical systems of large dimensions.

5.10 Local Linearization Methods for Strong / Weak Solutions of SDEs

Global linearization techniques are well known [Ramachandra and Roy 2001, Roy and Ramachandra 2001a,b, Iyengar and Roy 1998a, 1998b] for solving deterministic non-linear ODEs. Ozaki [1985, 1994] and Biscay et al. [1996] localized the technique for numerical solutions of non linear SDEs. The work by Ozaki is developed mostly on a heuristic framework in the form of multivariate autoregressive time series without the involvement of stochastic Ito–Taylor expansion whilst later work by Biscay et al. provided a theoretical basis to the formulations of Ozaki. We describe here a version of the local linearization scheme [Roy 2001, 2004] which is an implicit semi-analytical integration method, called the locally transversal linearization (LTL) method. The basic idea of the method lies in the assertion that at a chosen time instant, the LTL solution transversely intersects (see Fig. 5.15 below) the non linear solution manifold at that particular point or cross-section in the state space where the solution vector is sought.

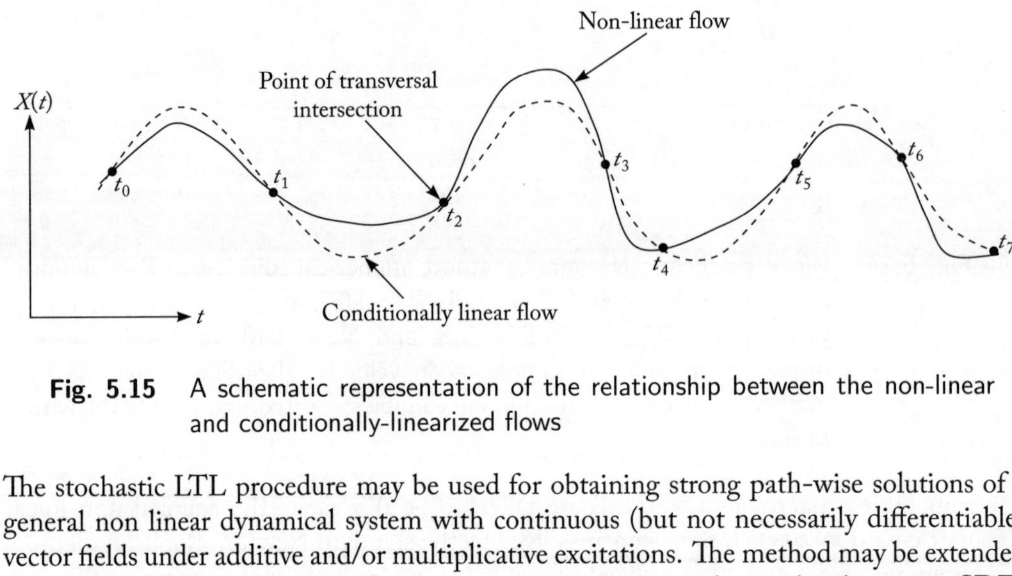

Fig. 5.15 A schematic representation of the relationship between the non-linear and conditionally-linearized flows

The stochastic LTL procedure may be used for obtaining strong path-wise solutions of a general non linear dynamical system with continuous (but not necessarily differentiable) vector fields under additive and/or multiplicative excitations. The method may be extended [Roy 2004, Saha and Roy 2007] to arrive at low or higher order weak solutions to SDEs and in particular noise-driven non linear mechanical oscillators of engineering interest. The advantage of the numeric–analytical LTL scheme lies in the fact that it effectively combines the versatility of numerical schemes with the elegance and computational ease of analytical schemes.

5.10.1 LTL-based schemes

Continuing with mechanical oscillators, consider the m-dimensional MDOF system:

$$\ddot{X} + C\dot{X} + KX + A_{nl}\left(X,\dot{X},t\right) = P(t) + \sum_{k=1}^{n} Q_k\left(X,\dot{X},t\right)\dot{B}_k(t) \tag{5.167}$$

$X, \dot{X} \in \mathbb{R}^m$ are m-dimensional state (displacement and velocity) vectors, C and K are the $m \times m$ state/time-independent damping and stiffness matrices respectively, and A_{nl} is the non linear part of the drift vector. $P(t)$ is a deterministic external force vector. $Q_k\left(X,\dot{X},t\right), k = 1,2,\ldots,n$ are the diffusion coefficients of n independently evolving Wiener processes, $\{B_k(t), k \leq n\}$. We assume without a loss of generality that $Q_k\left(X,\dot{X},t\right)$ is decomposable as:

$$Q_k\left(X,\dot{X},t\right) = \mu_k(t) + \sigma_k(X,\dot{X},t) \tag{5.168}$$

μ_k is the additive (possibly time-dependent) part and σ_k the multiplicative (state-dependent) part of the kth diffusion vector. The system of second order SDEs (5.167) is, as usual, recast into the following incremental form of $2m$ first order SDEs:

$$dX_1 = X_2 dt \tag{5.169a}$$

$$dX_2 = (-A_l(X_1,X_2) - A_{nl}(X_1,X_2,t) + P(t))\,dt + \sum_{k=1}^{n}\left(\mu_k(t) + \sigma_k(X_1,X_2,t)\right)dB_k(t) \tag{5.169b}$$

where, $X_1 = X = \left\{X_{1j} : j = 1,2,\ldots,m\right\}^T$ and $X_2 = \dot{X} = \dot{X}_1 = \left\{X_{2j} : j = 1,2,\ldots,m\right\}^T$. $A_l(X_1,X_2,) = CX_2 + KX_1$ stands for the linear part of the drift vector. It is assumed that the drift and diffusion functions are measurable, Lipschitz continuous and with appropriate growth bounds. Thus the sample continuity of any realization of the non linear flow $\Psi_i(\omega, X_1(0), X_2(0))$ for any $\omega \in \Omega$ is assured provided that, $E\left\|X_1(0)\right\|$, $E\left\|X_2(0)\right\| < \infty$. Let Π_κ be the partition of the time interval $[0,T]$ over which integration needs to be performed with $t_\kappa = T$ and $h_i = t_i - t_{i-1}$. Now, the objective of the LTL technique is to replace the non linear SDEs (5.169) by a suitably chosen set of κ linear systems of SDEs wherein the ith linear system should, in a sense, be able to approximately represent the non linear flow over the ith time sub-interval $T_i = (t_{i-1}, t_i]$. Such a replacement is non-unique. However, the following system of conditionally linearized SDEs (5.169) form a possible LTL system corresponding to the SDEs over T_i:

$$d\overline{X}_1 = \overline{X}_2 dt \tag{5.170a}$$

$$d\overline{X}_2 = \left(-A_l\left(\overline{X}_1,\overline{X}_2\right) + P(t)\right)dt - \left(A_{nl}(X_{1,i},X_{2,i},t)\right)dt + \sum_{k=1}^{n}\left(\mu_k(t_i) + \sigma_k(X_{1,i},X_{2,i},t)\right)dB_k(t) \tag{5.170b}$$

The LTL-based equation in $\overline{\mathbf{X}}(t) = \left\{\overline{X}_1{}^T, \overline{X}_2{}^T\right\}^T$ as above is obtained by replacing the arguments in the non-linear parts of the drift and diffusion fields, the latter associated with multiplicative noise, using the still unknown vector, $\mathbf{X}(t_i) = \mathbf{X}_i := \left\{X_{1,i}{}^T, X_{2,i}{}^T\right\}^T$. The term 'transversal' is associated with the fact that the tangent operators associated with the non linear drift and diffusion terms in the original SDE (5.169a,b) are different from those in the linearized model in Eq. (5.170a,b). Nevertheless, the vector fields of the non-linear and transversely linearized equations are instantaneously identical at $t = t_i$ provided the following vector identities are enforced:

$$X_{1,i} = \overline{X}_{1,i}, X_{2,i} = \overline{X}_{2,i} \tag{5.171}$$

Given the mutually transversal local evolutions of the solutions $\Psi_t(\omega, \mathbf{X}_{i-1})$ and $\overline{\Psi}_t(\omega, \overline{\mathbf{X}}_{i-1})$ for $t \in (t_{i-1}, t_i]$ and for $\omega \in \Omega$ fixed, of the non linear and linearized SDEs respectively, enforcement of Eq. (5.171) roughly implies that the two locally transversal solution manifolds intersect at $t = t_i$. In other words, the discrete unknown vector $\mathbf{X}_i$ may be obtained as the point of equality of $\Psi_t(\omega, \mathbf{X}_{i-1})$ and $\overline{\Psi}_t(\omega, \overline{\mathbf{X}}_{i-1})$ at $t = t_i$. Towards this, first an analytical solution for $\overline{\mathbf{X}}_t$ in Eq. (5.170) may be conditionally constructed in terms of the unknown state vector $\mathbf{X}_i$ as:

$$\overline{\mathbf{X}}_t = \overline{\Psi}_t = [\phi(t, t_{i-1})] \begin{pmatrix} \overline{\mathbf{X}}_{i-1} + \int_{t_{i-1}}^{t} [\phi(s, t_{i-1})]^{-1} (\widehat{\mathcal{P}}(s) - A_{nl}(\mathbf{X}, s)) ds \\ + \int_{t_{i-1}}^{t} [\phi(s, t_{i-1})]^{-1} \sum_{k=1}^{n} \left(\widehat{\mu}_k(s) + \widehat{\sigma}_k(\mathbf{X}, s)\right) dB_k(s) \end{pmatrix} \tag{5.172}$$

Here $\widehat{\mu}_k(t) = \left\{\{\mathbf{0}\}^T, \{\mu_k(t)\}^T\right\}^T$ is the $2m$ dimensional additive diffusion vector obtained by augmenting $\mu_k(t)$ with an m-dimensional zero vector $\{\mathbf{0}\}$. $\widehat{\sigma}_k(\mathbf{X}, t) = \left\{\{\mathbf{0}\}^T, \{\sigma_k(\mathbf{X}, t)\}^T\right\}^T$ is the $2m$-dimensional conditionally additive version of the original multiplicative diffusion vector σ_k, similarly augmented by an m-dimensional zero vector $\{\mathbf{0}\}$. $\widehat{\mathcal{P}}(t) = \left\{\{\mathbf{0}\}^T, \{P(t)\}^T\right\}^T$ is the augmented deterministic force vector. $\phi(t, t_{i-1})$ is the fundamental solution matrix (FSM) corresponding to the linear drift field only and is given by:

$$\phi(t, t_{i-1}) = \exp\{[A](t - t_{i-1})\} \tag{5.173}$$

A is a $2m \times 2m$ matrix and is of the form:

$$A = \begin{bmatrix} [\mathbf{0}] & [I] \\ -[K] & -[C] \end{bmatrix} \tag{5.174}$$

$[\mathbf{0}]_{m\times m}$ and $[\mathbf{I}]_{m\times m}$ are respectively zero and identity matrices. The unknown solution vector $\mathbf{X}_i$ may be obtained by substituting the linearized solution in Eq. (5.172) in the constraint Eq. (5.171). This results in a system of $2m$ coupled non linear algebraic equations for as many unknowns to determine $\boldsymbol{X}_{1,i}$ and $\boldsymbol{X}_{2,i}$ at each $t = t_i$. Roots of the algebraic equations may not be unique and may be found using a Newton–Raphson search algorithm. A possible multiplicity of solutions is consistent with the fact that non-linear dynamical systems may undergo bifurcations and may thus have multiple solutions.

Note that, while the basic LTL scheme as outlined above, is inherently implicit, an explicit variant of a similar class of schemes is provided by the so called phase space linearization [Iyengar and Roy 1998a, 1998b, and Roy 2000a, 2000b].

While Eq. (5.170) represents a basic LTL scheme, one may have other and even higher order versions of the LTL scheme by suitably using implicit substitutions via backward EM or Newmark expansions. For example, one can replace $\boldsymbol{A}_{nl}(\mathbf{X},t)$ and $\sigma_k(\mathbf{X},t)$ in Eq. (5.170) by an implicit EM substitution. To this end, the state variables $\boldsymbol{X}_1(t)$ and $\boldsymbol{X}_2(t)$ (appearing as arguments of $\boldsymbol{A}_{nl}$ and $\boldsymbol{\sigma}_k$) are backward expanded for $t \in T_i$ as:

$$X_{1j}(t) = X_{1j,i} - \int_t^{t_i} X_{2j,i}\,ds \tag{5.175a}$$

$$X_{2j}(t) = X_{2j,i} - \int_t^{t_i} \left(-A_{lj,i} + P_{ij} - A_{nlj,i}\right) ds - \sum_{k=1}^{n} \int_t^{t_i} \left(\mu_{jk,i} + \sigma_{jk,i}\right) dB_k(s) \tag{5.175b}$$

For example, if the non linearity is a function of only the displacement component X_{11} in the form $A_{nl1}(\mathbf{X},t) = (X_{11})^3$, then we have by implicit EM substitution:

$$\begin{aligned} A_{nl1,i}(\mathbf{X},t) &= \left(X_{11,i} - \int_t^{t_i} X_{21,i}\,ds\right)^3 \\ &= \left(X_{11,i} - X_{21,i}(t_i - t)\right)^3 \end{aligned} \tag{5.176}$$

Substitution of such backward expansions in Eq. (5.170) gives the following modified linearized SDEs:

$$d\overline{\boldsymbol{X}}_1 = \overline{\boldsymbol{X}}_2 dt \tag{5.177a}$$

$$d\overline{\boldsymbol{X}}_2 = \left(-\boldsymbol{A}_l\left(\overline{\mathbf{X}}\right) + \boldsymbol{P}(t)\right) dt - \left(\boldsymbol{A}_{nl}^E(\mathbf{X},t)\right) dt + \sum_{k=1}^{n} \left(\boldsymbol{\mu}_k(t_i) + \boldsymbol{\sigma}_k^E(\mathbf{X},t)\right) dB_k(t) \tag{5.177b}$$

$\boldsymbol{A}_{nl}^{E}(\mathbf{X},t)$ and $\sigma_k^E(\mathbf{X},t)$ denote the new linearized non-linear drift and diffusion terms based on such backward EM substitutions. Note that both the basic and the new LTL versions in Eqs. (5.170) and (5.177), respectively, need no computation of derivatives of the vector fields.

In order to achieve as till higher order accuracy without having to compute derivatives, the functions $\boldsymbol{A}_{nl}(\mathbf{X},t)$ and $\boldsymbol{\sigma}_k(\mathbf{X},t)$ may be conditionally replaced by the implicit two-parameter Newmark expansions of $\boldsymbol{X}_1$ and $\boldsymbol{X}_2$ over $t \in T_i$. We refer to Roy and Dash [2002] and Saha and Roy [2007] for the details.

Example 5.10

Consider a five-parameter Duffing oscillator [Saha and Roy 2007] represented by the following SDEs in incremental form:

$$dX_1 = X_2 dt \tag{5.178a}$$

$$dX_2 = \left(-2\pi c_1 X_2 - 4\pi^2 c_2\left(1 + X_1{}^2\right)X_1 + 4\pi^2 c_3 cos 2\pi t\right)dt$$

$$+4\pi^2 c_4 dB_1(t) + 4\pi^2 c_5 X_1 dB_2(t) \tag{5.178b}$$

$c_i, i = 1,2,3,4,5$ are real constants. $B_1(t)$ and $B_2(t)$ are independent standard Weiner processes. With $\mathbf{X} = (X_1, X_2)^T$, the non linear drift term $A_{nl}(\mathbf{X},t)$ is:

$$A_{nl}(\mathbf{X},t) = 4\pi^2 c_2 X_1{}^3 \tag{5.179}$$

We obtain solutions by the basic LTL scheme in Eq. (5.170) and the higher order scheme in Eq. (5.177).

Solution

The LTL system consisting of the linearized equations (as in Eq. 5.170) over the interval $T_i = (t_{i-1}, t_i]$ is given by:

$$d\overline{X}_{1,i} = \overline{X}_{2,i} dt \tag{5.180a}$$

$$d\overline{X}_{2,i} = \left(-2\pi c_1 \overline{X}_{2,i} - 4\pi^2 c_2 \overline{X}_{1,i} + 4\pi^2 c_3 \cos 2\pi t_i\right)dt + \left(-4\pi^2 c_2 X_{1,i}{}^3\right)dt$$

$$+4\pi^2 c_4 dB_1(t) + 4\pi^2 c_5 X_{1,i} dB_2(t) \tag{5.180b}$$

The matrix $\boldsymbol{A}$ (Eq. 5.174) is given by:

$$\boldsymbol{A} = \begin{bmatrix} 0 & 1 \\ -4\pi^2 c_2 & -2\pi c_1 \end{bmatrix} \tag{5.181}$$

The FSM corresponding to the linear drift part is given by $\phi(t,t_{i-1}) = \exp\{[A](t-t_{i-1})\}$. To compute the matrix exponential in $\phi(t,t_{i-1})$, a crude way is to use a deterministic Taylor expansion followed by the retention of a few terms only:

$$[\phi(t,t_{i-1})] = [I] + [A]h + [A]^2\frac{h^2}{2} + [A]^3\frac{h^3}{2} + \ldots \tag{5.182}$$

A similar expansion may also be utilized to compute $[\phi(t,t_{i-1})]^{-1}$ in the form:

$$[\phi(t,t_{i-1})] = [I] - [A]h + [A]^2\frac{h^2}{2} - [A]^3\frac{h^3}{2} + \ldots \tag{5.183}$$

Here, first four terms cf the above expansions are retained in computing ϕ and ϕ^{-1}.

Basic LTL scheme – response computation

Since A_{nl} involves non-linearity in X_1 alone, it suffices, for the basic LTL scheme, to impose the constraint equation, $\mathbf{X} = \overline{\mathbf{X}}$ at each $t = t_i$, on the X_1 component only. The linearized $\overline{X}_1(t)$ at $t = t_i$ is obtainable as:

$$\begin{aligned}\overline{X}_{1,i} &= \phi_{11,i}\overline{X}_{1,i-1} + \phi_{12,i}\overline{X}_{2,i-1} + 4\pi^2\left(\phi_{11,i}\phi^{-1}_{12,i} + \phi_{12,i}\phi^{-1}_{22,i}\right)\\ &\left(c_2{X_{1,i}}^3 h + c_3\int_{t_{i-1}}^{t_i}\cos 2\pi s\, ds\right)\\ &+4\pi^2\left(c_4 + c_5X_{1,i}\right)\left(\phi_{11,i}\phi^{-1}_{12,i} + \phi_{12,i}\phi^{-1}_{22,i}\right)I_1^{(1)}\end{aligned} \tag{5.184}$$

Now if the transversal intersection condition (for displacement alone), i.e., $X_{1,i} = \overline{X}_{1,i}$ is imposed, one obtains a non-linear, scalar algebraic equation to solve for $X_{1,i}$. Once $X_{1,i}$ is obtained, the velocity variable $X_{2,i}$ may be readily obtained from the following explicit map:

$$\begin{aligned}\overline{X}_{2,i} &= \phi_{21,i}\overline{X}_{1,i-1} + \phi_{22,i}\overline{X}_{2,i-1}\\ &+4\pi^2\left(\phi_{21,i}\phi^{-1}_{12,i} + \phi_{22,i}\phi^{-1}_{22,i}\right)\left(c_2{X_{1,i}}^3 h + c_3\int_{t_{i-1}}^{t_i}\cos 2\pi s\, ds\right)\\ &+4\pi^2\left(c_4 + c_5X_{1,i}\right)\left(\phi_{21,i}\phi^{-1}_{12,i} + \phi_{22,i}\phi^{-1}_{22,i}\right)I_1^{(1)}\end{aligned} \tag{5.185}$$

Higher order LTL scheme–response computation

By the backward (implicit) EM expansion, the non-linear drift term $A^E_{nl}(\mathbf{X}_i,t_i)$ and the multiplicative diffusion term $\sigma^E_{nl}(\mathbf{X}_i,t_i)$ take the forms $(X_{1,i} - X_{2,i}h)^3$ and $(X_{1,i} - X_{2,i}h)$

respectively. In this case, the transversal intersection condition needs to be imposed for both the state variables resulting in two coupled non-linear algebraic equations in the unknowns $X_{1,i}$ and $X_{2,i}$. Solving these two equations yields the solution of the corresponding SDEs (5.117) at $t = t_i$.

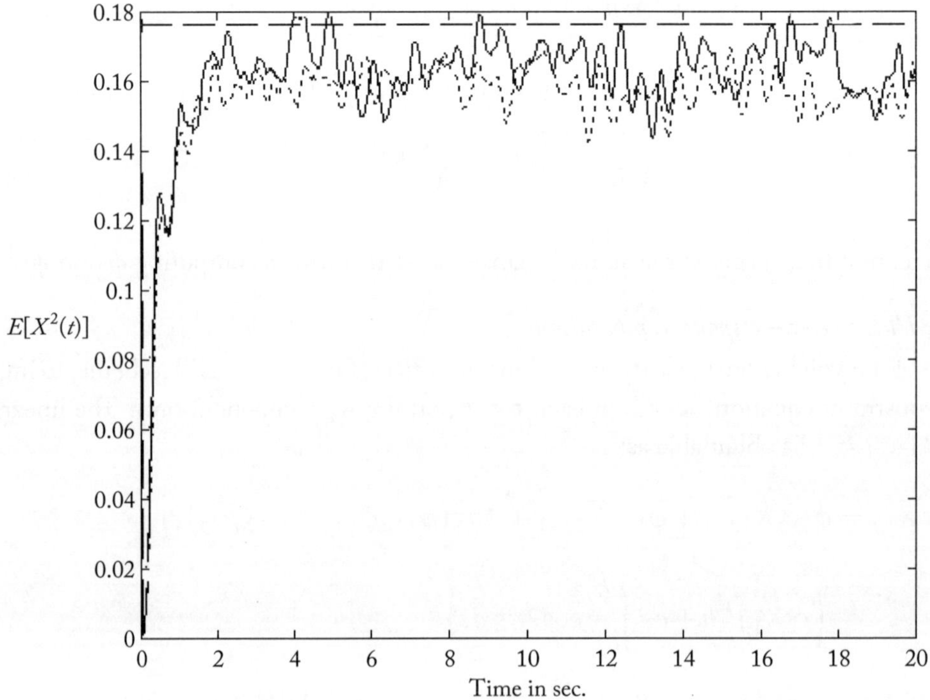

Fig. 5.16 Duffing oscillator in Example 5.10; case of additive noise only; sample approximation to $E[X^2(t)]$ by LTL schemes, $c_1 = 0.25,\ c_2 = 0.5,\ c_3 = 0,\ c_4 = 0.1,\ c_5 = 0,\ h = 0.01,\ N_e = 1000$, dark-line – basic LTL scheme (Eq. 5.170), dotted-line – higher order LTL scheme (Eq. 5.177), dashed-line – true stationary solution

Figure 5.16 shows the plot of $E[X^2(t)]$, the mean square displacement solution over $t \in [0,20]$ sempirically obtained with MC simulation using an ensemble size of $N_e = 1000$. The solution corresponds to additive noise only with $c_1 = 0.25,\ c_2 = 0.5,\ c_3 = 0,\ c_4 = 0.1$ and $c_5 = 0$. A time step size of $h = 0.01$ s is used in the simulation. The solution obtained from the higher order LTL scheme (Eq. 5.177) is also shown in Fig. 5.16. In this case, one can have the 'exact stationary solution' for $E[X^2(t)]$ via the analytically available two-dimensional *pdf* solving the reduced Fokker–Planck equation [Wang and Zhang 2000]. Specifically, the stationary *pdf* is given by:

$$f_{X_1,X_2}(x_1,x_2) = C\exp\left(-\frac{2\pi c_1}{2(4\pi^2 c_4)^2}\left(\frac{4\pi^2 c_2 x_1^2}{2} + \frac{4\pi^2 c_2 x_1^4}{4} + \frac{x_2^2}{2}\right)\right) \tag{5.186}$$

where, C is found by the normalization constraint $\int_{-\infty}^{\infty}\int_{-\infty}^{\infty} f_{X_1,X_2}(x_1,x_2)\,dx_1 dx_2 = 1$. The expectation of any deterministic function $\Psi(X_1,X_2)$ is then given by $E[\Psi(X_1,X_2)] = \int_{-\infty}^{\infty}\int_{-\infty}^{\infty} \Psi(x_1,x_2) f_{X_1,X_2}(x_1,x_2)\,dx_1 dx_2$, where an appropriate quadrature rule may be employed to numerically evaluate the integral to a desired precision. The true stationary solution for $E\left[X^2(t)\right]$ so evaluated is included in Fig. 5.16 for a comparison with the simulation results.

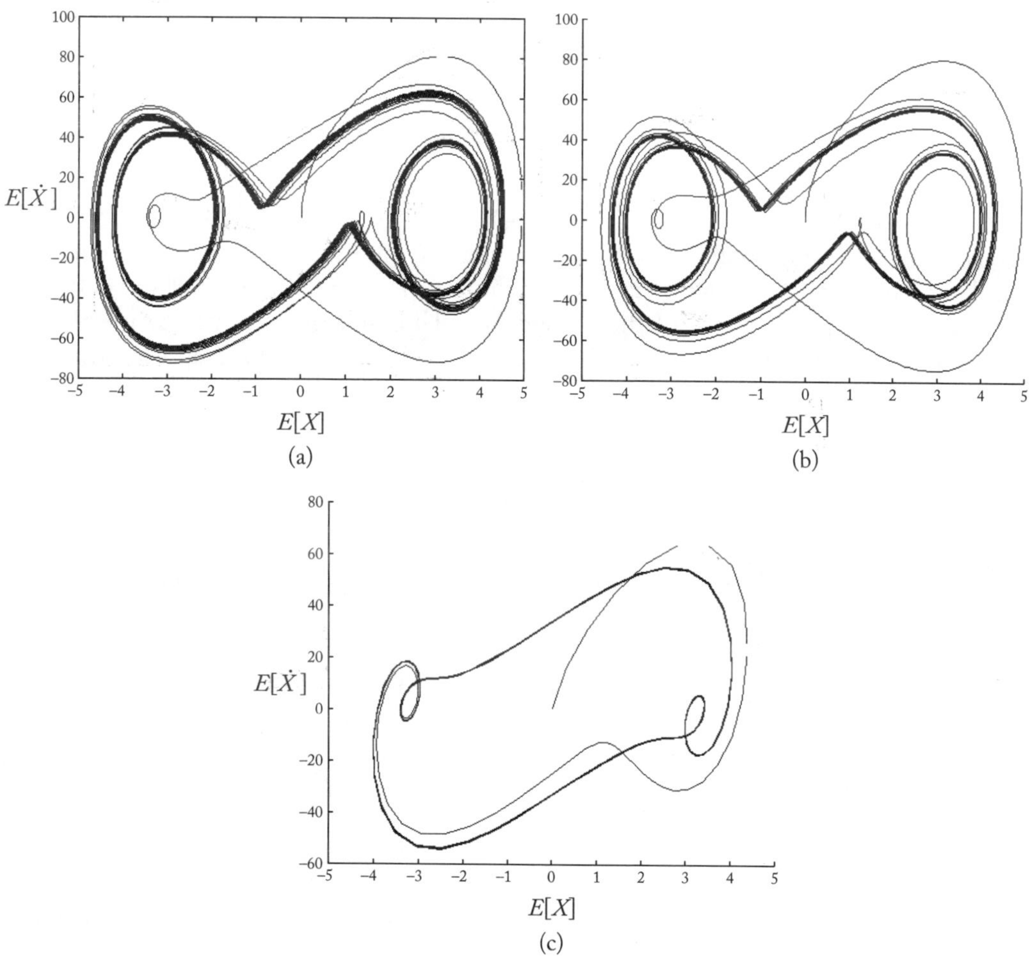

Fig. 5.17 Duffing oscillator in Example 5.10; harmonic force along with additive noise, phase plots by LTL schemes, $c_1 = 0.25$, $c_2 = 1$, $c_3 = 41$, $c_4 = 0.5$, $c_5 = 0$, $N_e = 1000$, (a) basic LTL scheme with $h = 0.001$, (b) higher order LTL scheme with $h = 0.001$ and (c) higher order LTL scheme with $h = 0.01$; note the unphysical distortion in the strange attractor

For the case of a harmonic (deterministic) force acting simultaneously with the additive noise and for certain choices of parameters, the oscillator exhibits chaotic response (characterized by an extreme sensitivity to initial conditions, i.e., exponential separation of two trajectories that start off with closely separated initial conditions). A chaotic trajectory never reaches a steady state in the usual sense, rather, as $t \to \infty$, it evolves within a region of the phase space called the strange attractor [Guckenheimer and Holmes 1983]. The phase portraits ($E[X_1]$ *vs.* $E[X_2]$ plots) obtained from MC simulations (with $N_e = 1000$) are shown in Fig. 5.17 for $c_1 = 0.25$, $c_2 = 1$, $c_3 = 41$, $c_4 = 0.5$ and $c_5 = 0$. The solution obtained by both the basic and higher order LTL schemes with $h = 0.001$ match closely. But with $h = 0.01$, while the quality of the solution by higher order scheme deteriorates (Fig. 5.17c), the solution by the basic LTL does not converge and may even overflow.

Figure 5.18 shows the time history plots of (sample approximations to) $E\left[X^2(t)\right]$ and $E\left[\dot{X}^2(t)\right]$ obtained by the higher order LTL scheme (Eq. 5.177) for the case of a harmonic (deterministic) forcing combined with a high-intensity multiplicative noise. The choice of oscillator parameters for this case is: $c_1 = 0.25$, $c_2 = 1$, $c_3 = 0.25$, $c_4 = 0$ and $c_5 = 0.4$. Once again, the performance of the higher order LTL scheme is satisfactory with $h = 0.001$ in contrast to the solution by the basic LTL scheme that blows up even with this apparently small step size.

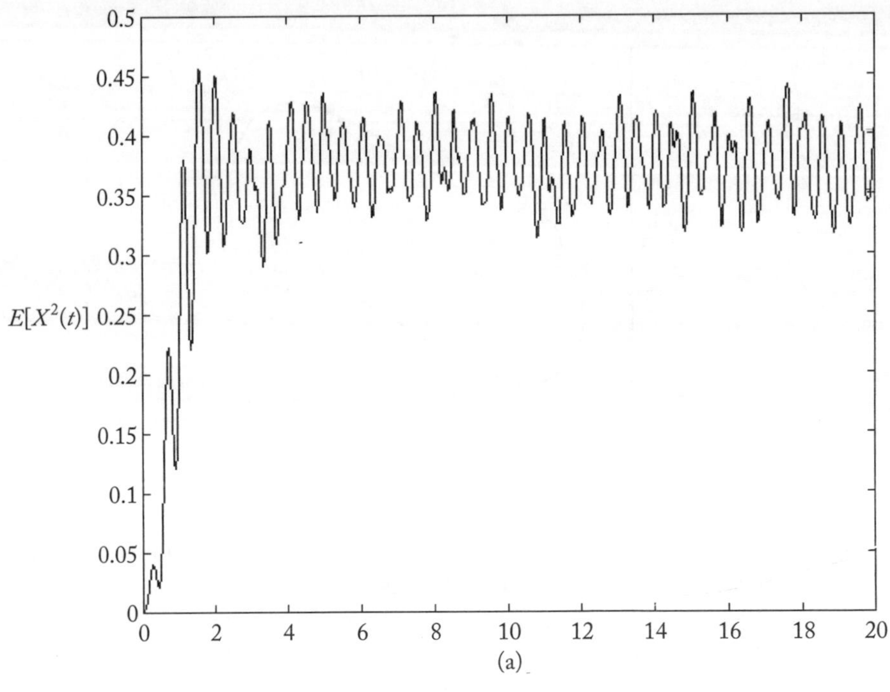

(a)

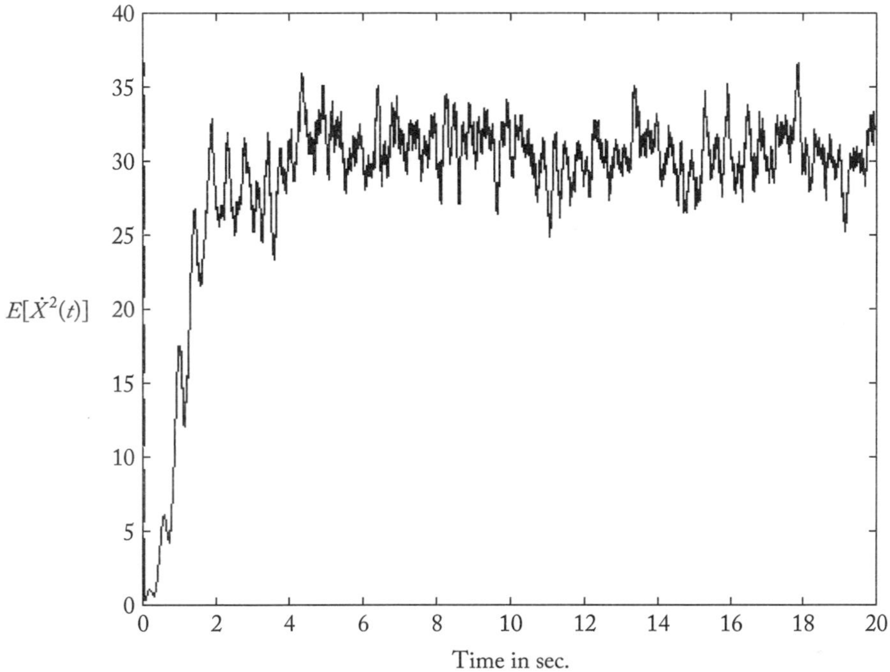

(b)

Fig. 5.18 Duffing oscillator in Example 5.10; high multiplicative noise along with deterministic harmonic force, time history plots by higher order LTL scheme with $h = 0.001$, $c_1 = 0.25$, $c_2 = 1$, $c_3 = 0.25$, $c_4 = 0$ and $c_5 = 0.4$, (a) sample approximation to $E[X^2(t)]$ and (b) sample approximation to $E[\dot{X}^2(t)]$

5.11 Concluding Remarks

As a follow-up to the theory of stochastic processes and Ito calculus developed in the earlier chapters, we have made a graded presentation in this chapter on the development of some numerical methods to integrate SDEs mainly driven by Brownian motions. Analogous to the dependence on the Taylor expansion of all numerical methods for integrating deterministic ODEs, Ito–Taylor expansion underlies the derivation of the explicit and implicit one-step integration methods for SDEs. The essential difference in having strong (path-wise) or weak (distribution-wise) solutions for SDEs has been highlighted along with the associated convergence theorems. The Ito–Taylor expansion, an asymptotic expansion derivable through repeated application of the Ito formula is derived and shown leading to numerical schemes of different orders of convergence. Some of the implementation issues, mainly of evaluation of multiple stochastic integrals (MSIs) involved in the truncated expansions, have been discussed. In this context, weak one-step approximate methods that typically replace MSIs by random variables,which are relatively more easily evaluated, are described and proved to have the same order of convergence. Implicit methods such as two-parameter stochastic Newmark methods (similar to the deterministic ones) have been derived using the Ito–Taylor forward and backward expansions and their application demonstrated in the context of a few typical (non linear) mechanical oscillators.

The numerical and semi-numerical (such as the numeric–analytical schemes of the LTL type) techniques presented in this chapter find numerous applications involving SDEs (within an MC set up) e.g., multi-dimensional dynamical systems under stochastic perturbations. Our endeavor in the next couple of chapters will be to demonstrate the ability and importance of numerical techniques in addressing a large class of estimation or system identification problems posed using the stochastic filtering theory.

Exercises

1. A non-linear (financial market) dynamic model of the interaction between liquidity and income [Semmler and Sieveking, 1993 – see Bibliography] is given by the following SDEs:

$$d\lambda = \lambda(\alpha - \beta\rho - \theta_1\lambda - g(\lambda,\rho))\,dt + \sigma_1 d\beta_1$$

$$d\rho = \rho(-\gamma + \delta\lambda - \theta_2\rho)\,dt + \sigma_2 d\beta_2$$

$\lambda(t)$ and $\rho(t)$ are the action variables denoting the liquidity and income normalized with respect to the capital stock. $\alpha,\ \beta,\ \theta_1, \gamma,\ \delta$ and θ_2 are scalar parameters. β_1 and β_2 are independent Brownian motions with σ_1 and σ_2 being their intensities respectively. $g(\lambda,\rho)$ is known as an acceleration term indicating regime changes (in terms of swings in liquidity and income) and is assumed of the form:

$$g(\lambda,\rho) = k(\lambda - \mu)(\rho - \nu), \text{if } 0 < \lambda < \mu < \lambda^* \text{ and } 0 < \rho < \nu < \rho^*$$

$$= 0, \quad \text{otherwise}$$

$k,\ \mu$ and ν are again scalar parameters. λ^* and ρ^* are threshold (steady state) values. Simulate the above SDEs by EM (explicit or implicit) method with the input values of the parameters: $\alpha = 0.1,\ \beta = 0.6,\ \theta_1 = 0.045, \gamma = 0.07,\ \delta = 0.7$ and $\theta_2 = 0.078$ and $k = 0.4,\ \mu = 0.06$ and $\nu = 0.08$. Further $\lambda^* = 0.082$ and $\rho^* = 0.161$.

2. Verify the stochastic chain rule for $V(t) = g(X(t)) = \sqrt{X(t)}$, with $X(t)$ governed by the SDE:

$$dX = (\alpha - X)\,dt + \beta\sqrt{X}dB, X(0) = X_0$$

(Hint: Use Ito's formula to derive the SDE for $V(t)$ as:

$$dV = \left(\frac{4\alpha - \beta^2}{8V} - \frac{1}{2}V\right)dt + \frac{1}{2}\beta dB, V(0) = \sqrt{X_0}$$

Solve the two SDEs by any of the numerical schemes to show that square of $V(t)$ closely match with $X(t)$.

3. Consider the following SDEs representing an FM demodulator (in a communication equipment):

$$dy_1(t) = -\left(c^{\frac{1}{3}}y_1 + \sin(y_2)\right)dt + \sqrt{c}\,(dB_1 - dB_2)$$

$$dy_2(t) = \left(\frac{1}{2}y_1 - \sin(y_2)\right)dt - \sqrt{c}\,dB_2$$

Use any of the numerical schemes described in the chapter to obtain the phase plane curves (y_1 *vs*. y_2 graphs) over the time interval $[0,10]$for i) $c = 0.001$ and ii) $c = 10$. Assume initial conditions $(y_{1,0} = 1, y_{2,0} = 1)$.

(*Notes*: The deterministic part of the model is characterized by the stable equilibria (fixed points corresponding to an unforced system): at $y_1 = 0,\ y_2 = 2\pi n,\ n = 0, \pm 1, \pm 2, \ldots$. Any trajectory starting in the domain of attraction of a stable equilibrium point remains in the domain of attraction – a state of `initial symmetry'. In the stochastic model, stochastic perturbations cause crossing of these domains. In other words, noise breaks the initial symmetry that may exist in physical systems.).

4. Another example (as in Exercise 3 above) exhibiting the phase transition—an exchange of stability between the possible steady states is the dynamics of a Duffing oscillator governed by the SDE $dX = X\left(\lambda t - X^2\right)dt + \mu dB$. The system has two equilibrium states in deterministic sense. If $\mu \gg \lambda$, the trajectories slip from $+\sqrt{\lambda t}$ to $-\sqrt{\lambda t}$ and vice versa before deciding one branch of the two steady states. Study this phenomenon by MC simulation by any of the numerical methods with $\lambda = 0.03$ and μ suitably varied in the range 0.1–0.75.

5. Analyze the higher order scheme in Eq. (5.70b) which is known as Milstein scheme to show that the strong and weak orders of convergence are both equal to unity. Test the scheme on the Black–Scholes SDE: $dX(t) = \lambda X dt + \mu X dB(t)$. Use $\lambda = 2$ and μ=0.01/0.1.

6. A strongly coupled Ornstein–Uhlenbeck model is governed by the SDEs:

$$dX_1 = (\alpha_1 X_1 + \beta_1 X_2)\,dt + \sigma_1 dB_1$$

$$dX_2 = (\alpha_2 X_1 + \beta_2 X_2)\,dt + \sigma_2 dB_2$$

where the parameters $\alpha_i, \beta_i, \sigma_i, i = 1, 2$ are real constants. The Brownian motions B_1 and B_2 are correlated with the correlation matrix given by:

$$\rho = \begin{bmatrix} 1 & r \\ r & 1 \end{bmatrix}$$

With $|r| \le 1$, use any of the numerical integration schemes to plot the trajectories of X_1 and X_2. Assume that $\alpha_i, \beta_i, \sigma_i, i = 1,2$ are all unity.

(Hint: Refer to Section 2.5.3, Chapter 2 for simulation of correlated normal random variables)

7. Solve the SDE: $dX = \alpha X(\beta - X)dt + \gamma X dB$ by i) EM scheme and the higher order scheme in Eq. (5.70d) and get a few trajectories with $\alpha = 2$, $\beta = 1$ and $\gamma = 0.25$. Also plot global error plots and obtain the strong and weak orders of convergence. Use time steps: 2^{-8}, 2^{-7}, 2^{-6}, 2^{-5} and 2^{-4} s. Assume the true solution to be the one obtained with a time step of 2^{-11} by the higher order scheme in Eq. (5.70d).

   (*Notes*: The SDE in this exercise is often used in population dynamics [Abundo 1991 - see Bibliography]. Extension to 2-dimensional case is straightforward and the effect of interactions among, say, different species makes an interesting study via a predator-prey model. For instance, the presence of noise is known to be responsible to the extinction of a species in the predator–prey model:

$$dX_1 = (a_{11}X_1 - a_{12}X_1X_2)\,dt + \sigma_{11}X_1^2 dB_1 + \sigma_{12}X_1X_2 dB_2$$

$$dX_2 = (a_{21}X_1 - a_{22}X_1X_2)\,dt + \sigma_{21}X_1X_2 dB_1 + \sigma_{22}X_2^2 dB_2$$

   $X_1(t)$ stands for the population of the prey and $X_1(t)$ of the predator. Figure below shows the phase plane plots corresponding to both the deterministic and stochastic cases. In the former case, a limit cycle type behavior – conservation of the species – is recognized and in the latter, noise extincts the population of the predator.)

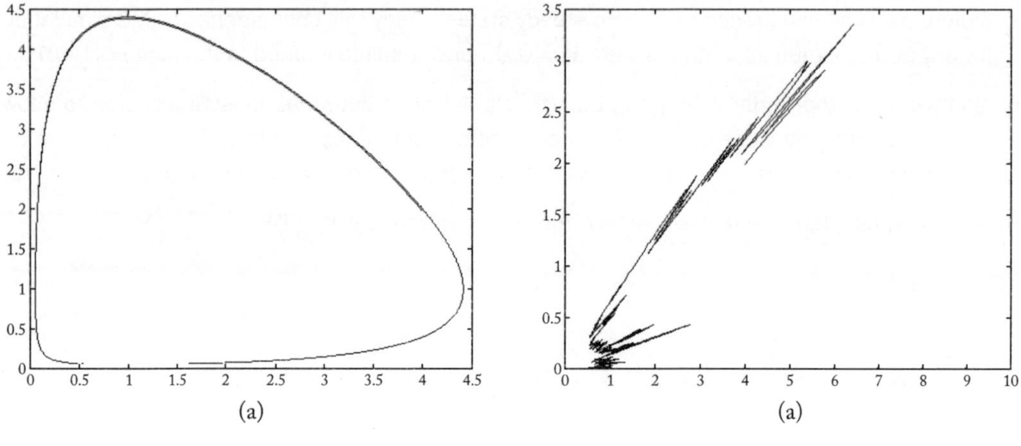

Fig. E.7 (a) deterministic case: limit cycle type behavior and (b) stochastic case: extinction of a species in the presence of noise

8. The following 3-parameter SDE represents a hardening single well (i.e., single equilibrium state – which is a zero state in the present case) non-linear Duffing oscillator:

$$dX_1 = X_2 dt$$

$$dX_2 = \left(-2\pi c_1 X_2 - 4\pi^2 c_2 (1 + X_1^2) X_1 + 4\pi^2 c_3 \cos 2\pi t\right) dt + \sigma dB$$

The oscillator exhibits a typical chaotic diffusion for higher amplitude values of the deterministic harmonic excitation [Roy 2001] without noise. Study the effect of noise on the chaotic attractor using the stochastic LTL-based schemes. Vary σ in the range $[2-100]$.

9. Consider the SDE of a double-well Duffing–Holmes oscillator under an additive noise:

$$dX_1 = X_2 dt$$

$$dX_2 = \left(-CX_2 - K(1 - X_1^2)\right) X_1 + \sigma dB$$

Use LTL-based numerical schemes to obtain the time histories of moments $E\left[X_1^2\right]$ and $E\left[X_2^2\right]$. Use the parameters: C =1, K =1 and $\sigma = 0.25$ and a time step $h = 0.1$s and the final time $T = 50$ s. Also plot the stationary *pdf* $f_{X_1,X_2}(x_1, x_2)$.

(*Notes*: The oscillator is expected to have a stationary bi-model *pdf* for large time \`t' under an additive noise of low intensity [Saha and Roy 2007]. The modes concentrate around the stable fixed points $(\pm 1, 0)$ in the phase space. The *pdf* may be constructed by binning the $X_1 - X_2$ space and approximating the *pdf* by the ratio of number of events falling inside each bin over the total and normalized to the bin volume.)

10. The following SDE represents a hardening Duffing oscillator under filtered white noise [Roy and Dash 2005]:

$$\ddot{x} + 2\pi c_1 \dot{x} + 4\pi^2 c_2 \left(1 + x^2\right) x = 4\pi^2 c_3 \cos 2\pi t \ + g(t)$$

$$\ddot{g} + 2\xi \overline{\omega} \dot{g} + \overline{\omega}^2 g = 4\pi^2 c_4 \dot{B}$$

Obtain the response histories and the possible chaotic phase plots by using the stochastic Newmark method. Use $c_1 = 0.25$ and $c_2 = 1.0$. The harmonic amplitude parameter c_3 can be varied over the interval $[0, 41]$ and the intensity c_4 of the additive white noise over the interval $[5-20]$. Assume the Newmark implicit parameters $\alpha = 0.5$ and ß $= 0.5$. Take the filter parameters as $\xi = 0.02$ and $\overline{\omega} = 10$.

11. Consider the SDEs 4.129 a–d corresponding to the MDOF system in Fig. 4.3 (of Chapter 4). Obtain the histories of the response moments, $E[X_1^2]$, $E[X_2^2]$, $E[X_1 X_2]$, $E[X_1 Y_1]$ $E[(X_2)(y_2)]$, $E[Y_1^2]$, $E[Y_2^2]$ and $E[(Y_1)(y_2)]$ by using stochastic Newmark method (Eq. 5.120). Assume the parameters as given in Example 4.18. Solve for the moments by the weak one-step approximation of the Newmark method (Section 5.9.3) also and compare the results with those obtained with MSIs.

12. Consider the hardening Duffing oscillator:

$$\ddot{x} + 2\pi c_1 \dot{x} + 4\pi^2 c_2 (1 + x^2) x = 4\pi^2 c_3 \cos 2\pi t + \sigma \dot{B}(t)$$

Show that the LTL-based schemes are numerically stable.

(*Notes*: Any one-step stochastic numerical scheme is said to be stochastically numerically stable if there exists a positive constant Δ_0 such that the following limiting condition is satisfied for each $\delta \in (0, \Delta_0)$ and for each $\varepsilon > 0$:

$$\lim_{(|x^\delta(t) - \hat{x}^\delta(t)| \to 0)} \sup_{(t_0 < t \le T)} P(|x^\delta(t) - \hat{x}^\delta(t)| \ge \epsilon) = 0$$

where, $x^\delta(t, \omega)$ and $\hat{x}^\delta(t, \omega)$ are time-discrete approximations obtained with maximum step size $\delta > 0$ with initial conditions x_0^δ and $\hat{x}_0^\delta$, respectively, and for a fixed $\omega \in \Omega$. T denote the right boundary of a finite time interval. In other words, if two solutions from the numerical scheme emerging from two closely separated initial conditions for the same realization of Brownian motion (implying that ω is fixed), stay close all along in forward time, the closeness serves as a pointer to the scheme being stochastically numerically stable. See Roy [2001] and Kloeden and Paten [1992] for further details.)

Notations

a	real constant
$\mathbf{A}$	amplification matrix in Eq. (5.44)
$\mathbf{A}_l$, $\mathbf{A}_{nl}$	linear and non-linear parts of the drift vector
c	damping coefficient in Eq. (5.40)
c_1, c_2	real constants
$\mathbf{C}$, $\overline{\mathbf{C}}$	damping matrices
$C^m(a,b)$	functions that, together with their first m derivatives, are continuous on (a, b)
$\mathfrak{E}$	error in a one-step approximation (Eq. 5.71)
$\mathbb{F}$	class of functions satisfying the inequality in Eq. (5.124)
$g(t, X(t)), g(t, \mathbf{X})$	scalar valued functions
h	time step
$I^{(1)}_{j_1 j_2 \cdots j_k}(h)$, $I^{(2)}_{j_1 j_2 \cdots j_k}(g, h)$	multi-stochastic integrals (MSIs)
k	stiffness coefficient in Eq. (5.40)

$\boldsymbol{K}, \overline{\boldsymbol{K}}$	stiffness matrices
L, L_0, L_{1j}	differential operators
m	dimension of an SDE / order of the truncated Ito-Taylor expansion
$\boldsymbol{M}$	mass matrix
n	dimension of the Brownian motion vector
N_e	ensemble size
p	order of accuracy of a numerical scheme
p_1, p_2	real constants (Eq. 5.6a)
$\mathcal{P}, \overline{\mathcal{P}}$	external deterministic force vectors
$\boldsymbol{Q}, \overline{\boldsymbol{Q}}$	diffusion coefficient matrices
R, R_1, R_2, R_3, R_4	remainder terms in the stochastic Ito-Taylor expansion
Z_{j1}, Z_{j2}	standard normal random variables
α,ß	real constants in stochastic implicit methods
β	strong order of convergence
γ	weak order of convergence
ε_s	strong measure of convergence (Eq. 5.2)
ε_w	weak measure of convergence (Eq. 5.2)
ζ, ζ_j	random variables
λ	real constant in Black-Scholes SDE (5.34)
$\lambda_{1,2}$	eigenvalues
μ	real constant in Black-Scholes SDE (5.34)
$\mu_k(t)$	additive part of diffusion vector
v_1, v_2	eigenvectors
ξ, ξ_j, ξ_{jk}	random variables
ρ	spectral radius
$\phi(t, t_{i-1})$	fundamental solution matrix
σ	additive noise intensity in Eq. (5.40)
υ	drift non-linearity parameter in Eq. (5.40)

Chapter **6**

Non-linear Stochastic Filtering and Recursive Monte Carlo Estimation

6.1 Introduction

With a reasonably broad exposition on SDEs and their solution techniques, mainly numerical, in the last chapter and an extensive reappraisal, in general, of stochastic processes in the preceding chapters, we are now ready to deal with a class of applications encountered in many fields of interest in science and engineering. For example, the problem of estimating the trajectory and the kinetic parameters of a target [Bar–Shalom and Li 1993, Blackman and Popoli 1999] is a classical engineering problem of signal analysis. Here, as in many other problems of a similar nature, the key goal is the estimation of system states of a stochastic dynamical system conditioned on noisy, and possibly sparse, observations of such states. Other such applications wherein the conditional state/parameter estimation is useful include structural health assessment/monitoring [Sarkar *et al.* 2012], regression analysis in climate modeling [Sokolov and Stone 1998], etc. For time–varying observations, the temporally evolving estimation problem is essentially one of stochastic filtering (see Fig. 6.1). A stochastic filter, as a modern tool for dynamical system identification [Banerjee 2009a, 2009b, 2009c], thus involves estimating the dynamically evolving states (system processes) and/or model parameters, declared as fictitiously evolving additional states, conditioned on an experimentally observed noisy data of known functions of the states till the current time.

Within a complete probability space, $(\Omega, \mathcal{F}, P)$, equipped with an increasing filtration $\{\mathcal{F}_t,\ 0 \le t \le T\}$ consisting of sub-σ-algebras of $\mathcal{F}$, a generic form of the process and observation models is typically represented by the following SDEs:

$$d\boldsymbol{X}_t = \boldsymbol{a}(t, \boldsymbol{X}_t)\, dt + \boldsymbol{\sigma}(t, \boldsymbol{X}_t)\, d\boldsymbol{B}_t \tag{6.1a}$$

$$dY_t = h(t, X_t)\,dt + dv_t \tag{6.1b}$$

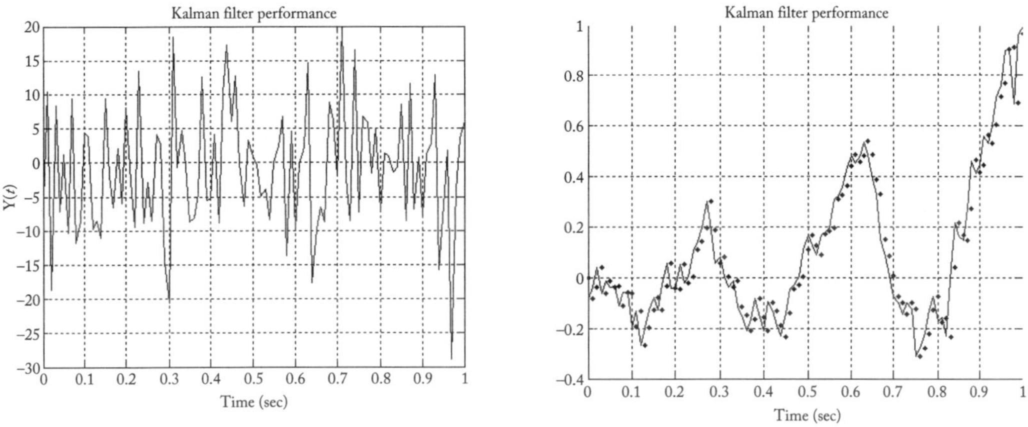

Fig. 6.1 Stochastic filtering; (a) time-varying observed data (with noise) and (b) time-evolutions of a sample of the process state X_t(dark line) and the filtered state $\widehat{X}_t$(dotted)

Here $X_t := X(t) \in \mathbb{R}^m$ is the hidden process state, which is only partly revealed by the noisy observation process $Y_t := Y(t) \in \mathbb{R}^q$ generating the sub-filtration, $\mathcal{F}_t^Y$. $a : \mathbb{R}^+ \times \mathbb{R}^m \to \mathbb{R}^m$ is the vector of (possibly non-linear) drift terms in Eq. (6.1a). The diffusion matrix, $\sigma : \mathbb{R}^+ \times \mathbb{R}^m \to \mathbb{R}^{m \times n}$ and the n-dimensional standard Brownian motion, $B \in \mathbb{R}^n$, together determine the process noise. $h : \mathbb{R}^+ \times \mathbb{R}^m \to \mathbb{R}^q$ is the vector of observation functions which may be non-linear. $v_t := v(t) \in \mathbb{R}^q$ a q-dimensional zero–mean P–Brownian motion representing the observation noise. It is assumed that the conditions (Chapter 4) for the existence of weak solutions to the above SDEs are satisfied. The central idea in a filtering problem is to find recursively at any time $t \in [0, T]$ the conditional distribution of the hidden process states given the sequence of observations till the current time t. In terms of the associated moments, of interest are typically the first two, of which the conditional mean vector $\widehat{X}_t = E[X(t) | \mathcal{F}_t^Y]$ typically defines the estimate and the second moment matrix $E[(X_t - \widehat{X}_t)(X_t - \widehat{X}_t)^T | \mathcal{F}_t^Y]$ defines the so called error covariance. Indeed, if the filtering distribution were Gaussian (which would be the case if the process/measurement models were linear and all the noises Gaussian), then the first two conditional moments as above would be adequate to define the distribution. Alternatively, within a Monte Carlo (MC) setup and a Bayesian update, one may solve the filtering problem by recursively determining an empirical approximation to the filtering distribution based on a finite ensemble of realizations called the particles. Denoting $X_i := X(t_i)$ and $Y_i := Y(t_i)$ with $t_i \in [0, T]$, the process states constitute a Markov process, i.e., $p(X_i | X_{0:i-1}) = p(X_i | X_{i-1})$, and the observations are typically assumed to be independent of the states*. Here $X_{0:i-1} := \{X_0, X_1, \ldots X_{i-1}\}^T$. From Bayes' theorem, it follows that:

$$
\begin{aligned}
p(X_i \mid Y_{0:i}) &= \frac{p(Y_{0:i}|X_i)\, p(X_i)}{p(Y_{0:i})} \\
&= \frac{p(Y_i, Y_{0:i-1}|X_i)\, p(X_i)}{p(Y_i, Y_{0:i-1})} \\
&= \frac{p(Y_i|Y_{0:i-1}, X_i)\, p(Y_{0:i-1}|X_i)\, p(X_i)}{p(Y_i|Y_{0:i-1})\, p(Y_{0:i-1})} \\
&= \frac{p(Y_i|Y_{0:i-1}, X_i)\, p(X_i|Y_{0:i-1})\, p(Y_{0:i-1})\, p(X_i)}{p(Y_i|Y_{0:i-1})\, p(Y_{0:i-1})\, p(X_i)} \\
&= \frac{p(Y_i|X_i)\, p(X_i|Y_{0:i-1})}{p(Y_i|Y_{0:i-1})}
\end{aligned}
\tag{6.2}
$$

$p(Y_i|Y_{0:i-1})$ in the denominator of the above equation is a normalization constant given by:

$$
p(Y_i|Y_{0:i-1}) = \int p(Y_i|X_i)\, p(X_i|Y_{0:i-1})\, dX_i \tag{6.3}
$$

$p(Y_i|X_i)$ is known as the likelihood function (which is also a Radon–Nikodym derivative; see Chapters 1 and 2) defined by the observation model (Eq. 6.1b) and the known statistics of Y_t. Equation (6.2) suggests that, in principle, the posterior (filtering) probability density $p(X_i \mid Y_{0:i})$ is obtainable recursively in two stages, viz. prediction and update. Suppose that the required *pdf* $p(X_{i-1} \mid Y_{0:i-1})$ at $t = t_{i-1}$ is available. Now, in the prediction stage, the prior *pdf* $p(X_i|Y_{0:i-1})$ may be obtained at $t = t_i$ using the system model via Chapman–Kolmogorov equation:

$$
p(X_i|Y_{0:i-1}) = \int p(X_i|X_{i-1})\, p(X_{i-1} \mid Y_{0:i-1})\, dX_{i-1} \tag{6.4}
$$

Here $p(X_i|X_{i-1})$ is the transition *pdf* of the state and is defined by the system Eq. (6.1a). At time t_i, when the observation (measurement) Y_i is available, one may update the prior *pdf* $p(X_i|Y_{0:i-1})$ via Bayes' theorem according to Eq. (6.2) to evaluate the posterior.

The recursive prediction and update formulas above form the basis to obtain an optimal Bayesian solution to the filtering problem in Eq. (6.1). Subject to the

---

*In order to remain consistent with the standard practice in the literature on particle filters, we have allowed for certain notational abuses here. Thus, p denotes a pdf and some arguments of p are written in terms of the random variables themselves, instead of their range.

mean-square measurability of the states, the implied optimality criterion is the minimum mean-squared error (MMSE) defined via the error covariance matrix, defined later. It is shown in Section 2.2.4, Chapter 2, that a conditional expectation is indeed a least mean square error estimator. Thus $E[.|\mathcal{F}_t^{\boldsymbol{Y}}]$ is an orthogonal projection operator on the subspace $L^2\left(\Omega, \mathcal{F}_t^{\boldsymbol{Y}}, P\right)$ and the estimate $\widehat{\boldsymbol{X}}_i = E[\boldsymbol{X}_i|\mathcal{F}_{t_i}^{\boldsymbol{Y}}]$ minimizes the mean square error between $\boldsymbol{X}_i$ and any $\mathcal{F}_{t_i}^{\boldsymbol{Y}}$-measurable approximation to $\boldsymbol{X}_i$. Here the mean square error is the trace of $\boldsymbol{P}_{i|i}$, given by $E[\left(\boldsymbol{X}_i - \widehat{\boldsymbol{X}}_i\right)^T \left(\boldsymbol{X}_i - \widehat{\boldsymbol{X}}_i\right)|\mathcal{F}_{t_i}^{\boldsymbol{Y}}]$. More generally, for any function, Φ, the best estimate in the mean square sense of $\Phi(\boldsymbol{X}_t)$ conditioned on $\boldsymbol{Y}_{s \le t}$ is given by the conditional expectation $\pi_t(\Phi) := E\left[\Phi(\boldsymbol{X}_t)|\mathcal{F}_t^{\boldsymbol{Y}}\right]$. These expectations are, as usual, with respect to the probability measure, P. $\pi_t(\Phi)$ is more generally referred to as the conditional distribution of $\boldsymbol{X}_t$ given the observations up to the time t, i.e., $\pi_t(\Phi) = \int_{\mathbb{R}} \Phi(\boldsymbol{x}_t)\, \pi_t(d\boldsymbol{x}), \Phi \in C^2$.

The Kalman filter posed in a time-discrete setting or its counterpart, the Kalman–Bucy filter, for time-continuous models [Kalman 1960, Kalman and Bucy 1961, Kailath 1981] yield exact closed form solutions to the filtering problem involving linear drift and observation functions along with Gaussian process and observation noises of the additive type. Except for a few special cases, the estimates in non-linear filtering are usually obtained either by approximate analytical schemes like extended Kalman filter (EKF) [Jazwinski 1970, IEEE 1983, Brown and Hwang 1997, Saha and Roy 2009, 2011], the unscented Kalman filter (UKF) [Julier and Uhimann 1997], or more appropriately via sequential Monte Carlo (SMC) techniques [Gordon *et al.* 1993, Doucet *et al.* 2000, Arulampalam *et al.* 2002, Sajeeb *et al.* 2007, 2010]. SMC schemes use an ensemble of weighted realizations (called particles) and, as briefly mentioned earlier, the particles provide an empirical approximation to the posterior *pdf.* The particle system evolves recursively using the process dynamics whilst assimilating the latest observation through generalized Bayes' theorem. A survey of such numerical schemes in the context of non-linear filtering may be found in Budhiraja *et al.* [2007].

6.2 Objective of Stochastic Filtering

In stochastic filtering, the focus is often on obtaining an expression for $\pi_t(\Phi)$ in terms of $\boldsymbol{Y}_{s \le t}$. $\pi_t(\Phi)$ is usually determined as a solution to an SDE called the filtering equation or the Kushner–Stratanovitch (KS) equation. Indeed, every filtering algorithm may be viewed as a scheme to obtain the state estimates, approximately or otherwise, based on the KS equation. A derivation of the KS equation is provided in the next section. The main tools required for the purpose are the familiar generalized Bayes' formula (Section 2.2.3, Chapter 2; also called the Kallianpur–Striebel [1968] formula in the filtering context), Girsanov's theorems on change of measures (Chapter 4) and Ito's formula (Chapter 4). We also provide an exposition on a host of filtering schemes based on the MC simulation route and discuss their performance as applied to a fairly wide class of non-linear filtering problems. These prominently include the problem of dynamical system parameter identification, pertaining mostly to mechanical oscillators.

6.3 Stochastic Filtering and Kushner–Stratanovitch(KS) Equation

For a simple exposition of the formulas involved, it is convenient to work with a scalar–valued function $\Phi(\boldsymbol{X})$ so that its estimate at time t is denoted by $\pi_t(\Phi(\boldsymbol{X}))$ or simply, $\pi_t(\Phi)$. According to the generalized Bayes' formula, if the probability measure $Q \gg P$ for some probability measure, Q, one has:

$$\pi_t(\Phi(\boldsymbol{X})) = E_P\left[\Phi(\boldsymbol{X}_t)|\mathcal{F}_t^{\boldsymbol{Y}}\right] = \frac{E_Q\left[\Phi(\boldsymbol{X}_t)\frac{dP}{dQ}|\mathcal{F}_t^{\boldsymbol{Y}}\right]}{E_Q\left[\frac{dP}{dQ}|\mathcal{F}_t^{\boldsymbol{Y}}\right]} \tag{6.5}$$

$\frac{dP}{dQ}$ is the Radon–Nikodym derivative (Chapter 1). In the context of stochastic filtering, we also refer to $\frac{dP}{dQ}$ as the weight applied to $\Phi(\boldsymbol{X}_t)$, the solution to the process dynamics. Under the new probability measure, Q, if the state (process) and the observation vectors are independent of each other, it is obvious that the conditional expectations under Q and hence $\pi_t(\Phi)$ can, in principle, be obtained by straight forward integration and in the filtering context by purely solving the process equation. This is indeed possible by an application of Girsanov's theorem. By the theorem, if Q is defined on $\mathcal{F}_t^{\boldsymbol{Y}}$ by:

$$\Lambda_t^{-1} = \frac{dQ}{dP} = \exp\left(-\int_0^t \boldsymbol{h}(s,\boldsymbol{X}_s)\,d\boldsymbol{v}_s - \frac{1}{2}\int_0^t \boldsymbol{h}^2(s,\boldsymbol{X}_s)\,ds\right), \tag{6.6}$$

then $\boldsymbol{Y}$ is a Wiener process under Q, independent of $\boldsymbol{X}$. It is also important to note that $\boldsymbol{X}$ continues to have the same law under Q. This is so since Λ_t^{-1} is a martingale with respect to P and $\mathcal{F}$ (Section 4.10.3, Chapter 4) and hence, $E_Q[g(\boldsymbol{X})] = E_P\left[g(\boldsymbol{X})\Lambda_T^{-1}\right] = E_P\left[g(\boldsymbol{X})E_P\left[\Lambda_T^{-1}|\mathcal{F}\right]\right] = E_P[g(\boldsymbol{X})]$ (since $E_P\left[\Lambda_T^{-1}|\mathcal{F}\right] = 1$) for every bounded measurable, g. Now if we define $\rho_t(\Phi) = E_Q\left[\Phi(\boldsymbol{X}_t)\Lambda_t|\mathcal{F}_t^{\boldsymbol{Y}}\right]$, where, $\Lambda_t = \frac{dP}{dQ}$ with Q as the reference measure, then:

$$\pi_t(\Phi) = \frac{\rho_t(\Phi)}{\rho_t(1)} \tag{6.7}$$

Λ_t is given by:

$$\Lambda_t = \exp\left(\int_0^t \boldsymbol{h}(s,\boldsymbol{X}_s)\,d\boldsymbol{Y}_s - \frac{1}{2}\int_0^t \boldsymbol{h}^2(s,\boldsymbol{X}_s)\,ds\right) \tag{6.8}$$

Λ_t is the solution of the SDE $d\Lambda_t = \Lambda_t\boldsymbol{h}(t,\boldsymbol{X}_t)\,d\boldsymbol{Y}_t$ with $\Lambda_0 = 1$ and is, in principle, a Q-martingale. Equation (6.7) is known as Kallianpur-Striebel formula. Note that under Q the observation $\boldsymbol{Y}_t$ itself behaves as a zero–mean Brownian motion (or an Ito integral, which is also a zero–mean martingale). $\rho_t(\Phi)$ is known as the unnormalized conditional distribution and satisfies the Zakai equation [1969] derived in the next section.

A difficulty in the numerical treatment of $\frac{dP}{dQ}$ is in the fact that it is generally a super-martingale, reducing to a (local) martingale only when $\boldsymbol{Y}_t - \int_0^t \boldsymbol{h}(s, \boldsymbol{X}_s)\, ds$ has the same law as $\boldsymbol{v}_t$, a zero mean martingale. This important point is further clarified by the following theorem [Klebaner 2005]:

Theorem: If $\mathcal{M}(t), 0 \leq t \leq T < \infty$ is a continuous local martingale (see Chapter 3 for a definition of local martingale) with $\mathcal{M}(0) = 0$, then its stochastic exponential $\varepsilon(\mathcal{M})$ is given by $\exp\left(\mathcal{M}(t) - \frac{1}{2}[\mathcal{M}, \mathcal{M}](t)\right)$ and is a continuous positive local martingale. Recall that $[\mathcal{M}, \mathcal{M}](t)$ is the quadratic variation of $\mathcal{M}(t)$. Consequently, $\varepsilon(\mathcal{M})$ is an integrable super-martingale and has finite non-increasing expectation.

Proof: Since $\boldsymbol{Y}_t$ is a Q-martingale, $\mathcal{M}(t) = \int_0^t \boldsymbol{h}(s, \boldsymbol{X}_s)\, d\boldsymbol{Y}_s$ is a Q-martingale and $[\mathcal{M}, \mathcal{M}](t) = \int_0^t \boldsymbol{h}^2(s, \boldsymbol{X}_s)\, ds$. To prove the above theorem, we need the following results on stochastic exponentials. If $\varepsilon(\mathcal{M}(t))$ is a stochastic exponential, then $d\varepsilon(\mathcal{M}) = \varepsilon(\mathcal{M})\, d\mathcal{M}(t)$ (see Appendix F). It is thus a stochastic integral with respect to $\mathcal{M}(t)$, i.e.:

$$\varepsilon(\mathcal{M}(t)) = 1 + \int_0^t \varepsilon(\mathcal{M}(s))\, d\mathcal{M}(s) \tag{6.9}$$

Since a stochastic integral with respect to a martingale (or a local martingale) is a local martingale, $\varepsilon(\mathcal{M})$ is a local martingale. Also it is positive. Now, suppose that τ_n is a localizing sequence. With $\varepsilon(\mathcal{M}(t \wedge \tau_n)) \geq 0$ one has by Fatou's lemma (Appendix C):

$$E\left[\lim_{n\to\infty} \inf \varepsilon(\mathcal{M}(t \wedge \tau_n))\right] \leq \lim_{n\to\infty} \inf E\left[\varepsilon(\mathcal{M}(t \wedge \tau_n))\right] \tag{6.10}$$

Since the limit exists, $\lim_{n\to\infty} \varepsilon(\mathcal{M}(t \wedge \tau_n)) = \varepsilon(\mathcal{M}(t))$, also implying that:

$$\lim_{n\to\infty} \inf \varepsilon(\mathcal{M}(t \wedge \tau_n)) = \varepsilon(\mathcal{M}(t)) \tag{6.11}$$

By the martingale property of $\varepsilon(\mathcal{M}(t \wedge \tau_n))$:

$$E\left[\varepsilon(\mathcal{M}(t \wedge \tau_n))\right] = E\left[\varepsilon(\mathcal{M}(0 \wedge \tau_n))\right] = E\left[\varepsilon(\mathcal{M}(0))\right] \tag{6.12}$$

It follows from Eq. (6.10) that $E[\varepsilon(\mathcal{M}(t))] \leq E[\varepsilon(\mathcal{M}(0))] < \infty$. Thus $\varepsilon(\mathcal{M}(t))$ is integrable.

By invoking Fatou's lemma for conditional expectations, we have for $s < t$:

$$E\left[\lim_{n\to\infty} \inf \varepsilon(\mathcal{M}(t \wedge \tau_n)) | \mathcal{F}_s\right] \leq \lim_{n\to\infty} \inf E\left[\varepsilon(\mathcal{M}(t \wedge \tau_n)) | \mathcal{F}_s\right] = E(\mathcal{M}(s \wedge \tau_n)) \tag{6.13}$$

Now, taking the limit as $n \to \infty$, we get the super-martingale property of $\varepsilon(\mathcal{M}(t))$:

$$E[\varepsilon(\mathcal{M}(t))|\mathcal{F}_s] \leq \varepsilon(\mathcal{M}(s)) \text{ a.s.} \tag{6.14}$$

The nature of difficulties in computing $\frac{dP}{dQ}$, which is the stochastic exponential $\varepsilon(\mathcal{M}(t))$, is thus obvious from the above theorem. Being a super-martingale, its mean may decrease to zero with time, rendering the ratio on the RHS of the Eq. (6.5) ill-behaved. This constitutes the root cause of the so-called degeneracy of weights in the implementation of SMC based (particle) filtering techniques and is further elaborated in a later section of this chapter.

6.3.1 Zakai equation

Suppose that Φ is C^2 with $E\left[\int_0^t \Phi^2(\boldsymbol{X}_s)\,ds\right] < \infty$, $E\left[\int_0^t \left|L_s(\Phi(\boldsymbol{X}_s))\right| ds\right] < \infty$, and $E\left[\int_0^t \left(\mathcal{M}_t^{\Phi}\right)^2 ds\right] < \infty$ where $L_s(.)$ is the backward Kolmogorov operator given by Eq. (4.256) (of Chapter 4) and $\mathcal{M}_t^{\Phi} = \int_0^t \frac{\partial(\Phi(\boldsymbol{X}_s))}{\partial x}\sigma(s,\boldsymbol{X}_s)\,d\boldsymbol{B}_s$. Considering the differential $d(\Phi(\boldsymbol{X}_t)\Lambda_t)$, one has:

$$d(\Phi(\boldsymbol{X}_t)\Lambda_t) = \Lambda_t d\Phi(\boldsymbol{X}_t) + \Phi(\boldsymbol{X}_t)\,d\Lambda_t + d[\Phi(\boldsymbol{X}_t),\Lambda_t] \tag{6.15}$$

$[\Phi(\boldsymbol{X}_t),\Lambda_t]$ is the quadratic co-variation term. The last term in the above equation has zero contribution as argued below. By Ito's formula for $\Phi(\boldsymbol{X}_t)$:

$$\Phi(\boldsymbol{X}_t) = \Phi(\boldsymbol{X}_0) + \int_0^t L_s(\Phi(\boldsymbol{X}_s))\,ds + \int_0^t \frac{\partial(\Phi(\boldsymbol{X}_s))}{\partial x}\sigma(s,\boldsymbol{X}_s)\,d\boldsymbol{B}_s$$

$$\Longrightarrow \mathcal{M}_t^{\Phi} = \Phi(\boldsymbol{X}_t) - \Phi(\boldsymbol{X}_0) - \int_0^t L_s(\Phi(\boldsymbol{X}_s))\,ds \tag{6.16}$$

is a martingale (see result 1 of Section 4.7.1, Chapter 4), which is independent of the exponential martingale $\Lambda_t = 1 + \int_0^t \Lambda_s \boldsymbol{h}(s,\boldsymbol{X}_s)\,d\boldsymbol{Y}_s$ since $\boldsymbol{Y}_t$ and $\boldsymbol{B}_t$ are independent. Accordingly, we have $[\Phi(\boldsymbol{X}_t),\Lambda_t] = 0$. Thus with $\Lambda_0 = 1$, we get from Eq. (6.15):

$$\Phi(\boldsymbol{X}_t)\Lambda_t = \Phi(\boldsymbol{X}_0) + \int_0^t \Lambda_s L_s(\Phi(\boldsymbol{X}_s))\,ds + \int_0^t \Lambda_s d\mathcal{M}_s^{\Phi}$$

$$+ \int_0^t \Phi(\boldsymbol{X}_s)\Lambda_s \boldsymbol{h}(s,\boldsymbol{X}_s)\,d\boldsymbol{Y}_s \tag{6.17}$$

By the aforesaid boundedness assumption, all the integrands on the RHS of the above equation are square integrable. We now take expectations on both sides with respect to Q conditioned on $\mathcal{F}_t^{\boldsymbol{Y}}$. Then the third term on the RHS of the last equation vanishes since $\boldsymbol{B}_t$ and $\boldsymbol{Y}_t$ are independent Wiener processes and thereby $E\left[\int_0^t G_s d\boldsymbol{B}_s | \mathcal{F}_s^{\boldsymbol{Y}}\right] = 0$ for any $\mathcal{F}_t$–adapted function, G. Thus, we get:

$$E_Q\left[\Phi(\boldsymbol{X}_t)\Lambda_t|\mathcal{F}_t^{\boldsymbol{Y}}\right] = E_Q\left[\Phi(\boldsymbol{X}_0)|\mathcal{F}_t^{\boldsymbol{Y}}\right] + E_Q\left[\int_0^t \Lambda_s L_s(\Phi(\boldsymbol{X}_s))\, ds|\mathcal{F}_s^{\boldsymbol{Y}}\right]$$

$$+E_Q\left[\int_0^t \Phi(\boldsymbol{X}_s)\Lambda_s \boldsymbol{h}(s,\boldsymbol{X}_s)\, d\boldsymbol{Y}_s|\mathcal{F}_s^{\boldsymbol{Y}}\right] \tag{6.18}$$

Since one can apply Fubini's theorem to stochastic integrals, the time integration and expectation operations may be interchanged, yielding the Zakai SDE, which is linear in $\rho_t(\Phi)$:

$$\rho_t(\Phi) = \rho_0(\Phi) + \int_0^t \rho_s(L_s(\Phi))\, ds + \int_0^t \rho_s(\Phi \boldsymbol{h})\, d\boldsymbol{Y}_s$$

$$\Longrightarrow d\rho_t(\Phi) = \rho_t(L_t(\Phi))\, dt + \rho_t(\Phi\boldsymbol{h})d\boldsymbol{Y}_t \tag{6.19}$$

Here we note that $\boldsymbol{X}_0$ and $\mathcal{F}_t^{\boldsymbol{Y}}$ are independent under Q and so $\rho_0(\Phi) = E_Q\left[\Phi(\boldsymbol{X}_0)|\mathcal{F}_t^{\boldsymbol{Y}}\right] = E_Q[\Phi(\boldsymbol{X}_0)] = E_P[\Phi(\boldsymbol{X}_0)]$. Refer to Liptser and Shiryaev [2001], Protter [2004], and Handel [2007a,b] for more details.

6.3.2 KS equation

The Zakai equation is an SDE satisfied by the unnormalized conditional distribution, $\rho_t(\Phi)$. Using this equation and the Kallianpur–Striebel formula in Eq. (6.7), one can derive the KS equation for $\pi_t(\Phi)$, the normalized distribution. To this end, we consider the differential $d\pi_t(\Phi)$:

$$d\pi_t(\Phi) = d\left(\frac{\rho_t(\Phi)}{\rho_t(1)}\right)$$

$$= \rho_t(\Phi)\, d\left(\frac{1}{\rho_t(1)}\right) + \frac{1}{\rho_t(1)} d\rho_t(\Phi) + d\left[\rho_t(\Phi), \frac{1}{\rho_t(1)}\right] \tag{6.20}$$

From Eq. (6.19):

$$\rho_t(1) = 1 + \int_0^t \rho_s(\boldsymbol{h})\, d\boldsymbol{Y}_s = 1 + \int_0^t \rho_s(1)\,\pi_s(\boldsymbol{h})\, d\boldsymbol{Y}_s$$

$$\Longrightarrow d\rho_t(1) = \rho_t(1)\,\pi_t(\boldsymbol{h})d\boldsymbol{Y}_t \tag{6.21}$$

which admits a solution given by:

$$\rho_t(1) = \exp\left(\int_0^t \pi_s(\boldsymbol{h})\, d\boldsymbol{Y}_s - \frac{1}{2}\int_0^t (\pi_s(\boldsymbol{h}))^2 ds\right) \tag{6.22}$$

It follows that:

$$d\left(\frac{1}{\rho_t(1)}\right) = -\frac{1}{(\rho_t(1))^2}\rho_t(1)\,\pi_t(\boldsymbol{h})\, d\boldsymbol{Y}_t + \frac{1}{(\rho_t(1))^3}(\rho_t(1))^2(\pi_t(\boldsymbol{h}))^2 dt$$

$$= -\frac{\pi_t(\boldsymbol{h})\, d\boldsymbol{Y}_t}{\rho_t(1)} + \frac{(\pi_t(\boldsymbol{h}))^2 dt}{\rho_t(1)} \tag{6.23}$$

Substituting the above equation in Eq. (6.20), we obtain:

$$d\pi_t(\Phi) = \rho_t(\Phi)\left(-\frac{\pi_t(\boldsymbol{h})\, d\boldsymbol{Y}_t}{\rho_t(1)} + \frac{(\pi_t(\boldsymbol{h}))^2 dt}{\rho_t(1)}\right) + \frac{\rho_t(L_s(\Phi))\, dt + \rho_t(\Phi\boldsymbol{h})d\boldsymbol{Y}_t}{\rho_t(1)} - \frac{\pi_t(\boldsymbol{h})\,\rho_t(\Phi\boldsymbol{h})dt}{\rho_t(1)}$$

$$= -\pi_t(\Phi)\,\pi_t(\boldsymbol{h})\, d\boldsymbol{Y}_t + \pi_t(\Phi)\,(\pi_t(\boldsymbol{h}))^2 dt + \pi_t(L_s(\Phi))\, dt + \pi_t(\Phi\boldsymbol{h})d\boldsymbol{Y}_t$$

$$-\pi_t(\boldsymbol{h})\,\pi_t(\Phi\boldsymbol{h})dt$$

$$= \pi_t(L_s(\Phi))\, dt + (\pi_t(\Phi\boldsymbol{h}) - \pi_t(\Phi)\,\pi_t(\boldsymbol{h}))\, d\boldsymbol{Y}_t$$

$$+\pi_t(\boldsymbol{h})\,(-\pi_t(\Phi\boldsymbol{h}) + \pi_t(\Phi)\,\pi_t(\boldsymbol{h}))\, dt$$

$$= \pi_t(L_s(\Phi))\, dt + (\pi_t(\Phi\boldsymbol{h}) - \pi_t(\Phi)\,\pi_t(\boldsymbol{h}))\,(d\boldsymbol{Y}_t - \pi_t(\boldsymbol{h})\, dt) \tag{6.24}$$

In interpreting the equations above, multiplication of a vector with another must be taken as their dot product so that the resulting quantity is a scalar. The KS equation in the integral form can be written as:

$$\pi_t(\Phi) = \pi_0(\Phi) + \int_0^t \pi_s(L_s(\Phi))\,ds + \int_0^t (\pi_s(\Phi\boldsymbol{h}) - \pi_s(\Phi)\,\pi_s(\boldsymbol{h}))\,d\boldsymbol{I}_s \quad (6.25)$$

where the process $\boldsymbol{I}_t = \boldsymbol{Y}_t - \int_0^t \pi_s(\boldsymbol{h})\,ds$ is called the innovation process. Whilst $\boldsymbol{Y}_t$ is a Wiener process under Q, $\boldsymbol{I}_t$ is a Wiener process under P adapted to the sub-filtration $\mathcal{F}_t^{\boldsymbol{Y}}$.

6.3.3 Circularity—the problem of moment closure in non-linear filtering problems

Given the process and observation in Eqs. (6.1a,b), the KS equation, Eq. (6.25), essentially provides an integral expression for the conditional expectation $\pi_t(\Phi)$ whose evolution, within a recursive time–discrete setting, may be written for any $t \in (t_i, t_{i+1}]$ as:

$$\pi_t(\Phi) = \pi_{t_i}(\Phi) + \int_{t_i}^t \pi_s(L_s(\Phi))\,ds$$
$$+\sum_{r=1}^{q}\int_{t_i}^t (\pi_s(\Phi h_r(s,\boldsymbol{X}(s))) - \pi_s(\Phi)\,\pi_s(h_r(s,\boldsymbol{X}(s))))\,dI_{r,s} \quad (6.26)$$

where, $\{I_{r,t}\} = \left\{Y_{r,t} - \int_{t_i}^t \pi_s(h_r(s,\boldsymbol{X}(s)))\,ds\right\} \in \mathbb{R}^q$. The purpose of stochastic filtering is to recursively update the predicted stochastic process $\Phi(\boldsymbol{X})$ to the filtered stochastic process $\widehat{\Phi} = \Phi(\widehat{\boldsymbol{X}})$ such that $I_{r,t}$ is reduced to a zero–mean Wiener process as $t \to \infty$ for each r. By choosing a family of functions, $\Phi_j(\boldsymbol{X}) = X_j, 1 \le j \le m$, one can find the estimate of the entire state vector $\boldsymbol{X}$. Suppose that the second term on the RHS of the KS equation, Eq. (6.26), is approximated as:

$$\int_{t_i}^t \pi_s(L_s(\Phi))\,ds \approx \int_{t_i}^t \pi_i(L_s(\Phi))\,ds$$
$$= \int_{t_i}^t E\left[L_s(\Phi)\,|\mathcal{F}_i^{\boldsymbol{Y}}\right]ds = E\left[\int_{t_i}^t (L_s(\Phi)\,ds)\,|\mathcal{F}_i^{\boldsymbol{Y}}\right] \quad (6.27)$$

where, $\pi_i(.) := \pi_{t_i}(.)$ and $\mathcal{F}_i^{\boldsymbol{Y}} := \mathcal{F}_{t_i}^{\boldsymbol{Y}}$. With this approximation (that helps uncoupling the prediction and updating stages over $(t_i, t_{i+1}]$), the first two terms on the RHS of Eq. (6.26) obtains the expectation $E[\Phi(\boldsymbol{X}(t))]$ according to the process dynamics given by (the solution to) Eq. (6.1a). The two terms constitute the familiar Dynkin's formula (see Eq. 4.128 of Chapter 4). Therefore, the source of random fluctuations is through the observation noise present in the last term on the RHS of Eq. (6.26). In other words, the KS equation involves averaging over paths generated by the process noise. However, a direct evaluation of the filtered estimate $\pi_t(\Phi)$ from the KS equation is severely hindered by the appearance of higher order estimates other than $\pi_t(\Phi)$ in the expression, $\pi_s(\Phi \boldsymbol{h}) - \pi_s(\Phi)\pi_s(\boldsymbol{h})$, thus leading to the so called problem of moment closure. Only when the drift coefficients $\boldsymbol{a}$ and $\boldsymbol{h}$ are linear in $\boldsymbol{X}$ and the equations are driven purely by additive (Gaussian) noises with $\boldsymbol{X}_0$ being a Gaussian random variable, one finds that it leads to an analytically feasible resolution to the moment closure problem, e.g., leading to the Kalman–Bucy filter [1961]. For example, if one considers a one-dimensional filtering problem defining (with some abuse of notation) $a(t,X) = Xa(t)$, $\sigma(t,X) = \sigma(t)$ and $h(t,X) = Xh(t)$, then $X(t)$ is Gaussian and, for $\Phi(X) = X$, the KS equation takes the following form:

$$\begin{aligned}\pi_t(X) &= \pi_0(X) + \int_0^t \pi_s(aX)\,ds \\ &\quad + \int_0^t \left(\pi_s\left(hX^2\right) - \pi_s(X)\pi_s(hX)\right)\left(dY_s - \pi_s(hX)\,ds\right) \end{aligned} \tag{6.28}$$

Assuming that the functions $a(t)$, $h(t)$ and $\sigma(t)$ are constants, one obtains:

$$\begin{aligned}\pi_t(X) &= \pi_0(X) + a\int_0^t \pi_s(X)\,ds \\ &\quad + h\int_0^t \left(\pi_s\left(X^2\right) - \left(\pi_s(X)\right)^2\right)\left(dY_s - h\pi_s(X)\,ds\right) \end{aligned} \tag{6.29}$$

Denoting the conditional mean $\pi_t(X)$ by μ_t and conditional variance $\pi_t\left(X^2\right) - \left(\pi_t(X)\right)^2$ by ${\beta_t}^2$, Eq. (6.29) simplifies to:

$$\begin{aligned}&\mu_t = \mu_0 + a\int_0^t \mu_s ds + h\int_0^t {\beta_s}^2\left(dY_s - h\mu_s ds\right) \\ &\Longrightarrow d\mu_t = a\mu_t dt + h{\beta_t}^2\left(dY_t - h\mu_t dt\right) \end{aligned} \tag{6.30}$$

Similarly one obtains for $\Phi(X) = X^2$:

$$\begin{aligned}\pi_t\left(X^2\right) &= \pi_0\left(X^2\right) + \int_0^t \left(\sigma^2 + 2a\pi_s\left(X^2\right)\right)ds \\ &\quad + h\int_0^t \left(\pi_s\left(X^3\right) - \pi_s\left(X^2\right)\pi_s(X)\right)\left(dY_s - h\pi_s(X)\,ds\right)\end{aligned}$$

$$\Longrightarrow d\pi_t\left(X^2\right)=\left(\sigma^2+2a\pi_t\left(X^2\right)\right)dt+h\left(\pi_t\left(X^3\right)\right.$$

$$\left.-\pi_t\left(X^2\right)\pi_t\left(X\right)\right)\left(dY_t-h\pi_t\left(X\right)dt\right) \tag{6.31}$$

With $d{\beta_t}^2=d\left(\pi_t\left(X^2\right)-\left(\pi_t\left(X\right)\right)^2\right)$, one has:

$$d{\beta_t}^2=d\pi_t\left(X^2\right)-\left(2\pi_t\left(X\right)d\pi_t\left(X\right)+d\left[\pi\left(X\right),\pi\left(X\right)\right]_t\right)$$

$$=\left(\sigma^2+2a\pi_t\left(X^2\right)\right)dt+h\left(\pi_t\left(X^3\right)-\pi_t\left(X^2\right)\pi_t\left(X\right)\right)\left(dY_t-h\pi_t\left(X\right)dt\right)$$

$$-2\pi_t\left(X\right)\left(a\pi_t\left(X\right)dt+h{\beta_t}^2\left(dY_t-h\pi_t\left(X\right)dt\right)\right)$$

$$-h^2{\beta_t}^4dt \tag{6.32}$$

$[\pi(X),\ \pi(X)]_t$ in Eq. (6.32) denotes the quadratic variation of $\pi(X)$. With $\pi_t\left(X^3\right)=E\left[X(t)^3|\mathcal{F}_t^Y\right]$ and X being a Gaussian random variable, one has, $\pi_t\left(X^3\right)=\left(\pi_t\left(X\right)\right)^3+3\pi_t\left(X\right){\beta_t}^2=3\pi_t\left(X\right)\pi_t\left(X^2\right)-2\left(\pi_t\left(X\right)\right)^3$. Substituting this expression for $\pi_t\left(X^3\right)$ in Eq. (6.32) and simplifying, one gets:

$$\frac{d{\beta_t}^2}{dt}=\sigma^2+2a{\beta_t}^2-h^2{\beta_t}^4 \tag{6.33}$$

The closed pair of Eqs. (6.30) and (6.33) constitute the linear Kalman–Bucy filter. Given the initial conditions $\pi_0\left(X\right)=\mu_0$ and ${\sigma_0}^2=Q_0$, the equations yield an unique solution [Bain and Crisan 2009] for $t\in\left[0,T\right]$. Equation (6.33) is known as the Riccati equation.

For non-linear filtering problems, one way to address the circularity problem is by adopting an MC strategy to approximately solve the filtering equation, wherein posterior distributions are empirically approximated over finite ensembles. A class of such MC filters [Manohar and Roy 2006], prominently the particle filters, will be described in Section 6.5.

6.3.4 Unnormalized conditional density and Kushner's theorem

Assume that the unnormalized conditional distribution $\rho_t\left(\Phi\right)$, satisfying Zakai Eq. (6.19), has a *pdf* $q_t(.)$ such that:

$$\rho_t\left(\Phi\right)=\int_{\mathbb{R}^m}\Phi\left(\boldsymbol{x}\right)\rho_t\left(d\boldsymbol{x}\right)=\int_{\mathbb{R}^m}\Phi\left(\boldsymbol{x}\right)q_t\left(\boldsymbol{x}\right)d\boldsymbol{x} \tag{6.34}$$

Then $q_t(\boldsymbol{x})$ satisfies the following stochastic PDE:

$$dq_t(\boldsymbol{x}) = L_t^*(q_t(\boldsymbol{x}))\,dt + \boldsymbol{h}d\boldsymbol{Y}_t q_t(\boldsymbol{x}) \tag{6.35}$$

with $q_0 \in \mathbb{R}$ being square integrable. L_t^* is the adjoint operator (the Fokker–Planck operator) given in Eq. (4.273) of Chapter 4.

Proof: From the Zakai equation, one gets:

$$\begin{aligned}\rho_t(\Phi) &= \int_{\mathbb{R}^m} \Phi(\boldsymbol{x})\,q_t(\boldsymbol{x})\,d\boldsymbol{x}\\ &= \int_{\mathbb{R}^m} \Phi(\boldsymbol{x})\,q_0(\boldsymbol{x})\,d\boldsymbol{x} + \int_0^t \left(\int_{\mathbb{R}^m} L_s(\Phi(\boldsymbol{x}))\,q_s(\boldsymbol{x})\,d\boldsymbol{x} \right) ds\\ &\quad + \int_0^t \left(\int_{\mathbb{R}^m} \Phi(\boldsymbol{x})\boldsymbol{h}q_s(\boldsymbol{x})\,d\boldsymbol{x} \right) d\boldsymbol{Y}_s\end{aligned} \tag{6.36a}$$

By integrating the second term by parts, one has:

$$\begin{aligned}\rho_t(\Phi) &= \int_{\mathbb{R}^m} \Phi(\boldsymbol{x})\,q_0(\boldsymbol{x})\,d\boldsymbol{x} + \int_0^t \left(\int_{\mathbb{R}^m} \Phi(\boldsymbol{x})L_s^*(q_s(\boldsymbol{x}))\,d\boldsymbol{x} \right) ds\\ &\quad + \int_0^t \left(\int_{\mathbb{R}^m} \Phi(\boldsymbol{x})q_s(\boldsymbol{x})\,\boldsymbol{h}d\boldsymbol{x} \right) d\boldsymbol{Y}_s\end{aligned} \tag{6.36b}$$

Applying Fubini's theorem and interchanging the integrations in the second and third terms of the RHS above, we obtain:

$$\begin{aligned}\rho_t(\Phi) &= \int_{\mathbb{R}^m} \Phi(\boldsymbol{x})\,q_0(\boldsymbol{x})\,d\boldsymbol{x} + \int_{\mathbb{R}^m} \Phi(\boldsymbol{x}) \left(\int_0^t L_s^*(q_s(\boldsymbol{x}))\,ds \right) d\boldsymbol{x}\\ &\quad + \int_{\mathbb{R}^m} \Phi(\boldsymbol{x}) \left(\int_0^t q_s(\boldsymbol{x})\,\boldsymbol{h}d\boldsymbol{Y}_s \right) d\boldsymbol{x}\\ &= \int_{\mathbb{R}^m} \left(q_0(\boldsymbol{x}) + \int_0^t L_s^*(q_s(\boldsymbol{x}))\,ds + \int_0^t q_s(\boldsymbol{x})\,\boldsymbol{h}d\boldsymbol{Y}_s \right) \Phi(\boldsymbol{x})\,d\boldsymbol{x}\end{aligned} \tag{6.37a}$$

By localization, one arrives at the identity:

$$q_t(\boldsymbol{x}) = q_0(\boldsymbol{x}) + \int_0^t L_s^*(q_s(\boldsymbol{x}))\,ds + \int_0^t q_s(\boldsymbol{x})\,\boldsymbol{h}d\boldsymbol{Y}_s \tag{6.37b}$$

This yields the required result in Eq. (6.35) for the unnormalized density $q_t(.)$ via Eq. (6.37b), which is also known as the adjoint Zakai equation.

In a similar manner, the Kushner theorem defines a *pdf* $p_t(\boldsymbol{x})$ for the conditional distribution $\pi_t(\Phi)$ of the KS equation, Eq. (6.25), i.e.:

$$\pi_t(\Phi) = \int_{\mathbb{R}^m} \Phi(\boldsymbol{x})\, p_t(\boldsymbol{x})\, d\boldsymbol{x} \tag{6.38}$$

The Kushner theorem states that $p_t(\boldsymbol{x})$ satisfies the stochastic PDE given by:

$$dp_t(\boldsymbol{x}) = L_t^*(p_t(\boldsymbol{x}))\, dt + p_t(\boldsymbol{x})(\boldsymbol{h} - \pi_t(\boldsymbol{h}))\, d\boldsymbol{I}_t \tag{6.39}$$

with $p_0 \in \mathbb{R}$ being square integrable.

Proof: From Eq. (6.25), it follows that:

$$\begin{aligned}\pi_t(\Phi) &= \int_{\mathbb{R}^m} \Phi(\boldsymbol{x})\, p_t(\boldsymbol{x})\, d\boldsymbol{x} \\ &= \int_{\mathbb{R}^m} \Phi(\boldsymbol{x})\, p_0(\boldsymbol{x})\, d\boldsymbol{x} + \int_0^t \left(\int_{\mathbb{R}^m} L_s(\Phi(\boldsymbol{x}))\, p_s(\boldsymbol{x})\, d\boldsymbol{x} \right) ds \\ &\quad + \int_0^t \left(\int_{\mathbb{R}^m} (\Phi(\boldsymbol{x})\boldsymbol{h} - \Phi(\boldsymbol{x})\pi_s(\boldsymbol{h}))\, p_s(\boldsymbol{x})\, d\boldsymbol{x} \right) d\boldsymbol{I}_s \end{aligned} \tag{6.40a}$$

By integrating the second term on the last RHS by parts, one has:

$$\begin{aligned}\pi_t(\Phi) &= \int_{\mathbb{R}^m} \Phi(\boldsymbol{x})\, p_0(\boldsymbol{x})\, d\boldsymbol{x} + \int_0^t \left(\int_{\mathbb{R}^m} \Phi(\boldsymbol{x})\, L_s^*(p_s(\boldsymbol{x}))\, d\boldsymbol{x} \right) ds \\ &\quad + \int_0^t \left(\int_{\mathbb{R}^m} (\Phi(\boldsymbol{x})\boldsymbol{h} - \Phi(\boldsymbol{x})\pi_s(\boldsymbol{h}))\, p_s(\boldsymbol{x})\, d\boldsymbol{x} \right) d\boldsymbol{I}_s \\ &= \int_{\mathbb{R}^m} \Phi(\boldsymbol{x})\, p_0(\boldsymbol{x})\, d\boldsymbol{x} + \int_{\mathbb{R}^m} \Phi(\boldsymbol{x}) \left(\int_0^t L_s^*(p_s(\boldsymbol{x}))\, ds \right) d\boldsymbol{x} \\ &\quad + \int_{\mathbb{R}^m} \Phi(\boldsymbol{x}) \left(\int_0^t (\boldsymbol{h} p_s(\boldsymbol{x}) - \pi_s(\boldsymbol{h})\, p_s(\boldsymbol{x}))\, d\boldsymbol{I}_s \right) d\boldsymbol{x} \\ \Longrightarrow \pi_t(\Phi) &= \int_{\mathbb{R}^m} \Phi(\boldsymbol{x}) \Bigg(p_0(\boldsymbol{x}) + \int_0^t L_s^*(p_s(\boldsymbol{x}))\, ds \end{aligned}$$

$$+\int_0^t \left(\boldsymbol{h}p_s(\boldsymbol{x}) - \pi_s(\boldsymbol{h})\,p_s(\boldsymbol{x})\right)d\boldsymbol{I}_s\Big)\,d\boldsymbol{x} \tag{6.40b}$$

Again, by localization, the following required identity is readily obtained:

$$p_t(\boldsymbol{x}) = p_0(\boldsymbol{x}) + \int_0^t L_s^*(p_s(\boldsymbol{x}))\,ds + \int_0^t \left(\boldsymbol{h}p_s(\boldsymbol{x}) - \pi_s(\boldsymbol{h})\,p_s(\boldsymbol{x})\right)d\boldsymbol{I}_s \tag{6.41}$$

Existence and uniqueness of the densities $p_t(x)$ and $q_t(x)$ depend on the nature of the filtering problem and certain (fairly general) regularity conditions. The interested reader may find more details in Kurtz and Ocone [1988], Florchinger and Schiltz [1996], and Yazigi [2010]. The normalized conditional density $p_t(x)$ can be expressed in terms of $q_t(\boldsymbol{x})$ as follows:

$$p_t(\boldsymbol{x}) = \frac{q_t(\boldsymbol{x})}{\int_{\mathbb{R}^m} q_t(\boldsymbol{z})\,d\boldsymbol{z}} \tag{6.42}$$

The validity of the above identity follows from the fact that $\int_{\mathbb{R}^m} p_t(\boldsymbol{x})\,d\boldsymbol{x} = 1$. It may also be verified from the Kallianpur–Striebel formula. Since, by the formula, $\pi_t(\Phi) = \frac{\rho_t(\Phi)}{\rho_t(1)}$, one has:

$$\pi_t(\Phi) = \int_{\mathbb{R}^m} \Phi(\boldsymbol{x})\,p_t(\boldsymbol{x})\,d\boldsymbol{x} = \frac{\int_{\mathbb{R}^m} \Phi(\boldsymbol{x})\,q_t(\boldsymbol{x})\,d\boldsymbol{x}}{\int_{\mathbb{R}^m} q_t(\boldsymbol{x})\,d\boldsymbol{x}} \Longrightarrow p_t(\boldsymbol{x}) = \frac{q_t(\boldsymbol{x})}{\int_{\mathbb{R}^m} q_t(\boldsymbol{z})\,d\boldsymbol{z}} \tag{6.43}$$

6.4 Non-linear Stochastic Filtering and Solution Strategies

As observed earlier, it is practically not feasible to have a finite dimensional (closed) representation for the non-linear filtering equation—the KS equation—which is a non-linear stochastic integro–differential equation (and not an SDE) in $\pi_t(\Phi)$. However, being a linear (parabolic) SPDE, the Zakai equation is generally looked upon as being less intractable and hence easier to work with. One has, for instance, the advantage of applying the classical PDE solution techniques to the Zakai equation. In particular, the equation may be directly approximated by a finite–dimensional projection scheme such as a Galerkin method [Germani and Piccioni 1984, Ahmed and Radaideh 1997] and the solution expressed as a finite combination of an orthogonal series. This yields a family of stochastic DEs which may be numerically solved and the approach may be quite effective for lower dimensional filtering problems. These so-called semi-analytical techniques of obtaining approximations to the Zakai equation also include those based on finite difference [Piccioni 1987, Gyöngy 1998], finite element [Walsh 2005] and spectral [Lototsky *et al.* 1997] methods.

6.4.1 Extended Kalman filter (EKF)

Extended Kalman filter [Jazwinski 1970] is an extension of the linear Kalman filtering (KF) theory to account for nonlinearity, but not non-Gaussianity, in stochastic filtering. Since a discussion on the EKF subsumes that on the KF, no separate discussion on the latter is included in this book. For the EKF [Judd 2003, Riadh 2006], the local linearization is the means to non-optimally deal with filtering problems with drift/diffusion nonlinearity. Apart from its inability to capture the non-Gaussian features in the estimation, the EKF is also limited by the dimensionality curse, as it typically needs elaborate tuning of the process noise covariance terms with increasing filter dimensions.

The EKF uses linearization tools in order to incorporate the traditional KF formulas for dealing with non-linear dynamic state estimation problems [Yun and Shinozuka 1980; Brown and Hwang 1997]. Consider, for example, the case of the state equation, Eq. (6.1a), with an additive noise and a deterministic forcing function, $\mathcal{P}(t) \in \mathbb{R}^m \times \mathbb{R}^+ \to \mathbb{R}$. Here, it is convenient to write the observation Eq. (6.1b) in an algebraic form (it is also a more practical portrayal over its SDE counterpart) and thus the process and observation equations are presently of the forms:

$$dX_t = a(t,X_t)\,dt + \mathcal{P}(t)\,dt + \sigma(t)\,dB(t) \tag{6.44a}$$

$$y_t = \hbar(t,X_t) + \sigma_y(t)B_y(t) \tag{6.44b}$$

$\hbar(t,X_t) : \mathbb{R}^+ \times \mathbb{R}^m \to \mathbb{R}^q$ is the observation vector function. With reference to Eq. (6.1b), $\hbar(t,X_t) = \int_0^t h(s,X_s)\,ds \in \mathbb{R}^q$. $B_y(t) \in \mathbb{R}^r$ an r-dimensional zero–mean P-Brownian motion. Given an initial estimate $\widehat{X}_0 := \pi_0(X) = E\left[X_0 | \mathcal{F}_0^y\right]$ of the state and error covariance $\widehat{P}_0 := E\left[\left(X_0 - \widehat{X}_0\right)\left(X_0 - \widehat{X}_0\right)^T | \mathcal{F}_0^y\right]$, the recursive state estimation follows the steps enumerated below:

Step 1. Begin with the filtered (updated) estimate $\widehat{X}_j$ and its predicted error covariance matrix $\widehat{P}_j$ for any $j = 0,1,\ldots,N$ with $t_N = T$

Step 2. Evaluate the predicted mean state $\widetilde{X}_{j+1}$ and error covariance matrix $\widetilde{P}_{j+1}$ by solving the following non-linear DE at $t = t_{j+1}$:

$$\dot{\widetilde{X}} = a\left(t,\widehat{X}\right) + \mathcal{P}(t) \tag{6.45}$$

$$\widetilde{P}_{j+1} = \Phi\left(t_{j+1},t_j\right)\widehat{P}_j\Phi^T\left(t_{j+1},t_j\right) + Q_{j+1} \tag{6.46}$$

Equation (6.45) may be numerically integrated, e.g., by the deterministic explicit Euler method. Alternatively, if $\nabla_j = \left[\frac{\partial a(t,X)}{\partial X}\right]_{X=\widehat{X}_j,t=t_j}$ is the $m \times m$ system Jacobian matrix

(Frechet derivative) of the non-linear function $\boldsymbol{a}(t,\boldsymbol{X})$, an approximate $\widetilde{\boldsymbol{X}}_{j+1}$ can be obtained from the linearized equation $\dot{\widetilde{\boldsymbol{X}}} = \nabla\widetilde{\mathbf{X}} + \mathcal{P}(t)$. $\boldsymbol{\Phi}\left(t,t_j\right)$, the state transition matrix of the linearized system appearing in Eq. (6.46), is given by:

$$\boldsymbol{\Phi}\left(t,t_j\right) = \exp\left[\left(\nabla_j\right)\left(t-t_j\right)\right] \tag{6.47}$$

$\boldsymbol{Q}_{j+1}$ is the process noise covariance matrix and is obtained as:

$$\boldsymbol{Q}_{j+1} = \iint_{t_j}^{t_{j+1}} \boldsymbol{\Phi}\left(t_{j+1},u\right)\sigma(u)E[\boldsymbol{B}(u)\boldsymbol{B}^T(v)]\sigma^T(v)\boldsymbol{\Phi}^T\left(t_{j+1},v\right)du\,dv \tag{6.48}$$

Evaluation of $\boldsymbol{Q}_{j+1}$, which can be a difficult task when the system dimension is large, may require a numerical integration tool, such as Gauss quadrature.

Step 3. Compute the optimum Kalman gain as:

$$\boldsymbol{K}_{j+1} = \widetilde{\boldsymbol{P}}_{j+1}\mathbb{H}^T{}_{j+1}\left[\mathbb{H}_{j+1}\widetilde{\boldsymbol{P}}_{j+1}\mathbb{H}^T_{j+1} + \boldsymbol{R}_{y,j+1}\right]^{-1} \in \mathbb{R}^{m\times q} \tag{6.49}$$

where, $\mathbb{H}_{j+1} = \left[\frac{\partial \boldsymbol{h}(t,X)}{\partial X}\right]_{X=\widetilde{X}_{j+1},t=t_{j+1}} \in \mathbb{R}^{q\times m}$ and $\boldsymbol{R}_{y,j+1} \in \mathbb{R}^{q\times q}$ is the covariance matrix of the observation noise at $t = t_{j+1}$.

Step 4. Obtain the updated estimate $\widehat{\boldsymbol{X}}_{j+1}$ and the associated covariance matrix $\widehat{\boldsymbol{P}}_{j+1}$ as:

$$\widehat{\boldsymbol{X}}_{j+1} = \widetilde{\boldsymbol{X}}_{j+1} + \boldsymbol{K}_{j+1}\left(\boldsymbol{y}_{j+1} - \boldsymbol{h}\left(t_{j+1},\widetilde{\boldsymbol{X}}_{j+1}\right)\right) \tag{6.50a}$$

$$\widehat{\boldsymbol{P}}_{j+1} = \left[\boldsymbol{I} - \boldsymbol{K}_{j+1}\mathbb{H}_{j+1}\right]\widetilde{\boldsymbol{P}}_{j+1} \tag{6.50b}$$

$\boldsymbol{y}_{j+1}$ in Eq. (6.50a) is the observation vector of the system states at $t = t_{j+1}$. Here $\boldsymbol{I}$ is an $m \times m$ identity matrix. Because of the linearization involved, the EKF-based estimate approximates the true mean square estimate corresponding to a strictly Gaussian posterior density function.

6.4.2 EKF using locally transversal linearization (LTL)

LTL based implicit methods of solving non-linear SDEs have been introduced in Chapter 5. The general idea of an LTL-based EKF [Ghosh *et al.* 2007] for stochastic filtering is to replace the non-linear vector field of the process Eq. (6.44a) by a time–invariant conditionally linearized vector field over a time step and then apply the Kalman filter estimation formulas at the right end of the time step. Consider the partition Π_N of the time interval $(0,T] \in \mathbb{R}$ with $t_N = T$ and $\Delta_{j+1} = t_{j+1} - t_j$. One now replaces the process equation by a suitably chosen set of N transversally linearized equations wherein the solution to any element of the set should be, in a sense, a representative of the associated non-linear flow over the j-th time step. Note that when the observation equation, Eq. (6.44b), is non-linear, N conditionally linearized algebraic equations must additionally be chosen.

Corresponding to Eq. (6.44), the following LTL system (with constant coefficients) may thus be written over $(t_j, t_{j+1}]$:

$$dX = A\left(t, \widehat{X}\right) X\, dt + \mathcal{P}(t)\, dt + \sigma(t)\, dB(t) \tag{6.51a}$$

$$y_t = H\left(t, \widehat{X}\right) X + \sigma_y(t)\, B_y(t) \tag{6.51b}$$

For convenience, no separate notations for the linearized approximations to the original vector states X_t and y_t are used. Since the estimate $\widehat{X}$ at time t is unknown, Eqs. (6.51a,b) may be viewed as conditionally linear. This however requires that the vector functions $a(t,X)$ and $\hbar(t,X)$ (in Eqs. (6.44a,b) are decomposable as:

$$a(t,X) = A(t,X)X, \hbar(t,X) = H(t,X)X \tag{6.52}$$

The LTL-based (EKF) filter employs steps similar to those of the conventional EKF (Eqs. 6.45-6.50) for state estimation. Thus, starting with the filter estimates $\widehat{X}_j$ and its error covariance matrix $\widehat{P}_j$ at time $t = t_j$, the predicted mean state $\widetilde{X}_{j+1}$ at $t = t_{j+1}$ is first obtained as an LTL-based solution of the non-linear process equation without the noise term. $\widetilde{P}_{j+1}$, the predicted error covariance matrix, is evaluated on the basis of Eq. (6.46) with $\Phi\left(t, t_j\right)$ given by:

$$\Phi\left(t, t_j\right) = \exp\left[A(t - t_j)\right] \tag{6.53}$$

Using $H_{j+1} = H\left(t_{j+1}, \widetilde{X}_{j+1}\right)$ as in Eq. (6.524), the Kalman gain matrix is obtained as:

$$K_{j+1} = \widetilde{P}_{j+1} H_{j+1}^T \left[H_{j+1} \widetilde{P}_{j+1} H_{j+1}^T + R_{y,j+1}\right]^{-1} \tag{6.54}$$

The updated estimate $\widehat{X}_{j+1}$ is determined using Eq. (6.50a). Finally thc updated error covariance matrix $\widehat{P}_{j+1}$ is evaluated from:

$$\widehat{P}_{j+1} = \left[I - K_{j+1} H_{j+1}\right] \widetilde{P}_{j+1} \tag{6.55}$$

Example 6.1

Consider a hardening Duffing oscillator under a deterministic harmonic driving force and a Gaussian white noise:

$$\ddot{X} + c\dot{X} + kX + \alpha X^3 = \mathscr{f} \cos \lambda t + \sigma W(t) \tag{6.56}$$

$\mathscr{f}$ is the amplitude of the deterministic forcing function. σ is the intensity of the process noise. We estimate $X(t)$ by EKF and LTL-based EKF using the observation Eq. (6.44b).

Solution

Results are obtained for two cases. One is with the observation $y(t)$ taken to be the displacement $X(t)$ and the other with $y(t)$ being the velocity $\dot{X}(t)$. Writing Eq. (6.56) in a state space form (with $X_1 := X$ and $X_2 := \dot{X}$), one has:

$$dX_1 = X_2 dt \tag{6.57a}$$

$$dX_2 = \left(-cX_2 - kX_1 - \alpha X_1^3 + \mathscr{f} \cos \lambda t\right) dt + \sigma dB(t) \tag{6.57b}$$

Estimation by EKF

With $\mathbf{X} = (X_1, X_2)^T, \widehat{\mathbf{X}} = (\widehat{X}_1, \widehat{X}_2)^T$, $\boldsymbol{a}(t, \mathbf{X}) = \left(X_2, -cX_2 - kX_1 - \alpha X_1^3\right)^T$ one has at any time $t_j \in (0, T]$:

$$\nabla_j = \left[\frac{\partial \boldsymbol{a}(t, \mathbf{X})}{\partial \mathbf{X}}\right]_{\mathbf{X}=\widehat{\mathbf{X}}_j, t=t_j} = \begin{bmatrix} 0 & 1 \\ -k - 3\alpha \hat{X}_{1,j}^2 & -c \end{bmatrix} \tag{6.58}$$

We have $\mathbb{H}_j = [1 \quad 0]$ if $\mathscr{h}(t, \mathbf{X}) = X_1(t)$, and $\mathbb{H}_j = [0 \quad 1]$ if $\mathscr{h}(t, \mathbf{X}) = X_2(t)$. With $\widehat{\mathbf{X}}_0 = \left(\widehat{X}_{1,0}, \widehat{X}_{2,0}\right)^T = (0, 0)^T$ and $\widehat{\mathbf{P}}_0 = 0.01\mathbf{I}_{2\times 2}$, the recursive filtering algorithm described in steps 1 – 5 (Eqs. 6.45 – 6.50) gives the estimates of the displacement and velocity as shown in Fig. 6.2 with a uniform time step $\Delta = 0.01$.

y_{j+1} in Eq. (6.50a) denotes the actually observed system state at discrete time instants. In this example, however, it is synthetically generated by first obtaining a solution $\mathbf{X}_{j+1}$ to the process dynamics (Eq. 6.44a) without the noise term) via any integration scheme (Chapter 5; typically a higher order scheme than the one used in the prediction step) and then corrupting it with a observation noise v_t. y_{j+1} is thus obtained from the equation:

$$y_{j+1} = \mathbb{H}_{j+1}\mathbf{X}_{j+1} + \sigma_{y,j+1} v_{j+1} \tag{6.59}$$

The observation noise variance is taken as 0.009.

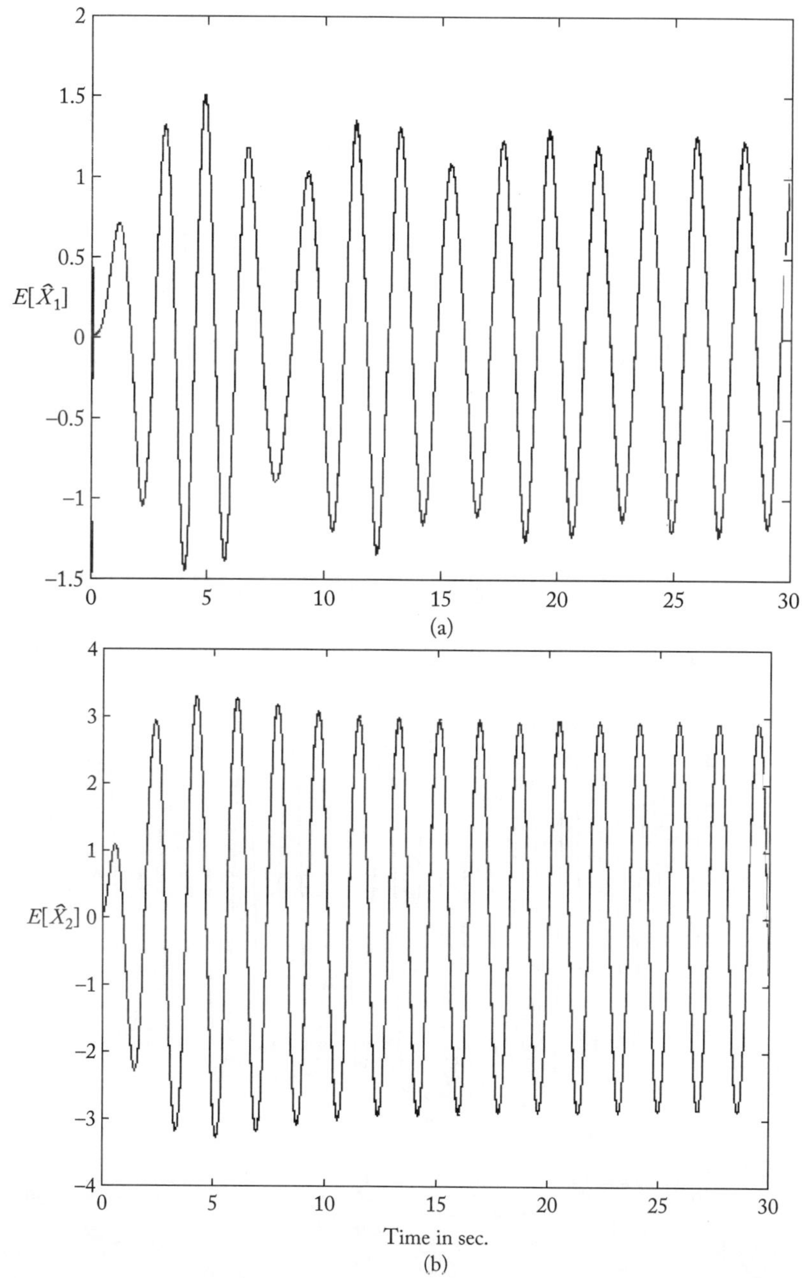

Fig. 6.2 Stochastic filtering by EKF; Example 6.1 – Duffing oscillator; $c = 0.05\ N\text{-}s/m$, $k = 4\ N/m$, $\omega = \sqrt{k}$; $\alpha = 7$ N/m^3, $f = 3$ N, $\lambda = 1.5\omega$, $T = 30$ s, $\Delta = 0.01\ s$, (a) filtered estimate $\hat{X}_1$ with measurement on displacement and (b) filtered estimate $\hat{X}_2$ with measurement on velocity; dark-line – filtered estimate, dashed-line – measured solution

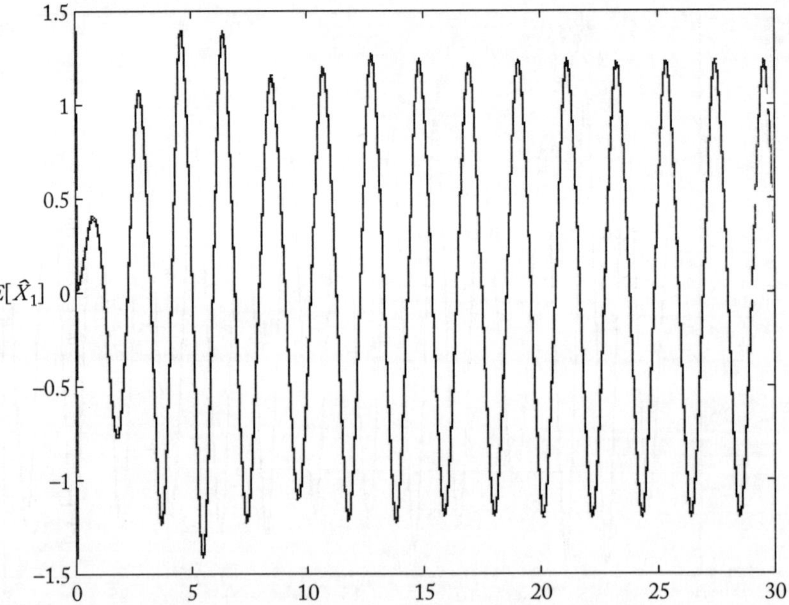

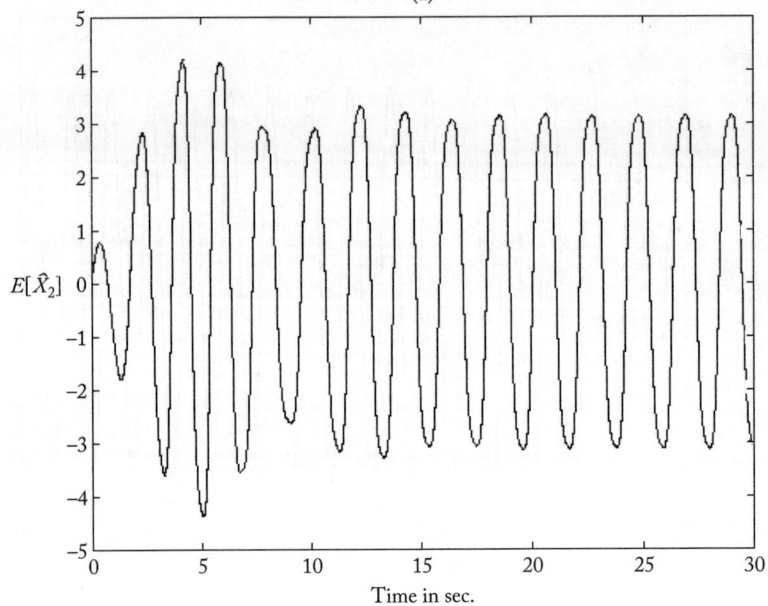

Fig. 6.3 Stochastic filtering by LTL-based EKF; Example 6.1 – Duffing oscillator; $c = 0.05\ N\text{-}s/m,\ k = 4\ N/m,\ \omega = \sqrt{k};\ \alpha = 7\ \text{N/m}^3,\ \not{f} = 3\ \text{N},\ \lambda = 1.5\omega,\ T = 30\,\text{s},\ \Delta = 0.01\ s$, (a) filtered estimate $\widehat{X}_1$ with measurement on displacement and (b) filtered estimate $\widehat{X}_2$ with measurement on velocity; dark-line – filtered estimate, dashed-line – measured solution

Estimation by LTL-based EKF

Here $A\left(t,\widehat{X}\right)$ in Eq. (6.52) is given by $\begin{bmatrix} 0 & 1 \\ \left(-K-\alpha \hat{X}_{1,j}^{2}\right) & -C \end{bmatrix}$. $H\left(t,\widehat{X}\right)=\begin{bmatrix}1 & 0\end{bmatrix}$ if the measurement is on displacement and $H\left(t,\widehat{X}\right)=\begin{bmatrix}0 & 1\end{bmatrix}$ if it is on velocity. With $\widehat{X}_0=\left(\hat{X}_{1,0},\hat{X}_{2,0}\right)^{\mathrm{T}}=(0,0)^T$ and $\widehat{P}_0=0.01\,\mathbf{I}_{2\times 2}$, the filtered estimates through the LTL-based EKF are shown in Fig. 6.3. As in the last case, during the simulations, y_{j+1} is generated by first obtaining an LTL-based solution X_{j+1} to the process equation (Eq. 6.56) without the noise term and then corrupting it with a measurement noise v_t whose variance is taken as 0.009.

6.4.3 EKF applied to parameter estimation

Prior to discussing MC filters, the more powerful of the filtering methods, we take a short detour to introduce the topic of parameter estimation, which is also popularly referred to as system identification. Stochastic filters, as a modern tool for this important class of problems, aim at estimating the model parameters appearing in the process equations by treating them (the parameters) as pseudo-stochastic processes (in contrast to the system or process states that naturally evolve as stochastic processes in accordance with the process dynamics) and conditioning them on the experimentally observed noisy data. System identification methods have their origin in the area of control [Alighanbari *et al.* 2005] of dynamical systems and are often posed within the framework of a Kalman filter.

In system identification problems considered in this book, the model parameters are declared as pseudo-states co-evolving with the original system states and thus one defines the augmented state vector as:

$$X=\left\{X_s^T,X_p^T\right\}^T \tag{6.60}$$

The subscript '*s*' refers to the system states and '*p*' to the parameter pseudo-states, the latter typically including stiffness, damping or forcing parameters or those appearing in the non-linear terms in the system model. Specifically, $X_s=\left\{Z^T,\dot{Z}^T\right\}^T$ where Z and $\dot{Z}$ are respectively the displacement and velocity vectors and $X_p\in\mathbb{R}^{n_p}$ where n_p is the dimension of the parameter vector. If the structural system model has '*m*' degrees of freedom (*dof* s), then $X\in\mathbb{R}^J$, where $J=2m+n_p$. The augmented process model is thus of the form:

$$dX_s(t)=\breve{a}\left(t,X_s(t),X_p(t)\right)dt+\breve{\mathcal{P}}(t)\,dt+\breve{\sigma}_s(t)\,dB_s(t) \tag{6.61a}$$

$$dX_p=0+\breve{\sigma}_p(t)\,dB_p(t) \tag{6.61b}$$

where, $\breve{a}:\mathbb{R}^+\times\mathbb{R}^{2m}\times\mathbb{R}^{n_p}\to\mathbb{R}^{2m}$, $\breve{\mathcal{P}}\in\mathbb{R}^{2m}$ and $\breve{\sigma}_s\in\mathbb{R}^{2m\times n}$ are respectively the drift vector, deterministic forcing vector and diffusion matrix. $B_s\in\mathbb{R}^n$ is a vector of independent standard Wiener processes. $\breve{\sigma}_p \in \mathbb{R}^{n_p \times n_p}$ is a diagonal matrix and $\boldsymbol{B}_p \in \mathbb{R}^{n_p}$ a vector of (Weiner) noise processes associated with the parameter evolution. $\boldsymbol{X}_p$ is thus assumed to evolve according to a zero drift SDE. The two equations, Eqs. (6.61a,b), may be combined to get:

$$d\boldsymbol{X}(t) = \boldsymbol{a}(t,\boldsymbol{X}(t))dt + \mathcal{P}(t)\,dt + \sigma(t)\,d\boldsymbol{B}(t) \tag{6.62}$$

with $\boldsymbol{a} : \mathbb{R}^+ \times \mathbb{R}^J \to \mathbb{R}^J$ and $\mathcal{P} \in \mathbb{R}^J$. $\sigma(t)$ is a $J \times (n + n_p)$ diffusion matrix and $\boldsymbol{B} \in \mathbb{R}^{n+n_p}$. Note that, with the augmentation of the states by the unknown parameters, Eq. (6.61a) becomes non-linear even if the system model were originally linear. The measurement equation is taken same as in Eq. (6.44b):

$$y_t = \int_0^t \boldsymbol{h}(s,\boldsymbol{X})\,ds + \boldsymbol{v}(t) = \boldsymbol{\hbar}(t,\boldsymbol{X}_t) + \sigma_y \boldsymbol{B}_y(t) \tag{6.63}$$

$\boldsymbol{\hbar}(t,\boldsymbol{X}) : \mathbb{R}^+ \times \mathbb{R}^J \to \mathbb{R}^q$ denotes the measurement vector function and $\boldsymbol{v}(t) := \boldsymbol{\sigma}_y \boldsymbol{B}_y(t) \in \mathbb{R}^q$ a q-dimensional Wiener process (the measurement noise) with zero mean and covariance matrix $\boldsymbol{R}_y = \boldsymbol{\sigma}_y^T \boldsymbol{\sigma}_y \in \mathbb{R}^{q \times q}$. Here $\boldsymbol{\sigma}_y$ is a diagonal matrix.

Given the process and measurement, Eqs. (6.62) and (6.63), one can proceed with the same recursive algorithm as in Eqs. (6.45)–(6.50) and implement the conventional EKF for combined state-parameter estimation. This is best illustrated with a simple example where the parameters of a Duffing oscillator are estimated using the algorithm.

Example 6.2

We consider a Duffing oscillator similar to the one in Example 6.1 and estimate the parameters - damping c, stiffness k and non-linear stiffness α- from a set of measurements related to the system states. The reference values for c, k and α are 0.9 N-s/m, 7 N/m and 3 N/m^3 respectively [Ghosh *et al.* 2007].

Solution

With $X_1 := X$ and $X_2 := \dot{X}$, it follows that $\boldsymbol{X}_s = \{X_1, X_2\}^T$. If we define $X_3 := c$, $X_4 := k$ and $X_5 := \alpha$, one has $\boldsymbol{X}_p = \{X_3, X_4, X_5\}^T$ and $\boldsymbol{X} = \{X_1, X_2, X_3, X_4, X_5\}^T$. Correspondingly, the augmented process model is given by:

$$dX_1 = X_2 dt \tag{6.64a}$$

$$dX_2 = \left(-X_3 X_2 - X_4 X_1 - X_5 X_1^3 + \mathcal{P}\cos\,\lambda t\right)dt + \sigma dB(t) \tag{6.64b}$$

$$dX_3 = \sigma_c dB_c(t) \tag{6.64c}$$

$$dX_4 = \sigma_k dB_k(t) \tag{6.64d}$$

$$dX_5 = \sigma_\alpha dB_\alpha(t) \tag{6.64e}$$

Here, one has:

$$\boldsymbol{a}(t,\boldsymbol{X}) = \left(X_2(t) \quad \left(-X_3X_2 - X_4X_1 - X_5X_1^3 + \mathcal{P}\cos\lambda t\right) \quad 0 \quad 0 \quad 0\right)^T \tag{6.65}$$

∇_j, the Jacobian matrix of the non-linear function, $\boldsymbol{a}(t,\boldsymbol{X})$, is obtained (at $t = t_j$) as:

$$\nabla_j = \begin{bmatrix} 0 & 1 & 0 & 0 & 0 \\ \left(-\hat{X}_{4,j} - 3\hat{X}_{5,j}\hat{X}_{1,j}^2\right) & -\hat{X}_{3,j} & \hat{X}_{2,j} & -\hat{X}_{1,j} & \hat{X}_{1,j}^3 \\ 0 & 1 & 0 & 0 & 0 \\ 0 & 1 & 0 & 0 & 0 \\ 0 & 1 & 0 & 0 & 0 \end{bmatrix} \tag{6.66}$$

If the measurement is on the displacement alone, we have $\hbar(t,\boldsymbol{X}) = X_1(t)$ and $\mathbb{H}_j = [1 \quad 0\; 0\; 0\; 0]$. If $\hbar(t,\boldsymbol{X}) = X_2(t)$, $\mathbb{H}_j = [0 \quad 1 \quad 0 \quad 0 \quad 0]$. Parameter estimation is performed for both the cases. The initial values for c, k and α are taken as 2, 5 and 5 respectively. The initial estimate $\widehat{\boldsymbol{X}}_0$ is taken as, $(0 \quad 0 \quad 2 \quad 5 \quad 5)^T$. The error covariance matrix, $\widehat{\boldsymbol{P}}_0$ is assumed to be the diagonal matrix, $diag\,[0.01 \quad 0.01 \quad 1 \quad 1 \quad 1]$. During the recursive procedure, the predicted mean state $\widetilde{\boldsymbol{X}}_{j+1}$ at $t = t_{j+1}$, is obtained by solving the linearized equation $\dot{\widetilde{\boldsymbol{X}}}_{j+1} = \nabla_j\widetilde{\boldsymbol{X}}_j + \mathcal{P}(t)$. $y_{j+1} := y(t_{j+1})$ (Eq. 6.63) is synthetically generated by first obtaining a solution to Eq. (6.64) with higher resolution and without the noise term (using the reference values for c, k and α) and then corrupting it with a realization of the measurment noise v_t. The measurement noise variance is taken 0.009. While the process noise intensity σ is taken as 0.001, those for the parameter states are taken as 0.00002 uniformly (with appropriate units). Figures 6.4 and 6.5 show the parameter estimation results for a uniform time step size of 0.001 s.

In the implementation of the EKF, accumulation and propagation of local errors may arise with a distinct possibility of divergence (the deviation of the estimated trajectory from the true/reference trajectory increasing with time), especially for filtering problems of higher dimensionality or if the recursive procedure starts with poor initial estimates [Shinozuka and Ghanem 1995, Grewal and Andrews 2001]. The alternative forms of EKF based on the LTL method or the multi-step transversal linearization method may perform somewhat better, even though the source of the problem remains (in the form of the updated covariance matrix via the Riccati equation losing its positive definiteness). The interested reader may refer to Ghosh et al. [2007] for results on state / parameter estimation of linear and non-linear structural systems using the higher order variants of EKF.

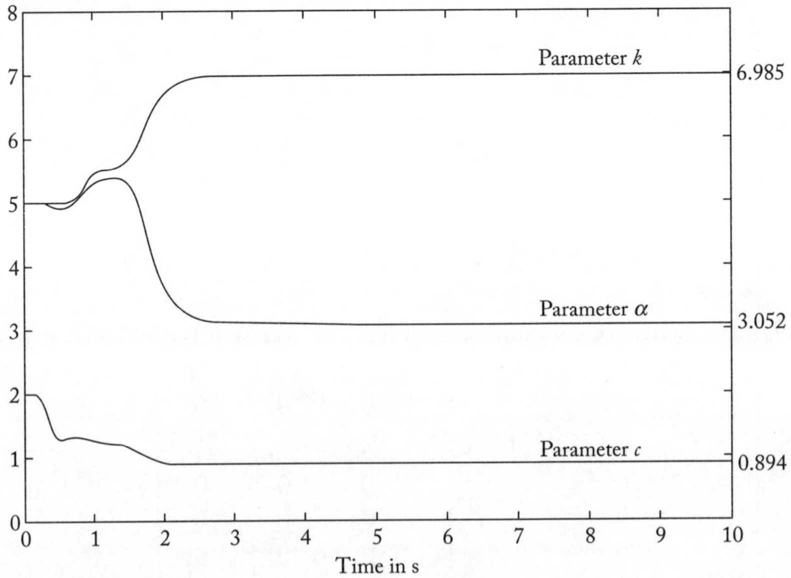

Fig. 6.4 Stochastic filtering by EKF; parameter estimation with measurements on displacement, Example 6.2 – Duffing oscillator; $\mathcal{P} = 4$ N, $\lambda = 3.5$ rad/s, reference values of parameters: $c = 0.9, k = 7$ and $\alpha = 3$

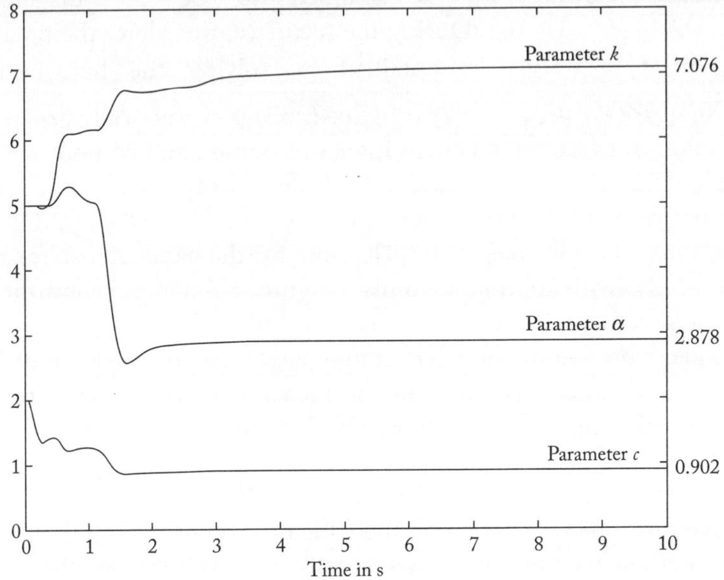

Fig. 6.5 Stochastic filtering by EKF; parameter estimation with measurements on velocity, Example 6.2 – Duffing oscillator; $\mathcal{P} = 4$ N, $\lambda = 3.5$ rad/s, reference values of parameters: $c = 0.9,\ k = 7$ and $\alpha = 3$

6.5 Monte Carlo Filters

Bayesian filters based on the MC approach are known to be more appropriate and versatile for non-linear/non-Gaussian estimation problems [Gordon *et al.* 1993, Tanizaki 1996, Doucet 1998, and Liu and Chen 1998]. One may find a comprehensive survey of the associated numerical methods in Crisan and Doucet [2002], Budhiraja et al. [2007], and Bain and Crisan [2009]. Of particular interest has been the development of the so-called particle filters via SMC techniques for state / parameter estimation problems in dynamical systems. While these methods have early origins during 1950s, most of the recent developments in this field have been prompted by the availability of inexpensive and fast computing power. The underlying concept is the choice of a set of sample points or particles representing an ensemble of system states. At the initial time, these are generated via an initial probability distribution of the states. The ensemble is then propagated through the given non-linear system model whilst incorporating the available information from measurements. The posterior *pdf* is empirically approximated by an ensemble of the system states. One can find a detailed review on particle filters and their applications in Doucet *et al.* [2000], Arulampalam *et al.* [2002], Ristic *et al.* [2004], Roychowdhury *et al.* [2013], and Doucet and Johnson [2008].

6.5.1 Bootstrap filter

While a number of PF algorithms are proposed in literature, the bootstrap filter is the simplest particle filter introduced by Gordon *et al.* [1993]. For a convenient exposition of the filter, the time–discretized process equation and the measurement equation are written in the form:

$$\boldsymbol{X}_i = \mathcal{G}(\boldsymbol{X}_{i-1}, \Delta\boldsymbol{B}_{i-1}) \tag{6.67a}$$

$$\boldsymbol{Y}_i = \mathcal{H}(\boldsymbol{X}_i, \boldsymbol{v}_i), i = 1, 2, ... \tag{6.67b}$$

Here $\mathcal{G} \in \mathbb{R}^m \times \mathbb{R}^n \to \mathbb{R}^m$ and $\mathcal{H} \in \mathbb{R}^m \times \mathbb{R}^q \to \mathbb{R}^q$. Within a recursive setup, suppose that, at $t = t_{i-1}$, one has a set of N_p particles $\boldsymbol{X}_{i-1}^{[j]}, j = 1, 2, \ldots, N_p$ from the available *pdf* $p(\boldsymbol{X}_{i-1} \mid \boldsymbol{Y}_{i-1})$. This presumes the availability of the initial *pdf* $p(\boldsymbol{X}_0|\boldsymbol{Y}_0) = p(\boldsymbol{X}_0)$. In a particle filter, we first propagate these samples through the process model and thus obtain the prior estimate of the states - the prediction step described below.

Prediction:

The transition *pdf* $p(\boldsymbol{X}_i|\boldsymbol{X}_{i-1})$ is defined by the process model (the solution to which is Markov) and known statistics of $\boldsymbol{B}_{i-1}$ as:

$$p(\boldsymbol{X}_i|\boldsymbol{X}_{i-1}) = \int p(\boldsymbol{X}_i|\boldsymbol{X}_{i-1}, \Delta\boldsymbol{B}_{i-1})\, p(\Delta\boldsymbol{B}_{i-1}|\boldsymbol{X}_{i-1})\, d(\Delta\boldsymbol{B}_{i-1}) \tag{6.68}$$

Further, the prior *pdf* is given by (see Eq. 6.4):

$$p\left(\boldsymbol{X}_i|\boldsymbol{Y}_{i-1}\right) = \int p\left(\boldsymbol{X}_i|\boldsymbol{X}_{i-1}\right) p\left(\boldsymbol{X}_{i-1} \mid \boldsymbol{Y}_{i-1}\right) d\boldsymbol{X}_{i-1} \tag{6.69}$$

Since $\Delta\boldsymbol{B}$ is independent of the current and past states, $p\left(\Delta\boldsymbol{B}_{i-1}|\boldsymbol{X}_{i-1}\right) = p\left(\Delta\boldsymbol{B}_{i-1}\right)$ and one has from Eq. (6.68):

$$p\left(\boldsymbol{X}_i|\boldsymbol{X}_{i-1}\right) = \int \delta(\boldsymbol{X}_i - \mathcal{G}(\boldsymbol{X}_{i-1}, \Delta\boldsymbol{B}_{i-1})\ ) p\left(\Delta\boldsymbol{B}_{i-1}\right) d\left(\Delta\boldsymbol{B}_{i-1}\right) \tag{6.70}$$

where, $\delta(.)$ is the Dirac delta operator. Equations (6.70) implies that the prior or predicted particles are given (at $t = t_i$) by:

$$\widetilde{\boldsymbol{X}}_i^{[j]} = \mathcal{G}(\boldsymbol{X}_{i-1}^{[j]}, \Delta\boldsymbol{B}_{i-1}^{[j]})\ , j = 1, 2, \ldots, N_p \tag{6.71}$$

Here $\Delta\boldsymbol{B}_{i-1}^{[j]}, j = 1, 2, \ldots, N_p$ are the samples drawn from the known *pdf* $p\left(\Delta\boldsymbol{B}_{i-1}\right)$ of the process noise.

Update:

At time t_i, the current measurement $\boldsymbol{Y}_i$ is available. The conditional density (likelihood function) $p\left(\boldsymbol{Y}_i|\widetilde{\boldsymbol{X}}_i\right)$is defined by the measurement model (Eq. 6.67b) and the known statistics of the measurement noise $\boldsymbol{v}_i$ as:

$$p\left(\boldsymbol{Y}_i|\widetilde{\boldsymbol{X}}_i\right) = \int \delta(\boldsymbol{Y}_i - \mathcal{H}(\widetilde{\boldsymbol{X}}_i, \boldsymbol{v}_i)\ ) p\left(\boldsymbol{v}_i\right) d\boldsymbol{v}_i$$

$$\Longrightarrow \boldsymbol{Y}_i = \mathcal{H}(\widetilde{\boldsymbol{X}}_i, \boldsymbol{v}_i) \tag{6.72}$$

With the likelihood function known, one may apply Bayes' rule (Eq. 6.2) to get the updated or posterior *pdf* as:

$$p\left(\boldsymbol{X}_i|\boldsymbol{Y}_i\right) = \frac{p\left(\boldsymbol{Y}_i|\widetilde{\boldsymbol{X}}_i\right) p\left(\widetilde{\boldsymbol{X}}_i|\boldsymbol{Y}_{i-1}\right)}{p\left(\boldsymbol{Y}_i|\boldsymbol{Y}_{i-1}\right)} \tag{6.73}$$

$p\left(\boldsymbol{Y}_i|\boldsymbol{Y}_{i-1}\right)$ in the denominator of the above equation is the normalization constant:

$$p\left(\boldsymbol{Y}_i|\boldsymbol{Y}_{i-1}\right) = \int p\left(\boldsymbol{Y}_i|\widetilde{\boldsymbol{X}}_i\right) p\left(\widetilde{\boldsymbol{X}}_i|\boldsymbol{Y}_{i-1}\right) d\widetilde{\boldsymbol{X}}_i \tag{6.74}$$

The normalization constant need not be necessarily known since the recursive procedure involves only the evaluation of $p\left(\boldsymbol{X}_i|\boldsymbol{Y}_i\right) \propto p\left(\boldsymbol{Y}_i|\widetilde{\boldsymbol{X}}_i\right) p\left(\widetilde{\boldsymbol{X}}_i|\boldsymbol{Y}_{i-1}\right)$. We evaluate the likelihood $p\left(\mathbf{Y}_i|\widetilde{\mathbf{X}}_i^{[j]}\right)$ of each prior sample $\widetilde{\mathbf{X}}_i^{[j]}$ and obtain a normalized weight $w_i^{[j]}$ for each $\widetilde{\mathbf{X}}_i^{[j]}$ as:

$$w_i^{[j]} = \frac{p\left(\mathbf{Y}_i|\widetilde{\mathbf{X}}_i^{[j]}\right)}{\sum_{k=1}^{N_p} p\left(\mathbf{Y}_i|\widetilde{\mathbf{X}}_i^{[k]}\right)} \tag{6.75}$$

where the likelihood vector $\boldsymbol{w}_i = \boldsymbol{w}_i\left(\widetilde{\mathbf{X}}_i\right)$ is represented by a set of discrete likelihoods $\left\{w_i^{[j]}\right\}$. Note that each scalar element of $\boldsymbol{w}_i$ may be identified with the Radon–Nikodym derivative $\frac{dP}{dQ}$ at $t = t_i$ (Eq. 6.5). One may resample N_p times from the discrete distribution $\left\{P\left(\widehat{\mathbf{X}}_i^{[j]} = \widetilde{\mathbf{X}}_i^{[j]}\right) = w_i^{[j]}\right\}$ to generate $\left\{\widehat{\mathbf{X}}_i^{[j]}, j = 1,2,\ldots,N_p\right\}$. The recursive algorithm (involving prior sampling and resampling) is repeated for all $i \in (1,N)$.

The formulation and the update procedure above are based on a result by Smith and Gelfand [1992]. The result can be stated thus:

Suppose that the samples $\left\{\widetilde{\mathbf{X}}^{[j]}, j = 1,2,\ldots,N_p\right\}$ are available from a continuous density function $g(\mathbf{X})$ and that new samples are required from a *pdf* proportional to $\psi(\mathbf{X})g(\mathbf{X})$, where $\psi(\mathbf{X})$ is a known function. The theorem states that a sample drawn with probability mass function $\psi(\widetilde{\mathbf{X}}^{[j]})/\sum_{k=1}^{N_p}\psi(\widetilde{\mathbf{X}}^{[j]})$ over $\left\{\widetilde{\mathbf{X}}^{[j]}\right\}$ tends in distribution to the required density as $N_p \to \infty$

Thus the central idea in the recursive algorithm is that the updated sample vector $\widehat{\mathbf{X}}_i$ is approximately distributed as per the *pdf* $p\left(\mathbf{X}_i|\mathbf{Y}_i\right)$. Here $g(\mathbf{X})$ is identified with the prior *pdf* $p\left(\mathbf{X}_i|\mathbf{Y}_{i-1}\right)$ and $\psi(\mathbf{X})$ with $p\left(\mathbf{Y}_i|\widetilde{\mathbf{X}}_i\right)$. During computation, resampling is performed by first drawing a random sample $u_j,\ j = 1,2,\ldots,N_p$ from a uniform distribution over $[0,1]$ and then selecting $\widehat{\mathbf{X}}_i^{[j]} = \widetilde{\mathbf{X}}_i^{[j]}$ if:

$$\sum_{k=1}^{j-1} w_i^{[k]} < u_j \leq \sum_{k=1}^{j} w_i^{[k]} \tag{6.76}$$

While the Bayesian bootstrap filter with the resampling strategy above is often referred to as the sequential importance resampling (SIR) filter, we henceforth use the coinage bootstrap filter to include the resampling step.

Example 6.3

We illustrate the performance of the bootstrap filter using a Van der Pol oscillator under an additive noise. The Van der Pol oscillator is a useful mathematical model for systems with self-excited limit cycle oscillations [Van der Pol 1920, Cartwright and Littlewood 1945]. It finds applications in science and engineering for its distinctive behavior under both deterministic and stochastic excitations. The governing SDE for the oscillator may formally be written as:

$$\ddot{X} + c\left(X^2 - 1\right)\dot{X} + kX = \sigma W(t) \tag{6.77}$$

$W(t)$ is a Gaussian white noise with σ being its intensity. c and k are the system parameters to be estimated along with the system states X and $\dot{X}$. The reference values for c and k are taken as 4 and 10, respectively.

Solution

With $X_1 := X$ and $X_2 := \dot{X}$, the process SDE (6.77) may be expressed in the incremental form:

$$dX_1 = X_2 dt \tag{6.78a}$$

$$dX_2 = \left(-c\left({X_1}^2 - 1\right)X_2 - kX_1\right)dt + \sigma dB(t) \tag{6.78b}$$

For parameter estimation, the system parameters are declared as purely diffusive stochastic processes evolving with time. Thus with $X_3 := c$ and $X_4 := k$, the process SDEs are augmented with:

$$dX_3 = \sigma_c dB_c(t) \tag{6.78c}$$

$$dX_4 = \sigma_k dB_k(t) \tag{6.78d}$$

where B_c and B_k are zero–mean independent Weiner processes with σ_c and σ_k being the corresponding diffusion coefficients. The process noise intensities σ, σ_c and σ_k are taken as 0.001, 0.5 and 0.5, respectively. The initial conditions for X_1 and X_2 are assumed to be 1.4 and 0, respectively. The augmented state vector is $\mathbf{X} = \{X_1, X_2, X_3, X_4\}^T$. The displacement X_1 is assumed to be measured and the standard deviation of the measurement noise is taken as 5% of the maximum absolute value of the displacement (without the measurement noise). The measurement noise v is assumed to be zero–mean Gaussian. The recursive procedure starts with the initial values of 2 and 8 for the augmented states X_3 and X_4, respectively. Figure 6.6 shows the phase plane plot of the estimated states obtained by the filter. The plot represents the self-excited limit cycle behavior of the unforced oscillator. The results correspond to a sample size of $N_p = 100$ with $T = 20$s and $\Delta = 0.01$ s. Figure 6.7 shows the estimates of c and k.

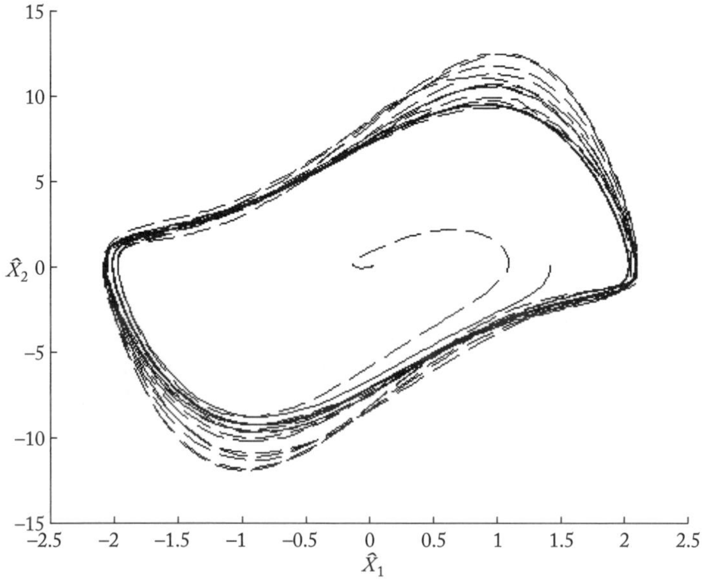

Fig. 6.6 Bootstrap filter applied to Van der Pol oscillator in Example 6.3; phase plane plot of the estimated states with measurements on displacement, initial values for c and k are 2 and 8, respectively, $T = 20$ s, $\Delta = 0.01$ s, $N_p = 100$, reference values of parameters: $c = 4$, $k = 10$

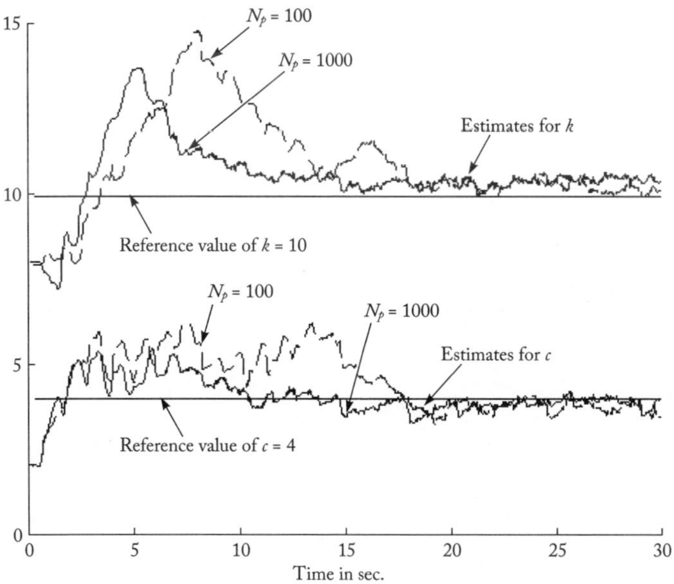

Fig. 6.7 Bootstrap filter applied to Van der Pol oscillator in Example 6.3, parameter estimates with measurements on displacement, initial values assumed for c and k are 2 and 8, respectively, $T = 20$ s, $\Delta = 0.01$ s

Computationally, the basic bootstrap (BS) filter suffers from a degeneracy of weights (as discussed earlier in Section 6.3). As iterations progress, one or a few of the particles may be the sole recipients of non-trivial and non-zero weights with the rest being assigned near-zero weights. This is ascribable mainly to the super-martingale character of the likelihood process. The number of truly distinct particles in the sample set may thus rapidly collapse as recursions progress. This may in turn lead to an early and spurious convergence, thereby resulting in a poor approximation to the posterior *pdf*. The performance of the filter may be improved by utilizing the notion of an effective sample size [Bergmann 1999]. For a given system dimension, the degeneracy problem is typically accentuated by a lower sample size. This is demonstrated in Fig. 6.8 wherein a relatively smaller sample size is seen to affect the filter performance, yielding premature convergence to a wrong solution. The results in the figure pertain to the Van der Pol oscillator in Example 6.3. An initial value (at $t = t_0$) of 20 (far away from the reference values) is assumed for both X_3 and X_4.

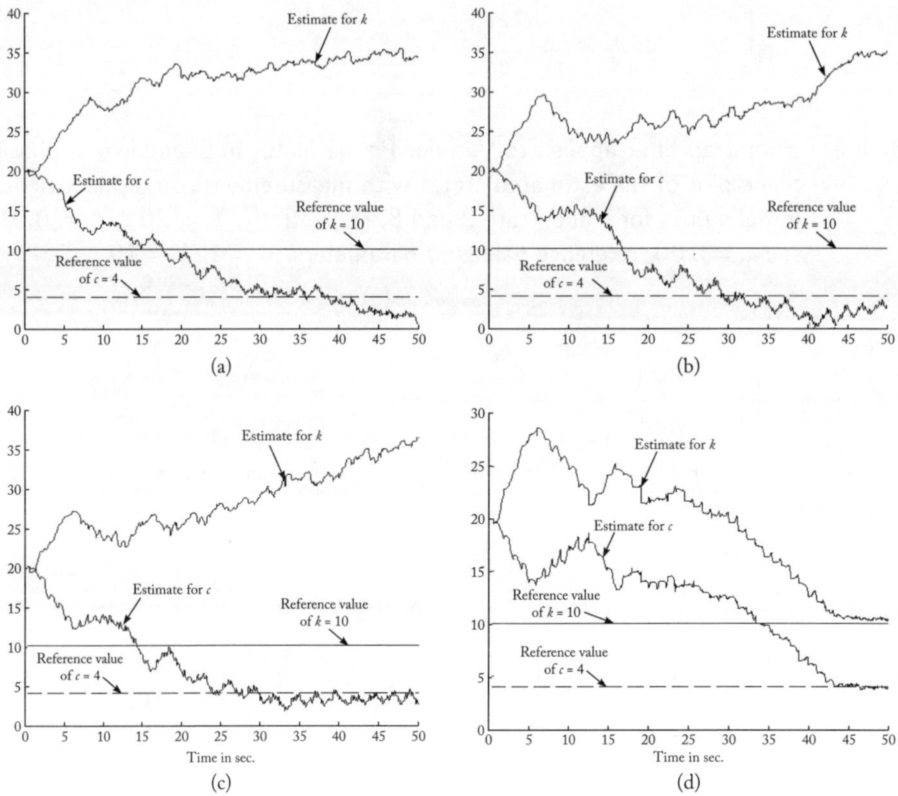

Fig. 6.8 Bootstrap particle filter—weight degeneracy owing to low sample size; parameter estimation for Van der Pol oscillator in Example 6.3, initial value of 20 assumed for both c and k; (a) $N_p = 100$ and (b) $N_p = 500$, (c) $N_p = 1000$ and (d) $N_p = 2000$; dark line – reference parameter k, dashed line – reference parameter c

Such degeneracy is most commonly encountered as the system dimension goes up, typically warranting an exponential increase in the required N_p [Silverman 1986]. An improvement to the BS is the sequential importance sampling (SIS) filter [Bruno 2003]. In the SIS filter, $\boldsymbol{w}_i$ in Eq. (6.75) is computed using the notion of importance sampling (Chapter 2). Suppose that $p(\boldsymbol{X}_i|\boldsymbol{Y}_i) \ll q(\boldsymbol{X}_i|\boldsymbol{Y}_i)$ and that it is easier to draw samples from $q(.)$ than from $p(.)$, then:

$$w_i \propto \frac{p(\boldsymbol{X}_i|\boldsymbol{Y}_i)}{q(\boldsymbol{X}_i|\boldsymbol{Y}_i)} \tag{6.79}$$

$q(\boldsymbol{X}_i|\boldsymbol{Y}_i)$ is the importance *pdf*. Suppose that it is factorized as:

$$q(\boldsymbol{X}_i|\boldsymbol{Y}_i) = q(\boldsymbol{X}_i \mid \boldsymbol{X}_{i-1}, \boldsymbol{Y}_i)\, q(\boldsymbol{X}_{i-1}|\boldsymbol{Y}_{i-1}) \tag{6.80}$$

Noting that $p(\boldsymbol{X}_i|\boldsymbol{Y}_i) \propto p(\boldsymbol{Y}_i|\boldsymbol{X}_i)\, p(\boldsymbol{X}_i|\boldsymbol{X}_{i-1})\, p(\boldsymbol{X}_{i-1} \mid \boldsymbol{Y}_{i-1})$ (see Eqs. (6.69) and (6.73)), one has:

$$w_i \propto \frac{p(\boldsymbol{Y}_i|\boldsymbol{X}_i)\, p(\boldsymbol{X}_i|\boldsymbol{X}_{i-1})\, p(\boldsymbol{X}_{i-1} \mid \boldsymbol{Y}_{i-1})}{q(\boldsymbol{X}_i \mid \boldsymbol{X}_{i-1}, \boldsymbol{Y}_i)\, q(\boldsymbol{X}_{i-1}|\boldsymbol{Y}_{i-1})}$$

$$= w_{i-1} \frac{p(\boldsymbol{Y}_i|\boldsymbol{X}_i)\, p(\boldsymbol{X}_i|\boldsymbol{X}_{i-1})}{q(\boldsymbol{X}_i \mid \boldsymbol{X}_{i-1}, \boldsymbol{Y}_i)} \tag{6.81}$$

Here the scalar w_i denotes the weight for a representative particle, $\boldsymbol{X}_i$. An optimal choice for the importance sampling density [Arulampalam *et al.* 2002] is $q(\boldsymbol{X}_i|\boldsymbol{Y}_i) = p(\boldsymbol{X}_i|\boldsymbol{X}_{i-1}, \boldsymbol{Y}_i)$ that minimizes the variance of the importance weight vector, $\boldsymbol{w}_i$. A common choice for the importance density from the point of easier implementation is the prior $p(\boldsymbol{X}_i|\boldsymbol{X}_{i-1})$ as in the BS. Then:

$$w_i = w_{i-1} p(\boldsymbol{Y}_i|\boldsymbol{X}_i) \tag{6.82}$$

Another feature of the SIS filter is the use of the resampling step selectively. This is done according to an effective sample size, $N_{p,\text{eff}}$ approximately given by:

$$N_{p,eff} \cong \frac{1}{\sum_{j=1}^{N_p} \left(w_i^{[j]}\right)^2} \tag{6.83}$$

The resampling step is implemented whenever $N_{p,\text{eff}}$ falls below a threshold. Here again, the BS may be seen as a special case of the SIS filter, where the resampling is adopted at every time step.

6.5.2 Auxiliary bootstrap filter

The auxiliary bootstrap (ABS) filter [Pitt and Shephard 1999] is a variant of the BS filter and adopts the SIS strategy of an importance density. The importance density adopted in the filter is $q(\mathbf{X}_i, k|\mathbf{Y}_i)$ from which the pairs $\left\{\mathbf{X}_i^{[j]}, k^{[j]}\right\}, j = 1,2,\ldots,N_p$ are sampled. Here k refers to an index of the particle at the previous time instant t_{i-1} and can be looked upon as an auxiliary variable. The procedure to get a representation for $q(\mathbf{X}_i, k|\mathbf{Y}_i)$ starts by applying Bayes' rule to the joint density, $p(\mathbf{X}_i, k|\mathbf{Y}_i)$:

$$p(\mathbf{X}_i, k|\mathbf{Y}_i) \propto p(\mathbf{Y}_i|\mathbf{X}_i)\, p(\mathbf{X}_i, k|\mathbf{Y}_{i-1})$$

$$= p(\mathbf{Y}_i|\mathbf{X}_i)\, p(\mathbf{X}_i|k, \mathbf{Y}_{i-1})\, p(k|\mathbf{Y}_{i-1})$$

$$= p(\mathbf{Y}_i|\mathbf{X}_i)\, p\left(\mathbf{X}_i|\mathbf{X}_{i-1}^{[k]}\right) w_{i-1}^{[k]} \tag{6.84}$$

The importance density is now defined as:

$$q(\mathbf{X}_i, k|\mathbf{Y}_i) \propto p\left(\mathbf{Y}_i|\mu_i^{[k]}\right) p\left(\mathbf{X}_i|\mathbf{X}_{i-1}^{[k]}\right) w_{i-1}^{[k]} \tag{6.85}$$

$\mu_i^{[k]}$ is some representation of $\mathbf{X}_i$ conditioned on $\mathbf{X}_{i-1}^{[k]}$, e.g., $\mu_i^{[k]} = E[\mathbf{X}_i|\mathbf{X}_{i-1}^{[k]}]$ or $\mu_i^{[k]} \sim p(\mathbf{X}_i|\mathbf{X}_{i-1}^{[k]})$, i.e., $\mu_i^{[k[j]]}$=$X_i^{[k[j]]}$, the jth predicted realization from the initial condition $X_{i-1}^{[k[j]]}, j = 1,2,\ldots,N_p$. By writing:

$$q(\mathbf{X}_i, k|\mathbf{Y}_i) = q(k|\mathbf{Y}_i)\, q(\mathbf{X}_i|k, \mathbf{Y}_i) \text{ and with} \tag{6.86a}$$

$$q(\mathbf{X}_i|k, \mathbf{Y}_i) := p\left(\mathbf{X}_i|\mathbf{X}_{i-1}^{[k]}\right), \tag{6.86b}$$

a comparison with Eq. (6.85) yields:

$$q(k|\mathbf{Y}_i) \propto p\left(\mathbf{Y}_i|\mu_i^{[k]}\right) w_{i-1}^{[k]} \tag{6.87}$$

Now with $q(\mathbf{X}_i, k|\mathbf{Y}_i)$ as defined in Eq. (6.85), the weights to be assigned to the sample $\left\{\mathbf{X}_i^{[j]}, k^{[j]}\right\}, j = 1,2,\ldots,N_p$ are obtained as:

$$w_i^{[j]} \propto \frac{p\left(\mathbf{X}_i^{[j]}, k^{[j]}|\mathbf{Y}_i\right)}{q\left(\mu_i^{[j]}, k^{[j]}|\mathbf{Y}_i\right)} = \frac{p(\mathbf{Y}_i|X_i^{[j]})\, p\left(\mathbf{X}_i^{[j]}|\mathbf{X}_{i-1}^{(k[j])}\right) w_{i-1}^{(k[j])}}{p\left(\mathbf{Y}_i|\mu_i^{[k[j]]}\right) p\left(\mathbf{X}_i^{[j]}|\mathbf{X}_{i-1}^{[k[j]]}\right) w_{i-1}^{[k[j]]}}$$

$$\Longrightarrow w_i^{[j]} \propto \frac{p(\mathbf{Y}_i|\mathbf{X}_i^{[j]})}{p\left(\mathbf{Y}_i|\mu_i^{[k[j]]}\right)} \tag{6.88}$$

Thus in the ABS filter, one resamples N_p times from the discrete distribution $\left\{P\left(\widehat{\mathbf{X}}_i^{[j]} = \widetilde{\mathbf{X}}_i^{[k[j]]}\right) = w_i^{[j]}\right\}$ to generate $\left\{\widehat{\mathbf{X}}_i^{[j]}, j = 1, 2, \ldots, N_p\right\}$. By accounting for the current measurement $\mathbf{Y}_i$ in the importance density, $q\left(\mathbf{X}_i, k|\mathbf{Y}_i\right)$, one expects the filter to perform better, e.g., by having more uniform weights prior to resampling. The performance of the ABS filter is illustrated in Fig. 6.9 in the context of parameter estimation of the Van der Pol oscillator in Example 6.3. The numerical values used in the computations are the same as those assumed earlier. In particular, to start the recursive procedure, initial values of 2 and 8 are assumed for X_3 and X_4 respectively. Result from the BS filter is also included in the figure for comparison.

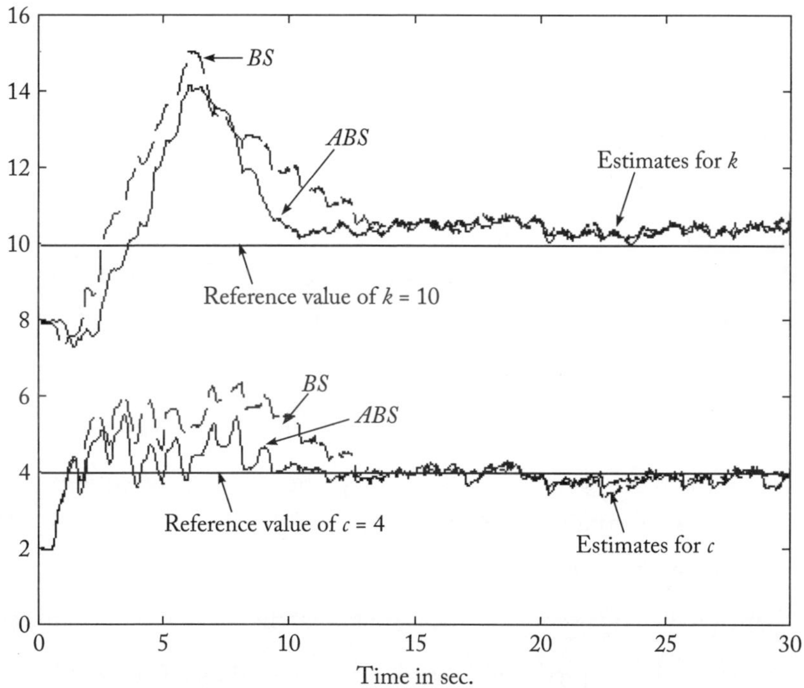

Fig. 6.9 ABS filter; parameter estimation, Van der Pol oscillator in Example 6.3, $T = 30$ s, $\Delta = 0.01$ s, $N_p = 500$

6.5.3 Ensemble Kalman filter (EnKF)

The Ensemble Kalman filter (EnKF) [Evensen 1994] is an MC filter combining the advantages of the analytical approach of a Kalman filter. The EnKF uses an ensemble of system states predicted through the process dynamics as in a PF, thus avoiding the

Gaussian closure, as in the EKF, at this stage. The additive nature of the gain-based update, whose form is the same as that of a KF, insures against possible degeneracy-weight collapse (and hence particle impoverishment)—often encountered with PFs. Thus, the EnKF, which may be viewed as an MC version of the KF, combines the closed-form features of the KF-based update with the PF-like simulation of the covariance. This makes it insensitive to process noise covariance, and this enables one to bypass the extensive tuning often needed in the EKF. The EnKF has found applications in higher dimensional filtering problems as in optical tomography [Raveendran *et al.* 2011, 2012], oceanographic and atmospheric modelling [Evensen 2003], where the infeasibility of storage-cum-manipulation of large state error covariance matrices and the need for extensive tuning of the noise covariance matrices have been the primary issues hindering the use of the KF.

Consider the dynamic state space model described by the non-linear process and measurement equations as in Eq. (6.44) ignoring the external forcing. Writing the process equation in a time-discretized form and the measurement equation in an algebraic form, we have at $t = t_i$:

$$\boldsymbol{X}_i = \boldsymbol{a}(\boldsymbol{X}_{i-1}) + \boldsymbol{\sigma}_{i-1}\Delta\boldsymbol{B}_{i-1} \tag{6.89a}$$

$$\boldsymbol{y}_i = \boldsymbol{h}(\boldsymbol{X}_i) + \boldsymbol{\sigma}_{y,i}\boldsymbol{B}_{y,i}, i = 1,2,... \tag{6.89b}$$

The notations remain the same as in Eq. (6.44). $\widehat{\boldsymbol{X}}_{i-1}$, the last updated (filtered) state at $t = t_{i-1}$, is used as the initial condition to obtain the predicted state $\widetilde{\boldsymbol{X}}_i$ at $t = t_i$. The sample mean vector of the predicted ensemble of states is obtained as $\overline{\widetilde{\boldsymbol{X}}}_i = \frac{1}{N_p}\sum_{j=1}^{N_p}\widetilde{\boldsymbol{X}}_i^{[j]}$. With $\widetilde{\boldsymbol{y}}_i = \boldsymbol{h}\left(\widetilde{\boldsymbol{X}}_i\right)$, the predicted ensemble of measurements, the mean vector of this ensemble is $\overline{\boldsymbol{y}}_i = \frac{1}{N_p}\sum_{j=1}^{N_p}\boldsymbol{h}\left(\widetilde{\boldsymbol{X}}_i^{[j]}\right)$. The predicted ensemble error covariance matrix $\widetilde{\boldsymbol{P}}_i$ is given by: The notations remain the same as in Eq. (6.44). $\widehat{\boldsymbol{X}}_{i-1}$, the last updated (filtered) state at $t = t_{i-1}$, is used as the initial condition to obtain the predicted state $\widetilde{\boldsymbol{X}}_i$ at $t = t_i$. The sample mean vector of the predicted ensemble of states is obtained as $\overline{\widetilde{\boldsymbol{X}}}_i = \frac{1}{N_p}\sum_{j=1}^{N_p}\widetilde{\boldsymbol{X}}_i^{[j]}$. With $\widetilde{\boldsymbol{y}}_i = \boldsymbol{h}\left(\widetilde{\boldsymbol{X}}_i\right)$, the predicted ensemble of measurements, the mean vector of this ensemble is $\overline{\boldsymbol{y}}_i = \frac{1}{N_p}\sum_{j=1}^{N_p}\boldsymbol{h}\left(\widetilde{\boldsymbol{X}}_i^{[j]}\right)$. The predicted ensemble error covariance matrix $\widetilde{\boldsymbol{P}}_i$ is given by:

$$\widetilde{\boldsymbol{P}}_i = \frac{1}{N_p - 1}\sum_{j=1}^{N_p}\left(\widetilde{\boldsymbol{X}}_i^{[j]} - \overline{\widetilde{\boldsymbol{X}}}_i\right)\left(\widetilde{\boldsymbol{X}}_i^{[j]} - \overline{\widetilde{\boldsymbol{X}}}_i\right)^T \tag{6.90}$$

The division by $N_p - 1$ ensures that $\widetilde{P}_i$ is the unbiased estimate of the $m \times m$ error covariance matrix. Introducing the prediction error anomaly or ensemble perturbation matrix (indicated by the superscript 'p'):

$$\mathbf{X}_i^p = \frac{1}{\sqrt{N_p - 1}}\left(\widetilde{X}_i^{[1]} - \overline{\widetilde{X}}_i, \widetilde{X}_i^{[2]} - \overline{\widetilde{X}}_i, \ldots \widetilde{X}_i^{[N_p]} - \overline{\widetilde{X}}_i\right) \tag{6.91}$$

we have:

$$\widetilde{P}_i = \mathbf{X}_i^p \mathbf{X}_i^{p\,T} \tag{6.92}$$

If we have a linear measurement operator $\hbar\left(\widetilde{X}_i\right) = H_i \widetilde{X}_i$ with $\mathbf{Y}_i^p = \frac{1}{\sqrt{N_p - 1}} H_i\left(\widetilde{X}_i - \overline{\widetilde{X}}_i\right) = H_i \mathbf{X}_i^p$, then the Kalman gain (Eq. 6.49) is obtained as:

$$\begin{aligned} K_i &= \widetilde{P}_i H_i^T \left[H_i \widetilde{P}_i H_i^T + R_{y,i}\right]^{-1} \\ &= \mathbf{X}_i^p \mathbf{X}_i^{p\,T} H_i^T \left[H_i \mathbf{X}_i^p \mathbf{X}_i^{p\,T} H_i^T + R_{y,i}\right]^{-1} \\ &= \mathbf{X}_i^p \mathbf{Y}_i^{p\,T} \left[\mathbf{Y}_i^p \mathbf{Y}_i^{p\,T} + R_{y,i}\right]^{-1} = \mathbf{X}_i^p \mathbf{Y}_i^{p\,T} S_i^{-1} \end{aligned} \tag{6.93}$$

where, $S_i = \mathbf{Y}_i^p \mathbf{Y}_i^{p\,T} + R_{y,i}$. Similar to $\mathbf{X}_i^p$, $\mathbf{Y}_i^p$ is the measurement anomaly matrix. The update for the system states including the augmented states (representing the system parameters in the case of parameter estimation), is obtained as (Eq. 6.50a):

$$\widehat{X}_i = \widetilde{X}_i + K_i\left(y_i - H_i \widetilde{X}_i\right) \tag{6.94}$$

Here y_i is the vector of actual measurements available at $t = t_i$. The error covariance update is given by:

$$\begin{aligned} \widehat{P}_i &= \left[I - K_i H_i\right] \widetilde{P}_i \\ &= \left[I - \mathbf{X}_i^p \mathbf{Y}_i^{p\,T} S_i^{-1} H_i\right] \mathbf{X}_i^p \mathbf{X}_i^{p\,T} \\ &= \mathbf{X}_i^p \left[I - \mathbf{Y}_i^{p\,T} S_i^{-1} H_i \mathbf{X}_i^p\right] \mathbf{X}_i^{p\,T} \\ &= \mathbf{X}_i^p \left[I - \mathbf{Y}_i^{p\,T} S_i^{-1} \mathbf{Y}_i^p\right] \mathbf{X}_i^{p\,T} \end{aligned} \tag{6.95}$$

Here $\mathbf{I}$ is an $m \times m$ identity matrix. Evaluation of $\boldsymbol{K}_i$ in Eq. (6.93) involves computation of an inverse for the $q \times q$ matrix $\boldsymbol{S}_i = \mathbf{Y}_i^p \mathbf{Y}_i^{p\,T} + \boldsymbol{R}_{y,i}$ which may pose difficulties for large-sized problems. In this regard, one may take note of the square root filter (SRF) [Livings *et al.* 2008] which is a variant of the EnKF involving a reduction in the computational effort. We write $\widehat{\boldsymbol{P}}_i$ similar to $\widetilde{\boldsymbol{P}}_i$ in Eq. (6.92) as:

$$\widehat{\boldsymbol{P}}_i = \mathbf{X}_i^u \mathbf{X}_i^{u\,T} \tag{6.96}$$

with $\mathbf{X}_i^u$ denoting the updated error anomaly matrix. From Eq. (6.95), a solution for $\mathbf{X}_i^u$ is written in the form:

$$\mathbf{X}_i^u = \mathbf{X}_i^p \boldsymbol{Z} \tag{6.97}$$

where $\boldsymbol{Z}$ can be treated as a square root matrix in the sense:

$$\boldsymbol{Z}\boldsymbol{Z}^T = \boldsymbol{I} - \mathbf{Y}_i^{p\,T} \mathbf{S}_i^{-1} \mathbf{Y}_i^p \tag{6.98}$$

Note that the matrix square root as defined above may not be unique since one can have $\mathbf{X}_i^u = \mathbf{X}_i^p \boldsymbol{Z}\boldsymbol{U}$ where $\boldsymbol{U}$ is any orthogonal matrix of size $N_p \times N_p$. Now, using the relationship $\boldsymbol{S}_i = \mathbf{Y}_i^p \mathbf{Y}_i^{p\,T} + \boldsymbol{R}_{y,i}$ in Eq. (6.93), one may arrive at the following identity:

$$\boldsymbol{I} - \mathbf{Y}_i^{p\,T} \mathbf{S}_i^{-1} \mathbf{Y}_i^p = \left[\boldsymbol{I} + \mathbf{Y}_i^{p\,T} \boldsymbol{R}_{y,i}^{-1} \mathbf{Y}_i^p\right]^{-1} \tag{6.99}$$

To show this, suppose that the LHS of the above equation is multiplied by $\boldsymbol{I} + \left[\mathbf{Y}_i^p\right]^T \boldsymbol{R}_{y,i}^{-1} \mathbf{Y}_i^p$. One may then write:

$$\left[\boldsymbol{I} - \mathbf{Y}_i^{p\,T} \mathbf{S}_i^{-1} \mathbf{Y}_i^p\right]\left[\boldsymbol{I} + \mathbf{Y}_i^{p\,T} \boldsymbol{R}_{y,i}^{-1} \mathbf{Y}_i^p\right]$$

$$= \boldsymbol{I} + \mathbf{Y}_i^{p\,T} \boldsymbol{R}_{y,i}^{-1} \mathbf{Y}_i^p - \mathbf{Y}_i^{p\,T} \mathbf{S}_i^{-1} \mathbf{Y}_i^p - \mathbf{Y}_i^{p\,T} \mathbf{S}_i^{-1} \mathbf{Y}_i^p \mathbf{Y}_i^{p\,T} \boldsymbol{R}_{y,i}^{-1} \mathbf{Y}_i^p$$

$$= \boldsymbol{I} + \mathbf{Y}_i^{p\,T} \boldsymbol{R}_{y,i}^{-1} \mathbf{Y}_i^p - \mathbf{Y}_i^{p\,T} \mathbf{S}_i^{-1} \mathbf{Y}_i^p - \mathbf{Y}_i^{p\,T} \mathbf{S}_i^{-1} \left[\mathbf{S}_i - \boldsymbol{R}_{y,i}\right] \boldsymbol{R}_{y,i}^{-1} \mathbf{Y}_i^p$$

$$\left(\text{since } \boldsymbol{S}_i = \mathbf{Y}_i^p \mathbf{Y}_i^{p\,T} + \boldsymbol{R}_{y,i}\right)$$

$$= \boldsymbol{I}$$

$$\Longrightarrow \boldsymbol{I} - \mathbf{Y}_i^{p\,T} \mathbf{S}_i^{-1} \mathbf{Y}_i^p = \left[\boldsymbol{I} + \mathbf{Y}_i^{p\,T} \boldsymbol{R}_{y,i}^{-1} \mathbf{Y}_i^p\right]^{-1} \tag{6.100}$$

showing the validity of the identity as in Eq. (6.99). This helps in avoiding the computation of S_i^{-1}in that getting $R_{y,i}^{-1}$ is easier than S_i^{-1} since $R_{y,i}$ is mostly diagonal. However, the task of computing the inverse on the entire RHS still remains and is simplified as follows. With $Y_i' := R_{y,i}^{-\frac{1}{2}} \mathbf{Y}_i^p$, we perform the eigenvalue decomposition of $\mathbf{Y}_i^{p\,T} R_{y,i}^{-1} \mathbf{Y}_i^p = Y_i' Y_i'^T = V \Theta V^T$ where Θ is the diagonal matrix of eigenvalues and V, the corresponding orthonormal eigenvector matrix. This yields:

$$I - \mathbf{Y}_i^{p\,T} S_i^{-1} \mathbf{Y}_i^p = \left(I + V\Theta V^T\right)^{-1} = V(I+\Theta)^{-1} V^T \tag{6.101}$$

This in turn (see Eq. 6.98) gives:

$$Z = V(I+\Theta)^{\frac{-1}{2}} \tag{6.102}$$

Since Θ is a diagonal matrix, it is easy to compute $(\mathbf{I}+\Theta)^{\frac{-1}{2}}$. Once Z is obtained, Eq. (6.97) gives $\mathbf{X}_i^u$ and hence $\widehat{P}_i$ from Eq. (6.96). The updated ensemble of states $\widehat{X}$ (see Eqs. (6.93) and (6.94)) in terms of Θ and V is obtainable from the following steps. The gain matrix K_i in Eq. (6.93) is first obtained as:

$$\begin{aligned}
K_i &= \mathbf{X}_i^p \mathbf{Y}_i^{p\,T} \left[\mathbf{Y}_i^p \mathbf{Y}_i^{p\,T} + R_{y,i}\right]^{-1} \\
&= \mathbf{X}_i^p Y_i'^T R_{y,i}^{\frac{1}{2}} \left(R_{y,i}^{\frac{1}{2}} Y_i' Y_i'^T R_{y,i}^{\frac{1}{2}} + R_{y,i}\right)^{-1} \\
&= \mathbf{X}_i^p Y_i'^T \left(Y_i' Y_i'^T + \mathrm{I}\right)^{-1} R_{y,i}^{-\frac{1}{2}} \\
&= \mathbf{X}_i^p Y_i'^T \left(V\Theta V^T + I\right)^{-1} R_{y,i}^{-\frac{1}{2}} \\
&= \mathbf{X}_i^p Y_i'^T V(I+\Theta)^{-1} V^T R_{y,i}^{-\frac{1}{2}} \\
&= \mathbf{X}_i^p \Theta^{\frac{1}{2}} (I+\Theta)^{-1} V^T R_{y,i}^{-\frac{1}{2}} (\text{ since } Y_i' Y_i'^T = V\Theta V^T)
\end{aligned} \tag{6.103}$$

Finally $\widehat{X}_i$ is computed using Eq. (6.94).

Example 6.4

We consider the Van der Pol oscillator in Example 6.3 for parameter estimation by the EnKF.

Solution

The reference parameter values for c and k are taken as 4 and 10, respectively. Figure 6.10 shows the performance of the EnKF along with that of ABS and BS filters. The sample size $N_p = 500$ is used for obtaining the parameter estimation results.

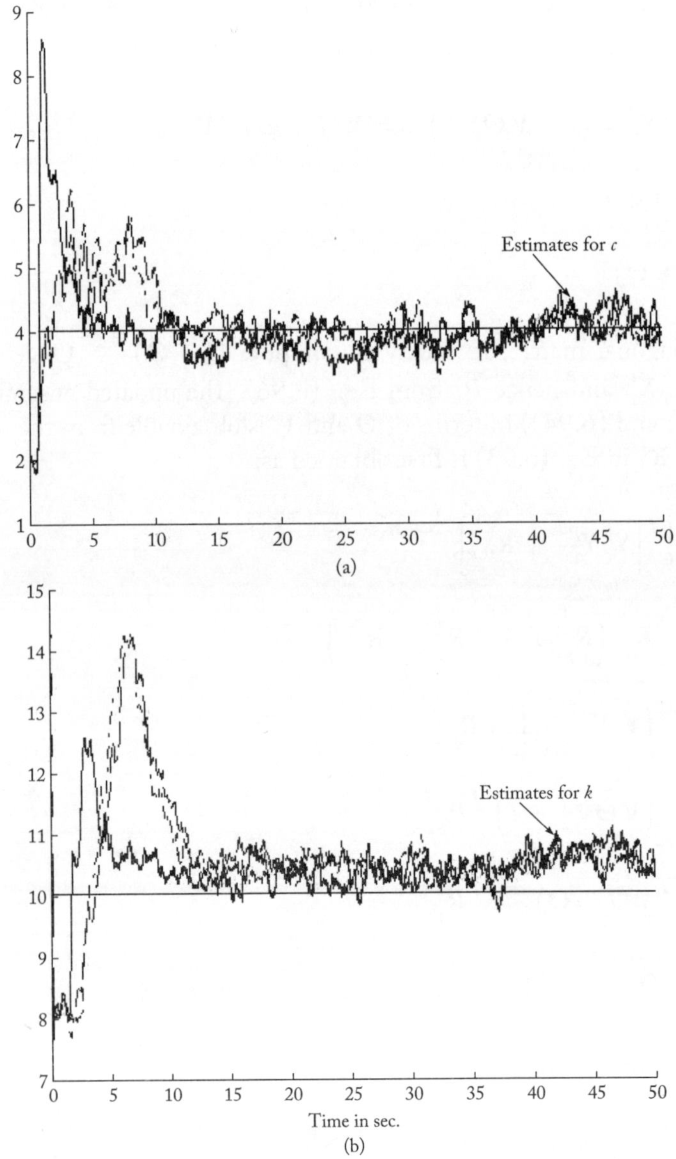

Fig. 6.10 Parameter estimation by EnKF; Van der Pol oscillator in Example 6.3, reference parameter values are: $c = 4$, $k = 10$; $T = 50$ s, $\Delta = 0.01$ s, $N_p = 500$, (a) estimates for c and (b) estimates for k, dark-line – EnKF, dashed-line – ABS filter, dash-dot line – BS filter

Example 6.5

To assess the performance of the EnKF for moderately large dimensional system identification problems, we consider an MDOF (multi-degree-of-freedom) shear frame [Clough and Penzien 1993] as shown in Fig. 6.11.

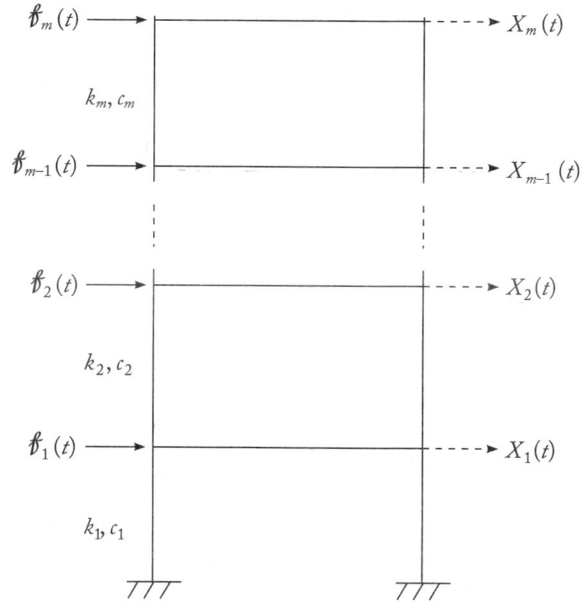

Fig. 6.11 An m–DOF shear frame model

Solution

Here the governing differential equation of the model (after pre-multiplying with the inverse of the original mass matrix denoted by $\boldsymbol{M}^*$) of the frame incorporating an additive Brownian diffusion term is formally represented as:

$$\ddot{\boldsymbol{X}} + \boldsymbol{C}\dot{\boldsymbol{X}} + \boldsymbol{K}\boldsymbol{X} = \mathcal{P}(t) + \sigma\dot{\boldsymbol{B}}(t) \tag{6.104}$$

Denoting by $\boldsymbol{C}^*, \boldsymbol{K}^*$, the original damping and stiffness matrices for the frame, $\boldsymbol{K} = [\boldsymbol{M}^*]^{-1}\boldsymbol{K}^*$ and $\boldsymbol{C} = [\boldsymbol{M}^*]^{-1}\boldsymbol{C}^*$ are the $m \times m$ (mass-normalized) stiffness and viscous damping matrices respectively. They are presently of the form:

$$\boldsymbol{K} = \begin{bmatrix} k_1 + k_2 & -k_2 & 0 & 0 & \dots & 0 & 0 & 0 \\ -k_2 & k_2 + k_3 & -k_3 & 0 & \dots & 0 & 0 & 0 \\ 0 & \cdot & \cdot & \cdot & & 0 & 0 & 0 \\ 0 & \cdot & \cdot & \cdot & & 0 & 0 & 0 \\ 0 & \cdot & \cdot & \cdot & & -k_{m-1} & k_{m-1} + k_m & -k_m \\ 0 & \cdot & \cdot & \cdot & & 0 & -k_m & k_m \end{bmatrix} \tag{6.105}$$

$$\boldsymbol{C} = \begin{bmatrix} c_1+c_2 & -c_2 & 0 & 0 & \dots & 0 & 0 & 0 \\ -c_2 & c_2+c_3 & -c_3 & 0 & \dots & 0 & 0 & 0 \\ 0 & \cdot & \cdot & \cdot & & 0 & 0 & 0 \\ 0 & \cdot & \cdot & \cdot & & 0 & 0 & 0 \\ 0 & \cdot & \cdot & \cdot & & -c_{m-1} & c_{m-1}+c_m & -c_m \\ 0 & \cdot & \cdot & \cdot & & 0 & -c_m & c_m \end{bmatrix} \tag{6.106}$$

If $\mathcal{P}^*(t)$ is the original deterministic force vector derived from an excitation acting at the supports, $\mathcal{P}(t) = [\boldsymbol{M}^*]^{-1}\mathcal{P}^*(t) \in \mathbb{R}^m$. The support excitation here is assumed to be harmonic, yielding uniform nodal force components of the form, $\mathscr{f}_j(t) = \mathscr{f}_0 \sin\lambda t,\ \forall j \in [1,m]$. Also, the mass-normalized noise intensity matrix σ is presently an $m \times m$ diagonal matrix. The damping matrix $\mathbf{C}$ is constructed based on Rayleigh's (proportional) damping [Clough and Penzien 1993], i.e., $\boldsymbol{C}^* = \alpha_1\boldsymbol{M}^* + \alpha_2\boldsymbol{K}^*$, $\alpha_1,\alpha_2 \in \mathbb{R}$ being constants chosen to obtain an appropriate damping mechanism for the example shear frame.

Prior to the augmentation with the parameter states, the state vector is given by, $\boldsymbol{X}_s = \left\{X_1,\dot{X}_1,X_2,\dot{X}_{2,\dots},X_m,\dot{X}_m\right\}^T$. The initial condition is $\boldsymbol{X}_s(t_0=0) = \mathbf{0} \in \mathbb{R}^{2m}$. Note that the combined state-parameter estimation problem is here a non-linear filtering problem (e.g., non-linear in the terms that contain the stiffness and damping parameters as they are considered as additional states), even though the process dynamics (Eq. 6.104) would have been strictly linear if the parameters were precisely known. Including the unknown (mass–normalized stiffness and damping) parameters as additional states, the augmented process state vector is given by $\boldsymbol{X} = \left\{\boldsymbol{X}_s^T,\boldsymbol{X}_p^T\right\}^T \in \mathbb{R}^{4m}$ with $\boldsymbol{X}_p \in \mathbb{R}^{2m}$ denoting the combined vector of stiffness $\{k_1,k_2,\dots,k_m\}^T$ and damping parameters, $\{c_1,c_2,\dots,c_m\}^T$. Incidentally, for this class of problems, the dimensions of the system states and parameters are the same.

Consider a 5-DOF shear frame model, i.e., $m = 5$ (yielding a 20-dimensional dynamical system for filtering purposes). The mass and stiffness parameters respectively are chosen as $m_1^* = m_2^* = \dots = m_5^* = 2.1012 \times 10^6$ Kg. and $k_1^* = k_2^* = \dots = k_5^* = 2.6125 \times 10^8$ N/m. Using the proportionality constants $\alpha_1 = 0.05$ and $\alpha_2 = 0.02$ in the Rayleigh damping mechanism yields the damping coefficients $c_1^* = c_2^* = \dots = c_5^* = 5.225 \times 10^6$ N-s/m. k_i^* and c_i^*, $i = 1,2,\dots,5$ constitute the parameter vector state $\boldsymbol{X}_p$. The specific values chosen for the stiffness and damping constitute the reference values for the parameters with which the synthetic measurements are generated. The shear frame being a stick model [Clough and Penzien 1993], the harmonic support excitation (in acceleration units) is transferred to the nodal DOFs with a uniform amplitude. Thus each nodal force so obtained is $\mathscr{f}_j(t) = 0.1\ \sin 2\pi t$, $j \in [1,5]$ with the excitation frequency being 1 Hz. Based on an eigenvalue analysis [Roy and Rao 2012], one can obtain the free vibration characteristics, e.g., the eigenvalues (or natural frequencies) and the corresponding eigenvectors (or mode shapes). Presently, the first two natural frequencies of the shear frame are about 10 and 86 rad/s. The estimates of stiffness parameters obtained via the EnKF are shown in Fig. 6.12. The estimates by the ABS are also included in the figure. The results by the ABS are found to deviate considerably from the reference values. The ensemble size is consistently taken as $N_p = 500$. For the results in Fig. 6.12, the dimension q of the measurement vector is 5. These correspond to the 5 displacement components X_1, X_2, X_3, X_4 and X_5 that are assumed to be available at each time instant t_i.

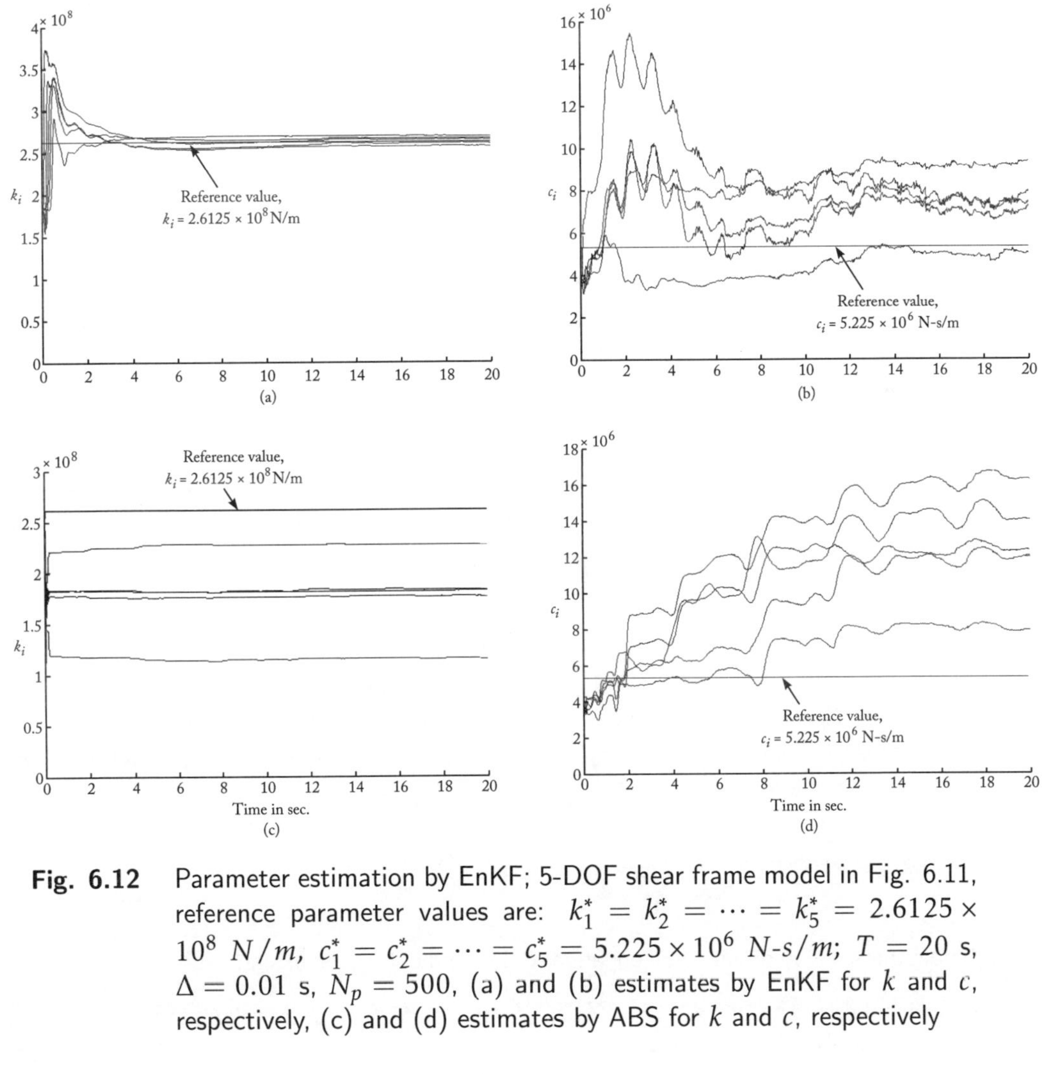

Fig. 6.12 Parameter estimation by EnKF; 5-DOF shear frame model in Fig. 6.11, reference parameter values are: $k_1^* = k_2^* = \cdots = k_5^* = 2.6125 \times 10^8\ N/m$, $c_1^* = c_2^* = \cdots = c_5^* = 5.225 \times 10^6\ N\text{-}s/m$; $T = 20$ s, $\Delta = 0.01$ s, $N_p = 500$, (a) and (b) estimates by EnKF for k and c, respectively, (c) and (d) estimates by ABS for k and c, respectively

6.6 Concluding Remarks

We have introduced the topic of non-linear stochastic filtering, wherein the key goal is to find, recursively in time, the conditional distribution of the hidden process states given the measurements available till that time. With the process and measurement models typically represented by SDEs, the Kushner–Stratonovich (KS) equation formally governs evolutions of the required estimates and it is derivable based on the generalized Bayes' formula. Closed-form solutions or a direct determination of the conditional moments (e.g., the estimates) using the KS equation are typically infeasible owing to the moment closure problem. However, if one allows for Gaussian closure, analytical or semi-analytical solutions are obtainable using variants of the EKF. Approximate non-Gaussian solutions are nevertheless possible via MC simulations and an exposition on a family such filtering schemes, including particle filters, is included along with limited numerical explorations of their performance against a few non-linear filtering problems. Even though particle filters, e.g., the bootstrap and auxiliary bootstrap filters, are not limited by a Gaussian closure, they, in general, use weight based updates and hence exhibit degeneracy of weights or weight collapse during computations leading sometimes to an erroneously premature convergence. That the degeneracy is related mainly to the super-martingale character of the likelihood process (the Randon–Nikodym derivative) has been highlighted. The EnKF—a hybrid scheme combining MC simulations with the additive update of the conditional mean via the KF formula that has the benefit of avoiding weight degeneracy—works better for relatively larger dimensional non-linear filtering problems. It is however not free from the approximations implied in the KF update. The filtering schemes considered in this Chapter are also not directly guided by the structure of the KS equation.

MC filters designed on the lines of the KS equation form the subject of the next chapter. Of specific interest are the variants of the so called KS filter, wherein the update terms are directly guided by the innovation integral in the KS equation and bear form-wise similarities with the update term containing the gain matrix in the EKF.

Exercises

1. Suppose that $X_t = I_{\tau \leq t}$ with τ being a non-negative random variable having the probability distribution, $F_\tau(t)$. The observation equation is governed by the SDE: $dY_t = X_t dt + d\nu_t$. Obtain an expression for $P(\tau \leq t | \mathcal{F}_t^Y)$ using Kallianpur–Sriebel formula (generalized Bayes' rule). Also show that:

$$\sigma_t(1) = \exp\left(\pi_s(f_\tau(t))\, dY_t - \frac{1}{2}(\pi_s(f_\tau(t)))^2 ds\right)$$

2. Obtain the solution of the Zakai equation (6.19) in the linear Gaussian case.

3. Consider the linear SDE: $dX_t = aX_t dt + dB_t$ where B is a Brownian motion and a is an unknown parameter. If a is a discrete random variable taking a finite number of values $\{a_1, a_2, \ldots, a_n\}$ with respective probabilities $\{p_1, p_2, \ldots, p_n\}$, obtain (i) a recursive formula for $\pi_t(j) := P(a \leq a_j | \mathcal{F}_t^X)$ via n-dimensional system of SDEs; (ii) find explicit solutions to the SDEs in (i) above; (iii) if $a \sim \mathcal{N}(0,1)$, clarify if the process $X(t)$ is Gaussian, $(X(t), a)$ Gaussian and if $X(t)\,|a$ is Gaussian.

4. Consider a finite state Markov chain X_t with values in, $\Omega = \{a_1, a_2, \ldots, a_n\}$. The transition intensities matrix is given by Λ and the initial distribution, p_0. If $I_t := I_{X_t = a_j}$ is the n-dimensional vector of indicator functions, a) show that the vector process $\mathcal{M}_t = I_t - I_0 - \int_0^t \Lambda^* I_t ds$ is a $\mathcal{F}_t^X$-martingale and b) find its variance, $E\left[\mathcal{M}_t \mathcal{M}_t^*\right]$.
5. For the process I_t in the Exercise 4 above, derive the filtering equation for the optimal linear estimate $\hat{I}_t := E\left[I_t | \mathcal{F}_t^Y\right]$ and the associated error covariance, where Y_t is given by the observation SDE: $dY_t = h(X_t)dt + dv_t$.

   a) Given the innovation process: $\tilde{B}_t = \int_0^t (dY_s - hX_s ds)$, show that with $0 \le s \le t$, $\tilde{B}_t$ possess the properties:

   b) $E\left[\tilde{B}_t | \mathcal{F}_s^Y\right] = \tilde{B}_s$ (martingale property) and

   c) $E\left[\left(\tilde{B}_t - \tilde{B}_s\right)^2\right] = t - s$

   Also derive the Kalman-Bucy equations, assuming that $E\left[X_t | \mathcal{F}_s^Y\right] = \int_0^t f(t,s) d\tilde{B}_s$ for some $f_{T,S}(t,s)$

6. Consider the following scalar linear Gaussian process and observation models governed by the SDEs:

$$X_{t+1} = aX_t + B_t, B_t \sim \mathcal{N}(0, \sigma_1)$$

$$Y_t = hX_t + \hat{B}_t, \hat{B}_t \sim \mathcal{N}(0, \sigma_2)$$

Given that $X_0 \sim \mathcal{N}(0,1)$. $a, h \in \mathbb{R}$. Find the conditional *pdf*s $f(X_{t+1}|X_t)$ and $f(Y_t|X_t)$. Implement the BS filter to find the optimal estimate of $X_t|Y_{1:t}$ for time $t = 1, 2, \ldots, T = 200$ sec with $a = 0.8$, $h = 0.5$, $\sigma_1 = 0.1$ and $\sigma_2 = 0.1$. Compare the estimate with a result from simulation of the model and plot the RMSE (root mean square error) $= \sqrt{\frac{1}{T}\sum_{t=1}^{T}\left(\hat{X}_{t|t}(N_p) - X_t\right)^2}$ as a function of the number of particles N_p used in the filter. $\hat{X}$ and X denote the estimated and true states respectively.

Notations

$c, c_1, c_2, \ldots, c_m$	damping parameters
$\boldsymbol{B}_Y(t)$	observation noise (a Weiner process)
$\boldsymbol{h}(t, \boldsymbol{X}_t)$	vector of observation functions in Eq. (6.1b)
$\boldsymbol{\hbar}(t, \boldsymbol{X}_t) = \int_0^t \boldsymbol{h}(s, \boldsymbol{X}_s) ds$	observation vector function (Eq. 6.44b)

$\mathbb{H}_t = \left[\frac{\partial \mathscr{h}(t,X)}{\partial X}\right] \in \mathbb{R}^{q\times m}$	a matrix (Eq. 6.49)
$\boldsymbol{I}_t = \boldsymbol{Y}_t - \int_0^t \pi_s(\boldsymbol{h})ds$	the innovation process
J	integer
$k, k_1, k_2, \ldots, k_m$	stiffness parameters
$\boldsymbol{K}_j$	optimum Kalman gain matrix at $t = t_j$ (Eqs. (6.49) and (6.93))
n_p	dimension of system parameters
N_p	sample size
$N_{p,eff}$	effective sample size
$p_t(.)$	normalized conditional density
$p(\boldsymbol{X}_i \mid \boldsymbol{X}_{i-1})$	transition pdf
$p(\boldsymbol{X}_i \mid \boldsymbol{Y}_{(0:i-1)})$	prior pdf
$p(\boldsymbol{X}_i \mid \boldsymbol{Y}_{(0:i)})$	posterior (filtering) pdf
$p(\boldsymbol{Y}_i \mid \boldsymbol{X}_i)$	likelihood function (which is also a Radon–Nikodym derivative)
$\frac{dP}{dQ}$	Radon–Nikodym derivative
$\widehat{\boldsymbol{P}}_j := E\left[(\boldsymbol{X}_j - \widehat{\boldsymbol{X}}_j)(\boldsymbol{X}_j - \widehat{\boldsymbol{X}}_j)^T \mid \mathcal{F}_j^{\boldsymbol{Y}}\right]$	error covariance matrix at $t = t_j$
$q_t(.)$	unnormalized probability density
$q(\boldsymbol{X}_i \mid \boldsymbol{Y}_i)$	importance pdf (Eq. 6.80)
Q	reference probability measure
$\boldsymbol{Q}_{j+1}$	process noise covariance matrix
$\boldsymbol{R}_y$	covariance matrix of the measurement noise
$\mathbf{v}_t \in \mathbb{R}^q$	a q- dimensional zero-mean P-Brownian motion
$\boldsymbol{V}$	orthonormal eigenvector matrix
$w_i(\widetilde{\boldsymbol{X}}_i)$	discrete likelihoods (Eq. 6.75)
$W(t)$	white noise
$\boldsymbol{X}_t$	process vector stochastic process

$\boldsymbol{X}_s, \boldsymbol{X}_p$	vector processes of system variables and parameters respectively
$\boldsymbol{X}_i^p, \boldsymbol{X}_i^u$	anomaly or ensemble perturbation' matrices
$\widetilde{\boldsymbol{X}}(t)$	predicted processes
$\widehat{\boldsymbol{X}}(t)$	filtered processes
$\boldsymbol{Y}_t$	observation vector stochastic process (Eq. 6.1)
$\boldsymbol{Z}$	square root matrix in Eq. (6.98)
α_1, α_2	real constants
$\beta_t^2 = \pi_t(X^2) - (\pi_t(X))^2$	the conditional variance
$\boldsymbol{\Theta}$	diagonal matrix of eigenvalues
$\boldsymbol{\Lambda}_t = \frac{dP}{dQ}$	Radon-Nikodym derivative (Eq. 6.8)
$\mu_t = \pi_t(X)$	the conditional mean
$\pi_t(.) = E\left[\Phi(\boldsymbol{X}_t) \mid \mathcal{F}_t^{\boldsymbol{Y}}\right] = \int_{\mathbb{R}} \Phi(\boldsymbol{X}_t)\pi_t(d\boldsymbol{x})$	conditional distribution of $\boldsymbol{X}_t$ given the observations up to the time t (Eq. 6.5)
$\rho_t(\Phi) = E_Q\left[\Phi(\boldsymbol{X}_t)\Lambda_t \mid \mathcal{F}_t^{\boldsymbol{Y}}\right]$	conditional distribution of $\Phi(\boldsymbol{X}_t)\Lambda_t$
σ	noise intensity
Φ	function of $\boldsymbol{X}_t$, a stochastic process
$\boldsymbol{\Phi}(t, t_j)$	state transition matrix of the linearized system (Eq. 6.47)
Δ	time step size

Chapter 7

Non-linear Filters with Gain-type Additive Updates

7.1 Introduction

The curse of dimensionality haunts the KF/EKF and the PF schemes. In the case of PFs, for instance, the number of particles needed for their successful implementation increases exponentially with increasing system dimension [Silverman 1986]. A failure to meet this condition almost invariably results in weight degeneracy wherein only one particle, with unit weight, survives in the MC simulation. Hence, a severe impediment to the numerical implementation of a PF appears by way of the requirement of disproportionately large ensemble sizes for higher dimensional non-linear filtering problems. Many of the PF schemes are not directly guided by the structure of the KS equation. The KS equation, though posing such numerical bottlenecks as circularity, is important in that it formally governs evolutions of the required estimates while solving state estimation and parameter identification problems in non-linear dynamical systems. We introduce, in this chapter, a family of MC filters that are more explicitly guided by the innovation integral in the KS equation. Our focus will be on the KS filter and its variants wherein the update terms are formed by directly manipulating the innovation integral in the KS equation.

7.2 Iterated Gain-based Stochastic Filter (IGSF)

The IGSF [Raveendran *et al.* 2014a] is one non-linear stochastic filtering scheme where an iterative evaluation of Kalman-like gain matrix within an MC scheme is incorporated into the algorithm. This is in fact motivated by the form of the parent KS equation, consistent with the view that every filtering algorithm may be interpreted as a scheme to determine the state estimates, approximately or otherwise, based on the KS equation. The filter is shown to be applicable to relatively higher dimensional dynamic system identification problems. The other feature is that it retains the simplicity of implementation of the EnKF. Over a given time step, the filter aims at iteratively updating the gain information whilst conforming to the form or structure of the non-linear KS equation. For an iterative evaluation of a Kalman-like gain $\mathcal{K}_i^{(l)}$ (i being the temporal recursion step and l the iteration index for a fixed i), the initial guess $\mathcal{K}_i^{(0)}$ is computed via an ensemble square root filter (SRF) (Livings *et al.* 2008). Unlike the PFs wherein information on the currently available measurement is incorporated as weights tagged to particles in the update stage, the IGSF uses the current measurement repetitively in the iterated evaluation of $\mathcal{K}_i^{(l)}$ such that the associated updates are in the form of additive corrections to the particles.

7.2.1 IGSF scheme

For a scalar-valued function, $\Phi(\boldsymbol{X})$, its estimate at time t is denoted by, $\pi_t(\Phi(\boldsymbol{X}))$ or simply, $\pi_t(\Phi)$. For a typical MDOF model of a mechanical oscillator with m degrees of freedom, one can determine the estimate of the augmented state vector $\boldsymbol{X} = \left\{\boldsymbol{X}_s^T, \boldsymbol{X}_p^T\right\}^T$ (see Eq. 6.60) by choosing, for instance, a family of functions, $\Phi_k(\boldsymbol{X}) = X_k$, $1 \le k \le J = 2m + n_p$. Recall that $\boldsymbol{X}_s \in \mathbb{R}^{2m}$ is the system state vector consisting of the displacement and velocity vectors and $\boldsymbol{X}_p \in \mathbb{R}^{n_p}$ the parameter vector. The augmented process model is governed by Eq. (6.62). The KS equation (Eq. 6.26) essentially provides an integral expression for evaluating $\pi_t(\Phi)$:

$$\pi_t(\Phi) = \pi_{t_i}(\Phi) + \int_{t_i}^{t} \pi_s(L_s(\Phi))\, ds + \sum_{r=1}^{q} \int_{t_i}^{t} \left(\pi_s(\Phi h_r(s, \boldsymbol{X}(s))) - \pi_s(\Phi)\, \pi_s(h_r(s, \boldsymbol{X}(s)))\right) dI_{r,s} \tag{7.1}$$

where, $\boldsymbol{I}_t$ is the so-called innovation vector given by $\boldsymbol{I}_t = \left\{I_{r,t}\right\} = \left\{Y_{r,t} - \int_{t_i}^{t} \pi_s(h_r(s, \boldsymbol{X}(s)))\, ds\right\} \in \mathbb{R}^q$ and L_t is the backward Kolmogorov operator given by Eq. (4.256).

Recall that we aim at recursively updating the predicted stochastic process $\Phi(\widetilde{\boldsymbol{X}})$ to the filtered stochastic process $\Phi(\widehat{\boldsymbol{X}})$ such that $I_{r,t}$ is reduced to a zero-mean noise process (an Ito integral) as $t \to \infty$ for each r. This is generally achieved by designing the temporal recursion such that the innovation process is reduced to a zero-mean martingale. Approximating the second term on the RHS of the KS equation (see also the discussion in Section 6.3.3, Chapter 6) as $\int_{t_i}^{t} \pi_s(L_s(\Phi))\, ds \approx \int_{t_i}^{t} \pi_i(L_s(\Phi))\, ds$ helps in uncoupling the prediction and updating stages over $t \in (t_i, t_{i+1}]$. Thus, the first two terms on the RHS of Eq. (7.1) obtain the expectation, $E_P\left[\Phi(\widetilde{\boldsymbol{X}})\right]$, involving the predicted process that solves the augmented SDE. The only source of stochastic variations is, then, through the measurement noise present in the third term on the RHS. In other words, the KS equation involves averaging over paths generated by the process noise. Nevertheless, in the MC filtering setup adopted here, no such averaging is performed while generating the sample realizations of the predicted process $\widetilde{\boldsymbol{X}}(t)$ *en route* to obtaining sample realizations of the filtered process, $\widehat{\boldsymbol{X}}(t)$. Thus, in relating the present scheme with (the above approximation to) the KS equation, one may interpret $\Phi(\widetilde{\boldsymbol{X}})$ to correspond to the first two terms on the RHS of Eq. (7.1) plus a noise term—an Ito integral with respect to the process noise $\boldsymbol{B}(t)$. The third term on the RHS involves a sum of integrals over weighted innovations. Clearly, each scalar innovation $I_{r,t}$ is a representative of the prediction error if the predicted process, $\widetilde{\boldsymbol{X}}(t)$, is used to compute the measurement function, $h_r(t,\widetilde{\boldsymbol{X}}(t))$, and thus the coefficient weights, $\mathcal{K}_{r,t} = (\pi_t(\Phi h_r(t,\boldsymbol{X}(t))) - \pi_t(\Phi)\,\pi_t(h_r(t,\boldsymbol{X}(t))))$. These terms that, upon integration, build up the appropriate row vector of the gain matrix, drive $I_{r,t}$ to zero–mean Wiener processes. The closure problem (i.e., appearance of estimates other than $\pi_t(\Phi)$ in the expressions for $\mathcal{K}_{r,t}$) and the need to obtain $\mathcal{K}_{r,t}$ continuously for $t_i < t \le t_{i+1}$ (or at least at a set of points in $(t_i, t_{i+1}]$) are a couple of reasons that hinder the development of a numerically accurate and efficient implementation of the KS equation. Nevertheless, these limitations are also suggestive of a possible iterative scheme for repetitively updating the gain information as well as the innovation process for a given time step form t_i to t_{i+1} in order to enhance the filter performance. Accordingly the iterative updates for $\pi_t(\Phi)$ would be denoted via the set, $\left\{\pi_t^{(\Gamma)}(\Phi) : 0 \le \Gamma \le \Gamma_{\max}\right\}$; one ideally desires $\pi_t^{(\Gamma)}(\Phi) \to \pi_t(\Phi)$ as the iteration parameter $\Gamma \to \infty$. The IGSF is implemented in two stages. In the first stage, an initial estimate, $\pi_t^{(0)}(\Phi)$, is found via an ensemble SRF (square root filter—see Section 6.5.3, Chapter 6). In the second stage, iterative updates, $\pi_t^{(\Gamma)}(\Phi), \Gamma > 0$, are obtained, in line with the discussion above. The SRF uses realizations drawn from (empirical approximations to) the prediction or posterior distribution with a fixed ensemble size. The first stage of the algorithm consists of the conventional propagation and initial update steps. Let $\widehat{\boldsymbol{X}}_i = \left\{\widehat{\boldsymbol{X}}_i^{[j]}, 1 \le j \le N_p\right\} \in \mathbb{R}^{(2m+n_p)\times N_p}$ be an ensemble consisting of N_p realizations of the last available filtered states at $t = t_i$ with a sample mean $\overline{\widehat{\boldsymbol{X}}}_i$ and sample covariance $\widehat{\Sigma}_i^{\boldsymbol{X}}$ corresponding to the filtering density at $t = t_i$. The predicted solution $\widetilde{\boldsymbol{X}}(t), t \in (t_i, t_{i+1}]$ obtained by integrating the augmented SDE (6.62) generates a set of predicted particles at any time t. Thus the sample predicted (ensemble averaged) mean vector at $t = t_{i+1}$ is given by:

$$\overline{\widetilde{\boldsymbol{X}}}_{i+1} = \frac{1}{N_p}\sum_{j=1}^{N_p} \widetilde{\boldsymbol{X}}_{i+1}^{[j]} \tag{7.2}$$

As with the EnKF (Section 6.5.3, Chapter 6), we use the measurement Eq. (6.89b) with $\boldsymbol{\hbar}(t,\boldsymbol{X}_t) = \int_0^t \boldsymbol{h}(s,\boldsymbol{X}_s)\,ds \in \mathbb{R}^q$ and form the so called prediction error (anomaly) and the predicted measurement anomaly matrices are now respectively constructed as:

$$\mathbf{X}_{i+1}^{p} = \frac{1}{\sqrt{N_p - 1}}\left(\left(\widetilde{X}_{i+1}^{[1]} - \overline{\overline{X}}_{i+1}\right), \left(\widetilde{X}_{i+1}^{[2]} - \overline{\overline{X}}_{i+1}\right), \ldots, \left(\widetilde{X}_{i+1}^{[N_p]} - \overline{\overline{X}}_{i+1}\right)\right) \tag{7.3a}$$

$$\mathbf{Y}_{i+1}^{p} = \frac{1}{\sqrt{N_p - 1}}$$

$$\left(\begin{array}{c}\left(\mathcal{h}\left(t_{i+1}, \widetilde{X}_{i+1}^{[1]}\right) - \mathcal{h}\left(t_{i+1}, \overline{\overline{X}}_{i+1}\right)\right), \left(\mathcal{h}\left(t_{i+1}, \widetilde{X}_{i+1}^{[2]}\right) - \mathcal{h}\left(t_{i+1}, \overline{\overline{X}}_{i+1}\right)\right), \ldots, \\ \left(\mathcal{h}\left(t_{i+1}, \widetilde{X}_{i+1}^{[N_p])}\right) - \mathcal{h}\left(t_{i+1}, \overline{\overline{X}}_{i+1}\right)\right)\end{array}\right) \tag{7.3b}$$

The initial update step, providing the zeroth iterate to start the iterative updating, proceeds as follows. The IGSF filtering step takes in the current measurement and generates the updated particles at $t = t_{i+1}$. The jth filtered particle is thus given by:

$$\widehat{X}_{i+1}^{[j](0)} = \widetilde{X}_{i+1}^{[j]} + \mathcal{K}_{i+1}^{(0)}\left(\mathcal{Y}_{i+1} - \mathcal{h}\left(t_{i+1}, \widetilde{X}_{i+1}^{[j]}\right)\right) \tag{7.4}$$

with the usual convention $\mathcal{Y}_{i+1} := \mathcal{Y}(t_{i+1})$, the superscript 0 indicating the zeroth update. The initial gain matrix is defined by (see Eq. 6.93):

$$\mathcal{K}_{i+1}^{(0)} = \mathbf{X}_{i+1}^{p} \mathbf{Y}_{i+1}^{p\ T}\left[\mathbf{Y}_{i+1}^{p} \mathbf{Y}_{i+1}^{p\ T} + \boldsymbol{R}_{\mathcal{Y},i+1}\right]^{-1} \tag{7.5}$$

where, $\boldsymbol{R}_{\mathcal{Y},i+1} = \boldsymbol{\sigma}_{\mathcal{Y},i+1}\boldsymbol{\sigma}_{\mathcal{Y},i+1}{}^{T}$. The second stage involves the crucial repetitive updates on the zeroth filtered particles via iterations on the iteration parameter Γ. For further exposition of the algorithm, it is convenient to indicize Γ as $0 = \Gamma_0 < \Gamma_1 < \cdots < \Gamma_{n_\Gamma} = \Gamma_{\max}$, with $n_\Gamma + 1$ denoting the maximum number of iterations, so that, $\pi_{t_i}^{\Gamma_l}(.) := \pi_i^{(l)}(.)$. An inspection of the KS equation (Eq. 7.1), along with the approximation to the second term on the RHS noted immediately afterwards, also reveals that such iterative updates merely affect only the third term on the RHS whilst leaving the prediction part of the estimate (i.e., the first two terms on the RHS) unaffected. Accordingly, considering a vector–valued process, $\boldsymbol{\Phi}(X)$, a generic form of the iterative updates (averaged over process noise) may be written as:

$$\overline{\pi}_{i+1}^{(l+1)}(\boldsymbol{\Phi}(X)) = \overline{\psi}_i(t_{i+1}) + \mathcal{K}_{i+1}^{(l)}\left(\mathcal{Y}_{i+1} - \mathcal{h}\left(t_{i+1}, \overline{\pi}_{i+1}^{(l)}\left(\widehat{X}\right)\right)\right) \tag{7.6a}$$

where, $\overline{\pi}_{i+1}^{(l+1)}(\boldsymbol{\Phi}(X)) = \frac{1}{N_p}\sum_{j=1}^{N_p} \boldsymbol{\Phi}\left(\widehat{X}_{i+1}^{(l+1)[j]}\right) \cong E_P\left[\boldsymbol{\Phi}\left(\widehat{X}_{i+1}^{(l+1)}\right)\right]$ denotes the vector of $(l+1)^{\text{th}}$ iterative approximation to the vector estimate $\boldsymbol{\Phi}\left(\widehat{X}\right) \in \mathbb{R}^J$ at $t = t_{i+1}$. In describing the IGSF and its variants in the next section, $\overline{\pi}_i^{(l)}$ always denotes the sample averaging (not the usual expectation) operator for the lth iteration. The overbar in the first term on the RHS of Eq. (7.6a) carries the same meaning of empirical averaging on the prediction as in Eq. (7.2). Thus, $\overline{\psi}_i(t_{i+1})$ is the empirical average of the prediction $\mathbf{\Phi}(\widetilde{X})$ at $t = t_{i+1}$ (via Eq. 6.62) obtained by the time marching map $\psi_i(t) := \mathbf{\Phi}(\widetilde{X}_{t_i, \widehat{X}_i}(t))$ with the last updated particle $\widehat{X}_i$ at $t = t_i$ as the initial condition for $\widetilde{X}$. $\mathcal{K}_{i+1}^{(l)}$ is the lth iterative gain matrix. The generic map ψ may, for instance, be obtained from the locally linearized solution of the augmented process SDE. For notational simplicity and consistency, we may at times use overbars even to denote the empirical estimates and hence Eq. (7.6a) may be recast as:

$$\overline{\mathbf{\Phi}}\left(\widehat{X}_{i+1}^{(l+1)}\right) = \overline{\psi}_i(t_{i+1}) + \mathcal{K}_{i+1}^{(l)}\left(\mathcal{Y}_{i+1} - \hbar\left(t_{i+1}, \overline{\widehat{X}}_{i+1}^{(l)}\right)\right) \tag{7.6b}$$

A particle based representation of the above takes the form:

$$\mathbf{\Phi}\left(\widehat{X}_{i+1}^{[j](l+1)}\right) = \psi_i^{[j]}(t_{i+1}) + \mathcal{K}_{i+1}^{(l)}\left(\mathcal{Y}_{i+1} - \hbar\left(t_{i+1}, \widehat{X}_{i+1}^{[j](l)}\right)\right), j \in [1, N_p] \tag{7.7}$$

with $\left\{\widehat{X}_{i+1}^{[j](l+1)}\right\}$ denoting the ensemble of filtered particles at t_{i+1} following the $(l+1)^{\text{th}}$ update. The computation of the gain matrix $\mathcal{K}_{i+1}^{(l)}$ requires the following definitions for the so called updated prediction and measurement anomaly matrices:

$$\mathbf{X}_{i+1}^{p,(l)} = \frac{1}{\sqrt{N_p - 1}}\left(\left(\widehat{X}_{i+1}^{[1](l)} - \overline{\widehat{X}}_{i+1}\right), \left(\widehat{X}_{i+1}^{[2](l)} - \overline{\widehat{X}}_{i+1}\right), \ldots, \left(\widehat{X}_{i+1}^{[N_p](l)} - \overline{\widehat{X}}_{i+1}\right)\right) \tag{7.8a}$$

$$\mathbf{Y}_{i+1}^{p,(l)} = \frac{1}{\sqrt{N_p - 1}} \left(\begin{array}{c} \left(\hbar\left(t_{i+1}, \widehat{X}_{i+1}^{[1](l)}\right) - \hbar\left(t_{i+1}, \overline{\widehat{X}}_{i+1}\right)\right), \left(\hbar\left(t_{i+1}, \widehat{X}_{i+1}^{[2](l)}\right) - \hbar\left(t_{i+1}, \overline{\widehat{X}}_{i+1}\right)\right), \ldots, \\ \left(\hbar\left(t_{i+1}, \widehat{X}_{i+1}^{[N_p](l)}\right) - \hbar\left(t_{i+1}, \overline{\widehat{X}}_{i+1}\right)\right) \end{array}\right) \tag{7.8b}$$

The updated gain matrix $\mathcal{K}_{i+1}^{(l)}$ is then computable as:

$$\mathcal{K}_{i+1}^{(l)} = \mathbf{X}_{i+1}^{p,(l)}\left(\mathbf{Y}_{i+1}^{p,(l)}\right)^T\left(\mathbf{Y}_{i+1}^{p,(l)}\left(\mathbf{Y}_{i+1}^{p,(l)}\right)^T + \mathbf{R}_{\mathcal{Y},i+1}\right)^{-1} \tag{7.9}$$

where the convention $\mathbf{X}_{i+1}^{p,(0)} = \mathbf{X}_{i+1}^{p}$ and $\mathbf{Y}_{i+1}^{p,(0)} = \mathbf{Y}_{i+1}^{p}$ is adopted. At the end of the iterations, we denote the updated particles as $\widehat{X}_{i+1}^{[j],(n_T)} := \widehat{X}_{i+1}^{[j]}$ before carrying them over to the next time step.

Example 7.1

The same MDOF shear frame model, shown in Fig. 6.11 (of Chapter 6) with $m = 20$, is chosen to assess the performance of the IGSF for a combined system-parameter identification.

Solution

The governing SDE is described in Eq. (6.104) incorporating an additive Brownian diffusion term. The stiffness and damping matrices $\boldsymbol{K}$ and $\boldsymbol{C} \in \mathbb{R}^{m \times m}$ are given in Eqs. (6.105 and 6.106). All the stiffness and damping terms are proposed to be estimated. With a stick model adopted for the shear frame, the reference values for the stiffness and damping terms are respectively taken as $k_1^* = k_2^* = \cdots = k_{20}^* = 2.6125 \times 10^8$ N/m and $c_1^* = c_2^* = \cdots = c_5^* = 5.225 \times 10^6$ Ns/m. Thus the number of parameters, $n_p = 40$ and the dimension of the filtering problem is $J = 2m + n_p = 80$. The Rayleigh damping mechanism [Clough and Penzien 1993], is assumed (with the proportionality constants $\alpha_1 = 0.05$ and $\alpha_2 = 0.02$) while generating the damping matrix. As in the last example, the support excitation is assumed to be harmonic, yielding uniform nodal force components of the form, $\mathfrak{f}_j(t) = \mathfrak{f}_0 \sin \lambda t$, $j \in [1, m]$. A forcing amplitude $\mathfrak{f}_0 = 0.1$ N and the excitation frequency $\lambda = 2\pi$ rad/s are assumed in the computations.

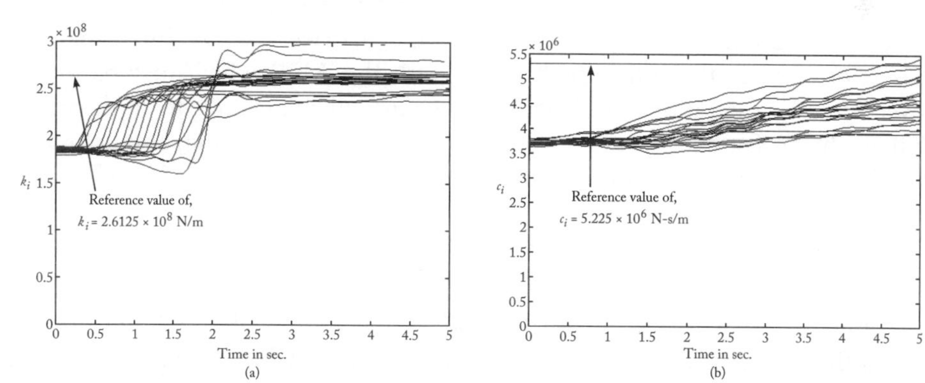

Fig. 7.1 Parameter estimation by IGSF; 20-DOF shear frame model in Fig. 6.11, reference parameter values are: $k_1^* = k_2^* = \cdots = k_{20}^* = 2.6125 \times 10^8$ N/m, $c_1^* = c_2^* = \cdots = c_{20}^* = 5.225 \times 10^6$ N-s/m, $T = 5$ s, $\Delta = 0.01$ s, $N_p = 400$, $n_\Gamma = 10$, (a) stiffness estimation (b) damping estimation

The original system state vector is $\boldsymbol{X}_s = \{X_1, \dot{X}_1, X_2, \dot{X}_2, \ldots, X_m, \dot{X}_m\}^T$. The initial condition is, $\boldsymbol{X}_s(t_0 = 0) = \boldsymbol{0} \in \mathbb{R}^{2m}$. The measurement is synthetically generated via Eq. 6.63 by adding a noise intensity of 3% to all the elements of the displacement vector, $(X_1, X_2, \ldots, X_m)^T$. An ensemble size of $N_p = 400$ is adopted in the MC simulation and the maximum number n_Γ of inner iterations is taken as 10 with $T = 5$s and $\Delta = 0.01$s The estimates of stiffness and damping parameters by the IGSF are shown in Fig. 7.1.

7.3 Improved Versions of IGSF

The modified versions of the IGSF (described in the last section) are due to Raveendran *et al.* [2013, 2014b]. These algorithms introduce an explicit non-Gaussian representation of the filtering density and an improved exploration of the process state space during the iterated updating stage. As in the basic IGSF, the inner iterations over a given time-step here are aimed at driving the innovation term to a zero–mean random variable measurable with respect to measurement filtration. The first variant employs non-Gaussian representations of the prediction and filtering densities provided through Gaussian sums. Additionally, the Gaussian sum filter bank [Sorensen and Alspach 1971] also helps exploring the phase space of the state variables better and the consequent diversity in the particles enables easier adaptation of the process dynamics with the measured variables. This version of IGSF is hereinafter termed the IGSF bank. As in the basic IGSF scheme, while the initial guess $\mathcal{K}_i^{(0)}$ is evaluated on similar lines as in ensemble square root filter (SRF), the iterations require repetitive computations of gain-like coefficient matrices, $\mathcal{K}_i^{(l)}$, consistent with the non-linear KS equation. In addition to the non-Gaussian representations of the densities through Gaussian sums, one may develop the iterative and additive update through an annealing-type parameterization using an artificial diffusion parameter (ADP) leading to the second variant of the IGSF. Specifically in this version, the iterations in the update stage require ADP-parameterized repetitive computations of the gain-like coefficient matrices, $\mathcal{K}_i^{(l)}$. This variant of the IGSF is referred to as the IGSF bank with ADP. The ADP, which may be lowered to zero over successive iterations at a much faster rate (allowing even for a discontinuous scheduling) than is feasible with the conventional simulated annealing, also helps enhance the so called mixing property [Rosenblatt 1956] of the iterative update kernels.

7.3.1 Gaussian sum approximation and filter bank

Gaussian sum approximation makes use of a weighted sum of Gaussian densities (the Gaussian mixture model) to approximate a *pdf*, wherein a tractable solution is arrived at from the sufficient statistics [Sorensen and Alspach 1971]. With $\mathcal{N}(\boldsymbol{\mu}, \Sigma^{1/2})$ denoting a normal density with the mean vector $\boldsymbol{\mu}$ and the covariance matrix Σ (assumed to be non-singular), the Gaussian mixture approximation theorem [Anderson and Moore 1979] suggests that, if the covariance matrices associated with temporally localized particle diffusion have relatively small norms, then the filtering and prediction densities may be adequately approximated as Gaussian mixtures. The following lemma from [Anderson and Moore 1979] , reproduced here for completeness, and the remarks that follow elucidate the underlying principle.

Lemma: Any probability density $p(\boldsymbol{x}), \boldsymbol{x} \in \mathbb{R}^J$ can be approximated as closely as desired in the space $L_1(\mathbb{R}^J)$ by a weighted mixture of the form

$$p_{N_G}(\boldsymbol{x}) = \sum_{n=1}^{N_G} w_n \mathcal{N}(\boldsymbol{\mu}^{(n)}, [\Sigma^{(n)}]^{1/2}) \tag{7.10}$$

for some integer N_G, positive scalars w_n with $\sum_{n=1}^{N_G} w_n = 1$, mean vectors $\boldsymbol{\mu}^{(n)}$ and positive definite matrices $\Sigma^{(n)}$, so that $\int_{\mathbb{R}^J} \left|p(\boldsymbol{x}) - p_{N_G}(\boldsymbol{x})\right| d\boldsymbol{x} < \varepsilon$ for any given $\varepsilon > 0$.

Now suppose that the process and measurement noise vectors in Eqs. (6.62) and (6.63) are Gaussian with zero mean and covariances $\boldsymbol{R_X} = \boldsymbol{\sigma\sigma}^T$ and $\boldsymbol{R_{\mathcal{Y}}} = \boldsymbol{\sigma_{\mathcal{Y}}\sigma_{\mathcal{Y}}^T}$, respectively. Let the prediction density $p\left(\widetilde{\boldsymbol{X}}_{i+1}|\mathcal{Y}_{1:i}\right) = \mathcal{N}\left(\boldsymbol{\mu}_{i+1|i}, \boldsymbol{R}^{1/2}_{\boldsymbol{X},i+1|i}\right)$ be Gaussian, where the subscripts $i+1|i$ in $\boldsymbol{\mu}_{i+1|i}$ and $\boldsymbol{R}_{\boldsymbol{X},i+1|i}$ indicate that the current measurement $\mathcal{Y}_{i+1}$ is yet to be assimilated within these statistical quantities. Then given $\boldsymbol{h}(.)$, $\boldsymbol{\mu}_{i+1|i}$ and $\boldsymbol{R}_{\mathcal{Y},i+1|i}$, the filtering density $p\left(\widehat{\boldsymbol{X}}_{i+1}|\mathcal{Y}_{1:i+1}\right) = c_{i+1}p\left(\widetilde{\boldsymbol{X}}_{i+1}|\mathcal{Y}_{1:i}\right)p\left(\mathcal{Y}_{i+1}|\widetilde{\boldsymbol{X}}_{i+1}\right)$ converges uniformly to the Gaussian density $\mathcal{N}\left(\boldsymbol{\mu}_{i+1|i+1}, \boldsymbol{R}^{1/2}_{\boldsymbol{X},i+1|i+1}\right)$ as $\boldsymbol{R}_{\boldsymbol{X},i+1|i} \to \boldsymbol{0}$ (in terms of a suitable matrix norm). Here c_{i+1} is the normalizing constant. Moreover, if the last filtering density $p\left(\widehat{\boldsymbol{X}}_i|\mathcal{Y}_{1:i}\right) = \mathcal{N}(\boldsymbol{\mu}_{i|i}, \boldsymbol{R}^{1/2}_{\boldsymbol{X},i|i})$ is Gaussian, then given $\boldsymbol{a}(.)$, $\boldsymbol{\mu}_{i|i}$ and $\boldsymbol{R_X}$, the prediction density $p\left(\widetilde{\boldsymbol{X}}_{i+1}|\mathcal{Y}_{1:i}\right) = \int p\left(\widetilde{\boldsymbol{X}}_{i+1}|\widehat{\boldsymbol{X}}_i\right)p\left(\widehat{\boldsymbol{X}}_i|\mathcal{Y}_{1:i}\right)d\widehat{\boldsymbol{X}}_i$ converges uniformly to the Gaussian density $\mathcal{N}(\boldsymbol{\mu}_{i+1|i}, \boldsymbol{R}^{1/2}_{\boldsymbol{X},i+1|i})$ as $\boldsymbol{R}_{\boldsymbol{X},i|i} \to \boldsymbol{0}$. In the above expressions, the subscripts $i+1|i+1$ denote quantities associated with the filtered solutions at time t_{i+1}. Instead of the *pdf* s, this theorem may be more generally stated in terms of the associated characteristic functions as well.

The theorem above (for a proof, see Anderson and Moore, 1979) implies that, with covariance matrices having relatively small norms, the filtering and prediction densities may be adequately approximated as Gaussian mixtures. Based on this approximation, one can construct a bank of, say, N_G Gaussian mixands wherein, at the start of recursion, i.e., at $t = 0$, equal number of particles are drawn from each Gaussian *pdf* in the mixand (i.e., the set of N_p particles is split into N_G subsets, each containing n_G particles drawn from each mixand) so as to populate the ensemble. This enables a tagging of the subsets of particles with the appropriate terms in the Gaussian sum all through the recursion/iteration stages. Also, the mixands are assigned equal weights $w_0^n := w^n(t_0) = \frac{1}{N_G}$, $n \le N_G$ to begin with.

7.3.2 Filtering strategy

The prediction and zeroth update stage is common to both IGSF bank and IGSF bank with ADP. Let $\left\{{}^n\widehat{\boldsymbol{X}}_i^{[j]} : 1 \le j \le n_G\right\}$ be the nth sub-ensemble, consisting of n_G realizations of the Gaussian random variable associated with the nth mixand ($n \in [1, N_G]$) in the Gaussian sum approximation of the last filtering density at $t = t_i$. Let the sample mean and sample covariance of this sub-ensemble be respectively denoted by, ${}^n\overline{\widehat{\boldsymbol{X}}}_i$ and ${}^n\widehat{\Sigma}_i$. For arriving at the predicted particles starting with ${}^n\widehat{\boldsymbol{X}}_i$ as the initial condition, numerical integration of the process SDE (6.62) may be accomplished through any available numerical/semi-analytical scheme (see Chapter 5). The predicted solution ${}^{n}\widetilde{\mathbf{X}}_{i+1}$ at t_{i+1} corresponding to the nth mixand thus generates a sub-ensemble of predicted particles $\left\{{}^{n}\widetilde{\mathbf{X}}_{i+1}^{[j]}:1 \le j \le n_G\right\}$ whose sample mean vector is given by:

$$ {}^{n}\overline{\widetilde{\mathbf{X}}}_{i+1} = \frac{1}{n_G}\sum_{j=1}^{n_G} {}^{n}\widetilde{\mathbf{X}}_{i+1}^{[j]} \tag{7.11} $$

In order to describe the zeroth update scheme, the prediction anomaly and predicted measurement anomaly matrices (corresponding to the nth mixand) are respectively defined as:

$$ {}^{n}\mathbf{X}_{i+1}^{p} = \frac{1}{\sqrt{n_G - 1}}\left(\left({}^{n}\widetilde{\mathbf{X}}_{i+1}^{[1]} - {}^{n}\overline{\widetilde{\mathbf{X}}}_{i+1}\right), \left({}^{n}\widetilde{\mathbf{X}}_{i+1}^{[2]} - {}^{n}\overline{\widetilde{\mathbf{X}}}_{i+1}\right), \ldots, \left({}^{n}\widetilde{\mathbf{X}}_{i+1}^{[N_p]} - {}^{n}\overline{\widetilde{\mathbf{X}}}_{i+1}\right)\right) \tag{7.12a} $$

$$ {}^{n}\mathbf{Y}_{i+1}^{p} = \frac{1}{\sqrt{n_G - 1}}\left(\left(\boldsymbol{h}\left(t_{i+1}, {}^{n}\widetilde{\mathbf{X}}_{i+1}^{[1]}\right) - \boldsymbol{h}\left(t_{i+1}, {}^{n}\overline{\widetilde{\mathbf{X}}}_{i+1}\right)\right), \left(\boldsymbol{h}\left(t_{i+1}, {}^{n}\widetilde{\mathbf{X}}_{i+1}^{[2]}\right)\right.\right. $$

$$ \left.\left. - \boldsymbol{h}\left(t_{i+1}, {}^{n}\overline{\widetilde{\mathbf{X}}}_{i+1}\right)\right), \ldots, \left(\boldsymbol{h}\left(t_{i+1}, {}^{n}\widetilde{\mathbf{X}}_{i+1}^{[N_p]}\right) - \boldsymbol{h}\left(t_{i+1}, {}^{n}\overline{\widetilde{\mathbf{X}}}_{i+1}\right)\right)\right) \tag{7.12b} $$

The above are used in generating the zeroth iterate, which is input to the iterative updating procedure outlined in the next section. The zeroth update step, fashioned after the currently adopted particle-based form of the EnKF, takes in the current measurement and generates the updated particles at $t = t_{i+1}$ via:

$$ {}^{n}\widehat{\mathbf{X}}_{i+1}^{[j](0)} = {}^{n}\widetilde{\mathbf{X}}_{i+1}^{[j]} + {}^{n}\mathcal{K}_{i+1}^{(0)}\left(\boldsymbol{y}_{i+1} - \boldsymbol{h}\left(t_{i+1}, {}^{n}\widetilde{\mathbf{X}}_{i+1}^{[j]}\right)\right), j \in [1, n_G], n \in [1, N_G] \tag{7.13} $$

Here ${}^{n}\mathcal{K}_{i+1}^{(0)}$ is the initial gain matrix:

$$ {}^{n}\mathcal{K}_{i+1}^{(0)} = {}^{n}\mathbf{X}_{i+1}^{p}\left({}^{n}\mathbf{Y}_{i+1}^{p}\right)^{T}\left[{}^{n}\mathbf{Y}_{i+1}^{p}\left({}^{n}\mathbf{Y}_{i+1}^{p}\right)^{T} + \boldsymbol{R_y}\right]^{-1} \tag{7.14} $$

Now, if $\overline{\pi}_{i+1}^{(0)}\left(\Phi\left({}^{n}\mathbf{X}\right)\right)$ denotes the sample mean with N_G elements of the nth filtered *pdf* component in the Gaussian sum following the zeroth iterate as above, then one may write $\overline{\pi}_{i+1}^{(0)}\left(\Phi\left({}^{n}\mathbf{X}\right)\right) \cong E\left[\Phi\left({}^{n}\widehat{\mathbf{X}}_{i+1}^{(0)}\right)\right]$.

7.3.3 Iterative update scheme for IGSF bank

The second stage involves repetitive updates over the zeroth filtered particles via iterations. The iterative updates entail, for every given time t_{i+1}, an iterating index $l = 1,2,\ldots$ that may be taken as defining a discrete iteration parameter set $1,2,...,n_\Gamma$ with n_Γ denoting the number of iterations. Let $\pi_t^{(l)}(.)$ denote the sample estimate (sub-ensemble wise) at the end of the lth iterate at time t. Then the iterative updates at t_{i+1}, upon sample averaging over process noise, may be written (similar to Eq. 7.6a) as:

$$\overline{\pi}_{i+1}^{(l+1)}\left(\boldsymbol{\Phi}\left({}^n\mathbf{X}\right)\right) = {}^n\overline{\psi}_i\left(t_{i+1}\right) + {}^n\mathcal{K}_{i+1}^{(l)}\left(y_{i+1} - \boldsymbol{h}\left(t_{i+1}, \overline{\pi}_{i+1}^{(l)}\left({}^n\mathbf{X}\right)\right)\right) \tag{7.15}$$

${}^n\mathcal{K}_{i+1}^{(l)}$ is the lth iterate of the gain matrix corresponding to the nth mixand in the Gaussian sum and ${}^n\overline{\psi}_i\left(t_{i+1}\right) := \boldsymbol{\Phi}\left({}^n\widetilde{\mathbf{X}}_{t_i,{}^n\widehat{\mathbf{X}}_i}(t)\right)$. The uncoupled nature of the prediction and iterative update, as adopted here, is reflected in the fact that the latter affects only the third term on the RHS of the KS Eq. (7.1) whilst leaving unaltered the prediction part of the estimate (i.e., the first two terms on the RHS of the equation).

We finally write the corresponding particle-wise version applicable to the nth sub-ensemble (similar to Eq. 7.7) as:

$$\boldsymbol{\Phi}\left({}^n\widehat{\mathbf{X}}_{i+1}^{[j](l+1)}\right) = {}^n\psi_i^{[j]}\left(t_{i+1}\right) + {}^n\mathcal{K}_{i+1}^{(l)}\left(\boldsymbol{y}_{i+1} - \boldsymbol{h}\left(t_{i+1}, {}^n\widehat{\mathbf{X}}_{i+1}^{[j](l)}\right)\right), j \in [1, n_G] \tag{7.16}$$

Note that $\{{}^n\widehat{\mathbf{X}}_{i+1}^{[j](l)}\}$ denotes the the lth iterated sub-ensemble of filtered particles at t_{i+1}. Now defining the nth sub-ensemble wise updated state and measurement anomaly matrices, ${}^n\mathbf{X}_{i+1}^{p,(l)}$ and ${}^n\mathbf{Y}_{i+1}^{p,(l)}$, of the lth iteration similar to those in Eq. (7.12), we evaluate the iterated gain matrix ${}^n\mathcal{K}_{i+1}^{l}$ (appearing in Eq. 7.16) as:

$${}^n\mathcal{K}_{i+1}^{(l)} = {}^n\mathbf{X}_{i+1}^{p,(l)}\left({}^n\mathbf{Y}_{i+1}^{p,(l)}\right)^T\left({}^n\mathbf{Y}_{i+1}^{p,(l)}\left({}^n\mathbf{Y}_{i+1}^{p,(l)}\right)^T + \boldsymbol{R_y}\right)^{-1} \tag{7.17}$$

where,

$${}^n\mathbf{X}_{i+1}^{p,(0)} := {}^n\mathbf{X}_{i+1}^{p} \text{ and } {}^n\mathbf{Y}_{i+1}^{p,(0)} := {}^n\mathbf{Y}_{i+1}^{p}.$$

The mixand weight is then updated and normalized using the particles ${}^n\widehat{\mathbf{X}}_{i+1}^{[j],(n_\Gamma)}$, available after the last, i.e., the n_Γ^{th} iteration, as:

$$^n\tilde{w}_{i+1} = {}^n w_i \frac{\mathcal{N}\left(E\left[\boldsymbol{h}\left(t_{i+1}, {}^n\widehat{\mathbf{X}}_{i+1}^{(n_\Gamma)}\right)\right],\ \left({}^n\mathbf{Y}_{i+1}^{p,(n_\Gamma)}\left({}^n\mathbf{Y}_{i+1}^{p,(n_\Gamma)}\right)^T + \boldsymbol{R}_y\right)^{1/2}\right)}{\sum_n^{N_G} \mathcal{N}\left(E\left[\boldsymbol{h}\left(t_{i+1}, {}^n\widehat{\mathbf{X}}_{i+1}^{(n_\Gamma)}\right)\right],\ \left({}^n\mathbf{Y}_{i+1}^{p,(n_\Gamma)}\left({}^n\mathbf{Y}_{i+1}^{p,(n_\Gamma)}\right)^T + \boldsymbol{R}_y\right)^{1/2}\right)},$$

$$n = 1, 2, \ldots, N_G \tag{7.18}$$

This is followed by weight normalization:

$$^n w_{i+1} = \frac{^n\tilde{w}_{i+1}}{\sum_{n=1}^{N_G} {}^n\tilde{w}_{i+1}} \tag{7.19}$$

Also, by convention, the last updated particles corresponding to the nth mixand in the Gaussian sum are renamed as $\left\{{}^n\widehat{\mathbf{X}}_{i+1}^{[j](n_\Gamma)} := {}^n\widehat{\mathbf{X}}_{i+1}^{[j]}\right\}$ before carrying them over to the next time step. Estimations may now be performed based on the empirical posterior *pdf* so obtained. Specifically, the sample estimate of the state vector at $t = t_{i+1}$ is given by:

$$E\left[\widehat{\mathbf{X}}_{i+1}\right] = \sum_{n=1}^{N_G} {}^n w_{i+1} E\left[{}^n\widehat{\mathbf{X}}_{i+1}\right] \tag{7.20}$$

It may be noted that the IGSF bank reduces to the basic IGSF for $N_G = 1$.

7.3.4 Iterative update scheme for IGSF bank with ADP

Compared to the update in Eq. (7.15), an additional parameter α_l is introduced while evaluating the iterative updates at t_{i+1} as:

$$\overline{\pi}_{i+1}^{(l+1)}\left(\boldsymbol{\Phi}\left({}^n\mathbf{X}\right)\right) = {}^n\overline{\boldsymbol{\psi}}_i\left(t_{i+1}\right) + \left(1 + \beta_l\right) {}^n\boldsymbol{\mathcal{K}}_{i+1}^{(l)}\left(\boldsymbol{y}_{i+1} - \boldsymbol{h}\left(t_{i+1}, \overline{\pi}_{i+1}^{(l)}\left({}^n\mathbf{X}\right)\right)\right) \tag{7.21}$$

All the other notations in the equation above remain the same as in Eq. (7.15). β_l is the iteration-dependent ADP that may be likened to the annealing parameter typically used with simulated annealing (SA) applied to a Markov chain [Lam and Delosme 1988]. Unlike the SA where the temperature is recursively reduced to zero whilst evolving a single Markov chain, the sequence $\{\beta_l | \beta_{l+1} \le \beta_l\}$ is used in the filter to evolve an ensemble of $N_p = n_G N_G$ Markov chains in the iteration parameter, so that, for a given $t = t_{i+1}$, the chains proceed in a controlled way to finally arrive at an ensemble that drives the innovation process $\boldsymbol{I}_{i+1}^{(l)} := \boldsymbol{y}_{i+1} - \boldsymbol{h}\left(t_{i+1}, \pi_{i+1}^{(l)}\left({}^n\mathbf{X}\right)\right)$ to a zero–mean Brownian martingale. Analogous to the SA, one must then supplement the iterative scheme with β_1, the initial ADP, and an appropriate schedule to drive β_l to zero over successive iterations. However, given that an MC scheme supplemented with a Gaussian sum representation is in itself a means to explore the phase space, the conservative or even geometric annealing schedules, typically used with standard SA algorithms or MCMC filters [Andrieu and Doucet 2000] and involving a very large number of iterations, need not be adhered to here. Indeed, as the numerical experiments confirm, considerable flexibility with the scheduling of the ADP (as well as in the choice of β_1) is possible with the present filter. Specifically, an exponentially decaying schedule may be given by $\beta_{l+1} := \frac{\beta_l}{\exp l}$, with β_1 determined in a problem–specific manner through a few trial runs (alternatively, it may be computed as the empirical average of an instantaneously defined cost functional evaluated at the particle locations corresponding to the zeroth update). The scheduled sequence of the ADP $\{\beta_l\}$ is intended to provide an additional handle in controlling the mixing property of the iterative update kernels and ensure that the process variables visit every finite subset of the phase space of interest sufficiently frequently [Rosenblatt 1956]. For this filter, the particle based version of Eq. (7.21) for the nth mixand takes the form:

$$\boldsymbol{\Phi}\left({}^{n}\widehat{\boldsymbol{X}}_{i+1}^{[j](l+1)}\right) = {}^{n}\overline{\boldsymbol{\psi}}_{i}^{[j]}(t_{i+1}) + (1+\beta_l)\,{}^{n}\mathcal{K}_{i+1}^{(l)}\left(\boldsymbol{y}_{i+1} - \boldsymbol{h}\left(t_{i+1}, {}^{n}\widehat{\boldsymbol{X}}_{i+1}^{[j](l)}\right)\right),$$

$$j \in [1, n_G] \tag{7.22}$$

The remaining steps involving the evaluation of the updated gain matrix ${}^{n}\mathcal{K}_{i+1}^{(l)}$ and the mixand weights ${}^{n}\tilde{w}_{i+1}$ and the estimation of the state vector are the same as those in the IGSF bank described earlier.

Example 7.2

We consider the same 80-dimensional filtering problem of Example 6.5 of Chapter 6 for parameter estimation by the IGSF bank and the IGSF bank with ADP. The problem corresponds to the 20-DOF shear frame model (Fig. 6.11 of Chapter 6) which is governed by the process SDE (6.104).

Solution

With the system properties characterized by the stiffness and damping matrices as in Eqs. (6.105) and (6.106) and the same reference parameters as in Example 6.5, the state–parameter estimation is performed with $N_p = 400$, $h = 0.01$ s. and $T = 5$ s. The number of Gaussian mixands $N_G = 10$ and the number of iterations $n_\Gamma = 10$ are adopted for both versions of the modified IGSF. For the IGSF bank with ADP, the initiating ADP $\beta_1 = 3$ is used.

In addition to the higher dimensional nature of the problem, wherein particle filters often fail to work, low-intensity measurement noises contribute to yet another performance barrier, needed nevertheless if small changes in the estimates were to be detected. Given the perceived ability of these filters to work with low measurement noise levels (a common scenario when using sophisticated measuring devices), the observation data—herein consisting only of the noisy displacement *dof*s—is synthetically generated by adding a very low noise intensity (less than 5%) to all the elements of the displacement vector. The estimation results are shown in Fig. 7.2. While applying the IGSF bank with ADP, β_1 is kept constant till the $n_\Gamma{}^{\text{th}}$ (last but one) iterative update and abruptly reduced to zero in the $(n_\Gamma + 1)^{\text{th}}$ (last) iterate.

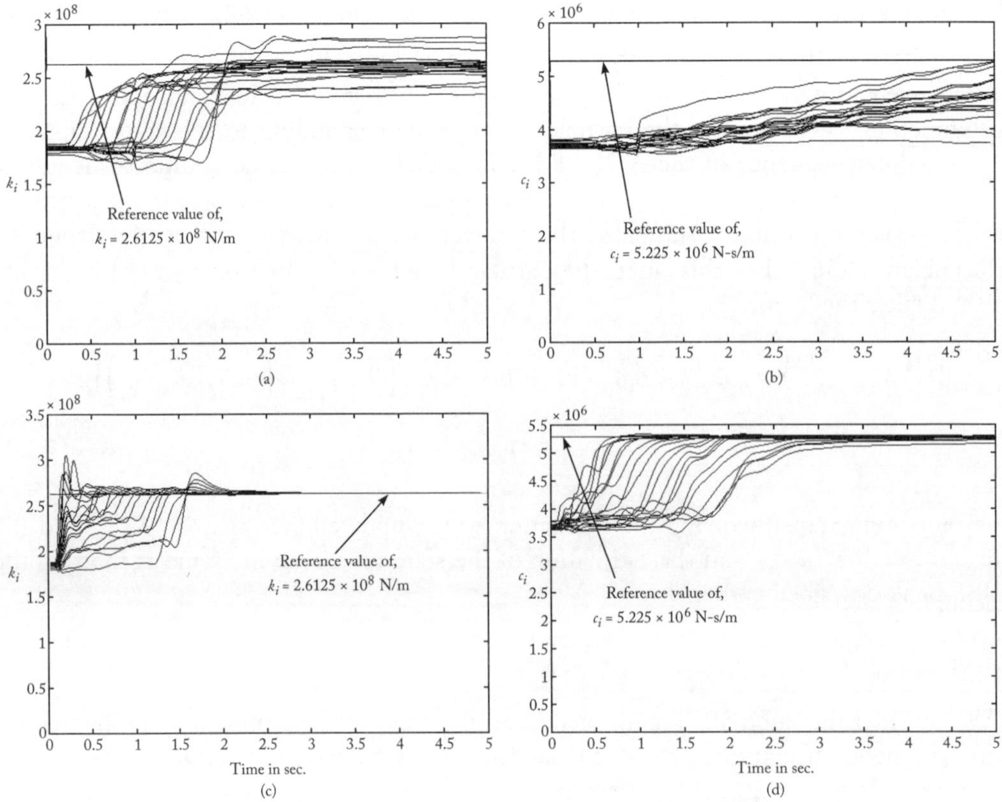

Fig. 7.2 Parameter estimation; 20-DOF shear frame model in Fig. 6.11, reference parameter values are: $k_1^* = k_2^* = \cdots = k_{20}^* = 2.6125 \times 10^8$ N/m, $c_1^* = c_2^* = \cdots = c_{20}^* = 5.225 \times 10^6$ N-s/m, $T = 5$ s, $\Delta = 0.01$ s, $N_p = 400$, (a) and (b) estimates by IGSF bank for k and c, respectively, and (c) and (d) estimates by IGSF bank with ADP for k and c, respectively

7.4 KS Filters

In this section, we will mostly work out the filter using a generic scalar valued function, $\Phi(\boldsymbol{X})$. The KS equation (Eq. 7.1) gives the evolution of $\pi_t(\Phi)$ via a stochastic integral expression. However, owing to the moment closure problem for non-linear, non-Gaussian dynamical systems, the KS equation cannot generally be reduced to stochastic PDEs for $\pi_t(\Phi)$ so that they could be numerically integrated. The particle based simulations in most SMC methods help avoid the Gaussian closure problem. But the EnKF, though applicable to relatively higher dimensional filtering problems, does retain the closed-form features of the KF-based update. Indeed, the IGSF and its variants, whilst retaining the advantages of the EnKF, try to capture the non-Gaussian nature of the prior and posterior densities by an iterative evaluation of KF-like gain matrix within an MC set-up.

The KS filter [Sarkar *et al.* 2014, Teresa *et al.* 2014] is a particle based approach that more closely mimics evolutions of the estimates through the KS equation and implements a non-linear gain-based particle update, which is additive in nature. The departure from a KF-type update, whilst retaining its additive character, is the key feature of the KS filter. The aim of the update here is to approximate the innovation integral in the KS equation and this is implemented through an annealing-type inner iterative procedure (as in the IGSF bank with ADP).

7.4.1 KS filtering scheme

Adopting the same partition Π_N of the time interval $(0,T] \in \mathbb{R}$ defined as $0 = t_0 < t_1 < \ldots < t_i < \ldots < t_N = T$, the estimate $\pi_t(\Phi)$ of $\Phi(\boldsymbol{X}_t)$ over a generic time step, $t \in (t_i, t_{i+1}]$ satisfies the KS equation (Eq. 7.1). As in the IGSF, an important ingredient of the method is a simplification of the second term on the RHS of the KS equation given by:

$$\int_{t_i}^{t} \pi_s(L_s(\Phi))\, ds \approx \int_{t_i}^{t} \pi_i(L_s(\Phi))\, ds = \int_{t_i}^{t} E_P\left[L_s(\Phi)|\mathcal{F}_i^{\boldsymbol{Y}}\right] ds \tag{7.23}$$

where, $\pi_i(.) := \pi_{t_i}$ and $\mathcal{F}_i^{\boldsymbol{Y}} := \mathcal{F}_{t_i}^{\boldsymbol{Y}}$. As observed earlier, this approximation helps in uncoupling the prediction and updating stages. By way of a maximal assimilation of $\boldsymbol{Y}_t$, the current observation, the aim of the filter is to drive $\Delta \boldsymbol{I}_t := \boldsymbol{I}_t - \boldsymbol{I}_i$ to a Brownian increment at the end of the filtering step over $(t_i, t_{i+1}]$, where $\boldsymbol{I}_i := \boldsymbol{I}_{t_i}$. Note that the KS equation is arrived at after averaging over the diffusion paths corresponding to the process noise, $\boldsymbol{B}(t)$. The first two terms on the RHS of the KS equation recover Dynkin's formula for the predicted mean $E_P(\Phi(\boldsymbol{X}_t|\boldsymbol{X}_{t_i} := \boldsymbol{X}_i))$ according to the process dynamics of Eq. (6.62). In the updating stage, as the current observation $\boldsymbol{Y}_t$ is available, the innovation vector $\boldsymbol{I}_t$ may be treated as a pseudo-Markov process $\boldsymbol{I}_t(\vartheta)$ in an artificially introduced (time-like) iteration parameter ϑ and the aim is to drive $\Delta \boldsymbol{I}_t(\vartheta) := \boldsymbol{I}_t(\vartheta) - \boldsymbol{I}_i$ weakly to a zero–mean Brownian increment via inner recursions over ϑ for a given $t \in (t_i, t_{i+1}]$, e.g. at $t = t_{i+1}$. In order to boost the mixing property of the associated transition kernel, the ϑ-recursion, also referred to as the inner iteration, is accomplished by multiplying the innovation integral (the last term on the RHS of the KS equation by $1 + \beta(\vartheta)$, where $1 + \beta(\vartheta)$ is a scalar annealing-type parameter (ATP) that is made to approach zero as iterations progress so as to ensure consistency with the original form of the KS equation.

Prediction

In view of the simplified form of the KS equation as above, the prediction SDE for $\Phi(\boldsymbol{X}_t)$, enabling particle-based simulation, is obtainable through Ito's formula applied to the scalar-valued function $\Phi(\boldsymbol{X}_t)$ where $\boldsymbol{X}_t$ follows the process SDE (6.62). The integral form of the prediction equation over $t \in (t_i, t_{i+1}]$ is:

$$\Phi(\boldsymbol{X}_t) = \Phi(\boldsymbol{X}_i) + \int_{t_i}^{t} L_s(\Phi(\boldsymbol{X}_s))ds + \int_{t_i}^{t} \nabla\Phi(\boldsymbol{X}_s).\sigma(\boldsymbol{X}(s))d\boldsymbol{B}_s \tag{7.24}$$

$\{\nabla\Phi(\boldsymbol{X})\}_\zeta := \frac{\partial\Phi(\boldsymbol{X}_s)}{\partial x_\zeta}$, $\boldsymbol{X} \in \mathbb{R}^J$, is the ζth element of the gradient vector $\nabla\Phi(\boldsymbol{X}_t)$. Equation (7.24) (or rather its SDE form) can be recursively integrated by any numerical scheme (Chapter 5) thus producing the predicted ensemble, $\left\{\Phi\left(\widetilde{\boldsymbol{X}}_{i+1}^{[j]}\right)\right\}_{j=1}^{N_p}$, N_p being the ensemble size. Specifically by choosing an appropriate set of such scalar functions, $\boldsymbol{\Phi} = \left\{\Phi_\zeta(\boldsymbol{X}) = X_\zeta : \zeta \in [1, J]\right\}$, one gets the ensemble of predicted states $\left\{\widetilde{\boldsymbol{X}}_{i+1}^{[j]}\right\}_{j=1}^{N_p}$.

Iterated updates

The iterative update for the KS filter based on a ϑ-parameterization, presently realized through the discrete sequence $\{\vartheta_l : l \in [0, \kappa - 1]\}$ (with $\vartheta_{l+1} > \vartheta_l \; \forall l$) at $t = t_{i+1}$ is given by

$$\Phi\left(\widehat{\boldsymbol{X}}_{i+1}^{(l+1)}\right) = \Phi\left(\widetilde{\boldsymbol{X}}_{i+1}\right) + (1 + \beta_l)\, U_{i+1}^{(l)} \tag{7.25a}$$

Here, $\widehat{\boldsymbol{X}}_{i+1}(\vartheta_{l+1}) := \widehat{\boldsymbol{X}}_{i+1}^{(l+1)}$ and $\beta_l = \Delta\vartheta_l = \vartheta_{l+1} - \vartheta_l$ and

$$U_{i+1}(\vartheta_l) := U_{i+1}^{(l)} = \sum_{r=1}^{q} \left\{\overline{\pi}_{i+1}\left(\Phi\left(\widehat{\boldsymbol{X}}_{i+1}^{(l)}\right) h_r\left(t_{i+1}, \widehat{\boldsymbol{X}}_{i+1}^{(l)}\right)\right)\right.$$

$$\left. -\overline{\pi}_{i+1}\left(\Phi\left(\widehat{\boldsymbol{X}}_{i+1}^{(l)}\right)\right)\overline{\pi}_{i+1}\left(h_r\left(t_{i+1}, \widehat{\boldsymbol{X}}_{i+1}^{(l)}\right)\right)\right\} \Delta I_{r,i+1}^{(l)} \tag{7.25b}$$

where, $\overline{\pi}_{i+1}(.) := \frac{1}{N_p}\sum_{j=1}^{N_p} (.)_{i+1}^{[j]}$ is the ensemble mean operator and $\Delta I_{r,i+1}^{(l)} := \left\{\Delta Y_{r,i+1} - h_r\left(t_{i+1}, \widehat{\boldsymbol{X}}_{i+1}^{(l)}\right)\Delta t_i\right\}$, $\Delta Y_{r,i+1} = Y_r(t_{i+1}) - Y_r(t_i)$ and $\Delta t_i = t_{i+1} - t_i$. Note that $h_r(.,.)$in the laste equation is the rth component of the vector $\boldsymbol{h}$, appearing in the SDE form of observation Eq. (6.1b). An alternative form of this update, wherein the initial update corresponding to $l = 0$ and $\beta_0 = 0$ is added to the prediction term before the subsequent updates with $\beta_l \neq 0$ for $l > 0$ take effect, is given by:

$$\Phi\left(\widehat{\boldsymbol{X}}_{i+1}^{(l+1)}\right) = \Phi\left(\widehat{\boldsymbol{X}}_{i+1}^{(1)}\right) + (1 + \beta_l)\, U_{i+1}^{(l)}, l = 1, 2, \ldots, \kappa - 1 \tag{7.26}$$

Here $\Phi\left(\widehat{\boldsymbol{X}}_{i+1}^{(1)}\right) = \Phi\left(\widetilde{\boldsymbol{X}}_{i+1}\right) + U_{i+1}^{(0)}$ and $\widehat{\boldsymbol{X}}_{i+1}^{(0)} := \widetilde{\boldsymbol{X}}_{i+1}$. $\widehat{\boldsymbol{X}}_{i+1}^{(l+1)}$ denote the $(l+1)^{\text{th}}$ inner-iterated update of $\widetilde{\boldsymbol{X}}_{i+1}$ conditioned on $\mathcal{F}_{i+1}^{\boldsymbol{Y}}$. In Eq. (7.26), $\beta_l := \vartheta_{l+1} - \vartheta_l > 0$ for $l > 0$ is the ϑ - varying ATP which is analogous to temperature in a simulated annealing (SA) scheme [Kirkpatrick *et al.* 1983]. Unlike the SA where the temperature is recursively reduced to zero whilst evolving a single Markov chain, the sequence $\{\beta_l | \beta_{l+1} \le \beta_l\}$ is used in the KS filter to evolve an ensemble of N Markov chains in ϑ so that, for a given t, the chains proceed in a controlled way to finally arrive at an ensemble that drives the pseudo-process $U_t(\vartheta)$ to a zero–mean Brownian motion in ϑ. The incorporation of β_l in the update equation should therefore be supplemented with an appropriate choice of $\beta_1 \gg 0$ to begin the inner iterations and, for a given filtering problem, β_1 should typically be arrived at through a few trial runs of the filter.

The sequence, $\{\beta_l\}$, whose elements for l close to one are relatively higher, helps drive the predictions in the non-Newton (derivative-free) direction given by the vector $U_{i+1}^{(l)}$. The choice of the annealing schedule, ensuring $\beta_l \to 0$ as $l \to \infty$, is governed by two important factors, namely the computational speed and an effective exploration of the state space to attain the desired solution. A conservative schedule, useful for SA schemes involving a single Markov chain, is given by [Lam and Delosme 1988] $\frac{1}{\beta_{l+1}} = \frac{1}{\beta_l} + \frac{\varkappa \rho(\beta_l)}{2\rho^3(\beta_l)}$, where $\varkappa < 1$ is a user-defined parameter. $\rho(\beta_l) := E\left[\left({\boldsymbol{I}_{i+1}^{(l)}}^T \boldsymbol{I}_{i+1}^{(l)} - {\boldsymbol{I}_{i+1}^{(l-1)}}^T \boldsymbol{I}_{i+1}^{(l-1)}\right)^2\right]$ is the variance of an incremental energy-like term and $\rho(\beta_l)$ is the standard deviation of ${\boldsymbol{I}_{i+1}^{(l)}}^T \boldsymbol{I}_{i+1}^{(l)}$, where $\boldsymbol{I}_{i+1}^{(l)} := \boldsymbol{I}_{i+1}(\vartheta_l)$. Although this schedule requires a large number of inner iterations to reduce β_l to zero, it does improve results for the parameter estimation problems considered here. Nevertheless, since an ensemble-based formulation provides an additional means of exploring the phase space of $\boldsymbol{X}_t$ through the particles, a more non-conservative schedule, e.g. $\beta_{l+1} = \frac{\beta_l}{\exp l}$ (which may not even qualify as a proper annealing schedule), appears to be more appropriate for the KS filter. Ideally, an appropriate stopping criterion should be used to fix κ so as to ensure that $U_{i+1,l}$ is indeed a zero–mean discrete Brownian motion (random walk) for $k \ge \kappa - 1$. Towards this, one may in principle employ the Kolmogorov–Smirnov test [Justel *et al.* 1997] in order to assess if the ensemble of realized observation errors $\left\{\boldsymbol{Y}_{i+1} - \boldsymbol{h}\left(t_{i+1}, \widehat{\boldsymbol{X}}_{i+1}^{[j](\kappa)}\right) : j \in [1, N_p]\right\}$ indeed correspond to the known (Gaussian) density of the random variable, $\boldsymbol{B}_{\boldsymbol{Y},i+1}$, the observation noise at t_{i+1}; see Eq. (6.44 b). However, since the number of iterations required to pass this test could be computationally prohibitive, the numerical illustrations in this work are based on a fixed value of κ assigned in a problem specific manner through a few trial runs whilst satisfying the constraint $\beta_{\kappa-1} \approx 0^+$.

A further modification to the above particle based scheme, once the inner iterations are over, may be effected by way of relaxing the approximation (Eq. 7.23) to the second term on the RHS of the KS equation. In doing this, the convention $\widehat{\boldsymbol{X}}_{i+1} := \widehat{\boldsymbol{X}}_{i+1}^{(\kappa)}$ is no longer applicable and the finally computed filtered state $\widehat{\boldsymbol{X}}_{i+1}$ is given by:

$$\Phi\left(\widehat{\boldsymbol{X}}_{i+1}\right) = \Phi\left(\widehat{\boldsymbol{X}}_{i}\right) + \frac{1}{2}\left[L_i\left(\Phi\left(\widehat{\boldsymbol{X}}_{i}\right)\right) + L_{i+1}\left(\Phi\left(\widehat{\boldsymbol{X}}_{i+1}^{(\kappa)}\right)\right)\right]\Delta t_i$$

$$+\nabla\Phi\left(\widehat{\boldsymbol{X}}_{i}\right).\left(\sigma\left(\widehat{\boldsymbol{X}}_{i}\right)\Delta\boldsymbol{B}_i\right) + U_{i+1}^{(\kappa)} \tag{7.27a}$$

On the other hand, if one were to iteratively correct the second term using the inner updates, $\widehat{\boldsymbol{X}}_{i+1}^{(l)}, l < \kappa$, large random fluctuations would possibly occur in $L_{i+1}(\Phi)$, the term that encapsulates the physical laws governing the system dynamics, owing to the typically large values of β_l for small l. This would render the numerical solutions prone to overflows and thus destroy the filtering accuracy. Another way of avoiding such fluctuations is to start updating the so-called prediction term only for $l \geq l_{\max}$ (with $l_{\max} < \kappa$) such that $\beta_l \ll 1$ for $l \geq l_{\max}$. In that case, one may write the updates corresponding to the inner iterations for $l \geq l_{\max}$ as:

$$\Phi\left(\widehat{\boldsymbol{X}}_{i+1}^{(l+1)}\right) = \Phi\left(\widehat{\boldsymbol{X}}_{i+1}^{(l)}\right) + (1+\beta_l)\,U_{i+1}^{(l)} \tag{7.27b}$$

For $l < l_{\max}$, Eq. (7.26) remains applicable. The proof of the convergence of the inner iterations is given in [Sarkar *et al.* 2014].

That the solution via the KS filter converges approximately to that of the KS equation may be intuitively demonstrated as follows. Using Eq. (7.27a) with $\beta_\kappa \approx 0$ and upon ensemble averaging over process noise, one may write as $\Delta t_i \to 0$:

$$\overline{\pi}_{i+1}\left(\Phi\left(\widehat{\boldsymbol{X}}_{i+1}\right)\right) \approx \overline{\pi}_i\left(\Phi\left(\widehat{\boldsymbol{X}}_{i}\right)\right) + \overline{\pi}_{i+1}\left(L_{i+1}\left(\Phi\left(\widehat{\boldsymbol{X}}_{i+1}^{(\kappa)}\right)\right)\right)\Delta t_i$$

$$+\overline{\pi}_{i+1}\left(\nabla\Phi\left(\widehat{\boldsymbol{X}}_{i}\right).\left(\sigma\left(\boldsymbol{X}_i\right)\Delta\boldsymbol{B}_i\right)\right) + \overline{\pi}_{i+1}\left(U_{i+1}^{(\kappa)}\right) \tag{7.28a}$$

where, $\Delta\boldsymbol{B}_i = \boldsymbol{B}_{i+1} - \boldsymbol{B}_i$ and

$$\overline{\pi}_{i+1}\left(U_{i+1}^{(\kappa)}\right) = \sum_{r=1}^{q}\left\{\overline{\pi}_{i+1}\left(\Phi\left(\widehat{\boldsymbol{X}}_{i+1}^{(\kappa)}\right)h_r\left(t_{i+1},\widehat{\boldsymbol{X}}_{i+1}^{(\kappa)}\right)\right) - \overline{\pi}_{i+1}\left(\Phi\left(\widehat{\boldsymbol{X}}_{i+1}^{(\kappa)}\right)\right)\overline{\pi}_{i+1}\right.$$

$$\left.\left(h_r\left(t_{i+1},\widehat{\boldsymbol{X}}_{i+1}^{(\kappa)}\right)\right)\right\}\Delta I_r^{(k-1)} \tag{7.28b}$$

with $\Delta I_r^{(k-1)} = \Delta Y_{r,i+1} - \overline{\pi}_{i+1}\left(h_r\left(t_{i+1},\widehat{X}_{i+1}^{(\kappa)}\right)\right)\Delta t_i$. The term $\overline{\pi}_{i+1}\left(\nabla\Phi\left(\widehat{X}_i\right).\right.$ $\left.\left(\sigma\left(X_i\right)\Delta B_i\right)\right)$ is the empirical expectation of an explicit Euler approximation to the Ito integral $\int_{t_i}^{t_{i+1}} \nabla\Phi\left(X_s\right).\sigma\left(X_s\right)dB_s$, which is a zero–mean martingale and hence vanishes reducing Eq. (7.28a) to:

$$\overline{\pi}_{i+1}\left(\Phi\left(\widehat{X}_{i+1}\right)\right) \approx \overline{\pi}_i\left(\Phi\left(\widehat{X}_i\right)\right)+\overline{\pi}_{i+1}\left(L_{i+1}\left(\Phi\left(\widehat{X}_{i+1}^{(\kappa)}\right)\right)\right)\Delta t_i+\overline{\pi}_{i+1}\left(U_{i+1}^{(\kappa)}\right) \quad (7.29)$$

Equation (7.29) is indeed a discrete and empirical approximation to Eq. (7.1), the original KS equation. The algorithm for implementing the KS filter is given in Table 7.1 for a vector-valued function, $\mathbf{\Phi}(X)$. However, for clarity of exposition, we consider $\mathbf{\Phi}(X) = \left\{\Phi_\zeta\left(X\right), \zeta \in [1,J]\right\} = X$.

Table 7.1 KS filtering scheme—alogrthmic steps

Step 1. Discretize the time interval of interest, say, $[0,T]$, using a partition $\Pi_N = \{0 = t_0, t_1, \ldots, t_N = T\}$ and $t_{i+1} - t_i = \Delta t$ uniformly for $i = 0,1,\ldots,N-1$. Choose an ensemble size, N_p.

Step 2. Generate the ensemble of initial conditions $\left\{X_0^{[j]}\right\}_{j=1}^{N_p}$ for the state vector. For each discrete time instant, t_{i+1}, $i = 0,1,\ldots,\ N-1$, the following steps are carried out.

Step 3. Prediction: Using $\left\{\widehat{X}_i^{[j]}\right\}_{j=1}^{N_p}$, the update available at the last time instant t_i (with the convention that $\left\{\widehat{X}_0^{[j]}\right\}_{j=1}^{N_p} = \left\{X_0^{[j]}\right\}_{j=1}^{N_p}$), propagate each particle to the current time instant t_{i+1} using an explicit EM approximation to Eq. (6.62), i.e.,

$$\widetilde{X}_{i+1}^{[j]} = \widehat{X}_i^{[j]} + a\left(t_i\ ,\widehat{X}_i^{[j]},\right)\Delta t + \mathcal{P}\left(t_i\right)\Delta t + \sigma\left(\widehat{X}_i^{[j]}\right)\Delta B_i, j = 1,2,\ldots,N_p \quad (7.30)$$

Step 4. Initial update:

$$\widehat{X}_{i+1}^{[j](1)} = \widetilde{X}_{i+1}^{[j]} + G_{i+1}(\Delta Y_i - h\left(t_{i+1}, \widetilde{X}_{i+1}^{[j]}\right)\Delta t) \quad (7.31a)$$

Evaluate each column of the G_{i+1} matrix, i.e., $g_{r,i+1} := G_{i+1}\left(\cdot,r\right)$, $r = 1,2,\ldots,q$ as:

$$g_{r,i+1} = \overline{\pi}\left(h_r\left(t_{i+1}, \widetilde{X}_{i+1}\right)\widetilde{X}_{i+1}\right) - \overline{\pi}\left(h_r\left(t_{i+1}, \widetilde{X}_{i+1}\right)\right)\overline{\pi}\left(\widetilde{X}_{i+1}\right) \quad (7.31b)$$

$G_{i+1} = \left[g_{1,i+1}, g_{2,i+1}, \ldots, g_{q,i+1}\right]$. Recall that $\overline{\pi}\left(\cdot\right)$ is the empirical expectation of the corresponding ensemble of particles.

Step 5. Iterated updates: Set $l=1$ and select $\beta_l,\ \kappa,\ l_{\max}$. Then update each particle as:

$$\widehat{\boldsymbol{X}}_{i+1}^{[j](l+1)} = \widehat{\boldsymbol{X}}_{i+1}^{[j](l)} + (1+\beta_l)\,\boldsymbol{G}_{i+1}^{(l)}\left(\boldsymbol{\Delta Y}_i - \boldsymbol{h}\left(t_{i+1}, \widehat{\boldsymbol{X}}_{i+1}^{[j](l)}\right)\Delta t\right) \tag{7.32a}$$

However, during the initial stages of inner iterations (i.e., for $l \le l_{\max}$), particles may be updated via the following map to avoid possible numerical oscillations in the updated solution:

$$\widehat{\boldsymbol{X}}_{i+1}^{[j](l+1)} = \widehat{\boldsymbol{X}}_{i+1}^{[j](1)} + (1+\beta_l)\,\boldsymbol{G}_{i+1}^{(l)}(\boldsymbol{\Delta Y}_i - \boldsymbol{h}\left(t_{i+1}, \widehat{\boldsymbol{X}}_{i+1}^{[j](l)}\right)\Delta t) \tag{7.32b}$$

Evaluate each column of the $\boldsymbol{G}_{i+1}^{(l)}$ matrix, i.e., $\boldsymbol{g}_{r,i+1}^{(l)} := \boldsymbol{G}_{i+1}^{(l)}(\cdot, r),\ r = 1,2,\ldots,q$ as:

$$\boldsymbol{g}_{r,i+1}^{(l)} = \overline{\pi}\left(h_r\left(t_{i+1}, \widehat{\boldsymbol{X}}_{i+1}^{(l)}\right)\widehat{\boldsymbol{X}}_{i+1}^{(l)}\right) - \overline{\pi}\left(h_r\left(t_{i+1}, \widehat{\boldsymbol{X}}_{i+1}^{(l)}\right)\right)\overline{\pi}\left(\widehat{\boldsymbol{X}}_{i+1}^{(l)}\right) \tag{7.33}$$

Step 6. Set $l = l+1$. If $l < \kappa$, set $\beta_{l+1} = \frac{\beta_l}{\exp(l+1)}$, and go to step 5; else, if $i < N$, go to step 3 with $i = i+1$; else terminate the algorithm.

The inner iteration to implement the non-linear additive update (in steps 4 and 5 of the algorithm above) is interpretable as a variant of stochastic Picard's iteration—a tool often used to prove existence and uniqueness of solutions to non-linear SDEs. In other words, the inner iterations involved in the KS filter are a means to asymptotically secure the true solution of the KS equation subject to the time discretization and sampling errors. Indeed the uniqueness of the true solution (i.e., the conditional distribution) corresponding to the filtering problem, described by the KS equation, is ensured under the (somewhat strong and further relaxable) assumption of uniform Lipschitz continuity of the drift and diffusion coefficients via the work of Kurtz and Ocone [1988]. They pose the problem as one of the filtered martingale and prove uniqueness by extending the original theory of Stroock and Varadhan [1979] - see also the discussion on martingale problem in Chapter 4. It may be worthwhile to understand the approximations involved in arriving at the filtered estimate. First, the explicit EM method used to generate the predicted solution provides a source of integration errors whose weak local order is $O(\Delta t)$ (Chapter 5). Additionally, integration errors also accrue owing to the iterative updates, which are consistent with the EM-based treatment of a diffusion term. Finally, one must also account for the MC error arising due to the empirical representations of the associated probability distributions over a finite ensemble. An interested reader may refer to Sarkar *et al.* [2014] for details on these error estimates involved in the KS filter.

Example 7.3

Consider the 80-dimensional filtering problem corresponding to the 20-DOF shear frame model in Fig. 6.11 of Chapter 6 for parameter estimation by the KS filter.

Solution

The augmented system model is governed by Eq. (6.104) with Eqs. (6.105) and (6.106) representing the stiffness and damping matrices of the shear frame. The proportionality constants $\alpha_1 = 0.05$ and $\alpha_2 = 0.02$ as used (as in the earlier Examples 7.1 and 7.2) are utilized in generating the viscous damping matrix, $\boldsymbol{C}$. The nodal forcing vector, $\mathcal{P}(t) = \left\{ f_j(t), j \in [1,J] \right\}$, is deterministic and is assumed to be uniformly harmonic with $f_j(t) = 0.1 \sin 2\pi t$. The state–parameter estimation is performed with $N_p = 400$ with time step $\Delta = 0.01$ s and $T = 5$ s A typical value of $\kappa = 10$ is used in the numerical computations (see Fig. 7.3).

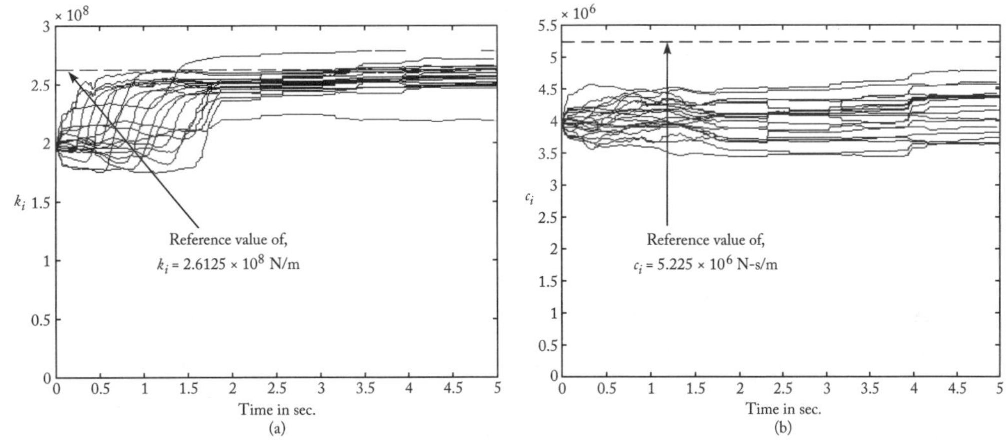

Fig. 7.3 Parameter estimation by KS filter; 20-DOF shear frame model in Fig. 6.11 of Chapter 6, reference parameter values are: $k_1^* = k_2^* = \cdots = k_{20}^* = 2.6125 \times 10^8$ N/m, $c_1^* = c_2^* = \cdots = c_{20}^* = 5.225 \times 10^6$ N-s/m, $T = 5$ s, $\Delta = 0.01$ s, $N_p = 400$, (a) stiffness estimation and b) damping estimation

7.5 EnKS Filter—a Variant of KS Filter

Many SMC methods, which attempt MC simulations based on averaging over the characteristics (i.e., the sample paths provided by the process and measurement dynamics) following a conditional Feynman–Kac formula [Milstein and Tretyakov 2009] and typically leading to a weighted particle system [Crisan and Lyons 1999] , have been tried to approximate the KS equation. However, as noted before, most of these methods are not free from the curse of weight collapse, even for moderately large filter dimensions. Building upon the idea of KS filter, the EnKS filter [Sarkar and Roy 2014] , whilst closely following the evolutions of the estimates based on the KS equation, implements the non-linear and strictly additive particle updates without an imperative necessity for the costly inner iterations. This development is based on a sequence of manipulations of the update term, so as to introduce an additional layer of numerical dispersion in the gain-like coefficient of the innovation. While inner iterations, as in the previous version of the filter, may still be utilized with some improvement in the estimate, the non-iterative form of the filter yields solutions that could be quite accurate even for large dimensional non-linear filtering problems.

7.5.1 EnKS filtering scheme

The second term on the RHS of the KS equation is first replaced by the approximation in Eq. (7.23). Then the first two terms (referred to as the prediction component) on the RHS of the KS equation, so approximated, recover Dynkin's formula for the predicted mean $E_P(\Phi(\boldsymbol{X}_t)|\boldsymbol{X}_{t-1:} := \boldsymbol{X}_{i-1})$ according to the process dynamics of Eq. (6.62). In motivating the EnKS filter, a particle based representation of Eq. (7.1) may be conceived of by putting back, in the prediction component, the diffusion term for the process dynamics. As a first step in deriving the particle-based additive update, as the current measurement, $\boldsymbol{Y}_t$ (say $t > t_i$), is available, an MC setting for solving Eq. (7.1) may be set up as:

$$\overline{\pi}_t(\Phi) = \overline{\pi}_{t_i}(\Phi) + \int_{t_i}^{t} \overline{\pi}_s(L_s(\Phi))\,ds + \sum_{r=1}^{q}\int_{t_i}^{t}(\overline{\pi}_s(\Phi h_r) - \overline{\pi}_s(\Phi)\overline{\pi}_s(h_r))\,d\overline{I}_{r,s} \tag{7.34}$$

where, $d\overline{\boldsymbol{I}}_t := \left\{d\overline{I}_{r,t}\right\} = \left[\sigma_{\boldsymbol{Y}}\sigma_{\boldsymbol{Y}}{}^T\right]^{-1}\{d\boldsymbol{Y} - \overline{\pi}_t(\boldsymbol{h}_r)\,dt\}$ is the incremental innovation process. $\sigma_{\boldsymbol{Y}_t}\sigma_{\boldsymbol{Y}_t}{}^T$ denotes the observation noise covariance matrix at time t. Let $\boldsymbol{\Phi} := (\Phi_1, \Phi_2, \ldots, \Phi_J)^T$ be the J -dimensional vector valued function to be estimated via the filtering technique with $\Phi_j(\boldsymbol{X}_t) \in C^2$ for $j \in [1, J]$. A particle based representation of Eq. (7.34) may be written in the matrix form as (for $t > t_i$):

$$\boldsymbol{\Phi}_t = \boldsymbol{\Phi}_i + \int_{t_i}^{t}\boldsymbol{\Psi}_s ds + \int_{t_i}^{t} d\boldsymbol{\Xi}_s + \frac{1}{N_p}\int_{t_i}^{t}\left\{\boldsymbol{\Phi}_s\boldsymbol{H}_s^T - \overline{\boldsymbol{\Phi}}_s\overline{\boldsymbol{H}}_s^T\right\}\left[\sigma_{\boldsymbol{Y}_s}\sigma_{\boldsymbol{Y}_s}{}^T\right]^{-1} \{d\mathbf{Y}_s - \boldsymbol{H}_s ds\} \tag{7.35}$$

where, $\boldsymbol{\Phi}_t := \left[\boldsymbol{\Phi}_t^{[1]}, \boldsymbol{\Phi}_t^{[2]}, \ldots, \boldsymbol{\Phi}_t^{[N_p]}\right] \in \mathbb{R}^{J\times N_p}$, $\boldsymbol{\Phi}_i := \boldsymbol{\Phi}_{t_i}$,

$$\boldsymbol{\Psi}_t := \left[L\left(\boldsymbol{\Phi}_t^{[1]}\right), L\left(\boldsymbol{\Phi}_t^{[2]}\right), \ldots, L\left(\boldsymbol{\Phi}_t^{[N_p]}\right)\right] \in \mathbb{R}^{J\times N_p},$$

$$d\boldsymbol{\Xi}_t := \left[\nabla\boldsymbol{\Phi}_t^{[1]}.\sigma_t^{[1]}d\boldsymbol{B}_t^{[1]}, \nabla\boldsymbol{\Phi}_t^{[2]}.\sigma_t^{[2]}d\boldsymbol{B}_t^{[2]}, \ldots, \nabla\boldsymbol{\Phi}_t^{[N_p]}.\sigma_t^{[N_p]}d\boldsymbol{B}_t^{[N_p]}\right] \in \mathbb{R}^{J\times N_p},$$

$$\boldsymbol{H}_t := \left[\boldsymbol{h}_t^{[1]}, \boldsymbol{h}_t^{[2]}, \ldots, \boldsymbol{h}_t^{[N_p]}\right] \in \mathbb{R}^{q\times N_p},$$

$$\overline{\boldsymbol{\Phi}}_t := [\overline{\pi}(\boldsymbol{\Phi}_t), \overline{\pi}(\boldsymbol{\Phi}_t), \ldots, \overline{\pi}(\boldsymbol{\Phi}_t)] \in \mathbb{R}^{J \times N_p},$$

$$\overline{\boldsymbol{H}}_t := [\overline{\pi}(\boldsymbol{h}_t), \overline{\pi}(\boldsymbol{h}_t), \ldots, \overline{\pi}(\boldsymbol{h}_t)] \in \mathbb{R}^{q \times N_p}$$

and

$$d\mathbf{Y}_t := [d\boldsymbol{Y}_t, d\boldsymbol{Y}_t, \ldots, d\boldsymbol{Y}_t] \in \mathbb{R}^{q \times N_p}.$$

Note that the identical (column) vector elements of $\overline{\boldsymbol{\Phi}}_t$ and $\overline{\boldsymbol{H}}_t$ are respectivelythe sample–averaged vectors corresponding to $\left\{\boldsymbol{\Phi}_t^{[j]}\right\}$ and $\left\{\boldsymbol{h}_t^{[j]}\right\}$, $j = 12, \ldots, N_p$ at the current time t. For the rest of the derivation, it is convenient to recast Eq. (7.35) by adding and subtracting $\overline{\boldsymbol{\Phi}}_t \boldsymbol{H}_t^T$ in the fourth term on the RHS as:

$$\boldsymbol{\Phi}_t = \boldsymbol{\Phi}_i + \int_{t_i}^{t} (\boldsymbol{\Psi}_s ds + d\boldsymbol{\Xi}_s) + \frac{1}{N_p} \int_{t_i}^{t} \left\{ \begin{array}{c} \boldsymbol{\Phi}_s \boldsymbol{H}_s^T - \overline{\boldsymbol{\Phi}}_s \boldsymbol{H}_s^T \\ + \overline{\boldsymbol{\Phi}}_s \boldsymbol{H}_s^T - \overline{\boldsymbol{\Phi}}_s \overline{\boldsymbol{H}}_s^T \end{array} \right\} \left[\sigma_{\boldsymbol{Y}_s} \sigma_{\boldsymbol{Y}_s}{}^T \right]^{-1}$$

$$\{d\mathbf{Y}_s - \boldsymbol{H}_s ds\} \tag{7.36}$$

A major hindrance in using Eq. (7.36), an MC based representation of Eq. (7.1), continues to be the problem of circularity in that the expectation of $\boldsymbol{\Phi}_t$ needs information on that of $\boldsymbol{\Phi}_t \boldsymbol{H}_t^T$, an higher order expectation. This impedes a direct solution of the set of non-linear equations. Starting with Eq. (7.36), an algorithm is devised to circumvent this problem.

7.5.2 EnKS filter—a non-iterative form

The EnKS filter constitutes a two-stage strategy. First, for a given $t \in \{t_i, t_{i+1}]$ in the current time interval, the process SDE (corresponding to the first two terms on the RHS of Eq. (7.36) is weakly solved using a numerical integration technique described in Chapter 5. In the second stage, an MC-based additive update term is derived using the third term on the RHS of Eq. (7.36). Although an explicit EM scheme is considered here for numerical integration of the process SDEs, a more accurate/stable stochastic integration scheme may be used. Presently, using the explicit EM-based integration, the recursive prediction-update filtering strategy that enables arriving at an empirical filtered distribution at time $t \in \{t_i, t_{i+1}]$ is described below. In all the numerical work, however, we set, $t = t_{i+1}$.

Prediction

$$\widetilde{\boldsymbol{\Phi}}_t = \boldsymbol{\Phi}_i + \boldsymbol{\Psi}_i (t - t_i) + \Delta \boldsymbol{\Xi}_i \tag{7.37}$$

where, $\boldsymbol{\Psi}_i := \boldsymbol{\Psi}_{t_i}$,

$$\Delta\Xi_i := \Delta\Xi_{t_i} = \left[\nabla\boldsymbol{\Phi}_i^{[1]}\sigma_i^{[1]}\left(\boldsymbol{B}_t^{[1]} - \boldsymbol{B}_i^{[1]}\right), \nabla\boldsymbol{\Phi}_i^{[2]}\sigma_i^{[2]}\left(\boldsymbol{B}_t^{[2]} - \boldsymbol{B}_i^{[2]}\right), \ldots, \right.$$
$$\left. \nabla\boldsymbol{\Phi}_i^{[N_p]}\sigma_i^{[N_p]}\left(\boldsymbol{B}_t^{[N_p]} - \boldsymbol{B}_i^{[N_p]}\right)\right],$$

$\sigma_i := \sigma_{t_i}$ and $\boldsymbol{B}_i := \boldsymbol{B}_{t_i}$.

Additive Update

A time–discretized MC-form of the update equation, motivated by an EM approximation to the third term on the RHS of Eq. (7.36), may be written as:

$$\boldsymbol{\Phi}_t = \widetilde{\boldsymbol{\Phi}}_t + \frac{1}{N_p}\left\{ \begin{array}{c} \left(\widetilde{\boldsymbol{\Phi}}_t - \overline{\widetilde{\boldsymbol{\Phi}}}_t\right)\widetilde{\boldsymbol{H}}_t^T \\ +\overline{\widetilde{\boldsymbol{\Phi}}}_t\left(\widetilde{\boldsymbol{H}}_t^T - \overline{\widetilde{\boldsymbol{H}}}_t^T\right) \end{array} \right\}\left[\sigma_{\mathbf{Y}_t}{}^T\sigma_{\mathbf{Y}_t}\right]^{-1}\left\{\Delta\mathbf{Y}_t - \widetilde{\boldsymbol{H}}_t(t - t_i)\right\} \tag{7.38}$$

Here $\widetilde{\boldsymbol{H}}_t := \left[\widetilde{\boldsymbol{h}}_t^{[1]}, \widetilde{\boldsymbol{h}}_t^{[2]}, \ldots, \widetilde{\boldsymbol{h}}_t^{[N_p]}\right]$ is the predicted ensemble of the measurement drift vectors. From the observation dynamics, we define $\Delta\mathbf{Y}_t := (t - t_i)\widetilde{\mathbf{Y}}_t = \Delta t\widetilde{\mathbf{Y}}_t$ and recast Eq. (7.38) as:

$$\boldsymbol{\Phi}_t = \widetilde{\boldsymbol{\Phi}}_t + \frac{1}{N_p}\left\{ \begin{array}{c} \left(\widetilde{\boldsymbol{\Phi}}_t - \overline{\widetilde{\boldsymbol{\Phi}}}_t\right)\widetilde{\boldsymbol{H}}_t^T\Delta t \\ +\overline{\widetilde{\boldsymbol{\Phi}}}_t\Delta t\left(\widetilde{\boldsymbol{H}}_t^T - \overline{\widetilde{\boldsymbol{H}}}_t^T\right) \end{array} \right\}\left[\sigma_{\mathbf{Y}_t}{}^T\sigma_{\mathbf{Y}_t}\right]^{-1}\left\{\widetilde{\mathbf{Y}}_t - \widetilde{\boldsymbol{H}}_t\right\} \tag{7.39}$$

where, $\widetilde{\mathbf{Y}}_t := [\mathcal{Y}_t, \mathcal{Y}_t, \ldots, \mathcal{Y}_t] \in \mathbb{R}^{q \times N_p}$. Here script $\mathcal{Y}_t$ stands for the vector of measurements used in the algebraic form of the observation Eq. (6.44b). The algebraic form is indeed the more practical way of modeling the observations as stated in Chapter 6. Note that the non-linear filtering (KS) Eq. (7.34) inherently demands SDE structures for both the system process and measurement models. Thus 'italicized' $\boldsymbol{Y}_t$ can be construed as a fictitious process so that the measurement over $(t_i,\ t_{i+1}]$ is $\Delta\boldsymbol{Y}_t \approx \mathcal{Y}_t\Delta_t, \Delta_t = t - t_i$. For small $\Delta t_i := t_i - t_{i-1}, \boldsymbol{h}(.,.)$ and $\boldsymbol{\hbar}(.,.)$ may be treated as identical. Recall that while the ensemble $\widetilde{\boldsymbol{\Phi}}_t = \left[\widetilde{\boldsymbol{\Phi}}_t^{[1]}, \widetilde{\boldsymbol{\Phi}}_t^{[2]}, \ldots, \widetilde{\boldsymbol{\Phi}}_t^{[N_p]}\right]$ contains the predicted particles, $\overline{\widetilde{\boldsymbol{\Phi}}}_t$ is constructed using N_p identical column vectors of the ensemble–averaged prediction, $\overline{\pi}_t(\widetilde{\boldsymbol{\Phi}})$. In the initial stages of time evolution, when the innovation process could have a significant drift component owing to the measurement-prediction mismatch (i.e., a significant departure from a zero–mean martingale), the gain-type coefficient matrix should be such (e.g. having a large norm) that the sample space is better explored. Keeping in mind that the temporal gradients of the evolving estimates have sharper variations in this regime, one way to construct the gain-type coefficient matrix would be to incorporate information on these gradients through the previous estimates. Thus $\widetilde{\boldsymbol{H}}_t \Delta t$ and $\overline{\widetilde{\boldsymbol{\Phi}}}_t \Delta t$ may be approximated as:

$$\widetilde{\boldsymbol{H}}_t \Delta t \approx \widetilde{\boldsymbol{H}}_t t - \overline{\widetilde{\boldsymbol{H}}}_i t_i - \Delta \overline{\widetilde{\boldsymbol{H}}}_t t \tag{7.40a}$$

$$\overline{\widetilde{\boldsymbol{\Phi}}}_t \Delta t = \overline{\widetilde{\boldsymbol{\Phi}}}_t t - \overline{\widetilde{\boldsymbol{\Phi}}}_i t_i \tag{7.40b}$$

where Ito's formula has been used in writing the approximation in Eq. (7.40a). Since this modification incorporates the filtered estimates at the previous time instant, t_i, it should be helpful in expediting the filter convergence especially in regions of sample space far away from the converged solution. Thus we get:

$$\boldsymbol{\Phi}_t = \widetilde{\boldsymbol{\Phi}}_t + \frac{1}{N_p} \left\{ \begin{array}{c} \left(\widetilde{\boldsymbol{\Phi}}_t - \overline{\widetilde{\boldsymbol{\Phi}}}_t\right)\left(\widetilde{\boldsymbol{H}}_t^T t - \overline{\widetilde{\boldsymbol{H}}}_i^T t_i - \Delta \overline{\widetilde{\boldsymbol{H}}}_t^T t\right) \\ + \left(\overline{\widetilde{\boldsymbol{\Phi}}}_t t - \overline{\widetilde{\boldsymbol{\Phi}}}_i t_i\right)\left(\widetilde{\boldsymbol{H}}_t^T - \overline{\widetilde{\boldsymbol{H}}}_t^T\right) \end{array} \right\} \left[\sigma_{\boldsymbol{Y}_t}^T \sigma_{\boldsymbol{Y}_t}\right]^{-1} \left\{\widetilde{\boldsymbol{Y}}_t - \widetilde{\boldsymbol{H}}_t\right\} \tag{7.41}$$

It may now be observed that, once the converged filtered estimate is available, the squared noise intensity term ${\sigma_{\boldsymbol{Y}_t}}^T \sigma_{\boldsymbol{Y}_t}$ may be replaced by the innovation covariance matrix, $E_p\left[\left((\boldsymbol{Y}_t - \boldsymbol{h}_t) - \pi_t(\boldsymbol{Y}_t - \boldsymbol{h}_t)\right)\left((\boldsymbol{Y}_t - \boldsymbol{h}_t) - \pi_t(\boldsymbol{Y}_t - \boldsymbol{h}_t)\right)^T\right]$. However, away from the converged solution and especially during the initial stages of the filter evolution, the last covariance matrix (e.g., its norm) would typically be rather large. Incorporation of this term, in lieu of ${\sigma_{\boldsymbol{Y}_t}}^T \sigma_{\boldsymbol{Y}_t}$ in Eq. (7.41) would have the effect of artificially increasing the measurement noise in the initial stages, thereby enabling the diffusion term in the system process model to explore the search space relatively more efficaciously. In the MC setup that we have adopted, the innovation covariance matrix may be empirically approximated as:

$$E_p\left[\left((\boldsymbol{Y}_t - \boldsymbol{h}_t) - \pi_t(\boldsymbol{Y}_t - \boldsymbol{h}_t)\right)\left((\boldsymbol{Y}_t - \boldsymbol{h}_t) - \pi_t(\boldsymbol{Y}_t - \boldsymbol{h}_t)\right)^T\right]$$

$$= \frac{1}{N_p - 1}\left(\widetilde{\boldsymbol{H}}_t^T - \overline{\widetilde{\boldsymbol{H}}}_t^T\right)\left(\widetilde{\boldsymbol{H}}_t - \overline{\widetilde{\boldsymbol{H}}}_t\right) \tag{7.42}$$

Finally, introducing a scalar parameter, $0 \le \alpha \le 1$, $\left[{\sigma_{\boldsymbol{Y}_t}}^T \sigma_{\boldsymbol{Y}_t}\right]^{-1}$ in Eq. (7.41) is replaced by $\left[\alpha \frac{1}{N_p - 1}\left(\widetilde{\boldsymbol{H}}_t^T - \overline{\widetilde{\boldsymbol{H}}}_t^T\right)\left(\widetilde{\boldsymbol{H}}_t - \overline{\widetilde{\boldsymbol{H}}}_t\right) + (1-\alpha)\left[{\sigma_{\boldsymbol{Y}_t}}^T \sigma_{\boldsymbol{Y}_t}\right]\right]^{-1}$. Equation (7.41) thus takes the form:

$$\boldsymbol{\Phi}_t = \widetilde{\boldsymbol{\Phi}}_t + \frac{1}{N_p}\left\{\begin{array}{c}\left(\widetilde{\boldsymbol{\Phi}}_t - \overline{\overline{\boldsymbol{\Phi}}}_t\right)(\widetilde{\boldsymbol{H}}_t^T t - \overline{\overline{\boldsymbol{H}}}_i^T t_i - \Delta\overline{\overline{\boldsymbol{H}}}_t^T t) \\ +\left(\overline{\overline{\boldsymbol{\Phi}}}_t t - \overline{\overline{\boldsymbol{\Phi}}}_i t_i\right)\left(\widetilde{\boldsymbol{H}}_t^T - \overline{\overline{\boldsymbol{H}}}_t^T\right)\end{array}\right\}$$

$$\left[\alpha\frac{1}{N_p-1}\left(\widetilde{\boldsymbol{H}}_t^T - \overline{\overline{\boldsymbol{H}}}_t^T\right)\left(\widetilde{\boldsymbol{H}}_t - \overline{\overline{\boldsymbol{H}}}_t\right) + (1-\alpha)\left[\sigma_{\mathbf{Y}_t}{}^T\sigma_{\mathbf{Y}_t}\right]\right]^{-1}\left\{\widetilde{\mathbf{Y}}_t - \widetilde{\boldsymbol{H}}_t\right\} \tag{7.43}$$

A more concise form of the update equation may be written as:

$$\boldsymbol{\Phi}_t = \widetilde{\boldsymbol{\Phi}}_t + \widetilde{\boldsymbol{G}}_t\left\{\widetilde{\mathbf{Y}}_t - \widetilde{\boldsymbol{H}}_t\right\} \tag{7.44}$$

where,

$$\widetilde{\boldsymbol{G}}_t := \frac{1}{N_p}\left\{\begin{array}{c}\left(\widetilde{\boldsymbol{\Phi}}_t - \overline{\overline{\boldsymbol{\Phi}}}_t\right)(\widetilde{\boldsymbol{H}}_t t - \overline{\overline{\boldsymbol{H}}}_i t_i - \Delta\overline{\overline{\boldsymbol{H}}}_t t) \\ +\left(\overline{\overline{\boldsymbol{\Phi}}}_t t - \overline{\overline{\boldsymbol{\Phi}}}_i t_i\right)\left(\widetilde{\boldsymbol{H}}_t^T - \overline{\overline{\boldsymbol{H}}}_t^T\right)\end{array}\right\}\left[\begin{array}{c}\alpha\frac{1}{N_p-1}\left(\widetilde{\boldsymbol{H}}_t^T - \overline{\overline{\boldsymbol{H}}}_t^T\right)\left(\widetilde{\boldsymbol{H}}_t - \overline{\overline{\boldsymbol{H}}}_t\right) \\ +(1-\alpha)\left[\sigma_{\mathbf{Y}_t}{}^T\sigma_{\mathbf{Y}_t}\right]\end{array}\right]^{-1} \tag{7.45}$$

In line with the traditional stochastic filtering terminology, the update term may be thought of as an innovation term $\boldsymbol{I}_t := \left\{\widetilde{\mathbf{Y}}_t - \widetilde{\boldsymbol{H}}_t\right\}$, weighted by the gain-type coefficient matrix $\widetilde{\boldsymbol{G}}_t$. An algorithm describing the computational features of the non-iterative EnKS is provided below.

Table 7.2 EnKS filtering scheme, non-iterative version

Step 1. Discretize the time interval of interest, say, $(0,T] \in \mathbb{R}$, using a partition $\Pi_N = \{t_0, t_1, \ldots, t_N\}$ such that $0 = t_0 < t_1 < \ldots < t_i < \ldots < t_N = T$ and $t_{i+1} - t_i = \Delta t_i (= \frac{T}{N}$ if the step size is chosen uniformly for $i = 0,1,\ldots,N-1))$. Choose an ensemble size, N_p. Generate the ensemble of initial conditions, $\left\{\boldsymbol{\Phi}^{[j](0)}\right\}_{j=1}^{N_p}$, or equivalently, $\left\{\boldsymbol{X}^{[j](0)}\right\}_{j=1}^{N_p}$, for the system state vector. For each discrete time instant, t_i, $i = 1,2,\ldots,N-1$, execute the following steps.

Step 2. Prediction

Using $\left\{\boldsymbol{\Phi}_i^{[j]}\right\}_{j=1}^{N_p}$, the update available at the last time instant t_i, propagate each particle discretely to the current time instant, t_{i+1}, using any appropriate integration scheme for SDEs, e.g., an explicit Euler–Maruyama (EM) approximation, applied to the process Eq. (6.62):

$$\widetilde{\boldsymbol{\Phi}}_{i+1}^{[j]} = \boldsymbol{\Phi}_i^{[j]} + L\left(\boldsymbol{\Phi}_i^{[j]}\right)\Delta t_i + \overline{\boldsymbol{\Phi}}_i^{[j]}\sigma_i^{[j]}\left(\boldsymbol{B}_{i+1}^{[j]} - \boldsymbol{B}_i^{[j]}\right), j = 1,2,\ldots,N_p \tag{7.46}$$

Step 3. Using $\left\{\widetilde{\boldsymbol{\Phi}}_{i+1}^{[j]}\right\}_{j=1}^{N_p}$, compute $\left\{\widetilde{\boldsymbol{X}}_{i+1}^{[j]}\right\}_{j=1}^{N_p}$. This step is trivial if $\boldsymbol{\Phi}(\boldsymbol{X})$ is the identity function, $\boldsymbol{\Phi}(\boldsymbol{X}) = \boldsymbol{X}$. Using $\left\{\widetilde{\boldsymbol{X}}_{i+1}^{[j]}\right\}_{j=1}^{N_p}$, compute $\left\{\widetilde{\boldsymbol{h}}_{i+1}^{[j]}\right\}_{j=1}^{N_p} = \left\{\boldsymbol{h}(\widetilde{\boldsymbol{X}}_{i+1}^{[j]})\right\}_{j=1}^{N_p}$. Also construct $\widetilde{\boldsymbol{\Phi}}_{i+1} := \left[\widetilde{\boldsymbol{\Phi}}_{i+1}^{[1]}, \widetilde{\boldsymbol{\Phi}}_{i+1}^{[2]}, \ldots, \widetilde{\boldsymbol{\Phi}}_{i+1}^{[N_p]}\right]$, $\overline{\overline{\boldsymbol{\Phi}}}_{i+1} := \left[\overline{\pi}\left(\widetilde{\boldsymbol{\Phi}}_{i+1}\right), \overline{\pi}\left(\widetilde{\boldsymbol{\Phi}}_{i+1}\right), \ldots, \overline{\pi}\left(\widetilde{\boldsymbol{\Phi}}_{i+1}\right)\right]$, $\widetilde{\boldsymbol{H}}_{i+1} := \left[\widetilde{\boldsymbol{h}}_{i+1}^{[1]}, \widetilde{\boldsymbol{h}}_{i+1}^{[2]}, \ldots, \widetilde{\boldsymbol{h}}_{i+1}^{[N_p]}\right]$ and $\overline{\overline{\boldsymbol{H}}}_{i+1} := \left[\overline{\pi}_{i+1}\left(\widetilde{\boldsymbol{h}}\right), \overline{\pi}_{i+1}\left(\widetilde{\boldsymbol{h}}\right), \ldots, \overline{\pi}_{i+1}\left(\widetilde{\boldsymbol{h}}\right)\right]$

Step 4. Additive update

Choose $\alpha \in [0,1]$ and update each particle as

$$\widehat{\boldsymbol{\Phi}}_{i+1}^{[j]} = \widetilde{\boldsymbol{\Phi}}_{i+1}^{[j]} + \widetilde{\boldsymbol{G}}_{i+1}\left(\boldsymbol{Y}_{i+1} - \widetilde{\boldsymbol{h}}_{i+1}^{[j]}\right), j = 1, 2, \ldots, N_p \tag{7.47}$$

where,

$$\widetilde{\boldsymbol{G}}_{i+1} := \frac{1}{N_p}\left\{\begin{array}{c}\left(\widetilde{\boldsymbol{\Phi}}_{i+1} - \overline{\overline{\boldsymbol{\Phi}}}_{i+1}\right)\left(\widetilde{\boldsymbol{H}}_{i+1}^{T} t_{i+1} - \overline{\overline{\boldsymbol{H}}}_{i}^{T} t_i - \Delta\overline{\overline{\boldsymbol{H}}}_{i+1}^{T} t_{i+1}\right) \\ + \left(\overline{\overline{\boldsymbol{\Phi}}}_{i+1} t_{i+1} - \overline{\overline{\boldsymbol{\Phi}}}_{i} t_i\right)\left(\widetilde{\boldsymbol{H}}_{i+1}^{T} - \overline{\overline{\boldsymbol{H}}}_{i+1}^{T}\right)\end{array}\right\}$$

$$\left[\alpha \frac{1}{N_p - 1}\left(\widetilde{\boldsymbol{H}}_{i+1}^{T} - \overline{\overline{\boldsymbol{H}}}_{i+1}^{T}\right)\left(\widetilde{\boldsymbol{H}}_{i+1} - \overline{\overline{\boldsymbol{H}}}_{i+1}\right) + (1-\alpha)\left[\sigma_{\boldsymbol{Y}_{i+1}}{}^{T}\sigma_{\boldsymbol{Y}_{i+1}}\right]\right]^{-1} \tag{7.48}$$

Step 5. If $i < N$, go to step 3 with $i = i + 1$, else terminate the algorithm.

Using a filtered martingale problem setup (See Section 4.12, Chapter 4 for brief notes on martingale problem), the existence and uniqueness of a posterior distribution satisfying the KS equation has been proved in [Kurtz 1998] under very general conditions on the drift and diffusion fields of the system process and measurement SDEs. Since Ito's theory is adopted in EnKS in interpreting the weak solutions of the SDEs, slightly more restrictive conditions, e.g. Lipschitz continuity and linear growth bound, are applicable to the drift and diffusion terms. The proof on convergence of the filtered estimate by the EnKS filter is given in Appendix G [also see Sarkar and Roy 2014].

7.5.3 EnKS filter—an iterative form

An iterative version of the EnKS filter is possible (as in the case of the KS filter) with the iterations providing an attractive, and sometimes indispensable, tool for an update procedure involving nonlinearities, e.g., those present in the system process and/or measurement models. While the additive nature of particle updates in the EnKS eliminates the curse of particle collapse, an iterative form could additionally help achieve faster convergence of the measurement-prediction mismatch to a zero–mean martingale. In other words, using an inner layer of iterations, one could, in a sense, attempt a maximal utilization of the current measurement within the particle update before moving over to the next time step. In order to provide an additional boost to the mixing property of the update kernel, an annealing-type non-decreasing sequence of scalar multipliers, $\{\beta_0, \beta_1, \ldots, \beta_{\kappa-1}\}$, is artificially applied to the iterated sequence of gain-weighted innovation terms. Here κ denotes the number of inner iterations at the current time t. The sequence $\{\beta_0, \beta_1, \ldots, \beta_{\kappa-1}\}$ is so chosen that $\beta_\kappa \to 1$ as $\kappa \to \infty$ (to approach the original update term) and $\beta_k \leq \beta_{k+1}$ for $k \in [0, \kappa - 1)$. Effect of the annealing type term is similar to the temperature (herein proportional to $1/\beta_k$) in a standard simulated annealing (SA) scheme. The added advantage we have over SA based schemes, is that, unlike SA where a single Markov chain is evolved and $1/\beta_k$ is slowly reduced to unity, in the present setting, an ensemble of pseudo-chains is propagated simultaneously, for a given t, allowing us to have a much faster and flexible reduction. Presently an exponential increment, $\beta_k = \exp(k + 1 - \kappa), k = 0, \ldots, \kappa - 1$ which is a more non-conservative schedule, is adopted. At a given time t, to initiate the inner iteration, an initial annealing term is required, for which, typically a large value of $1/\beta_k$ is taken. Although the filtered estimate asymptotically improves with increasing number of κ, at a given t, keeping in mind the computational feasibility, only a few iterations (say, $\kappa = 10$) may be prescribed for a sufficiently large class of problems. With the convention $\left\{\boldsymbol{\Phi}_t^{[j](0)}\right\}_{j=1}^{N_p} := \left\{\widetilde{\boldsymbol{\Phi}}_t^{[j]}\right\}_{j=1}^{N_p}$ and $\left\{\boldsymbol{h}_t^{[j](0)}\right\}_{j=1}^{N_p} := \left\{\widetilde{\boldsymbol{h}}_t^{[j]}\right\}_{j=1}^{N_p}$, a generic form of the iterative update may be given as:

$$\boldsymbol{\Phi}_t^{(l+1)} = \boldsymbol{\Phi}_t^{(l)} + \beta_l \boldsymbol{G}_t^{(l)}\left(\mathbf{Y}_t - \boldsymbol{H}_t^{(l)}\right), l = 1, 2, \ldots, \kappa - 1 \tag{7.49}$$

where,

$$\boldsymbol{G}_t^{(l)} := \frac{1}{N_p}\left\{\begin{array}{c}\left(\widetilde{\boldsymbol{\Phi}}_t^{(l)} - \overline{\overline{\widetilde{\boldsymbol{\Phi}}}}_t^{(l)}\right)\left(\widetilde{\boldsymbol{H}}_t^{(l)^T} t - \overline{\overline{\widetilde{\boldsymbol{H}}}}_i^{(l)^T} t_i - \Delta\overline{\overline{\widetilde{\boldsymbol{H}}}}_t^{(l)^T} t\right) \\ + \left(\overline{\overline{\widetilde{\boldsymbol{\Phi}}}}_t^{(l)} t - \overline{\overline{\widetilde{\boldsymbol{\Phi}}}}_i t_i\right)\left(\widetilde{\boldsymbol{H}}_t^{(l)^T} - \overline{\overline{\widetilde{\boldsymbol{H}}}}_t^{(l)^T}\right)\end{array}\right\}$$

$$\times\left[\alpha \frac{1}{N_p - 1}\left(\widetilde{\boldsymbol{H}}_t^{(l)^T} - \overline{\overline{\widetilde{\boldsymbol{H}}}}_t^{(l)^T}\right)\left(\widetilde{\boldsymbol{H}}_t^{(l)} - \overline{\overline{\widetilde{\boldsymbol{H}}}}_t^{(l)}\right) + (1 - \alpha)\left[\sigma_{\mathbf{Y}_t}{}^T \sigma_{\mathbf{Y}_t}\right]\right]^{-1}, \tag{7.50a}$$

$$\overline{\overline{\widetilde{\boldsymbol{\Phi}}}}_t^{(l)} = \left[\overline{\pi}_t^{(l)}(\boldsymbol{\Phi}), \overline{\pi}_t^{(l)}(\boldsymbol{\Phi}), \ldots, \overline{\pi}_t^{(l)}(\boldsymbol{\Phi})\right], \overline{\pi}_t^{(l)}(\boldsymbol{\Phi}) := \frac{1}{N_p}\sum_{j=1}^{N_p} \boldsymbol{\Phi}_t^{[j](l)} \tag{7.50b}$$

$$\boldsymbol{H}_t^{(l)} := \left[\boldsymbol{h}_t^{[1](l)}, \boldsymbol{h}_t^{[2](l)}, \ldots, \boldsymbol{h}_t^{[N_p](l)}\right] \text{ and } \overline{\boldsymbol{H}}_t^{(l)}$$

$$:= \left[\overline{\pi}_t^{(l)}(\boldsymbol{h}), \overline{\pi}_t^{(l)}(\boldsymbol{h}), \ldots, \overline{\pi}_t^{(l)}(\boldsymbol{h})\right] \tag{7.50c}$$

An algorithm for the iterative EnKS is described below.

Table 7.3 EnKS filtering scheme, iterative version

Steps 1 to 3. Same as steps 1 to 3 of the algorithm of the non-iterative version in Table 7.2. Set iteration counter $l = 1$ and choose κ. Compute $\beta_0 = \exp(1-\kappa)$.

Step 4. Iterative update: Using $\left\{\boldsymbol{\Phi}_{i+1}^{[j](l-1)}\right\}_{j=1}^{N_p}$ compute $\left\{\boldsymbol{X}_{i+1}^{[j](l-1)}\right\}_{j=1}^{N_p}$.

Now, generate

$\left\{\boldsymbol{h}_{i+1}^{[j](l-1)}\right\}_{j=1}^{N_p} := \left\{\boldsymbol{h}\left(\boldsymbol{X}_{i+1}^{[j](l-1)}\right)\right\}_{j=1}^{N_p}$, $\boldsymbol{\Phi}_{i+1}^{l-1} = \left[\boldsymbol{\Phi}_{i+1}^{[1](l-1)}, \boldsymbol{\Phi}_{i+1}^{[2](l-1)}, \ldots, \boldsymbol{\Phi}_{i+1}^{[N_p](l-1)}\right]$,

$\overline{\pi}_{i+1}^{(l-1)}(\boldsymbol{\Phi}) = \frac{1}{N_p}\sum_{n=1}^{N_p} \boldsymbol{\Phi}_{i+1}^{[j](l-1)}$, $\overline{\boldsymbol{\Phi}}_{i+1}^{l-1} = \left[\overline{\pi}_{i+1}^{(l-1)}(\boldsymbol{\Phi}), \overline{\pi}_{i+1}^{(l-1)}(\boldsymbol{\Phi}), \ldots, \overline{\pi}_{i+1}^{(l-1)}(\boldsymbol{\Phi})\right]$,

$\boldsymbol{H}_{i+1}^{(l-1)} = \left[\boldsymbol{h}_{i+1}^{[1](l-1)}, \boldsymbol{h}_{i+1}^{[2](l-1)}, \ldots, \boldsymbol{h}_{i+1}^{[N_p](l-1)}\right]$, $\overline{\pi}_{i+1}^{(l-1)}(\boldsymbol{h}) = \frac{1}{N_p}\sum_{n=1}^{N_p} \boldsymbol{h}_{i+1}^{[j](l-1)}$

and

$\overline{\boldsymbol{H}}_{i+1}^{(l-1)} = \left[\overline{\pi}_{i+1}^{(l-1)}(\boldsymbol{h}), \overline{\pi}_{i+1}^{(l-1)}(\boldsymbol{h}), \ldots, \overline{\pi}_{i+1}^{(l-1)}(\boldsymbol{h})\right]$.

Update each particle according to Eq. (7.49) as $\boldsymbol{\Phi}_i^{[j](l)} = \boldsymbol{\Phi}_i^{[j](l-1)} + \beta_l \boldsymbol{G}_i^{(l-1)}\left(\boldsymbol{Y}_i - \boldsymbol{h}_i^{[j](l-1)}\right)$, $j = 1, 2, \ldots, N_p$ where $\boldsymbol{G}_i^{(l-1)}$ is given in Eq. (7.50 a).

Step 6. Set $l = l+1$. If $l < \kappa$, set $\beta_l = \exp(l+1+\kappa)$ and go to step 2; else if $i < N$, go to step 3 with $i = i+1$; else terminate the algorithm.

Step 5. If $i < N$, go to step 3 with $i = i+1$, else terminate the algorithm.

Based on the notion of stochastic Picard's iterations often used to prove the existence and uniqueness of solutions to SDEs, the existence and uniqueness of solution (i.e., the conditional posterior distribution) via the proposed iterative scheme may be demonstrated [Sarkar and Roy 2014] . The relevant theorems with proofs are provided in Appendix G.

Example 7.4

The 80-dimensional filtering problem corresponding to the 20-DOF shear frame model (Fig. 6.11) is considered for state–parameter estimation by the EnKS filter.

Solution

The same stiffness and damping matrices in Eqs. (6.105 and 6.106) represent the augmented system model governed by Eq. (6.104). The stiffness and damping coefficients a well as the displacement and velocity states are estimated with the measurements available on all the displacement components. The (transverse) nodal forcing vector $\mathcal{P}(t) := \left\{ \mathscr{f}_i(t), i \in [1, 20] \right\}$ is assumed to be stochastic and the forcing at the ith floor is given by $\mathscr{f}_i(t) = 500 \exp(-t) |\xi| \cos(5t)$ where $\xi \approx \mathcal{N}(0, 1)$. Low measurement noise intensity (less than 1 %) is considered here for all the 20 components of the measured displacement vector. The state-parameter estimation is performed by the non-iterative form of the KS filter with $N_p = 800$ with time step $\Delta = 0.01$ s and $T = 10$ s The reference stiffness and damping parameters (used to integrate the system process *en route* to the generation of synthetic data by perturbing the computed solutions with appropriate noise) at each degree of freedom is taken as $k_i^* = 2.6125E08$ N/m and $c_i^* = 5.225E06$ N-s/m where $i \in [1, 20]$. The estimation results by the EnKS filter are shown in Fig. 7.4. A typical prescribed value of $\alpha = 0.8$ (Eq. 7.43) is used in the computations while the method is found to perform well for other values in the interval $(0, 1)$.

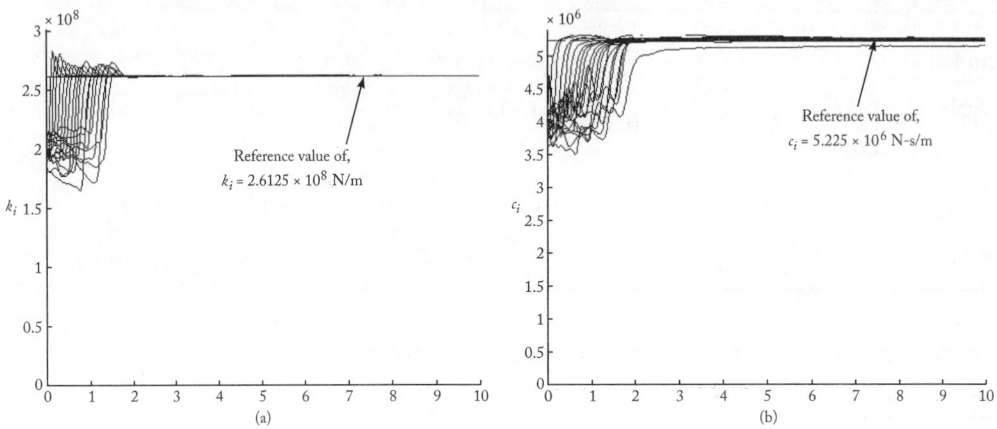

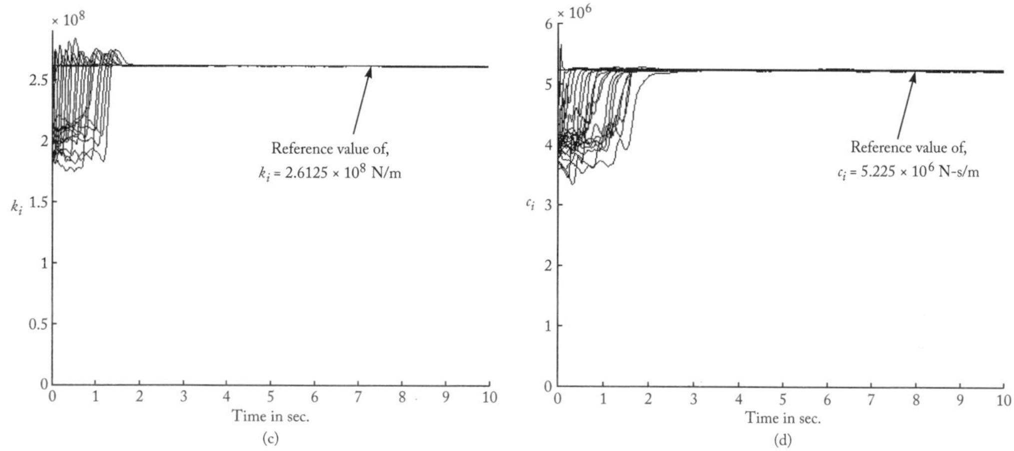

Fig. 7.4 Parameter estimation by the EnKS filter; 20-DOF shear frame model in Fig. 6.11, reference parameter values are: $k_1^* = k_2^* = \cdots = k_{20}^* = 2.6125 \times 10^8$ N/m, $c_1^* = c_2^* = \cdots = c_{20}^* = 5.225 \times 10^6\ N\text{-}s/m$; $T = 10$ s, $\Delta = 0.01$ s, $N_p = 800$, $\alpha = 0.8$ (a) and (b) estimation by non-iterative EnKS for k and c, respectively, and (c) and (d) estimation by iterative EnKS for k and c, respectively

Example 7.5

We consider dynamical system identification via a 200-dimensional non-linear filtering problem corresponding to a 50-DOF structural system.

Solution

For the state–parameter estimation, the system model in Fig. 6.11 (corresponding to a 50-storied shear frame with uncertain damping and stiffness parameters) is considered. The process dynamics equation is of the same form as in Eq. (6.104) with the forcing function $\mathcal{P}(t)$ being a random forcing vector as in Example 7.4 but with $i \in [1, 50]$. The stiffness and damping matrices are given in Eqs. (6.105 and 6.106). The viscous damping matrix $\boldsymbol{C}$ is obtained by assuming a Raleigh-type damping mechanism and using the proportionality constants $\alpha_1 = 0.02$ and $\alpha_2 = 0.05$. The aim is to estimate the stiffness and damping coefficients as well as the velocity and displacement states, conditioned only on all the measured displacement components. The augmented state vector is given by $\boldsymbol{X} := \left\{\left\{X_1, \dot{X}_1, \ldots, X_{50}, \dot{X}_{50}\right\}, \{k_1, k_2, \ldots, k_{50}\}, \{c_1, c_2, \ldots, c_{50}\}\right\}^T \in \mathbb{R}^{200}$. Particle filters are likely to diverge or collapse to a single particle for such a large dimensional state-parameter estimation problem with sparse data. Moreover, even for low dimensional problems, they may perform poorly with very low-intensity measurement noises, currently employed to reduce random fluctuations in the estimates due to large variance in the measurement noise. To demonstrate the performance of the EnKS filter with low measurement noise levels (possible with sophisticated measuring devices), very low measurement noise intensity (less than 1 %) is considered here for all the 50 components of the measured displacement vector. An ensemble size $N_p = 800$ and time step $\Delta = 0.01$ s with $T = 10$ s are taken for the state–parameter estimation. The parameter α is chosen to be 0.8.

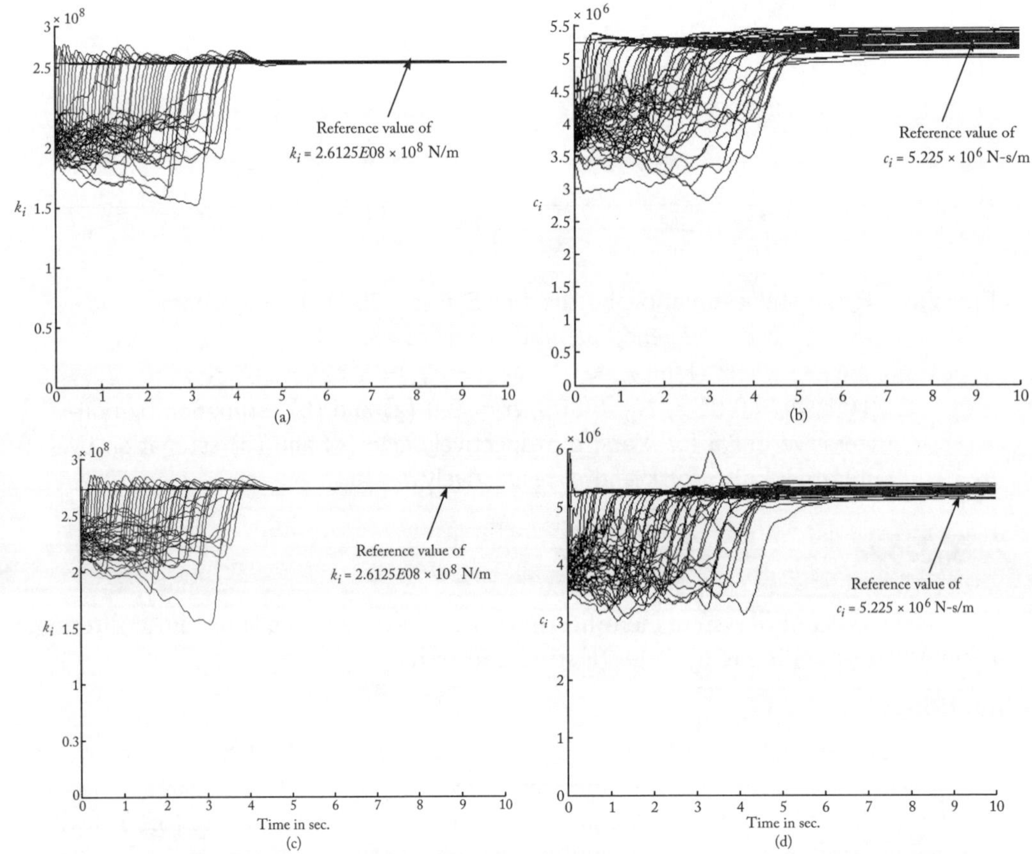

Fig. 7.5 Parameter estimation by the EnKS filter; 50-DOF shear frame model in Fig. 6.11, reference parameter values are: $k_1^* = k_2^* = \cdots = k_{50}^* = 2.6125 \times 10^8$ N/m, $c_1^* = c_2^* = c_{50}^* = 5.225 \times 10^6 N\text{-}s/m$; $T = 10$ s, $\Delta = 0.01$ s, $N_p = 800$, $\alpha = 0.8$ (a) and (b) estimation by non-iterative EnKS for k and c, respectively, and (c) and (d) estimation by iterative EnKS for k and c, respectively

The reference stiffness parameters (used to integrate the system process for the generation of synthetic data by perturbing the computed solutions with appropriate noise) at each degree of freedom are taken as $2.6125E08$ N/m as in the earlier Examples. Similarly the reference damping parameter at each floor is chosen as $5.225E06$ N-s/m. Figure 7.5 shows the performance of the non-iterative EnKS filter for the fairly large dimensional filtering problem considered here.

Example 7.6

We consider the shear frame with 20 *dof*s (as in Fig. 6.11) and utilize the state–parameter estimation by EnKS for damage detection.

Solution

The model for the shear frame, albeit of a lower dimension corresponding to 20 *dof*s is adopted for the damage detection problem. As in the last two examples, the nodal forces are taken to be stochastic. Thus, $\mathcal{P}(t) = \left\{f_i(t)\right\}_{i=1}^{20}$ is a random forcing vector whose i^{th} element, $f_i(t) = 500\ \exp(-t)\,|\xi|\cos(5t)$ where $\xi \approx \mathcal{N}(0,1)$ denotes the transverse loading at the i th storey. While the aim remains to estimate the stiffness and damping coefficients along with the velocity/displacement states (conditioned on only the measured displacements of the floors), the reference stiffness parameter k_{10} is kept at a slightly lower value vis-à-vis the rest $\{k_i | i \in [1,20]\} \setminus \{k_{10}\}$. This is a simple representation of a slightly reduced load-carrying capacity at the 10th floor, thereby indicating incipient damage/degradation. Specifically, we take $k_{10} = 0.98*2.6125E08$ N/m even as the rest of the stiffness parameters are maintained at $2.6125E08$ N/m. The reference value for the damping coefficients c_i^*, $i \in [1,20]$ remains at $5.225E06$ N–s/m. Here the augmented state vector is 80-dimensional and, as in the last example, low noise intensity (< 1%) is applied in generating the data. With $T = 10$ s, an ensemble size of $N_p = 800$ and time step $= 0.01$ s are taken for the filter run. In the computations, the parameter α is taken as 0.8. Fig. 7.6 shows the estimation results. The incipient damage introduced in the model is fairly detected by the filter (see the graph of the estimates in Fig. 7.6a and Fig. 7.6c being indistinguishable from that of the reference value of the stiffness parameter, k_{10}).

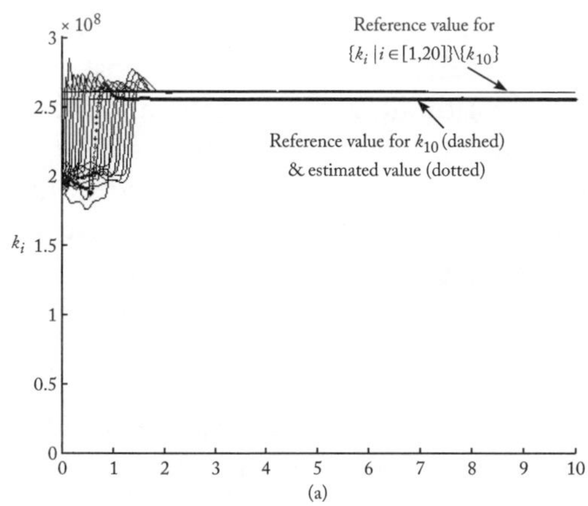

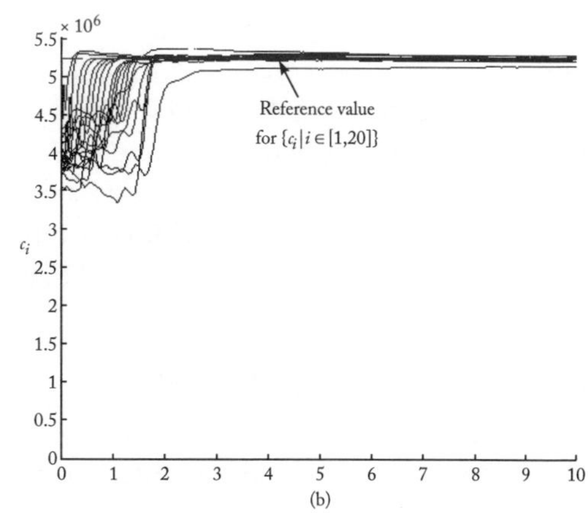

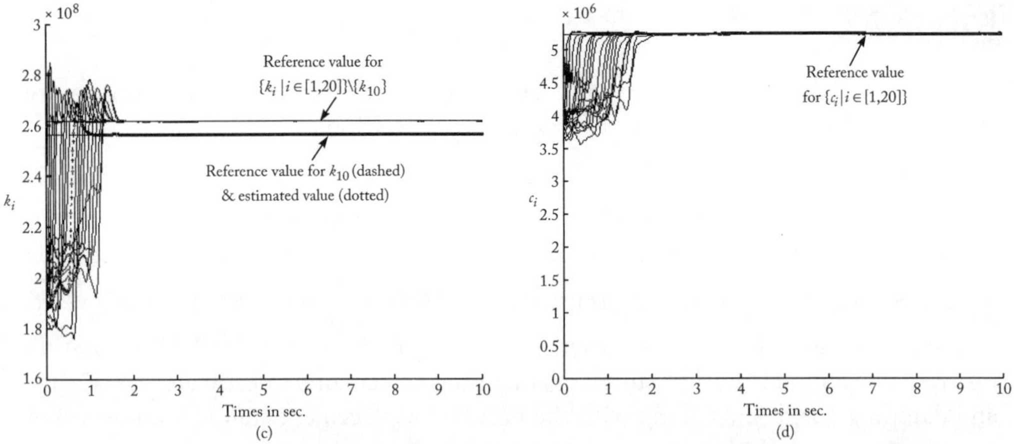

Fig. 7.6 Damage detection by parameter estimation using the EnKS filter; 20-DOF shear frame model in Fig. 6.11, reference parameter values are: $k_1^* = k_2^* = \cdots = k_{20}^* = 2.6125 \times 10^8\ N/m$, $c_1^* = c_2^* = \cdots = c_{20}^* = 5.225 \times 10^6\ N\text{-}s/m$; $T = 10$ s, $\Delta = 0.01$ s, $N_p = 1000$, $\alpha = 0.8$ (a) and (b) estimation by non-iterative EnKS, and (c) and (d) estimation by iterative EnKS

An interesting alteration to the setup for filtering problems (or the optimization problems to be covered in the last couple of chapters of this book), could be in the form of Poisson-type measurements. This would enable a more general yet a more practical modeling of the measurement equation. A key aspect of the development is the filter update scheme, derived from an ensemble approximation of the time-discretized non-linear filtering equation, modified to account for Poisson-type measurements. Specifically, the additive update through a gain-like correction term, empirically approximated from the innovation integral in the filtering equation, eliminates the problem of particle collapse. Though not covered here, we refer to Venugopal *et al.* [2016] for details on the scheme and specific algorithmic aspects of its applications related to non-linear dynamical systems. Finally , while we have covered only a small class of system identification problems as an application of stochastic filtering, readers may refer to Venugopal *et al,* [2016] for an application of the KS and EnKS filters to a class of inverse problems of interest in medical diagnostic imaging.

7.6 Concluding Remarks

In this treatise, we have viewed the KS equation as the parent filtering equation providing formal solutions to non-linear filtering problems. A set of recent particle filters broadly fashioned after the structure of the KS equation has been presented. The MC-adapted forms of the additive update, which attempts at driving the measurement-prediction misfit to a zero–mean martingale, are functionally analogous to non-Newton (i.e., derivative-free) directional information in arriving at the filtered solution. The additive update also eliminates the bottleneck of particle impoverishment, often engendered by the conventional weight-based implementation of the change of measures. In our understanding, the KS and EnKS filters provide for a couple of practicable numerical schemes for applications to reasonably large dimensional non-linear filtering problems.

A word on the possible use of Doob's h–transform (see Appendix D) in a better filtering scheme for parameter identification problems would be in order. As shown in Appendix D (Example D.1) based on a space–time trick on Doob's h–transform, the Brownian motion processes, which appear in the process SDEs corresponding to the parameters that are, say, strictly positive (or whose values may be bounded within an interval), could be restricted to remain positive (or within the stated intervals) by suitably added drift terms. This way of doing filtering, using the change of measure trick again, will not only be computationally faster, but numerically more accurate.

The rest of this book is mainly devoted to showcasing the power and near-universality of KS-type equations in the context of problems that go well beyond the direct purview of stochastic filtering. To start with, in the chapter to follow, we use this in improving the accuracy of numerically integrated solutions to SDEs by the concept of change of measure.

Notations

c_i, c_i^*	damping coefficients
$\boldsymbol{C} \in \mathbb{R}^m \times m,$	the damping matrix
$g_{r,i}^{(l)} := \boldsymbol{G}_{i+1}^{(l)}(.,r)-$	the rth column of the $\boldsymbol{G}_i^{(l)}, r = 1,2,\ldots,\boldsymbol{q}$
$\boldsymbol{G}_i^{(l)}$	$j \times \boldsymbol{q}$ matrix at l^{th} iteration (Equation (7.32)a)
$\boldsymbol{h}(t,\boldsymbol{X}_t)$	$\in \mathbb{R}^q$ vector of observation (measurement) functions (Eq. 6.1b)
$\boldsymbol{\hbar}(t,\boldsymbol{X}_t)$	$\in \mathbb{R}^q$ vector of observation (measurement) functions in algebraic form (Eq. 6.44b)
$\widetilde{\boldsymbol{H}}_t$	predicted ensemble of the measurement drift vectors
$\boldsymbol{I}_t$	$= \{I_{r,t}\} = \{Y_{r,t} - \int_{t_i}^{t} \pi_s(h_r(s,\boldsymbol{X}(s)))ds\} \in \mathbb{R}^{\boldsymbol{q}}$, the innovation vector
$J = 2m + N_p,$	dimension of the combined state- parameter vector of a typical MDOF system
k_i, k_i^*	stiffness coefficients
$\boldsymbol{K} \in \mathbb{R}^{m\times m},$	the stiffness matrix
$\mathcal{K}_t$	gain matrix
m	dimension of system variables
n_p	dimension of system parameters

n_Γ	number of iterations
N_G	number of Gaussian mixands
N_p	sample size
$\boldsymbol{R_Y} = \boldsymbol{\sigma_Y \sigma_Y^T}$	the measurement noise covariance matrix
${}^n\widetilde{w}_i$	mixand weights
$\boldsymbol{X} = \{\boldsymbol{X}_S^T\ \boldsymbol{X}_p^T\}^T$	$\in \mathbb{R}^{2m+n_p}$, a vector of stochastic processes representing augmented state variables
$\widetilde{\boldsymbol{X}}$	predicted state vector process
$\widehat{\boldsymbol{X}}$	updated state vector process
$\mathbf{X}^p$	prediction error (anomaly) matrix
${}^n\widehat{\mathbf{X}}_i$	sample mean of the sub-ensemble associated with the nth mixand
$\boldsymbol{Y}_t$	observation vector stochastic process
$\mathbf{Y}^p$	prediction measurement (anomaly) matrix
α	a scalar parameter, $0 \le \alpha \le 1$
α_1, α_2	proportionality constants, $\in \mathbb{R}$
α_l	artificial diffusion parameter at l^{th}iteration
β_l	annealing type parameter at l^{th} iteration
Γ	iteration parameter
$\overline{\pi}(\cdot)$	empirical expectation of the corresponding ensemble of particles
$\Phi_i(\boldsymbol{X}), i = 1, 2, ..., J$	scalar-valued function of $\boldsymbol{X}$
$\boldsymbol{\Phi}$	$:= (\Phi_1, \Phi_2, ..., \Phi_J)^T$, a J-dimensional vector valued function in EnKS scheme
$\boldsymbol{\Phi}_t$	$[\boldsymbol{\Phi}_t^{[1]}, \boldsymbol{\Phi}_t^{[2]}, \ldots, \boldsymbol{\Phi}_t^{[N_p]}] \in \mathbb{R}^{J \times N_p}$, a matrix with each column being a J-dimensional vector valued function of $\boldsymbol{X}$
$\nabla\Phi(\boldsymbol{X}_t)$	gradient vector
ψ_i	an explicit time marching map at time $t = t_i$
ϑ	iteration parameter
${}^n\widehat{\Sigma}_i$	sample covariance of the sub-ensemble associated with the nth mixand

Chapter 8

Improved Numerical Solutions to SDEs by Change of Measures

8.1 Introduction

Obtaining accurate numerical solutions to SDEs has been the focus of numerous studies owing to their relevance in many fields of engineering and science [Platen 1987, Kloeden and Platen 1992, Milstein 1995, Higham *et al.* 2002]. In the context of stochastic filtering (Chapters 6 and 7), it was seen that the process model is often a set of non-linear SDEs and imprecise integration techniques for these SDEs may precipitate significant numerical errors in the predicted particle locations leading to possible degradation in the filter performance. Determining solutions—strong or weak—of stochastically driven non-linear oscillators by direct numerical integration of the associated SDEs has been dealt with in Chapters 4 and 5. In particular, a universal framework for integration schemes is provided in Chapter 5 through an MC approach. There it has been shown that the Ito–Taylor expansion, which is based on an iterated Ito's formula, helps to construct integration schemes for SDEs. The possibility of developing higher order numerical integration schemes, e.g., the Milstein method [Milstein 1995], numeric–analytical techniques of LTL (locally transversal linearization) type [Roy 2000, 2001, 2004] and the stochastic Newmark method [Roy 2006], has been demonstrated along with the estimation of the order of accuracy of the higher order numerical integration schemes. However, unlike ordinary DEs, deriving higher order numerical schemes for SDEs is generally hindered by the difficulty of computing higher order MSIs. On the other hand, avoidance of the higher order MSIs, which implies retaining fewer terms in the hierarchical stochastic Taylor's approximation used to construct the integration scheme, naturally achieves relatively lower order accuracy. Most of the lower order explicit schemes (for instance, the explicit version of the EM scheme) may lose stability, for instance in the case of stiff SDEs, thus requiring impracticably low time steps to get stable solutions. Thanks to its computational expedience and ease of implementation, a lower order scheme would be ideal, were it not for a loss of integration accuracy. Motivated by this, we describe in this chapter an approach [Raveendran *et al.* 2013a, Raveendran *et al.* 2013b, Sarkar and Roy 2016] that aims at recovering the lost accuracy of a lower order scheme—the EM scheme to wit—using a change of measures.

The key role played by a change of measures and the associated Radon–Nikodym derivative was made evident in some of the previous chapters—from performing the simple task of numerically evaluating integrals by MC simulations to the development of stochastic filtering strategies (Chapters 6 and 7). It is further exploited in Chapters 9 and 10 (to follow) leading to a robust numerical scheme for global optimization.

First we briefly introduce the Girsanov linearization method (GLM) [Saha and Roy 2007] involving the concept of change of measures to solve SDEs under additive stochastic excitations. The method forms the basis for the numerical treatment of SDEs using a change of measures to achieve improved accuracy [Raveendran *et al.* 2013a, Sarkar and Roy 2016].

Girsanov linearization method (GLM)

Taking cue from the LTL schemes [Roy 2001, 2004] discussed in Chapter 5, the GLM [Saha and Roy 2007] uses a locally linearized replacement of the non-linear drift terms followed by a Girsanov transformation of measures to weakly correct for the linearization error. This correction, a Radon–Nikodym derivative and hence multiplicative in nature, satisfies a scalar SDE that is exactly solvable in terms of the linearized solution. Being a Radon–Nikodym derivative (the likelihood ratio), the correction process is an exponential martingale over bounded time intervals, subject to Novikov's condition. We outline the integration scheme in the context of non-linear mechanical oscillators (i.e. those with displacement $\boldsymbol{X}_1$ and velocity $\boldsymbol{X}_2$ states), even though the basic idea is applicable more generally. Consider a system of $2m$ first order non-linear SDEs in the following incremental form:

$$dX_{1j} = X_{2j}dt \tag{8.1a}$$

$$dX_{2j} = A_j(t,\boldsymbol{X})\,dt + \sum_{k=1}^{n} \sigma_{jk}(t)\,dB_k(t), j = 1,2,\ldots,m \tag{8.1b}$$

where, $\boldsymbol{X} := \left\{\boldsymbol{X}_1^T, \boldsymbol{X}_2^T\right\}^T \in \mathbb{R}^{2m}$ is the vector of state variables with $\boldsymbol{X}_1, \boldsymbol{X}_2 \in \mathbb{R}^m$, $\{B_k(t) : k = 1,2,\ldots,n\}$ is an n-dimensional vector of independently evolving zero–mean standard Wiener processes and $\boldsymbol{\sigma} = \left[\sigma_{jk} : j = 1,2,\ldots,m \text{ and } k = 1,2,\ldots,n\right] \in \mathbb{R}^{m\times n}$ the corresponding diffusion matrix. We assume that the drift vector $\boldsymbol{A} := \{A_j : j = 1,2,\ldots,m\}^T$ can be decomposed into two constituent parts as $A_j = A_{j\,l} + A_{j\,nl}$ (i.e., $\boldsymbol{A} = \boldsymbol{A}_l + \boldsymbol{A}_{nl}$), where $\boldsymbol{A}_l$ denotes the linear part of the vector field that should remain as it is in the linearized equations and $\boldsymbol{A}_{nl}$ the non-linear part. Equations (8.1) may represent, for example, an m-dimensional MDOF non-linear mechanical oscillator in the state space. Typically, the linear part $A_{j\,l}$ is of the form:

$$A_{j\,l}(t,\boldsymbol{X}) = -\sum_{k=1}^{m} C_{jk}X_{2k} - \sum_{k=1}^{m} K_{jk}X_{1k} + f_j(t) \tag{8.2}$$

$C, K \in \mathbb{R}^{m\times m}$ are the damping and stiffness matrices and $f(t) = \{f_j(t) : j = 1, 2, \ldots, m\}^T \in \mathbb{R}^m$ the deterministic force vector. (Note that in writing Eqs. (8.1) and (8.2), the system mass matrix is taken as an identity matrix). To ensure boundedness of the solution vector $\boldsymbol{X}$ and uniqueness in a weak sense, it is assumed that the drift and diffusion vectors, $\boldsymbol{A}$ and $\boldsymbol{\sigma}_k$ are measurable and satisfy the following bounds:

$$\left\|\boldsymbol{A}(t,\boldsymbol{X})\right\| + \left\|\sum_{k=1}^{n} \boldsymbol{\sigma}_k(t)\right\| \le C_1(1 + \|\boldsymbol{X}\|) \tag{8.3a}$$

$$\left\|\boldsymbol{A}(t,\boldsymbol{X}) - \boldsymbol{A}(t,\boldsymbol{Y})\right\| \le C_2(\|\boldsymbol{X} - \boldsymbol{Y}\|) \tag{8.3b}$$

where, $\boldsymbol{Y} \in \mathbb{R}^{2m}$, $C_1, C_2 \in \mathbb{R}^+$. Let the initial conditions be mean square bounded, i.e., $E\left[\left\|\boldsymbol{X}(t_0)\right\|^2\right] < \infty$ (without a loss of generality, it is presently treated as deterministic). Consider the partition Π_N of the time interval $(0, T] \in \mathbb{R}$ with $t_N = T$ and $h_i = t_i - t_{i-1}$. Let $T_i = (t_{i-1}, t_i]$ with $i \in \mathbb{Z}^+$. Assume T to be deterministic even though it could be a stopping time. For expositional convenience without a loss of focus on the main issues, we assume a uniform time step $h_i = h \forall i$ unless otherwise stated.

One is interested in computing the weak response, i.e., expectations of the form:

$$\nu = E[g(t,\boldsymbol{X})] \tag{8.4}$$

where, $g(.)$ denotes some real-valued scalar function. When we utilize MC simulations to evaluate the expectation, ν is replaced by its sample–mean formula:

$$\nu \approx \overline{\nu} = \frac{1}{N_p}\sum_{k=1}^{N_p}\left[g(t,\boldsymbol{X}^{[k]})\right] \text{ for large } N_p \tag{8.5}$$

where, $\boldsymbol{X}^{[k]}(t)$ denotes the kth sample of $\boldsymbol{X}(t)$ and N_p the ensemble size. Since it suffices to evaluate $\overline{\nu}$ using weak solutions of Eq. (8.1), the equation is first locally linearized and then a measure transformation based on Girsanov's theorem (Chapter 4) is used to correct for the linearization errors and thus obtain accurate estimates of the expectation. By local linearization (Section 5.10, Chapter 5), the governing SDEs are replaced by a system of linearized ones so the ith linearized SDE is taken to serve as a weak replacement for the non-linear SDE over T_i.

Briefly, the Girsanov theorem implies that if the drift coefficient of an Ito process (with a non-degenerate diffusion coefficient) is altered to an extent (such that the original and altered drift coefficients satisfy Novikov's condition—Eq. (4.240b) of Chapter 4), then the law of the process would not be altered drastically. Indeed, the law of the drift-modified process will remain absolutely continuous with respect to that of the original process and we can explicitly compute the Radon–Nikodym derivative.

While linearization of non-linear oscillators using the Girsanov theorem could be non-uniquely performed, an obvious choice is to simply remove $\boldsymbol{A}_{nl}$ from Eq. (8.1). If this route is adopted, then the transformed (linearized) SDEs over T_i take the form:

$$d\tilde{X}_{1j} = \tilde{X}_{2j}dt \tag{8.6a}$$

$$d\tilde{X}_{2j} = A_{j\,l}\left(t,\widetilde{X}_t\right)dt + \sum_{k=1}^{n} \sigma_{jk}(t)\left\{\left[\sigma^{-1}(t)\right]_{jk} A_{j\,nl}\left(t,\widetilde{X}_t\right)dt + dB_k(t)\right\},$$

$$j = 1,2,\ldots,m \tag{8.6b}$$

Here $\widetilde{\boldsymbol{X}}(t) := \left\{\widetilde{\boldsymbol{X}}_1^T, \widetilde{\boldsymbol{X}}_2^T\right\}^T \in \mathbb{R}^{2m}$ denotes the linearized solution. Equation (8.6) is subject to initial conditions $\left\{\widetilde{\boldsymbol{X}}(t_{i-1})\right\}_{T_i} = \left\{\widetilde{\boldsymbol{X}}(t_{i-1})\right\}_{T_{i-1}}$ for $i > 1$ and $\left\{\widetilde{\boldsymbol{X}}(t_0)\right\}_{T_1} = \widetilde{\boldsymbol{X}}(t_0)$ for $i = 1$. The term within the curly brackets on the RHS of Eq. (8.6b) may be taken as $d\tilde{B}_k(t)$. $\left\{\tilde{B}_k(t) : k \in [1,n]\right\}$ is another vector of n-dimensional Brownian motion process (that absorbs the non-linear part, $\boldsymbol{A}_{nl}$ of the drift field) by the Levy's characterization of Brownian motion (Chapter 4 and Appendix D) under a new probability measure Q. Note that $\widetilde{\boldsymbol{B}}(t)$ as well as $\widetilde{\boldsymbol{X}}(t)$ make sense when restricted to T_i. To compensate for the removal of non-linear drift terms, the linearized SDEs are augmented with a one-dimensional correction process $\Lambda(t)$ (i.e., the Radon–Nikodym derivative over T_i) that is a stochastic exponential:

$$d\Lambda(t) = \sum_{j=1}^{m}\sum_{k=1}^{n}\left[\sigma^{-1}(t)\right]_{jk} A_{j\,nl}\left(t,\widetilde{X}_t\right)d\tilde{B}_k(t) \tag{8.7}$$

subject to the initial condition known from $\Lambda(t)$ of T_{i-1} with $\Lambda(t_0) = 1$. An analytical solution to the linearized SDEs corresponding to Eq. (8.6) is available irrespective of the dimension $2m$ (see Eq. 5.172 of Chapter 5) and this solution is independent of $\Lambda(t)$. The solution for $t \in T_i$ is here rewritten (for a ready reference):

$$\widetilde{\boldsymbol{X}}_t = [\phi(t,t_{i-1})]\begin{pmatrix} \widetilde{\boldsymbol{X}}_{i-1} + \int_{t_{i-1}}^{t} [\phi(s,t_{i-1})]^{-1}\boldsymbol{F}(s)\,ds \\ + \int_{t_{i-1}}^{t} [\phi(s,t_{i-1})]^{-1} \sum_{k=1}^{n} \sigma_k(s)\,d\tilde{B}_k(s) \end{pmatrix} \tag{8.8}$$

where, $\boldsymbol{F}(t) = \left\{\{0,0,\ldots 0\}, \boldsymbol{f}^T\right\}^T \in \mathbb{R}^{2m}$ and $\boldsymbol{\phi}(t,t_{i-1}) = \exp\{[\boldsymbol{A}_l](t-t_{i-1})\}$ is the fundamental solution matrix (FSM) corresponding to the linear drift field with $\boldsymbol{A}_l = \begin{bmatrix} [\boldsymbol{0}] & [\boldsymbol{I}] \\ -[\boldsymbol{K}] & -[\boldsymbol{C}] \end{bmatrix} \in \mathbb{R}^{2m\times 2m}$, $\boldsymbol{0}$ being an $m\times m$ null matrix and $\boldsymbol{I}$ an $m\times m$ identity matrix. The Radon–Nikodym derivative $\Lambda(t)$ is a strictly positive random process—an exponential martingale—expressible as :

$$\Lambda(t) = \Lambda(t_{i-1})\exp\left[\sum_{j=1}^{m}\sum_{k=1}^{n}\int_{t_{i-1}}^{t}\left[\sigma^{-1}(s)\right]_{jk} A_{j\,nl}\left(s,\widetilde{\boldsymbol{X}}_s\right)d\tilde{B}_k(s) - \frac{1}{2}\int_{t_{i-1}}^{t}\left[\sigma^{-2}(s)\right]_{jk}\left(A_{j\,nl}\left(s,\widetilde{\boldsymbol{X}}_s\right)\right)^2 ds\right] \tag{8.9}$$

The required expectations in terms of the linearized solution may now be obtained as: $E\left[g\left(t,\boldsymbol{X}_t\right)\right] = E[\Lambda(t)g(t,\widetilde{\boldsymbol{X}}_t)]$, subject to a localized form of Novikov's restriction on $\boldsymbol{A}_{nl}$, i.e., $E\left[\exp\left(\frac{1}{2}\int_{t_{i-1}}^{t}\boldsymbol{A}_{nl}^2 ds\right)\right] < \infty$ for $t\in T_i$.

Numerical instabilities may occur in the computed moment histories, even if one chooses a very small step size. Reasons for this may essentially be traced to the super-martingale character of $\Lambda(t)$. Thus, even if one starts with $\Lambda(t_{i-1}) = 1$, $E\left[\Lambda(t_i)\right]$ may fall below 1 owing to numerical errors and finiteness of ensemble sizes. This is especially so if the drift nonlinearity is such that it affects the solution significantly away from that of the linear SDE. Moreover, these errors would also propagate fast in time, as explained in the following. Restricting attention to the interval T_i, we note that the argument of $\Lambda(t)$ may be expanded in an Ito–Taylor's series (each term of which is a zero–mean Gaussian Ito integral) so that $\Lambda(t)$ essentially admits an expression in the form of an infinite product of log-normal (exponentiated Gaussian) random variables. We may thus write, $\Lambda(t) = \mathcal{E}_t\prod_{k=1}^{i-1}\Lambda(t_k)$, where, $\mathcal{E}_t$ is the stochastic exponential that appears as the coefficient of $\Lambda(t_{i-1})$ in Eq. (8.9). Thus, numerical errors in the evaluation of $\Lambda(t)$ may quickly build up leading generally to underflows and sometimes overflows (see Fig. 8.1).

In order to circumvent such numerical difficulties, one can modify the linear SDEs [Saha and Roy 2007] so that they provide solutions as close as possible to those of the non-linear SDEs, at least locally (i.e., within a given sub-interval T_i). This would in turn reduce the oscillations (away from 1.0) in the sample paths of $\Lambda(t)$ to a considerable extent.

We now discuss two lower order numerical integration methods [Raveendran *et al.* 2013a, Sarkar and Roy 2016] that show better accuracy via a change-of-measure-based strategy.

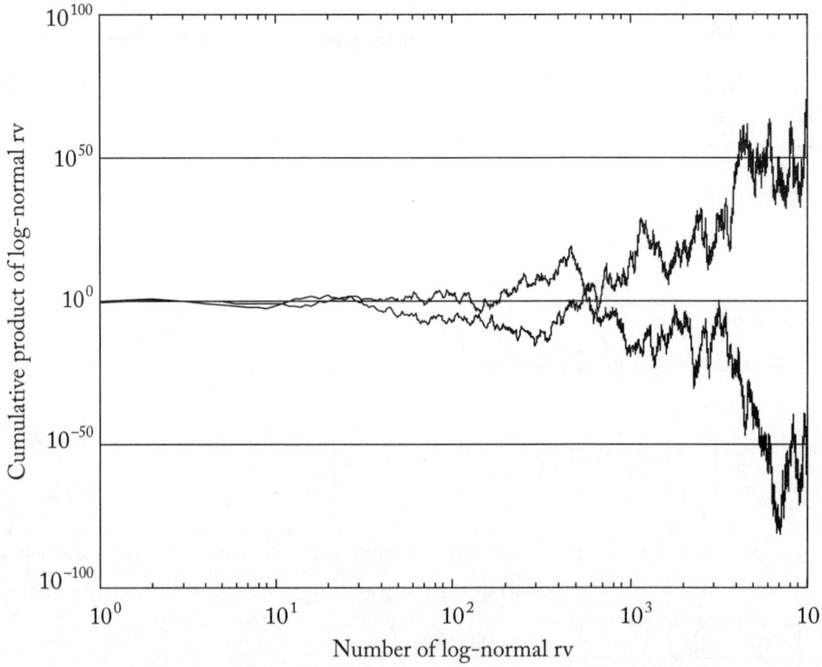

Fig. 8.1 Cumulative product of lognormal random variables leading to underflows or overflows

8.2 Girsanov Corrected Linearization Method (GCLM)

Let the linearized approximation to the non-linear term $\boldsymbol{A}_{nl}\left(t,\widetilde{\boldsymbol{X}}_t\right)$ over T_i be given by, $\widetilde{\boldsymbol{A}}_{nl}(t_{i-1},\widetilde{\boldsymbol{X}}_{i-1})$. The linearized SDEs (8.1) over T_i are recast as:

$$d\tilde{X}_{1j} = \tilde{X}_{2j}dt \tag{8.10a}$$

$$d\tilde{X}_{2j} = \left(A_{j\,l}\left(t,\widetilde{\boldsymbol{X}}_t\right) + \tilde{A}_{j\,nl}\left(t_{i-1},\widetilde{\boldsymbol{X}}_{i-1}\right)\right)dt$$

$$+\sum_{k=1}^{n}\sigma_{jk}(t)\left\{\left[\sigma^{-1}(t)\right]_{jk}\left(A_{j\,nl}\left(t,\widetilde{\boldsymbol{X}}_t\right)-\tilde{A}_{j\,nl}\left(t_{i-1},\widetilde{\boldsymbol{X}}_{i-1}\right)\right)dt + dB_k(t)\right\},$$

$$j = 1,2,\ldots,m \tag{8.10b}$$

As in Eq. (8.6), the term within the curly brackets on the RHS of the last equation may be taken as $d\tilde{B}_k(t)$. By Levy's characterization, $\left\{\tilde{B}_k(t)\,|k\in[1,n]\right\}$ is another n-dimensional vector Brownian motion that absorbs the linearization error $\left(\boldsymbol{A}_{nl}-\widetilde{\boldsymbol{A}}_{nl}\right)$ under a new probability measure Q.

As per Girsanov's theorem (Section 4.10, Chapter 4), if P and Q are equivalent, then there exists a unique $\mathcal{F}$-measurable Radon–Nikodym derivative Λ^{-1} so that, $dQ = \Lambda^{-1} dP$, where, $E_Q(\Lambda) = 1$. Let $\left|\varepsilon_{j\ nl}\left(t, \widetilde{\boldsymbol{X}}_t, \widetilde{\boldsymbol{X}}_{i-1}\right)\right| = \left|A_{j\ nl}\left(t, \widetilde{\boldsymbol{X}}_t\right) - \tilde{A}_{j\ nl}\left(t, \widetilde{\boldsymbol{X}}_{i-1}\right)\right|$ denote a formal measure of the $\mathcal{F}_t$-predictable linearization error for the jth drift component of $\widetilde{\boldsymbol{X}}_2$.

The Radon–Nikodym derivative is computable as:

$$\Lambda(t) = \frac{dP}{dQ}$$

$$= \exp\left[\sum_{j=1}^{m}\sum_{k=1}^{n}\int_{t_{i-1}}^{t}\left[\sigma^{-1}(s)\right]_{jk}\varepsilon_{j\ nl}\left(s, \widetilde{\boldsymbol{X}}_s, \widetilde{\boldsymbol{X}}_{i-1}\right)d\tilde{B}_k(s)\right.$$

$$\left.-\frac{1}{2}\int_{t_{i-1}}^{t}\left[\sigma^{-2}(s)\right]_{jk}\left(\varepsilon_{j\ nl}\left(s, \widetilde{\boldsymbol{X}}_s, \widetilde{\boldsymbol{X}}_{i-1}\right)\right)^2 ds\right] \tag{8.11}$$

subject to $\Lambda(0) = 1$. With increasing order of linearization and decreasing time-step size, $\left|\varepsilon_{j\ nl}\right|$ may become smaller so that $\Lambda(t)$ approaches 1, at least in principle. With numerical difficulties that afflict explicit evaluation of $\Lambda(t)$, the method uses the form of $\Lambda(t)$ to adopt rejection sampling/re-sampling strategy on a finite ensemble of GCLM-based linearized trajectories and thus directly compute sample expectations. If the functional form of a target density f is known up to a multiplicative constant, then one can simulate f from a simpler density g (called the instrumental density) with an accept-reject algorithm making use of the fundamental theorem of simulation [(Section 2.5, Chapter 2), Robert and Casella 2004]. Rejection sampling [Handschin and Mayne 1969] is useful when the upper bound of underlying density is known. Assume that there exists a known constant $C < \infty$ such that $f(x) < Cg(x)$ for every x. Then the sampling procedure is given as follows.

1. Generate a uniform random variable $u \sim U(0,1)$;
2. draw a sample $x \sim g(x)$;
3. if $< \frac{f(x)}{Cg(x)}$, return x; else go to Step 1.

The acceptance probability is inversely proportional to C. The choice of C is critical; if C is too small, the samples are not reliable because of the low rejection rate and if C is too large, the algorithm will be inefficient since the acceptance rate is too low. Before applying rejection sampling, the Ito integrals in the argument of the exponential in $\Lambda(t)$ are first written in terms of integrals with respect to t. Thus define:

$$\beta_k\left(t, \widetilde{\boldsymbol{X}}_t\right) = \sum_{j=1}^{m}\left[\sigma^{-1}(t)\right]_{jk}\varepsilon_{j\ nl}\left(t, \widetilde{\boldsymbol{X}}_t, \widetilde{\boldsymbol{X}}_{i-1}\right) \tag{8.12}$$

Using integration by parts, we have:

$$\int_{t_{i-1}}^{t} \beta_k d\tilde{B}_k = \int_{t_{i-1}}^{t} d\left(\beta_k \tilde{B}_k\right) - \int_{t_{i-1}}^{t} \tilde{B}_k d\beta_k - \int_{t_{i-1}}^{t} d\beta_k d\tilde{B}_k \tag{8.13}$$

Dropping the integration limits for notational convenience, one may write:

$$\begin{aligned}\int \beta_k d\tilde{B}_k &= \beta_k \tilde{B}_k\big|_{t_{i-1}}^{t} - \tilde{B}_k^2 \frac{\partial \beta_k}{\partial \widetilde{\boldsymbol{x}}_2}^T \sigma_k\big|_{t_{i-1}}^{t} - \int \frac{\partial \beta_k}{\partial \widetilde{\boldsymbol{x}}_2}^T \sigma_k ds - \int \tilde{B}_k \frac{\partial \beta_k}{\partial \widetilde{\boldsymbol{x}}_1}^T \widetilde{\boldsymbol{X}}_2 ds \\ &- \int \tilde{B}_k \frac{\partial \beta_k}{\partial \widetilde{\boldsymbol{x}}_2}^T \widetilde{\boldsymbol{A}} ds - \frac{1}{2}\int \tilde{B}_k \left(\left[\frac{\partial^2 \beta_k}{\partial \widetilde{\boldsymbol{x}}_2^2}\right]\sigma_k\right)^T \sigma_k ds \\ &+ \int \left(\tilde{B}_k^2 \left[\frac{\partial^2 \beta_k}{\partial \widetilde{\boldsymbol{x}}_1 \partial \widetilde{\boldsymbol{x}}_2}\right]\sigma_k\right)^T \widetilde{\boldsymbol{X}}_2 ds + \int \left(\tilde{B}_k^2 \left[\frac{\partial^2 \beta_k}{\partial \widetilde{\boldsymbol{x}}_2^2}\right]\sigma_k\right)^T \widetilde{\boldsymbol{A}} ds \\ &+ \int \left(\tilde{B}_k \left[\frac{\partial^2 \beta_k}{\partial \widetilde{\boldsymbol{x}}_2^2}\right]\sigma_k\right)^T \sigma_k ds + \int \left(\tilde{B}_k^2 \left[\frac{\partial^2 \beta_k}{\partial \widetilde{\boldsymbol{x}}_2^2}\right]\sigma_k\right)^T \sigma_k d\tilde{B}_k \end{aligned} \tag{8.14}$$

See Appendix H for a derivation of Eq. (8.14). Here the drift coefficient in Eq. (8.10b) is $\widetilde{\boldsymbol{A}} = \left\{A_{j\,l}\left(t,\widetilde{\boldsymbol{X}}_t\right) + \tilde{A}_{j\,nl}\left(t,\widetilde{\boldsymbol{X}}_{i-1}\right),\ j = 1,2,\ldots m\right\}^T$. The last term on the RHS of Eq. (8.14) can again be integrated by parts to obtain an expression corresponding to the next higher order. However, for a special case of $\beta_k = \beta_k(\widetilde{\boldsymbol{X}}_1)$ being a function of $\widetilde{\boldsymbol{x}}_1$ alone, the Ito integral on the RHS of the above equation drops off and thus the expression becomes closed. Now, for each $k \in [1,n]$ define the martingale:

$$\Lambda_k(t) = \exp\left(\int_{t_{i-1}}^{t} \beta_k d\tilde{B}_k - \frac{1}{2}\int_{t_{i-1}}^{t} \beta_k^2 ds\right) \tag{8.15}$$

where, $\int_{t_{i-1}}^{t} \beta_k d\tilde{B}_k$ is given by the RHS of Eq. (8.14) with the last (stochastic) integral removed. Recall that, in the special case of the non-linear drift terms being functions of $\boldsymbol{x}_1$ alone, one has an exact (and closed) expression for $\int_{t_{i-1}}^{t} \beta_k d\tilde{B}_k$. By clubbing the remaining integral terms separately from the rest in Eq. (8.14), one may write $\Lambda_k(t) = \Lambda_k^1(1)\Lambda_k^2(2)$ with:

$$\Lambda_k^{(1)}(t) = \exp\left(-\int_{t_{i-1}}^{t} \psi_k\left(\widetilde{\boldsymbol{X}}_t, \widetilde{\boldsymbol{X}}_{i-1}, \tilde{B}_k(s)\right) ds\right) \tag{8.16a}$$

$$\Lambda_k^{(2)}(t) = \exp \sum_{j=1}^{m} \left(\sigma_{jk}\right)^{-1} \varepsilon_{j\ nl} \tilde{B}_k|_{t_{i-1}}^{t} \tag{8.16b}$$

where, $\psi_k\left(\widetilde{\boldsymbol{X}}_t, \widetilde{\boldsymbol{X}}_{i-1}, \tilde{B}_k(t)\right) = \frac{d\left(\sum_{j=1}^{m}\left[\sigma^{-1}(t)\right]_{jk}\varepsilon_{j\ nl}\tilde{B}_k\right)}{dt}$. Also denote $\psi = \sum_{k=1}^{n} \psi_k$. Now the Radon–Nikodym derivative takes the product form:

$$\Lambda(t) = \prod_{k=1}^{n} \Lambda_k(t) \tag{8.17}$$

While the weight $\Lambda^{(2)}(t) = \prod_{k=1}^{n} \Lambda_k^{(2)}(t)$ can be computed explicitly, this is not so with $\Lambda^{(1)}(t) = \prod_{k=1}^{n} \Lambda_k^{(1)}(t)$ which still contains stochastic integrals. It is intended to bypass evaluation of these integrals via rejection sampling. Specifically, the aim is to perform rejection sampling with candidates from the linearized solution $\widetilde{\boldsymbol{X}}(t)$ with the law Q in order to sample from P. For this, a binary indicator I is generated such that $\left(I = 1|\widetilde{\boldsymbol{X}}(t)\right) = \exp\left(-\int_{t_{i-1}}^{t} \psi ds\right)$, where, $\widetilde{\boldsymbol{X}}(t) \sim Q$ is a sample linearized trajectory simulated only at a finite set of instances within T_i. The sample path is accepted if $I = 1$, i.e., with probability $\exp\left(-\int_{t_{i-1}}^{t} \psi ds\right)$. Being a function of the linearization error, $\varepsilon_{j\ nl}$, one anticipates that $\left\|\psi\right\|_{L^2(Q)}$ can be made (arbitrarily) small either by increasing the linearization order or reducing the step size $\Delta_i = t_i - t_{i-1}$. In particular, the method assumes that $0 \leq \psi \leq \frac{1}{\Delta_i}$. Hence the boundedness requirements over a finite interval $[0, T]$, including Novikov's condition, are numerically enforced via the above constraint on ψ. This is one of the stabilizing aspects of the method.

Events of probability $\int_{t_{i-1}}^{t} \psi ds$ can be obtained by drawing random points $(S, R) \sim U[T_i \times \left(0, \frac{1}{\Delta_i}\right)]$ so that the events $\{\psi(S) \geq R\}$ will have the desired probability. For extending this concept to an event of probability, $\exp\left(-\int_{t_{i-1}}^{t} \psi ds\right)$, the following theorem, adapted from Beskos and Roberts [2005], is used.

Theorem: Let $\widetilde{\boldsymbol{X}}(t) \sim Q$ be a sample linearized solution of Eq. (8.1) over T_i. Let $\tau = \left(V_j, W_j\right)_{j\geq 1}$ be a sequence of i.i.d. points from $U[T_i \times \left(0, \frac{1}{\Delta_i}\right)]$. Consider $u \sim U[0,1]$ and assume that $\widetilde{\boldsymbol{X}}, \tau$ and u are independent. For convenience, denote $\psi\left(\widetilde{\boldsymbol{X}}_t, \widetilde{\boldsymbol{X}}_{i-1}, B_r(t)\right) = \psi(\widetilde{\boldsymbol{X}}_t)$ and define the following events:

$$\Gamma_0^i = \Omega$$

$$\Gamma_r^i = \left\{\left(\psi(\widetilde{\boldsymbol{X}}(V_1)) \geq W_1\right), \ldots, \left(\psi(\widetilde{\boldsymbol{X}}(V_{r-1})) \geq W_{r-1}\right), \left(\psi(\widetilde{\boldsymbol{X}}(V_r)) \geq W_r\right), u \leq \frac{1}{r!}\right\}$$

$$r = 1, 2, \ldots. \tag{8.18}$$

Consider the sequence of events $(\Xi_r)_{r \geq 1}$ defined as:

$$\Xi_{2r+1} = \left(\Gamma_0^i - \Gamma_1^i\right) \cup \left(\Gamma_2^i - \Gamma_3^i\right) \cup \cdots \cup \left(\Gamma_{2r}^i - \Gamma_{2r+1}^i\right), r = 0, 1, \ldots \tag{8.19a}$$

$$\Xi_{2r+2} = \left(\Gamma_0^i - \Gamma_1^i\right) \cup \left(\Gamma_2^i - \Gamma_3^i\right) \cup \cdots \cup \left(\Gamma_{2r}^i - \Gamma_{2r+1}^i\right) \cup \Gamma_{2r+2}^i, r = 0, 1, \ldots \tag{8.19b}$$

where, $D - F = D \cap F^c$ for any set $F \subseteq D$. Then:

(1) $(\Xi_{2r+1})_{r \geq 0}$, $(\Xi_{2r+2})_{r \geq 0}$ are sequences of increasing and decreasing events, respectively, with

$$\Xi_{2\kappa+1} \subseteq \Xi_{2\lambda+2}, \text{ for any } \kappa, \lambda \in \{0, 1, \ldots\} \tag{8.20a}$$

and

$$P\left(\cap_0^\infty \Xi_{2r+2} - \cup_0^\infty \Xi_{2r+1}\right) = 0 \tag{8.20b}$$

(2) Let $\Xi = \cup_0^\infty \Xi_{2r+1}$. If I is a binary indicator such that I =1 when Ξ occurs and 0 otherwise, then

$$P\left(I = 1 | \widetilde{X}(t)\right) = \exp\left(-\int_{t_{i-1}}^{t} \psi(\widetilde{X}_t) ds\right). \tag{8.21}$$

Suppose that $2m$ iterations have occurred without a decision for I, i.e., E_{2r} happened and Ξ_{2r-1} did not. From Eqs. (8.18) and (8.19), one has:

$$\Xi_{2r} = \Xi_{2r-1} \cup \Gamma_{2r}^i \text{ and } \Xi_{2r+1} = \Xi_{2r} - \Gamma_{2r+1}^i = \Xi_{2r-1} \cup (\Gamma_{2r}^i - \Gamma_{2r-1}^i) \tag{8.22}$$

To reach a similar conclusion about Ξ_{2r+1}, it suffices to check if the sub-event of Γ_{2r}^i i.e.,

$$\Gamma_{2r}^i - \Gamma_{2r+1}^i = \Gamma_{2r}^i \cap \left\{\psi\left(\widetilde{X}(V_{r+1})\right) < W_{r+1} \text{ or } u > \frac{1}{(2r+1)!}\right\} \tag{8.23}$$

has occurred or not. The same interpretation applies for an even numbered iteration also.

The proof of this theorem, skipped here, may be readily adapted from Beskos and Roberts [2005]. Over T_i, the acceptance probability $\exp\left(-\int_{t_{i-1}}^{t} \psi(\widetilde{X}) ds\right)$ for the linearized solution $\widetilde{X}(t)$ is evaluated by Eq. (8.21). On realizing $\widetilde{X}(t)$ for an ordered sequence $\Sigma_i := \left\{t_{i-1} < t^{(j)} < t_i | j \in [1, r]\right\} \cup \{t_i\}$ of time instances within T_i, we need to ensure that $\psi\left(\widetilde{X}\left(t^*\right)\right) \geq 0\ \forall t^* \in \Sigma_i$. For this, let $\eta := \min\left\{\psi(\widetilde{X}(t^*)) \geq 0 \forall t^* \in \Sigma_i\right\}$ such that, $\psi\left(\widetilde{X}\right) \equiv \psi\left(\widetilde{X}\right) + \mu > 0$, where $\mu > \left|\eta\right|$ for all $t^* \in \Sigma_i$. A weak correction for this modification is included in the weight, $\Lambda^{(2)}\left(t\right) \equiv \Lambda^{(2)}\left(t\right) \times \exp\left(-\int_{t_{i-1}}^{t} \mu ds\right)$. $\equiv$ is used to imply the replacement of the LHS by the RHS. Now we can apply rejection sampling to check for the acceptance of the path (which is available in continuous time via the linearized solution). Upon acceptance, it is assigned the weight, $\Lambda^{(2)}\left(t\right)$. This enables applying the standard re-sampling procedure to modify the ensemble of the accepted paths. This completes the correction introduced by the Radon–Nikodym derivative.

8.2.1 Algorithm for GCLM

The algorithm for implementation of the GCLM is given below.

1. Select an ensemble size, N_p.
2. Discretize the time interval of interest $[0,T]$ into N time-steps: $[0,T] = \bigcup_{i=1}^{N} T_i$ and set the initial condition, $\widetilde{X}\left(t_0\right) = X_0$. Over every time step, i.e., for each, $i \in [1,N]$, the following steps are recursively followed over all the samples in the ensemble.
3. Generate the set $\Sigma_i \backslash \{t_i\}$ of r random time instances within T_i (uniformly distributed in length, (T_i)), sorted in the ascending order, i.e., $\Sigma_i \backslash \{t_i\} = \left\{t_{\gamma_1}, t_{\gamma_2,\ldots,} t_{\gamma_r}, t_i = t_{\gamma_{r+1}}\right\}$. Also let $t_{\gamma_0} = t_{i-1}$.
4. Draw $\nu \sim [0,1]$; set $k = 0$.
5. For $l = 1,\ldots,r+1$, repeat the following steps.

   a) Draw $u \sim \bigcup[0, \frac{1}{\Delta_i}]$; set $k = k+1$; obtain the Brownian increment $B\left(t_{\gamma_l}\right) - B\left(t_{\gamma_{l-1}}\right)$ and thus compute $\psi(\widetilde{X}(t_{\gamma_l}))$;
   b) If $\psi(\widetilde{X}(t_{\gamma_l})) < u$ or $\nu > \frac{1}{k!}$, then
      if k is even, reject the computed $\psi(\widetilde{X}(t_{\gamma_l}))$ and go to 5a;
      if k is odd, accept $\psi(\widetilde{X}(t_{\gamma_l}))$;
      else go to 5a.

6. Resample the N_p accepted sets, $\left\{\widetilde{X}(t^* \in \Sigma_i)\right\}$, with the particle $\widetilde{X}(t^*)$ assigned the weight $\Lambda^{(2)}\left(t^*\right)$, and use these re-sampled skeletal trajectories (or the associated continuous forms of the linearized solutions) to evaluate the sample moments.

Example 8.1

We consider the case of an additively driven, SDOF, hardening Duffing (HD) oscillator, modeled through the following SDE:

$$\ddot{X} + C\dot{X} + KX + \alpha X^3 = \sigma B(t) \tag{8.24}$$

$X(t)$ is the oscillator response process. K, α and C are positive real constants representing the stiffness, the intensity of nonlinearity, and the coefficient of viscous damping, respectively. $B(t)$ is a standard Brownian motion, and σ the intensity of noise. Equation (8.24) is subject to the deterministic initial condition $\boldsymbol{X}_0 = \left(X_0, \dot{X}_0\right)^T$ and the state vector $\boldsymbol{X}$ is represented as $\boldsymbol{X} = (X_1 = X, X_2 = \dot{X}_1 = \dot{X})^T$.

Solution

With a lower order explicit Ito–Taylor expansion of the non-linear drift term to locally linearize the SDE over T_i, one has (Eq. 8.10):

$$d\tilde{X}_1(t) = \tilde{X}_2(t)dt \tag{8.25a}$$

$$d\tilde{X}_2(t) = \left(-C\tilde{X}_2(t) - K\tilde{X}_1(t) - \alpha\tilde{X}_{1,i-1}^3\right)dt$$
$$+\sigma\left\{\sigma^{-1}\left(\alpha\left(\tilde{X}_1(t)\right)^3 - \alpha\tilde{X}_{1,i-1}^3\right)dt + dB(t)\right\} \tag{8.25b}$$

Denote the expression within the curly brackets on the RHS of the last equation by $d\tilde{B}(t)$ which is another Brownian motion that absorbs the linearization error under a new probability measure, Q. The solution to the linearized state, $\widetilde{\boldsymbol{X}} = \left(\tilde{X}_1, \tilde{X}_2\right)^T$ for $t \in T_i$ may be written as (see Eq. 8.8):

$$\widetilde{\boldsymbol{X}}_t = \exp\left(\boldsymbol{A}_l(t - t_{i-1})\right)\begin{pmatrix} \widetilde{\boldsymbol{X}}_{i-1} + \int_{t_{i-1}}^{t} \exp\left(-\boldsymbol{A}_l(s - t_{i-1})\right)\widetilde{\boldsymbol{A}}_{nl}ds \\ +\int_{t_{i-1}}^{t} \exp\left(-\boldsymbol{A}_l(s - t_{i-1})\right)\boldsymbol{G}d\tilde{B}(s) \end{pmatrix} \tag{8.26}$$

where, $\boldsymbol{A}_l = \begin{bmatrix} 0 & 1 \\ -K & -C \end{bmatrix}$, $\widetilde{\boldsymbol{A}}_{nl} = \left(0 \quad -\alpha\tilde{X}_{1,i-1}^3\right)^T$ and $\boldsymbol{G} = (0 \quad \sigma)^T$. The Radon–Nikodym derivative is given by:

$$\Lambda(t) = \exp\left\{\int_{t_{i-1}}^{t}\left(-\sigma^{-1}\left(\alpha\left(\tilde{X}_1(s)\right)^3 - \alpha\tilde{X}_{1,i-1}^3\right)\right)d\tilde{B}(s)\right.$$
$$\left.-\frac{1}{2}\int_{t_{i-1}}^{t}\left(\sigma^{-1}\left(\alpha\left(\tilde{X}_1(s)\right)^3 - \alpha\tilde{X}_{1,i-1}^3\right)\right)^2 ds\right\} \tag{8.27}$$

Following the discussion in the previous section, the above expression can be reduced to:

$$\Lambda(t_i) = \exp\left\{-\sigma^{-1}\alpha(\tilde{X}_{1,i}^3 B_i - \tilde{X}_{1,\gamma}^3 B_\gamma - \tilde{X}_{1,\gamma}^3(B_i - B_\gamma))\right\}\exp\left(-\int_{t_\gamma}^{t_i}\psi(s)\,ds\right) \tag{8.28}$$

where, $B_\gamma := B_{t_\gamma}, \tilde{X}_{1,\gamma} := \tilde{X}_1(t_\gamma)$. $B_t \equiv B_t - B_{t_\gamma}$ is the Brownian increment with $\mathcal{N}(0,(t-t_\gamma)^{1/2})$ distribution, $t_\gamma \geq t_{i-1}(t_\gamma \in \Sigma_i)$ and

$$\psi(t) = -3\sigma^{-1}\alpha\left(\tilde{X}_1(t)\right)^2\tilde{X}_2(t)B_t + \frac{1}{2}\sigma^{-1}\left\{\alpha\left(\left(\tilde{X}_1(t)\right)^3 - \tilde{X}_{1,\gamma}^3\right)\right\}^2 \tag{8.29}$$

This enables us to evaluate $\psi(t)$ and $\Lambda(t)$ at any time instant. Note that, in Eq. (8.27), the weight, $\Lambda(t_i) = \exp\left\{-\sigma^{-1}\alpha\left(\tilde{X}_{1,i}^3 B_i - \tilde{X}_{1,\gamma}^3 B_\gamma - \tilde{X}_{1,\gamma}^3\left(B_i - B_\gamma\right)\right) - \mu(t_i - t_\gamma)\right\}$, which can be directly evaluated, is used for re-sampling.

If the next higher order linearization of the non-linear drift term over T_i is considered, it yields the following linearized SDEs:

$$d\tilde{X}_1(t) = \tilde{X}_2(t)dt \tag{8.30a}$$

$$d\tilde{X}_2(t) = \left(-C\tilde{X}_2(t) - K\tilde{X}_1(t) - \alpha\left(\tilde{X}_{1,i-1} + \tilde{X}_{2,i-1}(t - t_{i-1})\right)^3\right)dt$$

$$+\sigma\left\{-\sigma^{-1}\left(\alpha\left(\tilde{X}_1(t)\right)^3 - \alpha\left(\tilde{X}_{1,i-1} + \tilde{X}_{2,i-1}(t - t_{i-1})\right)^3\right)dt + dB(t)\right\} \tag{8.30b}$$

The solution $\widetilde{\boldsymbol{X}}$ is given by Eq. (8.26) with $\boldsymbol{A}_l = \begin{bmatrix} 0 & 1 \\ -K & -C \end{bmatrix}$, $\widetilde{\boldsymbol{A}}_{nl} = \left(0 \;\; -\alpha\left(\tilde{X}_{1,i-1} + \tilde{X}_{2,i-1}(t - t_{i-1})\right)^3\right)^T$ and $\boldsymbol{G} = (0 \quad \sigma)^T$.

Writing $\psi(t_i)$ and $\Lambda(t_i)$ as:

$$\psi(t_i) = -3\sigma^{-1}\alpha\tilde{X}_{1,i}^2\tilde{X}_{2,i}B_i + 3\sigma^{-1}\alpha\left(\tilde{X}_{1,\gamma} + \tilde{X}_{2,\gamma}(t_i - t_\gamma)\right)^2\tilde{X}_{2,\gamma}B_i$$

$$+\frac{1}{2}\left\{\sigma^{-1}\alpha\left(\tilde{X}_{1,i}^3 - \left(\tilde{X}_{1,\gamma} + \tilde{X}_{2,\gamma}(t - t_\gamma)\right)^3\right)\right\}^2 + \mu \tag{8.31}$$

$$\Lambda(t_i) = \exp\left[-\sigma^{-1}\alpha\left\{\tilde{X}_{1,i}^3 B_i - \tilde{X}_{1,\gamma}^3 B_\gamma - \left(\tilde{X}_{1,\gamma} + \tilde{X}_{2,\gamma}(t_i - t_\gamma)\right)^3 B_i - \tilde{X}_{1,\gamma}^3 B_\gamma\right\}\right.$$

$$\left. -\mu(t_i - t_\gamma)\right] \tag{8.32}$$

we have the canonical Radon–Nikodym derivative $\Lambda(t_i) \equiv \Lambda(t_i)\exp\left(-\int_{t_{i-1}}^{t_i}\widetilde{\psi}(s)ds\right)$.

Figure 8.2 shows that $\psi_{\max}(k) := \max\left\{\psi(\widetilde{\boldsymbol{X}}(t,\omega_k)|t \in [0,T]\right\}$ is indeed bounded by $[0, \frac{1}{\Delta_i}]$ for both the lower and higher order linearization for the HD oscillator with C = 5, K =100, α=100, σ=5, Δ= 0.01, $\widetilde{\boldsymbol{X}}_0 = \{0,0\}^T$ and $N_p = 2000$.

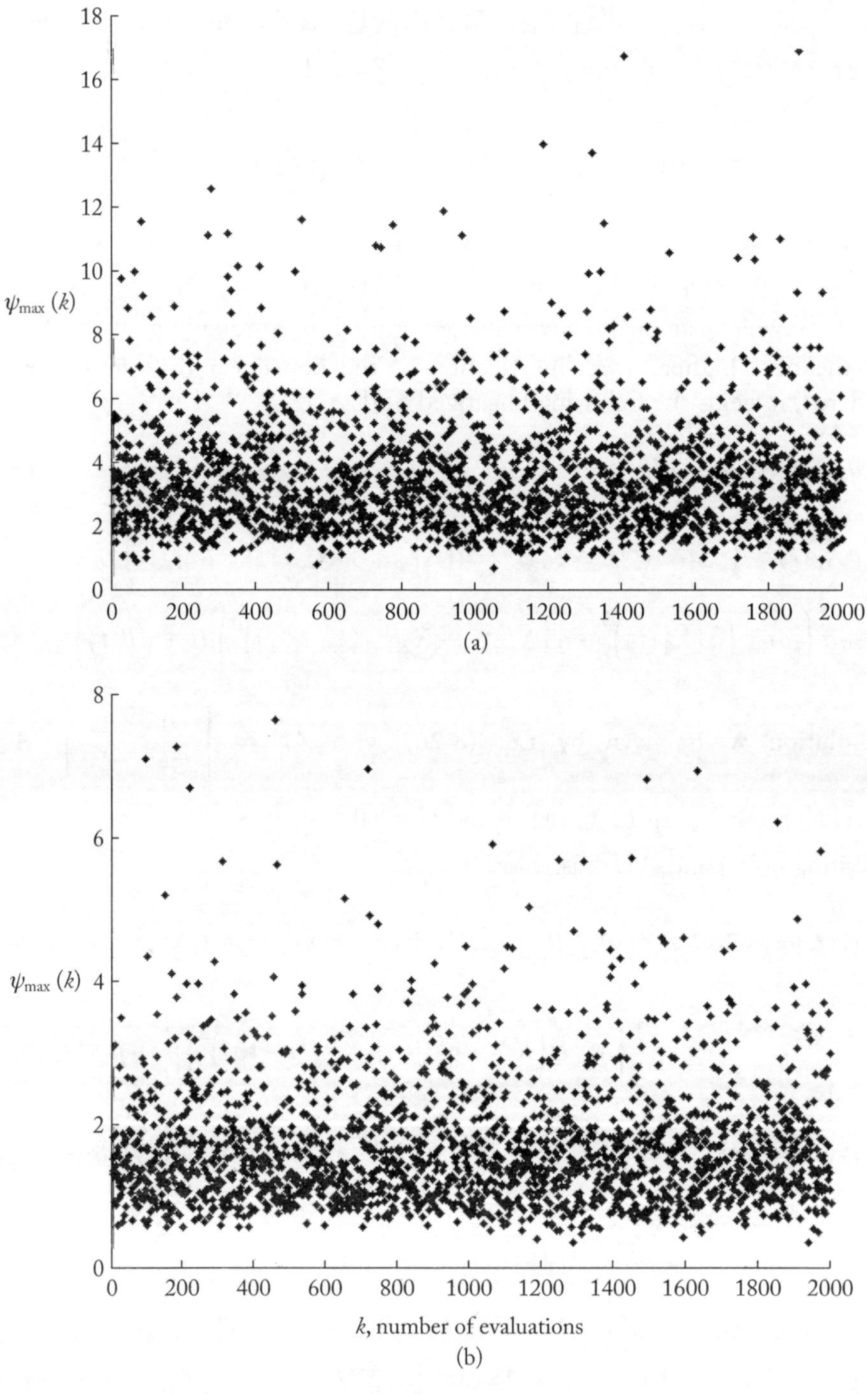

Fig. 8.2 HD oscillator: $C = 5$, $K = 100$, $\alpha = 100$, $\sigma = 5$, $\Delta = 0.01$, $\widetilde{\boldsymbol{X}}_0 = \{0, 0\}^T$, $N_p = 2000$; plots of $\psi_{\max}(k)$, (a) lower order linearization—Eq. (8.25) and (b) higher order linearization—Eq. (8.30)

The sample-averaged time histories of $E[X_1^2]$ and $E[X_2^2]$ and the acceptance rates (i.e., the ratio of the number of paths accepted to the total number of such paths generated) in the rejection sampling step for the two variants of the GCLM are shown in Fig. 8.3 along with the exact stationary limits for these moments. The exact stationary (joint) density function is available in this case through the solution of the reduced Fokker–Planck equation [Wang and Zhang 2000] and is given by:

$$p(x_1, x_2) = D \exp\left\{-\frac{Cx_2^2}{\sigma^2} - C\int_0^{x_1}\left(\frac{Ks + \alpha s^3}{0.5\sigma^2}\right)ds\right\} \tag{8.33}$$

The constant D is evaluated using the normalization constraint, $\iint_{-\infty}^{\infty} p(x_1, x_2)\, dx_1 dx_2 = 1$. The expectation of any integrable function $\chi(X_1, X_2)$ may then be found as, $E[\chi(X_1, X_2)] = \iint_{-\infty}^{\infty} \chi(x_1, x_2)\, p(x_1, x_2)\, dx_1 dx_2$. This integral is computed through numerical quadrature. While computed results via both the linearized versions are close to the known analytical solutions, the acceptance rates in the rejection sampling/re-sampling step of higher order linearization are typically higher than that of lower order one, as anticipated (Fig. 8.3c). Lower acceptance rates in the former case indicate that it takes longer to populate the sample set. Figure 8.4 shows the second moment histories of displacement and velocity components where the stiffness parameter K is 10 times the nonlinearity parameter α and the additive noise intensity ($\sigma = 0.5$) is lower.

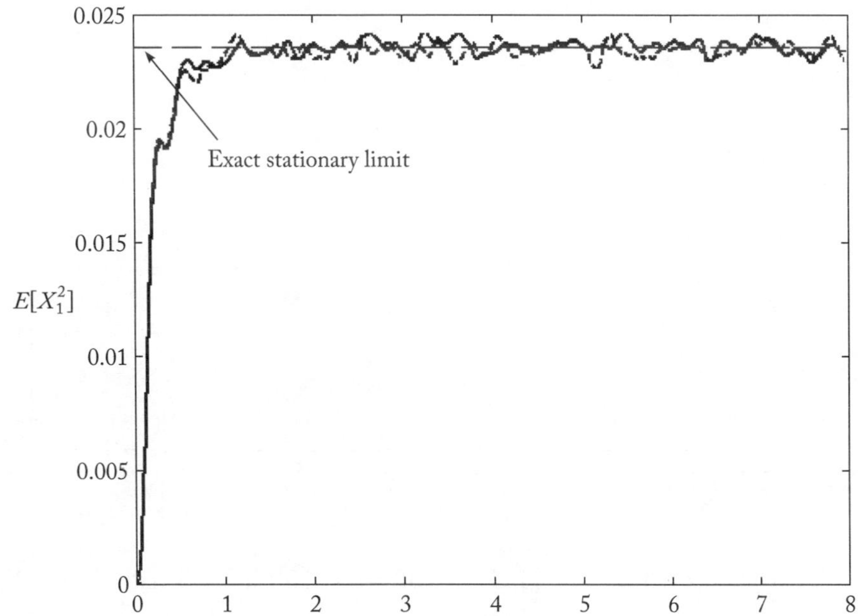

(a)

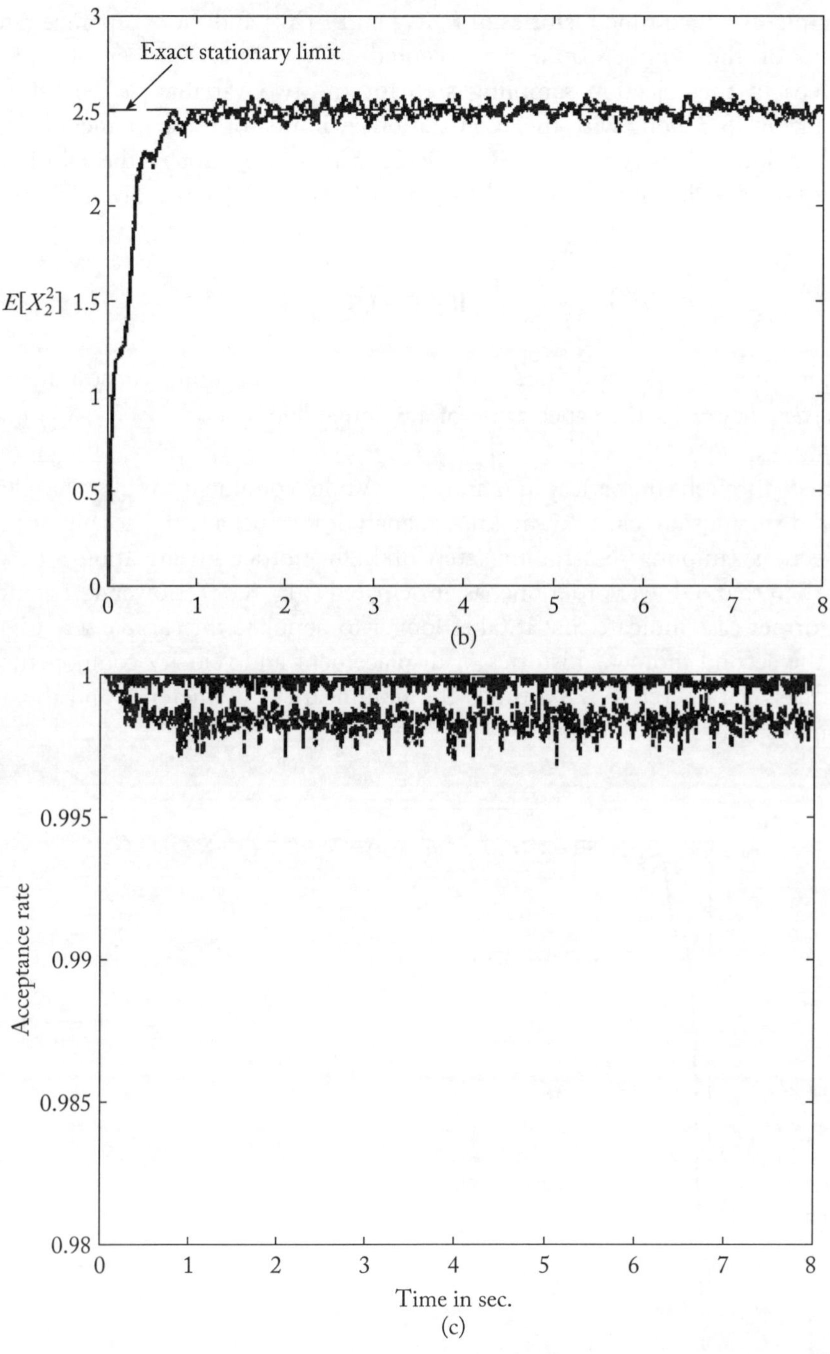

Fig. 8.3 HD oscillator: $C = 5$, $K = 100$, $\alpha = 100$, $\sigma = 5$, $\Delta = 0.01$, $\widetilde{\boldsymbol{X}}_0 = \{0, 0\}^T$, $N_p = 10000$; second moment histories for lower (dashed-line) and higher order (solid-line) linearizations, (a) sample–averaged $E[X_1^2]$, (b) sample–averaged $E[X_2^2]$ and (c) acceptance rate

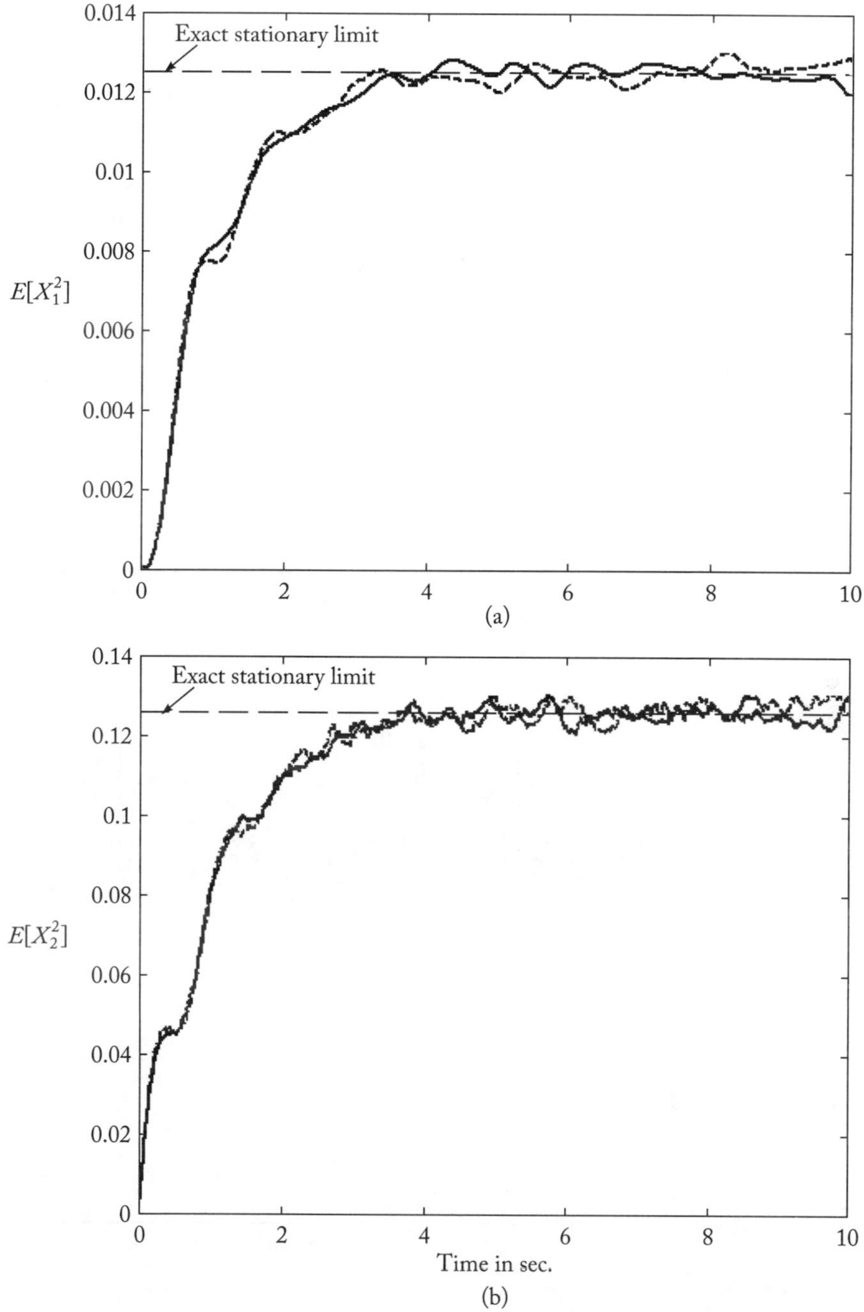

Fig. 8.4 HD oscillator: $C = 1$, $K = 10$, $\alpha = 1$, $\sigma = 0.5$, $\Delta = 0.01$, $\widetilde{\boldsymbol{X}}_0 = \{0, 0\}^T$, $N_p = 5000$; second moment histories for lower (dashed-line) and higher order (solid-line) linearizations, (a) sample–averaged $E[X_1^2]$ and (b) sample–averaged $E[X_2^2]$

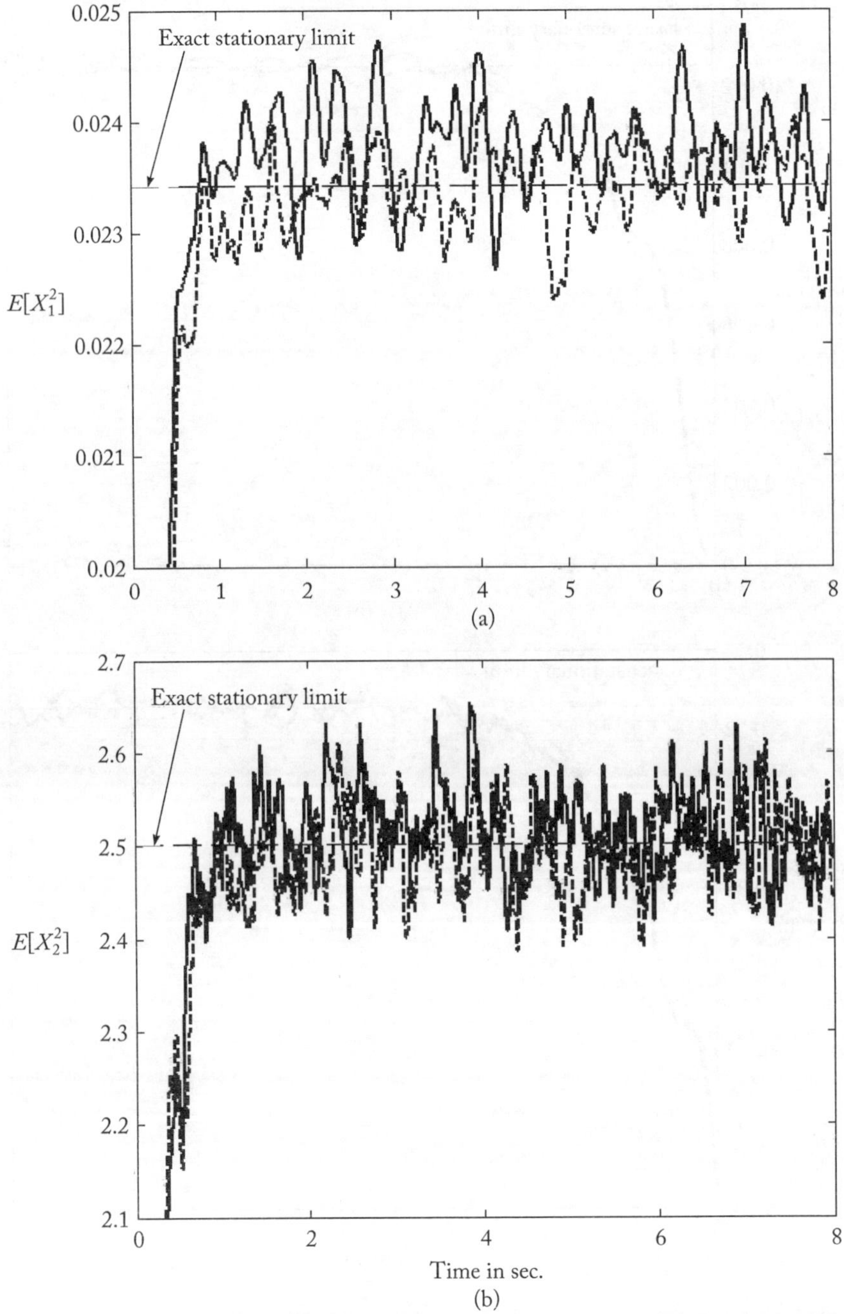

Fig. 8.5 HD oscillator: $C = 5$, $K = 100$, $\alpha = 100$, $\sigma = 5.0$, $\Delta = 0.01$, $\widetilde{\boldsymbol{X}}_0 = \{0,0\}^T$, $N_p = 5000$; second moment histories without Girsanov correction (solid-line) and lower order linearization (dashed-line), (a) sample–averaged $E[X_1^2]$ and (b) sample–averaged $E[X_2^2]$

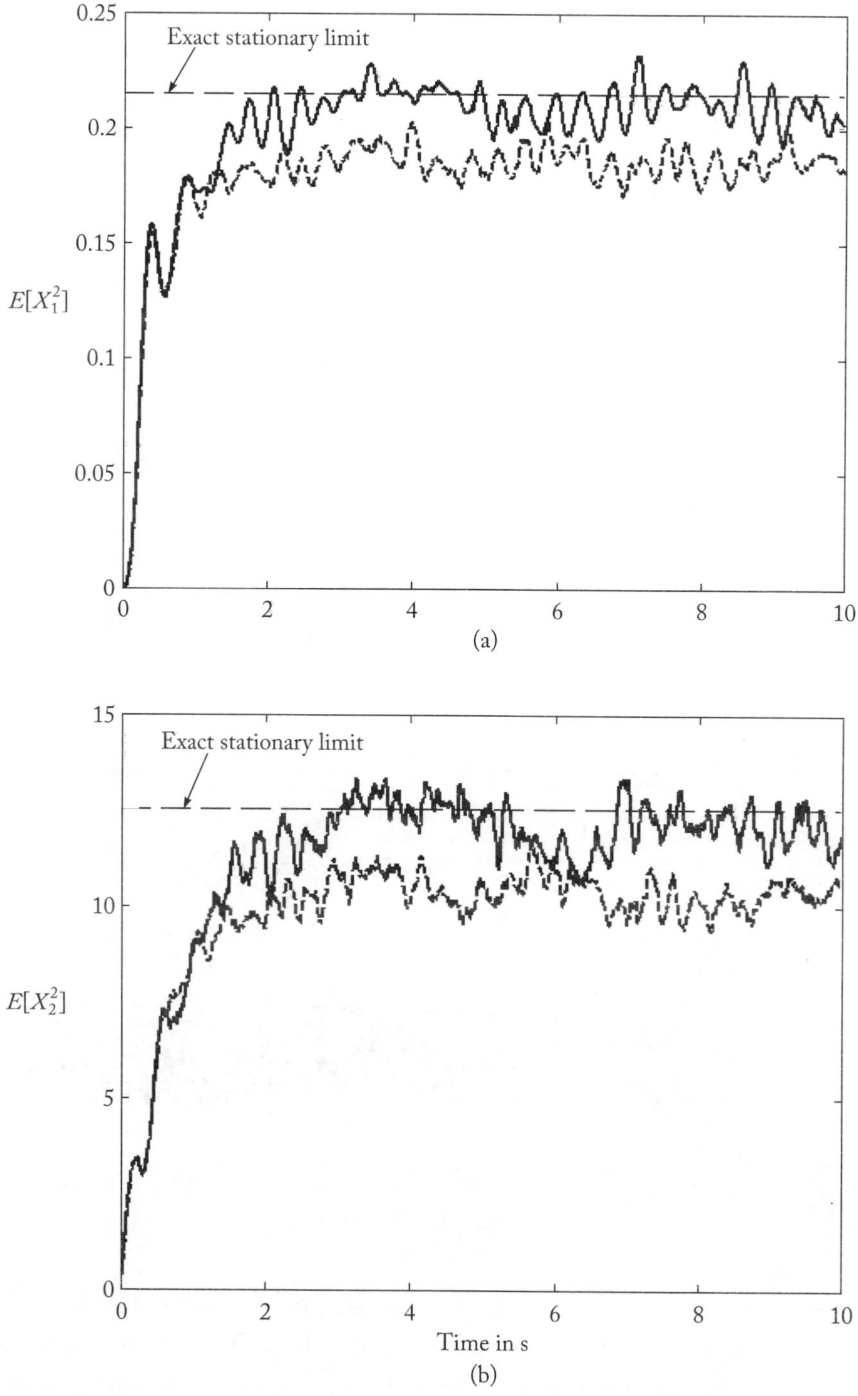
0.25
Exact stationary limit
0.2
0.15
$E[X_1^2]$
0.1
0.05
0
0
2
4
6
8
10
(a)
15
Exact stationary limit
10
$E[X_2^2]$
5
0
0
2
4
6
8
10
Time in s
(b)

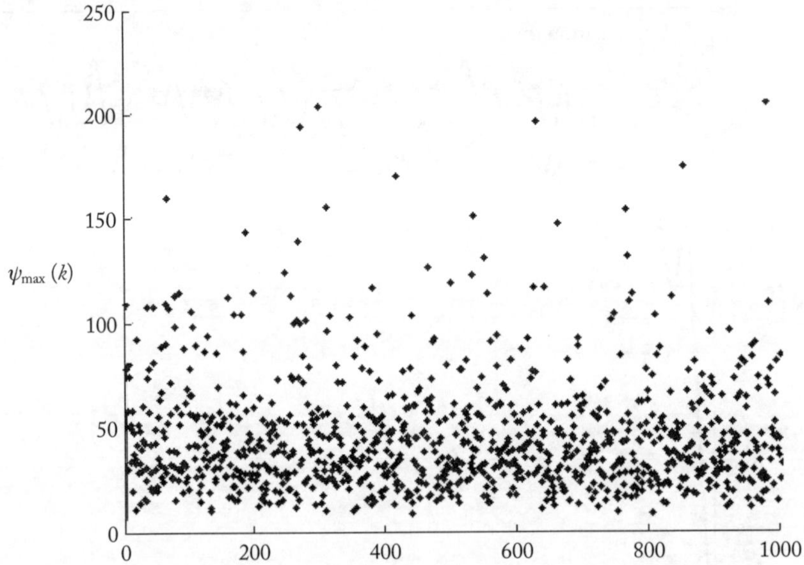

(c)

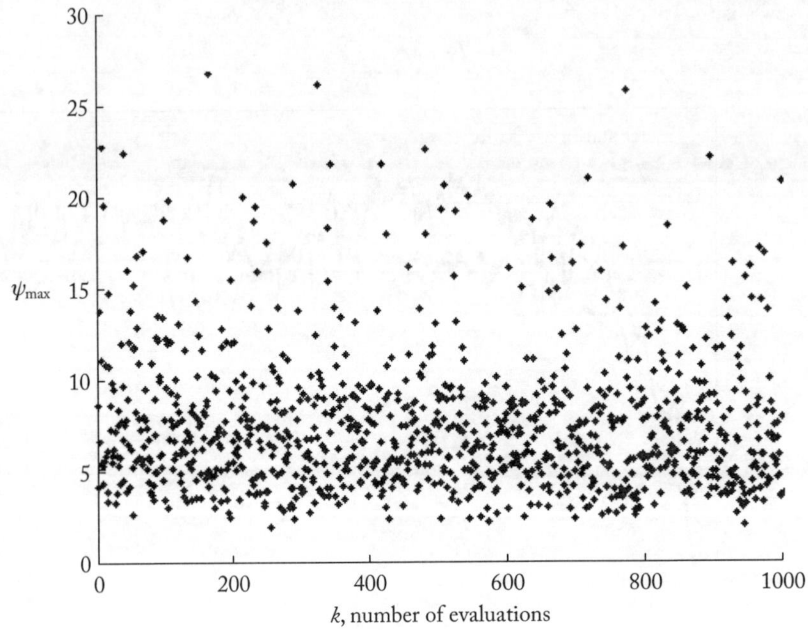

(d)

Fig. 8.6 HD oscillator: $C = 1$, $K = 10$, $\alpha = 100$, $\sigma = 5.0$, $\Delta = 0.005$, $\widetilde{X}_0 = \{0, 0\}^T$, $N_p = 1000$; second moment histories, (a) sample–averaged $E[X_1^2]$ and (b) sample–averaged $E[X_2^2]$, dashed-line—lower order and solid-line—higher order linearization, (c) $\psi_{\max}(k)$ for lower order linearization and (d) $\psi_{\max}(k)$ for higher order linearization

The contribution of Girsanov correction in improving the lower order linearized solution is illustrated in Fig. 8.5 through the plotted second moment histories of X_1 and X_2. Figure 8.6 shows the second moment histories in a case where $\psi_{\max}$ exceeds the permissible bound $\frac{1}{\Delta_i}$ in the case of lower order linearization while $\psi_{\max} < \frac{1}{\Delta_i}$ in the higher order linearization case. Thus, while the former solution is inaccurate, the other remains close to the exact stationary solution for the chosen, Δ_i. Here we have fixed a sufficiently small uniform time-step size $\Delta_i = \Delta$ *a-priori* for the numerical work so as to satisfy this constraint. The ideal strategy would however be to adaptively choose Δ_i a-posteriori (for each i) so as to impose, during run–time, the bound, $\psi_{\max}(k) < \frac{1}{\Delta_i}$. On another front, the displacement second–moment histories in Fig. 8.7 demonstrate that the correction through the Girsanov transformation becomes more conspicuous as the time step size increases (without violating the bound on $\psi_{\max}$).

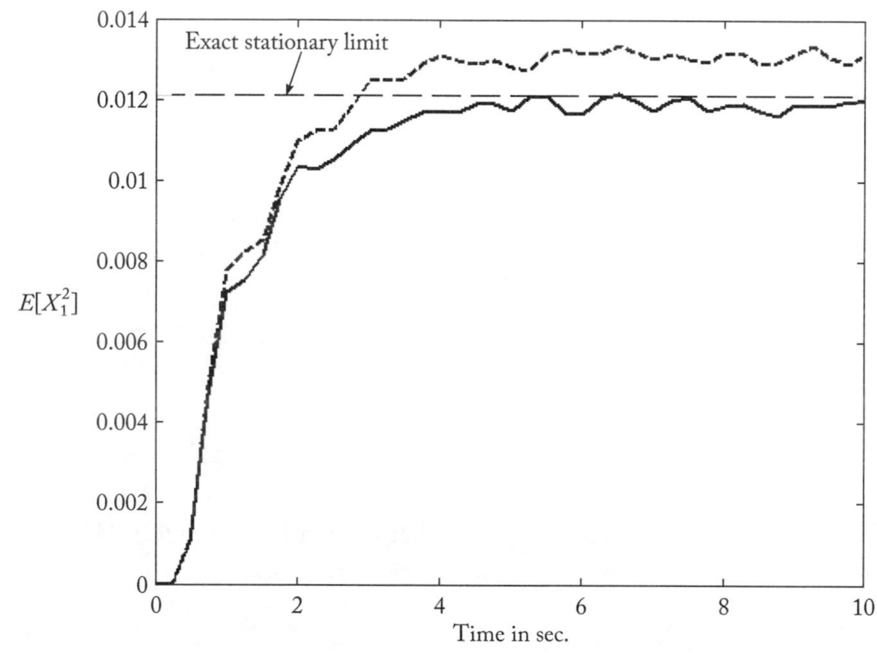

Fig. 8.7 HD oscillator: $C = 1$, $K = 10$, $\alpha = 10$, $\sigma = 0.5$, $\Delta = 0.25$, $\widetilde{X}_0 = \{0,0\}^T$, $N_p = 10000$; second moment history of sample–averaged $E[X_1^2]$ via GCLM with lower order linearization (solid-line) and without Girsanov correction (dashed-line)

Example 8.2

We apply the GCLM to a 2-*dof* non-linear oscillator under additive noises. The relevant equations of motion are as follows:

$$\ddot{X}_1 + C_1\dot{X}_1 + (K_1 + K_2)X_1 - K_2X_2 + \alpha X_1^3 = \sigma_1 B_1(t) \tag{8.34a}$$

$$\ddot{X}_2 + C_2\dot{X}_2 + (K_2 + K_3)X_2 - K_2X_1 = \sigma_2 B_2(t) \tag{8.34b}$$

Solution

Here the 4-dimensional state space vector is, $\mathbf{X} := \{X_{11} = X_1, X_{12} = \dot{X}_1, X_{21} = X_2, X_{22} = \dot{X}_2\}^T$, consisting of the displacement and velocity vectors $\mathbf{X}_1 := \{X_{11}, X_{21}\}^T$ and $\mathbf{X}_2 := \{X_{21}, X_{22}\}^T$, respectively. Note that the double subscript is used here, for convenience, to denote a vector ($\mathbf{X}$) with displacement and velocity components, and not a matrix. The incremental form of the above SDEs can be written as:

$$dX_{11}(t) = X_{12}(t)dt \tag{8.35a}$$

$$dX_{12}(t) = -\left(C_1X_{12}(t) + (K_1 + K_2)X_{11}(t) - K_2X_{21}(t) + \alpha(X_{11}(t))^3\right)dt$$
$$+\sigma_1 dB_1(t) \tag{8.35b}$$

$$dX_{21}(t) = X_{22}(t)dt \tag{8.35c}$$

$$dX_{22}(t) = -(C_2X_{22}(t) + (K_2 + K_3)X_{21}(t) - K_2X_{11}(t))dt + \sigma_2 dB_2(t) \tag{8.35d}$$

The GCLM formulation is quite similar to that written for the SDOF HD oscillator in Example 8.1. Thus, the lower order linearized SDEs are:

$$d\tilde{X}_{11}(t) = \tilde{X}_{12}(t)dt \tag{8.36a}$$

$$d\tilde{X}_{12}(t) = -\left(C_1\tilde{X}_{12}(t) + (K_1 + K_2)\tilde{X}_{11}(t) - K_2\tilde{X}_{21}(t) + \alpha\left(\tilde{X}_{11,i-1}\right)^3\right)dt$$
$$+\sigma_1\left\{-\sigma_1^{-1}\alpha\left(\left(\tilde{X}_{11}(t)\right)^3 - \left(\tilde{X}_{11,i-1}\right)^3\right)dt + dB_1(t)\right\} \tag{8.36b}$$

$$d\tilde{X}_{21}(t) = \tilde{X}_{22}(t)dt \tag{8.36c}$$

$$d\tilde{X}_{22}(t) = -\left(C_2\tilde{X}_{22}(t) + (K_2 + K_3)\tilde{X}_{21}(t) - K_2\tilde{X}_{11}(t)\right)dt + \sigma_2 dB_2(t) \tag{8.36d}$$

Solution for the linearized state $\widetilde{\boldsymbol{X}} := \left\{\tilde{X}_{11}, \tilde{X}_{12}, \tilde{X}_{21}, \tilde{X}_{22}\right\}^T$ may be obtained from Eq. (8.8), where $\boldsymbol{A}_l = \begin{bmatrix} 0 & 1 & 0 & 0 \\ -(K_1+K_2) & -C_1 & K_2 & 0 \\ 0 & 0 & 0 & 1 \\ K_2 & 0 & -(K_2+K_3) & -C_2 \end{bmatrix}$ and $\widetilde{\boldsymbol{A}}_{nl} = \left(0 \quad -\alpha\left(\tilde{X}_{11,i-1}\right)^3 \quad 0 \quad 0\right)^T$. Finally, the Radon–Nikodym derivative is given by:

$$\Lambda(t) = \exp\left\{\int_{t_{i-1}}^{t} \left[-\sigma_1^{-1}\left\{\alpha\left(\tilde{X}_{11}(t)\right)^3 - \alpha\left(\tilde{X}_{11,i-1}\right)^3\right\}\right] d\tilde{B}_1(s)\right.$$
$$\left. -\frac{1}{2}\int_{t_{i-1}}^{t} \left[-\sigma_1^{-1}\left\{\alpha\left(\tilde{X}_{11}(t)\right)^3 - \alpha\left(\tilde{X}_{11,i-1}\right)^3\right\}\right]^2 ds\right\} \tag{8.37}$$

Details on reducing the above to its canonical form are omitted for brevity.

Figure 8.8 shows a few second-moment histories for the 2-DOF oscillator, obtained through both lower and higher order linearizations for a reasonably large ensemble size (so as to reduce sampling errors). The favorable comparisons of the

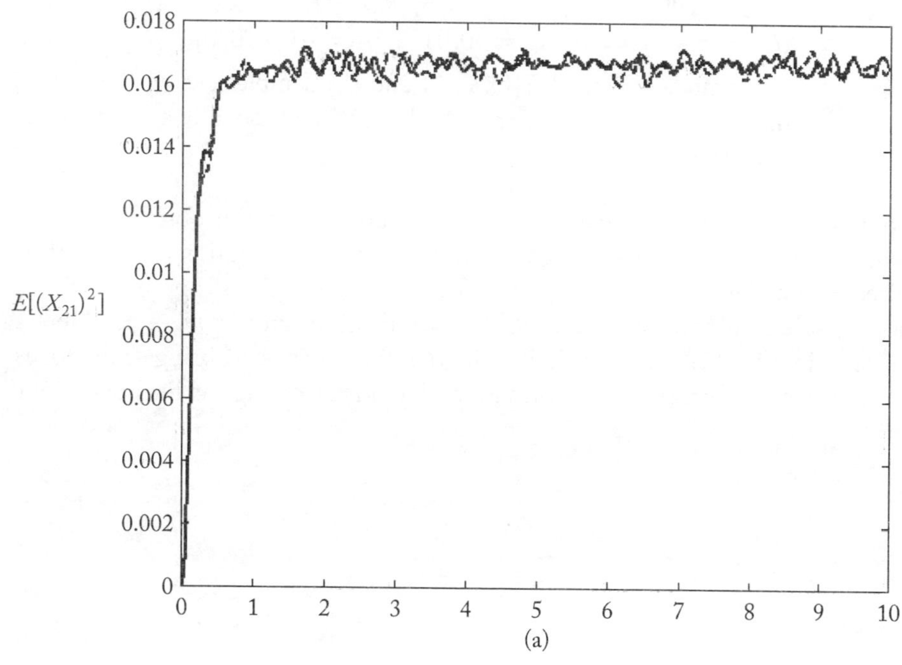

(a)

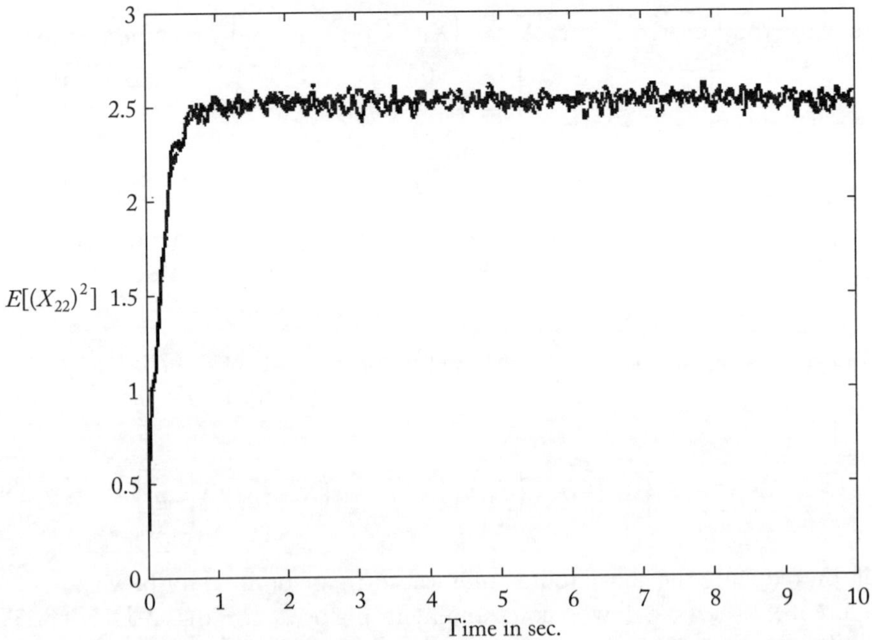

(b)

Fig. 8.8 2-dof oscillator: $C_1 = C_2 = 5$, $K_1 = K_2 = K_3 = 100$, $\alpha = 100$, $\sigma_1 = 5$, $\sigma_2 = 5$, $\Delta = 0.01$, $\widetilde{X}_0 = \{0,0,0,0\}^T$, $N_p = 10000$; (a) sample–averaged $E[(X_{21})^2]$ and (b) sample–averaged $E[(X_{22})^2]$, dashed-line—lower order linearization and solid-line—higher order linearization

computed solutions via the two linearizations along with the observation that the adopted step–size is small enough to restrict the linearization error within the required bound indicate that the GCLM-based solutions approach the exact stationary solutions without a perceptible discretization error as $m \to \infty$. Here again (as shown in Fig. 8.9), an admissibly higher time step size brings into focus the important role the Girsanov transformation plays in correcting the linearized solution.

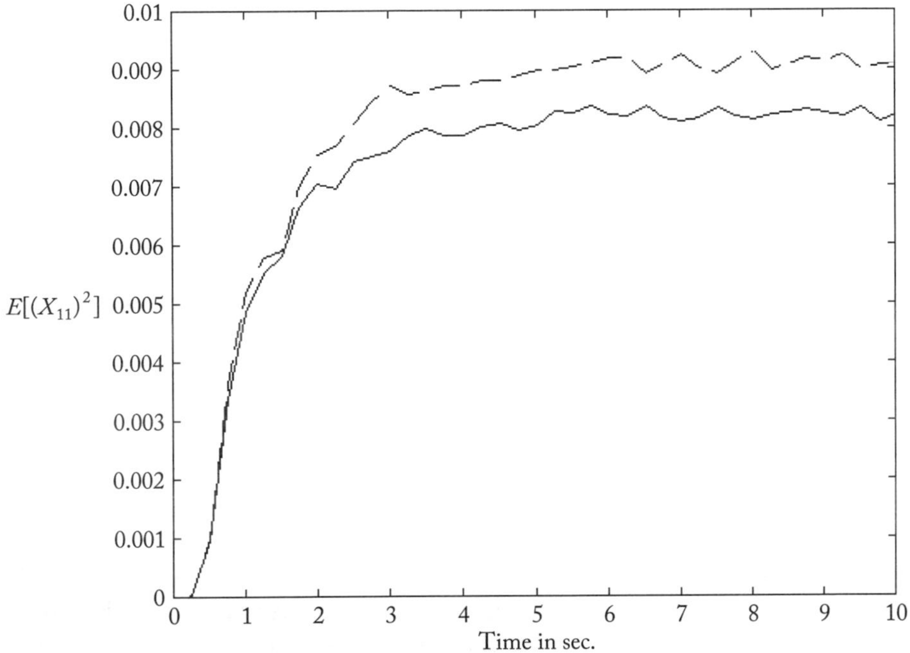

Fig. 8.9 2-dof oscillator: $C_1 = C_2 = 1$, $K_1 = K_2 = K_3 = 10$, $\alpha = 10$, $\sigma_1 = 0.5$, $\sigma_2 = 0.5$, $\Delta = 0.25$, $\widetilde{\boldsymbol{X}}_0 = \{0,0,0,0\}^T$, $N_p = 10000$; time history of sample–averaged $E[(X_{11})^2]$ via GCLM with lower order linearization (dashed-line) and without Girsanov correction (solid-line)

8.3 Girsanov Corrected Euler–Maruyama (GCEM) Method

The GCLM described in the previous section is somewhat elaborate and cumbersome. Here the linearized drift error correction, a Radon–Nikodym derivative obtainable as a stochastic exponential and expressible in terms of the linearized system response, appears as a factor defining the fitness or likelihood or weight distribution over an ensemble of realizations of the linearized solution at any given time. Brute-force strategies in exploiting the correction factor face a couple of numerical hurdles. While the first one is owing to its behaviour as a super-martingale, the second hurdle is in the accurate treatment of the stochastic integral terms in the correction factor. Numerical inaccuracy in computing the factor typically propagates in time and manifests in the so-called divergence, wherein all but one weight tends to zero. Indeed, as suggested in [Raveendran *et al.* 2013b], an ideal approach would be in the form of a scheme that does away with sampling of linearized solutions based on the weights and incorporates the factor through an additive correction free of stochastic exponentials.

In view of the popular use of the explicit EM method as a numerical integrator for SDEs, GCEM aims at incorporating an additive correction term, based on a change of measures, to the EM-based approximate numerical solution so that an improved weak solution is obtainable despite the inherently poor local convergence order of the EM method. Here the additive correction term is so realized that it does not require any exponentials to be computed. The aim is thus to lay out a framework, wherein the weak correction is nearly insensitive to sampling fluctuations. Following the definition of an appropriate error process, the essential idea is to project, through a change of measures, the EM-based approximately integrated process on the filtration generated by the error. Such an approach has its parallel in the non-linear stochastic filtering theory (Chapters 6 and 7). There, the observation-prediction mismatch was considered as an innovation process with the aim being to determine the integrated states (e.g., displacement and velocity states for mechanical oscillators) such that the innovation is rendered a zero–mean Brownian motion (or an Ito integral) characterizing the observation noise. In GCEM, on the other hand, the integration error process itself may be looked upon as the innovation, driven to an independently introduced zero–mean Brownian motion or Ito integral. This makes possible accessing parts of the rich theory of stochastic filtering, and specifically the Kushner–Stratonovich (KS) or non-linear filtering equations, to obtain the additive correction term within an MC setting. Recall that one may also contrast this with most particle filtering techniques, offering weighted particle based MC approximations to the non-linear filtering equations, as they encounter the problem of particle collapse rendering all but one of the particles unfit (i.e., with low weights) as the simulation progresses.

8.3.1 Additively driven SDEs and the GCEM method

Under a complete probability space $(\Omega, \mathcal{F}, P)$ supplied with an increasing filtration, $\{\mathcal{F}_t,\ 0 \le t \le T\}$, many systems of practical engineering interest may be modeled (possibly following semi-discretization via Galerkin-type projection in case the governing equations are in the form of stochastic partial differential equations) with the set of SDEs of a generic form:

$$d\boldsymbol{X}_t = \boldsymbol{a}(t, \boldsymbol{X}_t)\,dt + \boldsymbol{\sigma}(t)\,d\boldsymbol{B}_t, \boldsymbol{X}_0 := \boldsymbol{Z} \tag{8.38}$$

Here $\boldsymbol{X}_t(\omega) = \left\{X_{j,t}(\omega) : 1 \le j \le m\right\}^T : \mathbb{R}_+ \times \Omega \to \mathbb{R}^m$ is the state vector process, $\boldsymbol{a} = \left\{a_j\right\} : \mathbb{R}^+ \times \mathbb{R}^m \to \mathbb{R}^m$ is the vector of non-linear drift terms (possibly non-smooth) and $\boldsymbol{\sigma}(t) = \left\{\sigma_{jk}(t) : 1 \le k \le n\right\} : \mathbb{R}^+ \times \mathbb{R}^m \to \mathbb{R}^{m \times n}$ is the diffusion coefficient matrix (assumed to be state independent, which implies that only additive noises are being considered). $\boldsymbol{B}_t \in \mathbb{R}^n$ is an n-dimensional standard P-Brownian motion. $\boldsymbol{Z}$ is an $\mathcal{F}_0$-measurable random variable independent of the σ-algebra generated by $\boldsymbol{B}_s,\ s \ge 0$ and is square–integrable, i.e., $E\left[\left|\boldsymbol{Z}^2\right|\right] < \infty$. It is also assumed that the drift and diffusion terms satisfy the necessary conditions for a unique t-continuous weak solution.

Due to the general absence of analytical solutions, Eq. (8.38) is solved using a numerical integrator, the EM method to wit. The idea is to weakly correct for the error accrued in the numerical integration. Note that the method we consider here need not necessarily be applied with the EM method alone; many other numerical integrators for SDEs may be corrected this way. Before applying the scheme, the time interval of interest $[0,T]$ is discretized into N (uniform) sub-intervals with each time step taken as $\Delta t := T/N$. At the time instant, $t_i = i\Delta t$, the approximated process ${}^e\boldsymbol{X}_i := {}^e\boldsymbol{X}(t_i)$ is determined from the EM map:

$$ {}^e\boldsymbol{X}_i = {}^e\boldsymbol{X}_{i-1} + \boldsymbol{a}\left(t_{i-1}, {}^e\boldsymbol{X}_{i-1}\right)\Delta t + \sigma(t)\Delta \boldsymbol{B}_i \tag{8.39} $$

where, $\sigma \Delta \boldsymbol{B}_i \sim \mathcal{N}\left(\boldsymbol{0}, \left(\Sigma_i^B\right)^{1/2}\right)$ is a zero–mean Gaussian vector random variable with the covariance matrix $\Sigma_i^B = diag\left(\sum_{k=1}^{n} \int_{t_{i-1}}^{t_i} \left(\sigma_{jk}(t)\right)^2 dt : j \in [1,m]\right)$. The left superscript e in ${}^e\boldsymbol{X}_i$ denotes the EM-based approximation. In the special case of σ being a constant matrix (i.e., with time–invariant entries), the EM map reduces to:

$$ {}^e\boldsymbol{X}_i = {}^e\boldsymbol{X}_{i-1} + \boldsymbol{a}\left(t_{i-1}, {}^e\boldsymbol{X}_{i-1}\right)\Delta t + \sigma \Delta \boldsymbol{B}_{i-1}, \Delta \boldsymbol{B}_{i-1} := \boldsymbol{B}_i - \boldsymbol{B}_{i-1} \tag{8.40} $$

Note that the equation above is derivable through appropriately truncated stochastic Taylor expansions of the second and third terms on the RHS of the following integral form of the SDE (8.38) over $(t_{i-1}, t_i]$:

$$ \boldsymbol{X}_i = \boldsymbol{X}_{i-1} + \int_{t_{i-1}}^{t_i} \boldsymbol{a}\left(s, \boldsymbol{X}_s\right) ds + \int_{t_{i-1}}^{t_i} \sigma_s d\boldsymbol{B}_s \tag{8.41} $$

It is well known that the orders of local convergence of the EM method are quite low, with a strong convergence of $O(\sqrt{\Delta t})$ and weak convergence of $O(\Delta t)$ only [Section 5.2.1, Chapter 5]. In what follows, even though we assume σ to be constant, the method readily generalizes to time-variant diffusion coefficients.

8.3.2 Weak correction through a change of measure

An important starting point of the correction scheme is the observation that the EM map in Eq. (8.40) may be looked upon as the exact solution of the (locally) linearized SDE (henceforth called the EM–SDE) for $t \in (t_{i-1}, t_i]$:

$$ d\left({}^e\boldsymbol{X}_i\right) = \boldsymbol{a}\left(t_{i-1}, {}^e\boldsymbol{X}_{i-1}\right) dt + \sigma d\boldsymbol{B}_i \tag{8.42} $$

Based on the anomalies of the drift fields in Eqs. (8.38) and (8.42), one could define a vector valued error process $\mathcal{E}_t \in \mathbb{R}^n$ over $(t_{i-1}, t_i]$ such that:

$$ \sigma \mathcal{E}_t := \boldsymbol{a}\left(t, \boldsymbol{X}_t\right) - \boldsymbol{a}\left(t_{i-1}, \boldsymbol{X}_{i-1}\right) \tag{8.43} $$

This enables rewriting the original SDE (8.38) under $(\Omega, \mathcal{F}, P)$ as:

$$ d\boldsymbol{X}_t = \boldsymbol{a}\left(t_{i-1}, \boldsymbol{X}_{i-1}\right) dt + \sigma\left(\mathcal{E}_t dt + d\boldsymbol{B}_t\right) \tag{8.44} $$

Attempts have been made [Raveendran *et al.* 2013b] to effect a transformation of measures $P \to Q$ aimed at removing the drift term $\sigma\mathcal{E}_t dt$ from the above equation, so that the representation of the original SDE (8.38) under $(\Omega, \mathcal{F}, P)$ takes the following form over $(t_{i-1}, t_i]$

$$dX_t = a(t_{i-1}, X_{i-1})\,dt + \sigma d\widetilde{B}_t, \tag{8.45}$$

The above has a structure identical to the EM–SDE (8.42). Here $\widetilde{B}_t$ is a standard zero mean Q-Brownian motion independent of Z. Under P however, $d\widetilde{B}_t = \mathcal{E}_t dt + dB_t$ has the error term $\mathcal{E}_t$ driving its drift field. The EM map (Eq. 8.40) (with ΔB_i replaced by $\Delta\widetilde{B}_i$), i.e., the solution to SDE (Eq. 8.45), may then be numerically simulated to yield a finite ensemble of Q-realizations within an MC setup. However, the acceptability (or fitness) of these trajectories will depend on the Radon–Nikodym derivative $\Lambda_i = \frac{dP}{dQ}(t_i) = \exp\left(\int_{t_{i-1}}^{t_i} \mathcal{E}_s . d\widetilde{B}_s - \frac{1}{2}\int_{t_{i-1}}^{t_i} \|\mathcal{E}_s\|^2 ds\right)$. Λ_t is the solution to the scalar SDE $d\Lambda_t = \Lambda_t \mathcal{E}_t . d\widetilde{B}_t$. Thus, using the EM-based solution Eq. (8.45) at t_i, the corrected solution (in P-distribution) of Eq. (8.38) at this instant may be arrived at by the empirically evaluated distribution of the Q-ensemble of $\Lambda_i X_i$, which in turn ensures that the incremental error term $\Delta\widetilde{B}_t = \Delta(\mathcal{E}_t + B_t)$ behaves like a P-Brownian increment ΔB_t. In this scheme, Λ_i may be interpreted as defining a fitness distribution over the ensemble of P realizations of ${}^{\varepsilon}X_i$, the solution to the EM–SDE (8.42). However, being a stochastic exponential, the Radon–Nikodym derivative Λ_t behaves as a super martingale and hence precipitates the weight degeneracy problem as time recursion proceeds.

In order to address the problem, we consider a framework to incorporate the fitness factor Λ_t as an additive correction (or a curing) term, as in the KS or EnKS filters in Chapter 7. This should enable driving the relatively more unfit realizations of ${}^{\varepsilon}X_i$ to higher fitness regions in the solution space and hence bypasses any need for rejection sampling. First, given the EM–SDE (8.42), we denote the filtration generated by the n-dimensional, re-defined error process $\mathcal{E}_t := a(t, X_t) - a(t_{i-1}, X_{i-1})$ as $\mathcal{F}_t^e := \{\sigma(\mathcal{E}_s) | s \le t\}$, where $\sigma(\mathcal{E}_s)$ here denotes the σ-algebra corresponding to the random variable $\mathcal{E}_s$ for a given $s \le t$ (symbol σ in σ-algebra should not be confused with the elements of the diffusion matrix $\boldsymbol{\sigma}$). The essential idea is now to recognize that a weakly corrected form of ${}^{\varepsilon}X_t$, i.e., the solution to the EM–SDE, is obtainable via the conditional expectation, $\pi_t(X) := E_P[{}^{\varepsilon}X_t | \mathcal{F}_t^e]$, which renders the corrected solution $\pi_t(X)$ measurable with respect to the filtration $\mathcal{F}_t^e$. Note that while $\mathcal{F}_t^e \subset \mathcal{F}_t^X = \sigma(Z) \cup \mathcal{F}_t^B$ ($\mathcal{F}_t^X$ and $\mathcal{F}_t^B$ being the filtrations generated by the true solution X_t and B_t, respectively), $\mathcal{F}_t^e \not\subset \mathcal{F}_t^{{}^{\varepsilon}X}$, the filtration corresponding to the EM-based process ${}^{\varepsilon}X_t$. In the special case of X_t (and hence ${}^{\varepsilon}X_t$) being mean–square integrable, $\pi_t(X)$defines a projection of ${}^{\varepsilon}X_t$ on $\mathcal{F}_t^e$. At this stage, one faces the difficulty of characterizing $\mathcal{F}_t^e$. It is here that a change of measure $P \to Q$ may be usefully exploited. Specifically, under Q, we demand $\mathcal{E}_t$ to behave as a zero–mean martingale, independent of B_t and Z. We write

$\pi_t(\boldsymbol{X}) = \dfrac{E_Q[\Lambda_t{}^{\mathfrak{e}}\boldsymbol{X}_t|\mathcal{F}_t^e]}{E_Q[\Lambda_t|\mathcal{F}_t^e]}$ and tailor Q with a view to rendering $\mathcal{E}_t$ a zero–mean martingale. In other words, the following constraint must be satisfied a.s. under Q:

$$\boldsymbol{0} = -\mathcal{E}_t dt + \rho d\widetilde{\boldsymbol{B}}_t \tag{8.46}$$

where, $\widetilde{\boldsymbol{B}}_t$ is a standard $\mathcal{F}$-measurable n-dimensional Brownian motion independent of $\boldsymbol{B}_t$ and $\boldsymbol{Z}$ and ρ an $n\times n$ diffusion matrix, which may be chosen in a convenient and numerically expedient manner (viz. as a strictly diagonal matrix). Thus the associated Radon–Nikodym derivative takes the form:

$$\Lambda_t := \frac{dP}{dQ}(t) = \exp\left(\int_{t_{i-1}}^{t_i} \left(\rho^{-1}\mathcal{E}_s\right)^T d\widetilde{\boldsymbol{B}}_s - \frac{1}{2}\int_{t_{i-1}}^{t_i} \left\|\rho^{-1}\mathcal{E}_s\right\|^2 ds\right) \tag{8.47}$$

An aspect of the above framework is its similarity with that of non-linear filtering with Eqs. (8.42) and (8.46) playing the roles of the so-called process and observation equations (the measurement being an a.s. zero process). Derivation of the rest of the algorithm therefore is somewhat parallel to that of the non-linear filtering theory (Chapters 6). Given the apparent similarity with filtering theory, we provide below the governing equation for the scalar–valued process $\pi_t(\phi) := \pi_t(\phi(\boldsymbol{X}))$, (see Section 6.3.2 , Chapter 6 for a detailed derivation):

$$\pi_t(\phi) = \pi_{i-1}(\phi) + \int_{t_{i-1}}^{t} \pi_s(L_s(\phi))\, ds$$

$$+ \int_{t_{i-1}}^{t} \left(\pi_s\left(\boldsymbol{L}_s^1(\phi)\right) - \pi_s(\phi)\, \pi_s\left(\rho^{-1}\mathcal{E}_s\right)^T\right).d\boldsymbol{I}_s \tag{8.48}$$

where, $L_t(\phi) := \phi'(\boldsymbol{X}_t)^T \boldsymbol{a}(t_{i-1}, \boldsymbol{X}_{i-1}) + \frac{1}{2}\sum_{j,k=1}^{m}\sum_{l=1}^{n}\left(\frac{\partial^2\phi}{\partial x_j \partial x_k}\right)_t \sigma_{jl}\sigma_{kl}$ is the generator corresponding to the EM–SDE (8.42) and $\boldsymbol{L}_t^1(\phi) := \phi(\boldsymbol{X}_t)\left(\rho^{-1}\mathcal{E}_t\right)^T$, $d\boldsymbol{I}_t := \left\{d\widetilde{\boldsymbol{B}}_t - \pi_t\left(\rho^{-1}\mathcal{E}_t\right)dt\right\}$ is the incremental innovation term. Here $\phi'(\boldsymbol{X})$ and $\phi''(\boldsymbol{X})$ are, respectively, the first derivative (a vector) and the second derivative (a matrix) of $\phi(\boldsymbol{X})$ with respect to the vector, $\boldsymbol{X}$.

The first two terms on the RHS of Eq. (8.40) constitute Dynkin's formula determining the approximate mean $E[\phi({}^{\mathfrak{e}}\boldsymbol{X}_t)|\boldsymbol{X}_{i-1}]$ based on the EM–SDE (i.e., averaged $\phi({}^{\mathfrak{e}}\boldsymbol{X}_t)$ based on the EM map 8.40). In the MC setup, the approximate $\phi({}^{\mathfrak{e}}\boldsymbol{X}_t)$ is directly obtainable as a finite ensemble of EM-based trajectories /particles via Eq. (8.40). These EM-approximated set of particles is then weakly corrected by appropriately incorporating the last term on the RHS of Eq. (8.48). A step-wise implementation of the scheme, recursively implemented in time, is indicated below.

Step 1: Generating the EM-based approximated particles:

With $\phi(\boldsymbol{X})$ a chosen scalar function, the particle based EM approximation at $t = t_i$ is:

$$ {}^{\mathfrak{c}}\phi_i^{[j]} = \phi_{i-1}^{[j]} + L_{i-1}\left(\phi_{i-1}^{[j]}\right)\Delta t + \phi_{i-1}^{[j]}\sigma\Delta\boldsymbol{B}_{i-1} \tag{8.49} $$

where, ${}^{\mathfrak{c}}\phi_i := \phi({}^{\mathfrak{c}}\boldsymbol{X}_i)$ and the square bracketed superscript denotes the jth particle. In a recursive scheme, the initial condition ϕ_{i-1}, i.e., the first term on the RHS of Eq. (8.49), is assumed to be available in the weakly corrected form (hence the omission of the left superscript $\mathfrak{c}$). This step thus provides us with the EM-approximated particle set, $\left\{({}^{\mathfrak{c}}\phi_i)^{[j]}\right\}_{j=1}^{N_p}$.

Step 2: Weak additive correction

$$ \phi_i{}^{[j]} = ({}^{\mathfrak{c}}\phi_i)^{[j]} + G_i{}^{[j]} \tag{8.50} $$

where,

$$ G_i := \left\{\overline{\pi}_i\left(\boldsymbol{L}_i^1\left({}^{\mathfrak{c}}\phi_i\right)\right) - \overline{\pi}_i\left({}^{\mathfrak{c}}\phi_i\right)\overline{\pi}_i\left(\rho^{-1}\left({}^{\mathfrak{c}}\mathcal{E}_i\right)\right)^T\right\}.\left\{\Delta\widetilde{\boldsymbol{B}}_{i-1} - \left(\rho^{-1}\left({}^{\mathfrak{c}}\mathcal{E}_i\right)\right)\Delta t_{i-1}\right\} \tag{8.51} $$

Here ${}^{\mathfrak{c}}\mathcal{E}_t := \boldsymbol{a}(t, {}^{\mathfrak{c}}\boldsymbol{X}_t) - \boldsymbol{a}(t_{i-1}, \boldsymbol{X}_{i-1})$ is the EM-approximated error process and $\overline{\pi}_i(.) = \frac{1}{N_p}\sum_{j=1}^{N_p}(.)^{[j]}$ is the sample mean operator at time t_i which, owing to the finiteness of the ensemble, is an approximation to the exact population mean.

Note that, just as the solution to the SDE (8.42) is given by the EM map (8.40), similar such linearized SDEs (and hence appropriate error process, $\mathcal{E}_t$) may be readily constructed for maps corresponding to many other stochastic integration scheme as well.

8.4 Numerical Demonstration of GCEM Method

Toward a numerical demonstration of how the above weak correction scheme works, a few non-linear stochastically driven dynamical systems are considered. As a first example, the hardening Duffing oscillator under additive stochastic excitation with constant diffusion coefficients is considered, emphasizing the effect of the weak correction with reasonably large integration time step, Δt. As the next example, a non-smooth system involving an oscillator under discontinuous support is taken up. In the third example, numerical integration of a spatially discretized version of a one dimensional (1-D) Burger's equation is considered. Specifically arising from the semi-discrete formulation based on a mesh-free approximation, the large dimensional non-linear set of differential equations in time, supplemented with additive noise terms that presumably account for modeling/semi-discretization errors, is solved with the GCEM scheme.

Example 8.3

A 1-D hardening Duffing oscillator:

Under $(\Omega, \mathcal{F}, P)$, the system dynamics of the 1-D hardening Duffing (HD) oscillator under additive noise can be written as:

$$\ddot{X} + C\dot{X} + KX + \alpha X^3 = \mathcal{P}\cos\lambda t + \sigma\dot{B}(t) \tag{8.52}$$

with the initial condition $X(t=0) = 0$ and $\dot{X}(t=0) = 0$. The parameters $C, K, \alpha, \mathcal{P}$ and λ are taken as 4, 100, 100, 4 and 2 π, respectively.

Solution

Defining $X_1 = X$, $X_2 = \dot{X}$, and the vector $\mathbf{X} = (X_1, X_2)^T$, the state-space representation of Eq. (8.52) in the incremental form may be written as:

$$dX_1(t) = X_2(t)dt \tag{8.53a}$$

$$dX_2(t) = \left(-CX_2(t) - KX_1(t) - \alpha(X_1(t))^3 + \mathcal{P}\cos\lambda t\right)dt + \sigma dB(t) \tag{8.53b}$$

The EM-based time marching map for the above SDEs is:

$${}^{\varepsilon}X_{1,i} = X_{1,i-1} + X_{2,i-1}\Delta t \tag{8.54a}$$

$${}^{\varepsilon}X_{2,i} = X_{2,i-1} + \left(-CX_{2,i-1} - KX_{1,i-1} - \alpha(X_{1,i-1})^3 + \mathcal{P}\cos\lambda t_{i-1}\right)\Delta t$$

$$+\sigma\Delta B_{i-1} \tag{8.54b}$$

with $t \in (t_{i-1}, t_i]$. In the absence of exact solutions, reference solutions are presently obtained through EM simulations of Eq. (8.53) with a very small time step size $\Delta t = \Delta = 0.0001$s and very low additive noise intensity (dotted lines in Figs 8.10a,b). Indeed, in order to ensure that such reference solutions make sense, we choose such response regimes (e.g., one-periodic response in the absence of noise and once the transients die out) that are known to remain largely unaffected (in the weak sense) owing to the introduction of additive noise or small variations in the noise intensity. In order to assess the effect of the weak correction, an ensemble of trajectories ($N_p = 100$) are simulated using the EM-map with a reasonably large $\Delta = 0.05$s and then weakly corrected before computing the sample averages of the displacement and velocity components. Noise intensity is taken as $\sigma = 0.01$. The Radon–Nikodym derivative, from which the additive correction term may be derived, is of the form:

$$\Lambda_t = \exp\left(\int_{t_{i-1}}^{t} \mathcal{E}_s d\hat{B}_s - \frac{1}{2}\int_{t_{i-1}}^{t} \mathcal{E}_s{}^2 ds\right) \tag{8.55}$$

where

$$\mathcal{E}_t = \left(-CX_2(t) - KX_1(t) - \alpha(X_1(t))^3 + \mathcal{P}\cos\lambda t\right) \\ -(-CX_{2,i-1} - KX_{1,i-1} - \alpha X_{1,i-1}^3 + \mathcal{P}\cos\lambda t_{i-1}) \quad (8.56)$$

As observed from Fig. 8.10, the weakly corrected solutions (black lines), given in terms of $E_P[X_1(t)|\mathcal{F}_t^e]$ or $E_P[X_2(t)|\mathcal{F}_t^e]$ as the case may be, are nearly indistinguishably close to their respective references. This may be well contrasted with the regular EM solutions (dash–dot lines) that tend to diverge as time progresses.

Example 8.4

A non-smooth oscillator:

Non-smooth dynamical systems are ubiquitous in engineering dynamics, some useful applications being in oil drilling, rotor dynamics, etc. [Santos and Savi 2009], where the nonlinearity arises either from friction or intermittent contact of some parts of the system. The system dynamics of one such oscillator with intermittent contact has two phases, the first describing the dynamics when the moving body does not make a contact with the support and the other depicting the motion after a contact is made with the support. At the non-smooth contact region, the gap between the support and the body (in its static equilibrium state) is taken as $\mathcal{G}$ (see Fig. 8.11).

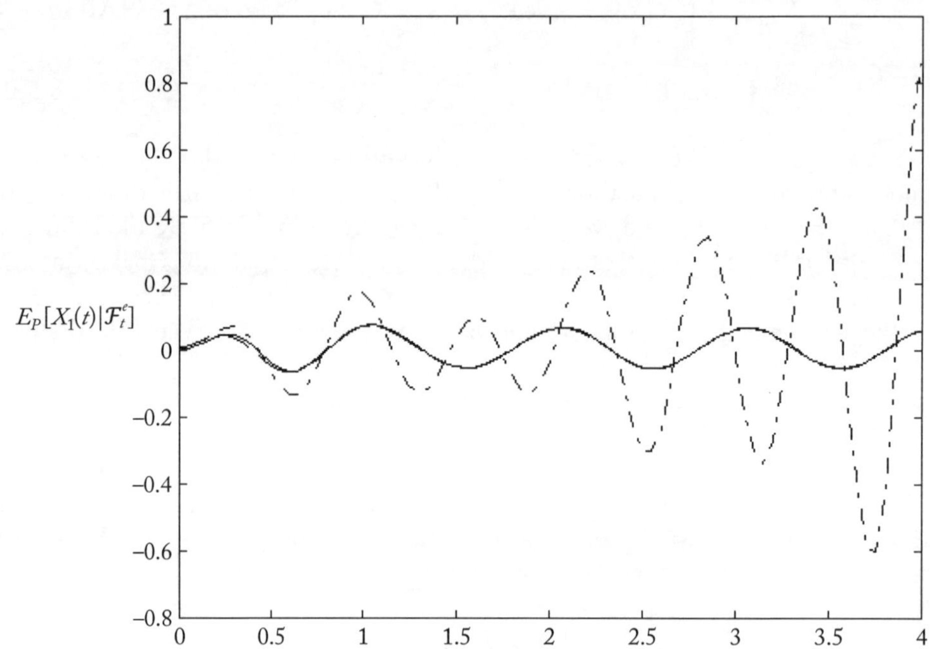

(a)

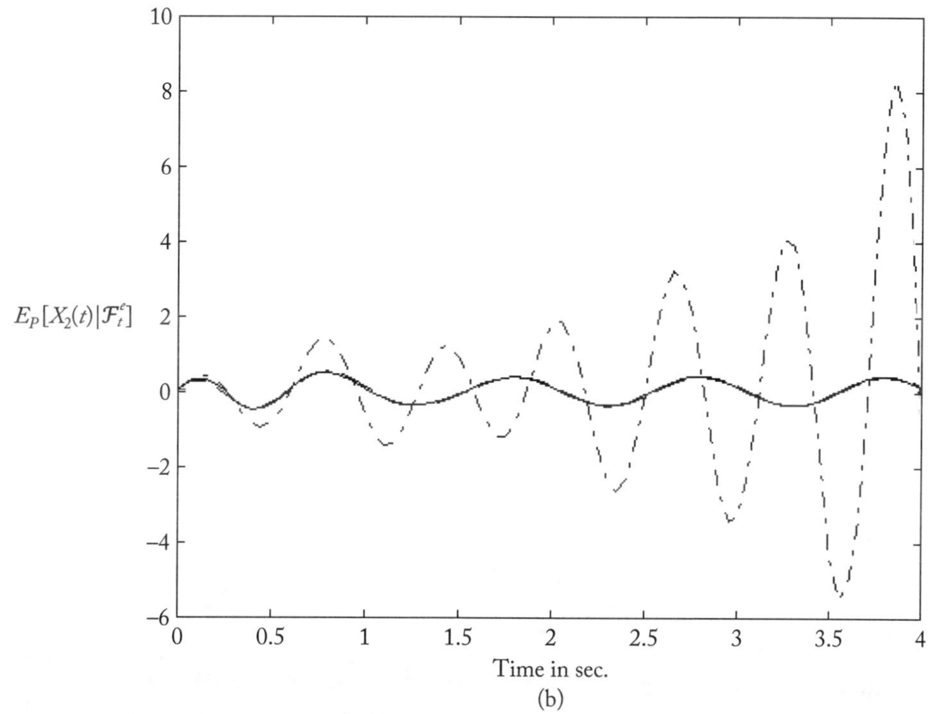

Fig. 8.10 HD oscillator in Example 8.3: $C = 4$, $K = 100$, $\alpha = 100$, $\mathcal{P} = 4$, $\lambda = 2\pi \frac{rad}{s}$ $\sigma = 0.01$; (a) sample mean displacement and (b) sample mean velocity profiles by GCEM method, dash-dot line—solution by (uncorrected) EM method with $\Delta = 0.05$, black line—corrected solution with $\Delta = 0.05$, dashed-line—reference solution with $\Delta = 0.0001$

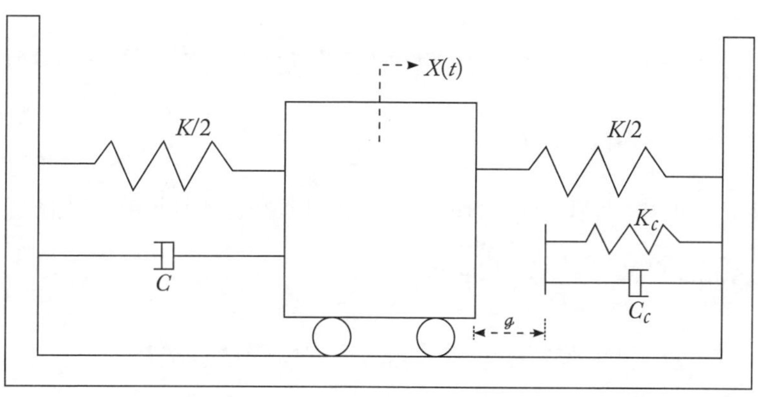

Fig. 8.11 A schematic diagram of the non-smooth oscillator

Solution

The two phases of the oscillator may be represented in terms of contact force $\mathscr{f}_c(t)$ as follows:

$$\text{no contact}: \mathscr{f}_c(t) = 0, X(t) < \mathscr{g} \tag{8.57a}$$

$$\text{with contact}: \mathscr{f}_c(t) = -\left(K_c X(t) + C_c \dot{X}(t)\right) < 0, X(t) \geq \mathscr{g}, \tag{8.57b}$$

The contact in the present problem is characterized through a linear elastic element leading to the restitution force given by, $K_c(X(t) - \mathscr{g})$. Finally the governing dynamics of the system may be written as:

$$\text{no contact}: \ddot{X}(t) + C\dot{X}(t) + KX(t) = \mathcal{P}\cos\lambda t + \sigma\dot{B}(t) \tag{8.58a}$$

$$\text{with contact}: \ddot{X}(t) + C\dot{X}(t) + KX(t) + \left(C_c\dot{X}(t) + K_c(X(t) - \mathscr{g})\right)$$

$$= \mathcal{P}\cos\lambda t + \sigma\dot{B}(t) \tag{8.58b}$$

The initial conditions are taken as $X(t=0) = 0.1$ and $\dot{X}(t=0) = 0.1$. As before, defining $X_1 = X, X_2 = \dot{X}$, the state-space representation of Eq. (8.58) may be written in incremental form as:

$$\text{no contact}: dX_1(t) = X_2(t)dt$$

$$dX_2(t) = (-CX_2(t) - KX_1(t) + \mathcal{P}\cos\lambda t)\,dt + \sigma dB(t) \tag{8.59a}$$

$$\text{with contact}: dX_1(t) = X_2(t)dt$$

$$dX_2(t) = \left(-(C + C_c)X_2(t) - (K + K_c)X_1(t) + K_c\mathscr{g} + \mathcal{P}\cos\lambda t\right)dt$$

$$+\sigma dB(t) \tag{8.59b}$$

Under $(\Omega, \mathcal{F}, P)$, the EM-map corresponding to Eq. (8.59) is given by:

$$\text{no contact}: {}^{\mathfrak{e}}X_{1,i} = X_{1,i-1} + X_{2,i-1}\Delta t$$

$${}^{\mathfrak{e}}X_{2,i} = X_{2,i-1} + (-CX_{2,i-1} - KX_{1,i-1} + \mathcal{P}\cos\lambda t_{i-1})\Delta t$$

$$+\sigma\Delta B_{i-1}, \tag{8.60a}$$

$$\text{with contact : } {}^{c}X_{1,i} = X_{1,i-1} + X_{2,i-1}\Delta t$$

$${}^{c}X_{2,i} = X_{2,i-1} + \left(-(C + C_c)X_{2,i-1} - (K + K_c)X_{1,i-1} + K_c\mathcal{G} + \mathcal{P}\cos\lambda t_{i-1}\right)\Delta t + \sigma\Delta B_{i-1} \quad (8.60b)$$

A reference solution (dashed lines in Fig. 8.12) for the problem is constructed through the EM–map (Eq. 8.60) with a small time step size $\Delta t = \Delta = 0.0001$s. While the ensemble size for the MC simulation is consistently kept at $N_p = 100$, a relatively large step size $\Delta = 0.01$ is chosen to implement the weak correction. Noise intensity is taken as $\sigma = 0.01$. As evidenced from Fig. 8.12, the weakly corrected EM solutions (black lines) under GCEM provide a substantively improved accuracy vis-à-vis the conventional EM solutions (dash-dot lines). The weakly corrected solutions are nearly indistinguishably close to their respective references.

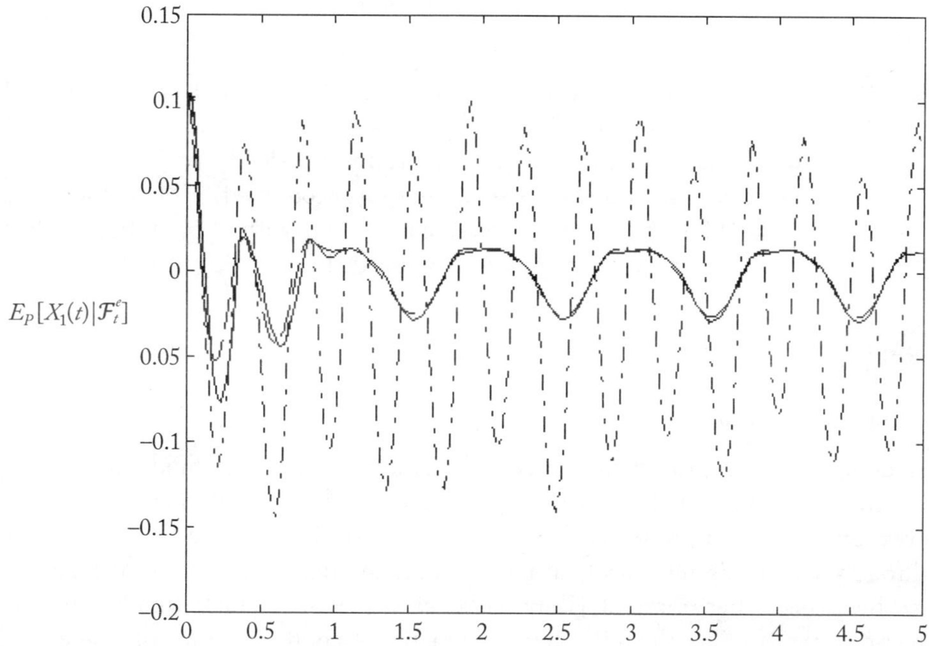

(a)

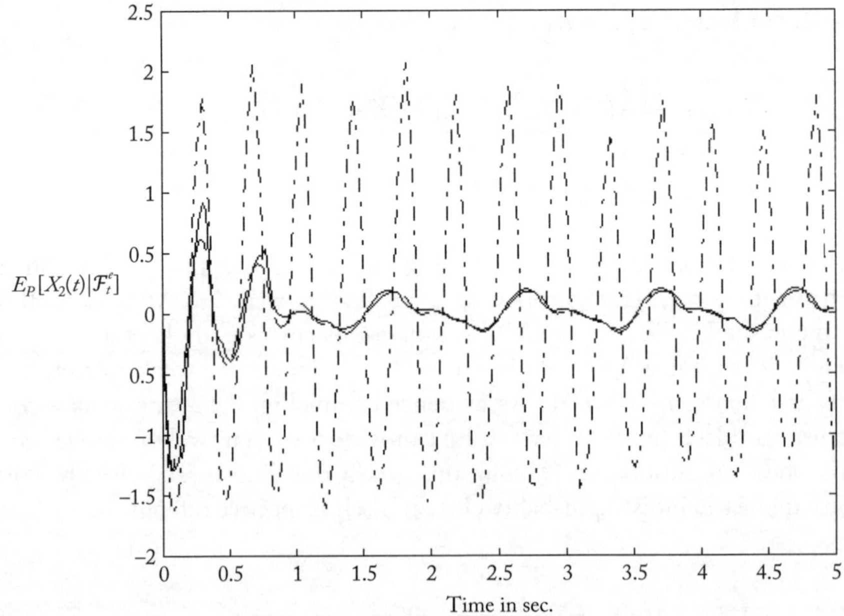

Fig. 8.12 The non-smooth oscillator in Example 8.4: $C = 4$, $K = 100$, $\mathcal{P} = 4$, $C_c = 5$, $K_c = 200$, $\lambda = 2\pi \frac{\text{rad}}{s}$, $\sigma = 0.01$, $\mathcal{G} = 0.0045$; (a) sample mean displacement and (b) sample mean velocity profiles by GCEM method, dash-dot line—solution by (uncorrected) EM method with $\Delta = 0.03$, black line—corrected solution with $\Delta = 0.03$, dashed-line—reference solution with $\Delta = 0.0001$

Example 8.5

1-D Burger's equation:

Besides serving as a prototypical model for studying turbulent fluid flow, Burger's equation is also used for demonstrating the efficacy of computational algorithms for convection dominated problems. One way to numerically simulate Burger's equation is through a semi-discrete formulation, wherein the operator may be discretized in space by a mesh-free method [Belytschko *et al.* 1994], which (unlike the finite element method) offers the advantage of functional approximations of higher order smoothness uniformly over the spatial domain of interest [Shaw and Roy 2008]. In the present case, a 1-D viscous (parabolized) Burger's equation is considered. The governing dynamics of the field variable $u(t,y) \in \mathbb{R}$ defined over the domain $[0,l]$ and evolving over time t, is given as:

$$\frac{\partial u(t,y)}{\partial t} + u(t,y)\frac{\partial u(t,y)}{\partial y} = \nu \frac{\partial^2 u(t,y)}{\partial y^2} \tag{8.61}$$

with initial condition $u(0,y) = \sin\frac{\pi y}{l}$ and boundary conditions $u(t,0) = 0$ and $u(t,l) = 0$. $\nu = 0.01$ is the viscosity parameter. The discretization starts with the weak formulation of Eq. (8.61)

$$\int_0^l W\left(\frac{\partial u(t,y)}{\partial t} + u(t,y)\frac{\partial u(t,y)}{\partial y} - \nu\frac{\partial^2 u(t,y)}{\partial y^2}\right)dy = 0 \tag{8.62}$$

where, W is a compactly supported weight (or test) function. For semi- discretization, a variables-separated form for $u(t,y) = \aleph(y)\hat{u}(t)$ is assumed, where $\aleph(y) \in \mathbb{R}^{1\times n_d}$ is a matrix containing mesh-free shape functions (Shaw and Roy 2007) evaluated at y, $n_d = 25$ defines the (state space) dimension of the semi-discrete system (which is identical to the number of discretization points or nodes in the spatial domain) and $\hat{u}(t) \in \mathbb{R}^{n_d}$ is the state vector containing the nodal values of $u(t,y)$. The shape functions are currently constructed using the element free Galerkin (EFG) method [Belytschko *et al.* 1994]. Since EFG typically incorporates a symmetric Galerkin formulation, shape functions and weight functions are chosen from the same function space. The n_d-dimensional semi-discrete system, in the form of a coupled system of first order non-linear ordinary differential equations (ODEs) is written as:

$$\boldsymbol{A}\dot{\widehat{\boldsymbol{u}}}(t) = \boldsymbol{b}(t)$$

where, $\boldsymbol{A} = \int_0^l \aleph^T(y)\aleph(y)\,dy$ is the coefficient matrix and

$$\boldsymbol{b}(t) = -\int_0^l \aleph^T(y)\left\{(\aleph(y)\widehat{\boldsymbol{u}}(t))\left(\frac{\partial\aleph(y)}{\partial y}\widehat{\boldsymbol{u}}(t)\right) - \nu\frac{\partial^2\aleph(y)}{\partial y^2}\widehat{\boldsymbol{u}}(t)\right\}dy \tag{8.63}$$

is the source vector. Such mathematical descriptors for most physical systems are typically subject to modeling as well as numerical (e.g., semi-discretization) errors. A natural setting to account for modeling uncertainty in the original Burger's equation would have been a space–time white noise over $[0,T]\times\mathbb{R}$, canonically associated with which is the $L^2(\mathbb{R})$- cylindrical Brownian motion (see, for instance, van Neerven [2008] and Riedle [2011] for a detailed exposition on cylindrical Wiener processes). However, we here take recourse to a simpler mathematical alternative and thus incorporate the effect of the modeling uncertainties via a set of classical Brownian motions in time in the semi-discretized equations. This yields the following formal system of non-linear SDEs:

$$\boldsymbol{A}\dot{\widehat{\boldsymbol{u}}}(t) = \boldsymbol{b}(t) + \boldsymbol{\sigma}\dot{\boldsymbol{B}}(t) \tag{8.64}$$

The proposed weak correction scheme is now applied to the EM-based solution of Eq. (8.64). Since Eq. (8.64) is only a formal representation involving $\boldsymbol{B}(t)$, a more appropriate incremental form of the SDEs is written as:

$$d\widehat{\boldsymbol{u}}(t) = \boldsymbol{A}^{-1}\boldsymbol{b}(t)\,dt + \boldsymbol{A}^{-1}\sigma d\boldsymbol{B}(t) \tag{8.65}$$

The EM–SDE corresponding to Eq. (8.65) over $t \in (t_{i-1}, t_i]$ is:

$$d\widehat{\boldsymbol{u}}(t) = \boldsymbol{A}^{-1}\boldsymbol{b}(t_{i-1})\,dt + \boldsymbol{A}^{-1}\sigma d\boldsymbol{B} \tag{8.66}$$

For reference, EM-simulated solutions of Eq. (8.65) with a very small time step, $\Delta = 0.0001$s (and a large ensemble size, $N_p = 5000$) is considered (dotted line in Fig. 8.13). For the weak correction, however, the EM-based simulations are performed with $\Delta = 0.05$s and $N_p = 100$. Noise intensity is uniformly taken as $\sigma = 0.1$. As observed from Fig. 8.13, the corrected solution (solid line) offers a close match with the reference and this may be contrasted with a somewhat poorer quality of the regular EM solutions (dashed line).

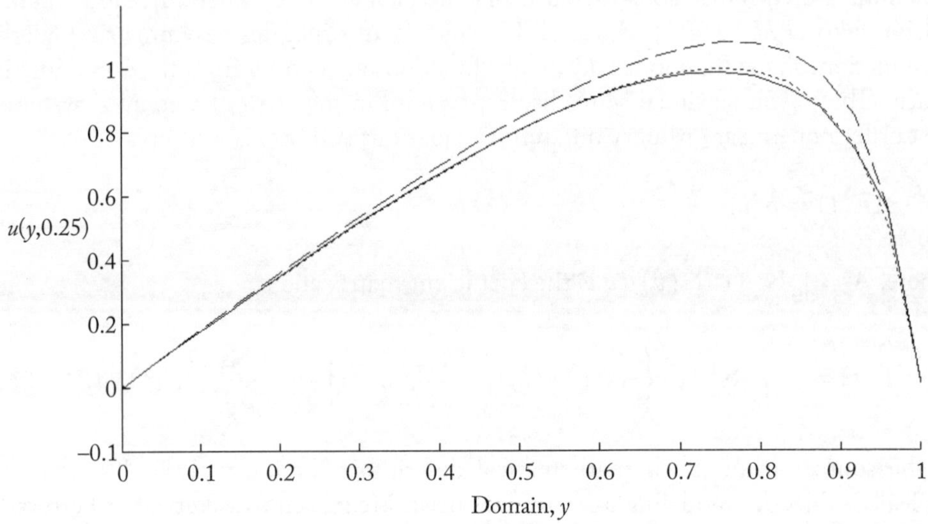

Fig. 8.13 1-D Burger's equation in Example 8.5: $\nu = 0.01$; sample mean solution profile at time $t = 0.25$ s, dotted-line—reference solution, dashed-line—EM solution without correction and solid-line—EM solution with correction by GCEM method

8.5 Concluding Remarks

We have described a plan to improve the numerical accuracy of, particularly, lower order explicit / semi-analytical integration schemes for SDEs using a change of measures so as to enable accurate weak simulations of the associated diffusion processes, especially for larger time step sizes. These schemes can be looked upon as improved versions of the lower order integration schemes presented in Chapter 5. The first one—the GCLM—an improvisation over the LTL scheme, is an MC approach for weakly correcting the linearization errors. It uses a combination of rejection sampling and re-sampling over an ensemble of locally linearized trajectories of additively driven non-linear oscillators. However, the requirement that the integrand in the canonical Radon–Nikodym derivative, which depends on the linearization error, must be bounded by the inverse of the time step imposes a restriction on the adoption of arbitrarily large step-sizes. Another scheme, the GCEM, is a weak correction scheme fashioned after the stochastic filtering strategy described in Chapters 6 and 7. It is general enough and offers the possibility of improved solutions as well as freedom from the necessity of very small time step sizes. Though used here to correct the EM scheme, the methodology does admit possible application with several other numerical integration schemes.

Having grasped the vital role that the notion of change of measure may play in deriving accurate numerical integrators for SDEs, we proceed to the last couple of chapters of this book that deal with schemes for stochastic optimization, emphasizing in particular a few recent ones that continue to exploit change of measures.

Notations

$\boldsymbol{A}(t,\boldsymbol{X})$	$:= \{A_j(t,\boldsymbol{X}), j = 1,2,\ldots,m\}$, a non-linear drift vector
$\boldsymbol{A}_c \in \mathbb{R}^{2m\times 2m}$	a matrix
$\boldsymbol{A}_l(t,\boldsymbol{X})$	linear part of the drift vector field $\boldsymbol{A}$
$\boldsymbol{A}_{nl}(t,\boldsymbol{X})$	non-linear part the drift vector field $\boldsymbol{A}$
$\tilde{\boldsymbol{B}}_t$	standard zero mean Q-Brownian motion
$C,\ C_1,\ C_2,\ C_c$	damping coefficients
$\boldsymbol{C} \in \mathbb{R}^{m\times m}$	damping matrix
$\mathscr{f} \in \mathbb{R}^m$	deterministic force vector
$\mathscr{f}_c(t)$	contact force in Example 8.4
$g(t,\boldsymbol{X})$	real-valued function
$\mathscr{g}$	gap between the support and the body in Example 8.4
$K,\ K_1,\ K_2,\ K_3,\ K_c$	stiffness coeffcients
$\boldsymbol{K} \in \mathbb{R}^{m\times m}$	stiffness matrix
m	dimension of the state variables, $\boldsymbol{X}_1, \boldsymbol{X}_2$
n_d	dimension of the semi-discrete system (Example 8.5)

$\aleph(y)$	mesh-free shape functions
$\mathcal{P}$	amplitude of the harmonic function (Eq. 8.52)
Q	probability measure
$u(t,y)$	field variable in Example 8.5
$X_{1j}, X_{2j}, j = 1,2,\ldots,m$	vector of state variables (stochastic processes)
$\boldsymbol{X}$	$:= \{X_{1j}, X_{2j}, j = 1,2,\ldots,m\}^T \in \mathbb{R}^{2m}$, a vector of state variables in Equation (8.1)
$\widetilde{\boldsymbol{X}}_i$	linearized solution to Eq. (8.1)
y	space variable in Example 8.5
α	non-linearity parameter
$\beta_k(t,\boldsymbol{X})$	function of $\boldsymbol{X}$
$\mathcal{E}_t \in \mathbb{R}^n$	vector valued error process (Eq. 8.43)
$\varepsilon_{nl}(t,.,.)$	linearization error
Δ	time step size
λ	frequency parameter
ν	viscosity parameter in Example 8.5
$\hat{\pi}_i(.) = \frac{1}{N_p}\sum_{j=1}^{N_p}(.)^{[j]}$	ensemble mean operator
$\phi(t,t_{i-1})$	$= \exp\{[A](t-t_{i-1})\}$, fundamental solution matrix (FSM)
$\Gamma_r, \Xi_r,$	events
$\boldsymbol{\varrho}$	$n \times n$ diffusion matrix

Chapter 9

Evolutionary Global Optimization via Change of Measures: A Martingale Route

9.1 Introduction

The efficacy of the concept of change of measures was demonstrated in the last few chapters in the context of non-linear stochastic filtering—a tool that also has considerable scientific usefulness in developing numerical schemes for system identification problems. This chapter also concerns an application of the same notion leading to a paradigm [Sarkar *et al.* 2014] on global optimization problems, wherein solutions are guided mainly through derivative-free directional information computable from the sample statistical moments of the design (state) variables within a MC setup. Before the ideas on this approach are presented in some detail, it is advisable to first focus on some of the available methodologies/strategies for solving such optimization problems.

In most cases of practical interest, the cost or objective functional, whose extremization solves the optimization problem, could be non-convex, non-separable and even non-smooth. Here separability means that the cost function can be additively split in terms of the component functions and the optimization problem may actually be split into a set of sub-problems. An optimization problem is convex if it involves minimization of a convex function (or maximization of a concave function) where the admissible state variables are in a convex set. For a convex problem, a fundamental result is that a locally optimal solution is also globally optimal. The classical methods [Fletcher and Reeves 1964, Fox 1971, Rao 2009] that mostly use directional derivatives are particularly useful in solving convex problems (Fig. 9.1). Non-convex problems, on the other hand, may have many local optima, and choosing the best one (i.e., the global extremum) could be an extremely hard task. In global optimization, we seek, in the design or state or parameter space, the extremal locations of nonconvex functions subject to (possibly) nonconvex constraints. Here the objective functional could be multivariate, multimodal and even non-differentiable, which together precludes applying a gradient-based Newton–step whilst solving the optimization problem.

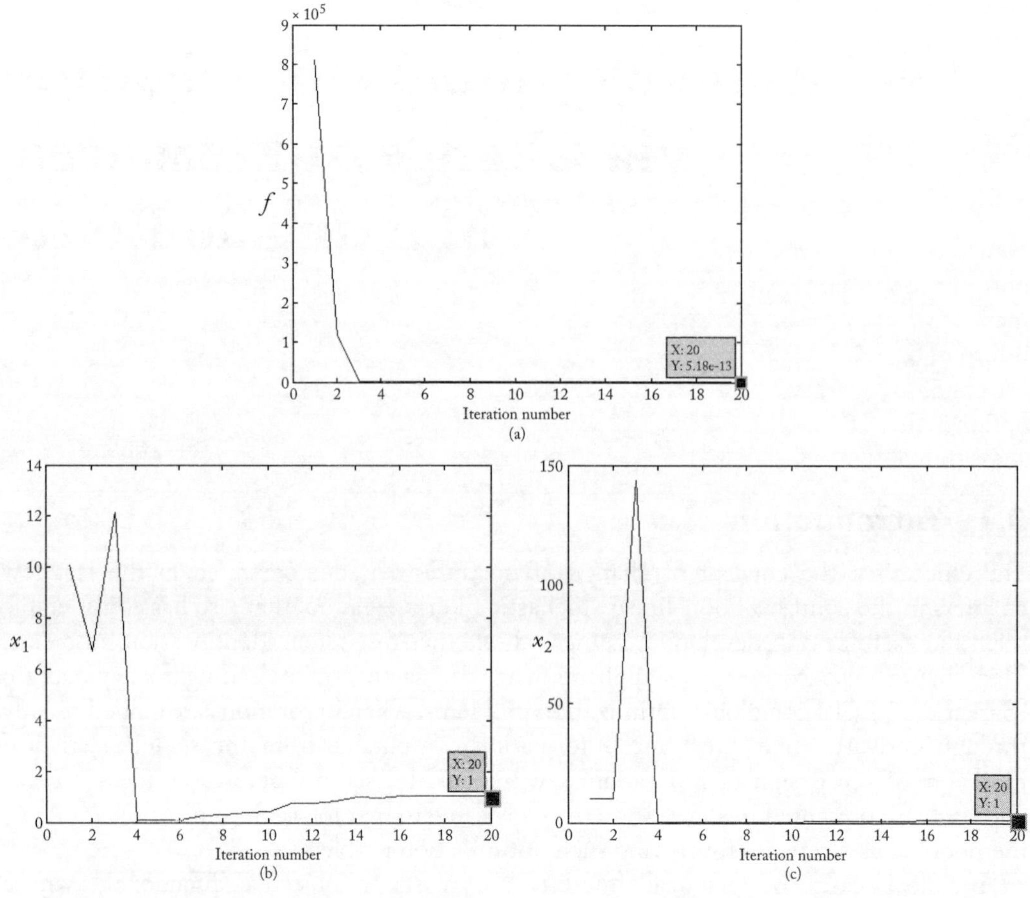

Fig. 9.1 Classical method of optimization: Davidon–Fletcher-Powell method [Davidon 1959, Fletcher and Powell 1963]; Rosenbrock function: $f(x_1,x_2) = 100\left(x_1^2 - x_2\right)^2 + (x_1 - 1)^2$—a unimodal convex function; evolution of (a) cost function, (b) variable x_1, (c) variable x_2, optimum point: $x_1 = 1, x_2 = 1$

Global Optimization

The need for global optimization is ubiquitous over a broad range of disciplines, from dynamical systems modeling and design in science and engineering to drug design and delivery. Most popularly available schemes for global optimization are essentially based on a stochastic search of the design space using diffusive noise processes, with applications ranging from VLSI circuit design [Edwards *et al.* 1999] to optimization of transportation routes [Powell 1986] to determination of protein structures [Zou *et al.* 1997, More and Wu 1999] in computational biology. In these problems, a possible manifestation of nonlinearity could be in the form of non-separability of the system parameters or the design variables. In fact, the filtering problems discussed in the last chapter could be looked upon as a small subset of problems in optimization, where the aim was to arrive at the states/parameters that minimize a data-prediction misfit metric. In contrast to finding a local extremal point, attainment of the global extremum of the objective functional within the search space, the domain of definition of the design variables, may be further challenged by a large dimensionality of the search space.

The research opportunities created by the ineffectiveness or inapplicability of a Newton-type search have been the fertile ground for numerous proposals for global optimization, many of which employ stochastic (i.e., random evolutionary) search with meta-heuristic origin [Holland 1975]. Indeed, in the absence of a generic directional information (equivalent to the Gateaux or directional derivative in the search for a local extremum of a sufficiently smooth cost functional); a direct search appears to be the only option left. In this regard, evolutionary stochastic search techniques (e.g., the genetic algorithm (GA) [Goldberg 1989]) have proved very effective compared to their gradient-based deterministic counterparts or even the derivative-free deterministic search methods (e.g., the pattern search methods [Hooke and Jeeves 1961, Nelder and Mead 1965] – see Fig. 9.2). This is in fact true [Fletcher 1987, Chong and Zak 2004] even for cases involving sufficiently smooth, yet multimodal, objective/ cost functionals, wherein the use of directional derivatives may be inadequate in obtaining the global optimum.

The ability of a meta-heuristic method in attaining the global optimum within a finite time may only be numerically demonstrated in a problem-specific manner, even though it typically offers no guarantee of reaching the global solution. Some of the notable schemes in this genre include variants of the GA [Sotiropoulos *et al.* 1997], ant colony optimization [Dorigo and Birattari 2010], simulated annealing (SA) [Van Laarhoven and Aarts 1987], particle swarm optimization (PSO) [Kennedy and Eberhart 1995], differential evolution (DiEv) [Storn and Price 1997] and, arguably, the covariance matrix adaptation evolution strategy (CMA–ES) [Hansen and Ostermeier 1996] though the last scheme may not necessarily be branded as meta-heuristic. While a stochastic route often facilitates an effective search, the posing of the original problem could itself be deterministic, wherein the aim would merely be to arrive at the design point that is the global extremum of an (or a collection of) objective functional(s), possibly subject to a set of prescribed constraints. To get some idea about how these meta-heuristic evolutionary stochastic schemes work, a short essay on some of these evolutionary methods is provided below.

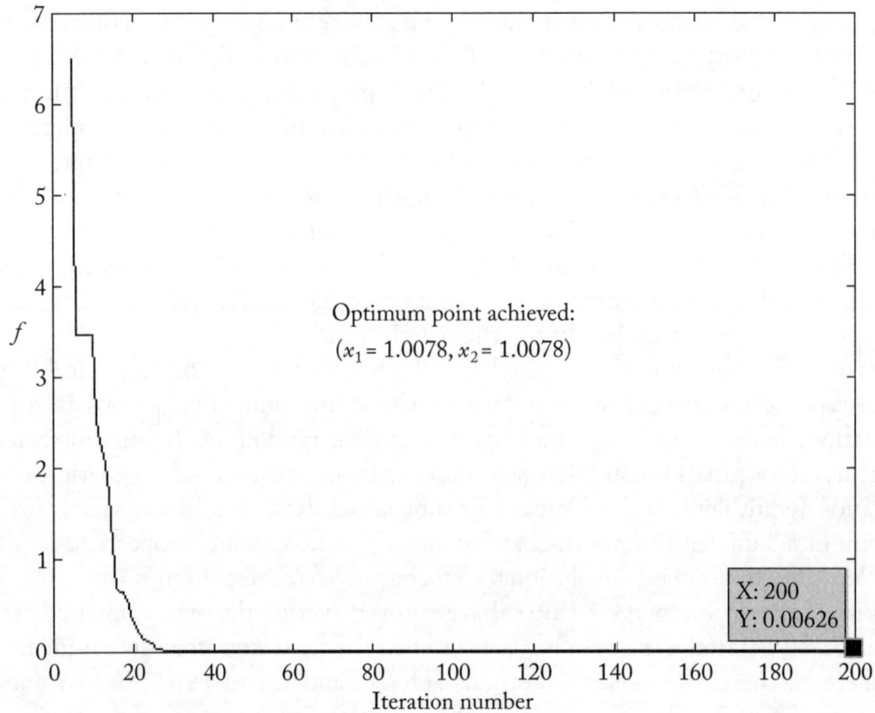

Fig. 9.2 Optimization by derivative-free deterministic search; Rosenbrock function: $f(x_1, x_2) = 100\left(x_1^2 - x_2\right)^2 + (x_1 - 1)^2$, solution by pattern search method [Hookes and Jeeves 1961]

Genetic Algorithm

The genetic algorithm (GA) [Holland 1975; Goldberg 1989] is an evolutionary global optimization scheme that is meant to closely mimic the genetic evolution of a species. That is to say, it is supposed to follow the biological processes with the ability to get adapted to the environment in the consecutive generations, where a notion of competitiveness in this environment is provided by the objective functional. The adaptation process is mainly accomplished in the form of genetic inheritance from parents to offspring and through the so-called survival of the fittest. In the GA, each element (member) in the population is called a chromosome, which is constructed using design variables, herein referred to as genes (Table 9.1). For the subsequent update, chromosomes are assigned a measure of competitiveness or goodness in terms of their fitness based on the objective functional. Like most evolutionary schemes, GA also relies upon *exploration–exploitation trade-off* which is implemented using three operators: crossover, mutation and selection. While crossover and mutation impart variations in the chromosomes, the selection operator is used to choose chromosomes for the subsequent population. Chromosomes with higher fitness values have higher chances of being selected than those with lower fitness values. For crossover, pairs of parents, which participate in mating to produce offspring, are chosen based on some probability. To impart additional layers of variation in the population, the mutation operator is applied to a randomly chosen chromosome to alter one or more of its genes (again based on a specified probability) (Table 9.1).

Table 9.1 GA scheme; crossover and mutation operations

Consider a 2-dimensional optimization problem, i.e., the number of design variables or parameters, $m = 2$. Initialize N_p, the population size.

Suppose $\boldsymbol{P}_i = \left(X_1^{[i]}, X_2^{[i]}\right)$ and $\boldsymbol{P}_j = \left(X_1^{[j]}, X_2^{[j]}\right) i, j \in \left(1, N_p\right)$, are two parent members (chromosomes) of the population and they are represented in binary form (with each parameter by, say, an 8-bit string).

i^{th} chromozome - $\boldsymbol{P}_i$: $\underbrace{1\ 0\ 0\ 0\ 1\ 0\ 0\ 1}_{X_1^{[i]}}\ \ \underbrace{0\ 1\ 1\ 1\ 1\ 1\ 1\ 0}_{X_2^{[i]}}$

j^{th} chromozome - $\boldsymbol{P}_j$: $\underbrace{0\ 1\ 0\ 0\ 0\ 1\ 0\ 0}_{X_1^{[j]}}\ \ \underbrace{1\ 1\ 1\ 0\ 0\ 1\ 1\ 1}_{X_2^{[j]}}$

Each bit represents a gene in the chromosome.

(a) single point crossover operation:

$\boldsymbol{P}_i$: 1 0 0 0 1 ¦ 0 0 1 0 1 1 1 1 1 1 0 parent 1

$\boldsymbol{P}_j$: 0 1 0 0 0 ¦ 1 0 0 1 1 1 0 0 1 1 1 parent 2

↓ set of genes selected for crossover

$\boldsymbol{C}_i$: 1 0 0 0 1 ¦ 1 0 0 1 1 1 0 0 1 1 1 offspring 1

$\boldsymbol{C}_j$: 0 1 0 0 0 ¦ 0 0 1 0 1 1 1 1 1 1 0 offspring 2

(b) multi point crossover operation:

$\boldsymbol{P}_i$: 1 0 0 0 1 ¦ 0 0 1 0 ¦ 1 1 1 ¦ 1 1 1 0 parent 1

$\boldsymbol{P}_j$: 0 1 0 0 0 ¦ 1 0 0 1 ¦ 1 1 0 ¦ 0 1 1 1 parent 2

↓ sets of genes selected for crossover

$\boldsymbol{C}_i$: 1 0 0 0 1 ¦ 1 0 0 1 ¦ 1 1 1 ¦ 0 1 1 1 offspring 1

$\boldsymbol{C}_j$: 0 1 0 0 0 ¦ 0 0 1 0 ¦ 1 1 0 ¦ 1 1 1 0 offspring 2

(c) mutation operation (assuming single point crossover)

C_i:	1	0	0	0	1	1	0	0	1	1	1	0	0	1	1	1	offspring 1
C_j:	0	1	0	0	0	0	0	1	0	1	1	1	1	1	1	0	offspring 2
↓												↑ gene selected randomly for mutation					
C_i:	1	0	0	0	1	1	0	0	1	1	1	1	0	1	1	1	offspring 1
C_j:	0	1	0	0	0	0	0	1	0	1	1	0	1	1	1	0	offspring 2

The selection operator that completes the basic structure of the GA is used to select the survived chromosomes, again based on some probability, for the next generation (see Fig. 9.3 for the optimization of the Rosenbrock function by the GA). If the value of the objective functional is known at the global extremum, then a stopping criterion may be based on the attainment of that value within a prescribed range. Otherwise, the algorithm must be stopped after a specified maximum number of iterations. Table 9.2 enumerates the steps defining the stochastic search scheme of the GA.

Table 9.2 Optimization by GA—the stochastic search algorithm

Step 1.	Generate the initial population The genes that code the design variables may be represented by Boolean, floating point/string variables or integers or a combination of these. The initial population is generated by randomly selecting the genes of each member (chromosome). Each chromosome stands for a candidate (realized) solution of the problem.
Step 2.	Find the fitness value (based on the objective functional) of each chromosome (or element) of the initial population.
Step 3.	Divide the population into sets of parents by a random combination among the members of the population.
Step 4.	Crossover operation: Get the offspring by exchanging the corresponding genes in the parent members by single point or multi-point crossover (see Table 9.1). The crossover points are randomly selected.
Step 5.	Mutation operation: Randomly change the value of a gene in each member of the offspring. The member and the gene to be changed are randomly selected.
Step 6.	Select the new generation (i.e., retain or reject the parent chromosomes) based on the fitness value of the offspring (new chromosomes).
Step 7.	Go to step 3 and iterate till the stopping criterion is satisfied.

For relatively better performance, there are variants of the GA [Srinivas and Patnaik 1994, Smith and Fogarty 1997] involving, for example, self-adaptation of the key parameters—cross-over and mutation rates. GA has been applied to a diverse range of optimization problems [Haupt 1995, Back 1996] including structural optimization [Goldberg and Samtani 1986, Doorly *et al.* 1996, Eby *et al.* 1999], control systems optimization [Krishnakumar and Goldberg 1992] and machine learning [Michalewicz 1996]. As a global optimization tool, GA is particularly well-suited for parallelism [Alba and Tomassini 2002] due to its capability of simultaneously exploring the search space in multiple directions.

Simulated Annealing

Annealing is the well-known metallurgical process of heating up a solid and then cooling it slowly until crystallization takes place. At very high temperatures, the atoms of the solid have higher energy and as the temperature is reduced in a controlled fashion, the energy reduces, until a state of minimum energy is achieved. In the context of optimization, the simulated annealing (SA) emulates this process. Based on this original idea of the SA, which dates back to the middle of the 20th century, Metropolis *et al.* [1953] introduced an algorithm to arrive at the equilibrium of a collection of atoms at a given temperature. This pioneering technique inspired Kirkpatrick and Vecchi [1983] to incorporate it as an optimization tool. SA begins at a very high temperature where the design variables are allowed to assume a wide range of variation. As the algorithm progresses, temperature is allowed to fall. This restricts the degree to which the variables are allowed to vary. This often leads the algorithm to a better solution, just as a metal achieves a 'better' crystal structure through the actual annealing process. Thus, as long as the temperature is decreasing, changes are produced yielding successively better solutions and finally giving rise to an optimum set of variables when the (absolute) temperature is close to zero.

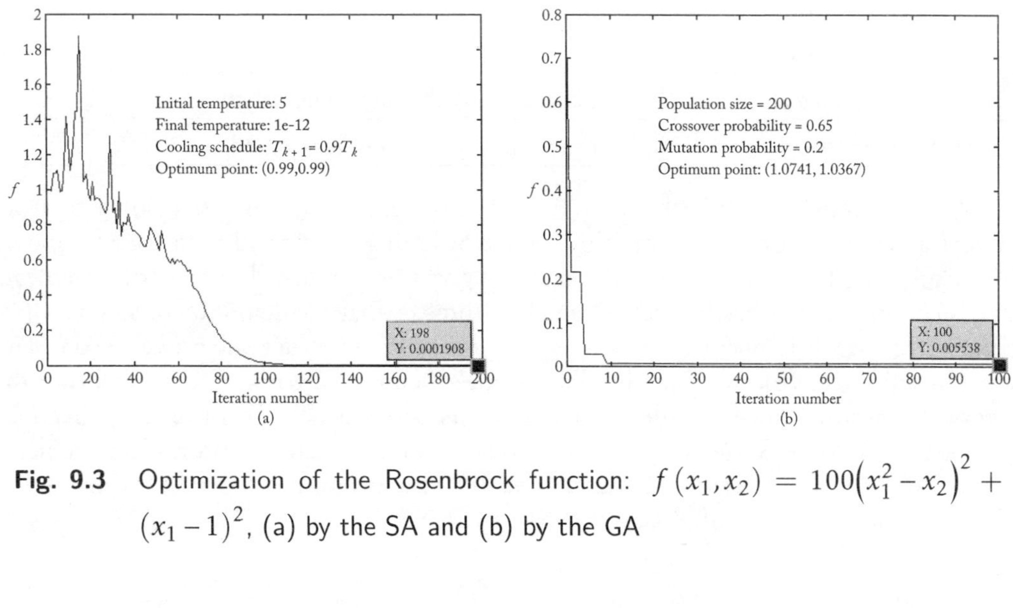

Fig. 9.3 Optimization of the Rosenbrock function: $f(x_1, x_2) = 100\left(x_1^2 - x_2\right)^2 + (x_1 - 1)^2$, (a) by the SA and (b) by the GA

The basic methodology of the SA algorithm is to generate at each iteration, a trial set in neighborhood of the current candidate. Next, it decides, based on an acceptance probability (that depends on the difference between their fitness values), whether the current candidate should be replaced by the trial set. In case of the SA, theoretically, an asymptotic convergence that is controlled by a so called cooling schedule to an optimal solution is guaranteed. The main control parameter in the cooling schedule is the temperature T. The role of T is to keep the probability of accepting a new move high in the earlier stages of iterations and force it towards zero asymptotically. In particular, the initial temperature is generally chosen sufficiently high so as to avoid a local minimum but small enough not to move off the global minimum. A proper cooling schedule (typically based on a logarithmic reduction) is needed for the SA algorithm to attain the convergence to the desired optimum in finite time. Table 9.3 details the relevant steps in the SA algorithm and Fig. 9.3 shows results by both SA and GA on Rosenbrock function.

Table 9.3 The SA algorithm

Step 1.	Initialize the current set of design variables, specify the initial temperature, the annealing (cooling) schedule and the acceptance probability criterion.
Step 2.	Calculate the current cost (fitness value), $f^{[i]}$.
Step 3.	Move to a new set of design variables and calculate the new cost, $f^{[j]}$.
Step 4.	Based on the change in the cost due to the new move, accept or reject the new variables; (i) accept the new cost and replace the current set of variables by the new one if $f^{[j]} - f^{[i]} \leq 0$ or $\exp\left(\frac{f^{[j]}-f^{[i]}}{T}\right) > U(0,1)$, a random number from a uniform distribution, (ii) reject otherwise.
Step 5.	Reduce the temperature according to the annealing schedule.
Step 6.	Repeat steps $3-5$ until the stopping criterion is reached.

As observed from Table 9.3, the main feature of the SA algorithm is the ability to avoid being trapped in a local minimum. This is done by letting the algorithm to accept not only better solutions but also worse solutions with a given probability. The main disadvantage, that is common in stochastic local search algorithms, is that the definition of some control parameters (initial temperature, cooling rate, etc.) is somewhat subjective [Wong and Constantinides 1998] and must be done from an empirical basis. This means that the algorithm must be tuned in order to maximize its performance. The diverse applications [Chibante 2010] of SA include inverse problems or parameter identification problems [Silva Neto and Özişik 1994; Souza *et al.*, 2007] and optimal control systems design [Grimble and Johnson 1988, Ogata 1995].

Particle Swarm Optimization (PSO)

The basic structure in these and many such other meta-heuristic optimization schemes typically lacks scientific rigor, grounded as it is in intuitive reasoning drawn from social or biological observations. Particle swarm optimization (PSO), originally proposed by Kennedy and Eberhart [1995], is another global optimization scheme that mimics the behavior of social organisms, e.g., a swarm of insects such as ants, termites, bees and wasps, a flock of birds, school of fish, etc. Each candidate, denoting a bee in a colony or a bird in a flock etc., moves in a generally random way, guided only by its own intelligence and the collective or group intelligence of the swarm. For example, if a candidate finds a good path to food, the rest of the swarm will also be able to follow the path instantly even if their location is far away in the swarm. In the PSO, each candidate, whose motion is characterized by position and velocity, wanders around in the design space and remembers the best position it has discovered. The particles (i.e., the candidates) communicate information of good positions to one another and adjust their individual positions and velocities accordingly. Note that optimization methods based on swarm intelligence are called behaviorally (socially) inspired algorithms in contrast with the GA schemes, which are biologically inspired approaches.

A striking feature of the PSO is that the search space is explored by a combination of the swarm's previous best and the individuals' previous best positions. At the heart of the flock or swarm behavior are three driving factors (which may, in part, contradict each other—a reflection of the exploration–exploitation trade-off): (1) cohesion- stick together, (2) adhesion- do not come too close, and (3) alignment- follow the general heading of the flock. To summarize, the PSO is derived following the model below.

(1) When a bird locates a target or food (i.e., an available extremum of the objective functional), it instantaneously shares the information with all other birds (non-local interaction across space-separated particles).

(2) All other birds tend to come towards the target or food.

(3) However, in doing so, each bird exercises its own intelligence consistent with its past memory.

Thus the model performs a random search, restricted by the conditions above, in the design space for the extremum of the objective functional, which, with progressing iterations, should approach the global extremum. Table 9.4 briefly highlights the main features of the algorithm.

c_1 and c_2 are also known as the cognitive and social parameters, respectively. Fine tuning of these parameters and the inertia weight w improves the performance of the PSO [Clerc and Kennedy 2002]. While a larger value of the inertia weight may help in global exploration initially, a smaller value applies to local exploration (e.g., near the final stages when much of the design space is already explored). This parameter thus has functional similarities with the temperature parameter in the SA.

Table 9.4 PSO—the algorithm

	Initialize an m-dimensional search space, and generate N_p particles (N_p realizations on sets of m design variables). The jth particle is defined by the m-dimensional position and velocity vectors denoted by $\boldsymbol{X}^{[j]} = \left(X_1^{[j]}, X_2^{[j]}, \ldots, X_m^{[j]}\right)^T$ and $\boldsymbol{V}^{[j]} = \left(V_1^{[j]}, V_2^{[j]}, \ldots, V_m^{[j]}\right)^T$, $j = 1, 2, \ldots, N_p$, respectively. Evaluate the cost function for the N_p particles and start the iteration counter that varies from 1 to $k_{\max}$
Step 1.	At any iteration, say $k \in (1, k_{\max})$, pick $\boldsymbol{p}_{\text{best}}^{[j]}$, the best position vector of the jth particle and pick $\boldsymbol{g}_{\text{best}}$, the best position vector of the swarm over all the past iterations, i.e., 1 to $k-1$.
Step 2.	Update the particles as: $\boldsymbol{V}^{[j]}(k+1) = w\boldsymbol{V}^{[j]}(k) + c_1 r_1 \left(\boldsymbol{p}_{\text{best}}^{[j]} - \boldsymbol{X}^{[j]}\right) + c_2 r_2 \left(\boldsymbol{g}_{\text{best}} - \boldsymbol{X}^{[j]}\right)$ (9.1a) $\boldsymbol{X}^{[j]}(k+1) = \boldsymbol{X}^{[j]}(k) + \boldsymbol{V}^{[j]}(k+1)$ (9.1b)
	$w \in \mathbb{R}^+$– inertia weight (or weight parameter on the previous velocity of the particle), r_1 and r_2-random numbers uniformly distributed in $[0,1]$ and c_1 and c_2 - learning factors of a particle—the first from the knowledge of its own success $\boldsymbol{p}_{\text{best}}^{[j]}$ and the second from that of the best position $\boldsymbol{g}_{\text{best}}$ of the swarm.
Step 3.	Repeat steps 1 and 2 till a stopping criterion is satisfied.

The PSO has been extensively used [Engelbrecht 2006] for many scientific and engineering applications. Improved versions of the scheme [Jiao *et al.* 2008, Hu *et al.* 2012] include those having an adaptivity to change w and c_1 and c_2 for better convergence of the scheme.

Example 9.1

We utilize the PSO scheme for a system identification problem (herein posed deterministically unlike a stochastic filtering problem). We identify the parameters of a Van der Pol oscillator (see Example 5.3 of Chapter 5 with no process noise) by posing it as an optimization problem. The governing DE for the oscillator may be written as:

$$\ddot{X} + c\left(X^2 - 1\right)\dot{X} + kX = 0 \tag{9.2}$$

c and k are the system parameters to be identified. Assume that the ideal system response under an assumed initial condition is known (implying no observation noise). To obtain this reference solution, the values for the parameters c and k are taken as 4 and 10, respectively.

Solution

With the state vector, $\boldsymbol{X} = \left(X_1 = X, X_2 = \dot{X}_1 = \dot{X}\right)^T$, the process Eq. (9.2) may be expressed in the state space form:

$$\frac{dX_1}{dt} = X_2 \tag{9.3a}$$

$$\frac{dX_2}{dt} = -c\left({X_1}^2 - 1\right)X_2 - kX_1 \tag{9.3b}$$

Let $\boldsymbol{\theta} := (c,k)^T$ be the vector of system parameters to be identified. Under a time discrete setting, $\{t_0,\ldots,t_N\}$, the objective functional may be defined as the sum of the squared error between the reference (observed) system response and the currently computed one:

$$f\left(\boldsymbol{\theta}^a\right) = \sum_{i=0}^{N} \left(\boldsymbol{X}_i - \boldsymbol{X}_i^a\right)^T \left(\boldsymbol{X}_i - \boldsymbol{X}_i^a\right) \tag{9.4}$$

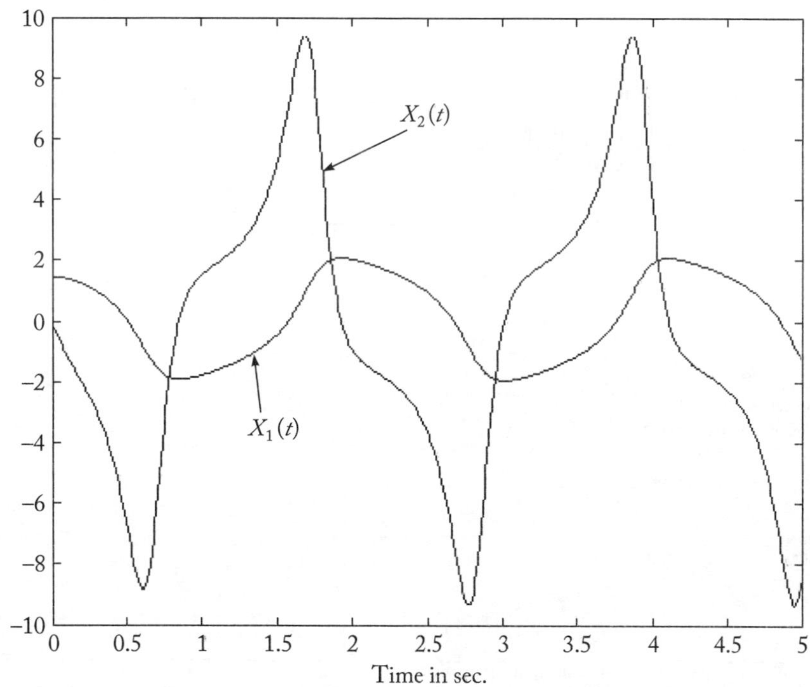

Fig. 9.4 Van der Pol oscillator Eq. (9.2), reference (observed) solution $\{X_1(t), X_2(t), t \in (0,\ 5\ \text{s})\}$ with $c = 4$, $k = 10$ and time step $\Delta t = 0.01$ s

where, $\boldsymbol{X}_i^a = (X_{1,i}^a, X_{2,i}^a)^T, i = 1,2,..N$ is the current solution to Eq. (9.2), when $\boldsymbol{\theta}$ is replaced by the current parameter vector $\boldsymbol{\theta}^a$. The reference time histories of displacement, $X_1(t)$ and velocity, $X_2(t)$ obtained by numerical integration of Eq. (9.2) using $c = 4$ and $k = 10$ and with the initial condition ($X_1(0) = 1.4$, $X_2(0) = 0$) are shown in Fig. 9.4.

The initial guesses of the parameters c and k are sampled from a uniform distribution with the constraint $(1,10)^T \le \boldsymbol{\theta}^a = (\theta_1^a = c, \theta_2^a = k)^T \le (10,20)^T$. In this example, the parameter or design space is 2-dimensional, i.e., $m = 2$ and the number of particles $N_p = 100$. We take the cognitive and social parameters as $c_1 = c_2 = 2$. The inertia weight w is reduced from $w_{\max} = 0.9$ to $w_{\min} = 0.2$ with increasing iteration k such that $w_k = w_{\max} - \frac{k}{k_{\max}}(w_{\max} - w_{\min})$. $k_{\max}$, the maximum number of iterations, is taken to be 500. Numerical results of optimization by the PSO are shown in Figs 9.5 and 9.6. Results using the GA and SA are also included in the figures.

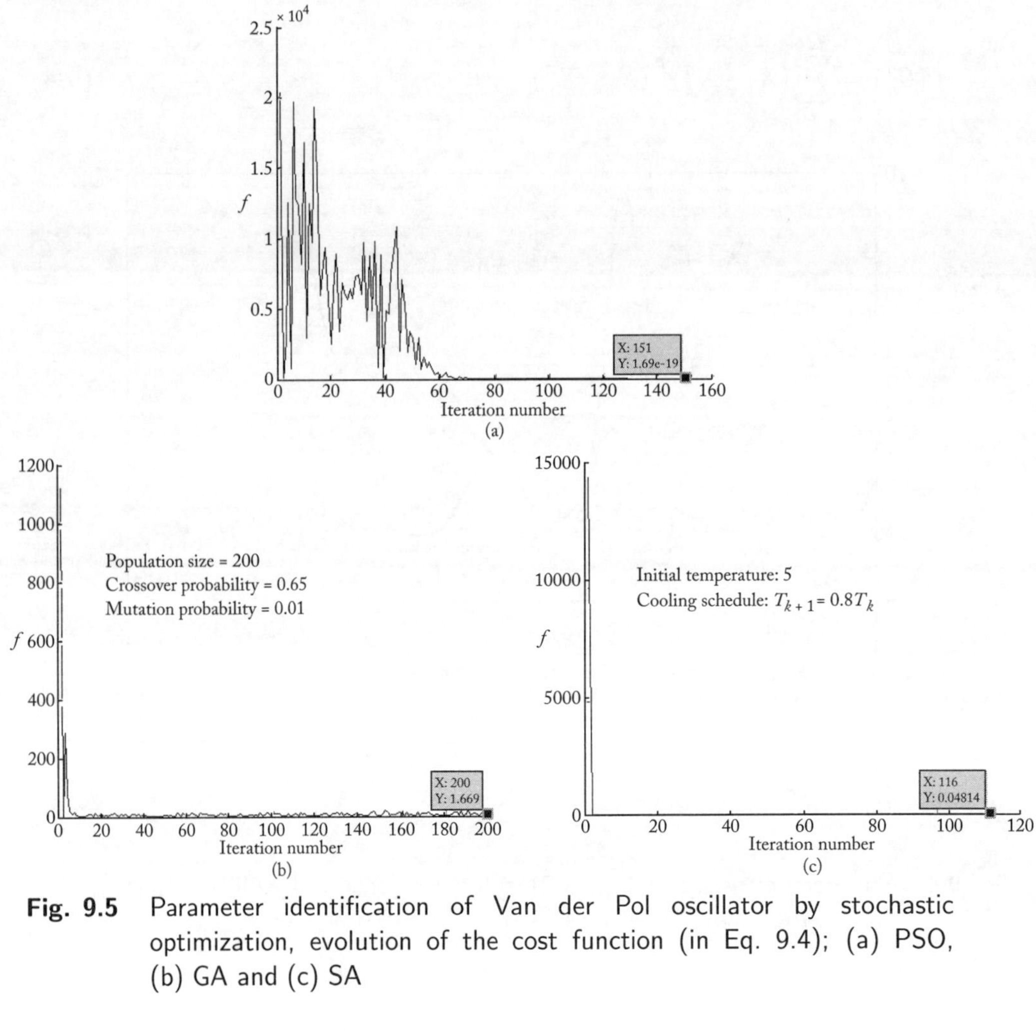

Fig. 9.5 Parameter identification of Van der Pol oscillator by stochastic optimization, evolution of the cost function (in Eq. 9.4); (a) PSO, (b) GA and (c) SA

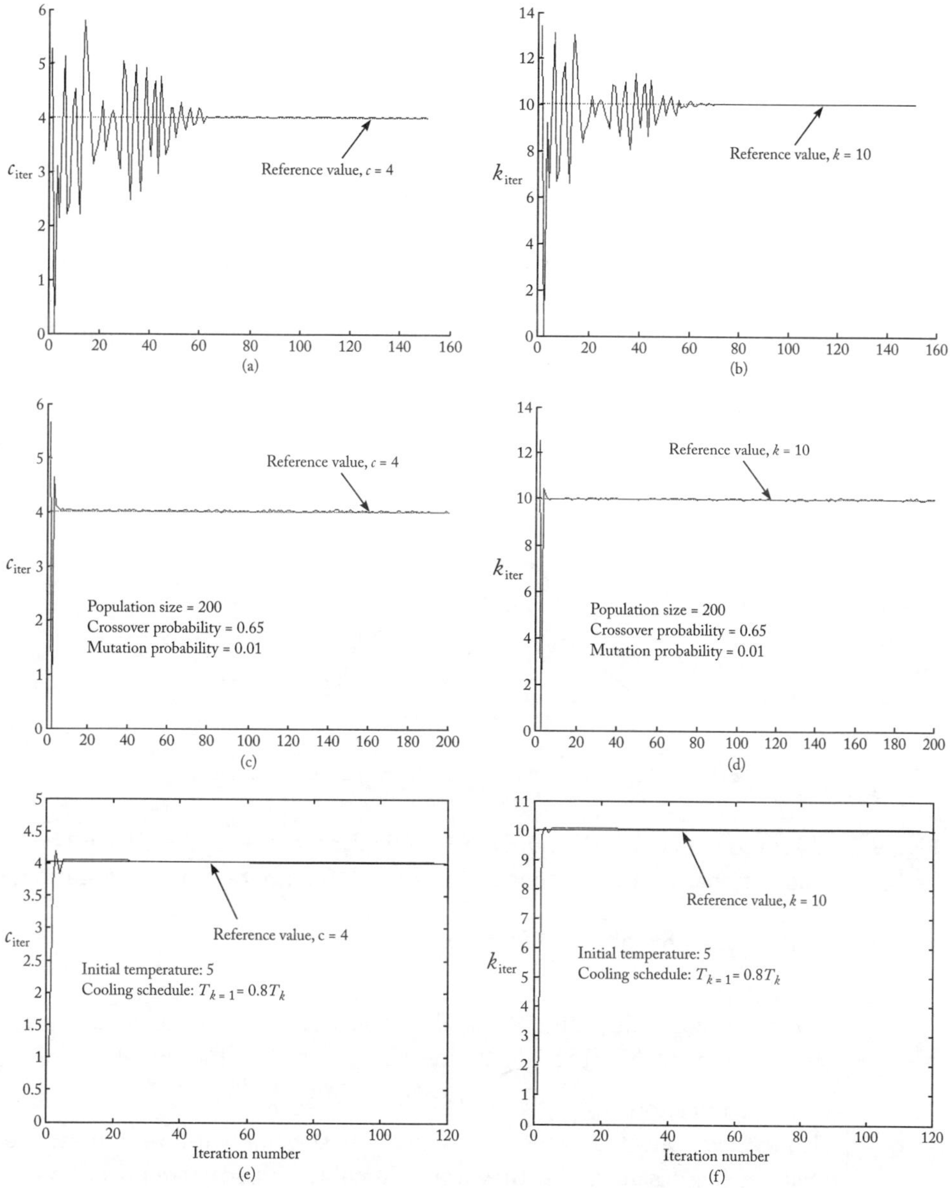

Fig. 9.6 Parameter identification of Van der Pol oscillator by stochastic optimization, evolution of the parameters c and k; (a) and (b) PSO, (c) and (d) GA, and (e) and (f) SA

Differential Evolution

Differential evolution (DiEv) [Storn and Price 1997] is another evolutionary optimization strategy, which relies on parallel direct search, working with a randomly chosen set of candidates. Like the GA, the DiEv is also characterized by strategic operations that are similar to mutation, cross-over and selection. The DiEv generates new candidates by mutation which, in the context of this method, is defined as adding the weighted difference between two randomly chosen particles to a randomly chosen third one. Figure 9.7 illustrates the mutation operation. The mutated vector's elements are then mixed with those of another randomly chosen vector, the target solution, to yield the so-called trial solution. Such a mixing operation is similar to the crossover used in the GA. The trial solution is accepted for the next population if it improves the fitness value over the previous one. This step is referred to as selection. To ensure that all the available particles take part in the exploration, each particle is forced to assume the role of the target solution once (the random choice of the target is thus in the set of particles that have not yet been chosen as targets). Table 9.5 details the salient features of the algorithm.

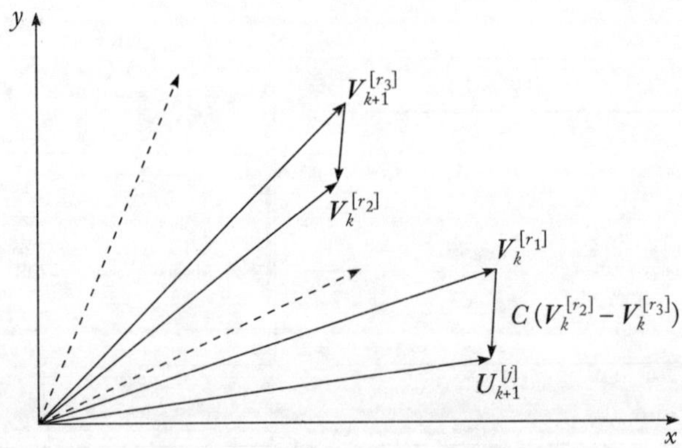

Fig. 9.7 Mutation operation in DiEv at the end of the kth iteration in a 2-dimensional parameter space; (see Table 9.5 for details on the notation)

Table 9.5 Salient features of the DiEv algorithm

	Given the cost function $f(.)$ and a specified m-dimensional parameter space, initialize N_p, the population size. Generate the initial set of N_p parameter vectors, $\mathbf{V}_0^{[j]}$, $j = 1,2,\ldots,N_p$, each of size m (the vector elements are selected randomly so that whilst being within the prescribed limits of the design space, they are reasonably well scattered over it). Start iterations, $k = 1,2,\ldots,k_{\max}$.
Step 1.	Generate the new set of N_p vectors by the mutation operation: $$\mathbf{U}_{k+1}^{[j]} = \mathbf{V}_k^{[r_1]} + C\left(\mathbf{V}_k^{[r_2]} - \mathbf{V}_k^{[r_3]}\right), \; j = 1,\, 2,\ldots,N_p \qquad (9.5)$$ r_1, r_2 and r_3 (different from each other and the running index j) are random integers drawn from a discrete uniform distribution on $[1{:}N_p]$. C is a scalar constant.

Step 2.	Perform the cross-over operation: $$\mathcal{T}^{[j]}_{r,(k+1)} = U^{[j]}_{r,(k+1)}\ if\, p_r \le q \text{ or } r = \varsigma_{\text{rand}}$$ $$= V^{[j]}_{r,k} \text{ otherwise, } j = 1,2,\ldots,N_p,\ r = 1,2,\ldots,m \quad (9.6)$$ where, p_r is a random number uniformly distributed over $[0,1]$ and q a user-specified real number in the interval $[0,1]$. ς_{rand} is a randomly chosen integer among $(1,m)$.
Step 3.	The selection operation: the final vectors at the $(k+1)^{\text{th}}$ iteration are determined as: $$\boldsymbol{V}^{[j]}_{k+1} = \mathcal{T}^{[j]}_{k+1} \text{ if } f(\mathcal{T}^{[j]}_{k+1}) < f(\boldsymbol{V}^{[j]}_{k})$$ $$\boldsymbol{V}^{[j]}_{k+1} = \boldsymbol{V}^{[j]}_{k} \text{ otherwise} \quad (9.7)$$
Step 4.	Repeat steps $1-3$ till a stopping criterion is satisfied.

Other, possibly more efficacious, variants of the DiEv have been reported [Wang *et al.* 2011] that differ from the basic version of Table 9.5 by way of how the mutation and cross-over operations are performed. The basic version may generally be denoted by DiEv/rand/1/bin. '/rand/' indicates that $\boldsymbol{V}^{[r_1]}_{k}$ in Eq. (9.5) is randomly selected for performing the mutation. '/1/' indicates the number of difference vectors used during the mutation. In Eq. (9.5), only one difference vector $\left(\boldsymbol{V}^{[r_2]}_{k} - \boldsymbol{V}^{[r_3]}_{k}\right)$ is used. '/bin/' indicates that independent binomial experiments decide the crossover in Eq. (9.6). A typical variant may be DiEv/best/2/bin wherein the mutation operation is executed over the best population vector, i.e.:

$$\boldsymbol{U}^{[j]}_{k+1} = \boldsymbol{V}^{[best](k)} + C\left(\boldsymbol{V}^{[r_1]}_{k} - \boldsymbol{V}^{[r_2]}_{k}\right) + D\left(\boldsymbol{V}^{[r_3]}_{k} - \boldsymbol{V}^{[r_4]}_{k}\right),\ j = 1,2,\ldots,N_p \quad (9.8)$$

and two difference vectors are involved in the mutation operation. $\boldsymbol{V}^{[\text{best}]}_{k}$ is the best population vector (based on the fitness value) at the kth iteration among the vectors $\boldsymbol{V}^{[j]}_{k}$, $j = 1,2,\ldots,N_p$. While the choice of the parameters, N_p, C and q is problem dependent, a possible guidance may be provided as $5m \le N_p \le 10m$, $C = 0.5$ and $q = 0.1$. Different population vector strategies and control parameter settings with a possible adaptivity have been studied [Fan and Lampinen 2003, Brest *et al.* 2006, Mallipeddi and Suganthan 2008] for gauging and improving the performance of DiEv with respect to many test cost functions that include multi-modal and hybrid composite functions.

Example 9.2

We consider the same system identification problem defined in Example 9.1. We identify the damping and stiffness parameters c and k of the Van der Pol oscillator by the DiEv. The system response (Fig. 9.4) under a specified initial condition ($X(0) = 1.4$ and $\dot{X}(0) = 0$) is assumed to be known. With reference values for the parameters c and k takes as 4 and 10, respectively, the optimization results by the DiEv are shown in Fig. 9.8.

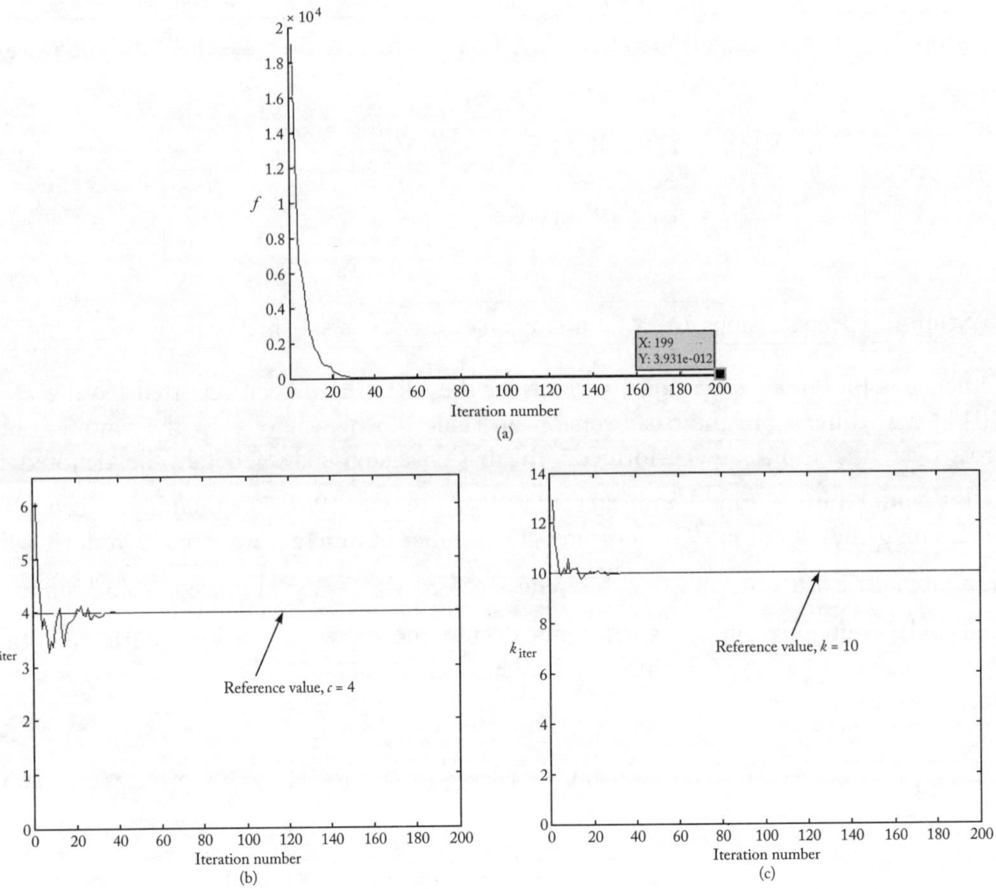

Fig. 9.8 Parameter identification of Van der Pol oscillator by DiEv; $N_p = 20$, control parameters: $C = 2$ and $q = 0.1$; evolutions of (a) cost function f, (b) damping parameter c and (c) stiffness parameter k

Covariance matrix adaptation evolution strategy (CMA–ES)

The CMA–ES, yet another evolutionary optimization strategy, is a derivative-free, robust local search performer [Hansen and Ostermeier 1996, Auger and Hansen 2005a and 2005b, Hansen 2007]. In the standard CMA–ES, the basic idea is to generate particles at each iteration k by sampling a multivariate normal distribution, $\mathcal{N}(\mathfrak{m}_k, \sigma_k)$, where, $\mathfrak{m}_k \in \mathbb{R}^m$ is the vector of mean values and σ_k^2, the covariance matrix of the vector, $\boldsymbol{X}_k^{[j]} \in \mathbb{R}^m$, $j = 1, 2, ..N_p$ of design variables (parameters). At each iteration, the scheme updates the particles as:

$$\boldsymbol{X}_{k+1} \sim \mathfrak{m}_k + s_k \mathcal{N}(\boldsymbol{0}, \mathcal{C}_k^{\frac{1}{2}}) \tag{9.9}$$

$\mathcal{C}_k = \sigma_k^2 \in \mathbb{R}^{m \times m}$ and s_k is known as an overall standard deviation or the step size at the iteration k. $\boldsymbol{0} \in \mathbb{R}^m$ is the zero–mean vector. Eq. (9.9) is equivalent to sampling $\boldsymbol{X}_{k+1} \sim \mathcal{N}(\mathfrak{m}_k, s_k \mathcal{C}_k^{\frac{1}{2}})$. Starting with an initial population $\boldsymbol{X}_0^{[j]} \sim U(\boldsymbol{a}, \boldsymbol{b})$, $j = 1, 2, \ldots, N_p$ where $\boldsymbol{a}$ and $\boldsymbol{b}$ are the specified limits defined on the m-dimensional parameters space, the goal is to evaluate $\mathfrak{m}_k$ and $\mathcal{C}_k$ and to fix the step size, s_k. To this end, the sampled particles $\boldsymbol{X}_k^{[j]}$, $j = 1.2, ..N_p$ are first ranked from below according to their fitness, i.e., $f\left(\boldsymbol{X}_k^{[1:N_p]}\right) \leq f\left(\boldsymbol{X}_k^{[2:N_p]}\right) \leq \ldots \leq f\left(\boldsymbol{X}_k^{[N_p:N_p]}\right)$ where $\boldsymbol{X}_k^{[l:N_p]}, l = 1, 2 \ldots, N_p$ is the lth best particle in the ensemble, $\{\boldsymbol{X}_k^{[j]}, j = 1.2, ..N_p\}$. Then:

$$\mathfrak{m}_k = \sum_{l=1}^{n_p} w_l \boldsymbol{X}_k^{[l:N_p]} \tag{9.10}$$

where, $n_p \leq N_p$ and w_l, $l = 1, 2, .., n_p$ are the positive weight coefficients. $\mathfrak{m}_k$ is thus a weighted average of the n_p selected particles from the sample $\boldsymbol{X}_k^{[j]}$, $j = 1.2, ..N_p$ with $w_l = \frac{1}{n_p}$.

$\mathcal{C}_k$ is evaluated from a hybrid combination of rank-n_p and rank-one updates. The rank-n_p update, as the name suggests, is again a weighted sum:

$$\mathcal{C}_k = \sum_{l=1}^{n_p} w_l \boldsymbol{Y}_k^{[l:N_p]} \boldsymbol{Y}_k^{[l:N_p]^T} \text{ with } \boldsymbol{Y}_k^{[l:N_p]} = \frac{\boldsymbol{X}_k^{[l:N_p]} - \mathfrak{m}_k}{s_k} \tag{9.11}$$

To get a reliable estimate for $\mathcal{C}_k$, information from the previous iterations is also incorporated. With $\mathcal{C}_0 = \boldsymbol{I}$, the update becomes:

$$\mathcal{C}_k = (1 - c_r)\mathcal{C}_{k-1} + c_r \sum_{l=1}^{n_p} w_l \boldsymbol{Y}_k^{[l:N_p]} \boldsymbol{Y}_k^{[l:N_p]^T} \tag{9.12}$$

$c_r \leq 1$ is the learning rate for updating the covariance matrix. The rank-one update is the result of updating the covariance matrix in the iteration sequence using only a single step as:

$$\mathcal{C}_k = (1 - c_1)\mathcal{C}_{k-1} + c_1 \boldsymbol{Y}_k^{[l:N_p]} \boldsymbol{Y}_k^{[l:N_p]^T} \tag{9.13}$$

c_1 is the learning rate for the rank-one update. In the above updates—Eqs. (9.12) and (9.13)—the sign information of $\boldsymbol{Y}_k^{[l:N_p]}$ is lost as $\boldsymbol{Y}\boldsymbol{Y}^T = -\boldsymbol{Y}(-\boldsymbol{Y}^T)$. In order to exploit the sign information, the so-called evolution path $\boldsymbol{p}_k^c \in \mathbb{R}^m$ is introduced and determined by elaborate adaptation formulae [Hansen 2007]:

$$\boldsymbol{p}_k^c = (1 - c_c)\boldsymbol{p}_{k-1}^c + \sqrt{c_c(2 - c_c)\mu_{\text{eff}}} \frac{\mathfrak{m}_k - \mathfrak{m}_{k-1}}{s_{k-1}}, \text{ with } \mu_{\text{eff}} = \frac{1}{\sum_l w_l^2} \tag{9.14a}$$

$$\boldsymbol{p}_k^s = (1 - c_s)\boldsymbol{p}_{k-1}^s + \sqrt{c_s(2 - c_s)\mu_{\text{eff}}} \mathcal{C}_{k-1}^{-\frac{1}{2}} \frac{\mathfrak{m}_k - \mathfrak{m}_{k-1}}{s_{k-1}} \tag{9.14b}$$

and

$$s_k = s_{k-1} \exp \frac{c_s}{d_s} \left(\frac{\left\| \boldsymbol{p}_k^s \right\|}{\left\| E\left[\mathcal{N}(\boldsymbol{0}, \boldsymbol{I})\right] \right\|} - 1 \right) \tag{9.14c}$$

$d_s \approx 1$, a damping parameter. c_c and c_s are the learning rates for cumulation corresponding to the rank-one update and the step-size control respectively. $\left\| E\left[\mathcal{N}(\boldsymbol{0}, \boldsymbol{I})\right] \right\|$ is the Euclidean norm of an $\mathcal{N}(\boldsymbol{0}, \boldsymbol{I})$ distributed random vector and $\approx \sqrt{m}$. Starting with $\boldsymbol{p}_0^c = \boldsymbol{0}$, the rank-one update in Eq. (9.13) is modified to:

$$\mathcal{C}_k = (1 - c_1)\mathcal{C}_{k-1} + c_1 \boldsymbol{p}_k^c \boldsymbol{p}_k^{c\,T} \tag{9.15}$$

Combining rank-n_p and rank-one updates, the final CMA update of the covariance matrix is obtained as:

$$\mathcal{C}_k = (1 - c_1 - c_r)\,\mathcal{C}_{k-1} + c_1 \boldsymbol{p}_k^c \boldsymbol{p}_k^{c\,T} + c_r \sum_{l=1}^{n_p} w_l \boldsymbol{Y}_k^{[l:N_p]} \boldsymbol{Y}_k^{[l:N_p]^T} \tag{9.16}$$

With the above updates at each iteration, sampling by Eq. (9.9) is equivalent to adopting a quadratic model of the objective function at each search point in the parameter space similar to the approximation of the inverse Hessian matrix in quasi-Newton methods (e.g., DFP method [Davidon 1959; Fletcher and Powell 1963], BFGS method [Broyden 1969; Fletcher 1970; Goldfarb 1970; Shanno 1970]). In fact, given $-\mathbf{H}^{-1}\nabla f(\boldsymbol{X}_i)^T$ as the search direction in a quasi-Newton method (Fig. 9.9) with $\nabla f(\boldsymbol{X}_i)$ denoting the gradient vector of $f(\boldsymbol{X})$ and $\mathbf{H}$ the Hessian matrix, the CMA–ES approximates the inverse Hessian matrix by $\mathcal{C}_k$.

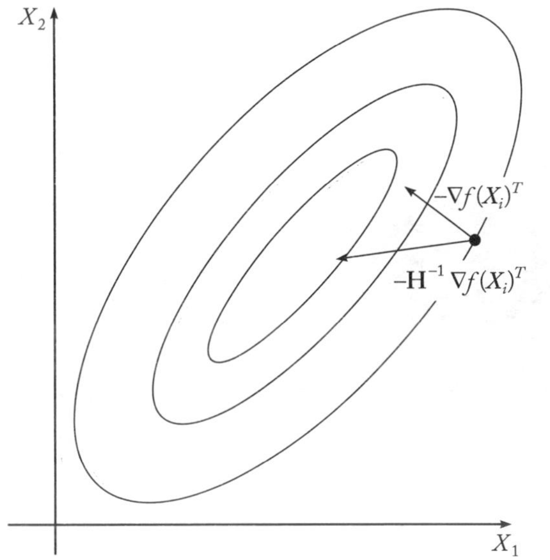

Fig. 9.9 Contours of (a 2-dimensional) objective function; $-\nabla f(\boldsymbol{X}_i)^T$ defines the search direction in steepest descent method and $-\boldsymbol{H}^{-1}\nabla f(\boldsymbol{X}_i)^T$ yields the search direction in Newton and quasi-Newton methods

Figure 9.10 shows the optimization result for the Rosenbrock function by the CMA–ES. The performance of the method is further discussed in Chapter 10 in the context of benchmark functions [Tang *et al.* 2010, Finck *et al.* 2014].

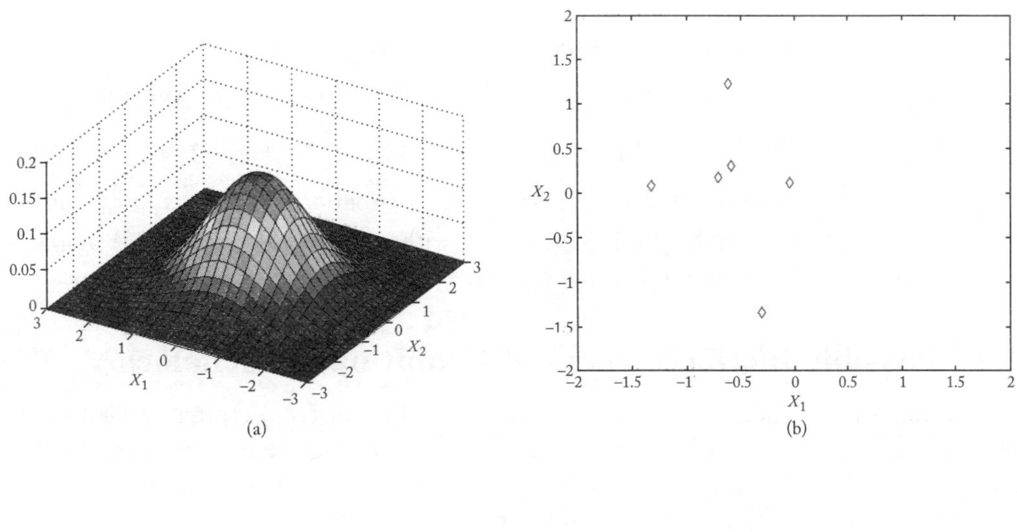

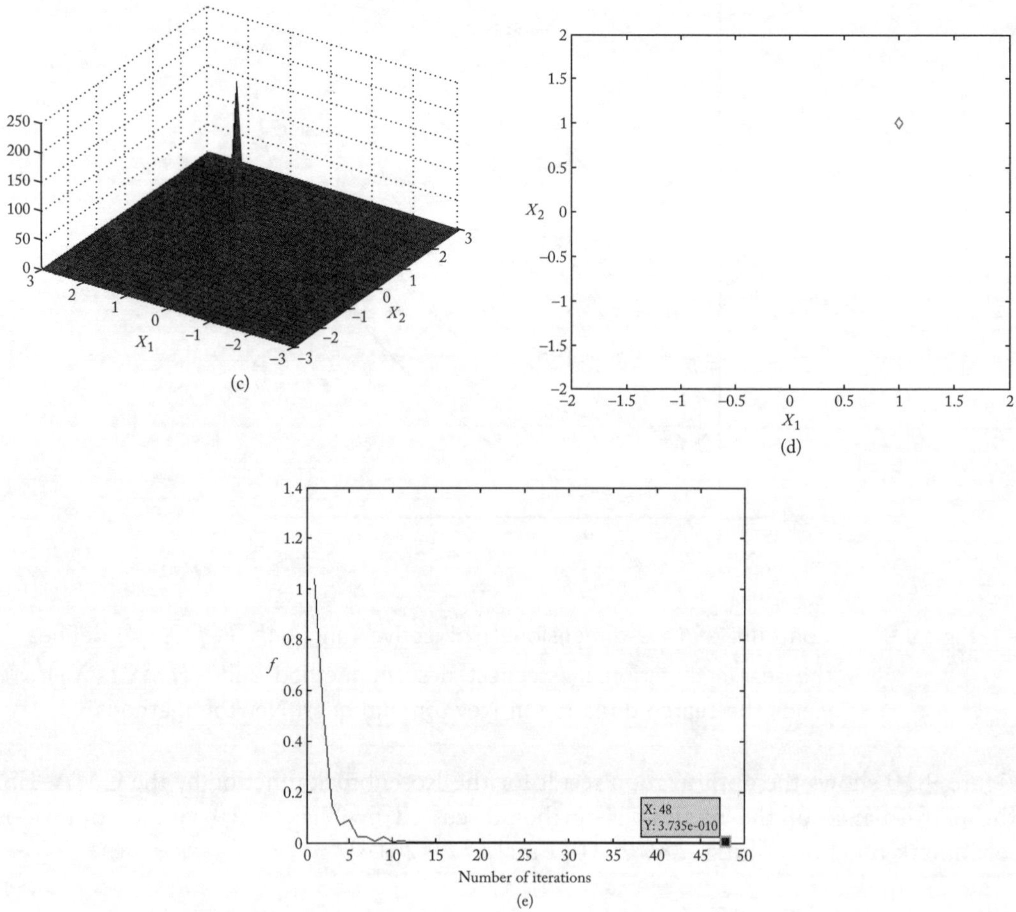

Fig. 9.10 Optimization of Rosenbrock function by the CMA–ES: $f(x_1, x_2) = 100\left(x_1^2 - x_2\right)^2 + (x_1 - 1)^2$, $N_p = 6 = n_p$; (a) & (b) sampling distribution $\mathcal{N}(\mathfrak{M}_0, s_0 \mathcal{C}_0^{\frac{1}{2}})$ and population $\boldsymbol{X}_1 \in \mathbb{R}^{N_p}$ at the start of iteration; (c) & (d) sampling distribution $\mathcal{N}(\mathfrak{M}_{50}, s_{50} \mathcal{C}_{50}^{\frac{1}{2}})$ and population $\boldsymbol{X}_{50} \in \mathbb{R}^{N_p}$ (merging with the optimal point (1,1)) at the final iteration; (e) evolution of the objective function with iterations

9.2 Possible Ineffectiveness of Evolutionary Schemes

Most evolutionary schemes depend on a random (diffusion type) scatter applied to the available candidates or particles and some criteria for selecting the new particles. Despite the wide adoption of a few evolutionary methods of the heuristic/meta-heuristic origin [Glover and Kochenberger 2003], the underlying justification is often based on sociological or biological metaphors [Goldberg 1989, Kennedy and Eberhart 1995,

Dorigo and Birattari 2010] that are hardly founded on a sound probabilistic basis even though a random search forms a key ingredient of the algorithm. Be that as it may, the popular adoption of these schemes is not only due to the algorithmic simplicity, but mainly because of their effectiveness in treating many Np-hard (Appendix I) optimization problems. Here a contrast may be drawn with a relatively poorer performance of some of the well-grounded stochastic schemes, e.g., simulated annealing [Van Laarhoven and Aarts 1987], stochastic tunneling [Wenzel and Hamacher 1999], etc. One anticipates that the notion of random evolution, built upon MC sampling of particles, should efficiently explore the search space, though at the cost of possibly slower convergence [Renders and Flasse 1996] in contrast to gradient based methods. Like some of the filtering schemes described in Chapter 6, some evolutionary optimization strategies, e.g., the GA, adopt multiplicative weight based strategy by assigning to each particle a weight. The weights, used to update the particles via selection of the best-fit individuals for subsequent evolution, may be considered functionally analogous to the derivatives used in gradient based approaches—after all a weight is a Radon–Nikodym derivative. Even though a weight based approach aims at reducing a measure of misfit between the available best and the rest within a finite population of particles, we have already seen that they bring with them the curse of weight collapse, wherein all but one particle tend to receive zero weights as iterations progress [Arulampalam *et al.* 2002]. As with non-linear filtering, this problem can be arrested, in principle, by exponentially increasing the ensemble size (number of particles), a requirement that can hardly be met in practice [Snyder *et al.* 2008]. Evolutionary schemes that replace the weight-based multiplicative approach by adopting additive updates for the particles succeed in eliminating particle degeneracy. Evolutionary schemes like the DiEv, PSO, etc. that are generally known to perform better than the GA utilize such additive particle update strategies without adversely affecting the craziness or randomness in the evolution. However, none of these methods obtains the additive term, which may be looked upon as the stochastic equivalent of derivative-based directional information, in a rigorous or rational manner—an issue addressed below.

The global optimization as described in the ensuing sections of this chapter is predicated on the premise that the basic idea behind the derivation of the KS equation (Chapter 6) to solve the non-linear stochastic filtering problems could be generalized to solve global optimization problems as well.

9.3 Global Optimization by Change of Measure and Martingale Characterization

The MC setup in Sarkar *et al.* [2014] for solving global optimization problems is based on rationally deriving an additive correction, once more largely exploiting a change of measures. In an attempt at framing a probabilistic setting that incorporates a rigorously derived directional update of the additive type, the problem of optimization is first posed as a *martingale problem* [Kurtz 1998] in the sense of Stroock and Varadhan [1972], which must however be randomly perturbed to facilitate global search. See Section 4.12,

Chapter 4 for brief notes on the martingale problem. Specifically, the local extremization of a cost functional is posed as a martingale problem realized through the solution of an integro–differential equation, which is randomly perturbed [Freidlin *et al.* 2012], so that the global extremum is attained as the perturbation vanishes asymptotically. The first martingale problem, whose solution essentially obtains a local extremum, involves an innovation (error) function, which is viewed as a stochastic process parametered by the iterations and must be driven to a zero–mean martingale.

The martingale structure of the innovation roughly implies that, by small perturbations of the argument vector, i.e., the design variables, the mean of the computed cost functional does not change and hence the argument vector corresponds to a local extremum. It may be noted that, even though the original extremization problem is posed within a deterministic setting, the cost functional as well as its argument vector are treated as stochastic diffusion processes. In order to realize a zero–mean martingale structure for the innovation, the particles are modified based on a change of measures adopted through an additive gain–type update strategy (on similar lines as the KS or the EnKS filters in Chapter 7). Thus each particle from the available population is iteratively guided by an additive correction term so as to locally extremize the given cost functional. The gain coefficient, which is a replacement for and a generalization over the Frechet derivative of a smooth cost functional, provides for an efficacious directional search without requiring the functional to be differentiable. In order to accomplish the global search, an annealing-type update and a random perturbation strategy are incorporated into the scheme which together aim at ensuring against a possible trapping of particles in local extrema.

9.4 Local Optimization as a Martingale Problem

In this section, the functional extremization posed as a martingale problem, which also includes a generic way of satisfying a given set of constraints, is described. However, before adopting a stochastic framework, a few general remarks on the expected functional features of this evolutionary optimization scheme are in order.

- The iterative solution is a random variable (defined on the search space) over every iteration.
- Thus, along the iteration axis, the solution process is considered a stochastic process, whose mean should evolve over iterations to the optimal solution.
- Upon convergence, the mean should be a constant, i.e., iteration–invariant. The random fluctuation about the converged mean is thus a zero–mean stochastic process (of the Ito integral type) and may be characterized as noise.

In posing the global optimization problem within a stochastic framework, a complete probability space $(\Omega, \mathcal{F}, P)$ is first adopted, within which the solution to the given optimization problem must exist as an $\mathcal{F}$-measurable random variable. Here Ω, known as the population set to remain consistent with the optimization parlance, necessarily includes all possible candidates or particles from which a set of randomly chosen candidates is evolved along the iteration axis τ. The introduction of τ as a positive monotonically increasing function in $\mathbb{R}$ is required to characterize the evolution of the solution as a stochastic process. A necessary aspect of the evolution is a random initial scatter provided to the particles so as to search the sample space for solutions that extremize the cost functional while satisfying the posed constraints, if any. Since the extremal points, global or local, on the cost functional may not be known *a priori*, the particles are updated iteratively (i.e., along τ) by conditioning on the evolution history of the so called extremal cost process based on the available candidates. This clearly has a parallel in the stochastic filtering method described in the last chapter. Thus denoting by $\mathcal{N}_\tau$ the filtration based on the increasing family of sub σ-algebras generated by the so called extremal (or rather the available extremal) cost process, defined as a stochastic process such that, for any $\xi \in [0,\tau]$, the mean of the associated random variable is given by the available best cost functional across the particles at the iteration step denoted by ξ. In case there are equality constraints, $\mathcal{N}_\tau$ also contains the history for the best realized constraint until τ. An elegant way of treating inequality constraints is possible by exploiting Doob's h–transform (see Appendix D); but such a scheme is outside the scope of this book.

Consider a multivariate multimodal cost functional $f(\boldsymbol{X}) : \mathbb{R}^m \longmapsto \mathbb{R}$ which is non-linear in $\boldsymbol{X} = \{X_r\}_{r=1}^m \in \mathbb{R}^m$ where m is the dimension of the parameter space. A point $\boldsymbol{X}^*$ needs to be found such that, $f(\boldsymbol{X}^*) \le f(\boldsymbol{X}), \forall \boldsymbol{X} \in \mathbb{R}^m$. The vector variable $\boldsymbol{X}$ is evolved in τ, the iteration parameter, as a stochastic process, thus allowing for the parameterization $\boldsymbol{X}_\tau := \boldsymbol{X}(\tau)$. Since there may not be any inherent physical dynamics in its evolution, $\boldsymbol{X}_\tau$ may be given a zero–mean random perturbation in Ω over every iteration. Specifically, since the number of iterations is a finite integer, one discretizes τ as $\tau_0 < \tau_1 < \cdots < \tau_N$ and evolves $\boldsymbol{X}_\tau$ for every τ increment. Thus, for the ith iteration with $\tau \in (\tau_{i-1}, \tau_i]$, $\boldsymbol{X}_\tau$ may be thought of as governed by the following SDE:

$$d\boldsymbol{X}_\tau = d\boldsymbol{B}_\tau \tag{9.17}$$

Here $\boldsymbol{B}_\tau$ is a zero–mean Brownian motion with covariance matrix, $\boldsymbol{R}_g = \boldsymbol{g}\boldsymbol{g}^T \in \mathbb{R}^{n\times n}$, where $\boldsymbol{g} \in \mathbb{R}^{n\times n}$ is the intensity matrix with its $(j,k)^{\text{th}}$ element denoted as g_{jk}. Note that, if, from physical considerations, the design variables in the vector $\boldsymbol{X}$ are known to lie within certain intervals or subsets of the real line (for instance, all of these parameters may perhaps be known to be positive), then Doob's h–transform may again be used to modify Eq. (9.17) with an appropriate drift term; see Appendix D, Example D.1. Though an elegant and useful strategy, it is not pursued further in the current edition of this book. The discrete τ-marching map for Eq. (9.17) may be written as:

$$\boldsymbol{X}_i = \boldsymbol{X}_{i-1} + \Delta\boldsymbol{B}_{i-1}, \Delta\boldsymbol{B}_{i-1} := \boldsymbol{B}_i - \boldsymbol{B}_{i-1} \tag{9.18}$$

We also denote the extremal cost process, generating the filtration, $\mathcal{N}_\tau$, as $\hat{f}_\tau$. Within a deterministic setting, a smooth cost functional $f(\boldsymbol{X})$ tends to become stationary (in the sense of a vanishing first variation) as $\boldsymbol{X}$ approaches an extremal value, $\boldsymbol{X}^*$. In the stochastic setup adopted here, a counterpart of this scenario would be that any conditioning of the future process $\boldsymbol{X}_\xi$ on the currently available filtration $\mathcal{N}_\tau$ generated by the extremal cost process $\hat{f}_s$, $s \le \tau$, identically yields the random variable $\boldsymbol{X}_\tau$ itself, i.e., $E[\boldsymbol{X}_\xi | \mathcal{N}_\tau] = \boldsymbol{X}_\tau$. Using the Markov structure of $\boldsymbol{X}_\tau$, one may thus postulate that a necessary and sufficient condition for the local extremization of $f_\tau := f(\boldsymbol{X}_\tau)$ is to require that $E\left[\boldsymbol{X}_\xi | \hat{f}_\tau\right] = \boldsymbol{X}_\tau$. Interestingly, this characterization endows $\boldsymbol{X}_\tau$ with the martingale property with respect to the extremal cost filtration, $\mathcal{N}_\tau$, i.e., once locally extremized, any future conditional mean of $\boldsymbol{X}_\tau$ on the cost filtration remains iteration invariant. An equivalent way of stating this is through an innovation process defined as $\hat{f}_\tau - f_\tau$, wherein local extremization would require driving the process $\boldsymbol{X}_\tau$ to an extremal process, $\boldsymbol{X}_\tau^*$ such that $\hat{f}_\tau - f(\boldsymbol{X}_\tau^*)$ becomes a zero mean martingale, e.g., a zero–mean Brownian motion or, more generally, an Ito integral (or, in a τ-discrete setting, a zero–mean random walk). In this way, it is possible to pose the determination of the local extrema as a martingale problem, a setup originally conceptualized by Stroock and Varadhan [1972] to provide a general setting for analyzing solutions to SDEs. If the optimization scheme is effective, one anticipates, in a weak sense, that $\boldsymbol{X}_\tau \to \boldsymbol{X}_\tau^*$ with increasing iterations, i.e., as $\tau \to \infty$. Moreover, as the noise intensity associated with $\boldsymbol{X}_\tau$ approaches zero, $\boldsymbol{X}_\tau^* \to \boldsymbol{X}^*$. However, since a strictly zero noise intensity is infeasible within the stochastic setup, the solution in a deterministic setting may be thought of as a degenerate version (with a Dirac measure) of that obtained through the stochastic scheme.

Note that a ready extension of this approach is possible with multi-objective cost functions, which would require building up the innovation as a vector–valued process. Similarly, the approach may also be adapted for treating equality constraints, wherein zero–valued constraint functions may be used to construct the innovation processes. The easy adaptability of the current setup for multiple cost functionals or constraints may be viewed as an advantage over many other evolutionary optimization schemes, wherein a single cost functional needs to be constructed based on all the constraints, a feature that may possibly engender instability or inaccuracy for a class of large dimensional problems.

9.5 The Optimization Scheme—Algorithmic Aspects

Driving $\boldsymbol{X}_\tau$ to an $\mathcal{N}_\tau$-martingale, or forcing the innovation process to a zero–mean martingale, may be accomplished expeditiously through a change of measures. In evolutionary algorithms, e.g., the GA, the accompanying change of measures is iteratively attained by assigning weights or fitness values to the current set of particles and subsequently selecting those with higher fitness, say, via rejection sampling. In an effort to explore the sample space better, these methods often use steps like crossover and mutation. These steps may be considered as biologically inspired safeguards against an intrinsic limitation of a weight-based approach, which tends to diminish all the weights but one to zero—the problem of particle impoverishment (see Chapter 6; Section 6.5.3), thereby leaving the realized particles nearly identical and thus precipitating premature convergence to a wrong solution. A more efficacious strategy to resist such degeneracy in the realized population (i.e., the ensemble) may be devised such that $\boldsymbol{X}_\tau$ at $\tau = \tau_i$ is updated using a purely additive term, derived through a Girsanov change of measures

$P \to Q$ (see Section 4.10, Chapter 4 for a detailed exposition on the Girsanov transformation), so as to ensure that the innovation or error process, originally described using measure P, becomes a zero–mean martingale under the new measure, Q. Recall that a basic ingredient in effecting this change of measure is the Radon–Nikodym derivative $\Lambda_t := \frac{dP}{dQ}$ (assuming absolute continuity of Q with respect to P and vice versa), a scalar-valued random variable also called the likelihood ratio that weighs a particle $\boldsymbol{X}_\tau$ as $\boldsymbol{X}_\tau \Lambda_t$. In executing the search for local extrema, the scheme obtains the additive particle update by expanding $\boldsymbol{X}_\tau \Lambda_t$ using Ito's formula (Section 4.4.2, Chapter 4; also see the derivation of the KS equation in Section 6.3.2, Chapter 6). An immediate advantage of using the additive update would be that the particles with lower weights are never killed, but rather corrected to become more competitive by being driven closer to the local optima. In direct analogy with Taylor's expansion of a smooth functional, where the first order term is based on Newton's directional derivative, the current version of the additive correction term may be thought of as a non-Newton directional term that drives the innovation process to a zero–mean martingale and a precise form of this term is derived in the section to follow. As noted before, the innovation process may be a vector; such a possibility may be usefully exploited while addressing possible numerical instability for a class of optimization problems—simply split a single cost functional into several and then drive each correspondent innovation into a zero–mean martingale so as to ensure that the innovation corresponding to the original cost functional is also rendered a zero–mean vector-valued martingale. For a simpler exposition, the building blocks of the method are presented using a single cost functional only. The formulation is readily adaptable for vector innovation processes of any finite dimension.

Within a τ-discretized framework and with $\tau \in (\tau_{i-1}, \tau_i]$, the evolution of $\boldsymbol{X}_\tau$ follows Eq. (9.17). The innovation used for the local search is given by:

$$\hat{f}_\tau - f(\boldsymbol{X}_\tau) = \Delta\eta_\tau \tag{9.19}$$

where, $\hat{f}_\tau \in \mathbb{R}$ is the extremal cost process; f the cost functional, possibly non-smooth, and $\Delta\eta_\tau = \eta_\tau - \eta_{i-1} \in \mathbb{R}$ a P-Brownian increment representing the diffusive fluctuations. Since the derivation of the integro–differential equation (analogous to the KS equation) for the local search is conveniently accomplished by imparting to Eq. (9.16) the explicit form of an SDE, an $\mathfrak{N}_\tau$-measureable process $\check{f}_\tau = \check{f}(\tau)$ may be constructed to arrive at the following incremental form:

$$\Delta\check{f}_\tau := \hat{f}\Delta\tau = f(\boldsymbol{X}_\tau)\Delta\tau + \Delta\eta_\tau\Delta\tau, \tag{9.20}$$

Since the τ-axis is entirely fictitious with the only requirement that $\tau(t)$ is a strictly increasing function of time, $\Delta\tau_i := \tau_i - \tau_{i-1}$ is taken to be a small increment. Hence, replacing $\Delta\tau$ with $\Delta\tau_i$, Eq. (9.20) may be recast as:

$$\Delta\check{f}_\tau = f(\boldsymbol{X}_\tau)\Delta\tau + \Delta\eta_\tau\Delta\tau_i \tag{9.21}$$

which is essentially correspondent to the SDE:

$$d\check{f}_\tau = f(\mathbf{X}_\tau)\,d\tau + \Delta\tau_i d\eta_\tau \tag{9.22}$$

Note that the replacement of $\Delta\tau$ by $\Delta\tau_i$ merely modifies the intensity of the noise process $\Delta\eta_\tau$ in Eq. (9.19) and does not interfere with the basic goal of driving the innovation to a zero–mean martingale. Indeed the form of SDE (9.22) implies that the diffusion coefficient $\Delta\tau_i$ is an order smaller relative to the drift coefficient, $f(\mathbf{X}_\tau)$. However, since η_τ is not a standard Brownian motion, it is more convenient to rewrite Eq. (9.22) as:

$$d\check{f}_\tau = f(\mathbf{X}_\tau)\,d\tau + \rho_\tau d\check{B}_\tau \tag{9.23}$$

Here $\check{B}_\tau$ is a standard P-Brownian motion and ρ_τ is a more general form of (scalar-valued) noise intensity that may explicitly depend on τ. However for multi-objective optimization problems that involve more than one cost functional and an n-dimensional vector Brownian motion, $\check{\mathbf{B}}, \boldsymbol{\rho}_\tau \in \mathbb{R}^{\mathfrak{n}\times n}$ is an intensity matrix and hence its inverse would be written as $\boldsymbol{\rho}_\tau^{-1}$. $\mathfrak{n}$ is the dimension of the multi-objective cost functional. A multi-dimensional form of Eq. (9.23) may thus be written as:

$$d\widetilde{\boldsymbol{f}}_\tau := \boldsymbol{\rho}_\tau^{-1} d\check{\boldsymbol{f}}_\tau = \boldsymbol{h}(\mathbf{X}_\tau)\,d\tau + d\check{\mathbf{B}}_\tau \tag{9.24}$$

where, $\boldsymbol{h}(\mathbf{X}_\tau) := \boldsymbol{\rho}_\tau^{-1}\boldsymbol{f}(\mathbf{X}_\tau)$. While it may be possible, or even desirable, to replace the Brownian (or diffusion-type) noise term above by others (e.g., a Poisson martingale or a sub- or super-diffusive noise), such modifications are not central to the basic idea and are not considered in this book. SDE (9.24) is assumed to satisfy the standard existence criteria (Chapter 4) so that it is at least weakly solvable. The $\mathfrak{N}_\tau$-measurable locally optimal solution may now be identified with the conditional mean, $E[\mathbf{X}_\tau|\mathfrak{N}_\tau]$, a measure-valued process characterized through the conditional distribution of $\mathbf{X}_\tau$ given the extremal cost process until τ. Considering a new measure Q under which $\mathbf{X}_\tau$ from Eq. (9.17) satisfies the constraint Eq. (9.19), the conditional mean may be represented by the generalized Bayes' formula as:

$$\pi_t(\mathbf{X}) := E_P[\mathbf{X}_\tau|\mathfrak{N}_\tau] = \frac{E_Q[\mathbf{X}_\tau\boldsymbol{\Lambda}_\tau|\mathfrak{N}_\tau]}{E_Q[\boldsymbol{\Lambda}_\tau|\mathfrak{N}_\tau]} \tag{9.25}$$

where the expectation $E_Q[.]$ is taken with respect to the new measure Q and Λ_τ is the likelihood given by:

$$\Lambda_\tau = \exp\left(\int_{\tau_{i-1}}^{\tau} \boldsymbol{h}_s.d\widetilde{\boldsymbol{f}}_s - \frac{1}{2}\int_{\tau_{i-1}}^{\tau} \boldsymbol{h}_s^2 ds\right) \tag{9.26}$$

Note that $\widetilde{f}_\tau$ is a Q-Brownian motion. Using Ito's expansion as in the case of the KS equation in Chapter 6, the incrementally additive updates on $\boldsymbol{X}_\tau$ to arrive at the local extrema is derivable as (see also Appendix I):

$$d\pi_\tau(\boldsymbol{X}) = (\pi_\tau(\boldsymbol{X}f) - \pi_\tau(\boldsymbol{X})\pi_\tau(f))\left(\rho_\tau\rho_\tau^T\right)^{-1}\left(d\check{f}_\tau - \pi_\tau(f)\,d\tau\right) \tag{9.27}$$

$d\check{f}_\tau - \pi_\tau(f)\,d\tau$ is the incremental innovation process which is driven to a zero–mean martingale. An equivalent integral representation of Eq. (9.27), referred to as the extremal equation, is:

$$\pi_\tau(\boldsymbol{X}) = \pi_{i-1}(\boldsymbol{X}) + \int_{\tau_{i-1}}^{\tau} (\pi_s(\boldsymbol{X}f) - \pi_s(\boldsymbol{X})\pi_s(f))\left(\rho_s\rho_s^T\right)^{-1}\left(d\check{f}_s - \pi_s(f)\,ds\right) \tag{9.28}$$

As with the KS equation, recall that the appearance of the unknown terms $\pi_t(f)$ and $\pi_t(\boldsymbol{X}f)$ on the RHS prevents Eq. (9.27) or (9.28) to be considered as an SDE in $\pi_\tau(\boldsymbol{X})$. In particular, a necessarily non-linear dependence of the cost functional f on $\boldsymbol{X}_\tau$ and the consequent non-Gaussianity of $\pi_\tau(f)$ would prevent writing the latter in terms of $\pi_\tau(\boldsymbol{X})$, thereby leading to the so called closure problem in solving for $\pi_\tau(\boldsymbol{X})$.

9.5.1 Discretization of the extremal equation

While a direct solution of Eq. (9.27) or (9.28) yields the local extrema in principle, exact/analytical solutions may be ruled out owing to the circularity inherent in the closure problem. Motivated by the MC filters (especially the EnKS filter considered in Chapter 7, Section 7.5), an MC scheme may be developed for a numerical treatment of Eq. (9.27) or (9.28) as well. Thus a two stage strategy, viz. prediction and update, may apparently be considered, even though, as we will soon see, the prediction step could be entirely eliminated in a variant of the final scheme. As with most evolutionary search schemes, a random exploration cum prediction may first be undertaken over the ith iteration, i.e., over $(\tau_{i-1}, \tau_i]$ based on Eq. (9.17). With N_p denoting the ensemble size, one realizes N_p predicted particles or MC candidates, $\left\{\boldsymbol{X}_\tau^{[j]}\right\}_{j=1}^{N_p}$ that must be updated based on Eq. (9.28). In developing an MC-based numerical solution to Eq. (9.28), a sample-averaged version of the equation is first written as:

$$\pi_\tau^{N_p}(\boldsymbol{X}) = \pi_{i-1}^{N_p}(\boldsymbol{X}) + \int_{\tau_{i-1}}^{\tau} \left(\pi_s^{N_p}(\boldsymbol{X}f) - \pi_s^{N_p}(\boldsymbol{X})\pi_s^{N_p}(f)\right)\left(\rho_s\rho_s^T\right)^{-1}$$

$$\left(d\check{f}_s - \pi_s^{N_p}(f)\,ds\right) \tag{9.29}$$

$\pi^{N_p}(.) := \frac{1}{N_p}\sum_{j=1}^{N_p}(.)^{[j]}$ is the sample-averaged or empirical approximation to the conditional mean $\pi(.)$. To avoid any confusion with notations, recall that, in Chapter 7, we had denoted π^{N_p} by $\overline{\pi}$. A candidate-wise representation of Eq. (9.29) may be given by:

$$X_\tau = X_{i-1} + \frac{1}{N_p}\int_{\tau_{i-1}}^{\tau}\left\{X_s F_s^T - \overline{X}_s \overline{F}_s^T\right\}\left[\rho_s\rho_s^T\right]^{-1}\left(d\check{F}_s - F_s ds\right) \tag{9.30}$$

where, $X_\tau := \left[X_\tau^{[1]}, X_\tau^{[2]}, \ldots, X_\tau^{[N_p]}\right] \in \mathbb{R}^{m\times N_p}$, $F_\tau := \left[f_\tau^{[1]}, f_\tau^{[2]}, \ldots, f_\tau^{[N_p]}\right] \in \mathbb{R}^{n\times N_p}$, $\overline{X}_\tau := \pi_\tau^{N_p}(X)r \in \mathbb{R}^{m\times N_p}$, $\overline{F}_\tau := \pi_\tau^{N_p}(f)r \in \mathbb{R}^{n\times N_p}$ and $d\check{F}_s := d\check{f}_s r \in \mathbb{R}^{n\times N_p}$. $r = \{1, 1, \ldots, 1\} \in \mathbb{R}^{N_p}$ is an N_p-dimensional row vector with all entries 1.

Note that the second term on the RHS of Eq. (9.30) is the update/correction term that has its parallel in the directional derivative term in gradient-based updates involving a smooth functional. For solving Eq. (9.30), a τ-discrete numerical scheme based on EnKS-like MC approximation to Eq. (9.28) is used. The scheme then involves the following two steps.

Prediction

The predicted candidate set, $\widetilde{X}_\tau = \left[\widetilde{X}_\tau^{[1]}, \widetilde{X}_\tau^{[2]}, \ldots, \widetilde{X}_\tau^{[N_p]}\right]$ at $\tau \in (\tau_{i-1}, \tau_i]$ is generated using the EM discretized map:

$$\widetilde{X}_\tau = X_{i-1} + \Delta B_\tau \tag{9.31}$$

where, $\Delta B_\tau := \left[\Delta B_\tau^{[1]}, \Delta B_\tau^{[2]}, \ldots, \Delta B_\tau^{[N_p]}\right]$, $\Delta B_\tau^{[j]} = B_i^{[j]} - B_\tau^{[j]}$, $j = 1, 2, \ldots, N_p$

Additive Update

The predicted candidates are improved using the correction term as in Eq. (9.30) based on an EM approximation to the integral. In both prediction and update, explicit EM schemes have been used merely for expositional convenience, even though other integration schemes could be applicable, especially for integrating the correction integral. The discrete update equation is given by:

$$X_\tau = \widetilde{X}_\tau + \frac{1}{N_p}\left(\widetilde{X}_\tau \widetilde{F}_\tau^T - \overline{\widetilde{X}}_\tau \overline{\widetilde{F}}_\tau^T\right)\left[\rho_\tau\rho_\tau^T\right]^{-1}\left(\Delta\check{F}_\tau - \widetilde{F}_\tau \Delta\tau\right) \tag{9.32}$$

Here $\Delta\check{F}_\tau = \Delta\check{f}_\tau r$, $\widetilde{F}_\tau := \left[\widetilde{f}_\tau^{[1]}, \widetilde{f}_\tau^{[2]}, \ldots, \widetilde{f}_\tau^{[N_p]}\right]$ and the predicted candidates, $\widetilde{X}_\tau$ are used to compute $\widetilde{f}_\tau := f(\widetilde{X}_\tau)$. Also note that $\overline{\widetilde{F}}_\tau^T := \pi_\tau^{N_p}\left(\widetilde{f}\right)r$ and $\overline{\widetilde{X}}_\tau := \pi_\tau^{N_p}\left(\widetilde{X}\right)r$. It is convenient to recast Eq. (9.32) as:

$$\boldsymbol{X}_\tau = \widetilde{\boldsymbol{X}}_\tau + \frac{1}{N_p}\left\{\left(\widetilde{\boldsymbol{X}}_\tau - \widetilde{\widetilde{\boldsymbol{X}}}_\tau\right)\widetilde{\boldsymbol{F}}_\tau^T + \widetilde{\widetilde{\boldsymbol{X}}}_\tau\left(\widetilde{\boldsymbol{F}}_\tau - \widetilde{\widetilde{\boldsymbol{F}}}_\tau\right)^T\right\}\left[\rho_\tau\rho_\tau^T\right]^{-1}\left(\Delta\check{\boldsymbol{F}}_\tau - \widetilde{\boldsymbol{F}}_\tau\Delta\tau\right) \tag{9.33}$$

From Eq. (9.20), recall that $\Delta\check{f}_\tau := \widehat{f}_\tau\Delta\tau$ and thus Eq. (9.33) may be rearranged as:

$$\boldsymbol{X}_\tau = \widetilde{\boldsymbol{X}}_\tau + \frac{1}{N_p}\left\{\left(\widetilde{\boldsymbol{X}}_\tau - \widetilde{\widetilde{\boldsymbol{X}}}_\tau\right)\left(\widetilde{\boldsymbol{F}}_\tau^T\Delta\tau\right) + \left(\widetilde{\widetilde{\boldsymbol{X}}}_\tau\Delta\tau\right)\left(\widetilde{\boldsymbol{F}}_\tau - \widetilde{\widetilde{\boldsymbol{F}}}_\tau\right)^T\right\}\left[\rho_\tau\rho_\tau^T\right]^{-1}\left(\widehat{\boldsymbol{F}}_\tau - \widetilde{\boldsymbol{F}}_\tau\right) \tag{9.34}$$

where, $\widehat{\boldsymbol{F}}_\tau = \widehat{f}_\tau\boldsymbol{r} \in \mathbb{R}^{m\times N_p}$. When the candidate solutions are far away from the local extrema, the correction terms should be large such that the candidates traverse more in the search space. In such cases, as in the initial stages of evolution, the innovation process may be far from behaving like a zero–mean martingale, i.e., it may have a significant drift component and the gain-like coefficient matrix in the correction term should enable better exploration of the search space. (e.g., by having a large norm). Since the estimates in this regime may have sharper gradients in τ, one way to modify the coefficient matrix in Eq. (9.34) would be to incorporate information on these gradients through the previous estimates. In doing so, $\widetilde{\boldsymbol{F}}_\tau^T\Delta\tau$ and $\widetilde{\widetilde{\boldsymbol{X}}}_\tau\Delta\tau$ are replaced, respectively, by the following approximations:

$$\widetilde{\boldsymbol{F}}_\tau^T\Delta\tau \cong \widetilde{\boldsymbol{F}}_\tau^T\tau - \widetilde{\widetilde{\boldsymbol{F}}}_{i-1}^T\tau_{i-1} - \Delta\widetilde{\widetilde{\boldsymbol{F}}}_\tau^T\tau \tag{9.35a}$$

$$\widetilde{\widetilde{\boldsymbol{X}}}_\tau\Delta\tau \cong \widetilde{\widetilde{\boldsymbol{X}}}_\tau\tau - \widetilde{\widetilde{\boldsymbol{X}}}_{i-1}\tau_{i-1} \tag{9.35b}$$

Note that we have used Ito's formula while approximating $\widetilde{\boldsymbol{F}}_\tau^T\Delta\tau$ in Eq. (9.35a). Using Eqs. (9.35a,b), Eq. (9.34) may be modified as:

$$\boldsymbol{X}_\tau = \widetilde{\boldsymbol{X}}_\tau + \frac{1}{N_p}\left\{\begin{array}{c}\left(\widetilde{\boldsymbol{X}}_\tau - \widetilde{\widetilde{\boldsymbol{X}}}_\tau\right)\left(\widetilde{\boldsymbol{F}}_\tau^T\tau - \widetilde{\widetilde{\boldsymbol{F}}}_{i-1}^T\tau_{i-1} - \Delta\widetilde{\widetilde{\boldsymbol{F}}}_\tau^T\tau\right)\\ +\left(\widetilde{\widetilde{\boldsymbol{X}}}_\tau\tau - \widetilde{\widetilde{\boldsymbol{X}}}_{i-1}\tau_{i-1}\right)\left(\widetilde{\boldsymbol{F}}_\tau - \widetilde{\widetilde{\boldsymbol{F}}}_\tau\right)^T\end{array}\right\}\left[\rho_\tau\rho_\tau^T\right]^{-1}\left(\widehat{\boldsymbol{F}}_\tau - \widetilde{\boldsymbol{F}}_\tau\right) \tag{9.36}$$

It may also be observed that, once the conditional mean is obtained in the sample-averaged form, the innovation noise covariance $\rho_\tau\rho_\tau^T$ should satisfy:

$$\begin{aligned}\rho_\tau\rho_\tau^T &\cong \pi_\tau^{N_p}\left(\left(\widehat{f}_\tau - \widetilde{f}_\tau\right)\left(\widehat{f}_\tau - \widetilde{f}_\tau\right)^T\right)\\ &= \frac{1}{N_p - 1}\left(\left(\widehat{\boldsymbol{F}}_\tau - \widetilde{\boldsymbol{F}}_\tau\right) - \left(\widehat{\widetilde{\boldsymbol{F}}}_\tau - \widetilde{\widetilde{\boldsymbol{F}}}_\tau\right)\right)\left(\left(\widehat{\boldsymbol{F}}_\tau - \widetilde{\boldsymbol{F}}_\tau\right) - \left(\widehat{\widetilde{\boldsymbol{F}}}_\tau - \widetilde{\widetilde{\boldsymbol{F}}}_\tau\right)\right)^T \end{aligned} \tag{9.37a}$$

One anticipates that, away from the local extrema, the (norm of the) RHS of the equation above should typically be relatively large. Accordingly, in order to impart higher diffusion to the local search in the initial stages wherein the innovation could have a significant drift component, $\boldsymbol{\rho}_\tau \boldsymbol{\rho}_\tau^T$ may be recast as:

$$\boldsymbol{\rho}_\tau \boldsymbol{\rho}_\tau^T = \alpha \frac{1}{N_p - 1}\left(\left(\widehat{\mathbf{F}}_\tau - \widetilde{\mathbf{F}}_\tau\right) - \left(\widehat{\overline{\mathbf{F}}}_\tau - \widetilde{\overline{\mathbf{F}}}_\tau\right)\right)\left(\left(\widehat{\mathbf{F}}_\tau - \widetilde{\mathbf{F}}_\tau\right) - \left(\widehat{\overline{\mathbf{F}}}_\tau - \widetilde{\overline{\mathbf{F}}}_\tau\right)\right)^T$$
$$+ (1-\alpha)\boldsymbol{\rho}_\tau \boldsymbol{\rho}_\tau^T \tag{9.37b}$$

Here $0 < \alpha < 1$(typically $\alpha = 0.8$ is taken in the numerical illustrations provided later on). Finally, then, Eq. (9.36) may be written as:

$$\mathbf{X}_\tau = \widetilde{\mathbf{X}}_\tau + \frac{1}{N_p}\left\{\left(\widetilde{\mathbf{X}}_\tau - \widetilde{\overline{\mathbf{X}}}_\tau\right)\left(\widetilde{\mathbf{F}}_\tau^T \tau - \widetilde{\overline{\mathbf{F}}}_{i-1}^T \tau_{i-1} - \Delta \widetilde{\overline{\mathbf{F}}}_\tau^T \tau\right) + \left(\widetilde{\overline{\mathbf{X}}}_\tau \tau - \widetilde{\overline{\mathbf{X}}}_{i-1}\tau_{i-1}\right)\left(\widetilde{\mathbf{F}}_\tau - \widetilde{\overline{\mathbf{F}}}_\tau\right)^T\right\}$$
$$\left[\alpha \frac{1}{N_p - 1}\left(\left(\widehat{\mathbf{F}}_\tau - \widetilde{\mathbf{F}}_\tau\right) - \left(\widehat{\overline{\mathbf{F}}}_\tau - \widetilde{\overline{\mathbf{F}}}_\tau\right)\right)\left(\left(\widehat{\mathbf{F}}_\tau - \widetilde{\mathbf{F}}_\tau\right) - \left(\widehat{\overline{\mathbf{F}}}_\tau - \widetilde{\overline{\mathbf{F}}}_\tau\right)\right)^T + (1-\alpha)\boldsymbol{\rho}_\tau \boldsymbol{\rho}_\tau^T\right]^{-1}$$
$$\left(\widehat{\mathbf{F}}_\tau - \widetilde{\mathbf{F}}_\tau\right) \tag{9.38}$$

In a more concise form, the update equation is thus of the form:

$$\mathbf{X}_\tau = \widetilde{\mathbf{X}}_\tau + \widetilde{\mathbf{G}}_\tau\left(\widehat{\mathbf{F}}_\tau - \widetilde{\mathbf{F}}_\tau\right) \tag{9.39}$$

where the gain-like update coefficient matrix is given by:

$$\widetilde{\mathbf{G}}_\tau = \frac{1}{N_p}\left\{\left(\widetilde{\mathbf{X}}_\tau - \widetilde{\overline{\mathbf{X}}}_\tau\right)\left(\widetilde{\mathbf{F}}_\tau^T \tau - \widetilde{\overline{\mathbf{F}}}_{i-1}^T \tau_{i-1} - \Delta \widetilde{\overline{\mathbf{F}}}_\tau^T \tau\right) + \left(\widetilde{\overline{\mathbf{X}}}_\tau \tau - \widetilde{\overline{\mathbf{X}}}_{i-1}\tau_{i-1}\right)\left(\widetilde{\mathbf{F}}_\tau - \widetilde{\overline{\mathbf{F}}}_\tau\right)^T\right\}.$$
$$\left[\alpha \frac{1}{N_p - 1}\left(\left(\widehat{\mathbf{F}}_\tau - \widetilde{\mathbf{F}}_\tau\right) - \left(\widehat{\overline{\mathbf{F}}}_\tau - \widetilde{\overline{\mathbf{F}}}_\tau\right)\right)\left(\left(\widehat{\mathbf{F}}_\tau - \widetilde{\mathbf{F}}_\tau\right) - \left(\widehat{\overline{\mathbf{F}}}_\tau - \widetilde{\overline{\mathbf{F}}}_\tau\right)\right)^T + (1-\alpha)\boldsymbol{\rho}_\tau \boldsymbol{\rho}_\tau^T\right]^{-1} \tag{9.40}$$

In the update strategy described in Eq. (9.39), while some measures are already taken for an exploration of the search space, we may still improve it by adopting an annealing type approach. The idea is to provide larger diffusion intensity to the update term in the initial stages of evolution and reduce it as the candidates approach the global optimum (a proper search scheme for the global optimum is presented in the next section). The annealing-type coefficient β_τ (with $1/\beta_\tau$ interpreted as the annealing temperature) here appears as a scalar factor multiplying the update term so that the update equation becomes:

$$\mathbf{X}_\tau = \widetilde{\mathbf{X}}_\tau + \beta_\tau \widetilde{\mathbf{G}}_\tau\left(\widehat{\mathbf{F}}_\tau - \widetilde{\mathbf{F}}_\tau\right) \tag{9.41}$$

Typically, β_τ starts with a small positive (near-zero) value at $\tau = \tau_0$ (implying that the initial temperature $1/\beta_0$ is high) and gradually increases to $\beta_{\max}$ at $\tau_{\max}$, the end of the iterations. In the numerical simulations presented here, $\beta_{\max}$ is taken as 2. β_τ is analogous to the annealing parameter found in the annealed Markov chain Monte Carlo (MCMC) simulation which we do not discuss here (see, for instance Gamerman and Lopes [2006]). Although the annealing temperature in the MCMC framework is prescribed to be reduced very slowly (e.g., logarithmically), $1/\beta_\tau$ in the present scheme may be reduced much faster (e.g., exponentially) in view of the fact that the job of an efficient exploration of the search space is already attended to, in part, through an appropriate construction of the coefficient matrix $\widetilde{\boldsymbol{G}}_\tau$. Indeed, while virtually any form of β_τ as an increasing function of τmay be prescribed, we presently use $\beta_{\tau_i} :=\beta_i = 1 - \frac{1}{\exp(i-1)}$ within a τ-discrete setting.

Coalescence and Scrambling: Schemes for Global Search

In any global optimization scheme, a major challenge is to eliminate the possibility of the candidate solutions getting trapped in local extrema. Although it seems possible to avoid the local traps through an exploitation of the innovation noise, whose intensity could be tuned using the annealing-type factor, β_τ, such an approach could be quite inefficient and prone to wild fluctuations. Indeed, as a local extremal point is approached, the strength (or norm) of the update term (in analogy with a directional derivative term in a gradient-based search) would be small and consequently the sensitivity of the update term to variations in β_τ would also be poor. This makes choosing β_τ difficult (e.g., necessitating β_τ to be too small for the search scheme to be efficient) and renders the annealing-type scheme computationally less effective for the global search. Moreover, smaller β_τ also implies larger diffusion and hence poorer convergence. A more effective way is probably to randomly perturb the candidates, so that they do not get trapped in the local wells. One such perturbation scheme, a combination of two approaches that are herein referred to as coalescence and scrambling, may be accomplished via yet another martingale problem and a perturbation kernel as explained below. Before going into the details of coalescence, we recall that the local extremization, posed as a martingale problem, requires that the jth candidate or particle should be updated following Eq. (9.41) as:

$$\boldsymbol{X}_i^{[j]} = \widetilde{\boldsymbol{X}}_i^{[j]} + \widetilde{\boldsymbol{C}}_i^{[j]}, 1 \le j \le N_p \tag{9.42}$$

where, $\widetilde{\boldsymbol{C}}_i^{[j]} := \beta_i \widetilde{\boldsymbol{G}}_i \left(\widehat{\boldsymbol{f}}_i - \widetilde{\boldsymbol{f}}_i^{[j]} \right)$. Being essentially a scheme for local extremization with the annealing-type term playing a poor role in the global search, the global update strategy must be efficiently equipped to prevent the evolving candidates from getting trapped at the local wells, i.e., the bad points. In doing so, the basic idea here is to provide several layers of random perturbation to the particles, whose intensity may be ideally thought of as being indexed by a positive integer l such that the perturbation vanishes as $l \to \infty$. Within the τ-discrete setting, we start with the prediction $\widetilde{\boldsymbol{X}}_i$ and denote ${}^l\Delta\widehat{\boldsymbol{X}}_i = {}^l\boldsymbol{X}_i - {}^l\widetilde{\boldsymbol{X}}_i$ as the randomly perturbed, l-indexed increment so that its limiting dynamics is provided by the process ${}^{\infty}\Delta\widehat{\mathbf{X}}_i \rightarrow \Delta\mathbf{X}_i$ as the random perturbations vanish asymptotically. During the ith iteration, the perturbed increment is arrived at using two transitional increments, ${}^{l}\mathcal{U}_i$ and ${}^{l}\mathcal{V}_i$ representing populations corresponding to the two perturbation operators, say $\mathbf{T}1$ and $\mathbf{T}2$, respectively. While the operator $\mathbf{T}1$ corresponds to the combined local search and coalescence operations, $\mathbf{T}2$ denotes the scrambling operation. Details of both coalescence and scrambling operations will be provided shortly. Specifically, the transitions may be written as, ${}^{l}\Delta\mathbf{X}_i \underset{\mathbf{T1}}{\longrightarrow} {}^{l}\mathcal{U}_i \underset{\mathbf{T2}}{\longrightarrow} {}^{l}\mathcal{V}_i \underset{\mathbf{T3}}{\longrightarrow} {}^{l}\Delta\widehat{\mathbf{X}}_i$ where $\mathbf{T3}$ is a selection operator, commonly used with most evolutionary optimization schemes in some form or the other. Here ${}^{l}\Delta\widehat{\mathbf{X}}_i$ is the finally obtained solution increment at τ_i which goes as input to the next iteration, i.e., ${}^{l}\mathbf{X}_i := {}^{l}\widetilde{\mathbf{X}}_i + {}^{l}\Delta\widehat{\mathbf{X}}_i$ and ${}^{l}\Delta\mathbf{X}_i = {}^{l}\widetilde{\mathbf{X}}_i - {}^{l}\mathbf{X}_{i-1}$ is the predicted increment. Ideally, one may start the iterations with a small l (i.e., high perturbation intensity) and gradually increase l with progressing iterations. However, in the present implementation of the scheme, the perturbation intensity is kept uniformly small all through the iterations. One can manage to take this latitude as the method arguably has superior convergence features vis-à-vis many existing evolutionary optimization algorithms, as demonstrated, to an extent, in the next Section. In view of this and for notational ease, the left superscript l is removed from the variables in the discussion to follow (provided that the perturbed nature of the variables is clear from the context). The operators are now defined below.

Operator $\mathbf{T}1$: Local search and coalescence

The operation for the local search has already been described and quantitatively captured through Eq. (9.42). Thus, in the following, we discuss the operation of coalescence. This perturbation is motivated by the observation that the *pdf*, if it exists, associated with the converged measure $\pi_\tau(.)$ should be unimodal, with its only peak located at the global extremum. Thus, when convergence to the global extremum occurs, all the candidate solutions should coalesce at the globally optimum point, except for a zero–mean noisy scatter around the latter. Ideally, for the sake of optimization accuracy, the noisy scatter should also have a low intensity. Once the global optimization scheme converges, the noisy scatter should then behave as a zero–mean martingale as a function of τ and with a unimodal transitional *pdf*. A zero–mean Brownian motion, which has a Gaussian *pdf*, is one such martingale. Clearly, such a property does not hold away from the global optimum, where the *pdf* should be multi-modal with a peak at every local extremum detected by the algorithm.

Now, we wish to make the above argument a little more precise and thus obtain a scheme to force a coalescence of candidates. Consider the update of the jth candidate, $\mathbf{X}_\tau^{[j]}$. A measure of the random scatter around $\mathbf{X}_\tau^{[j]}$ could be defined as $\delta_\tau^{[j]} = \mathbf{X}_\tau^{[\varsigma_1(j)]} - \mathbf{X}_\tau^{[j]}$, where, $\varsigma_1(j)$ denotes a random permutation on the indexing set $\{1, N_p\} \setminus \{j\}$ based on a uniform measure. Our goal is to drive $\delta_\tau^{[j]}$ to a zero–mean vector Brownian increment, $\Delta\eta_\tau^c$, with intensity matrix, ρ_τ^c (typically diagonal), which may be chosen uniformly for all j. Coalescence then implies that, in the limit of the noise term in

$\delta_\tau^{[j]}$ approaching zero, all the candidates would tend to merge into a single particle at the global extremum. Thus, similar to the innovation, $\widehat{\boldsymbol{f}}_\tau - \boldsymbol{f}(\boldsymbol{X}_\tau)$ on the LHS of Eq. (9.19), one treats $\boldsymbol{X}_\tau^{[\varsigma_1(j)]} - \boldsymbol{X}_\tau^{[j]}$ as yet another innovation process, whose characterization upon convergence would be of the form, $\boldsymbol{X}_\tau^{[\varsigma_1(j)]} - \boldsymbol{X}_\tau^{[j]} = \Delta\boldsymbol{\eta}_\tau^c$. Since $\Delta\boldsymbol{\eta}_\tau^c$ provides for an additional layer of randomness that imparts to each candidate $\boldsymbol{X}_\tau^{[j]}$ the structure of a stochastic process, the extremal cost filtration Ω_τ may now be suitably expanded to include the sub-filtration generated by $\Delta\boldsymbol{\eta}_s^c$ for $s \le \tau$ in order to remain theoretically consistent. Including the coalescence innovation within our search process, Eq. (9.42) may be modified as:

$$\boldsymbol{X}_i^{[j]} = \widetilde{\boldsymbol{X}}_i^{[j]} + \widetilde{\boldsymbol{D}}_i^{[j]}, \quad 1 \le j \le N_p \tag{9.43a}$$

where,

$$\widetilde{\boldsymbol{D}}_i^{[j]} := \beta_i \widetilde{\boldsymbol{G}}_i \widetilde{\boldsymbol{I}}_i^{[j]}, \qquad \widetilde{\boldsymbol{I}}_i^{[j]} := \begin{pmatrix} \widehat{\boldsymbol{f}}_i - \widetilde{\boldsymbol{f}}_i^{[j]} \\ \widetilde{\boldsymbol{X}}_i^{[\varsigma_1(j)]} - \widetilde{\boldsymbol{X}}_i^{[j]} \end{pmatrix} \tag{9.43b}$$

and we recall that over-tildes indicate either the predicted candidates or functions evaluated using the predicted candidates, as appropriate. Allowing for a convenient notational abuse, we have retained the same notation for the gain-like update coefficient matrix $\widetilde{\boldsymbol{G}}_i$ in Eq. (9.43b) (used earlier in the local update Eq. 9.40), which may now be computed as:

$$\widetilde{\boldsymbol{G}}_\tau = \frac{1}{N_p}\left\{\left(\widetilde{\boldsymbol{X}}_i - \overline{\widetilde{\boldsymbol{X}}}_i\right)\left(\widetilde{\boldsymbol{F}}_i^T \tau_i - \overline{\widetilde{\boldsymbol{F}}}_{i-1}^T \tau_{i-1} - \Delta\overline{\widetilde{\boldsymbol{F}}}_i^T \tau_i\right) + \left(\overline{\widetilde{\boldsymbol{X}}}_i \tau_i - \overline{\widetilde{\boldsymbol{X}}}_{i-1}\tau_{i-1}\right)\left(\widetilde{\boldsymbol{F}}_i - \overline{\widetilde{\boldsymbol{F}}}_i\right)^T\right\}$$

$$\left[\alpha \frac{1}{N_p - 1}\left(\left(\widehat{\boldsymbol{F}}_i - \widetilde{\boldsymbol{F}}_i\right) - \left(\overline{\widehat{\boldsymbol{F}}}_i - \overline{\widetilde{\boldsymbol{F}}}_i\right)\right)\left(\left(\widehat{\boldsymbol{F}}_i - \widetilde{\boldsymbol{F}}_i\right) - \left(\overline{\widehat{\boldsymbol{F}}}_i - \overline{\widetilde{\boldsymbol{F}}}_i\right)\right)^T + (1-\alpha)\boldsymbol{\gamma}_i\boldsymbol{\gamma}_i^T\right]^{-1} \tag{9.44}$$

where, $\boldsymbol{\gamma}_i\boldsymbol{\gamma}_i^T := \begin{bmatrix} \boldsymbol{\rho}_i\boldsymbol{\rho}_i^T & \boldsymbol{0} \\ \boldsymbol{0} & \boldsymbol{\rho}_i^c\left(\boldsymbol{\rho}_i^c\right)^T \end{bmatrix}$.

Since an additional innovation process, $\widetilde{\boldsymbol{X}}_\tau^{[\varsigma_1(j)]} - \widetilde{\boldsymbol{X}}_\tau^{[j]}$, is included in $\widetilde{\boldsymbol{I}}_i^{[j]}$, we have, $\widetilde{\boldsymbol{F}}_\tau \in \mathbb{R}^{(\mathrm{n}+m)} \times N_p$, $\overline{\widetilde{\boldsymbol{X}}}_\tau \in \mathbb{R}^{m \times N_p}$, $\overline{\widetilde{\boldsymbol{F}}}_\tau \in \mathbb{R}^{(\mathrm{n}+m)\times N_p}$ and $\widehat{\boldsymbol{F}}_\tau \in \mathbb{R}^{(\mathrm{n}+m)\times N_p}$. $\boldsymbol{\rho}_i^c\left(\boldsymbol{\rho}_i^c\right)^T$ may be chosen to be a diagonal matrix $\vartheta\boldsymbol{I}_{m\times m}$ with $\boldsymbol{I}$ denoting an identity matrix and $\vartheta \in \mathbb{R}^+$ and $\ll 1$. Thus, $\boldsymbol{\gamma}_i\boldsymbol{\gamma}_i^T \in \mathbb{R}^{(\mathrm{n}+m)\times(\mathrm{n}+m)}$. Here the jth column of $\widetilde{\boldsymbol{F}}_i$ is given as $\begin{pmatrix} \widehat{\boldsymbol{f}}_i - \widetilde{\boldsymbol{f}}_i^{[j]} \\ \widetilde{\boldsymbol{X}}_i^{[\varsigma_1(j)]} - \widetilde{\boldsymbol{X}}_i^{[j]} \end{pmatrix}$. Let a is a positive real number; define the closest integer smaller than a by $\lfloor a \rfloor$. Then the integer valued perturbation parameter for the local extremization cum coalescence step may be identified as $l = \left\lfloor \left\| \left[\gamma_i \gamma_i^T \right]^{-1} \right\| \right\rfloor$. While a proof is not provided here, the convergence and uniqueness of the iterative increment through this step, for a non-decreasing sequence of l converging to a limit point l^* which may be large yet finite or infinity, may be shown based on the work of Stroock and Varadhan [1972].

Operator **T**2*: Scrambling*

Similar to **T1**, the second perturbation operator is also based on random perturbations of the candidates in the population Ω. Granted that the gain-like coefficient matrix in the local update Eq. (9.42) is a derivative-free stochastic counterpart to the Frechet derivative, the term $\widetilde{\boldsymbol{C}}_i^{[j]}$ may be considered the equivalent of the directional derivative term responsible for updating the jth particle. Consequently, around any local extremum, the $L^2(P)$ norm $\left\| \widetilde{\boldsymbol{C}}_i^{[j]} \right\|$ is likely to be small. This may render further updates of the jth particle trivially small, leading to a possible stalling of the local scheme. In order to move out of these local traps, the basic idea here is to swap the gain-weighted directional information for the jth particle with that of another randomly selected one. This random perturbation at τ_i is accomplished by replacing the update equation (9.43a) by the following option:

$$\boldsymbol{X}_i^{[j]} = \widetilde{\boldsymbol{X}}_i^{[j]} + \widetilde{\boldsymbol{D}}_i^{[\varsigma_2(j)]}, \tag{9.45a}$$

or

$$\boldsymbol{X}_i^{[j]} = \widetilde{\boldsymbol{X}}_i^{[\varsigma_2(j)]} + \widetilde{\boldsymbol{D}}_i^{[j]}, 1 \le j \le N_p \tag{9.45b}$$

where, $\widetilde{\boldsymbol{D}}_i^{[j]}$ in Eq. (9.45b) is the update vector computed for the jth candidate via Eq. (9.43b). In Eq. (9.45a), $\widetilde{\boldsymbol{D}}_i^{[\varsigma_2(j)]} = \beta_i \widetilde{\boldsymbol{G}}_i \widetilde{\boldsymbol{I}}_i^{[\varsigma_2(j)]}$ is the correction vector originally computed for the $\varsigma_2(j)^{\text{th}}$ candidate similar to Eq. (9.43b) but with $\widetilde{\boldsymbol{I}}_i^{[\varsigma_2(j)]} = \begin{pmatrix} \widehat{\boldsymbol{f}}_i - \widetilde{\boldsymbol{f}}_i^{[\varsigma_2(j)]} \\ \widetilde{\boldsymbol{X}}_i^{[\varsigma_1(j)]} - \widetilde{\boldsymbol{X}}_i^{[\varsigma_2(j)]} \end{pmatrix}$; ς_2 being a random permutationon the integer set $\left\{1, N_p\right\} \backslash \{j\}$. Such perturbation may be described by a discrete probability kernel p_l on $\left[1, N_p\right] \times \left[1, N_p\right]$ such that:

$$\sum_{i \in \left[1, N_p\right]} p_l(i,j) = 1 \ \forall j \in \left[1, N_p\right] \tag{9.46}$$

Clearly, as $l \to \infty$, the matrix $p_l(i,j)$ should ideally approach the identity matrix, i.e., $p_l(i,j) \to \delta_{ij}$, where δ_{ij} is the Kronecker delta. However, since the coalescence step ensures that all particles crowd around the mean of a unimodal *pdf* with progressing iterations, directional scrambling across the set of such converged particles should not affect the numerical accuracy of the estimated global extremum. In other words, in a practical implementation of the scheme, $p_l(i,j)$ need not strictly approach the identity matrix for large l.

Operator **T**3*: Selection*

Use of diffusion-based random perturbations during exploration may sometimes result in bad candidates. This necessitates a selection step wherein candidates for the next iteration (say, the ith iteration) are chosen based on some selection criteria effected by a selection or fitness function $g(v|\mathbf{X}_{i-1})$, $v \in \Omega$. A general construction of the function, which corresponds to a Markov transition kernel on Ω and is conditioned on the ensemble of particles in the last iteration, should satisfy the following properties:

$$\text{(a) } g\left(v = \mathbf{X}_i^{[j]} \,\middle|\, \mathbf{X}_{i-1} = \mathbf{X}_{i-1}^{[k]}\right) = 0, \text{ if } j \neq k;\ \forall j,k \in \left[1, N_p\right] \tag{9.47a}$$

$$\text{(b) } g\left(v = \mathbf{X}_i^{[j]} \,\middle|\, \mathbf{X}_{i-1} = \mathbf{X}_{i-1}^{[j]}, f(\mathbf{X}_i^{[j]}) \geq f(\mathbf{X}_{i-1}^{[j]})\right) = \nu \ \forall j \in \left[1, N_p\right], \text{ where } \nu \in (0,1] \tag{9.47b}$$

$$\text{(c) } g\left(v = \mathbf{X}_{i-1}^{[j]} \,\middle|\, \mathbf{X}_{i-1} = \mathbf{X}_{i-1}^{[j]}, f(\mathbf{X}_i^{[j]}) < f(\mathbf{X}_{i-1}^{[j]})\right) = \nu \tag{9.47c}$$

The updated j^{th} particle $\mathbf{X}_i^{[j]}$ in the above clauses is computed using Eq. (9.45) which combines all the three operations of local extremization, coalescence and scrambling. Here the integer-valued perturbation parameter l may be identified with $l = \left\lfloor \frac{1}{1-\nu} \right\rfloor$. In the current numerical implementation of the scheme, we consistently take $\nu = 1$, which corresponds to l being infinity across all iterations and implies that the selection procedure does not involve any random perturbation.

9.5.2 Pseudo codes

The performance of the evolutionary optimization scheme described above will shortly be illustrated numerically. Prior to that, the following pseudo codes for the global optimization scheme are provided for better clarity. The reader is also urged to carefully go through the MATLAB codes given in the accompanying web site.

Table 9.6 Pseudo code 1

Step 1.	Discretize the τ-axis, say over $[\tau_{\min}, \tau_{\max}]$ using a partition $\{\tau_0 = \tau_{\min}, \tau_1, \ldots, \tau_N = \tau_{\max}\}$ such that $\tau_0 < \cdots < \tau_N$ and $\tau_i - \tau_{i-1} = \Delta\tau_i (= \frac{1}{N}$ if a uniform step size is chosen for $i = 0, \ldots, N-1)$. We assign $\tau_0 = 1$ and adopt $\Delta\tau$ small ($\leq 10^{-7}$). Choose an ensemble size N_p.
Step 2.	Generate the ensemble of initial population $\left\{\mathbf{X}_0^{[j]}\right\}_{j=1}^{N_p}$ for the solution vector. For each discrete $\tau_i, i = 1, \ldots, N-1$, execute the following steps.

Step 3.	Prediction Using$\left\{X_{i-1}^{[j]}\right\}_{j=1}^{N_p}$, the last update (or the initial population for $i = 1$) available at τ_{i-1}, obtain the predicted candidates using $\widetilde{X}_i^{[j]} = X_{i-1}^{[j]} + \Delta B_i^{[j]}$, $j = 1,2,\ldots,N_p$
Step 4.	Additive update Choose $\alpha \in \{0,1]$ (the parameter in Eq. 9.29b or 9.44); a typically prescribed value would be $\alpha \approx 0.8$, even though the method also performs well for other values in the interval indicated. Update each particle using Eq. (9.45a) or (9.45b). The annealing-type parameter, $\beta_i = 1 - \frac{1}{\exp(i-1)}$. The gain-like update coefficient matrix $\widetilde{G}_i$ is given by Eq. (9.44).
Step 5.	If $f\left(X_i^{[j]}\right) \geq f\left(X_{i-1}^{[j]}\right)$, $j = 1,2,\ldots,N_p$, then retain $X_i^{[j]}$ as the updated particle; else set $X_i^{[j]} = X_{i-1}^{[j]}$
Step 6.	If $i < N$, then go to step 3 with $i \equiv i+1$; else terminate the algorithm.

It may be noted that, while the prediction step of Eq. (9.31) appears to be helpful in exploration, this step may not be practically useful, especially as it does not exploit any directional information in exploring. Hence, the global search may be expedited by dropping the prediction step altogether from the *pseudo code* 1. Such a modification would typically mean that the number of evaluations of the cost functional is reduced by half. The modified pseudo-code (named *pseudo code* 2) involving no prediction step is provided below.

Table 9.7 Pseudo code 2

Steps 1 and 2.	These are the same as those of pseudo code 1
Step 3.	Additive update This step is similar to Step 4 of *pseudo code 1* except that the over-tildes over the variables are no longer needed and $\widetilde{X}_i^{[j]}$ may be replaced by $X_{i-1}^{[j]}$. Choose $\alpha \in \{0,1]$ and update each particle according to Eq. (9.45): $X_i^{[j]} = X_{i-1}^{[j]} + D_i^{[\varsigma_2(j)]}, \text{or} X_i^{[j]} = X_i^{[\varsigma_2(j)]} + D_i^{[j]}, 1 \leq j \leq N_p$ (9.48)

	where, $\boldsymbol{D}_i^{[\varsigma_2(j)]} = \beta_i \boldsymbol{G}_i \boldsymbol{I}_i^{[\varsigma_2(j)]}$ with $\boldsymbol{I}_i^{[\varsigma_2(j)]} = \begin{pmatrix} \widehat{\boldsymbol{f}}_{i-1} - \boldsymbol{f}_{i-1}^{[\varsigma_2(j)]} \\ \boldsymbol{X}_{i-1}^{[\varsigma_1(j)]} - \boldsymbol{X}_{i-1}^{[\varsigma_2(j)]} \end{pmatrix}$ and $\boldsymbol{D}_i^{[j]} := \beta_i \boldsymbol{G}_i \boldsymbol{I}_i^{[j]}$ with $\boldsymbol{I}_i^{[j]} := \begin{pmatrix} \widehat{\boldsymbol{f}}_{i} - \boldsymbol{f}_{i}^{[j]} \\ \boldsymbol{X}_{i}^{[\varsigma_1(j)]} - \boldsymbol{X}_{i}^{[j]} \end{pmatrix}.$ ς_1 and ς_2 are random permutations on the integer set $\{1, N_p\} \backslash \{j\}$. The gain-like update coefficient matrix $\boldsymbol{G}_i$ is: $\boldsymbol{G}_i = \frac{1}{N_p}\Big\{\big(\boldsymbol{X}_i - \overline{\boldsymbol{X}}_i\big)\big(\boldsymbol{F}_i^T \tau_i - \overline{\boldsymbol{F}}_{i-1}^T \tau_{i-1} - \Delta\overline{\boldsymbol{F}}_i^T \tau_i\big) + \big(\overline{\boldsymbol{X}}_i \tau_i - \overline{\boldsymbol{X}}_{i-1}\tau_{i-1}\big)\big(\boldsymbol{F}_i - \overline{\boldsymbol{F}}_i\big)^T\Big\}$ $\Big[\alpha \frac{1}{N_p - 1}\Big(\big(\widehat{\boldsymbol{F}}_i - \boldsymbol{F}_i\big) - \big(\widehat{\overline{\boldsymbol{F}}}_i - \overline{\boldsymbol{F}}_i\big)\Big)\Big(\big(\widehat{\boldsymbol{F}}_i - \boldsymbol{F}_i\big) - \big(\widehat{\overline{\boldsymbol{F}}}_i - \overline{\boldsymbol{F}}_i\big)\Big)^T + (1-\alpha)\,\gamma_i \gamma_i^T\Big]^{-1}$ (9.49)
Step 4.	Same as Step 5 of ***pseudo code 1***.
Step 5.	If $i < N$, go to Step 3 with $i \equiv i + 1$; else terminate the algorithm.

9.6 Some Applications of the Pseudo Code 2 to Dynamical Systems

The following two examples pertain to the reconstruction of system parameters [Sun *et al.* 2010] in the chaotic response regimes of non-linear dynamical systems. This could be a tricky problem, especially given the locally exponential separation of trajectories starting with closely separated initial conditions within the strange attractor [Holmes 1979]. Consider the following generic form of a deterministic dynamical system:

$$\dot{\boldsymbol{X}}_t = \boldsymbol{\psi}(\boldsymbol{X}_t, t, \boldsymbol{\theta}) \tag{9.50}$$

where, $\boldsymbol{X}_t$ is the state space representation of the system response vector (state vector) at time t. $\boldsymbol{\psi}$ is the non-linear vector field and $\boldsymbol{\theta}$ the unknown parameter vector of dimension n_p. Let $\boldsymbol{X}_i^a := \boldsymbol{X}^a(t_i)$ be the solution to Eq. (9.50) at time t_i, $i = 1, 2, \ldots, N$when $\boldsymbol{\theta}$ is replaced by an assumed parameter vector $\boldsymbol{\theta}^a$. $f(\boldsymbol{\theta}^a) = \sum_{i=1}^N \big(\boldsymbol{X}_i - \boldsymbol{X}_i^a\big)^T \big(\boldsymbol{X}_i - \boldsymbol{X}_i^a\big)$ (Eq. 9.4) needs to be minimized using a global optimization strategy to arrive at a converged estimate of $\boldsymbol{\theta}$. $\boldsymbol{X}_t$ denotes the true solution, i.e., it solves Eq. (9.50) when the vector $\boldsymbol{\theta}$ is known. The *pseudo-code* 2 is used in obtaining the optimization results and a comparison is made with those of PSO.

Example 9.3

The Lorenz oscillator (Lorenz 1963)

The Lorenz oscillator which is the first known chaotic dynamical system, is a low dimensional model for the convective motion of fluid. The governing DE is given as:

$$\dot{\boldsymbol{X}}_t = \begin{Bmatrix} \theta_1 (Y_t - X_t) \\ \theta_2 X_t - Y_t - X_t Z_t \\ X_t Y_t - \theta_3 Z_t \end{Bmatrix} \tag{9.51}$$

Here $\boldsymbol{X}_t := \{X_t, Y_t, Z_t\}^T$ and $\boldsymbol{\theta} := \{\theta_1, \theta_2, \theta_3\}^T$. While constructing the reference solution for purposes of validation, the parameter vector is taken as $\boldsymbol{\theta} := \{10 \quad 28 \quad 8/3\}^T$, which corresponds with the chaotic behavior of the oscillator. Eq. (9.51) is numerically solved over $t \in (0, 0.3\ \mathrm{s}]$ using the 4th order Runge–Kutta method with time step $\Delta t = 0.01 = \Delta$. The optimization problem is solved by *pseudo-code* 2 and the PSO method. The initial guesses for the parameters are randomly generated, based on a uniform distribution over $\{-10 \quad -10 \quad 0\}^T \leq \left\{\theta_1^a \quad \theta_2^a \quad \theta_3^a\right\}^T \leq \{51 \quad 60 \quad 40\}^T$. For both the methods, the ensemble size is 30. From Figs 9.11(a-f), the better convergence features of *pseudo-code* 2 is evident. Indeed, the substantively reduced numerical fluctuations via *pseudo-code* 2 may be largely attributed to the additive update that, owing to the encoded directional information, enables a better exploration–exploitation trade-off. The error norms in the parameters estimated by *pseudo-code* 2 are consistently lower by several orders vis-à-vis those based on the PSO.

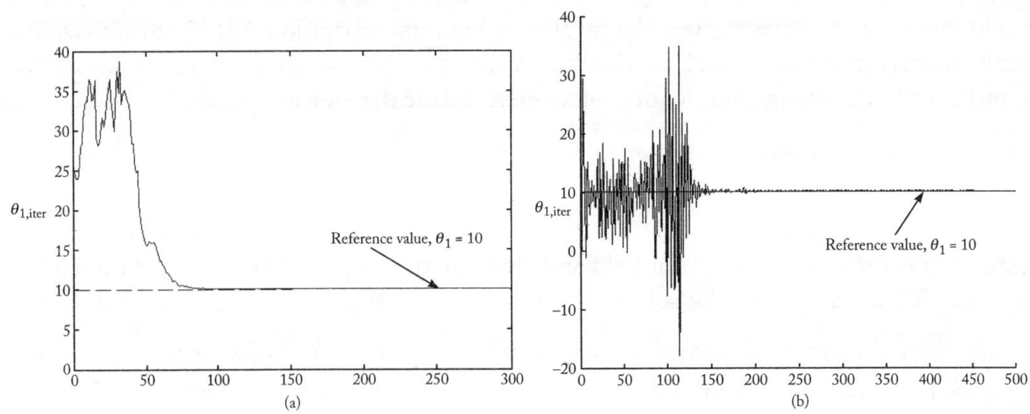

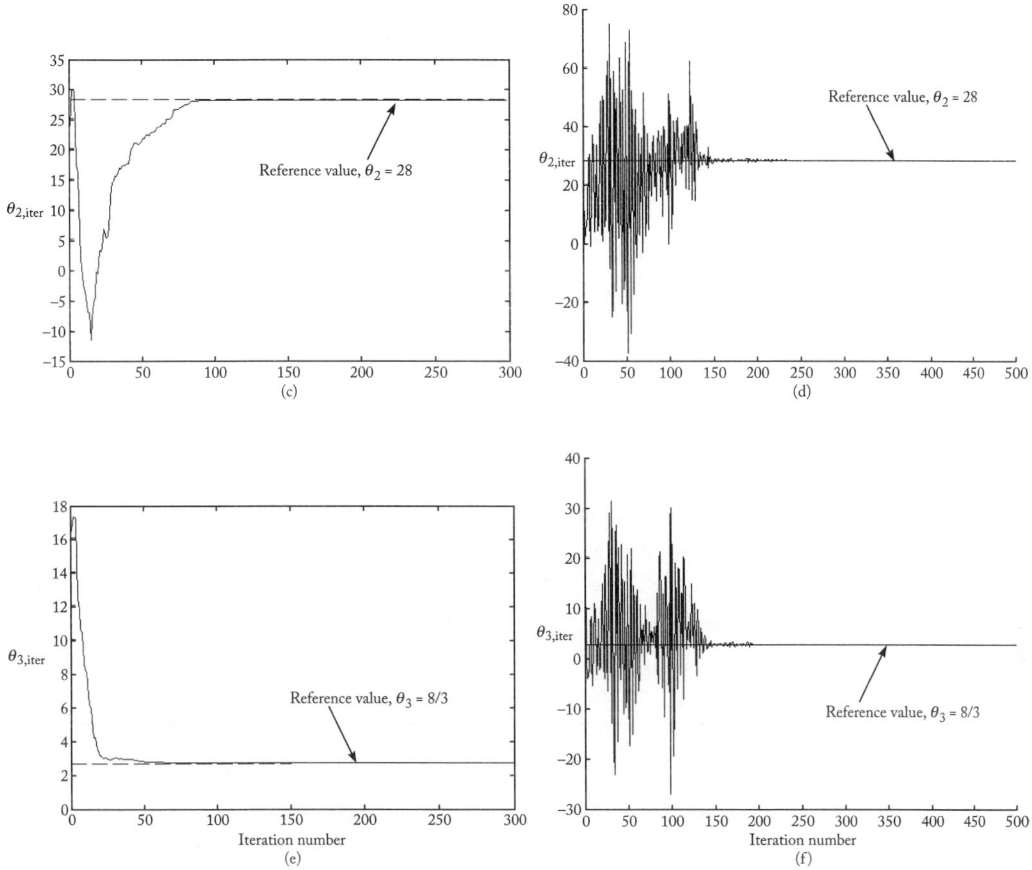

Fig. 9.11 Lorenz oscillator (Example 9.3)—identification of the system parameters in chaotic regimes; $N_p = 30$, (a), (c) and (e) results by the pseudo-code 2 and (b), (d) and (f) results by the PSO

The convergence of the *pseudo-code* 2 is shown in Fig. 9.12 (for the three parameters of the Lorenz oscillator) with different ensemble sizes: $N_p = 10, 20$, and 30. Fairly good convergence is achieved even with N_p as low as 10, though there are larger fluctuations in the initial stages of iteration.

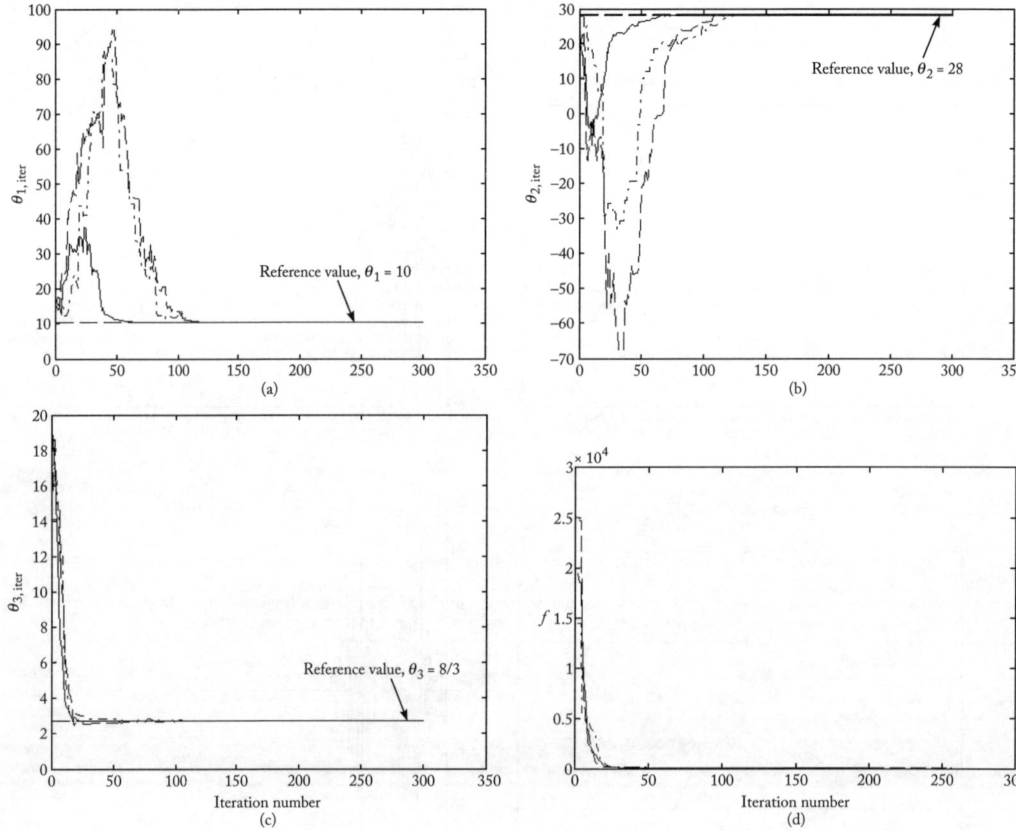

Fig. 9.12 Lorenz oscillator (Example 9.3)—identification of the system parameters in chaotic regimes; results by the pseudo-code 2, (a)–(c) evolution of the three parameters, θ_1, θ_2 and θ_3 and d) evolution of the objective function, $f(\boldsymbol{\theta}^a)$, dash-dot line—$N_p = 10$, dashed-line—$N_p = 20$, solid-line—$N_p = 30$

In obtaining the results in Figs 9.11 and 9.12, the updates at any iteration, τ_i, are obtained by Eq. (9.1), with $\boldsymbol{D}_i^{[\varsigma_2(j)]} := \beta_i \boldsymbol{G}_i \boldsymbol{I}_i^{[\varsigma_2(j)]}$, $j = 1, 2, \ldots, N_p$ where the innovation vector is $\boldsymbol{I}_i^{[\varsigma_2(j)]} := \left\{ \begin{array}{c} \hat{f}_{i-1} - f_{i-1}^{[\varsigma_2(j)]} \\ \boldsymbol{X}_{i-1}^{(\varsigma_1(j))} - \boldsymbol{X}_{i-1}^{[\varsigma_2(j)]} \end{array} \right\}$. In the absence of prior knowledge on the mean of the extremal cost function, $\hat{f}_{i-1}$ is taken at each iteration as the (available) minimum of the cost function.

Example 9.4

Chen's oscillator [Chen and Ueta 1999]

As another illustration of parameter identification (by optimization) in the chaotic regime, we consider Chen's oscillator [Chen and Ueta 1999] given by:

$$\dot{\boldsymbol{X}}_t = \begin{Bmatrix} \theta_1 (Y_t - X_t) \\ (\theta_3 - \theta_1) X_t + \theta_3 Y_t - X_t Z_t \\ X_t Y_t - \theta_2 Z_t \end{Bmatrix} \tag{9.52}$$

As before, $\boldsymbol{X}_t := \{X_t \quad Y_t \quad Z_t\}^T$ is the state vector and $\boldsymbol{\theta} := \{\theta_1 \quad \theta_2 \quad \theta_3\}^T$, the vector of unknown parameters. Here $\boldsymbol{\theta} := \{35 \quad 3 \quad 28\}^T$ is the reference parameter set that corresponds to Eq. (9.52) exhibiting chaotic response. The reference solution is generated by numerically integrating this equation with the (reference) parameters as input. As with the last example, numerical simulations of Eq. (9.52), required in the global search, are over $t \in (0.0.3 \text{ s}]$ using a 4th order Runge–Kutta method with time step $\Delta t = 0.01 = \Delta$. Initial guesses of the parameters are sampled from a uniform distribution over $\{-10 \quad -10 \quad 0\}^T \le \left\{\theta_1^a \quad \theta_2^a \quad \theta_3^a\right\}^T \le \{51 \quad 60 \quad 40\}^T$. The numerical results in Figs 9.13a–f (obtained with $N_p = 30$) support similar observations as in the case of the Lorenz oscillator regarding the performance of the *pseudo-code* 2. Equation (9.48) is used in obtaining the updates at each iteration.

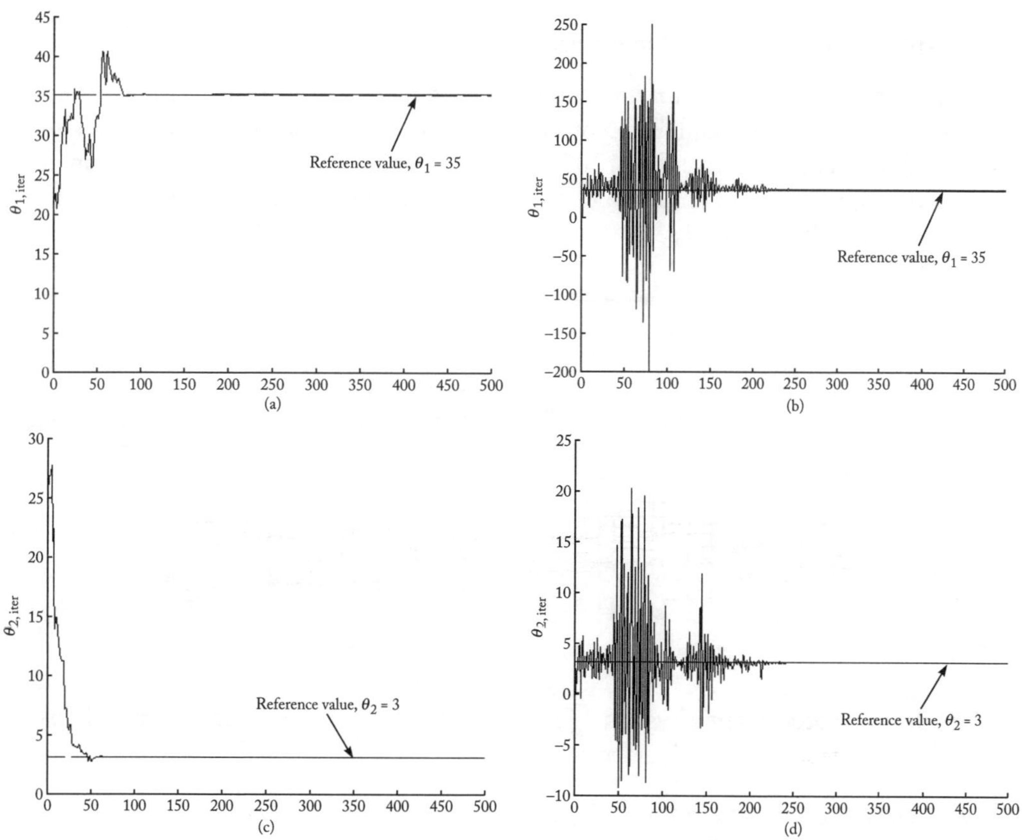

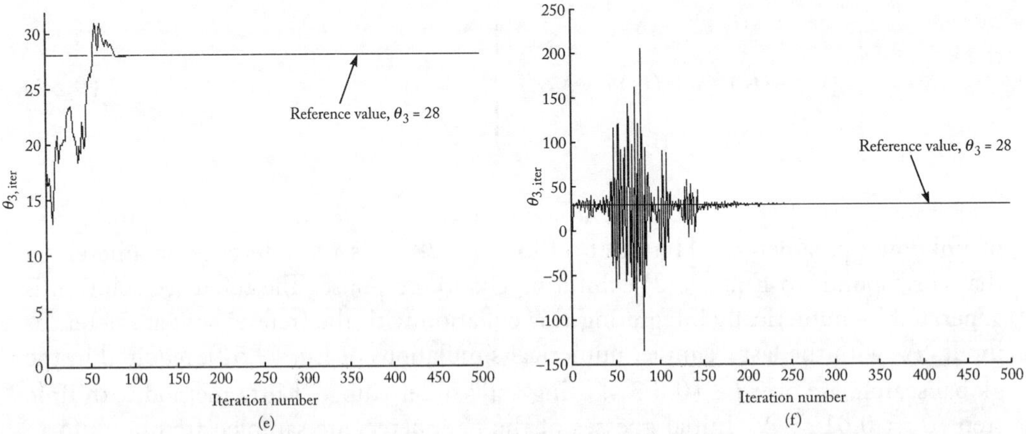

Fig. 9.13 Chen's oscillator—identification of the system parameters in the chaotic regime; $N_p = 30$ (a), (c) and (e) results by the pseudo-code 2 and (b), (d) and (f) results by the PSO

Example 9.5

Rotor dynamics and structural identification:

The inverse problem of identifying unknown parameters in a rotor bearing systems [Rajan *et al.* 1987, Diewald and Nordmann 1990, Wang and Shih 1990, Choi and Yang 2000, Assis and Steffen 2003] (based on any experimental measurement, assumed noise-free) is of importance in the study of their stability analysis, critical speeds and unbalance responses.

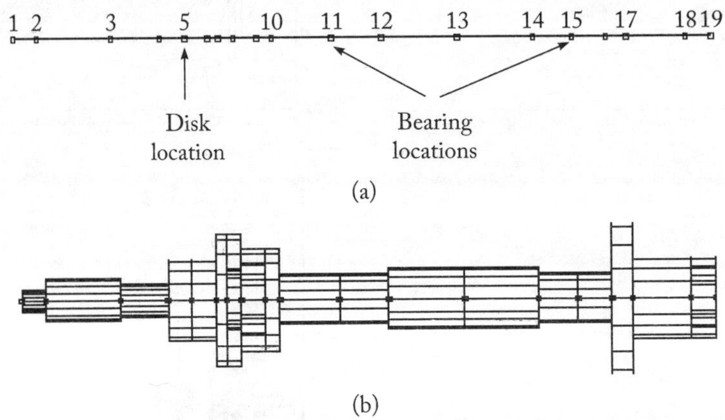

Fig. 9.14 Rotor bearing system. (a) FE model with total dofs $m = 76$ and (b) geometric configuration

In this example, a flexible rotor shaft [Nelson and McVaugh 1976] of length 0.3581 meter is considered. It is of a non-uniform circular cross-section and is supported on hydrodynamic bearings. It carries a concentrated disk at a distance of 0.0889 m from the left end. Figure 9.14 shows the geometric configuration and Table 9.8 contains the relevant geometric details. The FE (Finite Element) model of the rotor bearing system using beam elements is also shown in Fig. 9.14.

For details on the finite modeling of the rotor bearing systems, the reader may refer to the following literatures [Nelson and McVaugh 1976, Zorzi and Nelson 1977, Lalanne and Ferraris 1998, and Roy and Rao 2012]. Each beam element is represented by two nodes with each node carrying two transverse and two rotational *dof*s. Out of the 4 *dofs* per node, the first two stand for the two transverse displacements and the other two for the rotational ones. Thus, for the system in Fig. 9.14 the total number of *dof*s denoted by m is 76. The shaft has a disk element (with added mass/inertia) attached at node 5. The system equation of motion is of the form:

$$\boldsymbol{M}\ddot{\boldsymbol{X}} + (\boldsymbol{C} + \boldsymbol{G})\dot{\boldsymbol{X}} + \boldsymbol{K}\boldsymbol{X} = \mathbb{F}(t) \tag{9.53}$$

$\boldsymbol{M}, \boldsymbol{K}$ and $\boldsymbol{C}$ are, respectively, the system mass, stiffness and viscous damping matrices of size, $m \times m$. $\boldsymbol{X}(t)$ is the $m \times 1$ displacement response vector. The bearing stiffness and damping matrices (absorbed in $\boldsymbol{K}$ and $\boldsymbol{C}$) are, respectively, denoted by $\boldsymbol{K}_b$ and $\boldsymbol{C}_b$ (each of size 2×2) and they may be unsymmetric. $\boldsymbol{G}$ is the skew-symmetric matrix arising out of the Coriolis effect. In this example, the two bearings shown at nodes 11 and 15 are considered to be isotropic [Zorzi and Nelson 1977] with no cross-coupling terms so that:

$$\boldsymbol{K}_b = \begin{bmatrix} K_{11} & 0 \\ 0 & K_{22} \end{bmatrix}, \boldsymbol{C}_b = \begin{bmatrix} C_{11} & 0 \\ 0 & C_{22} \end{bmatrix} \tag{9.54}$$

Since the stiffness and damping coefficient matrices are symmetric, the only source of asymmetry in the system equations of motion is the Coriolis effect due to the shaft rotation λ. $\mathbb{F}(t)$ is the taken to be the excitation force vector due to a possible mass imbalance on the shaft. An imbalance in the form of eccentricity is assumed at the disk element (node 5 in the FE model).

Table 9.8 Rotor bearing system—shaft geometry

Element number	Axial distance from left end in m.	Inner radius in m.	Outer radius in m.	Element number	Axial distance from left end in m.	Inner radius in m.	Outer radius in m.
1	0.0		0.0051	10	0.1346		0.0127
2	0.0127		0.0102	11	0.1651		0.0127

5	0.0889		0.0203	14	0.2667		0.0127
3	0.0508		0.0076	12	0.1905		0.0152
4	0.0762		0.0203	13	0.2286		0.0152
6	0.1016		0.0330	15	0.2870		0.0127
7	0.1067	0.0152	0.0330	16	0.3048		0.0381
8	0.1141	0.0178	0.0254	17	0.3150		0.0203
9	0.1270		0.0254	18	0.3454	0.0152	0.0203

Hence, the two non-zero elements $\mathbb{F}_{17}$ and $\mathbb{F}_{18}$ of the vector $\mathbb{F}(t) \in \mathbb{R}^m$ are of the form $A \sin(\lambda t + \Phi)$. Here the subscripts 17 and 18 in $\mathbb{F}_{17}$ and $\mathbb{F}_{18}$ refer to the *dof*s corresponding to the two transverse directions at the node 5. The force amplitude is $A = m_d \varepsilon \lambda^2$ where ε is the eccentricity of the disk and m_d is the mass of the disk. In the computational work, m_d and ε are assumed to be 1.4 kg and 0.001 mm, respectively. The unbalance response is obtained by numerically solving Eq. (9.53) with the reference parameters $K_{11} = K_{22} = 4.378\text{E}07$ N/m and $C_{11} = C_{22} = 2.627\text{E}03$ N-s/m. λ varies in the range of 1500–1800 rad/s. The unbalance responses at the two bearing locations (Fig. 9.15) obtained in the frequency domain [Nelson and McVaugh 1976] are used in the optimization scheme to construct the cost function. Given the reference solution in Fig. 9.15, identification of the unknown bearing parameters—K_{11}, K_{22}, C_{11} and C_{22}—is posed as an inverse problem, presently solved via the global optimization route.

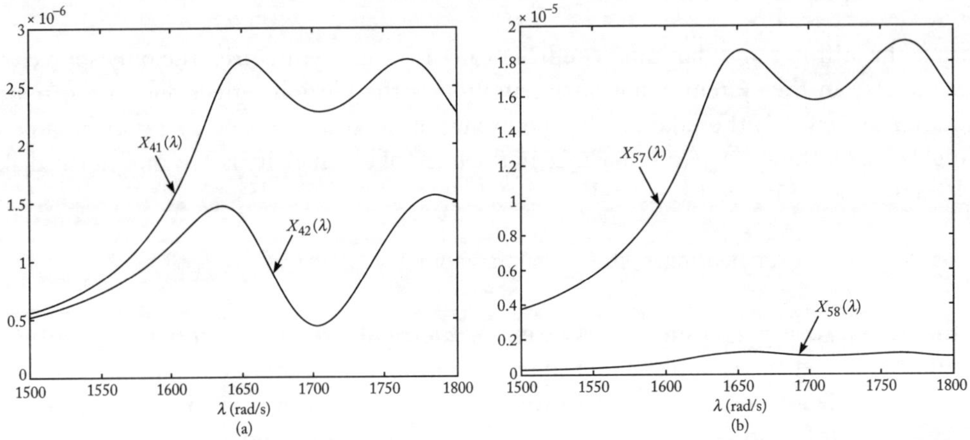

Fig. 9.15 The (reference) unbalance response of the rotor bearing system (Fig. 9.14); (a) responses in the two transverse directions (*dofs* 41, 42) at the bearing location at node 11, (b) responses in the two transverse directions (*dofs* 57, 58) at the bearing location at node 15

With $\boldsymbol{\theta}^a = (K_{11}, K_{22}, C_{11}, C_{22})^T$, the initial guesses are randomly generated, based on a uniform distribution over

$$(1E07,\ 1E07,\ 1E03,\ 1E03)^T \leq (\theta_1^a\ \ \theta_2^a\ \ \theta_3^a\ \ \theta_4^a)^T \leq (6E07,\ 6E07,\ 4E03,\ 4E03)^T \quad (9.55)$$

In the process of parameter identification, $f(\boldsymbol{\theta}^a) = \sum_{i=1}^{n_f} \left(\mathbf{X}_i - \mathbf{X}_i^a\right)^T \left(\mathbf{X}_i - \mathbf{X}_i^a\right)$ (as defined in Eq. 9.4), $\mathbf{X}(\lambda) = (X_{41}(\lambda), X_{42}(\lambda), X_{57}(\lambda), X_{58}(\lambda))^T$. n_f denotes the number of discrete frequency points on the λ-axis. The subscripts 41 and 42 to X in the vector $\mathbf{X}$ stand for the two transverse *dof* s at node 11 and subscripts 57 and 58 for the *dof* s at node 15. The ensemble size N_p is taken as 30. With maximum number of iterations being 200, the numerical result for the evolving cost function obtained by the *pseudo-code* 2 is presented in Fig. 9.16. The ability of the optimization scheme in reconstructing the bearing parameters (stiffness and damping) is manifest in Fig. 9.17.

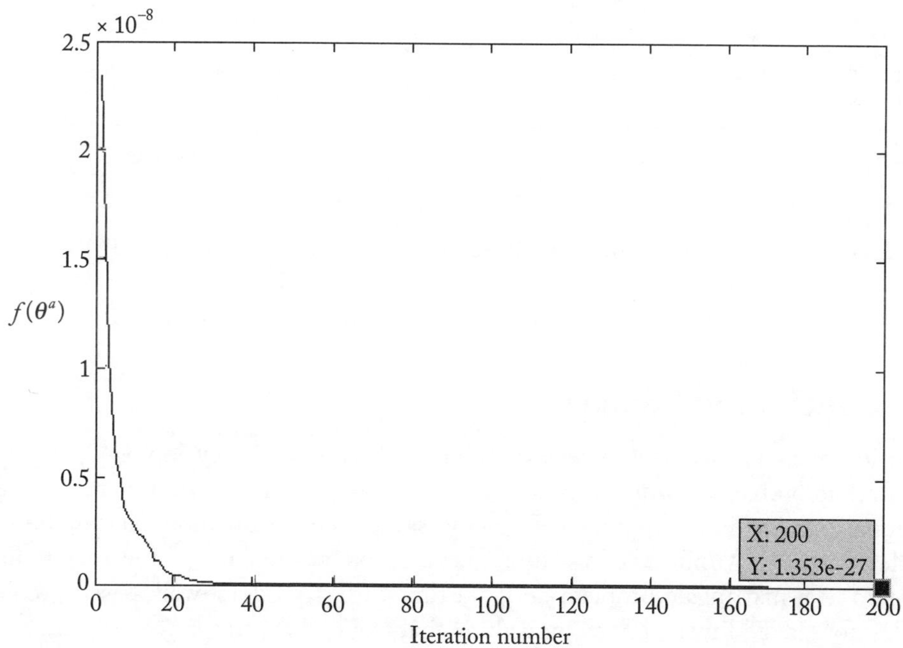

Fig. 9.16 Parameter identification of the rotor bearing system (in Fig. 9.14); evolution of $f(\boldsymbol{\theta}^a)$

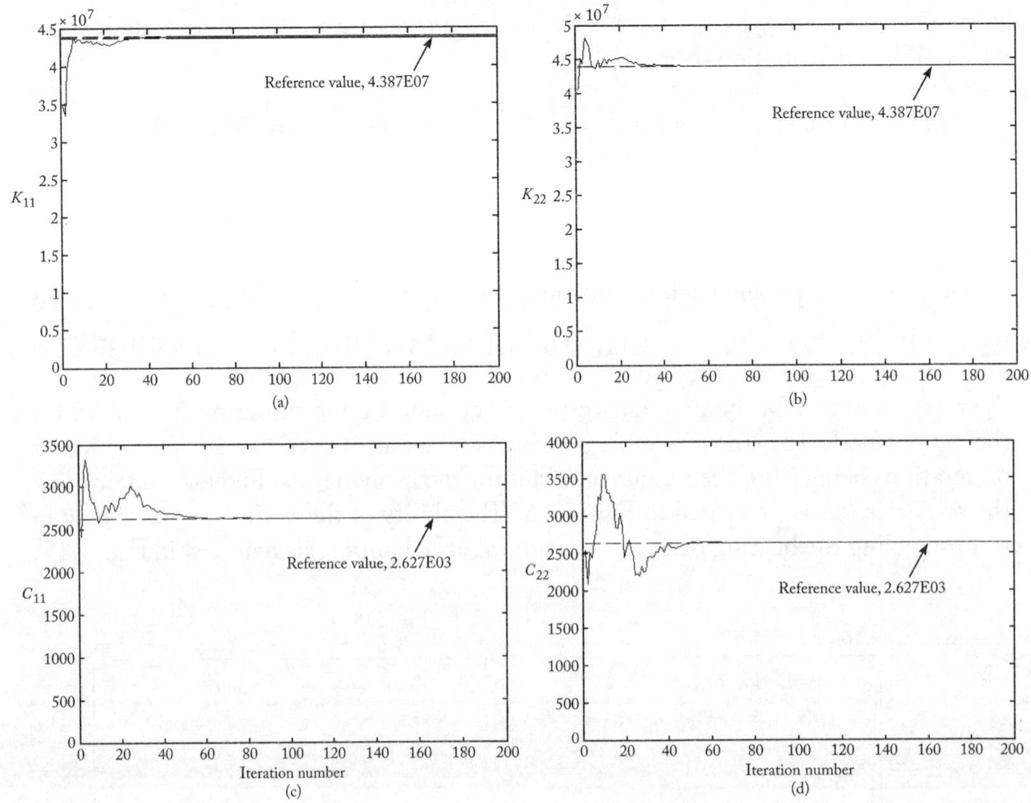

Fig. 9.17 Parameter identification of the rotor bearing system (in Fig. 9.14); $N_p = 30$, evolution of (a) stiffness parameter K_{11}, (b) stiffness parameter K_{22}, (c) damping parameter C_{11}, and (d) damping parameter C_{22}

9.7 Concluding Remarks

The notion of change of measures is exploited as a basic tool in developing algorithms for optimization problems, wherein solutions are guided mainly through derivative-free directional information computable from the sample statistical moments of the design variables within a Monte Carlo setup. The method poses part of the search for the extremum as a martingale problem and, starting with a random–walk based prediction, provides for an additive update procedure that could be interpreted as a non-trivial generalization over the directional derivative based update, classically employed for local extremizations of a smooth cost functional. By way of a further aid to the global search, i.e., in order to preempt possible stalling at local extrema, layers of random perturbation of the martingale problem, herein called the scrambling and coalescence strategies, are described. As demonstrated in the numerical work on minimizing suitable cost functionals *en route* to the parameter recovery of a couple of dynamical systems, the method marks an improvement over some of the relatively more well-known evolutionary stochastic search schemes. In the global optimization scheme described in this chapter, choices of scrambling and coalescence strategies are highly non-unique. We look into this aspect in some detail in the next chapter. We also show that it is also possible to improve the performance of the scheme by means of a sub-structuring of the design space. This and an exploration of different scrambling and coalescence strategies that potentially renders the presently adopted approach more competitive is the subject of our last chapter.

Notations

$\check{B}_t$	vector Brownian motion (Eq. 9.24 – measurement like equation)
c	damping coefficient
c_1, c_2	learning factors of a particle in PSO
c_r	learning rate in CMA-ES
C	damping matrix (Example 9.5)
C_b	bearing damping matrix
$\boldsymbol{C}_i, \boldsymbol{C}_j, i, j \in (1, N_p)$	two off-springs (in GA)
C_{11}, C_{22}	bearing damping coefficients
$\mathcal{C}_k$	covariance matrix in CMA-ES at iteration k^{th} iteration
$\widetilde{\boldsymbol{C}}_i^{[j]}, \widetilde{\boldsymbol{D}}_i^{[j]}$	update vectors for the jth particle at i^{th} iteration
$f(.)$	cost (objective) function
$f^{[i]}$	fitness (objective function) value for i^{th} particle
$f(\boldsymbol{\theta}^a)$	fitness (objective function) value at the current parameter vector $\boldsymbol{\theta}^a$ (Eq. 9.4)
$\boldsymbol{f}$	vector of cost (objective) functions
$\check{f}_\tau$	measurable process (Eq. 9.19)
$\mathbb{F}(t)$	excitation force vector in Example 9.5
$\mathbb{F}_{17}, \mathbb{F}_{18}$	elements of the force vector $\mathbb{F}(t)$ at 17th and 18th *dofs*
g_{best}	best position vector (in PSO) in the swarm
G	gyroscopic matrix (Example 9.5)

$\widetilde{\boldsymbol{G}}_\tau$	gain-like update coefficient matrix (Eq. 9.39 and 9.44)
$\boldsymbol{h}(\boldsymbol{X}_\tau)$	process of drift coefficients (Eq. 9.24)
$\boldsymbol{H}$	Hessian matrix
$\widetilde{\boldsymbol{I}}_i^{[j]}$	innovation vector for j^{th} particle at j^{th} iteration
k	stiffness coefficient
$\boldsymbol{K}$	stiffness matrix (Example 9.5)
$\boldsymbol{K}_b$	bearing stiffness matrix
K_{11}, K_{22}	bearing stiffness coefficients
	an integer. the dimension of a multi-objective cost functional f
m_d	mass of the disk in Example 9.5
$\boldsymbol{M}$	mass matrix
$\mathfrak{M}_k$	vector of mean values in CMA–ES
$\mathfrak{N}_\tau$	filtration
$\boldsymbol{p}_k^c, \boldsymbol{p}_k^s$	parameters at k^{th} iteration in CMA–ES
$\boldsymbol{p}_{\text{best}}^{[j]}$	best position vector (in PSO) for each jth particle
$\boldsymbol{P}_i, \boldsymbol{P}_j$	parent members (chromosomes) of the population (in GA)
r_1, r_2	random numbers uniformly distributed in $[0,1]$ (in PSO)
$\boldsymbol{r} = \{1,1,\ldots,1\} \in \mathbb{R}^{N_p}$	N_p-dimensional row vector with entries 1
s_k	overall standard deviation or the step size at k^{th} iteration in CMA-ES
T	temperature (control parameter in the cooling schedule of SA)
$\mathcal{T}_k$	vector after cross-over operation in DiEv
$\boldsymbol{U}_k$	parameter vector after mutation operation in DiEv
$\boldsymbol{V}_k$	velocity vector (in PSO), initial vector in DiEv
w	weight parameter (inertia weight) in PSO
w_l	positive weight coefficients for l^{th} particle in CMA–ES
$\left(X_1^{[j]}, X_2^{[j]}, \ldots, X_m^{[j]}\right)$, $j \in (1, N_p)$	parameters in a m-dimensional optimization problem

$\boldsymbol{X}^{[j]}$	position vector in PSO, vector of parameters
$\widetilde{\boldsymbol{X}}_\tau$	predicted candidates
$\boldsymbol{Y}_k$	normalized random variable
α	positive real constant
β_τ	annealing-type coefficient (Eq. 9.41)
$\delta_\tau^{[j]}$	random scatter around $X_\tau^{[j]}$
ε	eccentricity of the disk element in Example 9.5
η_t	Brownian motion
$\boldsymbol{\theta}, \boldsymbol{\theta}^a$	vectors of system parameters
$\theta_1, \theta_2, \theta_3$	parameters of Lorenz and Chen oscillators in Eqs 9.51 and 9.52, respectively
λ	harmonic frequency in rad/s
$\boldsymbol{\rho}_\tau$	noise intensity matrix
$\zeta_1(j), \zeta_2(j)$	random permutation on the indexing set $\{1, N_p\}/\{j\}$
σ_k^2	covariance matrix at k^{th} iteration in CMA–ES
τ	index for iteration
ψ	non-linear vector field

Chapter **10**

COMBEO—A New Global Optimization Scheme By Change of Measures

10.1 Introduction

In the last chapter, we have laid down a procedure to solve an optimization problem by first posing it as a martingale problem (see Section 4.12, Chapter 4), whose solution may lead to a local extremization of the cost functional. The stochastic search is next guided to reach global maximum by random perturbation strategies—coalescence and scrambling—specifically devised for the purpose. To realize a single reliable scheme that satisfies the diverse and conflicting needs of an optimization problem defined in terms of multi-cost functions under prescribed constraints is a tough task [Fonseca and Fleming 1995, Deb 2001]. This chapter addresses precisely this issue and considers some modifications to the skeletal optimization approach considered in the last chapter so as to impart greater flexibility with which the innovation process may be designed in the presence of conflicting demands *en route* to the detection of the global extremum. The efficiency of the global search basically relies upon the ability of the algorithm to explore the search space whilst preserving some directionality that helps in quickly resolving the nearest extremum. The development of the modified setup, referred to as COMBEO (Change **O**f **M**easure **B**ased **E**volutionary **O**ptimization), recognizes the near impossibility of a specific optimization scheme performing uniformly well across a large class of problems. Recognition of this fact had earlier [Hart *et al.* 2005, Vrugt and Robinson 2007] led to a proposal of an evolutionary scheme that simultaneously ran different optimization methods for a given problem with some communications built amongst the updates by the different methods. We herein similarly aim at combining a few of the basic ideas for global search used with some well-known optimization schemes under a single unified framework propped up by a sound probabilistic basis.

In a way to be explained in the sections to follow, the ideas (or their possible generalizations) behind some of the existing optimization methods may sometimes be readily included in the present setting—COMBEO—often by just tweaking the innovation process and attempting to incorporate the best practices of some of the available stochastic search methods.

10.2 COMBEO—Improvements to the Martingale Approach

Solution to a global optimization problem [Sarkar *et al.* 2015] requires that the posterior probability distribution associated with the converged solution be unimodal irrespective of the likely multi-modality of the cost function. Ideally, then, the converged measure $\pi_t(.)$ should correspond to a Dirac measure $\delta_{\boldsymbol{X}_{\min}}$ where $\boldsymbol{X}_{\min}$ denotes the global optimum. The unimodality constraint on the probability density may be imposed by coalescence, i.e., by forcing the particles to evolve according to a Brownian martingale whose mean is given by the available global extremum (Chapter 9). In contrast, coalescence may be ineffective or inappropriate in pulling out the particles stuck in local traps, a possibility rendered likely owing to the very small updates imparted to particles around the extremal (local or global) values. A stalling of the search scheme may be circumvented by scrambling, i.e., by randomly swapping the updates of two particles. The efficiency of the global search in the EnKS-type optimization scheme described in Chapter 9 (Section 9.5) and Tables 9.6 and 9.7, basically relies on the choice of the innovation term and the random perturbation strategies—coalescence and scrambling. Noting that these strategies are highly non-unique and based on the observed performance of the scheme against a few optimization problems in the last chapter, we consider a few substantial modifications of the optimization algorithm as incorporated in COMBEO. The approach is a graded one in that an attempt is made, via suitable illustrations, to bring out the specific ramifications of each modification on the original algorithm (the *pseudo code* 2 in Table 9.7).

10.2.1 Improvements to the coalescence strategy

Since coalescence is posed as a martingale problem, the associated update with any additional modification should share the same generic structure as Eq. (9.43a) of Chapter 9. In this context, a real strength of the coalescence step lies in the non-unique choice of the innovation vector—a feature that enables one to organize a powerful global search in the presence of conflicting costs. Thus one may also borrow some basic concepts from different existing global optimization schemes and adapt them within the present coalescence framework. For example, the personal and global information used in the PSO may be incorporated here by constructing the innovation vector as (in line with the *pseudo code* 2 in Table 9.7):

$$\boldsymbol{I}_i^{[\varsigma_2(j)]} = \left\{ \begin{array}{c} {}^{P}\boldsymbol{X}_{i-1}^{[\varsigma_1(j)]} - \boldsymbol{X}_{i-1}^{[\varsigma_2(j)]} \\ \\ {}^{G}\boldsymbol{X}_{i-1} - \boldsymbol{X}_{i-1}^{[\varsigma_2(j)]} \end{array} \right\} \in \mathbb{R}^{2m}, j = 1, 2, \ldots, N_p \tag{10.1}$$

Here ${}^{P}\boldsymbol{X}_{i-1}^{[j]} \in \mathbb{R}^m$ is the personal best location corresponding to the jth particle based on its evolution history till τ_{i-1}. ${}^{G}\boldsymbol{X}_{i-1} \in \mathbb{R}^m$ is the available best location among all the particles in the ensemble at τ_{i-1} over the same evolution history and thus $\boldsymbol{I}_i^{[j]} \in \mathbb{R}^{2m}$. $\varsigma_1(j)$ and $\varsigma_2(j)$ are random permutations on the integer set $\{1, N_p\}\backslash\{j\}$. Then the jth candidate or particle is updated as (see Eq. 9.48):

$$\boldsymbol{X}_i^{[j]} = \boldsymbol{X}_{i-1}^{[j]} + \boldsymbol{D}_i^{[\varsigma_2(j)]}, \boldsymbol{D}_i^{[\varsigma_2(j)]} = \beta_i \boldsymbol{G}_i \boldsymbol{I}_i^{[\varsigma_2(j)]} \tag{10.2}$$

$\boldsymbol{G}_i$ is the same as in Eq. (9.49) and is rewritten hereunder for a ready lookup

$$\boldsymbol{G}_i = \frac{1}{N_p}\left\{\left(\boldsymbol{X}_i - \overline{\boldsymbol{X}}_i\right)\left(\boldsymbol{F}_i^T\tau_i - \overline{\boldsymbol{F}}_{i-1}^T\tau_{i-1} - \Delta\overline{\boldsymbol{F}}_i^T\tau_i\right) + \left(\overline{\boldsymbol{X}}_i\tau_i - \overline{\boldsymbol{X}}_{i-1}\tau_{i-1}\right)\left(\boldsymbol{F}_i - \overline{\boldsymbol{F}}_i\right)^T\right\}$$

$$\left[\alpha\frac{1}{N_p - 1}\left(\left(\widehat{\boldsymbol{F}}_i - \boldsymbol{F}_i\right) - \left(\widehat{\overline{\boldsymbol{F}}}_i - \overline{\boldsymbol{F}}_i\right)\right)\left(\left(\widehat{\boldsymbol{F}}_i - \boldsymbol{F}_i\right) - \left(\widehat{\overline{\boldsymbol{F}}}_i - \overline{\boldsymbol{F}}_i\right)\right)^T + (1-\alpha)\gamma_i\gamma_i^T\right]^{-1} \tag{10.3}$$

$$\text{where, } \gamma_i\gamma_i^T := \begin{bmatrix} \rho_i^c\left(\rho_i^c\right)^T & \boldsymbol{0} \\ \boldsymbol{0} & \rho_i^c\left(\rho_i^c\right)^T \end{bmatrix}.$$

We recall from Chapter 9 that $\boldsymbol{X}_\tau := \left[\boldsymbol{X}_\tau^{[1]}, \boldsymbol{X}_\tau^{[2]}, \ldots, \boldsymbol{X}_\tau^{[N_p]}\right] \in \mathbb{R}^{m \times N_p}$. Since the innovation vector, $\boldsymbol{I}_i \in \mathbb{R}^{2m \times N_p}$, one has $\boldsymbol{F}_\tau, \overline{\boldsymbol{F}}_\tau \in \mathbb{R}^{2m \times N_p}$. m is the dimension of $\boldsymbol{X}$, the parameter set. N_p is the ensemble size, $\pi^{N_p}(.) := \frac{1}{N_p}\sum_{j=1}^{N_p}(.)^{[j]}$. Also, $\boldsymbol{\rho}^c(\boldsymbol{\rho}^c)^T \in \mathbb{R}^{m \times m}$ is the covariance matrix (generally diagonal) corresponding to each of ${}^{P}\boldsymbol{X}_{i-1}$ and ${}^{g}\boldsymbol{X}_{i-1}$ (see discussion on *Operator* **T1**: *Local search and coalescence*- Chapter 9, Section 9.5.1). In contrast to the original PSO (Table 9.4), the gain-type coefficient matrix here is founded on a probabilistic basis and hence its matrix structure enables iteration-dependent differential weighting of the scalar components of the innovation vector, thereby yielding faster convergence without losing the exploratory efficacy of the global search. Indeed, it is well recognized that the use of three parameters (e.g., the cognitive and social learning factors and the inertia weight) in the original PSO may lead to solutions with sensitive dependence upon the chosen parameters. It may even be possible to incorporate within the current setup the basic ideas driving some of the augmented PSO schemes, which attempt at removing some of the shortfalls of the original PSO [Sun *et al.* 2010].

Figure 10.1 shows a comparison of results obtained by (i) using Eq. (9.48) of the original *pseudo code* 2 and (ii) employing the modified coalescence (based on the PSO) in Eq. (10.1). Figure 10.1d shows the evolution of $f(\boldsymbol{\theta}^a)$, the cost functional as defined in Eq. (9.4). Convergence is achieved by the modified scheme with N_p as low as 10 as against $N_p = 30$

used in Example 9.3. In this modified coalescence strategy, information on the personal best of the jth particle (till τ_i) is included in the innovation vector $\boldsymbol{I}_i^{\varsigma_2(j)}$. Thus, the vector for scrambling of particles is taken as $\boldsymbol{I}_i^{[\varsigma_2(j)]} := \left\{ {}^{P}\boldsymbol{X}_{i-1}^{[\varsigma_1(j)]} - \boldsymbol{X}_{i-1}^{[\varsigma_2(j)]} \right\}$. For the present example, inclusion of ${}^{g}\boldsymbol{X}_{i-1}$ into the innovation vector is found to have no added advantage. In obtaining the results, the prediction step is avoided as in the *pseudo code* 2 (Table 9.7 of Chapter 9).

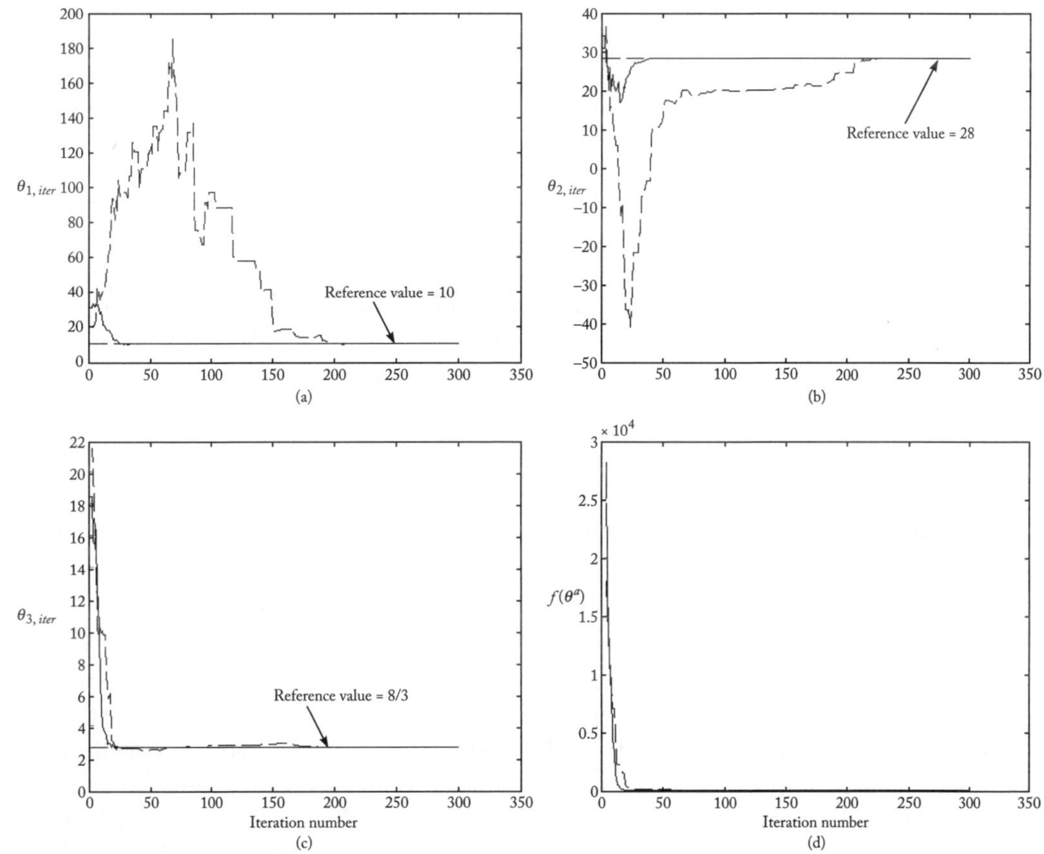

Fig. 10.1 Lorenz oscillator (see Example 9.3) revisited—identification of the system parameters in the chaotic regime; $N_p = 10$, dashed-line—results by Eq. 9.48; solid-line—results by the modified coalescence (based on the PSO) in Eq. (10.1).

10.2.2 Improvements to scrambling and introduction of a relaxation parameter

The scrambling as defined in Eq. (9.48) enables the particle updates to move out of the local traps and thus prevents a possible stalling of the algorithm. For instance, the scrambling there in involves swapping of the gain–weighted directional information for the jth particle with that of another randomly selected one. This scheme may be found to be promising in achieving better convergence with only a few particles (see Examples 9.3 to 9.5).

Here one may also borrow a basic idea from the DiEv to improve over the scrambling strategy. Probably at the cost of a somewhat slower convergence for a class of problems, a more effective modification of the scrambling strategy may be contemplated with a view to ensuring that the particles explore even more in the search space. Noting that every particle is an m-dimensional vector, one may execute the scrambling operation separately (m times) for the individual scalar components of the particles instead of swapping the update terms for the particles as a whole. This improvement, enabling element crossovers across different particles, allows the particles to assume larger variations (or diversity, to borrow from the usual parlance) and prompt them to explore more. With $r \in [1, m]$, the resulting update equation is then given by:

$$X_{r,i}^{[j]} = X_{r,i-1}^{[j]} + D_{r,i}^{[\varsigma_2(j)]}, D_{r,i}^{[\varsigma_2(j)]} = \beta_i \boldsymbol{G}_{r,i} . \boldsymbol{I}_{r,i}^{[\varsigma_2(j)]} \tag{10.4}$$

where, for instance, $X_{r,i}^{(j)}$ denotes the updated r^{th} scalar element of the jth particle at τ_i and

$$\boldsymbol{I}_{r,i}^{[\varsigma_2(j)]} = \left\{ \begin{array}{c} \hat{f}_{i-1} - f_{i-1}^{[\varsigma_2(j)]} \\ \\ X_{r,i-1}^{[\varsigma_1(j)]} - X_{r,i-1}^{[\varsigma_2(j)]} \end{array} \right\}, \; j = 1, 2, \ldots, N_p.$$

Note that the innovation vector here is retained in the same form as in Eq. (9.48) and $\boldsymbol{G}_{r,i}$ is the rth row vector of the matrix $\boldsymbol{G}_i$ Eq. (10.3).

Relaxation parameter

Improvement may further be had by a so called relaxation strategy (taking cue from the cross-over operation in DiEv – Table 9.5), wherein the resulting update in Eq. (10.2) or (10.4) is accepted with a probability < 1. Even with the suggested modifications to coalescence and scrambling, a search scheme that always chooses to update the particles might quickly collapse to a local trap owing to the highly directed and coalesced search imposed by the martingale problems. Such a scenario is prevented by the so-called relaxation, i.e., by assigning positive probabilities to the event of retaining particles without updates. With p_r denoting the relaxation parameter (or the inertia factor), this move ensures that the particles are updated with a probability, $(1 - p_r) < 1$. Results obtained for the same Lorenz oscillator (of Example 9.3) with the new scrambling procedure along with the relaxation approach are shown in Fig. 10.2.

Results in Figs 10.1 and 10.2 are indicative of the efficiency of the two modifications incorporated into the original coalescence and scrambling strategies. The additional relaxation is included in the algorithm for a selective choice of the update at each iteration. It is significant that convergence is realized with these modifications, again with $N_p = 10$ (as against $N_p = 30$ used in Example 9.3).

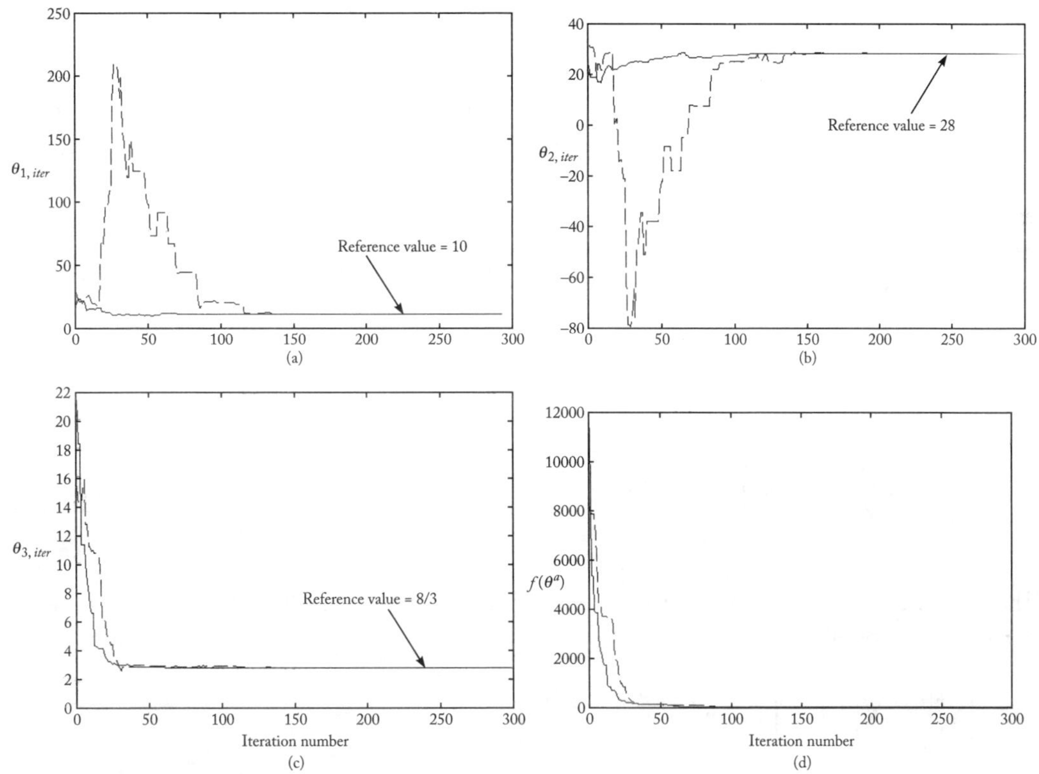

Fig. 10.2 Lorenz oscillator (see Example 9.3) revisited—identification of the system parameters in the chaotic regime; $N_p = 10$, $p_r = 0.9$, dashed-line—results by Eq. (9.48); solid-line—results by the modified scrambling strategy (based on DiEv) in Eq. (10.4)

10.2.3 Blending

If coalescence attempts at imposing the unimodality constraint, scrambling tries to avoid possible stalling of the search scheme. These mutually competing goals of the two strategies may result either in a premature collapse of the solutions to a fictitious point or a drift of solutions to infeasible regions. The coalescence-scrambling dilemma is reflective of the need for yet another (perhaps more) layer of randomness that would simultaneously ensure that the particles are repelled from the local valleys and, once at the supposed global extremum, do not drift apart. The idea for blending originated from the crossover technique in the GA [Holland 1975] wherein an offspring is created by the fusion of two individual particles. Even so, a blended particle at any iteration is generated by a linear combination of the original particle and its update rather than as a fusion of the two original particles as in the GA, where the term original refers to particles before the update.

$$\mathbf{X}_i^{b[j]} = w_i^{[j]}\mathbf{X}_{i-1}^{[j]} + \left(1 - w_i^{[j]}\right)\mathbf{X}_i^{[j]}, j = 1, 2, \ldots, N_p \tag{10.5}$$

where, $\mathbf{X}_i^{[j]}$ is the update obtained from Eq. (10.4). $\mathbf{X}_i^{b[j]}$ is the update by blending at the ith iteration. The weight vector, $\boldsymbol{w}_i$ is calculated based on a suitably defined fitness function, $\varkappa_i$, given below.

$$\varkappa_i^{[j]} = \sqrt{\left(\hat{f}_{1,i} - f_1(\mathbf{X}_i^{[j]})\right)^2 + \left(\hat{f}_{2,i} - f_2(\mathbf{X}_i^{[j]})\right)^2 + \cdots + \left(\hat{f}_{\mathfrak{m},i} - f_{\mathfrak{m}}(\mathbf{X}_i^{[j]})\right)^2} \tag{10.6a}$$

Equation (10.6a) applies to an $\mathfrak{m}$-dimensional multi-cost functional optimization where, $f = (f_1, f_2, \ldots, f_{\mathfrak{m}})^T$. We define:

$$\hat{w}_i^{[j]} = \left(\sum_{l=1}^{N_p} \varkappa_i^{[l]} w_i^{[l]}\right) - \varkappa_i^{[j]} w_i^{[j]} \tag{10.6b}$$

The weights $\boldsymbol{w}_i$ are obtained by normalizing $\widehat{\boldsymbol{w}}_i$ to sum to 1. Equation (10.5) may be rearranged to obtain:

$$\left|\mathbf{X}_i^{b[j]} - \mathbf{X}_{i-1}^{[j]}\right| = \left(1 - w_i^{[j]}\right)\left|\mathbf{X}_i^{[j]} - \mathbf{X}_{i-1}^{[j]}\right|$$

$$\Longrightarrow \left|\mathbf{X}_i^{b[j]} - \mathbf{X}_{i-1}^{[j]}\right| \le \left|\mathbf{X}_i^{[j]} - \mathbf{X}_{i-1}^{[j]}\right| \tag{10.7}$$

The above inequality and the imposition of the coalescence step render $\left|\mathbf{X}_i^{b[j]} - \mathbf{X}_{i-1}^{[j]}\right|$ a converging sequence, i.e., given $\delta > 0\ \exists i \in \mathbb{N}$ such that for $i' > i$, $\left|\mathbf{X}_{i'}^{b[j]} - \mathbf{X}_{i'-1}^{[j]}\right| < \delta$ a.s. One may thus draw the following conclusions.

(a) At any iteration, i, let the lth and m^{th} particles be the ones with the highest and lowest values of $\varkappa$; then $w_i(l) \le w_i(m)$ and $P\left(\mathbf{X}_i^b(m) = \mathbf{X}_{i-1}(m)\right) \ge P\left(\mathbf{X}_i^b(l) = \mathbf{X}_{i-1}(l)\right)$.

(b) Upon convergence, blending chooses the original particle with higher probability.

Example 10.1

Rotor dynamics and optimization of the unbalance response with constraints on critical speeds:

The inverse problem of identifying unknown parameters in a rotor bearing system has been handled in Chapter 9 by posing it as an optimization problem. Here we directly optimize the unbalance response of the rotor bearing system (in Fig. 9.14) given some constraints on the rotor critical speeds [Rajan *et al.* 1987, Shiau and Chang 1993]. The geometrical details of the rotor shaft are listed in Table 9.8. Figure 9.14 shows the FE (Finite element) model of the rotor bearing system using beam elements [Zorzi and Nelson 1977]. The finite element method (FEM) involves a semi-discretization of the governing equations, presently using 18 beam elements with the number of nodes = 19. Given the rotor bearing system with the original shaft geometry in Table 9.8 and with isotropic bearings of stiffness values, $K_{11} = K_{22} = 4.378\text{E}07$ N/m, and damping values, $C_{11} = C_{22} = 2.627\text{E}03$ N-s/m [Zorzi and Nelson 1977], the first two critical speeds denoted by $\check{s}_1$ and $\check{s}_2$ are around 1650 and 1765 rad/s. These are identifiable from Fig. 10.3 wherein the unbalance responses at the two bearings are plotted. These responses are obtained numerically upon solving Eq. (9.53) in the frequency domain (Roy and Rao 2012) for rotation speed, $\lambda \in [500 - 2500]$ using the bearing parameters values as noted. It may be noted that, in general, the job of semi-discretization, if done through a mesh-free method [Shaw and Roy 2008; Shaw *et al.* 2008], would generally require less number of nodes and thus might lead to considerable saving in the computational overhead in the implementation of the optimization code.

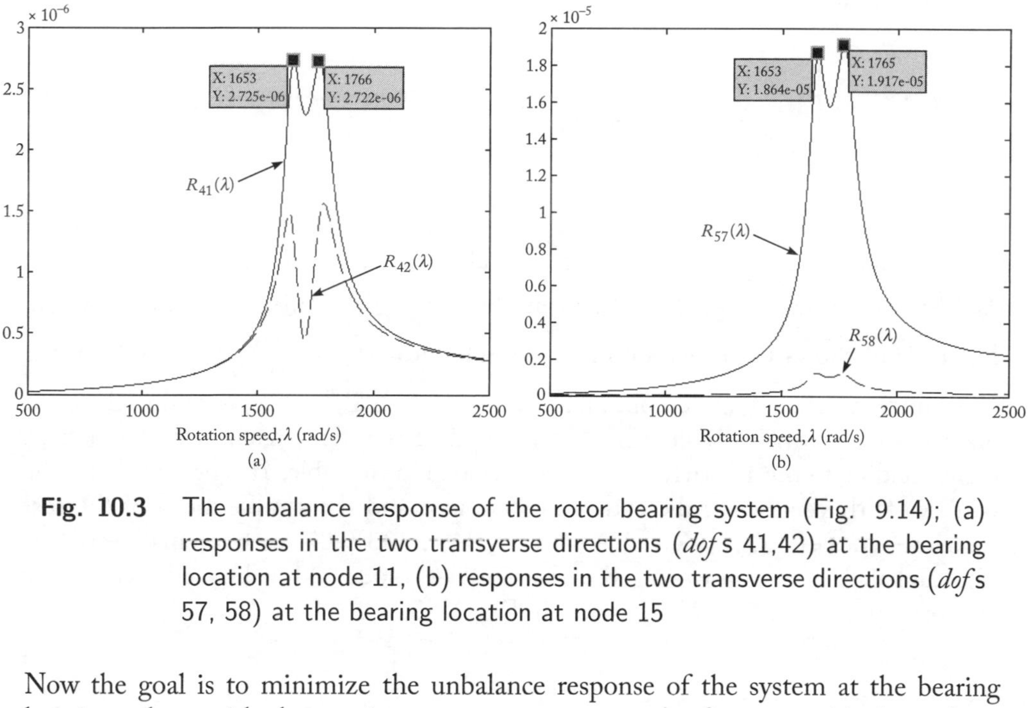

Fig. 10.3 The unbalance response of the rotor bearing system (Fig. 9.14); (a) responses in the two transverse directions (*dof*s 41,42) at the bearing location at node 11, (b) responses in the two transverse directions (*dof*s 57, 58) at the bearing location at node 15

Now the goal is to minimize the unbalance response of the system at the bearing locations along with the requirement to separate out the first two critical speeds so that they fall outside (e.g., straddle) the range 1000 – 2000 rad/s.

Solution

The shaft of the rotor bearing system is non-uniform as noticed from the plot in Fig. 9.14. The design variables are the shaft diameters collected in the vector, $\boldsymbol{X} := (X_1, X_2, \ldots, X_{15})^T$. The diameters of elements 7, 8 and 18 are kept out of $\boldsymbol{X}$, since the shaft has ring-type cross-sections for these elements with prescribed inner and outer radii (see Table 9.8). While these variations may also be incorporated into the set $\boldsymbol{X}$ of design variables, they are kept out of the purview in the present illustration. Thus $m = 15$ (i.e., each particle in COMBEO will be a 15-dimensional parameter vector). The optimization is aimed at minimizing the maximum peak response of $R_{41}(\lambda)$ (corresponding to the bearing location at node 11) under the aforementioned requirements on the first two critical speeds. We formulate the cost function $f(\boldsymbol{X})$ as:

$$f(\boldsymbol{X}) = \max_{500 \le \lambda \le 2500} R_{41}(\lambda) + (\check{s}_1 - 1000)^2 + (2000 - \check{s}_2)^2 \tag{10.8}$$

with the additional constraints $0.004 \le X_r^j \ge 0.05$ m., $r = 1,2,...,m, j = 1,2,\ldots,N_p$. The maximum number of iterations is $N = 200$ and the ensemble size $N_p = 10$.

Suppose that one wishes to optimize only the unbalance response, $\max_{500 \le \lambda \le 2500} R_{41}(\lambda)$, without any constraint on the critical speeds. The innovation used in this exercise is:

$$\boldsymbol{I}_{r,i}^{[\varsigma_2(j)]} = \begin{Bmatrix} \hat{f}_{i-1} - f_{i-1}^{[\varsigma_2(j)]} \\ X_{r,i-1}^{[\varsigma_1(j)]} - X_{r,i-1}^{[\varsigma_2(j)]} \end{Bmatrix}, \; r = 1,2,\ldots,m, \; j = 1,2,\ldots,N_p \tag{10.9}$$

The first component $\hat{f}_{i-1} - f_{i-1}^{[\varsigma_2(j)]}$ in $\boldsymbol{I}_{r,i}^{[\varsigma_2(j)]}$ signifies a greedy search and the second the coalescence plus scrambling at the element level of each particle. Here $f^{[\varsigma_2(j)]} = \max_{500 \le \lambda \le 2500} R_{41}^{[\varsigma_2(j)]}(\lambda)$ at the bearing at node 11 at any iteration. Figure 10.4 shows the evolution of the sample mean $\overline{f}_i = \frac{1}{N_p}\sum_{j=1}^{N_p} f_i^{[j]} \in \mathbb{R}$ of the cost function. The computed optimal solution of $\boldsymbol{X} \in \mathbb{R}^{m \times N_p}$ representing the shaft diameters is shown in Table 10.1. At the end of iteration, realized cost functions corresponding to the 10 particles are also included in the table. It is observed that the particles fairly coalesce at the end of N iterations and the sample mean of the cost function—$\max_{500 \le \lambda \le 2500} R_{41}(\lambda)$—is around 0.2333E-07 with a small variance. The unbalance responses at the bearing at node 11 obtained using the end-of-iteration solution, $\boldsymbol{X}$ are shown in Fig. 10.5.

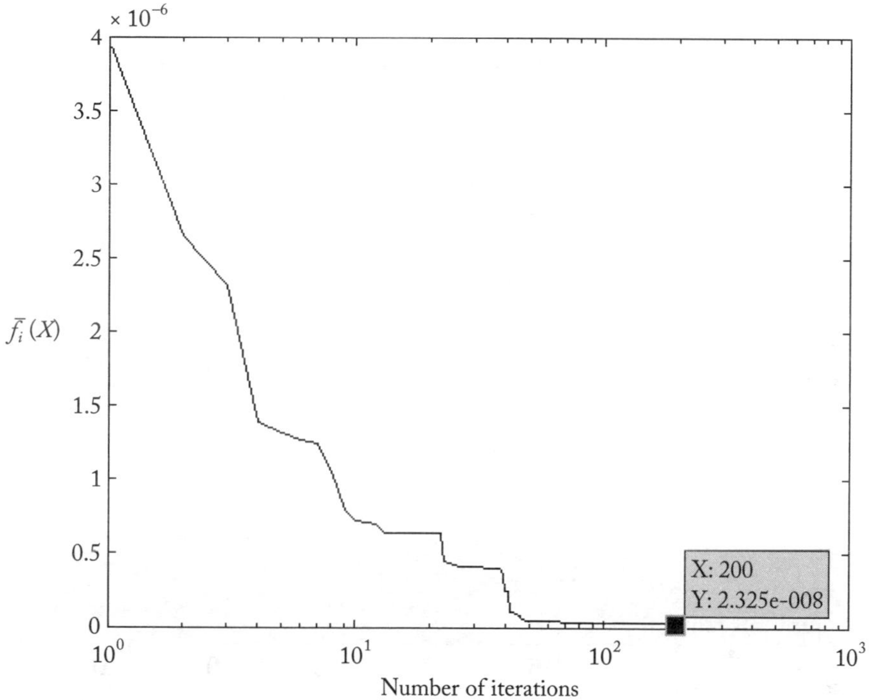

Fig. 10.4 Minimization of the cost function $f(\boldsymbol{X}) = \max_{500 \le \lambda \le 2500} R_{41}(\lambda)$ by COMBEO, $m = 15$, $N_p = 10$, $p_r = 0.9$, $N = 200$, evolution of the sample mean $\overline{f}_i(\boldsymbol{X})$ with iterations

Table 10.1 Minimization of the cost function $f(\boldsymbol{X}) = \max_{500 \le \lambda \le 2500} R_{41}(\lambda)$ by COMBEO; result on the shaft diameters $X_k^{[j]}$, $k = 1,2,\ldots,m = 15$ and $j = 1,2,\ldots,N_p = 10$ at the last (200th) iteration

Design variable (diameter in m)	Particle number, j				
	1	**2**	**3**	**4**	**5**
X_1	0.0060	0.0044	0.0043	0.0053	0.0061
X_2	0.0068	0.0051	0.0065	0.0049	0.0042
X_3	0.0443	0.0445	0.0431	0.0464	0.0438
X_4	0.0482	0.0489	0.0460	0.0494	0.0491
X_5	0.0499	0.0500	0.0491	0.0487	0.0496
X_6	0.0495	0.0494	0.0481	0.0437	0.0478

X_7	0.0496	0.0447	0.0496	0.0491	0.0499
X_8	0.0486	0.0488	0.0491	0.0498	0.0483
X_9	0.0086	0.0086	0.0086	0.0087	0.0087
X_{10}	0.0174	0.0201	0.0199	0.0170	0.0151
X_{11}	0.0181	0.0216	0.0199	0.0193	0.0191
X_{12}	0.0442	0.0187	0.0234	0.0208	0.0196
X_{13}	0.0441	0.0427	0.0369	0.0460	0.0499
X_{14}	0.0184	0.0218	0.0092	0.0141	0.0163
X_{15}	0.0071	0.0121	0.0081	0.0086	0.0090
$f(\mathbf{X}) = \max_{500 \le \lambda \le 2500} R_{41}(\lambda)$	0.2333 E-07	0.2341 E-07	0.2363 E-07	0.2330 E-07	0.2321 E-07

Design variable (diameter in m)	**Particle number, j**				
	6	**7**	**8**	**9**	**10**
X_1	0.0054	0.0052	0.0067	0.0043	0.0045
X_2	0.0050	0.0049	0.0059	0.0061	0.0055
X_3	0.0405	0.0489	0.0442	0.0448	0.0447
X_4	0.0497	0.0495	0.0499	0.0500	0.0497
X_5	0.0498	0.0485	0.0481	0.0489	0.0493
X_6	0.0445	0.0485	0.0497	0.0461	0.0480
X_7	0.0474	0.0482	0.0475	0.0497	0.0499
X_8	0.0498	0.0498	0.0491	0.0496	0.0498
X_9	0.0086	0.0088	0.0087	0.0087	0.0087
X_{10}	0.0167	0.0179	0.0180	0.0174	0.0162
X_{11}	0.0179	0.0176	0.0175	0.0199	0.0185
X_{12}	0.0237	0.0267	0.0371	0.0422	0.0400
X_{13}	0.0312	0.0391	0.0486	0.0479	0.0453
X_{14}	0.0190	0.0137	0.0139	0.0209	0.0198
X_{15}	0.0064	0.0100	0.0060	0.0081	0.0112
$f(\mathbf{X}) = \max_{500 \le \lambda \le 2500} R_{41}(\lambda)$	0.2338 E-07	0.2302 E-07	0.2333 E-07	0.2301 E-07	0.2286 E-07

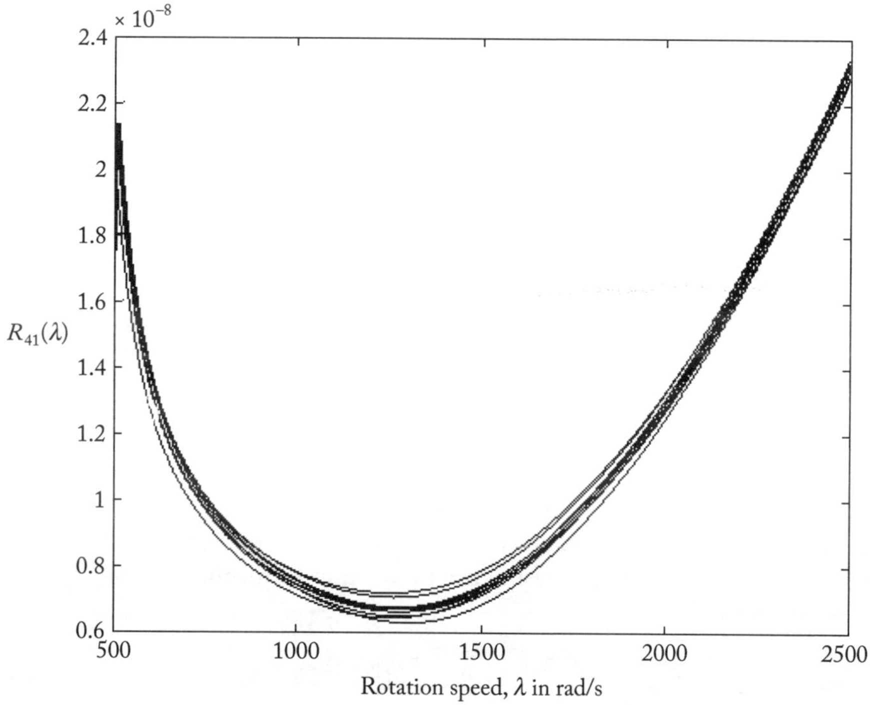

Fig. 10.5 Minimization of the cost function $f(\boldsymbol{X}) = \max_{500 \le \lambda \le 2500} R_{41}(\lambda)$ by COMBEO; $m = 15$, $p_r = 0.9$, unbalance responses $R_{41}(\lambda)$ obtained with end-of-iteration $\boldsymbol{X}^{[j]}$, $j = 1, 2, \ldots, N_p$

We now perform the optimization by COMBEO considering the cost function defined in Eq. (10.8), which also includes two terms incorporating the required separation of the first two critical speeds. In fact one may alternatively use multi-objective optimization, wherein each term in $f(\boldsymbol{X})$ would represent an independent cost function, and invoke Pareto optimality to arrive at an optimum solution [Lobato *et al.* 2012]. In the stochastic optimization setting, we may define Pareto optimality as follows. A solution $\boldsymbol{X}^*$ is Pareto optimal if there exists no other $\boldsymbol{X}$ such that:

$$E\left[f_i\left(\boldsymbol{X}^*\right) - f_i(\mathrm{X})\right] \ge 0, \forall\, i \tag{10.10}$$

with the strict inequality valid for at least one i. In this context, one may draw analogy from game theory [Neumann and Morgenstern 1947, Nash 1950, Osborne 2004] and refer to Nash equilibria which are the outcomes of non-cooperative strategies of the players. These equilibria are the local equilibrium solutions of individual objective functions in a multi-objective optimization. In contrast, Pareto optimum is the solution obtained by a cooperative (non-dominating) strategy of the constituent players. It is the result from a joint strategy so that no other strategy is better, in a weak sense, for all the players and strictly better for at least one individual player.

Thus in the context of multi-objective optimization by COMBEO, $\Delta f_i := f_i(\boldsymbol{X}^*) - f_i(\boldsymbol{X})\ \forall\ i$, a stochastic process parametered by the iterations, must be driven to a zero–mean super–martingale, i.e., $E\left[\Delta f_i(\tau) \,\middle|\, \mathcal{F}_\xi\right] \le \Delta f_i(\xi) \forall\ i$ a.s. for every $\xi \le \tau$ (perhaps with a martingale structure approaching in the limit of the number of iterations tending to infinity). Accordingly, the innovation $\boldsymbol{I}_{r,i}^{[\varsigma_2(j)]}$ (in Eq. 10.4) is chosen to be:

$$\boldsymbol{I}_{r,i}^{[\varsigma_2(j)]} = \begin{Bmatrix} X_{r,i-1}^{[\varsigma_1(j)]} - X_{r,i-1}^{[\varsigma_2(j)]} \\ \check{s}_{1,i-1}^{[\varsigma_2(j)]} - 1000 \\ 2000 - \check{s}_{2,i-1}^{[\varsigma_2(j)]} \end{Bmatrix}, \; j = 1, 2, \ldots, N_p \tag{10.11}$$

While the first component $X_{r,i-1}^{[\varsigma_1(j)]} - X_{r,i-1}^{[\varsigma_2(j)]}$ in $\boldsymbol{I}_{r,i}^{[\varsigma_2(j)]}$ denotes coalescence plus scrambling at the element level of each particle, the second and third ones represent a greedy search aimed at satisfying the constraints on the first two critical speeds, i.e., to move them away from the range 1000-2000 rad/s. Using the innovation so formulated, the update at each iteration is obtained as:

$$X_{r,i}^{[j]} = X_{r,i-1}^{[j]} + D_{r,i}^{[\varsigma_2(j)]}, D_{r,i}^{[\varsigma_2(j)]} = \beta_i \boldsymbol{G}_{r,i} . \boldsymbol{I}_{r,i}^{[\varsigma_2(j)]} \tag{10.12}$$

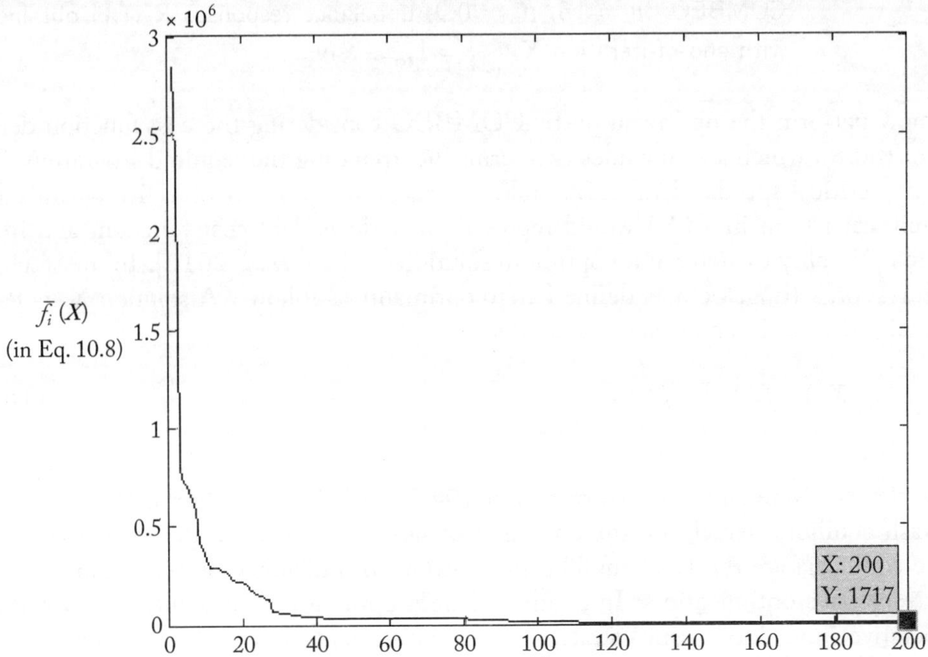

(a)

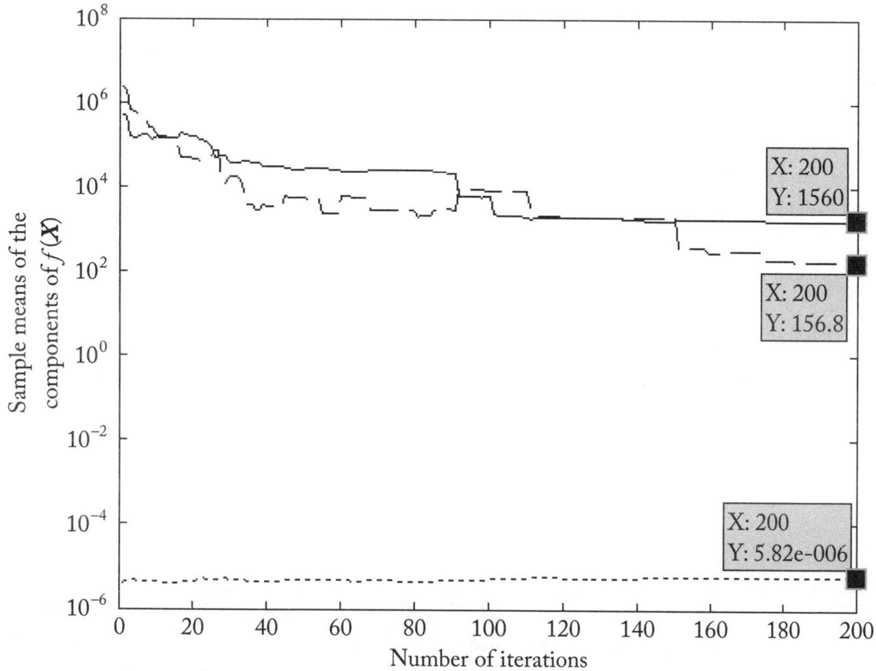

(b)

Fig. 10.6 Minimization of the cost function $f(\boldsymbol{X})$ in Eq. (10.8) by COMBEO: $m = 15$, $N_p = 10$, $p_r = 0.9$, $N = 200$, evolutions of the sample means of (a) $f(\boldsymbol{X})$ and (b) component costs with iterations; dotted-line—$\max_{500\le\lambda\le2500} R_{41}(\lambda)$, dashed-line—$(\check{s}_1 - 1000)^2$ and dark-line—$(2000 - \check{s}_2)^2$

Figure 10.6a shows the optimization result on the cost function $f(\boldsymbol{X})$ in Eq. (10.8) in terms of its sample mean. The sample means of the three component costs constituting $f(\boldsymbol{X})$ are shown in Fig. 10.6b. Evolving means of the first two critical speeds are shown in Fig. 10.7. That the required status of the two critical speeds at the end of $N = 200$ iterations is secured is evident from the figure. The unbalance responses $R_{41}(\lambda)$ at the left bearing obtained with the end-of-iteration $\boldsymbol{X} \in \mathbb{R}^{m\times N_p}$ are plotted in Fig. 10.8. All the N_p particles show the desired separation of the critical speeds attended to by an ample reduction of the response in this range of rotation speed. The optimal solution of $\boldsymbol{X} \in \mathbb{R}^{m\times N_p}$ at the end of iterations is shown in Table 10.2 along with the cost functions evaluated using each particle.

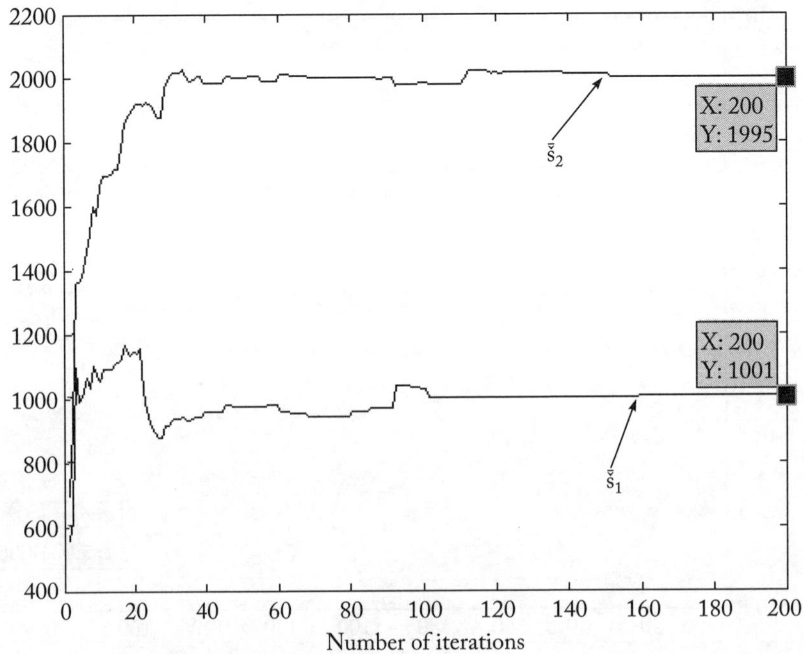

Fig. 10.7 Minimization of the cost function $f(X)$ in Eq. (10.8) by COMBEO: $m = 15, N_p = 10, p_r = 0.9, N = 200$; evolutions of the sample means of the first two critical speeds $\check{s}_1$ and $\check{s}_2$ with iteration

Table 10.2 Minimization of the cost function $f(X)$ in Eq. (10.8) by COMBEO; result on the shaft diameters $X_k^{[j]}$, $k = 1,2,\ldots,m = 15$ and $j = 1,2,\ldots,N_p = 10$ at the last (200th) iteration

Design variable (diameter in m)	Particle number, *j*				
	1	**2**	**3**	**4**	**5**
X_1	0.0050	0.0062	0.0045	0.0129	0.0068
X_2	0.0049	0.0105	0.0366	0.0123	0.0270
X_3	0.0334	0.0334	0.0469	0.0312	0.0066
X_4	0.0218	0.0473	0.0381	0.0341	0.0376
X_5	0.0191	0.0102	0.0072	0.0108	0.0058
X_6	0.0046	0.0044	0.0420	0.0043	0.0107

X_7	0.0063	0.0160	0.0141	0.0350	0.0237
X_8	0.0410	0.0371	0.0328	0.0430	0.0433
X_9	0.0409	0.0207	0.0456	0.0322	0.0289
X_{10}	0.0332	0.0453	0.0072	0.0481	0.0269
X_{11}	0.0066	0.0159	0.0135	0.0124	0.0468
X_{12}	0.0209	0.0152	0.0272	0.0491	0.0491
X_{13}	0.0183	0.0228	0.0206	0.0156	0.0146
X_{14}	0.0231	0.0482	0.0271	0.0224	0.0314
X_{15}	0.0169	0.0383	0.0468	0.0241	0.0112
$f(\mathbf{X}) = \max_{500\leq\lambda\leq2500} R_{41}(\lambda)$	7.222 E-06	4.036 E-06	6.777 E-06	4.182 E-06	2.377 E-06

Continued...

Design variable (diameter in m)	Particle number, *j*				
	6	**7**	**8**	**9**	**10**
X_1	0.0047	0.0282	0.0108	0.0049	0.0221
X_2	0.0063	0.0273	0.0159	0.0298	0.0371
X_3	0.0272	0.0464	0.0293	0.0114	0.0230
X_4	0.0412	0.0372	0.0071	0.0092	0.0205
X_5	0.0227	0.0448	0.0201	0.0053	0.0189
X_6	0.0072	0.0052	0.0048	0.0423	0.0053
X_7	0.0195	0.0111	0.0158	0.0477	0.0265
X_8	0.0406	0.0391	0.0340	0.0471	0.0384
X_9	0.0332	0.0083	0.0170	0.0486	0.0040
X_{10}	0.0404	0.0468	0.0451	0.0226	0.0249
X_{11}	0.0309	0.0281	0.0268	0.0431	0.0305
X_{12}	0.0472	0.0478	0.0312	0.0089	0.0120
X_{13}	0.0119	0.0104	0.0464	0.0348	0.0392
X_{14}	0.0355	0.0317	0.0489	0.0415	0.0377
X_{15}	0.0092	0.0473	0.0226	0.0076	0.0338
$f(\mathbf{X}) = \max_{500\leq\lambda\leq2500} R_{41}(\lambda)$	2.626 E-06	8.599 E-06	4.069 E-06	3.8316 E-06	1.3144 E-05

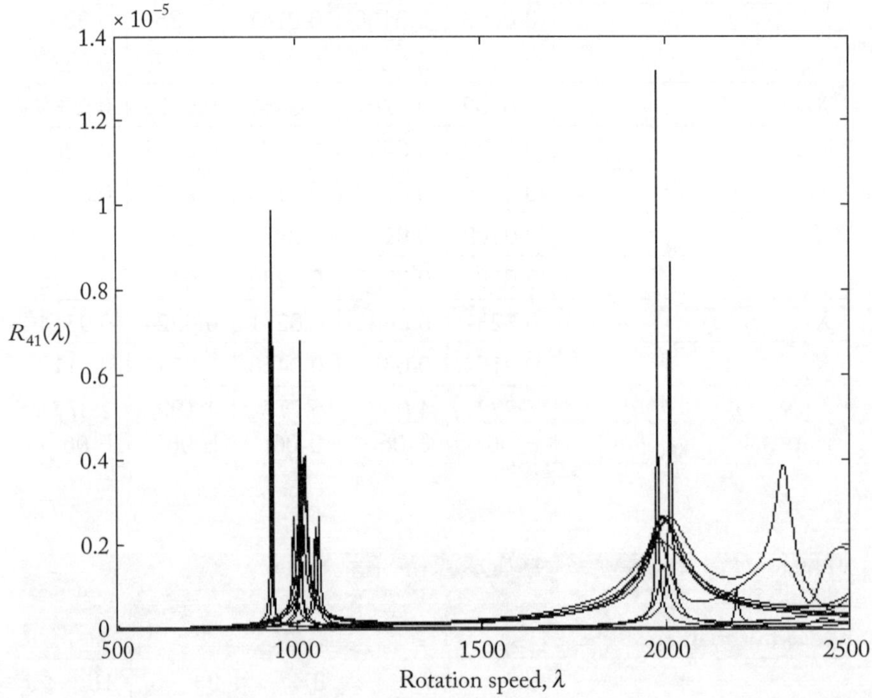

Fig. 10.8 Minimization of the cost function $f(\boldsymbol{X})$ in Eq. (10.8) by COMBEO: $m = 15$, $p_r = 0.9$, $N_p = 10$, $N = 200$; unbalance response $R_{41}(\lambda)$ obtained with end-of-iteration $\boldsymbol{X}^{[j]}$, $j = 1, 2, \ldots, N_p$

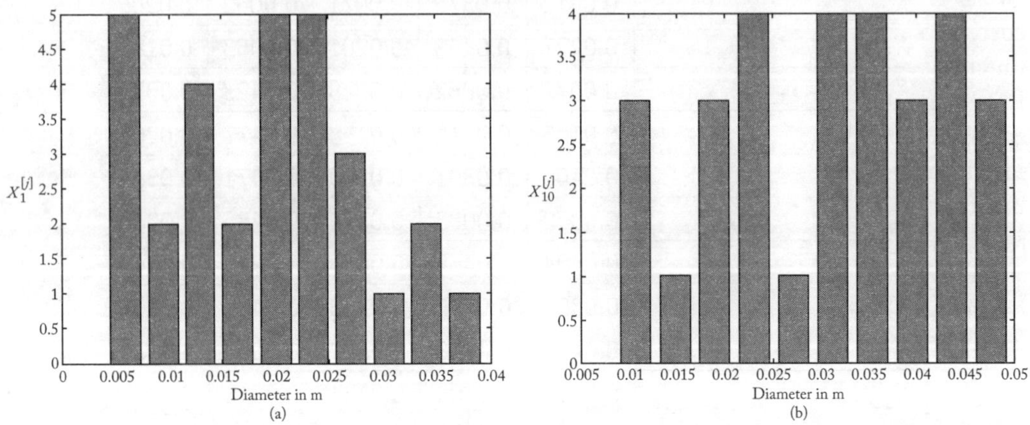

Fig. 10.9 Minimization of the cost function $f(\boldsymbol{X})$ in Eq. (10.8) by COMBEO: $m = 15$, $p_r = 0.9$, $N_p = 30$, $N = 200$, histograms corresponding to (a) $X_1^{[j]}, j = 1, 2, \ldots, N_p$—shaft diameter of 1st element and (b) $X_{10}^{[j]}, j = 1, 2, \ldots, N_p$—shaft diameter of 10th element of the rotor bearing system (see FE model in Fig. 9.14)

Given that coalescence and scrambling are two schemes imposing conflicting demands on COMBEO, the final (i.e., end-of-iteration) solution (see Table 10.2) may indicate no coalescence of the particles as such; thereby retaining multi-modality in the final posterior density (see Fig. 10.9). The last figure presents histograms typifying the multi-modality exhibited by some of the parameters (here the shaft diameters) at the conclusion of the last iteration. For clarity, results obtained with $N_p = 30$ are utilized to draw the histograms in the figure. The particles, upon convergence, may coalesce into separate groups in the parameter space, thereby yielding multi modal posteriors. In such cases, the solution vector corresponding to the best particle should be reported, rather than the empirical average over the set of particles which is likely to be erroneous. Thus, the particles of $\boldsymbol{X}$ recorded in Table 10.2 merely indicate feasible multiple solutions for the shaft geometry. While going through these examples, the reader is strongly advised to also carefully study the MATLAB codes provided in the companion web site.

10.3 COMBEO Algorithm

The improvements to the local search and random perturbation approaches, as described in the previous sections, provide one with a global optimization tool, the COMBEO with its roots founded on the KS approach and endowed with potentially numerous options easy to be incorporated in order to possibly meet a suite of conflicting demands either at a system analysis or design stage. For instance, incorporation of the innovation term in Eq. (9.19), which is equivalent to the term, $\widehat{f}_i - f_i$ in Eq. (9.43b) or (9.48), yields a greedy algorithm that helps optimization of functions that may be inherently unimodal or separable. Despite a possibly faster convergence, it is likely to be a poor performer in the global search. On the other hand, to facilitate (i) a more exhaustive global search at the cost of a substantively slower convergence, or (ii) system optimization where the cost functions are generally implicit in $\boldsymbol{X}$ of large dimension, one may adopt either a PSO-type innovation as in Eq. (10.1) or a DiEv-like scrambling as in Eq. (10.4). In addition, a relaxation strategy and/or a blending scheme respectively mimicking the roles of selection in the DiEv (Eq. 9.6) and cross-over operation in the GA (Table 9.2) may be incorporated into the COMBEO. It is thus appropriate at this stage to combine all the presented ideas in a single pseudo code that could readily offer a balance of convergence speed with exploratory efficiency depending on the nature of the problem at hand. In having an objective assessment of the strengths and demerits of different search tools, the versatility of the COMBEO is brought to focus in Table 10.3.

Table 10.3 Brief description of COMBEO algorithm

<table>
<tr><td rowspan="2">Step 1.</td><td colspan="4">Initialize m, the dimension of the optimization problem. Choose an ensemble size N_p. Fix convergence tolerance ε and the allowable maximum number of iterations it_max.</td></tr>
<tr><td colspan="4">Initialization of parameters</td></tr>
<tr><td></td><td>$\alpha \in [0,1]$ (parameter in Eq. (9.37b)- a typical value is 0.8)</td><td>$\beta \in [0,1]$ (annealing type parameter adopted from SA-Eq. 10.2 or 10.4)</td><td>$p_r \in [0,1]$ (relaxation parameter adopted from DiEv)</td><td>$w_0^{[j]} = \frac{1}{N_p}, j = 1,2,\ldots,N_p$ (weights in Eq. 10.5 – strategy adopted from GA)</td></tr>
<tr><td rowspan="2">Step 2.</td><td colspan="4">Generate the initial ensemble of particles, $\boldsymbol{X}_0 = \left\{X_0^{[j]}\right\}_{j=1}^{N_p}$ based on an initial, possibly uniform distribution in the admissible search space $\in \mathbb{R}^m$. The search space should be bounded, e.g., $[lb, ub]^m$ where lb and ub denote the lower and upper bounds.
Evaluate the (initial) cost function vector $\boldsymbol{f}_0 \in \mathbb{R}^{\mathfrak{m}\times N_p}$, where $\mathfrak{m}$ is the dimension of the multi-objective function</td></tr>
<tr><td colspan="4">Set iteration counter $i = 1$</td></tr>
<tr><td>Step 3.</td><td colspan="4">Formulate the innovation matrix, $\boldsymbol{I}_i$
Options:

i) greedy search:

$$\boldsymbol{I}_i = \left(\widehat{\boldsymbol{f}}_{i-1} - \boldsymbol{f}_{i-1}\right) \in \mathbb{R}^{\mathfrak{m}\times N_p} \tag{10.13}$$
ii) coalescence of candidate solutions (this requires generation of random index $\varsigma_1[j]$from the integer set $\left\{1,2,\ldots,N_p\right\}\backslash\{j\}$):

$$\boldsymbol{I}_i^{[j]} = \left(\boldsymbol{X}_{i-1}^{[\varsigma_1(j)]} - \boldsymbol{X}_{i-1}^{[j]}\right) \in \mathbb{R}^m, j = 12,\ldots,N_p \tag{10.14}$$
iii) scrambling with coalescence - particle-wise / element-wise (this requires generation of another random index $\varsigma_2[j]$ from the integer set $\left\{1,2,\ldots,N_p\right\}\backslash\{j\}$):</td></tr>
</table>

(a) particle-wise:

$$\boldsymbol{I}_i^{[\varsigma_2(j)]} = \boldsymbol{X}_{i-1}^{[\varsigma_1(j)]} - \boldsymbol{X}_{i-1}^{[\varsigma_2(j)]} \in \mathbb{R}^m, j = 12,\ldots,N_p \tag{10.15}$$

(b) element-wise:

$$\boldsymbol{I}_{r,i}^{[\varsigma_2(j)]} = X_{r,i-1}^{[\varsigma_1(j)]} - X_{r,i-1}^{[\varsigma_2(j)]} \in \mathbb{R}, r = 1,2,..,mj = 12,\ldots,N_p \tag{10.16}$$

iv) PSO-based coalescence:

$$\boldsymbol{I}_i^{[\varsigma_2(j)]} = \left\{ \begin{array}{c} {}^{P}\boldsymbol{X}_{i-1}^{[\varsigma_1(j)]} - \boldsymbol{X}_{i-1}^{[\varsigma_2(j)]} \\ {}^{G}\boldsymbol{X}_{i-1} - \boldsymbol{X}_{i-1}^{[\varsigma_2(j)]} \end{array} \right\} \in \mathbb{R}^{2m}, j = 12,\ldots,N_p \tag{10.17}$$

(from Eq. 10.1)

Note: Innovation may be formed by any combination of the options noted above. For example, the vector for greedy search along with coalescence and scrambling:

$$\boldsymbol{I}_i^{[\varsigma_2(j)]} = \left\{ \begin{array}{c} \widehat{\boldsymbol{f}}_{i-1} - \boldsymbol{f}_{i-1}^{[\varsigma_2(j)]} \\ \boldsymbol{X}_{i-1}^{[\varsigma_1(j)]} - \boldsymbol{X}_{i-1}^{[\varsigma_2(j)]} \end{array} \right\} \in \mathbb{R}^{n+m}, j = 12,\ldots,N_p \tag{10.18}$$

Step 4 Additive update – here the prediction step is avoided as this may not expedite the global search.

(a) *For coalescence alone*:

$$\boldsymbol{X}_i^{[j]} = \boldsymbol{X}_{i-1}^{[j]} + \boldsymbol{D}_i^{[j]},\ j = 1,2,\ldots,N_p, \tag{10.19}$$

where

$$\boldsymbol{D}_i^{[j]} = \beta_i \boldsymbol{G}_i \boldsymbol{I}_i^{[j]}$$

(b) *For coalescence with scrambling*:

$$\boldsymbol{X}_i^{[j]} = \boldsymbol{X}_{i-1}^{[j]} + \boldsymbol{D}_i^{[\varsigma_2(j)]},\ j = 1,2,\ldots,N_p \text{ particle-wise} \tag{10.20}$$

where

$$D_i^{[\varsigma_2(j)]} := \beta_i G_i I_i^{[\varsigma_2(j)]}$$

$$X_{r,i}^{[j]} = X_{r,i-1}^{[j]} + D_{r,i}^{[\varsigma_2(j)]},$$

$$j = 1,2,\ldots,N_p,\ r = 1,2,\ldots,m\text{- element-wise} \tag{10.21}$$

where

$$D_{r,i}^{[\varsigma_2(j)]} := \beta_i G_{r,i} I_i^{[\varsigma_2(j)]}$$

(c) *For blending*:

$$X_i^{b[j]} = w_i^{[j]} X_{i-1}^{[j]} + \left(1 - w_i^{[j]}\right) X_i^{[j]},\ j = 1,2,\ldots,N_p \tag{10.22}$$

w_i are evaluated from Eq. (10.6) for the next iteration. For blending along with coalescence and without or with scrambling, one first needs to obtain the respective updates $X_i^{[j]}$ in (a) or (b) of step 4 above before obtaining $X_i^{b[j]}$ from Eq. (10.22).

Updating option using relaxation parameter (adapted from DiEv):

a) without blending:
The updates in Eqs. (10.19)–(10.21) are accepted if $c = U(0,1) < 1 - p_r$, otherwise $X_i^{[j]} = X_{i-1}^{[j]},\ j = 1,2,\ldots,N_p$. If element-wise scrambling is involved, the updates are accepted or rejected with this strategy implemented at the element level.

b) with blending: If $c_1 = U(0,1) < 1 - p_r$, then:
If $c_2 = U(0,1) < p_{\text{rand}}$, where $p_{\text{rand}} \in [0,1]$; (one choice for p_{rand} is 0.5), obtain the update $X_i^{[j]}$ according to Eqs. (10.19)– (10.21).
Else obtain the update $X_i^{b[j]}$ according to Eq. (10.22) and set $X_i^{[j]} = X_i^{b[j]}$ for use in the next iteration.

Else, retain the original particle as the update, i.e., $X_i^{[j]} = X_{i-1}^{[j]}$.

Step 5.	If $f\left(X_i^{[j]}\right) \leq f\left(X_{i-1}^{[j]}\right)$, $j = 1,2,\ldots,N_p$, then retain $X_i^{[j]}$ as the updated particle; else set $X_i^{[j]} = X_{i-1}^{[j]}$
Step 6.	For blending (if required) in the next iteration step, update the weights w according to Eq. (10.6). For PSO-based coalescence strategy, update the personal and global information for use in the next iteration.
Step 7.	Calculate the empirical mean $\overline{X}_i = \frac{1}{N_p}\sum_{j=1}^{N_p} X_i^{[j]} \in \mathbb{R}^m$ and mean of the cost function by $\overline{f}_i = \frac{1}{N_p}\sum_{j=1}^{N_p} f_i^{[j]} \in \mathbb{R}^n$.
Step 8.	If $i < N$, then go to step 3 with $i \equiv i+1$; else terminate the algorithm.
Notes:	1. The gain matrix G_i is similar to the expression in Eq. (10.3). The dimensions of the variables involved in G_i depend on the corresponding innovation I_i. 2. The parameter β_i (annealing-type) in the jth update vector $D_i^{(j)}$ may be taken as $1 - \frac{1}{\exp(i-1)}$. Another choice is $\widehat{\beta_i}\zeta_i$ where $\widehat{\beta_i}$ is a specified scalar constant and ζ_i is a uniform random number between 0 and 1.

10.3.1 Some benchmark problems and solutions by COMBEO

At this stage, we assess the performance of COMBEO against a few benchmark minimization problems. The efficiency of a global optimization strategy is largely dependent on its ability to search all the promising search regions. The level of complexity in solving a given problem typically increases with increase in the system dimension as the multimodal nature of a cost function varies considerably with the dimension m of the design space. For instance, the Rastrigin function has roughly 10^m local optima [Tang *et al.* 2010]. Even though Rosenbrock's function $f(x_1,x_2) = 100\left(x_1^2 - x_2\right)^2 + (x_1 - 1)^2$ is unimodal for $m = 2$, it is highly multimodal in still higher dimensional search space. Difficulty in finding the global optimum may also be dependent on the specific characteristics of the cost function, which goes to explain the varied degree of complexity in solving different optimization problems of the same dimension. Further, for a given dimension, the difficulty levels may vary across problems owing to specific characteristics such as the degree of separability [Tang *et al.* 2010] of the cost function. Depending on the level of complexity, optimization problems could be categorized as separable, m-non-separable and non-separable. If the problem is separable or partially separable, i.e., if the cost function can be additively split in terms of component functions each of which is expressible in terms of just one element or a small subset of elements of the design variable vector, X, the original problem may actually be decoupled into a set of simpler sub-problems. Each sub-problem, involving one scalar variable or a small set of scalar variables, may be solved separately, e.g., by a simple line search for one scalar variable. In between the two extreme cases i.e., separable and non-separable, are the $\mathfrak{m}$-non-separable functionals ($\mathfrak{m}$ being the maximum number of scalar design variables appearing in the descriptions of the component cost functions) that correspond to partially separable problems. The nomenclature *separability* bears a similar meaning with *epistasis* in biology.

The following benchmark functions are considered to test COMBEO (see Table 10.4). They constitute the test suite that was released for the CEC 2010 special session and competition on large-scale global optimization [Tang *et al.* 2010]. Results obtained by the DiEv and the CMA–ES (Chapter 9) are also included in the table for a comparison.

Group I. Separable Functions:

(F1) Shifted Elliptic Function

(F2) Shifted Rastrigin's Function

(F3) Shifted Ackley's Function

Group II. Single-group $\mathfrak{m}$ *-non-separable Functions*

(F4) Single-group Shifted and $\mathfrak{m}$ -rotated Elliptic Function

(F5) Single-group Shifted and $\mathfrak{m}$ -rotated Rastrigin's Function

(F6) Single-group Shifted $\mathfrak{m}$ -rotated Ackley's Function

(F7) Single-group Shifted $\mathfrak{m}$ -dimensional Schwefel's function

(F8) Single-group Shifted $\mathfrak{m}$ -dimensional Rosenbrock's Function

Group III. $\frac{m}{2\mathfrak{m}}$*-group* $\mathfrak{m}$*-non-separable Functions*

(F9) $\frac{m}{2\mathfrak{m}}$-group Shifted and $\mathfrak{m}$-rotated Elliptic's Function

(F10) $\frac{m}{2\mathfrak{m}}$ -group Shifted and $\mathfrak{m}$ -rotated Rastrigin's Function

(F11) $\frac{m}{2\mathfrak{m}}$-group Shifted and $\mathfrak{m}$ -rotated Ackley's Function

(F12) $\frac{m}{2\mathfrak{m}}$ -group Shifted and $\mathfrak{m}$ -rotated Schwefel's Function

(F13) $\frac{m}{2\mathfrak{m}}$ -group Shifted and $\mathfrak{m}$ -rotated Rosenbrock's Function

Group IV. $\frac{m}{\mathfrak{m}}$*-group* $\mathfrak{m}$ *-non-separable Functions*

(F14) $\frac{m}{\mathfrak{m}}$ -group Shifted and $\mathfrak{m}$ -rotated Elliptic's Function

(F15) $\frac{m}{\mathfrak{m}}$-group Shifted and $\mathfrak{m}$ -rotated Rastrigin's Function

(F16) $\frac{m}{\mathfrak{m}}$-group Shifted and $\mathfrak{m}$ -rotated Ackley's Function

(F17) $\frac{m}{\mathfrak{m}}$-group Shifted and $\mathfrak{m}$ -rotated Schwefel's function

(F18) $\frac{m}{\mathrm{m}}$-group Shifted and m -rotated Rosenbrock's function

Group V. Non-separable Functions

(F19) Shifted Schwefel's function

(F20) Shifted Rosenbrock's function

The above functions are constructed using the following basic functions.

B1. The Sphere Function: $F_{\text{sphere}}(\boldsymbol{x}) = \sum_{j=1}^{m} x_j^2$

B2. The Elliptic Function: $F_{\text{elliptic}}(\boldsymbol{x}) = \sum_{j=1}^{m} \left(10^6\right)^{\frac{j-1}{n-1}} x_j^2$

B3. The Rotated Elliptic Function: $F_{\text{rot\_elliptic}}(\boldsymbol{x}) = F_{\text{elliptic}}(\boldsymbol{z})$, $\boldsymbol{z} = \boldsymbol{x} * \boldsymbol{M}$ (**M** being a orthogonal matrix) with $\boldsymbol{x}$ being a row vector.

B4. Schwefel's Problem 1.2: $F_{\text{schwefel}}(\boldsymbol{x}) = \sum_{k=1}^{m} \left(\sum_{j=1}^{k} x_j\right)^2$

B5. Rosenbrock's Function: $F_{\text{rosenbrock}}(\boldsymbol{x}) = \sum_{j=1}^{m-1} \left\{100\left(x_j^2 - x_{j+1}\right)^2 + \left(x_j - 1\right)^2\right\}$

B6. Rastrigin's Function: $F_{\text{rastrigin}}(\boldsymbol{x}) = \sum_{j=1}^{m} \left\{x_j{}^2 - 10\cos 2\pi\, x_j + 10\right\}$

B7. Rotated Rastrigin's Function: $F_{\text{rot\_rastrigin}}(\boldsymbol{x}) = F_{\text{rastrigin}}(\boldsymbol{z})$,$\boldsymbol{z} = \boldsymbol{x} * \boldsymbol{M}$ (**M** being a orthogonal matrix) with $\boldsymbol{x}$ being a row vector.

B8. Ackley's Function :

$$F_{\text{ackley}}(\boldsymbol{x}) = -20\exp\left(-0.2\frac{1}{n}\sqrt{\sum_{j=1}^{m} x_j^2}\right) - \exp\left(\frac{1}{m}\sum_{j=1}^{m}\cos\left(2\pi x_j\right)\right) + 20 + \exp(1)$$

B9. Rotated Ackley's Function: $F_{\text{rot\_ackley}}(\boldsymbol{x}) = F_{\text{ackley}}(\boldsymbol{z})$, $\boldsymbol{z} = \boldsymbol{x} * \boldsymbol{M}$ (**M** being a orthogonal matrix) with $\boldsymbol{x}$ being a row vector.

Explicit expressions for the functions, F1–F20 are given below. Let $\boldsymbol{z} = \boldsymbol{x} - \boldsymbol{o}$, $\boldsymbol{o} \in \mathbb{R}^m$ denotes a shifted global optimum and $\mathfrak{r}(1:\mathrm{m})$ the random permutation of the vector $\{1, 2, \ldots, \mathrm{m}\}$.

(1) $F1(\boldsymbol{x}) = F_{\text{elliptic}}(\boldsymbol{z})$

(2) $F2(\boldsymbol{x}) = F_{\text{rastrigin}}(\boldsymbol{z})$

(3) $F3(\boldsymbol{x}) = F_{\text{ackley}}(\boldsymbol{z})$

(4) $F4(\boldsymbol{x}) = F_{\text{rot\_elliptic}}(\boldsymbol{z}(\mathfrak{r}(1:\mathrm{m}))) \times 10^6 + F_{\text{elliptic}}(\boldsymbol{z}(\mathfrak{r}(\mathrm{m}+1:m)))$

(5) $F5(\boldsymbol{x}) = F_{\text{rot\_rastrigin}}(\boldsymbol{z}(\mathrm{r}(1:\mathrm{m}))) \times 10^6 + F_{\text{rastrigin}}(\boldsymbol{z}(\mathrm{r}(\mathrm{m}+1:m)))$

(6) $F6(\boldsymbol{x}) = F_{\text{rot\_ackley}}(\boldsymbol{z}(\mathrm{r}(1:\mathrm{m}))) \times 10^6 + F_{\text{ackley}}(\boldsymbol{z}(\mathrm{r}(\mathrm{m}+1:m)))$

(7) $F7(\boldsymbol{x}) = F_{\text{schwefel}}(\boldsymbol{z}(\mathrm{r}(1:\mathrm{m}))) \times 10^6 + F_{\text{sphere}}(\boldsymbol{z}(\mathrm{r}(\mathrm{m}+1:m)))$

(8) $F8(\boldsymbol{x}) = F_{\text{rosenbrock}}(\boldsymbol{z}(\mathrm{r}(1:\mathrm{m}))) \times 10^6 + F_{\text{sphere}}(\boldsymbol{z}(\mathrm{r}(\mathrm{m}+1:m)))$

(9) $F9(\boldsymbol{x}) = \sum_{k=1}^{\frac{m}{2\mathrm{m}}} \left\{ \begin{array}{c} F_{\text{rot\_elliptic}}(\boldsymbol{z}(\mathrm{r}(k-1)\mathrm{m}+k\mathrm{m}))) \times 10^6 \\ +F_{\text{elliptic}}\left(\boldsymbol{z}\left(\mathrm{r}\left(\frac{m}{2}+1:m\right)\right)\right) \end{array} \right\}$

(10) $F10(\boldsymbol{x}) = \sum_{k=1}^{\frac{m}{2\mathrm{m}}} \left\{ \begin{array}{c} F_{\text{rot\_rastrigin}}(\boldsymbol{z}(\mathrm{r}(k-1)\mathrm{m}+k\mathrm{m}))) \times 10^6 \\ +F_{\text{rastrigin}}\left(\boldsymbol{z}\left(\mathrm{r}\left(\frac{m}{2}+1:m\right)\right)\right) \end{array} \right\}$

(11) $F11(\boldsymbol{x}) = \sum_{k=1}^{\frac{m}{2\mathrm{m}}} \left\{ \begin{array}{c} F_{\text{rot\_ackley}}(\boldsymbol{z}(\mathrm{r}(k-1)\mathrm{m}+k\mathrm{m}))) \times 10^6 \\ +F_{\text{ackley}}\left(\boldsymbol{z}\left(\mathrm{r}\left(\frac{m}{2}+1:m\right)\right)\right) \end{array} \right\}$

(12) $F12(\boldsymbol{x}) = \sum_{k=1}^{\frac{m}{2\mathrm{m}}} \left\{ \begin{array}{c} F_{\text{schwefel}}(\boldsymbol{z}(\mathrm{r}(k-1)\mathrm{m}+k\mathrm{m}))) \times 10^6 \\ +F_{\text{sphere}}\left(\boldsymbol{z}\left(\mathrm{r}\left(\frac{m}{2}+1:m\right)\right)\right) \end{array} \right\}$

(13) $F13(\boldsymbol{x}) = \sum_{k=1}^{\frac{m}{2\mathrm{m}}} \left\{ \begin{array}{c} F_{\text{rosenbrock}}(\boldsymbol{z}(\mathrm{r}(k-1)\mathrm{m}+k\mathrm{m}))) \times 10^6 \\ +F_{\text{sphere}}\left(\boldsymbol{z}\left(\mathrm{r}\left(\frac{m}{2}+1:m\right)\right)\right) \end{array} \right\}$

(14) $F14(\boldsymbol{x}) = \sum_{k=1}^{\frac{m}{\mathrm{m}}} F_{\text{rot\_elliptic}}(\boldsymbol{z}(\mathrm{r}(k-1)\mathrm{m}+k\mathrm{m})))$

(15) $F15(\boldsymbol{x}) = \sum_{k=1}^{\frac{m}{\mathrm{m}}} F_{\text{rot\_rastrigin}}(\boldsymbol{z}(\mathrm{r}(k-1)\mathrm{m}+k\mathrm{m})))$

(16) $F16(\boldsymbol{x}) = \sum_{k=1}^{\frac{m}{\mathrm{m}}} F_{\text{rot\_ackley}}(\boldsymbol{z}(\mathrm{r}(k-1)\mathrm{m}+k\mathrm{m})))$

(17) $F17(\boldsymbol{x}) = \sum_{k=1}^{\frac{m}{\mathrm{m}}} F_{\text{schwefel}}(\boldsymbol{z}(\mathrm{r}(k-1)\mathrm{m}+k\mathrm{m})))$

(18) $F18(\boldsymbol{x}) = \sum_{k=1}^{\frac{m}{\mathrm{m}}} F_{\text{rosenbrock}}(\boldsymbol{z}(\mathrm{r}(k-1)\mathrm{m}+k\mathrm{m})))$

(19) $F19(\boldsymbol{x}) = F_{\text{schwefel}}(\boldsymbol{z})$

(20) $F20(\boldsymbol{x}) = F_{\text{rosenbrock}}(\boldsymbol{z})$

Table 10.4 Benchmark functions [Tang *et al.* 2010], performance of COMBEO vs the DiEv and the CMA–ES methods of optimization

Objective functional	COMBEO		DiEv		CMA–ES	
	Number of iterations	Mean error norm	Number of iterations	Mean error norm	Number of iterations	Mean error norm
F1	2.657×10^3	ε	2.892×10^3	ε	4.592×10^3	ε
F2	10^4	ε	3.822×10^3	ε	*it_max*	–*
F3	2.876×10^3	ε	3.017×10^3	ε	619	ε
F4	3.085×10^3	ε	12×10^5	ε	7.064×10^3	ε
F5	1.281×10^4	ε	*it_max*	–	*it_max*	–
F6	2.05×10^3 $(N_p=50)$	ε	*it_max* $(N_p=50)$	–	*it_max* $(N_p=50)$	–
F7	3.549×10^3	ε	2.577×10^3	ε	3.656×10^3	ε
F8	3.573×10^4	ε	2.481×10^4	ε	5.361×10^3	ε
F9	5.384×10^4	ε	*it_max*	–	5.957×10^3	ε
F10	3.902×10^4	ε	*it_max*	–	*it_max*	–
F11	8.162×10^4 $(N_p=100)$	ε	*it_max* $(N_p=100)$	–	*it_max* $(N_p=100)$	–
F12	1442	ε	4.508×10^3	ε	457	ε
F13	2.623×10^5	ε	1.035×10^5	ε	1.413×10^5	ε
F14	8.045×10^5	ε	*it_max*	–	7.103×10^3	ε
F15	1.7222×10^5	ε	2.758×10^4	ε	*it_max*	–
F16	2.5187×10^4	ε	1.2069×10^4	ε	*it_max*	–
F17	2092	ε	7.166×10^3	ε	501	ε
F18	*it_max*	–	*it_max*	–	4.332×10^4	ε
F19	1.095×10^6	ε	4.103×10^5	ε	1.273×10^3	ε
F20	*it_max*	–	*it_max*	–	7.269×10^3	ε

* meaning 'no convergence'

In the numerical work reported in Table 10.4, the number of design variables is $m = 40$, the relaxation parameter $p_r = 0.9$ and the cost function tolerance $\varepsilon = 10^{-5}$. For the three schemes, the program is terminated if the number of iteration exceeds the maximum allowed threshold, which is presently adopted as $it\_max = 4\times10^6$. COMBEO is able to solve most of the problems except F18 and F20. A small ensemble size of $N_p = 10$ is used for all the cases except for the functions F6 and F11 for which convergence is achieved by COMBEO with $N_p = 50$ and $N_p = 100$ respectively. Convergence is not attainable in 8 cases by the DiEv and in 7 cases by the CMA–ES (as observed from Table 10.2) even with $N_p = 100$.

10.4 Further Improvements to COMBEO

The dimensionality curse, which besets many stochastic search schemes including most stochastic filters (described in Chapters 6 and 7) with the necessity of an exponentially exploding ensemble size, comes in the way of solving optimization and inverse problems with a large number of unknowns. To an extent, the particle degeneracy problems encountered by weight-based schemes are circumvented in the additive–update strategies that attempt at healing the bad particles instead of eliminating them altogether. However, even with such an approach, the quality of solutions is highly dependent on the ensemble size, N_p. Here an increase in the system dimension m has to be inevitably accompanied with an increase in N_p to effect a proper exploration of the state space. This increase is likely to be orders of magnitude higher as opposed to a linear increase owing to the slow convergence rate, $\frac{1}{\sqrt{N_p}}$ [Chopin 2004], of the MC simulation. A possible amelioration could be by splitting the original problem into smaller parts and solve for each lower dimensional component separately. A divide and solve approach is justified on a fairly accurate assumption that a larger ensemble is required to solve the original problem compared to the locally-split problem to achieve the same level of accuracy.

10.4.1 State space splitting (3S)

We set the number of sub-problems/sub-structures, N_s, with each sub-structure corresponding to an $\frac{m}{N_s}$ -dimensional state (parameter) vector. If m is not divisible by N_s, then the first $N_s - 1$ sub-structures will have parameters equal to the integer part of $\frac{m}{N_s}$ and the remaining parameters constitute the last sub-structure. In general, let $n_s^{(k)}$ be the size of each kth sub-structure. Denote by $\boldsymbol{X} = \left(\boldsymbol{Y}^{(1)}, \boldsymbol{Y}^{(2)}, \ldots \boldsymbol{Y}^{(k)}, \ \ldots, \boldsymbol{Y}^{(N_s)}\right)^T$ with $\boldsymbol{Y}^{(k)}$ being the kth sub-structure. Following element-wise scrambling as in Eq. (10.4), each sub-structure is updated sequentially at i^{th} iteration as follows:

$$Y_{r,i}^{(k)[j]} = Y_{r,i-1}^{(k)[j]} + U_{r,i}^{(k)[\varsigma_2(j)]}, k = 1,2,\ldots,N_s, j = 1,2,\ldots,N_p \tag{10.23}$$

where, r takes $n_s^{(k)}$ values that correspond to the kth sub-structure and is contained in the parameter set $\{1,2,\ldots,m\}$. $Y_{r,i}^{(k)[j]}$ denotes the updated r^{th} component (parameter) of the kth sub-structure and the jth particle. Also, $\boldsymbol{U}_i^{(k)[\varsigma_2(j)]} = \beta_i \boldsymbol{G}_i^{(k)} \boldsymbol{I}_i^{(k)[\varsigma_2(j)]}$ with $\boldsymbol{I}_i^{(k)[\varsigma_2(j)]} = \left\{ \begin{array}{c} \widehat{\boldsymbol{f}}_{i-1} - \boldsymbol{f}_{i-1}^{[\varsigma_2(j)]} \\ \boldsymbol{Y}_{i-1}^{(k)[\varsigma_1(j)]} - \boldsymbol{Y}_{i-1}^{(k)[\varsigma_2(j)]} \end{array} \right\} \in \mathbb{R}^{\left(N_{\rm n} + n_s^{(k)}\right)}$, $j = 1,2,\ldots,N_p$, is the update vector corresponding to the kth sub-structure. β_i is an annealing type parameter as in Eq. (10.21).

Here $\boldsymbol{f}_{i-1}^{[j]} = \boldsymbol{f}_{i-1}\left(\boldsymbol{Y}_{i-1}^{[j]}\right)$ and $\boldsymbol{Y}_{i-1}^{[j]} = \left\{\boldsymbol{Y}_{i-1}^{(1)[j]}, \boldsymbol{Y}_{i-1}^{(2)[j]}, \ldots, \boldsymbol{Y}_{i-1}^{(k)[j]}, \ldots, \boldsymbol{Y}_{i-1}^{(N_s)[j]}\right\}$. Evident from the expression for $\boldsymbol{I}_i^{(k)}$ (and specifically of $\boldsymbol{Y}_{i-1}^{[j]}$) is the fact that the innovation vector for the kth sub-structure involves the entire set of state variables rather than the $n_s^{(k)}$ components belonging to it. It is computed by assimilating the last update information of the other $k-1$ sub-structures. The gain matrix $\boldsymbol{G}_i^{(k)} \in \mathbb{R}^{n_s^{(k)} \times (N_{\mathbb{m}} + n_s^{(k)})}$ takes into account an $N_{\mathbb{m}}$-dimensional multi-cost functional and is given by Eq. (10.3) with appropriate changes in the dimensions of the matrices involved due to the sub-structuring.

By following Eqs. (10.5) and (10.6), the blended update could be similarly written as:

$$\boldsymbol{Y}_i^{b(k)[j]} = \boldsymbol{w}_i^{(k)[j]} \boldsymbol{Y}_{i-1}^{(k)[j]} + \left(1 - \boldsymbol{w}_i^{(k)[j]}\right) \boldsymbol{Y}_i^{(k)[j]}, k = 1,2,\ldots,N_s, j = 1,2,\ldots,N_p \quad (10.24)$$

where the weights $\boldsymbol{w}_i^{(k)}$ are obtained by normalizing $\hat{w}_i^{(k)[j]} = \sum_{l=1}^{N_p} \varkappa_i^{(k)[l]} w_{i-1}^{(k)[l]} - \varkappa_i^{(k)[j]} w_{i-1}^{(k)[j]}$ to sum to 1. Moreover, the fitness value is calculated as:

$$\varkappa_i^{(k)[j]} = \sqrt{\left(\hat{f}_{1,i} - f_1(\boldsymbol{Y}_i^{(k)[j]})\right)^2 + \left(\hat{f}_{2,i} - f_2(\boldsymbol{Y}_i^{(k)[j]})\right)^2 + \cdots + \left(\hat{f}_{\mathbb{m},i} - f_{\mathbb{m}}\left(\boldsymbol{Y}_i^{(k)[j]}\right)\right)^2} \quad (10.25)$$

Remarks:

(i) Although the update—regular or blended—itself is done separately for each sub-structure, the information on correlation between the sub-structures is brought in by incorporating the last available information from the rest within the update of one.

(ii) The update at any iteration is not limited to the state variable components, $\boldsymbol{Y}^{(\cdot)}$, belonging to a sub-structure but involves the entire set of state variables, $\boldsymbol{Y}$. This could be contrasted with the localization techniques used in [Evensen 2003].

(iii) The finally updated particle at the ith iteration is obtained by concatenating the individual updates, i.e., $\boldsymbol{Y}_i^{[j]} = \left(\boldsymbol{Y}_i^{(1)[j]}, \boldsymbol{Y}_i^{(2)[j]}, \ldots, \boldsymbol{Y}_i^{(N_s)[j]}\right)^T \in \mathbb{R}^m$ for updates in Eq. (10.23) and $\boldsymbol{Y}_i^{b[j]} = \left(\boldsymbol{Y}_i^{b(1)[j]}, \boldsymbol{Y}_i^{b(2)[j]}, \ldots, \boldsymbol{Y}_i^{b(N_s)[j]}\right)^T \in \mathbb{R}^m$ for updates with blending in Eq. (10.24).

The current implementation of the 3S requires $O(N_s \times N_p)$ functional evaluations at each recursion of the algorithm in contrast with $O(N_p)$ evaluations that would have sufficed if one were to solve for $\mathbf{X}$ without splitting. Nevertheless, the extra functional evaluations may be justified given that the ensemble size N_p required to solve the original problem would almost always be far higher than $N_s \times N_p$. This is supported by the numerical evidence wherein some of the functional minimization problems of dimension 40 could not be solved by the CMA–ES even with ensemble sizes as high as 1000 whereas the presently outlined scheme with $N_s = 2$, 4 etc. could solve them with $N_p = 20$. Another advantage of the 3S is that it allows for a repetitive, and hence more comprehensive assimilation of the data owing to the inner iterations in the split-updates. This feature is especially handy while solving higher-dimensional inverse problems (posed as problems in optimization) with sparse data [Venugopal *et al.* 2016]. Using inner iterations in the original problem would have required a far higher computational overhead. Note, however, that the performance of the 3S scheme may deteriorate as the number of sub-structures is increased beyond a threshold for a given dimension. This could be owing to the inadequate data communication between the sub-structures as they grow in number. Thus, for best results, a balance must be struck between the number of sub-structures and the ensemble size.

Table 10.5 COMBEO–3S scheme—Algorithm of COMBEO with 3S scheme incorporated

Step 1.	As in step 1 of Table 10.3, initialize m, N_p, and the blending coefficients $w_0^{[j]} = \frac{1}{N_p}$, $j = 1,2,\ldots,N_p$. Fix ε and it_max. Generate the initial ensemble of particles, $\left\{\mathbf{X}_0^{[j]}\right\}_{j=1}^{N_p} \in \mathbb{R}^m$. As per the 3S scheme, choose the number of sub-structures N_s. Find $n_s^{(k)}$, $k = 1,2,\ldots,N_s$, i.e., if m is divisible by N_s, then $n_s^{(k)} = \frac{m}{N_s}$, else each of the $N_s - 1$ sub-structures will have parameters equal to the integer part of $\left(\frac{m}{N_s}\right)$ components and the remaining parameters form the last sub-structure. This implies that $\mathbf{X}_0 = \left[\mathbf{Y}_0^{(1)}, \mathbf{Y}_0^{(2)}, \ldots \mathbf{Y}_0^{(k)}, \ldots, \mathbf{Y}_0^{(N_s)}\right]^T \in \mathbb{R}^{m \times N_p}$. Calculate $f\left(\mathbf{X}_0^{(k)}\right) \in \mathbb{R}^{n \times N_p}$ and set $\widehat{f}$=min$(f) \in \mathbb{R}^n$. Also calculate the initial fitness values: $$\varkappa_0^{(k)[j]} = \sqrt{\sum_{l=1}^{N_f}\left(\hat{f}_l - f_l(\mathbf{Y}_0^{(k)[j]})\right)^2}, k = 1,2,\ldots,N_s, \quad j = 1,2,\ldots,N_p \qquad (10.26)$$ Set the inertia factor (relaxation parameter) p_r. *Set the iteration counter* $i = 1$.

Continued...

Step 2.	For implementation of the coalescence and scrambling strategies, generate $\varsigma = \left(\varsigma_l[1], \varsigma_l[2], \ldots, \varsigma_l\left(N_p\right)\right)$, $l = 1, 2$.
Step 3.	Set the counter for sub-structure, $k = 1$.
Step 4.	Set the counter for particles, $j = 1$.
Step 5.	If $c_1 = U(0,1) < 1 - p_r$, then If $c_2 = U(0,1) < p_{\text{rand}}$, where $p_{\text{rand}} \in [0,1]$; (one choice for p_{rand} is 0.5). Obtain the update $\boldsymbol{Y}_i^{(k)[j]}$ according to Eq. (10.23) and denote the update by $\mathbb{Y}_i^{(k)[j]} \in \mathbb{R}^{n_s^{(k)}}$; else obtain the update (with blending) according to Eq. (10.24) and set $\mathbb{Y}_i^{(k)[j]} = \boldsymbol{Y}_i^{b(k)[j]} \in \mathbb{R}^{n_s^{(k)}}$; else retain the original particle as the update, i.e., $\mathbb{Y}_i^{(k)[j]} = \boldsymbol{Y}_{i-1}^{(k)[j]} \in \mathbb{R}^{n_s^{(k)}}$.
Step 6.	Set $j \equiv j + 1$. If $j \leq N_p$, go to step 5; else go to step 7.
Step 7.	Construct $\boldsymbol{X}_i^{(k)[j]} = \left[\mathbb{Y}_i^{(1)[j]}, \mathbb{Y}_i^{(2)[j]}, \ldots \mathbb{Y}_i^{(k)[j]}, \boldsymbol{Y}_i^{(k+1)[j]} \ldots, \boldsymbol{Y}_i^{(N_s)[j]}\right]^T \in \mathbb{R}^m$. Set $k \equiv k + 1$. If $k \leq N_s$, go to step 4; else go to step 8.
Step 8.	Equate $\boldsymbol{Y}_i^{(k)[j]} = \mathbb{Y}_i^{(k)[j]}$, $k = 1, 2, \ldots, N_s$ and construct $\boldsymbol{X}_i^{[j]} = \left[\boldsymbol{Y}_i^{(1)[j]}, \boldsymbol{Y}_i^{(2)[j]}, \ldots, \boldsymbol{Y}_i^{(N_s)[j]}\right]^T \in \mathbb{R}^m$ for $j = 1, 2, \ldots, N_p$ and calculate the empirical mean $\overline{\boldsymbol{X}}_i = \frac{1}{N_p}\sum_{j=1}^{N_p} \boldsymbol{X}_i^{[j]} \in \mathbb{R}^m$ and mean of the cost function by $\overline{\boldsymbol{f}}_i = \frac{1}{N_p}\sum_{j=1}^{N_p} \boldsymbol{f}_i^{[j]} \in \mathbb{R}^{\text{n}}$.
Step 9.	Set $i \equiv i + 1$. If $i < it\_max$, go to step 2; else terminate the algorithm.

10.4.2 Benchmark problems

We assess the performance of COMBEO–3S scheme against a few benchmark minimization problems. The first set is the same 20 benchmark functions (F1-F20) considered earlier in Table 10.4. Results obtained by COMBEO–3S scheme are listed in Table 10.6. The number of design variables, $m = 40$, the ensemble size, $N_p = 20$ and the tolerance, $\varepsilon = 10^{-5}$. The relaxation parameter, $p_r = 0.9$. The 3S scheme takes four sub-structures $N_s = 4$ with $n_s^{(k)} = 10$, $k = 1, 2, \ldots, N_s$ to minimize these functions, all of which have their global minimum at 0. It is noted that except for F14, the COMBEO–3S is able to locate the global minimum for all the functions to the desired level of accuracy with an ensemble size as small as $N_p = 20$ and within far lower number of iterations. The scheme could solve F18 and F20 for which earlier version of COMBEO (without 3S) failed to converge (see Table 10.4).

Table 10.6 Performance of the COMBEO–3S scheme against cost functions F1–F20; $m = 40$, $N_p = 20$, $N_s = 4$, $\varepsilon = 10^{-5}$, $it\_max = 4\times10^5$

Cost Function	Number of Iterations	Error	Cost Function	Number of Iterations	Error
F1	709	ε	F11	212709	ε
F2	5438	ε	F12	855	ε
F3	721	ε	F13	76824	ε
F4	4505	ε	F14	*it_max*	934.6
F5	6419	ε	F15	145270	ε
F6	921	ε	F16	3426	ε
F7	614	ε	F17	1301	ε
F8	21387	ε	F18	*it_max*	0.0058
F9	218671	ε	F19	15050	ε
F10	46073	ε	F20	22476	ε

The second set of 24 functions (IF1–IF24) have been taken from the document on black-box optimization benchmarking [Finck *et al.* 2014]. The functions are categorized as:

Group I. Separable Functions:

(IF1) sphere function

(IF2) ellipsoidal function

(IF3) Rastrigin function

(IF4) Büche–Rastrigin function

(IF5) linear slope

Group II. Functions with low/moderate conditioning

(IF6) attractive sector function

(IF7) step ellipsoidal function

(IF8) original Rosenbrock function

(IF9) rotated Rosenbrock function

Group III. Functions with high conditioning and unimodal

(IF10) ellipsoidal function

(IF11) discus function

(IF12) bent cigar function

(IF13) sharp ridge function

(IF14) different powers function

Group IV. Multimodal functions with adequate global structure

(I15) Rastrigin function

(IF16) Weierstrass function

(IF17) Schaffers F7 function

(IF18) Schaffers F7 function, moderately ill-conditioned

(IF19) composite Griewank–Rosenbrock function F8F2

Group V. Multimodal functions with weak global structure

(IF20) Schwefel function

(IF21) Gallagher's Gaussian 101-me peaks function

(IF22) Gallagher's Gaussian 21-hi peaks function

(IF23) Katsuura function

(IF24) Lunacek bi–Rastrigin function

For this 2nd set of benchmark functions, it_max is set to 4×10^5 and $N_s = 2$. Table 10.7 reports the results of the COMBEO–3S scheme. Results by the CMA–ES are also included in the Table. The difficulty level of these problems is indeed high compared to the ones in Table 10.6. Nevertheless, except for some cases, the COMBEO–3S scheme could either solve the problem completely or arrive at errors that, though higher than the stipulated tolerance, remained relatively smaller. For the COMBEO–3S scheme, the error gradually decreases in most cases as iterations progress, revealing the possibility that it might reach the global optimal value asymptotically as the iterations are continued further. The showcasing of the performance of the COMBEO (with or without the 3S scheme) with respect to the two sets of benchmark functions (in Tables 10.4 and 10.6) also bears out the fact that, despite its many advantages already discussed, there is still considerable scope in improving its performance. Finally, we emphasize, once more, that the reader should carefully go through the MATLAB optimization codes available in the companion web site.

Table 10.7 Performance of COMBEO–3S scheme against INRIA cost functions IF1-IF24; $m = 40,\ N_p = 20,\ N_s = 2,\ \varepsilon = 10^{-5},\ p_r = 0.9,\ it\_max = 4 \times 10^5$

Cost Function	Target (optimum) Function Value	COMBEO–3S		CMA–ES	
		Number of Iterations	Error	Number of Iterations	Error
IF1	79.48	384	ε	216	ε
IF2	−209.88	572	ε	3611	ε
IF3	−462.09	4955	ε	*it_max*	6.28×10^1
IF4	−462.09	6517	ε	*it_max*	9.9×10^1
IF5	−9.21	84	ε	976	ε
IF6	35.9	4282	ε	1198	ε
IF7	92.94	*it_max*	2.89	*it_max*	1.58×10^1
IF8	149.15	15623	ε	4300	ε
IF9	123.83	*it_max*	3.33×10^{-2}	4026	ε
IF10	−54.94	*it_max*	1.19×10^2	3761	ε
IF11	76.27	91480	ε	2115	ε
IF12	−621.11	*it_max*	8.07	1712	ε
IF13	29.97	*it_max*	2.7×10^{-1}	*it_max*	9.9×10^{-1}
IF14	−52.35	*it_max*	6.84×10^{-5}	1643	ε
IF15	1000	*it_max*	1.7×10^2	*it_max*	6.29×10^1
IF16	71.35	*it_max*	1.95	*it_max*	2.48
IF17	−16.94	*it_max*	2.34×10^{-3}	*it_max*	1.58×10^{-1}
IF18	−16.94	*it_max*	4.13×10^{-1}	*it_max*	6×10^{-1}
IF19	−102.55	*it_max*	3.74	*it_max*	9.54
IF20	−546.5	*it_max*	3.15×10^{-1}	*it_max*	2.45
IF21	40.78	21965	ε	*it_max*	6.2×10^1
IF22	−1000	*it_max*	1.96	*it_max*	2.43
IF23	6.87	*it_max*	1.31	*it_max*	5.67
IF24	102.61	*it_max*	1.96×10^2	*it_max*	3.89×10^2

10.5 Concluding Remarks

The basic scheme of COMBEO, the evolutionary global optimization developed in the last chapter and mainly woven around change of measures, has been fitted with additional appurtenances with a view to making it more effective and viable for practical problems in science and engineering. Additional improvements in the global search have been attempted by suitably extending the scrambling and coalescence steps. Of significance is also the related attempt at uncovering an underlying thread that possibly relates the basic ideas driving the update strategies in a few other well-known optimization schemes. For instance, several random perturbation strategies, e.g., the coalescence step treated as a martingale problem, are explored and it is shown that a variant of the coalescence step may mimic a PSO-like search. A search as in DiEv, on the other hand, may be organized through a form of scrambling. Better performance also ensues, at least in some cases, with a sub-structuring principle—the 3S scheme. With numerical work on a host of higher dimensional benchmark cost functions of the separable, partially separable and non-separable types and comparisons with a few existing evolutionary methods prominently the DiEv and the CMA–ES, the performance features of the schemes are brought out to an extent.

Before concluding this chapter and the book, we wish to stress on the importance of Doob's h–transform (see Appendix D) in possible future modifications and restructuring of the COMBEO. The central point is that, after a suitable normalization, a stochastically evolving and strictly positive cost functional, upon extremization, could be interpreted as a Radon–Nikodym derivative that naturally induces a change of measures in the design space. Upon exploiting Doob's h–transform, one may then uniquely derive an appropriate drift term to modify the SDEs for the design variables such that a local, or quasi-local, search could be organized even more efficaciously. Of course, Doob's h–transform may also be used in parallel to weakly restrict the search to within certain feasible regions of the design space. This is possible, in a weak sense, despite using Brownian motions whose variations are unbounded (see Example D.1 in Appendix D). The authors look forward to sharing the details of some of these exciting ideas with our readers in a future edition of this book.

Notations

c_1, c_2	uniformly distributed random numbers
C_{11}, C_{22}	bearing damping coefficients
$\boldsymbol{D}_i^{[j]}$	update at i^{th} iteration to system variable vector corresponding to j^{th} particle
$f(\boldsymbol{X})$	objective (cost) function
$\boldsymbol{f}(\boldsymbol{X})$	vector cost function

$\boldsymbol{G}_i$	gain-like update coefficient matrix at i^{th} iteration
$\boldsymbol{I}_t$	innovation process
K_{11}, K_{22}	bearing stiffness coefficients
$\mathbb{m}$	dimension of the vector cost functional f
$n_s^{(k)}$	size of kth substructure
N	number of iterations
N_s	number of sub-structures
p_r	relaxation parameter
$R_{41}(\lambda)$	maximum unbalance response at 41st dof corresponding to the transverse direction at the rotor bearing (Fig. 9.14)
$\check{s}_1, \check{s}_2$	rotor critical speeds in Example 10.1
$\boldsymbol{U}_i^{(k)[j]}$	vector of updates for k^{th} substructure of jth particle at i^{th} iteration
w_i	vector of weights used in blending operation
$\boldsymbol{X}_0^{[j]}$	initial ensemble of particles
$\boldsymbol{X}_i^{[j]}$	vector of system variables corresponding to to j^{th} particle at i^{th} iteration
${}^P\boldsymbol{X}_i^{[j]} \in \mathbb{R}^m$	personal best vector corresponding to j^{th} particle at i^{th} iteration (PSO)
${}^G\boldsymbol{X}_i \in \mathbb{R}^m$	best location among all the particles at i^{th} iteration (PSO)
$Y_{r,i}^{(k)[j]}$	updated r^{th} component (parameter) that is in the kth sub-structure and belongs to the jth particle
$\boldsymbol{Y}_i^{b(k)[j]}$	blended update (Eq. 10.24)
β_i	annealing-type coefficient at i^{th} iteration
ε	tolerance value
λ	rotation speed in Example 10.1
γ_i	innovation noise covariance matrix at i^{th} iteration
$\boldsymbol{\varkappa}_i$	fitness function (Eq. 10.6a)

Appendix A (Chapter 1)

1. Topological space

A topology on a set X is a collection $\mathcal{T}$ of subsets of X with the following properties:

i) $\phi, X \in \mathcal{T}$ (A.1a)

ii) The union of the elements of any sub-collection of $\mathcal{T}$ is in $\mathcal{T}$ (A.1b)

iii) The intersection of the elements of any finite sub-collection of $\mathcal{T}$ is in $\mathcal{T}$ (A.1c)

where, ϕ is a null set. A set X for which a $\mathcal{T}$ is specified, is called a topological space. A topological space is often referred to as an ordered pair $(X, \mathcal{T})$ consisting of a set X and a topology $\mathcal{T}$ on X.

2. Open and closed sets

A subset $A \subseteq \mathbb{R}$ is open if $\exists \varepsilon > 0$ such that $(x - \varepsilon, x + \varepsilon) \subseteq A$, $\forall x \in A$. In higher dimensions, we may consider an open ball, $\mathcal{B}_r(x_0) = \{x : |x - x_0| < r\}$ with centre x_0 and radius $r > 0$ in $\mathbb{R}^n$. These open sets are measurable. For $n = 1$, the measure is denoted by the length of the interval spanned by the open set and for $n = 2$ and $n = 3$ it is the area and volume spanned by the corresponding open sets.

A subset $B \subseteq \mathbb{R}$ is closed if $A = B^c$ is open where the superscript c denotes complementation. Note that the null set ϕ and the universal set Ω of a topological space are both open and closed.

3. Metric space

A set M is termed a metric space if for every pair of elements $x,\ y \in M$ there is defined a real number $d(x,y)$ called the 'distance' (or a metric) between x and y satisfying the following axiomatic properties:

i) $d(x,y) > 0$ if $x \neq y, d(x,y) = 0$ iff $x = y$ (A.2a)

ii) $d(x,y) = d(y,x)$ for every x and y (symmetry) (A.2b)

iii) $d(x,y) \leq d(x,z) + d(y,z)$ for every z in M (triangle inequality) (A.2c)

A metric space is usually denoted by (M, d). $\mathbb{R}$, $\mathbb{R}^2$ and $\mathbb{R}^3$ are examples of metric spaces with the Euclidean distance between any two points in these spaces being a metric.

4. Cover of a set

'Cover of a set' is commonly associated with sets in a topology (see item 1 above). A cover of a set A in a topological space $(X, \mathcal{T})$ is a collection of sets in X, whose union contains A. In other words, an indexed family $C = \left\{B_{j\in\mathbb{N}}\right\}$, is a cover of A if:

$$A \subseteq \cup_j B_j \tag{A.3}$$

If C is a collection of open subsets of X, then C is said to be an open cover of A. If the index set is finite, then $\cup_j B_j$ is called a finite cover of A. A subcover of a cover C of A is a sub-collection S of C such that S covers A. In this context, a set A is called compact if every open cover of A has a finite subcover.

5. Proof for the properties of a distribution Function

The properties in Eq. (1.66) are restated below:

i) $0 \leq F_X(x) \leq 1, \ \forall x \in \mathbb{R}$ (A.4a)

ii) $F_X(x)$ is a non-decreasing function, i.e., $F_X(x) \leq F_X(y)$ whenever $x \leq y$ (A.4b)

iii) $\lim_{x\to\infty} F_X(x) = 1$ and $\lim_{x\to-\infty} F_X(x) = 0$ (A.4c)

iv) $F_X(x)$ is right continuous, i.e., for each $x \in \mathbb{R}$, $F_X(x) = \lim_{h\downarrow 0} F_X(x+h)$ (A.4d)

Proof:

i) $$F_X(x) = P(X \leq x) \in [0,1] \ \forall \ x \in \mathbb{R} \tag{A.5}$$

ii) For any $x \leq y$, we have $\{\omega : X(\omega) \leq x\} \subseteq \{\omega : X(\omega) \leq y\}$. Hence we have:

$$F_X(x) = P(\{\omega : X(\omega) \leq x\}) \leq P(\{\omega : X(\omega) \leq y\}) = F_X(y) \tag{A.6}$$

iii) Let $A_n = \{\omega : X(\omega) \leq -n\}$, for some $n \in \mathbb{N}$. Then $A_1 \supseteq A_2 \supseteq \ldots$ and $\cap_n A_n = \phi$. Hence we have:

$$F_X(-n) = P(A_n) \to P(\phi) = 0 \text{ as } n \to \infty \tag{A.7}$$

Similarly, let $B_n = \{\omega : X(\omega) \leq n\}$, for any $n \in \mathbb{N}$. Then $B_1 \subseteq B_2 \subseteq \ldots$ and $\bigcup_n B_n = \Omega$. Hence we have:

$$F_X(n) \leq P(B_n) \to P(\Omega) = 1 \text{ as } n \to \infty \tag{A.8}$$

(iv) Since $F_X(x)$ is non-decreasing, it is enough to prove that $\lim_{n\to\infty} F_X\left(x+\frac{1}{n}\right) = F_X(x)$. By definition:

$$F_X\left(x+\frac{1}{n}\right) - F_X(x) = P\left(\left\{\omega : X(\omega) \le x + \frac{1}{n}\right\}\right) - P(\{\omega : X(\omega) \le x\}) \quad \text{(A.9)}$$

In terms of inverse measure, the LHS of the above equation is equal to:

$$P\left(X^{-1}\left(-\infty, x+\frac{1}{n}\right]\right) - P\left(X^{-1}(-\infty, x]\right) = P\left(X^{-1}\left(x, x+\frac{1}{n}\right]\right) \quad \text{(A.10)}$$

Therefore:

$$\begin{aligned}\lim_{n\to\infty}\left\{F_X\left(x+\frac{1}{n}\right) - F_X(x)\right\} &= P\left(\cap_{n\in\mathbb{N}} X^{-1}\left(x, x+\frac{1}{n}\right]\right) \\ &= P\left(X^{-1}\left(\cap_{n\in\mathbb{N}}\left(x, x+\frac{1}{n}\right]\right)\right) \\ &= P\left(X^{-1}(\phi)\right) = P(\phi) = 0 \end{aligned} \quad \text{(A.11)}$$

Hence the result.

6. Continuity of a function

A function $f(x)$ is called continuous at a point $x = x_0$ if the increment of $f(x)$ is small over a small interval containing the point. That is:

$$\Delta f(x) = f(x) - f(x_0) \to 0 \text{ as } \Delta x = x - x_0 \to 0 \quad \text{(A.12)}$$

The above definition may also be stated as:

$$\lim_{x\to x_0} f(x) = f(x_0) \quad \text{(A.13)}$$

An alternative and more elegant definition for continuity (Fig. A.1) of $f(x)$ at a point $x = x_0$ is that for every $\varepsilon > 0$ there exists a $\delta > 0$ such that:

$$|x - x_0| < \delta \Rightarrow \left|f(x) - f(x_0)\right| < \varepsilon \quad \text{(A.14)}$$

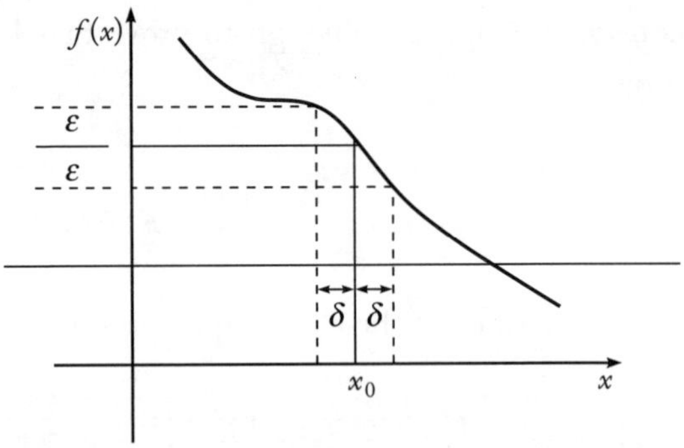

Fig. A.1 Definition for continuity of a function $f(x)$

Right-continuity and left-continuity

A function $f(x)$ is called right (left) continuous at a point $x = x_0$ if $f(x)$ approaches $f(x_0)$ when x approaches x_0 from the right (left). That is:

$$\lim_{x \downarrow x_0} f(x) = f(x_0) (\lim_{x \uparrow x_0} f(x) = f(x_0)) \tag{A.15}$$

Often we denote $\lim_{x \downarrow x_0} f(x)$ by $f(x_0+)$ and $\lim_{x \uparrow x_0} f(x)$ by $f(x_0-)$ so that we call $f(x)$ right continuous if $f(x_0+) = f(x_0)$ and left continuous if $f(x_0-) = f(x_0)$. A function is continuous if it is both right and left continuous.

Types of discontinuities

A point x is called a discontinuity of the first kind or a 'jump point' if both limits $f(x+)$ and $f(x-)$ exist and are not equal. The jump at this discontinuous point x is equal to $f(x+) - f(x-)$. Any other discontinuity is known as a discontinuity of the second kind.

Unit step function or Heaviside step function (Eq. 1.68) is an example of discontinuity of the first kind. $F(x) = \sin(\frac{1}{x})$ has a discontinuity at $x = 0$ and is of the second kind since the limits from the right or the left do not exist.

Uniformly continuous functions

Let $I = [a, b]$ be an interval. A function $f : I \to R$ is uniformly continuous on I if for each $\varepsilon > 0$ there exists $\delta > 0$ such that $|f(x) - f(x_0)| < \varepsilon$ for all $x, x_0 \in I$ that satisfy $|x - x_0| < \delta$.

While continuity can be defined at a point, the uniform continuity is a property of a function over an interval (or a set). Further, for a function to be continuous, the parameter δ depends on ε and x_0 whereas, for a uniformly continuous function, it is possible, for each $\varepsilon > 0$, to find one $\delta > 0$ for all points $x_0 \in I$.

Example A.1

Consider $f(x) = x^2$ on $I = [0,1]$. Then:

$$\left|f(x) - f(x_0)\right| = \left|x^2 - {x_0}^2\right| = \left|(x + x_0)(x - x_0)\right|$$

$$\leq 2|x - x_0| \forall x,\ x_0 \in [0,1] \tag{A.16}$$

If $|x - x_0| < \delta$, then $\left|f(x) - f(x_0)\right| < 2\delta$. Hence for a given ε, it suffices to define $\delta = \frac{\varepsilon}{2}$ so that $\left|f(x) - f(x_0)\right| < \varepsilon$ whenever $|x - x_0| < \delta$ and $f(x)$ is continuous. Since δ depends only on ε and not on x_0, $f(x)$ is uniformly continuous.

Example A.2

Let $f(x) = \frac{1}{x}$ and $I = (0,1]$. Then:

$$\left|f(x) - f(x_0)\right| = \frac{|x_0 - x|}{|x||x_0|} \tag{A.17}$$

For an arbitrary fixed x_0, let $0 < \delta < x_0$. Then $|x - x_0| < \delta$ gives $x > x_0 - \delta > 0$. Thus we have:

$$\left|f(x) - f(x_0)\right| < \frac{\delta}{(x_0 - \delta)\, x_0} \tag{A.18}$$

For a given ε, Eq. (A.18) gives $\delta = \varepsilon {x_0}^2/(1 + \varepsilon x_0)$. We find that f is continuous on the semi-closed interval $(0,1]$. Since δ depends on both ε and x_0, f is not uniformly continuous [Reddy 1998].

Absolutely continuous function

Absolute continuity of a function is a smoothness property and is stronger than continuity and uniform continuity. A function $f(x)$ is absolutely continuous on an interval $[a,b] \in \mathbb{R}$ if for each $\varepsilon > 0$, there exists a $\delta > 0$ such that $\sum_{i=1}^{n} \left|f(b_i) - f(a_i)\right| < \varepsilon$ for a finite collection of non-overlapping sub-intervals $\{[a_i, b_i],\ 1 \leq i \leq n\}$ in the interval $[a,b]$, satisfying the condition $\sum_{i=1}^{n} (b_i - a_i) < \delta$.

7. The non-zero jumps of the distribution function $F_X(x)$ are a countable set

Let E be the set of non-zero jumps of $F_X(x)$. If E_n is the subset of E so that:

$$E_n = \left\{a \in E :' \text{jump of } F_X(x)\, at\ a' \geq \frac{1}{n}\right\} \tag{A.19}$$

Let $a_1, \ldots, a_k \in E_n$ with $k \le n$ and $a_1 < \cdots < a_k$. Select $a_0 < a_1$ so that we have :

$$0 \le F_X(a_0) \le F_X(a_k) \le 1 \text{ and } F_X(a_j) - F_X(a_{j-1}) \ge \frac{1}{n} \tag{A.20}$$

This leads to:

$$F_X(a_k) - F_X(a_0) = \sum_{j=1}^{k} \left(F_X(a_j) - F_X(a_{j-1})\right) \ge \frac{k}{n} \tag{A.21}$$

But $F_X(a_k) - F_X(a_0) \le 1$ so that we find that $k \ngtr n$. Hence E_n is either empty or cannot have more than n elements. Since $E = \cup_n E_n$, it follows that E is countable.

8. Variation of a function

Variation of a function $f(x)$ over an interval $[a,b]$ is defined as:

$$V_f([a,b]) = \sup \sum_{i=1}^{n} \left|f(x_i) - f(x_{i-1})\right| \tag{A.22a}$$

where the supremum is taken over the partitions $\pi_n := \{x_0 < x_1 < \cdots < x_n\}$ with $x_0 = a$ and $x_n = b$. Since the sum increases with n, the variation of a function is indeed given by:

$$V_f([a,b]) = \lim_{\Delta x_n \to 0} \sum_{i=1}^{n} \left|f(x_i) - f(x_{i-1})\right| \tag{A.22b}$$

where $\Delta x_n = \max_{1 \le i \le n}(x_i - x_{i-1})$. $V_f(x) := V_f([0,x])$ is a non-decreasing function of x. $f(x)$ is of bounded (finite) variation, if $V_f(x) < \infty$ for all x.

9. Lebesgue integrable function

Given a set $E \subset \mathbb{R}$, a Borel σ-algebra $\mathcal{B}(\mathbb{R})$ and a finite measure μ, we define the Lebesgue integral:

$$I = \int_E g \, d\mu \tag{A.23}$$

of any function g on E. We say that $g(x) : E \to \mathbb{R}$ is Lebesgue integrable if and only if given $\varepsilon > 0$ there exists *simple* functions u and v on E such that $u \le g \le v$ and $\int_E (v-u)\, d\mu < \varepsilon$ with μ being a Lebesgue measure. Note that a function z: $E \to \mathbb{R}$ is called *simple* if z is measurable, takes on only a finite number of values, say, $\alpha_k, k = 1,2,...,n$ and always expressible as $\sum_k \alpha_k I_{A_k}$ where I_{A_k} are indicator functions (Eq. 1.13) and $A_k = \{x \in E : g(x) = \alpha_k\}$.

Thus, the Lebesgue integral I is given by:

$$\int_E g\, d\mu = \sup\left\{\int_E \phi d\mu \ : \phi \text{ is simple on } E \text{ such that } \phi \le g\right\} \tag{A.24}$$

In this context, we can define the lower and upper Lebesgue integrals to be:

$$\text{lower} \int_E g\, d\mu = \sup\left\{\int_E u d\mu \ : u \text{ is a simple function on } E \text{ such that } u \le g\right\} \tag{A.25a}$$

$$\text{upper} \int_E g\, d\mu = \inf\left\{\int_E v d\mu \ : v \text{ is a simple function on } E \text{ such that } g \le v\right\} \tag{A.25b}$$

We say that the function $g(x)$ is Lebesgue integrable if lower and upper Lebesgue integrals are the same. The Lebesgue integrals of u and v in Eqs. (A.25a,b) are similar to Reimann (or Darbaux) lower and upper sums defined for Reimann integrable functions.

10. X is a random variable on a measurable space (Σ, Υ, μ) if and only if it is a simple function or a limit of simple functions

If X is a non-negative random variable, $\exists$ a sequence of simple functions $\{Y_n\}$ that satisfy the following properties:

i) $Y_n(\omega) \ge 0$ (A.26a)

ii) $\cdots \le Y_n(\omega) \le Y_{n+1}(\omega) \le \ldots$ (A.26b)

so that $X(\omega) = \lim_{n\to\infty} Y_n(\omega)$

Proof:

Define:

$$Y_n(\omega) = \sum_{k=1}^{n2^n} \frac{k-1}{2^n} I_{(k-1)/2^n \le X < k/2^n}(\omega) + n I_{X>n}(\omega) \tag{A.27}$$

$\{Y_n(\omega)\}$ constitutes a set of simple functions for $n = 1, 2, \ldots$ with $\left|Y_n(\omega)\right| \le n$. Further, $0 \le Y_n(\omega) \le Y_{n+1}(\omega) \le X(\omega) \forall \omega$. For $X(\omega) \le n$, $\left|X(\omega) - Y_n(\omega)\right| \le \frac{1}{2^n}$. If $X > n$ and $X(\omega) = \infty$, $Y_n(\omega) = n$ that tends to ∞ as $n \to \infty$. Thus, the limit is the random variable $X(\omega)$.

In the general case (allowing $X(\omega)$ to be possibly negative), we use the fact that $X(\omega) = X^+(\omega) - X^-(\omega)$ where X^+ and X^- are non-negative functions and follow the arguments above.

11. Marginal densities of a joint normal random variable

The *pdf* of a 2-dimensional joint (and correlated) normal random variable $X = (X_1, X_2)$ is given by Eq. (1.137) and is reproduced below:

$$f_{X_1X_2}(x_1,x_2) = \frac{1}{2\pi\sigma_1\sigma_2\sqrt{(1-\rho^2)}}\exp\left[-\frac{1}{2(1-\rho^2)}\left\{\left(\frac{x_1-m_1}{\sigma_1}\right)^2 -2\rho\left(\frac{x_1-m_1}{\sigma_1}\right)\left(\frac{x_2-m_2}{\sigma_2}\right)+\left(\frac{x_2-m_2}{\sigma_2}\right)^2\right\}\right] \tag{A.28}$$

The marginal densities of the two random variables X_1 and X_2 are obtained as:

$$f_{X_1}(x_1) = \int_{-\infty}^{\infty} f_{X_1X_2}(x_1,x_2)\,dx_2 \text{ and } f_{X_2}(x_2) = \int_{-\infty}^{\infty} f_{X_1X_2}(x_1,x_2)\,dx_1 \tag{A.29}$$

Before we find explicit expressions for the two densities, we rewrite the expression $\left(\frac{x_1-m_1}{\sigma_1}\right)^2 - 2\rho\left(\frac{x_1-m_1}{\sigma_1}\right)\left(\frac{x_2-m_2}{\sigma_2}\right)+\left(\frac{x_2-m_2}{\sigma_2}\right)^2$ as:

$$\left\{\rho^2\left(\frac{x_1-m_1}{\sigma_1}\right)^2 - 2\rho\left(\frac{x_1-m_1}{\sigma_1}\right)\left(\frac{x_2-m_2}{\sigma_2}\right)+\left(\frac{x_2-m_2}{\sigma_2}\right)^2\right\}+(1-\rho^2)\left(\frac{x_1-m_1}{\sigma_1}\right)^2$$

$$=\left(\frac{x_2-m_2}{\sigma_2}-\rho\left(\frac{x_1-m_1}{\sigma_1}\right)\right)^2+(1-\rho^2)\left(\frac{x_1-m_1}{\sigma_1}\right)^2 \tag{A.30}$$

Now we have:

$$f_{X_1}(x_1) = \frac{1}{2\pi\sigma_1\sigma_2\sqrt{(1-\rho^2)}}\int_{-\infty}^{\infty}\exp\left[-\frac{1}{2(1-\rho^2)}\left\{\left(\frac{x_2-m_2}{\sigma_2}-\rho\left(\frac{x_1-m_1}{\sigma_1}\right)\right)^2 +(1-\rho^2)\left(\frac{x_1-m_1}{\sigma_1}\right)^2\right\}\right]dx_2$$

$$=\frac{1}{\sqrt{2\pi}\sigma_1}\exp\left\{-\frac{1}{2}\left(\frac{x_1-m_1}{\sigma_1}\right)^2\right\}\left[\frac{1}{\sqrt{2\pi}\sigma_2\sqrt{(1-\rho^2)}}\right.$$

$$\int_{-\infty}^{\infty} \exp\left\{-\frac{1}{2(1-\rho^2)}\left(\frac{x_2-m_2}{\sigma_2}-\rho\left(\frac{x_1-m_1}{\sigma_1}\right)\right)^2\right\} dx_2\Bigg] \qquad (A.31)$$

By suitable transformation of variables, the term within square brackets can be shown to be unity and hence we obtain the result for the marginal density of X_1 as in Eq. (1.138a).

If we write the expression $\left(\frac{x_1-m_1}{\sigma_1}\right)^2 - 2\rho\left(\frac{x_1-m_1}{\sigma_1}\right)\left(\frac{x_2-m_2}{\sigma_2}\right)+\left(\frac{x_2-m_2}{\sigma_2}\right)^2$ as:

$$\left\{\left(\frac{x_1-m_1}{\sigma_1}\right)^2 - 2\rho\left(\frac{x_1-m_1}{\sigma_1}\right)\left(\frac{x_2-m_2}{\sigma_2}\right)+\rho^2\left(\frac{x_2-m_2}{\sigma_2}\right)^2\right\}+(1-\rho^2)\left(\frac{x_2-m_2}{\sigma_2}\right)^2$$

$$=\left(\frac{x_1-m_1}{\sigma_1}-\rho\left(\frac{x_2-m_2}{\sigma_2}\right)\right)^2+(1-\rho^2)\left(\frac{x_2-m_2}{\sigma_2}\right)^2 \qquad (A.32)$$

and proceed on similar lines as in the case of X_1, we obtain the result for the marginal density of X_2 as in Eq. (1.138b).

12. Correlation coefficient, $\rho = \frac{\sigma_{12}}{\sigma_1\sigma_2}$

The *pdf* of a two-dimensional joint (and correlated) normal random variable $\boldsymbol{X} = (X_1, X_2)^T$ is given by Eq. (1.137) and is reproduced below:

$$f_{X_1X_2}(x_1,x_2) = \frac{1}{2\pi\sigma_1\sigma_2\sqrt{(1-\rho^2)}}\exp\left[-\frac{1}{2(1-\rho^2)}\left\{\left(\frac{x_1-m_1}{\sigma_1}\right)^2\right.\right.$$

$$\left.\left.-2\rho\left(\frac{x_1-m_1}{\sigma_1}\right)\left(\frac{x_2-m_2}{\sigma_2}\right)+\left(\frac{x_2-m_2}{\sigma_2}\right)^2\right\}\right] \qquad (A.33)$$

The cross-covariance between the normal random variables X_1 and X_2 is given by:

$$\sigma_{12} = E\left[(x_1-m_1)(x_2-m_2)\right] = E\left[X_1X_2\right] - E\left[X_1\right]E\left[X_2\right]$$

$$= E\left[X_1X_2\right] - m_1m_2 \qquad (A.34)$$

$E\left[X_1X_2\right]$ is known as cross-correlation and is defined as:

$$E\left[X_1X_2\right] = \int_{-\infty}^{\infty}\int_{-\infty}^{\infty} x_1x_2 f_{X_1,X_2}(x_1,x_2)\,dx_1dx_2 \qquad (A.35)$$

Before the above integral is evaluated, we rewrite the expression $\left(\frac{x_1 - m_1}{\sigma_1}\right)^2 - 2\rho\left(\frac{x_1 - m_1}{\sigma_1}\right)\left(\frac{x_2 - m_2}{\sigma_2}\right) + \left(\frac{x_2 - m_2}{\sigma_2}\right)^2$ as:

$$\left\{\left(\frac{x_1 - m_1}{\sigma_1}\right)^2 - 2\rho\left(\frac{x_1 - m_1}{\sigma_1}\right)\left(\frac{x_2 - m_2}{\sigma_2}\right) + \rho^2\left(\frac{x_2 - m_2}{\sigma_2}\right)^2\right\} + (1-\rho^2)\left(\frac{x_2 - m_2}{\sigma_2}\right)^2$$

$$= \left(\frac{x_1 - m_1}{\sigma_1} - \rho\left(\frac{x_2 - m_2}{\sigma_2}\right)\right)^2 + (1-\rho^2)\left(\frac{x_2 - m_2}{\sigma_2}\right)^2 \tag{A.36}$$

Now we have:

$$E\left[X_1X_2\right] = \frac{1}{2\pi\sigma_1\sigma_2\sqrt{(1-\rho^2)}}\int_{-\infty}^{\infty}\int_{-\infty}^{\infty} x_1x_2$$

$$\exp\left\{-\frac{1}{2(1-\rho^2)}\left\{\left(\frac{x_1 - m_1}{\sigma_1} - \rho\left(\frac{x_2 - m_2}{\sigma_2}\right)\right)^2 + (1-\rho^2)\left(\frac{x_2 - m_2}{\sigma_2}\right)^2\right\}\right\} dx_1 dx_2$$

$$= \int_{-\infty}^{\infty}\left\{\int_{-\infty}^{\infty} \frac{x_1}{\sqrt{2\pi}\sigma_1\sqrt{(1-\rho^2)}}\right.$$

$$\left.\exp\left(-\frac{1}{2(1-\rho^2)\sigma_1^2}\left(x_1 - m_1 - \rho\sigma_1\left(\frac{x_2 - m_2}{\sigma_2}\right)\right)^2\right) dx_1\right\} \frac{1}{\sqrt{2\pi}\sigma_2} x_2$$

$$\exp\left\{-\frac{1}{2}\left(\frac{x_2 - m_2}{\sigma_2}\right)^2\right\} dx_2 \tag{A.37}$$

The inner integral (within curly braces) with respect to dx_1 is identified as the mean of the normal distribution with mean $m_1 + \rho\sigma_1\left(\frac{x_2 - m_2}{\sigma_2}\right)$ and variance $\left(1-\rho^2\right)\sigma_1^2$ and therefore is equal to the mean itself, i.e. $m_1 + \rho\sigma_1\left(\frac{x_2 - m_2}{\sigma_2}\right)$. Hence the last equation can be written as:

$$E\left[X_1X_2\right] = \int_{-\infty}^{\infty}\left(m_1 + \rho\sigma_1\left(\frac{x_2 - m_2}{\sigma_2}\right)\right)\frac{1}{\sqrt{2\pi}\sigma_2} x_2 \exp\left\{-\frac{1}{2}\left(\frac{x_2 - m_2}{\sigma_2}\right)^2\right\} dx_2$$

$$= m_1 \frac{1}{\sqrt{2\pi}\sigma_2}\int_{-\infty}^{\infty} x_2 \exp\left\{-\frac{1}{2}\left(\frac{x_2-m_2}{\sigma_2}\right)^2\right\} dx_2 + \frac{\rho\sigma_1}{\sqrt{2\pi}\sigma_2}\int_{-\infty}^{\infty}\left(\frac{x_2-m_2}{\sigma_2}\right)x_2$$

$$\exp\left\{-\frac{1}{2}\left(\frac{x_2-m_2}{\sigma_2}\right)^2\right\} dx_2$$

$$= m_1 \frac{1}{\sqrt{2\pi}\sigma_2}\int_{-\infty}^{\infty} x_2 \exp\left\{-\frac{1}{2}\left(\frac{x_2-m_2}{\sigma_2}\right)^2\right\} dx_2 + \frac{\rho\sigma_1}{\sigma_2}\left\{\frac{1}{\sqrt{2\pi}\sigma_2}\int_{-\infty}^{\infty}(x_2-m_2)^2\right.$$

$$\left.\exp\left\{-\frac{1}{2}\left(\frac{x_2-m_2}{\sigma_2}\right)^2\right\} dx_2\right\} + \frac{\rho\sigma_1 m_2}{\sqrt{2\pi}\sigma_2}\int_{-\infty}^{\infty}(x_2-m_2)\exp\left\{-\frac{1}{2}\left(\frac{x_2-m_2}{\sigma_2}\right)^2\right\} dx_2 \tag{A.38}$$

and the part of the second term $\left\{\frac{1}{\sqrt{2\pi}\sigma_2}\int_{-\infty}^{\infty}(x_2-m_2)^2 \exp\left\{-\frac{1}{2}\left(\frac{x_2-m_2}{\sigma_2}\right)^2\right\} dx_2\right\} = E\left[(X_2-m_2)^2\right] = \sigma_2^2$.

We thus have the last term being zero and the part of the second term within curly braces is the variance $\sigma_2^2 = E\left[(X_2-m_2)^2\right]$ of X_2. Also, recognizing that in the first term, $\frac{1}{\sqrt{2\pi}\sigma_2}\int_{-\infty}^{\infty} x_2 \exp\left\{-\frac{1}{2}\left(\frac{x_2-m_2}{\sigma_2}\right)^2\right\} dx_2 = m_2$, we have:

$$E[X_1X_2] = m_1 m_2 + \frac{\rho\sigma_1\sigma_2^2}{\sigma_2} = m_1 m_2 + \rho\sigma_1\sigma_2 \tag{A.39}$$

Substituting the above result in Eq. (A.34) we have $\rho = \frac{\sigma_{12}}{\sigma_1\sigma_2}$.

13. System reliability and reliability index

Here we provide a brief account of the method of computing the reliability index for a system involving a set of random parameters and a random environment (e.g. external loading). We refer [Ang and Tang 1984, Melchers 1999, Maymon 1998, Nowak and Collins 2000, Manohar and Gupta 2003] for a more comprehensive treatment of the subject.

Suppose that $\boldsymbol{X} = (X_1, X_2, \ldots, X_n)^T$ represents the system states including the loading effects. If $g(\boldsymbol{X}) := g(X_1, X_2, \ldots, X_n)$ is taken as the performance function, $g(\boldsymbol{X}) = 0$ defines the limit state function representing the failure surface which is, in general, nonlinear in $\boldsymbol{X}$. The probability of failure is then defined as:

$$P\,(\text{failure}) = \int_{g(\boldsymbol{X})<0} f_{\boldsymbol{X}}\,(\boldsymbol{x})\,d\boldsymbol{x} \tag{A.40}$$

$f_{\boldsymbol{X}}\,(\boldsymbol{x})$ is the joint *pdf* of the state vector $\boldsymbol{X}$. The above integral is multi-dimensional and is difficult to evaluate - analytically or otherwise.

For example, in the 2-dimensional linear case (Fig. A.2), letting the loading $T = X_1$ and the system mechanical impedance, $S = X_2$, the performance function is given by:

$$g\,(X_1, X_2) = g\,(S, T) = S - T \tag{A.41}$$

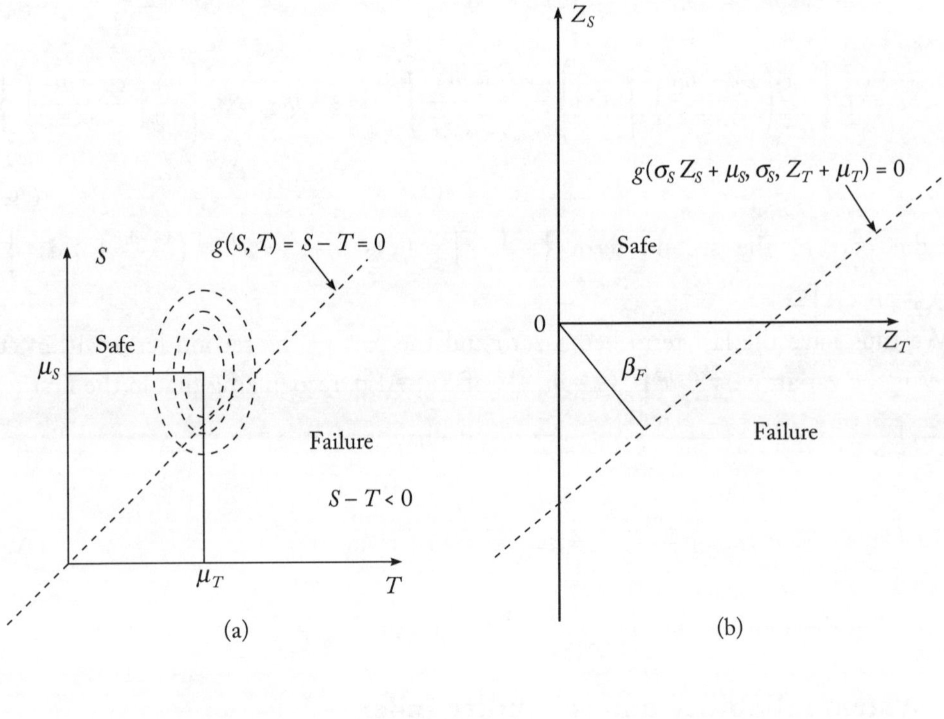

Fig. A.2 System reliability in a two-dimensional case; (a) failure surface in S and T (normal random variables) and (b) failure surface in reduced variates, Z_S and Z_T —standard normal variables

Suppose that the probability distributions of S and T are known with respective means μ_S and μ_T and variances σ_S^2 and σ_T^2. In case S and T are (uncorrelated) normals, $M := S - T$ is also normal. Now, $M < 0$ defines system failure (see Fig. A.3) whose probability is given by:

$$P\,(\text{failure}) = \int_{M<0} f_M\,(m)\,dm = P\,(M < 0) = F_M(0)$$

$$\Longrightarrow F_M(0) = P\left(\frac{M-\mu_M}{\sigma_M} + \frac{\mu_M}{\sigma_M} < 0\right)$$

$$\Longrightarrow P\left(\frac{M-\mu_M}{\sigma_M} < -\frac{\mu_M}{\sigma_M}\right) = P\left(Z_M < -\frac{\mu_M}{\sigma_M}\right), Z_M \text{ being standard normal}$$

$$= \Phi(-\frac{\mu_M}{\sigma_M}) \tag{A.42}$$

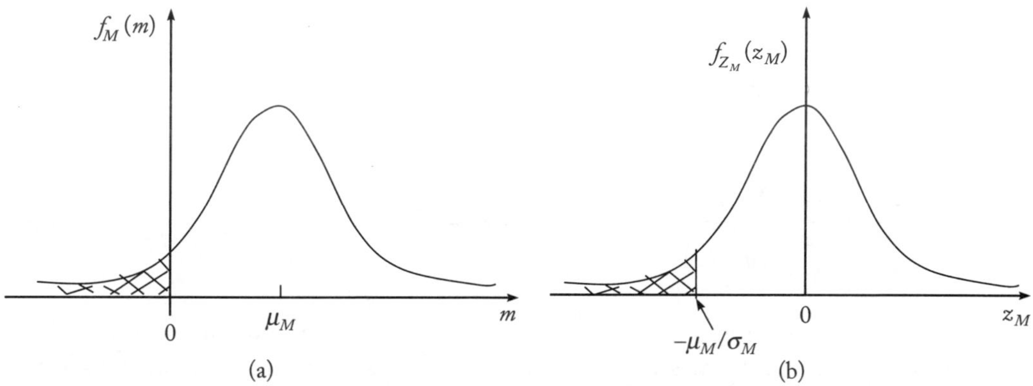

Fig. A.3 (a) pdf of limit state function $M = S - T$ and probability of failure $P(M < 0)$ shown by the hatched area and (b) pdf of limit state function in terms of reduced variate $Z_M = \frac{M-\mu_M}{\sigma_M}$ and probability of failure $P\left(Z_M < -\frac{\mu_M}{\sigma_M}\right)$ shown by the hatched area

Here $\Phi(.)$ is the probability distribution function of a standard normal variable. Thus, if $\beta_F := \frac{\mu_M}{\sigma_M}$, the probability of failure is given by $\Phi(-\beta_F)$ and β_F is known as the reliability index. In terms of reduced variates Z_S and Z_T defined as $Z_S = \frac{S-\mu_S}{\sigma_S}$ and $Z_T = \frac{T-\mu_T}{\sigma_T}$, the first two moments of M are given by $\mu_M = \mu_S - \mu_T$ and $\sigma_M^2 = \sigma_S^2 + \sigma_T^2$. Therefore, the probability failure is given by:

$$\Phi\left(-\frac{\mu_M}{\sigma_M}\right) = \Phi\left(-\frac{\mu_S - \mu_T}{\sqrt{\sigma_S^2 + \sigma_T^2}}\right) \tag{A.43}$$

Thus $\beta_F = \frac{\mu_S-\mu_T}{\sqrt{\sigma_S^2+\sigma_T^2}}$. Now, the limit state function $S-T = 0 \Longrightarrow Z_S\sigma_S - Z_T\sigma_T + \mu_S - \mu_T = 0$ is a straight line (Fig. A.2) in the space of reduced variates. It may be observed that the perpendicular distance from the origin to the straight line is $\frac{\mu_S-\mu_T}{\sqrt{\sigma_S^2+\sigma_T^2}}$ which leads to the result that the reliability index β_F is the minimum distance from the origin of the standard normal space to the surface of the limit state function g.

A generalization of the above result is as follows.

If $g(\boldsymbol{X})$ is a function of uncorrelated normal variables, the reliability index β_F is the least distance from the origin of the space of reduced variates to the hypersurface, $g\left(\sigma_{X_1}Z_1+\mu_{X_1},\sigma_{X_2}Z_2+\mu_{X_2},\ldots,\sigma_{X_n}Z_n+\mu_{X_n}\right)=0$, where $Z_i=\frac{X_i-\mu_{X_i}}{\sigma_{X_i}}$, $i=1,2,\ldots,n$. This implies that if Z_i^*, $i=1,2,..,n$ is the point on the hypersurface nearest from origin, $\beta_F=\left(\sum_{i=1}^n\left(Z_i^*\right)^2\right)^{1/2}$. The point $(Z_1^*,Z_2^*,..Z_n^*)$ is known as the most probable failure point or the design point.

The probability of failure is thus obtainable through computation of β_F instead of evaluation of the multi-dimensional integral in Eq. (A.40). If $g(\boldsymbol{X})$ is linear, say, expressed as $a_0+\sum_i a_iX_i$, a multi-dimensional hyperplane, we immediately have

$$\beta_F=\frac{\mu_g}{\sigma_g}=\frac{a_0+\sum_i a_i\mu_{X_i}}{\sqrt{\sum_i\left(a_i\sigma_{X_i}\right)^2}}\text{ and }P\left(\text{failure}\right)=\Phi(-\beta_F) \tag{A.44}$$

where, μ_g and σ_g are the mean and standard deviation of $g(\boldsymbol{X})$. μ_g and σ_g can be expressed in terms of μ_{X_i} and σ_{X_i}, the means and standard deviations of the normal random variables $X_i, i=1,2,\ldots$ If $g(\boldsymbol{X})$ is non-linear and is linearized by Taylor expansion, Eq. (A.62) may be at most a first-order approximation to β_F. The linearization may be carried out around the mean value vector of $\boldsymbol{X}$. It also amounts to approximating the non-normal $\boldsymbol{X}$ as normal. Further, with design point $\left\{Z_i^*\right\}_{i=1,2,\ldots}$ taken as corresponding to the mean values, there is an invariance problem: $\beta_F=\frac{\mu_g}{\sigma_g}$ is dependent on the specific form of $g(\boldsymbol{X})$. To avoid the invariance problem, the design point (on the failure surface) must be obtained in a different way. To elaborate, let us consider the following different cases.

Case A) *X* is normal (uncorrelated):

Z_i^*, $i=1,2,\ldots$ may be obtained by constrained minimization of the function $\sum_{i=1}^n\left(Z_i\right)^2$ using any optimization methods under the constraint $g\left(\sigma_{X_1}Z_1+\mu_{X_1},\sigma_{X_2}Z_2+\mu_{X_2},\ldots,\sigma_{X_n}Z_n+\mu_{X_n}\right)=0$. See Chapter 9 of this book for an account of evolutionary stochastic search techniques for global optimization.

Case B) *X* consists of non-normal random variables (uncorrelated):

In this case, one can transform each variable to a normal variable by Eq. (1.168), formulate $g\left(\boldsymbol{X}\right)$ in terms of the corresponding reduced variates and then get β_F by a constrained optimization procedure. Alternatively, an iterative procedure [Rackwitz and Fiessler 1978] may be followed to obtain Z_i^*, $i=1,2,..$ and then β_F. In the last method, the non-normals are treated as if they are normals at the beginning of each step and thus the form $g(\boldsymbol{X}) = g\left(\sigma_{X_1} Z_1 + \mu_{X_1}, \sigma_{X_2} Z_2 + \mu_{X_2}, \ldots, \sigma_{X_n} Z_n + \mu_{X_n}\right)$ remains unchanged. Table A.1 describes the steps involved in this iterative procedure.

Table A.1 Rackwitz and Fiessler iteration algorithm to obtain β_F (for non-normal $\boldsymbol{X}$)

	Assume that the marginal distributions $F_X(.)$ and $f_X(.)$ and first two moments μ_X and σ_X of the vector random variable $\boldsymbol{X}$ are available.
Step 1.	Start the iteration counter. Choose an initial $\boldsymbol{X}$ (one choice could be the mean value vector).
Step 2.	Assuming that $\boldsymbol{X}$ is normal, get the reduced variates (standard normals): $Z_i = \frac{X_i - \mu_{X_i}}{\sigma_{X_i}},\ i = 1, 2, \ldots$
Step 3.	Form the limit state function: $g\left(\sigma_{X_1} Z_1 + \mu_{X_1}, \sigma_{X_2} Z_2 + \mu_{X_2}, \ldots\right) = 0$ (A.45)
Step 4.	Minimize $\beta_F = \sum_{i=1}^{n} (Z_i)^2$ (with Eq. A.45 as a constraint) by any optimization method.
Step 5.	If $Z_i^*, i = 1, 2, \ldots$ is the solution from Step 4, determine new μ_X and σ_X (this step is applicable for the non-normal components of $\boldsymbol{X}$ only).
Step 6.	Repeat Steps 2 to 5 to update β_F and finally obtain the probability of failure by evaluating $\Phi(-\beta_F)$

In the above iterative procedure, Step 4 requires an optimization technique to get $Z_i^*, i = 1, 2, \ldots$ If the dimension of the problem permits, Lagrangian multiplier method may be conveniently used and in that case one may formulate the problem as:

$$\text{Minimize } G = \sum_{i=1}^{n} (Z_i)^2 - \lambda g\left(\sigma_{X_1} Z_1 + \mu_{X_1}, \sigma_{X_2} Z_2 + \mu_{X_2}, \ldots,\right) \tag{A.46}$$

Forming the derivatives $\frac{\partial G}{\partial \lambda}, \frac{\partial G}{\partial Z_i}, i = 1, 2, \ldots$ and equating them to zero, one needs to solve the resulting set of non-linear algebraic equations to obtain the design point at each iteration.

Another issue of the iterative procedure that may need elaboration is the requirement to solve for new μ_X and σ_X in Step 5 of Table A.1. For example, for $i = 1$ (the first component, which is assumed to be non-normal), we first obtain:

$$X_1^* = \sigma_{X_1} Z_1^* + \mu_{X_1} \tag{A.47}$$

Denoting the (unknown) new mean and standard deviation by $\mu^n_{X_1}$ and $\sigma^n_{X_1}$ and writing the (corresponding) new reduced variate as $Z^n_1 = \frac{X^*_1-\mu^n_{X_1}}{\sigma^n_{X_1}}$, we use the transformation in Eq. (1.168) to write:

$$z^n_1 = \Phi^{-1}\left(F_{X^*_1}(x^*_1)\right) \Longrightarrow \Phi\left(\frac{x^*_1-\mu^n_{X_1}}{\sigma^n_{X_1}}\right) = F_{X^*_1}(x^*_1) \tag{A.48}$$

Differentiating the above equation on both sides leads to:

$$\frac{1}{\sigma^n_{X_1}}\varPhi^{-1}\left(\frac{x^*_1-\mu^n_{X_1}}{\sigma^n_{X_1}}\right) = f_{X^*_1}(x^*_1) \Longrightarrow \sigma^n_{X_1} = \frac{\varPhi^{-1}\left(\frac{x^*_1-\mu^n_{X_1}}{\sigma^n_{X_1}}\right)}{f_{X^*_1}(x^*_1)} \tag{A.49}$$

$\varPhi$ denotes the *pdf* of standard normal. Eqs. (A.48) and (A.49) together imply that we equate the CDF and *pdf* of the equivalent normal distribution to those of the non-normal distribution at x^*_1. Knowing $\sigma^n_{X_1}$ from Eq. (A.49), $\mu^n_{X_1}$ is determined from Eq. (A.48) as:

$$\mu^n_{X_1} = x^*_1 - \sigma^n_{X_1}\left[\Phi^{-1}\left(F_{X^*_1}(x^*_1)\right)\right] \tag{A.50}$$

Case C) *X* with correlated non-normal components
In this case, these components need to be transformed to correlated standard normals by Nataf transformation (Section 1.11.4) and further transformed to uncorrelated ones by diagonalization of the associated covariance matrix (Section 1.11.3) before attempting to obtain β_F.

References

Ang, A. H. S. and W. H. Tang. 1984. *Probability Concepts in Engineering Planning and Design, vol. II: Decision, Risk and Reliability.* NY: John Wiley and Sons Ltd. NY.

Maymon, G. 1998. *Some engineering Applications in Random Vibrations and Random Structures.* Progress in Astronautics and Aeronautics. vol.178. Virginia: AIAA.

Manohar, C. S. and S. Gupta. 2003. 'Modelling and Evaluation of structural Reliability: Current Status and Future Directions'. in *Research Advances in Structural Engineering, edited by K S. Jagadish and R N. Iyengar.* Golden Jubilee Publications of Department of Civil Engineering, IISc. Bangalore. (India: Universities Press).

Melchers, R. E. 1999. *Structural Reliability, Analysis and Prediction.* 2nd ed. UK: John Wiley and Sons Ltd.

Nowak, A. S. and K. R. Collins. 2000. *Reliability of Structures.* NY: McGraw-Hill Inc.

Rackwitz, R. and B. Fiessler. 1978. 'Structural Reliability under Combined Random Load Sequences'. *Computers and Structures* 9: 489–94.

Reddy, D. B. 1998. *Introductory Functional Analysis with Applications to Boundary Value Problems and Finite Elements.* NY: Springer-Verlag.

Appendix B (Chapter 2)

1. Cauchy–Schwarz inequality

Cauchy–Schwarz inequality is a classical inequality that arises in different branches of mathematics—functional spaces, probability and statistics, real and complex analysis, numerical analysis, qualitative theory of differential equations. One case where the inequality is often used is given below.

If $\{a_i\}$ and $\{b_i\}$, $i = 1, 2, .., n$ are sequences of real numbers, then we have the following Cauchy–Schwarz inequality:

$$\sum_{i=1}^{n} a_i^2 \sum_{i=1}^{n} b_i^2 \leq \left(\sum_{i=1}^{n} a_i b_i \right)^2 \tag{B.1}$$

Proof:

$$\sum_{i=1}^{n} \sum_{j=1}^{n} \left(a_i b_j - a_j b_i \right)^2 = \sum_{i=1}^{n} a_i^2 \sum_{j=1}^{n} b_j^2 + \sum_{i=1}^{n} b_i^2 \sum_{j=1}^{n} a_j^2 - 2 \sum_{i=1}^{n} a_i b_i \sum_{j=1}^{n} b_j a_j$$

$$= 2 \sum_{i=1}^{n} a_i^2 \sum_{i=1}^{n} b_i^2 - 2 \left(\sum_{i=1}^{n} a_i b_i \right)^2 \tag{B.2}$$

Since the left hand side expression is positive, we have the inequality:

$$\sum_{i=1}^{n} a_i^2 \sum_{i=1}^{n} b_i^2 - \left(\sum_{i=1}^{n} a_i b_i \right)^2 \geq 0 \tag{B.3}$$

Hence the result in Eq. (B.1).

2. System reliability and reliability index by Rosenblatt transformation

We refer to the brief notes provided in Appendix A on computation of reliability index β_F and associated probability of failure P_F for a system with random parameters (including those of external loading) correlated or uncorrelated. The case of correlated parameters is discussed in Appendix A (Case C) where, $\boldsymbol{X}$ contains correlated non-normal components. In this case, one can also use Rosenblatt transformation as an alternative method to evaluate β_F and P_F.

By Rosenblatt transformation (see Eq. (2.39) of Chapter 2), we transform the correlated non-normals $\boldsymbol{X}$ to $\boldsymbol{Z}$, the independent standard normals and evaluate the reliability index β_F in the space of these standard normals. Starting from an initial approximate $\boldsymbol{X}_0$ and corresponding $\boldsymbol{Z}_0$, an iterative procedure due to Hohenbichler and Rackwitz [1981] can be adopted to obtain β_F and P_F. The iterative procedure is listed in Table B.1 below.

Table B.1 Hohenbichler and Rackwitz iteration algorithm to obtain β_F (for correlated non-normal $\boldsymbol{X}$)

	Assume that the joint distribution and first two moments μ_X and σ_X of the vector random variable $\boldsymbol{X}$ are available. Let n be the dimension of $\boldsymbol{X}$.
Step 1.	Start the iteration counter. Choose an initial $\boldsymbol{X}_0$ (one choice could be the mean value vector). Obtain $\boldsymbol{Z}_0$ by Rosenblatt transformation in Eq. (2.39).
Step 2.	At each iteration k, obtain the Jacobian matrix $J_k = \left[\frac{\partial z}{\partial x}\right]_{\boldsymbol{X}_k, \boldsymbol{Z}_k}$ using Eq. (2.39).
Step 3.	Form the limit state function $g(\boldsymbol{X}) = g\left(\sigma_{X_1} Z_1 + \mu_{X_1}, \sigma_{X_2} Z_2 + \mu_{X_2}, \ldots,\right) = g(\boldsymbol{Z})$. Find $g(\boldsymbol{X}_k)$. Here $g(\boldsymbol{Z}_k) = g(\boldsymbol{X}_k)$.
Step 4.	From the gradient vector $G(\boldsymbol{X}) = \left(\frac{\partial g}{\partial X_1}, \frac{\partial g}{\partial X_1}, \ldots, \frac{\partial g}{\partial X_n}\right)$, get $G(\boldsymbol{X}_k)$.
Step 5.	Obtain $G(\boldsymbol{Z}_k) = J_k^{-1} G(\boldsymbol{X}_k)$.
Step 6.	Evaluate the new failure point: $$\boldsymbol{Z}_k^* = \frac{1}{(G(\boldsymbol{Z}_k))^T G(\boldsymbol{Z}_k)} \left\{ (G(\boldsymbol{Z}_k))^T \boldsymbol{Z}_k - g(\boldsymbol{Z}_k) \right\} G(\boldsymbol{Z}_k) \quad \text{(B.4)}$$
Step 7.	Obtain the failure point in the space of original variables: $$\boldsymbol{X}_k^* = \boldsymbol{X}_k + J_k^{-1}(\boldsymbol{Z}_k^* - \boldsymbol{Z}_k) \quad \text{(B.5)}$$

Continued...

Step 8.	Get the reliability index, $\beta_F = \left(\left(Z_k^*\right)^T Z_k^*\right)^{1/2}$ and the $P_F = \Phi(-\beta_F)$.
Step 9.	Repeat steps 2–8 and continue the iterations till convergence is reached.

3. Convex functions

Let I be an open interval in $\mathbb{R}$. A function $g : I \to \mathbb{R}$ is convex on I, if $\forall x_1, x_2 \in I$, the graph of $g(x)$ lies below any of its chords (Fig. B.1). Thus, letting $x = \alpha x_1 + (1-\alpha)x_2$ with $0 \le \alpha \le 1$, we have the following inequality for a convex function:

$$g(x) \le \alpha g(x_1) + (1-\alpha) g(x_2) \tag{B.6}$$

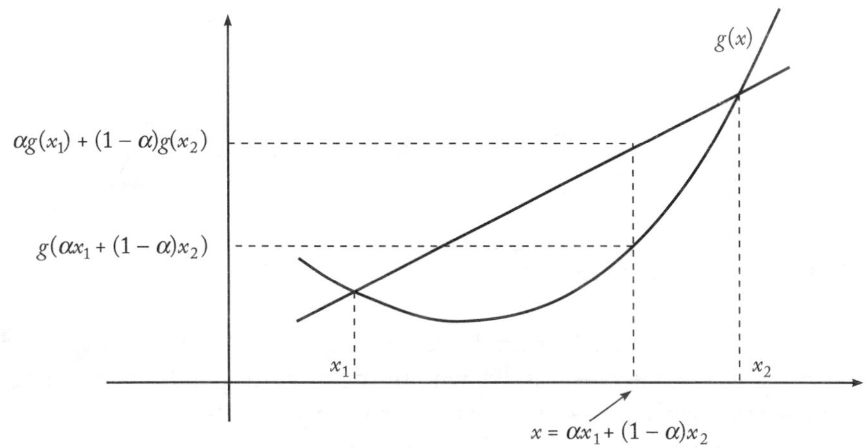

Fig. B.1 Convex function

The function $g(x)$ is strictly convex if the inequality is strict for $x_1 \neq x_2$.

4. Ising model and ferromagnetism

Ising model, originally proposed by Ising [1925], mathematically describes the phenomenon of ferromagnetism based on the principles of statistical mechanics. The model is helpful in identifying possible phase transitions in materials due to temperature variation. The model finds applications in diverse fields including quantum information theory, molecular biology, neuroscience, chemistry, and computer science. We provide in this note, only a brief description of the model with the aim being to highlight the role of MC simulation in arriving at the relevant physical (averaged) quantities such as magnetization and internal energy. A one dimensional Ising model, as solved by Ising himself, has no phase transition. A 2-dimensional model as studied by Peierls [1936] and Onsager [1944] exhibits the phase transition.

Electrons in an atom spin randomly about their own axis that points either up or down. Each electron constitutes a magnetic dipole. Spinning electrons in an atom produce a magnetic field. In non-magnetic materials, each dipole points in a random direction—parallel or anti-parallel—and the moments cancel one another. In ferromagnetic materials, the dipole moments locally align themselves in the same direction to produce a strongly non-zero magnetic field.

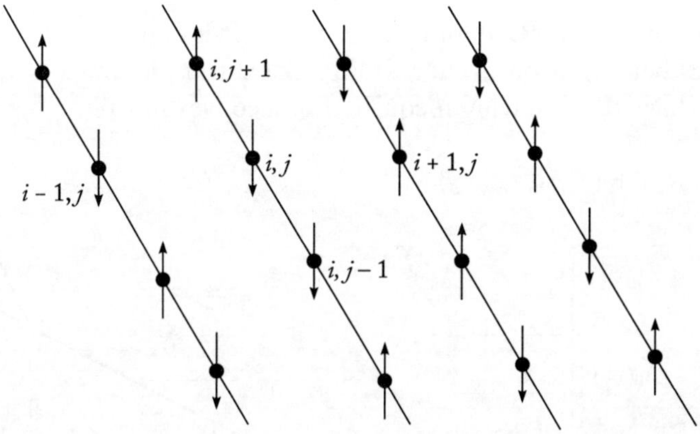

Fig. B.2 2-dimensional Ising model

Fig. B.2 shows a 2-dimensional square lattice Ising model with electron spins (dipoles) having only two possible states—up and down. The figure corresponds to one configuration of the system, wherein each spin is assumed to interact only with its nearest (local) neighbors on the lattice. We assign a discrete random variable S_k to each lattice location such that $S_k \in \{-1, +1\}$ represents a dipole (spin). If N is the size of the lattice (e.g. an $N \times N$ lattice), each spin configuration is defined by $\boldsymbol{S} = \{S_k\}$, $k = 1, 2, \ldots, N^2$, a sequence of Bernoulli random variables. Now, one can define the energy of each spin configuration (without an external field) in terms of an Hamiltonian:

$$H(\boldsymbol{S}) = -\sum_{\{i,j\}} J_{ij} s_i s_j \tag{B.7}$$

where, J_{ij} is a coupling parameter indicating the strength of interaction between the adjacent variables S_i and S_j. Here the interaction is restricted to the nearest neighbors (one above, one below, one to the right and one to the left as shown in Fig. B.2). Note that generalized models may include long-range interactions and more choices for spin direction. Further, in the present model, all the spin locations have the same number of neighbors ($=$ 4) due to the assumption of wrapping the model around.

The Ising model is classified as ferromagnetic if $J_{ij} > 0$, anti-ferromagnetic if $J_{ij} < 0$ and non-ferromagnetic if $J_{ij} = 0$. One may assume that all of the nearest neighbors $\{i, j\}$ have the same interaction strength. Setting $J_{ij} = J$, the Hamiltonian $H(\boldsymbol{S}) = -J \sum_{\{i,j\}} s_i s_j$.

For every spin configuration, we will assign a corresponding energy using the following rule. If two neighboring states are aligned, then we add $-J$ units of energy, and if they are anti-aligned add $+J$units of energy.

The configuration probability $P(\mathbf{S})$ (representing the probability of being in a thermal equilibrium state, at a given temperature, with configuration $\mathbf{S}$) is given by the Boltzmann distribution:

$$P(\mathbf{S}) = \frac{e^{-\beta H(\mathbf{S})}}{Z_\beta} \tag{B.8}$$

where, $\beta = (K_B T)^{-1}$ with K_B denoting the Boltzmann constant and T, the absolute temperature. $Z_\beta = \sum_{\mathbf{S}} e^{-\beta H(\mathbf{S})}$ is the normalization constant and is known as the partition function. One can derive all the thermo-dynamical quantities—internal energy, magnetization, magnetization susceptibility and specific heat—of interest from the partition function. For instance, the internal energy is given by $\frac{1}{Z_\beta}\sum_{\mathbf{S}} H(\mathbf{S}) e^{-\beta H(\mathbf{S})} = -\frac{\partial(\log Z_\beta)}{\partial\beta}$.

In equilibrium, the spins in a configuration $\mathbf{S}$, tend to align parallel in the same direction for ferromagnetic materials and in the opposite direction for non-ferromagnetic materials. The main goal in the study of the Ising model is to investigate the performance of the model in the vicinity of a critical temperature, denoted by T_c, where the model undergoes a phase transition, which assists in the quantitative characterization (e.g. bifurcation analysis) of material behavior as it goes through criticality. At $T = T_c$, spontaneous magnetization starts to appear and approaches one of two possible equilibrium states for low temperatures (for $T \ll T_c$). For higher temperatures ($T \gg T_c$), the material reaches a zero magnetic state (on average).

Since every spin site may have ± 1 spin, there are 2^{N^2} different configurations that are possible. The Ising model is thus difficult to evaluate numerically if N is large. This motivates simulating the model using MC Methods. One uses MC methods to compute various (averages of) quantities of interest as functions of temperature. The Metropolis algorithm described in Table B.2 is the most commonly used MC algorithm to compute these estimates.

Table B.2 Metropolis algorithm for Ising model

Step 1.	Initialize a configuration S_0 and select a temperature T. Start iterations $k = 1,2,\ldots$.
Step 2.	At any iteration k, pick a spin location on the lattice using a selection probability $= \frac{1}{N^2}$ and denote this configuration by $\mathbf{S}_k$. Compute the energy $H(\mathbf{S}_k)$ associated with $\mathbf{S}_k$.
Step 3.	Flip the value of the spin and calculate the new energy $H(\mathbf{S}_k^N)$.

Continued...

Step 4.	Retain the flipped value if the new energy is less.
Step 5.	If the new energy is more than the old one, accept the new one with probability $e^{-\beta(H(S_k^N)-H(S_k))}$.
Step 6.	Repeat steps 2–5 till equilibrium is attained or till the specified number of maximum steps k_{max} is reached.

5. Markov chains

A Markov chain is a sequence $\{X_n\}$ of random variables with the Markov property:

$$P(X_{n+1} = x_{n+1}|X_1 = x_1, X_2 = x_2, \ldots, X_n = x_n) = P(X_{n+1} = x_{n+1}|X_n = x_n) \quad \text{(B.9)}$$

Both the conditional probabilities in the above equation must be well defined through the associated Radon–Nikodym derivatives (see Sections 2.2.1 and 2.2.2, Chapter 2). Suppose that Markov chain is specified by the state space $S \subset \mathbb{Z}$. In Eq. (B.9), $(x_1, x_2, \ldots, x_n) \in S$ are the states of the system with each random variable $X_i : \Omega \to S$ where, Ω is the sample space. For instance, the weather conditions—cloudy, rainy and sunny—on a day may constitute Ω and S may be the sequence $\{1,2,3\}$.

If the chain is currently (at n^{th} step) in a state specified by x_i, it moves to the state x_j in the next (at $n+1^{\text{th}}$ step) with probability $P_{ij} := P\left(X_{n+1} = x_j|X_n = x_i\right)$. This probability does not depend on the states the chain was in before the current state according to the Markov property. P_{ij} is known as the one-step transition probability. The transition probability matrix $\boldsymbol{P}$ is characterized by the following features.

i) $P_{ij} \geq 0,\ i,j \in \mathbb{Z}$ and $\sum_{j \in I} P_{ij} = 1$ (i.e., each row is a distribution over S)

ii) The ij^{th} entry of the matrix $\boldsymbol{P}^n$ gives the probability that the Markov chain, starting in state x_i, will be in state x_j after n steps.

iii) If Þ_0 is the initial probability distribution of the states, the distribution of the states at the end of n steps is $\text{Þ}_n = \text{Þ}_0 \boldsymbol{P}^n$. This is the same as $\text{Þ}_n = (\boldsymbol{P}^n)^T \text{Þ}_0$ except that Þ_0 is a row vector in the former case and a column vector in the latter.

iv) The transition matrix $\boldsymbol{P}$ is regular if some power of the matrix contains all positive (non-zero) entries. A Markov chain is a regular Markov chain if its transition matrix $\boldsymbol{P}$ is regular.

v) If a Markov chain with a transition matrix $\boldsymbol{P}$ is regular, then there exists a unique probability vector Þ_v such that for any other probability vector Þ and for large n, $\text{Þ}\boldsymbol{P}^n = \text{Þ}_v$. The vector Þ_v is called the equilibrium vector or the limiting probability distribution (also called a stationary distribution) which is independent of the initial distribution.

vi) If a Markov chain with a transition matrix $\boldsymbol{P}$ is regular, then there exists a probability vector Þ_v such that $\text{Þ}_v \boldsymbol{P} = \text{Þ}_v$.

Example B.1

Suppose that $\Omega = \{\text{Cloudy, Rainy, Sunny}\}$ symbolizes the weather conditions on a day. With $S = \{1, 2, 3\}$, let the transition probability matrix P be:

$$P = \begin{bmatrix} 0.65 & 0.28 & 0.07 \\ 0.15 & 0.67 & 0.18 \\ 0.12 & 0.36 & 0.52 \end{bmatrix}$$

It is easy to verify that all powers of $\boldsymbol{P}$ will have positive (non-zero) entries. Thus, the Markov chain is regular Further, $\text{Þ}_v = (0.2864 \quad 0.4886 \quad 0.2250)$ is the equilibrium vector, i.e. the equilibrium distribution favors rainy weather.

An important Markov chain which is not regular is the absorbing Markov chain. In this case an element $P_{ii} = 1$ and all other entries in the ith row are zero. This implies while an absorbing state can be entered by any other non-absorbing state, it is not possible to leave an absorbing state once it is entered. Also, it is true that the system tends towards the absorbing state when $\boldsymbol{P}$ is raised to higher and higher powers. Another striking feature of an absorbing chain different from regular Markov chains is that the long term trend (raising $\boldsymbol{P}$ to higher powers) depends on the initial probability distribution, i.e., the equilibrium vector Þ_v is not unique. This property has more significance in the case where there is more than one absorbing state. Note that a Markov chain is a discrete analogue of a Markov process (Section 3.3.6, Eq. 3.42, Chapter 3).

References

Hohenbichler, M. and R. Rackwitz. 1981. 'Non-normal Dependent Vectors in Structural Safety'. *Jl. of EM Division., ASCE* 107(6): 1227–49.

Ising, Ernst. 1925. 'Beitrag zur Theorie des Ferromagnetismus'. *Zeitschrift fur Physik* 31: 253–58.

Onsager, L. 1944. 'Crystal Statistics. I. A Two-dimensional Model with an Order-Disorder Transition. *Phys. Rev.* 65: 117–49.

Peierls, R. 1936. 'On Ising's Model of Ferromagnetism'. *Mathematical Proceedings of the Cambridge Philosophical Society* 32(3): 477–81.

Appendix C (Chapter 3)

1. Monotone convergence theorem, Fatou's lemma and dominated convergence theorem

The above three are fundamental theorems for convergence of integrals.

Monotone convergence theorem:
Let $\{X_n\}$ be an increasing sequence of random variables on $(\Omega, \mathcal{F}, P)$. If X is another random variable on the same sample space, then:

$$0 \le X_n \uparrow X \Longrightarrow E[X_n] \uparrow E[X] \tag{C.1}$$

Proof:
By hypothesis, $X_n \le X_{n+1} \Longrightarrow X_j \le \lim_{n\to\infty} X_n = X$ for each j. Therefore $E[X_n] \le E[X] \Longrightarrow \lim_{n\longrightarrow\infty} E[X_n] \le E[X]$. It remains to prove the reverse inequality $\lim_{n\longrightarrow\infty} E[X_n] \ge E[X]$. To this end, let $Y_n^{(k)}$ be a sequence of simple random variables expressed in terms of a stair-case function (given below) such that $0 \le Y_n^{(k)} \uparrow X_k$ where X_k is a member of the given sequence $\{X_n\}$.

$$Y_n^{(k)} = \sum_{j\in\mathbb{N}} \left(\frac{j-1}{2^n}\right) I_{\frac{j-1}{2^n} < X_k \le \frac{j}{2^n}} + nI_{X_k > n} \tag{C.2a}$$

Likewise, Y_n is chosen to be a special sequence converging to X:

$$Y_n = \sum_{j\in\mathbb{N}} \left(\frac{j-1}{2^n}\right) I_{\frac{j-1}{2^n} < X \le \frac{j}{2^n}} + nI_{X > n} \tag{C.2b}$$

Since $Y_n^{(k)}$ is left continuous, $Y_n^{(k)} \uparrow Y_n$ as $k \to \infty$ such that $E\left[Y_n^{(k)}\right] \uparrow E[Y_n]$.

Now, choose any $\varepsilon > 0$ and fix n such that:

$$E[Y_n] \ge E[X] - \varepsilon \tag{C.3}$$

With this fixed n, choose k large so that:

$$E\left[Y_n^{(k)}\right] \geq E\left[Y_n\right] - \varepsilon \tag{C.4}$$

By combining the above two inequalities, we have:

$$E\left[X_k\right] \geq E\left[Y_n^{(k)}\right] \geq E\left[Y_n\right] - \varepsilon \geq E\left[X\right] - 2\varepsilon \tag{C.5}$$

Thus for any arbitrary ε, one can find sufficiently large k such that $E\left[X_k\right] \geq E\left[X\right] - 2\varepsilon$. This leads to the required result:

$$\lim_{k\longrightarrow\infty} E\left[X_k\right] \geq E\left[X\right] \tag{C.6}$$

Fatou's lemma:
Suppose that $\{X_n\}$ is a sequence of non-negative random variables. Then:

$$E\left[\lim\inf X_n\right] \leq \lim\inf E\left[X_n\right] \tag{C.7}$$

Proof: Let $Z_k = \inf_{n\geq k} X_n$. Then $\lim\inf_n X_n \uparrow \lim Z_k$. By monotone convergence theorem:

$$E\left[\lim\inf_n X_n\right] = \lim_k E\left[Z_k\right] \tag{C.8}$$

With $n \geq k$, $X_n \geq Z_k$. This leads to:

$$E\left[X_n\right] \geq E\left[Z_k\right]$$

$$\Longrightarrow E\left[Z_k\right] \leq \inf_{n\geq k} E\left[X_n\right] \tag{C.9}$$

In view of Eq. (C. 9), Eq. (C. 8) gives

$$E\left[\lim\inf_n X_n\right] \leq \lim_k \inf_{n\geq k} E\left[X_n\right]$$

$$= \lim\ \inf_n E\left[X_n\right] \tag{C.10a}$$

Hence the result.

A reverse Fatou's lemma states that if $\exists$ a Y such that $X_n \leq Y\ \forall n$ and $Y \in L^1$, then:

$$E\left[\lim\sup X_n\right] \geq \lim\sup E\left[X_n\right] \tag{C.10b}$$

To prove this, apply Fatou's lemma to $Y - X_n$.

Dominated convergence theorem

Let $\{X_n\}$ be an increasing sequence of random variables defined on $(\Omega, \mathcal{F}, P)$ and X a random variable on the same sample space such that $X_n \to X$ a.s. If the sequence $\{X_n\}$ is dominated by $Y \in L^1(\Omega, \mathcal{F}, P)$, i.e. $X_n(\omega) \leq Y(\omega),\ \forall \omega \in \Omega,\ n \in \mathbb{N}$ then:

$$X_n \to X \text{ in } L^1 \Longrightarrow E[X_n] \longrightarrow E[X] \tag{C.11}$$

Proof:

Given that $|X_n| \leq Y$. This implies that $E[|X_n|] \leq E[Y] < \infty$. So $X_n \in L^1 \forall n$. As $X_n \to X$ a.s. $|X| \leq Y \Longrightarrow X \in L^1$. Also $|X_n - X| \leq |X_n| + |X| \leq 2Y$. By reverse Fatou's lemma:

$$\lim\sup E[|X_n - X|] \leq E[\lim\sup |X_n - X|] = 0 \tag{C.12}$$

Hence:

$$\left|E[X_n] - E[X]\right| = \left|E[X_n - X]\right| \leq E[|X_n - X|] \longrightarrow 0. \tag{C.13}$$

Hence the result.

2. Borel–Cantelli Lemma-1 (BC1)

Let $(A_n : n \in \mathbb{N})$ be sequence of events such that $\sum_n P(A_n) < \infty$, then:

$$P(A_n, \text{ i.o.}) = 0 \tag{C.14}$$

where, $(A_n, \text{ i.o.})$ or (A_n infinitely often) is defined as:

$$(A_n, \text{ i.o.}) := \lim\ \sup A_n = \cap_m \cup_{n \geq m} A_n \tag{C.15}$$

Proof:

$$P(A_n, \text{ i.o.}) = \lim_{m \to \infty} P(\cup_{n \geq m} A_n) \leq \lim_{m \to \infty} \sum_{n \geq m} P(A_n) = 0 \tag{C.16}$$

3. Borel–Cantelli Lemma-2 (BC2)

Let $(A_n : n \in \mathbb{N})$ be sequence of independent events such that $\sum_n P(A_n) = \infty$. Then:

$$P(A_n, \text{ i.o.}) = P(\lim\ \sup A_n) = 1 \tag{C.17}$$

Proof:

$$(\lim\sup A_n)^c = \lim\inf A_n{}^c = \cup_m \cap_{n \geq m} A_n{}^c \tag{C.18}$$

Here, ($\liminf B_n$) for a sequence of events ($B_n : n \in \mathbb{N}$) means that the event B_n occurs eventually and is expressed as $\cup_m \cap_{n \geq m} B_n$. From Eq. (C. 18), we have:

$$P\left(\cap_{n \geq m} A_n{}^c\right) = \prod_{n \geq m} P\left(A_n{}^c\right) = \prod_{n \geq m} \left(1 - P(A_n)\right) \tag{C.19}$$

Since for $x \geq 0$, $1 - x \leq e^{-x}$:

$$\prod_{n \geq m} \left(1 - P(A_n)\right) \leq \exp\left(-\sum_{n \geq m} P\left(A_n\right)\right) = 0\left(\text{since} \sum_n P(A_n) = \infty\right) \tag{C.20}$$

so that $P\left(\lim \sup A_n\right)^c = 0 \Longrightarrow P(\lim \sup A_n) = 1$.

4. Kolmogorov's 0–1 law (proof by martingale theory)

Suppose that $\left\{X_1, X_2, \ldots, X_{n,} \ldots\right\}$ is a sequence of independent random variables. Define the tail σ-algebras $\mathcal{T}_n = \sigma\left(X_{n+1}, X_{n+2}, \ldots\right)$ and $\mathcal{T} = \cap_n \mathcal{T}_n$. Then, Kolmogorov's $0 - 1$ law states that all tail events have probability 0 or 1. That is, if $F \in \mathcal{T}$, $P(F) = 0$ or 1.

Proof: Since $F \in \mathcal{T}$, it is independent of $\mathcal{F}_n = \sigma(X_1, X_2, \ldots, X_n)$. Therefore:

$$E\left[I_F \mid \mathcal{F}_n\right] = E\left[I_F\right] = P(F) \tag{C.21}$$

But, $\mathcal{M}_n = E\left[I_F \mid \mathcal{F}_n\right]$ is a martingale since for $m < n$, $E\left[\mathcal{M}_n | \mathcal{F}_m\right] = \mathcal{M}_m$. Thus, by (forward) martingale convergence theorem (Section 3.4.4), we have:

$$\mathcal{M}_\infty = \lim_{n \to \infty} E\left[I_F \mid \mathcal{F}_n\right] = E\left[I_F \mid \mathcal{F}_\infty\right] \text{a.s.} \tag{C.22}$$

F being a tail event, it is $\mathcal{F}_\infty$-measurable and hence $E\left[I_F \mid \mathcal{F}_\infty\right] = I_F$ a.s. $\Longrightarrow P(F) = I_F$. This yields $P(F) = 0$ or 1.

5. To prove that $P(\tau < \infty) = 1$ where τ is a stopping time (hitting time) for simple random walk

Proof: Consider a sequence of independent non-negative random variables X_n, $n \in \mathbb{N}$ having an identical distribution with $P\left(X_j = 1\right) = P\left(X_j = -1\right) = \frac{1}{2}$. Let $\mathcal{F}_n = \sigma(X_1, X_2, \ldots, X_n)$ be the filtration generated by X_n. Define $S_0 := 0$, $S_n := \sum_{j=1}^n X_j$. Then the stochastic process S_n is also adapted to $\mathcal{F}_n$ since $\sigma\left(X_1, X_2, \ldots, X_n\right) = \sigma\left(S_1, S_2, \ldots, S_n\right)$. Let τ be a stopping time defined by:

$$\tau := \inf\{n : S_n = 1\} \tag{C.23}$$

Given a parameter $\theta \in \mathbb{R}$, we have:

$$E\left[e^{\theta X_j}\right] = \frac{1}{2}\left(e^{\theta} + e^{-\theta}\right) = \cosh\theta \tag{C.24}$$

It follows from the above that:

$$E\left[\text{sech}\ \theta\ e^{\theta X_n}\right] = 1,\ \forall n \tag{C.25}$$

With $\mathcal{M}_n^\theta := \prod_{j=1}^n \left(\text{sech}\ \theta\ e^{\theta X_j}\right) = (\text{sech}\ \theta\ )^n \prod_{j=1}^n \left(e^{\theta X_j}\right)$, one has (a.s.) for $n \geq 1$

$$E\left[\mathcal{M}_n^\theta | \mathcal{F}_{n-1}\right] = E\left[\mathcal{M}_{n-1}^\theta \text{sech}\theta\ e^{\theta X_n} | \mathcal{F}_{n-1}\right]$$

$$= \mathcal{M}_{n-1}^\theta E\left[\text{sech}\ \theta\ e^{\theta X_n} | \mathcal{F}_{n-1}\right]$$

(by properties of conditional expectation)

$$= \mathcal{M}_{n-1}^\theta E\left[\text{sech}\theta\ e^{\theta X_n}\right] = \mathcal{M}_{n-1}^\theta \tag{C.26}$$

(since X_n is independent of $\mathcal{F}_{n-1}$)

Thus $\mathcal{M}_n^\theta$ is a martingale. With the stopping time as defined in Eq. (C.23), the stopped martingale $\mathcal{M}_{\tau\wedge n}^\theta$ is given by:

$$\mathcal{M}_{\tau\wedge n}^\theta = (\text{sech}\theta\ )^{\tau\wedge n} e^{\theta S_{\tau\wedge n}} \tag{C.27}$$

and it follows:

$$E\left[\mathcal{M}_{\tau\wedge n}^\theta\right] = E\left[(\text{sech}\theta\ )^{\tau\wedge n} e^{\theta S_{\tau\wedge n}}\right] = 1,\quad \forall n \tag{C.28}$$

With $\theta > 0$, $e^{\theta S_{\tau\wedge n}}$ is bounded by e^θ so that $\mathcal{M}_{\tau\wedge n}^\theta$ is bounded by e^θ. Also, as $n \to \infty$, we have $\mathcal{M}_{\tau\wedge n}^\theta \to \mathcal{M}_\tau^\theta$. Now, the bounded convergence theorem gives in the limit as $n \to \infty$:

$$E\left[\mathcal{M}_\tau^\theta\right] = 1 = E\left[(\text{sech}\theta\ )^\tau e^\theta\right]$$

$$\Longrightarrow E\left[(\text{sech}\theta\ )^\tau\right] = e^{-\theta} \tag{C.29}$$

Now, as $\theta \downarrow 0$, $(\text{sech}\theta\ )^\tau \uparrow 1$ provided $\tau < \infty$, otherwise (if $\tau = \infty$), $(\text{sech}\theta\ )^\tau \downarrow 0$. By monotone convergence we have the result:

$$E\left[I_{\tau<\infty}\right] = 1 = P(\tau < \infty) \tag{C.30}$$

6. To prove $\lim_{t\to\infty}\sup_t B_t = \infty$ and $\lim_{t\to\infty}\inf_t B_t = -\infty$ a.s where B_t is a Brownian motion

Proof: Proving $\lim_{t\to\infty}\sup B_t = \infty$ is equivalent to showing $\sup\ B_t = \infty$ a.s., i.e.

$$P(\sup B_t = \infty) = 1.$$

By time-scaling property (Section 3.3.6, Chapter 3) of a Brownian motion:

$$\sup_t B_t = \sup_t \alpha^{-\frac{1}{2}} B(\alpha t)$$

$$\Longrightarrow \sup_t B_t = \sup_t n^2 B\left(\frac{t}{n}\right) = n^2 \sup_t B\left(\frac{t}{n}\right) = n^2 \sup_t B(t)$$

$$\Longrightarrow P(\sup_t B_t \leq n) = P\left(n^2 \sup_t B(t) \leq n\right) = P\left(\sup_t B(t) \leq \tfrac{1}{n}\right) \tag{C.31}$$

Appendix D (Chapter 4)

1. Continuous and discrete structural systems under stochastic input

(i) *Continuous structural systems (linear)*

For a linear continuous system [Meirovitch 1967, Roy and Rao 2012], the governing equation of motion is:

$$L\{Y(z,t)\} = F(z,t) \tag{D.1}$$

Here L is the partial differential operator that can be expressed as $L_s + m\frac{\partial^2}{\partial t^2} + c\frac{\partial}{\partial t}$ where L_s is a self-adjoint linear operator in spatial coordinate vector $z \in \sum$, the domain of the continuum. m and c are the distributed mass and damping parameters of the system. For instance, for a continuous beam of length l, $L_s Y = EI\frac{\partial^4 Y}{\partial z^4}$ with $z \in [0,l]$, E and I being the young's modulus and the moment of inertia respectively. $Y(z,t)$ is the field variable representing the system response, e.g. the transverse deflection of the beam. Since a continuous system is a distributed parameter system, the response Y is clearly of infinite dimension. $F(z,t)$ is the external loading function which is in reality a random field. Suppose that the input characteristics of $F(z,t)$ are known in terms of its mean $\mu_F(z,t)$ and autocorrelation $R_{FF}(z_1,z_2,t_1,t_2)$ (Chapter 3). At this stage, stationarity is not assumed for $F(z,t)$. Following the method of separation of variables and using the orthonormal eigenfunctions $\Phi_j(z)$, $j \in \mathbb{N}$ of the resulting Sturm–Liouville boundary value problem [Roy and Rao 2012], we express the system response as:

$$Y(z,t) = \sum_{j=1}^{\infty} q_j(t)\ \Phi_j(z) \tag{D.2}$$

$q_j(t)$, $j \in \mathbb{N}$ are known as the 'modal' responses each of which represents the response of an SDOF oscillator governed by the equation of motion (known also as a modal equation):

$$\ddot{q}_j(t) + 2\xi_j\omega_j\dot{q}_j(t) + \omega_j{}^2 q_j(t) = F_j(t), \tag{D.3a}$$

$$F_j(t) = \int_{\Sigma} \Phi_j(z) \frac{F(z,t)}{m} dz \tag{D.3b}$$

Here $F_j(t)$, $j \in \mathbb{N}$ are the generalized forces (modal forces). ω_j are the natural frequencies of the system and ξ_j the modal damping ratios. With zero initial conditions, the solution to Eq. (D.3a) is given by the convolution (Duhamel's) integral:

$$q_j(t) = \int_{-\infty}^{t} F_j(\tau) h_j(t-\tau) d\tau \tag{D.4}$$

$h_j(t)$ is referred to as the impulse response function of the jth modal equation (and hence that of an SDOF oscillator). $h_j(t) = 0$ for $t < 0$ and it is given by:

$$h_j(t) = \frac{e^{-\xi\omega_j t}}{m\omega_{dj}} \sin\omega_{d_j} t \text{ for } t \geq 0 \tag{D.5}$$

$\omega_{d_j} = \omega_j \sqrt{1-\xi_j^2}$ are known as the damped frequencies. In view of Eq. (D.3b), the means and autocorrelations of the modal forces $F_j(t)$ are related to $F(z,t)$ by:

$$\mu_{F_j}(t) = E\left[F_j(t)\right] = \int_{\Sigma} \Phi_j(z) \frac{E\left[F(z,t)\right]}{m} dz = \int_{\Sigma} \frac{\Phi_j(z)}{m} \mu_F(z,t) dz \tag{D.6a}$$

$$R_{F_jF_k}(t_1,t_2) = \int_{\Sigma}\int_{\Sigma} \Phi_j(z_1)\Phi_k(z_2) \frac{E\left[F(z_1,t_1)F(z_2,t_2)\right]}{m^2} dz_1 dz_2$$

$$= \int_{\Sigma}\int_{\Sigma} \frac{\Phi_j(z_1)\Phi_k(z_2)}{m^2} R_{FF}(z_1,z_2,t_1,t_2) dz_1 dz_2 \tag{D.6b}$$

The first and second order moments $E\left[q_j(t)\right]$ and $E\left[q_j(t_1)q_k(t_2)\right]$ of the modal responses are obtained from Eq. (D.3a):

$$E\left[q_j(t)\right] = \int_{-\infty}^{t} \mu_{F_j}(\tau) h_j(t-\tau) d\tau \tag{D.7a}$$

$$E\left[q_j(t_1)q_k(t_2)\right] = \int_{-\infty}^{t}\int_{-\infty}^{t} R_{F_jF_k}(\tau_1,\tau_2) h_j(t-\tau_1) h_k(t-\tau_2) d\tau_1 d\tau_2 \tag{D.7b}$$

The system response (in terms of the first and second order moments) is finally obtained by a superposition of the modal responses as in Eq. (D.2):

$$\mu_Y(z,t) = E[Y(z,t)] = \sum_{j=1}^{\infty} \Phi_j(z)\, E\left[q_j(t)\right] \tag{D.8a}$$

$$R_{YY}(z_1,z_2,t_1,t_2) = E[Y(z_1,t_1)\,Y(z_2,t_2)]$$

$$= \sum_{j=1}^{\infty}\sum_{k=1}^{\infty} E[q_j(t_1)\,q_k(t_2)]\,\Phi_j(z_1)\,\Phi_k(z_2) \tag{D.8b}$$

The last equation indicates that the system response moments depend on those of the modal responses, $q_j(\tau)$ and in turn on those of the input forcing function. Thus, knowing the input characteristics, the task finally reduces to determining the response of an SDOF oscillator (Eq. D.3a) under a stochastic input (see item iii below).

(ii) *Discrete structural systems (linear)*
Semi-discretization of a linear continuous system [Roy and Rao 2012] is often accomplished by the popular finite element method (FEM). Vast literature is available on FEM and readers can refer, for example, to Zienkiewicz [1977], Hughes [1987], Noor [1991] and Bathe [1996]. This leads to an MDOF system governed by matrix vector ordinary differential equations (ODEs) written hereunder:

$$\boldsymbol{M}\,\ddot{\boldsymbol{X}} + \boldsymbol{C}\dot{\boldsymbol{X}} + \boldsymbol{K}\,\boldsymbol{X} = \boldsymbol{F}(t) \tag{D.9}$$

Let N be the number of degrees of freedom (*dof*s) in the discrete model. $\boldsymbol{X}(t)$ is the response vector $(N \times 1)$ that corresponds to the unknown displacements at the nodal *dof*s. $\boldsymbol{M}, \boldsymbol{C}$ and $\boldsymbol{K}$ are respectively the $N \times N$ mass, damping and stiffness matrices which are generally symmetric. $\boldsymbol{F}(t)$ is the vector of nodal forces which are assumed to be stochastic, rendering Eq. (D.9) an SDE.

As in the case of a continuous system, the 'modal superposition method (Eq. D.2) can be utilized to solve the linear SDE (D.9) and obtain the system response in terms of its statistical moments, given similar quantities for the input forcing. However, the coupled system of Eqs. (D.9) is first reduced to a set of uncoupled SDOF systems driven by stochastic input, by assuming the system response in the form (similar to the one in Eq. D.2):

$$\boldsymbol{X}(t) = [\boldsymbol{\phi}]\{\boldsymbol{q}(t)\} \tag{D.10}$$

where, $[\boldsymbol{\phi}]$ is the eigenvector matrix of size $N \times m$ and $\boldsymbol{q}$ is the vector of modal responses. $\boldsymbol{\phi}$ is obtained by solving the eigenvalue problem:

$$\left[\boldsymbol{K} - \omega^2 \boldsymbol{M}\right]\boldsymbol{\phi} = \boldsymbol{0} \tag{D.11}$$

By the orthogonality properties of $\boldsymbol{\phi}$ with respect to $\boldsymbol{M}$ and $\boldsymbol{K}$, the system of Eqs. (D.9) reduces to a set of uncoupled modal equations of similar form as in Eq. (D.3). Rest of the solution procedure in this case also follows closely the one adopted for continuous systems. Whereas the eigenfunctions $\Phi_j(\boldsymbol{z})$ are employed for continuous systems, eigenvectors ϕ_j are presently used in the modal summation.

(iii) *Response mean and autocorrelation function in time domain of an SDOF oscillator under stochastic input*

For this purpose, let us take a typical oscillator (see Eq. D.3a):

$$\ddot{q}(t) + 2\xi\omega\dot{q}(t) + \omega^2 q(t) = F(t) \tag{D.12}$$

Eq. (D.4) gives the mean $E[q(t)]$ as:

$$\mu_q(t) := E[q(t)] = \int_{-\infty}^{t} E[F(\tau)]h(t-\tau)\,d\tau = \int_{-\infty}^{t} \mu_F(\tau)h(t-\tau)\,d\tau \tag{D.13}$$

The second moment $E[q(t_1)q(t_2)]$ is obtained as follows. Since one has:

$$q(t_1) = \int_{-\infty}^{t_1} F(\tau_1)h(t_1-\tau_1)\,d\tau_1,\quad q(t_2) = \int_{-\infty}^{t_2} F(\tau_2)h(t_2-\tau_2)\,d\tau_2 \tag{D.14}$$

it follows that:

$$\begin{aligned} R_{qq}(t_1,t_2) = E[q(t_1)q(t_2)] &= E\left[\int_{-\infty}^{t_1} F(\tau_1)h(t_1-\tau_1)\,d\tau_1 \int_{-\infty}^{t_2} F(\tau_2)h(t_2-\tau_2)\,d\tau_2\right] \\ &= \int_{-\infty}^{t_1}\int_{-\infty}^{t_2} E[F(\tau_1)F(\tau_2)]h(t_1-\tau_1)\;h(t_2-\tau_2)\,d\tau_1 d\tau_2 \\ &= \int_{-\infty}^{t_1}\int_{-\infty}^{t_2} R_{FF}(\tau_1,\tau_2)h(t_1-\tau_1)\;h(t_2-\tau_2)\,d\tau_1 d\tau_2 \end{aligned} \tag{D.15}$$

Eqs. (D.13) and (D.15) together describe the relation between the input and output characteristics of an SDOF oscillator.

If the input $F(t)$ is stationary, i.e., $\mu_F(t) = \mu_F$ (a constant) and $R_{FF}(t_1,t_2) = R_{FF}(t_2-t_1)$, the SDOF oscillator output characteristics in Eqs. (D.13) and (D.15) take the form:

$$\mu_q(t) = \mu_F \int_{-\infty}^{\infty} h(t-\tau)\,d\tau \tag{D.16a}$$

and

$$R_{qq}(t_1,t_2) = \int_{-\infty}^{\infty}\int_{-\infty}^{\infty} R_{FF}(\tau_2-\tau_1)\,h(t_1-\tau_1)\,h(t_2-\tau_2)\,d\tau_1 d\tau_2 \tag{D.16b}$$

From the above equations, it is seen that the oscillator output is non-stationary even if the input is stationary. In the specific case of a white noise input with intensity I, i.e., $R_{FF}(\tau_2-\tau_1) = I\delta(\tau_2-\tau_1)$, Eq. (D.16b) takes the form:

$$R_{qq}(t_1,t_2) = I\int_{-\infty}^{t_1}\int_{-\infty}^{t_2} \delta(\tau_2-\tau_1)h(t_1-\tau_1)\,h(t_2-\tau_2)\,d\tau_1 d\tau_2$$

$$= I\int_{-\infty}^{t} h(t_1-\tau_1)\,h(t_2-\tau_1)\,d\tau_1 \tag{D.17}$$

Substituting for the impulse response function $h_j(t) = \frac{e^{-\xi\omega t}}{\omega_d}\sin\omega_d t$ of the SDOF oscillator (Roy and Rao 2012), we evaluate the above integral as [Lin 1967, Yang 1986]:

$$R_{qq}(t_1,t_2) = \frac{I}{4\xi\omega^3}\,e^{-\xi\omega(t_2-t_1)}\left[\frac{e^{-2\xi\omega t_1}}{1-\xi^2}\left\{\xi^2\cos\omega_d(t_1+t_2)\right.\right.$$

$$\left.-\xi\sqrt{1-\xi^2}\,\sin\omega_d(t_1+t_2) - \cos\omega_d(t_2-t_1)\right\}$$

$$\left.+\left\{\cos\omega_d(t_2-t_1) + \frac{\xi}{\sqrt{1-\xi^2}}\sin\omega_d(t_2-t_1)\right\}\right] \tag{D.18}$$

With $t_1 = t_2 = t$, the non-stationary mean square response $R_{qq}(t,t) = E[q^2(t)]$ is:

$$R_{qq}(t,t) = \frac{I}{4\xi\omega^3}\left[e^{-2\xi\omega t}\left\{\frac{\xi^2}{1-\xi^2}\cos 2\omega_d t - \frac{\xi}{\sqrt{1-\xi^2}}\sin 2\omega_d t - \frac{1}{1-\xi^2}\right\}+1\right] \tag{D.19}$$

Fig. D.1 shows the time history of $R_{qq}(t,t)$. The parameter values $\omega = 10$ and $\xi = 0.02$ are used in the computation and the response history is computed over $t \in [0,10]$ sec. ($\cong$ 15 cycles of oscillation).

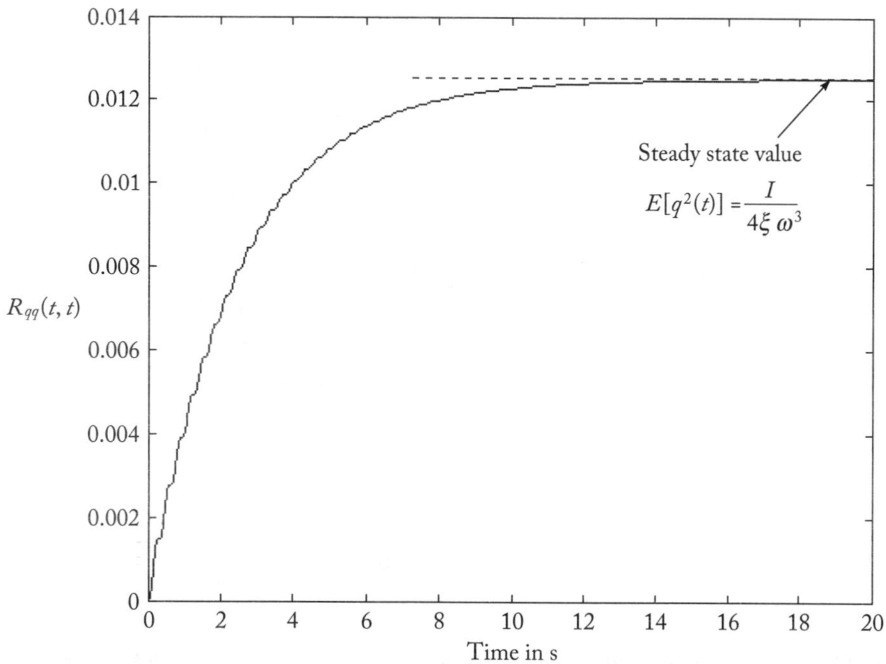

Fig. D.1 Non-stationary response $R_{qq}(t,t) = E\left[q^2(t)\right]$ of an SDOF oscillator (see Eq. D.19); $\omega = 10\frac{\text{rad}}{\text{s}}$, $\xi = 0.02$ and $I = 1.0$

For large times $(t_1, t_2 \to \infty)$ but with finite $t_2 - t_1 = \tau$, observe from Eq. (D.18) that $E[q(t_1)q(t_1+\tau)] = R_{qq}(t_1, t_1+\tau)$ attains stationarity, i.e., $\lim_{\substack{t_1,t_2 \to \infty \\ t_2 - t_1 \text{ finite}}} R_{qq}(t_1,t_2) \to R_{qq}(t_2 - t_1) = R_{qq}(\tau)$ and the stationary correlation is given by:

$$R_{qq}(\tau) = \frac{I}{4\xi\omega^3} e^{-\xi\omega|\tau|} \left\{ \cos\omega_d \tau + \frac{\xi}{\sqrt{1-\xi^2}} \sin\omega_d |\tau| \right\} \tag{D.20}$$

With $\tau = 0$, $R_{qq}(0)$ yields the stationary mean square response $E\left[q^2(t)\right] = \frac{I}{4\xi\omega^3}$. This is the same as the steady state value of $R_{qq}(t,t)$ in Fig. D.1.

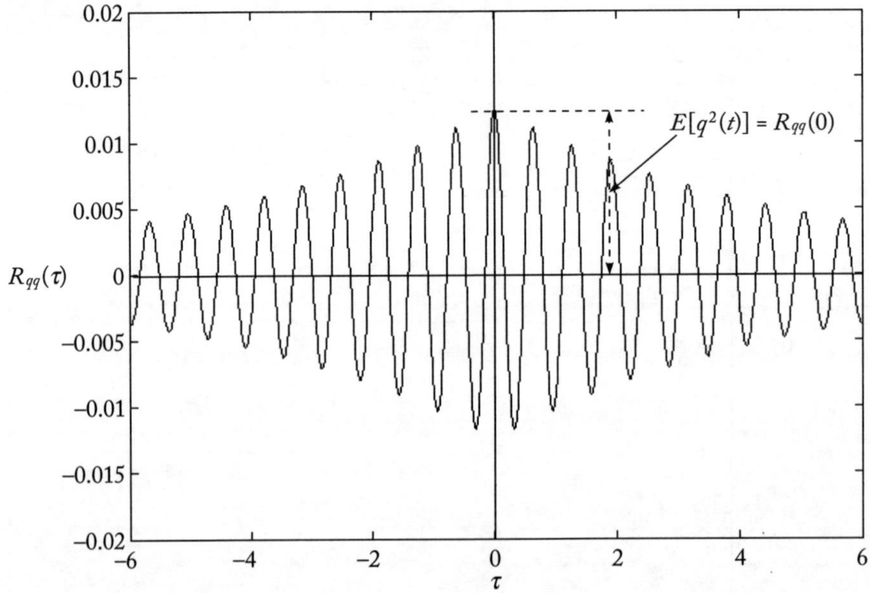

Fig. D.2 Stationary autocorrelation $R_{qq}(\tau) = E[q(t)q(t+\tau)]$ of an SDOF oscillator (see Eq. D.20) under white noise; $\omega = 10\frac{\text{rad}}{\text{s}}$, $\xi = 0.02$ and $I = 1.0$

2. Fokker-Planck equation (forward Kolmogororv pde)

According to the Fokker–Planck equation, the transition *pdf* $f(x;t|x_0;t_0)$ satisfies the initial value problem:

$$\frac{\partial f}{\partial t} = L_t^* f \text{ for } x_0, x \in \mathbb{R}, t > t_0, \lim_{t \to t_0} f(x;t|x_0;t_0) \to \delta(x - x_0) \tag{D.21}$$

where the forward Kolmogororv operator is given by (see Eq. 4.273 with $n = 1$ and $m = 1$):

$$L_t^* f = -\frac{\partial}{\partial x}(a(x,t)f) + \frac{1}{2}\frac{\partial^2}{\partial x^2}\left(\sigma^2(x,t)f\right) \tag{D.22}$$

Proof: Suppose that a scalar function $g(x)$ is sufficiently smooth and has a compact support. We have by Ito's formula (see Eq. 4.83):

$$g(X(t)) - g(x_0) = \int_{t_0}^{t} \left(a(s,X_s)\frac{\partial g(s,X_s)}{\partial x} + \frac{1}{2}\sigma^2(s,X_s)\frac{\partial^2 g(s,X_s)}{\partial x^2} \right) ds$$

$$+\int_{t_0}^{t}\sigma(s,X_s)\frac{\partial g(s,\mathbf{X}_s)}{\partial x}dB(s) \tag{D.23}$$

Taking conditional expectations on both sides of the above equation given $X(t_0)=x_0$ gives:

$$E\left[g(X(t)=x)|X(t_0)=x_0\right]=\int_{-\infty}^{\infty}g(x)f(x;t|x_0;t_0)\,dx$$

$$=g(x_0)+\int_{-\infty}^{\infty}\left(\int_{t_0}^{t}\left(a(s,x)\frac{\partial g(s,x)}{\partial x}+\frac{1}{2}\sigma^2(s,x)\frac{\partial^2 g(s,x)}{\partial x^2}\right)ds\right)$$

$$f(x;t|x_0;t_0)\,dx \tag{D.24}$$

The above equation is a weak form of the PDE (in integral form) that the transition *pdf* satisfies and $f(x;t|x_0;t_0)$ is the one that satisfies Eq. (D.24) for all test functions $g(x)$. Note that the expectation of the last term on the RHS of Eq. (D.23) is zero because of the zero mean property of an Ito integral. Differentiating both sides of the last equation with respect to t leads to:

$$\int_{-\infty}^{\infty}g(x)\frac{\partial f}{\partial t}dx=\int_{-\infty}^{\infty}a(t,x)\frac{\partial g(t,x)}{\partial x}f(x;t|x_0;t_0)\,dx$$

$$+\frac{1}{2}\int_{-\infty}^{\infty}\sigma^2(t,x)\frac{\partial^2 g(t,x)}{\partial x^2}f(x;t|x_0;t_0)\,dx \tag{D.25}$$

Integration by parts of the two integrals on the RHS of the above equation gives:

$$\int_{-\infty}^{\infty}a(t,x)\frac{\partial g(t,x)}{\partial x}f(x;t|x_0;t_0)\,dx=(af)g\Big|_{-\infty}^{\infty}-\int_{-\infty}^{\infty}g\frac{\partial(af)}{\partial x}dx \tag{D.26a}$$

$$\int_{-\infty}^{\infty}\sigma^2(t,x)\frac{\partial^2 g(t,x)}{\partial x^2}f(x;t|x_0;t_0)\,dx$$

$$=\left(\sigma^2 f\right)\frac{\partial g}{\partial x}\Big|_{-\infty}^{\infty}-g\frac{\partial\left(\sigma^2 f\right)}{\partial x}\Bigg|_{-\infty}^{\infty}+\int_{-\infty}^{\infty}g\frac{\partial^2\left(\sigma^2 f\right)}{\partial x^2}dx \tag{D.26b}$$

Substituting in Eq. (D.25) with the assumption that $f, \frac{\partial f}{\partial x} \to 0$ as $x \to \pm\infty$, we obtain:

$$\int_{-\infty}^{\infty} g(x)\frac{\partial f}{\partial t}dx = -\int_{-\infty}^{\infty} g(x)\left(\frac{\partial(af)}{\partial x} - \frac{1}{2}\frac{\partial^2(\sigma^2 f)}{\partial x^2}\right)dx$$

$$\Longrightarrow \int_{-\infty}^{\infty} g(x)\left(\frac{\partial f}{\partial t} + \frac{\partial(af)}{\partial x} - \frac{1}{2}\frac{\partial^2(\sigma^2 f)}{\partial x^2}\right)dx = 0 \tag{D.27}$$

As the test function $g(x)$ is arbitrary, we find that $f(x;t|x_0;t_0)$ satisfies the Fokker–Planck equation (D.22). For higher dimensions, the equation is:

$$\frac{\partial f}{\partial t} = -\sum_{i=1}^{m}\frac{\partial(a_i(\boldsymbol{x})f)}{\partial x_i} + \frac{1}{2}\sum_{i,j=1}^{m}\frac{\partial^2(\sigma_{ij}(\boldsymbol{x})f)}{\partial x_i \partial x_j} \tag{D.28}$$

where, $\boldsymbol{X}$ satisfies the SDE $d\boldsymbol{X}_t = \boldsymbol{a}(\boldsymbol{X}_t)\,dt + \boldsymbol{\sigma}(\boldsymbol{X}_t)\,d\boldsymbol{B}_t$. σ_{ij} is the diffusion term.

3. Levy's characterization of a Brownian motion

$X(t)$, a continuous stochastic process on a probability space $(\Omega, \mathcal{F}, P)$ with $X(0) = 0$ is a Brownian motion if and only if it is an $\mathcal{F}_t$-martingale having quadratic variation $[X,X](t) = t$.

Proof: To prove the 'if part' is straightforward. That is, if $X(t)$ is a Brownian motion, it is a continuous martingale (see Section 3.3.6, Chapter 3). Also, the quadratic variation of $X(t)$ is $[X,X](t) = t$ (Section 4.4.2, Chapter 4). For the 'only if part' (i.e., given $X(t)$ to be a martingale with $[X,X](t) = t$), it suffices to check $X(t)$ for the stationary and independent increment properties of a Brownian motion. To this end, consider the stochastic exponential of $X(t)$ (Section 4.6.3, Chapter 4):

$$Y(t) = \mathcal{E}(X)(t) := e^{X(t)-X(0)-\frac{1}{2}[X,X](t)} = e^{X(t)-\frac{1}{2}t}, \text{ with } Y(0) = 1 \tag{D.29}$$

Note that $Y(t)$ satisfies the SDE:

$$dY(t) = Y(t)\,dX(t) \tag{D.30}$$

For $Y(t)$ to be (an exponential) martingale, Novikov's condition in Eq. (4.240b) is to be satisfied. Here the condition is trivially satisfied. Utilizing the martingale property of $Y(t)$, we have for $0 \le s < t$:

$$E\left[e^{X(t)-\frac{1}{2}t}|\mathcal{F}_s\right] = e^{X(s)-\frac{1}{2}s}$$

$$\Longrightarrow E\left[e^{X(t)-X(s)}|\mathcal{F}_s\right] = e^{\frac{1}{2}(t-s)}$$

$$\Longrightarrow (X(t) - X(s)) \sim \mathcal{N}(0, t-s) \tag{D.31}$$

Also, the increment $X(t) - X(s)$ is independent of $\mathcal{F}_s$. Hence $X(t)$ is a Brownian motion.

4. Set of random variables $\{u(B_{t_1}, B_{t_2}, \ldots, B_{t_n}) \in C_0^\infty(\mathbb{R}^n), t_i \in [0,T], n = 1,2,\ldots\}$ is dense in $L^2(\mathcal{F}_T, P)$ - (the proof follows Øksendal 2003). Recall that $C_0^\infty(\mathbb{R}^n)$ is a space of infinitely differentiable functions with compact supports.

Proof: Let $\{t_i, i = 1,2,\ldots\}$ be a dense subset of $[0,T]$ and for $n = 1,2,\ldots$, let $\mathcal{H}_n$ be the σ-algebra generated by $B_{t_1}, B_{t_2}, \ldots, B_{t_n}$. As such, $\mathcal{H}_n \subset \mathcal{H}_{n+1}$ with $\mathcal{F}_T$ being the smallest σ-algebra containing all $\mathcal{H}_n$s. Now, consider a square-integrable $g \in L^2(\mathcal{F}_T, P)$. By martingale convergence theorem (Chapter 3, Section 3.4.4):

$$= E[g \mid \mathcal{F}_T] = \lim_{n\to\infty} E[g|\mathcal{H}_n] \tag{D.32}$$

The limit is pointwise a.e. (P) with the limit in $L^2(\mathcal{F}_T, P)$. Using Doob–Dynkin lemma (see the remark below), for each n and some Borel measurable function $g_n(B_{t_1}, B_{t_2}, \ldots, B_{t_n})$:

$$E[g \mid \mathcal{H}_n] = g_n(B_{t_1}, B_{t_2}, \ldots, B_{t_n}) \tag{D.33}$$

By the density argument, each $g_n(B_{t_1}, B_{t_2}, \ldots, B_{t_n})$ can be approximated in $L^2(\mathcal{F}_T, P)$ by functions $u_n(B_{t_1}, B_{t_2}, \ldots, B_{t_n})$ where $u_n \in C_0^\infty(\mathbb{R}^n)$. Hence we get the result.

Remark (*Doob–Dynkin lemma*): If $\boldsymbol{X}, \boldsymbol{Y} : \Omega \to \mathbb{R}^n$ are two random variables, then $\boldsymbol{Y}$ is $\mathcal{H}_X$-measurable if and only if there exists a Borel measurable function $\boldsymbol{g} \in \mathbb{R}^n \to \mathbb{R}^n$ such that $\boldsymbol{Y} = \boldsymbol{g}(\boldsymbol{X})$.

5. Doob's h-transform

Consider a stochastic process X_t on a filtered probability space $(\Omega, \mathcal{F}, \mathcal{F}_t, P_x)$ governed by a time–homogeneous SDE $dX(t) = a(X(t))dt + \sigma(X(t))dB(t)$. Let $f^t(y,x)$ be the transition density where the superscript t stands for a time interval. Then, with F^t denoting the homogeneous transition kernel, we define an associated operator $F^t h = \int_y f^t(y,x)h(y)\,dy$ satisfying the semi-group property $F^{t+s}h = F^t F^s h$ for non-overlapping intervals t and s. For an harmonic function $h(x)$ with $L_t h(x) = 0$, h is invariant under the transition kernel, i.e., $F^t h = h$. L_t is the generator of X_t. As discussed in Section 4.7.4, Chapter 4, a bounded harmonic function h is closely related to $\mathcal{I}$, the sigma–algebra generated by shift invariant functions $\mathfrak{H}$ having the property $\mathfrak{H} \circ \theta_t \equiv \mathfrak{H}$.

Here θ_t is a time shift operator $\theta_t : X_s \to X_{t+s}$. Doob's h–transform facilitates understanding the process X_t when it is conditioned path-wise on an event $A \in \mathcal{I}$ of positive (non-zero) probability. It is useful to think of A as a terminal event or one corresponding to the coffin state of the process X_t as $t \to \infty$ with $X_0 = x \in U$, the set of possible initial conditions. Let $h(x) = P_x(A)$ be the corresponding harmonic function (note that $h(X_t) = E_x[I_A|X_t] = E_x[I_A|\mathcal{F}_t]$ is an $\mathcal{F}_t$-martingale). Consider the set $U^* = \{x \in U \subset \mathbb{R} : h(x) > 0\}$, which is thus the initial set of values that reach A with non-zero probability. For $x \in U^*$, consider the conditioned path measure $P_x^* \ll P_x$ which is given by:

$$dP_x^* = \frac{I_A}{h(x)} dP_x \tag{D.34}$$

This leads to:

$$dP_x^*|_{\mathcal{F}_t} = E_x\left[\frac{I_A}{h(x)}|\mathcal{F}_t\right] dP_x|_{\mathcal{F}_t} = \frac{h(X_t)}{h(x)} dP_x|_{\mathcal{F}_t} \tag{D.35}$$

where $P_x|_{\mathcal{F}_t}$ is the restriction of P_x to $\mathcal{F}_t$. The statement of the Doob's h–transform now goes as follows. X_t is still a time homogeneous Markov process under the new probability law P_x^* with the transition kernel:

$$F^{*t}(dy,x) = \frac{h(y)}{h(x)} F^t(dy,x) \tag{D.36}$$

An outline of the proof: F^{*t} is a probability measure on U^* and so is P_x^*. Eq. (D.36) is just the generalized Bayes' rule:

$$P_x\left(X_{t+s} \in dy \mid A, \mathcal{F}_t\right) = \frac{P_x(A|X_{t+s} = y, \mathcal{F}_t) P_x(X_{t+s} \in dy|\mathcal{F}_t)}{P_x(A|\mathcal{F}_t)} = \frac{h(y)}{h(X_t)} F^s(dy, X_t) \tag{D.37}$$

The Markov property of the conditioned X_t under P_x^* is inherited from the unconditioned X_t:

$$E_x^*[g(X_{t+s})|\mathcal{F}_t] = h^{-1}(X_t) E_x[h(X_{t+s}) g(X_{t+s})|\mathcal{F}_t], P_x^*\text{–a.s., for } g \text{ bounded} \tag{D.38}$$

This may be proved as in the following steps. The RHS of the above equation is an $\mathcal{F}_t$-random variable. Let $B \in \mathcal{F}_t$ and one has:

$$E_x^*\left[h^{-1}(X_t) E_x[h(X_{t+s}) g(X_{t+s})|\mathcal{F}_t] I_B\right]$$

$$= E_x\left[h^{-1}(X_t) E_x[g(X_{t+s}) h(X_{t+s}) I_B|\mathcal{F}_t] \frac{h(X_t)}{h(x)}\right]$$

$$= E_x \left[g\left(X_{t+s}\right) I_B \frac{h\left(X_{t+s}\right)}{h(x)} \right]$$

$$= E_x^* \left[E_x^* \left[g\left(X_{t+s}\right) | \mathcal{F}_t \right] I_B \right] \tag{D.39}$$

Since $B \in \mathcal{F}_t$ is arbitrary, one can localize (i.e., remove the outer operator E_x^* from both sides of the last equation) to obtain $h^{-1}\left(X_t\right) E_x\left[h\left(X_{t+s}\right) g\left(X_{t+s}\right) | \mathcal{F}_t\right] = E_x^*\left[g\left(X_{t+s}\right) | \mathcal{F}_t\right]$.

Similar to $\frac{d}{dt} F^t h|_{t=0} = L_t h$, one has $\frac{d}{dt} F^{*t} h|_{t=0} = L_t^* h$. It follows from Eq. (D.36) that:

$$L_t^* \equiv h^{-1} L_t h \tag{D.40}$$

As an illustration, consider a scalar-valued C^2 function $g(\boldsymbol{X}_t)$ with $\boldsymbol{X}_t$ governed by the SDE:

$$d\boldsymbol{X}_t = \boldsymbol{a}\left(\boldsymbol{X}_t\right) dt + \boldsymbol{\sigma}(\boldsymbol{X}_t) d\boldsymbol{B}_t \text{ with } \boldsymbol{b} = \boldsymbol{\sigma}\boldsymbol{\sigma}^T \tag{D.41}$$

Applying the operator L_t^* on $g(\boldsymbol{X}_t)$, one obtains:

$$L_t^* g = \left(h^{-1} L_t h\right) g = h^{-1} L_t \left(hg\right)$$

$$= h^{-1} \left(\boldsymbol{a}.\left(g\boldsymbol{\nabla} h + h\boldsymbol{\nabla} g\right) + \frac{1}{2} \boldsymbol{b}\boldsymbol{\nabla}.\boldsymbol{\nabla}\left(hg\right) \right) dt$$

$$= gh^{-1} \left(\boldsymbol{a}.\boldsymbol{\nabla} h + \frac{1}{2} \boldsymbol{b}\boldsymbol{\nabla}.\boldsymbol{\nabla} h \right) dt + \boldsymbol{a}.\boldsymbol{\nabla} g\, dt + \frac{1}{2} \boldsymbol{b}\boldsymbol{\nabla}.\boldsymbol{\nabla} g\, dt + h^{-1} \boldsymbol{b}\boldsymbol{\nabla} h.\boldsymbol{\nabla} g dt$$

$$= \left(\boldsymbol{a}.\boldsymbol{\nabla} g + \frac{1}{2} \boldsymbol{b}\boldsymbol{\nabla}.\boldsymbol{\nabla} g \right) dt + h^{-1} \boldsymbol{b}\boldsymbol{\nabla} h.\boldsymbol{\nabla} g dt \tag{D.42}$$

where we have used $L_t h = 0$. ∇ is the gradient operator. The first term within the brackets on the last RHS is $L_t g$, leaving the correction component of the operator L_t^* as $\boldsymbol{b}\frac{\boldsymbol{\nabla} h}{h}.\boldsymbol{\nabla}$. Specifically, we have:

$$L_t^* \equiv L_t + \boldsymbol{b}\frac{\boldsymbol{\nabla} h}{h}.\boldsymbol{\nabla} \tag{D.43}$$

Example D.1

Suppose that the objective is to have a Brownian motion $X(t) = B(t)$ conditioned to remain within the interval $(0, \pi)$ up to time T, which may be a.s. bounded stopping time. To enable a treatment of the problem in a time–homogeneous setting, we enlarge the state space to include the time variable so that the latter is a pseudo-space variable. Thus $U^* = [0,\pi] \times [0,T]$ with an absorbing boundary $\delta U^* = (\{0,\pi\} \times (0,T]) \cup ((0,\pi) \times \{T\})$. The corresponding two-dimensional generator is $L_t = \frac{1}{2}\frac{\partial^2}{\partial x^2} + \frac{\partial}{\partial t}$. The conditioned X_t (denoted by X_t^*) shall stop when δU^* is encountered implying that $X_t = X_{t\wedge T}$ where $T = \inf\{t : X_t \in \delta U^*\}$ is indeed a stopping time.

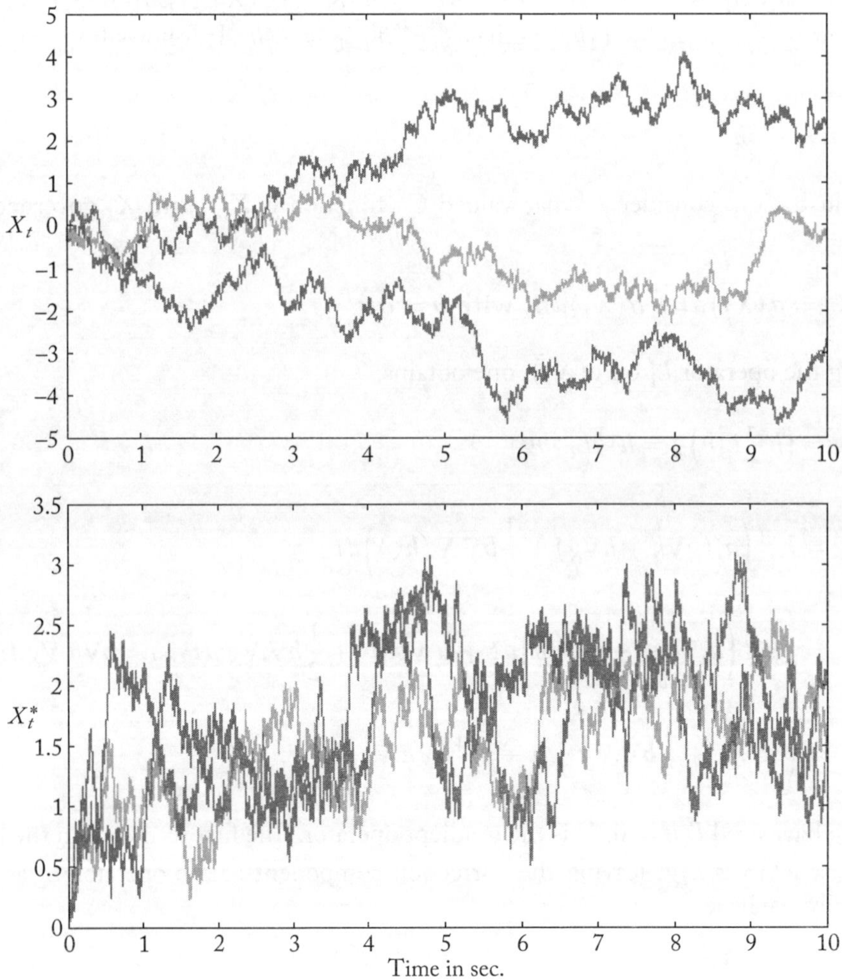

Fig. D.3 Doob's h-transform; a few realizations of (a) X_t and (b) the conditioned one, X_t^*

In view of Eq. (D.43), we have:

$$L_t^* = L_t + \frac{\nabla h}{h}\nabla, \nabla = \frac{d}{dx} \tag{D.44}$$

where, $h(x)$ can be obtained by solving $L_t h(x,t) = 0$ which is a two-dimensional heat equation in x and time variable $-t$. Using the initial (terminal) condition $h(x,T) = 1$ and the Dirichlet boundary conditions, $h(0,t) = h(\pi,t) = 0$, for $t \le T$, we get the solution in terms of a Fourier series expansion as $h(x,t) = \sum_{j=1,3,..}^{\infty} \frac{4}{j\pi} \exp\left(-j^2 (T-t)\right) \sin jx$. For demonstration, using only the first term corresponding to $j = 1$, we have $h(x,t) = \frac{4}{\pi} \exp(-(T-t)) \sin x$ so that $L_t^* = L_t + \frac{\cos x}{\sin x} \frac{d}{dx}$. Thus X_t^* satisfies the following SDE that contains an additional drift term:

$$dX_t^* = dB(t) + \frac{\cos X_t^*}{\sin X_t^*} dt \tag{D.45}$$

Fig. D.3 shows a few realizations of X_t and X_t^*.

We consider another example with the requirement of confining $X(t)$ to an arbitrary interval, say, to $[-\varepsilon, \varepsilon]$. This may be achieved by considering $dX(t) = \sigma(\varepsilon) dB(t)$ with $L_t = \frac{1}{2}\sigma^2(\varepsilon) \frac{\partial^2}{\partial x^2} + \frac{\partial}{\partial t}$. In this case, we obtain a Fourier series solution to $L_t h(x,t) = 0$ in the form $h(x,t) = \sum_{j=1,2,...}^{\infty} C_j \exp\left(-j^2 (T-t)\right) \cos \frac{j\pi x}{2\varepsilon}$ that satisfies the boundary conditions. Retaining only one term in the series and imposing $L_t h(x,t) = 0$ gives $\sigma(\varepsilon) = \frac{2\sqrt{2}\varepsilon}{\pi}$. Thus X_t^* satisfies the following SDE :

$$dX_t^* = \frac{2\sqrt{2}\varepsilon}{\pi} dB(t) + \tan \frac{\pi X_t^*}{2\varepsilon} dt \tag{D.46}$$

A simulation result suiting the intended requirement for $\varepsilon = 0.2$ is given in Fig. D.4.

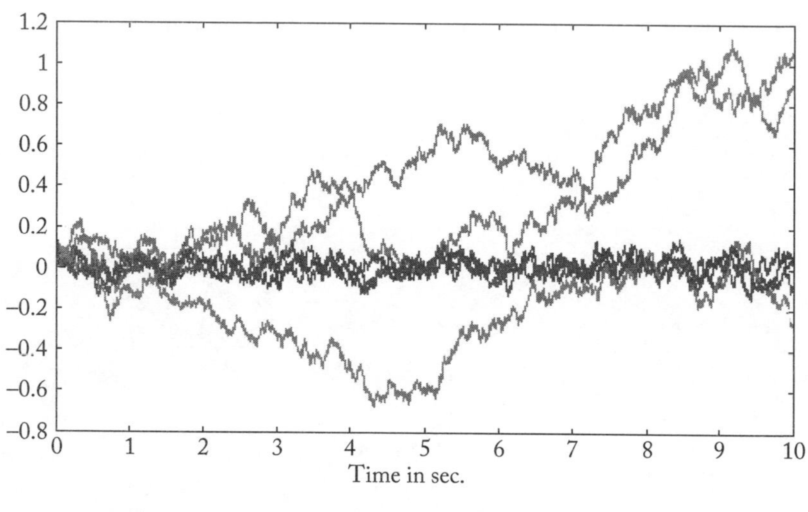

Fig. D.4 Doob's h-transform; a few (three) realizations of both X_t (in light black) and X_t^* (in dark black) for $\varepsilon = 0.2$

A formal, yet introductory, treatment of Doob's h-transform may be found in Revuz and Yor [1999] and Rogers and Williams [Volumes 1 and 2, 2000]. In any case, the astute reader, by now, may have realized that Doob's h–transform does the reverse job of Girsanov's theorem on change of measures. While Girsanov's theorem constructs the Radon–Nikodym derivative for a given change of drift in the SDE, Doob's h–transform starts with a given Radon–Nikodym derivative and constructs the associated change in drift in the SDE. It is precisely because of this that Doob's h–transform, if used properly, could be so effective.

References

Bathe, K. J. 1996. *Finite Element Procedures.* NJ: Prentice Hall.

Hughes, T. J. R. 1987. *The Finite Element Method.* Englewood Cliffs: Prentice Hall.

Lin, Y. K. 1967. *Probabilistic Theory of Structural Dynamics.* NY: McGraw-Hill.

Meirovitch, L. 1967. *Analytical Methods in Vibrations.* NY: Macmillan.

Noor, A. K. 1991. 'Bibliography of Books and Monographs on Finite Element Technology'. *Applied Mech. Rev.* 44(6): 307–17.

Øksendal, B. 2003. *Stochastic Differential Equations, An Introduction with Applications.* 6th ed. Heidelberg: Springer.

Revuz, D. and M. Yor. 1999. *Continuous Martingales and Brownian Motion.* 3rd ed. Springer.

Rogers, L. C. G. and D. Williams. 2000. *Diffusion, Markov Processes, and Martingales.* 1 & 2. 2nd ed. Cambridge: Cambridge Univ. Press.

Roy, D. and G. V. Rao. 2012. *Elements of Structural Dynamics : A New Perspective.* UK: John Wiley and Sons.

Yang, C. Y. 1986. *Random Vibration of Structures.* NY: John Wiley and Sons.

Zienkiewicz, O. C. 1977. *The Finite Element Method.* NY: McGraw-Hill.

Appendix E (Chapter 5)

1. Evaluation of MSIs

Computation of MSIs of the type $I^{(1)}_{j_1 j_2 \cdots j_k}(h)$, $k = 1, 2, \ldots$ in Eq. (5.72a) constitutes a crucial factor in maintaining the error orders of different one-step approximations to SDEs. Evaluations of some first level MSIs – particularly, $I^{(1)}_{j_1}(h)$, $I^{(1)}_{j_1 j_2}(h)$, j_1 and/or $j_2 = 0$ or $j_1 \neq 0$ and/or $j_2 \neq 0$ are already described in Section 5.7.1 of Chapter 5. In further exposition, the superscript '(1)' and argument 'h' are avoided for the sake of convenience. Also the lower limit $t = 0$ is assumed while determining the integrals. Eq. (5.76a) gives the MSIs involving single integrals, which are:

$$I_0 = \int_0^h ds = h \tag{E.1a}$$

$$I_j = \int_0^h dB_j(s) = B_j(h) - B_j(0) = B_j(h) = h^{\frac{1}{2}} Z_j \sim \mathcal{N}\left(0, h^{\frac{1}{2}}\right) \tag{E.1b}$$

where Z_j is the standard normal random variable $\mathcal{N}(0,1)$. For instance, these are involved in lower order Newmark schemes, wherein one needs to compute only I_j and $I_{j,0}$, $j = 1, 2, \ldots$. The two integrals $I_{j0} = \int_0^h \int_0^{s_1} dB_j(s_2)\, ds_1$ and $I_{0j} = \int_0^h \int_0^{s_1} ds_2 dB_j(s_1)$ are related by Eq. (5.81) and they are given by:

$$I_{j0} = \frac{1}{2} h^{\frac{3}{2}} \left(Z_{j1} + \frac{1}{\sqrt{3}} Z_{j2}\right),\ I_{0j} = \tfrac{1}{2} h^{\frac{3}{2}} \left(Z_{j1} - \tfrac{1}{\sqrt{3}} Z_{j2}\right) \tag{E.2}$$

where Z_{j1} and Z_{j2} are independent standard normal random variables. The case of I_{jj} involving a single Wiener process is given by Eq. (5.80):

$$I_{jj} = \frac{1}{2}\left(\left(I_j\right)^2 - h\right) \tag{E.3}$$

For higher order schemes, the multiple integrals such as $I_{j_1j_2}, I_{j_1j_20}, I_{j_100}, I_{0j_20}$, $I_{j_1j_2j_3}, I_{j_1j_2j_30}$ where $j_1, j_2, j_3 = 1, 2, \ldots$. need be modeled. In order to consistently generate these random variables numerically within a computer program, the following scaling is first considered [Roy and Dash 2005].

$$g_j(\theta) = \frac{B_j(\theta h)}{\sqrt{h}} \quad \text{with } B_j(0) = 0, \; 0 \le \theta \le 1 \tag{E.4}$$

Obviously, $g_j(\theta)$, $j = 1, 2, \ldots$ are standard Brownian motions with unit variance. Let the scaled k^{th} stochastic integral in terms of increments of $g_j(\theta)$ be denoted as:

$$\bar{I}_{j_1j_2\ldots j_k} = \int_0^1 \int_0^\theta \ldots \int_0^{\theta_{k-2}} dg_{j_1}(\theta_{k-1}) dg_{j_2}(\theta_{k-2}) \ldots dg_{j_k}(\theta) \tag{E.5}$$

where it is implied, as before, that $dg_0(\theta) = d\theta$. One thus has the following set of relations between the scaled and original multiple integrals:

$$I_j = h^{\frac{1}{2}} \bar{I}_j; \quad I_{j0} = h^{\frac{3}{2}} \bar{I}_{j0}; \quad I_{0j} = h^{\frac{3}{2}} \bar{I}_{0j};$$

$$I_{j_1j_20} = h^2 \bar{I}_{j_1j_20}; \quad I_{j_1j_2j_30} = h^{\frac{5}{2}} \bar{I}_{j_1j_2j_30}; \ldots \text{etc.} \tag{E.6}$$

for all $j_1, j_2, j_3 = 1, 2, \ldots$. Now, the task is to generate the third or higher level integrals approximately via a numerical scheme.

The Case of a single Wiener process

This particular case, involving only one Wiener process $B_1(t)$ is simpler to implement and is therefore dealt with first. It has been shown [Milstein 1995] that recursively using an expansion of the two-parameter Hermite n-polynomial, $H_n(\lambda s, \gamma B_1(s))$, followed by equating the like powers of λ and γ, one can arrive at the following exact expressions of the multiple integrals:

$$I_{11} = \frac{B_1^2(h) - h}{2}; \quad I_{110} = \frac{1}{2} \int_0^h B_1^2(s) ds - \frac{h^2}{4}; \quad I_{100} = h I_{10} - \int_0^h s B_1(s) ds$$

$$I_{010} = 2 \int_0^h s B_1(s) ds - h I_{10}; \quad I_{111} = \frac{B_1^3(h)}{6} - \frac{1}{2} h B_1(h)$$

$$I_{1110} = \frac{1}{6} \int_0^h B_1^3(s) ds - \frac{1}{2} \int_0^h s B_1(s) ds \tag{E.7}$$

Thus for an implementation of the higher order Newmark scheme, it suffices to approximately model the following basic MSIs: $A_1 = \int_0^h B_1^2(s)ds$; $A_2 = \int_0^h sB_1(s)ds$; and $A_3 = \int_0^h B_1^3(s)ds$. Expressed in terms of the standard Wiener process, $g_1(\theta) = B_1(\theta h)/\sqrt{h}$, $\theta = s/h \in [0,1]$, these integrals take the form:

$$A_1 = h^2 \int_0^1 g_1^2(\theta)d\theta = h^2 \bar{A}_1;\ A_2 = h^{5/2} \int_0^1 \theta g_1(\theta)d\theta = h^{5/2} \bar{A}_2$$

$$A_3 = h^{5/2} \int_0^1 g_1^3(\theta)d\theta = h^{5/2} \bar{A}_3 \tag{E.8}$$

Now the following four SDEs may be solved over $\theta \in [0,1]$ to determine these integrals:

$$d\bar{A}_0 = dg_1(\theta); d\bar{A}_1 = \bar{A}_0^2 d\theta; d\bar{A}_2 = \theta \bar{A}_0(\theta)d\theta; d\bar{A}_3 = \bar{A}_0^3(\theta)d\theta \tag{E.9}$$

The above equations are subject to initial conditions $\bar{A}_k(0) = 0; k = 0,1,2,3$. Moreover, the approximate numerical technique and the time step size h_1 to be used must be appropriately selected.

The case of multiple Wiener processes

A procedure similar to the case of a single Wiener noise input may be adopted in this case too. The only difference in this case is that a simplified recursive relationship between the multiple integrals based on Hermite expansions is not possible here. Thus all the multiple integrals of third and higher levels have to be determined by constructing a set of SDEs and solving the latter numerically. As in the case with a single Wiener process, these SDEs are formed in terms of increments of a set of standard (scaled) Wiener processes $g_k(\theta)$ (Eq. E.4) so that $0 \le \theta \le 1$, $k = 1,2,...,q$ $(q \ge 2)$.With a time step size h_1, the first level scaled integrals, $\bar{I}_k$, may be exactly generated using the map $(\bar{I}_k)_j = (\bar{I}_k)_{j-1} + \Delta_j g_k(h_1)$ with $\bar{I}_k(0) = (\bar{I}_k)_0 = 0$. Moreover, Eq. (E.2) may be used to exactly obtain the second level integrals $\bar{I}_{k0}$ and $\bar{I}_{0k}$. Next, one has the following set of SDEs for the approximate evaluations of other scaled multiple integrals over $[0,1]$ with a step size h_1 :

$$d\bar{I}_{j_1 j_2} = \bar{I}_{j_1} dg_{j_2}(\theta); d\bar{I}_{j_1 j_2 0} = \bar{I}_{j_1 j_2} d\theta; d\bar{I}_{j_1 00} = \bar{I}_{j_1 0} d\theta; d\bar{I}_{0 j_1 0} = \bar{I}_{0 j_1} d\theta;$$

$$d\bar{I}_{j_1 j_2 j_3} = \bar{I}_{j_1 j_2} dg_{j_3}(\theta); d\bar{I}_{j_1 j_2 j_3 0} = \bar{I}_{j_1 j_2 j_3} d\theta \tag{E.10}$$

subject to initial conditions

$$\bar{I}_{j_1 j_2}(0) = 0;\ \bar{I}_{j_1 j_2 0}(0) = 0;\ ...\ \bar{I}_{j_1 j_2 j_3 0}(0) = 0.$$

The following general relation between the scaled and original k^{th} level multiple integrals may be noted:

$$\bar{I}_{j_1 j_2 \dots j_k} = \int_0^1 \int_0^\theta \dots \int_0^{\theta_{k-2}} dg_{j_1}(\theta_{k-1}) dg_{j_2}(\theta_{k-2}) \dots dg_{j_k}(\theta)$$

$$I_{j_1 j_2 \dots j_k}(h, \{B_k(h) | k = 1, \dots, q\}) = h^{\sum_{m=1}^{k}(2-\bar{j}_m)/2} \bar{I}_{j_1 j_2 \dots j_k}(1, \{g_k(1) | k = 1, \dots, q\}) \quad \text{(E.11)}$$

where, $\bar{j}_m = 0$ if $j_m = 0$ else $\bar{j}_m = 1$. Now suppose that a stochastic Heun scheme be used to determine the scaled multiple integrals as:

$$(\bar{I}_{j_1 j_2})_i = (\bar{I}_{j_1 j_2})_{i-1} + 0.5((\bar{I}_{j_1})_{i-1} + (\bar{I}_{j_1})_i) \Delta_i g_{j_2}(h_1)$$

$$(\bar{I}_{j_1 j_2 0})_i = (\bar{I}_{j_1 j_2 0})_{i-1} + 0.5((\bar{I}_{j_1 j_2})_{i-1} + (\bar{I}_{j_1 j_2})_i) h_1$$

$$(\bar{I}_{j_1 00})_i = (\bar{I}_{j_1 00})_{i-1} + 0.5((\bar{I}_{j_1 0})_{i-1} + (\bar{I}_{j_1 0})_i) h_1$$

$$(\bar{I}_{0 j_1 0})_i = (\bar{I}_{0 j_1 0})_{i-1} + 0.5((\bar{I}_{0 j_1})_{i-1} + (\bar{I}_{0 j_1})_i) h_1$$

$$(\bar{I}_{j_1 j_2 j_3})_i = (\bar{I}_{j_1 j_2 j_3})_{i-1} + 0.5((\bar{I}_{j_1 j_2})_{i-1} + (\bar{I}_{j_1 j_2})_i) \Delta_i g_{j_3}(h_1)$$

$$(\bar{I}_{j_1 j_2 j_3 0})_i = (\bar{I}_{j_1 j_2 j_3 0})_{i-1} + 0.5((\bar{I}_{j_1 j_2 j_3})_{i-1} + (\bar{I}_{j_1 j_2 j_3})_i) h_1; \quad \text{(E.12)}$$

for all $j_1, j_2, j_3 \in [1, q]$. It must be noted that the above set of equations would have to be solved in the same hierarchical order as shown. For instance, to solve for $(\bar{I}_{j_1 j_2 0})_j$ using the second of the above set of equations, one must first solve for $(\bar{I}_{j_1 j_2})_i$ using the first equation. Note that it may be required to determine the appropriate step size h_1 in view of the original time step size h, so that the desired local error orders are formally maintained.

Finally, note that the following scheme may be adopted to generate Wiener increments, $\Delta_i B_k(s)$, $k = 1, 2, \dots, q$. To begin with, k sets of independent and $\mathcal{N}(0,1)$ random variables, $S_l = \{\overline{\omega}_1^{(l)}, \overline{\omega}_2^{(l)}, \dots, \overline{\omega}_i^{(l)}, \dots\}$, $l = 1, 2, .., k$, are generated. These Gaussian variables may be obtained from uniformly distributed pseudo-random variables in $[0, 1]$ via Box-Muller transformation. Next, the desired Wiener increments are generated via the scaling $\Delta_i B_l(h) = h^{\frac{1}{2}} \overline{\omega}_i^{(l)}$.

2. Orders of the MSIs (Eqs. 5.73 and 5.74)

In evaluating the orders of the MSIs, the following two results are utilized:

$$\text{a)}\ \left(E\left[I^{(1)}_{j_1 j_2 \cdots j_k}(h)\right]^2\right)^{1/2} = O\left(h^{\sum_{i=1}^{k}\frac{(2-\bar{j}_i)}{2}}\right)$$

where, $\bar{j}_i = 0$ if $j_i = 0$

$$= 1 \quad \text{if} \quad j_i \neq 0 \tag{E.13}$$

$$\text{b)}\ \left(E\left[I^{(2)}_{j_1 j_2 \cdots j_k}(g,h)\right]^2\right)^{\frac{1}{2}} \leq K\left(1 + E\left[\left|X(t)\right|^2\right]\right)^{\frac{1}{2}} h^{\sum_{i=1}^{k}\frac{(2-\bar{j}_i)}{2}}$$

$$\text{provided that}\left|L_{j_k} L_{j_{k-1}} \dots L_{j_1} g(t,x)\right| \leq K\left((1 + |x|^2\right)^{\frac{1}{2}} \tag{E.14}$$

Proof for a): The MSI of the type in Eq. (E.13) is:

$$I^{1}_{j_1 j_2 \cdots j_k}(h) = \int_t^{t+h} dB_{j_k}(s_1) \int_t^{s_1} dB_{j_{k-1}}(s_2) \dots \int_t^{s_{k-1}} dB_{j_1}(s_k) \tag{E.15}$$

Let us denote the order of smallness of $E\left[I^{(1)}_{j_1 j_2 \cdots j_k}(h)\right]^2$ by $p(j_1, j_2, \dots, j_k)$.

With $t = 0$ in the integral, we have:
if $j_k \neq 0$,

$$E\left[I^{(1)}_{j_1 j_2 \cdots j_k}(h)\right]^2 = E\left[\int_0^h dB_{j_k}(s_1) \int_0^{s_1} dB_{j_{k-1}}(s_2) \dots \int_0^{s_{k-1}} dB_{j_1}(s_k)\right]^2$$

$$= \int_0^h E\left[I^{(1)}_{j_1 j_2 \cdots j_{k-1}}(s_1)\right]^2 ds_1 \left(\text{since}\left(dB_{j_k}(s_1)\right)^2 = ds_1\right); \tag{E.16}$$

if $j_k = 0$, $dB_{j_k}(s_1) = ds_1$ and

$$E\left[I^{(1)}_{j_1 j_2 \cdots j_k}(h)\right]^2 = E\left[\int_0^h ds_1 \int_0^{s_1} dB_{j_k-1}(s_2) \dots \int_0^{s_{k-1}} dB_{j_1}(s_k)\right]^2$$

$$= E\left[\int_0^h I^{(1)}_{j_1 j_2 \cdots j_{k-1}}(s_1) ds_1\right]^2 \leq h \int_0^h E\left[I^{(1)}_{j_1 j_2 \cdots j_{k-1}}(s_1)\right]^2 ds_1 \tag{E.17}$$

The two Eqs. (E.16) and (E.17) together lead to the following recurrence relation:

$$p(j_1, j_2, \ldots, j_k) = p(j_1, j_2, \ldots, j_{k-1}) + \left(2 - \bar{j}_k\right) \tag{E.18}$$

which proves the claim in Eq. (E.13).

Proof for b): The MSI of the type in Eq. (E.14) is:

$$I^{(2)}_{j_1 j_2 \ldots j_k}(g,h)$$

$$= \int_t^{t+h} dB_{j_k}(s_1) \int_t^{s_1} dB_{j_{k-1}}(s_2) \ldots \int_t^{s_{k-1}} L_{j_k} L_{j_{k-1}} \ldots L_{j_1} g(s_{k-1}, X(s_{k-1})) dB_{j_1}(s_k) \tag{E.19}$$

The proof follows similar steps as in the case of (*a*). For instance, with $j_k \neq 0$:

$$E\left[I^2_{j_1 j_2 \ldots j_k}(g,h)\right]^2 = \int_0^h E\left[I^2_{j_1 j_2 \ldots j_{k-1}} g(s_1, X(s_1))\right]^2 ds_1 \tag{E.20}$$

If $j_k = 0$, $dB_{j_k}(s_1) = ds_1$ and therefore:

$$E\left[I^{(2)}_{j_1 j_2 \ldots j_k}(g,h)\right]^2 = E\left[\int_0^h I^{(2)}_{j_1 j_2 \ldots j_{k-1}} g(s_{k-1}, X(s_{k-1})) ds_1\right]^2$$

$$\leq h \int_0^h E\left[I^2_{j_1 j_2 \ldots j_{k-1}} g(s_{k-1}, X(s_{k-1}))\right]^2 ds_1 \tag{E.21}$$

Also, we have from the hypothesis in Eq. (E.14):

$$E\left[I^{(2)}_{j_1 j_2 \ldots j_{k-1}} g(s_{k-1}, X(s_{k-1}))\right]^2 \leq K\left((1 + E\left[\left|X(s_{k-1})\right|^2\right]\right)$$

$$\leq K\left((1 + E\left[\left|X(t)\right|^2\right]\right) \text{ for } t < s_{k-1} \tag{E.22}$$

Combining the statements in Eqs. (E.20–E.22) yields the result in Eq. (E.14).

References

Milstein, G. N. 1995. *Numerical Integration of Stochastic Differential Equations.* (originally published by Ural State University Press, Sverdlovsk, 1988). Dordrecht: Kluwer Academic Publishers, Springer. Science + Business Media .

Roy, D. and M. K. Dash. 2005. 'Explorations of a Family of Stochastic Newmark Methods in Engineering Dynamics. *Comput. Methods Appl. Mech. Engrg.* 194: 4758–96.

Appendix F (Chapter 6)

1. If $Y(t)$ is a stochastic exponential, then $dY(t) = Y(t)dX(t)$

In conventional calculus, the ODE $dy(t) = y(t)\,dt$ with the initial condition $y(0) = 1$ has a unique solution $y(t) = e^t$ which is an exponential. Similarly, if $dX(t)$ is a stochastic differential, then the SDE $dY(t) = Y(t)dX(t)$ with $Y(0) = 1$ has an a.s. unique solution:

$$Y(t) = e^{Z(t)} \tag{F.1}$$

with

$$Z(t) = X(t) - X(0) - \frac{1}{2}[X,X](t). \tag{F.2}$$

Drawing similarity with conventional calculus, $Y(t)$ is called a stochastic exponential and denoted by $\varepsilon(Y)$.

Proof for existence of the solution:
From Eq. (F.1), we have $d\varepsilon(Y) = d\left(e^{Z(t)}\right)$. Applying Ito's formula to $d\left(e^{Z(t)}\right)$ gives:

$$d\left(e^{Z(t)}\right) = e^{Z(t)}\left\{dZ(t) + \frac{1}{2}d[Z,Z](t)\right\} \tag{F.3}$$

With $Z(t)$ given by Eq. (F.2), $d[Z,Z](t)$ is given by:

$$d[Z,Z](t) = d[X,X](t) - d[X,[X,X](t)] \tag{F.4}$$

Since $X(t)$ is of finite variation and is continuous, $d[X,[X,X](t) = 0$ (Chapter 4) and Eq. (F.3) becomes:

$$d\left(e^{Z(t)}\right) = e^{Z(t)}\left\{dZ(t) + \frac{1}{2}d[X,X](t)\right\}$$

$$= e^{Z(t)}\left\{\left(dX(t) - \frac{1}{2}d[X,X](t)\right) + \frac{1}{2}d[X,X](t)\right\} = e^{Z(t)}dX(t)$$

$$\Longrightarrow d\varepsilon(Y) = e^{Z(t)}dX(t) = Y(t)\,dX(t) \quad \text{(F.5)}$$

This proves the existence of solution.

Proof for uniqueness of the solution

The solution to the SDE $dY(t) = Y(t)dX(t)$ with $Y(0) = 1$ in integral form is:

$$Y(t) = 1 + \int_0^t Y(s)\,dX(s) \quad (F.6)$$

If the SDE admits another solution, let it be denoted by $W(t)$. In such a case:

$$W(t) = 1 + \int_0^t W(s)\,dX(s) \quad \text{(F.7)}$$

By applying Ito's formula to $W(t)/Y(t)$, we have:

$$d\left(\frac{W(t)}{Y(t)}\right) = d(W(t)\,Y^{-1}(t))$$

$$= Y^{-1}(t)\,dW(t) + W(t)\,d\left(Y^{-1}(t)\right) + d\left[W, Y^{-1}\right](t)$$

$$= Y^{-1}(t)\,W(t)\,dX(t) + W(t)\left\{-Y^{-2}(t)\,dY(t) + Y^{-3}d[Y,Y](t)\right\}$$

$$-Y^{-2}(t)\,d[W,Y](t) \quad \text{(F.8)}$$

Note that,

$$d\left[W, Y^{-1}\right](t) = dW(t)\,dY^{-1}(t) = -Y^{-2}(t)\,dW(t)\,dY(t) = -Y^{-2}(t)\,d[W,Y](t).$$

Now,

$$[Y,Y](t) = \left[1 + \int_0^t Y(s)\,dX(s), 1 + \int_0^t Y(s)\,dX(s)\right](t)$$

$$= \left[\int_0^t Y(s)\,dX(s), \int_0^t Y(s)\,dX(s)\right](t)$$

$$= \int_0^t Y^2(s)\, d[X,X](s)$$

$$\implies d[Y,Y](t) = Y^2(t)\, d[X,X](t) \tag{F.9}$$

Similarly we have:

$$d[W,Y](t) = W(t)Y(t)d[X,X](t) \tag{F.10}$$

Substituting the above two results in Eq. (F.8) and writing $dY(t) = Y(t)\, dX(t)$, we have:

$$d\left(\frac{W(t)}{Y(t)}\right) = Y^{-1}(t)\, W(t)\, dX(t) + W(t)\left\{-Y^{-1}(t)\, dX(t) + Y^{-1} d[X,X](t)\right\}$$

$$-Y^{-1}(t)\, W(t) d[X,X](t)$$

$$= 0$$

$$\implies \frac{W(t)}{Y(t)} = \frac{W(0)}{Y(0)} = 1 \tag{F.11}$$

This shows that $W(t)$ and $Y(t)$ are identical a.s.

Appendix G (Chapter 7)

1. A proof on convergence of the filtered estimate by EnKS filter (non-iterative form, Chapter 7, Section 7.5.2)

In order to show that the proposed algorithm converges to the filtered estimate, we state and prove the following theorem [Sarkar and Roy 2014].

Statement of the theorem:

Let $\phi \in C_b^2(\mathbb{R})$. Assume that there exist constants $M_1, M_2 > 0$ such that

$$\left|b(t,x) - b(t,y)\right| + \left|f(t,x) - f(t,y)\right| \le M_1 \left|x - y\right| \tag{G.1a}$$

$$\left|b(t,x)\right| + \left|f(t,x)\right| \le M_2 \left|1 + |x|\right| \tag{G.1b}$$

and

$$E_P\left[|X_0|^2\right] < \infty \tag{G.2}$$

Assume additionally that h is a bounded and Lipschitz continuous function. Furthermore, we assume that ϕ is sufficiently smooth so that $\phi(x)$ and its derivatives satisfy an inequality of the form

$$\left|\phi(x)\right| \le M_3\left(1 + |x|^a\right) \tag{G.3}$$

for positive constants M_3, a. Then there exist constants $D' > 0$ and $D'' > 0$, independent of Δt_i, such that (we use $\bar{\phi} := \pi^e(\phi)$, $\bar{h} := \pi^e(h)$ for notational convenience; π^e denotes the conditional expectation operator for an EM-discretized argument; π'^e denotes the empirical averaging operator of the conditional expectation):

$$\left(E_P\left[\left|\pi(\phi) - \pi'^e(\phi)\right|^2\right]\right)^{\frac{1}{2}} \le D'(\Delta t)^{\frac{1}{2}}$$

$$+\frac{D''}{\sqrt{N}}\Delta t\left(|Y|+\|h\|\right)$$

$$\left\{\begin{array}{l}\left(\begin{array}{l}\left\|L\left((\phi-\bar{\phi})\left(ht-\bar{h}t+\bar{h}_{i-1}\Delta t\right)^T\right)\right\| \\ +\|\phi\|\|h\|+\left\|L\left((\bar{\phi}t-\bar{\phi}_{i-1}t_{i-1})\left(h-\bar{h}\right)^T\right)\right\|\end{array}\right)\left\|\left\{\begin{array}{l}\alpha\left(h-\bar{h}\right)^T\left(h-\bar{h}\right) \\ +(1-\alpha)\sigma^T\sigma\end{array}\right\}^{-1}\right\| \\ +\|\phi\|\|h\|\alpha\left(\Delta t\left\|L\left(\left(h-\bar{h}\right)\left(h-\bar{h}\right)^T\right)\right\|+\left\|\left(\left(h-\bar{h}\right)\left(h-\bar{h}\right)^T\right)\right\|\right)\left\|\left\{\begin{array}{l}\alpha\left(h-\bar{h}\right)^T\left(h-\bar{h}\right) \\ +(1-\alpha)\sigma^T\sigma\end{array}\right\}^{-1}\right\|^2\end{array}\right\}$$

$$+\frac{D''}{\sqrt{N}}\left(\left\|\tilde{G}(\phi)\right\|\left(\Delta t\left\|L(h)\right\|+\|h\|\right)+\left(\Delta t\left\|L(\phi)\right\|+\|\phi\|\right)\right) \tag{G.4}$$

For both the sake of conciseness and to indicate that the above inequality holds for any $t \in (t_{i-1}, t_i]$, we have removed the subscript '*i*' from ϕ, h, t, $\bar{h}$, Δt, $\tilde{G}$ and σ in the statement of the theorem. Even in the proof of the theorem given below, the subscript '*i*' must be assumed to be present whenever the above variables/operators appear as un-subscripted.

Proof of the theorem:

In the least square sense, the error in the filtered estimate due to approximations in time and finiteness of the ensemble may be denoted by $\left(E_P\left[\left|\pi_i(\phi)-\pi_i'^e(\phi)\right|^2\right]\right)^{\frac{1}{2}}$, where $(\cdot)^e$ and $(\cdot)'$ denote EM and ensemble approximations respectively. As a precursor to getting an error bound, this term is split into two parts, one corresponding to EM integration error and the other due to ensemble approximation:

$$\left(E_P\left[\left|\pi_i(\phi)-\pi_i'^e(\phi)\right|^2\right]\right)^{\frac{1}{2}} \le \underbrace{\left(E_P\left[\left|\pi_i(\phi)-\pi_i{}^e(\phi)\right|^2\right]\right)^{\frac{1}{2}}}_{\text{(term-1: time discretization error)}}$$

$$+ \underbrace{\left(E_P\left[\left|\pi_i{}^e(\phi)-\pi_i'^e(\phi)\right|^2\right]\right)^{\frac{1}{2}}}_{\text{(term-2: ensemble approximation error)}} \tag{G.5}$$

We first obtain an error bound for the time discretization (term-1 on the RHS of the last equation) through Lemma 1 below and for the ensemble approximation error (term-2 on the RHS of the equation) through Lemma 2 stated at the end of proof of Lemma 1.

Lemma 1: According to this lemma, one needs to first prove:
If the process has bounded moments of any order and $\phi \in C_b^2(\mathbb{R})$, then:

$$\left(E_P\left[\left|\pi_i(\phi) - \pi_i^e(\phi)\right|^2\right]\right)^{\frac{1}{2}} \leq D'(\Delta t_i)^{\frac{1}{2}} \tag{G.6}$$

Proof of Lemma 1
Using the conditional version of Jensen's inequality, we have:

$$\begin{aligned} E_P\left[\left|\pi_i(\phi) - \pi_i{}^e(\phi)\right|^2\right] &= E_P\left[\left|E_P\left[\phi_i - \phi_i{}^e\right] | \mathcal{F}_i^Y\right|^2\right] \\ &\leq E_P\left[E_P\left[\left|\phi_i - \phi_i^e\right|^2 | F_i^Y\right]\right] \\ &= E_P\left[\left|\phi_i - \phi_i^e\right|^2\right] \end{aligned} \tag{G.7}$$

Using the standard strong order of convergence of the EM method [Kloeden and Platen 1992], we obtain:

$$\left(E_P\left[\left|X_i - X_i^e\right|^2\right]\right)^{\frac{1}{2}} \leq D_1(\Delta t_i)^{\frac{1}{2}} \tag{G.8}$$

where, $D_1 > 0$ is a constant independent of Δt_i. In general, for $p \geq 1, D_2 > 0$, one can write:

$$E_P\left[\left|X_i - X_i^e\right|^{2p}\right]^{\frac{1}{2p}} \leq D_2(\Delta t_i)^{\frac{1}{2}} \tag{G.9}$$

Furthermore, we assume that ϕ is sufficiently smooth so that $\phi(x)$ and its derivatives satisfy an inequality of the form:

$$\left|\phi(x)\right| \leq D_3(1 + |x|^a) \tag{G.10}$$

for some constants $D_3, a > 0$. Hence we can write:

$$\left|\phi_i - \phi_i^e\right| \leq D_4\left(1 + |X_i|^a + \left|X_i^e\right|^a\right)\left|X_i - X_i^e\right| \tag{G.11}$$

where $D_4 > 0$. Then, using the Cauchy–Schwarz inequality, we have

$$E_P\left[\left|\phi_i - \phi_i^e\right|^{2p}\right] \leq (D_4)^{2p} E_P\left[\left(1 + |X_i|^a + \left|X_i^e\right|^a\right)^{2p} \left|X_i - X_i^e\right|^{2p}\right]$$

$$\leq (D_4)^{2p} \sqrt{E_P\left[\left(1+|X_i|^a+\left|X_i^e\right|^a\right)^{4p}\right]}\sqrt{E_P\left[\left|X_i - X_i^e\right|^{4p}\right]} \tag{G.12}$$

$$\leq D'(X_i)\left((\Delta t_i)^{\frac{1}{2}}\right)^{2p}$$

Hence, for $p = 1$, we get

$$\left(E_P\left[\left|\pi_i(\phi) - \pi_i^e(\phi)\right|^2\right]\right)^{\frac{1}{2}} \leq D'(\Delta t_i)^{\frac{1}{2}} \tag{G.13}$$

where, $D'(X_i) > 0$ is independent of Δt_i.

The proof of lemma 1 *is complete.*

Next, we consider the error due to the ensemble approximation (term-2 on the RHS of Eq. G.5) within a time–discretized framework. In a recursive setting, given the empirical filtered distribution of X_t at $t = t_{i-1}$, we may consider:

$$E_P\left[\left|\pi_{i-1}^e(\phi) - {\pi'_{i-1}}^e(\phi)\right|^2\right] \leq D_5 \frac{\|\phi\|^2}{N} \tag{G.14}$$

where, $\|\cdot\|$ denotes the supremum norm on $C_b(\mathbb{R}^n)$.

Lemma 2:

Assume that for any $\phi \in C_b^2(\mathbb{R}^n)$,

$$E_P\left[\left|\pi_{i-1}^e(\phi) - {\pi'_{i-1}}^e(\phi)\right|^2\right] \leq D_5 \frac{\|\phi\|^2}{N} \tag{G.15}$$

Then, for $D_5 > 0$

$$\left(E_P\left[\left|\pi_i'^e(\phi) - {\pi_i}^e(\phi)\right|^2\right]\right)^{\frac{1}{2}}$$

$$\leq \frac{D_{13}}{\sqrt{N}} \Delta t_i \left(|Y_i| + \|h_i\|\right)$$

$$\left\{\begin{array}{l}\left(\begin{array}{l}\left\|L\left((\phi-\bar{\phi})\left(ht-\bar{h}t+\bar{h}_{i-1}\Delta t_i\right)^T\right)\right\| \\ +\|\phi\|\|h\|+\left\|L\left((\bar{\phi}t-\bar{\phi}_{i-1}t_{i-1})\left(h-\bar{h}\right)^T\right)\right\|\end{array}\right)\left\|\left\{\begin{array}{l}\alpha\left(h-\bar{h}\right)^T\left(h-\bar{h}\right) \\ +(1-\alpha)\,\sigma_i^T\sigma_i\end{array}\right\}^{-1}\right\| \\ +\|\phi\|\|h\|\alpha\left(\Delta t_i\left\|L\left(\left(h-\bar{h}\right)\left(h-\bar{h}\right)^T\right)\right\|+\left\|\left(\left(h-\bar{h}\right)\left(h-\bar{h}\right)^T\right)\right\|\right)\left\|\left\{\begin{array}{l}\alpha\left(h-\bar{h}\right)^T\left(h-\bar{h}\right) \\ +(1-\alpha)\,\sigma_i^T\sigma_i\end{array}\right\}^{-1}\right\|^2\end{array}\right\}$$

$$+\frac{D_{13}}{\sqrt{N}}\left(\left\|\tilde{G}_i(\phi)\right\|\left(\Delta t_i\left\|L(h)\right\|+\|h\|\right)+\left(\Delta t_i\left\|L(\phi)\right\|+\|\phi\|\right)\right) \quad \text{(G.16)}$$

Proof:

Using Minkowski's inequality, we can write,

$$\left(E_P\left[\left|\pi_i'^e(\phi)-\pi_i^e(\phi)\right|^2\right]\right)^{\frac{1}{2}} \leq \left(E_P\left[\left|\left\{\tilde{\pi}_i'^e(\phi)+\tilde{G}'_i(\phi)\tilde{I}'_i(\phi)\right\}\right.\right.\right.$$

$$\left.\left.\left.-\left\{\tilde{\pi}_i'^e(\phi)+\tilde{G}_i(\phi)\tilde{I}_i(\phi)\right\}\right|^2\right]\right)^{\frac{1}{2}}$$

$$+\left(E_P\left[\left|\left\{\tilde{\pi}_i'^e(\phi)+\tilde{G}_i(\phi)\tilde{I}_i(\phi)\right\}-\left\{\tilde{\pi}_i^e(\phi)+\tilde{G}_i(\phi)\tilde{I}_i(\phi)\right\}\right|^2\right]\right)^{\frac{1}{2}}$$

$$=\left(E_P\left[\left|\tilde{G}'_i(\phi)\tilde{I}'_i(\phi)-\tilde{G}_i(\phi)\tilde{I}_i(\phi)\right|^2\right]\right)^{\frac{1}{2}}+\left(E_P\left[\left|\tilde{\pi}_i'^e(\phi)-\tilde{\pi}_i^e(\phi)\right|^2\right]\right)^{\frac{1}{2}} \quad \text{(G.17)}$$

where, $\tilde{\pi}_i{}^e(\phi) := \pi_i{}^e(\tilde{\phi})$, $\tilde{I}_i'(\phi) := Y_i - \tilde{\pi}_i'^e(h)$ and

$$\tilde{G}_i' = \frac{1}{N}\left\{\left(\tilde{\Phi}_i-\widehat{\tilde{\Phi}}_i\right)\left(\tilde{H}_i^T t_i-\widehat{\tilde{H}}_{i-1}^T t_{i-1}-\Delta\,\widehat{\tilde{H}}_i^T t_i\right)+\left(\widehat{\tilde{\Phi}}_i t_i-\widehat{\tilde{\Phi}}_{i-1}t_{i-1}\right)\left(\tilde{H}_i^T-\widehat{\tilde{H}}_i^T\right)\right\}$$

$$\left\{\alpha\frac{1}{N-1}\left(\tilde{H}_i-\widehat{\tilde{H}}_i\right)\left(\tilde{H}_i^T-\widehat{\tilde{H}}_i^T\right)+(1-\alpha)\,\sigma_i^T\sigma_i\right\}^{-1}$$

$$=\left\{\tilde{\pi}_i'^e\left((\phi-\bar{\phi})\left(ht-\bar{h}t+\bar{h}_{i-1}\Delta t_i\right)^T\right)\right.$$

$$\left.+\tilde{\pi}_i'^e\left((\bar{\phi}t-\bar{\phi}_{i-1}t_{i-1})\left(h-\bar{h}\right)^T\right)\right\}\left\{\alpha\tilde{\pi}_i'^e\left(\left(h-\bar{h}\right)\left(h-\bar{h}\right)^T\right)+(1-\alpha)\,\sigma_i^T\sigma_i\right\}^{-1} \quad \text{(G.18)}$$

As before, we sometimes replace the conditional expectation of a variable with an over-bar for conciseness. One may write similar expressions for $\tilde{G}_i$ and $\tilde{I}_i$ by appropriately replacing the ensemble approximation operator in $\tilde{G}_i'$ and $\tilde{I}_i'$.

Using Minkowski's inequality on the second term of the RHS of Eq. (G.17), we have

$$\left(E_P\left[\left|\tilde{\pi}'^e_i(\phi)-\tilde{\pi}^e_i(\phi)\right|^2\right]\right)^{\frac{1}{2}} \le \left(E_P\left[\left|\tilde{\pi}'^e_i(\phi)-\left(\pi'^e_{i-1}(\phi)+\pi'^e_{i-1}(L(\phi))\Delta t_i\right)\right|^2\right]\right)^{\frac{1}{2}}$$

$$+\left(E_P\left[\left|(\pi'^e_{i-1}(\phi)+\pi^e_{i-1}(L(\phi))\Delta t_i)-\tilde{\pi}^e_i(\phi)\right|^2\right]\right)^{\frac{1}{2}} \tag{G.19}$$

$$\le \Delta t_i\left(E_P\left[\left|\pi'^e_{i-1}(L(\phi))-\pi^e_{i-1}(L(\phi))\right|^2\right]\right)^{\frac{1}{2}}+\left(E_P\left[\left|\pi'^e_{i-1}(\phi)-\pi^e_{i-1}(\phi)\right|^2\right]\right)^{\frac{1}{2}}$$

$$\le D_6\left(\Delta t_i\frac{\|L(\phi)\|}{\sqrt{N}}+\frac{\|\phi\|}{\sqrt{N}}\right)$$

Towards getting a bound for the first term on the RHS of Eq. (G.17), the term is split as:

$$\left(E_P\left[\left|\tilde{G}'_i(\phi)\tilde{I}'_i(\phi)-\tilde{G}_i(\phi)\tilde{I}_i(\phi)\right|^2\right]\right)^{\frac{1}{2}} \le \left(E_P\left[\left|\tilde{G}'_i(\phi)\tilde{I}'_i(\phi)-\tilde{G}_i(\phi)\tilde{I}'_i(\phi)\right|^2\right]\right)^{\frac{1}{2}}$$

$$+\left(E_P\left[\left|\tilde{G}_i(\phi)\tilde{I}'_i(\phi)-\tilde{G}_i(\phi)\tilde{I}_i(\phi)\right|^2\right]\right)^{\frac{1}{2}} \tag{G.20}$$

$$\le (|Y_i|+\|h_i\|)\left(E_P\left[\left|\tilde{G}'_i(\phi)-\tilde{G}_i(\phi)\right|^2\right]\right)^{\frac{1}{2}}$$

$$+\left\|\tilde{G}_i(\phi)\right\|\left(E_P\left[\left|\tilde{\pi}^e_i(h)-\tilde{\pi}'^e_i(h)\right|^2\right]\right)^{\frac{1}{2}}$$

Considering the last factor of the first term on the RHS of Eq. (G.20), we have:

$$\left(E_P\left[\left|\tilde{G}'_i(\phi)-\tilde{G}_i(\phi)\right|^2\right]\right)^{\frac{1}{2}}$$

$$\leq \left(E_P \left[\left| \begin{array}{l} \left\{ \begin{array}{l} \tilde{\pi}'^e_i \left((\phi - \bar{\phi}) \left(ht - \bar{h}t + \bar{h}_{i-1} \Delta t_i \right)^T \right) \\ + \tilde{\pi}'^e_i \left((\bar{\phi}t - \bar{\phi}_{i-1} t_{i-1}) \left(h - \bar{h} \right)^T \right) \end{array} \right\} \\ - \left\{ \begin{array}{l} \tilde{\pi}^e_i \left((\phi - \bar{\phi}) \left(ht - \bar{h}t + \bar{h}_{i-1} \Delta t_i \right)^T \right) \\ + \tilde{\pi}^e_i \left((\bar{\phi}t - \bar{\phi}_{i-1} t_{i-1}) \left(h - \bar{h} \right)^T \right) \end{array} \right\} \end{array} \right|^2 \right] \right)^{\frac{1}{2}}$$

$$\left(E_P \left[\left| \left\{ \alpha \tilde{\pi}'^e_i \left(\left(h - \bar{h} \right) \left(h - \bar{h} \right)^T \right) + (1 - \alpha) \sigma_i^T \sigma_i \right\}^{-1} \right|^2 \right] \right)^{\frac{1}{2}}$$

$$+ \left(E_P \left[\left| \left\{ \tilde{\pi}^e_i \left((\phi - \bar{\phi}) \left(ht - \bar{h}t + \bar{h}_{i-1} \Delta t_i \right)^T \right) + \tilde{\pi}^e_i \left((\bar{\phi}t - \bar{\phi}_{i-1} t_{i-1}) \left(h - \bar{h} \right)^T \right) \right\} \right|^2 \right] \right)^{\frac{1}{2}}$$

$$\left(E_P \left[\left| \begin{array}{l} \left\{ \alpha \tilde{\pi}'^e_i \left(\left(h - \bar{h} \right) \left(h - \bar{h} \right)^T \right) + (1 - \alpha) \sigma_i^T \sigma_i \right\}^{-1} \\ - \left\{ \alpha \tilde{\pi}^e_i \left(\left(h - \bar{h} \right) \left(h - \bar{h} \right)^T \right) + (1 - \alpha) \sigma_i^T \sigma_i \right\}^{-1} \end{array} \right|^2 \right] \right)^{\frac{1}{2}} \quad \text{(G.21)}$$

Considering the first term on the RHS of the inequality (Eq. G.21), we get:

$$\left(E_P \left[\left| \begin{array}{l} \left\{ \tilde{\pi}'^e_i \left((\phi - \bar{\phi}) \left(ht - \bar{h}t + \bar{h}_{i-1} \Delta t_i \right)^T \right) + \tilde{\pi}'^e_i \left((\bar{\phi}t - \bar{\phi}_{i-1} t_{i-1}) \left(h - \bar{h} \right)^T \right) \right\} \\ - \left\{ \tilde{\pi}^e_i \left((\phi - \bar{\phi}) \left(ht - \bar{h}t + \bar{h}_{i-1} \Delta t_i \right)^T \right) + \tilde{\pi}^e_i \left((\bar{\phi}t - \bar{\phi}_{i-1} t_{i-1}) \left(h - \bar{h} \right)^T \right) \right\} \end{array} \right|^2 \right] \right)^{\frac{1}{2}}$$

$$\leq \left(E_P \left[\left| \tilde{\pi}'^e_i \left((\phi - \bar{\phi}) \left(ht - \bar{h}t + \bar{h}_{i-1} \Delta t_i \right)^T \right) - \tilde{\pi}^e_i \left((\phi - \bar{\phi}) \left(ht - \bar{h}t + \bar{h}_{i-1} \Delta t_i \right)^T \right) \right|^2 \right] \right)^{\frac{1}{2}}$$

$$+ \left(E_P \left[\left| \tilde{\pi}'^e_i \left((\bar{\phi}t - \bar{\phi}_{i-1} t_{i-1}) \left(h - \bar{h} \right)^T \right) - \tilde{\pi}^e_i \left((\bar{\phi}t - \bar{\phi}_{i-1} t_{i-1}) \left(h - \bar{h} \right)^T \right) \right|^2 \right] \right)^{\frac{1}{2}} \quad \text{(G.22)}$$

From Eq. (G.19) we write,

$$\left(E_P\left[\left\|\begin{array}{l}\left\{\tilde{\pi}'^e_i\left((\phi-\bar{\phi})\left(ht-\bar{h}t+\bar{h}_{i-1}\Delta t_i\right)^T\right)+\tilde{\pi}'^e_i\left((\bar{\phi}t-\bar{\phi}_{i-1}t_{i-1})\left(h-\bar{h}\right)^T\right)\right\}\\-\left\{\tilde{\pi}^e_i\left((\phi-\bar{\phi})\left(ht-\bar{h}t+\bar{h}_{i-1}\Delta t_i\right)^T\right)+\tilde{\pi}^e_i\left((\bar{\phi}t-\bar{\phi}_{i-1}t_{i-1})\left(h-\bar{h}\right)^T\right)\right\}\end{array}\right\|^2\right]\right)^{\frac{1}{2}}$$

$$\leq D_7\left(\begin{array}{l}\Delta t_i\dfrac{\left\|L\left((\phi-\bar{\phi})\left(ht-\bar{h}t+\bar{h}_{i-1}\Delta t_i\right)^T\right)\right\|}{\sqrt{N}}+\dfrac{\left\|\left((\phi-\bar{\phi})\left(ht-\bar{h}t+\bar{h}_{i-1}\Delta t_i\right)^T\right)\right\|}{\sqrt{N}}\\+\Delta t_i\dfrac{\left\|L\left((\bar{\phi}t-\bar{\phi}_{i-1}t_{i-1})\left(h-\bar{h}\right)^T\right)\right\|}{\sqrt{N}}+\dfrac{\left\|\left((\bar{\phi}t-\bar{\phi}_{i-1}t_{i-1})\left(h-\bar{h}\right)^T\right)\right\|}{\sqrt{N}}\end{array}\right) \tag{G.23}$$

where, $D_7 > 0$ is a constant. The last term of inequality Eq. (G.21) may be written as:

$$\left(E_P\left[\left\|\begin{array}{l}\left\{\alpha\tilde{\pi}'^e_i\left(\left(h-\bar{h}\right)\left(h-\bar{h}\right)^T\right)+(1-\alpha)\,\sigma_i^T\sigma_i\right\}^{-1}\\-\left\{\alpha\tilde{\pi}^e_i\left(\left(h-\bar{h}\right)\left(h-\bar{h}\right)^T\right)+(1-\alpha)\,\sigma_i^T\sigma_i\right\}^{-1}\end{array}\right\|^2\right]\right)^{\frac{1}{2}}$$

$$\leq\alpha\left(E_P\left[\left\|\begin{array}{l}\tilde{\pi}^e_i\left(\left(h-\bar{h}\right)\left(h-\bar{h}\right)^T\right)\\-\tilde{\pi}'^e_i\left(\left(h-\bar{h}\right)\left(h-\bar{h}\right)^T\right)\end{array}\right\|^2\right]\right)^{\frac{1}{2}}$$

$$\left(E_P\left[\left\|\begin{array}{l}\left\{\alpha\tilde{\pi}'^e_i\left(\left(h-\bar{h}\right)\left(h-\bar{h}\right)^T\right)+(1-\alpha)\,\sigma_i^T\sigma_i\right\}^{-1}\\\left\{\alpha\tilde{\pi}^e_i\left(\left(h-\bar{h}\right)\left(h-\bar{h}\right)^T\right)+(1-\alpha)\,\sigma_i^T\sigma_i\right\}^{-1}\end{array}\right\|^2\right]\right)^{\frac{1}{2}}$$

$$\leq D_8\alpha\left(\Delta t_i\frac{\left\|L\left(\left(h-\bar{h}\right)\left(h-\bar{h}\right)^T\right)\right\|}{\sqrt{N}}+\frac{\left\|\left(\left(h-\bar{h}\right)\left(h-\bar{h}\right)^T\right)\right\|}{\sqrt{N}}\right)$$

$$\left\|\left\{\alpha\left(h-\bar{h}\right)^T\left(h-\bar{h}\right)+(1-\alpha)\,\sigma_i^T\sigma_i\right\}^{-1}\right\|^2 \tag{G.24}$$

$D_8 > 0$ is a constant. Using Eqs. (G.21), (G.23) and (G.24), we may arrive at the following inequality:

$$\left(E_P\left[\left|\tilde{G}'_i(\phi)-\tilde{G}_i(\phi)\right|^2\right]\right)^{\frac{1}{2}}$$

$$\leq D_9\begin{pmatrix} \Delta t_i\frac{\left\|L\left((\phi-\bar{\phi})\left(ht-\bar{h}t+\bar{h}_{i-1}\Delta t_i\right)^T\right)\right\|}{\sqrt{N}} \\ +\frac{\left\|\left((\phi-\bar{\phi})\left(ht-\bar{h}t+\bar{h}_{i-1}\Delta t_i\right)^T\right)\right\|}{\sqrt{N}} \\ +\Delta t_i\frac{\left\|L\left((\bar{\phi}t-\bar{\phi}_{i-1}t_{i-1})\left(h-\bar{h}\right)^T\right)\right\|}{\sqrt{N}} \\ +\frac{\left\|\left((\bar{\phi}t-\bar{\phi}_{i-1}t_{i-1})\left(h-\bar{h}\right)^T\right)\right\|}{\sqrt{N}}\end{pmatrix}\left(E_P\left[\left\|\begin{Bmatrix}\alpha\tilde{\pi}'^e_i\left(\left(h-\bar{h}\right)\left(h-\bar{h}\right)^T\right) \\ +(1-\alpha)\sigma_i^T\sigma_i\end{Bmatrix}^{-1}\right\|^2\right]\right)^{\frac{1}{2}}$$

$$+D_9\left(E_P\left[\left\|\begin{Bmatrix}\tilde{\pi}^e_i\left((\phi-\bar{\phi})\left(ht-\bar{h}t+\bar{h}_{i-1}\Delta t_i\right)^T\right) \\ +\tilde{\pi}^e_i\left((\bar{\phi}t-\bar{\phi}_{i-1}t_{i-1})\left(h-\bar{h}\right)^T\right)\end{Bmatrix}\right\|^2\right]\right)^{\frac{1}{2}}$$

$$\alpha\begin{pmatrix}\Delta t_i\frac{\left\|L\left(\left(h-\bar{h}\right)\left(h-\bar{h}\right)^T\right)\right\|}{\sqrt{N}} \\ +\frac{\left\|\left(\left(h-\bar{h}\right)\left(h-\bar{h}\right)^T\right)\right\|}{\sqrt{N}}\end{pmatrix}\left\|\begin{Bmatrix}\alpha\left(h-\bar{h}\right)^T\left(h-\bar{h}\right) \\ +(1-\alpha)\sigma_i^T\sigma_i\end{Bmatrix}^{-1}\right\|^2 \tag{G.25}$$

$D_9 > 0$ is a constant. This implies:

$$\left(E_P\left[\left|\tilde{G}'_i(\phi)-\tilde{G}_i(\phi)\right|^2\right]\right)^{\frac{1}{2}}$$

$$\leq\frac{D_9}{\sqrt{N}}\begin{pmatrix}\Delta t_i\left\|L\left((\phi-\bar{\phi})\left(ht-\bar{h}t+\bar{h}_{i-1}\Delta t_i\right)^T\right)\right\| \\ +\left\|\left((\phi-\bar{\phi})\left(ht-\bar{h}t+\bar{h}_{i-1}\Delta t_i\right)^T\right)\right\| \\ +\Delta t_i\left\|L\left((\bar{\phi}t-\bar{\phi}_{i-1}t_{i-1})\left(h-\bar{h}\right)^T\right)\right\| \\ +\left\|\left((\bar{\phi}t-\bar{\phi}_{i-1}t_{i-1})\left(h-\bar{h}\right)^T\right)\right\|\end{pmatrix}\left\|\begin{Bmatrix}\alpha\left(h-\bar{h}\right)^T\left(h-\bar{h}\right) \\ +(1-\alpha)\sigma_i^T\sigma_i\end{Bmatrix}^{-1}\right\|$$

$$+\frac{D_9}{\sqrt{N}}\left\{\begin{array}{l}\left\|(\phi-\bar{\phi})\right\|\left\|ht-\bar{h}t+\bar{h}_{i-1}\Delta t_i\right\| \\ +\left\|(\bar{\phi}t-\bar{\phi}_{i-1}t_{i-1})\right\|\left\|h-\bar{h}\right\|\end{array}\right\}$$

$$\alpha\left(\begin{array}{l}\Delta t_i\left\|L\left(\left(h-\bar{h}\right)\left(h-\bar{h}\right)^T\right)\right\| \\ +\left\|\left(\left(h-\bar{h}\right)\left(h-\bar{h}\right)^T\right)\right\|\end{array}\right)\left\|\left\{\begin{array}{l}\alpha\left(h-\bar{h}\right)^T\left(h-\bar{h}\right) \\ +(1-\alpha)\,\sigma_i^T\sigma_i\end{array}\right\}^{-1}\right\|^2 \tag{G.26}$$

From Eqs. (G.20) and (G.26), we get:

$$\left(E_P\left[\left|\tilde{G}'_i(\phi)\,\tilde{I}'_i(\phi)-\tilde{G}_i(\phi)\,\tilde{I}_i(\phi)\right|^2\right]\right)^{\frac{1}{2}}$$

$$\leq\frac{D_{10}}{\sqrt{N}}\left(|Y_i|+\|h_i\|\right)$$

$$\left\{\begin{array}{l}\left(\begin{array}{l}\Delta t_i\left\|L\left((\phi-\bar{\phi})\left(ht-\bar{h}t+\bar{h}_{i-1}\Delta t_i\right)^T\right)\right\| \\ +\left\|\left((\phi-\bar{\phi})\left(ht-\bar{h}t+\bar{h}_{i-1}\Delta t_i\right)^T\right)\right\| \\ +\Delta t_i\left\|L\left((\bar{\phi}t-\bar{\phi}_{i-1}t_{i-1})\left(h-\bar{h}\right)^T\right)\right\| \\ +\left\|\left((\bar{\phi}t-\bar{\phi}_{i-1}t_{i-1})\left(h-\bar{h}\right)^T\right)\right\|\end{array}\right)\left\|\left\{\begin{array}{l}\alpha\left(h-\bar{h}\right)^T\left(h-\bar{h}\right) \\ +(1-\alpha)\,\sigma_i^T\sigma_i\end{array}\right\}^{-1}\right\| \\ +\left\{\begin{array}{l}\left\|(\phi-\bar{\phi})\right\|\left\|ht-\bar{h}t+\bar{h}_{i-1}\Delta t_i\right\| \\ +\left\|(\bar{\phi}t-\bar{\phi}_{i-1}t_{i-1})\right\|\left\|h-\bar{h}\right\|\end{array}\right\} \\ \alpha\left(\begin{array}{l}\Delta t_i\left\|L\left(\left(h-\bar{h}\right)\left(h-\bar{h}\right)^T\right)\right\| \\ +\left\|\left(\left(h-\bar{h}\right)\left(h-\bar{h}\right)^T\right)\right\|\end{array}\right)\left\|\left\{\begin{array}{l}\alpha\left(h-\bar{h}\right)^T\left(h-\bar{h}\right) \\ +(1-\alpha)\,\sigma_i^T\sigma_i\end{array}\right\}^{-1}\right\|^2\end{array}\right\}$$

$$+\frac{D_{10}}{\sqrt{N}}\left\|\tilde{G}_i(\phi)\right\|\left(\Delta t_i\left\|L(h)\right\|+\|h\|\right) \tag{G.27}$$

Finally, from Eq. (G.17), we get an error bound due to the ensemble approximation of the filtered estimate at time t_i:

$$\left(E_P\left[\left|\pi'^e_i(\phi)-\pi^e_i(\phi)\right|^2\right]\right)^{\frac{1}{2}}$$

$$\le \frac{D_{11}}{\sqrt{N}}\left(|Y_i| + \|h_i\|\right)$$

$$\left\{\begin{array}{l}\left(\begin{array}{l}\Delta t_i\left\|L\left((\phi-\bar{\phi})\left(ht-\bar{h}t+\bar{h}_{i-1}\Delta t_i\right)^T\right)\right\| \\ +\left\|\left((\phi-\bar{\phi})\left(ht-\bar{h}t+\bar{h}_{i-1}\Delta t_i\right)^T\right)\right\| \\ +\Delta t_i\left\|L\left((\bar{\phi}t-\bar{\phi}_{i-1}t_{i-1})\left(h-\bar{h}\right)^T\right)\right\| \\ +\left\|\left((\bar{\phi}t-\bar{\phi}_{i-1}t_{i-1})\left(h-\bar{h}\right)^T\right)\right\|\end{array}\right)\left\|\left\{\begin{array}{l}\alpha\left(h-\bar{h}\right)^T\left(h-\bar{h}\right) \\ +(1-\alpha)\sigma_i^T\sigma_i\end{array}\right\}^{-1}\right\| \\ +\left\{\begin{array}{l}\left\|(\phi-\bar{\phi})\right\|\left\|ht-\bar{h}t+\bar{h}_{i-1}\Delta t_i\right\| \\ +\left\|(\bar{\phi}t-\bar{\phi}_{i-1}t_{i-1})\right\|\left\|h-\bar{h}\right\|\end{array}\right\} \\ \alpha\left(\begin{array}{l}\Delta t_i\left\|L\left(\left(h-\bar{h}\right)\left(h-\bar{h}\right)^T\right)\right\| \\ +\left\|\left(\left(h-\bar{h}\right)\left(h-\bar{h}\right)^T\right)\right\|\end{array}\right)\left\|\left\{\begin{array}{l}\alpha\left(h-\bar{h}\right)^T\left(h-\bar{h}\right) \\ +(1-\alpha)\sigma_i^T\sigma_i\end{array}\right\}^{-1}\right\|^2\end{array}\right\}$$

$$+\frac{D_{11}}{\sqrt{N}}\left(\left\|\tilde{G}_i(\phi)\right\|\left(\Delta t_i\left\|L(h)\right\|+\|h\|\right)+\left(\Delta t_i\left\|L(\phi)\right\|+\left\|\phi\right\|\right)\right) \tag{G.28}$$

Here, $D_{12} > 0$ is a constant. Upon simplification of Eq. (G.28), we get:

$$\left(E_P\left[\left|\pi'^e_i(\phi)-\pi^e_i(\phi)\right|^2\right]\right)^{\frac{1}{2}}$$

$$\le \frac{D_{12}}{\sqrt{N}}\left(|Y_i| + \|h_i\|\right)$$

$$\left\{\begin{array}{l}\left(\begin{array}{l}\Delta t_i\left\|L\left((\phi-\bar{\phi})\left(ht-\bar{h}t+\bar{h}_{i-1}\Delta t_i\right)^T\right)\right\| \\ +\left\|\phi\right\|\|h\|\Delta t_i \\ +\Delta t_i\left\|L\left((\bar{\phi}t-\bar{\phi}_{i-1}t_{i-1})\left(h-\bar{h}\right)^T\right)\right\| \\ +\left\|\phi\right\|\|h\|\Delta t_i\end{array}\right)\left\|\left\{\begin{array}{l}\alpha\left(h-\bar{h}\right)^T\left(h-\bar{h}\right) \\ +(1-\alpha)\sigma_i^T\sigma_i\end{array}\right\}^{-1}\right\| \\ +\left\|\phi\right\|\|h\|\Delta t_i\alpha\left(\begin{array}{l}\Delta t_i\left\|L\left(\left(h-\bar{h}\right)\left(h-\bar{h}\right)^T\right)\right\| \\ +\left\|\left(\left(h-\bar{h}\right)\left(h-\bar{h}\right)^T\right)\right\|\end{array}\right)\left\|\left\{\begin{array}{l}\alpha\left(h-\bar{h}\right)^T\left(h-\bar{h}\right) \\ +(1-\alpha)\sigma_i^T\sigma_i\end{array}\right\}^{-1}\right\|^2\end{array}\right\}$$

$$+\frac{D_{11}}{\sqrt{N}}\left(\left\|\tilde{G}_i(\phi)\right\|\left(\Delta t_i\left\|L(h)\right\|+\|h\|\right)+\left(\Delta t_i\left\|L(\phi)\right\|+\left\|\phi\right\|\right)\right) \quad \text{(G.29)}$$

where, $D_{12} > 0$ is a constant. Simplification of the inequality in Eq. (G.29) yields:

$$\left(E_P\left[\left|\pi'^e_i(\phi)-\pi^e_i(\phi)\right|^2\right]\right)^{\frac{1}{2}}$$

$$\leq \frac{D''}{\sqrt{N}}\Delta t_i\left(|Y_i|+\|h_i\|\right)$$

$$\left\{\begin{array}{l}\left(\begin{array}{l}\left\|L\left((\phi-\bar{\phi})\left(ht-\bar{h}t+\bar{h}_{i-1}\Delta t_i\right)^T\right)\right\| \\ +\|\phi\|\|h\|+\left\|L\left((\bar{\phi}t-\bar{\phi}_{i-1}t_{i-1})\left(h-\bar{h}\right)^T\right)\right\|\end{array}\right)\left\|\left\{\begin{array}{c}\alpha\left(h-\bar{h}\right)^T\left(h-\bar{h}\right) \\ +(1-\alpha)\sigma_i^T\sigma_i\end{array}\right\}^{-1}\right\| \\ +\|\phi\|\|h\|\alpha\left(\Delta t_i\left\|L\left(\left(h-\bar{h}\right)\left(h-\bar{h}\right)^T\right)\right\|+\left\|\left(\left(h-\bar{h}\right)\left(h-\bar{h}\right)^T\right)\right\|\right)\left\|\left\{\begin{array}{c}\alpha\left(h-\bar{h}\right)^T\left(h-\bar{h}\right) \\ +(1-\alpha)\sigma_i^T\sigma_i\end{array}\right\}^{-1}\right\|^2\end{array}\right\}$$

$$+\frac{D''}{\sqrt{N}}\left(\left\|\tilde{G}_i(\phi)\right\|\left(\Delta t_i\left\|L(h)\right\|+\|h\|\right)+\left(\Delta t_i\left\|L(\phi)\right\|+\left\|\phi\right\|\right)\right) \quad \text{(G.30)}$$

$D'' > 0$ is a constant. Combining the proofs of Lemmas 1 and 2 above, we arrive at the error bound for the filtered estimate proving the statement of the theorem in Eq. (G.4).

2. Theorems 1 and 2 (stated below) and their proofs for existence and uniqueness of solution (i.e. the conditional posterior distribution) via the proposed iterative version of EnKS filter (Chapter 7, Section 7.5.3)

Theorem 1 *for existence of a solution*:
Suppose that,

a) $\phi_{i-1} \in L^2(P)$

b) $\phi_t \in C_b^2$

c) $\{\beta_0, \beta_1, \ldots\}$ is a Cauchy sequence and converges to 1.

d) $\mathrm{L}_t(\phi)$ and σ_t are Lipschitz continuous uniformly on $(t_{i-1}, t_i]$.

Then there exists a limiting solution ϕ_t for Eq. (7.49) as $k \to \infty$.

Proof:

The prediction equation is taken as:

$$\tilde{\phi}_t = \phi_{i-1} + \int_{t_{i-1}}^{t} \mathrm{L}(\phi_s)ds + \int_{t_{i-1}}^{t} \phi'_s f_s dB_s \quad \text{(G.31)}$$

$$\Rightarrow \left\|\tilde{\phi}_i\right\|_{2,(\nu_{t_i}\times P)} \leq \left\|\phi_{i-1}\right\|_{2,(\nu_{t_i}\times P)} + \left\|\int_{t_{i-1}}^{t_i} \mathrm{L}(\phi_s)ds\right\|_{2,(\nu_{t_i}\times P)} + \left\|\int_{t_{i-1}}^{t_i} \phi_s' f_s dB_s\right\|_{2,(\nu_{t_i}\times P)} \quad \text{(G.32)}$$

where, ν_t is a Lebesgue measure in t and $\|\cdot\|_{2,(\nu_{t_i}\times P)}$ is a standard matrix norm, e.g. the 2-norm [van Handel 2007]. The first term on the RHS of Eq. (G.32) satisfies $\left\|\phi_{i-1}\right\|_{2,(\nu_{t_i}\times P)} = \sqrt{\Delta t}\left\|\phi_{i-1}\right\|_{2,P} < \infty$. For the second term, using Jensen's inequality $\left(\left(t\int_0^t a_s ds\right)^2 \leq t^{-1}\int_0^t a_s^2 ds\right)$ and linear growth property of $\mathrm{L}(\phi_t)$, we may write:

$$\left\|\int_{t_{i-1}}^{t_i} \mathrm{L}(\phi_s)ds\right\|^2_{2,(\nu_{t_i}\times P)} \leq (\Delta t_i)^2\left\|\mathrm{L}(\phi)\right\|^2_{2,(\nu_{t_i}\times P)} \leq (\Delta t)^2\left(1+\left\|\phi\right\|_{2,(\nu_{t_i}\times P)}\right)^2 < \infty \quad \text{(G.33)}$$

Similarly, for the third term on the RHS of Eq. (G.32), using Ito's isometry and linear growth property of $\phi_t' f_t$, we may write:

$$\left\|\int_{t_{i-1}}^{t_i} \phi_s' f_s dB_s\right\|^2_{2,(\nu_{t_i}\times P)} \leq (\Delta t_i)\left\|\phi' f\right\|^2_{2,(\nu_{t_i}\times P)} \leq (\Delta t_i)\left(1+\left\|\phi\right\|_{2,(\nu_{t_i}\times P)}\right)^2 < \infty \quad \text{(G.34)}$$

Hence we conclude that $\tilde{\phi}_i \in L^2(\nu_{t_i}\times P)$. This paves the way for proving the existence of a limiting solution for the inner iteration at a given time, t_i. The iterated update equation corresponding to the conditioned process ϕ_t may be written as:

$$\phi_{t,k} = \phi_{t,k-1} + \beta_{k-1}G_{t,k-1}\left\{Y_t - h_{t,k-1}\right\},\ k = 1,2,... \quad \text{(G.35)}$$

For further work, the update function is denoted as:

$$\aleph_{t,k-1}(\phi) := \phi_{t,k-1} + \beta_{k-1}G_{t,k-1}\left\{Y_t - h_{t,k-1}\right\} \quad \text{(G.36)}$$

From the above equation, we may write:

$$\left\|\aleph_{t,k-1}(\phi)\right\|_{2,(\nu_{t_i}\times P)} \leq \left\|\phi_{t,k-1}\right\|_{2,(\nu_{t_i}\times P)} + \left\|\beta_{k-1}G_{t,k-1}\left\{Y_t - h_{t,k-1}\right\}\right\|_{2,(\nu_{t_i}\times P)} \quad \text{(G.37)}$$

Defining

$$\aleph_t(\phi) := \phi_t + G_t\{Y_t - h_t\} \quad \text{(G.38)}$$

we have:

$$\left\|\aleph_t(\phi)\right\|_{2,(\nu_{t_i}\times P)} \leq \left\|\phi_t\right\|_{2,(\nu_{t_i}\times P)} + \|G_t\|_{2,(\nu_{t_i}\times P)}\,\|\{Y_t - h_t\}\|_{2,(\nu_{t_i}\times P)} \quad \text{(G.39)}$$

Now, recalling that $\phi_t \in C_b^2$ and h satisfy linear growth property, we may conclude from Eq. (G.39) that $\left\|\aleph_t(\phi)\right\|_{2,(\nu_{t_i}\times P)} \leq \infty$ which implies that $\aleph_t$ indeed maps to an F_t^Y adapted process in $L^2(\nu_{t_i} \times P)$.

Next, we show that $\aleph_t$ is a continuous map, i.e., as $\left\|(\phi_{t,k-1} - \phi_t)\right\|_{2,(\nu_{t_i}\times P)} \to 0$ and $\beta_{k-1} \to 1$,

$$\left\|\aleph_{t,k-1}(\phi) - \aleph_t(\phi)\right\|_{2,(\nu_{t_i}\times P)} \to 0$$

Taking norm of $\aleph_{t,k-1}(\phi) - \aleph_t(\phi)$ we get:

$$\left\|\aleph_{t,k-1}(\phi) - \aleph_t(\phi)\right\|_{2,(\nu_{t_i}\times P)} \leq \left\|\phi_{t,k-1} - \phi\right\|_{2,(\nu_{t_i}\times P)}$$

$$+\left\|\beta_{k-1} G_{t,k-1}\left\{Y_t - h_{t,k-1}\right\} - G_t\{Y_t - h_t\}\right\|_{2,(\nu_{t_i}\times P)} \tag{G.40}$$

For expositional ease, we replace $\|\cdot\|_{2,(\nu_{t_i}\times P)}$ by $\|\cdot\|$. The second term on the RHS of Eq. (G.40) may be split as:

$$\left\|\beta_{k-1} G_{t,k-1}\left\{Y_t - h_{t,k-1}\right\} - G_t\{Y_t - h_t\}\right\| \leq \left|\beta_{k-1} - 1\right| \left\|G_{t,k-1}\left\{Y_t - h_{t,k-1}\right\}\right\|$$

$$+\left\|G_{t,k-1}\left\{Y_t - h_{t,k-1}\right\} - G_t\{Y_t - h_t\}\right\|$$

$$\leq \left|\beta_{k-1} - 1\right| \left\|G_{t,k-1}\left\{Y_t - h_{t,k-1}\right\}\right\|$$

$$+\left\|G_{t,k-1} - G_t\right\| \left\|\left\{Y_t - h_{t,k-1}\right\}\right\|$$

$$+\|G_t\| \left\|h_t - h_{t,k-1}\right\| \tag{G.41}$$

The factor $\left\|G_{t,k-1} - G_t\right\|$ in the second term on the RHS of Eq. (G.41) may be split as:

$$\left\|G_{t,k-1} - G_t\right\| \leq$$

$$\left\| \begin{array}{l} \frac{1}{N}\left\{ \begin{array}{l} \left(\Phi_{t,k-1}-\widehat{\Phi}_{t,k-1}\right)\left(H_{t,k-1}^{T}t-\widehat{H}_{i-1}^{T}\,t_{i-1}-\Delta\,\widehat{H}_{t,k-1}^{T}\,t\right) \\ +\left(\widehat{\Phi}_{t,k-1}\,t-\widehat{\Phi}_{i-1}\,t_{i-1}\right)\left(H_{t,k-1}^{T}-\widehat{H}_{t,k-1}^{T}\right) \end{array} \right\} \\ \quad \left\{\alpha\frac{1}{N-1}\left(H_{t,k-1}-\widehat{H}_{t,k-1}\right)\left(H_{t,k-1}^{T}-\widehat{H}_{t,k-1}^{T}\right)+(1-\alpha)\,\sigma_t^T\sigma_t\right\}^{-1} \\ -\frac{1}{N}\left\{\left(\Phi_t-\widehat{\Phi}_t\right)\left(H_t^Tt-\widehat{H}_{i-1}^{T}\,t_{i-1}-\Delta\,\widehat{H}_t^{T}\,t\right)+\left(\widehat{\Phi}_t\,t-\widehat{\Phi}_{i-1}\,t_{i-1}\right)\left(H_t^T-\widehat{H}_t^{T}\right)\right\} \\ \quad \left\{\alpha\frac{1}{N-1}\left(H_t-\widehat{H}_t\right)\left(H_t^T-\widehat{H}_t^{T}\right)+(1-\alpha)\,\sigma_t^T\sigma_t\right\}^{-1} \end{array} \right\|$$

$$\leq \left\| \begin{array}{l} \frac{1}{N}\left\{ \begin{array}{l} \left(\Phi_{t,k-1}-\widehat{\Phi}_{t,k-1}\right)\left(H_{t,k-1}^{T}t-\widehat{H}_{i-1}^{T}\,t_{i-1}-\Delta\,\widehat{H}_{t,k-1}^{T}\,t\right) \\ +\left(\widehat{\Phi}_{t,k-1}\,t-\widehat{\Phi}_{i-1}\,t_{i-1}\right)\left(H_{t,k-1}^{T}-\widehat{H}_{t,k-1}^{T}\right) \end{array} \right\} \\ -\frac{1}{N}\left\{\left(\Phi_t-\widehat{\Phi}_t\right)\left(H_t^Tt-\widehat{H}_{i-1}^{T}\,t_{i-1}-\Delta\,\widehat{H}_t^{T}\,t\right)+\left(\widehat{\Phi}_t\,t-\widehat{\Phi}_{i-1}\,t_{i-1}\right)\left(H_t^T-\widehat{H}_t^{T}\right)\right\} \end{array} \right\|$$

$$\left\|\left\{\alpha\frac{1}{N-1}\left(H_{t,k-1}-\widehat{H}_{t,k-1}\right)\left(H_{t,k-1}^{T}-\widehat{H}_{t,k-1}^{T}\right)+(1-\alpha)\,\sigma_t^T\sigma_t\right\}^{-1}\right\|$$

$$+\left\| \frac{1}{N}\left\{\left(\Phi_t-\widehat{\Phi}_t\right)\left(H_t^Tt-\widehat{H}_{i-1}^{T}\,t_{i-1}-\Delta\,\widehat{H}_t^{T}\,t\right)+\left(\widehat{\Phi}_t\,t-\widehat{\Phi}_{i-1}\,t_{i-1}\right)\left(H_t^T-\widehat{H}_t^{T}\right)\right\} \right\|$$

$$\left\| \begin{array}{l} \left\{\alpha\frac{1}{N-1}\left(H_{t,k-1}-\widehat{H}_{t,k-1}\right)\left(H_{t,k-1}^{T}-\widehat{H}_{t,k-1}^{T}\right)+(1-\alpha)\,\sigma_t^T\sigma_t\right\}^{-1} \\ -\left\{\alpha\frac{1}{N-1}\left(H_t-\widehat{H}_t\right)\left(H_t^T-\widehat{H}_t^{T}\right)+(1-\alpha)\,\sigma_t^T\sigma_t\right\}^{-1} \end{array} \right\| \tag{G.42}$$

From the first term on the RHS of Eq. (G.42) we take the following term and split it as:

$$\left\| \begin{array}{l} \frac{1}{N}\left\{ \begin{array}{l} \left(\Phi_{t,k-1} - \widehat{\Phi}_{t,k-1}\right)\left(H_{t,k-1}^{T} t - \widehat{H}_{i-1}^{T}\, t_{i-1} - \Delta \widehat{H}_{t,k-1}^{T}\, t\right) \\ + \left(\widehat{\Phi}_{t,k-1}\, t - \widehat{\Phi}_{i-1}\, t_{i-1}\right)\left(H_{t,k-1}^{T} - \widehat{H}_{t,k-1}^{T}\right) \end{array} \right\} \\ -\frac{1}{N}\left\{\left(\Phi_t - \widehat{\Phi}_t\right)\left(H_t^T t - \widehat{H}_{i-1}^{T}\, t_{i-1} - \Delta \widehat{H}_t^{T}\, t\right) + \left(\widehat{\Phi}_t\, t - \widehat{\Phi}_{i-1}\, t_{i-1}\right)\left(H_t^T - \widehat{H}_t^{T}\right)\right\} \end{array} \right\|$$

$$\leq \frac{1}{N}\left\|\left(\Phi_{t,k-1} - \widehat{\Phi}_{t,k-1}\right) - \left(\Phi_t - \widehat{\Phi}_t\right)\right\|\left\|\left(H_{t,k-1}^{T} t - \widehat{H}_{i-1}^{T}\, t_{i-1} - \Delta \widehat{H}_{t,k-1}^{T}\, t\right)\right\|$$

$$+\frac{1}{N}\left\|\left(\Phi_t - \widehat{\Phi}_t\right)\right\|\left\| \begin{array}{l} \left(H_{t,k-1}^{T} t - \widehat{H}_{i-1}^{T}\, t_{i-1} - \Delta \widehat{H}_{t,k-1}^{T}\, t\right) \\ -\left(H_t^T t - \widehat{H}_{i-1}^{T}\, t_{i-1} - \Delta \widehat{H}_t^{T}\, t\right) \end{array} \right\|$$

$$+\frac{1}{N}\left(\begin{array}{l} \left\|\left(\widehat{\Phi}_{t,k-1}\, t - \widehat{\Phi}_{i-1}\, t_{i-1}\right) - \left(\widehat{\Phi}_t\, t - \widehat{\Phi}_{i-1}\, t_{i-1}\right)\right\|\left\|\left(H_{t,k-1}^{T} - \widehat{H}_{t,k-1}^{T}\right)\right\| \\ + \left\|\left(\widehat{\Phi}_t\, t - \widehat{\Phi}_{i-1}\, t_{i-1}\right)\right\|\left\|\left(H_{t,k-1}^{T} - \widehat{H}_{t,k-1}^{T}\right) - \left(H_t^T - \widehat{H}_t^{T}\right)\right\| \end{array} \right)$$

$$\leq \frac{D_{13}}{N}\left\|(\phi_{t,k-1} - \phi_t)\right\|\left\|\left(H_{t,k-1}^{T} t - \widehat{H}_{i-1}^{T}\, t_{i-1} - \Delta \widehat{H}_{t,k-1}^{T}\, t\right)\right\|$$

$$+\frac{D_{13}}{N} t\left\|\left(\Phi_t - \widehat{\Phi}_t\right)\right\|\left\|(h_t - h_{t,k-1})\right\|$$

$$+\frac{1}{N}\left(t\left\|(\phi_{t,k-1} - \phi_t)\right\|\left\|\left(H_{t,k-1}^{T} - \widehat{H}_{t,k-1}^{T}\right)\right\| + \left\|\left(\widehat{\Phi}_t\, t - \widehat{\Phi}_{i-1}\, t_{i-1}\right)\right\|\left\|(h_t - h_{t,k-1})\right\|\right) \quad \text{(G.43)}$$

where the constant $D_{13} > 0$. From the last term on the RHS of Eq. (G.42) we take the following term and split it as:

$$\left\| \begin{array}{l} \left\{\alpha \frac{1}{N-1}\left(H_{t,k-1} - \widehat{H}_{t,k-1}\right)\left(H_{t,k-1}^T - \widehat{H}_{t,k-1}^T\right) + (1-\alpha)\,\sigma_t^T \sigma_t\right\}^{-1} \\ -\left\{\alpha \frac{1}{N-1}\left(H_t - \widehat{H}_t\right)\left(H_t^T - \widehat{H}_t^T\right) + (1-\alpha)\,\sigma_t^T \sigma_t\right\}^{-1} \end{array} \right\|$$

$$\leq \alpha \frac{1}{N-1}\left(\left\|\left\{\left(H_t - \widehat{H}_t\right)\left(H_t^T - \widehat{H}_t^T\right) - \left(H_{t,k-1} - \widehat{H}_{t,k-1}\right)\left(H_{t,k-1}^T - \widehat{H}_{t,k-1}^T\right)\right\}\right\|\right) \quad \text{(G.44)}$$

$$\left\|\left\{\alpha \frac{1}{N-1}\left(H_{t,k-1} - \widehat{H}_{t,k-1}\right)\left(H_{t,k-1}^T - \widehat{H}_{t,k-1}^T\right) + (1-\alpha)\,\sigma_t^T \sigma_t\right\}^{-1}\right\|$$

$$\left\|\left\{\alpha \frac{1}{N-1}\left(H_t - \widehat{H}_t\right)\left(H_t^T - \widehat{H}_t^T\right) + (1-\alpha)\,\sigma_t^T \sigma_t\right\}^{-1}\right\|$$

Splitting the first term on the RHS of Eq. (G.44):

$$\alpha \frac{1}{N-1}\left(\left\|\left\{\left(H_t - \widehat{H}_t\right)\left(H_t^T - \widehat{H}_t^T\right) - \left(H_{t,k-1} - \widehat{H}_{t,k-1}\right)\left(H_{t,k-1}^T - \widehat{H}_{t,k-1}^T\right)\right\}\right\|\right)$$

$$\leq \alpha \frac{1}{N-1}\left(\left(\left\|(H_t - H_{t,k-1})\right\| + \left\|\left(\widehat{H}_{t,k-1} - \widehat{H}_t\right)\right\|\right)\left\|\left(H_t^T - \widehat{H}_t^T\right)\right\|\right)$$

$$+\alpha \frac{1}{N-1}\left(\left\|\left(H_{t,k-1} - \widehat{H}_{t,k-1}\right)\right\|\left(\left\|\left(H_t^T - H_{t,k-1}^T\right)\right\| + \left\|\left(\widehat{H}_{t,k-1}^T - \widehat{H}_t^T\right)\right\|\right)\right)$$

$$\leq D_{14}\alpha \frac{1}{N-1}\left(\left\|(h_t - h_{t,k-1})\right\|\left\|\left(H_t^T - \widehat{H}_t^T\right)\right\| + \left\|\left(H_{t,k-1} - \widehat{H}_{t,k-1}\right)\right\|\left\|(h_t - h_{t,k-1})\right\|\right) \quad \text{(G.45)}$$

where the constant $D_{14} > 0$. From Eqs. (G.42–G.45), we have:

$$\left\|G_{t,k-1} - G_t\right\| \leq D_{15}$$

$$\left\{\begin{array}{l} \frac{1}{N}\left\|(\phi_{t,k-1}-\phi_t)\right\|\left\|\left(H_{t,k-1}^T t-\widehat{H}_{i-1}^T t_{i-1}-\Delta\widehat{H}_{t,k-1}^T t\right)\right\| \\ +\frac{1}{N}t\left\|\left(\Phi_t-\widehat{\Phi}_t\right)\right\|\left\|(h_t-h_{t,k-1})\right\| \\ +\frac{1}{N}\left(t\left\|(\phi_{t,k-1}-\phi_t)\right\|\left\|\left(H_{t,k-1}^T-\widehat{H}_{t,k-1}^T\right)\right\|+\left\|\left(\widehat{\Phi}_t\, t-\widehat{\Phi}_{i-1}\, t_{i-1}\right)\right\|\left\|(h_t-h_{t,k-1})\right\|\right) \end{array}\right\}$$

$$\left\|\left\{\alpha\frac{1}{N-1}\left(H_{t,k-1}-\widehat{H}_{t,k-1}\right)\left(H_{t,k-1}^T-\widehat{H}_{t,k-1}^T\right)+(1-\alpha)\,\sigma_t^T\sigma_t\right\}^{-1}\right\|$$

$$+D_{15}\left\|\frac{1}{N}\left\{\left(\Phi_t-\widehat{\Phi}_t\right)\left(H_t^T t-\widehat{H}_{i-1}^T t_{i-1}-\Delta\widehat{H}_t^T t\right)+\left(\widehat{\Phi}_t\, t-\widehat{\Phi}_{i-1}\, t_{i-1}\right)\left(H_t^T-\widehat{H}_t^T\right)\right\}\right\|$$

$$\left\{\begin{array}{l} \left(\alpha\frac{1}{N-1}\left(\left\|(h_t-h_{t,k-1})\right\|\left\|\left(H_t^T-\widehat{H}_t^T\right)\right\|+\left\|\left(H_{t,k-1}-\widehat{H}_{t,k-1}\right)\right\|\left\|(h_t-h_{t,k-1})\right\|\right)\right) \\ \left\|\left\{\alpha\frac{1}{N-1}\left(H_{t,k-1}-\widehat{H}_{t,k-1}\right)\left(H_{t,k-1}^T-\widehat{H}_{t,k-1}^T\right)+(1-\alpha)\,\sigma_t^T\sigma_t\right\}^{-1}\right\| \\ \left\|\left\{\alpha\frac{1}{N-1}\left(H_t-\widehat{H}_t\right)\left(H_t^T-\widehat{H}_t^T\right)+(1-\alpha)\,\sigma_t^T\sigma_t\right\}^{-1}\right\| \end{array}\right\} \quad \text{(G.46)}$$

Since h_t is Lipschitz continuous, from Eq. (G.46) we may conclude:

$$\left\|G_{t,k-1}-G_t\right\|\to 0$$

$$\text{as}\left\|(\phi_{t,k-1}-\phi_t)\right\|\to 0 \text{ and } \left|\beta_{k-1}-1\right|\to 0$$

which implies by Eqs. (G.40) and (G.41):

$$\left\|\aleph_{t,k-1}(\phi)-\aleph_t(\phi)\right\|\to 0 \tag{G.47}$$

$$\text{as}\left\|(\phi_{t,k-1}-\phi_t)\right\|\to 0\text{and }\left|\beta_{k-1}-1\right|\to 0.$$

Now consider the sequence $\left\{\phi_{t,k+1} = \aleph_t(\phi_k) = \aleph^{k-1}(\phi_{t,0})\right\}_{k=1}^{\infty}$ where the iterations are of the stochastic Picard's form. If, for a given t, $\left\{\phi_{t,k}\right\}_{k=1}^{\kappa}$ is a Cauchy sequence in $L^2(\nu_{t_i} \times P)$, it must converge to some F_t^Y process $\phi_t$in $L^2(\nu_{t_i} \times P)$, i.e. ϕ_t must be a fixed point in

$L^2(\nu_{t_i} \times P)$.

By continuity of $\aleph_t$, $\aleph_t(\phi_k) \to \aleph_t(\phi)$. But $\aleph_t(\phi_k) = \phi_{k+1} \to \phi_t$. Hence, $\aleph_t(\phi) = \phi_t$ which means that ϕ_t is the desired solution at time t arrived at through inner iterations. In other words, the aim of driving $\{Y_t - h_t\}$ to a zero-mean martingale is met. Hence, the remaining job is to show that $\left\{\phi_{t,k}\right\}$ is indeed a Cauchy sequence in $L^2(\nu_{t_i} \times P)$.

In the same way as the inequality Eq. (G.39) is arrived at, we may have:

$$\left\|\aleph_t(\phi) - \aleph_t(\chi)\right\| \le \left\|\phi_t - \chi_t\right\| + \left\|G_t(\phi)\{Y_t - h_t(\phi)\} - G_t(\chi)\{Y_t - h_t(\chi)\}\right\| \tag{G.48}$$

We may finally arrive at the following bound:

$$\left\|\aleph_t(\phi) - \aleph_t(\chi)\right\| \le D_{16}\left\|\phi_t - \chi_t\right\| \tag{G.49a}$$

where the constant $D_{16} > 0$. Iterating upon this bound (which parallels the derivation of a Taylor-like expansion), we have [van Handel 2007]:

$$\begin{aligned}
\left\|\aleph_{t,k}(\chi) - \aleph_{t,k}(\phi)\right\|^2 &= \int_{t_{i-1}}^{t_i} \left\|\aleph_{t,k}(\chi) - \aleph_{t,k}(\phi)\right\|_{2,P}^2 dt \\
&\le D_{16}^2 \int_{t_{i-1}}^{t_i} \left\|\aleph_{t,k-1}(\chi) - \aleph_{t,k-1}(\phi)\right\|_{2,(\nu_{s_1} \times P)}^2 ds_1 \\
&\le \ldots \le D_{16}^{2k} \int_{t_{i-1}}^{t_i} \int_{t_{i-1}}^{s_1} \cdots \int_{t_{i-1}}^{s_{k-1}} \left\|\chi - \phi\right\|^2 ds_k \ldots ds_1 \\
&\le \frac{D_{16}^{2k} (\Delta t_i)^k}{k!} \left\|\chi - \phi\right\|^2
\end{aligned} \tag{G.49b}$$

Specifically:

$$\sum_{k=0}^{\infty} \left\|\aleph_{t,k+1}(\phi_0) - \aleph_{t,k}(\phi_0)\right\|^2 \le \left\|\aleph_{t,1}(\phi_0) - \phi_0\right\|^2 \sum_{k=0}^{\infty} \frac{D_{16}^{2k} (\Delta t_i)^k}{k!} < \infty \tag{G.50}$$

From Eq. (G.50), we see that $\left\{\aleph_{t,k}(\phi_0)\right\}$ is indeed a Cauchy sequence in $L^2(\nu_{t_i} \times P)$.

Theorem 2:

The solution via the proposed iterated algorithm is unique.

Proof: Let us consider two solutions χ_t and ϕ_t, whose realizations are continuous paths, arrived at by the proposed iterative algorithm. If $\chi_t = \phi_t$ a.s. under the product measure, $\nu_{t_i} \times P$, then χ_t and ϕ_t are indistinguishable P a.s. Assume that $\chi_t, \phi_t \in L^2(\nu_{t_i} \times P)$. As in the last proof, we continue to replace $\|\cdot\|_{2,(\nu_{t_k} \times P)}$ by $\|\cdot\|$. Following Eq. (G.49b), we already have:

$$\|\chi - \phi\|^2 = \|\aleph_{t,k}(\chi) - \aleph_{t,k}(\phi)\|^2 \leq \frac{D_{16}^{2k}(\Delta t_i)^k}{k!}\|\chi - \phi\|^2 \tag{G.51}$$

It is evident from Eq. (G.51), $\|\chi - \phi\| \to 0$ with $k \to \infty$. This implies $\chi_t = \phi_t$ $(\nu_{t_i} \times P)$ a.s. One must however show that χ_t, given $\chi_0 = \phi_0 \in L^2(P)$, belongs to $L^2(\nu_{t_i} \times P)$. Once this is established, the proof is complete. Using Ito's formula:

$$\|\chi_t\|^2 = \|\chi_{i-1}\|^2 + \int_{t_{i-1}}^{t} \left(2\chi_s^T \chi_s' b(X_s, s) + \left\|\chi_s'' f(X_s, s)\right\|^2\right) ds$$

$$+ \int_{t_{i-1}}^{t} 2\chi_s^T \chi_s' f(X_s, s)\, dB_s \tag{G.52}$$

Taking expectations on both sides:

$$E_P\left[\|\chi_t\|^2\right] = E_P\left[\|\chi_{i-1}\|^2\right] + E_P\left[\int_{t_{i-1}}^{t} \left(2\chi_s^T \chi_s' b(X_s, s) + \left\|\chi_s'' f(X_s, s)\right\|^2\right) ds\right]$$

$$+ E_P\left[\int_{t_{i-1}}^{t} \left(2\chi_s^T \chi_s' f(X_s, s)\right) dB_s\right] \tag{G.53}$$

Using linear growth property and noting that the Ito integral,

$$E_P\left[\int_{t_{i-1}}^{t} \left(2\chi_s^T \chi_s' f(X_s, s)\right) dB_s\right] = 0,$$

we have:

$$E_P\left[\|\chi_t\|^2\right] \leq E_P\left[\|\chi_{i-1}\|^2\right] + D_{17} E_P\left[\int_{t_{i-1}}^{t} \left(2\|\chi_s\|(1 + \|\chi_s\|) + (1 + \|\chi_s\|)^2\right) ds\right] \tag{G.54}$$

where, $D_{17} > 0$. Eq. (G.54) implies:

$$E_P\left[\|\chi_t\|^2\right] \leq E_P\left[\|\chi_{i-1}\|^2\right] + D_{17}E_P\left[\int_{t_{i-1}}^{t} ((1 + \|\chi_s\|) + \|\chi_s\|)^2\, ds\right] \tag{G.55}$$

Eq. (G.55) implies:

$$E_P\left[\|\chi_t\|^2\right] \leq E_P\left[\|\chi_{i-1}\|^2\right] + D_{18}E_P\left[\int_{t_{i-1}}^{t} (1 + \|\chi_s\|)^2\, ds\right] \tag{G.56}$$

where, $D_{18} > 0$. Applying Minkowski's inequality and Fubini's theorem:

$$E_P\left[\|\chi_t\|^2\right] \leq E_P\left[\|\chi_{i-1}\|^2\right] + D_{19}\int_{t_{i-1}}^{t} E_P\left[1 + \|\chi_s\|^2\right] ds \tag{G.57}$$

where, $D_{19} > 0$. Finally applying Gronwall's lemma:

$$E_P\left[\|\chi_t\|^2\right] \leq E_P\left[\|\chi_{i-1}\|^2\right] + E_P\left[1 + \|\chi_{i-1}\|^2\right]\exp\left(D_{19}\Delta t\right) \tag{G.58}$$

From Eq. (G.58), the claim follows.

References

Additive Particle Updates in Non-Iterative and Iterative Forms'. *arXiv.org, article no:* 1402. 1253.

Kloeden, P. E. and E. Platen. 1992. *Numerical Solution of Stochastic Differential Equations.* Berlin: Springer-Verlag.

Sarkar, S. and D. Roy. 2014. 'An Ensemble Kushner–Stratonovich (EnKS) Non-linear Filter:

van Handel, R. 2007. *'Stochastic Calculus, Filtering, and Stochastic Control'.* Lecture Notes. AMS

Appendix H (Chapter 8)

1. Derivation of $\int_{t_{i-1}}^{t} \beta_r dB_r$ using integration by parts

By integration by parts, we have:

$$\int_{t_{i-1}}^{t} \beta_r dB_r = \int_{t_{i-1}}^{t} d(\beta_r B_r) - \int_{t_{i-1}}^{t} B_r d\beta_r - \int_{t_{i-1}}^{t} d\beta_r dB_r \tag{H.1}$$

For notational convenience the integration limits are dropped in the further explanation.

$$d\beta_r = \frac{\partial \beta_r}{\partial \hat{X}_1}^T d\hat{X}_1 + \frac{\partial \beta_r}{\partial \hat{X}_2}^T d\hat{X}_2 + \frac{1}{2}\left(\left[\frac{\partial^2 \beta_r}{\partial \hat{X}_2^2}\right] d\hat{X}_2\right)^T d\hat{X}_2$$

$$= \frac{\partial \beta_r}{\partial \hat{X}_1}^T \hat{X}_2 dt + \frac{\partial \beta_r}{\partial \hat{X}_2}^T \left\{\hat{A}dt + G_r dB_r\right\} + \frac{1}{2}\left(\left[\frac{\partial^2 \beta_r}{\partial \hat{X}_2^2}\right]\right.$$

$$\left.\left(\hat{A}dt + G_r dB_r\right)\right)^T \left(\hat{A}dt + G_r dB_r\right)$$

$$= \frac{\partial \beta_r}{\partial \hat{X}_1}^T \hat{X}_2 dt + \frac{\partial \beta_r}{\partial \hat{X}_2}^T \hat{A}dt + \frac{1}{2}\left(\left[\frac{\partial^2 \beta_r}{\partial \hat{X}_2^2}\right] G_r\right)^T G_r dt + \frac{\partial \beta_r}{\partial \hat{X}_2}^T G_r dB_r \tag{H.2}$$

Substituting the expression for $d\beta_r$ in second and third term on the RHS of Eq. (H.1), it follows:

$$\int B_r d\beta_r = \int B_r \frac{\partial \beta_r}{\partial \hat{X}_1}^T \hat{X}_2 dt + \int B_r \frac{\partial \beta_r}{\partial \hat{X}_2}^T \hat{A}dt + \int \frac{1}{2}\left(B_r\left[\frac{\partial^2 \beta_r}{\partial \hat{X}_2^2}\right] G_r\right)^T$$

$$G_r dt + \int B_r \frac{\partial \beta_r}{\partial \hat{X}_2}^T G_r dB_r \tag{H.3a}$$

$$\int d\beta_r dB_r = \int \frac{\partial \beta_r}{\partial \hat{X}_2}^T G_r dt \tag{H.3b}$$

The last integral on the RHS of Eq. (H.3a) can be written as:

$$\int B_r \frac{\partial \beta_r}{\partial \hat{X}_2}^T G_r dB_r = \int d\left(B_r^2 \frac{\partial \beta_r}{\partial \hat{X}_2}^T G_r\right) - \int B_r d\left(B_r \frac{\partial \beta_r}{\partial \hat{X}_2}^T G_r\right)$$
$$- \int d\left(B_r \frac{\partial \beta_r}{\partial \hat{X}_2}^T G_r\right) dB_r \tag{H.4}$$

where,

$$d\left(B_r \frac{\partial \beta_r}{\partial \hat{X}_2}^T G_r\right) = B_r \left(\left[\frac{\partial^2 \beta_r}{\partial \hat{X}_1 \partial \hat{X}_2}\right] G_r\right)^T \hat{X}_2 dt + B_r \left(\left[\frac{\partial^2 \beta_r}{\partial \hat{X}_2^2}\right] G_r\right)^T$$
$$\hat{A} dt + B_r \left(\left[\frac{\partial^2 \beta_r}{\partial \hat{X}_2^2}\right] G_r\right)^T G_r dB_r \tag{H.5}$$

and

$$\int d\left(B_r \frac{\partial \beta_r}{\partial \hat{X}_2}^T G_r\right) dB_r = \int B_r \left(\left[\frac{\partial^2 \beta_r}{\partial \hat{X}_2^2}\right] G_r\right)^T G_r dt \int d\left(B_r \frac{\partial \beta_r}{\partial \hat{X}_2}^T G_r\right) dB_r$$
$$= \int B_r \left(\left[\frac{\partial^2 \beta_r}{\partial \hat{X}_2^2}\right] G_r\right)^T G_r dt \tag{H.6}$$

Thus the required result follows:

$$\int \beta_r dB_r = \beta_r B_r\Big|_{t_{i-1}}^{t} - B_r^2 \frac{\partial \beta_r}{\partial \hat{X}_2}^T G_r\Bigg|_{t_{i-1}}^{t} - \int \frac{\partial \beta_r}{\partial \hat{X}_2}^T G_r dt - \int B_r \frac{\partial \beta_r}{\partial \hat{X}_1}^T \hat{X}_2 dt - \int B_r \frac{\partial \beta_r}{\partial \hat{X}_2}^T \hat{A} dt$$
$$-\frac{1}{2} \int B_r \left(\left[\frac{\partial^2 \beta_r}{\partial \hat{X}_2^2}\right] G_r\right)^T G_r dt + \int \left(B_r^2 \left[\frac{\partial^2 \beta_r}{\partial \hat{X}_1 \partial \hat{X}_2}\right] G_r\right)^T \hat{X}_2 dt + \int \left(B_r^2 \left[\frac{\partial^2 \beta_r}{\partial \hat{X}_2^2}\right] G_r\right)^T \hat{A} dt$$
$$+ \int \left(B_r \left[\frac{\partial^2 \beta_r}{\partial \hat{X}_2^2}\right] G_r\right)^T G_r dt + \int \left(B_r^2 \left[\frac{\partial^2 \beta_r}{\partial \hat{X}_2^2}\right] G_r\right)^T G_r dB_r \tag{H.7}$$

The last term in the expression can again be integrated by parts to obtain expressions corresponding to the next higher order.

Appendix I (Chapter 9)

1. Np hard optimization problems

The class of problems, P, Np, Np hard and Np complete, signify the complexity of computation (See Figure I.1). A formal definition of these classes can be found in [Garey and Johnson 1979, Leeuwen 1998]. The class P corresponds to those problems that are solvable in polynomial time. Suppose that an algorithm runs in $O(n)$ time where n is problem size, the run time is said to be a linear function of n. Similarly one can have an algorithm that runs in quadratic time $(O(n^2))$, cubic time $(O(n^3))$ and, in general, in polynomial time and we then say that the problems it solves are in class P.

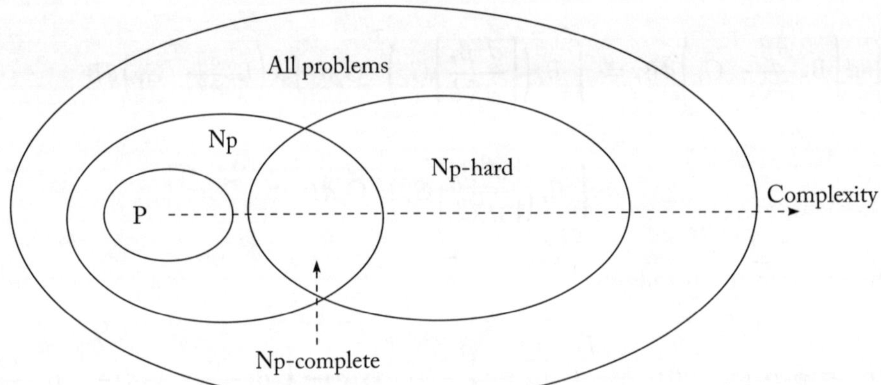

Fig. I.1 Class of problems—P, Np, Np-complete and Np-hard

There exist algorithms that do not run in polynomial time on regular computers, but run in polynomial time on a non-deterministic Turing machine [Martin 1997]. These programs belong to Np that stands for non-deterministic polynomial time. Equivalently, Np defines problems (decision problems) that can be verified in polynomial time. This does not necessarily mean that there is a polynomial-time way to find a solution. Intuitively Np is a class of decision problems where one can verify an answer quickly in reasonable time if one has been provided with a solution. For example, consider the

Hamiltonian path problem. The Hamiltonian path is a path - in an undirected or directed graph $G(V,E)$ where V denotes the vertices and E the edges - that visits each vertex exactly once. Assume that we have a solution path on hand. Verifying if it is correct can be performed in a polynomial time. Note that all problems in P are also in Np.

A problem $\mathcal{A}$ is Np-complete, if

(i) it is in Np, i.e. if it is polynomial-time verifiable, and

(ii) every problem in Np is reducible to $\mathcal{A}$ in polynomial time.

As an example, consider a travelling salesman problem (TSP) - the minimum Hamiltonian cycle of cost $\leq k$, where each edge has an associated weight. A Hamiltonian cycle is a simple cycle in a graph that visits all vertices exactly once before returning to the start vertex. One can verify in polynomial time if a cycle visits all vertices and has cost $\leq k$ and hence a TSP is Np. It is Np-complete since a known Np-complete problem such as a Hamiltonian cycle can be reduced to the TSP [Garey and Johnson 1979]. It is not known if every problem in Np can be solved and this is called the P versus Np problem [Lance 2009]. However, if any Np-complete problem can be solved, then every problem in Np can be solved because of condition (ii) appearing in the definition of Np-completeness.

A problem satisfying condition (ii) above is an Np-hard problem whether it satisfies the condition stated in (i) or otherwise. Np-hard problems are the hardest of all problems in Np. They belong to a class where, even when one has a solution, it cannot be verified in polynomial time. More precisely, a problem is called Np-hard if it is polynomial-time reducible but not necessarily polynomial-time verifiable. TSP where one needs to find the shortest distance (cycle) covered, is an example of an Np-hard problem. That is, given a weighted graph G, the task is to find the shortest cycle (an optimization problem) that visits every vertex. Finding the shortest cycle is obviously harder than determining if a cycle exists at all; so the TSP is Np-hard.

2. Derivation of Eq. 9.27 which reads as:

$$d\pi_\tau(\mathbf{X}) = (\pi_\tau(\mathbf{X}f) - \pi_\tau(\mathbf{X})\pi_\tau(f))\left(\rho_\tau\rho_\tau^T\right)^{-1}\left(d\check{f}_\tau - \pi_\tau(f)\,d\tau\right) \tag{I.1}$$

Note that a product of vectors above implies a vector dot product. For instance, the product $(\mathbf{X}f)$ is the dot product $(\mathbf{X}.f)$.

Solution

The gain based update equation for a bounded, at least twice differentiable function $\phi_\tau := \phi(\mathbf{X}_\tau)$ of $\mathbf{X}_\tau$ may be arrived at by expanding $\phi_\tau\mathbf{\Lambda}_\tau$, where $\tau \in (\tau_{i-1}, \tau_i]$, using Ito's formula:

$$d\left(\phi_\tau\mathbf{\Lambda}_\tau\right) = \phi_\tau d\mathbf{\Lambda}_\tau + d\phi_\tau\mathbf{\Lambda}_\tau + \left[d\phi_\tau, d\mathbf{\Lambda}_\tau\right] \tag{I.2}$$

[.] denotes the quadratic covariation. Here ϕ_τ is assumed to be a scalar function. With the observation SDE in the form of Eq. (9.24), a further expansion of Eq. (I.2) leads to:

$$d\left(\phi_\tau \Lambda_\tau\right) = \phi_\tau \Lambda_\tau h_\tau d\tilde{f}_\tau + \Lambda_\tau \phi_\tau' dX_\tau + \frac{1}{2}\Lambda_\tau \left[dX_\tau, \phi_\tau'' dX_\tau\right] \tag{I.3}$$

By explicitly writing the term $\left[dX_\tau, \phi_\tau'' dX_\tau\right]$ we get:

$$d\left(\phi_\tau \Lambda_\tau\right) = \phi_\tau \Lambda_\tau h_\tau d\tilde{f}_\tau + \Lambda_\tau \phi_\tau' dX_\tau + \frac{1}{2}\Lambda_\tau \sum_{j,k=1}^{m} \sum_{l=1}^{n} \left(\frac{\partial^2 \phi_\tau}{\partial x_j \partial x_k}\right)_\tau g_{jl} g_{jl} d\tau \tag{I.4}$$

The incremental form as in Eq. (I.4) may be given the following integral representation:

$$\phi_\tau \Lambda_\tau = \phi_{i-1}\Lambda_{i-1} + \int_{\tau_{i-1}}^{\tau} \phi_s \Lambda_s h_s d\tilde{f}_s$$

$$+ \int_{\tau_{i-1}}^{\tau} \Lambda_s \left(\phi_s' dB_s + \frac{1}{2} \sum_{j,k=1}^{m} \sum_{l=1}^{n} \left(\frac{\partial^2 \phi_\tau}{\partial x_j \partial x_k}\right)_s g_{jl} g_{jl} ds \right) \tag{I.5}$$

In deriving Eq. (I.5), Eq. (9.17) is made use of, where $g \in \mathbb{R}^{n \times n}$ is the intensity matrix with its $(j,k)^{th}$ element denoted as, g_{jk}. Taking conditional expectation with respect to $\mathfrak{N}_\tau$ under Q we get:

$$E_Q\left[\phi_\tau \Lambda_\tau | \mathfrak{N}_\tau\right] = E_Q\left[\phi_{i-1}\Lambda_{i-1} | \mathfrak{N}_\tau\right] + E_Q\left[\left\{\int_{\tau_{i-1}}^{\tau} \Lambda_s \phi_s h_s d\tilde{f}_s\right\} | \mathfrak{N}_\tau\right]$$

$$+ E_Q\left[\left\{\int_{\tau_{i-1}}^{\tau} \Lambda_s \phi_s' dB_s\right\} | \mathfrak{N}_\tau\right] + \frac{1}{2} E_Q\left[\left\{\int_{\tau_{i-1}}^{\tau} \sum_{j,k=1}^{m} \sum_{l=1}^{n} \left(\frac{\partial^2 \phi_\tau}{\partial x_j \partial x_k}\right)_s g_{jl} g_{jl} ds\right\} | \mathfrak{N}_\tau\right] \tag{I.6}$$

Using Fubini's theorem:

$$E_Q\left[\phi_\tau \Lambda_\tau | \mathfrak{N}_\tau\right] = E_Q\left[\phi_{i-1}\Lambda_{i-1} | \mathfrak{N}_\tau\right] + \int_{\tau_{i-1}}^{\tau} E_Q\left[\phi_s \Lambda_s h_s | \mathfrak{N}_s\right] d\tilde{f}_s$$

$$+ \int_{\tau_{i-1}}^{\tau} E_Q\left[\Lambda_s \phi_s' | \mathfrak{N}_s\right] dB_s + \frac{1}{2} \int_{\tau_{i-1}}^{\tau} E_Q\left[\sum_{j,k=1}^{m} \sum_{l=1}^{n} \left(\frac{\partial^2 \phi_\tau}{\partial x_j \partial x_k}\right)_s g_{jl} g_{jl} | \mathfrak{N}_\tau\right] ds \tag{I.7}$$

Noting that $\int_{\tau_{i-1}}^{\tau} E_Q\left[\boldsymbol{\Lambda}_s\phi_s'|\Omega_s\right]d\boldsymbol{B}_s = 0$ and for notational convenience denoting the un-normalized conditional expectation operator, $E_Q\left[(.)\boldsymbol{\Lambda}_\tau|\Omega_\tau\right]$ as $\rho_\tau(.)$, we arrive at the following equation:

$$\rho_\tau(\phi) = \rho_{i-1}(\phi) + \int_{\tau_{i-1}}^{\tau} \rho_s(\phi\boldsymbol{h})\,d\tilde{\boldsymbol{f}}_s$$

$$+\frac{1}{2}\int_{\tau_{i-1}}^{\tau}\rho_s\left(\sum_{j,k=1}^{m}\sum_{l=1}^{n}\left(\frac{\partial^2\phi}{\partial x_j\partial x_k}\right)_s g_{jl}g_{jl}\right)ds \tag{I.8}$$

Equation (I.8) finds its equivalence with the Zakai equation (Chapter 6). An incremental representation of Eq. (I.8) may be given as:

$$d\rho_\tau(\phi) = \rho_\tau(\phi\boldsymbol{h})\,d\tilde{\boldsymbol{f}}_\tau + \frac{1}{2}\rho_\tau\left(\sum_{j,k=1}^{m}\sum_{l=1}^{n}\left(\frac{\partial^2\phi}{\partial x_j\partial x_k}\right)g_{jl}g_{jl}\right)d\tau \tag{I.9}$$

In order to obtain the normalized conditional law, i.e. $\pi_\tau(\phi) = \frac{\rho_\tau(\phi)}{\rho_\tau(1)}$, it is expanded using Ito's formula as given below:

$$d\pi_\tau(\phi) = d\left(\frac{\rho_\tau(\phi)}{\rho_\tau(1)}\right) = \rho_\tau(\Phi)\,d\left(\frac{1}{\rho_\tau(1)}\right) + \frac{1}{\rho_\tau(1)}d\rho_\tau(\Phi) + d\left[\rho_\tau(\Phi), \frac{1}{\rho_\tau(1)}\right] \tag{I.10}$$

$d\left(\frac{1}{\rho_\tau(1)}\right)$ may be expanded as:

$$d\left(\frac{1}{\rho_\tau(1)}\right) = -\frac{1}{\rho_\tau^2(1)}d\rho_\tau(1) + \frac{1}{\rho_\tau^3(1)}\left[d\rho_\tau(1), d\rho_\tau(1)\right] \tag{I.11}$$

Putting $\phi = 1$in Eq. (I.9), we get an Ito expansion for $d\rho_\tau(1)$, which is given below:

$$d\rho_\tau(1) = \rho_\tau(\boldsymbol{h})\,d\tilde{\boldsymbol{f}}_\tau \tag{I.12}$$

Using Eq. (I.12) in Eq. (I.11):

$$d\left(\frac{1}{\rho_\tau(1)}\right) = -\frac{\pi_\tau(\boldsymbol{h})\,d\tilde{\boldsymbol{f}}_\tau}{\rho_\tau(1)} + \frac{(\pi_\tau(\boldsymbol{h}))^2 d\tau}{\rho_\tau(1)} \tag{I.13}$$

Using Eqs. (I.9) and (I.13) in Eq. (I.10) we get:

$$d\pi_\tau(\phi) = \pi_\tau(\phi \boldsymbol{h})\, d\tilde{\boldsymbol{f}}_\tau + \frac{1}{2}\pi_\tau\left(\sum_{j,k=1}^{m}\sum_{l=1}^{n}\left(\frac{\partial^2\phi}{\partial x_j \partial x_k}\right) g_{jl} g_{jl}\right) d\tau$$

$$+\left(-\pi_\tau(\phi)\,\pi_\tau(\boldsymbol{h})\, d\tilde{\boldsymbol{f}}_\tau + \pi_\tau(\phi)\,\pi_\tau^2(\boldsymbol{h})\, d\tau\right) - \pi_\tau(\phi\boldsymbol{h})\,\pi_\tau(\boldsymbol{h})\, d\tau \qquad (I.14)$$

Thus we arrive at the integro–differential equation describing the evolution of the normalized conditional estimate $\pi_\tau(\phi)$ given as:

$$d\pi_\tau(\phi) = \frac{1}{2}\pi_\tau\left(\sum_{j,k=1}^{m}\sum_{l=1}^{n}\left(\frac{\partial^2\phi}{\partial x_j \partial x_k}\right) g_{jl} g_{jl}\right) d\tau + \left(\pi_\tau(\phi\boldsymbol{h}) - \pi_\tau(\phi)\,\pi_\tau(\boldsymbol{h})\right)$$

$$\left(d\tilde{\boldsymbol{f}}_\tau - \pi_\tau(\boldsymbol{h})\, d\tau\right) \qquad (I.15)$$

Equation (I.15) has its parallel in the KS equation, which is well known in stochastic filtering. Since we are typically interested in the evolution of the conditional estimate of $\boldsymbol{X}_\tau$, i.e., ϕ is the identity function, Eq. (I.15) may be simplified as:

$$d\pi_\tau(\boldsymbol{X}) = \left(\pi_\tau(\boldsymbol{X}\boldsymbol{h}) - \pi_\tau(\boldsymbol{X})\,\pi_\tau(\boldsymbol{h})\right)\left(d\tilde{\boldsymbol{f}}_\tau - \pi_\tau(\boldsymbol{h})\, d\tau\right) \qquad (I.16)$$

Replacing $\boldsymbol{h}(\boldsymbol{X}_\tau)$ by $\rho_\tau^{-1}\boldsymbol{f}(\boldsymbol{X}_\tau)$ in Eq. (I.16), we get the required result in Eq. (I.1).

References

Garey, M. R. and D. S. Johnson. 1979. *Computers and Intractability: A Guide to the Theory of NP-Completeness.* NY: W. H. Freeman & Co.

Lance, F. 2009. 'The Status of the P versus NP problem'. *Commun. ACM.* 52(9): 78–86.

Leeuwen, Jan van, ed. 1998. *Handbook of Theoretical Computer Science. vol. A. Algorithms and complexity.* Amsterdam: Elsevier.

Martin, J. C. 1997. *Introduction to Languages and the Theory of Computation,* 2nd ed. NY: McGraw-Hill.

References

Chapter 1

Ang, A. H. S. and W. H. Tang. 1975. *Probability Concepts in Engineering Planning and Design, Vol. 1, Basic Principles*. NY: John Wiley and Sons Ltd.

Berger, J. 1985. *Statistical Decision Theory and Bayesian Analysis*. NY: Springer Verlag.

Feller, W. 1968. *An Introduction to Probability theory and its Applications, Volume* 1. 3rd ed. NY: John Wiley and Sons Inc.

Kolmogorov, A. N. 1933. *Grundbegriffe der Wahrscheinlichkeitsrechnung*. Berlin: Springer.

Kolmogorov, A. N. 1933. *Grundbegriffe der Wahrscheinlichkeitsrechnung*. Berlin: Springer. English translation: *Foundations of the Theory of Probability*. 1950. NY: Chelsea.

Laurendeau, N. M. 2005. *Statistical Thermodynamics: Fundamentals and Applications*. UK: Cambridge University Press.

Maymon, G. 1998. *Some engineering applications in random vibrations and random structures*. Progress in Astronautics and Aeronautics. Vol 178. USA: AIAA Inc.

Nataf, A. 1962. 'Determination des Distribution dont les Margessontdonnees'. *Comptesrendus de lacademie des sciences* 225: 42–43.

Papoulis, A. 1991. Probability, Random Variables and Stochastic Processes. 3rd ed. NY: McGraw-Hill Inc.

Pearson, K. 1900. 'On the Criterion that a Given System of Deviations from the Probable in the Case of a Correlated System of Variables is Such that it can be Reasonably Supposed to have Arisen from Random Sampling'. *Philosophical Magazine Series* 5, 50 (302): 157–75.

Rackwitz, R. and B. Fiessler. 1978. 'Structural Reliability under combined random load sequences'. *Computers and Structures* 9. 489–94.

Ross, S. 2002. *A First Course in Probability*. 6th ed. Singapore: Pearson Prentice Hall.

Rudin, W. 1976. *Principles of Mathematical Analysis*. 3rd ed. NY: McGraw-Hill.

Chapter 2

Ang, A. H. S. and W. H. Tang. 1984. *Probability Concepts in Engineering Planning and Design, (Vol. II:) Decision, Risk and Reliability*. NY: John Wiley and Sons Ltd.

Ash, R. B. 2008. *Basic Probability Theory*. NY: Dover Publications, Inc.

Bishop, C. M. 2006. *Pattern Recognition and Machine Learning*. Berlin: Springer.

Breiman, L. 1968. *Probability*. Reading: Addison-Wesley.

Candy, J. V. 2009. *Bayesian Signal Processing: Classical Modern and Particle Filtering Methods*. NJ: Wiley-IEEE Press.

Chaikin, P. M. and T. C. Lubensky. 2007. *Principles of condensed matter physics*. 4th ed. Cambridge: Cambridge University Press.

Dimov. L. T. 2008. *Monte Carlo Methods for Applied Scientists*. London: World Scientific.

Elishakoff, I. and Y. J. Ren. 2003. *Finite Element Methods for Structures with Large Stochastic Variations*. UK: Oxford University Press.

Field, E. H. 2007. 'A Summary of Previous Working Groups on California Earthquake Probabilities'. *Bulletin of the Seismological Society of America* 97(4): 1033–53.

Fisher, E. and M. N. Barber. 1972. 'Scaling Theory for Finite-Size Effects in the Critical Region'. *Phys. Rev. Lett.* 28: 1516.

Fishman, G. S. 1995. *Monte Carlo: Concepts, Algorithms, and Applications*. NY. USA: Springer.

Gentle, J. E. 2003. *Random Number Generation and Monte Carlo Methods*. 2nd ed. NY: Springer-Verlag.

Glasserman, P. 2003. *Monte Carlo Methods in Financial Engineering*. NY: Springer-Verlag.

Halmos, P. R. and L. J. Savage. 1949. 'Application of the Radon–Nikodym Theorem to the Theory of Sufficient Statistics'. *Ann. Math. Statist.* 20: 225–41.

Haykin, S. 1996. *Adaptive Filter Theory*. 3rd ed. NJ: Prentice-Hall.

Hohenbichler, M. and R. Rackwitz. 1981. 'Non-normal Dependent Vectors in Structural Safety'. *Jl. of EM division, ASCE*. 107(6): 1227–49.

Hubalek, F. and W. Schachermayer. 2001. 'The Limitations of No-Arbitrage Arguments for Real Options'. *Int. J. Theory. Appl. Finan.* 4(2): 361.

Ising, Ernst. 1925. 'Beitrag zur Theorie des Ferromagnetismus'. *Zeitschrift für Physik* 31: 253–58.

Kailath, T. 1981. *Lectures on Kalman and Wiener Filtering Theory*. NY: Springer-Verlag.

Knuth, D. 1997. *The Art of Computer Programming. Vol. 2: Seminumerical Algorithms*. 3rd ed. Massachusetts: Addison-Wesley.

Kroese, D. P., T. Taimre, and Z. I. Botev. 2011. *Handbook of Monte Carlo Methods*. NJ: John Wiley & sons.

Landau, D. P. 1976. 'Finite-size Behavior of the Ising square lattice'. *Phys. Rev. B*. 13: 2997.

Landau, D. P. and K. Binder. 2000. *A Guide to Monte Carlo Simulations in Statistical Physics*. Cambridge: Cambridge University Press. Cambridge, UK.

Manekar, M. A, S. Chaudhary, M. K. Chattopadhyay, K. J. Singh, S. B. Roy, and P. Chaddah. 2001. 'First-order Transition from Antiferromagnetism to Ferromagnetism in $Ce(Fe_{0.96}Al_{0.04})_2$'. *Physical Review* B 64(2.).

Matsumoto, M. and T. Nishimura. 1998. 'Mersenne Twister: A 623-Dimensionally Equidistributed Uniform Pseudorandom Number Generator'. *ACM Transactions on Modeling and Computer Simulation*. 8(1): 3–30.

Mitchell, T. M. 1997. *Machine Learning*, International Student Edition. UK: McGraw-Hill.

Neyman, J. 1935. 'Sur un teorema concernente le cosidette statistiche sufficient'. *Giorn. Ist. Ital. Att.* 6: 320–34.

Onsager, L. 1944. 'Crystal Statistics. I. A Two-dimensional Model with an Order-Disorder Transition'. *Phys. Rev.* 65: 117–49.

Papoulis, A. 1991. Probability, Random Variables and Stochastic Processes. 3rd ed. NY: McGraw-Hill Inc.

Peierls, R. 1936. 'On Ising's Model of Ferromagnetism'. *Mathematical Proceedings of the Cambridge Philosophical Society* 32(3): 477–48.

Rosenblatt, M. 1952. 'Remarks on a Multivariate Transformation'. *The Annals of Mathematical Statistics* 23(3): 470–72.

Ross. S. 2002. *A First Course in Probability*. 6th ed. Singapore: Pearson Prentice Hall.

Roy, D. and G. V. Rao. 2012. *Elements of Structural Dynamics: A New Perspective*. UK: John Wiley and Sons.

Rubinstein, R. Y. and D. P. Kroese. 2007. Simulation and the Monte Carlo Method. 2nd ed. NY: John Wiley and Sons.

Saha, N. and D. Roy. 2007. 'Variance-Reduced Weak Monte Carlo Simulations of Stochastically Driven Oscillators of Engineering Interest'. *Journal of Sound and Vibration* 305: 50–84.

Thomasian, A. 1969. *The Structure of Probability Theory with Applications*. NY: McGraw-Hill.

Thrun, S., W. Burgard, and D. Fox. 2005. *Probabilistic Robotics*. Cambridge, MA: MIT Press.

Tobochnik, J. and H. Gould. 2008. 'Teaching Statistical Physics by Thinking about Models and Algorithms'. *Am. J. Phys.* 76: 353–59.

Chapter 3

Bhattacharya, R. N. and E. C. Waymire. 2009. *Stochastic Processes with Applications*. Philadelphia: Siam publications.

Black, F. and M. Scholes. 1973. 'The Pricing of Options and Corporate Liabilities'. *J. Polit. Econ.* 81: 637–54.

Brzeźniak, Z. and T. Zastawniak. 1999. *Basic Stochastic Processes*. Springer Undergraduate Mathematics Series. London: Springer.

Cramér, H. and M. R. Leadbetter. 1967. *Stationary and Related Stochastic Processes*. NY: Wiley. Reprinted by Dover Publications. 2004.

Doob, J. 1953. *Stochastic Processes*. 1953. Wiley Series in Probability and Mathematical Statistics. NY: Wiley.

Ewald, J. B. D. 2002. *'Numerical Methods for Stochastic Differential Equations in the Geosciences'.* PhD. diss. Indiana University, Bloomington.

Gardner, W. A. 1990. *Introduction to Random Processes: With Applications to Signals and Systems.* 2nd ed. NY: McGraw-Hill.

Grimmett, G. R. and D. R. Stirzaker. 2001. *Probability and Random Processes.* 3rd ed. UK: Oxford University Press.

Hughes, B. D. 1996. *Random Walks and Random Environments.* UK: Oxford University Press.

Itô, K. and H. P. McKean Jr. 1996. *Diffusion Processes and Their Sample Paths.* Reprint of the 1974 Edition. Berlin: Springer-Verlag.

Karatzas, I. and S. E. Shreve. 1991. *Brownian Motion and Stochastic Calculus.* 2nd ed. NY: Springer-Verlag.

________. 1998. *Methods of Mathematical Finance.* NY: Springer-Verlag.

Klebaner, F. C. 2005. *Introduction to Stochastic Calculus with Applications.* 2nd ed. London: Imperial College Press.

L'evy, P. 1937. *Th'eories de L'Addition Al'eatories.* Paris: Gauthier-Villars.

Lawler, G. F. 1995. *Introduction to Stochastic Processes.* NY: Chapman and Hall.

Lin, Y. K. 1967. *Probabilistic Theory of Structural Dynamics.* NY: McGraw-Hill.

Mörters, P. and Y. Peres. 2010. *Brownian Motion.* Cambridge: Cambridge University Press.

Nagasawa, M. 2000. *Stochastic Processes in Quantum Physics.* Basel, Switzerland: Birkhauser Verlag.

Nigam, N. C. 1983. *Introduction to Random Vibrations.* Cambridge: MIT press.

Nigam, N. C. and S. Narayanan. 1994. *Applications of Random Vibrations.* New Delhi, India: Narosa Publications.

Øksendal, B. 2003. *Stochastic Differential Equations, An Introduction with Applications.* NY: Springer-Verlag.

Rogers, L. C. G. and D. Williams. 2000. *Diffusions, Markov Processes and Martingales, vol.* 1: *Foundations.* 2nd ed. UK: Cambridge University Press.

Ross, S. M. 1995. *Stochastic Processes.* 2nd ed. NY: John Wiley and Sons.

Roychowdhury, S., D. Roy, J. N. Reddy, and A. Srinivasa. 2016. 'Fluctuation Relation Based continuum model for thermoviscoplasticity in metals'. *Journal of the Mechanics and Physics of Solids* 96: 353–68.

Sarkar, S., R. C. Shubhankar, D. Roy, and R. M. Vasu. 2015. 'Internal noise-driven generalized Langevin equation from a nonlocal continuum model'. *Physical Review E.* (DOI: 10.1103/PhysRevE.00.002100)

Stroock, D. W. and S. R. S. Varadhan. 1979. *Multidimensional Diffusion Processes.* NY: Springer-Verlag.

Weiss, G. H. 1994. *Aspects and Applications of the Random Walk.* Amsterdam: North Holland Press.

Wiener, N. 1923. 'Differential Space'. *Jl. of Mathematical Physics* 2: 131–74.

Williams, D. 1991. *Probability with Martingales*. Cambridge: Cambridge University Press.

Yates, R. D and D. J. Goodman. 1999. *Probability and Stochastic Processes: A Friendly Introduction for Electrical and Computer Engineers*. NY: John Wiley and Sons.

Chapter 4

Björk, T. 2001. *Stochastic Calculus*. Lecture Notes for the Course Numerical Methods for Stochastic Differential Equations given in Stockholm and Lund. Sweden.

Blevins, R. D. 1990. *Flow Induced Vibrations*. Florida: Krieger Publishing Co.

Chatfield, C. 2004. *The Analysis of Time Series. An Introduction*. (originally published in 1975) 6th ed. Boca Raton, Fla: Chapman and Hall/CRC.

'Cramer, H. 1960. 'On Some Classes of Non-stationary Processes'. *Proc. 4th Berkeley Symposium Math. Stat. and Prob.* 2: 57.

Crooks, G. E. 1999. 'Excursions in Statistical Dynamics'. PhD thesis. Univ. of California. Berkeley, USA.

Friedman, A. 1975. *Stochastic Differential Equations and Applications*. vol I and II. NY: Academic Press.

Ge, H. and D. Q. Jiang. 2008. 'Generalized Jarzynski's Equality of Inhomogeneous Multidimensional Diffusion Processes'. *Jl. Stat. Phys.* 131: 675–89.

Handel, R. V. 2007. *Stochastic Calculus, Filtering, and Stochastic Control.* Lecture Notes.

Harrison, R. F. and J. K. Hammond. 1986. 'Analysis of the Non-stationary Response of Vehicles with Multiple Vehicles'. *Jl. Dynamic Sys. Meas. and Control, ASME*. 108: 69–73.

Howell, L. J. and Y. K. Lin. 1971. 'Response of Flight Vehicles to Non-stationary Atmospheric Turbulence'. *AIAA Jl.* 9: 2201–7.

Hummer, G. and A. Szabo. 2001. 'Free-energy Reconstruction from Non-equilibrium Single-molecule Pulling Experiments'. *Proc. Natl. Acad. Sci. USA*. 98(7): 3658–61.

Ito, K. 1951. On a Formula Concerning Stochastic Differentials. *Nagoya Mathematical Journal* 3: 55–65

Jarzynski, C. 1997. 'Equilibrium Free-energy Differences from Non-equilibrium Measurements: A Master-equation Approach'. *Phys. Rev.* E56: 5018–35.

Kanai, K. 1957. 'Semi-empirical Formula of Seismic Characteristic of Ground'. *Bulletin of Earthquake Research Institute, University of Tokyo*. 35: 309–25.

Karatzas, I. and S. E. Shreve. 1991. *Brownian Motion and Stochastic Calculus*. 2nd ed. NY: Springer-Verlag.

Kiureghian, D. and J. Crempien. 1989. 'An Evolutionary Model for Earthquake Ground Motion'. *Structural Safety* 6. 235–46.

Klebaner, F. C. 2005. *Introduction to Stochastic Calculus with Applications*. 2nd ed. London: Imperial College Press.

Liphardt, J., S. Dumont, S. B. Smith, I. Tinoco, and C. Bustamante. 2002. 'Equilibrium Information from Non-equilibrium Measurements in an Experimental Test of Jarzynski's Equality'. *Science* 296: 1832–35.

McKean, H. P. 1969. *Stochastic Integrals.* NY: Academic Press. reprinted in 2005 by American Mathematical Society.

Nigam, N. C. and S. Narayanan. 1994. *Applications of Random Vibrations.* New Delhi, India: Narosa Publications.

Øksendal, B. 2003. *Stochastic Differential Equations, An Introduction with Applications.* 6th ed. Heidelberg: Springer.

Priestley, M. B. 1965. 'Evolutionary Spectra and Non-stationary Processes'. *Stat. Soc. Ser. B (Methodol.) Series B.* 27: 204–37.

Priestley, M. B., 1967. 'Power Spectral Analysis of Non-stationary Random Processes'. *J. Sound Vib.* 6: 86–97.

Ramasubramanian, S., 2003. 'Diffusions and the Martingale Problem of Stroock and Varadhan'. In *Connected at Infinity: A Selection of Mathematics by Indians* edited by R. Bhatia. Texts and Readings in Mathematics. 25: 105–29. ND, India: Hindustan Book Agency.

Rogers, L. C. G. and D. Williams. 2000. *Diffusion, Markov Processes, and Martingales.* vol. 1 and 2. 2nd ed. Cambridge: Cambridge Univ. Press.

Roy, D. and G. V. Rao. 2012. *Elements of Structural Dynamics–A New Perspective.* UK: John Wiley and Sons.

Rudin, W. 1987. *Real and Complex Analysis.* 3rd ed. NY: McGraw-Hill.

Shinozuka, M. 1972. 'Monte Carlo Simulation of Structural Dynamics'. *Computers and Structures.* 2: 855–74.

Shinozuka, M. and C. M. Jan. 1972. 'Digital Simulation of Random Processes and its Applications'. *Jl. of Sound and Vib.* 25(1): 111–28.

Shinozuka, M. and Y. Sato. 1967. 'Simulation of Nonstationary Random Processes'. *Jl. Eng. Mech. Div., ASCE.* 93 (EM1): 11–40.

Stratonovich, R. L. 1966. 'A New Representation for Stochastic Integrals and Equations'. *SIAM J. Control.* 4: 362–71.

Stroock, D. W. and S. R. S. Varadhan. 1969. 'Diffusion Processes with Continuous Coefficients, I'. *Comm. Pure Appl. Math.* 22: 345–400.

__________. 1969. 'Diffusion Processes with Continuous Coefficients, II'. *Comm. Pure Appl. Math.* 22: 479–530.

__________. 1972. 'On the Support of Diffusion Processes with Application to the Strong Maximum Principle'. *Proceedings of the Sixth Bercley Symposium on Mathematical Statistics and Probability.* 3: 330–59.

Tazimi, H. 1960. 'A Statistical Method of Determining the Maximum Response of Structures during Earthquake'. *Proc. of* 2nd *World Conference on Earthquake Engineering.* Tokyo and Kyoto. Japan.

Yadav, D. and N. C. Nigam. 1978. 'Ground Induced Non-stationary Response of Vehicles'. *Jl. Sound and Vib.* 61(1): 117–26.

Zhang, R., Y. Yong, and Y. K. Lin. 1991. 'Earthquake Ground Motion Modeling II: Stochastic Line Source'. *Jl. Eng. Mech. Div., ASCE.* 117(9): 2133–48.

Zipfel, P. H. 2000. *Modeling and Simulation of aerospace vehicle dynamics.* USA: AIAA, Inc.

Chapter 5

Belytschko, T. and T. J. R. Hughes, eds. 1983. *Computational Methods for Transient Analysis.* Amsterdam: North Holland.

Biscay, R., J. C. Jimenez, J. J. Riera, and P. A. Valdes. 1996. 'Local Linearization Method for the Numerical Solution of Stochastic Differential Equations'. *Ann. Inst. Statist.Math.* 48(4). 631–44.

Burrage, K, P. M. Burrage, and T. Tian. 2004. 'Numerical Methods for Strong Solutions of Stochastic Differential Equations: An Overview'. *Proc. Roy. Soc. London.* A 460: 373–402.

Burrage, K., P. M. Burrage, and T. Mitsui. 2000. 'Numerical Solutions of Stochastic Differential Equations—Implementation and Stability Issues. *Jl. Comp. Appl. Math.* 125(1–2): 171–82.

Carrier, G G. and C. E. Pearson. 1991. *Ordinary Differential Equations.* SIAM Classics in Applied Mathematics vol. 6. Philadelphia: SIAM

Chung, J. and G. M. Hulbert. 1993. 'A Time Integration Algorithm for Structural Dynamics with Improved Numerical Dissipation: The Generalized α-Method'. *ASME. Jl. of Applied Mechanics* 60: 371–75.

Crandall, S. 1973. 'Correlations and Spectra of Non-linear System Response'. *Probl. Non-linear Vib.* 14: 39–53.

Gard, T. C. 1988. *Introduction to Stochastic Differential Equations.* NY: Marcel Dekker.

Gikhma, I. and A. V. Skorokhod. 1972. *Stochastic Differential Equations.* Berlin: Springer.

Guckenheimer, J. and P. Holmes. 1983. *Non-linear Oscillations, Dynamical Systems, and Bifurcations of Vector Fields.* NY: Springer-Verlag.

Higham, D. J. 1998. 'Mean-square and Asymptotic Stability of Numerical Methods for Stochastic Ordinary Differential Equations'. University of Strathclyde Mathematics Research Report 39.

Higham, D. J. 2001. 'An Algorithmic Introduction to Numerical Simulation of Stochastic Differential Equations'. *SIAM Review* 43(3): 525–46.

Higham, D. J., X. Mao, and A. Stuart. 2002. 'Strong Convergence of Euler-type Methods for Non-linear Stochastic Differential Equations'. *SIAM Jl. Numer. Anal.* 40: 1041–63.

Hilber, H. M., T. J. R. Hughes, and R. L. Taylor. 1977. 'Improved Numerical Dissipation for Time Integration Algorithm in Structural Dynamics'. *Int. Jl. of Earthquake Engineering and Structural Dynamics* 5: 283–93.

Ibrahim, R. A. 1978. 'Stationary Response of a Randomly Parametric Excited Non-linear System'. *Jl. Appl. Mech.* 45: 910–16.

Ibrahim, R. A. 1985. *Parametric Random Vibration.* Hertfordshire, NY: Research Studies Press. John Wiley and Sons.

Iwan W. D. and P. D. Spanos. 1978. 'Response Envelope Statistics for Non-linear Oscillators with Random Excitation'. *Jl. Appl. Mech.* 45: 170–74.

Iyengar, R. N. and D. Roy. 1998a. 'New Approaches for the Study of Non-linear Oscillators'. *Jl. Sound and Vib.* 211: 843–75.

__________. 1998b. 'Extensions of the Phase Space Linearization (PSL) Technique for Non-linear Oscillators'. *Jl. Sound and Vib.* 211: 877–906.

Iyengar, R. N. and P. K. Dash. 1978. 'Study of the Random Vibration of Non-linear Systems by the Gaussian Closure Technique'. *Jl. Appl. Mech.* 45. 393–99.

Kloeden, P. E. and E. Platen. 1992. *Numerical Solution of Stochastic Differential Equations.* Berlin: Springer-Verlag.

Lin, Y. K. 1967. *Probabilistic Theory of Structural Dynamics.* NY: McGraw-Hill.

Manohar, C. S. 1995. 'Methods of Non-linear Random Vibration Analysis'. *Sadhana* 20(2–4): 345–71.

Maruyama, G. 1955. 'Continuous Markov Processes and Stochastic Equations'. *Rendiconti del Circolo Matematica di Palermo* 4: 48–90.

Milstein, G. N. 1974. 'Approximate Integrations of Stochastic Differential Equations'. *Theory Probability Appl.* 19: 557–62.

__________. 1995. *Numerical Integration of Stochastic Differential Equations.* Originally published by Ural State University Press, Sverdlovsk. 1988. Dordrecht: Kluwer Academic Publishers, Springer Science+Business Media.

Miranda, I., R. M. Ferencz, and T. J. R. Hughes. 1989. 'An Improved Implicit–Explicit Time Integration Method for Structural Dynamics'. *Int. Jl. of Earthquake Engineering and Structural Dynamics* 18: 643–55.

Newmark, N. M. 1959. 'A Method of Computation for Structural Dynamics'. *ASCE. Jl. of Engineering Mechanics Division* 85, EM3: 67–94.

Ozaki, T. 1985. 'Non-linear Time Series Models and Dynamical Systems'. In *Handbook of Statistics.* vol 5, edited by E. J. Hannan, P. R. Krishnaiah, and M. M. Rao North HollandHolland: Elsevier Science.

__________. 1994. 'The Local Linearization Filter with Application to Non-linear System Identification'. In *Proceedings of the First US/Japan Conference on the Frontiers of Statistical Modeling: An Informational Approach*, edited by H. Bozdogan, 217–40. Boston: Kluwer Academic Publishers.

Ramachandra, L. S. and D. Roy. 2001. 'A New Method for Two-point Non-linear Boundary Value Problems in Solid Mechanics'. *Jl. Applied Mechanics* 68: 776–86.

Roberts, J. B. and P. D. Spanos. 1990. *Random Vibration and Statistical Linearization.* Chichester, UK: John Wiley and Sons.

Roy, D. 2000a. 'Phase Space Linearization for Non-linear Oscillators: Deterministic and Stochastic Excitations'. *J. Sound Vib.* 231: 307–41.

__________. 2000b. 'Explorations of the Phase–Space Linearization Method for Deterministic and Stochastic Non-linear Dynamical Systems'. *Non-linear Dynamics*. 23(3): 225–58.

__________. 2001. 'A New Numeric analytical Principle for Non-linear Deterministic and Stochastic Dynamical Systems'. 2001. *Proc. R. Soc. Lond. A.* 457: 539–66.

__________. 2003. 'A Weak Form of Stochastic Newmark Method with Applications to Engineering Dynamical Systems'. *Applied Mathematical Modelling* 27: 421–36.

__________. 2004. 'A Family of Lower and Higher Order Transversal Linearization Techniques in Non-linearStochastic Engineering Dynamics'. *Int. J. Numer. Methods Eng.* 61: 764–90.

__________. 2006. 'A Family of Weak Stochastic Newmark Methods for Simplified and Efficient MonteCarlo Simulations of Oscillators'. *Int. Jl. Numer. Methods Eng.* 67: 364–99.

Roy, D. and G. V. Rao. 2012. *Elements of Structural Dynamics: A New Perspective.* Singapore: John Wiley and Sons.

Roy, D. and L. S. Ramachandra. 2001a. 'A Semi-analytical Locally Transversal Linearization Method for Non-linear Dynamical Systems'. *Intl. Jl. of Numerical Methods in Engineering* 51: 203–24.

__________. 2001b. 'A Generalized Formulation Based on Locally Transversal Linearization for MDOF Non-linear Oscillators'. *Jl. of Sound and Vib.* 241: 653–79.

Roy, D. and M. K. Dash. 2002. 'A Stochastic Newark Method for Engineering Dynamical Systems'. *Jl. of Sound and Vib.* 249(1): 83–100.

__________. 2005. 'Explorations of a Family of Stochastic Newmark Methods in Engineering Dynamics'. *Comput. Methods Appl. Mech. Eng.* 194: 4758–96.

Saha, N. and D. Roy. 2007. 'Higher Order Weak Linearizations of Stochastically Driven Non-linear Oscillators'. *Proc. R. Soc. A.* 463: 1827–56.

Saito, Y. and T. Mitsui. 1996. 'Stability Analysis of Numerical Schemes for Stochastic Differential Equations'. *SIAM Jl. Numer. Anal.* 33: 2254–67.

Simmons, G. F. 1991. *Differential Equations with Applications and Historical Notes.* 2nd ed. McGraw-Hill, Inc.

Simmons, G. F. and S. G. Krantz. 2006. *Differential Equations–Theory, Technique, and Practice.* McGraw-Hill, Inc.

Spanos, P. D. 1981. 'Stochastic Linearization in Structural Dynamics'. *Applied Mechanics Reviews* 34(1): 1–8.

Wagner, W. and E. Platen. 1978. 'Approximation of Ito Integral Equations'. *Preprint, ZIMM, Akad. Wissenschaften.* DDR. Berlin.

Wang, R. and Z. Zhang. 2000. 'Exact Stationary Solutions of the Fokker–Planck Equation for Non-linear Oscillators under Stochastic Parametric and External Excitations'. *Nonlinearity* 13: 907–20.

Yang, C. Y. 1986. *Random Vibration of Structures.* NY: John Wiley and Sons.

Zhu, W. Q. 1988. 'Stochastic Averaging Methods in Random Vibration'. *Appl. Mech. Rev.* 41: 189–99.

Zienkiewicz, O. C. and R. L. Taylor. 1991. *The Finite Element Method. vol. 2.* UK: McGraw-Hill.

Chapter 6

Ahmed, N. U. and S. M. Radaideh. 1997. 'A powerful numerical technique solving Zakai equation for non-linear filtering'. *Dynamics and Control* 7: 293–303.

Alighanbari, M., A. Homaifar, and B. Sayarrodsari. 2005. 'Robust Adaptive Control and Parameter Estimation Using Multi Objective Evolutionary Algorithm'. *Proc. IEEE International Conference on Systems, Man and Cybernetics.* Hawaii.

Arulampalam, M. S., S. Maskell, N. Gordon, and T. Clapp. 2002. 'Tutorial on Particle Filters for Online Non-linear/Non-Gaussian Bayesian Tracking'. *IEEE Transactions on Signal Processing.* 50(2): 174–88

Bain, A. and D. Crisan. 2009. *Fundamentals of Stochastic Filtering.* NY: Springer.

Banerjee, B., D. Roy, and R. M. Vasu. 2009a. 'Efficient Implementations of a Pseudodynamical Stochastic Filtering Strategy for Static Elastography'. *Medical Physics.* 36. 3470–6.

__________. 2009b. 'A Pseudo-dynamic Sub-optimal Filter for Elastography under Static Loading and Measurements'. *Physics in Medicine and Biology* 54: 285–305.

__________. 2009c. 'A Pseudo-dynamical Systems Approach for a Class of Inverse Problems in Engineering'. *Proceedings of the Royal Society A* 465 (2105): 1561–79.

Bar-Shalom, Y. and X. R. Li. 1993. *Estimation and Tracking: Principles, Techniques and Software.* Boston, MA: Artech House.

Bergman, N. 1999. 'Recursive Bayesian Estimation: Navigation and Tracking Applications'. PhD Thesis. Linköping Univ. Linköping, Sweden.

Blackman, S. and R. Popoli. 1999. *Design and Analysis of Modern Tracking Systems.* Boston, MA: Artech House.

Brown, R. G. and P. Y. C. Hwang. 1997. *Introduction to Random Signals and Applied Kalman Filtering.* 2nd ed. NY: John Wiley and Sons Inc.

Bruno, M. G. S. 2003. 'Sequential Importance Sampling Filtering for Target Tracking in Image Sequences'. *IEEE Signal Processing Letters.* 10(8): 246–49.

Budhiraja, A., L. Chen, and C. Lee. 2007. 'A Survey of Numerical Methods for Non-linear Filtering Problems'. *Phys. D. 230* (1–2): 27–36.

Cartwright, M. L. and J. E. Littlewood. 1945. 'On Non-linear Differential Equations of the Second Order: I. The Equation $\ddot{y} - k(1 - y2)\dot{y} + y = b\lambda k cos(\lambda t + a)$; k large. J. *London Math. Soc.* 20: 180–89.

Clough, R. W. and J. Penzien. 1993. *Dynamics of Structures.* Mc-Graw Hill Inc.

Crisan, D. and A. Doucet. 2002. 'A Survey of Convergence Results on Particle Filtering Methods for Practitioners'. *IEEE Transaction on Signal Processing* 50(3): 736–46.

Doucet, A. 1998. 'On Sequential Monte Carlo Methods for Bayesian Filtering.' *Statistics and Computing* (2000) 10: 197–208.

Doucet, A. and A. M. Johansen. 2008. 'A Tutorial on Particle Filtering and Smoothing: Fifteen years later.' *In The Oxford Handbook of Non-linear Filtering*, edited by D. Crisan and B. Rozovsky. Oxford: Oxford University Press

Doucet, A., Godsill, S., and C. Andrieu. 2000. 'On Sequential Monte Carlo Sampling Methods for Bayesian Filtering'. *Statistics and Computing* 10(3): 197–208.

Evensen, G. 1994. 'Sequential Data Assimilation with a Non-linear Quasi-geostrophic Model using Monte Carlo Methods to Forecast Error Statistics'. *Jl. of Geophysics Res.* 99(c5): 10143–62.

Evensen, G. 2003. 'The Ensemble Kalman Filter: Theoretical Formulation and Practical Implementation'. *Ocean Dyn.* 53(4): 343–67.

Florchinger, P. and J. Schiltz. 1996. 'Smoothness of the Density for the Filter Under Infinite Dimensional Noise and Unbounded Observation Coefficients'. *In Proceedings of the 2nd Portuguese Conference on Automatic Control.* 115–18.

Germani, A., M. Piccioni, and A. Galerkin. 1984. 'Approximation for the Zakai Equation, System Modelling and Optimization' in *System Modelling and Optimization.* Berlin: Springer, pp. 415–23.

Ghosh, S. J., D. Roy, and C. S. Manohar. 2007. 'New Forms of Extended Kalman Filter via Transversal Linearization and Applications to Structural System Identification'. *Comput. Methods Appl. Mech. Eng.* 196: 5063–83.

Gordon, N. J., D. J. Salmond, and A. F. M. Smith. 1993. 'Novel Approach to Non-linear/Non-Gaussian Bayesian State Estimation'. *IEE Proc. F Radar Signal Process.* 140: 107–13.

Grewal, M. S. and A. P. Andrews. 2001. *Kalman Filtering: Theory and Practice Using Matlab.* NY: John Wiley and Sons Inc.

Gyöngy, I. 1998. 'Lattice Approximations for Stochastic Quasi-linear Parabolic Partial Differential Equations Driven by Space-time White Noise I'. *Potential Anal.* 9: 1–25.

Handel, R. V. 2007a.'Filtering, Stability and Robustness'. PhD Thesis. California Inst. of Tech. Pasadena, California.

__________. 2007b. 'Stochastic Calculus, Filtering and Stochastic Control'. Lecture Notes. California Inst. of Tech., Pasadena.

IEEE Trans. Automatic Control. 1983. *Special Issue on Applications of Kalman Filtering.* 28(3).

Jazwinski, A. H. 1970. *Stochastic Processes and Filtering Theory.* NY: Academic Press.

Judd, K. 2003. 'Non-linear State Estimation, Indistinguishable States and the Extended Kalman Filter'. *Physica D: Non-linear Phenomena.* 183: 273–81.

Julier, S. J. and J. K. Uhlmann. 1997. 'A New Extension of the Kalman Filter to Non-linear Systems'. In *Proc. AeroSense: 11th Int. Symp. Aerospace/Defense Sensing, Simulation and Controls.* 182–93.

Kailath, T. 1981. *Lectures on Wiener and Kalman Filtering. CISM Monographs.* No. 140. Berlin: Springer-Verlag.

Kallianpur, G. and C. Striebel. 1968. 'Estimation of Stochastic Systems: Arbitrary System Process with Additive White Noise Observation Errors'. *Ann. Math. Statist.* 39(3): 785–801.

Kalman, R. E. 1960. 'A New Approach to Linear Filtering and Prediction Problem'. *Jl. of Basic Engg., Trans. ASME, Series D.* 82: 34–45.

Kalman, R. E. and R. S. Bucy. 1961. 'New Results in Linear Filtering and Prediction Theory'. *Jl. of Basic Engg., Trans. ASME D.* 83: 95–107.

Klebaner, F. C. 2005. *Introduction to Stochastic Calculus with Applications.* 2nd ed. UK: Imperial College Press.

Kurtz, T. G. and D. L. Ocone. 1988. 'Unique Characterization of Conditional Distributions in Non-linear Filtering'. *The Annals of Probability* 16(1): 80–107.

Liptser, R. S. and A. N. Shiryaev. 2001. *Statistics of Random Processes,* II, Applications. Berlin: Springer-Verlag.

Liu, J. S. and R. Chen. 1998. 'Sequential Monte Carlo Methods for Dynamic Systems'. *Journal of the American Statistical Association.* 93: 1032–44.

Livings, D. M., S. L. Dance, and N. K. Nichols. 2008. 'Unbiased Ensemble Square Root Filters'. *Physica D.* 237: 1021–8.

Lototsky, S., R. Mikulevicius, and B. L. Rozovskii. 1997. 'Non-linear Filtering Revisited: A Spectral Approach'. *SIAM J. Control Optim.* 35: 435–61.

Manohar, C. S. and D. Roy. 2006. 'Monte Carlo Filters for Identification of Non-linear Structural Dynamical Systems'. *Sadhana* 31: 399–427.

Piccioni, M. 1987. 'Convergence of Implicit Discretization Schemes for Linear Differential Equations with Application to Filtering, Stochastic Partial Differential Equations and Applications'. *Trento Lect. Notes in Math.* 1236: 208–29

Pitt, M. and N. Shephard. 1999. 'Filtering via Simulation: Auxiliary Particle Filters'. *J. Amer. Statist. Assoc.* 94(446): 590–99.

Protter, P. 2004. 'Stochastic Integration and Differential Equations'. in *Applications of Mathematics,* 2nd ed. Berlin: Springer-Verlag.

Raveendran, T., S. Gupta, R. M. Vasu, and D. Roy. 2011. 'Pseudo-time Particle Filtering for Diffuse Optical Tomography'. *Journal of the Optical Society of America A.* 28: 2070–81.

__________. 2012. 'A Pseudo-time EnKF Incorporating Shape Based Reconstruction for Diffuse Optical Tomography'. *Medical Physics* 39: 1092–101.

Riadh, K. 2006. 'Different Aspects of Non-linear Stochastic Filtering Theory'. PhD Thesis. University of Mannheim, Germany.

Ristic, B., S. Arulampalam, and N. Gordon. 2004. *Beyond the Kalman Filter: Particle Filters for Tracking Applications.* MA: Artech House.

Roy, D. and G. V. Rao. 2012. *Elements of Structural Dynamics–A New Perspective.* UK: John Wiley and Sons Inc.

Roychowdhury, S., D. Roy, and R. M. Vasu. 2013. 'Variance Reduced Particle Filters for Structural System Identification Problems'. *Journal of Engineering Mechanics (ASCE)* 139: 210–18.

Saha, N. and D. Roy. 2009. 'Extended Kalman Filters Using Explicit and Derivative-Free Local Linearizations'. *Applied Mathematical Modelling* 33: 2545–63.

_________. 2011. 'A Derivative-free, 2-Stage Extended Kalman Filter'. *Journal of Engineering Mechanics (ASCE)* 137: 537–46.

Sajeeb, R., C. S. Manohar, and D. Roy. 2007. 'Use of Particle Filters in an Active Control Algorithm for Noisy Non-linear Structural Dynamical Systems'. *Journal of Sound and Vibration* 306: 111–35.

_________. 2010. 'A Semi-analytical Particle Filter for Identification of Non-linear Oscillators'. *Probablistic Engg. Mech.* 25 (1): 35–48.

Sarkar, S., T. Raveendran, D. Roy, and R. M. Vasu. 2012. 'Stochastic Filtering in Structural Health Assessment: Some Perspectives and Recent Trends'. *Recent Trends in Structural Health Monitoring and Assessment, (edited by A. Haldar).* World Scientific.

Shinozuka, M. and R. Ghanem. 1995. 'Structural System Identification II: Experimental Verification'. *J. Eng. Mech., ASCE.* 121 (2): 265–73.

Silverman, B. W. 1986. *Density Estimation for Statistics and Data Analysis.* London: Chapman and Hall.

Smith, A. F. M. and A. E. Gelfand. 1992. 'Bayesian Statistics Without Tears: A Sampling–Resampling Perspective'. *The Americal Statistician* 46(2): 84–8.

Sokolov, A. P. and P. H. Stone. 1998. 'A Flexible Climate Model for Use in Integrated Assessments'. *Clim. Dyn.* 14: 291–303.

Tanizaki, H. 1996. *Non-linear Filters: Estimation and Applications.* 2nd Ed. Berlin: Springer.

Van der Pol, B. 1920. 'A Theory of the Amplitude of Free and Forced Triode Vibrations'. *Radio Review* 1. 701–10 and 754–62.

Walsh, J. B. 2005. 'Finite Element Methods for Parabolic Stochastic PDE's'. *Potential Anal.* 23: 1–43.

Yazigi, A. 2010. 'Stochastic Filtering'. Master's thesis. Univ. of Helesinki, Finland.

Yun, C. B. and M. Shinozuka. 1980. 'Identification of Non-linear Dynamic Systems'. *Journal of Structural Engineering* 8(2): 187–203.

Zakai, M. 1969. 'On the Optimal Filtering of Diffusion Processes'. *Z. Wahrsch. Verw. Geb.* 11: 230–43.

Chapter 7

Anderson, B. D. O. and J. B. Moore. 1979. *Optimal Filtering.* Englewood Cliffs, NJ: Prentice-Hall.

Andrieu, C. and A. Doucet. 2000. 'Simulated Annealing for Maximum A Posteriori Parameter Estimation of Hidden Markov Models'. *IEEE Transactions on Information Theory* 46(3): 994–1004.

Clough, R. W. and J. Penzien. 1993. *Dynamics of Structures.* McGraw-Hill Inc.

Crisan, D. and T. Lyons. 1999. 'A Particle Approximation of the Solution of the Kushner–Stratonovich Equation'. *Probab. Theory Related Fields* 115: 549–78.

Justel, A., D. Peña, and R. Zamar. 1997. 'A Multivariate Kolmogorov-Smirnov Test of Goodness of Fit'. *Statistics and Probability Letters*. 35: 251–59.

Kirkpatrick, S., C. D. Gelatt, and M. P. Vecchi. 1983. 'Optimization by Simulated Annealing'. *Science, New Series*. 220(4598): 671–80.

Kurtz, T. G. 1998. 'Martingale Problems for Conditional Distributions of Markov Processes'. *Electron. J. Probab*. 3(9): 1–29.

Kurtz, T. G. and D. L. Ocone. 1988. 'Unique Characterization of Conditional Distributions in Non-linear Filtering'. *Ann. Probab*. 16: 80–107.

Lam, J. and J. M. Delosme. 1988. 'An Efficient Simulated Annealing Schedule: Derivation'. *Technical Report* 8816 *Yale Electrical Engineering Department. New Haven, Connecticut*.

Livings, D. M., S. L. Dance, and N. K. Nichols. 2008. 'Unbiased Ensemble Square Root Filters'. *Physica D*. 237: 1021–8.

Milstein, G. V. and M. V. Tretyakov. 2009. 'Solving Parabolic Stochastic Partial Differential Equations via Averaging Over Characteristics'. *Math. Comp*. 78: 2075–106.

Raveendran, T., D. Roy, and R. M. Vasu. 2013. 'A Scaled Unscented Transformation Based Directed Gaussian Sum Filter for Non-linear Dynamic System Identification'. *Probabilistic Engineering Mechanics* 34: 83–90.

__________. 2014a. 'Iterated Gain-based Stochastic Filters for Dynamic System Identification'. *Jl. of the Franklin Institute-Engineering and Applied Mathematics*. 351 (2, SI): 1093–111.

__________. 2014b. 'Iterated Stochastic Filters with Additive Updates for Dynamic System Identification: Annealing-type Iterations and the Filter Bank'. *Probablistic Engineering Mechanics*. 38 (SI): 77–87.

Rosenblatt, M. 1956. 'A Central Limit Theorem and a Strong Mixing Condition'. *Proc. N. A. S.* 42: 43–7.

Sarkar, S., S. R. Chowdhury, M. Venugopal, R. M. Vasu, and D. Roy. 2014. 'A Kushner–Stratonovich Monte Carlo Filter Applied to Non-linear Dynamical System Identification'. *Physica-D-Non-linear Phenomena* 270: 46–59.

Sarkar, S. and D. Roy. 2014. 'An Ensemble Kushner-Stratonovich (EnKS) Non-linear Filter: Additive Particle Updates in Non-Iterative and Iterative Forms'. *arXiv.org, article no:* 1402. 1253.

Silverman, B. W. 1986. *Density Estimation for Statistics and Data Analysis*. London: Chapman and Hall.

Sorensen, H. W. and D. L. Alspach. 1971. 'Recursive Bayesian Estimation using Gaussian Sums'. *Automatica* 7(4): 465–79.

Stroock, D. W. and S. R. S. Varadhan. 1979. *Multidimensional Diffusion Processes*. Berlin: Springer-Verlag.

Teresa, J., M. Venugopal, D. Roy, R. M. Vasu, and R. Kanhirodan. 2014. 'Diffraction Tomography from Intensity Measurements: a Novel Evolutionary Stochastic Search to Invert Experimental Data'. *Journal of the Optical Soc. Am. (A)* 31(5): 996–1006.

Venugopal, M., R. M. Vasu, and D. Roy. 2016. 'An Ensemble Kushner–Stratonovich–Poisson Filter for Recursive Estimation in Non-linear Dynamical Systems'. *IEEE Trans. on Automatic Control* 61(3): 823–28.

Venugopal, M. , van. Es. Peter, S. Manohar, D. Roy, and R. M. Vasu. 2016. 'Quantitative photoacoustic tomography by stochastic search: Direct recovery of the optical absorption field'. *Optics Letters* 41(18): 4202–5.

Chapter 8

Belytschko, T., Y. Y. Lu, and L. Gu. 1994. 'Element-free Galerkin Methods'. *IJNME* 37: 229–56.

Beskos, A. and G. O. Roberts. 2005. 'Exact Simulation of Diffusions'. *The Annals of Applied Probability* 15(4): 2422–44.

Handschin, J. E. and D. Q. Mayne. 1969. 'Monte Carlo Techniques to Estimate the Conditional Expectation in Multi-state, Non-linear Filtering'. *Int. Jl. Contr.* 9: 547–59.

Higham, D. J., X. Mao, and A. Stuart. 2002. 'Strong Convergence of Euler-type Methods for Non-linear Stochastic Differential Equations'. *SIAM Jl. Numer. Anal.* 40: 1041–63.

Kloeden, P. and E. Platen. 1992. *Numerical Solution of Stochastic Differential Equations.* Berlin: Springer.

Milstein, G. N. 1995. *Numerical Integration of Stochastic Differential Equations.* Kluwer. Dordrecht.

Platen, E. 1987. 'Derivative-free Numerical Methods for Stochastic Differential Equations'. *Lec Notes Control Inform Sci.* 96: 187–93.

Raveendran, T., D. Roy, and R. M. Vasu. 2013a. 'A Nearly Exact Reformulation of the Girsanov Linearization for Stochastically Driven Non-linear Oscillators'. *J. Appl. Mech.*. 065002: 20.

_________. 2013b. 'A Novel Filtering Framework Through Girsanov Correction for the Identification of Non-linear Dynamical Systems'. *Inverse Problems*. 21020.

Riedle, M. 2011. 'Cylindrical Wiener Processes'. *Séminaire de Probabilitiés XLIII*: 191–214.

Robert, C. P. and G. Casella. 2004. *Monte Carlo Statistical Methods.* NY: Springer.

Roy, D. 2000. 'Exploration of the Phase-Space Linearization Method for Deterministic and Stochastic Non-linear Dynamical Systems'. *Non-linear Dyn.* 23(3): 225–58.

_________. 2001. 'A Numeric-analytic Technique for Non-linear Deterministic and Stochastic Dynamical Systems'. *Proc. Roy. Soc. A.* 457: 539–66.

_________. 2004. 'A Family of Lower- and Higher-order Transversal Linearization Techniques in Non-linear Stochastic Engineering Dynamics'. *IJNME* 249: 764–90.

_________. 2006. 'A Family of Weak Stochastic Newmark Methods for Simplified and Efficient Monte Carlo Simulations of Oscillators'. *Int. J. Numer. Methods Eng.* 67: 364–99.

Saha, N. and D. Roy. 2007. 'The Girsanov Linearisation Method for Stochastically Driven Non-linear Oscillators'. *J. Appl. Mech.* 74: 885–97.

Santos, B. C. D. and M. A. Savi. 2009. 'Non-linear Dynamics of a Non-smooth Shape Memory Alloy Oscillator'. *Chaos Solutions and Fractals* 40: 197–209.

Sarkar, S. and D. Roy. 2016. 'Weakly Corrected Numerical Solutions to Stochastically Driven Non-linear Dynamical Systems'. *Applied Mathematical Modelling* 40(2): 859–70.

Shaw, A. and D. Roy. 2007. 'A NURBS-based Error Reproducing Kernel Method with Applications in Solid Mechanics'. *Computational Mechanics* 40: 127–48.

__________. 2008. 'A NURBS-based Parametric Method Bridging Mesh-free and Finite Element Formulations'. *CMES-Comp. Model Engg.* 26–1: 31–57.

Van Neerven, J. M. A. M. 2008. *Stochastic Evolution Equations, Open Courseware*, TU: Delft.

Wang, R. and Z. Zhang. 2000. 'Exact Stationary Solutions of the Fokker–Planck Eq. for Non-linear Oscillators under Stochastic Parametric and External Excitations'. *Nonlinearity* 13(3): 907–20.

Chapter 9

Alba, E. and M. Tomassini. 2002. 'Parallelism and Evolutionary Algorithms'. *IEEE Trans. Evol. Comput.* 6: 443–62.

Arulampalam, M. S., S. Maskell, N. Gordon, and T. Clapp. 2002. 'A Tutorial on Particle Filters for Online Non-linear/Non-Gaussian Bayesian Tracking'. *IEEE Trans. Signal Processing* 50: 174–88.

Assis, E. G. and V. Steffen Jr. 2003. 'Inverse Problem Techniques for the Identification of Rotor Bearing Systems'. *Inverse Problems in Science and Engineering* 11(1): 39–53.

Auger, A. and N. Hansen. 2005a. 'A Restart CMA Evolution Strategy with Increasing Population Size'. In *Proc. IEEE Congress on Evolutionary Computation.* (CEC'05) 1769–76

__________. 2005b. 'Performance Evaluation of an Advanced Local Search Evolutionary Algorithm'. In *Proc. IEEE Congress on Evolutionary Computation* (CEC 2005). 2. 1777–84.

Back, T. 1996. *Evolutionary Algorithms in Theory and Practice.* Oxford: Oxford University Press.

Brest, J., S. Greiner, B. Boskovic, M. Mernik, and V. Zumer 2006. 'Self-adapting Control Parameters in Differential Evolution: A Comparative Study on Numerical Benchmark Problems'. *IEEE Transactions on Evolutionary Computation* 10(6): 646–57.

Broyden, C. G. 1969. 'A New Double-rank Minimization Algorithm'. *Notices American Math. Soc.* 16: 670.

Chen, G. and T. Ueta. 1999. 'Yet Another Chaotic Attractor'. *International Journal of Bifurcation and Chaos* 9(07): 1465–66.

Chibante, R. 2010. *Simulated Annealing Theory with Applications.* Rijeka, Croatia: Sciyo.

Choi, B. G. and B. G. Yang. 2000. 'Optimal Shape Design of Rotor Shafts Using Genetic Algorithm'. *Jl. of Vibration and Control.* 6: 207–22.

Chong, E. K. P. and S. H. Zak. 2004. *An Introduction to Optimization*, 2nd ed. NY: John Wiley and Sons.

Clerc, M. and J. Kennedy. 2002. 'The Particle Swarm–Explosion, Stability, and Convergence in a Multidimensional Complex Space'. *IEEE Trans. Evol. Comput.* 6: 58–73.

Davidon, W. C. 1959. 'Variable Metric Methods for Minimization'. *Argonne National Labs Report*. ANL-5990.

Diewald, W. and R. Nordmann. 1990. 'Parameter Optimization for the Dynamics of Rotating Machinery'. *Proc. of 3rd. Int. Conf. on Rotor Dynamics*. 51–55.

Doorly, D., J. Peiro, and J. Oesterla. 1996. 'Optimization of Aerodynamic and Coupled Aerodynamic–Structural Design Using Parallel Genetic Algorithm'. *Proc. of 6th AIAA/NASA/ISSMD Symposium on Multi-disciplinary Analysis and Optimization*. 401–409.

Dorigo, M. and M. Birattari. 2010. 'Ant Colony Optimization'. In *Encyclopedia of Machine Learning*. 36–39.

Eby D., R. Averill, E. Goodman, and W. Punch. 1999. 'The Optimization of Flywheels Using an Injection Island Genetic Algorithm'. In *Evolutionary Design by Computers*, edited by P. Bentley. San Francisco: Morgan Kaufmann. 167–90.

Edwards, R. T., M. Cohen, G. Cauwenberghs, M. A. Vorontsov, and G. W. Carhart. 1999. 'Analog VLSI Parallel Stochastic Optimization for Adaptive Optics'. In *Learning on Silicon,* edited by G. Cauwenberghs and M. Bayoumi., Norwell MA: Kluwer Academic.

Engelbrecht, A. P. 2006. *Fundamentals of Computational Swarm Intelligence*. NY: John Wiley and Sons.

Fan, H. Y. and J. Lampinen. 2003. 'A Trigonometric Mutation Operation to Differential Evolution'. *Jl. of Global Optimization* 27: 105–29.

Finck, S., N. Hansen, R. Ros, and A. Auger. 2014. 'Real-Parameter Black-Box Optimization Benchmarking 2010: Presentation of the Noiseless Functions'. Technical Report 2009/2. Research Center PPE

Fletcher, R. 1970. 'A New Approach to Variable Metric Algorithms'. *The Computer Journal* 13(3): 317.

Fletcher, R. 1987. *Practical Methods of Optimization*. 2nd ed. NY. John Wiley and Sons.

Fletcher, R. and C. Reeves. 1964. 'Function Minimization by Conjugate Gradient'. *Computer Jl.* 7: 149–54.

Fletcher, R. and M. J. D. Powell. 1963. 'A Rapid Convergent Descent Method for Minimization'. *Computer Journal* 6(2): 163–68.

Fox, R. L. 1971. *Optimization Methods for Engineering Design*. Massachusetts: Addison-Wesley.

Freidlin, M. I., A. D. Wentzell, and J. Szücs. 2012. *Random Perturbations of Dynamical Systems*. NY: Springer-Verlag.

Gamerman, D. and H. F. Lopes. 2006. *Markov Chain Monte Carlo: Stochastic Simulation for Bayesian Inference,* 2nd ed. London: Chapman & Hall/CRC Texts in Statistical Science. 2nd Ed..

Glover, F. and G. Kochenberger, ed. 2003. *Handbook of Metaheuristics*. Netherlands: Kluwer Academic Publishers.

Goldberg, D. E. 1989. *Genetic Algorithms in Search, Optimization and Machine Learning*, 1st ed. Boston: Addison-Wesley Publishing Company.

Goldberg, D. E. and M. P. Samtani. 1986. 'Engineering Optimization via Genetic Algorithms'. *Proc. 9th Conf. on Electronic Comput., ASCE, NY.*471–82.

Goldfarb, D. 1970. 'A Family of Variable Metric Updates Derived by Variational Means'. *Math. Comp.* 24(109): 23–26.

Grimble, M. J. and M. A. Johnson. 1988. Optimal Control and Stochastic Estimation. vol. 1 and 2. Chichester: John Wiley and sons.

Hansen, N and A. Ostermeier. 1996. 'Adapting Arbitrary Normal Mutation Distributions in Evolution Strategies: The Covariance Matrix Adaptation'. *ICEC* 96. *IEEE Press*. 312–17.

Hansen, N. 2007. *The CMA Evolution Strategy: A Tutorial.* Berlin: Springer-Verlag

Haupt, R. L. 1995. 'An Introduction to Genetic Algorithms for Electromagnetics'. *IEEE AP-S Mag.* 37: 7–15.

Holland, J. H. 1975. 'Adaptation in Natural and artificial systems'. Internal Report. University of Michigan, Ann Arbor, MI.

Holmes, P. 1979. 'A Non-linear Oscillator with a Strange Attractor'. *Philosophical Transactions of the Royal Society A.* 292: 419–48.

Hooke, R and T. A. Jeeves. 1961. 'Direct Search Solution of Numerical and Statistical Problems'. *Jl. of the Association for Computing Machinery.* 8: 212–19.

Hu, M., T. Wu, and J. D. Weir. 2012. 'An Adaptive Particle Swarm Optimization with Multiple Adaptive Method'. *IEEE Trans. on Evol. Comput.* 17: 1–15.

Jiao, B., Z. Lian, and X. Gu. Aug. 2008. 'A Dynamic Inertia Weight Particle Swarm Optimization Algorithm'. *Chaos, Solitons Fractals* 37: 698–705.

Kennedy, J. and R. Eberhart. 1995. 'Particle Swarm Optimization'. *Proceedings of IEEE International Conference on Neural Networks* 4(2): 1942–48.

Kirkpatrick, S. and M. P. Vecchi. 1983. 'Optimization by Simulated Annealing'. *Science* 220(4598): 671–80.

Krishnakumar, K. and D. E. Goldberg. 1992. 'Control System Optimization Using Genetic Algorithms'. *Jl. of Guidance, Control and Dynamics* 15(3): 735–40.

Kurtz, T. G. 1998. 'Martingale Problems for Conditional Distributions of Markov Processes'. *Electron. Jl. Probab.* 3(9): 1–29.

Lalanne, M. and G. Ferraris. 1998. *Rotor Dynamics Prediction in Engineering.* NY: John Wiley and Sons.

Lorenz, E. N. 1963. 'Deterministic Nonperiodic Flow'. *Jl. of Atmospheric Sciences* 20(2): 130–41.

Mallipeddi, J. R. and P. N. Suganthan. 2008. 'Empirical Study on the Effect of Population Size on Differential Evolution Algorithm'. *Proc. IEEE Congr. Evol. Comput.* 3663–70.

Metropolis, N., A. W. Rosenbluth, M. N. Rosenbluth, A. H. Teller, and E. Teller. 1953. 'Equation of State Calculations by Fast Computing Machines'. *The Jl. of Chemical Physics* 21(6): 1087–92.

Michalewicz, Z. 1996. *Genetic Algorithms + Data Structures = Evolution Programs*, 3rd ed. Berlin: Springer–Verlag.

More, J. J. and Z. Wu. 1999. 'Distance Geometry Optimization for Protein Structures'. *Jl. of Global Optimization*. 15: 219–34.

Nelder, J. A and R. Mead. 1965. 'A Simplex Method for Function Minimization'. *Computer Jl.* 7: 308–13.

Nelson, H. D. and J. M. McVaugh. 1976.' The Dynamics of Rotor Bearing Systems using Finite Elements'. *Trans. ASME, J. Eng. Ind.* 98(2): 593–600.

Ogata, K. 1995. *Discrete-time Control Systems*, 2nd ed. NJ: Prentice-Hall, Inc.

Powell, W. B. 1986. 'A Stochastic Model of the Dynamic Vehicle Allocation Problem'. *Transportation Science* 20: 117–29.

Rajan, M., S. D. Rajan, H. D. Nelson, and W. J. Chen. 1987. 'Optimal Placement of Critical Speeds in Rotor-bearing Systems'. *Transactions of the ASME* 109: 152.

Rao, S. S. 2009. *Engineering Optimization: Theory and Practice*, 4th ed. NJ: John Wiley and Sons.

Renders, J. M. and S. P. Flasse. 1996. 'Hybrid Methods using Genetic Algorithms for Global Optimization'. *IEEE Transactions on Systems, Man, and Cybernetics: Part B. Cybernetics*. 26(2): 243–58.

Roy, D. and G. V. Rao. 2012. *Elements of Structural Dynamics–A New Perspective*. UK: John Wiley and Sons.

Sarkar, S., D. Roy, and R. M. Vasu. 2014. 'A Perturbed Martingale Approach to Global Optimization'. *Physics Letters A*. 378(38): 2831–44.

Shanno, D. F. 1970. 'Conditioning of Quasi–Newton Methods for Function Minimization'. *Math. Comp.* 24(111): 647–56.

Silva Neto, A. J. and M. N. Özişik. 1994. 'The Estimation of Space and Time Dependent Strength of a Volumetric Heat Source in a One-dimensional Plate'. *Int. J. Heat Mass Transfer* 37(6): 909–15.

Smith, J. and T. Fogarty. 1997. 'Operator and Parameter Adaptation in Genetic Algorithms'. *Soft Computing* 1(2): 81–7.

Snyder, C., T. Bengtsson, P. Bickel, and J. Anderson. 2008. 'Obstacles to High-Dimensional Particle Filtering'. *Monthly Weather Review* 136: 4629–40.

Sotiropoulos, D. G., E. C. Stavropoulos, and M. N. Vrahatis. 1997. 'A New Hybrid Genetic Algorithm for Global Optimization'. *Non-linear Analysis Theory, Methods and Applications* 30(7): 4529–38.

Souza, F. L., F. J. C. P. Soeiro, A. J. Silva Neto, and F. M. Ramos. 2007. 'Application of the Generalized External Optimization Algorithm to an Inverse Radiative Transfer Problem'. *Inverse Problems in Science and Engineering* 15(7): 699–714.

Srinivas, M. and L. Patnaik. 1994. 'Adaptive Probabilities of Crossover and Mutation in Genetic Algorithms'. *IEEE Trans. on Systems, Man and Cybernetics* 24(4): 656–67.

Storn, R. and K. Price. 1997. 'Differential Evolution—A Simple and Efficient Heuristic for Global Optimization Over Continuous Spaces'. *Jl. of Global Optimization* 11: 341–59.

Stroock, D. W. and S. R. Varadhan. 1972. 'On the Support of Diffusion Processes with Application to the Strong Maximum Principle'. *Proceedings of the Sixth Berkeley Symposium on Mathematical Statistics and Probability* (Univ. California, Berkeley, Calif. 1970/1971). 3: 330–59.

Sun, J., J. Zhao, X. Wu, W. Fang, Y. Cai, and W. Xu. 2010. 'Parameter Estimation for Chaotic Systems with a Drift Particle Swarm Optimization Method'. *Physics Letters A*. 374(28): 2816–22.

Tang, K., X. Li, P. N. Suganthan, Z. Yang, and T. Weise. 2010. 'Benchmark Functions for the CEC'2010 Special Session and Competition on Large-scale Global Optimization'. Nature Inspired Computation and Applications Laboratory, USTC, China, 153–177

Van Laarhoven, P. J. and E. H. Aarts. 1987. *Simulated Annealing*. Netherlands: Springer. 7–15

Wang, J. H. and F. M. Shih. 1990. 'Improve the Stability of Rotor Subjected to Fluid Leakage by Optimum Diameters Design'. *ASME, Jl. Vib. Acoust. Stress Reliab. Des.*112: 59–64.

Wang, Y., Z. X. Cai, and Q. F. Zhang. 2011. 'Differential Evolution with Composite Trial Vector Generation Strategies and Control Parameters'. *IEEE Transactions on Evolutionary Computation* 15(1): 55–66.

Wenzel, W. and K. Hamacher. 1999. 'A Stochastic Tunneling Approach for Global Minimization of Complex Potential Energy Landscapes'. *Physics Rev. Lett.* 82: 3003–7.

Wong, K. L. and A. G. Constantinides. 1998. 'Speculative Parallel Simulated Annealing with Acceptance Prediction'. *Electronics Letters* 34(3): 312–13.

Zorzi, E.S. and H. D. Nelson. 1977. 'Finite Element Simulation of Rotor Bearing Systems with Internal Damping'. *Transactions of the ASME, Journal of Engineering for Power* 99: 71–6.

Zou Z., R. H. Byrd, and R. B. Schnabel. 1997. 'A Stochastic/Perturbation Global Optimization Algorithm for Distance Geometry Problems'. *Jl. Global Optimization* 11: 91–105.

Chapter 10

Chopin N. 2004. 'Central Limit Theorem for Sequential Monte Carlo Methods and its Application to Bayesian Inference'. *Ann. Statist.* 32: 2385–411.

Deb, K. 2001. *Multi-Objective Optimization Using Evolutionary Algorithms*. 1st ed. UK: John Wiley.

Evensen, G. 2003. 'The Ensemble Kalman Filter: Theoretical Formulation and Practical Implementation'. *Ocean Dynamics* 53: 343–67, (doi: 10.1007/s10236-003-0036-9).

Finck, S., N. Hansen, R. Ros, and A. Auger. 2014. 'Real-Parameter Black-Box Optimization Benchmarking 2010: Presentation of the Noiseless Functions'. Technical Report 2009/21, Research Center PPE.

Fonseca, C. M. and P. J. Fleming. 1995. 'An Overview of Evolutionary Algorithms in Multi objective Optimization'. *Evolutionary Computation* 3(1): 1–16.

Hart, W. E., N. Krasnogor, and J. E. Smith. 2005. *Recent Advances in Memetic Algorithms*. Berlin: Springer.

Holland, J. H. 1975. 'Adaptation in Natural and Artificial Systems'. Internal Report. University of Michigan, Ann Arbor, MI.

Lobato, F. S., E. G. Assis, V. Steffen Jr, and A. J. da Silva Neto. 2012. 'Design and Identification Problems of Rotor Bearing Systems Using the Simulated Annealing Algorithm'. *In Simulated Annealing—Single and Multiple Objective Problems edited by Marcos de Sales Guerra Tsuzuki, InTech*, (http://dx.doi.org/10.5772/47833).

Nash Jr., J. F. 1950. 'Equilibrium Points in n-person Games'. *Proc. Nat. Acad. Sci. USA.* 36. 48–49, (doi: 10.1073/pnas.36.1.48).

Neumann, J. V. and O. Morgenstern. 1947. *Theory of Games and Economic Behavior.* Princeton, NJ, USA: Princeton University Press.

Osborne, M. J. 2004. *An Introduction to Game Theory.* NY: Oxford University Press.

Rajan, M., S. D. Rajan, H. D. Nelson, and W. J. Chen. 1987. 'Optimal Placement of Critical Speeds in Rotor-bearing Systems'. *Transactions of the ASME* 109: 152.

Roy, D. and G. V. Rao. 2012. *Elements of Structural Dynamics—A New Perspective.* UK: John Wiley and Sons.

Sarkar, S., D. Roy, and R. M. Vasu. 2015. 'A Global Optimization Paradigm Based on Change of Measure. *Royal Society Open Science* 2: 150123.

Shaw, A. and D. Roy. 2008. 'NURBS-based Parametric Mesh-free Methods'. *Computer Methods in Applied Mechanics and Engineering* 197(17–18): 1541–67.

Shaw, A., B. Banerjee, and D. Roy. 2008. 'A NURBS-based Parametric Method Bridging Mesh-free and Finite Element Formulations'. *Computer Modeling in Engineering and Sciences* 26(1): 31–57.

Shiau, T. N. and J. R. Chang. 1993. 'Multi-objective Optimization of Rotor-bearing System with Critical Speed Constraints'. *Journal of Engineering for Gas Turbines and Power* 115(2): 246–56.

Sun, J., W. Fang, and W. Xu. 2010. 'A Quantum-Behaved Particle Swarm Optimization with Diversity-Guided Mutation for the Design of two Dimensional IIR Digital Filters'. *IEEE Trans Circuits Systems II* 57 (2): 141–5.

Tang, K., X. Li, P. N. Suganthan, Z. Yang, and T. Weise. 2010. 'Benchmark Functions for the CEC, 2010 Special Session and Competition on Large-scale Global Optimization'. Nature Inspired Computation and Applications Laboratory, USTC, China, 153177

Venugopal, M., D. Roy, and R. M. Vasu. 2016. 'Stochastically Evolving Non-local Search and Solutions to Inverse Problems with Sparse Data'. *Probabilistic Engineering Mechanics* 46: 37–47.

Vrugt, J. A., and B. A. Robinson. 2007. 'Improved Evolutionary Optimization from Genetically Adaptive Multi Method Search'. *Proceedings of the National Academy of Sciences* 104(3): 708–11.

Zorzi, E. S. and H. D. Nelson. 1977. 'Finite Element Simulation of Rotor-bearing Systems with Internal Damping'. *Transactions of the ASME, Journal of Engineering for Power* 99: 71–76.

Bibliography

Abundo, M. 1991. 'A Stochastic Model for Predator–Prey Systems: Basic Properties, Stability and Computer Simulation'. *Jl. of Mathematical Biology* 29(6): 495–511.

Alspach, D. L. and H. W. Sorenson. 1972. 'Non-linear Bayesian Estimation Using Gaussian Sum Approximations'. *Automatic Control, IEEE Transactions on.* 17(4): 439–48.

Anderson, B. D. O. and J. B. Moore. 1979. *Optimal Filtering.* Englewood Cliffs: Prentice-Hall.

Ang, A. H. S. and W. H. Tang. 1984. *Probability Concepts in Engineering Planning and Design, vol. II: Decision, Risk and Reliability* . NY: John Wiley and Sons Ltd.

Bendat, J. S. and A. G. Piersol. 1986. *Random Data: Analysis and Measurement Procedures.* 2nd ed. NY: John Wiley and Sons Inc.

Berg, B. A. 2004. *Markov Chain Monte Carlo Simulations and Their Statistical Analysis (With Web-Based Fortran Code).* NJ: World Scientific.

Berzuini, C., N. G. Best, W. R. Gilks, and C. Larizza. 1997. 'Dynamic Conditional Independence Models and Markov Chain Monte Carlo Methods'. *Journal of the American Statistical Association* 92(440): 1403–12.

Bhattacharya, R. N, and E. C. Waymire. 2009. *Stochastic Processes with Applications.* Philadelphia: Siam Publications.

Billingsley, P. 1995. *Probability and Measure.* 3rd ed. NY: Wiley.

Carpenter, J., P. Clifford, and P. Fearnhead. 1999. 'Improved Particle Filter for Non-linear Problems'. *IEE Proceedings–Radar, Sonar and Navigation* 146(1): 2–7.

Casella, G. and R. L. Berger. 2002. *Statistical Inference.* Pacific Grove, CA: Duxbury.

Charles, K. and K. Herbert. 1980. *Thermal Physics.* W. H. Freeman.

Chorin, A. J. and X. Tu. 2009. 'Implicit Sampling for Particle Filters'. *Proc. Natl. Acad. Sci. USA* 106: 17249–54.

Cipra, B. A. 1987. 'An Introduction to the Ising Model'. *American Mathematics Monthly* 94: 937–59.

Cox, D. R. and D. Oakes. 1984. *Analysis of Survival Data.* London: Chapman & Hall.

Crisan, D. and B. Rozovskii. 2011. *The Oxford Handbook of Non-linearFiltering.* Oxford: Oxford University Press.

Crisan, D. and T. Lyons. 1999. 'A Particle Approximation of the Solution of the Kushner–Stratonovitch Equation'. *Probab. Theory Relat. Fields* 115: 549–78.

Daum, F. E. 1986. 'Exact Finite Dimensional Non-linear Filters'. *IEEE Trans. Automat. Contr.* 31: 616–22.

de Freitas, J. F. G. 1998. *'Bayesian Methods for Neural Networks'.* PhD. Dept. Engg., Cambridge.

Di Masi, G.B., M. Pratelli, and W. J. Runggaldier. 1985. 'An Approximation for the Non-linear Filtering Problem with Error Bound'. *Stochastics* 14(4): 247–71.

Doucet, A., 2001. *Sequential Monte Carlo Methods.* John Wiley and Sons Inc.

Durrett, R. 1991. *Probability: Theory and Examples.* USA: Wadsworth.

Elishakoff, I. and G. Falsone. 1993. 'Some Recent Developments in Stochastic Linearization Technique'. In *Computational Stochastic Mechanics*, Edited by A. Cheng and C. Y. Yang. London: Elsevier Applied Science. 175–94.

Elliott, R. J. and R. Glowinski. 1989. 'Approximations to Solutions of the Zakai Filtering Equation'. *Stochastic Analysis and Applications* 7(2): 145–68.

Evensen, G., 2009. *Data Assimilation: The Ensemble Kalman Filter.* Springer.

Gard, T. C. 1988. *Introduction to Stochastic Differential Equations.* NY: Marcel Dekker Inc.

Gardiner, C. W. 2004. *Handbook of Stochastic Methods.* 3rd ed. Springer.

Geweke, J. and H. Tanizaki. 1999. 'On Markov Chain Monte Carlo Methods for Non-linear and Non-Gaussian State-Space Models'. *Commun. Stat. Simul. C.* 28: 867–94.

Grigoriu, M. 2003. 'A Class of Models for Non-stationary Gaussian Processes'. *Probabilistic Engineering Mechanics* 18: 203–13.

Grigoriu, M., 2002. *Stochastic Calculus: Applications in Science and Engineering.* Boston, MA: Birkhäuser.

Grimmett, G. A. 2005. *Probability and Random Processes.* 3rd ed. Oxford: Oxford University Press.

Gubner, J. A. 2006. *Probability and Random Processes for Electrical and Computer Engineers.* NY: Cambridge University press.

Hald, A. 1998. *A History of Mathematical Statistics from* 1750 *to* 1930. USA: John Wiley and sons.

Hammerseley, J. M. and D. C. Handscomb. 1964. *Monte Carlo Methods.* London: Chapman & Hall.

Hansen, N. 2006. 'The CMA Evolution Strategy: a Comparing Review, Towards a New Evolutionary Computation'. in *Advances on Estimation of Distribution Algorithms.* Springer. 1769–76.

Hansen, N. 2006. 'Compilation of Results on the 2005 CEC Benchmark Function Set'. *Computational Laboratory (CoLab), Institute of Computational Science, ETH Zurich, Tech. Rep.*

Hansen, N. 2011.A 'CMA–ES for Mixed–Integer Non-linear Optimization'. *RR (Research Report)* -7751, INRIA, HAL archives Id: inria-00629689.

Hansen, N. and S. Kern. 2004. 'Evaluating the CMA Evolution Strategy on Multimodal Test Functions'. in *Lecture Notes in Computer Science: Parallel Problem Solving from Nature*-PPSN VIII. 3242. 282–291. Berlin, Heidelberg: Springer.

Hansen, N., S. D. Müller, and P. Koumoutsakos. 2003. 'Reducing the Time Complexity of the Derandomized Evolution Strategy with Covariance Matrix Adaptation (CMA–ES)'. *Evolutionary Computation* 11(1): 1–18.

Harrison, J. M. 1985. *Brownian Motion and Stochastic Flow Systems.* New York: John Wiley and &. Sons.

Hart, W. E., N. Krasnogor, and J. E. Smith. 2005. *Recent Advances in Memetic Algorithms.* Berlin: Springer.

Hasminskii, R. Z. 1980. *Stochastic Stability of Differential Equations.* Netherlands: Sijthoff and Noordhoff.

Higham, D. J. and N. J. Higham. 2000. *MATLAB Guide.* Philadelphia: SIAM.

Hoshiya, M. and E. Saito. 1984. 'Structural Identification by Extended Kalman Filter'. *Journal of Engineering Mechanics.* 110(12): 1757–70.

Hutzenthaler, M., A. Jentzen, and P. E. Kloeden. 2011. 'Strong and Weak Divergence in Finite Time of Euler's Method for Stochastic Differential Equations with Non-globally Lipschitz Continuous Coefficients'. *Proc. R. Soc. Lond. Ser. AMath. Phys. Eng. Sci.* 467. 1563–76.

IAEA (2012): *One year on: The Fukushima Nuclear Accident and its aftermath*, available at http://www.iaea.org/newscenter/news/2012/fukushima1yearon.html

Insua, D. R, F. Ruggeri, and M. P. Wiper. 2012. *Bayesian Analysis of Stochastic Process Models.* John Wiley and Sons.

Iserlas, A. 2008. *A First Course in the Numerical Analysis of Differential Equations.* UK: Cambridge University Press.

Ito, K. and B. Rozovskii. 2000. 'Approximation of the Kushner Equation for Non-linear Filtering'. *SIAM J. Control Optim.* 38(3): 893–915.

Ito, K. and H. McKean. 1974. *Diffusion Processes and Their Sample Paths.* Berlin: Springer-Verlag.

Jimenez, J.C. 2002. 'A Simple Algebraic Expression to Evaluate the Local Linearization Schemes for Stochastic Differential Equations'. *Appl. Math. Lett.* 15: 775–80.

Julier, S. J. and J. K. Uhlmann. 2004. 'Unscented Filtering and Non-linear Estimation'. *Proc. IEEE* 92: 401–22.

Kalbfleisch, J. D. and R. L. Prentice. 1980. *The Statistical Analysis of Failure Time Data.* NY: John Wiley & Sons.

Kallenberg, O. 2002. *Foundations of Modern Probability, Probability and its Applications.* 2nd ed. Springer-Verlag.

Kallianpur, G. 1980. *Stochastic Filtering Theory.* NY: Springer-Verlag Inc.

Kallianpur, G. and C. Striebel. 1968. 'Estimation of Stochastic Systems: Arbitrary System Process with Additive White Noise Observation Errors'. *The Annals of Mathematical Statistics* 39(3): 785–801.

Karlin, S. and H. Taylor. 1981. *A Second Course in Stochastic Processes*. Academic Press.

Karr, A. F. 1993. *Probability*. NY: Springer-Verlag.

Kishor, S.T. 1982. *Probability and Statistics with Reliability, Queuing and Computer Science Applications*. NJ: Prentice-Hall Inc.

Kloeden, P. and E. Platen. 1992. *Numerical Solutions of Stochastic Differential Equations*. Berlin: Springer.

Kloeden, P., E. Platen, and H. Schurz. 1994. *Numerical Solution of SDE through Computer Experiments*. Berlin: Springer.

Kotecha, J. H. and P. M. Djuric. 2003. 'Gaussian Sum Particle Filtering'. *IEEE Trans. on Signal Processing* 51: 2602–12.

Kushner, H. J. 1967. 'Dynamical Equations of Optimal Non-linear Filtering'. *J. Diff. Equations* 3: 179–90.

Landau, L. D. and E M. Lifshcitz. 1980. *Statistical Physics*. Oxford, UK: Pergamon Press.

Laurendeau, N. M. 2005. *Statistical Thermodynamics: Fundamentals and Applications*. UK: Cambridge University Press.

Li, J. and D. Xiu. 2009. 'A Generalized Polynomial Chaos Based Ensemble Kalman Filter with High Accuracy'. *Journal of Computational Physics* 228(15): 5454–69.

Liu, P. L. 1986. 'Multivariate Distribution Models with Prescribed Marginals and Covariances'. *Probabilistic Engineering Mechanics* 1(2): 105–12.

Loéve, M. 1977. *Probability Theory*. NY: Springer-Verlag.

Los Alamos Report. 2000. *'A Review of Criticality Accidents'*. LA-13638.

Müller, C. L., B. Baumgartner, and I. F. Sbalzarini. 2009. 'Particle Swarm CMA Evolution Strategy for the Optimization of Multi-Funnel Landscapes'. *IEEE Congress on Evolutionary Computation (CEC 2009)*. 2685–92.

MacGillivray, H. T. and R. J. Dodd. 1984. *Monte-Carlo Simulations of Galaxy Systems. Astrophysics and Space Science* 105(2): 331–37.

Manohar, C. S. and S. Gupta. 2003. 'Modelling and evaluation of structural reliability: Current status and future directions'. in *Research Advances in Structural Engineering* edited by K. S. Jagadish, and R. N. Iyengar. Golden Jubilee Publications of Department of Civil Engineering, IISc., Bangalore. India: Universities Press.

Mao, X. 1997. *Stochastic Differential Equations and Applications*. Chichester: Horwood.

Melchers, R. E. 1999. *Structural Reliability, Analysis and Prediction*, 2nd ed. UK: John Wiley and Sons Ltd.

Milstein, G. N. and M. V. Tretyakov. 2009. 'Monte Carlo Methods for Backward Equations in Non-linear Filtering'. *Adv. Appi. Prob.* 41: 63–100.

Monbet, V., P. Ailliot, and M. Prevesto. 2007. 'Survey of Stochastic Models for Wind and Sea State Time Series'. *Probab. Eng. Mech.* 22: 113–26.

Morton, J. K. and D. Mayers. 1994. *Numerical Solution of Partial Differential Equations*. Cambridge: Cambridge University Press.

Murphy, K. and S. Russell. 2001. 'Rao-Blackwellised Particle Filtering for Dynamic Bayesian Networks'. In *Sequential Monte Carlo Methods in Practice,* edited by Doucet, A., N. de Freitas, and N. Gordon. NY: Springer-Verlag.

Newland, D. E. 2005. *An Introduction to Random Vibrations, Spectral and Wavelet Analysis*, 3rd ed. Mineola, NY: Dover.

Nocedal, J. and S. J. Wright. 1999. *Numerical Optimization.* NY: Springer-Verlag Inc.

Nowak, A. S. 2000. *Reliability of Structures.* NY: McGraw-Hill Companies Inc.

Papoulis, A. and S. U. Pillai. 2002. *Probability, Random Variables and Stochastic Processes,* 4th ed. NY: McGraw-Hill.

Parsopoulos, K. E. and M. N. Vrahatis. 2004. 'On the Computation of All Global Minimizers Through Particle Swarm Optimization'. *IEEE Transactions on Evolutionary Computation* 8(3): 211–24.

Platen, E. 1999. *'An Introduction to Nnumerical Methods for Stochastic Differential Equations'* Acta Numer. 8: 197–246.

Platen, E. and L. Shi. 2008. 'On the Numerical Stability of Simulation Methods for SDEs'. *Quantitative Finance Research Centre Research Paper.* 234.

Press, W. H., S. A. Teukolsky, W. T. Vetterling, and B. P. Flannery. 2007. *Numerical Recipes. The Art of Scientific Computing,* 3rd ed. Cambridge: Cambridge University Press.

Raveendran, T., S. Sarkar, D. Roy, and R. M. Vasu. 2013. 'A Novel Filtering Framework through Girsanov Correction for the Identification of Non-linear Dynamical Systems'. *Inverse Problems* 29(6): 065002.

Revuz, D. and M. Yor. 1991. *Continuous Martingales and Brownian Motion.* Berlin: Springer-Verlag.

Rohatgi, V. K. 1976. *An Introduction to Probability Theory and Mathematical Statistics.* John Wiley and Sons.

Rozkosz, A. and L. Slomiliski. 1997. 'On Stability and Existence of Solutions of SDEs with Reflection at the Boundary'. *Stochastic Processes and their Applications* 68: 285–302.

Samuelson, Paul. A. 1965. 'Rational Theory of Warrant Pricing'. *Industrial Management Review* 6(2): 13.

Semmler, W. and M. Sieveking. 1993. 'Non-linear Liquidity–Growth Dynamics with Corridor–Stability'. *Jl. of Economic Behavior and Organization* 22: 189–208.

Shinozuka, M. and G. Deodatis. 1988. 'Stochastic Process Models for Ground Motion'. *Jl. Eng. Mech. Div., ASCE* 3(3): 114–23.

Shinozuka, M. and G. Deodatis. 1991. 'Simulation of Stochastic Processes by Spectral Representation'. *Applied Mechanics Review* 44(4): 191–203.

Shiryaev, A. N. 1995. *Probability.* New York: Springer.

Shreve, S. E. 2004. *Stochastic Calculus for Finance I: The Binomial Asset Pricing model, II: Continuous–Time models.* NY: Springer-Verlag.

Shynk, J. J. 2013. *Probability, Random Variables and Random Processes: Theory and Signal Processing Applications.* Hoboken, NY: John Wiley and Sons.

Snyder, C., T. Bengtsson, P. Bickel, and J. Anderson. 2008. 'Obstacles to High-Dimensional Particle Filtering'. *Mon. Wea. Rev.* 136: 4629–40.

Soong, T. T. and M. Grigoriu. 1993. *Random Vibration of Mechanical and Structural Systems.* Englewood Cliffs, NJ: P T R Prentice-Hall.

Talay, D. and L. Tubaro. 1990. 'Expansion of the Global Error for Numerical Schemes Solving Stochastic Differential Equations'. *Stoch. Anal. Appl.* 8: 483–509.

Tanizaki, H. 1999. 'On the Non-linear and Non-normal Filter Using Rejection Sampling'. *IEEE Trans. Automat. Contr.* 44: 314–19.

Taylor, H. M. and S. Karlin. 1998. *An Introduction to Stochastic Modeling,* 3rd ed. NY: Academic Press.

Tocino, A. and R. Ardanuy. 2002. 'Runge–Kutta Methods for Numerical Solution of Stochastic Differential Equations'. *Jl. Comp. Appl. Math.* 138: 219–41.

van Handel, R. 2007. 'Stochastic Calculus, Filtering, and Stochastic Control', Course Notes. *URL http://www. princeton. edu/~ rvan/acm217/ACM217. pdf.*

Vidyadhar, G. K. 2010. *Modeling and Analysis of Stochastic Systems,* 2nd ed. Chapman & Hall/CRC Texts in Statistical Science.

Von Mises, R. 1964. *Mathematical Theory of Probability and Statistics.* NY: John Wiley.

Von Neumann, J. and O. Morgenstern. 1944. *The Theory of Games and Economic Behavior.* Princeton, N.J: Princeton Univ. Press.

Vrugt, J. A., and B. A. Robinson. 2007. 'Improved Evolutionary Optimization from Genetically Adaptive Multi Method Search'. *Proceedings of the National Academy of Sciences* 104(3): 708–11.

Yates, J. R. D. and D. J. Goodman. 2005. *Probability and Stochastic Processes: A Friendly Introduction for Electrical and Computer Engineers,* 2nd ed. John Wiley and Sons.

Index

31